Food, Pharmaceutical and Bioengineering Division 2017

Core Programming Area at the 2017 AIChE Annual Meeting

Minneapolis, Minnesota, USA
29 October - 3 November 2017

Volume 1 of 2

ISBN: 978-1-5108-5802-2

Printed from e-media with permission by:

Curran Associates, Inc.
57 Morehouse Lane
Red Hook, NY 12571

Some format issues inherent in the e-media version may also appear in this print version.

Printed by Curran Associates, Inc. (2018)

AIChE
120 Wall Street, FL 23
New York, NY 10005-4020

Phone: (800) 242-4363
Fax: (203) 775-5177

www.aiche.org

Additional copies of this publication are available from:

Curran Associates, Inc.
57 Morehouse Lane
Red Hook, NY 12571 USA
Phone: 845-758-0400
Fax: 845-758-2633
Email: curran@proceedings.com
Web: www.proceedings.com

TABLE OF CONTENTS

VOLUME 1

VOLUME 2

Some abstracts in this book refer to images that were not made available by the publisher

497316 Sustained Supersaturation of Erlotinib Sdd Ternary Amorphous Systems

Abstract:

As more low-solubility drug candidates enter the pharmaceutical development pipeline, the need for solubilization-enhancing technologies is ever greater. In an amorphous dispersion, the solubility of a drug is enhanced by kinetically trapping the molecules into the higher-energy amorphous state, often realizing a 5-50x enhancement in solubility relative to the crystalline state. Spray dried dispersions (SDDs), containing amorphous drug and a concentration-enhancing matrix polymer are a scalable and mature technology which takes advantage of the amorphous drug solubility enhancement. Typically, an SDD must meet numerous performance metrics including rapid dissolution rate, sustainment of supersaturation, physical stability, chemical stability and manufacturability. We have successfully formulated over 1,000 low-solubility compounds as SDDs over the past two decades.

Despite this success, for some compounds it can be challenging to meet all of the performance requirements using only one excipient in the SDD system. In these cases, a multi-component amorphous system approach may warranted, in which two excipients are utilized, each tailored to separately achieve specific performance metrics. In this presentation, we discuss the case study of erlotinib, a rapidly-crystallizing compound which requires both fast dissolution and sustained supersaturation. In this case, erlotinib SDD is spray dried with a fast-dissolving matrix polymer (Eudragit L100) and formulated with a crystallization-inhibiting polymer (HPMC) to maximize both dissolution rate and sustainment. The combination of Eudragit L100 and HPMC succeeds by increasing the in vitro test AUC over ten-fold compared to binary SDD formulations.

488954 Connecting the Product-Process-Performance Interplay for Improved Understanding in Development of a Spray Dried Dispersion Drug Product

Abstract:

The pharmaceutical industry has faced a shift in the physicochemical properties of pre-clinical drug candidates rendering them less soluble and bioavailable. Spray drying has become an important unit operation in the pharmaceutical industry to produce spray dried amorphous dispersions (SDD) which can improve apparent solubility versus the crystalline drug as well as dissolution rate, thus improvingthe bioavailability of poorly soluble active pharmaceutical ingredients (API). The development of drug products consisting of spray dried solid dispersions is complex as it involves the need for strengthened understanding of the solid dispersion and drug product formulation attributes as they relate to enhancement of drug dissolution rate and maintenance thereof in the drug product. In contrast to standard drug particle erosion as main dissolution mechanism in conventional crystalline API formulation, solid dispersions can release drug in various types of species ranging from free drug, nano-particles, bile salt micelles, precipitated drug, and so on[1]. The characteristic dissolution mechanisms are impacted by API physiochemical properties and formulation as well as the process conditions utilized to make them.

This work will cover: 1. different mechanisms of dissolution and species formed for various APIs, 2. how the spray drying process influences dissolution and amorphous product stability, 3. utility of novel spectroscopic and other characterization techniques to study dispersion quality and performance, 4. influence of dispersion and drug product formulation properties on dissolution and tablet properties, 5. end to end considerations for ensuring how key quality and performance attributes are designed in and maintained in the drug product.

Spray dried dispersion particles and powders were manufactured on various scales ranging from single particle levitation, ProCepT 4M8, GEA mobile minor and GEA PSD-2 spray dryers. Multiple process conditions resulting in a range of drying rates were studied and will be presented. The resulting powders will be characterized by standard techniques such as modulated differential scanning calorimetry (mDSC) and x-ray powder diffraction (XRPD) to evaluate bulk amorphous dispersion uniformity. In several cases, additional techniques such as focused ion beam scanning electron calorimetry (FIB-SEM), x-ray photoelectron spectroscopy (XPS), and energy dispersive x-ray spectroscopy (EDX) were utilized to look a particle level homogeneity. Drug product formulations were made by standard tumble blenders, dry granulated, lubricated and compressed into tablets. Discriminating dissolution methodologies were used to study drug release from the dispersions and tablet formulations. Environmental stressing was conducted to look at stability and impact on drug release from tablets.

The choice of polymers and surfactants that make up the SDD composition was shown to have a significant impact on the amorphous quality, particle level uniformity, resulting phase stability and dissolution behavior. The choice of solvent composition and spray drying conditions was shown to have a modest to profound impact on these quality attributes. This was shown through dissolution studies coupled with human PK data as well as phase uniformity issues

discovered immediately following manufacture or after post-process stressing. The API has inherent physio-chemical properties that tend to give rise to characteristic speciation behavior during dissolution of SDD powder. The rate of release of overall drug into these entities was negatively impacted as the dispersion load in the tablet increases, tablet void fraction exceeded critical levels, and tablet disintegrant system was not properly designed.

In conclusion, spray dried dispersion based drug products require end to end study and understanding of the API behavior, formulation and process components, characterization methodology for amorphous quality and dissolution methodology and environmental stresses to be successful. Discriminating dissolution methodologies can help understand the specific mechanisms of dissolution enhancement and novel characterization methods can be useful in characterizing and ensuring maintenance of amorphous quality in the SDD and final drug product.

References:

1: D.T Friesen et al. Hydroxypropyl Methylcellulose Acetate Succinate-Based

Spray-Dried Dispersions: An Overview. Molecular Pharmaceutics. 2008. VOL. 5, NO. 6, pg. 1003â€"1019

495763 Strategically-Designed Polymer Synthons As Pharmaceutical Excipients in Oral Drug Delivery

Abstract:

Synthetic polymer excipients, equipped with tunable structures and properties, can enable more bioavailable, inexpensive, and safe oral drug formulations when rationally designed to solubilize active pharmaceutical ingredients in amorphous solid dispersions. However, the underlying polymer-drug interactions in these materials to control drug supersaturation generation and maintenance currently remain poorly understood. We present a new approach to discover specialized excipients for a highly-crystalline drugs, driven by molecular evolution-inspired ideas to rapidly construct polymer synthons (macromolecules with structural subunits correlating to desired noncovalent interactions, analogous to the retrosynthetic analysis taught in organic chemistry) using high-throughput synthesis and screening protocols. In this manner, the molar mass and chemical composition state spaces were sampled to identify an optimal excipient synthon among 70 copolymers to solubilize phenytoin. The leading system contained N-isopropylacrylamide (NIPAm, responsible for inhibiting phenytoin crystallization) and N,N-dimethylacrylamide (DMA, providing polymer hydrophilicity), and polymer-drug interactions were revealed using NOESY and DOSY NMR spectroscopy. Complementary all-atom molecular dynamic simulations corroborated experimental evidence of NIPAm synthons strongly binding to phenytoin. When administered to rats, we observed a record 23-fold increase in area-under-the-curve bioavailability over neat drug with excellent in vitro-in vivo correlations. Altogether, these discoveries demonstrate how existing synthesis tools in the polymer chemistry landscape can accelerate the translation of new drug candidates (or even failed compounds) into more effective and accessible therapeutic treatments.

497295 Pill Burden Reduction through Engineering the Mechanical Properties of Spray-Dried Amorphous Solid Dispersions

Abstract:

Drug candidates with poor oral bioavailability due to low aqueous solubility and dissolution rate are common in the pharmaceutical industry. One of the most mature and well understood techniques to improve oral absorption of drug candidates is to formulate the drug as a spray-dried amorphous solid dispersion (SDD). Given SDDs are diluted with excipients to, primarily, stabilize the amorphous form of the drug, a common challenge in SDD oral delivery of tablets is pill burden. One promising solution to reduce pill burden is engineering the SDD particles to replace the use of tablet fillers, which primarily attenuate undesirable mechanical and powder flow properties of crystalline drugs. Given that the mechanical and powder flow properties of SDDs can be significantly tuned during the spray drying process, the use of fillers can be substantially reduced. However, tuning the particle properties of SDDs is a hierarchical multidimensional optimization problem, and for that reason, a clear relationship between particle properties and physical and chemical state stability of the drug and in vivo performance is generally established prior to optimization of mechanical and powder flow properties. There have been several recent reviews on this subject, but there lacks the direct connection between spray-drying process conditions and their impact on final tablet formulation [1-2].

The aim of the current study is to establish relationships between spray drying process parameters and the resulting physical and mechanical properties of the SDD particles, and secondly, to use these relationships to formulate high SDD loaded (greater than 70%) immediate release tablets. Felodipine spray-dried with PVP VA64 was used as the model SDD system. The powder mechanical properties measured include compression profiles, SEM imaging, modified in-die compression analysis [3], BET surface analysis, and Hiestand indices.

Results showed that, for a given SDD formulation, specific surface area of the powder was controllable by tuning the particle size, particle wall thickness, and to a lesser extent, particle morphology through drying kinetics and atomization during spray drying. This resulted in SDDs that spanned a 10-fold difference in specific surface area. The resulting mechanical properties of these high surface area SDDs included significant increases in tensile strength and bonding index (BI) when compressed, due to larger bonding area between particles [4].

The material properties were strongly dependent on the drug loading, with higher drug loading SDDs having diminished compactibility compared to their placebo counterparts. These results suggest that, counter to selecting the highest drug loading SDD with acceptable physical stability and in vivo performance, a lower drug loading SDD formulation could allow for less filler in the final drug product, resulting in an overall higher loading within the tablet. Consequently, this also improves the physical stability robustness.

In addition, it was observed that, when compressed as the neat material, the compacts had varying degrees of anisotropy, which resulted in greater axial elasticity. High elastic recovery decreases the bonding area between particles, reducing the strength of the tablets and increasing the risk for capping and lamination. Therefore, decreasing the degree of anisotropy could reduce the risk of these tablet defects and should be considered when engineering SDD particles for integration into tablet formulations [5].

Overall, the powder properties resulted in various immediate release tablet formulations with as high as 80% concentration of SDD. It was demonstrated that the SDD could span the tabletability range of typical fillers depending on how it was spray dried, allowing it to â€œdo the jobâ€ of different types of fillers. The study ultimately concludes that rationally selecting spray-drying conditions to target specific particle properties, such as high surface area, high compactibility, and isotropic geometries, allows tablet formulations that can successfully decrease pill burden.

[1] Patel, B.B. et al. Revealing facts behind spray dried solid dispersion technology used for solubility enhancement. Saudi Pharm. J. 2015, 23 (4), 352-365.

[2] Singh, A. and Van den Mooter, G. Spray drying formulation of amorphous solid dispersions. Adv. Drug Deliv. Rev. 2015.

[3] Katz, J. M., Roopwani, R., Buckner, I. S. A Material-Sparing Method for Assessment of Powder Deformation Characteristics Using Data Collected During a Single Compression-Decompression Cycle. J. Pharm. Sci. 2013, 102 (10), 3687-3693.

[4] Osei-Yeboah, F., Chang, S., and Sun, C.C. A critical Examination of the Phenomenon of Bonding Area - Bonding Strength Interplay in Powder Tableting. Pharm Res. 2016, 33, 1126-1132.

[5] Sun, C.C. Microstructure of Tablet â€" Pharmaceutical Significance, Assessment, and Engineering. Pharm. Res. 2017, 34, 918-928.

500057 Reprocessing of Spray Dried Dispersions: Minimizing Resources in Process Development and Dealing with the Possible Failure

Abstract:

Reprocessing of spray dried dispersions: minimizing resources in process development and dealing with the possible failure

T. Porfirio[1,2*], I. Duarte[1], R. Ferreira[1], B. Henriques[1], J. Vicente[1], V. Semião[2]

[1] Hovione Farmaciência SA, Sete Casas, 2674-506 Loures, Portugal; *tporfirio@hovione.com

[2]LAETA, IDMEC, Mechanical Engineering Department, Instituto Superior Técnico, Universidade de Lisboa, Av. Rovisco Pais 1, 1049-001 Lisbon, Portugal

Spray drying is a well-established manufacturing process to produce amorphous solid dispersions to improve the bioavailability of poorly soluble drugs. However, the scale-up of the process is often considered a resource-intensive challenge requiring large quantities of expensive drugs [1]. Therefore, the possibility of reprocessing spray dried dispersions (SDD) is very appealing since it minimizes the material's requirements during the process development. Furthermore if the feasibility of the reprocessing steps is demonstrated during development it can be applied during commercial routine as risk mitigation strategy.

However, the reprocessing can have significant potential to adversely affect the identity, strength, quality, purity or other attribute of the drug and excipients [2]. The reprocessing should be performed on the cases where there is an improvement in one or more quality attributes without adverse effect on the others. The drive of the present work is to understand the effect of successive reprocessing steps on chemical and physical properties of spray dried solid dispersions using different stabilizing polymers commonly used in the production of amorphous solid dispersions.

Three different systems were produced and reprocessed (as shown in Figure 1) using itraconazole (ITZ) as model drug and HPMC E3, HPMCAS-MG and Copovidone as different stabilizing polymers. After each spray drying step a sample was collected for analytical characterization to assess the physicochemical attributes of spray dried material viz. particle size, bulk density and content of residual solvents, assay, and dissolution profile. The remaining spray dried product was reprocessed to prepare the feed solution of the subsequent spray drying step. This study carried out two reprocessing in order to understand when the material begins to present modifications in its properties.

Figure 1 – Scheme of the adopted method

All the tests were performed in a lab-spray dryer equipped with a two-fluid nozzle and using the same drying conditions (such as outlet temperature, drying gas flowrate, feed flowrate and atomization flowrate). A solvent mixture of dichloromethane and methanol was used to dissolve the material in a solid concentration of 5 wt.%.

Placebo tests with polymer-only were also performed to assess the physiochemical properties, such as molecular weight, monomer content and heavy metal levels of the stabilizing polymers. The same conditions were used as in the tests with active substance.

The use of different formulations and placebos tests will clarify the role of the polymer type on the SDD reprocessing.

Reference:

[1] Hallow, D., Mudryk, B., Braem, A., Tabora, J., Lyngberg, O. & Tummala, S., 2010. An Example of Utilizing Mechanistic and Empirical Modeling in Quality by Design. Journal of Pharmaceutical Innovation, 5(4), pp.193–203.
[2] Guidance for Industry - Drug Substance Chemistry, Manufacturing, and Controls Information, U.S. Department of Health and Human Services – Food and Drug Administration

494745 *In silico* Development of Amorphous Solid Dispersions for Optimal Performance and Stability

Abstract:

The focus of the present work lies on screening formulations for amorphous solid dispersions (ASDs) with a computational tool which allows a comprehensive set of virtual experiments to be performed without API expenditure and with a fast turnover and consequently to explore the design space in full prior to manufacturing prototypes in the laboratory.

ASDs of poorly water soluble active pharmaceutical ingredients (APIs) in hydrophilic polymeric matrices are a commonly used strategy to increase the solubility and dissolution rate of oral-dosage forms [1]. However, successful development of amorphous solid dispersions for optimal performance and stability depends on both the formulation (e.g. drug load, polymer system and surfactants) and on the manufacturing technology/process conditions. For example, adequate solvent mixtures used in the process offer the possibility of influencing particle morphology and drying rate [2] and have been reported to offer advantages in the solid state drug-polymer miscibility and on the stability of the ASDs [3,4]. On the other hand, excessively complex formulations can lead to undesirable phase separation which can compromise physical stability [5]. However, optimizing the formulation (e.g. drug load, polymer system and solubilizers) and the process (e.g. drying kinetics and droplet/particle size) requires extensive laboratory work, which is often compromised by time and cost constraints. Therefore the ability to perform virtual experiments is highly advantageous and cost-effective.

In this presentation we cover the mathematical modeling of the various concurrent physical phenomena that a solution of API and polymer (and sometimes other excipients) undergo prior to forming an amorphous solid dispersion, such as atomization, co-solvent evaporative dynamics, propensity for phase separation and shell formation [6,7]. We show a set of case studies of *in silico* screening benchmarked against data from solvent casting and spray drying experiments and also show a case-study where an excessively complex ASD formulation led to phase separation and compromised the stability of the amorphous intermediate.

Figure 1: Sketch of various stages throughout the droplet drying history and the concomitant propensity for phase separation and crystalization if the formulation and/or process are not adequately designed.

References

[1] Perrie Y., Rades T. "Pharmaceutics - Drug Delivery and Targeting", Pharmaceutical Press, 2010.

[2] Wulsten, E., Kiekens, F, van Dycke, F., Voorspoels, J. & Lee, G., "Levitated single-droplet drying: Case study with itraconazole dried in binary organic solvent mixtures", Int. J. Pharmaceutics, 2009, 378(1-2) pp. 116-121

[3] Janssens, S, Anné, M., Rombaut, P. & Van den Mooter, G., "Spray drying from complex solvent systems broadens the applicability of Kollicoat IR as a carrier in the formulation of solid dispersions", Eur. J. Pharm. Sci. 2009, 237, pp. 241-248.

[4] Paudel, A. & Van den Mooter, G., "Influence of Solvent Composition on the Miscibility and Physical

Stability of Naproxen/PVP K 25 Solid Dispersions Prepared by Cosolvent Spray-Drying", Pharm. Res. 2012, 29, pp. 251-270.

[5] Janssens, S., Nagels, S., de Armas, H.N., D'Autry, W., Van Schepdael, A. & Van den Mooter, G., "Formulation and characterization of ternary solid dispersions made up of Itraconazole and two excipients, TPGS 1000 and PVPVA 64, that were selected based on a supersaturation screening study", Eur. J. Pharm. Biopharm, 69:1, 2008

[6] Duarte, Í., Santos, J. L, Pinto, J. G & Temtem, M., "Screening methodologies for the development of spray-dried amorphous solid dispersions", Pharm. Res., 2015, 32 pp. 222 – 237, 2015.

[7] Prudic, A., Ji, Y. & Sadowski, G., "Thermodynamic Phase Behavior of API/Polymer Solid Dispersions", Mol. Pharmaceutics, 2014, 11(7), pp. 2294-2304

490777 Identification and Reconstruction of Pathways for Lignin Catabolism

Abstract:

Cellulosic biofuel production yields a substantial lignin byproduct stream that currently has few applications. Biological conversion of lignin compounds into chemicals and fuels has the potential to improve the economics of cellulosic biofuels. Microbial strains have been isolated with the ability to specifically degrade challenging lignin dimers, and the associated pathways could substantially improve the conversion of lignin into valuable biochemicals. However, no dimer degradation pathway has been characterized to the level necessary for reconstruction in a heterologous host, largely due to the combined challenges of fully identifying the relevant enzymes and then optimizing the resulting pathways for heterologous function. We have used a barcoded transposon library to rapidly identify entire catabolic pathways in genetically-intractable microbes such as the lignin-degrading strain *Novosphingobium aromaticivorans*. Simultaneously, we are reconstructing pathways for lignin catabolism in *Escherichia coli* to enable pathway characterization in a tractable and well-defined host. Evolutionary optimization of these heterologous pathways identified factors limiting pathway activity and will allow the design of characterized, optimized synthetic operons for lignin dimer degradation.

499050 Investigating New Pathways and a 'Funneling' Strategy to Enhance Production of Chemicals Derived from Shikimic Acid Biosynthesis

Abstract:

The engineering of non-natural pathways is a promising to approach to developing renewable and sustainable alternatives to many conventional petrochemicals. This is true of a multitude of useful chemicals that can be synthesized from precursors of native shikimic acid biosynthesis, including various aromatics and other molecules. For example, several recent studies have demonstrated how *Escherichia coli* can be engineered for the efficient production of muconic acid which, in turn, can be used as a precursor to adipic acid – a platform chemical used for the synthesis of many different polymers (e.g., nylon 6,6). To date, no fewer than five distinct pathways have been reported for muconic acid production from glucose, each involving different precursors of the shikimic acid pathway and enzyme chemistries. This unique flexibility (i.e., the ability to synthesize the same product via multiple different pathways) arises in part due to the fact that muconic acid is also a naturally-occurring metabolite that arises as a key intermediate in the beta-ketoadipate pathway, used by various microbes to degrade a range of aromatic compounds. Further exploiting this principle, here we demonstrate how three additional routes to muconic acid can be constructed by linking a series of recently-engineering pathways for phenol biosynthesis with its subsequent, partial degradation. The resultant phenol-dependent pathways support muconic acid production at up to 0.5 g/L in shake flasks. We also report a novel, four-step pathway for muconic acid production from endogenous chorismate via p-hydroxybenzoate which supports muconic acid titers of up to about 1 g/L while circumventing auxotrophic limitations experienced via the original, 3-hydroshikimate-derived route. Moreover, as inspired by the mechanisms of 'biological funneling' employed by soil microbes for the simultaneous degradation of diverse aromatic mixtures, we investigate the potential of a synthetic 'funneling' strategy for enhancing achievable muconic acid titers and yields. More specifically, by co-expressing multiple pathways that stem from different endogenous precursors before then converging on a common intermediate, muconic acid yields have been enhanced by as much as 40% relative to the best performing, single pathway control. In addition to maximizing the availability of precursors, this novel strategy offers a more holistic approach to balancing pathway performance and host fitness. Analogous 'funneling' approaches have similarly been extended to investigate the general utility of this strategy for improving the production of aromatic products, including phenol, catechol, and others.

498494 Engineering of ADP1 for Production of Vanillin from Lignin

Abstract:

Vanillin is one of the most widely used natural flavoring agents in the world. Its popularity exceeds the supply from traditional plant extraction. Therefore, chemical synthesis has been used to meet global demand. Although this has successfully improved vanillin availability and pricing, growing consumer preferences for "natural" ingredients have steered food and agriculture firms to instead rely on biologically derived vanillin in their products. This forces the need for a creative solution for vanillin products to be both natural and affordable. Biotechnological approaches have been used to meet this need, with glucose and ferulic acid used as starting materials for a biological conversion to vanillin to date. Here, we look to harness the bacteria ADP1's natural ability to degrade lignin monomers to produce vanillin from biomass. Our approach involves engineering ADP1 by deleting ADP1's natural vanillic acid and protocatechuate degradation pathways and by introducing a catechol o-methyltransferase to convert protocatechuate to vanillic acid. In addition, we look to introduce a decarboxylase enzyme to convert vanillic acid to vanillin, to potentially provide an affordable and renewable source of natural vanillin.

Dow AgroSciences

C4-C9 Alcohols Production By '+1' Pathway Engineered E. Coli: Evaluation of Vector Expression Systems and Fermentation Process Conditions

Authors: Mamatha Devarapalli, Paresh Sanghani, Chris Stowers, Sarah Delaplane, Ryan Hill and Devon Rosenfeld

Solutions for the Growing World

Renewable Approaches for C4-C9 Alcohol Production

- C4-C9 linear alcohols are commonly used as solvents and perfumes
- C4-C9 alcohols can also be reduced into alkanes and alkenes to serve as precursors for other molecules
- Petrochemical processes for medium-chain alcohol (C4-C9) production are costly due to high catalyst cost and poor selectivity
  - Odd chains are even more challenging
- Three main pathway options
  - Fatty acid synthesis
  - Reverse beta oxidation (pioneered by R. Gonzalez lab at Rice)
  - **'+1 Pathway'** (pioneered by J. Liao at UCLA)
    - Branch chain alcohols
    - **Linear alcohols**

Dow AgroSciences | 2

Background on the LeuABCD "+1 Pathway"

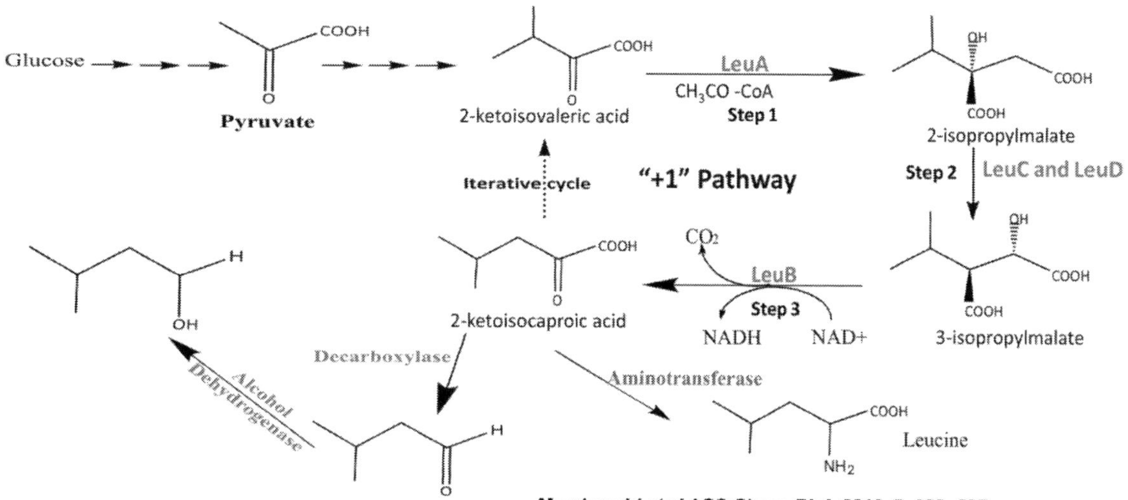

Marcheschi et al ACS Chem. Biol. 2012, 7, 689–697

Dow AgroSciences

3

"+1 Pathway' Pathway Development Strategy

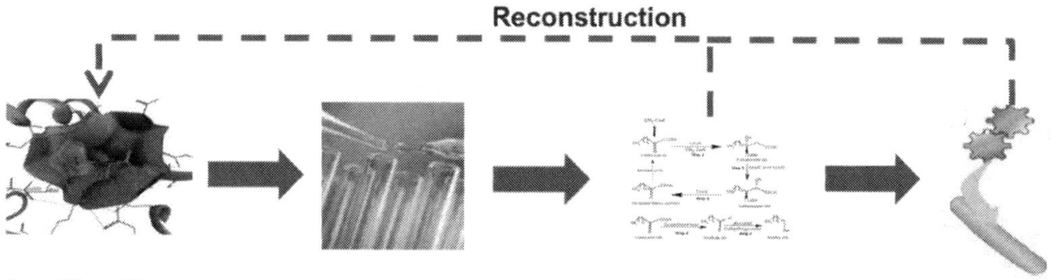

In silico Enzyme Design	*In vitro* Enzyme Characterization	*In vitro* Pathway Evaluation	*In vivo* Pathway Characterization
•Protein crystallography •Active site analysis	•Kinetics k_{cat}, k_M •Substrate specificity •pH dependence	•Substrate inhibition •Pathway kinetics •Carbon balance	•Titer •Overall yield •Byproduct analysis •Fermentation Optimization

Dow AgroSciences

4

Strain Engineering to Drive Flux Towards '+1 Pathway' LCAs

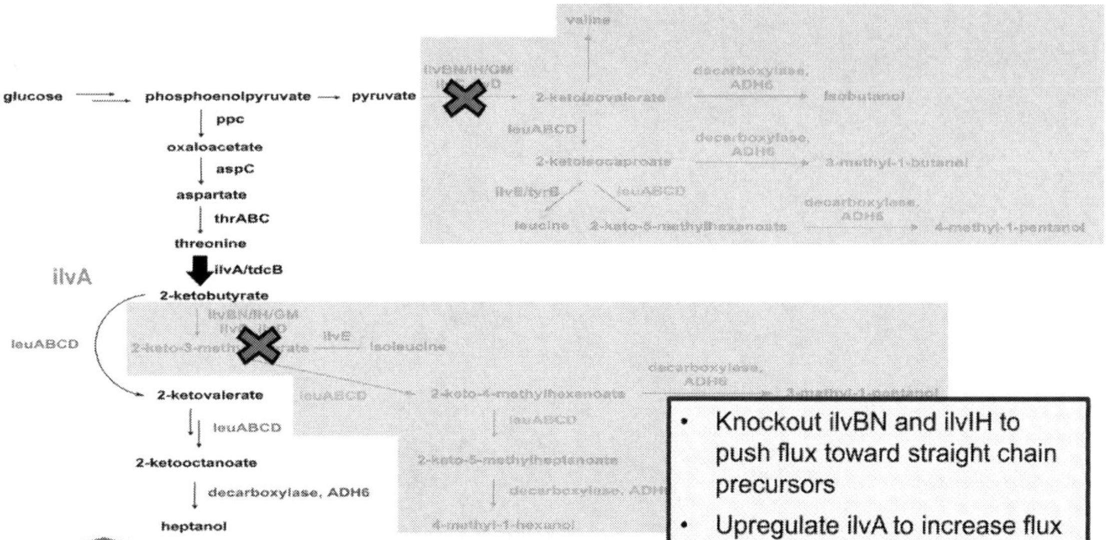

- Knockout ilvBN and ilvIH to push flux toward straight chain precursors
- Upregulate ilvA to increase flux toward 2-ketobutyrate

Dow AgroSciences

+1 Pathway Expression in Duet Vectors

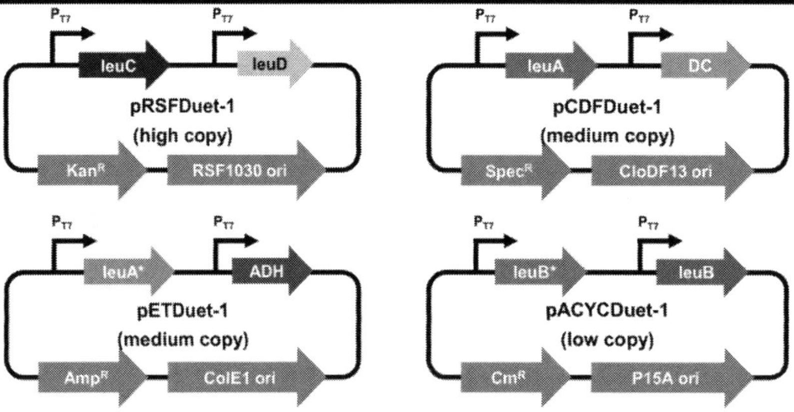

- Engineered leuA (leuA*), leuA, engineered leuB (leuB*), leuB, leuC, leuD, DC, and ADH6 in *E. coli* using the duet vector system
- Plasmid copy number can be varied to tune pathway performance

Dow AgroSciences

| 6

Modified LeuB Shifts Alcohol Distribution to Longer Chain Length

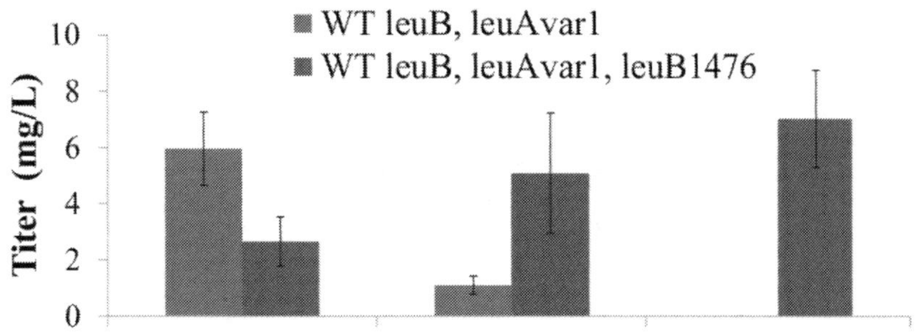

- Initial shake flasks experiments accumulated only 15 mg/L of long-chain alcohols
- Modified LeuB seemed to shift selectivity to longer chain lengths
- **Titers were far lower than compared to published results**

Dow AgroSciences | 7

Liao Construct Design

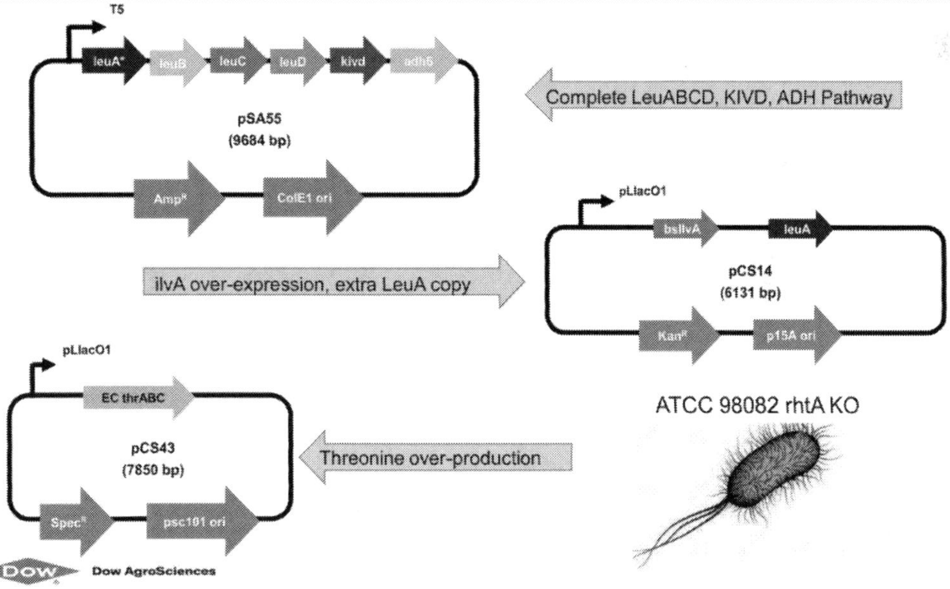

Dow AgroSciences | 8

Total Alcohol Titer

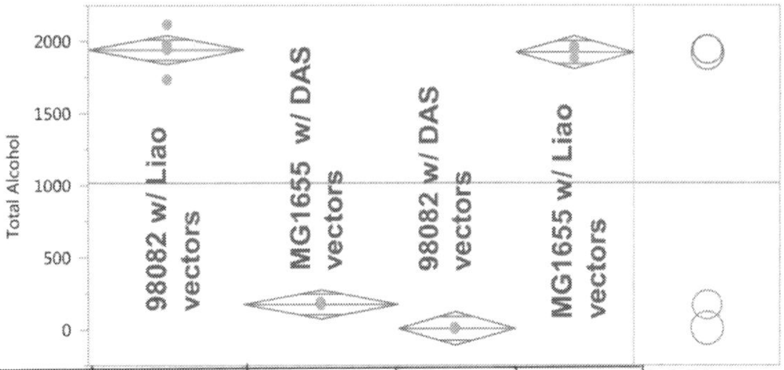

Strain	Liao	DAS	Liao	DAS
Vectors	Liao	DAS	DAS	Liao

Lower performance in DAS strains seem to be due to the expression strategy (vectors).

Dow AgroSciences | 9

Proteomics Evaluation of Liao vs. DAS Vector System

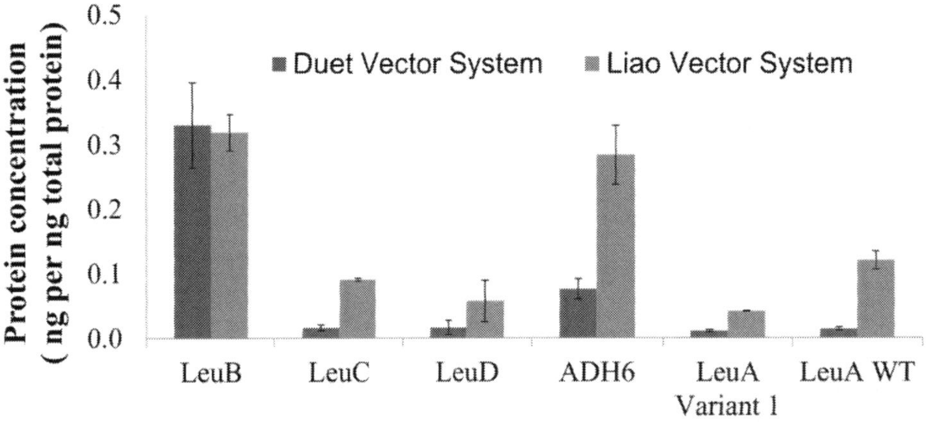

Surprisingly the Liao vector system had higher expression levels of several key enzymes in the pathway

Dow AgroSciences | 10

MG1655 3-vector vs. 2-vector Expression System

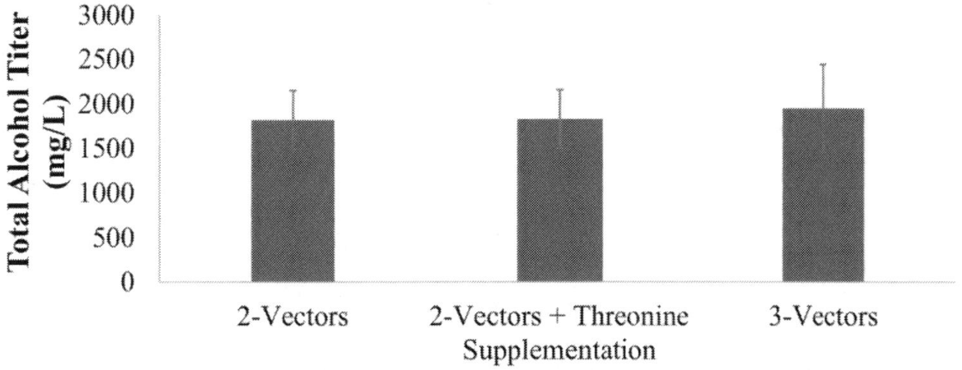

- Elimination of the ThrABC over-expression vector seemed to have little impact on performance.
- Threonine supplementation seemed to have no effect.
- No effect on alcohol distribution was observed.

Dow AgroSciences | 11

Is both LeuA and ilvA over-expression needed?

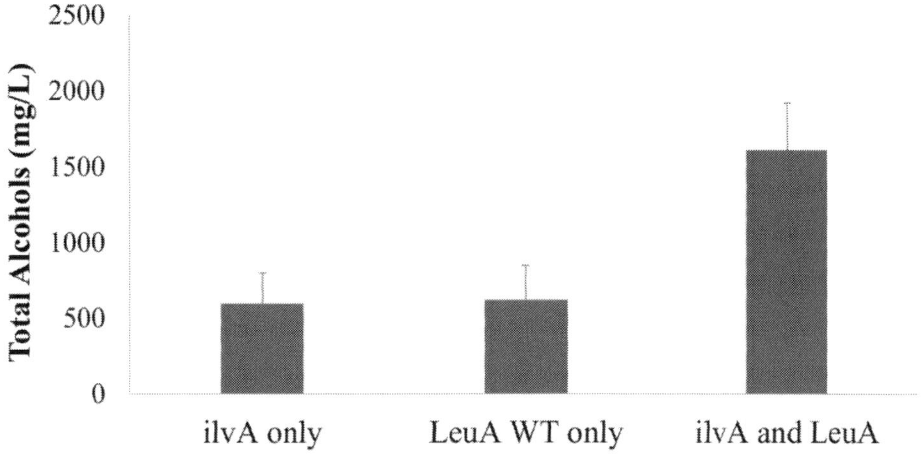

Both IlvA and LeuA are needed in order to achieve high alcohol titers

Dow AgroSciences | 12

Metabolomic Characterization Reveals Enzymatic Limitations

- Large amount of 2-KB and short-chain 2-alkyl malates detected with threonine feed
- Accumulation of 2-alkymalates is consistent with a LeuCD limitation
- Drop off in metabolite concentration at increased chain length could indicate LeuA limitation

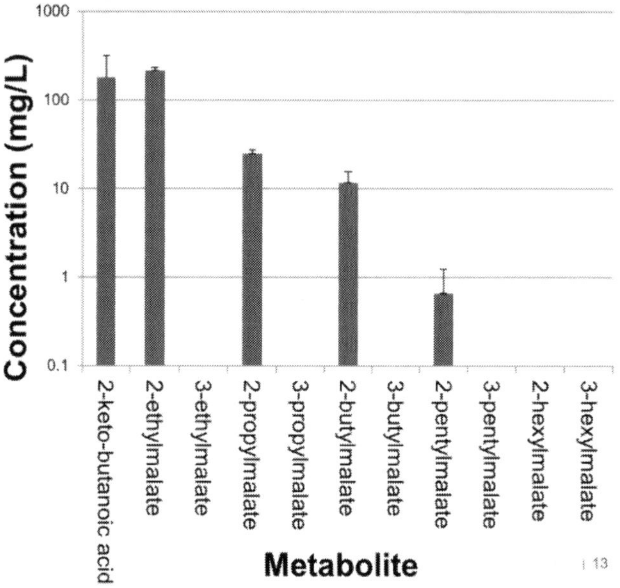

2-HM Feeding

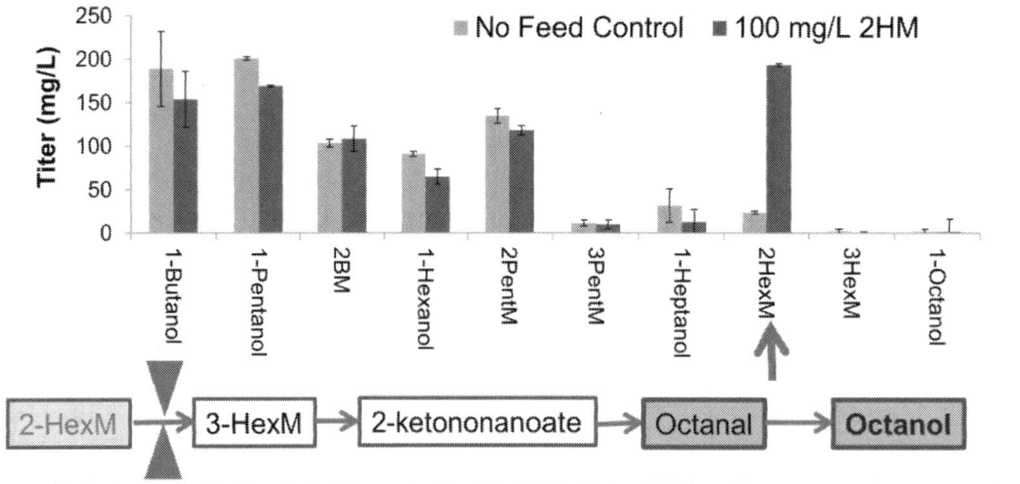

No conversion is observed when 2-HexM is fed indicating LeuCD bottleneck

3-HM Feeding

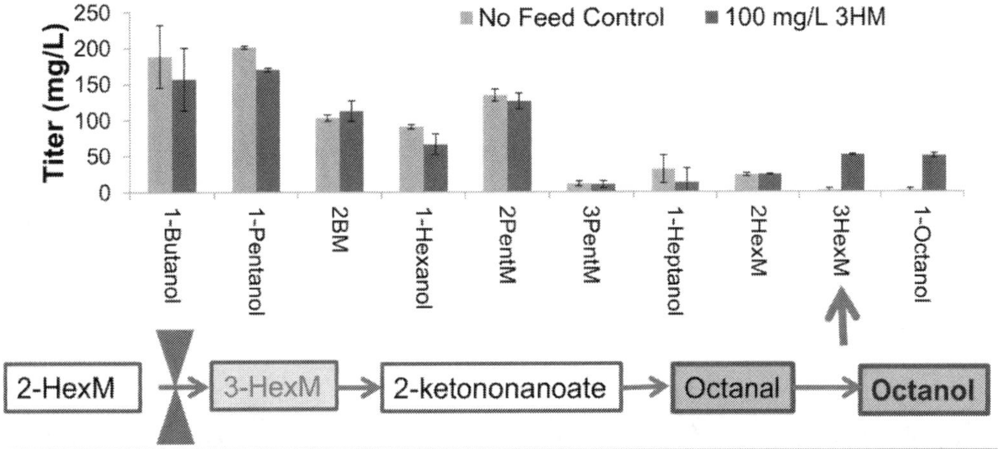

Half of the fed 3-HexM converted into octanol. Near perfect mass balance occurred between residual 3-HexM and octanol.

C8 Keto acid Feeding

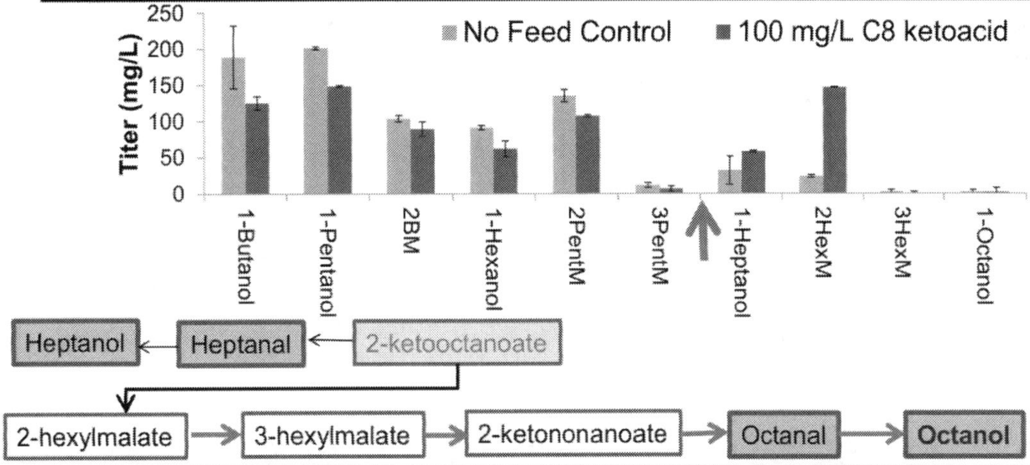

An increase in heptanol and 2-HM was observed when C8 keto acid was fed. Most of the carbon is trapped in 2-HM indicating LeuCD limitation

pH effect on alcohol production- Bioreactor studies

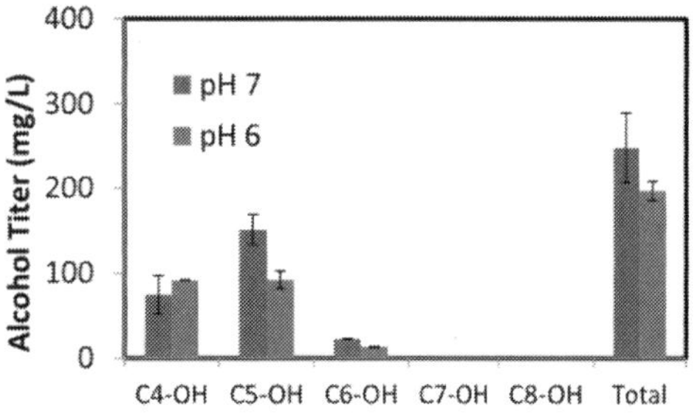

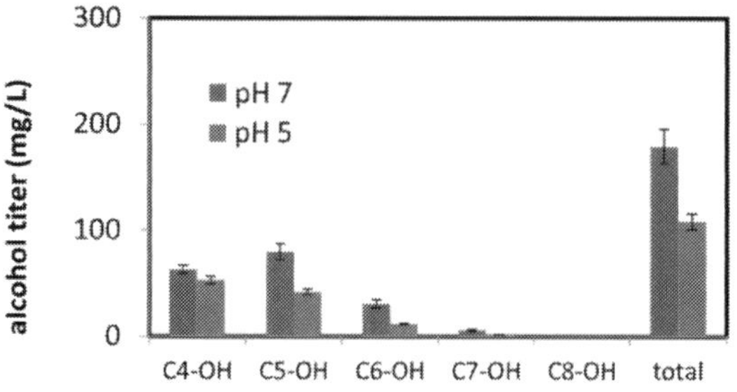

Alcohol production was higher at pH 7 followed by pH 6 and then pH 5

Effect of induction time and post induction temperature

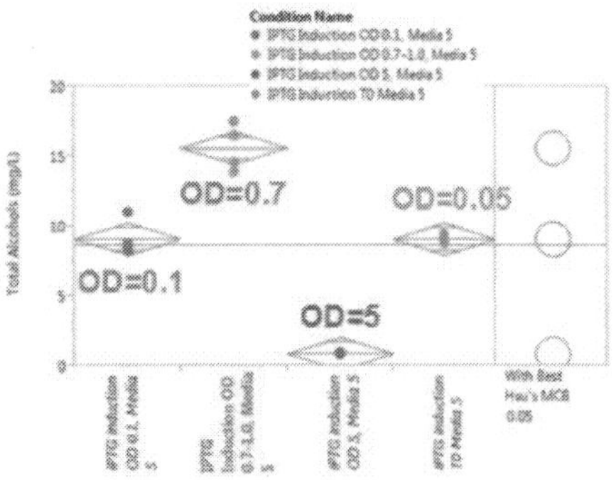

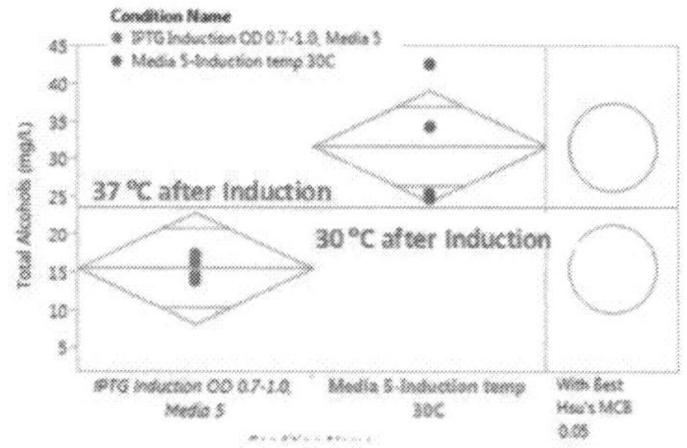

Lowering temperature to 30 °C post-induction and inducing the E.coli cultures at 0.7- 1.0 OD(600 nm) showed higher production of total alcohols

Conclusions

- '+1 Pathway' is functional for long-chain alcohols production
- Improved selectivity to longer chain lengths can be achieved by engineering LeuB, PPDC and LeuCD, but work remains to shift the alcohol distribution more towards C6-C9
- 'Omics' is a powerful tool to guide strain development and enzyme engineering.
  - ➢ Proteomics quickly demonstrated the underlying cause for performance differences in two expression systems
  - ➢ Metabolomics illuminated pathway bottlenecks and guided future enzyme engineering
- Fermentation conditions such as pH, induction timing and post-induction temperature has significant impact on the total alcohol production

Acknowledgements

- Mike Harris
- Scott Greenwalt
- Bryan Ward
- Matt Roach
- Matt Robinson
- Sowji Garapati
- Pam Torkelson
- Mark Mikola
- Allison Lutocka
- Samantha Hall

HTP Enzyme Characterization of LeuB

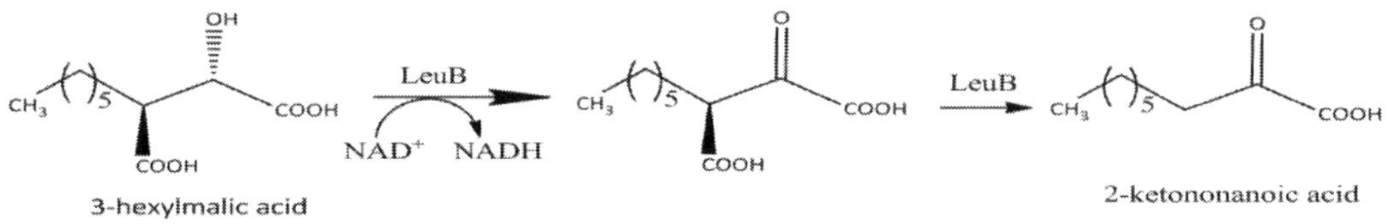

Substrates used for evaluating LeuB variants:

Substrate	Cycle Number	Product	Terminal Product
3-isopropylmalic acid (3-IPM)	1	2-ketoisocaproate (C6)	Isopentanol
3-butylmalic acid (3-BM)	3	2-ketoheptanoate (C7)	Hexanol
3-hexylmalic acid (3-HM)	5	2-ketononanoate (C9)	Octanol

Dow AgroSciences | 21

Activity of Wild Type LeuB Against Native and Non-native Substrates

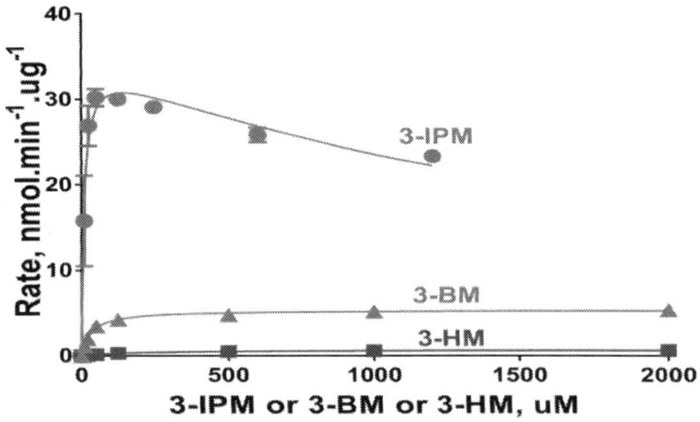

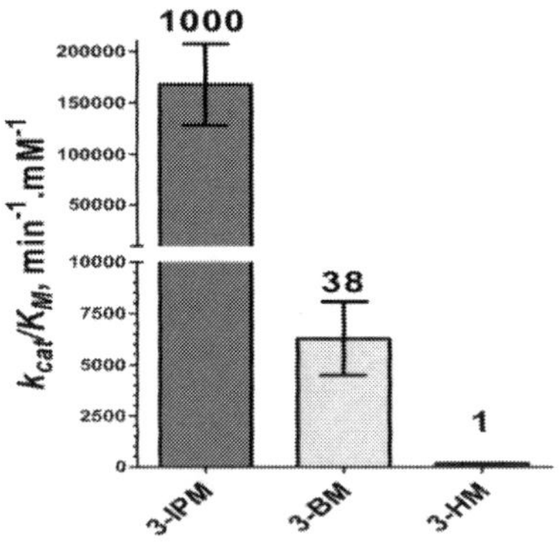

LeuB requires modification to improve activity towards longer substrates

Dow AgroSciences | 22

Active Site Model of LeuB (Homology Modeling)

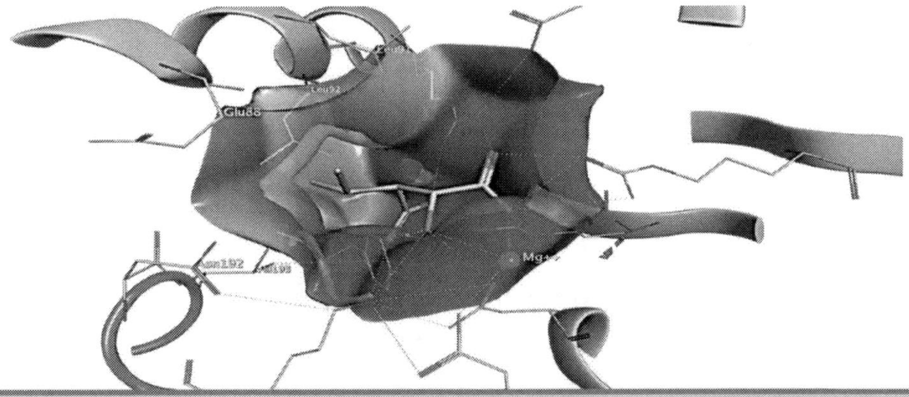

29 LeuB variants designed by rational engineering (P. Sanghani)

LeuB variants with improved activity against 3-HM

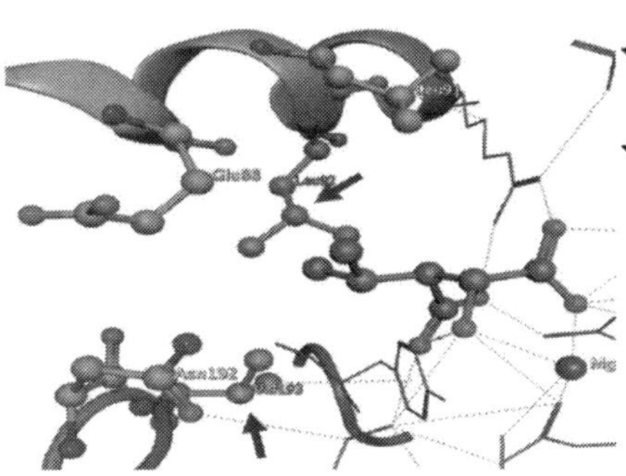

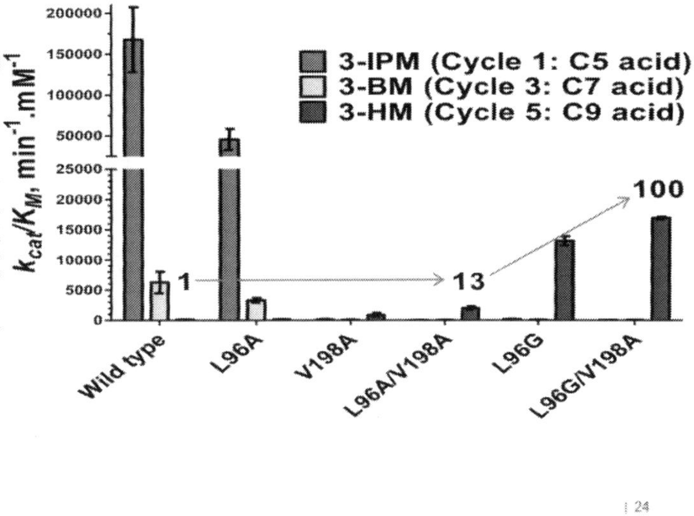

Pathway Validation in vitro Using Engineered Enzymes

- LeuB and decarboxylase enzymes engineered to improve selectivity to long chain length alcohols
- In vitro pathway evaluation methods developed
- Pathway performance evaluated at varying enzyme concentrations

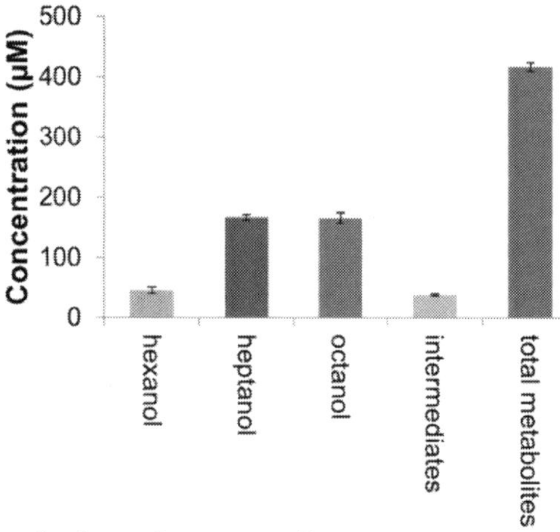

2-KB + Acetyl CoA + NADPH → Alcohols + Intermediates

Liao Results w/ the LeuACBD "+1 Pathway"

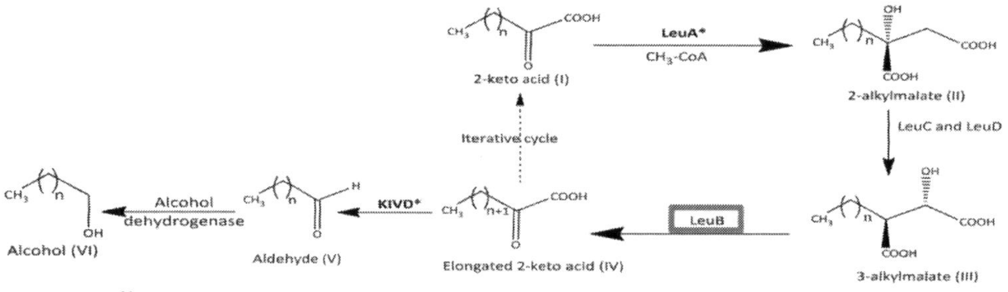

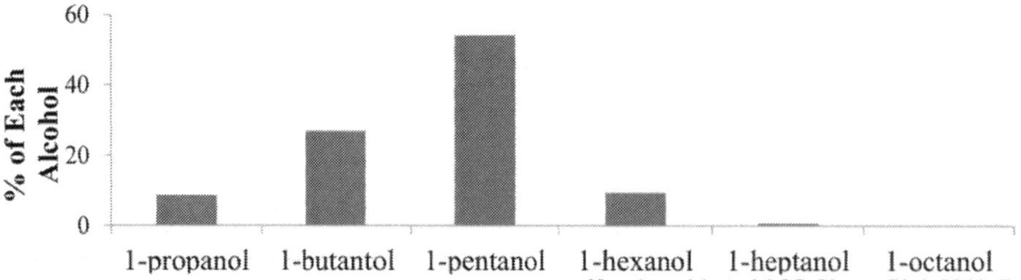

Marcheschi et al ACS Chem. Biol. 2012, 7, 689−697

Metabolomic Characterization of Pathway to Identify Pathway Bottleneck

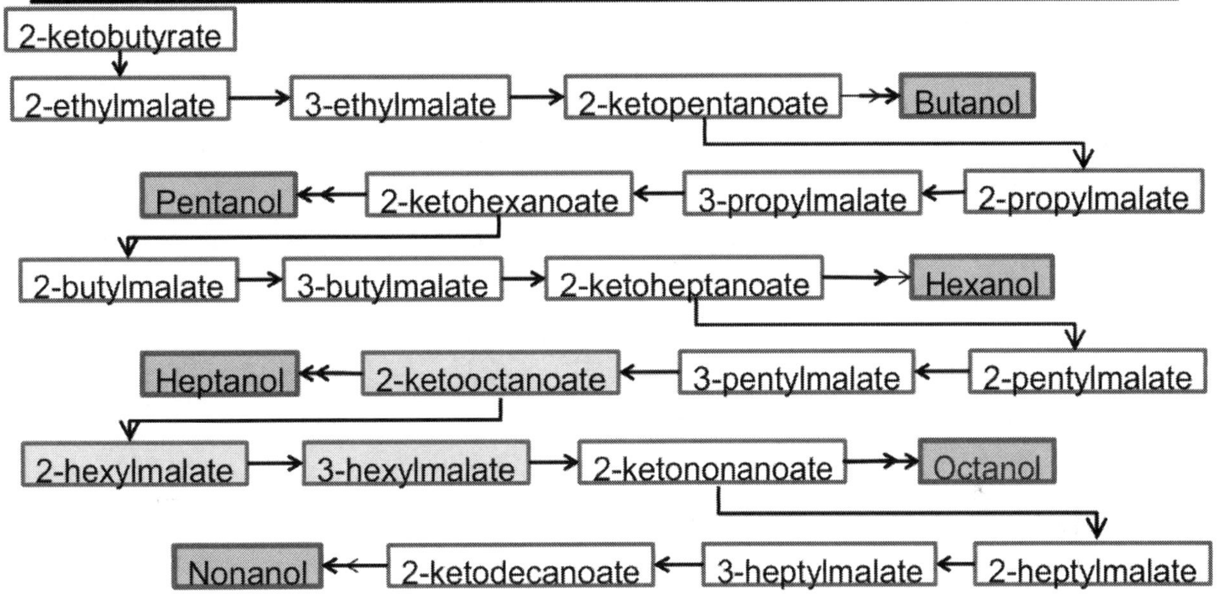

Dow AgroSciences

493164 Establishing a Platform *Escherichia coli* Strain to Generate Xylose-Derived Value-Added Products

Abstract:

Xylose is the most abundant C5 sugar in lignocellulosic biomass and also represents a source of carbon from non-edible feedstocks. Here, we report the construction of a platform *Escherichia coli* strain for the production of high value compounds via the nonphosphorylative xylose metabolism. Firstly, a 3,4-dihydroxybutanal over-producing pathway was constructed by efficient enzymes screening and host strain engineering. Then two 3,4-dihydroxybutanal dehydrogenases were identified to efficiently convert 3,4-dihydroxybutanal into 3,4-dihydroxybutyric acid (3,4-DHBA), which is the hydrolyzed form of 3-Hydroxy-Î^3-butyrolactone (3HBL). 3HBL is one of the top value-added building block for synthesis of various drugs and nutraceuticals. This novel 3,4-DHBA biosynthetic pathway produced 1.27 g/L of 3,4-DHBA in shake flasks, which is the highest titer reported so far. The application of this platform was further demonstrated by building an artificial pathway for biosynthesis of 1,4-butanediol (1,4-BDO). The 3,4-dihydroxybutanal over-producing strain was transformed into a 1,2,4-butanetriol (1,2,4-BTO) over-producing strain by over-expression of endogenous alcohol dehydrogenase, which enabled 1.5 g/L 1,2,4-BTO produced from xylose. Then the *Klebsiella oxytoca* diol dehydratase was engineered to achieve non-native catalysis of 1,2,4-BTO into 1,4-BDO by a series of rational protein engineering strategies. Those efforts enabled 209 mg/L 1,4-BDO produced via a novel metabolic route from xylose. This work demonstrates great potential for large-scale production of 3,4-DHBA and sets an example to build novel biosynthetic pathways via rational protein engineering.

500133 Understanding and Enhancing Bioconversion of Acetic Acid to Biodiesel Using *Yarrowia lipolytica*

Abstract:

Acetate is an inexpensive and renewable carbon source that can be generated from various sources such as pyrolysis of lignocellulosic biomass and anaerobic digestion of fermentable wastes. The oleaginous yeast *Yarrowia lipolytica* is a promising biocatalyst that can utilize acetate and convert it into triacylglycerides (TAGs), a precursor for green diesel. However, currently there is limited knowledge on the metabolism of *Y. lipolytica* when cultured on acetate. To develop a better understanding, we used sodium acetate as the sole carbon source and performed ^{13}C-MFA on two different strains as well as two fermentation phasesâ€"growth (nitrogen present) and lipid production (nitrogen depleted and cells no longer synthesize biomass). The resolved flux distributions demonstrated that the glyoxylate shunt pathway is constantly active in both strains and phases. Furthermore, the cells were able to divert an adequate amount of glyoxylate shunt flux to gluconeogenesis through the PEP-pyruvate node. In this way, the oxidative pentose phosphate pathway remained active even when the cells no longer divide to generate sufficient lipogenic NADPH for TAG accumulation. Based on these conclusions, we also demonstrated that the enhancement of acetate utilization and conversion into TAGs can be achieved by providing the cells with another substrate to increase flux through upper glycolysis.

504091 Metabolic Engineering of Bacteria for Production of Oleochemicals

Abstract:

Finding a sustainable alternative for todayâ€™s petrochemical industry is a major challenge facing chemical engineers and society at large. To be sustainable, routes for converting carbon dioxide and light into organic compounds for use as both fuels and chemical building blocks must be identified, understood, and engineered. Advances in metabolic engineering, synthetic biology, and other bioengineering disciplines have expanded the scope of what can be produced in a living organism. As in other engineering disciplines, synthetic biologists want to apply a general understanding of science (e.g. microbiology and biochemistry) to construct complex systems from well-characterized parts (e.g. DNA and protein). Once novel synthetic biological systems (e.g. enzymes for biofuel synthesis) are constructed, they must be engineered to function inside evolving cells without negatively impacting the hostâ€™s physiology.

In this talk, I will describe pathways for producing high-value commodity chemicals derived from fatty-acids and how my group and others have combined synthetic biology and systems biology to improve oleochemical production in bacteria using sustainable feedstocks. The talk will describe the critical regulatory points in native fatty acid metabolism, strategies for deregulating the pathway, and alternatives that by-pass it altogether. I will highlight the use of heterologous plant and bacterial enzymes to alter the chain length distribution of products from common long-chain molecules to higher-value medium-chain analogs. I will also describe our collaborative efforts with the Maranas lab at Penn State to use computational tools to redesign the substrate specificity of enzymes used to produce medium chain oleochemicals. Here, we have used the IPRO software to identify a small (~60) library of thioesterase mutants that resulted in a new variant that produces 2-fold more fatty acid with over 50% C_8 chain lengths, molecules not produced by the native enzyme.

496260 Addressing the Downstream Challenges of Viral Therapy for Cancer Treatment – a Study of Rhabdoviral Vector Purification

Abstract:

Oncolytic viruses (OVs) are a class of cancer therapies that are currently going through clinical trials on their way to full approval. For example, the world's first clinical trial using a combination of two viruses to kill cancer cells and stimulate an immune response is currently being conducted. Existing downstream purification processes use a combination of chromatography and membrane-based processes to remove both process-related (e.g. cell debris, host cell proteins, and nucleic acids) and product-related impurities (e.g. aggregated and empty virus particles). One of the main challenges during viral vector manufacturing is the mandatory final sterile filtration step. The primary issue is the severe membrane fouling, as indicated by a dramatic and uncontrollable transmembrane pressure increase associated with the low recovery of the desired final product - recent studies reporting up to 80% losses in the total vector titer. We have evaluated the fouling behaviour of commercially available sterile filters with various membrane morphologies using a Rhabdoviral vector.

As a potential improvement in the downstream process, we have investigated the use of monolithic column chromatography. This technology features a continuous stationary phase with a network of interconnected pores that has been previously shown to be ideal for the production of 'large' biomolecules. The monolithic column (1 mL capacity) made of polymethacrylate copolymers functionalized with quaternary ammonium groups (BIA Separations) was evaluated using NGC Quest Plus chromatography system (BioRad). The flow rate and buffer composition conditions used for the monolith chromatography were optimized in terms of the recovery of the Rhabdoviral vector. At a relatively high flow rate of 4 ml/min, the purified Rhabdoviral vector sample was purified using a step-elution salt concentration gradient with the majority of impurities washed out at lower salt concentrations. To investigate the fouling propensity of the different biomolecules that were collected in each chromatographic fraction, a series of small-scale (13mm polycarbonate filter holder with the filter area of 0.5 cm^2) sterile filtration studies were conducted. The resulting TMP profiles for each fraction (at constant flux) were analyzed using various pore plugging/blocking models. Our current hypothesis is that by increasing removal of the process- and product-related impurities using the monolithic column, we can improve the sterile filtration process performance in terms of throughput and product recovery that is needed for the large-scale production of oncolytic viruses.

497681 A Small Molecule Based Affinity Membrane Spin Column for Antibody Purification

Abstract:

Antibodies have extraordinary specificity and affinity to antigens, which in turn makes them important candidates to be used in numerous applications including clinical, academic and industrial, while new implementations are continuously being explored in academic research. A major contributor to the cost of antibody applications is mostly linked to the downstream production process, or more specifically the usage of Protein A (or G) affinity columns for purification of antibodies. These columns are expensive and have short life cycles with several obstacles that preventing them to be used repeatedly. Here, we design a novel affinity spin column for the purification of monoclonal and polyclonal antibodies from complex protein mixtures such as cell culture media of hybridomas and ascites fluids. This method utilizes the nucleotide-binding site (NBS) that is located on the Fab variable domain of immunoglobulins to enable capturing of antibody molecules on a membrane affinity column via a small molecule, which has a moderate binding affinity to the NBS. A ring structured small molecule was attached to the regenerated cellulose membrane in order to target the NBS for capturing antibodies on the spin column. Antibody capture was accomplished by running of samples in an equilibration buffer (20 mM TRIS pH 7.0) through the spin column at 500 rpm and elution of the antibody was achieved by running a mild elution buffer (0.5 M KCl in 20 mM TRIS pH 7.0). Several pharmaceutical antibodies were tested on the spin column, and the results indicate that the column efficiency for antibodies was >95%, with a purity level of >95%. Furthermore, small molecule based affinity column has a strong binding specificity towards antibodies, and has a dynamic binding capacity of ~48 μg/mg which is greater than any current protein A (or G) spin columns. The quality and the capacity of this small molecule affinity membrane spin column method is further evaluated for a number of parameters such as: injection concentrations, volumes, number of membranes, antibody/protein-contaminant combinations, and column reusability and stability. This study provides a good tool for fast antibody purification at a small scale.

486714 Removal of Cyanobacterial Toxins Using Polymeric Nanoparticles

Abstract:

Communities are faced with significant threats to health by the supplied water from lakes or reservoirs that develop algal bloom. During a bloom, cyanobacteria release the bacterial cell wall component endotoxin, and other toxins into the water, which enter the bloodstream upon ingestion causing an overwhelming inflammatory response in humans. Water treatment plants currently use a series of standard cleansing procedures to diminish cyanotoxins, however, finding an effective and affordable technique remains a challenge. Here, we propose a new approach based on biocompatible polymeric NPs in particulate, and membrane forms with a previously proven capability to achieve improved effectiveness and efficiency in removing cyanotoxins from water. Removal of cyanotoxins were assessed using nanometer-sized poly-e-caprolactone (PCL) NPs with specifically different diameters and shapes as synthesized in the PIâ€™s laboratory. Membranes were constructed incorporating PCL NPs of various diameters and shapes to create a corrugated surface for effective cyanotoxin removal from contaminated water.

486597 Increased Production of Yersiniabactin (Ybt) and the Anthranilate Analog through Media Optimization and Ybt Metal BindingÂ Characterization

Abstract:

Yersiniabactin (Ybt) is a mixed nonribosomal peptide-polyketide natural product that binds a wide range of metals with the potential to impact processes requiring metal retrieval and removal. The heterologous production of Ybt and an associated anthranilate analog are increased through systematic screening and optimization of culture medium components. Specifically, a Plackett-Burman design-of-experiments methodology was used to screen 22 components and to determine those contributing most to siderophore production. L-cysteine, L-serine, glucose, and casamino acids significantly contributed to the production of both compounds. Ybt and the anthranilate analog titers were increased to 867 ± 121 mg/L and 16.6 ± 0.3 mg/L, respectively, an increase of ~3,800% and ~7,900% relative to production in M9 medium.

The range of metals that Ybt forms a complex with were determined by using mass spectrometry. Using LC-MS the analysis of water samples exposed to aqueous and surface-localized Ybt, quantitative assessment of binding was completed with metals that included palladium, magnesium, and zinc. In total, Ybt showed affinity for 10 metals. Next, Ybt-modified XAD-16N resin (Ybt-XAD) was utilized to quantify the degree of metal binding and metal removal abilities, showing a rank order of Fe>Ga>Ni>Cu>Crâ‰ˆZn>Co>Pd>Mg>Al. Batch tests were conducted using Ybt-XAD for the treatment of wastewater from a local precious metal plating company, showing selective removal of nickel from the aqueous effluent.

499390 Applications of Magnetically Responsive Micro/Mesoporous Carbon Adsorbents in Biorefining

Abstract:

Inhibitory compounds commonly arise at multiple points in the biorefining pipeline, including with respect to both toxic byproducts in the feedstock as well as the final target product itself. In both cases, if not effectively removed, their presence can have deleterious effects on the microbial biocatalyst and, in turn, its overall performance. Here we demonstrate how a series of novel, magnetically responsive micro/mesoporous carbon powder (MMCP) adsorbents can be employed for the facile and efficient removal of inhibitory bioproducts and byproducts alike. In addition to large surface areas (>1400 m^2/g) promoting high adsorption capacities, said MMCPs also uniquely include magnetic Ni nanoparticles that enable the direct and facile magnetic retrieval of spent adsorbents from the culture broth, bypassing the need for column configurations (e.g., packed or expanded bed). As a preliminary demonstration of concept, the use of MMCPs for *in situ* removal of n-butanol has been studied under model conditions. Although it represents a promising biofuel, n-butanol production by conventional batch fermentation is limited as a result of its cytotoxic effects. Butanol loading capacities of a family of MMCPs (with 4-10%wt. Ni content) were found to not be significantly impacted by Ni content and, furthermore, performed comparably to other, previously-studied commercial resins and activated carbons. Magnetic retrieval of the mesoporous powder is dependent on the Ni content, with up to 89% retrieval achieved in 6 min with 10%wt. Ni. Desorption studies using retrieved adsorbents demonstrated an average of 93% recovery of the total adsorbed butanol. Kinetic studies, meanwhile, indicate that neither the effective adsorption or desorption rates should constitute a bottleneck with respect to the future development of semi-continuous butanol fermentation processes using MMCPs. Finally, the further utility of MMCPs for the upstream removal of toxic byproducts derived from the pretreatment of lignocellulose has been investigated. Furfural, for example, can be readily removed using MMCPs that are subsequently retrieved from the culture. Said pretreatment steps improve the fermentability of biomass-derived sugar mixtures, supporting improved production of ethanol and lactate by engineered *Escherichia coli*.

501387 Improving Enzymatic Synthesis of β-Lactam Antibiotics By in-Situ Crystallization

Abstract:

Improving enzymatic synthesis of β-lactam antibiotics by in situ crystallization

Matthew A. McDonald*, Lukas Bromig*, Andreas S. Bommarius*, Martha A. Grover*, and Ronald W. Rousseau*

**School of Chemical & Biomolecular Engineering, Georgia Institute of Technology, Atlanta, Georgia*

Keywords: reactive crystallization, β-lactam antibiotics, penicillin G acylase

In 2010 the two most common classes of β-lactam antibiotics, the penicillins and cephalosporins, accounted for nearly 60% of world antibiotic consumption, with delivery of over 20 billion doses of penicillin-derived antibiotics.[1] Traditionally, β-lactam antibiotics have been made batch-wise in a process requiring several chemically intensive steps, such as amine protection by benzyl chloroformate and deprotection by hydrogenation with palladium catalysts, resulting in large amounts of potentially contaminated waste.[2] The contaminated waste poses two threats: the spread of antibiotic resistance in wastewater treatment plants[3] and inadvertent exposure of sensitive individuals to a sever allergen. A 2012 study found that 92 out of 11,761 patients at Mount Sinai Hospital in New York City suffered anaphylaxis (a rapid, potentially deadly allergic reaction) from penicillin, a prevalence of 0.7%.[4]Enzymatic synthesis of β-lactam antibiotics can help to mitigate these risks by operating in mild aqueous conditions. The enzyme used industrially for this purpose is penicillin G acylase (PGA); it also catalyzes the hydrolytic degradation of the antibiotic (termed secondary hydrolysis), making the desired product an intermediate. In-situ product isolation can prevent the enzyme from degrading the antibiotic.

We use ampicillin and cephalexin as a model penicillin and cephalosporin, respectively. PGA catalyzes the condensation of 6-aminopenicillanic acid (6-APA) and phenylglycine methyl ester (PGME) into ampicillin and 7-aminodesacetoxy-cephalosporanic acid (7-ADCA) and PGME into cephalexin. Ampicillin and cephalexin have lower saturation concentrations than 6-APA and 7-ADCA, respectively, making it possible to supersaturate the solution with respect to the antibiotic. The supersaturation can then be used to crystallize pure ampicillin and cephalexin, isolating it from PGA.[5]Degradation of ampicillin and cephalexin by PGA yields phenylglycine, a sparingly soluble contaminant. PGA also hydrolyzes the highly soluble PGME into phenylglycine (termed primary hydrolysis). To maintain a pure solid phase phenylglycine must not be allowed to reach its saturation concentration. Typical crystallization processes utilize evaporation, cooling, and other chemical process to generate supersaturation. In this process, PGA generates the supersaturation necessary for the separating phase change (from solute to crystal).

To better understand the mechanism and quantitatively predict the process behavior, a model of the enzyme kinetics, including the synthesis, primary, and secondary hydrolysis reactions, was developed. Existing kinetic models for ampicillin and cephalexin synthesis by PGA do not consider the effects of pH value on the reaction despite the fact that pH is known to affect the enzyme selectivity and activity. The pH value is shown to decrease with increasing conversion because of the generation of an acid group in phenylglycine. We have improved on previous models such that our model considers the effect of pH and removal of species from solution via crystallization.[6]

For ampicillin the new process showed maximum improvements in yield and selectivity of 33% and 49% over the non-crystallizing control, respectively, all while maintaining a strict purity constraint. However, these two maxima cannot be simultaneously achieved; instead selectivity is optimized at a lower pH value around 6 where the enzyme is suspected to bind 6-APA more strongly, and yield improves at a higher pH value around 7 where catalytic activity is improved but less specific.[6]Preliminary results show cephalexin enjoys similar benefits from reactive crystallization.

More recently we have explored continuous reactive crystallization to synthesize β-lactams because it may enable greater efficiency and access to different chemistries. To operate a continuous process the stability and longevity of the enzyme must be characterized. We have determined the temperature activity profile and the empirical parameter T^{60}_{50}, the temperature at which 50% of the initial activity degrades within one hour, of an industrially relevant PGA, Assemblase® from DSM-Sinochem. . Temperature ramp studies were also used to determine catalyst longevity (in the form of total turnover number, TTN). Continuous operation could increase conversion over batch by tailoring the conditions to optimize reactive crystallization; for example, a plug flow reactor/crystallizer can operate with excess 6-APA or 7-ADCA relative to PGME to hinder hydrolysis of the PGME while a recycle stream could act to seed the plug flow apparatus to slow antibiotic hydrolysis via crystallization. The aforementioned model is combined with enzyme stability data to map the large design space opened by continuous operation.

References

[1] Gelbrand, H., Miller-Petrie, M., Pant, S., Gandra, S., Levinson, J., Barter, D., White, A., Laxminarayan, R., Ganguly, N. K., and Kariuki, S. (2015) The State of the World's Antibiotics 2015, *Wound Healing Southern Africa 8*, 30-34.

[2] Giordano, R. C., Ribeiro, M. P., and Giordano, R. L. (2006) Kinetics of β-lactam antibiotics synthesis by penicillin G acylase (PGA) from the viewpoint of the industrial enzymatic reactor optimization, *Biotechnology advances 24*, 27-41.

[3] Deschamps, E., Vasconcelos, O., Lange, L., Donnici, C. L., Silva, M. C. d., and Sales, J. A. (2012) Management of effluents and waste from pharmaceutical industry in Minas Gerais, Brazil, *Brazilian Journal of Pharmaceutical Sciences 48*, 727-736.

[4] Albin, S., and Agarwal, S. Prevalence and characteristics of reported penicillin allergy in an urban outpatient adult population, 6 ed., pp 489-494, OceanSide Publications, Inc.

[5] Encarnación-Gómez, L. G., Bommarius, A. S., and Rousseau, R. W. (2016) Reactive crystallization of β-lactam antibiotics: strategies to enhance productivity and purity of ampicillin, *Reaction Chemistry & Engineering 1*, 321-329.

[6] McDonald, M. A., Bommarius, A. S., and Rousseau, R. W. (2017) Enzymatic reactive crystallization for improving ampicillin synthesis, *Chemical Engineering Science 165*, 81-88.

490473 Separation of Fatty Acids and Fatty Acid Esters Using Nanoporous Polymeric Membranes

Abstract:

Membrane separations are typically inexpensive and can be scaled to industrial quantities, but the development of membranes that can separate organic chemicals lag behind those that separate gasses and water. Part of the challenge in this field is that the polymeric membranes swell in the presence of organic chemicals and the swelling impacts their abilities to separate chemicals. Despite these challenges, the potential applications of membranes that separate organic liquids and solids is impressive. For instance, worldwide over 200M tons of vegetable, tall, and fish oils are isolated each year, and their fatty acids are biorenewable, safe feedstocks for the chemical community. Unfortunately, the purification of fatty acids using traditional methods such as distillation, selective precipitation, and chromatography are energy intensive or not readily scaled to separate unsaturated fatty acids. Commercial applications of fatty acids requires starting with a mixture of fatty acids that can vary based on which season the crops were grown and the plant source for the oils.

We recently developed a series of membranes based on polyepoxies that can separate some of the key fatty acids found in vegetable and fish oils. These membranes were synthesized by the addition of diamines and di or triepoxides that react to yield highly cross-linked membranes. The selectivity of the membranes can be tuned by the selection of the epoxide and diamines used in their synthesis. These membranes were able to separate saturated and unsaturated fatty acids.

499012 A Microfluidic Model of Hemostasis Sensitive to Platelet Function and Coagulation

Abstract:

Background

Hemostasis is the physiological process by which a vascular injury is sealed by the formation of a blood clot, or thrombus, while maintaining vessel patency. Several in vitro and in vivo models of intravascular thrombus formation have been developed. These models are relevant to thrombotic events such as those following atherosclerotic plaque rupture. However, relatively few models of extravascular thrombus formation exist, and these are relevant to bleeding diatheses and trauma.

Objective

The objective of our study is to develop an in vitro model of extravascular thrombus formation. The model should show the dependence of hemostasis on platelet function and coagulation in order to plug a hole in a model vessel under physiologic pressure gradients.

Methods

Microfluidic Device Design and Fabrication

The master template for the device was prepared with photoresists (KMPR 1010 and KMPR 1050) to define a two-layer device with heights of 20 µm and 50 µm. Polydimethylsiloxane (PDMS) was molded off of these masters and covalently bonded to glass using standard soft lithography procedures. The device was designed in the shape of the letter 'H' where the outer two vertical channels represent the vascular and extravascular compartments (10 mm long X 100 µm wide X 50 µm high), respectively. The vertical channels are connected by a horizontal channel (150 µm long X 50 µm wide X 20 µm high) representing a hole in the vessel wall, which we refer to hereafter as the injury channel. Type I fibrillar collagen, recombinant human tissue factor (TF), or a combination of the two were adsorbed to the walls of the injury channel. Bovine serum albumin (2%, BSA) in HEPES buffered saline (HBS), collagen, TF or collagen-TF solutions were allowed to adsorb on the walls of the injury channel at 4 °C for 12 h.

Blood Collection and Preparation

Blood was collected from healthy donors by venipuncture into vacutainer tubes containing 3.2% sodium citrate. Platelet counts and hematocrit were measured and recorded for each donor. 960 µL aliquots of blood collected in sodium citrate was combined in tubes with 40 µL of Alexa-555 labeled fibrinogen (final concentration 56 µg/mL) which was added to visualize fibrin deposition. Platelets were labeled with the lipophilic dye DiOC6 (1 µM final concentration). Labeled platelets were incubated at 37 °C in the presence or absence of 12 nM anti-FVIII antibody or 100 µM the P2Y12 antagonist 2-MeSAMP for 15 minutes prior to the assay.

Device Operation

Pressure was controlled for the blood, recalcification buffer, and wash buffer (3.2% sodium citrate in HBS) independently by applying a pressure to the headspace of their respective reservoirs using a pressure-based flow controller. Blood was recalcified in the ratio of 9:1 (citrated whole blood:recalcification buffer) using a herringbone mixer. The output from the herringbone mixer was connected to the blood channel of the extravascular injury device. The pressures of the blood and recalcification reservoirs were held constant at 10 kPa. The wash buffer reservoir was connected to the extravascular channel by. The pressure in the wash buffer was initially set to 3.5 kPa to drive a small amount of wash buffer through the injury channel while the blood channel was filled with recalcified blood driven by 10 kPa of pressure in the blood and recalcification reservoirs. The pressure in the wash reservoir was then reduced to 1.75 kPa so that the blood passes from the blood channel into the injury channel and out into the extravascular channel. Thrombus formation was monitored through an inverted microscope equipped with a 16 bit CCD camera using relief contrast and epifluorescence microscopy with a 20X objective for up to 45 minutes. Time to closure was defined as the first time that no red blood cells pass through the injury channel for 5 seconds.

Results

Type I collagen and TF and their effect on the time to closure of the injury channel are considered. Channels coated with BSA did not support platelet adhesion or significant fibrin deposition. This suggests that velocity gradients at the entrance and exit of the injury channel do not promote thrombus formation in the absence of agonist. Type I collagen alone supports platelet adhesion and fibrin accumulation in the near-wall-region underneath platelet aggregates. The initial thrombus formation is dictated by the position of collagen fibers in the injury channel and not by fluid dynamics in this model. There was no consistent location for initial platelet adhesion in the channel for collagen or collagen-TF surfaces. However, full closure is not achieved over 45 minutes, as the thrombus approaches closure, platelet aggregates embolize. TF alone leads to initial fibrin formation in the corners of the injury channel followed by platelet adhesion to the deposited fibrin. The time to closure of the horizontal channel is 15.8 ± 2.0 minutes. Type I collagen and TF together support platelet accumulation and fibrin formation with a time to occlusion of 7.5 ± 1.6 minutes. On the combined collagen-TF surface, platelets first adhere and aggregate to form a plug in the injury channel that stops the loss of blood into the extravascular compartment. Fibrin was observed within the dense platelet plug, particularly in the near-wall region. Following cessation of flow, fibrin rapidly polymerizes across the entire horizontal channel. These acquired results demonstrate that collagen and TF act synergistically to form a stable thrombus. Moreover, both primary hemostasis (formation of a platelet plug) and secondary hemostasis (stabilization by fibrin) are necessary.

Normal whole blood treated with a function blocking anti-human FVIII antibody did not form stable thrombi. Initial platelet accumulation was similar to that for the collagen-TF surface with control blood. However, as the thrombus approached closure large aggregates of platelets were shed. This process was repeated several times over the duration of the experiment. After 30 minutes there was small increase in platelet accumulation and blood continued to leak through the channel for up to 45 minutes. Fibrin accumulation was significantly reduced compared to controls. Moreover, fibrin accumulation was limited to the near-wall region where large platelet aggregates were attached likely protecting coagulation products from dilution by flow.

Normal whole blood treated with the P2Y12 antagonist 2-MeSAMP resulted in a prolonged time to closure (15.2 ± 3.4 minutes) compared to control blood on collagen-TF surfaces. The platelet plug was larger in area than in controls, typically filling the entire horizontal channel, and took longer to accumulate than for the untreated blood. Thrombi appeared less compact and contained less fibrin, potentially due to reduced platelet activation, and were less mechanically stable than in control blood as indicated by the displacement of large platelet aggregates in the direction of flow. Importantly, following closure as defined the arrest of red blood cells moving across the injury channel, we continued to observe leakage of plasma indicating a porous thrombus. No such leakage was observed in controls.

Conclusions

Hemostasis is a function of blood flow, platelet function, and coagulation. The described microfluidic model of hemostasis depends on all three of these phenomena. A thrombus forms in a small side channel to stop blood loss from one channel into another in a process that better represent hemostasis than similar models of intravascular thrombus formation. The presented geometry provides two unique features compared to conventional flow models. First, there are two paths for blood flow, namely; through the injury channel or the blood channel. Both are controlled independently. When the injury channel occludes, the flow rate through the blood channel remains unchanged. Moreover, plasma proteins and blood cells are continuously delivered to the periphery of the thrombus. Second the normal force is smaller in an extravascular injury compared intravascular injury for comparable pressure gradients and channel geometries. The sizes of the channels are similar to the microvasculature in humans. Most bleeds that occur in genetic bleeding disorders such as hemophilia, von Willebrand disease, or platelet disorders take place in the microvasculature. The wall shear rate in the blood channel and the pressure drop across the injury channel are comparable to those found in the arterioles. Normal whole blood in a collagen-TF coated injury channel recreates primary and secondary hemostasis with the initial formation of a platelet plug that is then stabilized with fibrin. The time to closure of the injury depends on both FVIII and autocrine signaling through the ADP receptor P2Y12, with deficits in either resulting in prolonged closure times and unstable or leaky thrombi. We described an experimental platform for studying hemostasis with a focus on collagen-TF induced thrombus formation within a model vascular wall. In the current configuration, the platform could be extended to study the effect of extravascular coagulation and the geometry of the extravascular space.

References

R. M. Schoeman, K. Rana, N. Danes, et al.. A Microfluidic Model of Hemostasis Sensitive to Platelet Function and Coagulation. *Cellular and Molecular Bioengineering.* 10:3-15, 2017. DOI: 10.1007/s12195-016-0469-0

M. Lehmann, A. M. Wallbank, K. A. Dennis, A. R. Wufsus, K. M. Davis, K. Rana, and K. B. Neeves. On-chip recalcification of citrated whole blood using a microfluidic herringbone mixer. *Biomicrofluidics.* 9, 064106:1-13, 2015. DOI: 10.1063/1.4935863

497169 Microfluidic Platforms for Aging Studies

Abstract:

Aging is becoming a major challenge in todayâ€™s society. Lifespan has been dramatically increased; yet serious age-associated diseases significantly impact the quality of life at older age. Better understanding the aging process should shed light on disease prevention and on relevant environmental factors that could significantly improve quality of life at older age. Aging studies, however, pose significant technical challenges. These require long experimental length-scales, while acquiring longitudinal information is typically unfeasible.To address this problem, we have developed microfluidic platforms to perform aging studies in *C. elegans*.

The multicellular organism *C. elegans* has been widely studied as a model for multiple fundamental biological processes, ranging from gene silencing to apoptosis, metabolism, and neurogenesis. In particular, *C. elegans* has had a large impact in aging research. We have developed microfluidic platforms tailored for studying aging of the nervous system. This platform allows for lifelong longitudinal tracking of the same *C. elegans* population. Dedicated compartments allow lifelong culture and high-resolution live imaging of cellular (i.e., neurons) and subcellular (i.e., synapses) features. Importantly, this platform enables performing aging studies without traditionally used drugs that prevent reproduction. We are also developing microfluidic arrays to perform genetic screens on late-onset phenotypes. These arrays enable lifelong tracking of individual animals, while collecting progeny and thus allow performing forward genetic screens on late-onset (post-reproductive) phenotypes. These systems are part of a toolbox for quantitative aging research in *C. elegans* and should allow better understanding thus far elusive questions in longevity and age-associated disease.

496402 Microfluidic Platform to Decode Mechanotransduction Mechanisms in Developing Organs

Abstract:

Mechanical forces are critical but poorly understood inputs into organ development and wound healing. Calcium ions are critical second messengers in cells for integrating environmental and mechanical cues, but the regulation of calcium signaling is poorly understood in developing epithelial tissues. Here we report a chip-based regulated environment for micro-organs (REM-Chip) that enables systematic investigations of the crosstalk between an organâ€™s mechanical stress environment and biochemical signaling under genetic and chemical perturbations with concurrent live imaging. This method enabled us to define the essential conditions for generating organ-scale intercellular calcium waves in *Drosophila* wing discs that are also observed *in vivo* during organ development. We discovered that mechanically induced intercellular Ca^{2+} waves require fly extract growth serum as a chemical stimulus. Using the REM-Chip, we demonstrate the release of mechanical compression rather than the initial application is sufficient to initiate intercellular Ca^{2+} waves. The Ca^{2+} response depends on the pre-stress intercellular Ca^{2+} activity and not on the magnitude or duration of mechanical stimulation applied. Mechanically induced intercellular calcium waves rely on IP_3R mediated calcium induced calcium release and propagation through gap junctions. Thus, intercellular calcium waves in developing epithelial tissues may be a consequence of stress dissipation observed during organ growth and wound healing.

493503 3D Biomimetic Model for Cellular Invasion: A Versatile Platform to Examine 3D Physiological Invasion in Angiogenesis and Pathological Tumor Cell Invasion

Abstract:

Cell invasion is an essential and highly regulated process. Cells invade to vascularize tissues, to form tissues and to respond to inflammation at injured sites. Unfortunately, cell invasion is also involved in numerous pathological conditions such as in invasive tumors. Much of our understanding of cell invasion has been from traditional 2D cell cultures and from cells that are randomly dispersed inside 3D hydrogels. However, such methods also have drawbacks as they often lack tissue architecture, tissue-tissue interfaces, gradient of cytokines, and fluid flow, which may all affect cell invasion [1]. Microfluidic platforms, therefore, offer a possibility to overcome some of these limitations. Microfluidic devices are often fabricated with photolithographic etching to introduce patterned structures and tissue compartments with perfused fluidic chambers. However, because of the photolithographic fabrication, fluidic chambers presented in microfluidic devices sometimes lack the architecture of a tissue compartment. For instance, a widely accepted feature of fluidic chamber is a rectangular fluidic channel with one side exposed to hydrogel, such as collagen to mimic tissue interface while the other three surfaces are either made from glass or PDMS. As a result, cells inhabited inside these channels are constrained to only interact with the microenvironment through one surface interface whereas cells in vivo such as endothelial cells in blood vessels or epithelial cells in ducts are fully surrounded by the extracellular matrices. Additionally, these artificial hard surfaces from glass or PDMS may in turn influence cellular invasion.

To address these limitations, we engineered a microfluidic platform that contain 2 perfusable hollow cylindrical channels that are fully embedded within a 3D extracellular matrix to mimic ductal compartments within a tissue [2]. Soluble factors may be added into one of the channels to generate a gradient across the interstitium matrix between the two channels. Various cell types may also be seeded into these channels to create a basic structure of a ductal compartment. For instance, we utilize this microfluidic platform to model physiological cellular invasion in angiogenesis, a process where new blood vessels form from existing blood vessels. We seeded endothelial cells into one of the channels to form a rudimentary blood vessel. When a gradient of angiogenic factors was administered into the system, endothelial cells began to extend their protrusion and invaded into the interstitium to form 3D angiogenic sprouts. These angiogenic sprouts contain features that resemble in vivo sprouts such as tip and stalk cells, formation of lumen, and formation of branches. The sprouts ultimately develop into perfusable blood vessels. To capture pathological cellular invasion, we seeded pancreatic cancer cells inside one of the channels. Surprisingly, when the pancreatic cancer duct was placed juxtaposed to the rudimentary blood vessel, the pancreatic tumor cells began to invade into the tissue towards the engineered blood vessel. Upon reaching the blood vessel, the pancreatic tumor cells migrated along the blood vessels and invaded into the blood vessel, a phenomenon commonly observed in human pancreatic cancer patients [3]. All together, we demonstrate that our microfluidic platform is a versatile and important platform that enables us to observe and capture multiple processes of cellular invasion in both angiogenesis and tumor cell migration.

[1] Galie, P.A., **Nguyen, D.H.**, Choi, C.K., Cohen, D.M., Janmey, P.A., Chen, C.S. (2014). â€œFluid shear stress threshold regulates angiogenic sproutingâ€. **Proceedings of the National Academy of Sciences**. 111(22): 7968-7973

[2] **Nguyen, D.H.**, Stapleton, S.C., Yang, M.T., Cha, S.S., Choi, C.K., Galie, P.A., Chen, C.S. (2013). *Biomimetic model to reconstitute angiogenic sprouting morphogenesis in vitro*. **Proceedings of the National Academy of Sciences**. 110(17): 6712-6717.

[3] **Nguyen, D.H.**, Esak, L., Alimperti, S., Wong, A., Eyckmans, J., Stanger, B.Z., Chen, C.S. (2017). *Pancreatic ductal adenocarcinoma replaces endothelium during tissue invasion*. **Manuscript in revision**.

500551 "Lymphatics-on-a-Chip" to Reconstitute Lymphatic Drainage Function and Lymphedema

Abstract:

Lymphatic vessels (LVs), comprised of lymphatic endothelial cells (LECs) are central for tissue fluid homeostasis, immune surveillance, and tumor metastasis [1]. Given that LVs serve as a route for tumor spreading, LV interactions with tumor cells were studied using tumor secretome analysis, biochemistry, MRI imaging, bioinformatics, and mouse models to understand paracrine loops in lymphatic tumor metastasis and to provide new therapeutic strategies [2-5]. Moreover, novel anti-lymphangiogenic peptides were developed and treated in mouse models, showing reduced metastases in the lungs and the lymph nodes [6-10]. Apart from the lymphatic functions in cancers, more general understanding of lymphatic transport of fluid, lipids, proteins, and immune cells is important in investigation of the vast majority of LV-related diseases, such as lymphedema, obesity, heart failure, autoimmune diseases, inflammation, and infections. All of these disorders are closely linked to poor lymphatic transport.

Among many lymphatic diseases, "Lymphedema (LE)", featured by abnormal tissue swelling, is the most common lymphatic disorder, influencing 150 million individuals worldwide. In physiological condition, normal LVs properly drain excess interstitial fluid that is leaked from blood capillaries; and the drained lymph fluid travels through the lymphatic system to go back to the blood circulation through the subclavian veins. These recirculating processes maintain interstitial fluid homeostasis in normal tissues. However, any failures in the LV drainage trigger abnormal fluid accumulation in the tissues, causing "swelling". Currently, mechanisms of LV drainage in normal or LE condition are poorly understood, thus therapeutic options for LE treatment are limited. There is no clinically available drug for LE, and widely practiced conservative therapies, such as massage and compression garments are palliative.

One of the major obstacles to better understanding and curing LE is a lack of appropriate experimental tools for evaluating lymphatic drainage function. Lymphatic drainage is the most crucial lymphatic function, directly linked to LE and influenced by numerous biological factors. Although animal models have contributed to major discoveries in the field, isolating and controlling biological factors that may contribute to LE in the animal models is challenging. Majorities of in vitro models that have attempted to solve these problems were two-dimensional (2D) cell culture models on a plastic dish or a transwell, which have not successfully recapitulated lymphatic cell behaviors in 3D environment as in in vivo settings. As such, detailed, mechanistic investigation of LE has remained limited. Thus, 3D biomimetic vascular models would be useful to study blood and lymphatic biology in many different contexts [11-13].

To study lymphatic drainage function and LE pathogenesis in 3D in vitro, we built a new biomimetic "human lymphatics-on-a-chip" model system by fabricating a microfluidics-based device that includes a poly-dimethyl siloxane (PDMS) housing and two parallel micro-channels within 3D collagen matrix. Employing physiological luminal flow in the micro-channels, human primary dermal lymphatic endothelial cells (LECs) seeded in the channels formed lumenized and perfusable LVs within 3D collagen, exhibiting physiologically relevant LV structure and function. For example, interstitial flow containing lymph molecules, such as albumins, fatty acids, phospholipids, and particles, demonstrated that LVs much better drain the lymph molecules than blood vessels (BVs) do. Surprisingly, we observed that LVs exhibited dilated lumens under interstitial flow, however, BVs were squeezed in the same condition. Immunostaining revealed that LVs have weak tight junctions and jagged adherens junctions with specialized portal-like structures to promote fluid transport. The dynamic LV lumens under interstitial pressure, weaker junctions, and preformed portals have been reported in mouse models and human lymphatic tissues, so we concluded that we have a functionally relevant LV model in 3D.

Based on the normal lymphatics-on-a-chip model, inflammatory cytokine treatment and Podoplanin (Pdpn) deficiency significantly impaired lymphatic drainage function by abnormally tightening LV junction. This is a notable finding because most of the inflammation conditions cause vessel leakages by loosening vascular junction. Mechanistically, we discovered that one of the integrin families is a new therapeutic target for LE. This integrin is normally inactivated, however, becomes highly activated in LE conditions, including both cytokine treated- and Pdpn deficient groups. The activated integrin dramatically tightens lymphatic junction, delays fluid, lipid, cell transport through the LVs, and ultimately impairs lymphatic drainage rate. Employing multiple inhibitors of this integrin and the integrin-related pathways, we could normalize pathological lymphatic junction, restoring almost 100% of lymphatic drainage function. Now, we are in validating these findings in mouse LE models.

In conclusion, our "lymphatics-on-a-chip" model recapitulates physiological lymphatic drainage function and LV structure in nature. Moreover, the "lymphatics-on-a-chip" provides an important in vitro platform for revealing previously unappreciated disease mechanisms of LE. Together, these data suggest that the "lymphatics-on-a-chip" may serve as a new platform for mechanistic studies and drug screening for lymphatics associated disorders.

REFERENCES (*Equal contribution)

[1] **Lee E**, Pandey NB, Popel AS, "Crosstalk between cancer cells and blood endothelial and lymphatic endothelial cells in tumour and organ microenvironment", **Expert Rev Mol Med** 17:e3, **2015**

[2] **Lee E**, Pandey NB, Popel AS, "Lymphatic endothelial cells support tumor growth in breast cancer", **Sci Rep**, 4:5853, **2014**

[3] **Lee E**, Fertig EJ, Jin K, Sukumar S, Pandey NB, Popel AS, "Breast cancer cells condition lymphatic endothelial cells within pre-metastatic niches to promote metastasis", **Nat Commun,** 5:4715, **2014**

[4] **Lee E**, Pandey NB, Popel AS, "Pre-treatment of mice with tumor-conditioned media accelerates metastasis to lymph nodes and lungs: a new spontaneous breast cancer metastasis model", **Clin Exp Metastasis,** 31(1):67-79, **2014**

[5] Fertig EJ*, **Lee E***, Pandey NB, Popel AS, "Analysis of gene expression of secreted factors associated with breast cancer metastases in breast cancer subtypes", **Sci Rep,** 5:12133, **2015**

[6] **Lee E*,** Rosca EV*, Pandey NB, Popel AS, "Small peptides derived from somatotropin domain-containing proteins inhibit blood and lymphatic endothelial cell proliferation, migration, adhesion and tube formation", **Int J Biochem Cell Biol,** 43:1812-21, **2011**

[7] **Lee E*,** Kim YS*, Bae SM, Kim SK, Jin S, Chung SW, Lee M, Jeon OC, Park RW, Kim IS, Byun Y, Kim SY, "Polyproline-type helical-structured low-molecular weight heparin (LMWH)-taurocholate conjugate as a new angiogenesis inhibitor", **Int J Cancer**, 124: 2755-65, **2009**

[8] **Lee E,** Koskimaki JE, Pandey NB, Popel AS, "Inhibition of lymphangiogenesis and angiogenesis in breast tumor xenografts and lymph nodes by a peptide derived from transmembrane protein 45A", **Neoplasia,** 15(2):112-24, **2013**

[9] Koskimaki JE*, **Lee E*,** Chen W, Rivera CG, Rosca EV, Pandey NB, Popel AS, "Synergy between a collagen IV mimetic peptide and a somatotropin-domain containing peptide as angiogenesis and

lymphangiogenesis inhibitors", **Angiogenesis,** 16(1):159-70, **2013**

[10] **Lee E**, Lee SJ, Koskimaki JE, Han Z, Pandey NB, Popel AS, "Inhibition of breast cancer growth and metastasis by a biomimetic peptide", **Sci Rep**, 4:7139, **2014**

[11] Choi, D., Park, E., Jung, E., Seong, Y.J., Yoo, J., **Lee, E.**, Hong, M., Lee, S., Ishida, H., Burford, J., Peti-Peterdi, J., Adams, R.H., Srikanth, S., Gwack, Y., Chen, C.S., Vogel, H.J., Koh, C., Wong, A., Hong, Y.K., "Laminar flow downregulates Notch activity to promote lymphatic sprouting", **J Clin Invest,** doi:10.1172/JCI87442. **2017**

[12] **Lee E**, Song HHG, Chen CS, "Biomimetic on-a-chip platforms for studying cancer metastasis", **Curr Opin Chem Eng**, 11:20-7, **2016**

[13] Nguyen DHT*, **Lee E***, Alimperti SA, Wong A, Eyckmans J, Stanger BZ, Chen CS, "Pancreatic ductal adenocarcinoma replaces endothelium during tissue invasion", In revision **2017**

ACKNOWLEDGEMENTS

This work was supported in part by grants from the National Institutes of Health (EB00262, UH3EB017103, UC4DK104196). Esak Lee acknowledges financial support from LE&RN postdoctoral grant from Lymphatic Education and Research Network (LE&RN), and BU-CTSI grant (TL1TR001410) from the National Center for Advancing Translational Sciences at the National Institutes of Health.

496496 An Optically Controlled Microphysiological System for the Heart-Brain Axis

Abstract:

An Optically Controlled Microphysiological System for the Heart-Brain Axis

Jonathan Soucy[1], Tess Torregrosa[1], Abigail Koppes[1], Nasim Annabi[1,2], Ryan Koppes[1]

[1]Chemical Engineering, Northeastern University, Boston, MA

[2]Harvard-MIT Division of Health Sciences and Technology, Massachusetts Institute of Technology, Cambridge, MA

Introduction:

While, there are more than 700,000 people in the United States suffer a heart attack annually, the underlying cellular mechanisms that lead to cardiac dysfunction remains poorly understood[1]. Previously, animal models have been used to investigate cardiovascular disease, but because of their inherent complexities, variability, and failure to translate to human physiology, *in vitro* alternatives must be developed[2]. Towards this goal, we aim to recapitulate the human cardiac sympathovagal balance *in vitro* to study the underlying cellular changes of cardiac tissue in response to ANS modulation.

In vivo, the heart responds to environmental changes by controlling cardiac output via the ANS. In the case of deteriorating cardiac function, the sympathetic nervous system (SNS) will innervate the cardiac tissue, while parasympathetic nervous system (PSNS) will withdraw to maintain resting heart rate[3]. However, as the branches of the ANS work to restore cardiac output, the overall balance of neural input leans towards the sympathetic. This understanding of the cardiac ANS pathophysiology has led to development of beta-blocker therapies to inhibit SNS activity. Conversely, the role of the PSNS in heart failure not completely understood and is an area of ongoing research. Preliminary cardiac ANS studies have shown that increased cardiac PSNS activity can reduced risk of arrhythmias, potentially leading to vagal stimulation as a new therapeutic strategy[4]. Yet, mimicking this imbalance in vitro for investigating the underlying cellular mechanisms has not been demonstrated.

Towards developing innervated muscle organ systems on a chip, microfluidic devices have been used for compartmentalized co-culture of skeletal muscle cells with neurons[5], but only recently have applied to cardiac systems[2]. These neural cardiac co-culture devices can be improved upon by culturing cardiomyocytes (CM) in a physiologically relevant three-dimensional (3D) environment as well as the addition of means for real-time, bi-directional control/output of neural and cardiac function. Optogenetics allows for optical control of specific cell populations via genetic modification with light-sensitive ion channels, microbial rhodopsins[6]. In contrast to electrical stimulation, optogenetics enables temporally precise excitatory and inhibitory control of neural activity in distinct cell types expressing light-gated ion channels and pumps, such as channelrhodopsin 2 (ChR2) and halorhodopsin (NpHR)[7]. Furthermore, optogenetics is compatible with simultaneous neural recording due to the decoupling of electrical and optical signals. Independent, multiplexing modulation of multiple cell types can be accomplished with opsins possessing distinct absorption spectra such as channelrhodopsin and Chrimson[8]. Therefore, there is are advantages for utilizing an optogenetic toolbox to control SNS and PSNS activity independently without concern for confounding, non-specific stimulation or confounding influence on monitored electrical activity.

Here, we aim to design a custom chip to mimic an *in vivo* cardiac ANS and allow for optical control of neural activity. *In vitro* models provide numerous benefits over their *in vivo* counterparts by reducing variability, lowering cost, and by allowing for the inclusion of knockout transgenic cell lines used to study specific genes. Developing a functional model of the cardiac ANS will allow for an improved fundamental understanding of the relationship between the cholinergic PSNS neurons, the adrenergic SNS neurons. Additionally, by studying the cellular responses to physiological relevant changes, new therapeutic targets may be identified to treat cardiovascular disease.

Methods:

A custom microfluidic chip made of PDMS with three independent channels, separated by microposts for capillary separation of the central cardiac laden 3D hydrogel (Fig 1). This microfluidic chip will be mounted onto a 60-channel multielectrode array (MEA, Multichannel Systems) to enable both the stimulation and monitoring each cell's electrical potential.

Pure populations of CM and cardiac fibroblast (CF) from the heart, cholinergic neurons from the intrinsic cardiac ganglia, and adrenergic neurons from the superior cervical ganglia from neonatal rat pups will be isolated for the in vitro platform. Using isolated cells, we will encapsulate varying ratios of CM and CF within a photocrosslinkable gelatin based hydrogel, gelatin methacrylate (GelMA)[9] *in situ* using a prototype chip design and in micropatterned gel channels. We used an in-house photocrosslinking system with 405nm LEDs to polymerize GelMA in the presence of lithium phenyl-2,4,6-trimethylbenzoylphosphinate (LAP, BioBots). The use of this system resulted in CM viability significantly greater than systems using UV light; however, beating was only observed when both CM and CF were present within the gel.

Light sensitive neuron populations will be produced via transfection with adeno-associated viruses carrying either Chronos-GFP or ChrimsonR-dTomato (UNC Vector Core), to allow for independent optical control of SNS or PSNS populations. Transfection will be validated via positive expression of appropriate fluorescent reporter protein.

In addition to monitoring electrical activity, to enable quantification of cell beating activity, we developed a custom MATLAB code to calculate beating on a cell-by-cell basis using video microscopy. Furthermore, we developed a novel method to quantify the synchronicity of CM beating (Fig. 1B), based on an algorithm originally developed to investigate neuronal activity[10].

Results and Discussion:

We have successfully fabricated the PDMS microfluidic platform on a 64-channel microelectrode array for real-time monitoring of both neural and muscle populations. Primary populations for cardiomyocytes, fibroblasts, as well as adrenergic, and cholinergic neurons were successfully isolated from neonatal rat pups. Independent transfection of the SNS and PSNS populations was validated with robust expression of fluorescent reporter proteins in Tyrosine Hydroxylase or Acetyl Choline positive neurons. Furthermore, seeding of our platform showed spontaneous neurite ingrowth into the cardiomyocyte laden hydrogel.

Our visual method for quantifying cardiomyocyte contraction demonstrated good correlation to contraction monitored on the MEA. Our method was able to detect changes in beating and coordinated contraction induced via the addition of Isoproterenol, a beta-adrenergic agonist (Fig. 1C). Further, using our visual quantification method of cardiomyocyte beating over long-term culture, we observed that the degree of synchrony increased to a steady value after 5 days. However, for measurements from longer culture times great than 2 weeks, there were increased arrhythmias. Nevertheless, immunocytochemistry showed that CM phenotype was maintained throughout the study.

References:

1 Loffredo, F. S., Nikolova, A. P., Pancoast, J. R. & Lee, R. T. Heart failure with preserved ejection fraction: molecular pathways of the aging myocardium. *Circ. Res.* **115**, 97-107, (2014).

2 Oiwa, K. *et al.* A device for co-culturing autonomic neurons and cardiomyocytes using microfabrication techniques. *Integr Biol (Camb)* **8**, 341-348, (2016).

3 Ripplinger, C. M., Noujaim, S. F. & Linz, D. The nervous heart. *Prog. Biophys. Mol. Biol.* **120**, 199209, (2016).

4 Vanoli, E. *et al.* Vagal stimulation and prevention of sudden death in conscious dogs with a healed myocardial infarction. *Circ. Res.* **68**, 1471-1481 (1991).

5 Uzel, S. G. *et al.* Microfluidic device for the formation of optically excitable, three-dimensional, compartmentalized motor units. *Sci Adv* **2**, e1501429, (2016). 6 Deisseroth, K. Optogenetics. *Nat Methods* **8**, 26-29, (2011).

7 Boyden, E. S., Zhang, F., Bamberg, E., Nagel, G. & Deisseroth, K. Millisecond-timescale, genetically targeted optical control of neural activity. *Nat. Neurosci.* **8**, 1263-1268, (2005).

8 Klapoetke, N. C. *et al.* Independent optical excitation of distinct neural populations. *Nat Methods* **11**, 338-346, (2014).

9 Saini, H., Navaei, A., Van Putten, A. & Nikkhah, M. 3D cardiac microtissues encapsulated with the co-culture of cardiomyocytes and cardiac fibroblasts. *Adv Healthc Mater* **4**, 1961-1971, (2015).

10 Kreuz, T., Mulansky, M. & Bozanic, N. SPIKY: a graphical user interface for monitoring spike train synchrony. *J. Neurophysiol.* **113**, 3432-3445, (2015).

505030 Contracting 3D Printed Microtissues: Solid and Fluid Instabilities

Abstract:

Living cells are often dispersed in extracellular matrix (ECM) gels like collagen and Matrigel as minimal tissue models. Generally, large-scale contraction of these constructs is observed, in which the degree of contraction and compaction of the entire system correlates with cell density and ECM concentration. The freedom to perform diverse mechanical experiments on these contracting constructs is limited by the challenges of handling and supporting these delicate samples. Here, we present a method to create simple cell-ECM constructs that can be manipulated with significantly reduced experimental limitations. We 3D print mixtures of cells and ECM (collagen-I) into a 3D growth medium made from jammed microgels. With this approach, we design microtissues with controlled dimensions, composition, and material properties. We also control the elastic modulus and yield stress of the jammed microgel medium that envelops these microtissues. Similar to well-established bulk contraction assays, our 3D printed tissues contract. By contrast, the ability to create high aspect ratio objects with controlled composition and boundary conditions allows us to drive these microtissues into different regimes of physical instability. For example, a contracting tissue can be made to buckle as a whole or break up into droplets, depending on composition, size, and shape. These new instabilities may be employed in tissue engineering applications to anticipate the physical evolution of tissue constructs under the forces generated by the cells within.

504124 Adaptable Biomaterials to Maintain Stemness of Neural Progenitor Cells

Abstract:

While neural progenitor cells (NPCs) and their progeny have significant therapeutic promise, the difficulty and cost of expanding a large number of NPCs remains a significant barrier to widespread clinical use. Recently, 3D hydrogels have been proposed as *in vitro* culture platforms for the expansion of stem cell populations to overcome the space limitations of 2D culture. However, very little is known about which 3D material properties are required to maintain NPCs in an undifferentiated state for expansion. It is well-established that matrix stiffness modulates stemness in strongly adherent stem cells, including mesenchymal stem cells and muscle satellite cells, but the impact of stiffness on stemness maintenance in less contractile stem cells such as NPCs is not well known. We present data demonstrating that 3D matrix stiffness does not correlate with the maintenance of NPC stemness over a broad range of matrix mechanical properties (E~0.5-50 kPa). In contrast, matrix degradability strongly correlated with the expression of NPC stem markers and NPC proliferation in two different biomaterial systems. Our results have identified matrix remodeling as a previously unknown requirement for maintenance of NPC stemness in 3D hydrogels and suggest that adaptable biomaterials will be useful for expansion of therapeutically relevant numbers of NPCs.

495947 Radiation-Induced Changes in the Extracellular Matrix Alter the Invasiveness of Heterogeneous Tumors

Abstract:

Breast cancer metastasis remains a significant clinical problem linked to reduced patient survival. The role of the extracellular matrix (ECM) in metastasis, however, is not well understood. We hypothesize that the irradiated ECM influences tumor and immune cell behavior. In this study, we characterize the effects of radiation on heterogeneous tumor cell migration to evaluate how tumor-stromal interactions modulate metastasis after therapy. This work represents the first step toward elucidating how changes in the ECM after radiation contribute to tumor cell metastasis.

The effect of radiation on tumor-ECM changes was studied using an orthotopic breast cancer model. Nude mice were inoculated with heterogeneous mixtures of luciferase- and mCherry-labeled 4T1 or luciferase- and GFP-labeled 67NR murine breast cancer cells in the mammary fat pad (MFP) in ratios of 0:100 10:90, 50:50, 90:10, and 100:0. 4T1 cells exhibit a highly metastatic phenotype while 67NR cells are tumorigenic but not metastatic. Either primary tumors or contralateral MFPs were irradiated to a dose of 20 Gy with a 250 kVp cabinet x-ray machine when tumors were palpable. Cell migration was monitored with bioluminescence imaging 10 d after irradiation. In addition, lungs, livers, tumors, and MFPs were harvested for confocal imaging and flow cytometry to examine metastatic cell compositions. Flow cytometry was also performed on dissociated irradiated and control tissues to characterize immune cell populations. Of particular interest were CD11b+F4/80+ macrophages. Normal and irradiated MFPs were then decellularized and used as hydrogel scaffolds for studying cell-cell interactions. Scanning electron microscopy was performed to determine the structure of the control and irradiated MFPs.

The presence of 4T1 cells in the primary tumor directly enhanced the metastatic capacity of 67NR cells. With only 10% of 4T1 cells in the initial inoculated population, we observed 67NR metastases to the lung and liver. Different fluorophores allowed the visualization of metastatic cells as aggregates. Radiation of primary tumors enhanced circulating tumor cell levels, and irradiation increased tumor cell migration to normal tissues. Macrophage infiltration increased both in metastatic lesions and in tissues after irradiation. In decellularized MFP hydrogels, 67NR cells developed an invasive morphology when cultured with 4T1 cells, and this effect was enhanced in the irradiated MFP. The fibrous structure of the MFPs was also altered after irradiation.

This study establishes that tumor heterogeneity determines metastatic spread and underscores the importance of *in vivo* models that reflect the variation of naturally occurring tumors. We found that the ECM modulates tumor cell migration. The increase of macrophage infiltration in metastases as well as in irradiated tissues indicates a crucial role of the immune system in tumor progression and metastasis. Our work suggests that the ECM may facilitate tumor cell invasion and metastasis following radiotherapy. Future studies will utilize these results to engineer improved *in vitro* tumor microenvironment models to investigate the multifaceted physical, chemical, and biological cues that influence cancer metastasis.

511561 Creating Biomaterial Gradients for Musculoskeletal Interfacial Tissue Engineering

Abstract:

Tissue engineering is an ever-growing field, which aims to advance current treatments for injuries and disease, through a combination of biomaterials, cells, and signaling molecules to restore tissue function. Of particular interest, musculoskeletal impediments are the second most frequent disability and affect over 1.7 billion people worldwide. Musculoskeletal tissues, such as bone, tendon, and cartilage tissue, are important for load bearing and everyday movement. Additionally, musculoskeletal tissues are heterogeneous, where these tissues connect and form chemical and physical gradients responsible for transferring load from one tissue type to another (i.e. tendon-bone interface). Electrospinning is a common technique for fabricating nanofibrous scaffolds that can mimic the structure of the native extracellular matrix (ECM). However, current electrospinning techniques do not easily allow for the replication of the chemical and physical gradients present in musculoskeletal interfacial tissues. In this work, magnetic electrospinning was utilized to produce a tunable fiber alignment gradient that mimics the physical gradient in the tendon-bone interface, where the tendon ECM is highly aligned and the bone ECM is not aligned. Magnetic electrospinning uses permanent magnets to create a magnetic field, where the electrospun fibers will align in the direction of the magnetic field and transition to unaligned fibers as the field weakens. As a proof of concept, electrospinning was performed using poly(caprolactone). Permanent magnet strips were selectively adhered onto a rotating collection mandrel to spatially control the size of the aligned fiber compartment of the scaffold. Fiber alignment percent was quantified using scanning electron microscopy and bright field images of the electrospun fibers as a function of scaffold length and with different magnet configurations. Results successfully show a high degree of fiber alignment on the surface and very close to the magnets with decreasing alignment at distances away from the magnet. Furthermore, by placing multiple magnets next to each other in â€œseriesâ€ the length of the highly aligned compartment increases. Future research will combine this technique with offset electrospinning to independently control gradients in fiber alignment (via magnetic field) and chemistry (via degree of offset between electrospinning solutions) to better mimic the tendon-bone interfacial tissue.

499047 Rapid Generation and Simultaneous Detection of Biomimetic Oxygen Concentration Gradients *in vitro*

Abstract:

Introduction: Oxygen homeostasis is critical for the functioning of multicellular organisms. Deficiency of oxygen or hypoxia can lead to several pathological conditions such as ischemia, tumorigenesis and drug resistance. Most studies utilize specialized O_2 incubators to generate singular oxygen concentrations that vary significantly from the physiological conditions where hypoxic gradients exist within the tissue. Current microfluidic technology using polydimethylsiloxane-based (PDMS) devices enables generation of such oxygen concentration gradients, but yield low-to-moderate spatial resolution and mostly involve tedious device assembly. We have developed a novel and simplistic approach of reproducibly and rapidly generating stable biomimetic oxygen gradients with high spatial resolution and integrated detection capability. The microfluidic split and recombine strategy utilizing O_2-rich and O_2-depleted media allows generation of prolonged dissolved oxygen (DO) gradients while an underlying sensor layer allows real-time detection of DO gradients generated. A 3-sided glass coating prevents multi-directional diffusion of ambient oxygen. Device functionality for cell based hypoxia studies was demonstrated through viability analysis of immortalized mammary epithelial cells (MCF-12A) in hypoxic environments.

Materials and Methods: PDMS devices were fabricated using standard lithography techniques. A 3-sided glass coating (cured sol-gel) was deposited on the inner walls of the microfluidic channels to prevent diffusion of ambient oxygen into the media as well as underlying sensor layer. Temperature and reaction time of the curing process were characterized to attain an optimal and reproducible glass-coating. Platinum (ii) octaethylporphyrin ketone in polystyrene (PtOEPK/PS) matrix was spin-coated onto a 1x3 inch glass slide and used as the sensor layer for oxygen detection. A thin PDMS-membrane was spin-coated on top of the sensor layer to inhibit PtOEPK mediated cytotoxicity and facilitate cell growth. Calibration for both gaseous and DO were conducted prior to oxygen gradient generation experiments. MCF-12A cells were then seeded in a multi cell-outlet device and subjected to hypoxic gradient by flowing regular (21% O_2) and deoxygenated (0% O_2) media at 1.5μLmin^{-1} for 8 hours incubation period.

Results and Discussion: The average glass coating thickness progressively increases with both temperature and time. Heating at 80^{0}C for 60s produced an optimal glass thickness of approximately 7μm with lower frequency of cracks and no unpolymerized spaces. Compared to uncoated devices, the coated devices offer an increased sensitivity by 20%. Linear Stern-Volmer relationship was maintained for both types of devices, the coated devices exhibiting higher I_0:I_{100} ratio in all cases. Stable oxygen concentration gradients with high spatial resolution were generated over various flow rates, ranging from 1μLmin^{-1} to 100μLmin^{-1}. The slope of the oxygen gradient was constant, irrespective of flow rate. Viability analysis in the multi cell-outlet device showed increased mortality of MCF-12A cells with increasing hypoxic stress, with approximately 10% live cells being viable at 0% oxygen conditions. Since the chambers are physically separated, the cells are not able to migrate to oxygen rich areas.

Conclusion: Our proposed technique enables convenient generation and simultaneous detection of biomimetic oxygen gradients *in vitro* for relatively long periods. The multi cell-outlet design mimics the functionality of several specialized O_2 incubators at once. To further explore wide-scale applicability of this platform, the activity of hypoxia-activated prodrugs (HAP) in cancer cells under hypoxic gradient is currently being investigated.

Acknowledgments: Funding for this project was provided by the National Science Foundation Award # 1642794 and 1645195.

499196 Evaluation of a Chitosan-Gelatin Thermogelling Hydrogel As a Bioprinter Ink Using an Inexpensive Platform

Abstract:

Tissue engineering focuses on producing tissues and organs *in vitro*, to meet the demand for transplantable tissue, using tissue templates called scaffolds.Â Cell population of scaffolds is random.Â Cells infiltrate from the surrounding environment into the scaffold and their arrangement, distribution, and orientation are either random, or factors of the scaffoldâ€™s properties (i.e. porosity). However, tissue morphologies are non-random and show distinct arrangement.Â Bioprinting, 3D printing of cells encapsulated in carrier inks, will allow cells to be printed similar to natural tissue morphologies.

Hydrogels are overwhelmingly used for bioprinting due to their dual phase of liquid and solid. Current hydrogels require damaging crosslinking agents (ultraviolet, pH or chemical) to gel.Â In contrast, gelatin-chitosan hydrogels gel when the temperature is raised to body temperature (37°C). However, when using an injectable, thermogelling hydrogel a number of factors must be considered. These factors include the preparation method, gelation characteristics, bed characteristics, feed rate, and layer height. We obtained a $700 commercial 3D printer and modified it to print from a sterile syringe for $100. Custom print codes were generated in Python and then script was rendered using a software provided by the manufacturer.

We investigated a 2%, 3%, and 4% (wt/v) 1:1 chitosan-gelatin (CG) hydrogel. We investigated various solution preparation steps (centrifugation, mixing, and degassing) that could help in obtaining consistent printability and fiber formation. Since gelation depends on the temperature of the print surface, we developed temperature profiles of the bed via IR imaging and grid-based assessment using thermocouples. We used agarose to create a uniform print surface to maintain a constant gap with the needle tip. We evaluated the effect of feed rate and needle height on fiber size, the distribution of cells post-printing, and the viability of cells printed in media. We found that the fiber size decreased from 760µm, to 243µm as the feed rate increased from 10cm/min to 100cm/min, increased needle height may decrease fiber size to an optimal minimum. Distinct fiber morphology was observed for the optimum height. Finally, we were able to encapsulate well distributed cells and show cells printed in media with our processing parameters exhibited excellent viability and no contamination after 5 days. Overall we printed 3D, sterile, cell laden structures with an inexpensive bioprinter. Temperature played the greatest role in fiber size, followed by feed rate. These findings will help in printing cell-laden structures at expected to lead to more functional tissues.

502909 Engineered Peptide Modified Hydrogel Platform for Biomanufacturing of Human Pluripotent Stem Cells

Abstract:

Human pluripotent stem cells (hPSCs) have the potential to revolutionize healthcare and have the capacity to differentiate into all mature cell types, making them attractive candidates for cell therapy based regenerative medicine. However, such commercial and clinical applications of hPSCs will require its mass production, necessitating an integrated, robust bioprocess for its scalable culture. Current approaches utilize suspension culture of hPSC aggregates, which maintains cell-cell contact and is highly scalable. However, controlling aggregate size, multi-aggregate coalescence, and hydrodynamic shear stress when cultured in a bioreactor is difficult. Hence it is challenging to control the heterogeneity in hPSC propagation, which is reflected in heterogeneity in hPSC differentiation, thereby compromising its functional maturation.

In contrast to the current approaches, we propose to develop a transformative approach which allows high viability and controlled propagation of single cell hPSCs. We hypothesize that ensuring single cell survival of hPSCs will be critical for its homogenous propagation. In addition, propagating the hPSCs within hydrogel capsule will protect the cells from bioreactor hydrodynamics, thereby ensuring controlled scale-up. We are thus designing peptide-conjugated alginate capsules to meet the biomanufacturing need of hPSCs. The designed peptides will ensure single cell survival and propagation, while alginate capsule will protect from bioreactor hydrodynamics.

hPSC survival and proliferation critically depends on the tight cell-cell contact, the absence of which results in ROCK pathway activation, actomyosin hyperactivation, and ultimately dissociation induced apoptosis. We have designed synthetic peptides which mimics cell-cell contact in hPSCs, promoting cell viability and pluripotency. Synthetic peptides which mimic the bioactive regions of full proteins are low cost and easily produced. Four synthetic peptides were designed and were first evaluated in 2D planar alginate substrates. hPSCs attached to the peptide modified alginate, and attachment increased as function of peptide concentration for each of the four peptide conjugated substrates. Substrates supported single cell attachment and viability of hPSCs in planar culture. Single cell hPSCs grew into colonies upon continued culture on modified substrates, supporting hPSC propagation. After propagation, hPSCs retained high levels of pluripotency comparable to parallel cultures on Matrigel, confirmed by high levels of *OCT4* and *Nanog* gene and protein expression. Maintenance of pluripotency was dependent on peptide length and sequence, and in cases showed higher expression compared to Matrigel controls. Additionally, hPSC retained differentiation potential after propagation in the peptide conjugated alginate, comparable to cell differentiated on Matrigel. Cells were chemically induced to the definitive endoderm stage, and showed high levels of *SOX17* and *FOXA2* gene expression. Similar to propagation, differentiation was dependent on peptide length and sequence. In parallel, we have also determined properties of alginate capsules to support hPSC propagation and differentiation. We are currently integrating the identified peptides with the alginate capsules, to facilitate culture in the bioreactor setting. This will allow for large scale biomanufacturing of hPSCs within the hydrogel with high viability, while protecting hPSCs from aggregation or shear stress.

501317 Substrate Surface Tension and Marangoni Effects Mediate Cell-Cell Coordination in Migration and Multicellular Assembly

Abstract:

Cell motion plays a critical role in many biological processes, including multicellular assembly and wound healing. It is still largely unclear through which mechanisms cells communicate with each other in coordinated motions during migration and assembly. We hypothesize that surface tension of the extracellular matrix conveys messages from cell to cell during collective processes. We find that cells can spread and migrate efficiently on viscous liquid silicone materials that have high surface tension. We observe that cells on these materials migrate through a highly collective, swarming-type motility with enhanced mutual spatiotemporal correlations compared to those on the substrates with lower ratios of surface tension to elastic modulus. Enhanced cell-cell correlation results from various mechanical interactions, including cell-matrix contractile forces and intermolecular attractive forces (surface tension) at the interface between extracellular materials and the aqueous cell culture solution. Marangoni flow is driven by surface tension gradients at the interface between two phases. In our case, extracellular materials and the aqueous cell culture solution have different surface tensions. Thus, we propose that the interface between these two phases undergoes convective shear motion driven by the gradient of surface tensions generated by cell contractility on the substrate surface. Together, our results suggest that cells on viscous substrates with relatively high surface tension coordinate their motions through the convective shear flow of the interface.

491710 Nitrogen Recovery and Biocrude Formation from Hydrothermal Liquefaction of Protein

Abstract:

Nitrogen Recovery and Biocrude Formation from the Hydrothermal Liquefaction of Protein

James D. Sheehan and Phillip E. Savage

Hydrothermal liquefaction is a thermochemical process that uses water at elevated temperatures (200-370°C) and pressures (10-22 MPa) to transform wet biomass into an energy dense "biocrude" oil and aqueous-phase nutrient coproducts.[1–3] The biocrude can be upgraded and co-refined with petroleum crudes into liquid fuels while the N- and P-containing nutrient coproducts can be recovered and recycled to cultivate additional biomass (e.g., microalgae). N is an essential element for cultivating microalgae and its demand can be met in part by recycling recovered N-containing nutrients from HTL, which are mostly derived from microalgal proteins. Proteins are major biochemical constituents of microalgae and contain approximately 90% of microalgal N.[4]The N atoms in the biomass proteins are ideally recovered as nutrients, however N-containing compounds can also partition into biocrude, which lowers the fuel quality and reduces the heating value of biocrude. Thus, the sustainability of HTL can be highly dependent on proteins and their transformation into valuable nutrient coproducts or detrimental N-containing compounds that partition into biocrude.

We performed HTL of bovine serum albumin (BSA), a well-defined model protein, and observed its transformation into valuable nutrient coproducts (e.g. amino acids, ammonia) and biocrude. We used temperatures ranging from 200 °C to 400 °C and holding times between 30 s to 1 hour. BSA readily decomposed during HTL into smaller peptides. The presence of peptides with molecular weights exceeding 5 kDa was no longer observable at temperatures at 250°C or higher. At 200°C, BSA aggregated into aqueous- and organic solvent-insoluble solids, in yields up to 80 wt%, that eventually decomposed into peptides which persevered in the hydrothermal media. The yields of amino acids were less than 6 wt% under all conditions, however, at temperatures of 350°C or higher the yields of primary and secondary amines in the aqueous-soluble products were upwards of 50 wt% and over 40% of N was recovered as ammonia, a preferred N-source for cultivating microalgae. The highest yields of biocrude were between 17 and 24 wt% and were achieved at temperatures ranging 300 to 400°C and holding times of 1 to 10 min. Mild (200°C) and severe (400°C, >10 min) conditions resulted with biocrude yields less than 5 wt%. This presentation will also include a discussion of the reaction pathways, kinetics, and mechanisms and thereby breaks new ground by connecting operationally defined lumped products (e.g., biocrude) with molecular-level chemistry for HTL of proteins.

References

(1) Peterson, A. A.; Vogel, F.; Lachance, R. P.; Fröling, M.; Antal, Jr., M. J.; Tester, J. W. Thermochemical Biofuel Production in Hydrothermal Media: A Review of Sub- and Supercritical Water Technologies. *Energy Environ. Sci.* **2008**, *1*, 32–65.

(2) Toor, S. S.; Rosendahl, L.; Rudolf, A. Hydrothermal Liquefaction of Biomass: A Review of Subcritical Water Technologies. *Energy* **2011**, *36*, 2328–2342.

(3) López Barreiro, D.; Prins, W.; Ronsse, F.; Brilman, W. Hydrothermal Liquefaction (HTL) of Microalgae for Biofuel Production: State of the Art Review and Future Prospects. *Biomass and Bioenergy* **2013**, *53*, 113–127.

(4) Becker, E. *Microalgae Biotechnology and Microbiology*; Cambridge University Press: New York, 1994.

491087 Bio-Succinic Acid Production from Tartaric Acid

Abstract:

Fossil fuels are currently the primary source of energy and chemicals because of their high availability and energy content.[1] However, the depletion of fossil fuels and environmental problems such as greenhouse gas emissions have triggered the development of biomass-based technologies, as biomass is the only source of renewable carbon.[2] Those new technologies also bring novel value added chemicals by utilizing the functional groups which are already present in biomass molecules.[3] Succinic acid ($C_4H_6O_4$), a four-carbon dicarboxylic acid, is recognized by the Department of Energy (DOE) as one of the top biomass-derived platform chemicals.[4] Its main applications include replacement of maleic anhydride as the C4 building block and synthesis of biodegradable and biobased polymers (Scheme 1).[4] The traditional method for succinic acid production relies on fossil fuels. Fermentation has been proposed as a renewable alternative, yet low product concentration and high concentration of organic acid are still drawbacks.[5,6]

Scheme 1. Succinic acid synthesis and applications

We developed a novel one-step catalytic route to produce bio-succinic acid using biomass-derived tartaric acid (Scheme 1), in high yields. Tartaric acid, which is naturally produced in grapes, can be recovered from winery waste streams as a byproduct of wine industry.[7] A liquid-phase system under hydrogen atmosphere comprised of molybdenum oxide catalyst supported on carbon black (MoO_x/BC) is effective in catalyzing C-O bond cleavage of tartaric acid. The role of solvent was investigated to gain a better understanding of the reaction system. The MoO_x/BC performance under different pre-reduction temperatures combined with characterization data such as temperature programmed reduction (TPR), X-ray powder diffraction (XRD), X-ray photoelectron spectroscopy (XPS) reveals that the high catalyst activity is correlated with the formation of specific Mo oxidation states. Finally, recyclability studies and structural characterization of the catalyst after reaction indicate that MoO_x/BC remains active upon reuse.

References

(1) Beerthuis, R.; Rothenberg, G.; Shiju, N. R. *Green Chem.* **2014**, *17*, 1341–1361.

(2) Ruppert, A. M.; Weinberg, K.; Palkovits, R. *Angew. Chemie - Int. Ed.* **2012**, *51* (11), 2564–2601.

(3) Corma, A.; Iborra, S.; Velty, A. *Chem. Rev.* **2007**, *107* (6), 2411–2502.

(4) Carlson, A.; Coggio, B.; Lau, K.; Mercogliano, C.; Millis, J. *Chemicals and Fuels from Bio-Based Building Blocks*, **2016**, 173–190.

(5) Cheng, K. K.; Zhao, X. B.; Zeng, J.; Wu, R. C.; Xu, Y. Z.; Liu, D. H.; Zhang, J. A. *Appl. Microbiol. Biotechnol.* **2012**, *95* (4), 841–850.

(6) Bechthold, I.; Bretz, K.; Kabasci, S.; Kopitzky, R.; Springer, A. *Chem. Eng. Technol.* **2008**, *31* (5), 647–654.

(7) Oliveira, M.; Duarte, E. **2016**, *10* (1), 168–176.

495881 1,6-Hexanediol Synthesis from Cellulose

Abstract:

1,6-hexanediol (HDO) is a high value commodity chemicals ($4,400/ton) used as a polyester or polyurethane component traditionally produced from adipic acid. In this presentation we discuss a three step catalytic route to produce HDO from cellulose[1] that includes: i) hydrolysis/dehydration of cellulose into LGO and HMF with sulfuric acid, ii) hydrogenation/hydrogenolysis of LGO and HMF into THFDM over Pd/Si-Al_2O_3 [2] and iii) hydrogenolysis of THFDM into HDO over a bifunctional Pt-acid catalyst. The overall HDO yield is 50%. In this presentation we will discuss the catalytic chemistry of each of these steps and elucidate the relationship between the catalyst structure and performance. We use cellulose, white birch and corn stover as feedstocks.

The THFDM is converted into 1,2,6 hexanetriol and then HDO in the final step through a dehydration-hydrogenation mechanism on a bifunctional Pt-based catalysts in over 90% HDO selectivity (70% HDO yield). We will show that the active catalytic acid phase for the dehydration, a BrÃ¸nsted acid site, is generated on the oxophilic support through hydrogen spillover from the Pt.

Â

Reference:

1 J. He, K. Huang, K. J. Barnett, S. Krishna, D. M. Alonso, Z. J. Brentzel, S. P. Burt, T. Walker, W. F. Banholzer, C. T. Maravelias, I. Hermans, J. A. Dumesic and G. W. Huber, *Faraday Discuss.*, 2017.

2 S. H. Krishna, D. J. McClelland, Q. A. Rashke, J. A. Dumesic and G. W. Huber, *Green Chem.*, 2017, **19**, 1278â€“1285.

497020 High-Efficiency Photo-Electrochemical Chloro-Alkali Production

Abstract:

The increased penetration of photovoltaic (PV) technologies has led to significant challenges on the management of the electrical grid. This has made evident the need for energy storage technologies that can bridge the temporal mismatch between electricity production and demand imposed by intermittent solar power sources. Alternatively, the implementation of PVs for large-scale off-grid chemical processes could further accelerate the deployment of clean energy technologies. Electrochemical processes, such as the ones implemented for chlor-alkali production, account for more than 1% of the global energy consumption and could be adapted to directly operate with solar energy sources. In this study, we demonstrate a solar chloro-alkali reactor that operates at unprecedented solar-to-chemical efficiencies as high as 25.1%. The reactor implements high efficiency multi-junction InGaP/GaAs/InGaAsNSb solar cells within a novel planar solar concentrator which maintains a $\pm 40^o$ acceptance angle with minimal tracking requirements. The concentrated PV system provides the required potential ($V_{OC} = 3.16$ V) to drive a customized electrochemical cell fabricated via additive manufacturing. The electrochemical cell is composed of a Nickel cathode and a dimensionally stable anode based on a Titanium substrate coated with an Iridium and Ruthenium oxide blend. The two electrodes are separated by a Nafion® 117 membrane which allows for the collection of nearly pure H_2 and Cl_2 gas streams as well as a NaOH enriched aqueous stream in the cathode side. Under operation conditions the faradaic efficiency towards chlorine production of the anode was measured to be between 96-99%. The demonstration presented here provides insights and foundational design principles for the implementation of solar chemical processes. Related developments for the production of solar-Hydrogen and solar-textiles will be discussed.

500094 Gold Nanoparticle Catalysis: Colloidal Versus Supported Heterogeneous Catalysis and Methods for Colloidal Nanoparticle Recovery

Abstract:

Gold nanoparticles (AuNPs) have attracted enormous attention due to their unique catalytic activities. Potential advantages of colloidal AuNPs over supported AuNPs is the greater catalytic activity and selectivity. Colloidal catalysts also catalyze reactions under mild conditions and are very effective for chiral catalysis. Recovery of colloidal catalysts involve tedious methods like pH or temperature induced aggregation, solvent extraction, etc. The advantage of supported catalyst lies in its reusability and recovery by simple methods like centrifugation, filtration, sedimentation, etc. Supported catalysts are preferred over colloidal catalyst in industries because they can be used in a continuous reactor system. For colloidal systems, surface functionalization with ligands is a prerequisite to prevent aggregation. Surface passivation causes significant reduction in catalytic activity and selectivity. For supported catalysts on metal oxide supports, a certain portion of the active surface area of the catalyst is lost due to the immobilization process and simultaneously new edges at metal-metal oxide interfaces are created. In this work, we have attempted to characterize the catalytic activity and active sites of both colloidal and supported catalysts systems by employing a ligand adsorption based method in aqueous media. An in-depth study of the effect of nanoparticle stabilizing ligands such as thiolated polyethylene glycol (PEG) and 11-mercaptoundecanoic acid (MUA) on the nanoparticle catalytic activity has been carried out and we have quantified the loss of active surface sites due to the ligand stabilization. Decreasing PEG chain length and increasing surface coverage of PEG on AuNP reduces the catalytic activity. Moreover, the functionalization of AuNP with MUA completely passivates the surface and the AuNPs show no catalytic activity. Although, high MUA coverage on the AuNP are detrimental to the catalytic activity, it renders the AuNPs easily recoverable from the solution by pH induced aggregation and redispersion. To maintain both catalytic properties and reversible pH induced recovery, we have considered partial surface coverages of the AuNP with MUA. This work also reports the synthesis of thiolated poly (acrylic acid) (PAA) functionalized AuNPs and explored its application as a recoverable catalyst where the PAA provides pH responsive dispersibility in aqueous media. Thus, PAA-AuNPs are easily and completely recovered from the reaction mixture and reused in subsequent reactions. It was found that the AuNP-PAA catalyst was highly active and reusable with modest reductions in the reaction rate with up to four catalyst recycles. We have also supported â€˜bareâ€™ colloidal AuNPs (trisodium citrate stabilized) on metal oxides and quantified the loss of active surface area. We report the comparison of loss of surface area due to ligand stabilization and metal oxide supports and compare the effectiveness and recoverability of both system of catalysts. A direct consequence of being able to characterize active surface area of both colloidal and supported system simultaneously is the ability to measure the effect of the support on the catalytic activity of the AuNPs. We have quantified the metal oxide support effect on the catalytic activity of AuNPs for various metal oxide-AuNP catalysts.

500274 Isosorbide Production from Aqueous Sorbitol Solution over Silica-Aluminas

Abstract:

Isosorbide is an important chemical for the production of fuels, plastics, and pharmaceuticals. Current isosorbide production is from plant biomass. Most current studies have been conducted in non-aqueous systems or pure sorbitol. However, processing plant biomass usually involves water. Thus, isosorbide production in aqueous systems is more practical. In the present study, we investigated the conversion of aqueous sorbitol to isosorbide using amorphous silica-alumina (ASA) with various SiO_2/Al_2O_3 ratios. The isosorbide yield of ~ 56% was achieved at 220ËšC and 22 h. Surface area/pore volume, densities of acid sites, and total acidity of ASA from Brunauer-Emmett-Teller (BET), Fourier transform infrared spectroscopy (FT-IR)-pyridine adsorption and NH_3-Temperature Programmed Desorption (TPD), respectively, were found to be responsible for this high isosorbide selectivity in the aqueous system.

498148 Simple One-Step Reaction to Achieve Cross-Linked Polymer and Polymer Composites: Models for Rubber Tires That Can Recycled Multiple Times with Robust Recovery of Properties Related to Cross-Link Density

Abstract:

Conventional polymer networks cannot be recycled into high-value products because of permanent, covalent crosslinks. Rubber tires illustrate well the issues ranging from major economic loss to environmental problems that arise with spent, crosslinked polymer. Here, we will discuss a simple, one-step reaction strategy using nitroxide-mediated polymerization (NMP) that we developed to synthesize recyclable crosslinked polymer with dynamic covalent bonds from any monomers or polymers that contain carbon-carbon double bonds that are amenable to free radical polymerization. (See Advanced Materials 2016, 28, 6746.) Using this reaction strategy, we have designed network polymers using polybutadiene and styrene monomer as models for recyclable tires, yielding full property recovery after multiple melt-reprocessing steps. No previous study has achieved direct formation of recyclable networks from commercial monomer/polymer without presynthesis or prefunctionalization and demonstrated full recovery of crosslinks after multiple reprocessing steps. We have likewise demonstrated the effectiveness of this strategy with commercial natural (uncrosslinked) rubber, leading to crosslinked natural rubber that can be effectively recycled and reprocessed multiple times with full recovery of properties associated with crosslink density. Thus, the reaction strategy can be employed with exactly the same polymers currently used in the tire industry and with reaction temperatures comparable to those employed in tire molding operations. As a result, this strategy provides for green reaction engineering not with the use on biobased feedstock but instead by effective, high-value recycling of synthetic materials that have been highly engineered for demanding uses and that are unlikely to be replaced by biobased feedstock. Importantly, we have also shown that this strategy can work well with polymer composites, e.g. networks incorporating carbon black filler exactly of the type used in rubber tires. Additionally, we have also used one-step NMP to produce polymer networks from monomers with very good reprocessabiltiy and relative uniformity of chains between crosslinks. Examples of other novel material development using simple, one-step NMP will also be discussed, time permitting.

502687 Developing Improved Enzymes for Industrial Biocatalysis

Abstract:

Pharmaceutical agents typically contain multiple functional groups with a few stereocenters that must be synthesized in high chemical purity. The field of Biocatalysis uses enzymes to perform highly selective chemical transformations. These enzymes are mainly evolved around a specific metabolic or chemical transformation and in many cases screening panels of these enzymes results in just enough activity for small-scale delivery and proof-of-concept experiment, not large-scale manufacturing. To overcome this challenge, we utilize the power of high-throughput rapid screening technologies and machine-learning algorithms during reaction and protein engineering to develop an optimized, highly selective biochemical transformation that enables large-scale manufacturing. Recent success from our laboratories will be discussed.

495941 Technoeconomic Optimisation, Antisolvent Selection and Comparative Environmental Evaluation for Continuous Paracetamol Crystallisation

Abstract:

Pharmaceutical production has long relied solely on batch processing, which offers many benefits including equipment flexibility and extensive know-how of a mature technology. Nevertheless, in an environment of growing R&D expenditure and an ever-increase drive towards sustainability, strong research interest is evident towards Continuous Pharmaceutical Manufacturing (CPM) (Gutmann et al., 2015). Continuous production can achieve higher yields at the required high purities purities, better heat and mass transfer, decreased processing times, and improved efficiency and reliability. The multifacetted research effort in developing CPM technologies includes demonstrations of novel continuous synthesis routes (Baumann and Baxendale, 2015), the development of cutting-edge microscale flow devices (Jensen, 2017), advances in downstream product formulation (Ierapetritou et al., 2016), and demonstrations of full end-to-end continuous production lines for approved pharmaceutical formulations (Mascia et al., 2013).

The need for cost-effective R&D methodologies brings process modelling and simulation to the forefront of detailed, knowledge-based process option evaluation. Furthermore, they are also pivotal towards evaluating alternative design parameters for existing or newly developed processes (Benyahia et al., 2012) and for developing control strategies, including model-predictive control (MPC) (Lakerveld et al., 2013; Mesbah et al., 2015).

Our recent publications have illustrated the use of systematic process modelling, plantwide simulation and detailed costing for technoeconomic evaluation of CPM processes (Jolliffe and Gerogiorgis, 2016). Moreover, we have employed nonlinear optimisation on the basis of rigorous process models to determine optimal design and operating parameters for key product separation operations (Jolliffe and Gerogiorgis, 2017a, 2017b). The present paper focuses on systematically evaluating a range of solvents, antisolvents and operating conditions for the crystallisation of paracetamol.

The solvents and antisolvents studied are those that are relatively benign and are most preferable to use, and for which thorough experimental solubility data have been compiled (Granberg and Rasmuson, 1999, 2000; Hojjati and Rohani, 2006). These substances are namely water, toluene, methanol, ethanol, 1-propanol, 2-propanol, 1-butanol, acetone, acetonitrile and ethyl acetate. Where data are not available, rigorous thermodynamic estimation tools (such as the UNIFAC and NRTL activity coefficient prediction methods) have been employed. The impact of crystallisation temperature and antisolvent addition ratio are also considered. The cases are formulated into a unified nonlinear optimisation (NLP) framework to determine optimal design variables and process conditions. In addition to technical metrics for evaluation such as product recovery, the sustainability of the different cases has been evaluated by monitoring quantitative green chemistry metrics, such as the E-factor. The use of the solvent and antisolvent in plug flow and Continuous Oscillatory Baffled Crystallisers (COBC) has relied on published models thereof, because these novel unit operations have been shown to offer promising scalability and excellent heat and mass transfer control (McGlone et al., 2015; Su et al., 2015). The solved optimisation cases illustrate the relative benefits and drawbacks of different combinations of solvent and antisolvent, in both technical and sustainability terms. Accordingly, the considered crystalliser designs and unit volume/size variations underscore the benefits of COBC technology, and its potential for production-scale implementation.

LITERATURE REFERENCES

Baumann, M., Baxendale, I.R., 2015. The synthesis of active pharmaceutical ingredients (APIs) using continuous flow chemistry. Beilstein J. Org. Chem. 11, 1194–1219.

Benyahia, B., Lakerveld, R., Barton, P.I., 2012. A plant-wide dynamic model of a continuous pharmaceutical process. Ind. Eng. Chem. Res. 51, 15393–15412.

Granberg, R.A., Rasmuson, Å.C., 2000. Solubility of paracetamol in binary and ternary mixtures of water + acetone + toluene. J. Chem. Eng. Data 45, 478–483.

Granberg, R.A., Rasmuson, Å.C., 1999. Solubility of paracetamol in pure solvents. J. Chem. Eng. Data 44, 1391–1395.

Gutmann, B., Cantillo, D., Kappe, C.O., 2015. Continuous-flow technology—A tool for the safe manufacturing of active pharmaceutical ingredients. Angew. Chem. Int. Ed. 54, 6688–6728.

Hojjati, H., Rohani, S., 2006. Measurement and prediction of solubility of paracetamol in water–isopropanol solution. Part 1. Measurement and data analysis. Org. Process Res. Dev. 10, 1101–1109.

Ierapetritou, M., Muzzio, F., Reklaitis, G., 2016. Perspectives on the continuous manufacturing of powder-based pharmaceutical processes. AIChE J. 62, 1846–1862.

Jensen, K.F., 2017. Flow chemistry—Microreaction technology comes of age. AIChE J. 63, 858–869. doi:10.1002/aic.15642

Jolliffe, H.G., Gerogiorgis, D.I., 2017a. Technoeconomic optimisation of a conceptual flowsheet for continuous separation of an analgaesic Active Pharmaceutical Ingredient (API). Ind. Eng. Chem. Res., in press.

Jolliffe, H.G., Gerogiorgis, D.I., 2017b. Technoeconomic optimisation and comparative environmental evaluation of continuous crystallisation and antisolvent selection for artemisinin recovery. Comput. Chem. Eng., in press.

Jolliffe, H.G., Gerogiorgis, D.I., 2016. Plantwide design and economic evaluation of two Continuous Pharmaceutical Manufacturing (CPM) Cases: ibuprofen and artemisinin. Comput. Chem. Eng. 91, 269–288.

Lakerveld, R., Benyahia, B., Braatz, R.D., Barton, P.I., 2013. Model-based design of a plant-wide control strategy for a continuous pharmaceutical plant. AIChE J. 59, 3671–3685.

Mascia, S., Heider, P.L., Zhang, H., Lakerveld, R., Benyahia, B., Barton, P.I., Braatz, R.D., Cooney, C.L., Evans, J.M.B., Jamison, T.F., Jensen, K.F., Myerson, A.S., Trout, B.L., 2013. End-to-End continuous manufacturing of pharmaceuticals: integrated synthesis, purification, and final dosage formation. Angew. Chem. Int. Ed. 52(47): 12359–12363.

McGlone, T., Briggs, N.E.B., Clark, C.A., Brown, C.J., Sefcik, J., Florence, A.J., 2015. Oscillatory Flow Reactors (OFRs) for continuous manufacturing and crystallization. Org. Process Res. Dev. 19, 1186–1202.

Mesbah, A., Paulson, J.A., Lakerveld, R., Braatz, R.D., 2015. Plant-wide model predictive control for a continuous pharmaceutical process, in: 2015 American Control Conference (ACC), pp. 4301–4307.

Su, Q., Nagy, Z.K., Rielly, C.D., 2015. Pharmaceutical crystallisation processes from batch to continuous operation using MSMPR stages: Modelling, design, and control. Chem. Eng. Process. Process Intensif. 89, 41–53.

498830 Metric-Driven Process Development of a Late Stage Drug Substance

Abstract:

Batch chemical manufacture of a pharmaceutical intermediate typically includes three common elements: reaction, work-up, and isolation. At the start of development, resources are often divided among these elements to enable work in parallel. While such practice allows investigations to proceed quickly, there is a risk that choices made for each element are not necessarily optimal for processing nor environmentally sound when examining the process in its entirety. This occurs because the criterion for a successful reaction differs for work-up and isolation. For example, consider solvent selection. It is often desirable to choose a solvent system that affords homogenous reaction conditions, becomes immiscible during work-up to remove water-soluble waste, and ultimately proves insoluble for the product yet highly soluble for impurities. This presentation will contextualize these challenges and share experimental practices through a case study of a process to synthesize a late-stage active pharmaceutical ingredient (API). The process described was constrained to deliver a specific crystal form and of course meet the general criteria of high impurity purge and high yield. Material properties of the API â€" a BCS Class I material â€" as well as of the reagents further challenged development by requiring a thorough understanding of the impact of both pH and the presence of water. A combination of automated and manual experiments streamlined work and serves as a platform for future development. Several metrics such as greenness, cost, overall cycle time, and process mass intensity were used to evaluate the options and guide development decisions.

501507 Scalable Continuous Synthesis of Enantiomerically Pure Amine APIs in an Enzyme Membrane Reactor

Abstract:

Chiral amine groups were found in 11% of the 100 highest selling drugs in 2015. Notable examples include oseltamivir phosphate (sold as Tamiflu®), used for treating influenza symptoms, dexmethylphenidate (sold as Focalin®), an ADHD medication, and sitagliptin (sold as Januvia®), an anti-type II diabetes medicine. Amine dehydrogenases (AmDHs), which were first developed in 2012[1], catalyze the synthesis of a chiral amine through reductive amination of a prochiral ketone with ammonia as the nitrogen source and NADH as cofactor, to produce an *(R)*-amine, with NAD^+ and water as by-products. To date, the largest scale at which AmDHs have been used was in a 100 mL batch process with an isolated yield of 213 mg of amine product. Because of the high cost of NADH, formate dehydrogenase (FDH) as a cofactor regeneration enzyme to convert NAD^+ to NADH. Atom economy for this process is extremely high.

We present the first example of a continuous process for chiral amine synthesis with AmDHs, employing an enzyme membrane reactor (EMR). The EMR consists of a continuously stirred reaction volume with a semipermeable UF membrane at the outlet which traps macromolecules within the reactor, but allows the small molecule substrates and products to pass[2-3]. Conversion of substrates is measured by following the change in the optical rotation of the outlet stream with a polarimeter. This rapid, online measurement allows for constant control of conversion, which is crucial for process optimization.
In this presentation, we outline the initial process development for this new reactor system for the synthesis of a model chiral amine compound. This begins with the characterization of the kinetic parameters of both enzymes which will determine rates and inhibition patterns. After a stable operating point is determined, the space-time yield (STY) can be determined. As we examine the enzyme deactivation behavior over a period of multiple days, the total turnover number (TTN) for both enzymes is obtained and will be reported. Data from these experiments will be used to optimize STY and TTN. Lastly, we will analyze continuous amine production via EMR with Green Chemistry metrics.

1. Abrahamson, M. J.; Vazquez-Figueroa, E.; Woodall, N. B.; Moore, J. C.; Bommarius, A. S., Development of an Amine Dehydrogenase for Synthesis of Chiral Amines. *Angewandte Chemie-International Edition* **2012,** *51* (16), 3969-3972.

2. Bommarius, A. S.; Drauz, K.; Gunther, K.; Knaup, G.; Schwarm, M., L-methionine related L-amino acids by acylase cleavage of their corresponding N-acetyl-DL-derivatives. *Tetrahedron-Asymmetry* **1997,** *8* (19), 3197-3200.

3. Kragl, U.; Kruse, W.; Hummel, W.; Wandrey, C., Enzyme engineering aspects of biocatalysis: Cofactor regeneration as example. *Biotechnology and Bioengineering* **1996,** *52* (2), 309-319.

Development of a Biocatalytic Cascade for the Deracimization of a Racemic Alcohol

Shane Grosser[1], Jeff Moore[2], Amude Kassim[2], Rob Leighty[3], and Wenyong Chen[1]

[1]Process Research and Development, Merck & Co., Inc., Rahway, NJ USA

[2]Chemical Biotechnology and Structure, Merck & Co., Inc., Rahway, NJ USA

[3]Global Vaccines and Biologics Commercialization, Merck & Co., Inc., Kenilworth, NJ USA

This submission is focused on the development and lab-, kilo- and pilot- scale demonstration of a biocatalytic deracimization process for the generation of an optically pure chiral alcohol, the key precursor to a novel dimeric catalyst used in the endgame of an active API synthesis. In particular, the team was tasked with developing a biocatalytic cascade capable of performing the requisite oxidation and reduction chemistry, followed by establishing a process to purify the chiral alcohol from the resulting complex reaction media.

What is particularly novel about this chemistry is that it uses two different ketoreductase enzymes, one which is S-selective, and another which is R-selective, to stereoselectively oxidize the S-enantiomer of the racemic alcohol substrate to the corresponding ketone, and then to stereoselectively reduce that ketone back to the desired R-enantiomer. In addition to the two ketoreductase enzymes, this process also uses two different enzymatic cofactor recycling systems for maintenance of the NAD^+ and $NADP^+$ co-factors in the appropriate oxidation states. The orthogonal nature of these recycling systems is afforded through an aerobic oxidation of NADH back to NAD^+ by NADH oxidase (NOX) for the first reaction, and the reduction of $NADP^+$ to NADPH by phosphite dehydrogenase (PDH) in the second reaction. In the absence of fully optimized enzyme variants with perfect selectivity for NAD^+ vs. $NADP^+$, this system is executed as a sequential catalytic cascade. As both ketoreductases are capable of running in either the oxidation or the reduction directions, successful conversion to >98%ee is controlled by both a staged charging approach, as well as control of the cofactor recycling systems. For example, following complete conversion of the undesired enantiomer to the ketone during the oxidation step, the oxidation's zinc dependent ketoreductase enzyme is deactivated by the addition of EDTA, a powerful zinc chelator. Since the second ketoreductase enzyme is not zinc dependent, it is unaffected by the presence of residual EDTA and proceeds without issue once the reduction enzyme, and it's recycling enzyme PDH, have been charged.

Additional obstacles overcome in the development of this process center around the initial oxidation reaction and its catalyst recycling system. The primary oxidant in the NAD^+ recycle happens to be molecular oxygen, delivered through compressed air to the aqueous reaction. As the moderately slow rate of O_2 delivery tends to impose a mass transfer limitation for this step, precise control at the lab scale, as well as thoughtful considerations of equipment geometry and process execution were critical in ensuring a robust process across scales. Towards that end, significant effort was placed on the refinement and characterization of a reactor setup with line of sight to a commercial synthetic chemistry reactor, including online monitoring and automated control of reaction performance via air flow monitoring and control, dissolved oxygen monitoring and control, as well as reaction component monitoring (via in-situ fluorescence spectroscopy and online LC sampling.) Employing these reactor designs, and further leveraging data-rich experimentation practices, the team developed a reaction fingerprint for both the oxidation and reduction reactions, successfully demonstrating each of them across easymax, optimax, 20L and 200 gal reactors. This includes refinement of kLa regression methodologies across scales for the O_2 dependent oxidation reaction, as well as reaction profiling using online HPLC. This body of information was critical in ensuring appropriate reactor configuration and process execution and has even enabled real-time modifications to the pilot scale equipment configuration and operational parameters based on at-line profiling data.

Enzyme removal is frequently a challenge in free solution biocatalysis, particularly in instances where the product possesses significant aqueous phase solubility. This was certainly true for this chemistry, which presented a number of such challenges regarding protein/product purification. Ultimately, the team developed and implemented a robust protein/product separation, including a novel kosmotropic enzyme precipitation and extraction regimen. Specifically, acetonitrile was added to the aqueous end of reaction mixture, followed by the kosmotropic salt potassium phosphate. This induced a number of phase separation phenomena, which were exploited to enact a robust purification. First, the addition of the potassium phosphate effectively salted-out the protein, rendering a precipitated protein phase which could be readily filtered for enzyme removal. Secondly, the salt also helped to induce a phase split between the otherwise miscible acetonitrile/water mixture, and lastly, the salt also helped to shuttle the highly water soluble product into the organic layer, thereby permitting extractive workup away from the rest of the reaction byproducts. From this point, moderately standard techniques, such as carbon treatment for color removal, and a distillative crystallization could be employed resulting in a robust, scalable, and highly portable process for the deracimization of the racemic alcohol.

497589 Design of New Reactors to Produce Levan Biopolymer through Enzymatic Catalysis: Kinetics and Mass Transfer

Abstract:

Biopolymers are rising in importance during last years because of its potential application in biomedicine and therapeutics. Most of them are usually produced by microbial fermentation and that implies some drawbacks such as sterilization control, high-cost downstream, problems with uniformity of different parameters, etc. One of the most promising biopolymers is levan due to its amphiphilic properties. However, its production has been only focused on microorganisms and its value on market is too high so as to continue researching on its potential applications.

In this work; levan enzymatic way was explored by using the enzyme involved in the process, levansucrase from *Bacillus subtilis*. Levan synthesis by biocatalysis was studied and adjusted to a kinetic model which explains the behavior. Moreover, a complete analysis of reaction conditions was carried out by studying temperature, agitation speed, and pH. Furthermore, the differences between hydrolysis reaction and transfructosylation reaction were analyzed so as to a better understanding of catalytic mechanism. Results show that hydrolysis is clearly the limiting step and for that reason, the use of fructose instead of sucrose is presented as a successful alternative in order to improve yield.

Moreover, levansucrase was immobilized in two supports (a packed bead reactor and a monolithic reactor) so as to a better use and recycling of the enzyme. The effect of immobilization in production rate was also studied and a model of mass transfer through both reactors was also proposed. Mass transfer model shows not limiting transfer phenomena in external transport for both reactors, and, concerning internal transport, monolith reactor shows successful results with an effectiveness factor greater than 0.9. Additionally, the enzyme activity when it was recovered from monolith reactor was studied and it shows a low reduction in catalysis rate compared with the native one.

These results can allow reducing its high value in market and a greater implementation in biomedicine and therapeutics.

UCL ENGINEERING

Towards the optimal design of a minimal set of clinical tests for the identification and characterization of von Willebrand disease

Beatrice Taverna[1,2], Alessandra Casonato[3], Fabrizio Bezzo[1], **Federico Galvanin**[2]

[1] Department of Industrial Engineering, University of Padova, Padova, Italy

[2] Department of Chemical Engineering, University College London, London, United Kingdom

[3] Department of Cardiologic, Thoracic and Vascular Sciences, University of Padova Medical School, Padova, Italy

AIChE Annual Meeting
October 29 - November 3, 2017,
Minneapolis, MN, USA

UCL ENGINEERING
Change the world

UCL ENGINEERING

Outline

- **Introduction**
  - Von Willebrand Disease (VWD)
  - Clinical tests for VWD diagnosis
  - Pharmacokinetic (PK) model of VWD
  - Project objectives
- **Definition of a minimal set of clinical tests for VWD diagnosis**
  - Model-based information analysis
  - Response surface methodology (RSM)
  - First level of modification of the PK model
  - Second level of modification of the PK model
- **Results**
- **Final remarks**

AIChE Annual Meeting
October 29 - November 3, 2017,
Minneapolis, MN, USA

2

UCL ENGINEERING
Change the world

UCL ENGINEERING

Von Willebrand disease

Von Willebrand Disease (VWD) is one of the most common bleeding disorders seen in humans[1]

- **Qualitative or quantitative deficiency of Von Willebrand Factor (VWF)**
- **Alteration of VWF** in the coagulation process

VWF is a *multymeric glycoprotein* which:

✓ Transports and protects coagulation factor VIII
✓ Mediates platelet adhesion, aggregation and thrombus growth.

VWD symptoms:

- Sporadic nosebleeds, excessive bleeding from small lesions in skin, menorrhagia

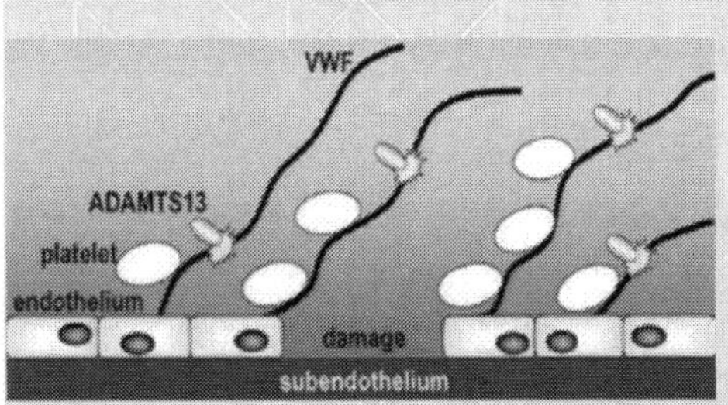

VWD diagnosis is complex: diverse types and variants of VWD

- Type 1
- Type 2A
- Type 2B
- Vicenza

→ *Pharmacokinetic (PK) models have been recently proposed for disease classification and characterization ("model-based diagnosis")*[2]

[1]Lillicrap, D. (2007). *Thromb. Res.*, **120**, S11-S16.
[2]Castaldello, C., Galvanin, F., Casonato, A., Padrini, R., Barolo, M., & Bezzo, F. (2017). Canadian Journal of Chemical Engineering (in press)

UCL ENGINEERING

Clinical trials for VWD diagnosis

The **diagnosis of von Willebrand disease** is based on the following tests:

- **VWF:Ag test**, to evaluate the number of VWF antigens in the blood stream (**overall VWF quantity**)
- **VWF:CB test**, to analyse the ability of VWF in binding with collagen (**quantity of high molecular weight multimers**)
- **DDAVP test, invasive**: repetition of blood samples [0, 15, 30, 60, 120, 180, 240, 480, 24 h] after administration of a drug (desaminopressin), forcing the release of VWF from the endothelial cells

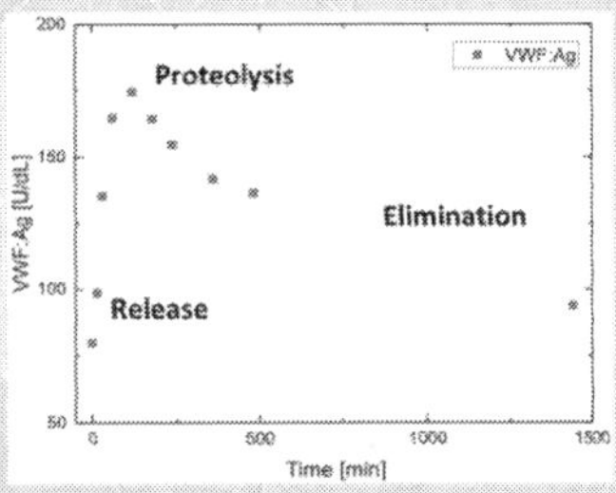

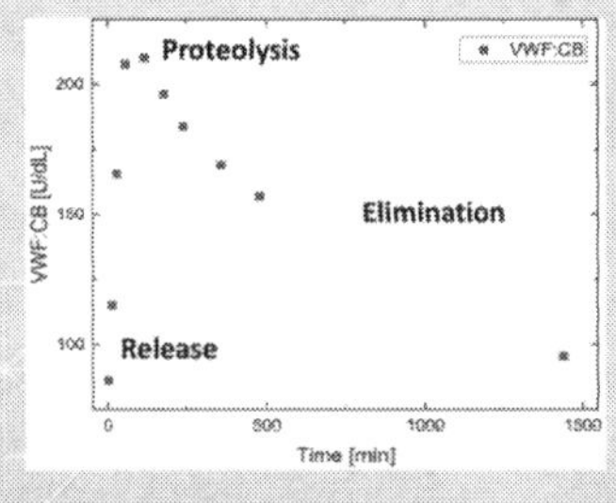

Base Mechanisms of VWF Distribution

VWF RELEASE → **VWF PROTEOLYSIS** → **VWF CLEARANCE**

secretion of super ultra-large multimers (SUL)

proteolysis of **SUL** to smaller species operated by a specific enzyme ( )

elimination from plasma (independent by multimer size)

UCL ENGINEERING

Pharmacokinetic models of VWD

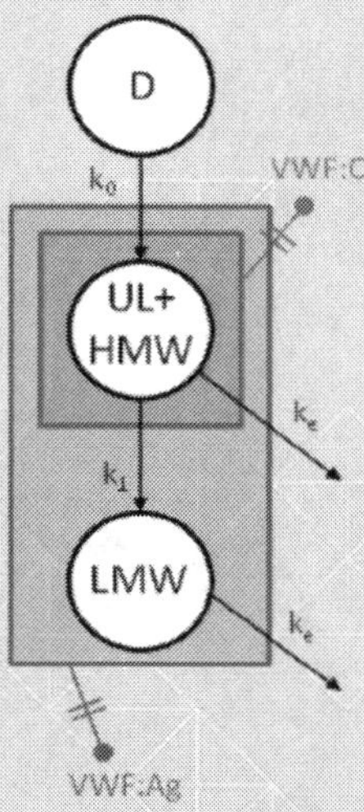

Two compartment model[1,2] described by a system of *differential and algebraic equations (DAEs)* to characterize the mechanism of

- VWF release **(k_0)**
- VWF proteolysis **(k_1)**
- VWF elimination **(k_e)**

$$\frac{dx^{UL+HMW}}{dt} = k_0 D\, e^{-k_0(t-t_{max})} - k_1\left(x^{UL+HMW} - x_b^{UL+HMW}\right) - k_e\left(x^{UL+HMW} - x_b^{UL+HMW}\right)$$

$$\frac{dx^{LMW}}{dt} = k_1\left(x^{UL+HMW} - x_b^{UL+HMW}\right) - k_e\left(x^{LMW} - x_b^{LMW}\right)$$

$$y^{AG} = \frac{x^{UL+HMW} + x^{LMW}}{V_d} \qquad y^{CB} = \frac{x^{UL+HMW}}{V_d}$$

DDAVP test needs to be carried out to achieve a **statistically satisfactory estimation** of the disease metabolic parameters: release (k_0), proteolysis (k_1), elimination (k_e).

[1]Galvanin, F., Barolo, M., Padrini, R., Casonato, A., Bezzo, F. (2014). *AIChE Journal*(60), 1718-1727.
[2]Ferrari, M., Galvanin, F., Padrini, R., Barolo, M., Bezzo, F., Casonato, A. (2017). Thromb. Haemostasis, (under revision)

UCL ENGINEERING Change the world

UCL ENGINEERING

Basal clinical trials

Main Problem ➡ **DDAVP** is a long (at least <u>24 h</u>) and **invasive** non-routine test.

The scientific community is pushing researchers to achieve model identification (i.e. quantify the PK parameters of release (k_0), proteolysis (k_1), elimination (k_e)) using *basal (static) clinical trials* only → can be carried out through a **<u>conventional blood test</u>**

Basal tests considered in this study:

- **VWFpp**: to quantify VWF elimination from the blood stream
- **VWF:Ag**: to evaluate the number of VWF antigens in the blood stream
- **VWF:CB**: to analyse the ability of VWF in binding with collagen.

Two important *physiological quantities* are derived:

- VWFpp ratio expressed as **VWFpp/VWF:Ag**
- VWF:R defined as **VWF:CB/VWF:Ag**

UCL ENGINEERING

Project objectives

Challenging goal of the project ➡ *To reproduce the dynamic profiles of VWF:Ag and VWF:CB* in a 24 h DDAVP test, by identifying the pharmacokinetic model parameters (k_e, k_0, k_1) exploiting **basal clinical trials and/or a time-reduced DDAVP**

Model modification has been carried out considering the following cohort of subjects:

- **20 VWD patients** (type 2B and Vicenza)
- **42 normal subjects** (HnonO, HO)

Information content analysis of the pharmacokinetic model → Definition of basal state correlations with response surface methodology → First-level of model modification → Second-level of model modification

Understand which pharmacokinetic model parameters (k_e, k_0, k_1) **require the longest time for being identified** by the DDAVP test with the highest information.

Information content analysis steps

1. **Sensitivity analysis**
2. **Fisher information analysis**

UCL ENGINEERING

Pharmacokinetic models: information analysis

PK models → System of Differential and Algebraic Equations (DAEs)

$$\mathbf{f}(\dot{\mathbf{x}}(t), \mathbf{x}(t), \mathbf{u}(t), \boldsymbol{\theta}, t) = 0$$

$$\hat{\mathbf{y}}(t) = \mathbf{g}(\mathbf{x}(t))$$

- $\mathbf{x}$ = **VWF multimeric concentrations (UL+HMW, LMW)**
- $\mathbf{u}$ = drug administration (DDAVP, fixed in this case)
- $\hat{\mathbf{y}}$ = measurements of VWF:Ag/VWF:CB
- $\boldsymbol{\theta}$ = set of PK parameters to be estimated

Design vector

$$\boldsymbol{\varphi} = [\, \mathbf{t}^{sp}, \tau]^T$$

- $\mathbf{t}^{sp}$ **set of time instants** at which the measured variables are sampled
- τ the **duration of clinical test**

Fisher information matrix (FIM)

$$\mathbf{H}_\theta(\boldsymbol{\theta}, \varphi) = \mathbf{H}_\theta^0 + \sum_{k=1}^{n_{sp}} \sum_{i=1}^{N_y} \sum_{j=1}^{N_y} s_{ij} \left[\frac{\partial \hat{y}_i(t_k)}{\partial \theta_l} \frac{\partial \hat{y}_j(t_k)}{\partial \theta_m}^T \right]_{l,m=1...N_\theta}$$

INFO from time domain (Dynamic Sensitivities)

Optimal design for improving parameter estimation

$$\varphi = \arg\max_{\varphi \in D} \{\psi[\mathbf{H}_\theta(\boldsymbol{\theta}, \varphi)]\}$$

Fisher Information metric (trace, determinant, eigenvalues)

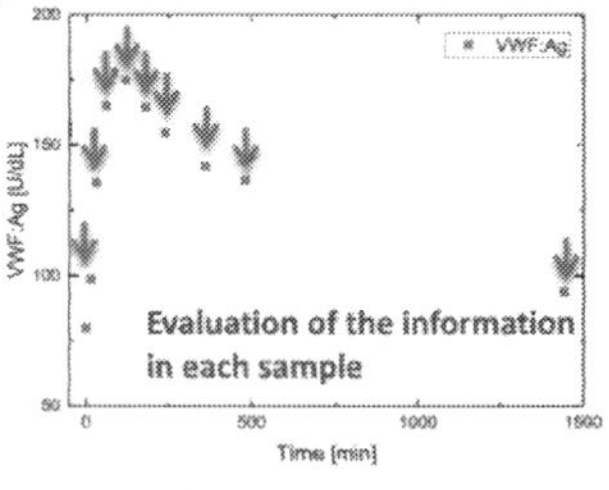

UCL ENGINEERING

Results from sensitivity analysis

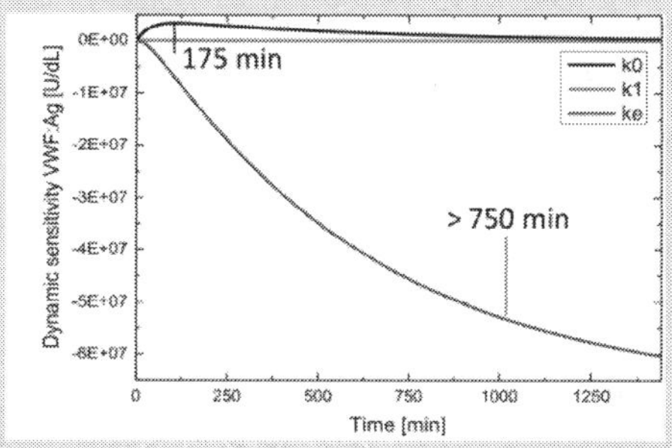

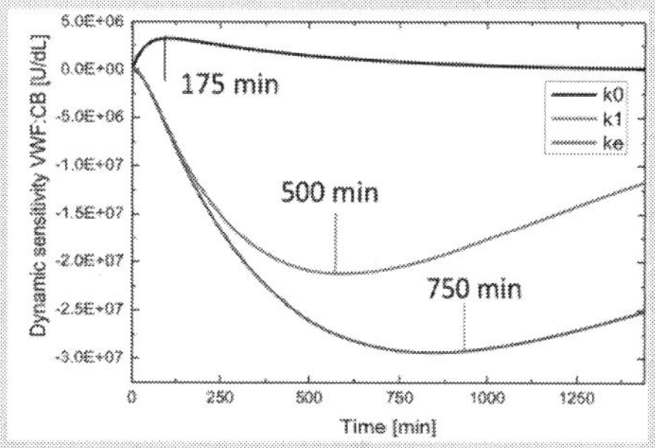

- *Release constant k_0*: reaches the *sensitivity peak around at 175 min* (approx. 3 hours)
- *Proteolytic constant k_1*: reaches the *sensitivity peak around 500 min* (VWF:CB response)
- *Elimination constant $k_{e:}$* requires a **long time** for reaching the sensitivity peak for all the categories of subjects
- Sensitivity analysis gives only a partial idea of the information. The overall information can by quantified by the Fisher Information Matrix (FIM)

UCL ENGINEERING

Fisher information analysis

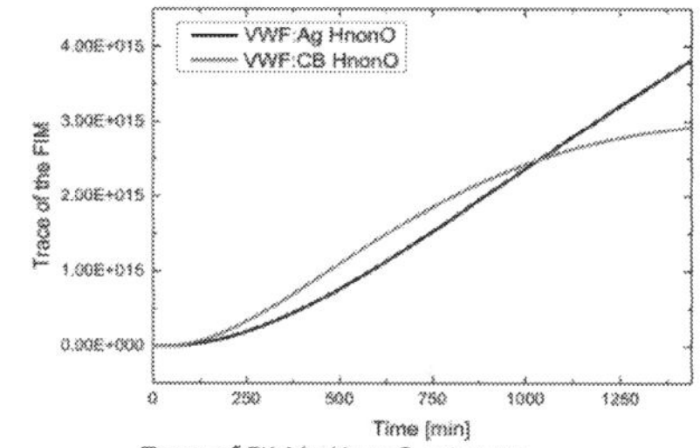

Trace of FIM in HnonO category.

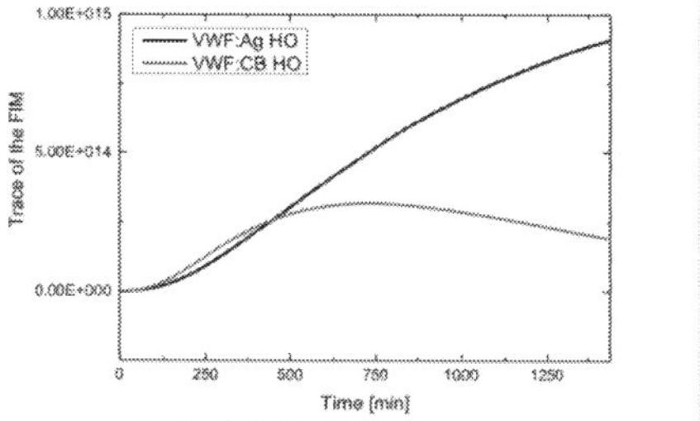

Trace of FIM in HO category.

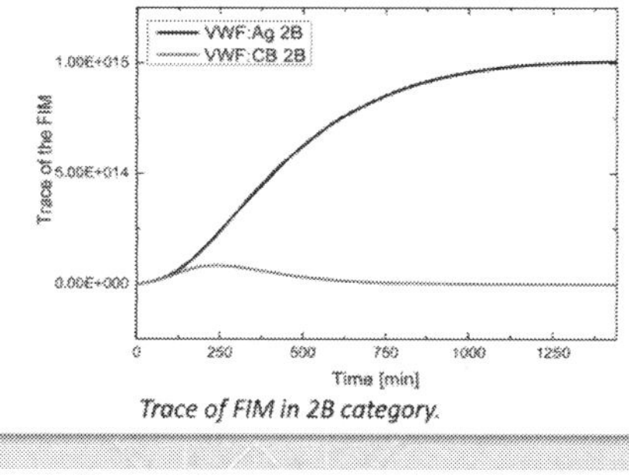

Trace of FIM in 2B category.

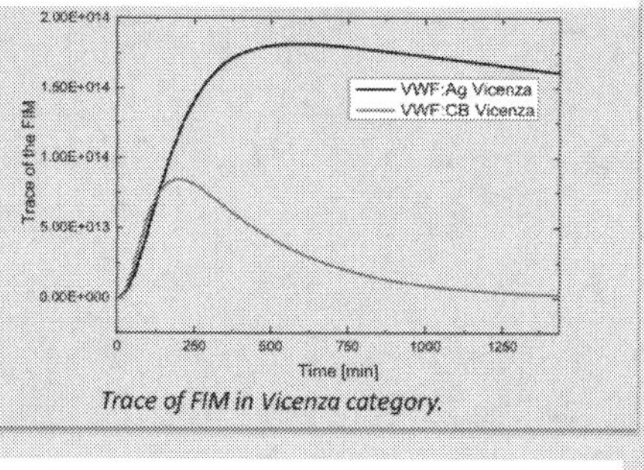

Trace of FIM in Vicenza category.

- Trace of FIM <u>does not reach a maximum in the information content</u> (except for Vicenza subjects)
- This behaviour is mostly determined by the **dynamic sensitivity of the elimination parameter k_e**

UCL ENGINEERING

Response surface methodology (RSM)

- Response Surface Methodology (RSM) is a technique used to define black-box regression models[1]
- Key aspect to reduce the DDAVP test execution time → **development of basal state correlation for the elimination kinetic constant k_e**
- RSM allows to *investigate potential correlations between input and output variables* in systems where no information is given on what is happening inside

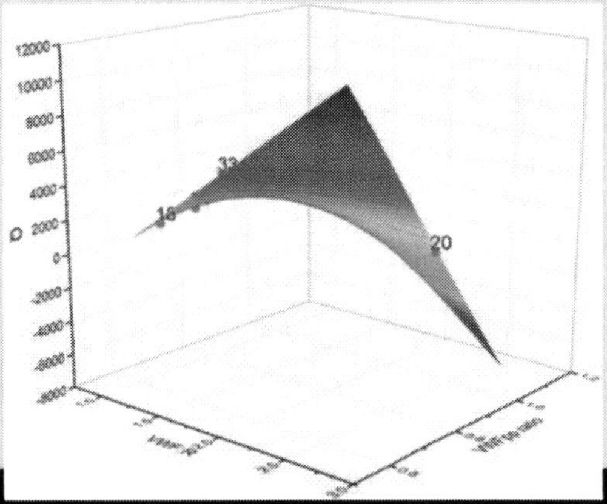

- The objective is to *assume an analytical form of the response surface (linear/quadratic, w/o interactions)* which manages to approximate the experimental data.
- Mathematically, in our case, the workspace is 3D, whereas the hyper-surface is 2D.

[1]Box, G. E. P. and Wilson, K.B. (1951) On the Experimental Attainment of Optimum Conditions (with discussion). Journal of the Royal Statistical Society Series B 13(1):1–45.

11 UCL ENGINEERING Change the world

UCL ENGINEERING

Response surface for the elimination constant k_e

- Interaction with the clinicians has been fundamental in the definition of the correlation
- The relation for the elimination constant k_e is mathematically expressed as:

$$k_e = A + B \cdot \text{VWFpp ratio} + C \cdot \text{VWF:R} + D \cdot \text{VWFpp ratio} \cdot \text{VWF:R}$$

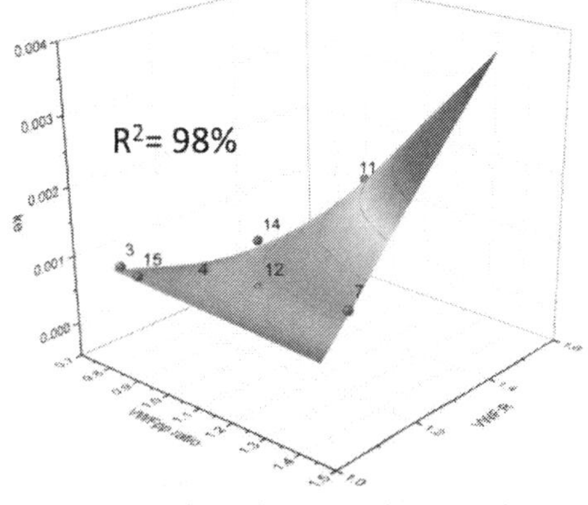

Linear response model surface with interactions for HnonO subjects.

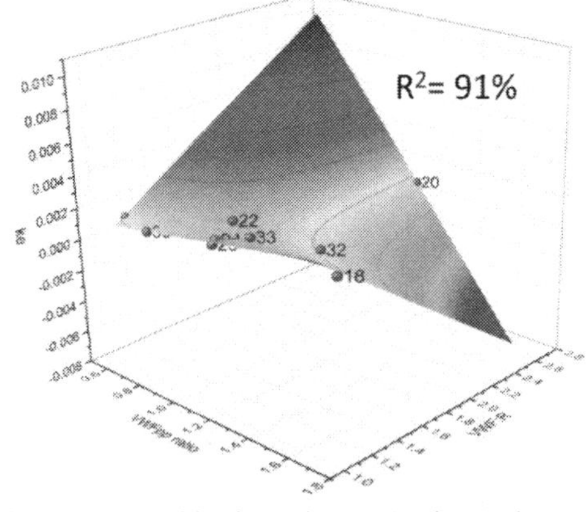

Linear response model surface with interactions for HO subjects.

AIChE Annual Meeting
October 29 - November 3, 2017,
Minneapolis, MN, USA

12 UCL ENGINEERING Change the world

UCL ENGINEERING

Response surface for the elimination constant k_e

- Interaction with the clinicians has been fundamental in the definition of the correlation
- The relation for the elimination constant k_e is mathematically expressed as:

$$k_e = A + B \cdot \text{VWFpp ratio} + C \cdot \text{VWF:R} + D \cdot \text{VWFpp ratio} \cdot \text{VWF:R}$$

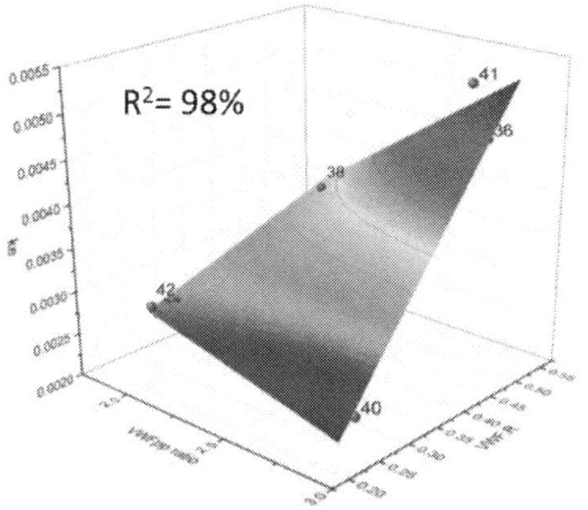

Linear response model surface with interactions for 2B subjects.

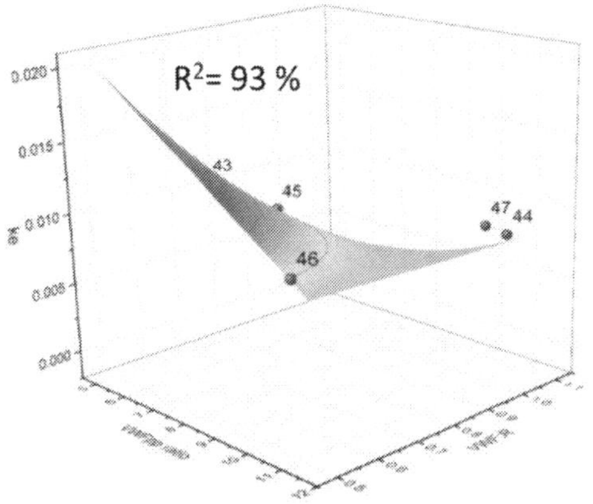

Linear response model surface with interactions for Vicenza subjects.

UCL ENGINEERING

First-level of model modification

Original pharmacokinetic model
+
new basal state RSM correlation
for the **elimination kinetic constant k_e**

First-level of model modification

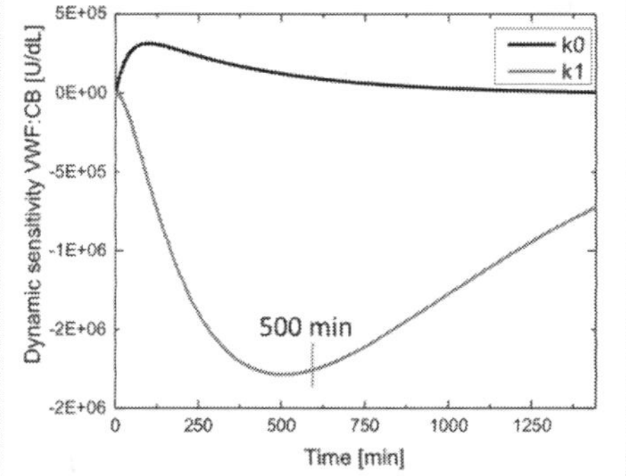

Comparison between the calculated and estimated k_e.

	k_e calculated by the first-level of modification [h^{-1}]	k_e estimated through the original model [h^{-1}]	relative error
Hnon0	6.73E-04	7.04E-04	4.6%
H0	0.00186	0.00152	9.8%
2B	0.00333	0.00323	3.1%
Vicenza	0.00840	0.00818	2.7%

The relative error is less than 10%

- The time to reach the information peak is **halved** compared to that required by original model
- *This result is valid for all the VWD categories*
- Still *500 minutes are required* to identify the proteolytic constant k_1 with the maximum information.

UCL ENGINEERING

Response surface for the proteolytic constant k_1

- **Significant effort** to establish a correlation on k_1 ➡ quality of the fitting from standard clinical trials VS Model complexity
- The relation for the proteolytic constant k_1 is mathematically expressed as:

$$k_1 = A + B \cdot \mathrm{VWF{:}Ag} + C \cdot \mathrm{VWFpp\ ratio} + D \cdot \mathrm{VWF{:}Ag} \cdot \mathrm{VWFpp\ ratio}$$

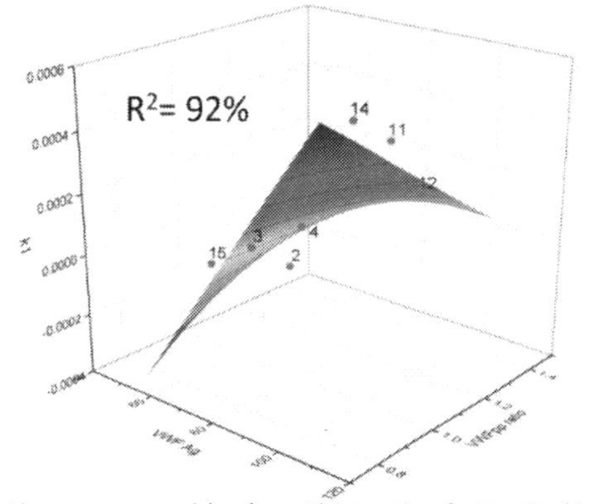

Linear response model surface with interactions for HnonO subjects.

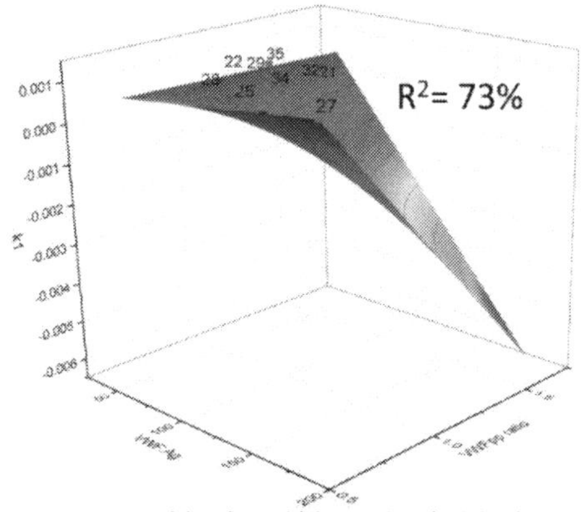

Linear response model surface with interactions for HO subjects.

AIChE Annual Meeting
October 29 - November 3, 2017,
Minneapolis, MN, USA

UCL ENGINEERING
Change the world

UCL ENGINEERING

Response surface for the proteolytic constant k_1

- **Significant effort** to establish a correlation on k_1 ➡ quality of the fitting from standard clinical trials VS Model complexity
- The relation for the proteolytic constant k_1 is mathematically expressed as:

$$k_1 = A + B \cdot \mathrm{VWF{:}Ag} + C \cdot \mathrm{VWFpp\ ratio} + D \cdot \mathrm{VWF{:}Ag} \cdot \mathrm{VWFpp\ ratio}$$

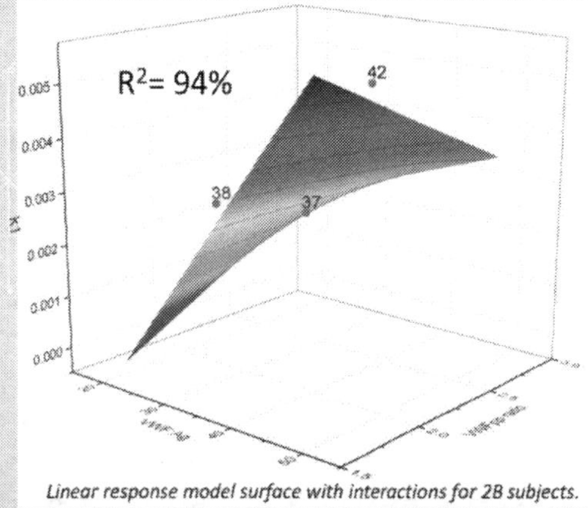

Linear response model surface with interactions for 2B subjects.

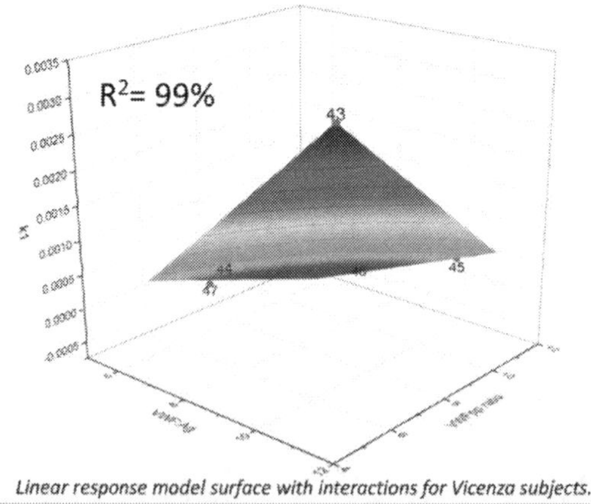

Linear response model surface with interactions for Vicenza subjects.

AIChE Annual Meeting
October 29 - November 3, 2017,
Minneapolis, MN, USA

UCL ENGINEERING
Change the world

UCL ENGINEERING

Second-level of model modification

Original pharmacokinetic model
+
new basal state correlation
for the elimination kinetic constant k_e
+
new basal state correlation
for the proteolytic kinetic constant k_1

→ **Second-level of model modification**

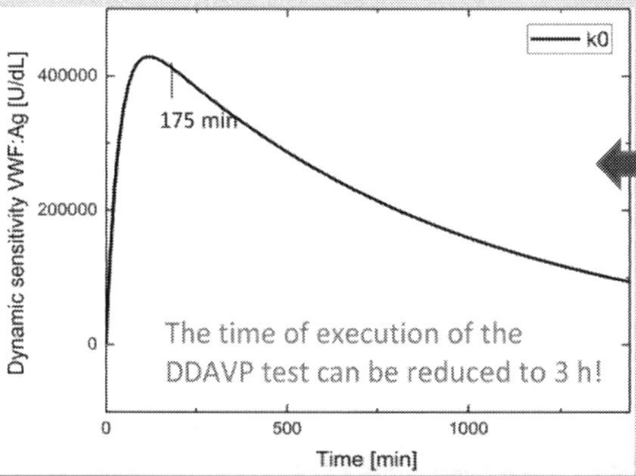

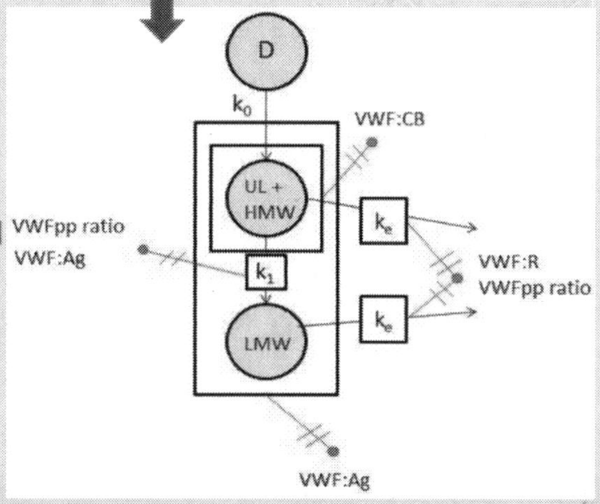

UCL ENGINEERING

Second-level of model modification: validation

The second-level of model modification has been validated considering patients (1, 19, 39, 50) that **do not belong to the original pool**

The execution time of the DDAVP test reduced to 3h → **Only the release kinetic constant k_0 needs to be estimated**

Estimated values of the release constant k_0 for patients 1, 19, 39 and 50.

	1		19		39		50	
	24 h DDAVP	3 h DDAVP	24 h DDAVP	3 h DDAVP	24 h DDAVP	3 h DDAVP	24 h DDAVP	3 h DDAVP
k_0 [min^{-1}]	0.02570	0.02487	0.04430	0.03832	0.02620	0.02592	0.03690	0.03632

The 3h DDAVP allows to correctly estimate k_0 as with the 24h DDAVP

Patient 1

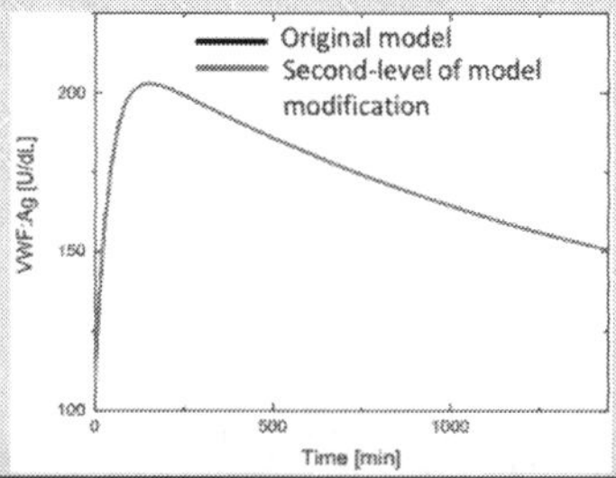

Patient 19

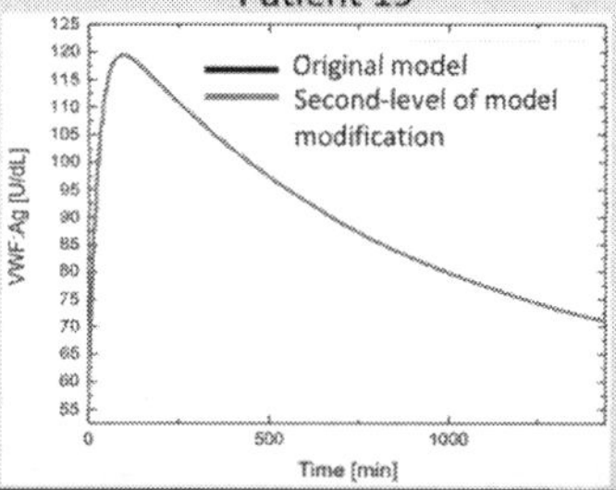

The result is remarkable
↓
the profiles do overlap

Comparison between the simulated profiles of VWF:Ag response with original and second-level of modification of the pharmacokinetic model.

UCL ENGINEERING

Final remarks

UCL

Main Achievements

- Exploitation of the information content of **basal clinical tests**
- Significant reduction of the execution time of the DDAVP test **from 24 h to 3 h**
- Definition of basal state correlations for the values of
  - The proteolytic kinetic constant k_1
  - The elimination kinetic constant k_e

→ **Reduced PK model**

Further investigations

- **Model-based design** of the 3 h DDAVP sampling protocol
- Third-level of modification of the pharmacokinetic model → k_0

  → only partially satisfactory (uncertainty on release definition)

AIChE Annual Meeting
October 29 - November 3, 2017,
Minneapolis, MN, USA

19 UCL ENGINEERING Change the world

UCL ENGINEERING

Final remarks

UCL

Current research activities

- Validation of the reduced PK model using data from additional subjects affected by VWD
- Model generalisation and extension to type 1 variants of VWD
- Development of an open-source software for model-based VWD diagnosis

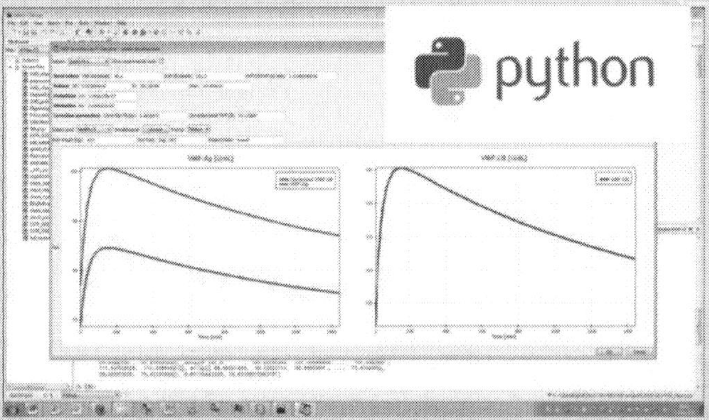

AIChE Annual Meeting
October 29 - November 3, 2017,
Minneapolis, MN, USA

20 UCL ENGINEERING Change the world

UCL ENGINEERING

Acknowledgements

Thank you for your attention

AIChE Annual Meeting
October 29 - November 3, 2017,
Minneapolis, MN, USA

21 UCL ENGINEERING
Change the world

499991 Experimental Verification of Endothelial Cell Aquaporin-1 Expression's Influence on Sub-Endothelial Intima Thickness and Relevance to Early Atherosclerosis

Abstract:

Experimental Verification of Endothelial Cell Aquaporin-1 Expression's Influence on Sub-Endothelial Intima Thickness and Its Relevance to Early Atherosclerosis

Roman Yakobov, Klaudio Mitri, Deyvi Tenemaza, Ruven Pinkhasov, Diana Yakobov, Jiah Lee, Kung-ming Jan and David S. Rumschitzki

Atherosclerosis is perennially among the top two killers in the US and in all Western countries. It begins decades before symptoms arise by the infiltration of low density lipoprotein (LDL) cholesterol from the blood across the arterial endothelium into the arterial wall, where it can bind to extracellular matrix (ECM) in the subendothelial intima (SI) wall region. This SI lipid accumulation appears to trigger a cascade of events that can lead to early lesion formation. A large (~22nm diameter) aggregate, LDL crosses the endothelium via the junctions around rare (~1/(2000-5000) cells in rat aorta) endothelial cells (EC) that have temporarily (~1h) widened junctions with their neighbors. Our group has shown that this transport is advection-dominated, via a slow (~30 nm/s) transmural pressure difference-driven transmural flow of the liquid portion of the blood through these leaky junctions. Naturally this transmural pressure difference and flow also act across the normal endothelium and its tight junctions. We have shown, both theoretically and experimentally, that this flow passes not just paracellularly through the EC-EC tight junctions, but a portion (~1/3 of the total, but in this case pure H_2O) also passes transcellularly via endothelial cell aquaporin-1 (AQP1) water channel membrane proteins. This water-and-plasma flow is of great interest because it can wash unbound LDL from the SI, dilute the local SI LDL concentration and, since binding to ECM is a chemical reaction, potentially strongly affect LDL-ECM binding kinetics. Thus more EC AQP1 could mean slower LDL-ECM binding and slower progression to lesions.

Interestingly our group has observed that the aortic SI region is mostly (>95%) void, in contrast to the aortic media (~42% void) and that transmural pressure can compress the SI, causing the ECs to block SI fenestrae and significantly increase transmural flow resistance; this drastically lowers transmural flow. Our theory predicts that increased/decreased EC AQP1 expression and activity raises/lowers the critical transmural pressure needed to accomplish this SI compression and concomitant increase in flow resistance. Since under normal conditions, the SI is compressed, if this theory is correct, increased ECAQP1 expression has the potential to decompress the SI, decrease flow resistance and thus increase SI LDL dilution in the physiological range. We have already shown that an increase/decrease in functioning EC AQP1 indeed increases/decreases the transmural pressure at which flow resistance drops. What is missing is to show that changes in functioning EC AQP1s accomplish these changes in flow resistance via a corresponding shift in the critical pressure for SI compression. This is the subject of this paper, i.e., to directly test these predictions experimentally.

We anaesthetize two rats each for each set of conditions. These sets of conditions include treatment with either (a sub-toxic level of) $HgCl_2$, which reversibly blocks EC AQP1 channels, Forskolin (which increases EC AQP1 expression) or a blank control solution each at one of four different transmural pressures. We then cannulate the ratâ€™s carotid artery and attach it to a constant pressure reservoir at one of four different pressures. We stop the ratâ€™s heart, fix the vessel *in situ*, extract, further fix and embed the aorta in Epon, section (60-100 nm) for transmission electron microscopy (TEM), stain to increase contrast, view the sections under TEM and analyze the sections for SI thickness using a custom-written Matlab code. We analyze 100-200 *random* sections per set of conditions (i.e., treatment and pressure). We plot the results in the form of SI thickness vs pressure for each of the three treatments and compare with theoretical predictions. The results are a very strong test of our theory and tie together theory and experiments on both SI thickness, and on aortic wall hydraulic conductivity (the inverse of its flow resistance) as functions of transmural pressure and functioning EC AQP1 numbers.

496756 Mathematical Modeling and Data Analytics Using WBC Populations for the Prognosis and Diagnosis of Acute Coronary Syndrome

Abstract:

The complete blood count (CBC) is one of the most common clinical tests, integral to the diagnosis, treatment, and monitoring of almost all diseases because it provides a simple high-level assessment of the health of a patientâ€™s hematologic and immunologic systems by reporting an estimate of the current number of each type of blood cell circulating per unit volume blood. Increases or decreases in the countsÂ of different cell types may indicate anemia, infection, malignancy or more. Most routine CBCs involve high-resolution and high-throughput single-cell measurements of the morphology of tens of thousands of blood cells, providing single-cell details of morphology, protein concentration, or other characteristics. This additional high-throughput information is not utilized for clinical purposes, which results in the squander of precious data. These single-cell characteristics reflect states of maturation, activation, production, destruction, and the perturbation of those processes in different disease conditions. If we can infer these states and their rates of change from routine blood counts, we can diagnosis disease earlier and more precisely.

Here we develop a mathematical model of white blood cell (WBC) population dynamics inspired by cellular mechanisms for this purpose. We first show that this model can be useful to distinguish healthy individuals from those with a range of acute disease processes, and we then show how the model can improve the risk-stratification of patients being evaluated for acute coronary syndrome. Instead of using one model parameter for the purpose of diagnosis and disease stratification, we use advanced machine learning techniques to build a robust, cross-validated classifier that can help with the classification of disease etiology in patients having the same external manifestation of symptoms. We can show how patients indistinguishable based on the CBC indices alone can be classified using the parameters depicting the population dynamics of WBC morphology. Our study demonstrates how mechanistic modeling of existing clinical data can realize the vision of precision medicine in a cost-effective way.

UNIVERSITY OF MICHIGAN

Role of Red Blood Cell Deformability on Cellular Dynamics in Blood Flow

Mario Gutierrez

10/29/2017

Background – Red Blood Cells

Erythrocytes
(**Red Blood Cells**)

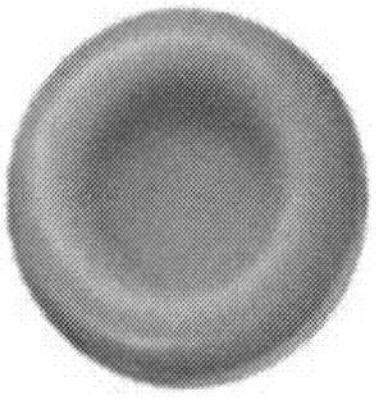

Flexible

Diameter: ~8 µm

Count per mL: ~5x10^9

R. A. Rhoades and D. R. Bell, Medical Phisiology: Principles for Clinical Medicine. Lippincott Williams & Wilkins, 2012.
Mebius, R. and G. Kraal. Nat Immunol 2005, 5, 606-616.

Background - Blood Composition

	Erythrocytes (**Red Blood Cells**)	Thrombocytes (**Platelets**)	Leukocytes (**White Blood Cells**)
	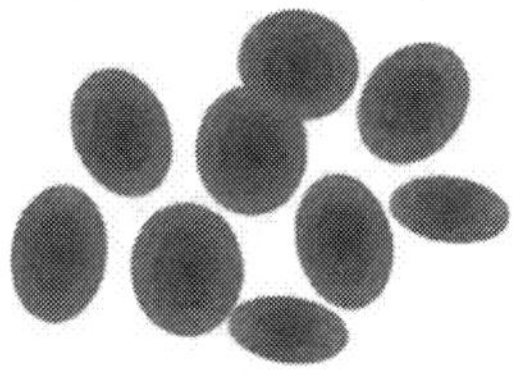	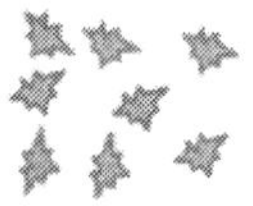	
Diameter:	~8 µm	~3 µm	~12 µm
Count per mL:	~5×10^{9}	~3×10^{8}	~7×10^{6}
RBC Ratio:	1 : 1	17	700
Shear Elastic Modulus:	2×10^{-12} dyn/cm	2×10^{-11} dyn/cm	1.15×10^{-9} dyn/cm
Deformability Ratio:	1 : 1	10	600

Background – Margination Phenomenon

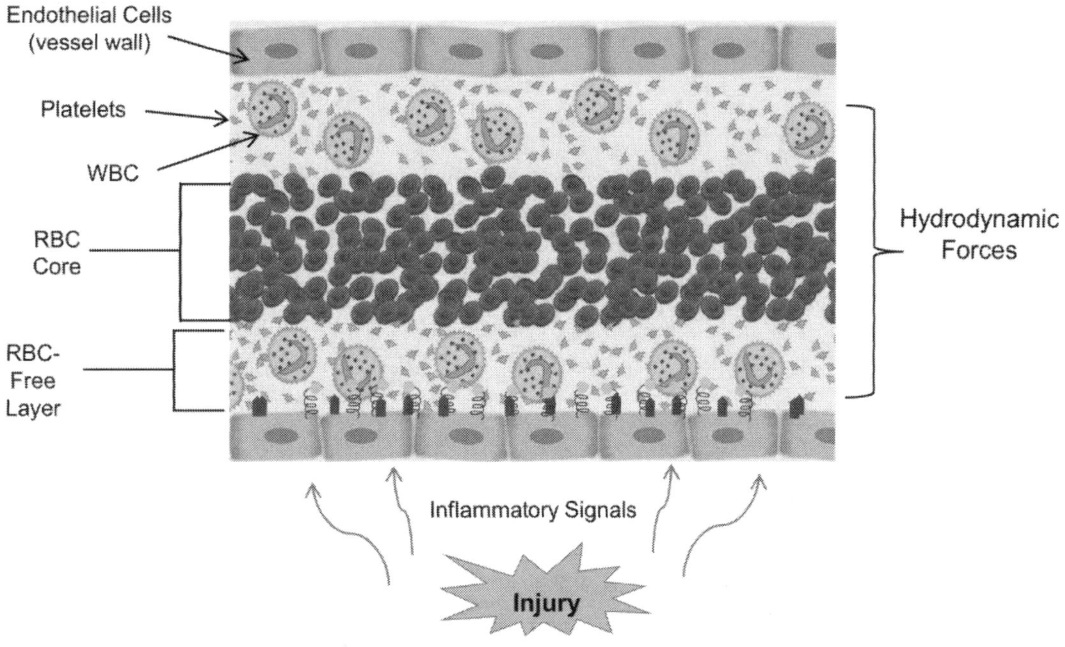

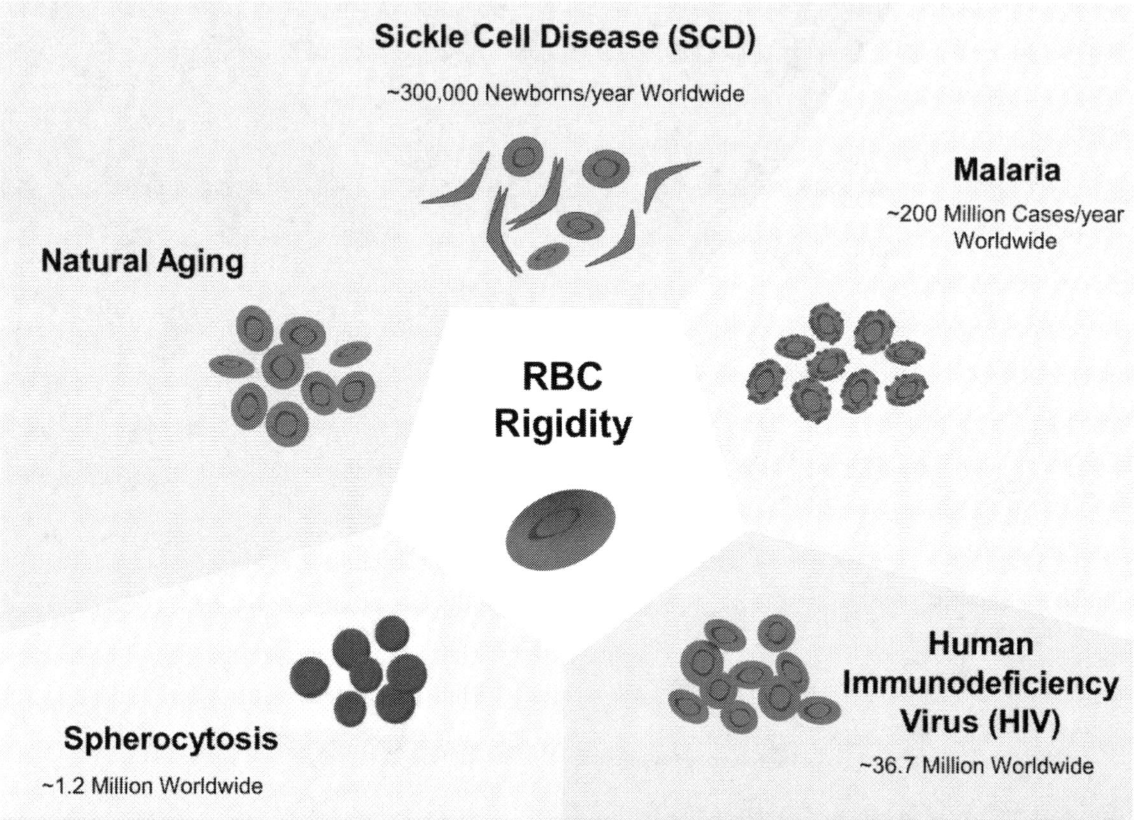

D. C. Rees, et al., "Sickle-cell disease.," Lancet, Dec. 2010.
G. A. Athanassiou, et al., "Red blood cell deformability in patients with human immunodeficiency virus infection,"
R. Khecuriani, et al., "Deformability of red blood cells and human aging.," May 2010.

How does RBC Rigidity Impact Blood Segregation ?

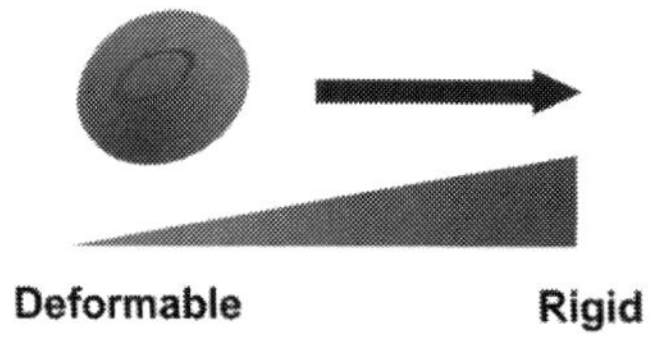

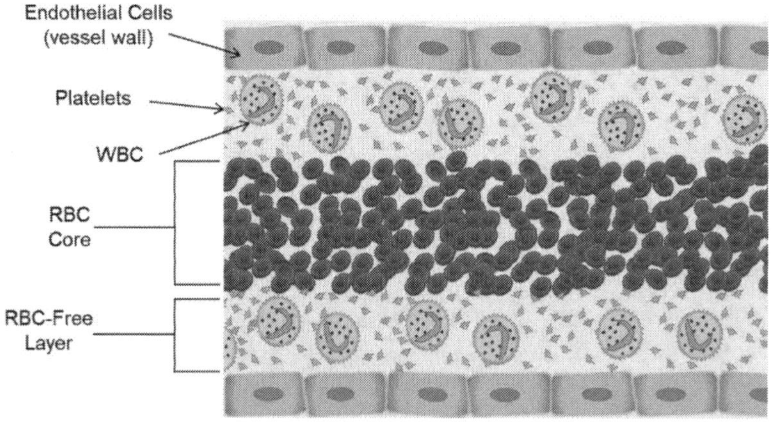

Cantat et al. "Lift Force and Dynamical Unbinding of Adhering Vesicles under Shear Flow"," 1999
A. Kumar et al., "Mechanism of margination in confined flows of blood and other multicomponent suspensions," ,2012.

Results – Quantify Degree of Rigidity

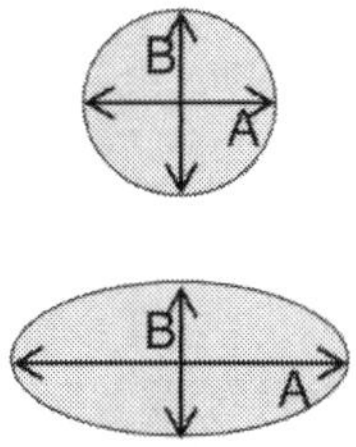

Ektacytometry

$$EI = \frac{A - B}{A + B}$$

Elongation Index

tert – Butyl Hydroperoxide

(TBHP)

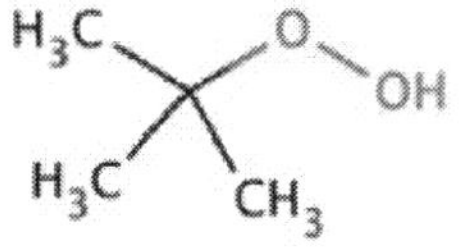

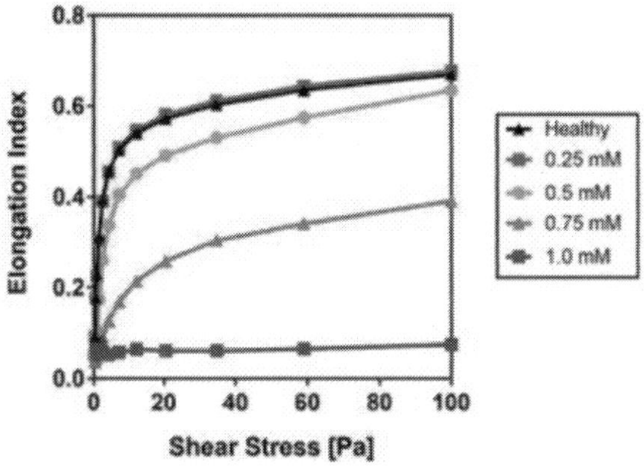

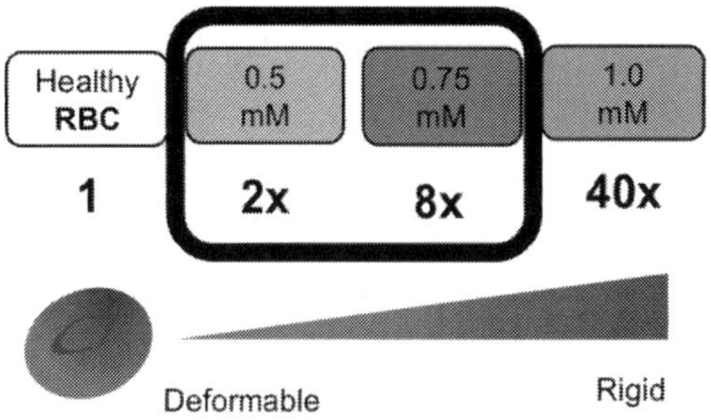

in vitro Flow Chamber Assays

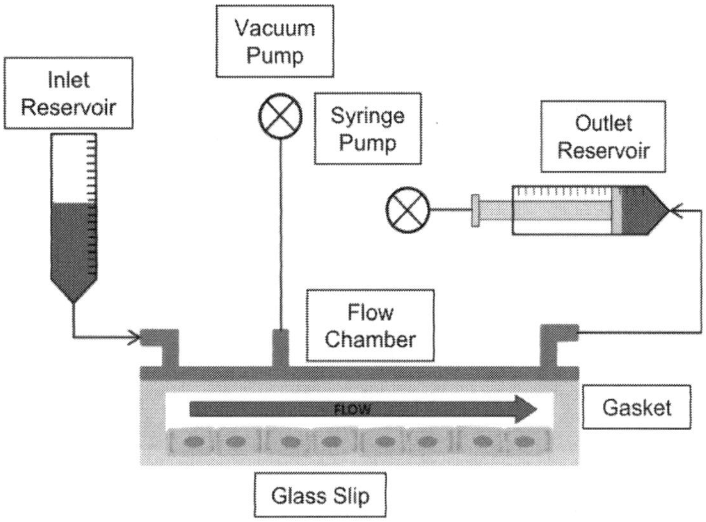

$$\gamma_w = \frac{6Q}{h^2 w} \quad [s^{-1}]$$

Shear:	**Relevance:**
1,000 s^{-1}	Capillaries / Arterioles
500 s^{-1}	Venules
200 s^{-1}	Large Veins

Results – Effect of Degree of Rigidity

Conditions:
- 1,000 s^{-1}
- 1.0 mM TBHP Rigid RBCs

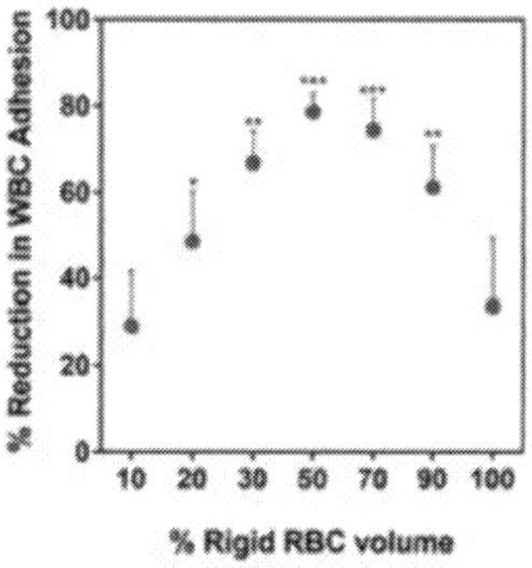

Healthy

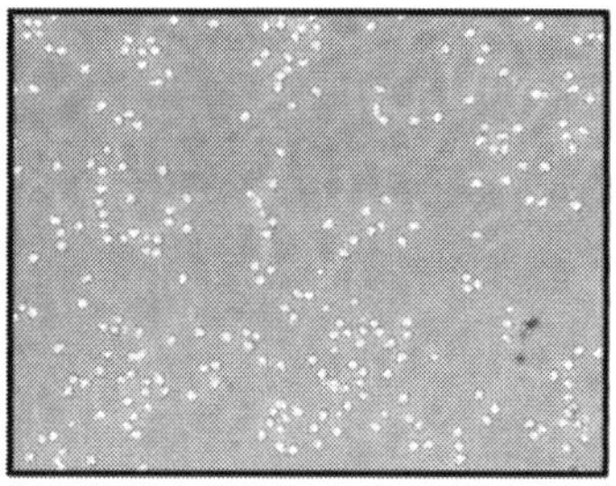

~50% Rigid RBC vol.

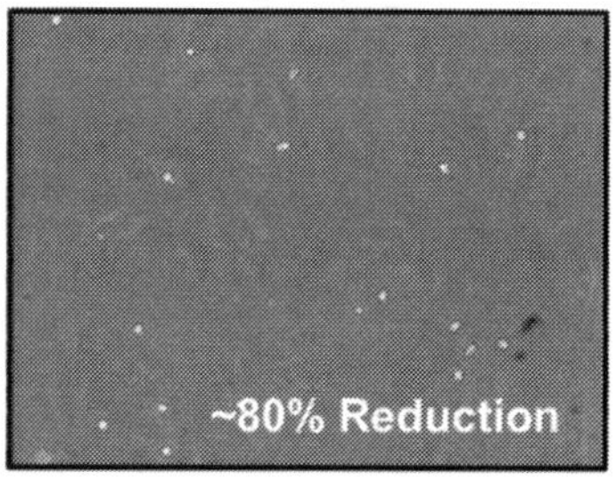

* p<0.1
** p<0.01
*** p<0.001

Results – Effect of Degree of Rigidity and Shear

1,000 s^{-1}

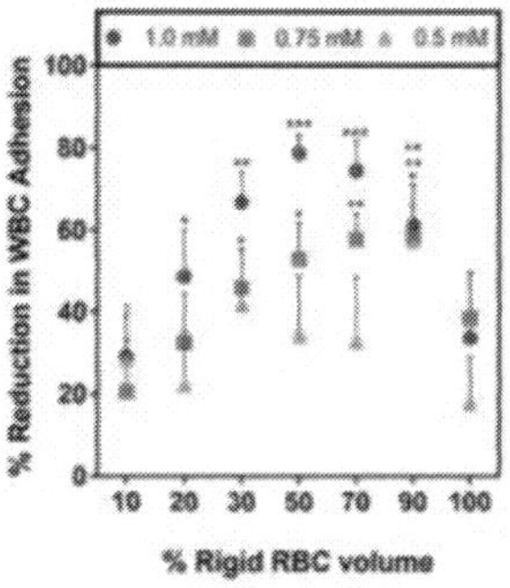

500 s^{-1}

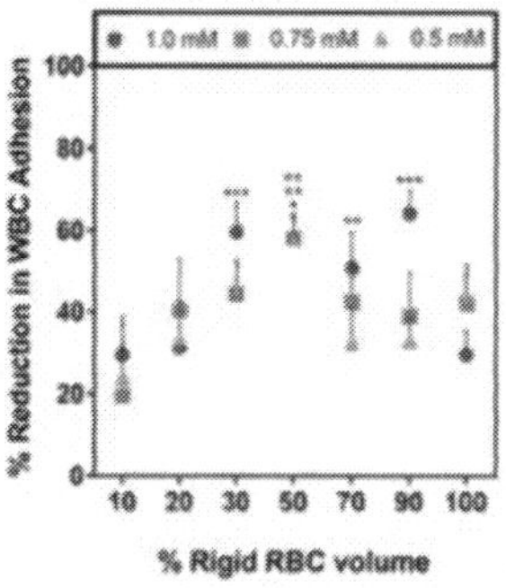

200 s^{-1}

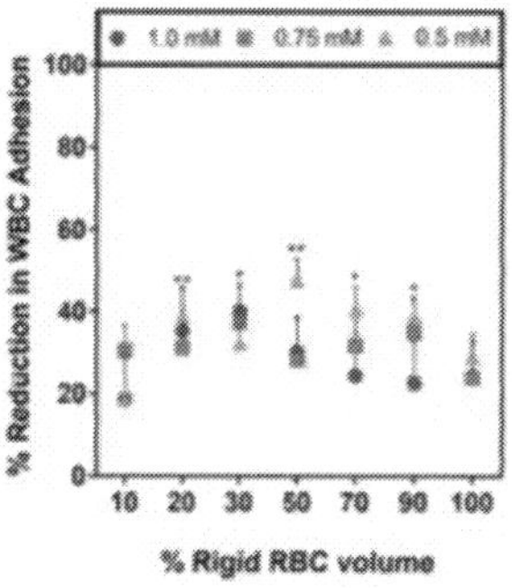

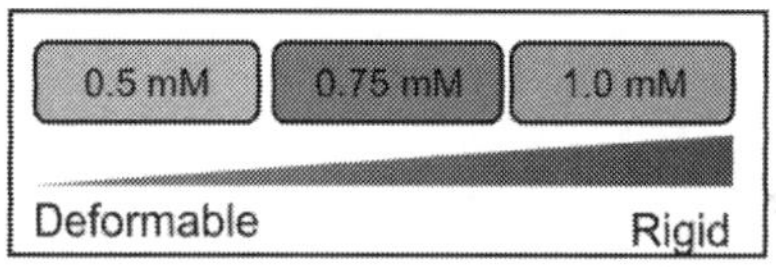

* p<0.1
** p<0.01
*** p<0.001

Potential Disease Relevance

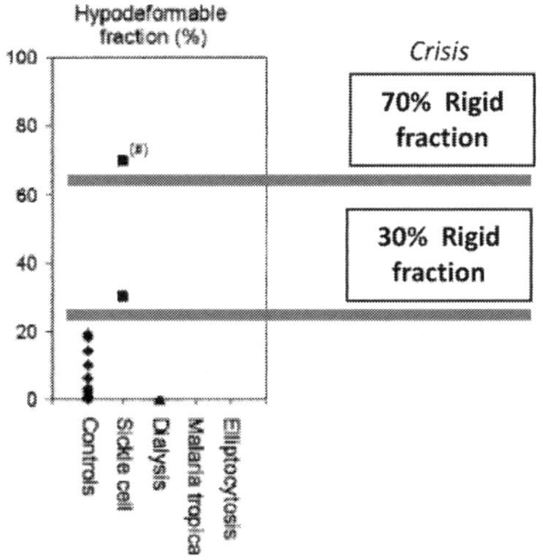

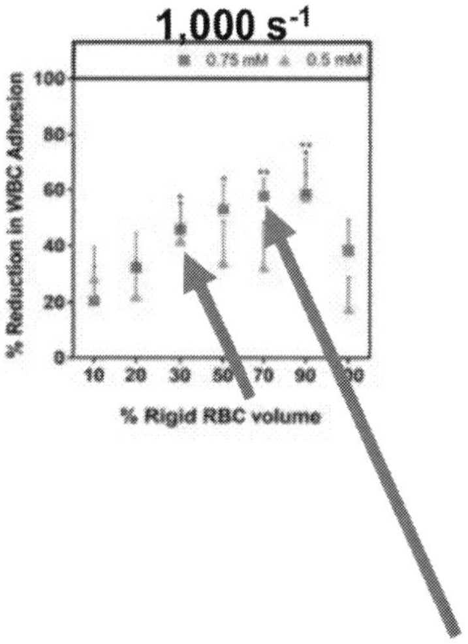

SS, Severe Crisis

+ Infected Blood

• Parasitized Cells

What is going on with the RBC Core?

Manuscript in preparation. Eniola Lab.

Methods – RBC Confocal Distributions

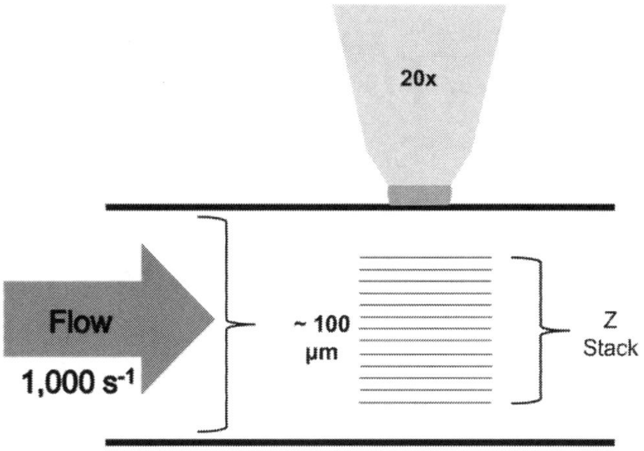

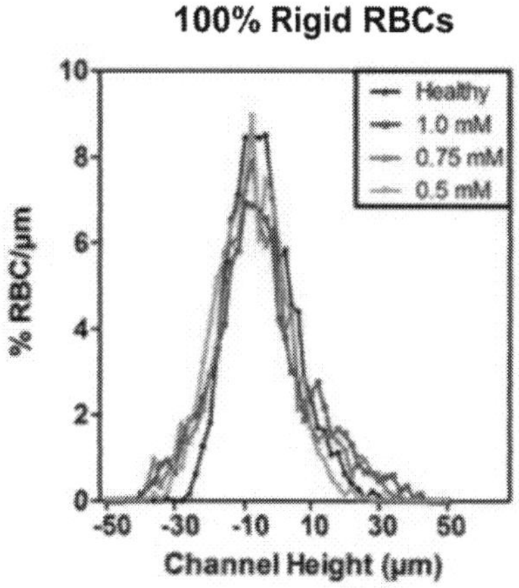

Manuscript in preparation. Eniola Lab.

Results – RBC Confocal Distributions

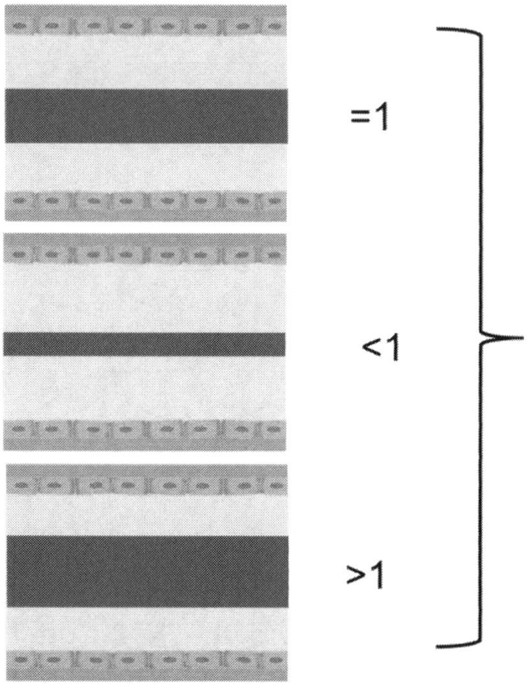

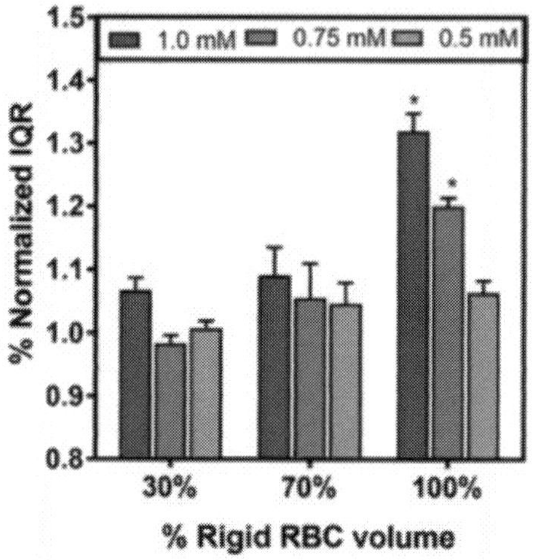

* p<0.1
** p<0.01
*** p<0.001

Manuscript in preparation. Eniola Lab.

Major Observations/Implication

Observations:

1) WBC Adhesion is reduced in the presence of Rigid RBCs
2) WBC Adhesion Reduction does not linearly correlate to % Rigid RBC Concentration increase
3) WBC Reduction is more pronounced in higher shears
4) Rigid RBCs change RBC core size

Implications:

1) **Reduction of WBCs in the presence of Rigid RBCs can relate to poor immune response**
2) **Combination of Rigid and Healthy RBCs producing maximum adhesion reduction can potentially explain different Immune response in patients**
3) **Venules are more susceptible to WBC Adhesion Reduction**
4) **RBC Core enlargement can be a factor in poor immune response**

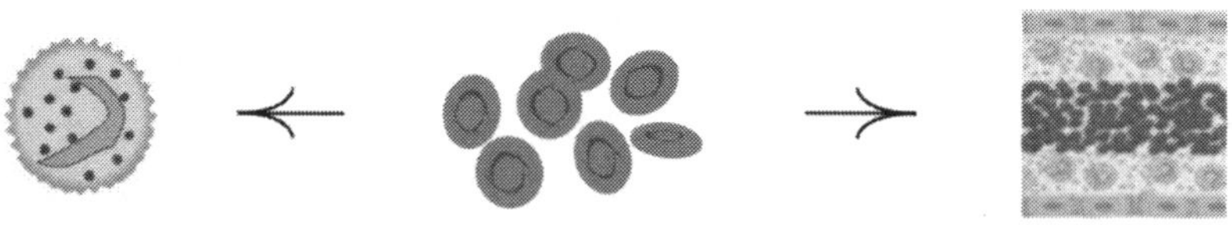

Acknowledgments

People

Research Advisor: Dr. Eniola-Adefeso

Eniola Research Group

Funding

National Science Foundation

Graduate Research Fellowship

Manuscript in preparation. Eniola Lab.

COLLEGE OF ENGINEERING
CHEMICAL ENGINEERING
UNIVERSITY OF MICHIGAN

14

500409 Model-Based Novel Strategy for Individualized Treatment of Sickle Cell Disease with Hydroxyurea

Abstract:

Sickle Cell Disease (SCD) is a hereditary disease caused by a single nucleotide mutation in the β subunit of hemoglobin producing abnormal, rigid, and sickle shaped red blood cells. Current treatment method involves daily oral dosage of hydroxyurea over a patient's entire lifetime. Hydroxyurea is known to induce fetal hemoglobin synthesis that inhibits hemoglobin polymerization and thus relieves the symptoms associated with the disease. The major side effect of the drug is the inhibition of the enzyme ribonucleotide reductase, which causes bone marrow toxicity.

The main challenges associated with hydroxyurea are substantial interpatient variability in pharmacokinetic (PK) and pharmacodynamic (PD) profiles, cytotoxicity, lack of an effective biomarker to predict treatment efficacy and consequently lack of optimal dosing regimen for individual patients. The current biomarkers in use are fetal hemoglobin (HbF) and mean cell volume (MCV) of red blood cells (RBCs) to monitor treatment progression. A semi-mechanistic compartmental PK model is developed, which includes enzymatic reaction that converts hydroxyurea into nitric oxide and its derivatives. The PD model captures the biochemical and morphological changes the cell populations are undergoing during bone marrow differentiation. A population balance model approach is used where cells are compartmentalized according to their functionalities in hematopoietic process. All the cellular properties are lumped into two internal coordinates, cell maturation age and cell volume. The model predicts the volume distribution of reticulocytes and RBCs, and concentrations of leukocytes, reticulocytes and erythrocytes of SCD patients under treatment with hydroxyurea. Through our studies, we established that MCV of reticulocytes – an immediate precursor of erythrocytes – is a better biomarker as its dynamics changes on a faster timescale compared to HbF and MCV of RBCs. Further, global sensitivity analysis is done using Sobol's method to identify sensitive parameters for obtaining individual patient model. Once the individual patient PK, PD trajectory is predicted, nonlinear model predictive control (NMPC) is applied to determine optimal dosage that bounds the model predictions within target values.

Another major issue with the lifelong treatment of hydroxyurea is non-compliance. It is difficult to differentiate non-compliance from treatment inefficacy, as non-complying patients might mislead clinicians into believing that treatment at the current dose level is insufficient. In this case, the dose might be increased but once the patient starts complying, the excess dose might prove to be harmful. The PK-PD model incorporates non-compliance to predict response and optimal dose for a non-complying patient. The current model is being tested on data collected from Riley Hospital for Children in Indianapolis.

499529 M2-Mediated Influenza Virus Budding and Scission: From Basic Principles to Control

Abstract:

Viral release of influenza is believed to occur independently of the usual cellular machinery of endosomal sorting complexes required for transport (ESCRT); instead, scission of the extruding bud is facilitated by the M2 protein.[1] The shape of the M2 transmembrane domain is distinctly conical and the assumed mechanism by which scission proceeds is that M2 aggregates in the budding neck to cause constriction that eventually pinches off the enveloped virus.

To investigate this hypothesis, we have constructed a minimal physical model consisting of a coarse-grained lipid bilayer membrane[2] whose geometry is shaped to a budding neck by an analogue of the viral particle contents. In this way, we can model the fascinating phenomenon of budding in a generic, physiologically inspired theoretical framework on the relevant *in vivo* length scale. Our model is used to delineate molecular mechanisms in enveloped virus release and predict M2 aggregate topologies in various representative stages of the budding process, allowing us to propose novel antiviral strategies.

REFERENCES

[1] Rossman *et al*. Cell 141, pp. 902-913, 2010

[2] Brannigan *et al*. Phys. Rev. E 72, 011915, 2005

501743 Transport of Molecular Motors on Networks of Cytoskeletal Filaments

Abstract:

Molecular motor proteins use the energy from ATP hydrolysis to perform work and facilitate a variety of cellular processes. Motors of the myosin family move along actin filaments, and their directed motion facilitates processes including vesicle and organelle transport, cytoplasmic streaming, and cell division. The organization of actin filaments can vary both within a cell and between different cells. Thus, it is important to understand how the organization of the cytoskeletal network influences the cellular-scale transport of motors.

In this work, we use stochastic, particle-based computer simulations to investigate the motion of molecular motors on networks of filaments. We vary properties including the density and length of filaments, the organization of filaments, and the density of motors. We are particularly interested in the effects of filament bundling on transit times for a motor to cross a spatial domain. Actin bundling is common in cells, and has been implicated in cytoplasmic streaming, in which myosin motors induce bulk flow of the cytosol in plant and fungal cells. In our model, motors can bind to and unbind from filaments, undergo diffusive motion in the cytoplasm, and undergo directed motion when bound to a filament. We generate spatially resolved stochastic simulation trajectories using the Gillespie algorithm. To characterize the transport of motors, we measure the first passage times for motors to traverse spatial domains of varying sizes. At intermediate filament densities, the mean first passage time decreases with increasing number of filaments, and for a fixed number of filaments, shorter first passage times are promoted by longer filament lengths and bundling of filaments. We find that spatiotemporal correlations and rebinding of motors to filaments in the bundle facilitate the faster transport, even with a reduced bulk density of filaments. We further explore the effects of excluded volume at higher densities, which can produce density heterogeneities and introduce traffic jams on filaments. We introduce analytical calculations to help interpret our simulation results at low motor densities.

496181 Pathway Analysis of Combined *in Utero* exposure to Heavy Metals and Phthalates and Its Association with Child Psychomotor Development

Abstract:

The exposome represents the totality of exposures from conception onwards, simultaneously identifying, characterizing and quantifying the exogenous and endogenous exposures and modifiable risk factors that predispose to and predict diseases throughout a person's life span. Exposome came as a complement to the human genome; although decoding the human genome increased our understanding of the underlying causes of disease, the genome explains only a percentage of the population burden. Current evidence indicates that environmental factors are equally or even more important than genetic predisposition alone. What is actually critical is the interaction of environmental factors with biological systems. Towards understanding better the causal links among the genome, the environment and disease, unraveling the exposome implies that both environmental exposures and genetic variation are reliably measured simultaneously.

Exposome studies will require novel tools to address the complexity of emerging environmental health issues. Critical for success will be the ability to bring together existing geospatial, environmental, health and socioeconomic data, and to collect new high resolution data using innovative environmental micro-sensors, remote sensing or other community and omics/systems biology based approaches to describe the exposome for e.g. endocrine disruption-related syndromes and sex-related changes (menopause), neurodevelopment, neurodegenerative or respiratory diseases. Mapping the entire lifecycle of an individual may not be necessary if critical lifetime events where an individual's geospatial lifeline crosses a noteworthy environmental event are recognized and understood. Whereas exposure during all life stages may entail adverse effects, children, pregnant women and the elderly are particularly susceptible. Thus, focus will have to be on susceptibility windows during growth (including pregnancy) and development, and on the unequal distribution of the burden of epigenetically active food and environment-related disease to vulnerable populations such as the young, elderly, socio-economic disadvantaged, gender and ethnic minorities.; thus these population sub-groups have been the focus of this study.

Large-scale exposome projects like HEALS (Health and Environment-wide associtations via Large population Studies) try to reverse the paradigm of "nature versus nurture" and adopt one defined by complex and dynamic interactions between DNA sequence, epigenetic DNA modifications, gene expression and environmental factors that all combine to influence disease phenotypes. Data collected in on-going epidemiological studies across the world involving mother/infant pairs, children, or adults including the elderly have been analysed to evidence relevant environmental exposure/health outcome associations. These associations aid in designing pilot surveys using an integrated approach, where the combined set of selected biomarkers of exposure, effects and individual susceptibility results in integrated risk assessment. Environmental conditions that can affect the epigenome of an individual include both external and internal factors. Individual behaviors such as smoking and alcohol consumption, physical activity, dietary intake, temperature changes and stress are external factors that have been proposed to have a long-term influence on epigenetic modifications. However, it is possible that small defects in transmitting epigenetic information through successive cell divisions, or maintaining it in differentiated cells, accumulate in a process that could be considered as an "epigenetic drift" associated with aging.

In the frame of HEALS we have been applying the exposome connectivity paradigm on several pre-existing cohorts re-analysing biobanked samples using a full multi-omics platform coupling transcriptomics and metabolomics and applying integrative bioinformatics and exposome-wide association algorithms to draw links between combined exposures to metals and endocrine disrupters and metabolic pathway dysregulation, as well as between metabolic pathway perturbations and clinically observed phenotypes of neurodevelopmental disorders such as problems in linguistic, motor development and cognitive capacity.

The children (n=148) who are subjects of this study were tested at year 1 and 2 of their life using the Bailey battery of neurodevelopmental testing. Their mothers had been exposed to metals and/or endocrine disrupting compounds during the 2nd and 3rd trimester of pregnancy in Lodz, Poland. Biobanked plasma and urine samples from the mothers were analysed using a combination of LC-MS/MS ToF and NMR. The metabolomics data were further analysed using the MZMine and Genespring bioinformatics platforms to identify the metabolic pathways that seemed to be the most perturbed from the exposure to the mixture of phthalates and metals that was the primary objective of this study. The results were summarized in correlation globe that revealed the most statistically significant associations between the exposure variables as measured using human biomonitoring (in this case, exposure to metals such as Pb, Se, Cu, and tobacco smoke measured by the cotinine levels in blood).

A second correlation globe analysis provided an overview of the observed statistically significant associations between neurodevelopmental health outcomes as measured in terms of the Bayley test scores for different types of neurodevelopmental endpoints (motor development, cognitive function, linguistic skills, verbal expression) and the main metabolic pathways that are perturbed in the sampled population. Coupling the outcomes of the two correlation globes, several key metabolic pathways that are linked to exposure to metals and phthalates on the one hand and to perturbations in normal child neurodevelopment at 1 and 2 years of age on the other have been identified. Exposome-wide associations of metabolic pathway perturbations with adverse outcomes regarding neurological development during early life revealed two main mechanisms through which co-exposure to phthalates and metals during pregnancy (in particular during the second and third trimester) may hamper the normal neurological development of newborns and infants.

One key mechanism is perturbation of oxidative phosphorylation leading to disruption in mitochondrial respiration. The latter has been clinically associated with problems in motor development. ATP overproduction inhibits the activity of one of the enzymes involved in glycolysis, thus blocking sugar catabolism. Disruption in oxidative phosphorylation in the mitochondria is also observed after exposure to metals such as Pb or Hg. This could lead to reduction in mitochondrial energy production during fetal development, affecting negatively its neurodevelopment. Moreover, the overproduction of reactive oxygen species (ROS) due to the presence of metals inside the cell induces oxidative stress. The latter favors the production of pro-oxidant species that may lead to oxidative damage. Redox homeostasis may be disrupted due to exposure to xenobiotics. Overexpression of superoxide dismutase (SOD)-coding genes will result in increased quantities of SOD; the latter is linked to neurodevelopmental disorders due to the overproduction of ROS. Verbal development is linked to the presence of glutathione peroxidase (GPx3) during pregnancy and the presence of GPx1 in the umbilical cord. Finally, motor development is associated with the presence of Se at birth. A possible putative mechanism that explains this is the fact that Se is bound to selenoproteins, which protect the cell from ROS-induced destruction.

Our results are corroborated by previous studies that have shown changes in lipid concentration in the brain after exposure to phthalates, when the former are linked to oxidative stress. Additional metabolic pathways the induction of which during the third trimester of pregnancy is statistically associated with clinically observed neurodevelopmental disorders in the first two years of the child's life based on our results include the following:

- citrulline nitric oxide cycle,

- urea cycle,
- cysteine biosynthesis/homocysteine degradation,
- lysine degradation II and
- purine ribonucleoside degradation to ribose 1 phosphate

So far no experimental validation of the induction of these specific pathways and neurotoxicity exists in vitro or in vivo in the literature. Our current results indicate that in vitro testing coupled to targeted metabolomics on metabolically active relevant cell lines could be a viable way forward towards providing the mechanistic evidence of the observed exposome-wide associations.

Two are the main characteristics that stand out in this first-of-a-kind study:

(a) The fact that we have chosen not to consider factors such as smoking behavior or socio-economic status as confounders. Rather, we were interested in capturing the overall metabolome perturbation and associate it with (i) the totality of environmental exposure factors we could quantify; and (ii) clinically observed neurodevelopmental disorders. The scope of the study was to investigate how the perinatal and early-life exposome affected child neurodevelopment.

(b) Exposome-based associations of clinical or sub-clinical health outcomes with the early-life external and internal exposome supports the elucidation of the mechanisms through which xenobiotics interact with and eventually perturb cell metabolism to induce specific pathways of toxicity in infants and young children.

The connectivity approach to environmental health builds upon the connectivity across different biological scales in a systems biology approach to elucidate the mechanisms underlying the environmental burden of disease. Adopting a data-driven paradigm that is guided by systems biology principles this approach couples comprehensiveness in exposome and health associations and biological plausibility. Thus, it can unravel causal links between environmental exposures, genetic predisposition and early biological markers of adverse health outcomes without needing to revert to very large population sizes.

501302 Functional Graphenes Oxide Coating with Speek Coaxial Electrospun Fibers for Proton Exchange Membranes

Abstract:

Proton exchange membrane fuel cell (PEMFC) is considered to be clean and efficient energy storage and conversion in the 21th century. As one of the key components of PEMFC, proton exchange membrane (PEM) with properties of cost-effectiveness, high conductivity and low fuel crossover attracts much attention. Many efforts have been done on the inorganic hybrid PEMs for the good barrier or hygroscopic abilities of inorganic fillers as GO, ZrP, SiO_2 etc. While poor compatibility between inorganic and organic components becomes the most challenging issue.

In this study, the inorganic/organic compatibility is improved by coaxial electrospinning method. Encapsulation of sulfonated organosilane functionalized graphene oxide (SSi-GOs) in the core of sulfonated poly (ether ether ketone) (SPEEK) nanofibers prevents the inorganic component from dropping off, and thus facilitates high compatibility, conductivity and single cell performance. Strong electric force induces the alignment of SSi-GOs sheets along the electrospun fiber, which is evidenced by the spindle-like morphology of the SSi-GOs containing fibers through TEM. As comparison, SSi-GOs cure up and float between the hybrid electrospun fibers due to the lack of restriction. With the increase of SSi-GO content, water absorption of the membranes increases to a maximum value at around 2.5 wt% and then decreases, while the swelling ratio keeps decreasing to as low as 10% in 70°C water. The sulfonated organosilane functional groups on SSi-GO contribute to better ionic aggregations, as a result, the co-electrospun membranes exhibit high conductivity of around 160mS cm^{-1} with 2.5 wt% SSi-GO content, 1.4 and 1.5 folds of that with 2.5wt% SSi-GO hybrid electrospun and without SSi-GO electrospun membranes, respectively. Better single cell performances, including higher power density and lower hydrogen crossover current density are achieved with the co-electrospun membranes comparing with the hybrid electrospun membranes and even Nafion membranes.

496081 Performance of a Submerged Anaerobic Membrane Bioreactor (AnMBR) at Long HRT and High Solid Concentration: Focusing on Membrane Fouling and Effluent Quality

Abstract:

Performance of a submerged anaerobic membrane bioreactor (AnMBR) at long HRT and high solid concentration: focusing on membrane fouling and effluent quality

Yifru Waktole Berkessa[1, 2], Binghua Yan[1*], Ming Tan[1], Yang Zhang[1*]

[1] Lab of Waste Valorization and Water Reuse, Qingdao Institute of Bioenergy and Bioprocess Technology, Chinese Academy of Sciences, 189 Songling Road, Laoshan District, Qingdao 266101, PR China

[2] University of Chinese Academy of Sciences, No.19A Yuquan Road, Beijing 100049, PR China

Corresponding author e-mail: zhangyang@qibebt.ac.cn, yanbh@qibebt.ac.cn

Aerobic membrane bioreactor is proved to be efficient for wastewater treatment and its application is widespread. However, it has drawbacks of high energy consumption and high solid yield. Interests in anaerobic membrane bioreactor (AnMBR) have increased in the last three decades due to its low energy consumption, low sludge production and biogas generation. Despite its merits membrane fouling is bottleneck to AnMBR technology and this study aimed at assessing treatment performance and mitigate membrane fouling by using newly designed submerged AnMBR under high MLSS concentration (22.30 g/l) and long HRT (47days). Submerged AnMBRs (two) ran in parallel with three phase separator placed between bulk sludge and membrane module. Microfiltration (0.40μm) membrane modules with effective total filtration area of $0.14m^2$ were installed. Results showed steady COD removal (average COD removal of >98%), high specific flux and mean biogas generation of 0.29 LCH_4/g $COD_{removed}$. The later could be due to active methanogens and syntrophic bacteria abundant in the sludge as confirmed from high-throughput sequencing of 16s rRNA targeting bacterial and archaeal community. Effluent quality analysis showed the effluent soluble microbial products were dominated by proteins compared to carbohydrates. Besides from specific ultraviolet absorbance (SUVA) analysis, effluent from both reactors had low aromaticity with SUVA <1 (L¡¤mg^{-1}¡¤M^{-1}) except for the first ten days. Furthermore, mean total resistance of both reactors (2.23E+09 m^{-1}) and fouling rate (4.0E+08m^{-1}¡¤day^{-1}) values over course of operation time were low compared to literature values (usually >E+10 for both resistance and fouling rate) which could suggest the newly designed AnMBR mitigated fouling and produce effluent with better quality.

Key words: Anaerobic membrane bioreactor; mixed liquor suspended solid; membrane fouling; membrane flux

Fig. 1 Influent COD and COD removal over operation period

Fig. 2 Membrane permeability over operation period

Acknowledgement: Financial support of the Shandong Provincial Natural Science Foundation (ZR2016bb06) and ¡°100 Talents Program¡± of Chinese Academy of Sciences (CAS).

496031 Mixed Matrix Membranes for Butanol Recovery Based on Zeolites As Macro-Crosslinkers

Abstract:

Mixed matrix membranes for butanol recovery based on zeolites as macro-crosslinkers

Song Hu[1], *Peiyong Qin*[1,*]

[1]National energy R&D center for biorefinery, Beijing University of Chemical Technology, Beijing, China

Biobutanol produced by biomass fermentation is critical for the development of sustainable and renewable biofuel. Mixed matrix membrane (MMM) is considered to be a promising choice for butanol recovery from fermentation via pervaporation.

In previous work, methods for fabrication of MMM only focus on improving fillers¡¯ physical compatibility and ignore the potential of chemical bridging between polymer and fillers, thus it is still hard to obtain a sufficiently thin selective membrane layer with high particles loading.

Herein, we report a method for the fabrication of an efficient membrane for butanol recovery using micron-sized particles of surface modified silicalite-1 as macro-crosslinkers within a polydimethylsiloxane (PDMS) matrix (as shown in Fig. 1). In this method, silicalite-1 of nano- and micron-sized particles were synthesized and surface modified with a range of alkyloxysilanes in the fabrication of mixed matrix membranes (MMMs) in order to minimize the undesirable voids. Moreover, silicalite-1 particles in MMMs have a double function which not only provided transport channels of high selectivity, but also reacted with the hydroxyl groups of PDMS during membrane formation. The highest separation performance was obtained by incorporating micron-sized n-propyltrimethoxysilane (PTMS)-modified silicalite-1 into the PDMS membrane at a loading of 63 wt. %. A high separation factor of 72.9 and a normalized total flux of 1156 g/m^2h were obtained in separating n-butanol from a 1.5 wt. % aqueous solution at 55 ^oC. Compared to the pure PDMS membrane, the separation factor and normalized butanol flux were increased by 83.6 % and 19.7 %, respectively.

Fig. 1 Schematic diagram of the strategies for improving the interactions and eliminating the undesirable voids, and rheology of membranes in curing process.

Additionally, MCM-41 (mesoporous) was also used as fillers in fabricating of mixed matrix membrane via the same method. Compared to the pure PDMS membrane, the total flux increased by 40 % when using PTMS-modifiled MCM-41 as macro-crosslinker.

492577 What Causes Fuel Cell Membranes to Dry out?

Abstract:

The requirements for feed humidification in Polymer Electrolyte Membrane (PEM) fuel cells was examined with one and two dimensional PEM fuel cell (PEMFC) reactors. Fuel cells were built with flow patterns that operated as 1D, stirred tank reactors or 2D, plug flow reactors. The fuel cell were operated as autohumidified fuel cells with bone dry feeds. Autohumidified fuel cell operation is sustained in 1D at temperatures of 40-80°C provided the production of water (1/2 current) exceeds the convection of water vapor exiting the fuel cell (saturation water vapor mole fraction × vapor molar flow rate). Detailed water balances for the one dimensional reactor demonstrate that in autohumidified fuel cell operation water crosses from the cathode to the anode such that the ratio of the water activity at the anode to that at the cathode is approximately 0.8. The water activity difference between the anode and cathode sets an upper limit of electro-osmotic drag < 0.1. Membrane dehydration occurs when the convective water vapor flow from the anode and cathode exceeds water produced by reaction. Membrane dehydration is more significant with higher temperature, higher gas feed rates. Membrane dehydration is more severe in 2-dimensional fuel cells with long thin gas flow channels because of greater gas flow velocity convecting more water vapor. Autohumidified operation with moderate current density (~ 0.5 $A.cm^{-2}$) was sustained in a 2D H_2/O_2 fuel cell with small stoichiometric excess. Sustained autohumidified operation in a 2D H_2/air was only possible with low current densities (<0.1 $A.cm^{-2}$).

Polyarylester nanofiltration membranes with solvent resistance

Yan Fu, Ayang Zhou, Jinli Zhang, Wei Li (*)

Collaborative Innovation Center of Chemical Science and Chemical Engineering (Tianjin), School of Chemical Engineering &Technology, Tianjin University, Tianjin 300350, People's Republic of China.

Nanofiltration has been widely used to concentrate, purify and remove small pollutants as a low-energy separation process without phase transition. As one of the important varieties, organic solvent nanofiltration (OSN) membrane have received more attention because of its stability in organic solvent media[1-3]. Most OSN membranes are hydrophilic polyimide membranes formed by interfacial polymerization (IP) using amines as aqueous monomers. To date, there are few hydrophobic TFC OSN membranes prepared via IP due to the absence of suitable monomers. Lignin and its derivatives exhibit lower solubilities in many organic solvents[4], and consist of typical phenolic structures which could react with trimesoyl chloride (TMC). Therefore, we are motivated to employ lignin derivatives involving ammonium lignosulfonate (AL), vanillic alcohol (VA) and guaiacol (GA) to develop hydrophobic OSN membranes (illustrated in Scheme 1 and 2), and study its separation performance in organic solvents including THF, DMSO, DMF DMAc and acetonitrile (ACN).

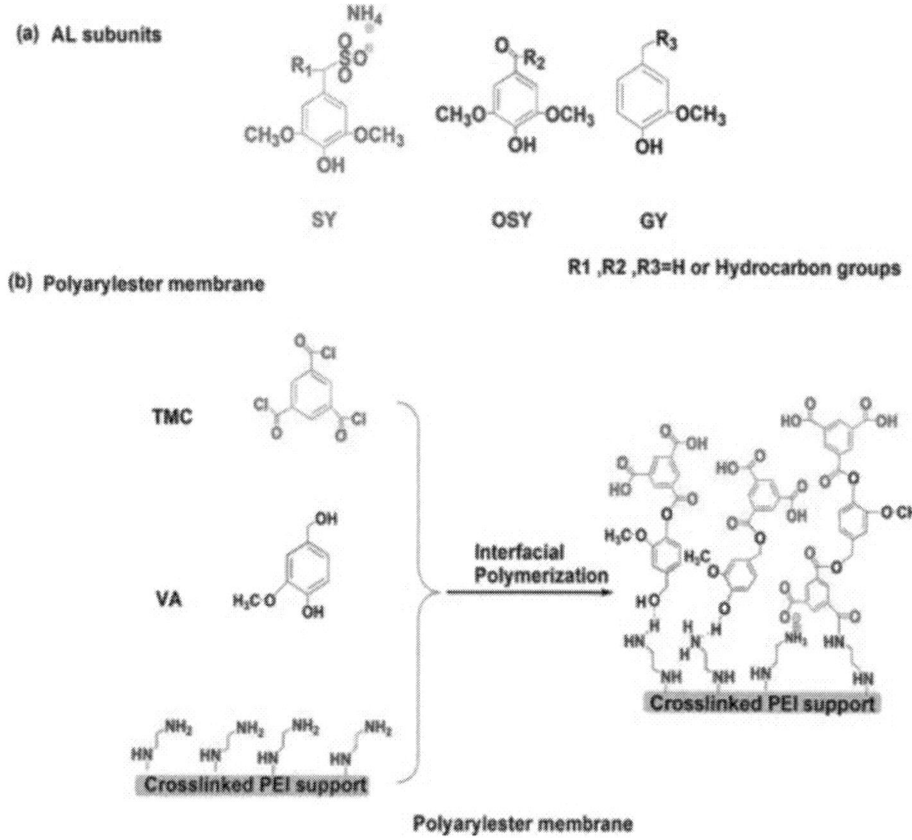

Scheme 1. (a) Three subunit structures of AL; (b) Polyarylester NF membrane obtained via the interfacial polymerization between VA and TMC on the cross-linked polyetherimide support.

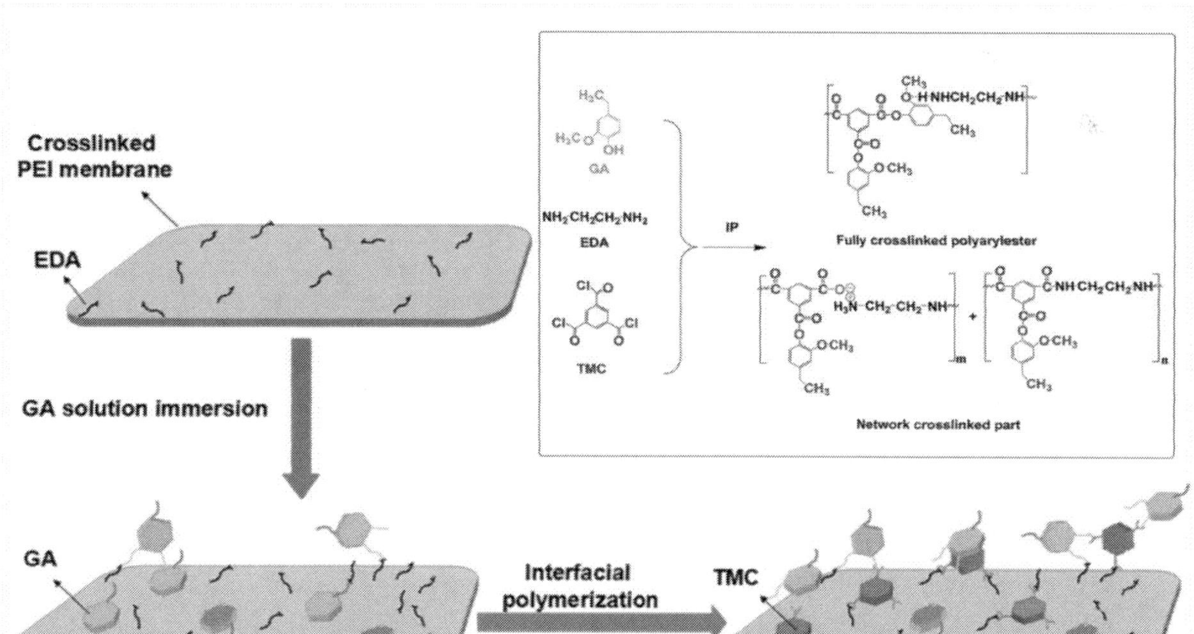

Scheme 2. Schematic diagram of preparation process of NF-GA membrane.

Fig. 1 shows the performance of the membranes with different concentrations of AL reacting with 0.06% (w/v) TMC to separate a DMSO solution containing 0.01 g L^{-1} BB. The AL-TMC NF membranes show superior organic solvent resistance, the optimal NF-2AL membrane has a permeance of 0.97 L m^{-2} h^{-1} MPa^{-1} and a 98% rejection for Brilliant Blue R-250 (BB, 825 g mol^{-1}) in DMSO.

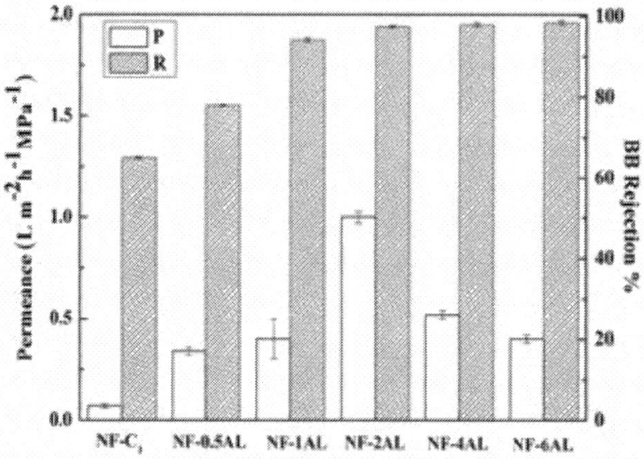

Fig. 1. The performance of membranes with different AL concentrations (0, 0.5%, 1%, 2%, 4%, 6% (w/v)) for a 0.01g·L^{-1} BB in DMSO solution at 1 MPa.

To enhance the permeance, we adopted VA as the monomer, which has the molecular structure similar to AL, to synthesize the solvent resistant NF membrane (VA-TMC). As shown in **Fig. 2**, the VA-TMC NF membrane exhibits high DMSO permeance about 21 L m^{-2} h^{-1} MPa^{-1}, the comparable BB rejection of 89%.

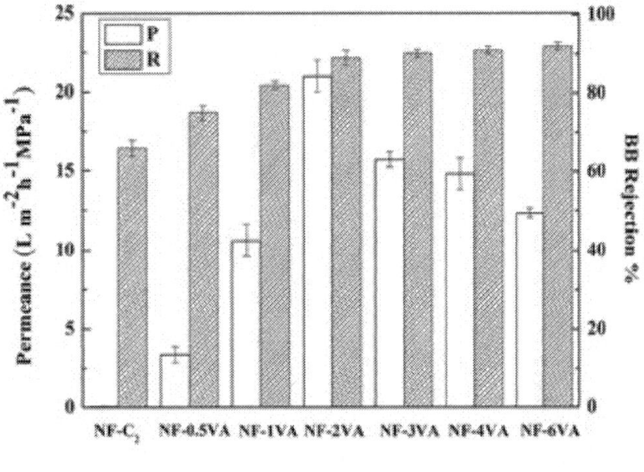

Fig. 2 The performance of membranes with different VA concentrations (0, 0.5%, 1%, 2%, 3%, 4%, 5% (w/v)) for a 0.01gL $^{-1}$ BB in DMSO solution at 1 MPa.

Fig. 3 illustrates the morphologies of NF-2VA, NF-4VA and NF-6VA. These membranes have an asymmetric structure composed of a finger-like porous support layer and a top skin layer. The skin layer thickness is 198, 278 and 322 nm for NF-2VA, NF-4VA and NF-6VA, respectively. The hydrophilicity was evaluated by measuring the static water contact angles. The water contact angle is 54±3° for NF-C_2, 76±2° for NF-2VA, 81±2° for NF-4VA, and 83±1° for NF-6VA. The increase in the contact angle with the amount of VA suggests that the polyarylester membranes have relatively high hydrophobicities, which can be attributed to the phenyl-enriched surfaces. According to ATR-FTIR we can conclude that the contents of ester groups and methoxyl groups can be enhanced by increasing the VA concentration. Based on the solid-state ^{13}C NMR spectra shown in **Fig 4**, the peaks at 172.8 (carbonyl carbon of acyl chloride), 165 (ester) and 135 ppm (the typical peak of TMC) confirm that VA reacts with TMC to form an ester bonded structure in the product of VA-TMC.

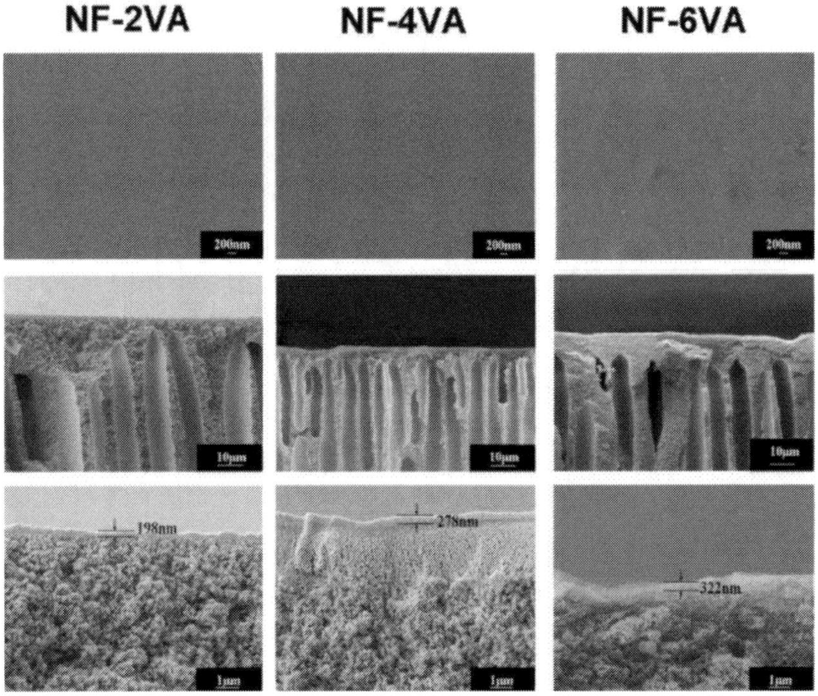

Fig. 3 SEM images of the NF-2VA, NF-4VA and NF-6VA membranes.

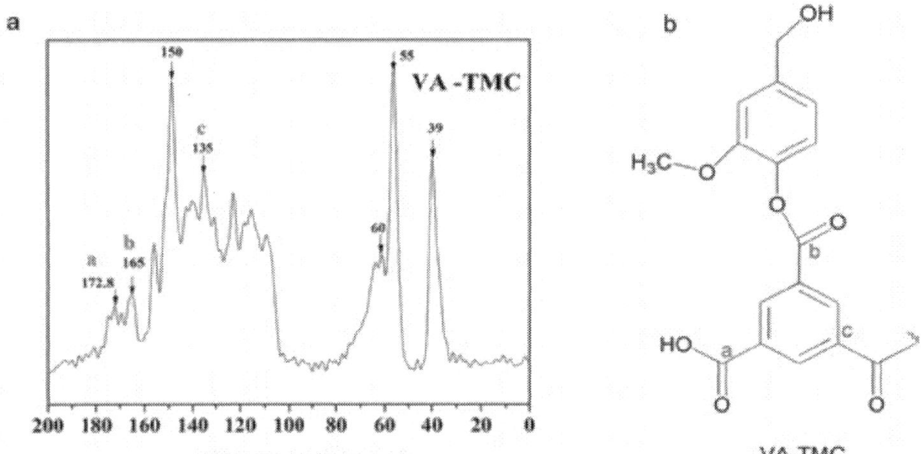

Fig. 4 (a) The solid-state ^{13}C NMR spectrum of the product from the reaction of VA and TMC in the same ratio as the top layer of the NF-2VA membrane; (b) The structure of the polyarylester product, VA-TMC.

Fig. 5 contains the performance of the membranes with different GA concentrations when separating a 0.01 g L^{-1} Rose Bengal (RB) in THF, ACN, NMP, DMSO and DMF solutions. For the membranes prepared via IP between GA and

TMC, the NF-3GA has a permeance of 73.05 L $m^{-2} h^{-1} MPa^{-1}$ and 99% rejection of RB in THF, a permeance of 7.01 L $m^{-2} h^{-1} MPa^{-1}$ and 87% rejection of RB in DMSO, a permeance of 6.49 L $m^{-2} h^{-1} MPa^{-1}$ and 91% rejection of RB in NMP, a permeance of 10.9 L $m^{-2} h^{-1} MPa^{-1}$ and 75% rejection of RB in DMF, a permeance of 9.5 L $m^{-2} h^{-1}$ MPa^{-1} and 100% rejection of RB in ACN. The solvent permeance for GA-TMC membrane increases with decreasing polarity of the solvent: for THF, a permeance as high as 73.05 L $m^{-2} h^{-1} MPa^{-1}$ was obtained. The permeance decreases in the order DMSO<NMP<DMF<ACN<THF.

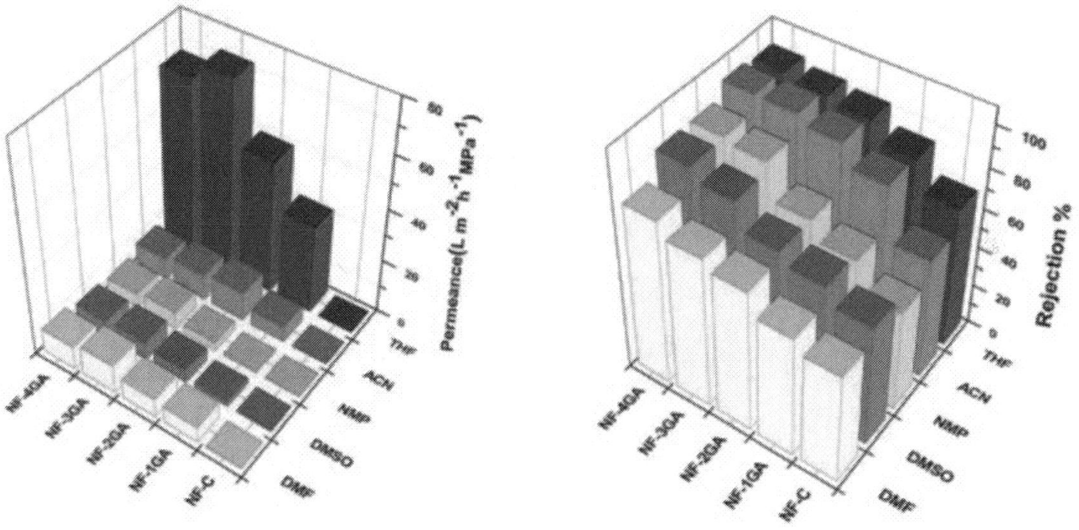

Fig. 5 The performance of membranes with different GA concentrations (0, 1%, 2%, 3%, 4%(w/v)) for a 10 mg·L^{-1}RB in THF, ACN, NMP, DMSO and DMF solution at 1 MPa.

Fig. 6 presents the morphology the surface and cross-sectional morphologies of NF-2GA, NF-3GA and NF-4GA. All the membranes have an asymmetric structure composed of a finger-like porous support layer and a top skin layer. The thickness of the top-layer remarkably increases from 83 to 208 nm with the loading of GA from 2to 4g. For GA-TMC, the water contact angle increases with the amount of GA indicates that the polyarylester membranes have higher hydrophobicity. According to ATR-IR shown in **Fig. 7**, the calculated intensity ratio of I_{1235}/I_{1650} increases in the following order: NF-2GA < NF-3GA < NF-4GA, while the ratio of I_{1036}/I_{1650}

increases as the order of NF-2GA < NF-3GA < NF-4GA. Therefore, we can conclude that the contents of ester groups and methoxyl groups can be enhanced by increasing the GA concentration.

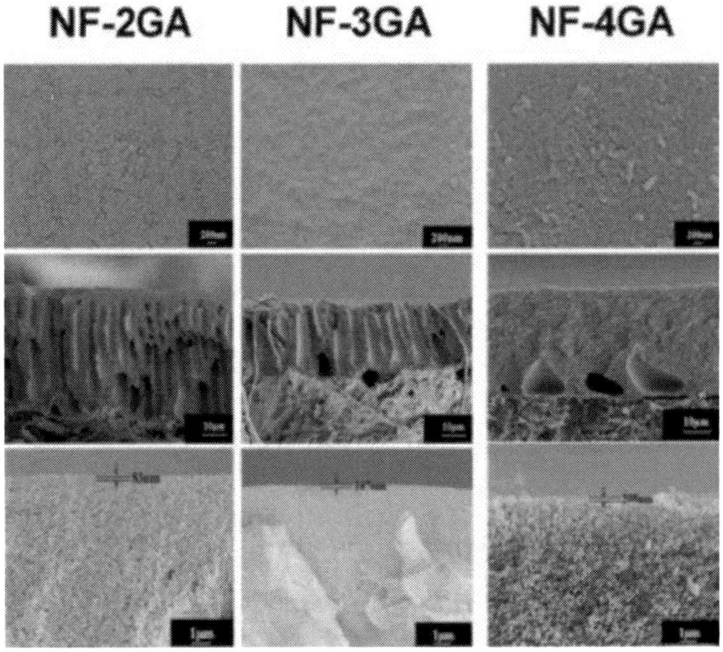

Fig. 6 SEM images of the NF-2GA, NF-3GA and NF-4GAmembranes.

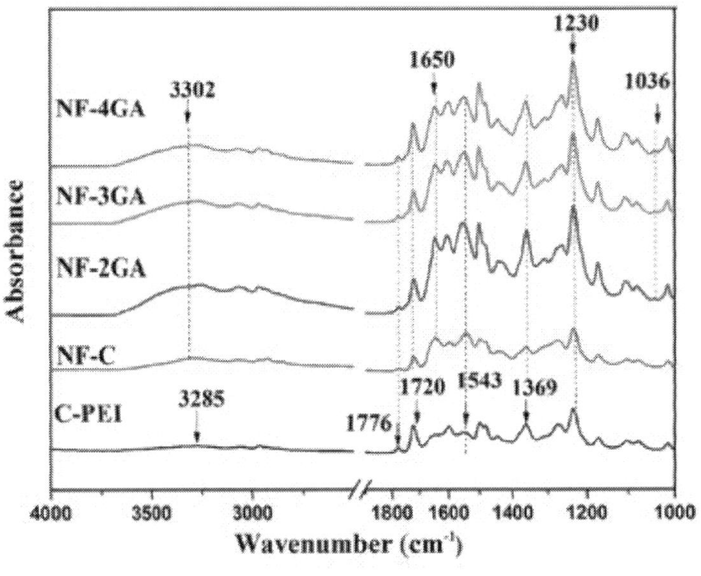

Fig. 7 The ATR-FTIR spectra of the C-PEI, NF-C, NF-2GA, NF-3GA and NF-4GA membranes.

In summary, a TFC NF membrane with solvent resistance was synthesized via interfacial polymerization using AL and TMC as monomers. The separation performance of the membranes was assessed in various organic solvents including THF, DMSO, DMAc, and DMF; the membranes showed good rejection but low permeance. To enhance the permeance, we adopted VA and GA, which possesses similar molecular structures to AL subunits, to synthesize NF membranes. VA-TMC NF membrane exhibits high DMSO permeance about 21 L m^{-2} h^{-1} MPa^{-1}, the comparable BB rejection of 89% and a rejection of 81% for clindamycin phosphate. Moreover, GA-TMC NF membrane has a maximum permeance of 73.05 L m^{-2} h^{-1} MPa^{-1} and 99% rejection of RB in THF. These results demonstrate the polyarylester nanofiltration membrane could be effective as a recovery technique for drug residues from organic solvents in the pharmaceutical industry. Interfacial polymerization involving other lignin derivatives can be further employed in the fabrication of new OSN membranes.

References

(1) Marchetti, P.; Jimenez Solomon, M.F.; Szekely, G.; Livingston, A.G. (2014) Molecular separation with organic solvent nanofiltration: a critical review. *Chemical reviews*, 114(21): 10735-10806.

(2) Gohil, J.M.; Ray, P. (2017) A review on semi-aromatic polyamide TFC membranes prepared by interfacial polymerization: Potential for water treatment and desalination. *Separation and Purification Technology*, 181: 159-182.

(3) Werth, K.; Kaupenjohann, P.; Skiborowski, M. (2017) The potential of organic solvent nanofiltration processes for oleochemical industry. *Separation and Purification Technology*, 182: 185-196.

(4) Duval, A.; Lawoko, M. (2014) A review on lignin-based polymeric, micro- and nano-structured materials. *Reactive and Functional Polymers*, 85: 78-96.

Acknowledgement

This work was supported by the National Key Research and Development Program (2016YFC1201503), NSFC (21621004) and the Program for Changjiang Scholars and Innovative Research Team in University (IRT_15R46).

495387 Bipolar Membrane Electrodialysis for Cleaner Production of Amino/Organic Acids

Abstract:

A bipolar membrane (BPM) is a kind of composition membrane that at least consists of a layered ion-exchange structure with a cationic selective layer and an anionic selective layer. Bipolar membrane electrodialysisis considered as the-state-of-art technology for separation and conversion of amino/organic acids. This technology has considerable promise for environmentally friendly production of organic/amino acids which is able to operate without the need for significant chemicalsand energy input. In this study, the production of three amino acids (methylglycine, dimethylglycine and trimethylglycine) with similar structure but different alkyl groups was investigated usingbipolar membranes electrodialysis. Effects of feed concentration, current density and cell configuration on product yield were intensively analyzed in terms of ion dimensions, intrinsic transport velocities, ion concentration, ion orientation, and the interaction between amino acids and membranes.Results indicated that the optimal recovery efficiency was achieved at current density of 20 mA/cm^2 and feed concentration of 6%. Meanwhile, the optimal energy consumptions and current efficiencies were 2.3kWh/kg and 78% for methylglycine, 2.49 kWh/kg and 69.5% for dimethylglycine, and 3.52kWh/kg and 39.6 % for trimethylglycine, respectively. Additionally, it was found that the product yield ratio for three tested amino acids followed the order: methylglycine>dimethylglycine>trimethylglycine. Naturally, it was clearly shown that the separation and purification of amino acids using bipolar membranes were significantly affected by intrinsic molecular characteristics and molecular volume

487044 Cross-Linked Mixed Matrix Membranes for Organic Solvent Nanofiltration (OSN) Consisting of P84 Polyimide and Carboxyl-Functionalized Multi-Walled Carbon Nanotubes

Abstract:

To design solvent-resistant nanofiltration membranes from polyimide materials, one must cross-link the polyimide membranes in order to improve the stability and resistance of the membrane in the harsh environments but incorporate porous fillers to open up the channels for a better solvent transport. A moderate annealing is also needed to manipulate the pore size for a better solute rejection. We have fabricated mixed matrix membranes (MMMs) consisting of carboxyl-functionalized multi-walled carbon nanotubes (MWCNTs-COOH) and P84 polyimide in this study for organic solvent nanofiltration applications, such as pharmaceutical, with the aid of cross-linking by 1, 6-hexanediamine (HDA). The incorporation of hydrophilic functionalized carbon nanotubes into P84 not only improves liquid sorption and transport but also enhances membrane porosity and pore size. Consequently, the permeances of water, ethanol, and isopropanol across the MMMs increases with an increase in MWCNTs-COOH loading up to 0.075 wt. %. However, a higher loading of carbon nanotubes may result in agglomeration and reduce the separation performance. The cross-linked MMM comprising 0.05 wt. % MWCNTs-COOH has a rejection of 85% to rose bengal molecules of 1017.65 Dalton with an ethanol permeance of about 9.6 (L/m^2 h bar) at 5 bar. A higher transmembrane pressure or a lower feed concentration show a better separation performance for the MMMs. Interestingly, the rejection of rose bengal in isopropanol solutions is higher than that in ethanol solutions (i.e., 99 vs. 85%) even though the former has a lower solvent flux than the latter. After thermal annealing at 150°C in a 3/1 EG/PEG400 (weight ratio) solution, the resultant membranes show very improved rejections to small dyes. The cross-linked and then annealed MMM comprising 0.05 wt. % MWCNTs-COOH has a rejection of almost 100% to Safranin O dye molecules of 350.85 Dalton with an ethanol flux of about 4.6 LMH at 20 bar.

501516 Modifying IgG Glycan Profile Using a High-Throughput Multi-Gene Assembly Platform

Abstract:

Protein N-glycosylation is a post-translational modification that is crucial for protein folding, biological activity, half-life and stability. It is one of the most complex post-translational modifications in that it involves a complex network of enzymes that may act on identical substrates to generate many different glycosylation patterns. A drawback of protein expression in mammalian cells is the presence of heterogeneous populations of the same protein due to differences in protein N-glycosylation. The complex nature of glycosylation makes it a good target for control using genetic engineering approaches. While pipelines exist for the generation of multi-gene constructs they are limited in their utility. Utilizing the latest DNA synthesis and assembly technology we developed a platform for combinatorial assembly of multi-gene constructs for use in altering glycosylation of immunoglobulin G (IgG) produced in chinese hamster ovary cells (CHO). A kinetic based model of N-glycosylation was used to predict N-glycosylation genes responsible for generating targeted glycosylation profiles of IgG. Expression of selected glycosylation genes from multi-gene constructs led to altered glycan profiles of IgG demonstrating the efficacy of our platform and validating predictions of our model.

501762 Synthetic Regulation of Sporulation Genes for Tunable Display of Proteins on Bacterial Spore Surface

Abstract:

Bacteria endospores, with multiple protein layers encasing and protecting the genetic material in the core, are the most robust type of cells found in nature. A few proteins on the outmost layer of the spore have been used as anchors to display proteins such as enzymes and vaccines. These displayed proteins benefit from the protective protein layer and gain enhanced robustness. In addition, they are assembled to the spore surface during sporulation and no further purification or immobilization steps are needed, which significantly reduces the production costs. In all, bacterial spores are an attractive platform for protein display. However, the assembly of spores is delicately regulated and even the outermost layer consists of over 70 different proteins. How these proteins are interacted and organized is elusive. Currently display of a protein on the spore surface relied on a few abundant surface proteins, the expression level of which is completely regulated by native sporulation regulators. In order to display multiple enzymes on the same spore surface and regulate their relative abundance, a tunable expression system in response to non-sporulation signals are needed. In this work, we designed tunable expression systems that integrate the sporulation regulatory elements and binding sites for non-sporulation regulators, including transcription factors LacI and TetR, as well as a series of dCas9:sgRNA complexes targeting at various sites of the spore surface gene. Using GFP and mCherry as reporters, the expression level of the spore surface anchor was evaluated under various induction conditions using the wildtype as a control. Most of the designed synthetic systems show a wide range of expression levels without impeding the integrity of the spores. These synthetic regulation systems allow the fine tuning of displayed proteins on the spore surface, which is critical to optimize a multi-protein pathway.

495026 Kinetically-Controlled RNA Aptamer Biosensors for High-Throughput Production Phenotype Screening

Abstract:

A major bottleneck in the biological production of biochemicals from renewable feedstocks is the lack of high throughput tools to screen for small molecule production in the generation, optimization, and industrialization of microbial strains. We have developed a new computational pipeline to design kinetically-controlled RNA based sensors that enable the rapid implementation of high-throughput screens for secreted biochemicals and production phenotypes. The biosensors combine in vitro selected RNA aptamers for molecular recognition with DNA strand displacement for signal generation, and require only standard laboratory equipment such as a qPCR machine or a plate reader. Currently, the RNA biosensor platform has a throughput of at least 128 HPLCs, and because it is not hardware-centric, can screen the same number of samples at a fraction of the cost. We validated our new approach by the construction of 38 novel devices for the aromatic amino acid derivative *p*-amino-phenylalanine (*p*-AF), a precursor in an engineered styrene derivative production pathway. We have developed a set of *p*-AF biosensors with EC_{50} values between 1 - 0.5 mM, ideal for screening production of *p*-AF producing *E.coli* strains with titers as low as 100 mg/L. In this presentation, I will discuss these results and on-going efforts to demonstrate industrially relevant synthetic biology applications including production titer screens of genetic variants, biocatalysis validation of heterologous pathway enzymes, and media optimization for increased productivity and industrialization of microbial strains.

502900 Modcell: A Multiobjective Strain Design Platform for Modular Cell Engineering

Abstract:

Metabolic engineering has enabled the use of microbial cell factories for industrial production of biochemicals. However, developing an optimal strain for synthesis of one product with the conventional strategy is laborious and expensive. To accelerate and reduce the cost of strain engineering, we develop the modular cell (MODCELL) design principle by exploiting the modular organization of metabolic networks and combinatorial possibilities of metabolic modules that enable the synthesis of a large space of biochemicals. Using the multiobjective optimization methods, we develop novel algorithms to implement the MODCELL design for genome-scale metabolic networks. We demonstrate MODCELL to design the biomass-degrading *Clostridium thermocellum* modular cell for combinatorial synthesis of biochemicals, e.g., alcohols and bioesters from lignocellulosic biomass. We envision the MODCELL will provide a useful platform for modular cell engineering.

495599 Optimization Driven Top-Down Synthesis of Genome Minimized Strains for Bioproduction

Abstract:

Genome minimized strains are obtained by removing genome segments associated with genes or processes either detrimental or unneeded under bioproduction conditions. They offer advantages as production chassis [1, 2] by reducing transcriptional cost, eliminating competing functions and limiting unwanted regulatory interactions. Genetically minimized strains have been experimentally explored before, demonstrating beneficial characteristics [3]. Existing approaches for identifying stretches of DNA to remove are largely *ad hoc* based on information on presumably dispensable regions through experimentally determined non-essential genes and comparative genomics. As more sophisticated genome editing tools (e.g., CRISPR) are becoming commonplace, the need for a computational aid that will help sequentially minimize genomes consistent with a set of performance criteria is becoming more pressing. Here we introduce a versatile genome reduction algorithm that implements a mixed integer linear program (MILP) to iteratively identify the largest contiguous sequences within the genome that can be deleted without affecting the organismâ€™s growth rate or other desired attributes. Known essential genes or genes that cause significant fitness or performance loss can be flagged and their deletion is thus prohibited. This optimization-based procedure also accounts for promoter regions ensuring that retained genes will be properly transcribed. The algorithm avoids the deletion of synthetic lethal pairs by maintaining biomass production even upon the simultaneous deletion of multiple stretches. In addition, we assess the possibility of removing even larger stretches of DNA if only one or two essential genes are within the deleted sequence. These deleted essential genes can then be re-inserted into the genome in a different location. We applied the algorithm to design minimized *E. coli* strains and found that we were able to recapitulate the long deletions identified in previous experimental studies [3]. Moreover, we discovered alternative combinations of deletions which have not been explored *in vivo*. Identification of large DNA stretches allows us to significantly reduce the time needed to perform deletions during each cycle of the genome reduction process.

[1] Lian, Jiazhang, et al. "Design and construction of acetyl-CoA overproducing *Saccharomyces cerevisiae* strains." *Metabolic engineering* 24 (2014): 139-149.

[2] Vickers, Claudia E., et al. "Grand challenge commentary: Chassis cells for industrial biochemical production." *Nature chemical biology* 6.12 (2010): 875.

[3] Juhas, Mario, et al. "*Bacillus subtilis* and *Escherichia coli* essential genes and minimal cell factories after one decade of genome engineering." *Microbiology* 160.11 (2014): 2341-2351.

504106 Genome-Scale Engineering: A New Frontier in Metabolic Engineering

Abstract:

Advances in reading, writing and editing genomes have greatly expanded our ability to reprogram biological systems at the resolution of a single nucleotide and on the scale of a whole genome. Such capacity has drastically accelerated the cycle of design, build and test in synthetic biology to construct organisms as cell factories for synthesis of fuels and chemicals. In this presentation, I will highlight our recent work in the development of new molecular parts and tools for genome-scale engineering and our attempt in automating the design, build and test cycle. Examples include but are not limited to: (a) design and construction of the CRISPR-Cas system for multiplexed genome engineering in Saccharomyces cerevisiae and Streptomycetes and for genome-scale engineering in S. cerevisiae; (b) design and construction of the RNA interference machinery for genome-scale engineering in S. cerevisiae; and (3) design and development of a biological foundry called Illinois Biological Foundry for Advanced Biomanufacturing (iBioFAB) for protein and cellular engineering.

501691 Targeting Chemoresistance in Bioengineered 3D Tumor Models: Rationally-Designed Combination Therapies Informed By Physical Stress and Heterocellular Communication

Abstract:

Tumor heterogeneity and drug resistance to conventional therapies remain major causes of treatment failure, recurrence and dismal survival rates for patients with advanced stage cancers. A range of cellular, architectural, and physical cues in the tumor microenvironment influence the intrinsic and acquired resistance mechanisms that lead to treatment failure. These cues include physical forces such as hydrodynamic shear stress and communication with heterocellular stromal partners, which remain understudied as determinants of tumor heterogeneity and variability in treatment response. Strategies that leverage photodynamic therapy (PDT) a photochemistry-based biophysical treatment modality to regionally target and prime stubborn tumor populations may be essential to overcoming key barriers to durable cancer management while minimizing toxicity from traditional agents. A multi-faceted approach is needed to evaluate and optimize PDT-based combination therapies, including the development of bioengineered 3D models that integrate cues such as physical forces and heterotypic cellular communication. Here the impact of hydrodynamic stress and stromal biology is evaluated in the context of ovarian cancer (OvCa). Metastatic OvCa spreads predominantly via flushing of ascites along preferential fluidic pathways and communicates with local microenvironment, including the extracellular matrix (ECM) and TECs, to form peritoneal implants. A microfluidic model that supports 3D tumor growth was developed to investigate the role of fluidic stress on the heterogeneity of metastatic OvCa. The motivation for this study was based on clinical observations that the most stubborn tumors are often found in regions such as the peritoneal gutter, a common site of resistance and recurrence, and a region that is subjected to fluidic stress from ascites. Tumor nodules cultured under flow showed increased epithelial-mesenchymal transition (EMT) compared to non-flow 3D cultures. Molecular and morphological changes consistent with EMT included a transcriptionally-regulated significant decrease in E-cadherin, a significant increase in vimentin, and significant decrease in fractal dimension, a metric adapted to quantify spindle-like morphology. A concomitant significant post-translational upregulation of epidermal growth factor receptor (EGFR) expression and activation was seen under flow. The impact of heterotypic communication between TECs and OvCa cells was investigated in a 3D model. Tumors grown in the presence of TECs were differentially susceptible to chemotherapy and benzoporphyrin derivative (BPD)-based PDT and showed increased heterogeneity in treatment response in the presence of endothelial cells. The potential value of using bioengineered models to guide customized, rationally-designed PDT-based combination regimens will be presented.

500875 Co-Administration of the Tumor-Penetrating Peptide Irgd Improves the Therapeutic Efficacy of paclitaxel in a 3D Air-Grown Lung Cancer Model

Abstract:

Current cancer research relies on two-dimensional (2D) cell culture, which does not recapitulate *in vivo* conditions in their entirety. Due to this unnatural microenvironment, 2D cell assays often provide unpredictable and contrasting results in comparison to animal studies and clinical trials, and contributes to the low success rates and high costs associated with drug development. Three-dimensional (3D) spheroids are better representative of cancer cells that tend to form solid tumors *in vivo*. By exhibiting intrinsic physiological and morphological characteristics similar to tumor tissue, 3D spheroids closely mimic *in vivo* biology and are characterized for having hypoxic cells and a necrotic core, which often gives them high resistance to the treatments.

3D cell culture platforms are increasingly being used in cancer research and drug development since they mimic avascular tumors *in vitro*. In the present study, we focused on the development of air-grown multicellular spheroids (MCS) to mimic *in vivo* tumors for understanding lung cancer biology and improving treatment methods. 3D MCS were formed using A549 lung adenocarcinoma cells, comprising cellular heterogeneity with different proliferative and metabolic gradients. The growth kinetics, morphology, and 3D structure of the MCS were characterized by brightfield, fluorescent, and scanning electron microscopy. In a drug resistance study MCS demonstrated significant decrease in the growth when the tumor penetrating peptide iRGD and paclitaxel (PTX) were co-administered as compared to PTX alone. It was also found that when treated with both iRGD and PTX, A549 MCS exhibited an increase in apoptosis and decrease in clonogenic survival capacity as compared to PTX treatment alone. This study demonstrated that co-administration of iRGD resulted in the improvement of tumor penetration ability of PTX in an *in vitro* A549 3D MCS model. In addition, this is the first time an air-grown lung cancer tumor spheroid model has been developed and evaluated.

Keywords: Lung cancer; Spheroids; Tumor-penetrating peptide iRGD; Paclitaxel, A549 lung adenocarcinoma

501166 Capturing Microenvironmental Regulation of Metastatic Dormancy and Recurrence

Abstract:

Disseminated tumor cells (DTCs) undergo varying periods of dormancy in ectopic tissue sites before developing overt metastatic tumors. Accumulating evidence suggests that metastatic dormancy and recurrence of DTCs is regulated by intrinsic genetic instability and close interaction with the surrounding microenvironment. However, the mechanism by which DTCs enter and escape dormancy remains largely uncertain due to the lack of experimental model systems that can capture the initial activity of extremely rare DTCs. Here, we introduce a tissue-engineered approach to capture the critical events regulated by the extrinsic tissue microenvironments on DTCs with high experimental control and fidelity. We have developed human soluble factor available and richly vascularized microenvironments by subdermally implanting human bone marrow stromal cell pre-seeded scaffolds into an immunodeficient NSG mouse. Humanized tissue analogues recruited circulating human tumor cells released from physiologically relevant orthotopic prostate tumor xenograft. Tail-vein delivery of human peripheral blood mononuclear cells further increased cellular complexity. Established an early stage humanized metastatic niches were serially transplanted into a naïve NSG mouse to continuously monitor evolution of mature metastatic microenvironment. Our approach successfully recapitulated the heterogeneous phenotypes, dormant and aggressive, of DTCs and demonstrated human stromal and immune cell derived niche regulation. We envision that the controlled, analytical, and transplantable features of tissue-engineered humanized tumor microenvironments will be an invaluable tool for studying rare DTC biology, potentially leading to anti-metastatic strategies that target the tumor microenvironment.

493493 3D Organotypic Cancer Model: Pancreatic Ductal Adenocarcinoma Replaces Endothelium during Tissue Invasion

Abstract:

Although the diagnosis and treatment of cancer in its earliest stages has significantly improved outcomes in many tumors, survival rates in patients with tumors that have spread to distant sites remain dismal. As such, the vast majority of cancer mortalities stem from metastasis and its complications. Metastasis describes the process where tumor cells escape from the tumor primary sites to enter the blood stream, and arrive at other distant organs to form secondary tumors. Among the many metastatic cancers, pancreatic ductal adenocarcinoma (PDAC) over the past decades has remained one of the deadliest metastatic diseases owing to the high incidence of metastasis of pancreatic cancer cells. Remarkably, patients with PDAC exhibits a high degree of vascular invasion and abundant circulating tumor cells in the bloodstream, which suggest that PDAC cells are highly efficient to gain access to the blood stream. Despite this, the interactions between PDAC cells and blood vessels are poorly understood, partly due to the lack of *in vitro* models to enable close examination of tumor cells and endothelial cells at the blood vessel interface.

Here, we describe an organotypic PDAC-on-a-chip culture model to examine vascular invasion in PDAC. The model features two hollow cylindrical channels embedded within a three-dimensional matrix [1]. One of the channels was lined with endothelial cells to mimic blood vessel while the other was seeded with PDAC cells to mimic a tumorous pancreatic duct [2]. In this model, PDAC cells from the tumorous pancreatic duct invaded through the intervening matrix and into the blood vessel lumen, induced endothelial cell apoptosis, and ultimately replaced the endothelial cells. A phenomenon we termed vascular replacement. We later confirmed the observation of vascular replacement by tumor cells in our organotypic model by using an *in vivo* model of ectopic PDAC growth. Interestingly, PDAC human patients have also exhibited vascular invasion and tumor-like cells, which populate the lumen side of the blood vessels in previous reports. This confirms the clinically relevant observation of vascular replacement in our 3D organotypic model. Additionally, using small molecule inhibitors and CRISPR technology, we identified the Activin-ALK7 pathway as a critical mediator of vascular replacement by PDAC. This tumor-on-a-chip model thus provides an important *in vitro* platform for revealing previously unappreciated mechanisms by which tumors interact with vessels.

[1] **Nguyen, D.H.,** Stapleton, S.C., Yang, M.T., Cha, S.S., Choi, C.K., Galie, P.A., Chen, C.S. (2013). *Biomimetic model to reconstitute angiogenic sprouting morphogenesis in vitro*. **Proceedings of the National Academy of Sciences**. 110(17): 6712-6717.

[2] **Nguyen, D.H.,** Esak, L., Alimperti, S., Wong, A., Eyckmans, J., Stanger, B.Z., Chen, C.S. (2017). *Pancreatic ductal adenocarcinoma replaces endothelium during tissue invasion*. **Manuscript in revision**.

496781 Photothermal Therapy of Bladder Cancer Via Targeted Single-Walled Carbon Nanotubes

Abstract:

Single-walled carbon nanotubes (SWNTs) are unique in that they strongly absorb near-infrared (NIR) light, while biological systems have very low levels of absorption of NIR light. This project focuses on targeting the SWNTs using conjugation with the human annexin A5 protein. Annexin A5 is a monomeric protein that binds with high affinity to phophatidylserine (PS) in phospholipid bilayers. PS is the most abundant anionic phospholipid of the plasma membrane and is tightly segregated to the internal side of the plasma membrane in most cell types. Anionic phospholipids are largely absent from the external surfaces of resting mammalian cells under normal conditions. PS is expressed on the external surface of tumor cells and also of endothelial cells that line the blood vessels in tumors, but it is not expressed on the outside surface of the vascular endothelium in normal organs. The SWNTs serve as an NIR target resulting in tumor heating and ablation.

This project focuses on the photothermal therapy treatment of superficial bladder cancer. Approximately 80% of bladder cancer patients have recurrent tumors starting with stage I and quickly progressing towards more muscle invasive and metastatic malignancies with each recurrence. The high incidence of reappearance is believed to be due to the residual tumor left behind which leads to a high patient to cost ratio. The current form of treatment is a transurethral resection for lower grade tumors and cystectomy for higher grade masses. Novel treatments are needed for eliminating all bladder tumor cells with a minimum number of treatments, ideally one.

In the study of this therapy for treating bladder cancer, *in vitro* binding tests confirmed a strong binding affinity of SWNT-annexin A5 to MB49 mouse and J82 human bladder cancer cells. Combining SWNT-annexin A5 heating via NIR light confirmed significant cell death as compared to untreated controls for both cell lines. The *in vitro* tests provided statistically significant validation for the potential of this targeted ablation therapy. *In vivo* testing on immune competent mice was conducted to confirm the efficacy of this treatment even further. Immunohistochemistry for the annexin A5 component of the conjugate was conducted 24 hours post nanotube delivery. SWNT-annexin A5 binding was seen only on bladder cancer tissue and no non-specific binding was detected on normal urothelium. A biodistribution study followed by FT-Raman analysis confirmed no non-specific accumulation of SWNT-annexin A5 in any organs. Using a 360° diffuser to deliver NIR light to the inside of mouse bladders, NIR power tolerance tests confirmed that no healthy tissue damage occurred at 50 J/cm^2. *In vivo* therapy testing of MB49 cells implanted in the bladder of immune competent mice confirmed a promising modality with no detectable tumor presence 24 hours after treatment. Survival studies have confirmed a significant increase in median survival. Studies are ongoing to assess rate of recurrence.

496783 Sprouty2 Regulates Signaling and Phenotypic Responses of Glioblastoma Cells to DNA Damaging Agents and Receptor Kinase Inhibitors

Abstract:

Cell signaling mechanisms govern tumor response to approved and investigational therapeutics, including DNA damaging agents and targeted therapeutics such as EGFR and MET kinase inhibitors. Understanding these mechanisms quantitatively is key to improving patient survival, especially in tumors characterized by a high degree of chemoresistance. We previously identified the protein Sprouty2 (SPRY2) as a possible tumor promoter in glioblastoma multiforme (GBM). This discovery was unexpected because SPRY2 functions as a tumor suppressor in other tumor types. Specifically, we found that SPRY2 knockdown reduced proliferation of human GBM cells and antagonized the growth of subcutaneous GBM tumor xenografts in mice. Here, we explored the role of SPRY2 in modulating the response of glioblastoma cells to two different classes of cancer therapeutics: DNA damaging agents, temozolomide and carboplatin, and targeted inhibitors of oncogenic receptor kinases, Gefitinib + PHA665752 (G+P). SPRY2 knockdown increased GBM cell line response to both classes of therapeutics. Thus, the normal expression of SPRY2 in glioblastoma cells antagonizes cell-autonomous responses that would otherwise lead to increased cancer cell death, pointing to SPRY2 as a broad driver of glioblastoma resistance to therapy. We went on to survey dynamic alterations in the DNA damage response, proliferation, and survival signaling pathways. We then modeled the relationships between cell death, cell cycle, and cell signaling using partial least squares regression (PLSR), a data-driven computational modeling approach. Our analysis revealed that a core subset of kinase-driven signaling processes, spanning multiple distinct signaling pathways that are common to the cellular response to both classes of therapeutics, is responsible for explaining how SPRY2 broadly modulates glioblastoma response to therapeutics. In combination with the original DNA damaging and receptor kinase inhibition treatments, we then inhibited these key kinase driven pathways, and observed the expected perturbations in cellular phenotypes. Co-inhibiting kinases identified by the PLSR model to contribute to variability between SPRY2 expressing versus knockdown cells, enhanced cell death and cell cycle arrest. These findings begin to elucidate the mechanism by which SPRY2 may act as a tumor promoter in glioblastoma cells. Ongoing studies aim to identify the molecular SPRY2 interaction partners that are upstream of these resistance mechanisms. Ultimately, this work has the potential to guide the rational combination of several available classes of therapeutics in order to improve glioblastoma patient survival.

497956 Function and Regulation of Regenerating Proteins in Pancreatic Cells

Abstract:

Regenerating islet proteins (Regs) comprise a family of 16-20 kDa polypeptides which have been implicated in diverse processes including proliferation, differentiation and prevention of apoptosis in normal pancreas development, diabetes, pancreatitis and pancreatic cancer [1]. Patients with pancreatic ductal adenocarcinoma exhibit significantly higher levels of Reg1A than healthy subjects. Yet, the exact mechanisms underlying Reg protein, expression, function and regulation remain unknown. We focused on a particular member of the Reg family, namely Reg1, which is involved in pancreatic tissue development and cancer. To that end, we have been successful in engineering the production of mouse Reg1 and human REG1A proteins in yeast to facilitate our experimentation. Exposure of cultured cells to the recombinant proteins caused the increase in proliferation thereby confirming the Reg1 activity. We also hypothesized that Reg1 targets cyclin D1 which controls cellular growth. Treatment of pancreatic MIN6 beta-cells resulted in increased expression of cyclin D1 as determined by quantitative PCR and a luciferase-based assay using a cyclin D1 promoter construct. The expression of microRNAs (miRNAs) implicated in pancreatic cancer was profiled aiming at discovering regulators of the expression of Regs. This led to the identification of miRNAs, some of which are differentially expressed in pancreatic acinar and islet cells as well as in pancreatic adenocarcinoma cell lines. *In silico* identification of miRNAs targeting Reg1 were combined with experiments. Following miRNA overexpression in acinar cells, we observed a reduction in mouse *Reg1* expression corroborating our findings at the protein level (western blotting). Our findings are a step toward uncovering the function(s) of Reg proteins and the regulation of their expression in the context of normal and pathological states, especially in pancreatic cancer.

498331 *In vitro* Model of Breast Cancer Cell Dormancy Under Hypoxia-Mimicking Microenvironments Using Cobalt Chloride

Abstract:

In vitro model of breast cancer cell dormancy under hypoxia-mimicking microenvironments using cobalt chloride

Hak Rae Lee[1], Faith Leslie[1], Pedram Motallebnejad[1], and Samira M. Azarin[1]

1. Department of Chemical Engineering and Materials Science, University of Minnesota

The majority of cancer-related deaths among breast cancer patients result from metastatic disease that originates from disseminated tumor cells (DTCs). Clinical observations have indicated that some DTCs undergo a period of dormancy before they start colonization at the metastatic site, with this latent period sometimes lasting years to decades. Cancer cells can enter dormant states to survive unfavorable environmental factors such as metabolic or oxidative stress. While hypoxia has been well-studied in various tumor microenvironments, its role in DTC dormancy is not yet fully understood, in part due to a lack of well-established *in vitro* and *in vivo* models. Hypoxic conditions under conventional hypoxic chambers are extremely unstable and cannot be maintained during characterization outside the chamber since normoxic response is quickly established. Cobalt chloride ($CoCl_2$) is a hypoxia-mimetic agent which stabilizes hypoxia inducible factor 1-alpha (HIF1Î±), a major regulator protein of hypoxia signaling. Thus, we sought to develop an *in vitro* breast cancer dormancy model under hypoxia-mimicking microenvironments using $CoCl_2$. Western blot analysis demonstrated that addition of $CoCl_2$ to MCF-7 breast cancer cells was sufficient to increase HIF1Î± expression, and the level of HIF1Î± expression could be modulated by the dose and duration of $CoCl_2$ treatment. Immunofluorescence staining of HIF1Î± and glucose transporter 1 (Glut1), a key target protein of HIF1 signaling, also demonstrated that both proteins were upregulated and stabilized under $CoCl_2$ treatment. Growth curve analysis showed that $CoCl_2$ treatment led to inhibition of cell growth, with the cell number remaining constant for up to 24 days. The inhibition in cell growth by $CoCl_2$ treatment matched the growth inhibition observed in a hypoxic chamber control, indicating that the effects matched those of true hypoxia. Furthermore, the $CoCl_2$-treated cells exhibited additional hallmarks of dormancy, such as reduced expression of Ki67, a marker of actively cycling cells, and accumulation of cells in the G0/G1 phase of the cell cycle, as determined by flow cytometry following propidium iodide staining. Importantly, longitudinal tracking of cell growth and Ki67 status showed that these non-proliferative breast cancer cells resumed their growth and became Ki67-positive again after removal of $CoCl_2$, indicating the cells under hypoxia-mimicking conditions were quiescent and not senescent. Collectively, the results demonstrate that $CoCl_2$ treatment can effectively mimic dormancy under hypoxic conditions *in vitro* with high stability, thereby enabling robust characterization of dormant cells under hypoxia-mimicking conditions *in vitro*.

509826 Automated Systems for Screening, Kinetics, and Optimization of Chemical Synthesis and On Demand Production

Abstract:

Over the two past decades, small-scale continuous synthesis has matured from simple demonstration examples to applications in pharmaceuticals and fine chemicals. The field has moved beyond single transformations to continuous multistep synthesis of active pharmaceutical ingredients (APIs) by incorporating in-line workup techniques. Integration of on-line measurements of reactant flows, reactor temperature, and outlet concentrations with feedback control systems has enabled automated optimization of reaction yields as well as determining kinetic information. The present contribution starts with automated screening and optimization of chemical reaction in microsystems and continues with integration of small-scale reactors and separators in on-demand continuous synthesis of APIs. Automated continuous/oscillating droplet flow systems are demonstrated for simultaneous optimization of discrete process variables (e.g., catalyst, ligands, and solvents) and continuous conditions (temperature and reagent concentrations). Coupling the droplet reactor with a light source converts the system into an automated microfluidic platform for exploring and optimizing visible-light photoredox catalysis.

As an example of process integration for on demand manufacturing, we present a plug-and-play, reconfigurable, refrigerator-sized manufacturing platform for on-demand synthesis of common pharmaceuticals, e.g. ciprofloxacin. This flexible system is capable of complex multi-step synthesis, in-line purification, post-synthesis work-up and formulation. Multistep synthesis occurs at elevated temperatures and pressures to enhance reaction rates, and the resulting residence times are a few minutes, in contrast to the multiple hour-long synthesis typically needed for batch. Typical production rates are grams/hour sufficient to produce thousands of doses per day of common pharmaceuticals. Finally, we discuss opportunities for integration of machine learning for synthesis planning with fully automated chemical synthesis systems.

509827 Can Laboratory Automation Finally Fulfill the Promise of Revolutionizing Pharmaceutical Discovery?

Abstract:

From Bruce Merrifield's pioneering work in the automation of peptide synthesis in the 60's to the boom and bust of combinatorial chemistry in the 90's, laboratory automation technology for chemical synthesis in early stage drug discovery research has played a role of both hero and villain. This lecture will unearth insights into this rich history and will detail Lilly's research experience with discovery-scale laboratory automation. A specific emphasis will be on our more than a decade of experience with our one of a kind remote access fully integrated Automated Synthesis Laboratory (ASL). Discussion of the key learnings, successes, and challenges in our ASL experience will serve as the backdrop for a bold new vision for the future of remote access, fully integrated laboratory science. The audience will learn about Lilly's ongoing investment in the construction of the Lilly Life Sciences Studio (L2S2), which we believe will significantly change pharmaceutical research in both industry and academia.

509828 Emerging Technology: A Key Enabler for Modernizing Pharmaceutical Manufacturing and Advancing Product Quality

Abstract:

Over the past decade, there has been substantial progress towards modernizing pharmaceutical manufacturing. However, pharmaceutical manufacturing continues to confront a number of challenges, which result in unacceptably high occurrence of product quality issues. Nearly two thirds of all drug shortages relate to quality issues surrounding product manufacturing or facilities. These quality issues expose patients to unnecessary risk and negatively impact public health. Working with industry to encourage the modernization of pharmaceutical manufacturing technology is a mutually beneficial approach to increase manufacturing efficiency, improve overall drug quality, avoid shortages, and prepare for future regulatory challenges.

To encourage the adoption of innovative approaches in pharmaceutical manufacturing and to prepare for reviews and inspections involving technology for which there is little experience, the Agency established the Emerging Technology Program. This program features a cross-functional Emerging Technology Team (ETT) which works directly with industry to identify and resolve potential scientific and policy issues that may impact technologies new to the pharmaceutical industry. In a relatively short period of time, the Emerging Technology Program has facilitated the approvals of the first 3D-printed drug product, the first drug product made via continuous manufacturing, and the first switch to a continuous manufacturing process for an approved drug product.

Advancements in regulatory science are essential to FDA's mission and enable the ETT to issue clear recommendations to sponsors about the risks and benefits of products that involve new technologies. FDA's research program aims to address any knowledge and experience gaps to support these needed improvements. The knowledge gained from the internal and sponsored research will inform policy, review, and inspection activities, ensuring that FDA regulatory policies reflect state-of-the-art manufacturing science.

In this talk, FDA's emerging technology program is described including a summary of industry interactions and observed trends. FDA's manufacturing science research programs are highlighted e.g. continuous manufacturing, 3D printing, and process modeling to demonstrate how advances in regulatory science support the emerging technology program.

500280 Construction of Genetic Devices By Engineering Transcriptional Interference

Abstract:

Different levels of gene expression are often obtained by the action of transcription factors in the *cis*â€'regulatory region. Transcriptional interference is a mechanism where a transcriptional process is negatively affected by the presence of neighboring proteins on the same DNA region. Therefore, Transcriptional Interference offers a promising mechanism for building elements for the design of tunable gene regulation. Here we use tandem regulatory elements in the *cis*-regulatory region that are susceptible to Transcriptional Interference. By tweaking the characteristics of each regulatory element, the extent of Transcriptional Interference on each other can be engineered thus modifying aspects of the gene expression response such as overall regulatory range, sharpness of the induction curve for each inducer as well as sensitivity to each of the inducers. Such behaviors are successfully described by a mathematical model of Transcriptional Interference, presenting the opportunity to both understand how naturally occurring systems integrate multiple inputs exerted by tandem transcription factors and how this type of systems can be engineered *de novo*.

498332 Optogenetic Control of Gene Expression for Metabolic Engineering

Abstract:

Metabolic engineering to produce chemicals faces the following challenges: (1) biosynthetic pathways of interest frequently compete with essential metabolic pathways, (2) desired compounds are often toxic to the cells engineered to produce them, and (3) imbalances in enzyme levels can greatly affect the performance of engineered metabolic pathways. To address these challenges, we introduce a new technological platform that utilizes light-sensitive proteins to control gene expression, and achieve unprecedented control over engineered metabolic pathways. This platform includes optogenetic circuits to control metabolic valves for essential pathways that compete with pathways of interest, as well as biosynthetic pathways of toxic compounds. To avoid light shading effects of high cell density fermentations that would limit the applicability of optogenetics to metabolic engineering, we have developed systems that respond to very low light levels, as well as inverter circuits that repress toxic pathways in the light, during the growth phase (when cell densities are low), and induce them in the dark during the production phase (when cell densities are high). There are many advantages of using light to control gene expression for metabolic engineering, compared to established methods that use chemical inducers. Light is inexpensive relative to many chemical inducers, non-toxic, and less likely to produce off-target effects. Also, our optogenetic circuits are tunable, and reversible, which allows for real-time corrections that are not available to chemical inducers. Furthermore, this technology enables a periodic operation of fermentations, using light pulses to maximize product formation. With this new platform, we have surpassed the maximum yields, titers, and productivities of desired products reported in the literature. Our approach significantly departs from existing strategies to control gene expression for metabolic engineering, and raises the possibility of optimizing engineered strains by simultaneously balancing biosynthetic pathways for products of interest, while keeping in check essential pathways that compete with product formation, and rates of toxic product formation.

499206 Establishment of Artificial Dynamic Regulatory Network and its Application in Metabolic Engineering

Abstract:

In the past decades, metabolic engineering endows microorganisms with an ability to produce new chemicals. However, a big challenge of metabolic engineering remains on how to maximize the metabolic flux through exogenous pathways while maintaining the balanced state of the microbial hostâ€™s native metabolism. Dynamic pathway regulation is a promising alternative to balance endogenous metabolism and the expression of exogenous pathways. However, the current studies mainly focused on dynamically regulating the expression of exogenous pathway genes. Few research was reported to dynamically control the endogenous competing pathways, especially the essential pathways for cell growth to rewire the metabolic flux towards the target compunds. To address this issue, in this study, we designed and established an artificial dynamic regulatory network and demonstrated its effectiveness in improving the biosynthesis of a widely-used platform chemical muconic acid in *Escherichia coli*. The system was based on a promoter-regulator system in combination with interference antisense RNA responsive to the generated muconic acid and achieved simultaneous up- and down-regulation in an orthogonal manner. With this system, the titer of muconic acid was elevated to the highest level so far for this pathway. This artificial dynamic regulatory network provided a valuable framework to autonomously regulate the desirable pathway genes and the native competing pathway genes responsive to the changing cellular physiological state for optimal microbial production.

500008 Using Promoter Architecture to Guide Engineering of Strong Fatty Acid Inducible Hybrid Promoters in *Yarrowia Lipolytica*

Abstract:

Recently, there has been significant improvements in the genetic toolbox for the industrial oleaginous microbe, *Y. lipolytica*. These tools range from CRISPR-Cas9 mediated genome editing to libraries of hybrid promoters made from strong upstream activating sequences for tunable strength; however, strong inducible promoters remain undeveloped. In addition to tunable strength, having control of timing of gene expression can improve cellular efficiency by separating the growth and production phase. Inducible regulation is particularly important when certain products are toxic and inhibit growth, or when requiring the ability to switch on production at stationary phase. Here, we demonstrate the development of the only known and strongest fatty acid inducible promoter system in *Y. lipolytica*. The hybrid promoter has nearly 50-fold induction strength relative to glucose with expression ranging from 2 to 10-fold higher than the commonly used native inducible acyl CoA oxidase (POX2) promoter. Additional engineering is accomplished with tandem repeats of an approximately 200 bp fatty acid inducible element to create a tunable hybrid promoters with a significantly larger dynamic expression range than the native POX2 promoter. We explore the mechanism of regulation by investigating epigenetics associated with inducible regulation in the POX2 promoter. Libraries of promoter strengths are accomplished by manipulating the different core functional elements of the promoter system. While repressed in the absence of fatty acids, the hybrid promoter lacks catabolite repression in the presence of either glucose or glycerol in fatty acid containing media. This inducible genetic switch can be can be strongly activated at stationary phase with low concentrations of a wide range of fatty acids ranging from oleic acid to eicosapentaenoic acid. The activity of the inducible promoter correlates well with changes in intracellular fatty acid pools, showing strong promise in application as a tool for strain engineering.

492923 Design and Characterization of Orthogonal Pol II gRNA Expression Systems for dCas9 Transcriptional Repression Networks in *S. Cerevisiae*

Abstract:

Robust, modular, and orthogonal genetic control systems have applications in metabolic pathway regulation and optimization, the creation of whole cell biosensors, and cell-fate decisions for therapeutics and tissue engineering. These systems require an interoperable biological framework, a library of orthogonal characterized parts, and an understanding of inherent retroactivity.Â In recent work [Gander, M. W. et al. Nat. Commun. 8, 15459 doi:10.1038/ncomms15459 (2017)] we adapted the nuclease-inactive bacterial CRISPR/dCas9 system to create a functionally complete genetic NOR logic gate where a single output may be targeted to one of two inputs using only gRNA. The NOR gate input stage is a pol II promoter with two unique target sites (pGRR) which drives the expression of computationally-designed self-cleaving-ribozyme flanked gRNA (RGRs). These ribozymes mediate post-transcriptional processing to remove the 5â€™ cap and poly-A tail nuclear export signals. With minimal characterization these NOR gates were composed to create five two input boolean logic functions and a six layer repression cascade. Despite the overall success, impedance mismatching of uncharacterized components and strong retroactive effects resulted in failed logic circuits and unexpected signal gain in longer cascades. To characterize these components and investigate sources of retroactivity, we have developed titratable 1 layer (gRNA repressing - reporter protein) and 2 layer (gRNA repressing - gRNA repressing - reporter promoter) repression cascade framework. We have found that improper ribozyme processing leads to a degradation of signal with increasing cascade size and that gRNA input sensitivity decreases with increasing cascade size. Surprisingly, gRNA input sensitivity is recovered with decreased dCas9 expression - implying that a major source of retroactivity is caused by differences in gRNA binding affinity for the dCas9. In this presentation, we will discuss the computational methods used to design functional RGR devices at the level of RNA, computational methods for selecting orthogonal components, how retroactivity affect the behavior of concurrent gRNA expression, and ongoing experiments to harness gRNA expression as a platform for building biosensors and dynamically responsive genetic control systems.

504107 Uncovering How RNA Molecules 'Make Decisions' on the Fly: Towards Understanding and Engineering Cotranscriptional RNA Folding

Abstract:

RNAs are emerging as a powerful substrate for engineering gene expression and cellular behavior since they are now known to control almost all aspects of gene expression. As with all biomolecules, RNA function is intimately related to its structure, since RNA can adopt structures that selectively modulate gene expression. Central questions in biology and bioengineering then are: How do RNAs fold inside cells?; and How can we engineer these folds to control gene expression? In this talk, I will present our work at the interface of these two questions and share results that are beginning to uncover design principles for understanding natural RNAs and engineering RNAs for an array of applications.

I will start by presenting our work on engineering RNA molecular switches that control transcription. The desire to uncover design principles for engineering these RNAs motivates our development of SHAPE-Seq, a technology that couples chemical probing with next-generation sequencing and that helps characterize RNA structures on an 'omics' scale. I will then describe our exciting recent developments in using SHAPE-Seq to help break open one of the frontiers of RNA structure function relationships by uncovering at nucleotide resolution how RNAs fold cotranscriptionally. Specifically, I will highlight new data on uncovering the ligand-dependent folding pathways of riboswitches, and how we are beginning to use these datasets to computationally reconstruct cotranscriptional folding pathways. This new ability is allowing us to ask deep questions about how RNA molecules make regulatory decisions 'on the fly' during the dynamic process of transcription. By probing the fundamental processes of RNA folding and function, these studies are expected to greatly aid RNA engineering.

503235 Overcoming Oxygen Supply Limitations in Islet Transplantation

Abstract:

Islet transplantation is an approved treatment for diabetes in much of the world but not yet in the USA. Recipients must be immunosuppressed, which has motivated extensive study of islet micro- and macro-encapsulation approaches. Encapsulated islets are separated from the bloodstream and receive oxygen by diffusion from the nearest source, and oxygen supply limitations can have deleterious effects on islet viability and function. More than 30 years ago, Martin Yarmush suggested these issues needed investigation because islets in the pancreas are exposed to high blood perfusion rates with arterial blood having pO2 in excess of 100 mmHg. Studies revealed that insulin secretion rate was increasingly comprised as pO2 decreased to low values that nevertheless were substantially higher than required for maintenance of viability. Ensuing studies have endeavored to quantify the effects of oxygen supply limitations and to develop solutions to overcome this problem. These efforts are reviewed with emphasis given to recent results with exogenous oxygen supply that give promise to maintenance of viability and function of implanted macroencapsulation devices over extended periods of time.

505281 The Amyloid Regulatory Network Hypothesis

Abstract:

"Amyloid" is a general term describing protein aggregates with several physicochemical features in common: fibrillar morphology, predominantly cross b-sheet secondary structure, and protease resistance. Deposits of aggregated protein are linked to several degenerative disorders including Huntington's, Alzheimer's, Parkinson's, and prion diseases. These deposits may be extracellular (e.g., in Alzheimer's) or intraneuronal (e.g., in Huntington's). Despite the shared fibrillar morphology, the proteins involved have no sequence homology or common native fold, and the specific cellular sub-types affected in each disorder are unique. Numerous in vitro and animal studies provide support for the 'amyloid cascade hypothesis' – that aggregation is causally linked to cellular toxicity. At one time it was believed that the fibrils were neurotoxic. Increasingly, it has become apparent that these amyloidogenic proteins also self-assemble into a variety of soluble nonfibrillar oligomeric species, and that these oligomeric species may be the most damaging to cells.

Intrinsically disordered proteins such as b-amyloid or a-synuclein must partially fold in order to aggregate. On the other hand, proteins with a stable native fold undergo partial unfolding prior to aggregation. One could argue that amyloidogenesis is attributable to a too-small difference between the stability of the native state (whether intrinsically disordered or stably folded), and the stability of the alternatively folded aggregated state. Although the native proteins are conformationally diverse, aggregates from different proteins are conformationally convergent. These structural features uniquely arise during the course of self-assembly of amyloidogenic proteins and are not present in the native unaggregated state. Given this, we argue that aggregation of amyloid proteins is a structure-based, rather than a sequence-based, homotypic (self) association. Furthermore, since structure rather than sequence is the common feature shared by amyloid aggregates, it seems reasonable to imagine that heterotypic (non-self) as well as homotypic association of amyloidogenic proteins is possible. In this talk I will present our hypothesis that, whereas homotypic association leads to amyloid aggregates, heterotypic association of amyloidogenic proteins may regulate and inhibit amyloid fibril formation. We will present three case studies to illustrate these ideas. These studies include: (1) a comparison of polyasparagine versus polyglutamine aggregation pathways, and a biological motivation for the different pathways, (2) the competing roles of cystatin C as an inhibitor of beta-amyloid aggregation and an inhibitor of the enzyme that degrades beta-amyloid and (3) the design of anti-amyloid peptides that mimic the structure of the amyloidogenic protein transthyretin.

505285 Linking Cancer and Metabolism via Isotope Labeling and Network Analysis

Abstract:

The metabolic engineerâ€™s toolbox, comprising, among others, stable isotope tracers, pathway flux estimation and analysis, pathway identification, and pathway kinetics and regulation, has long been used to elucidate and quantify pathways primarily in the context of engineering microbes for product overproduction. These tools are also well-suited for analyzing the physiology of tumor cells as they commonly undergo metabolic rewiring during oncogenesis to meet the biosynthetic needs required for aggressive growth. Metabolism is now emerging as a new space of potential therapeutic interest in cancer, so it is important to identify effective targets through fundamental understanding of the metabolic pathways preferentially used by tumor cells. To this end, these tools are increasingly finding use in cancer biology due to their unparalleled ability to quantify intracellular metabolism of mammalian cells and in particular the role of glucose and glutamine that must be efficiently directed to contribute carbon and nitrogen for biomass and cofactor production. An overview of this approach, highlighting a systems-theoretic analysis of pathways and bioreaction networks will be presented. In particular, we will focus on implications of the Warburg effect, the role of reductive carboxylation and IDH activity and the interplay between glucose and glutamine uptake as sources of carbon skeletons, energy and reducing equivalents required for tumor cell growth. These results suggest new targets for cancer therapy some of which will be addressed in the presentation.

501646 Modeling-Guided Engineering Enables Efficient Limonene Production in Cyanobacteria

Abstract:

Terpenes are plant-derived secondary metabolites with diverse industrial applications. Many terpene molecules are potential candidates for pharmaceutical, nutraceutical, and biofuel applications. To produce high titer of terpenes in microbes, high carbon flux toward terpene pathway is desired. In cyanobacteria, terpene precursors are synthesized from the MEP (2-*C*-methyl-D-erythritol 4-phosphate) pathway. The inherent low carbon flux toward this secondary metabolic pathway, and the still elusive regulation of carbon partition among carbon catabolic pathways are key limitations impeding high titer of terpene production. In our study, computational modeling-guided metabolic engineering led to efficient limonene production in the cyanobacterium *Synechococcus elongatus*. The downstream terpene synthase was determined as a key metabolic flux controlling node in the MEP-derived terpene biosynthesis. Through synthetic biology design, we achieved >100 fold increase in limonene productivity, and a significant productivity increase compared to that achieved through stepwise metabolic engineering. The strong limonene carbon sink led to enhanced carbon flux into the MEP pathway, and potentially saturated the native MEP flux. The limonene productivity change along bacterial growth also indicates that MEP pathway is highly regulated, and constantly competing carbon source with sugar metabolism. Through various â€˜-omicsâ€™ studies, we also determined flux controlling points in the MEP pathway. Further identifying upstream regulation point that controls MEP enzyme expression will bring important insight into carbon partition between primary and secondary metabolic pathways, and provide valuable targets for metabolic engineering efforts.

500902 Engineering an Environmentally-Isolated Strain of *Bacillus megaterium* for Biofuel Production and Recovery Under Supercritical CO_2

Abstract:

Using a targeted bioprospecting approach by sampling fluid from a deep subsurface supercritical carbon dioxide ($scCO_2$) well, a strain of *Bacillus megaterium* was isolated that is rarely able to demonstrate consistent, robust growth in the presence of $scCO_2$. Due to the broad microbial lethality and solvent chemistry of $scCO_2$, we hypothesize that a dual-phase reactor of growth media and $scCO_2$ will simultaneously provide a sterile growth environment and the capacity to continuously strip off strain-produced biofuels (short chain alcohols) to alleviate product toxicity. Additionally, $scCO_2$ is a sustainable, labile solvent that can be separated from desired products through depressurization, leaving these products at high concentrations and providing advantages over other co-solvent extraction systems. We have developed a transformation protocol for our strain of *B. megaterium* and evaluated a xylose-inducible promoter under $scCO_2$, which was found to have similar expression compared to anaerobic cultures. We engineered the $scCO_2$ tolerant strain to produce isobutanol by introducing a two-enzyme (2-ketoisovalerate decarboxylase (KivD) and alcohol dehydrogenase (Adh)) pathway. Combining our recombinant biofuel strain with $scCO_2$ culturing, isobutanol production was observed for cultures under scCO2 as well as was extracted into the $scCO_2$ phase. Currently we are analyzing the transcriptome for our *B. megaterium* strain, which we expect will provide hypotheses for $scCO_2$ tolerance as well as aid in further metabolic engineering of this unique host. Additionally, we have developed an integrated process model for culturing and product extraction to analyze the energetics of producing and separating isobutanol using $scCO_2$. We have found process conditions that result in comparable if not better energetics than existing *in situ* extraction techniques such as gas stripping. Lastly, we are working on building a genome scale model of our strain of *Bacillus* to incorporate our transcriptome data, creating metabolic pathways to make fuels directly from common carbon substrates, and developing genomic integration/knockout protocols to enhance the metabolic engineering of this unique host.

501212 Development and Application of a Crispri System for the Syngas-Fermenting MicrobeÂ *Clostridium Ljungdahlii*

Abstract:

Clostridium ljungdahlii has recently emerged as an attractive candidate for the bioconversion of synthesis gas (CO, CO_2, H_2) to a variety of fuels and chemicals through the Wood-Ljungdahl (WL) pathway. However, the lack of robust tools for generating gene deletions and inducible knockdowns in this organism limits the abiltity to both divert carbon flux away from acetate toward higher-value products, and to assign gene function in the poorly annotated genome. To address this need, here we develop a CRISPRi system for *C. ljungdahlii* that allows for the controlled knockdown of specific genes using the deactivated Cas9 nuclease (dCas9) and a single-guide RNA (sgRNA). qRT-PCR and enzymatic assays confirmed the ability of the system to regulate key genes with proposed roles in energy conservation and carbon flux. Screening of tightly controlled promoters for dCas9 expression reduced background downregulation, and multiplexing of sgRNAs through the use of a BioBricks assembly method allowed the simultaneous regulation of multiple gene targets in a hight-throughput manner. As a proof of concept to test the ability of the CRISPRi system to divert flux to heterologous product synthesis in an engineered strain, plasmids were constructed that enabled 3-hydroxybutyrate (3HB) synthesis, and production was assessed in strains co-transformed with various CRISPRi plasmids. The results provide interesting insights into the hitherto opaque metabolic and regulatory networks of *C. ljungdahlii*, and serve as a useful tool to identify gene targets for improving heterologous product yield.

498054 Cyanobacterial Glycogen Production and Hydrolysis for Production of Media for Industrial Bioprocessing

Abstract:

Cyanobacteria, a branch of photosynthetic microorganisms, as well as terrestrial plants can convert sunlight and carbon dioxide into stored carbohydrates, such as starch and glycogen. Cyanobacterial glycogen production could offer an alternate source of carbohydrates, as compared to terrestrial plants, for biocatalytic processes to produce value-added chemicals. Cyanobacteria offer the potential for higher annual yield than terrestrial plants, and they need neither fresh water nor arable land to grow robustly. This work demonstrates an increased glycogen accumulation in the cyanobacterium *Synechococcus* PCC 7002 through metabolic engineering, and the use of a growth-associated screen for glycogen production. Evaluation of these strains grown in wastewater nutrient sources yielded improved glycogen productivity. Further, optimization of cyanobacterial hydrolysate produced from glycogen rich strains of PCC 7002 supported growth of *Escherichia coli* for production of value-added chemicals. Finally, a techno-economic analysis of proposed industrial processes for producing cyanobacterial hydrolysate has shown that through improvements in biomass titers and separation techniques, such a process could compete with corn-based sugar feedstocks.

499048 Deciphering Cyanobacterial Phenotypes for Fast Photoautotrophic Growth

Abstract:

Synechococcus elongatus UTEX 2973 is the fastest growing cyanobacterium characterized to date. Its genome was found to be 99.8% identical to *Synechococcus elongatus* 7942, yet it grows twice as fast. Microbial physiological studies mainly focus on genome/transcription/protein levels, but a cell's fluxes and metabolome represent the actual function outputs across multiple levels. Current genome-to-phenome mapping is still poorly performed for nonmodel organisms. Even for species with identical genomes, cell phenotypes can be strikingly different. To understand *Synechococcus* 2973's fast-growth phenotype, isotopically nonstationary metabolic flux analysis (INST-MFA), biomass compositional analysis, gene knockouts, and metabolite profiling were performed on both strains under various growth conditions. The *Synechococcus* 2973 flux maps demonstrate strong Calvin cycle, photorespiration, and pyruvate kinase activity, but minimal flux through malic enzyme and oxidative pentose phosphate pathways. Moreover, central metabolite pool sizes under optimal conditions were found to be lower, while under suboptimal light the cell accumulates central metabolites. In addition, *Synechococcus* 2973 shows similar flux ratios to *Synechococcus* 7942, but exhibited greater carbon assimilatory and photorespiratory flux, less accumulation of glycogen, and potentially metabolite channeling. Finally, *Synechococcus* 2973 has weak flux through a linear TCA pathway

and small pool sizes of acetyl-CoA/TCA intermediates under all growth conditions. This study employed experimental dynamic flux analysis methodologies to obtain new insights into optimal photoautotrophic metabolisms. The outcomes indicate a highly effective metabolism in *Synechococcus* 2973 compared to other model cyanobacteria. Metabolite pool sizes indicate energy metabolism, rather than carbon fixation pathways, constrains fast cyanobacterial growth under our experimental conditions. Together, the strain's many features, such as glycogen accumulation, carbon assimilatory and metabolite channeling, result in increased biomass growth and support a photosynthetic platform to produce valuable products from its sugar phosphate pathways.

504979 Metabolite Cross-Feeding Drives the Symbiotic Growth of *Chlorella vulgaris* and Heterotrophic Microbes

Abstract:

Climate change due to rising atmospheric greenhouse gases is one of the most threatening problems facing humanity. Methods for capturing greenhouse gasses and producing biofuels have gained much attention. Microalgae and cyanobacteria can be used to capture carbon dioxide from point-source emitters and produce biofuels at high levels. Open bioreactors are needed to economically grow these photoautotrophs; however, heterotrophs often contaminate open bioreactors. The presence of heterotrophs can be advantageous in some cases, but frequently they are detrimental to the growth of photoautotrophs.

In this work, we show that *Chlorella vulgaris* UTEX #395 (a unicellular, non-motile, photosynthetic microalga) releases significant amounts of polysaccharides during photoautotrophic growth, termed â€œphotosynthateâ€. Using *C. vulgaris* photosynthate, we isolated heterotrophic microorganisms from soil samples that can grow on the released sugars and used ^{13}C-labeled photosynthate to confirm that soil microbes efficiently consumed it. In a co-culture with *C. vulgaris* the soil microbes significantly improved overall biomass production. Using pulse-chase tracer experiments in a co-culture of *C. vulgaris* and soil microbes, we demonstrated that soil microbes efficiently consumed photosynthate while *C. vulgaris* received carbon dioxide in return. The symbiotic exchange between the organisms where each partner benefited from the other was quantified using ^{13}C-tracers. The increased understanding of photoautotroph-heterotroph interactions obtained in this study may help to improve the possibility of economically viable large scale biofuel production from waste carbon dioxide sources.

497685 Using Synthetic Biology to Engineer Epistasis to Deter Bacterial Adaptation

Abstract:

Antibiotic resistance continues to be a major public health concern, with fewer new antibiotics being developed to combat increasingly resistant bacteria. Previous antibiotic resistance research has focused primarily on a "bottom up" approach – identifying specific genetic changes that confer resistance, and developing new reactionary therapies to circumvent these changes. A "top down" approach – understanding how such resistance emerges in the first place – would enable a more preventative strategy to fight resistance. To this end, work in our lab has focused on non-genetic factors such as changes at the epigenetic level that contribute to the acquisition of antibiotic resistance.

Here we present the use of deactivated CRISPR (Clustered Regularly Interspaced Short Palindromic Repeats) –Cas9 constructs to selectively inhibit or activate sets of multiple genes in *Escherichia coli*. Simultaneous gene expression perturbations induced in this fashion demonstrate synthetic engineering of epigenetic epistatic interactions – the combined impact of changes in two or more genes on the overall fitness of the cell. We systematically investigate all possible combinations of gene perturbations within two sets of genes. We show strong evidence that as more genes are perturbed, the cell's ability to adapt decreases as quantified by decreased fitness during antibiotic exposure and a slowed rate of increase in the Minimum Inhibitory Concentration of said antibiotic over time. This work demonstrates a novel approach to engineer control over the emergence of antibiotic resistance through synthetic induction of negative epigenetic epistatic interactions.

491654 Programmable Control of CRISPR-CAS9 Systems By Engineering Sgrna As Toehold-Switchable Riboregulators

Abstract:

Predictable control over gene expression is necessary for diverse applications in molecular biology, synthetic biology, and biotechnology. One of the most promising strategies to exert these types of control is the recently developed CRISPR interference (CRISPRi) and activation (CRISPRa) approach, which provides simple and highly effective RNA-based methods for targeted silencing and upregulation of transcription in bacterial and eukaryotic cells. While these current methods are capable of sequence-specific targeting, they lack the ability to integrate endogenous cellular signals as inputs for targeted regulation of gene expression. Here we present the design and construction of novel riboregulators by integrating toehold riboswitches into sgRNA scaffolds. These synthetic constructs are able to process specific information such as native metabolism and stress responses by taking advantage of highly predictable Watson-Crick base pairing mediated by RNA-RNA strand-displacement reactions. By exploiting the exceptional fidelity of these reactions, our toehold-sgRNA chimera can convert endogenous signals into cellular responses in the form of gene repression by CRISPRi. We demonstrate the programmability and adaptability of these engineered systems to control expression of multiple genes in *E. coli* with minimal cross-talks. These synthetic riboregulators represent a new class of RNA-based regulatory elements to facilitate targeted genetic perturbations and to guide cellular behaviors in a variety of context.

492796 Engineering Post-Translational Proofreading to Discriminate Non-Standard Amino Acids

Abstract:

One goal of engineers seeking to harness the advantages of biological processes is to expand the repertoire of biocompatible and bio-orthogonal chemistry. The use of non-standard amino acids (NSAAs) beyond the standard twenty amino acids is a salient example. NSAAs expand the structural and functional diversity of proteins ranging from catalysts to therapeutic agents. However, the low fidelity of existing engineered orthogonal translation systems (OTSs) hinders broader adoption of site-specific NSAA incorporation in protein engineering. Current methods to determine OTS fidelity require the low throughput and costly approaches of protein purification and mass spectrometry. To discriminate incorporation of desired NSAAs from incorporation of related standard amino acids into a target protein *in vivo*, we engineered the first example of a synthetic post-translational quality control system. We achieve Post-Translational Proofreading (PTP) by hijacking and modifying the N-end rule pathway of protein degradation for NSAAs. Under this scheme, desired NSAAs are targeted for incorporation at the N-terminus of a reporter protein, which is stabilized by correct incorporation and destabilized by misincorporation of undesired related standard amino acids. We show that PTP is compatible with a variety of useful reporter proteins. We further demonstrate that PTP is highly tunable by engineering variants of ClpS, the N-end rule discriminator protein, which distinguish incorporation of different types of NSAAs from one another. We illustrate the utility of PTP by incorporating this approach into a fluorescence activated cell sorting (FACS) pipeline for OTS evolution to achieve increased activity and selectivity for two different OTS families on desired NSAA substrates. Due to improved OTS fidelity resulting from this method, we improve the performance of *Escherichia coli* strains engineered for biocontainment.

493817 Developing Robust, Environmentally-Responsive Genetic Circuits for Real-World Applications

Abstract:

As synthetic biology transitions out of the lab and towards real-world applications, there is a need to develop a well-characterized toolbox of genetic sensors and regulators that can respond to extrinsic environmental conditions, instead of lab chemicals (e.g., IPTG). In addition, many sophisticated cellular functions require complex genetic circuits, which can lead to significant consumption of finite cellular resources and thus resource-coupled interference in predictable circuit behavior. To solve these problems, we have developed environmental sensors, RNA regulators, and feedback circuits. First, we have built and characterized sensors that respond to pH, temperature, light, oxygen levels, nitrogen source levels, or concentrations of diverse compounds found in a variety of environments. These environmental sensors have also been combined to construct multi-input logic gates (e.g., AND gates). Second, we have employed the CRISPR interference system (CRISPRi) and synthetic antisense RNAs (asRNAs) to repress or derepress target genes in a programmable manner. Importantly, we provided design rules for asRNA regulators, based on a comprehensive study of more than a hundred *de novo* designed asRNA regulators, enabling predictable implementation of these regulators in complex genetic circuits (e.g., NAND gates that respond to pH and temperature). Third, we have investigated the effects of resource competition on circuit behaviors. For the first time, we experimentally demonstrated that negative feedback can significantly reduce resource-coupled interference in complex circuits, allowing for robust circuit performance. These environmental sensors, RNA regulators, and feedback motif lay the groundwork for the implementation of robust, environmentally-responsive genetic circuits in metabolic engineering, environmental, and medical applications.

497204 RNA-Aptamer-in-Droplet (RAPID) High-Throughput Screening for Secretory Phenotypes

Abstract:

Synthetic biology tools have been used to rewire a wide range of microorganisms to create living foundries for the biorenewable production of high value chemicals. This engineering process often comprises a long and costly design-build-test cycle in which pools or libraries of very large numbers of variants must be constructed and tested to optimize performance. Currently, library generation capacity often greatly exceeds the rate of high throughput testing and screening. Overcoming this mismatch is especially challenging for secretory phenotypes, as the secreted product must be physically entrapped with the responsible producer cell and a biosensor module to enable identification and isolation of improved variants. Well plate assays are flexible and achieve this goal with limited throughput, but droplet microfluidic techniques can enable ultrahigh-throughput screening for secretory production. Here, we present an approach we term RNA Aptamer In Droplet (RAPID) screening, a method that greatly expands the scope and generality of high throughput microfluidic screening. Using *in vitro* transcribed RNA aptamers, we synthetically transduce target product concentration into fluorescence, allowing modular biosensor construction, fast measurement, and rapid screening of millions of variants in droplets. We apply the RAPID approach by enhancing secreted production of a small molecule metabolite (tyrosine) and a recombinant protein (streptavidin) from yeast. In these case studies, we isolate mutants with improvements exceeding 20-fold for metabolite production. Our proof-of-concept results demonstrate that RAPID screening affords a more general approach to massively enhancing screening power for secretory phenotypes.

501044 Enabling Microbial Cell Factories: Synthetic Biology Tools for Efficient Pathway Assembly and Integration

Abstract:

Advances in DNA sequencing has unveiled an immense treasure trove of biological parts and biosynthetic pathways encoded in genomic information, and the emergence of synthetic biology over the past decade has accelerated our ability use these components to systematically design microbial cell factories for the production of a wide variety of useful compounds. As our understanding of biological parts and systems improve, genetic circuit and pathway designs invariably become more complex. As a result, the sizes of DNA fragments encoding these designs are ever-increasing. However, the speed of uncovering genetic components and designing synthetic biological systems has far outstripped our ability to build and test these systems. Despite recent development in DNA synthesis methods, the quick and reliable assembly of individual DNA components into complete pathways remains one of the limiting factors in the implementation of the design-build-test cycle of synthetic biology. Moreover, large DNA fragments encoding the constructed pathways often suffer from genetic instability in production hosts, necessitating the need to integrate them into the host genome for genetic stability, which is a challenge as the integration efficiency of DNA fragment decreases precipitously with increasing DNA size.

Here, we present our recent progress in developing synthetic biology tools to address these challenges. We developed a scarless, sequence-independent, and enzyme free method of assembling PCR amplified DNA fragments, named Twin-Primer Assembly (TPA). TPA is capable of the assembly of up to a 31 kb plasmid with high fidelity. Even without the use of enzymes, the performance of TPA is on par or better with some of the best *in vitro* assembly methods currently available. The absence of an enzymatic assembly step in can simplify workflow, reduce human intervention and make TPA amendable to the automation process. To stably propagate the genetic pathways assembled and maximize their gene expression, we also a developed simple platform for high-efficiency, single-step, markerless, and multi-copy chromosomal integration of full biochemical pathways in *Saccharomyces cerevisiae*. This Di-CRISPR (delta integration CRISPR-Cas) platform is mediated by the use of CRISPR-Cas to generate double strand breaks at multiple delta sites in the *S. cerevisiae* genome, allowing simultaneous integration of multiple copies of linearized donor DNA containing large biochemical pathways. With the Di-CRISPR platform, we were able to attain highly efficient and markerless integration of large biochemical pathways and achieve an unprecedented 18-copy genomic integration of a 24 kb combined xylose utilization and (R,R)-2,3-butanediol (BDO) production pathway in a single step, thus generating a strain that was able to produce BDO directly from xylose. These tools developed provide simple and highly efficient methods that have the potential to streamline the central design-test-build tenet of synthetic biology and translate the product of this endeavour into industrially applicable microbial cell factories.

504105 Biology By Design: Emerging Tools in Cell-Free Synthetic Biology

Abstract:

Imagine a world in which we could adapt biology to manufacture any therapeutic, material, or chemical from renewable resources, both quickly and on demand. Industrial biotechnology is one of the most attractive approaches for addressing this need, particularly when large-scale chemical synthesis is untenable. Unfortunately, current approaches to engineering organisms remain costly and slow. This is because cells themselves impose limitations on biobased product synthesis. It is difficult to balance intracellular fluxes to optimally satisfy a very active synthetic pathway while the machinery of the cell is functioning to maintain reproductive viability. Further, chemical reactions take place behind a selective barrier, the cell wall, which limits sample acquisition, monitoring, and direct control. In addition, cells are adapted to a relatively simple chemical operating system (*i.e.*, a few common sugars, 20 amino acids), which presents researchers a limited set of accessible molecules with which to work. In this presentation, I will discuss my group's efforts to overcome these limitations and widen the aperture of the traditional model of biotechnology. In one direction, we seek to create a new cell-free framework that enables systematic optimization and debugging of biosynthetic pathways for metabolic engineering. In another area, we are catalyzing new directions for the synthesis of homogeneous glycoprotein therapeutics and vaccines. Our new paradigms for biochemical engineering are enabling a deeper understanding of why natureâ€™s designs work the way they do, as well as opening the way to novel biobased products that have been impractical, if not impossible, to produce by other means.

498481 Single Particle Tracking of Oncogenic Microvesicle Interactions with Planar, Supported Stem Cell Bilayers

Abstract:

Sustained angiogenesis is a hallmark of breast cancer. Previous data suggests that breast cancer cell-derived microvesicles promote the differentiation of adipose derived stem cells into myofibroblasts.[1] These myofibroblasts, in turn, stimulate tumor angiogenesis via increased proangiogenic factor section and extracellular matrix remodeling.[1] Thus, there appears to be a critical transfer of information from oncogenic microvesicles to stem cells that promotes their cellular transformation and the resulting biological outcome. However, the binding moieties of microvesicles and the mechanisms of triggering cargo release intra- and extracellularly remain unknown. Here, we use single particle tracking techniques to study the interactions of oncogenic microvesicles with stem cell surfaces. Because membrane binding and fusion are ubiquitous ways biomolecules are transported into cells and cellular compartments, our studies characterize these entry steps. A critical feature of this experiment is the development of planar stem cell bilayers inside microfluidic devices, a platform which is compatible with a variety of characterization tools and microscopy methods. With this platform, we are able to monitor the binding and fusion of breast cancer derived microvesicles to chemically-tunable, planar bilayers mimicking stem cell surfaces. In this presentation, we describe this new approach to studies of oncogenic microvesicle entry and delineate the effects of microvesicle-associated integrins and the properties of the extracellular environment (calcium, pH, *etc.*) on cancer microvesicle binding. Furthermore, single particle tracking was also used to define the mechanism of cancer microvesicle delivery of cargo via membrane fusion. The results indicate the impact of the microenvironment on the binding of cancer microvesicle and cargo delivery to adipose derived stem cells.

Reference

1. Song, Y. H.; Warncke, C.; Choi, S. J.; Choi, S.; Chiou, A. E.; Ling, L.; Liu, H.-Y.; Daniel, S.; Antonyak, M. A.; Cerione, R. A. Breast cancer-derived extracellular vesicles stimulate myofibroblast differentiation and pro-angiogenic behavior of adipose stem cells. *Matrix Biology* 2016.

499476 Label-Free Interference-Based Quantitative Study of Filopodia-like Structures in Cancer-Cells of Different Metastatic Potential

Abstract:

Identification of cells with high metastatic potential is critical for patient prognosis and treatment when performing single cell analysis, for instance, from a population of rare cells isolated from blood in a liquid-biopsy. However, methods for cancer cell identification can be time-consuming and costly because they typically involve the use of specific antibodies and biomarkers. Filopodia, "finger-like" plasma membrane protrusions used by cells to probe their environment, play a central role in cell adhesion and migration, and have emerged as important contributors to cancer metastasis; for instance, breast carcinomas with poor prognosis exhibit upregulated filopodia-associated genes. Although filopodia lengths could reach several micrometers, filopodia characterization is hampered by their small diameters, ~200-400 nm, typically requiring the use of fluorescence microscopy and even super-resolution techniques. Here, we study filopodia-like structures in cancer cells of different metastatic potential, such as PC3 and LNCaP, to identify characteristic behaviors that can be used for cancer cell identification. Cells are analyzed using a label-free interference-based technique, reflection interference contrast microscopy (RICM), where interferograms embed detailed topographical information of the cell-substrate interaction, down to the nanometer-scale. RICM's unique non-invasive "view from below" perspective directly reveals the presence filopodia-like structures in single cells as they settle and interact with a plasma-treated glass surface. The computational analysis of RICM interferograms, using fast custom-developed algorithms, facilitates label-free quantification of cell adhesion and filopodia dynamics. Results indicate that RICM-based analysis of the dynamics of cancer cells' filopodia-like structures enables discrimination between cell lines of high and low metastatic potential. These results illustrate RICM's potential for identification of highly metastatic cancer cells using the dynamics of filopodia-like structures as a label-free biomarker.

501000 A Genetically Encoded Toolbox for Glycocalyx Engineering: Tunable Control of Cell Adhesion, Survival, and Cancer Cell Behaviors

Abstract:

The glycocalyx is a coating of protein and sugar on the surface of all living cells. Dramatic perturbations to the composition and structure of the glycocalyx are frequently observed in aggressive cancers. However, tools to experimentally mimic and model the cancer-specific-glycocalyx remain limited. Here we develop a genetically encoded toolkit to engineer the chemical and physical structure of the cellular glycocalyx. By manipulating the glycocalyx structure, we are able to switch the adhesive state of cells from strongly adherent to fully detached. Surprisingly, we find that a thick and dense glycocalyx with high O-glycan content promotes cell survival even in a suspended state, characteristic of circulating tumor cells during metastatic dissemination. Our data suggest that glycocalyx-mediated survival is largely independent of receptor tyrosine kinase and mitogen activated kinase signaling. While anchorage is still required for proliferation, we find that cells with a thick glycocalyx can dynamically attach to a matrix scaffold, undergo cellular division, and quickly disassociate again into a suspended state. Together, our technology provides a needed toolkit for engineering the glycocalyx in glycobiology and cancer research.

499189 Quantification of the Effects of High Shear Stresses on Single Circulating Tumor Cells Using a Microfluidic Device

Abstract:

An important step in the metastatic progression of cancer is the migration of cancer cells from a primary tumor to distal location through intravasation, in which cancer cells travel through the bloodstream as circulating tumor cells (CTCs). While in the blood vessels, CTCs experience hemodynamic shear stresses at high levels (5-60 dyn/cm^2). Previous studies identified that exposing cancer cells to fluid shear stress (FSS) induced a phenotypic change causing the cells to be more metastatic. While a novel finding, this study was limited in its ability to control the FSS used to challenge the cells and was limited to only final measurements at the end of the experiment. The goal of this project was to address this limitation by developing a microfluidic device to isolate and trap single CTCs and then interrogate them with varying amounts FSS. The microfluidics traps were designed with a semi-circular geometry with a space in the middle to allow for optimal trapping and testing of FSS-induced cell deformability. COMSOL simulations were performed to optimize the fluid velocity profile of the device to ensure a high trapping efficiency. The simulations also identified the flow profiles needed to produce FSS on the cells that they typically encounter in the bloodstream. These flow profiles were used to test the phenotypic change in MDA-MB-231 cells trapped in the microfluidic device. Preliminary findings demonstrate the ability to efficiently trap single cells in the microfluidic device to test FSS-induced changes on cell deformability and viability allowing for a more accurate determination of cell phenotype. These results will provide new insight into the mechanisms of how CTCs survive in the bloodstream and if changes in FSS results in a more aggressive cancer phenotype.

489753 Investigating the Effects of Cold Atmospheric Plasma on Cervical Cancer

Abstract:

Evidence indicates that cold atmospheric plasma (CAP) may selectively induce apoptosis in cancerous cells. Due to these observed selective effects, the use of CAP is a promising new avenue for cancer therapy. Despite its potential impact, the effects of CAP in biological systems are not well understood. This project investigates the cellular and molecular responses induced by CAP treatment within cancerous (SiHa) and non-cancerous (Ect1/E6E7 and End1/E6E7) cervical cell lines, in vitro, in order to better understand the viability of CAP as a therapy for cervical cancer. Cervical cancer is the fourth most common cancer in women, and it was recently revealed that its impact has been historically grossly underestimated. Its prevalence and accessibility for topical treatment make cervical cancer an ideal target for CAP treatments.

Plasma is a state of ionized gas, often considered the fourth state of matter; it is produced using a strong, controlled electric field or extreme heat. CAP, however, is a subset of plasma that is not in thermodynamic equilibrium. This allows for electron temperatures that are necessary to achieve ionization, while the bulk plasma temperature remains below 40°C. Because of the non-thermal nature of this type of plasma, biological systems are able to be exposed to CAP treatments without heating-induced effects.

A floating electrode dielectric barrier discharge (FE-DBD) plasma pen was developed for this study in order to investigate the cellular and molecular responses induced by CAP in cervical cells. The CAP was produced with 98% argon, 2% oxygen carrier gas mixture at a flowrate of 2 L/m. Standard CAP treatment for this investigation was produced with 10kV, 0.1A, and the cells were treated for 30 seconds.

After exposure to standard CAP treatment, 71 ± 7% of cancerous cells underwent apoptosis as indicated by flow cytometric analysis with Annexin V and propidium iodide staining protocols (n=3). One hypothesized cause of CAP induced apoptosis is through CAP alteration of the cellular membrane, allowing for plasma produced ions, such as reactive oxygen and nitrogen species (RONS), to enter the cell more easily. Indication of membrane alteration was observed in the cancerous and non-cancerous cervical cell lines. All of the cell lines had altered levels and locations of uptake of FITC and quantum dots following CAP treatment as compared to carrier gas treatment controls. Altering membrane permeability may be an important mechanism through which CAP treatment increases intercellular exposure to apoptotic triggers. Thus, while altering the membrane permeability in a controlled and selective manner could allow CAP treatment to be used to target for delivery of imaging dyes and small molecule compounds.

In conclusion, developing a comprehensive understanding of CAP effects on cancerous and non-cancerous cells could lead to the development of effective therapeutic interventions for cervical cancer.

493101 Targeted and Controlled Combination Therapy Using siRNA and Resveratrol for Inducing Leukemic Cell Apoptosis

Abstract:

Title: Inducing leukemic cells death by targeted combination therapy of siRNA and Resveratrol

Abstract:

Introduction: Approximately every 3 minutes, a person is diagnosed with blood cancer accounts for almost 10.2% of new cancer cases in the US. Almost 171,550 people are diagnosed with leukemia, lymphoma, and myeloma in 2016. Many approaches have been explored including small interfering RNA (siRNA) and small chemicals. siRNA has been proposed with 84% protein suppression level, however, cell death levels have not been shown. On the other hand, Resveratrol, a polyphenol found in wine and grapes, induces ~60% apoptosis via inhibition of the sphingosine kinase 1 enzyme (SPHK1). Generally, leukemia treatment with small molecules, suggests that despite inhibiting certain pathways with high percentage, cells can still survive by alternative pathways, hence low cell apoptosis level are obtained. In addition, small molecule cancer therapeutics are also hindered by issues with pharmacokinetics. For example, resveratrol has a very short half live (3-5 hours) and its bioavailability is very low. In this study, we questioned whether a combination of siRNA with controlled release of resveratrol can induce higher cell death level. First, we explored the effect of combination of resveratrol and siRNA on K562 cell line derived from chronic myelogenous leukemia (CML). Using cell concentration of 2 x 10^5cells/ml of solution, and 40 µM Resveratrol, we explored forming coaxial hybrid fibers from hydrophobic poly(ε-caprolactone) (PCL) and hydrophilic gelatin (GT) in two different configurations, and the release of resveratrol at 37°C over five days.

Selected siRNA was a custom-synthesized BCR-ABL siRNA (5'-GCAGAGUUCAAAAGCCCTT-3'). First, 2 x 10^5K562 cells/mL of solution were exposed to a range of resveratrol concentrations (10, 20, 40, 80, and 160 µM) and a range of siRNA concentrations (12, 24, 36, 48, and 60 nM). Both conditions were analyzed for cell viability after incubation for 72 hours using Annexin V, PI staining and flow cytometry. Apoptotic cells level of 48% was achieved with resveratrol concentration of 40 µM. Apoptotic cell level of 47% was achieved with 36 nM siRNA concentration. A combination of both drugs was then investigated following the same methodology. From these studies, 40 µM resveratrol concentration and 36 nM siRNA concentration was selected in combination therapy. Also, similar concentration was used in forming the electrospun fibers. In addition, a factorial design of experiments was used to develop a surface plot using a mid-point concentration of 20 µM resveratrol and 18 nM siRNA. Results showed that addition of both molecules had an additive effect on apoptosis; nearly 72(±2.9)% cells and 25(±1.8)% (mid-point) were apoptotic under these conditions. Next, PCL and GT were dissolved separately in trifluoroethanol (TFE) and mixed together in 1:1 ratio for the hybrid layers. Fabricated fibers at the same settings and same solvent were characterized for their structural morphology, various components, viability and apoptosis of K562 cell over 72 hr. Resveratrol release from the fibers were also characterized by incubating samples in a buffer, similar to our previously publication [1]. Amount of resveratrol released at various times was assessed using HPLC.

These results show a promising potential of the approach.

[1] Khalf A, Madihally SV. Modeling the Permeability of Multiaxial Electrospun poly(ε-caprolactone)- Gelatin Hybrid Fibers for Controlled Doxycycline Release. Materials Science & Engineering C: Materials for Biological Applications. 76:161-170, 2017.

504947 Nanostructured Platforms to Evaluate Cell Migration-Microenvironment Interactions in Glioma

Abstract:

Research has established that two dimensional tissue culture models, such as tissue culture polystyrene, fail to adequately recapitulate conditions in the tumor microenvironment. Thus, there is substantial interest in developing new models to mimic unique chemical, mechanical, and topographical conditions *in vivo*. For example, hydrogels composed of collagen, hyaluronic acid, Matrigel, and peptide-modified synthetic materials, have all been employed to evaluate interactions of cancer cells with specific microenvironment cues. Similarly, electrospun fibers can mimic the topography of ordered anatomical structures, such as blood vessels and white matter tracts.

Here, we will discuss important caveats that we have uncovered in using and developing these models to evaluate cell response to mechanical cues in the microenvironment. First, we show that cells can sense the mechanical environment of the underlying support within an "edge effects" region close to the interface. Thus, in this region, cell responses may not reflect those corresponding to the intended hydrogel mechanical properties. Further, many tissues are viscoelastic; however, tissue characterization primarily focuses on the Young's or Elastic modulus, rather than the viscoelastic response. We show that tissues can have similar moduli, but very different Maxwell elasticity and viscoelasticity, which may influence cell behaviors. Finally, we show that shearing forces may be better predictors of cell behavior than tensile forces or Young's modulus. All of these factors should be considered in the design of biomaterials to mimic the cancer microenvironment.

As a model system, we will discuss hydrogels and electrospun fiber mats to mimic the brain microenvironment for gliomas, astrocytomas of the nervous system. Because the average patient survival is ~ 12-15 months with best of care, additional strategies for glioma are urgently needed. In addition, glioma is highly infiltrative, spreading to the contralateral hemisphere even with complete resection of the tumor-containing hemisphere. Gliomas typically travel along blood vessels and white matter tracts that provide ordered topographical features. These structures are well modeled by fibrous collagen hydrogels or electrospun nanofibers, whereas hyaluronic acid (HA) is the primary tissue component of brain. Thus, this presentation will describe collagen, HA, and collagen-HA hydrogels, as well as electrospun fiber mat models of glioma.

505286 Human *In Vitro* Models to Improve Preclinical Testing of Drugs

Abstract:

Human microphysiological or "Body-on-a-Chip" systems are powerful tools to assess the potential efficacy and toxicity of drugs in pre-clinical studies. Having a human based, multiorgan system, that emulates key aspects of human physiology can provide important insights to complement animal studies in the decision about which drugs to move into clinical trials. Our human surrogates are constructed using a low cost, robust "pumpless" platform. We use this platform in conjunction with "functional" measurements of electrical and mechanical activity of tissue constructs. Using a system with four or more organs we can predict the exchange of metabolites between organ compartments in response to various drugs and dose levels. We have constructed models incorporating barrier tissues such as GI tract, blood brain barrier, and skin with internal organs such as liver, cardiac, and neuromuscular junctions. We will provide examples of using the system to both predict the response of a target tissue as well as off target responses in other tissues/organs. We believe such models will allow improved predictions of human clinical response from preclinical studies.

505288 CAR-T Manufacturing: Delivering on the Promise of a Transformational Therapy

Abstract:

CAR-T cells have been proven clinically to be life-saving therapies for a number of hematological cancers, and offer further promise for additional hematological indications as well as for solid tumors. Commercializing these CAR-T therapies offers numerous challenges in all areas of drug development, including manufacturing and supply chain. Raw material procurement, patient tracking, cost-effective and robust manufacturing, process changes and comparability, process and assay validation, QC/QA release, manufacturing facility strategies, staffing models, and cold supply chain are among the many technical challenges that will be discussed in this presentation.

505289 The Importance of Thinking Big in Academia

Abstract:

One of the most important lessons I learned from my postdoctoral fellowship (2002-2004) with Maish Yarmush was to think big and to be confident in tackling new challenges. I joined the CEM after completing my PhD at Rutgers with a background in Protein Engineering. Soon I was surrounded by an array of diverse new projects, big ideas, and fearless enthusiasm. I have tried to embrace this spirit in my lab as we have become involved in a range of topics including enzymology, biomaterials, bioelectrocatalysis, therapeutic delivery, biophysics, and bioenergy. My time with Maish at the CEM helped to inspire me to think big and to constantly embrace new opportunities.

505290 Plasmonic Nanomaterials for Tissue Sealing and Radiation Dosimetry

Abstract:

The ability to modulate electromagnetic radiation has diverse applications in biotechnology and medicine. Plasmonic nanoparticles (e.g. gold or silver nanoparticles) can efficiently absorb and scatter incident radiation, which makes them useful in biological imaging and therapeutics. In this presentation, I will discuss the use of plasmonic nanoparticles in tissue sealing and radiation sensing and dosimetry. Gold and silver nanoparticles, were incorporated within polypeptide matrices, leading to the formation of laser-activated tissue-integrating sealant (LATIS) nanomaterials. Irradiation of these materials with non-ionizing laser light facilitated a photothermal response, which, in turn, resulted in rapid, fluid-tight sealing of ruptured intestinal tissue ex-vivo and incised skin in live mice. As a second application, we demonstrate the use of gold nanoparticles as indicators of ionizing radiation at doses used in clinical radiotherapy. Irradiation of colorless gold salt formulations with ionizing photons (X-rays) and protons resulted in the formation of gold nanoparticles which rendered a visible color change to the liquid dispersion or an encapsulating gel. The intensity of the color formed was a function of the radiation dose employed, and was used to calibrate this nanoscale dosimeter. This technology was further extended for the detection of ionizing radiation doses delivered to specific regions in a spatially resolved manner, which opens up possibilities to detect â€œhot spotsâ€ in tissues. The efficacy of this novel dosimetry device was evaluated using clinical anthropomorphic phantoms and in animals. Our studies demonstrate that plasmonic nanomaterials can further advance the use ionizing and non-ionizing radiation in diverse clinical applications including tissue sealing and repair, and cancer radiotherapy.

503670 Digital Alchemy for Assembly Engineeringc

Abstract:

Back in the day, Stuart Churchill was a pioneer in the use of computers for modeling chemical and thermal processes. Since that time, the role of computing in chemical engineering has become indispensable. In this talk, we present new computational methods we call "digital alchemy" that allow for predictive modeling of thermodynamic assembly, and permit the design of building blocks optimized for assembly into target structures with desired properties. In this approach, standard thermodynamic ensembles are generalized to include alchemical variables and their thermodynamic conjugates, allowing for rapid, real-time computational investigation of large, complex design spaces.

503671 Improved Algebraic, Numerical, and Graphical Representations in Fluid Mechanics

Abstract:

The concept of equivalents is applied to fully developed flows through straight channels, and two applications are introduced: the equivalents that produce equal rates of flow through channels of different cross-section, and the diminishment that occurs between the regimes of laminar and turbulent flow. The validity of the eddy diffusivity and the invalidity of the mixing-length as measures of the shear stress due to turbulence have perhaps never before been demonstrated so clearly. The possible choices of primary variables and the consequences are also examined. The objective of these explorations is an improved understanding of the basic fluid mechanics by teachers, students, and industrial practitioners.

503672 The Scaling of Turbulence Near the Wall and the Churchill Turbulent Flux Correlation: Insights with Lagrangian Simulations

Abstract:

The evaluation of mean velocity and mean temperature in wall-bounded turbulent flows has been based on the "classical" theory of turbulence and the law of the wall. The viscous wall parameters, namely the friction velocity, the friction temperature, and the kinematic viscosity of the fluid have been used to scale not only the mean values of velocity, but also other mean turbulent quantities close to the wall, including the Reynolds stresses and the turbulent velocity and temperature fluctuations. Stuart Churchill, as he questioned the validity of the Reynolds analogy between momentum and heat or mass transfer in his works [1], he realized that extending the scaling of momentum to the scaling of scalar quantities in turbulence was not a sound approach. He proposed a scaling of turbulence that depends on the turbulent momentum flux and the turbulent heat flux. In his AIChE Institute Lecture (in Miami, 1998), Churchill presented a summary of his work on this topic, and closed form solutions for the mean velocity and temperature [2]. The turbulent Prandtl number, the most significant parameter for the modeling of turbulent transport not only in theory but also in modern CFD commercial software packages, was also interpreted physically based on these ideas [3]. We have used direct numerical simulation results and results from Lagrangian simulations [4] for the transport of passive scalars in turbulent channel flows to explore Churchill's ideas on turbulent flow scaling and to develop an updated theory for the scaling of heat transport in turbulent flows [5-7]. His passion to remove empiricisms and approximations from the development of predictive correlations in turbulent transport has been an inspiration. In this talk, we will revisit the Churchill scaling in view of our simulation results, and will discuss the counterintuitive finding that wall turbulence can be used to separate rather than mix substances under special circumstances [8].

1. Churchill, S. W., 1996, "A Critique of Predictive and Correlative Models for Turbulent Flow and Convection," Ind. Eng. Chem. Res., 35, pp. 3122–3140.
2. Churchill, S.W., 2000, "Progress in the Thermal Sciences: AICHE Institute Lecture," AIChE Journal, 46(9), pp. 1704-1722.
3. Churchill, S. W., 2002, "A reinterpretation of the Turbulent Prandtl Number," Ind. Eng. Chem. Res., 41, pp. 6393–6401.
4. Papavassiliou, D.V, 2002, "Turbulent transport from continuous sources at the wall of a channel," Int. J. Heat Mass Transfer, 45(17), pp. 3571-3583.
5. Le, P.M., and Papavassiliou, D.V., 2008, "On the scaling of heat transfer using thermal flux gradients for fully developed turbulent channel and Couette flows," Int. Commun. Heat Mass Transfer, 35(4), pp. 404-412.
6. Srinivasan, C., and Papavassiliou, D. V., 2011, "Prediction of Turbulent Prandtl Number in Wall Flows With Lagrangian Simulations," Ind. Eng. Chem. Res., 50(15), pp. 8881–8891.
7. Srinivasan, C., and Papavassiliou, D.V., 2013, "Heat transfer scaling for wall bounded turbulent flows," Applied Mechanics Review 65(3), Art. 031002.

8. Nguyen, Q., Srinivasan, C., and Papavassiliou, D.V., 2015, "Flow induced separation in wall turbulence", Phys Rev E, 91, Art. 033019.

503674 Flow Boiling Using a Piranha Pin Fin Heat Sink

Abstract:

Flow Boiling using a Piranha Pin Fin Heat Sink

Corey Woodcock, Xiangfei Yu, Yoav Peles[1] and Joel Plawsky

Rensselaer Polytechnic Institute; [1]University of Central Florida

The *Piranha Pin Fin* (*PPF*) microdevice is a heat exchanger engineered to dissipate extreme thermal loads using either dielectric fluids such as HFE7000 or water. Fabrication of the device is CMOS compatible and the first generations of the device have been fashioned from silicon substrates. The PPF operates on the principle that heat transfer would be more efficient if heated fluids could be removed nearly as fast as they are created. This process would keep the driving force as high as possible. Thus the heat exchanger is built as two layers; a primary substrate that dissipates the bulk of the heat and shunts the heated fluids to a booster section that continues to heat the fluid and extends the total surface area of the device. Three generations of devices were fabricated. The first two generations were single-layer systems designed to help understand the operation of the primary layer and optimize the shape and location of the pin fins. The third generation was the first, two-layer system to be fabricated and tested.

Figure 1 Representative PPF array

Correlations for the heat transfer coefficient, h_{tp}, based on the Boiling Number and the Weber Number in the PPF design was developed and validated with Gen II. The discrepancy of heat transfer coefficient between predicted and measured values was within 20%. The ratio of two-phase to single-phase pressure drop was correlated with the Boiling Number (Bo) and the Jakob Number (Ja). The discrepancy of the two-phase multiplier between prediction and experimental measurements were within 12%.

The third generation *PPF* heat sinks were the first series to provide both a primary substrate populated with *PPF*s, as well as a booster substrate populated with solid micro pin fins. The substrates are micro-aligned and fusion-bonded to provide a full three dimensional (3D) architecture with bulk silicon properties. Experimental results from the Gen III *PPF* microdevices provide a proof of concept of the true *PPF* heat sink concept with fluid venting into multiple substrates. At low flowrates (G = 670 [kg/s/m^2]) the Gen III *PPF* heat sink has been shown to dissipate heat loads approaching 1.1 [kW/cm^2] at an exit quality of 95%, while at higher flowrates (G = 3,300 [kg/s/m^2]) the thermal limit of the heat sink remains unknown due to exceeding the available power supply at >1.4 [kW/cm^2].

503673 Transport Problems in the Spirit of Stuart Churchill for Teaching and Research at the University of Michigan

Abstract:

Stuart Churchillâ€™s solutions to important heat transport problems have greatly influence both pedagogy and research. The correlation of Churchill and Chu for the natural convection heat transfer to or from a vertical plate, over an astonishing range of Rayleigh numbers is based partly on a similarity solution of a type common in heat, mass, and momentum transfer for problems in chemical engineering. One such problem is asphaltene or wax transport and deposition onto the wall of an oil pipeline, which can be solved using similar methods, as will be discussed. In addition, such problems, and the wealth of their applications make for excellent motivations and examples in the teaching of graduate or undergraduate transport, as will be discussed in the talk.

500209 A Compartmentalized, Numerical Model of Continuous Blending

Abstract:

Here we present an empirically-derived, numerical model of a continuous blender capable of describing both the mass flow rate and residence time distribution of a given powder as a function of processing conditions. The approach discretizes the blending process in both time and space, describing a stepwise progression of powder through segmented portions of the equipment, which are defined primarily by the geometry of the mixing blades. The model defines three distinct blending regimes within the process: A conveying mode, where powder is rapidly transported from the blender inlet to outlet; a mixing mode, where intermixing of powder between nearby blender segments is more completely described; and a transition regime between the two. Further, we show that coupling the model with a given materials system on a specific blender provides an approach to identifying appropriate process ranges in the context of desired blending regimes. Moreover, the model is shown to serve as a useful analytical tool for defining design spaces in the context of QbD, affording greater mechanistic insights into blending behavior than a standard linear regression analysis of an orthogonal design of experiments.

491186 Impact of Raw Material and Blend Properties on the Screw Feeding, Continuous Blending and Tableting Unit Operation of an Integrated Continuous Direct Compression Platform

Abstract:

Continuous manufacturing will likely change the way how drug products are developed and commercialized. Although continuous processes simplify the development process by removing the need of multiple unit operations and scale up, they also bring in new challenges. Only limited resources are available early in the development stage and hence the formulation and process needs to be developed in a short time frame with only a limited quantity of active ingredient. This is particularly challenging for continuous processes because most of the equipment is designed at production scale. Drug product development for continuous processes requires therefore innovative strategies and simulation to maximize the information with a minimal amount of active. Understanding and modeling the impact of material and blend properties at each unit operation is therefore the key step to develop drug products of high quality with minimal resources.

Direct compression is by far the simplest and most cost efficient manufacturing route for oral solid dosage forms. Although direct compression is an inherently continuous technique, simple unit operations preceding tableting (i.e. weighing and blending) are historically performed batch wise. To enable continuous direct compression, the integration of continuous powder feeding units, continuous dry powder mixers and a tablet press is required.

This work aims to leverage raw material and blend properties with required settings for optimal feeding, blending and tableting in an integrated continuous direct compression platform. First, the raw material properties of a wide range of fillers for direct compression and active pharmaceutical ingredients (API) were determined. The characterization included determination of particle size and shape (laser diffraction, static image analysis, scanning electron microscopy), specific surface area, tapped, bulk and true density, flowability (drained angle of repose, dynamic angle of repose, flow function coefficient, basic flowability energy, specific energy, aeration index, permeability), compressibility, charge density, moisture content (loss-on-drying) and dynamic vapor sorption. Subsequently, principal component analysis was applied on the API and filler dataset. This allowed to select 6 APIs (acetaminophen micronized, powder and dense power, caffeine powder, metoprolol micronized and etravirine spray dried) and 5 DC fillers (Avicel PH 101 and 200, Emcompress, Pearlitol 100SD, Tablettose 80) with unique properties. These were combined together with magnesium stearate in 30 tertiary mixtures that were also characterized for their properties.

Continuous direct compression trials were performed with these formulations to evaluate the impact of blending (impeller speed and configuration), die filling (paddle wheel speed) and compression (main compression force) parameters on responses of the process (fill level, mean residence time and number of blade passes) and product (blend and content uniformity, tablet weight variability and tensile strength).

The hopper refill performance could be related to the flowability of the powder. Materials with a flow function coefficient below 2 (i.e. no powder flow) gradually reduced the apparent volume of the refill valve over time. At the feeding stage, the amount of material dispensed per screw revolution was related to the bulk and tapped density whereas the variability on the feeder flow rate was correlated with the flow function coefficient of the material. At the first or intensive mixing stage, the fluidization state of the powder bed depended on the process settings. The residence mass in the first continuous mixer was positively correlated with the bulk and tapped density of the blend when the mixer operated in the densely stirred powder bed flow regime. Additionally, more permeable blends resulted in lower residence masses as these blends fluidized more easily. At the second blending or lubrication stage, all blends were completely fluidized and as a result this response was not correlated with material properties. At the die filling stage of the tableting unit operation, the residence mass in the feed frame of the tablet press was a strong function of the bulk density of the blend (R^2=0.97) as the powder bed was in a densely stirred regime during processing. In contrast, the fill depth required for achieving target tablet weight (175 mg) was negatively correlated with the bulk density (R^2=-0.90). In contrast, tablet weight variability depended on the compressibility (R^2=0.68), unconfined yield strength (R^2=0.72) and flow function coefficient (R^2=0.56) of the blends. Overall, blends characterized by a low density and poor flowability yielded the highest tablet weight variability at low paddle speed.

The tabletability was strongly dependent on the selected filler. The cellulose based fillers resulted overall in tablets with the highest tensile strength. Additionally, Avicel PH101 showed better tabletability compared to PH200 driven by the higher bonding area of the smaller particles. However, the cellulose based fillers exhibited a high lubricant sensitivity. As the cellulose particles were subjected to a higher number of paddle wheel revolutions, the tensile strength decreased with 45% for some blends. Pearlitol 100SD also resulted in tablets of good strength and showed only a limited lubrication sensitivity. However, high ejection forces were observed with pearlitol 100 SD and higher levels of magnesium stearate (1.25 % w/w) were required to limit the ejection force during tableting. Formulations consisting of Tablettose 80 or Emcompress fillers with the elastically deforming acetaminophens resulted in relatively weak tablets.

The database will be further completed with blend and content uniformity data. Subsequently, advanced multivariate models such as T-PLS will be applied to link the API and Filler properties to the process parameters and all responses of the continuous direct compression process. These models are valuable tools to select excipient properties as a function of API properties during drug product development for continuous direct compression processes.

493141 Feeder Characterization and Model Development Accounting for Incoming Material Properties

Abstract:

In previous years, research focus has been placed in characterizing feeder performance and tooling selection using a formulation based approach [1-3]. This approach consists on studying feeder performance for a specific ingredient (e.g., excipient, lubricant, or API) based on its percentage in a formulation and a desired process flow rate [4, 5]. Although efficient for the early stages of continuous manufacturing, this approach does not provide process developers with sufficient information to ensure process performance on instances where process throughput is changed and/or material properties change from the ones measured during characterization. The formulation based approach also does not cut down significantly the amount of work needed to design new processes for new formulations. To address these issues a novel approach based on the characterization of feeders for a broad range of materials and flow rates (i.e., independent of formulation requirements) was performed by our group. The goal of this work is to develop models and tools that predict flow rate variability and feeding capacity over a broad range of ingredient material properties, which serve as inputs to the model.

In this presentation, we describe a characterization procedure applied to a GEA compact feeder using a series of materials relevant to the pharmaceutical industry. Our experimental plan involved flow rate characterization and Residence Time Distribution (RTD) characterization. Feeder flow rate characterization is performed by measuring the accuracy and variability of material dispensing as a function of volumetric throughput and feeder hopper fill level. Twenty (20) different materials were run in a GEA compact feeder. The materials span a broad range of material properties, as shown by study of our internal material property database using Principal Component Analysis (PCA). Eight (8) feeder configurations based on the screw size and gear box configuration were studied. At each feeder configuration, flow rate experiments were carried out in both volumetric and gravimetric flow. The objective was to understand the effect of gear box, screw size, and material properties on the outgoing flow of the feeder (i.e., feed factor deviation, mean flow rate, and standard deviation) to develop a model capable of predicting the behavior. The flow rate model, the flow rate data analysis, the procedure for calibration, and the overall flow rate model will be introduced in this presentation.

The RTD of the feeder was characterized for a variety of materials. Feeder RTD is important for two main reasons: (1) traceability of raw materials and (2) characterization of the dynamic response time to changes in incoming material properties. In both cases, the goal is to understand how long it takes for a new material (e.g., lot added due to refill or lots with new material properties) introduced into the feeder to "wash out" the material that was already in the feeder. The RTD for the GEA feeder was characterized using four (4) different materials (i.e., tracer-base combinations) and two (2) different flow rates. The aim was to understand if material properties had an impact on the RTD profile (i.e., mixing behavior) inside of the feeder. The results from the RTD profile were used to model the mixing behavior inside of the feeder using a tanks-in-series model with side compartments. Regressed coefficients were used to compare the mixing behavior between the materials and then correlated to those materials' properties.

References

1. Boukouvala, F., et al., *An integrated approach for dynamic flowsheet modeling and sensitivity analysis of a continuous tablet manufacturing process.* Computers & Chemical Engineering, 2012. **42**: p. 30-47.

2. Toebermann, J., .-C., Rosenkranz, J., Werther, J., Gruhn, G., *Block-oriented process simulation of solids processes.* Computers & Chemical Engineering, 2000. **23**: p. 1773-1782.

3. Hartage, E.U., Pogodda, M., Reimers, C., Schwier, D., Gruhn, G., Werther J., *Flowsheet simulation of solids processes.* KONA, 2006. **24**: p. 146-156.

4. Engisch, W.E. and F.J. Muzzio, *Method for characterization of loss-in-weight feeder equipment.* Powder Technology, 2012. **228**: p. 395-403.

5. Yang, S. and J.R.G. Evans, *Metering and dispensing of powder; the quest for new solid freeforming techniques.* Powder Technology, 2007. **178**: p. 56-72.

494769 Disintegration and Dissolution Modeling for Accelerated Drug Product Process Development

Abstract:

We present a methodology to streamline tableting process development complemented by a mathematical model of disintegration and dissolution. In a state-of-the-art tableting process development the main goal is to achieve an optimal tradeoff between mechanical strength, disintegration time, friability, dissolution profile, tablet weight uniformity and any other performance variable of interest. Excluding dissolution testing, the assessment of tensile strength, disintegration time, friability and tablet weight uniformity is based on simple and fast tests which can be performed during the tablet manufacturing and the results can be promptly used for the process optimization. In contrast, the dissolution assessment is laborious, time consuming and can only be reliably performed in analytical laboratories. Therefore, taking into account the dissolution profile in the optimization loop largely increases the development cycle-time which is often not feasible at early process development stages. However, this is particularly important when formulating amorphous drug product intermediates to optimize the release rate and supersaturation maintenance which are paramount for drug performance [1].

In the proposed methodology the early assessment of dissolution profile is performed with a first principles mathematical model which takes into account intrinsic material properties such as drug diffusivity in the polymer matrix and in the dissolution media [2], the polymer swelling/erosion rate [3] and whether the API is crystalline or is in an amorphous solid dispersion. The disintegration dynamics are handled with a Population Balance Equation [4] and are coupled to the dissolution model. Whenever direct assessment of intrinsic material properties is not possible, these parameters are obtained via curve-fitting the model to experimental dissolution data. The curve-fitting is performed using a particle swarm optimization algorithm, which is suitable for multivariate and multi-objective problems.

We present a case-study on the application of the methodology to the downstream process development of an amorphous solid dispersion which illustrates the advantages of complementing laboratory development with a computational tool in terms of process and drug performance without hindering the development cycle-time.

Figure 1: Dissolution profile of a tablet formulated from an amorphous solid dispersion intermediate obtained both experimentally and from the presented mathematical model.

References

[1] Henriques, J., Valente, P. & Winters, C. "Formulating Amorphous Solid Dispersions: Bridging Particle Engineering and Formulation", American Pharmaceutical Review, Nov/Dec. 2016.

[2] Wang, Y., Abrahamsson, B., Lindfors, L. & Brasseur, J.G. "Comparison and Analysis of Theoretical Models for Diffusion-Controlled Dissolution", Mol. Pharmaceutics, 9, 1052-1066, 2012.

[3] Siepmann, J., Kranz, H., Peppas, N. A., & Bodmeier, R. "Calculation of the required size and shape of hydroxypropyl methylcellulose matrices to achieve desired drug release profiles", Int. J. of Pharm., 201(2), 151–164, 2000.

[4] Wilson, D., Wren, S. & Reynolds G., "Linking Dissolution to Disintegration in Immediate Release

Tablets Using Image Analysis and a Population Balance Modelling Approach", Pharm. Res., 29:198-208, 2012.

490283 Modeling of Flow and Drying of Aqueous Polymer Coatings on Porous Pharmaceutical Tablets

Abstract:

The application of polymer coating films on the surface of tablet cores is crucial in the manufacture of solid dosage drugs. Understanding the coating film behaviour at the point of contact between the liquid solution and the porous tablet core can reduce the number of defective tablets being discarded and allow a better prediction of the optimal operating conditions for the coating process. In our work, to estimate the coating film attributes and uniformity, we developed a new mathematical model that defines and describes the spreading, absorption and evaporation of polymer solutions onto and from a tablet.

Several researchers (Weidner et al., 1995; Schwarz et al., 2001) have considered polymer coating film drying but have not accounted for changes in coating density owing to solvent evaporation. In our work, we considered that the film density increases as the solvent (water) evaporates from the surface of the tablet. The influence of solvent evaporation on important coating spreading parameters, such as the viscosity of the aqueous solution and the diffusion coefficient of the polymer in the solvent, was also investigated.

To better describe the behaviour of the film on the surface of the tablet we divided the overall process into three stages. For the first stage, we considered the coating solution impacting on the tablet surface and then spreading on the surface and absorbing into the tablet. We modelled the coating solution as a liquid of changing density and viscosity due to solvent (water) evaporation. The absorption during the first stage is affected by polymer particle retention in the tablet pores, the volume fraction of the latter not being necessarily uniform in the tablet. We assumed that the viscosity of the coating solution becomes infinite when the mass fraction of the polymer reaches a certain value (Weidner et al., 1995). At this point, the film spreading and absorption are hindered and the second stage commences. During the latter, we assumed that only solvent evaporation takes place from the tablet surface. This stage lasts until the polymer concentration reaches a new critical value, at which point a polymer particle crust develops at the surface of the coating film (Kadja and Bergeles, 2003). In the final stage, we considered the evaporation of the solvent (water) from beneath the coating crust (i.e., tablet porous core).

We derived an expression for the movement of the absorbed coating fluid inside the porous tablet based on a kinematic boundary condition of the free wetting front surface (Alleborn and Raszillier, 2004). The pressure and the vertical and radial velocities inside the porous substrate were calculated from the Laplace and Darcy equations, respectively. We also took into account the effect of the tablet porosity and permeability on the polymer particle retention.

We implemented our novel model in gPROMS, employing the Modelbuilder modelling platform (Process Systems Enterprise Ltd., 2015). The model estimated the water and polymer content inside the porous tablet core and the dry coating film position and height. The numerical results were in good agreement with experimental data taken from the literature (Schwartz et al. 2001; Möltgen et al. 2012). In future work, our approach could be extended to also predict the spreading of non-Newtonian coating liquids employed by the pharmaceutical industry as there are many different models that propose relationships between viscosity, stress and strain rate.

References:

Alleborn, N. and Raszillier, H. (2004) "Spreading and sorption of a droplet on a porous substrate", Chemical Engineering Science, 59(10), pp. 2071-2088.

Kadja, M., Bergeles, G., (2003) "Modelling of slurry droplet drying", Applied Thermal Engineering, 23(7) pp. 829-844.

Möltgen, C., Puchert, T., Menezes, J.C., Lochmann, D. and Reich, G. (2012) 'A novel in-line NIR spectroscopy application for the monitoring of tablet film coating in an industrial scale process', Talanta, 92, pp. 26–37.

Process Systems Enterprise Ltd. gPROMS User Guide. London, UK, December 2015.

Schwartz, L.W., Roy, R.V., Eley, R.R. and Petrash, S. (2001) "Dewetting patterns in a drying liquid film", Journal of Colloid and Interface Science, 234(2), pp. 363-374.

Weidner, D.E., Schwartz, L.W. and Eley, R.R. (1996) "Role of surface tension gradients in correcting coating defects in corners", Journal of Colloid and Interface Science, 179(1), pp. 66-75.

17AIChE Dr.Reddy's

Modeling and simulation efforts in scale up and characterisation of semi-solid dosage forms

AICHE Annual Meeting 2017

Saurav S Rath, Birendra K David and Ravi C Palaparthi
Process Engineering, **Dr. Reddy's Laboratories, Hyderabad**

Executive summary — Generic drug development

17AIChE Dr.Reddy's

Challenges faced by generic pharmaceutical industry

- ✓ Tight timelines of product development
- ✓ Market constraints
- ✓ Manufacturing capabilities
- ✓ Regulatory restrictions

Getting it **right the first time** demands detailed product & process understanding and implementing appropriate mechanistic modeling and Quality-by-design (QbD) approaches in product life cycle, to minimize downstream risks.

This work provides case studies in **topical dosage products**.

Introduction — Typical process flow and composition

17AIChE Dr.Reddy's

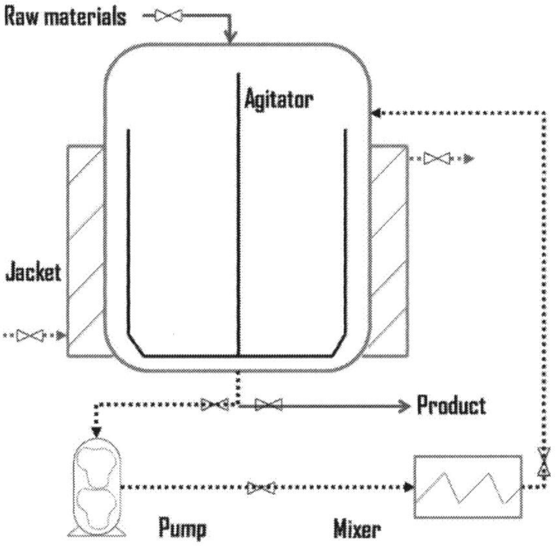

Raw materials	
Aqueous phase	**Oil phase**
Water, surfactant, suspending agent	Wax, fatty alcohol, anti-oxidant, **drug**

Unit operations

Heating, mixing, homogenization, cooling

Formulation

Emulsion, solution, suspension

WHY mechanistic modeling? — Mapping of CPPs, CMAs and CQAs

17AIChE Dr.Reddy's

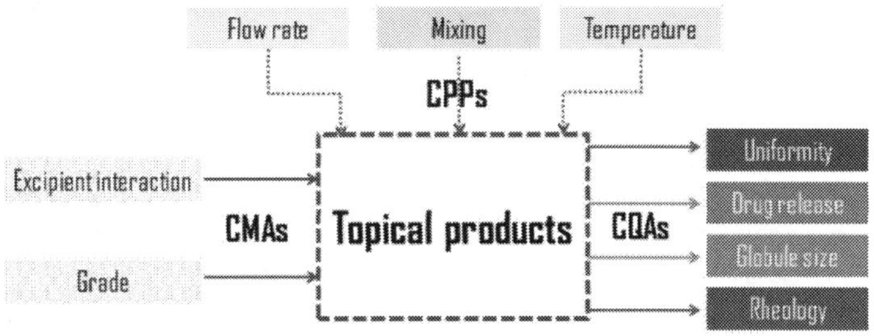

In practical sense, these QbD elements are inter-linked. For example in emulsion (O/W) systems, the non-Newtonian rheology varies not only with CPPs and CMAs but also is an implicit function of globule size.

Complex relation

Excipients + Process + Rheology → Globule size

This calls for various mechanistic models, to help predict the product behaviour.

Scale-up heuristics — Modeling and simulation capability

17AIChE Dr.Reddy's

Prediction of globule size distribution – **Population balance modeling (PBM)**

Drawbacks of general trend

- X Can't predict the distribution (d_{10}, d_{50} and d_{90})
- X Unsuitable for both internally and externally circulated mixers

Highlights of PBM

- ✓ Valid for both Newtonian and non-Newtonian fluid based systems
- ✓ Working for both laminar or turbulent flow regimes
- ✓ Can be used for all types of available high shear homogenizers
- ✓ What-if studies

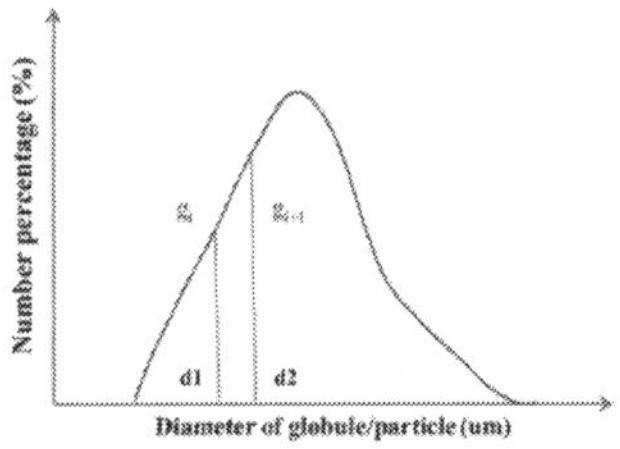

Number percentage distribution

Number of drops in [v,v+dv] per unit volume — *Breakage rate* — *Probability of forming daughter droplet (v) from mother (v')*

$$\frac{\partial n(v,t)}{\partial t} = -\,g(v)n(v,t) + \int_0^\infty \beta(v,v')g(v')n(v',t)dv'$$

$$^1g(d_p) = c_{1,b}\frac{\epsilon^{1/3}}{d_p^{2/3}}\exp\left(-\frac{c_{2,b}\gamma}{\rho_d\,\epsilon^{2/3}d_p^{5/3}}\right), where\;\; \varepsilon = [\mu\dot{\gamma}^2 v + Po\,(QN^2D^2\rho_c)]/V$$

[1]*Determination of breakage rates using single drop experiments, Matthias Kraume*

Case study – I — Predicting globule size distribution

17AIChE Dr.Reddy's

System – O/W emulsion

Inputs of PBM

- ✓ Shear and velocity profile in tank (CFD)
- ✓ Premix concentration (Microscopy or any relevant equipment)
- ✓ Geometry and process parameters

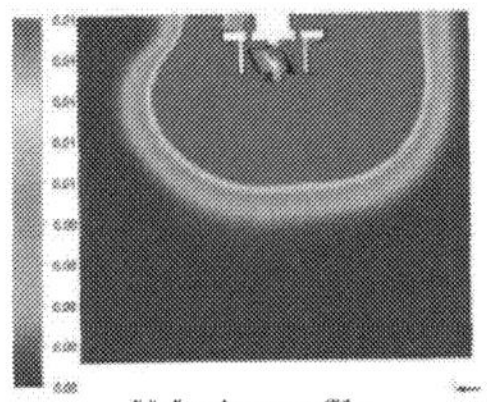

Velocity profile

What-if studies

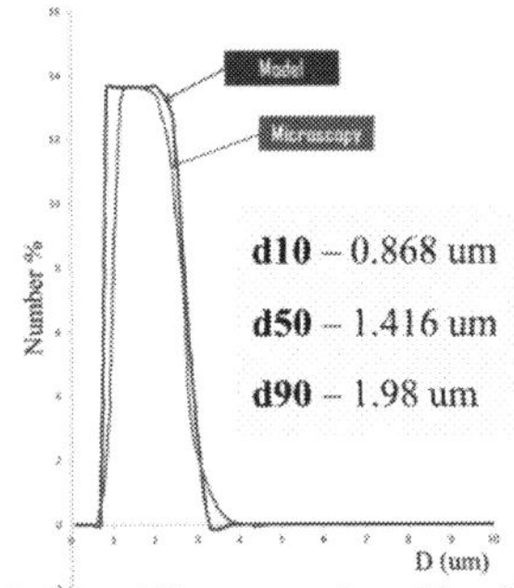

Model vs Microscopy data (Number %)

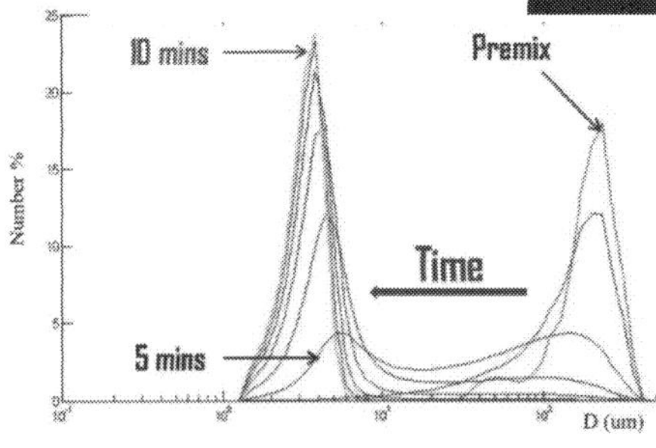

Shift of number % distribution with time

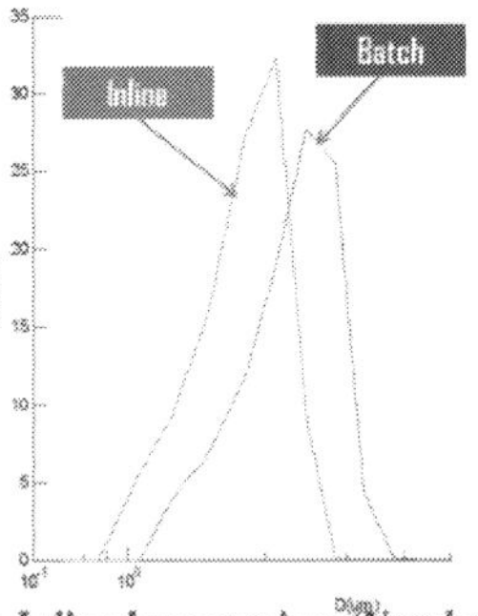

Batch vs Inline homogenizer (Number %)

Scale-up heuristics — Modeling and simulation capability

17AIChE Dr.Reddy's

Shear Strain – Scale-up correlation for homogenization

Drawback of general trend

X Doesn't take into account the effect of viscosity on other CQAs (GSD)

Overview of shear strain correlation

Case A: Shear < Shear $_{critical}$

With increase in shear strain, the apparent viscosity increases because of decrease in globule size. Order of (increase in viscosity due to increasing shear) >> Order of (decrease in viscosity due to thixotropic behaviour)

Case B: Shear > Shear $_{critical}$

With increase in shear strain, the globule size seizes to reduce and it reaches a plateau[2]. Increase RPM beyond max point → More shear imparted to the system →Viscosity reduces

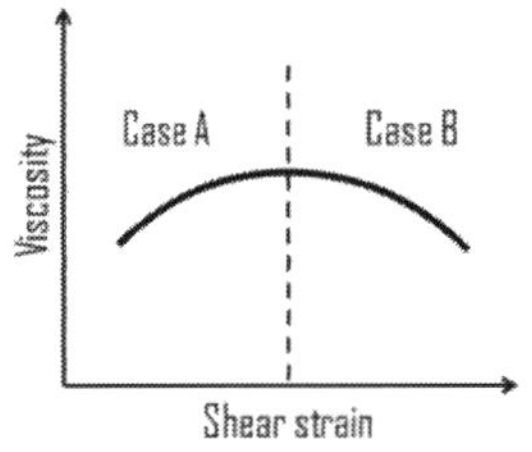

Shear rate

$$Shear\ strain = \beta.\left(\frac{N.D}{\delta}\right).t, \text{where}$$

D	m	Rotor diameter
δ	m	Rotor-stator gap
N	s^{-1}	Homogenizer speed
t	s	Homogenizer time

[2]*Deformation and breakup of drops in simple shear flows. Bruijn, de, R.A. and High shear mixers: A review of typical applications Jinli Zhang*

Case study – II — Estimating shear strain

17AIChE Dr.Reddy's

CQA [Viscosity ~ 20,000 cP (measured at 2/s) and average globule size ~ 3.9 – 4 um]

Given the CQA, the required shear ~ β x 110 **(Point A)**. Using the **algorithm**, homogenizer speed and time are calculated.

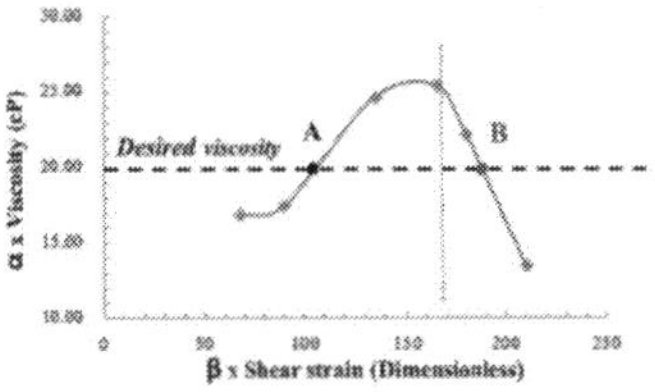

Variation of viscosity with shear

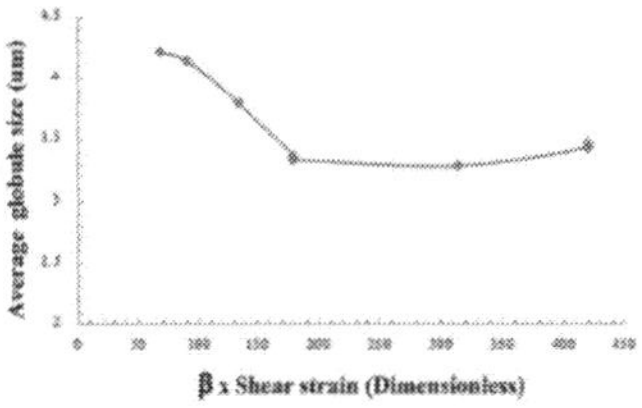

Variation of globule size with shear

Case Study – III — Mapping of shear and equipment

For shear sensitive products, it is imperative to know the possibilities by which shear can be applied to the system.

- ✓ Homogenizer (Size reduction and homogeneity)
- ✓ Lobe pump (Recirculation) ~ **5-10 % of homogenizer**
- ✓ Impeller (Uniformity of bulk) ~ 0.05-3 % of homogenizer
- ✓ Others – 0.01-0.05 % of homogenizer

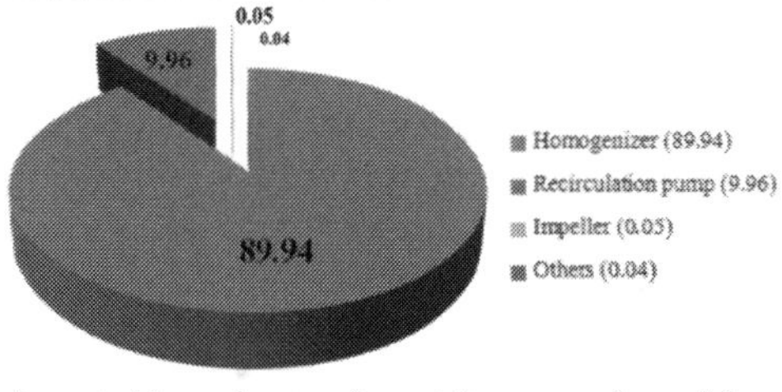

Shear imparted by various equipment for one such emulsion

Scale-up heuristics — Modeling and simulation capability

17AIChE Dr.Reddy's

Algorithm for calculation of homogenizer parameters

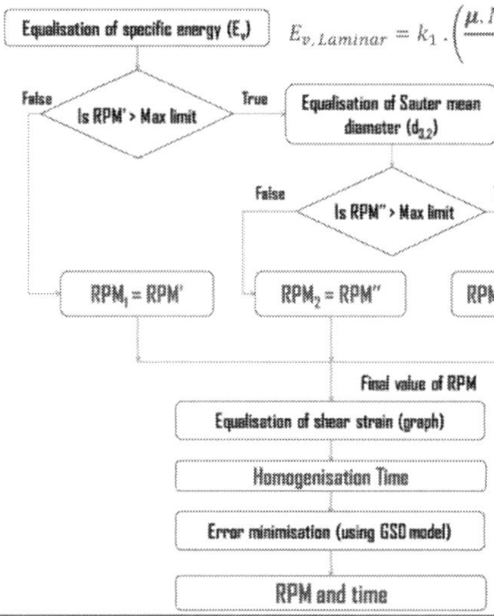

$$E_{v,Laminar} = k_1 \cdot \left(\frac{\mu . N^2 . D^3}{M}\right) . t$$

$$E_{v,Turbulent} = k_2 \cdot \left(\frac{N^3 . D^5}{M}\right) . t$$

$$d_{3,2} = k_3 . D . \left(\frac{\rho . N^2 . D^3}{\sigma}\right)^{-3/5}$$

Weber number

$$Shear = \beta . \left(\frac{N . D}{\delta}\right) . t$$

Glossary

D	(m)	Rotor diameter
δ	(m)	Rotor-stator gap
N	(s^{-1})	Homogenizer speed
t	(s)	Homogenizer time
ρ	(kg/m^3)	Density of phase
μ	(Pa.s)	Viscosity of phase
σ	(N/m)	Surface tension
M	(Kg)	Mass of phase
Q	(Kg/s)	Flow rate of phase
β	-	Correction factor

Case study- IV — Effect of cooling rate on product CQA

17AIChE Dr.Reddy's

System – O/W emulsion

Cooling of bulk can be controlled by modifying the following:

- ✓ Flow rate of coolant fluid
- ✓ Temperature of coolant fluid
- ✓ Mixing parameters

IV (b) Cooling time may not directly affect product viscosity.

IV (a) Fast cooling → Higher viscosity

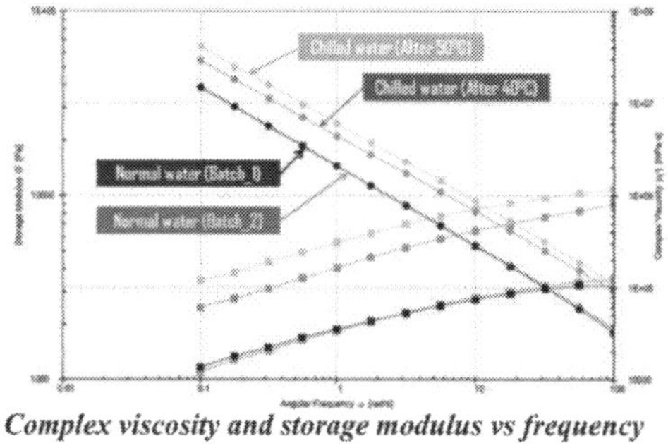

Complex viscosity and storage modulus vs frequency

Cooling rate	Average viscosity (cP) at $2s^{-1}$	
	Lab	Plant
Fast	82,312	100,000
Slow	81,800	69,000

⇨ The lab-scale data shows that cooling rate doesn't impact viscosity. In plant scale, the shear due to extra recirculation (>40 mins) resulted in drop of viscosity (shear-viscosity relation).

It is critical to optimise shear and cooling rate to attain desired CQA.

Scale-up heuristics — Modeling and simulation capability

17AIChE Dr.Reddy's

Algorithm for calculation of cooling time

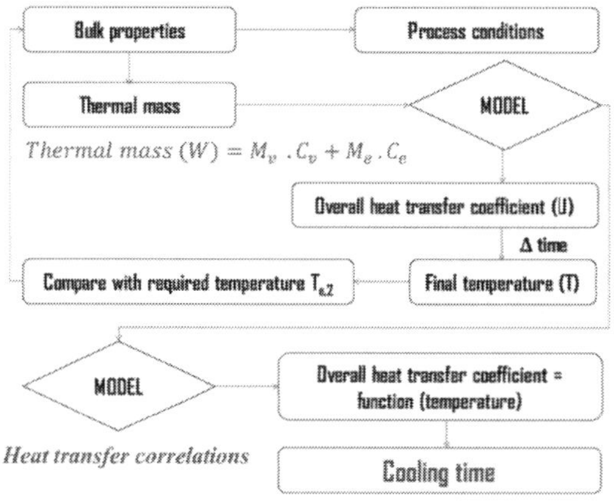

$$Thermal\ mass\ (W) = M_v \cdot C_v + M_e \cdot C_e$$

Heat transfer correlations

$$Cooling\ time\ (t) = W \cdot \frac{exp\left(\frac{u.a}{m.C}\right)}{m.C.\left(exp\left(\frac{u.a}{m.C}\right)-1\right)} \cdot ln\left(\frac{T_{e,1}-T_{w,1}}{T_{e,2}-T_{w,2}}\right)$$

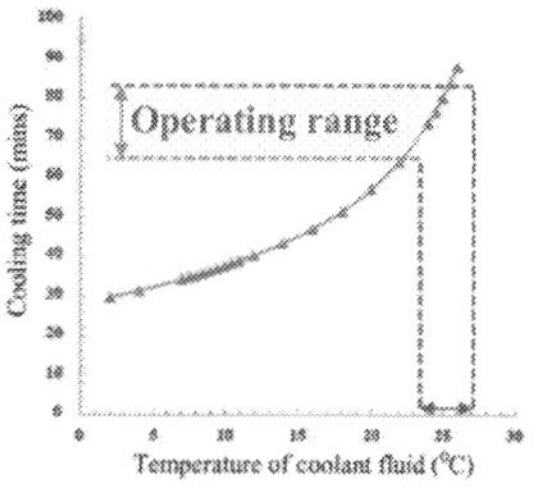

Time vs inlet fluid temperature

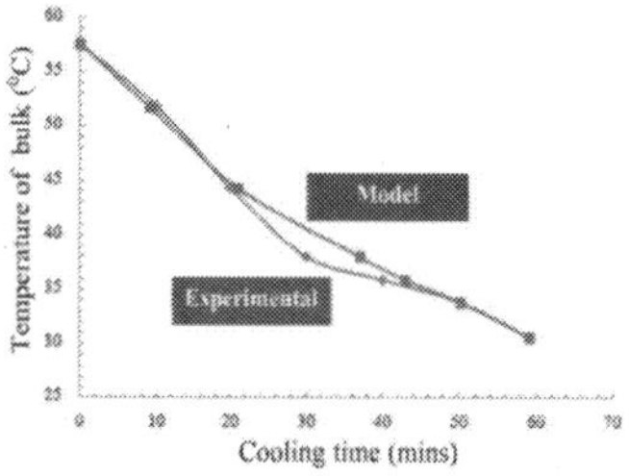

Model vs experiment (cooling time)

Glossary

u	W/(m².K)	Overall heat transfer coefficients
a	m²	Surface area
m	Kg/s	Flow rate of coolant
$T_{e,1}/T_{e,2}$	K	Initial/Final temperature of bulk
$T_{w,1}/T_{w,2}$	K	Inlet/Outlet temperature of coolant
M_e/M_v	Kg	Mass of bulk/Vessel
C_e/C_v	J/Kg.K	Specific heat capacity of bulk/Vessel

Scale-up heuristics — Modeling and simulation capability

17AIChE Dr.Reddy's

Correlations for mixing parameters – Effect on bulk uniformity

The mapping of right equipment (impeller, baffles and geometry) and process parameters is crucial for attaining homogeneity (*reducing dead zones*) in bulk.

- ✓ Minimum suspension velocity [N_{js}] – Suspensions
- ✓ Circulation velocity and solubility/miscibility time – Solutions

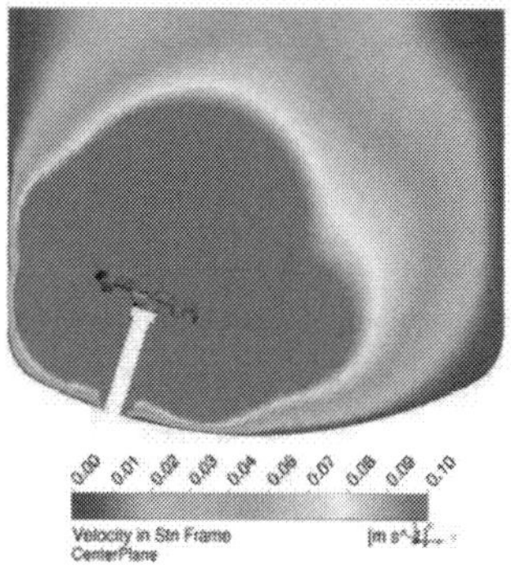

CFD analysis of mixing tank

Case study- V — Defining rate limiting step in mixing

$$Circulation\ velocity\ (V_s) = N_q.N.D^3/A$$

$$Circulation\ time\ (t) \sim 5L/V_s$$

Characteristic length

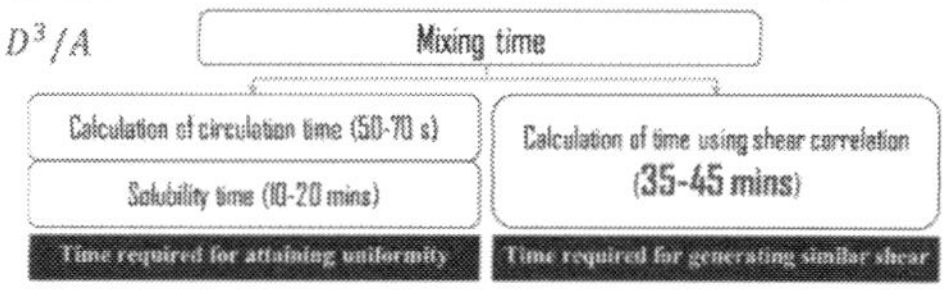

Glossary

N	s⁻¹	Impeller speed
N_q	-	Flow number
D	m	Impeller diameter
A	m²	Surface area
L	m	Characteristic length

Mixing parameters	200 kg	1000 kg	Correlations
Speed (RPM)	500	700-800	Power/volume
Time (mins)	15	40	Shear
Viscosity	**55,000**	57,000	Viscometer (at 2/s)

System properties

Ointment (Wax base with soluble API)

Density = 990 Kg/m³

k – 191.35 and n – 0.3 (Shear thinning)

Conclusion

Generic drug development

AIChE Dr.Reddy's

This work focusses on models obtained from computational fluid dynamics (CFD) coupled with population balance modeling (PBM) and constitutive models (like shear, energy density).

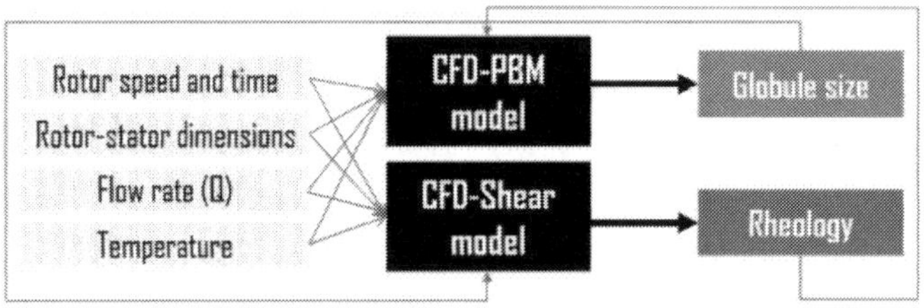

Defining, and scaling up the right manufacturing process with a given set of ingredients, to achieve the right product characteristics presents as a challenge to the process engineer.

Thank You

Back up slides

Batch vs Inline homogenizer

In case of inline homogenizer, the energy given into the system is more guided. Given the similar mixing efficiency for both batch and inline process, the amount of dead zones are more in case of batch process (CFD).

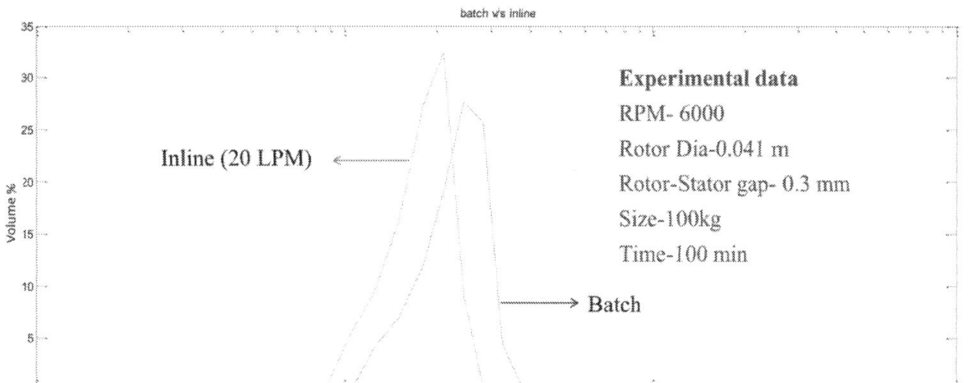

Scale-up heuristics — Modeling and simulation capability

17AIChE Dr.Reddy's

Prediction of globule size distribution – **Population balance modeling (PBM)**

Drawbacks of general trend

- X Can't predict the distribution (d_{10}, d_{50} and d_{90})
- X Unsuitable for both internally and externally circulated mixers

Highlights of PBM

- ✓ Valid for both Newtonian and non-Newtonian fluid based systems
- ✓ Working for both laminar or turbulent flow regimes
- ✓ Can be used for all types of available high shear homogenizers
- ✓ What-if studies

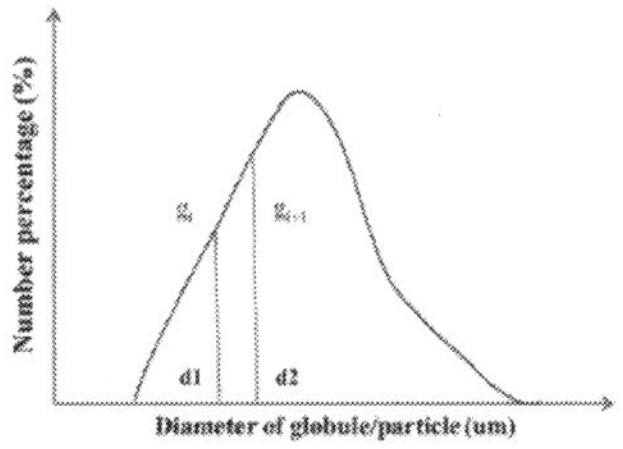

Number percentage distribution

Number of drops in [v,v+dv] per unit volume — *Breakage rate* — *Probability of forming daughter droplet (v) from mother (v')*

$$\frac{\partial n(v,t)}{\partial t} = -\,g(v)n(v,t) + \int_0^\infty \beta(v,v')g(v')n(v',t)dv'$$

$$^{1}g(d_p) = c_{1,b}\frac{\epsilon^{1/3}}{{d_p}^{2/3}}\exp\left(-\frac{c_{2\,b}\gamma}{\rho_d\,\epsilon^{2/3}{d_p}^{5/3}}\right), where\ \ \varepsilon = [\mu\dot{\gamma}^2 v + Po\,(QN^2D^2\rho_c)]/V$$

[1]Determination of breakage rates using single drop experiments, Matthias Kraume

Population balance modeling (Framework)

Method Approximations

i. Coalescence is neglected (Concentration of surfactant >> CMC)

ii. Material properties (ρ,σ) are assumed based on experiments and literature.

iii. GSD for premix was assumed to be that of the simply shaken mixture.

PBM Solution Approximations

i. Use of Trapezoidal rule for solving integral part of PBM equation.

ii. C1,C2 - constants in PBM equation are set using result of a particular product and need to be refined using parameter estimation.

iii. The prediction of GSD is done on a discrete basis where input is given in form of data points (nodes), accuracy of the system can be increased by including more nodes.

Population balance modeling (Framework)

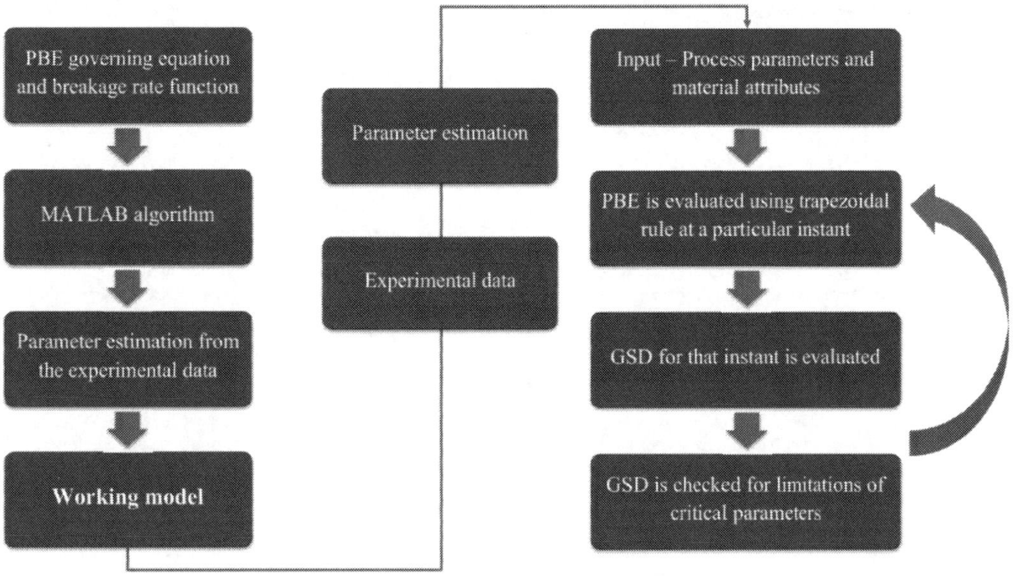

Effect of cooling rate (Literature)

Mixing speeds and cooling rates can impact of scale-up and influence physical properties of semisolids, such as its viscosity [1]

High cooling rate → High viscosity [2]

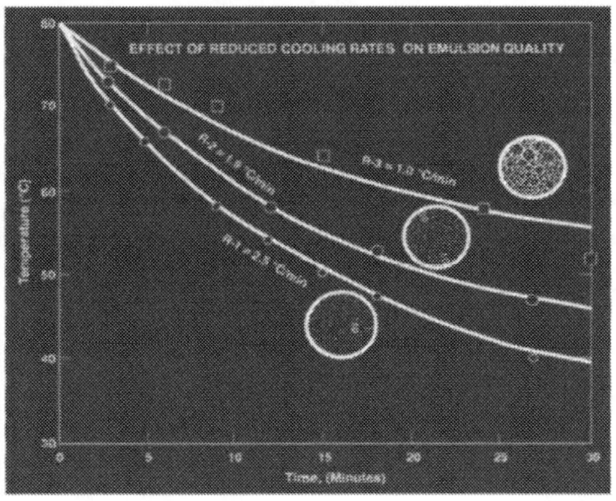

(1) Wessels P, Holz M, Erni F, Krummen K and Ogorka J. 1997. Statistical Evaluation of Stability Data of Pharmaceutical Products for Specification Setting. Drug Development and Industrial Pharmacy, 23(5): 427-439

(2) T. J. Lin, "Rheology Fundamentals and Applications in Cosmetic Formulation, Manufacturing and Quality Control" in The Chemistry and Manufacture of Cosmetics, Vol. 1 Basic Science (2000)

489812 Dissolution Kinetics Modelling of a BCS Class II Active Pharmaceutical Ingredient

Abstract:

Mechanisms of dissolution kinetics of active pharmaceutical ingredient (API) crystals have been extensively studied in the pharmaceutical domain, as these have a significant impact upon the bioavailability of drugs within the body. In such studies, the pharmaceutical dosage form of the API is typically suspended in a stirred vessel (in which the hydrodynamic conditions are well controlled), allowing the rate of dissolution to be measured using either UV or conductivity methods, providing an insight into the drug's behavior within the body. Dissolution kinetics can also be used to improve industrial crystallization processes, as the selective dissolution of fines could lead to narrower crystal size distributions (termed kinetic ripening). However, while the dissolution of APIs is of critical importance in the pharmaceutical industry, very few studies have been conducted to date which combine experimental methods with numerical simulations in order to achieve a comprehensive understanding of the principle influencing factors. This study attempts to address this shortcoming by examining the dissolution performance of Ibuprofen, a BCS Class II API, with PAT techniques experimentally and mathematical modelling.

The dissolution rate of Ibuprofen with different particle size distribution (PSD) and crystal habit was measured experimentally at different agitation rates. In order to simulate the environment encountered by the drug in the human body, all experiments were carried out in water at 37ÊšC using a phosphate buffer to maintain a pH of 7.20 (in accordance with U.S. Pharmacopoeia requirements). Different crystal habits were obtained by recrystallizing Ibuprofen in methanol and hexane respectively. Crystals generated using methanol as the solvent produce polyhedral crystals while those generated using hexane are needlelike. Different agitation conditions have been investigated while the effect of particle size upon the dissolution rate was examined by using four different size fractions (sub 150Âµm, 150-300Âµm, 300-500µm and 500-850µm).

Experiments were carried out in a 100mL, temperature controlled Easymax vessel. Agitation in the vessel was achieved using a stainless-steel pitched blade impeller. Solvent and crystals were added into the glass vessel and the rate of dissolution was *in-situ* measured using on-line UV probe to track the liquid phase Ibuprofen concentration, while focused beam reflectance measurement (FBRM) was applied to track the solid phase particle size by measuring chord length distribution (CLD) and particle vision measurement (PVM) was used to monitor the shape and structure of the crystals. The initial and final PSD of Ibuprofen were also measured *off-line* using a Malvern Mastersizer 2000.

As dissolution is the reverse process of crystal growth, it involves two steps: (1) surface reaction and disintegration of the surface species; (2) mass transfer of these species into the bulk solution across the diffusion layer. It is typically assumed that the overall process is limited by the slower step. Based on this theory, a kinetic model has been devised to describe the dissolution behavior of a polydisperse powder under non-sink conditions. Two rate limiting regimes, surface detachment controlling and mass transfer controlling, were derived respectively. This model was solved using numerical methods and combined with a fraction of the collected experimental data in order to obtain a shape

modulated dissolution rate and an overall correlation for the dissolution rate. The remaining experimental data was used to verify the accuracy of the model, demonstrating a very good agreement between theory and experiment. This dissolution kinetic model, which was derived from fundamental principles, is of great use to identify favorable crystal characteristics enabling higher rates of dissolution. It can be used to inform parameter selection during Quality by Design (QbD), for instance, the prediction of PSD of API crystals could be applied to guide the crystallization process for API manufacturing while the estimation of the API concentration changing with time provides useful information for formulation development.

Keywords: Dissolution kinetics; Mathematical modelling; API manufacturing and formulation

497278 ERK1/2-Driven Sprouty2 Expression Mediates Resistance to Receptor Tyrosine Kinase-Targeted Therapeutics in Glioblastoma

Abstract:

The efficacy of inhibitors of oncogenic kinases in glioblastoma multiforme (GBM) has been disappointing to date. In GBM and other malignancies, resistance to targeted therapeutics can arise through dynamic rewiring of cell signaling pathways responsible for tumorigenesis and cell survival. Here, we describe a signaling rewiring process in glioblastoma cells that promotes expression of Sprouty2 (SPRY2), which we recently identified as a driver of GBM tumor growth and resistance to kinase inhibitors. Specifically, we found that SPRY2 depletion reduced the ability of GBM cells to form colonies in soft agar, to form subcutaneous tumors in mice, and to resist co-inhibition of the EGFR and MET receptor tyrosine kinases. In the present study we found that, in a panel of GBM cell lines, SPRY2 expression was initially reduced in response to inhibition of EGFR and MET. However, 24-48 hours after receptor inhibition, resurgent SPRY2 expression was observed while EGFR and MET phosphorylation remained suppressed. Multivariate measurements of several kinases downstream of EGFR and MET revealed that phosphorylation of extracellular regulated kinase 1/2 (ERK1/2), which was initially suppressed in response to EGFR and MET inhibition, also displayed a resurgence that paralleled SPRY2 expression. This correlation between ERK phosphorylation and SPRY2 expression is consistent with the documented ability of ERK to regulate SPRY2 transcription. Antibody array analysis suggested that ERK reactivation is mediated by other RTKs activated in response to EGFR and MET inhibition. This analysis was supported further by measuring transcriptional differences in ligands responsible for activating these receptors. Suppression of the alternate RTKsâ€™ activity through inhibitors or siRNA-mediated knockdown augmented cell death response to EGFR and MET inhibitors. In addition to population-level studies, this effect was demonstrated with single-cell resolution using a fluorescence-based reporter of ERK activity and RNA FISH. Thus, our data support a model wherein resurgent ERK phosphorylation and resultant SPRY2 expression occur due to the ability of glioblastoma cells to activate alternative RTKs, suggesting that targeting multiple RTKs could be a way to overcome resistance to single RTK inhibitor treatments. To explore this hypothesis in tumor xenograft models, we recently engineered a luminescence reporter of ERK activity that enables intravital imaging of ERK activity. Using this new tool, we have demonstrated the improved ability of specific combinations of RTK inhibitors to maintain ERK activity suppression within the tumor.

497512 Single-Cycle Growth Kinetics of Infection By Zika Virus

Abstract:

The Zika virus (ZIKV) has become a public health concern because of its correlation between ZIKV infection and neural disorders in both fetuses and adults. In this study, we analyzed experimental ZIKV infection of cell cultures to characterize virus-cell adsorption, intracellular reproduction and particle degradation. Virus adsorption was tested at different multiplicity of infection (MOI), the ratio of added virus particles to cell numbers. Actual efficiencies of adsorption were lower than anticipated, owing in part to degradation of free virus particles and mass transfer limitations of virus particle transport to susceptible cells. The length of the latent period of ZIKV single-cycle growth, reflecting the duration of intracellular biosynthesis and particle assembly, depended on the MOI, and the average burst size or number of virus particles released per cell was approximately 300. These results provide the basic parameters of ZIKV infection single-cycle kinetics which will inform future models of multi-cycle infection spread. Further, the low efficiency of ZIKV absorption and high degradation rate will be important factors to consider as we develop quantitative and integrative approaches to better understand how ZIKV grows, spreads and ultimately contributes to the development of disease.

497635 Dynamic Modeling of Pancreatic Cancer Metabolism to Investigate Optimal Therapeutic Strategies

Abstract:

Pancreatic cancer is an aggressive disease with an overall 5-year survival rate of ~5%. Cancer cells in pancreatic tumors rely on glucose and glutamine to generate energy and building blocks for proliferation. Therapies that target tumor metabolism aim to impede the unique metabolic phenotype of the cancer cells; however, in order to develop effective treatment strategies, quantitative information about how specific reactions contribute to and control tumor metabolism is needed.

We have built a computational model that predicts the dynamics of glucose and glutamine metabolism in pancreatic cancer cells over time. The model predicts relevant metabolic information that is difficult to measure experimentally, including the time courses of the metabolite concentrations and the fluxes of each reaction [1]. The model also links metabolism to cancer cell growth and predicts the number of pancreatic cancer cells over time. We fit the model to published data obtained from experimental studies of pancreatic cancer cells and validated the model using a distinct set of measurements not included in the fitting. As a result of this work, we constructed a validated model of pancreatic cancer metabolism that predicts tumor growth in response to nutrient availability and utilization. To our knowledge, this is the first model of cellular metabolism that specifically predicts the dynamics of pancreatic cancer cells.

The model provides mechanistic insight into novel targets for inhibiting pancreatic cancer cell metabolism. We applied the model to predict the effects of inhibiting various enzymes in the metabolic network, identifying effective therapeutic targets to inhibit tumor metabolism impede cell growth. We specifically investigate strategies that target core metabolic enzymes individually and in combination. Thus, the model can be used to design novel therapies for impeding cancer cell proliferation, complementing *in vitro* and *in vivo* pre-clinical studies. We are building on this work to predict the dynamics of carbon utilization and labeling using isotope-labeled metabolomics experiments. Altogether, we demonstrate the importance of kinetic modeling to investigate novel therapeutic strategies for pancreatic cancer.

[1] Roy, M. and Finley, S.D., (2017) Computational model predicts the effects of targeting cellular metabolism in pancreatic cancer. *Front. Physiol.* 8:217. doi: 10.3389/fphys.2017.00217

487220 Regulation of Amyloidogenesis and Proteolysis in the Beta-Amyloid/Cathepsin B/Cystatin C Network

Abstract:

Cystatin C (CysC) is an important protease inhibitor of the central nervous system. Hereditary Cystatin C Amyloid Angiopathy (HCCAA) is a rare and fatal condition caused by misfolding, aggregation, and deposition of CysC variant L68Q. This single mutation enhances the unfolding rate of the monomer to a partially-unfolded intermediate state, allowing the protein to dimerize by domain-swapping. Since L68Q is prone to both domain-swapping and to fibrillogenesis, it has been hypothesized that CysC forms amyloid via propagated domain-swapping. We tested this idea by characterizing the stability of a domain-swapping-resistant CysC mutant, V57N. Contrary to the propagated domain-swapping hypothesis, V57N readily formed fibrils despite its inability to form domain-swapped dimers. Furthermore, we discovered that, at physiological conditions, CysC can self-associate into small meta-stable oligomers that retain the secondary structure of monomeric CysC and are not domain-swapped. These non-swapped oligomers exhibited some pre-amyloid characteristics, yet were more resistant to fibril formation than CysC monomers. We propose a new hypothesis that CysC amyloid aggregation initiates from the monomeric state and does not require domain-swapping. The Alzheimer's peptide, Aβ, is well known to associate into amyloid fibrils. There are a few reports suggesting that CysC interacts with Aβ and may inhibit Aβ fibril formation, although the published data are inconclusive. We used several biophysical tools to compare the efficacy of CysC monomers (mCys), domain-swapped dimers (dCys), and non-swapped oligomers (oCys) in inhibiting Aβ aggregation. Both mCys and dCys caused a modest reduction in Aβ fibrillation. Surprisingly, oCys was a much more potent inhibitor of Aβ fibrillation. We hypothesize that this strong interaction is a specific consequence of oCys pre-amyloid structure, suggesting that heterotypic interactions between pre-amyloid-like oligomers arising from different proteins may potently arrest further maturation into fibrils.

Because CysC can interact with Aβ and inhibit its aggregation, it has been speculated that CysC could be a neuroprotective agent in Alzheimer's, a hypothesis supported in some but not all transgenic animal studies. However, one of the primary functions of CysC is to inhibit the cysteine protease cathepsin B (catB). CatB is known to proteolytically degrade monomeric Aβ into harmless fragments, and may also digest preformed oligomers and protofibrils. This raises the question: what is the net outcome of the interactions of these three proteins, when CysC binds to Aβ, CatB degrades Aβ, and CysC inhibits CatB. Does CysC inhibit Aβ fibrillogenesis, or does CysC inhibit CatB, preventing the degradation of Aβ? In this way, the putative neuroprotective functions of CysC and CatB are antagonized by the other's presence. Further complicating the situation, monomeric CysC is metastable, and as described above, it can re-fold into domain-swapped dimers that no longer inhibit CatB, or it can associate into oligomers that retain CatB inhibitory activity but interact more strongly with Aβ.

We developed a mathematical model to simulate the complex interactions between Aβ, CatB and CysC. We measured CysC binding affinity with Aβ through a solution-based FRET assay. It has generally been believed that CatB activity is constrained to acidic cellular compartments and is drastically diminished upon secretion to neutral extracellular fluids. However, we monitored CatB proteolysis of Aβ by and discovered

that even at neutral pH, this reaction was fast enough to be physiologically relevant. We quantified enzymatic degradation rates of $A\beta_{42}$, $A\beta_{40}$, and CysC by CatB to determine k_{cat}/K_M values. We validated the model with experimental data, then used the model to define a set of conditions under which CysC and CatB concentrations critically affect the amount of free Aβ and the rate of Aβ amyloidogenesis. Several published studies have explored the relationship between CysC and Aβ in animal models, with researchers reaching different conclusions. Our analysis can be used to interpret these *in vivo* studies mechanistically, and to propose novel strategies for controlling amyloidogenesis and interfering with disease processes.

485439 Receptor Endoyctosis and Cell Memory in Growth Factor Signaling

Abstract:

Cells need to sense and respond to signals in a constantly changing environment. Responding to relative rather than absolute changes in the stimuli enables cells to remain sensitive over a wide range of background signal levels. Indeed, various biological systems ranging from bacterial chemotaxis to human sensory perception systems, such as vision, are known to respond to relative changes in their environment. With a combined modeling and experimental approach we identify a novel mechanism of relative sensing of extracellular ligands by mammalian signaling cascades. Specifically, we study the epidermal growth factor (EGF) induced signaling cascade and demonstrate that the pathwaysâ€™ immediate-early phosphorylation response can sense relative change in extracellular level of EGF. Using a theoretical model, we analytically explain the mechanism of relative sensing by deriving constraints on the network parameters and confirm their consistency with the values derived through fitting experimental phospho-proteomic data. Importantly, the derived constraints are characteristic of multiple other signaling cascades. This points to the potential generality of the discovered phenomenon of extracellular ligand relative sensing across many other signaling systems. Overall, our work presents a novel quantitative framework to understand how cells respond to fluctuating environments in vivo.

496588 Gut Microbiota-Derived Tryptophan Metabolites Modulate Inflammatory Response in Hepatocytes and Macrophages

Abstract:

Gut Microbiota-derived Tryptophan Metabolites Modulate Inflammatory Response in Hepatocytes and Macrophages

Smitha Krishnan[1], Yufang Ding[2], Maria Choi[1], Nima Saedi[3], Martin L. Yarmush[3], Arul Jayaraman[2], and Kyongbum Lee[1]

1. Department of Chemical and Biological Engineering, Tufts University, Medford, MA 02155

2. Department of Biomedical Engineering, Texas A&M University, College Station, TX 77843

3. Center for Engineering in Medicine, Massachusetts General Hospital, Boston, MA 02114

The human gastrointestinal (GI) tract harbors an estimated 10 to 100 trillion microorganisms, encompassing several hundred different species. Collectively referred to as the gut microbiota, the microbes perform a number of essential physiological functions, including the metabolism of complex carbohydrates [1], development of the immune system [2], and defense against pathogens [3]. Beyond the GI tract, gut-brain, gut-lung, and gut-liver associations have also been identified, highlighting the importance of the microbiota in host physiology. Alterations in the intestinal microbiota composition (i.e. dysbiosis) have been correlated to various diseases, including obesity [4, 5], diabetes [6], and cancer [7].

Non-alcoholic fatty liver disease (NAFLD) has emerged as the most prevalent chronic liver disease in Western countries. It is characterized by steatosis, or fat deposition in the liver, which can remain a benign condition. However, in 10-20% of patients, NAFLD progresses from steatosis to non-alcoholic steatohepatitis (NASH), which can lead to cirrhosis and liver cancer. Direct, potentially causal involvement of the microbiota in NASH was highlighted by Le Roy et al. [11], who reported that high fat diet (HFD)-fed germ-free (GF) mice inoculated with commensal bacteria from GI tract of HFD-fed hyperglycemic mice developed steatohepatitis, whereas HFD-fed GF mice inoculated with bacteria from HFD-fed, but normoglycemic mice only showed mild steatosis. This suggests that the composition of the intestinal microbiota, and hence their metabolic products, could determine whether NAFLD progresses to NASH. The study described in this abstract utilizes metabolomics in conjunction with biochemical assays to identify and characterize microbiota-dependent immunomodulatory metabolites that could modulate liver inflammation in steatosis.

In previous work [13], we identified a panel of aromatic amino acid (AAA)-derived metabolites whose levels in the mouse intestine depended on the presence of the microbiota. Furthermore, several of these metabolites were found to activate the aryl hydrocarbon receptor (AhR), a transcriptional regulator of host-microbiota interactions in the intestine. We compared the profiles of these metabolites in serum, cecum, and liver samples from 14-week old male C57BL/6J mice raised for 8 weeks either on HFD or low-fat diet (LFD). Untargeted LC-MS experiments found that several of the previously identified gut microbiota-dependent metabolites were significantly depleted in HFD mice compared to LFD mice (Figure 1a-c), including Ahr activators tryptamine (TA) and indole-3-acetate (I3A) (Figure 1d).

We next investigated the effects of TA and I3A on inflammatory responses in hepatocytes using both human (HepG2) and murine (AML12) cell lines. Cultured HepG2 and AML12 cells were treated with 10, 100, or 500 uM TA or I3A for 24 h, and then exposed to the pro-inflammatory cytokine TNF_ for 24 h in the presence of the metabolites. In both human and murine-derived hepatocytes, pretreatment with TA or I3A significantly attenuated TNF_-stimulated increases in intracellular fatty acid levels, with I3A exhibiting greater potency. Similarly, pretreatment with TA or I3A significantly attenuated TNF_Ðstimulated increases in intracellular bile acid levels. These effects were also observed when the hepatocytes were preloaded with lipid droplets prior to metabolite and cytokine treatment by growing the cells in fatty acid supplemented culture medium to mimic a steatotic state.

To confirm that the metabolites can activate the AhR pathway in hepatocytes, we applied TA and I3A to H1G1.1c3 cells expressing a stable enhanced green fluorescent protein (EGFP) reporter construct regulated by the AhR response element (AHRE) derived from the CYP1A1 promoter [14]. Consistent with our previous findings [13], 24 h exposure to TA or I3A dose-dependently increased GFP expression in the reporter cells, indicating ligand activation of the AhR in hepatocytes by these metabolites.

Finally, we investigated the effects of I3A and TA on macrophages, as polarization of resident macrophages towards a pro-inflammatory state is a key step in the progression of steatohepatitis. RAW 264.7 murine macrophages were treated with I3A and TA for 4 h, and then stimulated with palmitate for 18 h followed by LPS exposure for 6 h in the presence of the metabolites. Treatment with I3A significantly attenuated the induction of the pro-inflammatory cytokines TNFa, IL-1b and MCP-1 by palmitate and LPS in a dose dependent manner at both mRNA and protein levels. Unlike I3A, TA only decreased the gene expression of IL-1b. However, TA significantly decreased TNFa and MCP-1 expression at the protein level. Both I3A and TA significantly decreased bone marrow-derived macrophage (BMDM) migration towards MCP-1 in a dose dependent manner.

In summary, our results show that several microbiota-dependent metabolites derived from tryptophan are significantly depleted in a murine model of diet-induced liver steatosis, and that these metabolites can act directly on hepatocytes and macrophages to modulate inflammatory pathways. Our results demonstrate that these microbiota metabolites are ligands for the AhR. Taken together, our findings support the hypothesis that dysbiosis of the gut microbiota could predispose the liver to inflammation in diet-induced steatosis through an altered microbiota metabolite profile. Prospectively, insights into the mechanisms underlying the link between microbiota dysbiosis and NAFLD could provide novel strategies to treat or prevent the progression of fatty liver diseases through the use of probiotics or postbiotics.

Figure 1

[1] T. Arora and R. Sharma, "Fermentation potential of the gut microbiome: implications for energy homeostasis and weight management," Nutrition Reviews, vol. 69, pp. 99-106, 2011.
[2] L. V. Hooper, D. R. Littman, and A. J. Macpherson, "Interactions Between the Microbiota and the Immune System," Science, vol. 336, pp. 1268-1273, June 8, 2012 2012.
[3] N. Kamada, G. Y. Chen, N. Inohara, and G. Nunez, "Control of pathogens and pathobionts by the gut microbiota," Nat Immunol, vol. 14, pp. 685-690, 2013.
[4] J. Shen, M. S. Obin, and L. Zhao, "The gut microbiota, obesity and insulin resistance," Molecular Aspects of Medicine, vol. 34, pp. 39-58, 2013.
[5] S. Greenblum, P. J. Turnbaugh, and E. Borenstein, "Metagenomic systems biology of the human gut microbiome reveals topological shifts associated with obesity and inflammatory bowel disease," Proceedings of the National Academy of Sciences, vol. 109, pp. 594-599, January 10, 2012 2012.
[6] J. K. Nicholson, E. Holmes, J. Kinross, R. Burcelin, G. Gibson, W. Jia, et al., "Host-Gut Microbiota Metabolic Interactions," Science, vol. 336, pp. 1262-1267, June 8, 2012 2012.

[7] J. C. Arthur, E. Perez-Chanona, M. MŸhlbauer, S. Tomkovich, J. M. Uronis, T.-J. Fan, et al., "Intestinal Inflammation Targets Cancer-Inducing Activity of the Microbiota," Science, vol. 338, pp. 120-123, October 5, 2012 2012.
[8] R. E. Ley, F. Backhed, P. Turnbaugh, C. A. Lozupone, R. D. Knight, and J. I. Gordon, "Obesity alters gut microbial ecology," Proc Natl Acad Sci U S A, vol. 102, pp. 11070-5, Aug 2 2005.
[9] P. J. Turnbaugh, R. E. Ley, M. A. Mahowald, V. Magrini, E. R. Mardis, and J. I. Gordon, "An obesity-associated gut microbiome with increased capacity for energy harvest," Nature, vol. 444, pp. 1027-31, Dec 21 2006.
[10] S. G. Sheth, F. D. Gordon, and S. Chopra, "Nonalcoholic steatohepatitis," Ann Intern Med, vol. 126, pp. 137-45, Jan 15 1997.
[11] T. Le Roy, M. Llopis, P. Lepage, A. Bruneau, S. Rabot, C. Bevilacqua, et al., "Intestinal microbiota determines development of non-alcoholic fatty liver disease in mice," Gut, vol. 62, pp. 1787-94, Dec 2013.
[12] W. R. Wikoff, A. T. Anfora, J. Liu, P. G. Schultz, S. A. Lesley, E. C. Peters, et al., "Metabolomics analysis reveals large effects of gut microflora on mammalian blood metabolites," Proceedings of the National Academy of Sciences of the United States of America, vol. 106, pp. 3698-3703, Mar 2009.
[13] G. V. Sridharan, K. Choi, C. Klemashevich, C. Wu, D. Prabakaran, L. B. Pan, et al., "Prediction and quantification of bioactive microbiota metabolites in the mouse gut," Nat Commun, vol. 5, p. 5492, 2014.

[14] Nagy, Scott R., et al. "Development of a green fluorescent protein-based cell bioassay for the rapid and inexpensive detection and characterization of Ah receptor agonists." Toxicological Sciences 65.2 (2002): 200-210.

501131 Integrated Modeling Framework for Signaling, Transcription and Cell Fate and Their Clinical Significance

Abstract:

Abstract: Cellular behavior is regulated by complex signaling networks which in turn are influenced extracellular environment, membrane trafficking processes and in combination regulate the nuclear events leading to transcription. These processes operate at very different length and time scales and key information about them like rates of interactions, expression levels are often incomplete which makes mathematical modeling of these networks difficult. Here we develop a general approaches to integrate signaling models operating at different time scales and also a signaling and trafficking model operating at different timescales. To validate the models we performed extensive sensitivity analyses and studied the robustness and evolvability of these models. These were utilized in interpreting the experimental results, designing further experiments and making useful clinical predictions. This framework was applied to a core signaling module of Ras-MAPK, PI3K/AKT and JAK/STAT pathway which is a critical cellular network often altered in various cancers of lung, liver and prostate. We studied the interaction of this network with p53 mediated DNA damage response pathway and Rab8/Rab11 regulated late endocytic recycling pathways.

Acknowledgements: This Work was supported by European Commission Grant FP7-ICT-2011-9-600841 and NIH 1U54CA193417

509054 Solve this! Fundamental Approach to Problem Solving in Industrial Processes I (Invited Talks)

Abstract:

Solving complex plant problems is an everyday exercise that affects all projects and budgets, and it is very rarely introduced to students at any level. In fact, by the time young engineers are few years into working in industry, most have lost the engineering skills that they developed in school. We hope to encourage future engineers to retain those skills and learn the importance of plant problem-solving.

496057 Retron-Based Targeted Mutagenesis Enabling *in vivo* Continuous Evolution in *E. coli*

Abstract:

Envisioned as an accelerated imitation of Darwinian evolution, *in vivo* continuous evolution, when compared with traditional *in vitro* methods, is supposed to be a highly effective approach of engineering enzymes or pathways with tailor-made properties under few human intervention. A typical *in vivo* continuous evolution process contains iterative rounds of global mutagenesis and fitness-coupled selection. Generally used global mutagenesis methods, however, usually lead to error catastrophe or evolutionary escape. Targeted mutagenesis, which can largely improve the mutation rate of a target region against the relative steady genetic background, has raised much attention. Existing targeted mutagenesis tools mainly rely on mutagenic protein anchoring or orthogonal DNA polymerase-plasmid pairs, while lacking stringent targeting property and general applicability. Recently, a novel targeted mutagenesis tool based on yeast native retrotransposons, where diversity is generated in transcription level and recorded by reverse transcription and recombination, has opened a new direction.

In this work, we proposed a novel targeted mutagenesis tool based on an *E. coli* native retron cassette, which is able to produce ssDNAs encoded within it *in vivo* through transcription and reverse transcription processes. By inserting the target region sequence into the wild-type retron cassette and applying the error-prone T7 RNA polymerase, an ssDNA mutant library can be generated, and integrated chromosomally into the target area through single-stranded DNA recombination. As a proof of concept, the gene *sacB*, encoding levansucrase, was integrated into *E. coli* DH10B genome as a counter-selection marker for mutants. When the ssDNA mutant library targeting the *sacB* region was generated in this strain, a considerable mutation rate improvement over its genetic background as well as the wild-type strain can be observed through sucrose lethal selection. Details of this work will be given in the conference. In the light of its strong ability to construct mutant library of an artificially designated region with high throughput and low workload, we believe that the retron-based targeted mutagenesis, when integrated into *in vivo* continuous evolution, will be developed into a practical method in rapid and robust protein and metabolic engineering.

501814 Optogenetic Platform for the "on Demand" Production of Proteins in Biopharma and Biomedical Applications

Abstract:

Optogenetics, the use of light to trigger and control genetic expression, is a biological technique that is finding diverse application niches, and multiple versions of optogenetic control have been documented in recent literature.

Here, we describe a simple optogenetic platform for controlling protein expression in mammalian cell lines. Our platform uses two distinct vectors, GAV and Z-NEP. Briefly, GAV encodes for the constitutive expression of green fluorescent protein (GFP) and a fusion protein that contains a protein monomer. This last protein forms dimmers upon exposure to blue light, and dimers regulate the expression of the Z-NEP vector that encodes for a reporter (i.e., red fluorescent protein (RFP)) and/or other protein of interest (i.e., a monoclonal antibody).

The platform was tested in the CHO-S cells (the battle horse in the production of glycosylated recombinant proteins) and DU145 cells (a prostatic cancer cell line). The optogenetic control of the production of a monoclonal antibody of therapeutic interest (a biosimilar of infliximab) was demonstrated in CHO-S cells cultured in bottles and microfluidic devices. To our knowledge, this is the first report of application of optogenetics for the "on-demand" production of recombinant proteins of biopharmaceutical interest. In addition, we successfully triggered and controlled the expression of fluorescent proteins in DU145 cells exposed to blue light. We envision the control of protein expression in co-cultures of healthy and cancerous cells as a useful research and potentially therapeutic tool (i.e., controlled release mediated through light).

Our results suggest that the application of optogenetic tools to biotechnology and biomedical scenarios will enable a tight control of protein expression in organ-on-chip/lab-on-chip applications.

486576 Elucidating and Engineering the Role of Arabinogalactan Proteins in Somatic Embryogenesis

Abstract:

There is a critical need for efficient reproduction of high-value crops including timber crops, biofuel feedstocks and medicinal plants. Our work focuses on the loblolly pine (*Pinus taeda*), a critical crop to the timber industry, currently occupying 30 million acres of U.S. timberlands and making up one half of the U.S. standing pine volume. Somatic embryogenesis (SE) is anÂ *in vitro*Â cellular process that leads to the formation of a viable embryo from a single cell, and provides a means to create a clonal crop, enabling selection for plants with preferred characteristics (e.g., fast growth rate, disease resistance, superior wood quality). Our goal is to identify and manipulate predictors of embryo yield in the earliest stage of development to streamline SE and enable process commercialization. The primary limitation to large-scale implementation of SE is that the process is not fundamentally understood, leading to large variability in embryo yieldâ€"a number that cannot be determined until 3-4 months from SE initiation. Arabinogalactan proteins (AGPs) are a promising biomarker and have been implicated in other SE systems, providing many avenues for manipulation to improve yields. In this work, targeted treatments were applied to maintenance cultures to evaluate influence on yield. For example, we demonstrated that AGPs are influential on loblolly pine somatic embryo yield by inhibiting AGP function *in vitro* using the AGP-binding Î²-glucosyl Yariv reagent. We also introduced generic, exogenous AGPs (e.g., gum arabic, from the Acacia tree) to cultures, supplementing the endogenous AGPs to quantify the molecular response and effect on SE. We further extend this approach using novel characterization techniques to elucidate the role of AGPs in culture, including new colorimetric assays, ELISAs using antibodies with AGP subclass specificity and UPLC characterization and fractionation. Understanding the complex properties of AGPs and their influence on SE will enable a better fundamental understanding of SE and identification of targeted strategies to promote high embryo yield.

501850 A Segregated Kinetic Model for Antibody Producing Cell Lines

Abstract:

Monoclonal antibodies (MAbs) have extensive biomedical applications, account for nearly 40% of the sale of biologics in the U.S., and are produced in animal cell bioreactors at a variety of scales. Mammalian cell cultures have become the favored production hosts for MAbs, since microbial systems are not able to carry out the complex post-translational and functional modifications of these proteins, such as glycosylation. The efficiency and performance of these cell cultures require use of highly specialized culture media to enhance MAb yield in in vitro production, since the cells are subject to death and a considerable understanding of the variations in MAb productivity due to variations in culture conditions.

Cell death is a key issue in the bioreactor culture, as it decreases overall cell yield. Death of mammalian cells occurs mainly in two distinct forms: uncontrolled cell death (necrosis) and programmed cell death (apoptosis). Necrosis is a sudden and passive form of cell death, caused by cell membrane damage occurring upon extreme environmental stress, such as high shear stress on the cell membrane as a result of collision with the bioreactor walls or stirrers, exposure to high levels of toxic metabolites or any environmental insult. Apoptotic death can account for up to 80% of mammalian cell death in bioreactors, which restricts cell growth and viability and ultimately decreases cell yield. Therefore, apoptosis has been considered as a target to maintain high cell culture viability and increase antibody production in many prior research studies.

Based on metabolic flux analysis (MFA), a segregated modeling approach is presented in this work for MAb production by hybridoma cells. The viable cell population is segregated into
growing and apoptotic subpopulations to predict more precisely the effect of culture conditions on cell growth, cell death and target product formation. The MFA-based model is developed based on a set of key macro-reactions and considering effect of specific growth rate on production and consumption rates of extracellular nutrients and metabolites. The cell-specific kinetics of growth are expressed in terms of concentrations of glucose, glutamine, lactate, and ammonia. The integrated model predicts the temporal profiles of concentrations of essential nutrients, key metabolites and the cell populations for specific initial culture conditions for batch and fed-batch cultures. The segregation of the cell population enables the quantitative assessment of apoptotic state of the cell culture, impact of apoptosis and necrosis, and distribution of MAb production among growing and non-growing cells. The overall productivity is indirectly affected by the apoptotic state of the culture since the apoptotic state determines the overall cell viability. The model predictions compare well with the experimental data for many batch and extended batch experiments.

493551 Extracellular Production of Soluble Single-Chain Variable Fragment (scFv) Using Recombinant *E. coli* By Precisely Controlled Fed-Batch Culture with Do Stat

Abstract:

Single-chain variable fragment (scFv) antibodies are currently utilized as affinity reagents for diagnostics instead of full-length antibodies. ScFvs can be easily expressed in recombinant *E. coli,* however, there are some drawbacks such as low productivity and the formation of inclusion bodies. In this study, the extracellular production of soluble scFv using recombinant *E. coli* b was aimed by employing a fed-batch culture with DO-stat to realize the cost-effective production of soluble scFvs.

Five *E. coli* horboring four different scFv genes with periplasmic secretion signal(pelB leader) were constructed employing pET Expression System. Using the strains, the influence of temperature, culture media and host strains on scFv expression was evaluated and the induction timing was optimized. Based on these results, fed-batch cultures for scFv production were performed. Glucose feeding of the fed-fed batch system was carried out based on the DO-stat with PID control.

Fed-batch cultures with DO-stat were successfully performed and the cell concentration was increased to approximately 130 (OD600) during 100 hrs cultivation. During the fed-batch cultivation, soluble scFv was gradually excreted from cells and accumulated in the borth. Finally, 3.1g/L of soluble Anti-human IgG scFv and 4.8g/L of soluble Anti-CRP scFv were extracellularly produced during 100 hrs fed-batch cultivation. Total scFv production including inclusion bodies of scFvs expressed in the periplasm was 6.0g/L and 8.7g/L, respectively. These results will greatly contribute to achieve the cost-effective production of scFvs because the soluble scFV does not require the refolding step during purification process of scFvs.

500033 Investigating *Clostridium Carboxidivorans* P7 Metabolisms during Syngas Fermentations

Abstract:

Clostridium Carboxidivorans P7 is a model microorganism for industrial syngas fermentation. This study used ^{13}C tracers to delineate functional pathways and to profile metabolic responses to syngas substrates as well as mass transfer. Labeling of P7 proteinogenic amino acids suggests: deficient biosynthesis capability for some key amino acids (such as phenylalanine and leucine); active *Re-citrate* synthase to complete the TCA pathway (but no citramalate synthase detected); bicarbonate utilization via pyruvate:ferredoxin oxidoreductase and anaplerotic pathways. To investigate the effect of syngas mass transfer on cell metabolism, inverse dynamic labeling was performed in continuous-flow-stirred-tank reactors (i.e., P7 culture was prepared in a ^{13}C-medium then fed with unlabeled syngas feedstock). In tested conditions, P7 metabolized CO/CO_2 after 5-hour acclimation and the key metabolite Acetyl-CoA was labeled at the speed of 0.010~0.016 $hour^{-1}$ (flow rates ranging from 10mL/min to 20mL/min). Mass transfer was proved to be more influential on energy metabolism (cofactor generation) than carbon flux. However, syngas substrates showed strikingly slower rates percolating through central pathways, which demonstrate different sensitivity to substrates mass transfer conditions. Decrease of syngas flow rates from 20mL/min to 1 mL/min would not significantly hamper gaseous carbon utilization rates, but high flowrate facilitates NAD(P)H generation and increases ratio of ethanol in fermentation products.

503775 CFD Modeling of Bioreactor Mixing Properties

Abstract:

During scale-up/scale-down of therapeutic protein manufacturing processes, mammalian cell culture bioreactors from tens of milliliters to thousands of liters in size are utilized. Reactor design changes significantly across this large volume range, with aspects like vessel geometry, impeller type and number, and gas sparging setup varying considerably. This often makes mixing and gas transfer behavior difficult to predict, necessitating numerous trial-and-error experiments to choose control parameters like impeller RPM and gas sparge rate.Â This study investigates the use of Computational Fluid Dynamics (CFD) to model bioreactor behavior at different volume scales.

CFD modeling has the potential to describe bioreactor mixing properties such as mixing time, dissolved gas concentration, and shear stress, over a large operating parameter design space with minimal experimentation. In this presentation, we will present the use of the â€œANSYS CFXâ€ CFD solver to compare mixing properties of AMBR250 250-mL microbioreactors and 3-L bench top bioreactors. We will focus on the key scaling properties of power per volume ratio (P/V) and volumetric gas transfer constant ($k_L a$).

498170 Targeted Protein Therapeutics As Powerful Tools for Understanding and Overcoming Drug Resistance in Cancer

Abstract:

Over the past few decades, protein-based therapies have emerged as the gold standard for targeted, potent anticancer treatment, often in combination with small molecule-based drugs. Despite remarkable progress in development of these protein drugs, intrinsic and acquired tumor resistance substantially limits the breadth and depth of their clinical efficacy. The issue of resistance is often studied or addressed only after a drug fails to elicit complete responses in patients. Protein drugs can be engineered for superior target specificity and tumor selectivity, making them ideal not only as therapeutics, but also as probes for interrogating the target of interest to reveal mechanisms of cancer drug resistance. Here, we present a broadly applicable platform for mapping the pathways of resistance to a targeted protein therapy. The results of this screen were used to rationally design and evaluate drug combinations to overcome resistance to the targeted protein therapeutic.

For proof-of-concept of this approach, we used a multivalent pro-apoptotic protein composed of engineered protein scaffold domains targeting death receptor 5, a clinically relevant target for colorectal cancers (CRCs). This death receptor agonist (DRA) induces apoptosis at picomolar concentrations in many CRC lines. However, the clinical relevance of this drug is severely limited due to its failure to induce apoptosis in a significant number of cell lines. In this work, we show that the efficacy of a targeted therapy can be resurrected through systematic evaluation and understanding of its limitations. We used a CRISPR/Cas9 knockout screen to identify genes that confer DRA drug sensitivity to DRA-resistant cell lines. We then identified the small molecule drugs targeting pathways and proteins corresponding to these genetic drivers of resistance, and tested them in combination with the DRA *in vitro* to select synergistic combinations for *in vivo* validation. Our rational approach elegantly provides optimal protein-small molecule drug combinations that elicit a robust anticancer response, exhibit minimal toxicity, and combat drug resistance.

492606 Numerical Modelling Strategy for the Scale-up of Single-Use Bioreactors

Abstract:

Wave-mixed bioreactors have experienced increased importance in recent years in pharmaceutical processing. The low shear environment makes them particularly appealing for shear sensitive product. However, scale-up relationships between different sized mixing bags are not fully understood. It is advantageous to test the impact of the mixing process on high value product at a smaller scale before commencing production at large scale.

Numerical modelling, through the use of Computational Fluid Dynamics (CFD) was used to characterise the mixing times and hydrodynamic environment within a 500 mL and 50 L (production scale) mixing bag. The 50 L bag was characterised for a prescribed set of operating conditions (rocking speed and rocking angle of the wave bioreactor). Operating conditions for the small scale 500 mL mixing bag were identified that would generate the similar shear rates in both the 500 mL and 50 L mixing bag. Additionally, a scale-up relationship was developed. Validation experiments were then conducted at the 500 mL scale. This allows the high value product to be tested under the same shear environment at the 500 mL scale before processing at the 50 L scale.

493155 Challenges in Single-Use Bioprocessing Systems: Evaluating the Cytotoxicity of a Leachate from Plastic Single-Use Bioreactors

Abstract:

The rapid expansion of single-use systems for bioprocessing is largely due to the ease of use and availability in pre-sterilized forms. While plastics used in biopharmaceutical production have gained acceptance for regular use, particularly in pilot-scale or experimental facilities, the additives used as stabilizers or to improve the processing of these plastics have a tendency to migrate or leach over time. One such additive is Irgafos 168®, or tris(2,4-di-tert-butylphenyl)phosphite, which is used as an antioxidant stabilizer. When undergoing radiation treatment during sterilization this is converted to bis(2,4-di-tertibutylphenyl)phosphate, or bDtBPP. bDtBPP has been found to leach from single-use bioprocessing bags into growth media, and our work has focused on determining its impact on cell growth for cell lines used in biopharmaceutical production. Specifically, we selected a Chinese Hamster Ovary cell line (CHO-K1), which was grown in both adherent and suspended form, to determine LC50 values of bDtBPP and other compounds used in plastics. Specifically for bDtBPP, the LC50 values were found to be higher for adherent cells (where albumin likely masks some of the toxic effects), with a drop in viable cell density observed at lower concentrations than cell viability (% live). This second observation indicates that the bDtBPP impacts metabolism and cell reproduction at low concentrations prior to cell death being prominent, so this may impact production rates of biopharmaceuticals. Our work seeks to establish a protocol for evaluating cytotoxicity of leachates, while at the same time gaining a better understanding the toxic behavior of plastic additives (and their degradants) when used in single-use bioprocessing systems.

498111 Reincubation of Heat ShockedÂ *Pseudomonas aeruginosa* Biofilm

Abstract:

When bacteria colonize a medical implant surface, they form a biofilm which cannot be eradicated chemically. The current standard of care is surgical explantation of the device and surrounding tissue, with eventual reimplantation of a replacement device with twice the probability of infection. These infections are a $5 billion problem in the U.S. alone, impacting over 100,000 patients annually. By applying a localized thermal shock, we have shown that these biofilms can be eliminated, though the mechanism of elimination has been unclear. In the presence of antibiotics which have little effect on the biofilm by themselves, biofilms are eliminated by heat shocks which are too mild to have any noticeable effect in the absence of antibiotics. One hypothesis for this synergistic efficacy is that the heat shock triggers a change in bacterial phenotype, prompting the bacteria to disperse from the biofilm in the planktonic state, and thereby making them susceptible to antibiotic elimination.

Reincubation studies, however, indicate that biofilm elimination by thermal shock is not instantaneous. Biofilms were cultured to a steady-state population density of 10^7 colony forming units (CFU) per cm^2 and subjected to heat shocks of varying intensity (60, 70, or 80 Â°C for 1, 5, or 30 min). Their population density was quantified immediately following thermal shock, demonstrating populations reductions ranging from none (no statistically significant reduction) after mild heat shock to complete (no viable bacteria detected) after aggressive heat shocks. Reincubation of thermally shocked biofilms for 2, 4, 12, 24 and 96 hours followed by resuspension and plating showed a bimodal response. Following heat shock, biofilms slowly regrew, eventually reaching their original 10^7 CFU/cm^2 population densityâ€"if the population density immediately following these heat shocks was typically 10^3 CFU/cm^2 or above. However, if the population density immediately following these heat shocks was typically below 10^3 CFU/cm^2 (yet clearly detectable and quantifiable), the biofilms would continue dying off, with no viable CFU after a few hours, despite being able to form colonies when immediately resuspended and plated. This supports an opposing hypothesis that the bacteria do not flee the biofilm, but rather die in place, leaving a matrix full of toxic enzymes that proceed to kill off the surviving bacteria. These hypotheses are further examined by applying thermal shock in a flow cell system in which any fleeing bacteria are collected in the flowing fluid and quantified as a function of time.

Cosmetic and food ingredients produced by engineered microbes – the future of the industry

Congqiang Zhang [a,*], Xixian Chen [a], Heng-Phon Too [a,b]

*a. Biotransformation Innovation Platform (BioTrans), Agency for Science, Technology and Research (A*STAR), Singapore*

b. Department of Biochemistry, Yong Loo Lin School of Medicine, National University of Singapore, Singapore

*To whom correspondence may be addressed. zcqsimon@outlook.com, congqiang_zhang@biotrans.a-star.edu.sg

Natural flavors and fragrances are currently mainly supplied with plant extracts, which is susceptible to environmental factors, such as pests, disease and weather, price and political volatility and the logistical complexity of sourcing raw materials from farmers in remote location. And the recent changes in consumer perception and regulation have resulted in the dramatic shift from synthetic to natural ingredients in the flavors and fragrances sector. Consequently, there are growing interests in engineering microbes to produce the ingredients from renewable resources. Apocarotenoids, such as α-, β-ionone and retinol, contain many high value compounds in the food and cosmetic industries. Ionones, C13 apocarotenoids, have superior aromatic properties with extremely low odour threshold of less than 1 ppb in air. However, the supply of natural ionones and retinoids is severely limited by their extremely low abundance in nature. For instance, it requires 100 tons of raspberries, or 20 hectares of agricultural area, to produce merely 1 gram of α-ionone. As for retinoids, there is no natural source from plants. Here, an *Escherichia coli*-based modular system was developed to produce various apocarotenoids. A high lycopene-accumulating strain (~107 mg/L in flasks) was initially obtained by modular optimization of the heterologous mevalonate and lycopene biosynthetic pathways. The lycopene strain served as the chassis for the biosynthesis of different apocarotenoids. Incorporation of enzyme engineering approaches (N-terminal truncation and protein fusion) into modular metabolic engineering strategy significantly improved α-ionone production from 0.5 to 30 mg/L in flasks, and yielded 480 mg/L of α-ionone in fed-batch fermentation. By modifying apocarotenoid genetic module, this platform strain was successfully re-engineered to produce 32 mg/L and 500 mg/L of β-ionone in flask and bioreactor, respectively (> 80-fold higher than previously reported). Similarly, 33 mg/L of retinoids was produced in flask by reconstructing apocarotenoid module, demonstrating the versatility of the "plug-n-play" modular system. Collectively, this study highlights the efficacy of our approaches and sheds light on the bright future of microbial production of food and cosmetic ingredients.

INTRODUCTION

Apocarotenoids are a class of compounds derived from carotenoids by carotenoid cleavage oxygenases. Widely distributed in bacteria, fungi, plants and animals, apocarotenoids act as aroma and scent compounds (α- and β-ionone), photosensory pigment (bixin, crocin), hormones (abscisic acid) and signalling compounds (strigolactones) (Walter et al. 2010). Among various apocarotenoids, α- and β-ionone are two important aromatic compounds. α-ionone has a sweet and violet-like aroma with odour threshold of ~0.4 ppb (Larsen and Poll 1990). Its isomer, β-ionone, has a warm, woody, and violet aroma and an even lower odour threshold of ~0.007 ppb in air and 1 ppb in water (Beekwilder et al. 2014). Due to their significantly low odour threshold and pleasant smell, they are widely exploited in cosmetics and perfume industry (Lalko et al. 2007). Besides ionones, retinol (or vitamin A) is another commercially important apocarotenoid. Retinol plays an essential function in vision, bone development and skin health as antioxidants (O'Byrne and Blaner 2013). As active cosmetic ingredients and effective medicines for skin diseases, the retinol market size is estimated to be about 1.6 billion dollars (Jang et al. 2011).

Despite their high commercial values, supply of natural ionones and retinoids is severely limited by their extremely low abundance in nature. Ionones are present in sub-ppm level in many flowers and fruits, such as rose, sweet osmanthus, orris root and raspberry. For instance, it requires 100 tons of raspberries, or 20 hectares of agricultural area, to produce merely 1 gram of α-ionone. As for retinoids, there is no natural source from plants. Exceptionally low amount of retinoids exists in some animal-derived food, such as eggs and butter. Hence, the current supply through extraction from natural sources cannot meet the increasing demands for natural ionones and retinoids. Although these compounds could be chemically synthesized, apocarotenoids such as α-ionone have chiral centres, and synthetic ones are usually a mixture of different enantiomers. More importantly, consumers tend to prefer natural to synthetic flavours and thus natural ingredients have significantly higher prices (such as natural β-ionone sells ~$300/kg about 10 time higher than that of synthetic ones). With rapid developments in biotechnology, especially metabolic engineering, many valuable compounds can be produced in microbial cell factories in sustainable ways. In nature, α-ionone is synthesized from α- or ε-carotene. Retinoids and β-ionone are derived from β-carotene. Ionones are biosynthesized by the carotenoid cleavage dioxygenase enzyme (CCD including CCD1 or CCD4 (Lashbrooke et al. 2013)) in plants and retinal is produced by β-carotene 15,15' dioxygenase (BCDO, coding gene *blh*) (Martinez et al. 2007) in animals or bacteria (Fig. 1). As a subclass of isoprenoids, the biosynthesis of carotenoids and apocarotenoids are mainly derived from lycopene which is produced from two universal isoprenoid precursors, isopentenyl pyrophosphate (IPP) and dimethylallyl pyrophosphate (DMAPP). These precursors are produced by the 1-deoxy-D-xylulose 5-phosphate (DXP) and/or the mevalonate (MVA) pathway (Fig. 1).

The production of α-ionone was previously demonstrated in engineered *Escherichia coli*, but with a very low yield, ~336 ng/L (Lashbrooke et al. 2013). Although β-ionone was synthesized from glucose in *E. coli* (Simkin et al. 2004) and *Saccharomyces cerevisiae* (Beekwilder et al. 2014; López et al. 2015), the highest titer achieved was ~6 mg/L in bioreactor (López et al. 2015) due to the complexity of the biosynthetic pathway. Adapted from our previous modular metabolic engineering approach, experimental design-aided systematic pathway optimization (EDASPO) (Zhang et al. 2015), this study has constructed and systematically optimized a lycopene-overproducing *E. coli* strain using the heterologous MVA pathway. Extending the modular design, we introduced an additional module to facilitate the production of apocarotenoids. However, the ε-carotene and α-ionone titers were much lower than that of lycopene which was due to the poor expression of lycopene epsilon-cyclase (LCYe) and CCD1 enzyme. We thus applied protein engineering approaches (protein truncation and fusion partners), which markedly improved enzyme expression and product yields. Furthermore, the modular design enabled easy adaptation of the system to the β-ionone production by replacing LCYe with lycopene beta-cyclase (*crtY* gene) or to retinoid production by overexpressing β-carotene dioxygenase (BCDO, or *blh* gene) instead of *CCD1* with *crtY*, indicating that the modular platform developed here could function as a 'plug-n-play' platform to produce various apocarotenoids.

RESULTS

Optimization of lycopene strain as the chassis strain

It is very challenging to maintain high metabolic flux along a recombinant pathway that simultaneously overexpresses 13 enzymes in *E. coli* cells to produce apocarotenoids (Fig. 1). To overcome the issue, we divided the pathway into four different modules (Module 1, upstream mevalonate pathway including *hmgS-atoB-hmgR*; Module 2, downstream mevalonate pathway including *mevK*, *pmk, pmd and idi*; Module 3, lycopene biosynthetic pathway including *crtEBI and ispA*; Module 4, apocarotenoid biosynthesis module, including *LCYe-CCD1* or *crtY-CCD1* or *crtY-blh*).

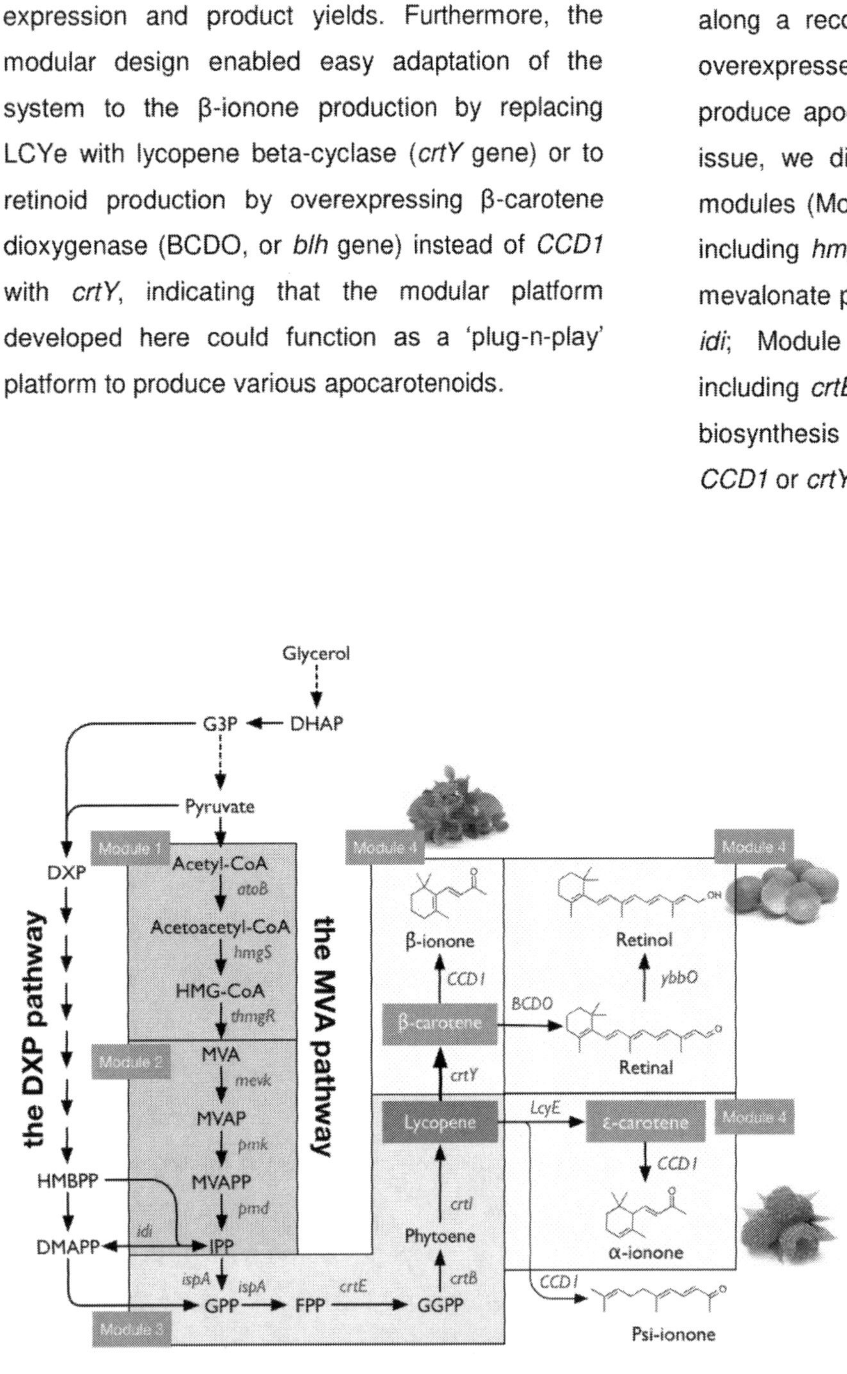

Fig. 1. Schematic diagram of biosynthetic pathway of α, β-ionone and retinoids. The biosynthetic pathway was grouped into four major modules: the MVA pathway (module 1 and 2), the lycopene pathway (module 3) and the apocarotenoid pathway (module 4). The genes expressed encode the following enzymes: *atoB*, Acetoacetyl-CoA thiolase; *hmgS*, HMG-CoA synthase; *thmgR*, truncated HMG-CoA reductase; *mevk*, mevalonate kinase; *pmk*, phosphomevalonate kinase; *pmd*, mevalonate pyrophosphate decarboxylase; *idi*, IPP isomerase; *ispA*, FPP synthase; *crtE*, GGPP synthase; *crtB*, phytoene synthase; *crtI*, phytoene desaturase; *LCYe*, lycopene epsilon-cyclase; *crtY*, lycopene beta-cyclase; *CCD1*, carotenoid cleavage dioxygenase; *BCDO* (or *blh*), β-carotene dioxygenase; *ybbO*, NADP+-dependent aldehyde reductase. Abbreviation for the compounds: G3P, D-glyceraldehyde-3-phosphate; DHAP, Dihydroxyacetone phosphate; HMG-CoA, 3-hydroxy-3-methyl-glutaryl-coenzyme A; MVA, mevalonate; MVAP, phosphomevalonate; MVAPP, diphosphomevalonate; IPP, isopentenyl pyrophosphate; DMAPP, dimethylallyl pyrophosphate; HMBPP, (E)-4-Hydroxy-3-methyl-but-2-enyl pyrophosphate; GPP, geranyl pyrophosphate; FPP, farnesyl pyrophosphate; GGPP, geranylgeranyl pyrophosphate. Dashed arrow indicates multiple enzymatic steps.

As lycopene is the common precursor for different apocarotenoids, we first optimized the first three modules to overproduce lycopene by EDASPO approach (Zhang et al. 2015). Briefly, the 3 modules were controlled by three different T7 promoter variants with different strength, and the lycopene yield was then correlated with the promoter strength of each module to locate the optimal transcriptional range for each module (Ternary diagram, Fig. 2). As a result, we identified the optimal range for each module and obtained an optimized strain (121 strain) that produced ~107 mg/L of lycopene in a shake flask, or 26,000 ppm, which is comparably to previously reported (Alper et al. 2005; Matthaus et al. 2014; Xie et al. 2015; Zhang et al. 2013) (Fig. 2). This lycopene-producing strain served as parental platform strain for further production of downstream carotenoids and apocarotenoids.

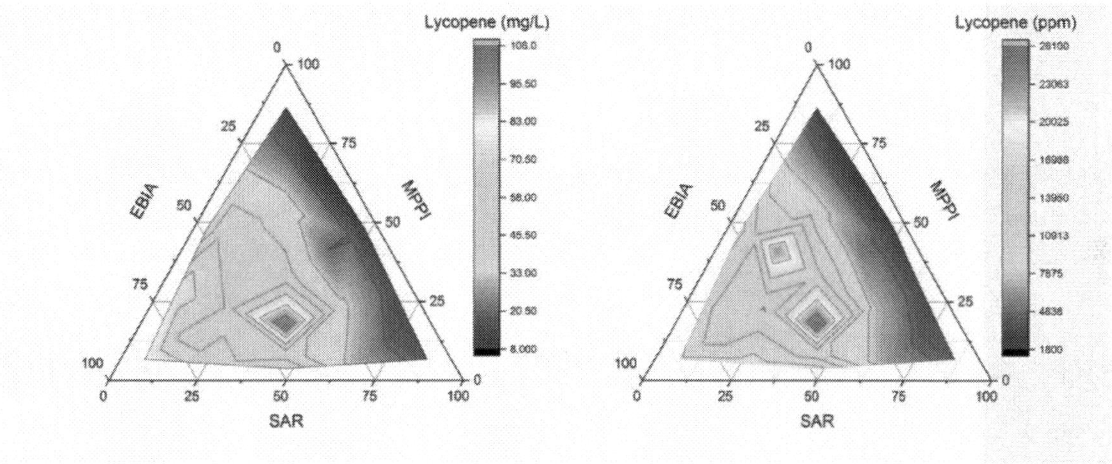

Fig. 2. Using EDASPO method to optimize lycopene production.

The number represents the different strength of T7 promoter variants, with 1, 2 and 3 representing approximately 92%, 37% and 16% of native T7 promoter strength, respectively (Zhang et al. 2015). Module 1, or SAR module, hmgS-atoB-hmgR; Module 2, or MPPI module, mevK, pmk, pmd and idi; Module 3, or EBIA module, lycopene biosynthetic pathway (crtEBI and ispA).

Construction and optimization of the ε-carotene strain

Although plants contain both the lycopene beta-cyclase (*crtY*) and lycopene epsilon-cyclase (*LCYe* gene), ε-carotene is known to be accumulated at very low levels in nature. To date, only the LCYe from *Lactuca sativa* (LsLCYe) is known to preferentially form the bicyclic ε-ring from lycopene (Cunningham and Gantt 2001). When we overexpressed the wild type LsLCYe enzyme in the lycopene strain (121 strain), a mixture of δ-carotene (carotene with an ε-ring at one end) and ε-carotene (carotene with two ε-rings at both ends) was produced, with more than 70% lycopene uncyclized (0.1 mM, Fig. 3). We initially optimized ε-carotene production by tuning the transcription levels and obtained highest amount of ε-carotene with moderate IPTG induction (0.03 mM). However, the highest ε-carotene titer was merely 18 mg/L (2500 ppm), corresponding to less than 30% in the total carotenoids (Fig. 3), suggesting that the activity of LsLCYe enzyme was insufficient.

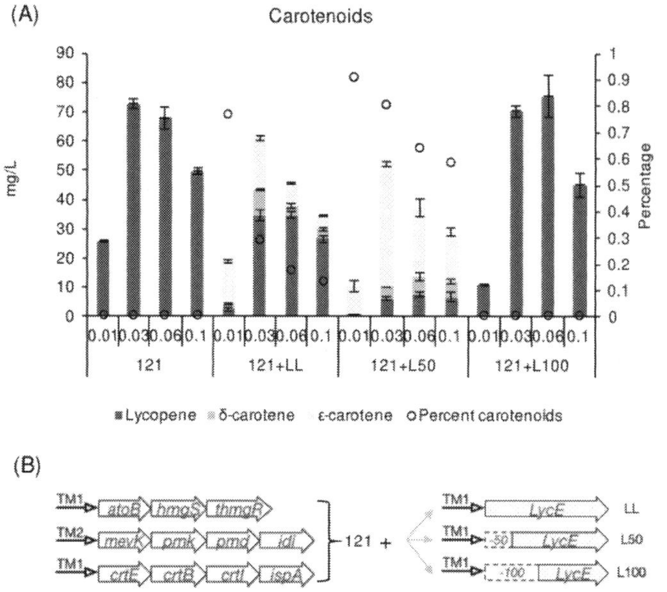

Fig. 3. Production of α-ionone in the ε-carotene-accumulating *E. coli* strain.

(A) Optimization of CCD1 with N-terminal fusion partners (FP), a figure insert with a different y-axis scale was supplemented to compare the ionone titers of strains 121-LO and 121-L50-O. (B) Intracellular carotenoids that remained uncleaved. (C) An illustration of promoters and regulated genes for different strains. The abbreviations are as follows. LO, LsLCYe and OfCCD1 were expressed in a polycistronic manner on the same plasmid; L50-O, LsLCYe enzyme with the first 50 amino acids removed (L50) and OfCCD1 were expressed in a polycistronic manner. L50-SUMO-O, L50 and SUMO-fused OfCCD1 were co-expressed. L50-MBP-O, L50 and MBP-fused OfCCD1 were co-expressed. L50-TrxA-O, L50 and TrxA-fused OfCCD1 were co-expressed.

Modifications of N-terminal residues involved in membrane interactions and signal peptide coding regions have previously been reported to enhance solubility and activity of recombinant protein (Biggs et al. 2016; Gopal and Kumar 2013). Therefore, we next examined if truncation of the N-terminus sequences involved in anchoring to membrane would increase the catalytic efficiency of LsLCYe. A series of N-terminal truncated LsLCYe was constructed by systematically removing 50 residues at a time using the truncate cross-lapping in vitro assembly method (Zou et al. 2013). The enzymatic expression level was increased by ~40% when the first 50 amino acids (Fig. 4A) were removed, with a concomitant greater than twofold increase in ε-carotene (>40 mg/L, 5500 ppm and 80% of final products) production (Fig. 2, 121+L50 strain). Remarkably, removing the first 100 residues (Fig. 2, 121+L100 strain) led to an increase in the protein concentration by ~140% (Fig. 4A and B). However, the cyclase activity was completely lost. In order to better understand how truncation affected LsLCYe expression, we compared the transcription level and mRNA secondary structure of truncated LsLCYe and its native form. The results indicated that increased expressions of truncated LsLCYe did not appear to be due to significant increases in mRNA expressions (Fig. 4C), (Goodman et al. 2013; Zadeh et al. 2011).

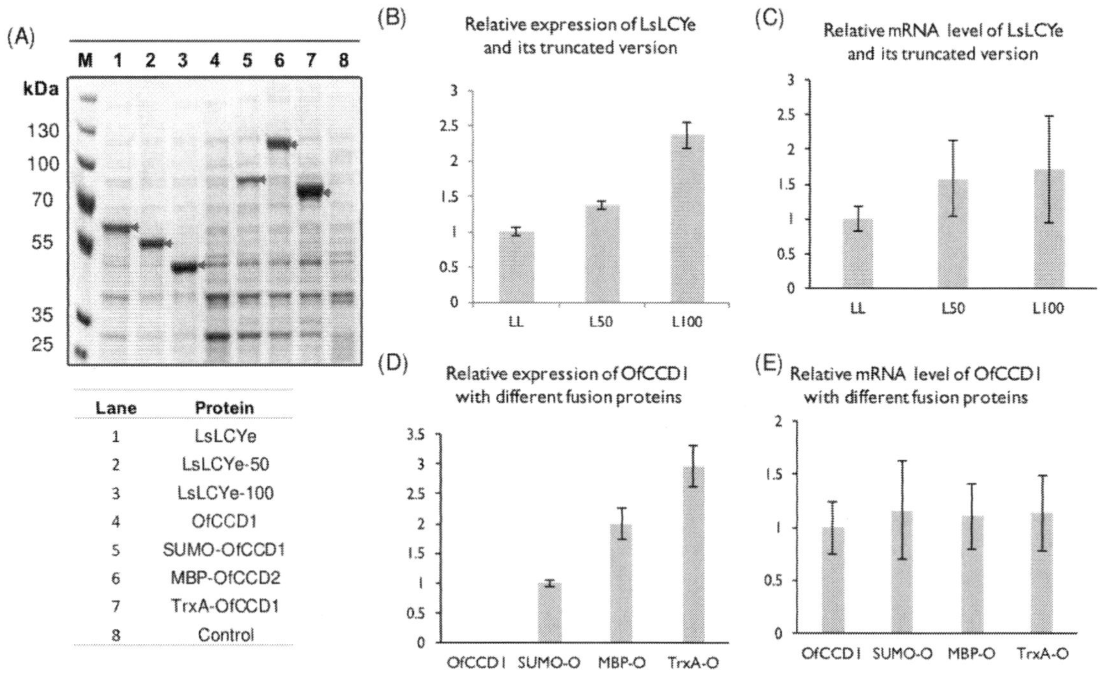

Fig. 4. Protein and mRNA expression analysis of *LsLCYe* and *OfCCD1* with N-terminal modifications. (A) SDS-PAGE gel image of the whole-cell protein expression of LsLCYe and OfCCD1 with various N-terminal modifications. The proteins of interest were indicated with blue arrow. Lane 1: E. coli cells with overexpressed LsLCYe. Lane 2: E. coli cells with overexpressed Δ50-LsLCYe (L50). Lane 3: E. coli cells with overexpressed Δ100-LsLCYe (L100). Lane 4: E. coli cells with overexpressed OfCCD1. Lane 5: E. coli cells with overexpressed SUMO-OfCCD1 (SUMO-O). Lane 6: E. coli cells with overexpressed MBP-OfCCD1 (MBP-O). Lane 7: E. coli cells with overexpressed TrxA-OfCCD1 (TrxA-O). Lane 8: E. coli cells without any overexpression (control). Quantitative analysis of the overexpressed protein yield was based on the band intensity on SDS PAGE gel. (B) LsLCYe expression levels with various N-terminal truncations. (C) The mRNA expression levels of LsLCYe and its truncated variants. (D) OfCCD1 expression with different fusion partners. (E) mRNA expression levels of OfCCD1 with different fusion proteins. The relative fold change of protein expression was normalized by the expression of LsLCYe (B and C) and SUMO-OfCCD1 (D and E). The results were the average of triplicates with standard error bar shown.

Systematic optimization of α-ionone production

LsLCYe with N-terminus 50 amino acids truncated (ΔN50- LsLCYe*)* was further explored for α-ionone production. It has been demonstrated previously that CCD1 from *Osmanthus fragrans* (*OfCCD1*) can cleave carotene at C9-C10 position to produce both the α- and β-ionone (Baldermann et al. 2010). Thus, we overexpressed OfCCD1 enzyme in ε-carotene strains (121-L and 121-L50 in Fig. 3). However, even with enhanced availability of ε-carotene in strain 121-L50-O, the amount of α-ionone detected was less than 500 μg/L (Fig. 5A, insert). As previously noted, lycopene and ε-carotene were the main intermediates accumulated in the LsLCYe and ΔN50-LsLCYe overexpressed strains, respectively (Fig. 5B). SDS-PAGE analysis showed that hardly any OfCCD1 protein was observed 24 h after induction (Fig. 4A and D), even though mRNA was well detected (Fig. 4E). The results indicated that OfCCD1 expression may be limited by insufficient translation, consistent with previous studies on other CCD1 overexpression (Ilg et al. 2009; Ilg et al. 2014).

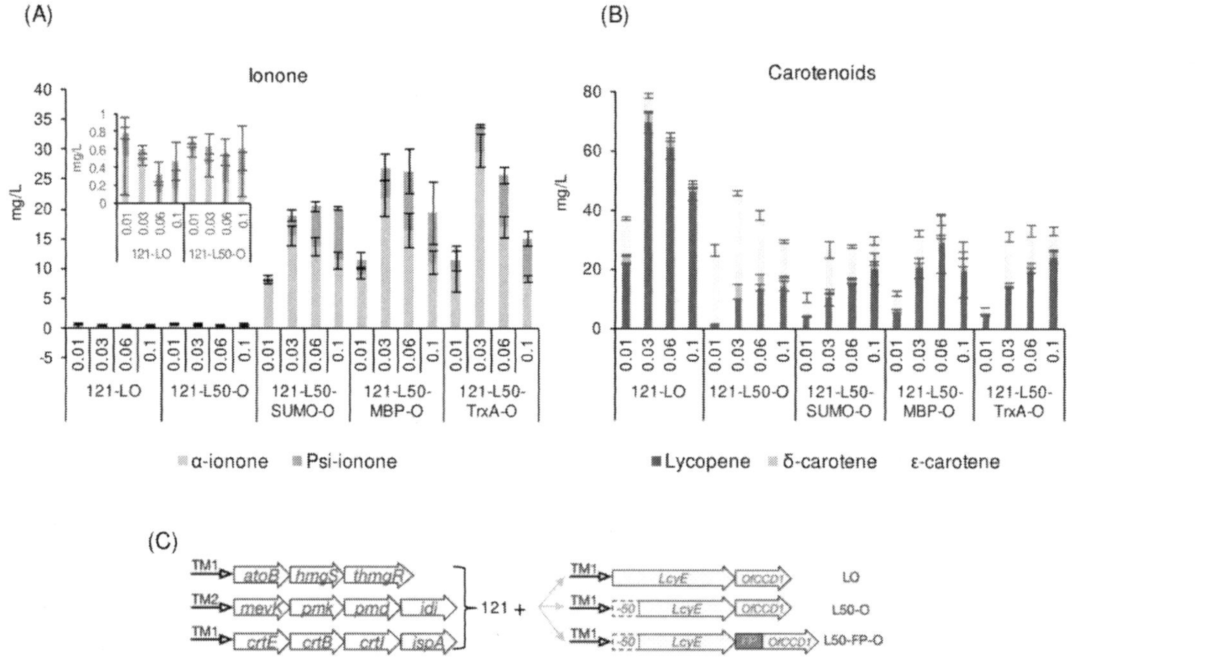

Fig. 5. Production of α-ionone in the ε-carotene-accumulating *E. coli* strain. (A) Optimization of CCD1 with N-terminal fusion partners (FP), a figure insert with a different y-axis scale was supplemented to compare the ionone titers of strains 121-LO and 121-L50-O. (B) Intracellular carotenoids that remained uncleaved. (C) An illustration of promoters and regulated genes for different strains. The abbreviations are as follows. LO, LsLCYe and OfCCD1 were expressed in a polycistronic manner on the same plasmid; L50-O, LsLCYe enzyme with the first 50 amino acids removed (L50) and OfCCD1 were expressed in a polycistronic manner. L50-SUMO-O, L50 and SUMO-fused OfCCD1 were co-expressed. L50-MBP-O, L50 and MBP-fused OfCCD1 were co-expressed. L50-TrxA-O, L50 and TrxA-fused OfCCD1 were co-expressed.

Literature suggested that fusion with glutathione-S-transferase (GST) would significantly increase expression and solubility of CCD1 in *E. coli* (Ilg et al. 2009). This knowledge prompted us to examine if fusion partners would improve the functional expression of *OfCCD1*. Three commonly used N-terminal fusion partners, small ubiquitin-like modifier (SUMO) protein (Panavas et al. 2009), maltose binding protein (MBP) (Sun et al. 2011) and thioredoxin (TrxA) (LaVallie et al. 1993), were tested. *OfCCD1* when fused with the three fusion partners significantly increased the titers of α-ionone (Fig. 5A). Among them, the strain with TrxA-fused *OfCCD1* (or TOfCCD1) was able to produce 29.7 mg/L α-ionone with low induction (0.03 mM IPTG), which is more than 50-fold improvement (Fig. 5A). To study the underlying mechanism of fusion partners, we further investigated protein expression by denaturing SDS-PAGE gel. Without fusion partner, OfCCD1 was barely detectable (Fig. 4A and D). In contrast, OfCCD1 enzymes fused with SUMO, MBP and TrxA were well expressed. The relative protein band intensities of SUMO-, MBP- and TrxA-fused OfCCD1 were well correlated with their corresponding α-ionone production in Fig. 5A. For example, TOfCCD1 expressing strain had the best expression and produced the highest amount of α-ionone. Interestingly, even with the highest ionone producer strain (Fig. 5, 121-L50-TrxA-O strain with 0.03mM IPTG), both ε-carotene and lycopene (~30 mg/L total carotenoids) were still accumulated to the similar levels as α-ionone (Fig. 5B), suggesting that α-ionone yield could be further improved. Hence, we managed to progressively improve the production of α-ionone from 500 μg/L to 29 mg/L.

Next, we tested the performance in fed-batch fermentation using the best α-ionone producing strain (121-L50-TrxA-O strain). As α-ionone is volatile, isopropyl myristate was used to entrap the compound during fermentation. As shown in Fig. 6A, about 480 mg/L of α-ionone was produced within 48h. In sum, we have achieved ~1000-fold increase in the α-ionone production from initial 500 μg/L to 480 mg/L. Thus, the α-ionone titer achieved in this study is about 1428 times higher than previously reported (Lashbrooke et al. 2013). In addition, the α-ionone produced in our strain is 100% pure (R)-(+)-(E)-alpha-ionone enantiomer. In contrast, chemically synthesized sample is half R and half S form mixture (Fig. 6B).

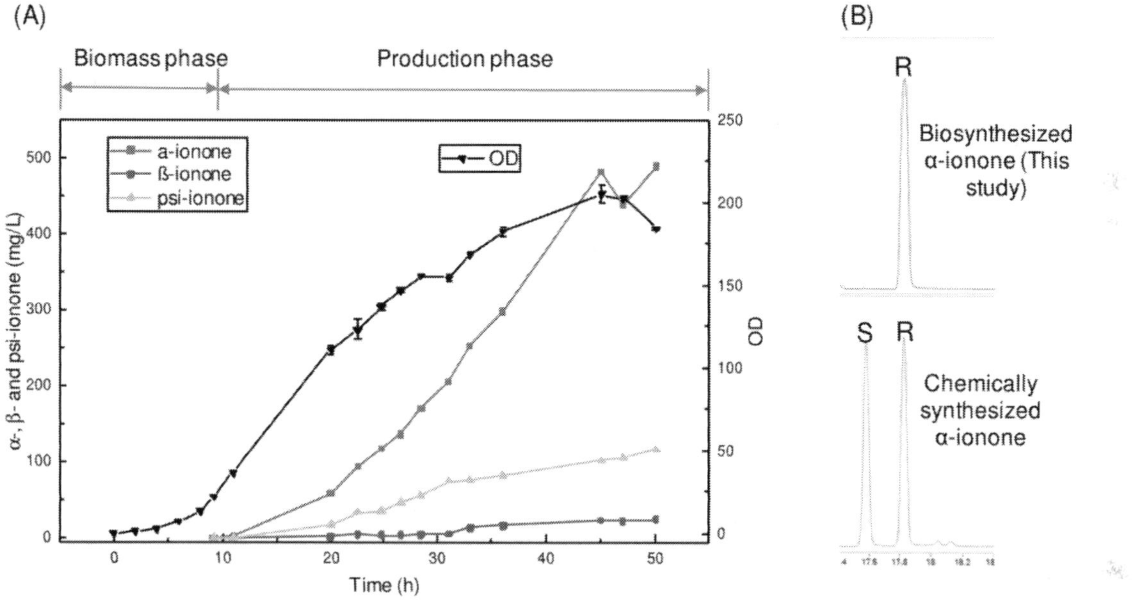

Fig. 6. Fed-batch fermentation of apocarotenoids. (A) Time-course profiles for ionones and biomass. (B) Chirality check of α-ionone produced in this study. Different from chemically synthesized α-ionone including 50% R- and 50% S-enantiomers, the ionone produced in our strain is 100% R- enantiomer.

Optimization of β-ionone production

In order to demonstrate the broader utility of this modular approach, we next produced other apocarotenoids by replacing the LCYe enzyme with the expression of lycopene beta cyclase (or *crtY*) from *Pantoea ananatis*. Unlike LsLCYe, the native lycopene beta cyclase was highly active and converted greater than 70% of lycopene into β-carotene (65 mg/L), with residual amount of lycopene (24 mg/L) (crtY strain in Fig. 7). The β-ionone production with wild-type *OfCCD1* expressing strain (Of strain) was 140 μg/L, corresponding to merely less than 1% conversion yield from β-carotene (Fig. 7). As the N-terminal TrxA fused *OfCCD1* (TOfCCD1) improved α-ionone (Fig. 5), we tested the possibility that it may similarly improve β-ionone production. Disappointingly,

TOfCCD1 expression did not result in a significant increase in the production of β-ionone (Fig. 7). Instead, we detected another major peak at 9.46 min in the GC chromatogram, which was predicted to be psi-ionone with the National Institute of Standards and Technology (NIST) mass spectrum library and further validated with authentic standard of psi-ionone. The expression of TOfCCD1 resulted in a 37-fold increase in psi-ionone production from ~0.4 to ~14.6 mg/L. Noteworthy, psi-ionone was also produced by the α-ionone production strain, although it was a minor product (Fig. 5A). Hence, it appeared that lycopene was also cleaved by TOfCCD1 and produced psi-ionone (Fig. 1), consistent with the observation that certain types of CCD1 could convert lycopene to psi-ionone or 6-methyl-5-hepten-2-one(Ilg et al. 2014; Lashbrooke et al. 2013).

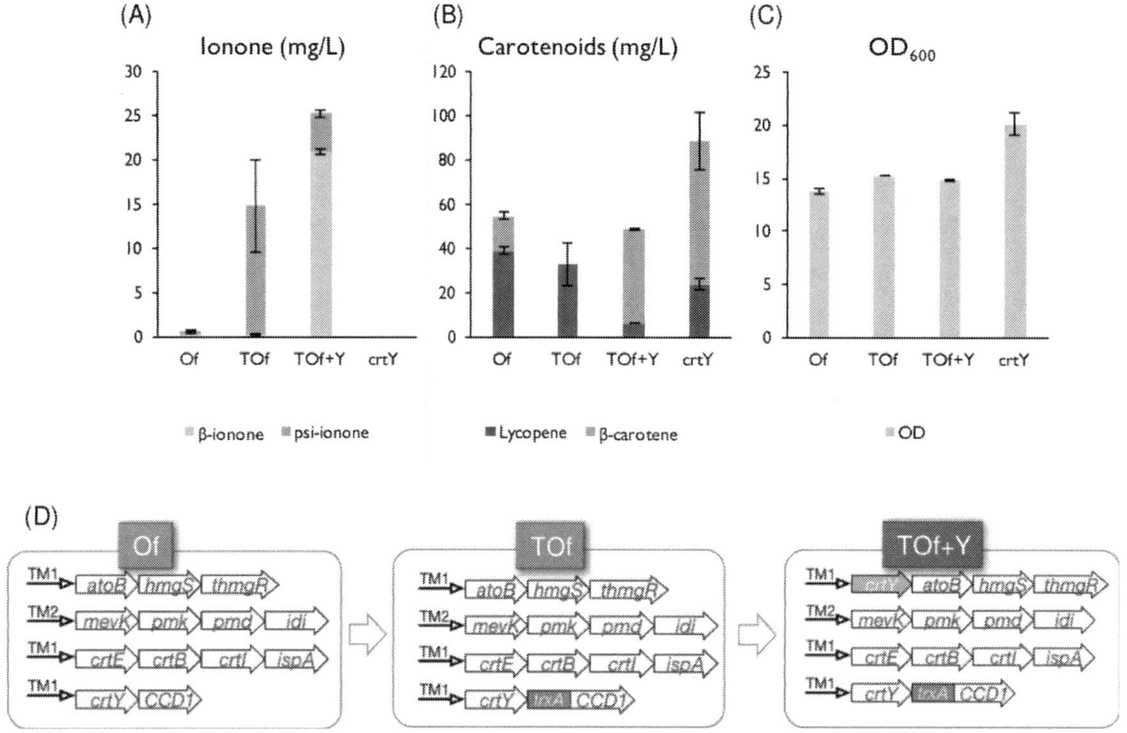

Fig. 7. Metabolic engineering strategies employed for the production of β-ionone. Comparison of wild-type OfCCD1, N-terminal modified OfCCD1 and supplementation of crtY. In Of, TOf and crtY strain, the operon/gene crtY-OfCCD1, crtY-TrxA-OfCCD1 and crtY were expressed, respectively. In TOf+Y strain, additional copy of crtY gene was supplemented in the p15A-spec vector. (A) Ionone production in engineered strain. Strain crtY was the control strain that did not express CCD1 gene thus only produced β-carotene. (B) Accumulated carotenoids in the engineered strain. (C) Biomass of the engineered strains. (D) An illustration of promoters and regulated genes for each strain.

As large amount of lycopene and lycopene-derived psi-ionone were produced in the TOf strain, we hypothesized that increasing the gene dosage of *crtY* (TOf+Y strain) may have two potential benefits. Firstly, it may increase the carbon fluxes from lycopene to β-carotene and hence increase β-ionone production. Secondly, the production of psi-ionone may be reduced by the decrease in the availability of intracellular lycopene. To test the hypothesis, we first inserted *crtY* into the first module of our system, *atoB-hmgS-hmgR*, resulting in a new operon, *crtY-atoB-hmgS-hmgR* (illustrated in Fig. 7D). As the plasmids used in the study are about 10 copies per cell, the insertion increased the crtY copy number from 10 to 20 copies per cell. Interestingly, gene dosage of *crtY* effectively resulted in more lycopene converted to β-carotene, hence, from undetectable amount of β-carotene (TOf strain, Fig. 7) to a significant yield of 42.4 mg/L (TOf+Y strain, Fig. 7). Consequently, the titer of β-ionone was significantly increased from 0.3 to 20.9 mg/L, and psi-ionone was decreased from 14.6 to 4.4 mg/L (Fig. 7), indicative of a higher carbon flux diverted into β-ionone production through β-carotene.

Although supplementation of *crtY* markedly improved β-ionone production, there was still 4.4 mg/L of psi-ionone produced. To maximize the β-ionone production and minimize by-product/intermediates formation, we screened for a CCD with higher selectivity for β-carotene. Another three CCD1 enzymes, from *Arabidopsis thaliana* (AtCCD1) (Schmidt et al. 2006), *Vitis vinifera* (VvCCD1) (Lashbrooke et al. 2013) and *Petunia hybrid* (PhCCD1) (López et al. 2015) were tested. With the expression of wild-type AtCCD1, VvCCD1 and PhCCD1, β-ionone was produced at 0.1, 13.0 and 27.0 mg/L, respectively, accompanied by 0.2, 14.8 and 4.8 mg/L psi-ionone, respectively (Fig. 8A). The data suggested that wild-type VvCCD1 and PhCCD1 had higher activities in *E. coli* than AtCCD1 and OfCCD1. In addition, PhCCD1 also had higher selectivity for β-carotene than VvCCD1 and TOfCCD1. Nevertheless, there were still 13.3 mg/L β-carotene and 13.8 mg/L lycopene accumulated in the PhCCD1 expressing strain (Fig. 8B). We next investigated the fusion of TrxA to PhCCD1 and supplementation of additional copy of *crtY*. With additional supplementation of *crtY* (Ph+Y strain), the β-ionone titer increased from 27 to 32.4 mg/L and psi-ionone titer decreased from 4.8 to 1.6 mg/L. Concomitantly, β-carotene production increased from 13.3 (Ph strain) to 24.3 (Ph+Y strain) mg/L and lycopene production decreased from 13.8 to 1.5 mg/L. However, the fusion of TrxA (TPh strain) merely resulted in a moderate increase in β-ionone production from 27 to 30.4 mg/L and a slight decrease in psi-ionone production from 4.8 to 3.2 mg/L. Furthermore, there was no synergistic benefit with the combination of *crtY* supplementation and TrxA fusion (TPh+Y strain, Fig. 8). Collectively, the results suggested that unlike OfCCD1, native PhCCD1 activity was sufficiently efficient in converting β-carotene to β-ionone. In sum, we improved the β-ionone production from initial 140 µg/L to 32.4 mg/L, a ~230-fold increase in tubes and flasks over a period of 24 h. Further in fed-batch fermentation, ~500 mg/L of β-ionone was produced with Ph+Y strain (Fig. 8C), which was about 3700-fold increase over that of the initial strain.

Production of retinoids using the "plug-n-play" system

Retinoids are biosynthesized by cleaving β-carotene at C15-C15' position. To further illustrate the utility of our "plug-n-play" system to produce other apocarotenoids, we replaced the CCD1 gene with two *blh* genes (BCDO enzyme, Fig. 1), from *Uncultured marine bacterium HF10_19P19* (*blh1*) (Martinez et al. 2007) and *Uncultured marine bacterium 66A03* (*blh2*) (Jang et al. 2015). Without any optimization of the upstream pathway, we produced 18.5 mg/L retinol and 5.1 mg/L retinal in

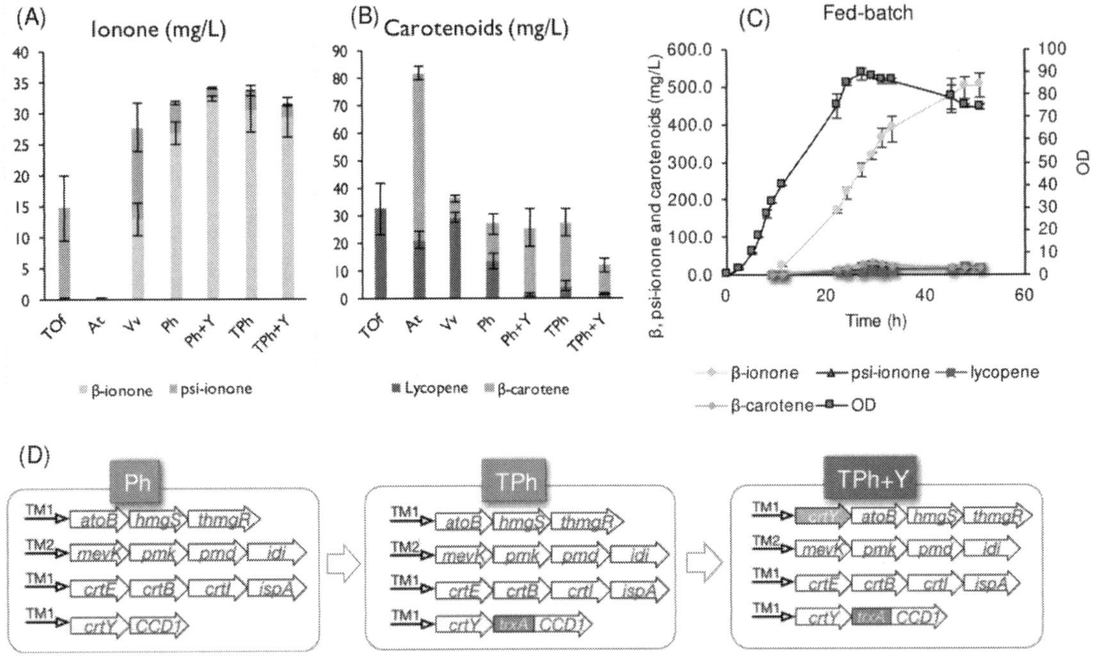

Fig. 8. Screening of CCDs with higher selectivity. (A) Ionone production using different CCDs. (B) Carotenoids accumulated in E. coli cells with different CCDs. In TOf, At, Vv, Ph, TPh strain, the operon/gene crtY-TrxA-OfCCD1, crtY-AtCCD1, crtY-VvCCD1, crtY-PhCCD1 and crtY-TrxA-PhCCD1 were expressed, respectively. In Ph+Y and TPh+Y strain, additional copy of crtY gene was supplemented in the p15A-spec vector. It was found that PhCCD1 had higher selectivity for β-carotene versus lycopene than VvCCD1 and OfCCD1, thus produced higher titer of ionone and lower amount of psi-ionone. (C) Fed-batch fermentation of β-ionone with strain TPh+Y. (D) An illustration of promoters and regulated genes for different strains.

the blh1 strain and 3.0 mg/L retinol and 1.1 mg/L retinal in blh2 strain (Fig. 9A). In blh1 strain, lycopene accumulated (28.2 mg/L) but no β-carotene was detected (Fig. 9B), leading us to hypothesize that retinoid production was limited by lycopene beta-cyclase instead of BCDO in blh1 strain. As expected, the co-expression of additional *crtY* gene (blh1+Y) increased retinoid production (from 23.6 to 33.2 mg/L). As for blh2 strain, retinoid production was likely to be limited by the activity of BCDO instead of lycopene beta-cyclase, as the increase in *crtY* gene dosage failed to improve the retinoid production despite that >90% lycopene was converted into β-carotene (Fig. 9). It is worth noting that in the retinoid-producing strains, the *ybbO* gene, predicted as an oxidoreductase in *E. coli* and validated experimentally with high retinol dehydrogenase activity (Jang et al. 2015) (Genbank Accession ID is ECD_00444, Fig. 1), was not overexpressed but greater than 80% of retinal was converted into retinol. While the underlying mechanism was unknown, it could be due to the relatively high intracellular expression of ybbO enzyme or other unknown endogenous proteins with retinol dehydrogenase activity. The retinoid and ionone results suggested that this "plug-n-play" system is versatile and will be useful for the production of many apocarotenoids which share a common upstream pathway (Fig. 9C).

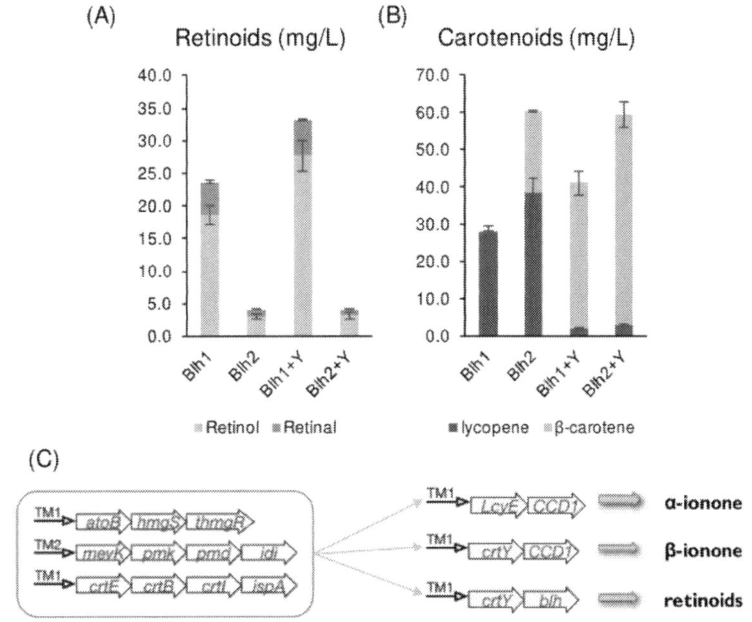

Fig. 9. Production of retinoids in the modularized system. (A) Retinoid production using different blh genes and increased gene copy of crtY. (B) Carotenoids accumulated in different E. coli cells. To produce other apocarotenoids, we could merely replace the CCD1 enzyme with other corresponding carotenoid cleavage oxygenases. Here, we used BCDO (blh) to produce retinoids. (C) A summary illustration of the modular system for the production of different apocarotenoids.

CONCLUSION

The recent changes in the regulation of natural ingredient labelling have resulted in the increase in demand of natural flavours and fragrances. Consequently, there are growing interests in engineering microbes to produce the ingredients from renewable resources. Particularly, many apocarotenoids are important aroma molecules. However, the complex metabolic characteristics of apocarotenoid pathway have hampered the development of highly efficient microbial processes. Here, modular metabolic engineering and enzyme engineering strategies were systematically applied to effectively minimize the metabolic burden imposed by overexpression of 13 enzymes and to overcome the challenge from critical enzymes of low activities. Together, the strategy enabled the development of a robust *E. coli* strain capable of producing unprecedented yields of α-ionone and β-ionone, demonstrating the great potential of using microbes in production of natural flavours and fragrances.

496433 Construction of a Zeaxanthin Biosynthetic Pathway As the Base for Crocin Synthesis inÂ Saccharomyces Cerevisiae

Abstract:

Crocin is a pharmacologically active carotenoid pigment mainly present in the stigmas of Crocus sativus. It has been well explored in experimental animal models of cognitive impairments, depression, anxiety and epilepsy.But extraction from plant is the only way for its production. Crocin synthetic pathway in Crocus sativus shown that four steps progressed from Î²-carotene to crocin. The intermediate productssÂ are zeaxanthin, crocetin dialdehyde, crocetin, and the final crocin.

Synthetic biology provides many novel approaches for microbiology that many plant intermediate products can be produce in microbe. Our study use correlation technique to construct a new system aimed to produce crocin.

Based on a high-level Î²-carotene production yeast strain from Yu Hongwei, our main work is construct a high-level zeaxanthin product strain. Î²-carotene hydroxylase(BCH), which is the key enzyme for the reaction, can be found in many species. We worked on 8 bch genes ,which were screened out and condon optimized. Then we set up the a best zeaxanthin biosynthesis system in yeast cell.

Of the next three steps to produce crocin, only one reaction was not clear about its enzyme and conditions. Our study also do some work to find out the undiscovered enzyme and try to assembly the other components for steps, to prepare for the crocin-producing yeast strain.

497597 Membrane Engineering in *Escherichia coli* to Enhance Production of Bio-Fuels and Chemicals

Abstract:

The production of advanced biofuels and chemicals has advanced substantially in the last decade. Methods for improving the performance of engineered microbes includes a wide range of strategies, from the overexpression of existing genes to the replacement of entire pathways. Experimental results and *in silico* models have shown that tolerance is often related to production. In other words, the economic viability of bio-based production systems is highly dependent on the performance of the engineered microbial cell factory. Low production titers of bio-products compromise their ability to economically compete with petroleum-based processes. We aim to increase microbial production of biofuels and biochemicals, such as ethanol, butanol, phenol and styrene, by increasing microbial robustness to these inhibitors. Particularly, the cell membrane is frequently damaged by accumulation of these chemicals, decreasing the production capacity of the microbial cell factory. Engineering of the membrane has been demonstrated both to increase tolerance towards these inhibitors and to enable increased production of biofuels and chemicals.

Previously, our lab demonstrated that membrane engineering could improve the production of both octanoic acid and styrene (Tan et al., 2016). Specifically, *Escherichia coli* was engineered to isomerize some of the membrane-associated cis-unsaturated fatty acids to the trans form via expression of the Cti enzyme from *Pseudomonas aeruginosa*. This engineering strategy enabled a tighter packing of phospholipids in the membrane, resulting in increased rigidity. Currently, the ability of this engineering strategy to improve tolerance and production of other important membrane-damaging compounds is being further explored. We are also tuning the saturated and unsaturated (S/U) fatty acid ratio, as well as other membrane properties. This strategy offers additional insight into compatibility of chemicals produced and the microbial cell factory. These strategies enable higher tolerance towards certain chemicals of interest, advanced biofuels and inhibitory compounds associated with lignocellulose treatment. An increase in tolerance will allow us to not only to enhance production, but also to decrease the associated unit operations for processing feedstock for advanced biofuels production. Thus, current economic limitations for biofuel production at large scale could be reduced.

497739 Cell-Free Production of Isobutanol

Abstract:

Limitations on biofuel production using cell culture include low product concentrations (<few vol%) due to product feedback inhibition, instability of cells, and lack of economic product recovery. In previously published studies, a novel cell-free immobilized enzyme system was developed in our group to overproduce the biofuel, isobutanol. Two enzymes, (keto-acid decarboxylase (KdcA) and alcohol dehydrogenase (ADH)) were immobilized and one enzyme, (formate dehydrogenase (FDH)), was free in solution, producing isobutanol titers much higher than previously reported with *S. cerevisiae*. Building on these previous studies, we have utilized mutagenesis to increase enzyme activity while immobilized. One enzyme (KdcA) was supplemented with the higher preforming 2-ketoisovalerate decarboxylase (KivD), and site-directed mutagenesis was conducted on two other enzymes (ADH, FDH) from previous publications to increase enzyme activity. We also immobilized FDH previously free in solution, creating a truly cell free system. Future work will combine the reaction system with a separation method.

499857 Systematic Carbon and Growth Analysis of a Promising Methanotroph Strain

Abstract:

Systematic carbon and growth analysis of a promising methanotroph strain

Kyle Stone, Q. Peter He, and Jin Wang

Department of Chemical Engineering, Auburn University, Auburn, AL, 36849, USA

Methanotrophs are widespread bacteria that use methane as their sole carbon and energy source[1]. In doing so the organisms, are not only able to reduce the global warming potential of natural gas leaks or effluent from coal ventilation, but also fulfill the role of a steady biocatalyst utilizing the 376-490 trillion Btu worth of biogas made in the US[2]. Due to recent developments in genetic studies, metabolic modeling, and downstream processing, these cells have been studied for commercial applications, including the production of single cell protein (SCP), lipids (for biodiesel), methanol, poly-ß-hydroxybutyrate (PHB), as well as waste treatment or bioremediation[3]. One strain in particular, *Methylomicrobium buryatense* 5GB1, has been identified as a promising biocatalyst because it is a fast growing organisms with the ability to generate a variety of products including lipids for biodiesel, organic acids, and ectoine. Since the organism is haloalkaliphilic the high pH and salt concentrations are inhibitory to many overpowering contaminants, as well [3-6].

In this work we provide a systematic approach in analyzing the carbon use and growth patterns of 5GB1 with an in house gas mixing system that safely creates a variety of headspace conditions with methane, oxygen, and nitrogen. Of particular interest are the yields of methane to biomass, to carbon dioxide, and to organic liquid products in batch cultures with balanced, methane limited, or oxygen limited headspaces. The approaches used accurately measure carbon products with an overall balance of 95%-102%, taking into account CO_2 that becomes far more soluble in alkaliphilic conditions. This solubility studied, illustrated the requirement of an effective Henry's constant to properly characterize the gas-liquid equilibrium of CO_2 and to have an accurate carbon balance.

The growth patterns observed thus far with the systematic comparison suggest that oxygen concentration amongst different oxygen:methane ratios plays a significant role on the optimal growth rate of this organism. Specifically, an excess of oxygen leads to a general decrease of growth rates in the cultures tested. Additionally, gas uptake rates are evaluated to further gauge when substrate limitation occurs and to use as an indicator of possible metabolic pathway changes. Overall the study contributes to the much needed knowledge base for successful methane fermentation strategies with this promising methanotroph.

1) Conrad R. 2009. The global methane cycle: recent advances in understanding the microbial processes involved.

Environ. Microbiol. Reports 1:285–292.

2) USDA. Biogas Opportunities Roadmap: 2014.

3) Kalyuzhnaya MG, Puri AW, Lidstrom ME. 2015. Metabolic engineering in methanotrophic bacteria. *Metab. Eng.* **29**:142–152.

4) Gilman A, Laurens LM, Puri AW, Chu F, Pienkos PT, Lidstrom ME. 2015. Bioreactor performance parameters for an industrially-promising methanotroph Methylomicrobium buryatense 5GB1. *Microb. cell factories* **14**.

5) Henard CA, Smith H, Dowe N, Kalyuzhnaya MG, Pienkos PT, Guarnieri MT. 2016. Bioconversion of methane to lactate by an obligate methanotrophic bacterium. *Sci. reports* **6**.

6) Strong PJ, Kalyuzhnaya M, Silverman J, Clarke WP. 2016. A methanotroph-based biorefinery: potential scenarios for generating multiple products from a single fermentation. *Bioresour. Technol.* **215.**

495619 Improved Production of Small Molecule Compound Though E. coli Metabolic System Optimization

Abstract:

Metabolic engineering provides an opportunity to access the small molecule compounds with superior benefit to disease therapy and environment protection but of low production and impurity in nature. Different from the normal biosynthesis pathway build-up in which multi plasmids bearing the specific genes are roughly introduced to a heterologous host as Escherichia coli, we enhanced and simplified biosynthesis pathway by integrating all these genes into only one plasmid and realizing the relative biosynthesis in the optimal host chosen from a number of candidates. The optimizaed synthesis system enables a consecutive gene translation, homogeneous production microenvironment and efficient material transportation which promise a highly-improved mechanism. Based on the study, we succeeded in boosting production of a plant-based anti-inflammatory agent salicylate 2-O-Î²-D-glucoside (SAG) and a nonribosomal peptide-polyketide siderophore Yersiniabactin (Ybt).

496246 Zn Recovery from Electroplating Sludge Using Stirred Tank Bioreactor

Abstract:

Electroplating sludge contains various harmful metals such Zn, Cu, Ni, Cr, Fe, etc which is produced from electroplating industries in India. The metals discarded into environment by electroplating industries are in huge amount. Hence, there is a need for the recovery of metals from sludge before it gets discharged into the environment. Bioleaching is one of the suitable method for the recovery of metals from solid waste. The present study deals with the recovery of Zn from electroplating in a lab scale stirred tank bioreactor.

A 20 L capacity stirred tank reactor was fabricated along with other accessories to conduct lab scale pilot plant studies for II batch electroplating sludge. The experiment was started with growing *T.ferroxidans* culture in 2 L of 9K media. The pH of the media was 2.3. The growth of *T.ferrooxidans* was observed within 2-3 days. Two liters well grown culture of *T.ferroxidans*was added in stirred tank along with 8 L of 9K media having 2.2% (w/v%) of ferrous sulfate (half strength). The pH of the media was adjusted to 6.0 in the beginning of the experiment and 1% (w/v%) of II batch electroplating sludge (100g) was added in stirred tank. Proper mixing was provided through stirrer to the entire mixture. Later, 2 ml of sample was collected in the regular interval of 2 days for the Zn analysis. The total duration of the experiment was 49 days. The experiments were conducted in duplicates and average value was plotted along with time.

It was observed that the concentration of Zn was increased significantly for initial days of bioleaching process and later it became steady. The final concentration of Zn in supernatant was observed as 190 ppm on 49th day of bioleaching process. From the material balance (as per XRF result), the amount of zinc in initial sludge sample and after 49th day of bioleaching process was obtained as 3649 mg and 1900 mg. This shows the 52.06% recovery of Zn from solid sludge. The results obtained in the lab scale stirred tank (52.06%) are in agreement with the results obtained from shake flask studies (56.17%). The final concentration of Zn in solid sludge was estimated as 0.919% through XRF analysis which also match with the results of shake flask studies (Table 6). These results indicate the feasibility of scaling up the bioleaching process from lab scale to pilot or large scale operation.

495628 Blocking Lactic Acid Pathway for Enhanced HA Production in *C. Glutamicum*

Abstract:

Hyaluronic acid (HA), containing D-glucuronic acid (GlcA) and N-acetyl-D-glucosamine (GlcNAc) as a disaccharide unit, plays multiple roles in human body and has been widely applied in cosmetics, surgery, medical, food industry and so on. According to the estimation, the market value of HA would exceed 10 billion US dollars until 2020. Presently, the main approach to produce HA is fermentation. With the development of bioengineering, *Streptococcus*or genetically engineered *Bacillus* could obtain HA with titer by 6-10 g/L.

Corynebacterium glutamicum (*C. glutamicum*) is a non-pathogenic, non-sporulating bacterium that generally recognized as safe (GRAS). It is an ideal alternative recombinant host for HA production. In our previous study, co-expressing HA synthase (HasA) from *S. equisimilis* and native UDP-glucose dehydrogenase gene (HasB) in engineered *C. glutamicum* successfully achieved 5-6 g/L HA titer in flask culture and 8-9 g/L HA titer in batch culture of 5L fermentor. In this work, a novel idea is further proposed to engineer the recombinant *C. glutamicum* for enhanced HA synthesis.

HA biosynthesis is an ATP consuming pathway. Every 2 mol glucose needs 5 mol ATPÂ equivalent to form a disaccharide unit. Lactate dehydrogenase (LDH) participates transforming 1 mol pyruvic acid to lactic acid, in which 2 mol ATPÂ equivalent was consumed. Blocking lactic acid pathway would enable more pyruvic acid flowing into TCA cycle and creating ATP. Therefore, a LDH-inactive *C. glutamicum* mutant (â^†*ldh*) was constructed by single cross over homologous recombination. An engineered HA producer, *C. glutamicum* â^†*ldh*-AB with HA synthesis operon (*hasA-hasB*) was further obtained through introducing *C. glutamicum- E. coli* shuttle plasmid pXMJ19-*ssehasA-hasB* (AB). In 5 L fermentor, the biomass and HA titer reached 14.7 g/L and 10.6 g/L, respectively, with 40g/L glucose as carbon source. To enhance the HA accumulation, glucose feeding strategy was further applied to keep residual glucose concentration higher than 8 g/L after 20 h. In the fed-batch culture, the maximal titer of HA achieved 21.6 g/L, which is the highest result reported so far. The M_w of the HA product consisted two part, 1.28 MDa (mass percentage 14.2%) and 290 KDa (mass percentage 85.8%), respectively.

496331 Intensification of Biosynthesizing Tyramine with Engineering Bacteria Expressing L-Tyrosine Decarboxylase By Permeabilization of Cell Membranes

Abstract:

It's important to develop whole cell biocatalysts with high cell-bound tyrosine decarboxylase (TDC) activity for the biosynthesis of tyramine with higher efficient and lower cost. A recombinant *Escherichia coli* (Origami(DE3)-pET21a-*tdc*) strain was constructed by expressing the TDC from *Lactobacillus brevis* CGMCC 1306 in this study. To increase the membrane permeability and cell-bound TDC activity of Origami(DE3)-pET21a-*tdc*, we treated the recombinant cells with nine different organic solvents. The results showed permeabilization with organic solvents increased cell-bound TDC activity as a function of their hydrophobicity, and solvents with high hydrophobicity (log *P*> 2.28) are effective in increasing cell-bound TDC activity of cells. Hexane was the most effective among the tested solvents, which could increase cell-bound TDC activity to 1.48 U•mg^{-1} under optimum condition (1% hexane, 10 min). In addition, treatment of the Origami(DE3)-pET21b-*tdc* under optimum permeabilized condition resulted in structural changes of the cell envelope, but did not substantially affect overall morphology of recombinant cells. By using 0.47 g•L^{-1} hexane-treated Origami(DE3)-pET21b-*tdc* as biocatalyst, 18 g•L^{-1} L-tyrosine was almost totally (>99%) converted to tyramine at pH 5.5 and 40 °C within 4.5 h. These results allow systematic selection of suitable organic solvents for improving cell-bound TDC activity of the recombinant *E. coli* cells and the whole-cell biocatalyst developed here showed great promising for industrial production of tyramine.

488516 Two-Step Biocatalytic Reaction Using Whole Cells for Efficient Production of Phenyllactic Acid from l-Phenylalanine

Abstract:

3-Phenyllactic acid (PLA) is a natural product with broad and effective antimicrobial activity that is often used as a preservative in the food industry and as feed additive to replace antibiotics in livestock feeds. In this study, a two-step biocatalytic process for production of PLA from l-phenylalanine was developed and investigated. In the first step, l-phenylalanine was converted to phenylpyruvic acid (PPA) by recombinant *Escherichia coli* cells expressing membrane-bound l-amino acid deaminases from *Proteus vulgaris*. In the second step, unpurified PPA was directly converted to PLA by recombinant *E. coli* cells co-expressing d-2-hydroxyisocaproate dehydrogenase from *Lactobacillus delbrueckii* subsp. *bulgaricus* and formate dehydrogenase from *Mycobacterium vaccae* N10. Through two-step cascade reactions, 121 mM PLA was produced from 150 mM l-phenylalanine with a yield of 81%. Thus, this process showed great production efficiency and is a promising method for industrial production of PLA.

490424 Evaluation of Culture Conditions on the Production of Antimicrobial Compounds Against *Staphylococcus aureus* from *Lactobacillus viridescens Â*

Abstract:

One of the major challenges facing the food industry today is to ensure the long life of food and food safety, due to the growing demands of the consumer market and enforcement agencies. Allied to this, the use of synthetic chemical compounds for food preservation and the consequences of its use has been discussed. Other highlighted topics are the use of chemical compounds in crops or grains to control and combat pests besides the alternatives for the treatment of new resistant bacteria to the usual treatments with known antibiotics. In this context, some natural compounds produced by bacteria which have an inhibitory effect against microorganisms associated with food contamination and diseases, have gained space in research and scientific studies. Most lactic acid bacteria species are capable to produce organic substances that are active against contaminating organisms and help in food preservation. One of the main limiting factors for the use of antimicrobial compounds on a large scale in the food industry or other fields is the need of further studies to identify new strains of bacteria which produce these compounds, together with the characterization and verification of safety for its use. In addition, the determination of ideal parameters for the cultivation and purification of these compounds, could help reduce the high costs involved in obtaining them, justify the need for further studies. In this context, this work evaluated the culture conditions for production of antimicrobial compounds by *Lactobacillus viridescens* and verified its potential of inhibition against the pathogenic microorganism *Staphylococcus aureus*. First, a culture of the microorganism *Lactobacillus viridescens* was carried out to determine its growth curve and to evaluate which phase of the growing curve shows the highest production of antimicrobial compounds. For this, three cultures were made in MRS broth (sharpe, de man and rugosa) at 36 °C for 54 h. During cultivation, samples were taken every 6 h for determination of the cell mass and inhibition generated by the antimicrobial compounds produced. Subsequently, the produced metabolites concentration through a freeze-drying process was performed. The produced cell mass was evaluated by the dry weight method. The evaluation of the potential inhibition against *Staphylococcus aureus* by the antimicrobial compounds was carried out by the agar plate diffusion method. According to the results, the microorganism *Lactobacillus viridescens* showed that at 6 h it was already in the exponential growth phase, which lasted 24 h, where the transition to the stationary growth phase occurred. It was verified that at the time of 24 h, largest halos of inhibition against the microorganism *Staphylococcus aureus* were obtained, indicating that at this time the maximum production of antimicrobial compounds should occur. After the determination of the growth curve, a study was carried out to evaluate the influence of physical and nutritional factors on the productivity of the antimicrobial compounds. This study was conducted according to the fractional factorial experimental design type (2^{6-2}) with four replicates at the central point, and the study carried out in duplicate. The selected physical factors for this study were the incubation temperature of cultivation and the initial pH of the culture medium. The selected nutritional factors were glucose concentration, lactose concentration, concentration of the yeast extract and sodium chloride. The culture medium used was MRS. After cultivation, each experimental point had was tested and its inhibition evaluated following the same methods used during the evaluation of the growth curve. As a result of this study, it was possible to verify that the temperature and sodium chloride concentration factors had a negative effect, statistically significant effect at a 95% confidence level. This means that the higher the concentration of sodium chloride and the higher the temperature, the lower inhibition zone generated against *Staphylococcus aureus*. The factors pH, glucose concentration, lactose concentration and yeast extract concentration had a positive effect on the inhibition halo formed, but none of these factors was statistically significant at a 95% confidence level.These results demonstrate the ability of lactic acid bacteria *Lactobacillus viridescens* to produce compounds with antimicrobial effect against the pathogen *Staphylococcus aureus*. The study also showed that the antimicrobial metabolites produced by *Lactobacillus viridescens* are produced at the end of their exponential growth phase, and the incubation temperature during cultivation as well as the concentration of sodium chloride in the culture medium are the main parameters associated with production of antimicrobial compounds by *Lactobacillus viridescens*.

489215 Engineering Gut Microbes to Treat Metabolic Disorders

Abstract:

Inborn errors of metabolism (IEM) are a family of more than 500 potentially lethal congenital genetic disorders that cumulatively affect 1 in 1000 newborns. In many IEMs, pathologies manifest as a result of improper metabolism of nutrients in food. In Phenylketonuria (PKU) for example, elevated levels of phenylalanine and the accumulation of aberrant metabolic intermediates in the system lead to acute and chronic toxicities. Resultantly, many disorders within this group are generally treated through lifelong nutritional management due to the lack of alternative and pharmacological options. Longitudinal studies have indicated that even with strict adherence to a diet of synthetic supplements, patients experience chronic issues like frailty, delayed growth, and intellectual disabilities. Recently, enzyme-replacement therapies (ERT) have demonstrated promise in pre-clinical and clinical settings by providing a metabolic sink for phenylalanine in PKU. As an enhancement to traditional ERT, we are developing a novel therapeutic for IEMs associated with amino acids by expressing metabolic enzymes in lactic acid bacteria (LAB) that natively colonize the human gastrointestinal (GI) tract. Starting with an enzyme under clinical development for PKU, phenylalanine ammonia-lyase (PAL), and by promoting the intestinal adhesion and colonization characteristics, the engineered LAB will intervene before amino acid absorption occurs in the small intestines during digestion. To engineer new enzymes with activities required for treating IEMs, we have developed a novel facile selection and screening methodology. This can potentially be utilized to enhance enzymatic properties or identify mutants with altered substrate specificity, creating a spectrum of PALs that can be used to treat IEMs associated with other amino acids. Here we describe the methodology, development, and optimization of this method. To characterize and engineer microbial adhesion to intestinal mucus, we developed a novel assay that is able to capture the quantitative and mechanistic binding thermodynamics of cells to mucus. We will discuss the development of this assay and its implementation for engineering improved mucus binding. The platform technologies discussed here will be instrumental in realizing microbiota-based therapeutics as an emerging and urgently-needed treatment for IEMs that currently have inadequate or no options.

494673 Engineering a Rubisco-Deletion Crypthecodinium Cohnii for Increased Growth and Lipid AccumulationÂ

Abstract:

The heterotrophic dinofagellate microalga *Crypthecodinium cohnii* is well known for its ability to accumulate lipids with a high fraction of polyunsaturated fatty acid, such as docosahexaenoic acid (DHA). However, it has been challenging for genetic modification in this valuable microorganism. In this study, we developed a new genetic transformation methodology for DHA-producing *C. cohnii* by evaluating suitable selected marker and optimizing transformation protocol. In addition, a gene potentially involved in CO_2 fixation under autotrophic condition, ribulose 1,5-bisphosphate carboxylase (*rubisco*), was selected as the target for construction of gene knock-down mutant, the results demonstrated that the constructs can be successfully inserted into the *C. cohnii* chromosome by homologous recombination, which was further confirmed by sequencing, decreased gene expression enzymatic activity, and Western blotting analysis. Interestingly, comparative analysis showed that the deletion of the *rubisco* gene promoted cell growth and increased lipid content of *C. cohnii* under heterotrophic condition when compared with the wild type. To explore the possible mechanism related to the enhanced growth and lipid accumulation in the *rubisco*-deletion mutants, a LC-MS based metabolomic analysis was conducted, and the results showed that the metabolites involved in energy metabolism were up-regulated, suggesting the deletion of the *rubisco* gene may contribute to the re-direction of more carbon or energy towards growth and lipid accumulation under heterotrophic condition.

490562 Systems Analysis and Engineering of Oleaginous Red Yeasts

Abstract:

Microbial lipids have been recognized as important renewable resources for biofuels and oleochemicals. There are some natural oleaginous yeasts capable of accumulating lipids to over 70% of their dry cell mass. In particular, the red yeast *Rhodosporidium toruloides* is superb for production of neutral lipids under various conditions. To understand the molecular bases of its oleaginicity, we sequenced the genome of *R. toruloides* and performed trans-omic analysis. Our results demonstrated that TAG accumulation under nitrogen-limited conditions is tightly connected with cellular processes related to lipogenesis, nitrogenous compounds recycling, macromolecules metabolism and autophagy. Under phosphorus-limited conditions, cells activate degradation of phosphorus-containing components including RNA and nucleic acids such as adenosine monophosphate, leading to reduced cell proliferation and enhanced lipid biosynthesis. Proteomic analysis identified 226 lipid droplet (LD)-associated proteins, and 77 of them changed abundance significantly along with lipid accumulation. The LD structural protein Ldp1 was demonstrated to facilitate giant LD formation in other yeasts. Furthermore, we have confirmed some of these data by genetic modification of *R. toruloides* using in-house developed tools. Taken together, our systems study revealed secrets behind yeast oleaginicity and provided rich information for engineering other cell factories.

Representative references

Zhu Z, et al. Nat. Commun. **2012**, 3,1112.

Lin X, et al. *FEMS Yeast Res.* **2014**, 14, 547–555.

Fischer M, et al. *Protein Sci. 2015, 24,* 987–995.

Zhu Z, et al. Eukaryot. Cell **2015**, 14, 252–264.

Wang Y, et al. *Appl. Biochem. Biotechnol.* **2016**, *180*, 1497–1507.

Liu H, et al. *FEMS Yeast Res.* **2017**, 17, doi: https://doi.org/10.1093/femsyr/fox017

497193 Considerations for Using Hammerhead-Based Riboswitches in the 5′-UTR to Control Genes in Bacteria

Abstract:

Riboswitches are small regulatory elements composed of RNA that have garnered much attention over the past decade due to their high selectivity. These single-stranded nucleic cids are generally composed of two functional domains: (i) an aptamer domain, which selectively binds a specific target molecule, and (ii) a functional domain or expression platform that controls whether the gene is turned "on" or "off" through allosteric means. The majority of natural riboswitches are located upstream of the genes they control, typically repressing expression of the genes they control when the ligand they recognize reaches a specific concentration. However, many synthetic riboswitches have been constructed that function in either the 5′-UTR or the 3′-UTR, depending on their mechanism.

The hammerhead ribozyme has been used as a riboswitch functional domain for ligand-dependent control of gene expression by linking an aptamer to one of its three arms. The most successful use of a hammerhead riboswitch thus constructed has been demonstrated in the 3′-UTR of eukaryotic organisms. In prokaryotic organisms, the 3′-UTR switch does not control gene expression well, and 5′-UTR switches have met with limited success. Here, we investigate various design considerations, such as use of an anti-RBS fragment upstream of the riboswitch, number of nucleotides between the riboswitch and gene, and speed of protein synthesis, to use hammerhead riboswitches placed in the 5′-UTR to control expression of the *lacZ* and *gfp* genes in the bacterium *Escherichia coli*. The results of this work will provide a groundwork for design of efficient 5′-UTR hammerhead riboswitches in prokaryotic organisms.

500476 Elucidation of Aromatic Metabolism Pathways in a Non-Model, Non-Conventional Oleaginous Yeast

Abstract:

Oleaginous yeasts have long been a target for developing industrial-scale biochemical processes due to their ability to accumulate high amounts of lipids, synthesize complex chemicals and proteins, and robustly metabolize diverse feedstocks. In parallel, interest in lignocellulosic biomass as a feedstock has grown. *Trichosporon oleaginosus*, previously known as *Cryptococcus curvatus*, is a non-model, non-conventional, oleaginous yeast that we have discovered both tolerates and metabolizes lignin-derived phenolics. We have characterized *T. oleaginosus* biomass and lipid accumulation while utilizing glucose, phenol, catechol, 4-hydroxybenzoic acid, and resorcinol as sole carbon sources under nitrogen rich and nitrogen starved conditions. Aromatic metabolism does not hinder lipid accumulation, but toxicity limits how much carbon can be dosed at a single time to 1 g/L for three of the four aromatics studied. As such, we explored different feeding strategies to overcome aromatic toxicity, and increased lipid accumulation by more than 85%. Since many recalcitrant waste streams are heterogeneous feedstocks, we also characterized growth and lipid accumulation when cells are cultured in dual carbon feedstocks, and found that feeding glucose and resorcinol at once improves lipid accumulation by 41% when compared to glucose as a sole carbon source. To date, the mechanism for aromatic metabolism in this yeast remains unknown. BLAST analysis suggests a potential putative pathway for metabolism of aromatic compounds, but non-intuitive qPCR results indicates the mechanism could involve a cryptic pathway or an alternative mechanism of aromatic metabolism. RNAseq data was utilized in effort to elucidate the pathways responsible for aromatic metabolism. In all, this work hopes to showcase this yeast as a potential model oleaginous yeast for aromatic metabolism, with future applications for valorization of recalcitrant feedstocks such as lignin and phenolic wastewater effluents.

489459 Engineering Metabolic Pathways By Using Standardized DNA Parts

Abstract:

Almost all metabolic engineering practices require plasmid construction. Currently, tens of thousands of research laboratories across the globe are constructing plasmids every day in an inefficient way â€“ researchers customize materials they need, pay commercial companies to synthesize them from scratch (oligos and/or genes), wait for the materials to be delivered, and assemble the materials in their own laboratories.

A more efficient way of constructing plasmids is through assembly of standardized DNA parts, which are from a user-defined library and readily available. Here, we developed a scarless DNA assembly method that can assemble up to 12 standardized DNA parts in any order. Our method does not need to avoid any forbidden restriction enzyme site, and is compatible with multiple-tier DNA assembly, which makes it to be advantageous to all existing methods that use standardized DNA parts (e.g. BioBricks assembly and BASIC). To demonstrate the usefulness of this method, we built a library of DNA parts for metabolic engineering applications, and used them to overproduce isoprenoid and aromatic compounds in commonly used species (*Escherichia coli, Bacillus subtilis* and *Saccharomyces cerevisiae*). In this era of CRISPR-Cas9, we also designed a large number of parts for this genome editing method, and used them to construct some plasmid-free microbes that have more stable fermentation characteristics in metabolic engineering applications. Through sharing and improving this library of DNA parts, our DNA assembly method would make it cheaper and faster to engineer microbial metabolic pathways for overproduction of valuable chemicals.

497370 Application of 13C Flux Analysis to Determine Impacts of Media Alterations on Industrial CHO Cell Metabolism

Abstract:

Industrial bioprocesses place extraordinary demands on the metabolism of host cells to meet the biosynthetic requirements for maximal growth and protein production. Identifying host cell metabolic phenotypes that promote high recombinant protein titer is a major goal of the biotech industry. ^{13}C metabolic flux analysis (MFA) provides a rigorous approach to quantify these metabolic phenotypes by applying stable isotope tracers to map the flow of carbon through intracellular metabolic pathways. We have conducted a series of ^{13}C MFA studies to examine the metabolic impacts of altering the composition of a proprietary chemically defined growth medium on CHO cell metabolism.

CHO cell cultures characteristically produce excess ammonia and lactate as byproducts, both of which are toxic at high concentrations. Whereas lactate is often consumed during stationary growth phase in CHO cell cultures, ammonia continues to accumulate in the extracellular media throughout the course of cell growth due mainly to glutamine catabolism. For CHO cells that utilize glutamine, rational media design can alleviate ammonia stress from the cell culture. However, manipulating carbon sources in the growth medium can also have negative effects on cellular metabolism such as decreased culture growth, viability, recombinant protein productivity, or longevity. This study highlights a rationally engineered cell culture medium that successfully reduces culture ammonia levels by 40% while maintaining the original metabolic phenotype.

First, the basal media developed in-house by Sanofi was chemically altered to cause CHO cells to produce significantly less ammonia byproduct. This low ammonia-producing media variant was experimentally developed by altering the ratio of carbon sources in the media to strategically reduce flux through metabolic pathways that result in ammonia production while supplementing complementary, non-ammonia producing pathways to balance metabolism. This altered media variant successfully decreased the ammonia concentration in industrial CHO while maintaining culture growth, viability, and specific productivity.

Parallel ^{13}C MFA studies were performed on IgG-producing CHO cells grown identically in three media variants: the basal control media, the low-ammonia media, and the low-ammonia media supplemented with basal ammonia levels. The latter media was used to control for any direct effects of changing ammonia concentrations on cellular metabolism. ^{13}C labeling studies utilizing [U-$^{13}C_5$]glutamine and [1,2$^{13}C_2$]glucose were carried out in parallel for each condition.

From the comparison of the ^{13}C flux analysis across the three media types, we have concluded that the media alterations did not have a significant impact on the intracellular metabolism of CHO cultures. This suggests that Sanofi can use their newly developed media formulation to decrease toxic ammonia buildup in IgG-producing CHO cell lines without significantly altering host metabolic phenotype or productivity.

500130 Use of an *Escherichia coli* pyruvate-Overproducing Platform Strain to Produce ÊŸ-Valine

Abstract:

Biocatalytically derived, carbon-neutral chemicals are becoming increasingly attractive over traditional catalytic methods, but are impeded by the slow, metabolic engineering design-build-test cycle. To expedite this process, a precursor-abundant strain to industrially relevant chemicals will enable high-throughput engineering of diverse biosynthesis pathways. To this end, the metabolic potential of *Escherichia coli* was assessed computationally and pyruvate identified as the optimal precursor to produce commercially relevant chemicals. Subsequently, a genome-scale metabolic model of *E. coli* was used to design strains that achieved astonishing pyruvate yields up to 95% of the theoretical maximum. Next, as a proof-of-concept, ÊŸ-valine was chosen as a target for overproduction to test the platform capabilities of a pyruvate-producing strain since it is an amino acid widely used as a nutritional supplement in several industries with a global demand of about 500 tons annually. Currently, a pyruvate-producing *E. coli* strain harboring a plasmid expressing ÊŸ-valine biosynthetic genes has been engineered to achieve 69% of the theoretical maximum yield, the highest reported for *E. coli*, as opposed to 39% for wild-type *E. coli* with the same plasmid, in approximately 22 hours of aerobic culturing in a defined, glucose minimal medium. This study demonstrates that a pyruvate platform strain is, indeed, a versatile tool to rapidly engineer strains to produce various chemicals.

500228 Engineering a Novel 3-Methyl-1-Butanol Biosynthetic Pathway in *Escherichia coli*

Abstract:

Production of branched-chain higher alcohols as gasoline substitutes via a coupled branched-chain amino acid biosynthesis pathway and Ehrlich degradation pathway has been well established. In this work, we constructed a novel pathway for production of isobutanol and 3-methyl-1-butanol (3MB) in *E.coli*. In addition, we explored its potential for specific 3MB biosynthesis using metabolic engineering, bioprospecting, and protein engineering approaches. Preferential production of 3MB over the one-carbon shorter byproduct, isobutanol, involves selective termination of carbon chain elongation at the desired chain length, which has been shown to represent a challenge in various engineered pathways. To overcome this challenge, we selected the long-chain favoring keto acid dehydrogenase complex (KDHC) as the termination enzyme. Subsequently, downstream modification enzymes aldehyde dehydrogenase (AldH) and alcohol dehydrogenases (ADH) with enhanced product selectivity were obtained using structure-guided protein engineered and bioprospecting, respectively. The flux distribution between the desired chain elongation step and the pre-mature termination step was also improved through titrating the expression levels of the two competing enzymes. Taken together, the optimized pathway yielded a ratio of 3:1 3MB to isobutanol, a 400% increase over the non-engineered pathway.

486585 Next Steps in Engineering *E. coli* Erythromycin Production

Abstract:

Erythromycin has been widely used to against bacterial infections. Instead of using the original producing host *Saccharopolyspora erythraea* (*S. erythraea*), here we harness *Escherichia coli* (*E. coli*) as an expression surrogate due to the relative ease of genetic manipulation. A recent version of the production system features five plasmids hosting large (~55 kb) heterologous expression pathway.Â In this work, weâ€™ve improved the strain stability by rearranging the expression pathway. Two bacterial artificial chromosomes(BAC) were employed rather than our previous five plasmids to reduce cell metabolic burden and improve efficiency. We also improved erythromycin titers by accumulating the common deoxysugar intermediate TDP-4-keto-deoxyglucose. Additionally, an *E. coli* multidrug efflux pump (MacAB) was overexpressed to further elevated erythromycin titer. Finally, six heterologously expressed deoxysugars were studied to biosynthesize erythromycin analogous in this study.

493052 Photocatalytic Production of a Jet Fuel Precursor Limonene

Abstract:

With the increased interest in developing more sustainable sources of fuels and chemicals to limit our carbon footprint, cyanobacteria are attractive platform strains for metabolic engineering efforts. Terpenoids are a class of chemicals that have received considerable attention for use as renewable drop-in biofuels [1]; they are attractive for fuel applications because they have a greater energy density and lower hygroscopicity than short chain alcohols, and the structural diversity to mimic the alkanes and aromatics of desired chain lengths for gasoline, diesel and jet fuels. Importantly, this means that terpenoids can be blended with petroleum-based fuels, and are compatible with existing transport fuel infrastructure. Limonene ($C_{10}H_{16}$) is a monocyclic terpenoid that has properties suitable as a precursor for next generation jet fuels, and as a fuel additive that enhances cold-weather performance [2,3]. Currently, limonene is extracted from citrus fruit peel as a byproduct of juice processing, and used in the flavor/fragrance industry and as an environmentally-friendly solvent and degreaser. The worldwide production of limonene in 2013 was over 70,000 tons [4]. However, the extraction methods from citrus peel are energy-intensive, and the seasonal and environmental factors that affect citrus fruit crops make the pricing and supply of limonene extremely volatile. The largest potential market for limonene is for jet fuel, which has a current demand of 1.53 million barrels per day in the United State (U.S. Energy Information Administration http://www.eia.gov/forecasts/steo/report/us_oil.cfm); which translates to an almost $60 billion market for jet fuel alone. We have chosen to engineer the fast growing cyanobacterium, *Synechococcus* sp. PCC 7002 (*Synechococcus* hereafter) as a platform for photosynthetic limonene production. *Synechococcus* has been identified as an excellent platform for biotechnological applications because of its fast growth and ability to thrive in salt water and changing light conditions. We have engineered *Synechococcus* to yield 4 mg limonene per L culture through heterologous expression of the *Mentha spicata* L-limonene synthase [5]. To date, our limonene-producing (LS) strain has the highest reported limonene yield among photosynthetic engineered organisms with comparable genetic manipulation. Further, we have ensured a high purity of limonene through careful selection of the *M. spicata* LS, which has high product specificity [6,7] We have also overcome another important challenge of working with cyanobacteria: stable and homoplastic integration of the LS gene into the *Synechococcus* genome. Cyanobacteria have several to hundreds of copies of their chromosome depending on the species and/or growth condition [8], therefore, a long selection and outgrowth period are required to ensure that all copies of the chromosome are homogenous. The next step of our research is to use ^{13}C-MFA to help develop a rationale engineering approach to further increase limonene titers. We will discuss the results of our flux analysis and compare how the MEP pathway in cyanobacteria differs from heterotrophic bacteria.

References

1 Keasling, J. D. Synthetic biology and the development of tools for metabolic engineering. *Metabolic Engineering* **14**, 189-195, doi:10.1016/j.ymben.2012.01.004 (2012).

2 Renninger, N. S., Ryder, J. A. & Fisher, K. J. **Jet fuel compositions and methods of making and using same**. (2011).

3 Tracy, N. I., Chen, D., Crunkleton, D. W. & Price, G. L. Hydrogenated monoterpenes as diesel fuel additives. *Fuel* **88**, 2238-2240, doi:10.1016/j.fuel.2009.02.002 (2009).

4 Ciriminna, R., Lomeli-Rodriguez, M., Cara, P. D., Lopez-Sanchez, J. A. & Pagliaro, M. Limonene: a versatile chemical of the bioeconomy. *Chemical Communications* **50**, 15288-15296, doi:10.1039/c4cc06147k (2014).

5 Davies, F. K., Work, V. H., Beliaev, A. S. & Posewitz, M. C. Engineering limonene and bisabolene production in wild type and a glycogen-deficient mutant of *Synechococcus* sp. PCC 7002. *Front Bioeng Biotechnol.*, doi:10.3389/fbioe.2014.00021 (2014).

6 Alonso, W. R., Rajaonarivony, J. I. M., Gershenzon, J. & Croteau, R. Purification of 4*S*-limonene synthase, a monoterpene cyclase from the glandular trichomes of peppermint (*Mentha* x *piperita*) and spearmint (*Mentha spicata*). *Journal of Biological Chemistry* **267**, 7582-7587 (1992).

7 Rajaonarivony, J. I. M., Gershenzon, J. & Croteau, R. Characterization and mechanism of (4*S*)-limonene synthase, a monoterpene cyclase from the glandular trichomes of peppermint (*Mentha* X *piperita*). *Archives of Biochemistry and Biophysics* **296**, 49-57, doi:10.1016/0003-9861(92)90543-6 (1992).

8 Griese, M., Lange, C. & Soppa, J. Ploidy in cyanobacteria. *FEMS Microbiology Letters* **323**, 124-131, doi:10.1111/j.1574-6968.2011.02368.x (2011).

499752 Metabolic Engineering ofÂ *Clostridium cellulovoran* for n-Butanol Production from Cellulose

Abstract:

n-Butanol can be used as a significant industrial chemical and potential superior biofuel. Especially, compared with ethanol, *n*-butanol can be used as an ideal substitute of gasoline because of its high energy density, low water solubility, and low vapor pressure. Unfortunately, bio-butanol production using the conventional acetone-butanol-ethanol (ABE) fermentation process is not economically feasible because of the co-production of other metabolites, including CO_2, acetone, acetate, butyrate, ethanol, and lactate resulting in a low butanol yield. In addition, starch and sugar as subtract are not suitable for the production of bulk biochemicals or biofuels, due to the high feedstock cost and impact on food supply. Lignocellulosic biomass is an abundant, cheap, and renewable source of carbon source, and thus is a desired feedstock for biofuels production. Conventional biorefinery of lignocellulosic biomass requires different operations, including feedstock pretreatment, cellulase/hemicellulase production, biomass enzymatic hydrolysis, sugar fermentation, and product recovery, which is not cost-effective due to the complex process and subsequently high equipment capital requirement. Consolidated bioprocessing (CBP) combines enzyme production, biomass hydrolysis, and sugar fermentation into one step, which will greatly simplify the process and dramatically reduce the equipment investment. In our previous research, *Clostridium cellulovorans*, a natural cellulose/hemicelluloses utilizing, non-natural butanol producing bacterium, was metabolically engineered for n-butanol production from cellulose by overexpressing *adhE2* gene from *Clostridium acetobutylicum*. However, this mutant only produced 1.42 g/L n-butanol and 1.60 g/L ethanol from microcrystalline cellulose in 10 days. Moreover, due to the plasmid DNAs were always digested by *C. cellulovorans* cell extract, current transformation efficiency is low, making metabolic engineering difficult. So, efficient transformation for developing better mutants are urgently required. In order to solve this problem, its restriction modification (RM) systems were analyzed and a new clostridium shuttle plasmid p83151d2 without restriction sites was constructed for its successful transformation by improved electroporation in this study. Ethanol and n-butanol production from cellulose were investigated by introduced *adhE1*, *bdhB*, *aor*, *thl*, *fnr*, and *fdh* with *adhE2* into *C. cellulovorans*, respectively. Moreover, adding artificial electron carrier such as methyl viologen (MV) in engineered *C. cellulovorans* redistributed the metabolic flux from acid production to alcohol production, leading to a high n-butanol titer of 5.2 g/L from cellulose. These results showed that *C. cellulovorans* is a promising CBP platform host for biofuel production from cellulosic biomass.

501727 Metabolism of the Pyrolytic Sugar Levoglucosan and Engineered Pyrolysate Tolerance in *Lactococcus lactis*

Abstract:

Lactococcus lactis, a gram-positive lactic acid bacterium, is an important organism in the dairy, chemical, and health industries. We have engineered this organism to consume the sugar levoglucosan, produced from pyrolysis of lignocellulosic biomass. Expression of the levoglucosan kinase gene from *Lipomyces starkeyi*, allows *L. lactis* to consume pure levoglucosan and levoglucosan-rich bio-oil as a sugar source, producing optically pure L-lactic acid. However, lignocellulosic pyrolysate contains several inhibitory phenolic compounds which limit the growth of bacteria under fermentation conditions. To promote growth of *L. lactis* in bio-oil, a genome-scale library was constructed from *L. lactis* subsp. cremoris NZ9000 to identify genes which contribute to growth in bio-oil and confer resistance to inhibitors. Furthermore, we discuss metabolic engineering strategies for in vivo conversion of L-lactate to other high-value chemicals.

500616 Efficient Energy Utilization in Carbon-Fixing *Moorella Thermoacetica*

Abstract:

Carbon dioxide is the main greenhouse gas that is produced in combustion and cellular respiration. On the other side of the carbon cycle, plants and algae reduce CO_2 using photosynthesis. Despite technological advances, industrial society emits CO_2 faster than the rate at which it can be eliminated, breaking balance in the carbon cycle. Metabolic CO_2 reduction requires energy in two forms: reducing power and ATP. Here we find that when glucose is the sole energy source, non-photosynthetic carbon-fixing anaerobe *Moorella thermoacetica* is limited by reducing power. Despite electron-limitation, we observe near-complete carbon utilization for acetate and biomass production. Using ^{13}C glucose and $^{12}CO_2$, we observe that CO_2 accounts for most carbons of non-aromatic amino acids and secreted acetate. In this talk, we will introduce and apply simultaneous ^{13}C and ^{18}O tracing to identify energy-efficient pathways that allow cells to proliferate in chronically nutrient- and energy-limited environments. Identification of metabolic limitations using similar isotope tracer strategies should be broadly useful for improving bioprocesses.

497740 Chiral Membranes for Enantiomer Separation

Abstract:

Most essential substances of living organisms are chiral in nature. Recently, enantiomeric separation of racemic mixtures has developed rapidly to meet the needs of the biotechnology, materials, agrochemicals, fragrances and pharmaceutical industries. Chiral membrane filtration is a faster, more efficient, less energy-intensive, easier to scale up and broadly applicable technique than crystallization and chromatography. An approach for separating enantiomers using modified regenerated cellulose membrane is presented. This novel method of coating membranes is useful recovering the desirable enantiomer of important drugs like propranolol, thalidomide, carnitine and methotrexate. A high selectivity towards a particular enantiomer was achieved after coating the membranes with a chiral and structural polymer. The membranes were characterized using Attenuated Total Reflectance Fourier Transform Infrared Spectroscopy and X-Ray Photoelectron Spectroscopy and Scanning Electron Microscopy. To determine selectivity and permeability, the membranes were evaluated via filtration of chiral feed streams containing amino acids and/or proteins with selectivities evaluated via LC-MS.

498766 The Introduction and Removal of β-Glucans in a Biological Purification Process – a Case Study

Abstract:

The Introduction and Removal of β-glucans in a Biological Purification Process – a Case Study

Adele Pearson, Robert G. Luo, Kent E. Göklen

GlaxoSmithKline

King of Prussia, PA, USA

Poster Abstract

β-glucans are polysaccharides of D-glucose monomers linked by β-glycosidic bonds. Because they pose an immunogenicity risk to patients, drug products should have low levels of β-glucans. They can be introduced into a process through multiple sources: 1) host cell impurity from a yeast cell line; 2) component of yeastolate which may be an additive in production media; 3) leachable from cellulosic depth filters; and 4) potential contaminant in raw materials (e.g. sucrose). In this study, β-glucans were found at unexpectedly high levels in batches of the product of a mAb purification process; levels were higher than seen earlier in the program. The investigation that led to an understanding of the phenomenon and the change that caused it will be presented. In addition, the studies evaluating removal of β-glucans by filtration, including filter screening and sizing, will also be presented.

498815 Numerical Evaluation of theÂ Two-PhaseÂ Fluid Dynamics in a Bench Scale BioreactorÂ Applied to Microalgae Cultivation

Abstract:

The cultivation of microalgae is applied in the production of food, biofuels, biochemicals, pharmaceuticals and fertilizers. Through the photosynthesis process, these algae convert water, carbon dioxide and light into oxygen, organic compounds and water. *Spirulina* is a prokaryote microalgae found in different habitats. *Spirulina* has a high content of macro and micronutrients, thus it can be used for human consumption, as well as presents pharmaceutical properties, including antioxidant, anti-inflammatory, antiallergic, anticancer and antibacterial activity. In the cultivation of *Spirulina* microalgae, some factors influence the production of biomass, such as aeration, composition of the culture medium, lighting, pH and temperature. The aeration of the culture medium keeps the cells in suspension, favoring the contact between microalga and substrate, and illumination of the cells, however, with too much air flow, cell disruption and crop compromise can occur. To help in the investigation of these parameters, the computational fluid dynamics (CFD) can be used, as this technique provides detailed and accurate results of the variables that interferes in microalgae production. In this study, the fluid dynamics behavior found in the cultivation of *Spirulina* sp. LEB 18, due to the air flow used for aeration, was numerically evaluated. The study was carried out for a bench scale photobioreactor with a capacity of 2 L, being 1.96 L of occupied by the liquid phase, with aeration of the medium through a duct in the center of the photobioreactor with air outlet at the base at a flow of 1 L/min. The computational fluid dynamics technique was used to estimate the turbulent two-phase flow in an axissimetric geometry using the OpenFOAM v4.1 code. The simulated phases consisted of air, fed through a central duct to the erlenmeyer, and microalgal medium formed by the Zarrouk medium and *Spirulina* sp. LEB 18 microalgae. The k-Omega SST model was used to predict the turbulent characteristics of the flow. Meshes with different refinement levels were elaborated, containing 13, 29 and 49.5 thousand control volumes. The simulation of these data provided results for the calculation of the grid convergence index (GCI), whose value indicated that the use of a mesh containing 49.500 control volumes is adequate to capture the variables of interest with precision. The transient simulations were conducted for approximately 0.6 seconds, from which the simulations were continued for a further 14.4 seconds, with calculation of mean values. Process parameters such as velocity, pressure, turbulent kinetic energy, dissipated turbulent kinetic energy, vorticity, shear stress and phase distribution were studied in the present study, verifying its suitability for microalgae cultivation.

492844 Polysorbate 80 Disposition Following Tangential Flow Filtration

Abstract:

Tangential Flow Filtration (TFF) is known to cause partial unfolding and aggregation of proteins, leading to poor product quality and potential patient safety concerns. While downstream controls are typically in-place, the use of a surfactant such as Polysorbate 80 (PS80) can prevent unfolding and aggregation by protecting sensitive proteins from stress during TFF and subsequent operations. However, PS80 has not been commonly used upstream of TFF because it is challenging to predict its disposition following TFF operations. The two following items highlight sources of this challenge: (1) PS80 is difficult to consistently filter because it can be present as a monomer, micelle, or larger non-spherical micelle. (2) PS80 distribution during and after TFF may be affected by protein-PS80 interactions, which would inhibit separation via TFF.

By better understanding the mechanisms effecting PS80 disposition after TFF, new insight can be gained as to how PS80 could be best used upstream of TFF to increase drug product quality while maintaining predictable output stream composition. For this study, several experiments were performed to evaluate the effect of key filtration and TFF parameters on PS80 disposition. One such experiment included centrifugal filtration or depth flow filtration to assist with the screening of parameters non-specific to TFF such as PS80 concentration, protein presence, and membrane molecular weight cut off (MWCO). These experiments showed that the main contributions to PS80 retention is protein presence and protein concentration. The potential presence of larger non-spherical micelles due to high PS80 concentration does not inhibit PS80 removal. This presentation will review different approaches used to study which factors impact PS80 retention and removal and how final PS80 compositions may be predicted when it is introduced upstream of TFF.

499888 Purification of a Pegylated Protein

Abstract:

PEGylation can bring along many challenges, one of which is the formation of process related impurities. Typically, size exclusion chromatography has been advantageous in removal of these impurities at small-scale, but this is not feasible for large-scale implementation. With the use of High Throughput Screens (HTS), several other purification methods were identified. This poster will highlight a challenging process in which large polymer-protein complexes were conjugated and subsequently purified downstream using HIC chromatography.

500563 The Role of Bacterial Outer Membrane Vesicles in Establishing an Ecological Niche

Abstract:

Gram negative bacteria constitutively secrete outer membrane vesicles (OMVs), which are small, membrane-bound, spherical vesicles that encapsulate a number of molecules including toxins and other virulence factors. These vesicles are hypothesized to play a role in the establishment of an ecological niche for the organism by delivering cargo to both eukaryotic and prokaryotic cells. In this project, we have studied the role of OMVs in the ability of an oral pathogen, *Aggregatibacter actinomycetemcomitans*, to colonize the host, by communicating with host cells as well as neighboring bacteria. Using dynamic light scattering (DLS) and scanning electron microscopy (SEM), we found that *A. actinomycetemcomitans* OMVs are approximately 85 nm in diameter and enriched in a protein toxin, leukotoxin (LtxA). LtxA is located entirely on the surface of the vesicle, bound to the negatively charged lipopolysaccharide (LPS) present on the surface of the vesicle through electrostatic interactions. We hypothesized that in this location, LtxA might act as a targeting agent, trafficking the OMVs to target cells in a cholesterol- and receptor-dependent manner. Because free LtxA must bind to both cholesterol and a specific receptor, lymphocyte function-associated antigen-1 (LFA-1) in order to kill target cells, we investigated the specific role of these two molecules in the binding and toxicity of *A. actinomycetemcomitans* OMVs. Our results, however, demonstrate that *A. actinomycetemcomitans* OMVs require neither cholesterol nor LFA-1 to induce toxicity. Additionally, we have discovered that unlike free LtxA, OMVs are not endocytosed by target cells; instead they fuse with the plasma membrane of the host cell to deliver their cargo directly into the cytosol. We have also characterized delivery of *A. actinomycetemcomitans* OMVs to other *A. actinomycetemcomitans* cells and found that the OMVs are efficiently taken up by the bacterial cells. The presence of OMVs appears to decrease the extent of bacterial growth, consistent with reports of OMVs containing quorum sensing molecules. We are currently investigating the delivery of OMVs to other periodontal inhabitants to study the role of OMVs in the establishment of the oral microbiome and biofilms. We anticipate that these results will facilitate the identification of new therapeutic targets aimed at preventing pathogen colonization within the oral microbiome.

497071 Engineering Glucose Binding Proteins with a Chemo-Enzymatic Tag for Glucose Detection in Exhaled Breath Condensates (EBC)

Abstract:

Despite recent advances in glucose biosensors and glucose detection, reliable measurement of glucose at the very low concentrations is highly desired. Glucose concentrations in non-invasive body fluids, such as saliva, sweat, and exhaled breath are only a fraction of those found in blood. In particular, we have found that exhaled breath condensate (EBC) contains a consistent fractional glucose concentration compared to blood glucose owing to the rapid glucose exchange between lung fluid and blood. However, this concentration is approximately 1/1500 of blood glucose concentration, as confirmed by our preliminary human trials conducted using a selective EBC condenser developed in our lab. While electrochemical sensors based on glucose oxidase (GOx) are the current gold standard, the lack of stability due to environmental interferences and the high dissociation constant of the enzyme prevent them from being able to resolve the minute changes in glucose at low concentrations [1]. The *E. coli* glucose binding protein (GBP) has a stronger and more specific affinity to glucose (K_D = 0.35 μM) compared to that of GOx (K_D = 20 mM) [2]. We propose to harness this protein to achieve a lower limit of detection and higher sensitivity via fluorescence-based competitive binding biosensor.

We take advantage of the competitive binding of glucose (K_D = 0.35 μM) and galactose (K_D = 1.4 μM) to GBP to develop a fluorescent biosensor that is quenched when bound to galactose-analog and fluorescent when bound to glucose. We engineer GBP to contain a biorthogonal azide tag which will allow conjugation to fluorophores, quantum dots, or another matrix of interest directly from cell lysate. This will enable the competitive detection of glucose via galactose-analog displacement upon interaction with EBC samples. Our chemoenzymatic tagging method [3] selectively and efficiently labels the N-terminus of the GBP with 12-azidodecanoic acid (12-ADA). We simultaneously develop a fluorescently conjugated galactose that quenches the fluorescence of the quantum dot when bound to GBP.

This presentation will discuss our initial results characterizing wt-GBP and 12-ADA-GBP binding to glucose and galactose. This work will also include assessing the binding capabilities of 12-ADA-GBP to TAMRA fluorophores via click chemistry directly from cell lysate. We find that our chemoenzymatic tagging method enables us to directly bind the 12-ADA-GBP from cell lysates containing overexpressed proteins to fluorophores of interest without any purification steps.

[1] Chen, C ., Q. Xie, D. Yang, H. Xiao, Y. Fu, Y. Tan, and S. Yao. "Recent advances in electrochemical glucose biosensors: a review." *RSC Adv.* vol. 3, pp. 4473-4491, 2013

[2] L. Tolosa and G. Rao, "The Glucose Binding Protein as Glucose Sensor," in *Glucose Sensing*, C. D. Geddes and J. R. Lakowicz, Eds. Boston, MA: Springer US, 2006, pp. 323–331.

[3] C. Kulkarni, M. Lo, J. G. Fraseur, D. A. Tirrell, and T. L. Kinzer-Ursem, "Bioorthogonal Chemoenzymatic Functionalization of Calmodulin for Bioconjugation Applications," *Bioconjug. Chem.*, vol. 26, no. 10, pp. 2153–2160, Oct. 2015.

497140 Engineering Novel "Designer" Glycopeptides in Planta As a Molecular Carrier for Directing the Accumulation of Recombinant Proteins/Enzymes

Abstract:

This project is aimed at leveraging a plant-specific, O-glycosylation process called hydroxyproline (Hyp)-O-glycosylation for de novo design and engineering of Hyp-O-glycosylated peptides (HypGPs) in plant cells that function as a molecular carrier in facilitating extracellular secretion of conjoined heterologous proteins/enzyme. Hyp-O-glycosylation involves post-translational Pro to Hyp and subsequent *O*-glycosylation, a modification that is unique to higher plants and green algae. Strategically designed Pro-rich HypGP modules engineered in plants can direct extensive *O*-glycosylation on each Pro residues, with HypGPs comprised of contiguous Pro residues, as in extensin-like "Ser-Pro-Pro-Pro-Pro" motifs are glycosylated with oligo-arabinosides; HypGPs comprised of non-contiguous Pro residues, as clustered in arabinogalactan protein (AGP)-like "Ser-Hyp-Ser-Hyp..." motifs, are glycosylated with highly branched arabinogalactan (AG) polysaccharides. In plant cell culture, dramatically increased secreted protein yields up to 1500-fold were detected when heterologous proteins, including a reporter protein GFP and many therapeutic proteins such as interferon α2, human growth hormone, interleukin-12 and stem cell factor, were expressed with a designer HypGP tag. Engineering a thermostable endoglucanase from *Acidothermus cellulolyticus* in tobacco plants with a designer HypGP tag could facilitate the deposition and stabilization of the enzyme in the cell wall matrix, leading to increased biomass saccharification efficiency by 3. 5-fold compared to wild type plants. The growth and biomass accumulation of the transgenic plants was not significantly impacted.

497945 Protein Detection Using Paper-Based Graphene Ink Biosensors from a Flexographic Proofer

Abstract:

A low-cost graphene ink-based biosensor was developed using a flexographic proofer with paper and non-paper substrates. Due to low-cost materials and an easy-to-follow device fabrication procedure, these biosensors are attractive as lab-on-a-chip devices for rapid diagnosis. These biosensors require no pumps for testing, and testing can be performed on a multimeter in a matter of minutes. A printing procedure was developed to fabricate biosensors using graphene inks from XG Sciences on which electrical resistivity measurements can be performed to detect bovine serum albumin (BSA) in phosphate buffer (PB). Various paper substrates including copy paper, chromatography-grade filter paper, and Tyvek were used with different size anilox rolls for printing. A polyethylene terephthalate (PET) substrate was used as a non-porous printing substrate for a control with protein detection tests. From electrical resistivity measurements on a Keithley 2400 SourceMeter, the graphene ink-copy paper biosensors were able to detect the presence of BSA in a PB solution. These biosensors can be further enhanced by functionalizing with anti-BSA to specifically bind and detect BSA in a non-ideal solution. Paper-based, graphene biosensors have the high potential to aid in the development of inexpensive biosensors that could prove useful in medical diagnostics in underdeveloped countries.

498937 Engineering of a Protein Probe for Alpha-Synuclein Detection

Abstract:

The aggregation of the amyloid, alpha-synuclein, has been implicated in the pathology of the neurodegenerative disorder, Parkinsonâ€™s Disease. Natively unstructured monomers of alpha-synuclein mis-fold and self-assemble to form prefibrillar oligomers, which are the most neurotoxic conformers. A system for the rapid and specific detection of these oligomeric intermediates would be beneficial towards uncovering the mechanism behind alpha-synuclein oligomerization, developing of therapeutic treatments, and the discovery of early diagnostic tools. Our group has developed a biomolecular probe, which retains the natural self-assembly of alpha-synuclein and incorporates a conformational fluorescence sensitivity to the biarsenical dye, FlAsH, in the presence of oligomers. A directed evolution approach was applied to optimize the probe and identify the best recombinant variants.

499167 Expression of Snake Antivenom Proteins with *B. Subtillis*

Abstract:

Bacillus subtillis is a promising organism for high expression levels and secretion of recombinant proteins. This project is exploring the use of *B. subtillis* as a potential host for a fab-like fragment of an opossum protein, DM43, that has been shown to be able to neutralize metalloproteinases from snake venom in mice. Another protein is also part of the project that is from the serum of the python (*Python reticulatus*). Upstream of the target protein gene sequence we have inserted a sequence that encodes for a piece of a cellulase that is expressed and secreted by *B. subtillis*. The presentation will discuss the expression strategy and the effectiveness of the expressed proteins as antivenom.

499200 Expression of Snake Antivenom Peptide Chain in *Pichia Pastoris*

Abstract:

The North American Opossum (*Didelphis virginiana*) is resistant to snake bite by US snakes from the crotalid family, such as the Western Diamondback Rattlesnake. The N-terminal peptide fragment of a serum protein from the opossum has been shown to be a potent inhibitor of venom toxins. The recombinant expression of the peptide is sought as a potential low-cost route to production of the antivenom peptide. We have expressed the peptide as a concatenated chain of peptides in *P. pastoris* and found good levels of the peptide chain secreted into the medium. The presentation will discuss the development of the *P. pastoris* expression host, the production process and the activity of the produced peptides.

496528 Computational Analysis of Solid Tumor Oxygenation Facilitated By Polymerized Human Hemoglobins

Abstract:

Purpose: In the U.S., over 60% of patients diagnosed with stage III and IV solid tumors will undergo chemo- and/or radio-therapy during the course of their treatment. A major constraint in chemo-and radio-therapeutic cancer treatment is inadequate oxygenation of solid tumors. Hypoxic conditions in the tumor microenvironment induce quiescence in cancer cells, which reduces the effectiveness of cancer therapies. Consequently, alleviating hypoxia in solid tumors is considered a promising target for improving the efficacy of anti-cancer therapeutics. Polymerized human hemoglobin (PolyhHb) can be transfused to increase solid tumor oxygenation and improve the efficacy of anti-cancer therapeutics. In this study, we analyzed the biophysical properties of synthesized PolyhHbs with low and high oxygen (O_2) affinity. By locking PolyhHb in the relaxed (R)-state, O_2 offloading at low O_2 tensions (<20 mmHg) may be increased, while O_2 offloading at high O_2 tensions (>20 mm Hg) is facilitated with tense (T)-state PolyhHb. Therefore, R-state PolyhHb may deliver significantly more O_2 to hypoxic tissues. Furthermore, transfusion of PolyhHbs may alter microvascular hemodynamics, which could improve O_2 transport into the tumor. We hypothesize that *in silico* models of the tumor microenvironment may be used to guide the dosage and type of PolyhHb as a function of tumor O_2 consumption and O_2 tension.

Methods: To yield low and high affinity PolyhHbs respectively, we first polymerized tense (T) and relaxed (R) state PolyhHb via glutaraldehyde. Clarification, purification, and concentration were each performed with tangential flow filtration. The diameter, cooperativity, O_2 tension at 50% saturation (p50), and rapid offloading kinetics of the synthesized PolyhHb were each analyzed. The resulting biophysical parameters were used to populate an O_2 diffusion model into the tumor tissue. Here, blood fluid flow was modeled with the Quemeda constitutive law and the radius of the red blood cell rich core was approximated from experimental data. Starling flow was modeled with Brinkmanâ€™s equation for flow through the blood vessel wall and tissue space. The O_2 equilibrium for all hemoglobin species was handled with the Hill Equation. In the simulation, we modeled PolyhHb delivery at various transfusion volumes. The inlet partial pressure of dissolved O_2 ($pO_{2,in}$) was varied from normal conditions (90 mm Hg) to extremely hypoxic conditions (1 mm Hg). The fluid velocity profiles, apparent viscosity through the lumen, wall shear stress, O_2 distributions in the tissue space, O_2 flux through the blood vessel, hypoxic volume were each analyzed. Sensitivity to environmental conditions were examined by varying the diameter of the arteriole, maximum rate of O_2 consumption, Michaelis-Menten Constant, and thickness of the tissue space.

Results: In general, we found that increasing the volume of PolyhHb transfusion decreased the apparent viscosity of blood in the arteriole. In addition, we found that PolyhHb transfusion decreased the wall shear stress at large arteriole diameters (> 20 Î¼m) but increased wall shear stress for small arteriole diameters (< 10 Î¼m). Both T- and R-state

PolyhHb transfusions may lead to elevated O_2 delivery at low $pO_{2,in}$. In addition, R-state PolyhHb transfusions may be more effective than T-state PolyhHb for O_2 delivery at similar transfusion volumes. R-state PolyhHb was less effective at maintaining O_2 delivery under normoxic conditions. The pO_2 pressure drop per unit length signifies that while the O_2 flux across the vessel wall is similar at high $pO_{2,in}$, there is significantly more O_2 lost at high pO_2s but more O_2 retained at low pO_2s. At low $pO_{2,in}$ the radius of the arteriole had the greatest effect on O_2 delivery. At high $pO_{2,in}$ the maximum rate of O_2 consumption had the greatest effect on O_2 delivery. Interestingly, R-state PolyhHb is much less sensitive to arteriole radius than T-state PolyhHb under hypoxic conditions (< 10 mm Hg).

Conclusions: Decreases in the apparent viscosity resulting from PolyhHb transfusion may result in significant changes in flow distributions throughout the tumor microcirculatory network. The difference in wall shear stress implies that PolyhHb may have a more significant effect on capillary beds through mechano-transduction. The increased O_2 flux and decreased pO_2 drop per unit length indicates that both PolyhHbs are suited to deliver O_2 under hypoxic conditions. However, the hypoxic volumes estimated here are inconsistent with the literature. This indicates that the assumptions in the Krogh tissue cylinder model may not adequately describe the tumor microenvironment. The system was more sensitive to changes in the tumor microenvironment than to the PolyhHb which indicates that a more descriptive model should be developed to match results from previous studies. Furthermore, the limited oxygenation results of this study may indicate that increases in oxygenation alone may not resolve tumor hypoxia. Therefore, future work should further evaluate the hemodynamic effects of PolyhHb in tumor capillary networks.

501790 Strategies for Quorum Sensing Inhibition inÂ *Staphylococcus aureus*

Abstract:

Many pathogenic bacteria regulate gene expression in response to cell density using quorum sensing. Quorum sensing is responsible for virulence in *Staphylococcus aureus* and requires the synthesis and exchange of compounds known as autoinducers. Disruption of the quorum sensing system, known as quorum quenching, can be achieved by degradation of autoinducers therefore making quorum quenching a potential non-antibiotic intervention against resistant pathogens. Here, we reconstitute *S. aureus* quorum signaling in non-pathogenic bacterial model hosts. We also present an inverter-type cellular circuit to facilitate high-throughput screening of potential quorum-quenching enzymes. Finally, we use directed evolution to drive mutations in enzymes towards increased performance in autoinducer degradation.

499260 Luminescent Nanoparticles for High-Throughput Microfluidic Droplet Barcoding

Abstract:

Molecularly-targeted therapeutics and personalized medicine have dramatically increased the prognosis of patients suffering from cancer. However, one major challenge facing cancer patients is the presence of drug-resistant tumor cells. This is further complicated by the high patient-to-patient variability due to the heterogeneous, personalized nature of cancer. As such, multiple therapeutic combinations have required the development of new analytical methods capable of multiplexed high-throughput screening (HTS) technologies that can quantify *ex vivo* samples like patient biopsies. The use of droplet microfluidic devices has garnered significant interest as a means to facilitate high-throughput, single cell analysis of heterogeneous populations like tumor biopsies. Additionally, the inclusion downstream trapping arrays can facilitate droplet collection and dynamic investigation of the single cell response. However, these devices are still limited in their to assess multiple inputs such as combinations of drugs or different doses. While some droplet barcoding techniques exist, they are limited by overlapping fluorescence spectra between barcoding fluorophores and biochemical fluorophores. In this work, rare earth (RE)-doped nanoparticles are coupled with droplet microfluidics for a novel approach at droplet barcoding. The Î²-hexagonal $NaYF_4$ nanoparticles are roughly 750 nm in diameter and are doped with rare-earth emitters (Tb^{3+}, Eu^{3+} or Dy^{3+}). RE dopants have been extensively employed in solid-state applications due to their stability and wide excitation/emission pathways, allowing for visible luminescence with UV and IR excitation. Here, we use three different dopants (Eu^{3+}, Tb^{3+}, and Dy^{3+}) incorporated into the nanoparticles which have a luminescence emission spectra in the red, green, and blue regions upon excitation with UV light. Using a droplet microfluidic trapping array, we have demonstrated that these particles are biologically inert and spectrally independent with common biological stains like GFP. Moreover, we have shown that the RE-doped nanoparticles do not interfere with fluorescent live and dead stains. To demonstrate the feasibility of these nanoparticles as droplet barcodes, we have co-encapsulated the nanoparticles with model cancer cell lines exposed to established therapeutic combinations to quantify concentration and time-dependent drug induced cell death. Thus, this work offers a novel antibody-free method for multiplexed droplet barcoding to facilitate high-throughput *ex vivo* screening of cancer therapeutics.

500039 Development of NIR-II Nanoparticle Contrast Agents for Photoacoustic Imaging

Abstract:

Developing technologies for cancer treatment and diagnosis is a vast yet very complex healthcare problem. New treatments are designed every day for each specific case of tumors, but cancer diagnosis technology has yet to keep up. This creates a situation where many treatments options are available but physicians are unable to select the best course of action due to limited information on tumor morphology. This lack of a fast, inexpensive, and safe method to image cancers remains the current bottleneck for effective therapies. The upcoming light-sound hybrid imaging modality, photoacoustic imaging (PA), is already recognized for being able to produce high resolution images without the need for expensive machinery and ionizing radiation. To maximize imaging depth and reduce signal attenuation, there is a drive to develop photoacoustic contrast agents that absorb in the Near Infrared II (900-2000nm) ranges. We here present the development of stable NIR-II active nanoparticles (NPs), utilizing the controlled self-assembly mechanism, Flash Nanoprecipitation (FNP). After screening potential NIR-II dyes, 10 distinct dye-encapsulated NPs were developed with sizes ranging from 20 to 180nm. These results shed light on the molecular requirements for nanoencapsulation into stable particles. NP sizes could be precisely tuned with co-core material, whereas the absorbance profiles remained constant. Thus, the addition of co-core material can be used to tune bulk properties and nanoparticle accumulation via EPR to tumors or clearance through the liver and spleen, while keeping the imaging properties constant. When kept frozen and in the dark, NIR-II dye NPs were determined to be stable for up to 7 days, and are thus suitable for long-term storage and use. While there was slight blue-shifts in the dyes' absorbance due to H-stacking after encapsulation, these particles exhibited strong NIR-II photoacoustic activity in the 1200-1300 nm region, where water and background signals are negligible. These particles represent the first successful incorporation of NIR-II active organic dye into NPs as contrast agents for PA imaging.

495493 Colorimetric Virus Detection Using Gold Nanoparticle Aggregation

Abstract:

Recent studies have shown that the plasmonic shift caused by the aggregation of gold nanoparticles (AuNPs) can be applied to virus detection. A typical biochemical detection involves either antibodies for capturing virus capsid proteins or complementary oligonucleotides for detecting virus nucleic acids. The antibodies and oligonucleotides are very expensive, have limited chemical stability, and can only detect one specific type of virus at a time. Therefore, the use of AuNPs red-shift to detect the presence of virus without the need of expensive antibody or oligonucleotides functionalization has drawn our interest.

Our model virus porcine parvovirus (PPV), can be non-specifically adsorbed to AuNPs to form a virus corona. The AuNPs coated in virus have a measurable aggregation when osmolyte is present as compared to AuNPs coated in model protein bovine serum albumin (BSA) and exposed to osmolyte. Osmolytes are natural, organic compounds that are found in many organisms. Our lab has shown that osmolytes preferentially flocculate both an enveloped and non-enveloped virus but not model proteins. In order to create a quick, sensitive, inexpensive and portable method for detecting the presence of viral particles, we are employing the preferential flocculation of virus particles with osmolytes in the presence of AuNPs.

To distinguish the aggregation of AuNPs with virus solutions from that with BSA, solutions at various ionic strengths and with different concentration of osmolyte solutions was explored. Glycine at 1 M can be used to distinguish the AuNP aggregation of coronas formed from PPV and BSA. Lower ionic strength buffer used for corona formation give the largest difference between BSA and PPV. In addition, the key parameters that increase the sensitivity and decrease the limit of detection of PPV will be discussed. This unique method can easily be transferred to other virus systems and be used in the detection of norovirus on cruise ships.

493242 Engineering Hepatitis B Viral-like Particles into Protein Delivery Vehicles

Abstract:

Nanodevices, which can specifically target diseased cells via molecular level signals and deliver high therapeutic payloads, have attracted considerable interest as next generation platforms for advanced disease therapeutics. A significant challenge in engineering these therapeutic nanodevices is incorporating a reliable release mechanism to enable the dispersion of the drug payload within the cellular environment after uptake of the nanodevice. The reducing environment of the cytosol has often been exploited to release molecules tethered through disulfide bonds; however, ability to form these reversible bonds *in vitro* is hindered by nondiscriminatory free cysteines and the formation of unwanted homodimeric byproducts. In nature, error in disulfide bond formation in proteins is abated by tertiary and quaternary structure, which stabilizes the correct orientation of the cysteine residues. Inspired by this, we have developed a genetically encoded protein module which uses affinity domains to drive the formation of disulfide bonds between the nanodevice and cargo. Furthermore, we have combined this technology with our previously developed Hepatitis B Virus-Like Particle (HBV VLP) nanocarrier platform to create a protein delivery device capable of selectively targeting cancer cells and releasing proteins into the cytosol.

497593 Production of Homogeneous Antibody-Drug Conjugates Using the Nucleotide Binding Site

Abstract:

Antibody-drug conjugates (ADCs) are a recent pinnacle of immunotherapeutic agents that enhance the antitumor effects of antibodies and reduce adverse systemic effects of potent drugs. To obtain an efficient ADC, a potent drug should be conjugated to an antibody while preserving antibody activity and high potency of the drug with an optimized drug-antibody ratio (DAR). Conventional ADCs are typically produced by conjugating a drug to an antibody through the side chains of either surface-exposed lysines or free cysteines generated through reduction of inter-chain disulfide bond. Many obstacles have been associated with the conventional drug conjugation methods to antibodies, such as aggregation of antibodies, high DAR yielding rapid *in vivo* clearance, heterogenous modifications causing weak performance of the ADC, and requirement of harsh chemical modification resulting in damaged antibodies. In this study, we describe a site-specific method for conjugation of drugs to monoclonal therapeutic antibodies using the nucleotide-binding site (NBS). We used a previously identified NBS ligand that has moderate binding affinity to the NBS located between heavy and light chains on the variable region of Fab fragment to conjugate a drug with a PEG linker to the antibody. We used the UV-NBS method that uses UV-energy to initiate a photo-cross-linking reaction between the NBS and the NBS-ligand, to obtain controlled site-specific conjugation of the drug to the antibody. We have assessed the effectiveness of the ADCs generated with the UV-NBS method utilizing a chemotherapeutic agent in combination with various pharmaceutical antibodies. We further characterized *in vitro* efficacy, potency, and the DAR of the ADCs as well as activity and sensitivity of the antibodies post-conjugation with the drugs.

499963 Arduino-Based POC System for the Diagnosis of Viral Diseases through on-Line Conductivity Measurement

Abstract:

The development of cost-efficient Diagnostic Point of Care (POC) systems is a research niche of high relevance; the recent pandemic/epidemic episodes associated with viral diseases (Influenza epi-centered in México 2009; Ebola in West Africa 2013-2015; and Zika in Latin America 2016) are tangible and cruel reminders of the need for portable, low-cost, and easy-to-use diagnostic systems that can effectively address epidemic episodes in remote or underprivileged areas. Here, we describe the development of a simple Arduino-based POC device for quantitative detection of viral pathogens. As a proof-of concept model, we demonstrate the detection, amplification, and quantification of Ebola virus (EBOV) genetic material.

We used Recombinase Polymerase Amplification (RPA), an isothermal amplification technique, for the specific amplification of a sequence of nucleotides that encodes a region of the Ebola GP protein (expressed in *Escherichia coli*). The use of an isothermal amplification technique, instead of PCR, greatly simplifies the complexity of the system. We show that simple temperature control at 37C in a small chamber allows the amplification of a DNA sequence (at a minimum concentration of 0.0322 ng/mL) in a short time (i.e., 10â€“20 minutes). An increase in resistivity of the amplification mix (from 10,000 to 600,000 ohms) within the chamber, associated with a decrease in the ionic strength related to the occurrence of the DNA/RNA amplification, is detected online by two electrodes connected to the Arduino.

Our platform exhibits the essential attributes of a POC system: (a) the use of small volumes, (b) low capital cost, (c) portability, (d) and a fast, accurate, and selective response.

497329 A Novel Diagnostic Liposomal Platform, Nanoallergen, for Clinical Evaluation of Epitope Immunogenicity of Peanut Allergen

Abstract:

Currently, several diagnostics have been developed for diagnosis of allergies such as Skin Prick Test (SPT) and ImmunoCAP specific IgE blood test. Although the existing diagnostics are able to determine whether patients are allergic to a suspected allergen, they are unable to determine the effect of specific epitopes present on allergen proteins on physiological allergic responses, leading to a high likelihood of inaccurate predictions of physiological allergy symptoms for patients. Since it has been shown that severe allergic symptoms are correlated with certain immunogenic epitopes, it is important to evaluate individual epitope immunogenicity for predictions of severe clinical allergic responses. In this regard, we developed a liposomal-based platform we call nanoallergens to evaluate epitope immunogenicity of the suspected allergen so as to determine which immunogenic epitopes play a critical role in severe allergic responses. Nanoallergens are a liposomal synthetic allergen that multivalently displays potential IgE binding epitopes of the allergen proteins. As experimental approaches, we synthesized IgE binding epitope mimetics derived from the major peanut allergen protein (Ara h 2) with a lipid conjugate, loaded them on the nanoallergen platform, and performed *in vitro* RBL cellular degranulation assays to evaluate epitope immunogenicity of the Ara h 2 protein for triggering degranulation responses using clinical samples of highly allergic patients. The nanoallergen study showed that certain immunogenic epitopes induced strong degranulation responses for the tested patient sera with clinical history of severe allergic symptoms, suggesting that nanoallergens allow for predictions of severe allergic reactions by correlating between immunogenic epitopes and severe allergic reactions. Furthermore, we performed basophil activation tests using clinical samples of a large population of severe allergic subjects to investigate whether the evaluated immunogenic epitopes by nanoallergen study are critical for severe physiological responses of peanut allergy patients. In conclusion, nanoallergens could potentially be used as a diagnostic tool for predictions of severe peanut allergic responses.

498003 The Discovery of Enzymatically Depolymerized Heparin Derivatives for the Treatment of Ulcerative Colitis

Abstract:

Heparin is a member of glycosaminoglycan and consists of variably sulfated repeating disaccharide units. Besides its well-known and fully exploited anticoagulant activity, heparin was found to possess many non-anticoagulant activities that were valuable for therapeutic use. Heparan sulfate(HS), very closely related in structure to heparin, is a ubiquitous glycosaminoglycan regulating a wide variety of biological activities including developmental processes, inflammation, coagulation, immunity and tumor metastasis. Based on the review of existing researches, the molecular size and sulfation pattern of heparin/HS both determine their non-anticoagulant efficacy. Accordingly, to maximizing the diversity of heparin structure, strategies of controllable enzymatic depolymerization using three types of heparinases with different substrate specificity and chemically selective modification of heparin molecule were combined. The enzyme kinetics of three heparinases towards different chemically modified heparins were studied, giving instructions to improve the process of enzymatically production of low molecular weight heparin derivatives. Therefore, a library of heparin derivatives with different molecular weights and sulfation patterns was designed and constructed. In spite of a heterogenous mixture of different polysaccharide chains, each species of the heparin library had distinct molecular weight distribution and defined structure, which is the fundamental prerequisite for structrue-activity relationship analysis.

The anti-inflammatory effects of the heparin derivatives were systematically screened using the animal model of DSS induced colitis in mice. The results indicated that an enzymatically depolymerized low molecular weight heparin derivative without anticoagulant activity had better colon-protective effects through the inhibition of cytokine secretion and macrophage recruitment. The degree of desulfation and de-anticoagulation of heparin is extremely important for the treatment of colitis in mice. Furthermore, enzymatically depolymerization of heparin with specific type of heparinase show benefits in ameliorating colitis in mice. The selected heparin derivatives also attenuated intestinal epithelial permeability according to the studies on a normal human colon mucosal epithelial cell line. Several molecular pathways involved in the ulcerative colitis with the treatment of the potent heparin derivative were depicted, and the structure factors of heparin influencing anti-colitis activity were thoroughly analyzed.

499879 Methods for Development and Characterization of DNA Polymerase Based Bio-Recorders

Abstract:

Biosensors engineered to respond to custom input functions are invaluable tools for reporting changes in the physiological states of cells. Such biosensors are genetically encoded, making their delivery and induction precisely controllable. Taking biosensing a step further, we intend to develop a DNA polymerase-based *in vivo* cation bio-recorder. Such a nano-scale device would have an altered error rate when the concentration of a cation of interest changes in its environment. We are currently using neural firing as the ideal system where we would employ such a genetically encoded bio-recorder, using it to write the temporal trace of neural activity as error patterns onto DNA molecules. This would enable us to uncover neural interactions with single neuron precision. Along those lines, here we report methods we have established for the development and characterization of such a device. We first used rational design to make the processivity of the distributive Y-family *Sulfolobus acidocaldarius* DNA polymerase, *Dbh,* suitable for a neural recording application. We then developed methods to characterize the fidelity of the engineered *Dbh* under neural firing conditions and compared it with the fidelity of the wild-type *Dbh*. We intend to heavily rely on the tools established here to reach the final goal of an engineered DNA polymerase with kinetic parameters (error rate and processivity) fitting their bio-recording application in a neural environment.

498121 Selection and Affinity Enhancement of Alpha-Synuclein-Specific Single Domain Antibody Using Experimental and Simulation Techniques

Abstract:

The ability to modulate protein-protein interactions is of great interest to both fundamental biological science and applied research such as drug design. While many studies have demonstrated the ability to use structure-based information to rationally improve protein-protein interactions, the rational design of target-specific binders is still an unsolved problem. In this context, single domain antibodies (VHHs) are promising designable molecules, given their small size, stability, and ability to bind surfaces of any shape. Alpha-Synuclein (AS) is a natively disordered protein, implicated in the pathogenesis of Parkinsonâ€™s disease (PD) and related neurological disorders. The Non-Amyloid Component (NAC) region of AS has been previously targeted for inhibiting the aggregation and/or reducing toxicity of AS. Various methods such as yeast surface display, phage display and, more recently, rational design approaches have been used for generating antibody fragments for this purpose. In this work, we have pursued a dual modeling-experimental approach for rationally designing an AS-specific VHH.

Starting from an immunized Camelid library against the NAC region of A53T mutant of AS (A53T), a single-round of a bacteria-based selection technique was used to obtain a NAC-specific VHH. Atomistic models of the VHH and the VHH-NAC binding were constructed using a combination of Molecular Dynamics techniques (Replica Exchange Molecular Dynamics, Umbrella Sampling) and numerical calculations (Finite-Difference Poisson-Boltzmann Equation). Furthermore, affinity enhancing mutations based on computational models were proposed and tested using experimental techniques such as ELISA and Surface Plasmon Resonance (SPR). Using this dual approach of prediction and experimental verification, we were able to enhance the affinity of the VHH by an order of magnitude. Hence, our work demonstrates the ability to develop a predictive in-silico model of binding for the VHH-NAC system. This approach can be especially powerful for targeting natively disordered and weakly immunogenic antigens such as AS, for which, crystal structures are not readily available and experimentally validated computational models can be leveraged for rational design and affinity enhancement.

499754 Impact of Linker Attachment Site on Structure and Dynamics of Enzymes

Abstract:

Immobilized enzymes have several applications in a broad range of industries, such as, food and beverages, production of biofuels, biosensors, textiles, and pharmaceuticals. Enzymes are immobilized on surfaces by covalent linkers to enhance their stability and consequently their lifetime. These stabilizing effects of enzyme immobilization often come at the expense of specific enzyme activity. This is due to the complex interplay of enzyme-linker-surface interactions, mass transfer limitations, and immobilization related enzyme unfolding. Accordingly, current immobilization methods are limited to certain enzyme-surface systems. An understanding of the molecular interactions between the enzyme-linker-substrate for a wide range of systems is therefore necessary to generalize immobilization approach.

We use combined experimental and simulation techniques to develop such a widely applicable immobilization strategy. We first explore the role of linker molecules on the catalytic activity of enzymes. Our hypothesis is that given identical linker attachment chemistry, the linker effect on enzyme activity is primarily related to the structural changes and steric hindrance caused by enzyme-linker interactions. In this regard, we perform all-atom molecular dynamics simulations of modified T4 lysozyme (WT*) and its cysteine mutants in the presence and absence of succinimidyl polyethylene glycol ester (SM (PEG)2) linker molecules. This combination of linker chemistry and cysteine mutants provides an absolute certainty on the linker attachment site. The simulations are performed on selected cysteine mutants that are experimentally observed to be either influenced or not influenced by linker attachment. We compare the structure and dynamics of the WT* and its variants in the presence and absence of linker molecules. In our presentation, we will discuss these results and comment on the strategies to attach linker molecules with less impact on structure and dynamics of enzymes.

491151 Building Disulfide Bonds Between Subunits to Improve the Stability of Nitrile Hydratase

Abstract:

Stability is an indispensable property of enzymes for industrial applications. Nitrile hydratase (NHase) with two subunits is a vital biocatalyst for commercial bioproduction of acrylamide. The bioconversion process shows many advantages such as mild reaction conditions, low energy consumption, high conversion rate and product purity. However, the NHase was not stable enough in the late catalytic process because of the high concentration acrylamide immersion. Enhancing stability of NHase is thus urgently required in industry. Generally, the tertiary structure of protein is maintained by noncovalent interactions such as hydrophobic interactions, hydrogen bonds and salt bridges, which are all relatively weak interactions. To significantly improve the stability of NHase, covalent disulfide bond is specifically highlighted in this study. By using the Disulfide by Design software, 9 disulfide bridges between the two subunits were designed. And corresponding mutant enzymes were constructed with genetic engineering tools. In the host of *Rhodococcus ruber* TH3, one NHase mutant, 23-307, displayed significant enhanced activity (3915 U/mL), 8.2% higher than the original one, although the other mutants completely lost their activities due to the conformation changes. The in vivo stabilizing effect of the *R. ruber* TH3/23-307 cells harboring the mutant NHase was assessed using both heat shock and acrylamide immersion experiments. While immersing the cells of *R. ruber* TH3/23-307 and the original *R. ruber* TH3/NHase at 60°C for 10 min, 52.5% of NHase activity remained in the mutant *R. ruber* TH3/23-307, but this value dropped to 29.4% in the original control. The deactivation behavior of *R. ruber* TH3/23-307 against acrylamide immersion was conducted in a solution with dynamically increased acrylamide concentrations. When the acrylamide concentration was increased to 150 g/L, the remaining NHase activity in *R. ruber* TH3/23-307 was as high as 92%, compared with 37% of that in the original *R. ruber* TH3/NHase. The hydration synthesis was further conducted to produce 500 g/L acrylamide with continuous acrylonitrile feeding. The bacteria were recovered through centrifugation for multiple batches of hydration after the acrylamide concentration reached 500 g/L. Results showed that the cells of *R. ruber* TH3/23-307 could complete three batches production of 500 g/L acrylamide, but the cells of original *R. ruber* TH3/NHase could only complete one batch.

499832 Engineering Bioresponsive Materials from Recombinant Oleosin

Abstract:

Oleosin is a tri-block structural protein that helps stabilizing oil bodies in plant cells. The wild type oleosin found in sunflower seeds is around 200 amino acids long and has a hydrophobic block in the center, with two hydrophilic arms at the N- and C- termini. This work explores the multiple applications of functionalized oleosin, with the help of molecular biology.

Targeting nano-particles for drug delivery has great potential for improving efficacy and reducing side effects from systemic toxicity. In particular, the RGD motif has been shown to effectively target the Î±$_v$Î²$_3$receptor, which is overexpressed in breast cancer cell lines. In addition, new developments in the assembly of materials afford the opportunity to expose cryptic targeting domains in tissue-specific microenvironments in which certain proteases are expressed. Here, recombinant oleosin are designed to combine the responsiveness to environmental proteases with specific targeting. Materials made recombinantly allow complete control over amino acid sequence, in which each molecule is identically functionalized. Previously, oleosin, a naturally occurring plant protein that acts as a surfactant, has been engineered to self-assemble into spherical micelles - a useful structure for drug delivery. In order to make oleosins that are locally activated to bind receptors, oleosin is genetically modified to incorporate the integrin-binding motif RGDS just behind a domain cleavable by thrombin. The resulting modified oleosin self-assembles into spherical micelles in aqueous environments, with the RGDS motif protected by the thrombin-cleavable domain. Upon the addition of thrombin, the RGDS is exposed and the binding of the spherical micelles to breast cancer cells is increased 4-fold. The strategy of combining tissue-specific protease cleavable domains with different adhesive domains is modular, wherein numerous different protease/receptor pairs could be combined to optimize targeting in specific diseases. Current work includes incorporating multiple ligand combinations such as PHSRN and RGDS, and enhancing cellular uptake by incorporating cell-penetrating peptides such as Tat-peptide.

500558 Exploiting the PAF Receptor to Target Infectious Diseases in the Lungs

Abstract:

Many bacterial pathogens that cross cell membranes in the lungs express lipooligosaccharides (LOS) laden with phosphorylcholine (CHoP) moieties. Our lab has demonstrated that these CHoP-containing LOS interact with the PAF receptor on lung cells. Previous students have explored using CHoP-containing LOS as targeting ligands for particles, but there is an inherent toxicity in these difficult-to-purify bacterial extracts. Currently, we are approaching this challenge by using CHoP-containing synthetic polymer coatings as a means of imbuing our particles with targeting to the PAF receptor for targeted uptake. We are investigating copolymer systems, so we need an easily-modified platform for study. Thus, we have designed our polymers with thiol end-groups to form self-assembled monolayers onto gold nanoparticles. By modifying chain length and ion content in these polymers we modify chain behavior and charge effects of these particle coatings. These are tools that we use to modify the way these particles interact with proteins and cells to enhance their ability to effectively carry drug and gene payloads to targeted areas. The development of systems that can target specific cell or tissues is a major goal in drug delivery, and biomaterials that minimally interact with proteins in the body are of additional value in a variety of medical devices.

501039 Interaction of Multiple Drops and Formation of Toroidal-Spiral Particles

Abstract:

In the development of drug delivery technologies for treating complex diseases, encapsulating multiple compounds and manipulating their sustained-release kinetics independently (for optimal therapeutic effect) can be challenging. We previously developed a fluid-dynamic technology based on multiple-drop interactions to achieve this goal. During sedimentation in a viscous liquid, polymeric droplets self-assemble into a reproducible and controllable toroidal-spiral (TS) structure, which could be solidified into solid particles by photo-initiated cross-linking of the polymer. The goal of encapsulating multiple drops featuring different physical properties generally requires complicated and time-consuming experimental iteration on the starting conditions, because all satellite drops (containing drugs) must catch up and coalesce simultaneously with the host drop that forms the surrounding matrix upon solidification. To guide laboratory trials and optimize conditions, a computational simulation is needed to accurately predict the configurations and interfaces of the multiple drop phases, especially when each drop comprises different physical properties. Our creeping-flow numerical simulations of drop sedimentation and interactions closely resemble the experimental shapes of single and multiple, interacting drops, and also reveal the internal interfaces and structures. Furthermore, understanding fundamental hydrodynamics of multiple drops could lead us to potential scale up production of TS particles and impact on mixing and printing in general.

498169 Formation of Aggregates in Perfluorocarbon-Based Oxygen Carriers WHEN Diluted on Plasma Expanders

Abstract:

Perfluorocarbon (PFC) based oxygen carriers supply the blood function of respiratory gases transport. PFC emulsions do not have a component that generate oncotic pressure, thus it is mixed with a plasma expander. However, it generates aggregation. The objective of this study was to evaluate the influence of plasma expanders and main electrolytes in the emulsion aggregation.

Three phases of experimentation were carried out. First, the effect of mixing the PFC emulsion with hydroxyethyl starch (HES) solution was evaluated. Second, the physicochemical properties of the mixture with different solutions of chlorine, phosphate, potassium and sodium were studied. Third, the interaction of bovine serum albumin (BSA) and human serum albumin (HSA) with components of the PFC emulsion was analyzed.

First, stronger aggregation was observed at higher concentrations of HES, which was attributed to the amount of electrolytes in the mixture. This was corroborated in the second phase, where it was determined that the electrolyte charge is the factor that affects the emulsion stability causing aggregates and particle size increase. Third, it was evidenced a different behavior of the emulsion in the presence of BSA and HSA, the second one being an aggregate forming substance.

In conclusion, it was determined that the plasma expanders and the electrolyte ions charge generate aggregates when they are diluted in PFC emulsion. Additionally, the chemical structure of the HSA interacts with the egg yolk lecithin structure (emulsifier) increasing the particle size. Further, it is known that electrostatic charges also destabilize plasma proteins, for this reason, our future work include studying the mixture of blood, plasma expander and PFC emulsion.

489387 Structural Studies of Protein Based Nanoparticle Synthesis

Abstract:

Protein and peptide directed synthesis of inorganic nanomaterials offers several advantages over traditional synthesis methods; however, despite being used to great effect, the fundamental mechanisms for how biomolecules control and drive nanomaterial synthesis is not well understood. We are using high-resolution structural biology techniques on proteins that are able to synthesize gold nanoparticles to investigate how biomolecules are able to influence inorganic nanomaterial nucleation and growth. In this presentation we will present our functional studies of this system as well as gold bound protein crystal structures we have determined of this system thus far.

492814 Probing the Influence of Sibling Proteins on Collagen I Fibrillogenesis and Denaturation

Abstract:

Native bone tissue is comprised of a mixture of collagen, non-collagenous proteins, and hydroxyapatite (HA). Additionally, the SIBLING (small integrin binding, N-linked glycoprotein) family of proteins is the primary group of non-collagenous proteins found in mineralized tissue. The SIBLING family of proteins have all been demonstrated to contain a cell binding RGD sequence, one or more HAP or calcium binding domains, and at least one collagen binding domain. There have been extensive studies to date examining the cell binding and mineralization induction capabilities of this family of proteins, but few if any studies on the role of these proteins in collagen fibrillogenesis. In this study, the influence of six SIBLING proteins and two proteoglycan modified SIBLING proteins on both collagen fibrillogenesis and denaturation are investigated. The results suggest that dentin phosphoprotein, dentin sialoprotein, and the proteoglycan modified dentin sialoprotein have the largest impacts on the native fibrillogenesis process. When combined with the existing literature, these results provide further insight into the natural biological function of the SIBLING family of proteins during bone development.

500611 Oxidative Modification of Peptoids Utilizing Bleach and TEMPO As Green Chemistry Catalysts for Protein Therapeutic Applications

Abstract:

Biotherapeutic drugs, derived from biological molecules such as proteins and DNA, are becoming an integral and exceptionally critical aspect of modern medicine. Compared to common pharmaceutical drugs, biotherapeutics are much larger in size and have greater target specificity, allowing them to treat many chronic diseases ranging from cancer to rheumatoid arthritis. The major issue with protein based therapeutics is that they readily undergo proteolysis, or enzymatic degradation, when administered through subcutaneous injections. Traditionally, biotherapeutic modification procedures have centered on the use of PEG derivatives. This process, called PEGylation, is unfavorable due to the increases in molecular weights of the proteins and the heterogeneous mixture of products formed. Instead of PEG derivatives, we propose peptoids with N- methoxyethylglycine (NMEG) side chains to decrease proteolysis. NMEG groups are more advantageous than PEG derivatives due to their low molecular weight and ability to form homogeneous products. Our work focuses on increasing the protease resistance of target biotherapeutic proteins by cross-linking a NMEG-5 peptoid to a cytochrome c via reductive amination. In the presence of a reducing agent, an imine bond is formed through the reduction of the peptoid's aldehyde group and cytochrome c's primary amine groups. Due to the expensive and unstable nature of commercially available aldehyde side chains, a green chemistry method, using only sodium hypochlorite (bleach) and 2,6,6-Tetramethylpiperidinoxy (TEMPO, free radical), oxidized the peptoid's hydroxyl group into the desired aldehyde for cross linkage. By incorporating the NMEG peptoid, we drastically increased the biotherapeutic's resistance to enzyme degradation, ultimately suggesting that NMEGylation is a capable method of modification to improve efficacy of biothereapeutic proteins.

501137 Resonant Soft X-Ray Scattering of Proteins in Solution

Abstract:

Protein structure is crucial for biological function and may suggest important parameters for the design of therapeutics and enzymes. We propose resonant soft X-ray scattering (RSOXS) as an approach to study proteins and other biological assemblies in solution. Calculation of the scattering contrast suggests that soft X-ray scattering is more sensitive than hard X-ray scattering, because of contrast generated at the absorption edges of constituent elements such as carbon, nitrogen, and oxygen. As a proof-of-principle, we have examined the structure of bovine serum albumin (BSA) in solution by RSOXS. We find that by varying incident X-ray energies, we are able to achieve higher scattering contrast near the absorption edge. From our RSOXS scattering result we are able to reconstruct the structure of BSA in three-dimensions. The RSOXS results also agree with hard X-ray experiments, including crystallographic data. Our study demonstrates the potential of RSOXS for studying protein structure in solution and highlights advantages of this technique over traditional X-ray scattering.

501824 Heterologous Expression of Highly Specific Antimicrobial Peptides in Probiotic E. coli

Abstract:

Antimicrobial peptides (AMPs) present an interesting alternative to conventional antibiotics because certain classes of AMP are very specific for particular genera of bacteria. Thuricin CD is an antimicrobial peptide that has been found to be specific to *Clostridium* and *Listeria* genera. Here, we demonstrate expression of this complex heterodimeric antimicrobial peptide in *E. coli* hosts and test the efficicacy against *Listeria innocua,* a non-infectious model organism. Efficacy is evaluated using a well diffusion assay and minimum inhibitory concentration assay. In addition , we study the transport mechanism and post-translational modification of Thuricin CD that influences the performance of mutants in our model.

495259 Engineered Bacterial Biosensor to Detect Endocrine Disruptors

Abstract:

Hormone disrupting compounds are pervasive. They are found in plastics, pesticides, and medications, among other sources. Though the full impact of these compounds has not been fully elucidated, these chemicals are linked to a variety of health problems, including diseases such as diabetes and cancer, obesity and infertility. We have developed a detection strategy based on electrochemical impedance spectroscopy to quantify endocrine disrupting compounds that is both fast and portable. Our method requires no specialized skills to perform, and, rather than responding to individual chemicals, this system reports the total estrogenic activity of a sample. We have engineered E. coli to surface-express a native estrogen receptor construct. This approach will be broadly applicable to the detection of chemically dissimilar classes of compounds that bind to a single receptor.

501710 Systems Biology Analysis of Natural Biomass Utilization Microbiomes for Biotechnology Applications

Abstract:

Many natural biomass utilization systems (NBUS) can efficiently degrade ligocellulosics. Typical NBUS include microbiomes for efficient biomass degradation, such as the microbiota in bovine rumen and herbivorous insect guts. We have carried out comparative systems biology analysis of multiple microbiome systems in insect guts and cattle rumen. Metagenome analysis of the gut symbionts of three different insects specialize on different food types was conducted as a means of comparing taxonomic and metabolic diversity of gut microbiomes to diet and life history of the insect hosts. A second goal was the discovery of novel biocatalysts for biorefinery applications. The comparative analysis revealed dramatic differences among the three insect species in the abundance and taxonomic composition of the symbiont populations present in the gut. The composition and abundance of symbionts was correlated with their previously identified capacity to degrade and utilize the different types of food consumed by their hosts. The metabolic reconstruction revealed that the gut metabolome of cutworms and grasshoppers was more enriched for genes involved in carbohydrate metabolism and transport than wood-feeding termite, whereas the termite gut metabolome was enriched for glycosyl hydrolase (GH) enzymes relevant to lignocellulosic biomass degradation. Moreover, we cloned and further characterized multiple biomass-degrading enzymes and established that the grasshopper symbiont enzymes were generally more efficient in biomass degradation than the homologous enzymes from cutworm symbionts. Together, these results demonstrated a correlation between the composition and putative metabolic functionality of the gut microbiome and host diet, and suggested that this relationship could be exploited for the discovery of symbionts and biocatalysts useful for biorefinery applications. In addition to insect guts, we have carried out comprehensive proteomics analysis of microbiomes in cattle rumen to study the molecular mechanisms of biomass deconstruction and utilization. We have included four treatments of high fiber, low fiber and medium fiber diet in the treatments in a 72 day feeding trial. Metagenome sequencing was first carried out to derive the gene models for proteomics data search. Enzyme assays revealed that the cellulytic enzyme activities are higher in the high fiber biomass treated time point. MudPIT-based shot-gun proteomics was carried out to characterize both the soluble and fiber portion of the rumen material. The proteome profiling highlighted that microbial diversity and enzyme profile change dramatically in response to the feeding content. Furthermore, cluster analysis and protein co-regulatory network modeling of protein abundance revealed that several groups of enzymes are coordinatively regulated in response to the high fiber biomass feeding. Each group of these coordinative enzymes includes multiple glycoside hydrolase (GH) families and functional category, indicating the synergistics toward biomass deconstruction. More importantly, the protein profiling correlates with the metagenome analysis in a way that some co-regulating enzymes were found in the same operon in metagneome sequences. The results highlighted both the regulation mechanisms and the reliability of proteomics data. Based on the network modeling, we have cloned and characterized multiple enzymes toward reverse design of a biorefinery process for efficient biomass utilization. Overall, the systems biology analysis revealed the potential mechanisms for efficient biomass degradation in various biomass-degrading microbiota, which can be used to guide the enzyme discovery and biorefinery design.

490010 Metabolic Pathway Engineering in Mammalian Cells through Kinetic Model Optimization

Abstract:

Fast growing cells, including stem cells, cancer cells, and many immortal cell lines, have a metabolic behavior known as aerobic glycolysis, characterized by a high rate of glycolysis and lactate production. This is in contrast to the metabolic behavior of many quiescent cells, which consists of a lower glucose consumption, low lactate production, and a larger reliance on oxidative phosphorylation for energy production. Altering the metabolic phenotype of aerobic glycolysis has been an active area of research in developing cancer therapies. In industrial cell culture, reducing lactate production resulting from rapid growth has been shown to enhance productivity. A novel strategy to alter cellular metabolic behavior may advance cancer therapy and cell culture bioprocessing.

The metabolic reaction network is highly nonlinear due to complex allosteric feedback and feedforward controls and different combinations of isoenzymes. A detailed kinetic metabolic model thus has a large parameter space with many possible combinations of enzyme kinetics and expression. This necessitates a method to intelligently search the parameter space for key changes which can be made to manipulate metabolism, as well as other biological reaction networks to specific ends.

A promising approach for deciding which enzymes and genes to experimentally manipulate is to embed a kinetic model into an optimization problem in order to target desirable modifications to metabolic behavior. We employed a kinetic model of metabolism which encompasses glycolysis, the pentose phosphate pathway, and the TCA cycle with the known allosteric regulations.

The model was solved using a local optimizer embedded in General Algebraic Modeling System (GAMS) to find the combinations of parameters that enable the target metabolic behavior with stated constraints. A multi objective optimization problem is also constructed to demonstrate the metabolic tradeoff between rates of glucose consumption and lactate production. Additionally, a penalty is imposed to changes from the original state incurred in the optimization process; the penalty increases with increasing magnitude of genetic alteration. This penalty also provides insight into the variables which have the most impact on the objective function. Varying the weight of this penalty allows for improvements in target metabolism to be balanced by the number of desired changes to the system, limiting the complexity of resulting experiments. The optimization schema was first applied to reduce lactate production while maintaining the energetic requirements for growth by allowing the expression level of multiple enzymes to vary. Then, the effect of knockdown or amplification of genes in the pathway were evaluated, and combinations of genetic alterations that can steer the metabolic behavior from the aerobic glycolysis type to the target type were identified.

Here we have demonstrated a method for rationally guiding cellular engineering through optimization. By identifying combinations of parameter changes which yield desirable outcomes, optimization of kinetic models can greatly reduce the effort required to engineer cell metabolism and provide deep insight into the reaction networks and their behavior.

499057 Comparative Transcriptomics Analysis Pipeline for a Customized CHO Microarray Platform

Abstract:

Comparative transcriptomics analysis, which has been enabled by microarray and emerging RNA-Seq profiling technologies, provides a powerful tool to understand different phenotypes in bioprocessing related cell lines at the gene expression level. To translate large transcriptomics data sets into biologically relevant information that can be leveraged for cell line development and process engineering, multiple tools that span the spectrum of raw data processing, gene expression analysis, pathway analysis, and data visualization are required. However, a seamless integration of these tools into an automated pipeline that is specifically designed for analyzing transcriptomics data from CHO cells, the leading cell line for commercial recombinant protein production, is lacking. Moreover, commercial software are usually not structured to conveniently enable CHO pathway analysis and interoperation. To address this gap, we have developed an automated pipeline in R which leverages public domain statistical methods to analyze gene expression data from a customized microarray platform for proprietary CHO cell lines. The pipeline not only tests the significance of difference between two conditions, but also identifies differentially expressed genes as well as differentially expressed pathways with multiple intuitive data analysis visualizations as outputs. The utility of this pipeline was demonstrated by comparing the transcriptome profiles of recombinant protein expressing CHO cells before and after a bioreactor process temperature shift. In addition, we used this pipeline to analyze microarray data where CHO cell lines with different recombinant protein expressing capacities were compared. Moreover, to justify the utility of the customized microarray based comparative analysis pipeline for CHO cells, we compared the outcome from microarray based analysis with the results from RNA-seq based analysis. With similar differential pathway analysis results and interpretation, the majority of differentially expressed genes identified in microarray analysis were also recognized as differentially expressed in the RNA-Seq analysis. We hope an automated microarray based comparative transcriptomics analysis pipeline such as the one presented in this study will shorten the time required from acquiring data to biological interpretation and can help accelerate the introduction of mechanism-driven approaches for cell line and bioprocess optimization.

494994 An Efficient Brownian Dynamics Approach for Modeling Multivalent Ligand-Receptor Assembly in the Cell Membrane

Abstract:

Multivalent ligand-receptor assembly is a key feature of many cell signal transduction systems. Nonetheless, the underlying mechanisms of ligand-receptor assembly in the plasma membrane of cells remain poorly understood. Current modeling and simulation techniques provide limited capabilities for the spatiotemporal modeling of multivalent species (molecules or complexes). A key barrier to modeling such systems is the multiscale nature of the problem. Multivalent molecules may interact via their subnanometer-scale binding features (domains or motifs). On the other hand, the time and spatial scales of assembly and diffusion of macromolecular species could be relatively large. In this work, we present a time-adaptive Brownian Dynamics (BD) simulation algorithm that enables efficient modeling of diffusion, reaction, and assembly of multivalent species. Combined with agent-based modeling, the BD approach provides a coarse-grained description of the structural details of multivalent molecules and their site-specific features. It enables motion of the multivalent molecules and their complexes due to lateral diffusion and rotation. The BD algorithm adaptively adjusts step sizes to capture the subnanometer scale site-specific interactions of the molecules and complexes, while enabling fast computation. We demonstrate these capabilities by developing models on multivalent antigen-receptor interactions in the plasma membrane of immune cells. Using the models, we provide analysis on the computational performance of the simulation approach both in the concentrated and dilute regimes. The models predict the distribution of receptor aggregate sizes in response to antigen molecules of variable valencies and structures. Consistent with experimental reports, the models also reveal and explain self-limiting antigen-receptor aggregation in the cell membrane.

496062 A Deep Learning Framework Decodes Coordination of Microbial Metabolism Under Genetic and Environmental Perturbations

Abstract:

One of the main goals of systems biology research is to decode the complex relationship between an organismâ€™s genotype and phenotype. A key method that has been used extensively towards this end is inferring regulatory structure and active regulatory events from biological (omics) data. Metabolism ties extracellular information with intracellular signals in a complex regulatory network made up of modular regulatory circuits built on a small set of metabolites. It has been hypothesized that metabolic regulation exists as a nested loop of interacting modules with local and global levels of regulation. Key challenges identified with omics data integration include sparse and disparate datasets, computational complexity, and difficulty in deducing biological meaning from the integration.

We present a novel framework, DeepOmics, that attempts to solve these challenges by uniquely integrating data-driven and constraint-based techniques. DeepOmics is a specially designed multi-hidden-layer neural network, in which each neuron and its associated weights and biases are designed to mimic actual biological regulation. Thus, DeepOmics goes beyond mere prediction of phenotypes but also seeks to uncover regulatory interactions between genes, gene products and the associated metabolic network. This is achieved by leveraging machine learning techniques of deep learning and transfer learning in conjunction with well-established constraint-based methods.

One key application of our framework, that distinguishes it from previous attempts, is the characterization of the efficiency of metabolic engineering strategies. DeepOmics builds on established computational strain design algorithms to uncover influential factors (such as medium composition, cultivation modes, pathway steps, and genetic engineering strategies) that are critical to the principal metrics of effectiveness: yield, titer, and rate. In this way DeepOmics enables efficient screening of metabolic engineering designs based on decades of data from published literature.

We have tested DeepOmics on the central metabolism of the microbe *Escherichia Coli*. Our framework recapitulates known biology and generates testable predictions about cellular regulatory interactions involved from genotype to phenotype. DeepOmics can provide valuable insights that will enhance applications in biotechnology, human health, and environmental remediation

References

1. Machado, D., Herrgard, M. J. & Rocha, I. Stoichiometric Representation of Gene-Protein-Reaction Associations Leverages Constraint-Based Analysis from Reaction to Gene-Level Phenotype Prediction. *PLoS Comput. Biol.* **12**, 1â€"24 (2016).
2. Gerosa, L. et al. Pseudo-transition Analysis Identifies the Key Regulators of Dynamic Metabolic Adaptations from Steady-State Data. *Cell Syst.* **1**, (2015).
3. Chubukov, V., Gerosa, L., Kochanowski, K. & Sauer, U. Coordination of microbial metabolism. *Nat. Rev.*

Microbiol. **12,** 327â€"40 (2014).

492961 Predicting Metabolic Disruptions Due to Heterologous Pathway Expression

Abstract:

Synthetic biology and metabolic engineering have been instrumental in transforming organisms such as *E. coli* and yeast to microbial factories to produce a wide range of commercially useful biomolecules by introducing non-native synthesis pathways in microbial hosts. While the general consensus has been, that enzymes are highly selective biocatalysts, recent evidences suggest that enzymes are often promiscuous. We hypothesize that such promiscuity in heterologous enzymes could promote metabolic disruptions in the expression host by catalyzing unwanted or unexpected reactions. Current computational synthesis methods like *PathPred* and *ProPath* can identify pathway(s) to direct host metabolites to desired end-products but cannot predict metabolic disruptions incurred by overexpression of recombinant pathways. In this work, we developed a computational method to assess the disruptive impact of synthesis pathways to microbial hosts. The steps in this predictive algorithm are as follows. First, we use PROXIMAL, an algorithm that predicts structural modifications to reactants by enzymes, generating potential byproducts that could be synthesized due to promiscuous activities of the heterologous enzymes. Second, generated products and associated reactions are incorporated into the underlying host network model as new nodes and edges. Third, yield is evaluated based on the augmented host. We quantify the metabolic disruption introduced as a change in maximum attainable product yield. We also explored the correlation of disruption with enzymatic choices and network topology. This work is novel as it is the first to examine the question of enzyme promiscuity in relationship to synthesis pathways.

493118 Maximum Entropy Prediction of Distributions for Stochastic Biochemical Reaction Networks with Oscillatory Dynamics

Abstract:

Ever since the Belousov-Zhabotinsky discovery, and with the molecular revolution in biology, an increasing number of experimental papers have been published on a variety of biological systems that exhibit oscillatory behavior (Novák and Tyson, 2008). In systems such as regulatory gene networks, cell cycle, and enzymatic processes, the number of molecules involved is often far from the thermodynamic limit. Therefore, the internal noise or the inherent stochasticity for such systems has to be taken into account while modeling these reactions (Gillespie, 2007). Although stochastic models based on the probabilistic approach of the Chemical Master Equation (CME) have been proposed, studies in the literature have been limited by the challenges of solving the CME and the lack of computational power to perform large-scale stochastic simulations.

An alternative approach through a series of mathematical transformations is to rewrite the CME in terms of probability moments (Smadbeck and Kaznessis, 2013). Since the dynamics of the lower order moments explicitly depends on the higher ones, the set of moment equations for nonlinear systems is underspecified. In this presentation, we show that the infinite set of stationary moment equations describing stochastic reactions networks with oscillatory behavior can be truncated and solved using maximization of the entropy of the distributions. This prediction is accomplished without any prior knowledge of the system dynamics and without imposing any biased assumptions on the mathematical relations among species involved (Constantino et al., 2016).

We explore the Brusselator and Schnakenberg models as study cases. These are hypothetical sets of chemical reactions that can produce limit cycle oscillations and provide a qualitative description of biochemical oscillators. The results from our numerical experiments compare with the distributions obtained from well-established kinetic Monte Carlo methods, such as the Stochastic Simulations Algorithm (SSA). They suggest that the accuracy of the prediction increases exponentially with the closure order chosen for the system. Hence, we conclude that maximum entropy models can be used as an efficient closure scheme alternative for moment equations to predict the non-equilibrium stationary distributions of stochastic chemical reactions with oscillatory dynamics.

References

Constantino, P.H., Vlysidis, M., Smadbeck, P., Kaznessis, Y.N. (2016) Modeling stochasticity in biochemical reaction networks. J. Phys. D: Appl. Phys. 49 093001; DOI: 10.1088/0022-3727/49/9/093001

Gillespie, D. T. (2007) Stochastic Simulation of Chemical Kinetics, Ann. Rev. Phys. Chem. 58:35-55; DOI: 10.1146/annurev.physchem.58.032806.104637

Novák, B., Tyson, J.J. (2008) Design Principles of Biochemical Oscillators. Nature Reviews Molecular Cell Biology, 9, 981-991; DOI: 10.1038/nrm2530

Smadbeck, P., Kaznessis, Y. N. (2013) A closure scheme for chemical master equations. PNAS, 110, 35:14261-65; DOI: 10.1073/pnas.1306481110

Agent-Based Modeling of a Mammalian Cell Culture Bioreactor with High Performance Computing to Predict Cell Behavior

Armour College of Engineering
ILLINOIS INSTITUTE OF TECHNOLOGY

AMGEN®

Robert Jackson[1], Mostafa Safdarnejad[1], Elif Seyma Bayrak[2], Tony Wang[2], Radu Georgescu[2], Myra Coufal[2], Chetan Goudar[2], Cenk Undey[2], Ali Cinar[1]
[1]Illinois Institute of Technology [2]Amgen, Inc.

Abstract

Knowledge of cell physiology is important for making process control decisions in the production of therapeutic proteins. In the current cell culture paradigm, analytical techniques used to obtain data requires substantial effort associated with sampling and off-line experimentation and analysis. An ideal alternative approach is decision making based on robust in-silico prediction of cell physiology that virtually eliminates experimental effort and offers a seamless and real-time approach for process control.

Agent-Based Modeling (ABM), is a novel modeling approach to address the complexity of systems that comprise heterogeneous interacting individuals. An agent is an autonomous computational entity that takes actions within its environment based on the conditions it perceives. We have integrated agent-based modeling of cells in a bioreactor with high-performance computing to leverage parallel processing that allows the ABM program to run faster, more efficiently, and with a higher capacity for the amount of cells that can be modeled. Results of this model will be discussed in this presentation.

Introduction

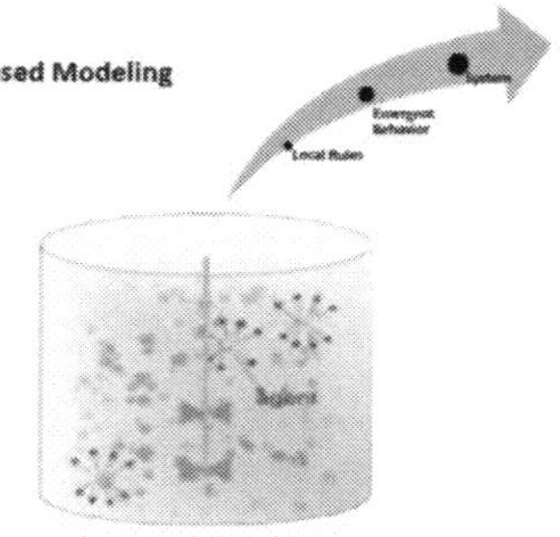

CHO Cell Agents in Bioreactor Environment

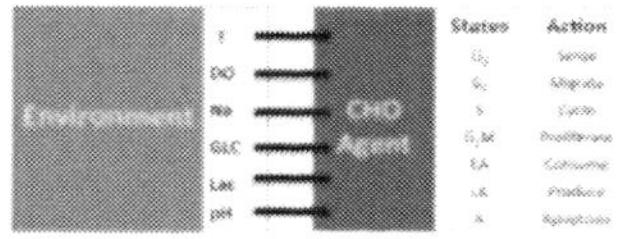

CHO Cell – Environment Interactions

Methods

The Repast (Recursive Porous Agent Simulation Toolkit) High Performance Computing libraries were used with C++ to implement the agent based modeling of the CHO cells in a bioreactor.
The simulations were done on a Linux Ubuntu 14.04 dual-boot Windows 10 machine with a Intel Core i5 3.2 GHz processor and 8 GB of RAM.

Agent Based Modeling was used to predict the cycle behavior and viable cell density of CHO cells growing in a laboratory-scale bioreactor for two different cell culture processes.

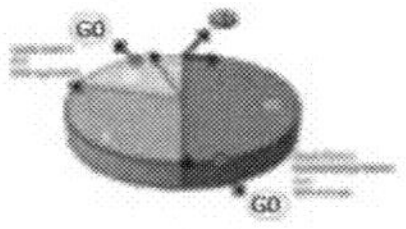

Cell Cycle

Environmental Condition	Oxygen	Glucose	Lactate/Na
Highly Favorable	High	High	Low
Favorable	High	Low	Low
Mild	Low	High	Low
Mild to Severe	Low	Low	Low
Severe	High	High	High
Severe	High	Low	High
Severe	Low	High	High
Severe	Low	Low	High

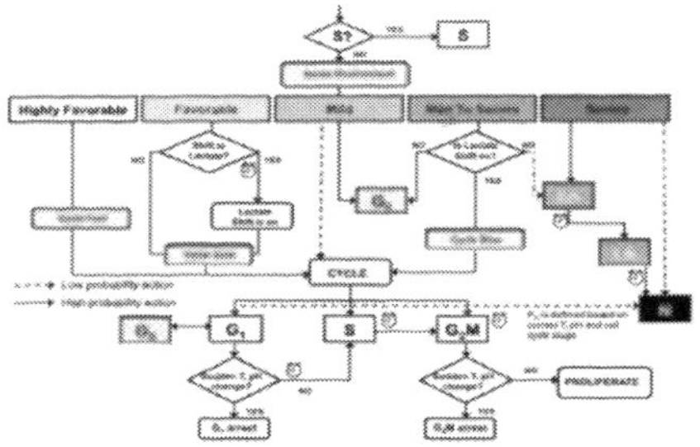

Agent Based Model Cell Cycle Rule Base (Bayrak et al. 2016)[1]

Results

Amgen Process A Cell Line 1 with JAVA and HPC Predictions

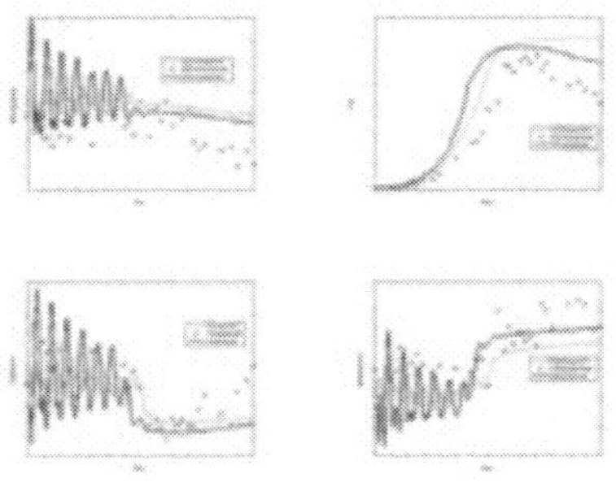

Amgen Process B Cell Line 2 with HPC Predictions

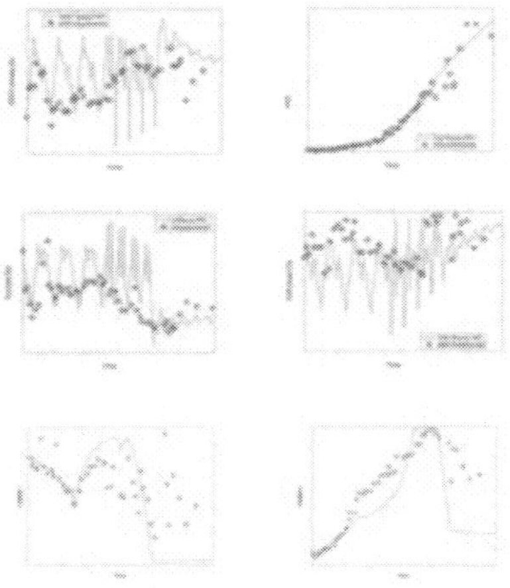

G0G1 phase %, S phase %, and G2M phase % refer to cell cycle phases with Growth 0 and Growth 1 combined into G0G1, S phase representing the DNA synthesis phase, and G2M representing Growth 2 Combined with Mitosis (Cell Division). VCD is viable cell density. Variable 1 and Variable 2 are Amgen proprietary.

Computing Performance Comparison for Process A		
Software	Repast Simphony	Repast HPC
Programming Language	JAVA	C++
Operating System	Windows 10	Linux Ubuntu 14.04
# of Processes	1	4
Initial Number of Cells	120	120
Time Ticks	1400	1400
Simulation Run Time	~2 hours	~1 minute

Conclusion

This work demonstrates the successful integration of an ABM of a cell culture bioreactor with high-performance computing.
Furthermore this high-performance ABM is still in good agreement with the experimental data, but it can predict the cell cycle behavior over 100 times faster through parallel processing, which will enhance real-time monitoring and control. This high-performance ABM has the capability of simulating more cells in less run-time.
This cell culture ABM will be used to improve industrial monitoring and control, resulting in improved production and quality of therapeutic proteins.

1. Elif S. Bayrak et al. IFAC-PapersOnLine 49-7 (2016) 200–205

Contact Information
Robert Jackson
rjacks4@hawk.iit.edu

496419 Maximizing P-Glycoprotein Expression and Transport in the Presence of Therapeutic Compounds

Abstract:

In 2004, a study of brain tissue obtained from elderly patients showed a local inverse correlation for the expression of P-glycoprotein (P-gp), a blood-brain barrier (BBB) transport protein, and the deposition of aggregated amyloid-β protein (Aβ), a hallmark of Alzheimer's disease (AD). The most common form of neurodegenerative disorder, AD affects over 5 million people and is the 6th leading cause of death in the United States. AD is characterized by an accumulation of Aβ in the brain. Currently under study as a therapeutic approach is the transport of excess Aβ out of the brain through the single endothelial cell thickness of the BBB. Our lab has shown that P-gp can transport oligomeric aggregates of Aβ. While monomeric Aβ is inert, oligomeric Aβ exhibits neurotoxicity and leads to the formation of Aβ fibrils that deposit as amyloid plaques, which are characteristic of AD brain. However, formation of oligomeric Aβ may be important to its clearance from the brain.

It has been recently discovered that the presence of certain forms of Aβ in the cells of the BBB can reduce P-gp levels through the ubiquitin-proteasome degradation pathway. These findings identify P-gp as a novel therapeutic target for AD. This study seeks to identify therapeutic compounds that will increase the expression levels of P-gp at the blood-brain barrier in order to mitigate the accumulation of Aβ by allowing for increased clearance of Aβ from the brain.

Madin-Darby Canine Kidney (MDCK) epithelial cells continuously express the multidrug resistance 1 (MDR1) gene responsible for the presence of P-gp. Cells were treated with three prospective therapeutic compounds: verapamil, caffeine, and rifampicin. Verapamil is a known P-gp substrate recently suggested to up-regulate P-gp expression and has exhibited a neuroprotective effect on AD model rats. Rifampicin, an antibiotic used to treat tuberculosis and leprosy, and caffeine have been shown in epidemiological studies to reduce dementia symptoms. Following treatment, alteration of P-gp expression was determined using fluorescent microscopy to visualize cell membrane location of P-gp and SDS-PAGE with Western blotting to quantify P-gp expression. A cell transport assay was used to quantify the transport of Aβ the treated MDCK cells. Results supporting increased P-gp expression and Aβ transport highlight the prospective therapeutic potential of these compounds.

507047 Understanding Preferential Consumption of Aromatic Compounds in *Acinetobacter baylyi* ADP1

Abstract:

Rapid depletion of fossil resources and an increasing threat of anthropogenic climate change are driving a need for renewable alternatives to petrochemicals. Lignin, a common waste product from paper and farming industries, offers a rich diversity of aromatic molecules that can be converted into biorenewable alternatives to petrochemicals. *Acinetobacter baylyi* ADP1 (referred to as ADP1) is an attractive candidate for lignin valorization due to its ability to metabolize a diverse array of lignin monomers and its natural propensity to incorporate foreign DNA into its genome. However, ADP1 exhibits strict substrate preference among these carbon sources through mechanisms that are incompletely understood. The presented work investigates possible mechanisms and motivations underlying the native carbon source preference of ADP1 primarily using metabolomics, an analytical method that enables the simultaneous profiling of concentrations of hundreds of metabolites within the cell. As the closest predictor of cellular phenotype, metabolomics is a powerful tool for understanding changes in phenotype. Combined with analytical methods for measuring metabolite uptake/secretion and protein expression, these tools provide a robust framework for elucidating the mechanisms causing substrate preference in ADP1. Once the native substrate preference of ADP1 is fully understood, this knowledge can be applied to engineer ADP1 for production of valuable products such as advanced biofuels.

508430 Single-cell Analysis for Identifying an Effective Combination Therapy for Melanoma

Abstract:

Targeted therapy that inhibits the BRAF (V600E) protein has shown impressive initial responses in metastatic melanoma patients. However, acquired drug resistance, derived from tumor heterogeneity, always limits its efficacy and compromises its treatment outcomes. Melanocyte to neural crest transition (MNT) has been observed to play crucial role in early stage adaptive drug tolerance, and the mathematic modeling indicates the transition is contributed from both stochastic phenotypic transition and non-genetic selection. Whole transcriptomic analysis enriched important signaling proteins that are involved in the resistance development process. Single cell proteomic analysis of those proteins was performed on patient-derived melanoma models to further study the early stage drug tolerance. This resolves how protein signaling network coordination rewires in the process, which cannot be resolved from proteomic analysis of bulk tumor tissue. Signaling network coordination analysis indicates that co-inhibiting the hubs of the network (MEK, NFKB-p65) could keep cells in the drug-susceptible stages. The combination therapy that co-inhibits mutant BRAF, MEK and NFKB-p65 together successfully arrested the MNT transition and induced a sustained tumor growth suppression in multiple patient-derived melanoma cell lines. This study provides us a brand-new way of using single cell proteomic analysis to infer effective combinations by resolving protein signaling network coordination that is inaccessible through traditional bulk-level proteomic measurement.

510217 Anaerobiosis revisited: yeast physiology under extremely low oxygen availability

Abstract:

The budding yeast *Saccharomyces cerevisiae* plays an important role in biotechnological applications, ranging from fuel ethanol to recombinant protein production. It is also used as a model organism to understand eukaryotic gene regulation and cellular physiology. *S. cerevisiae*â€™s ability to grow under anaerobic conditions is of particular interest for some industrial applications, in order to maximize product formation and decrease costs related to mixing and aeration. Despite the abundant amount of information and the industrial relevance, it is still not conclusive whether *S. cerevisiae* can grow under complete anaerobiosis, since the biosynthesis of lipids requires molecular O_2. Thus, we revisited the much-studied fermentative metabolism in *S. cerevisiae* using a set of well-defined experiments to clarify if it can indeed grow under complete anaerobiosis. We used an O_2 trap (*OTC-2, Agilent*) coupled to a N_2 stream (Certified; 5.2, Air Products, < 3 ppm O_2), to provide an inlet N_2 gas with less than 15 ppb O_2. Under such severe O_2 limitation in a chemostat cultivation without lipid supplementation, the biomass concentration dropped, but washout did not occur. By performing lipidomics, we noticed that the fraction of unsaturated fatty acids in cells dropped to 40.5% (w/w), as compared to 70% when cultivated with oleic acid/ergosterol. The same cultivation system was used to grow other yeast species (*Kluyveromyces lactis* and *K. marxianus*) and the results attained will be compared to those involving *S. cerevisiae*.

510218 Growth Kinetics and Modeling of 2, 3-Butylene Glycol Fermentation Using Carbon Monoxide

Abstract:

Two strict anaerobic species, C. autoethanogenum and C. ljungdahlii, have been shown to produce 2,3-butylene glycol (2,3-BG) in the stationary phase. 2,3-BG glycol is used as a fuel additive, polymers, and gas chromatograph applications. Currently, the largest source of 2,3-butylene glycol is through hydrolysis of 2,3-epoxybutane, and 2,3-epoxybutane largely comes from oil. It is desired to detour from oil as a primary source for a valuable substance.

These two species in the Clostridium genus can utilize carbon monoxide as the primary substrate source, which would make them useful in industrial applications. Therefore, intent of this research is to quantify the growth kinetics of these two organisms and formulate a kinetic model to predict the behavior of the said microorganisms in batch fermentation of 2,3-BG. Modified Monod model is utilized to describe the effects of substrate and end-products on cell growth, substrate utilization, and 2,3-BG production.

501836 Heterologous Reconsitutition of the Quorum Sensing System of Clostridium Difficile in Non-Pathogenic Hosts

Abstract:

It has recently been discovered that toxin secretion in hypervirulent strains of *Clostridium difficile*, a highly antibiotic-resistant opportunistic pathogen, is upregulated by a 2-component quorum sensing apparatus, leading to increased mortality in patients infected with these strains. In this work we reconstitute the quorum sensing system of *C. difficile* in non-pathogenic hosts. We discuss synthetic strategies to systematically isolate responsive regulators and development of high-throughput screens for detection of quorum sensing signaling molecules. Finally, we reconstitute a homologous system from the food-borne pathogen *Listeria monocytogenes* and examine implications of orthogonality and cross-talk between gram-positive QS in pathogens.

500391 Fine-Tuning of the Flavonoid Biosynthesis Pathway By Promoter Strategies

Abstract:

Promoter is one of the most important and basic tools to achieve diverse of synthetic biology goals. During the fine-tuning of metabolic or regulation networks, the limited number of well-characterized inducible promoters have made implementing complicated strategies difficult. In this study, 104 native promoter-5â€²-UTR complexes (PUTR) from *E. coli* were screened and characterized based on a series of RNA-seq data. The strength of the 104 PUTRs varied from 0.007% to 4630% of that of the P_{BAD} promoter in transcriptional level and from 0.1% to 137% in translational level. In order to further upregulate gene expression, a series of combinatorial PUTRs and cascade PUTRs were constructed by integrating strong transcriptional promoters with strong translational 5â€²-UTRs. Finally, two combinatorial PUTRs (P_{ssrA}-UTR_{rpsT} and P_{dnaKJ}-UTR_{rpsT}) and two cascade PUTRs ($PUTR_{ssrA}$-$PUTR_{infC\text{-}rplT}$ and $PUTR_{alsRBACE}$-$PUTR_{infC\text{-}rplT}$) were identified as having the highest activity. These engineered PUTRs are stable for the expression different genes, such as the red fluorescence protein gene and the Î²-galactosidase gene. These PUTRs could be used for the efficient static regulation of the flavonoids biosynthesis pathway combined with ePathBrick and high-throughput screening process. Besides, combination of the strong PUTRs and a P_{fdeR}-dependent dynamic regulation system, fine-tuning of the naringenin biosynthesis pathway was achieved. The static and dynamic regulatory elements of screening and transformation obtained here will provide technical and theoretical references for metabolic engineering of microorganisms for the constitutive expression of complicated metabolic pathways.

487562 Treatment of Autosomal Dominant Progressive Hearing Loss By *in vivo* Delivery of Genome Editing Agents

Abstract:

Although genetic factors contribute to almost half of prelingual deafness, treatment options for genetic deafness are limited. We developed a genome editing approach to target a dominantly inherited form of genetic deafness using the clustered, regularly interspaced, short palindromic repeats (CRISPR)-associated protein Cas9 (CRISPR-Cas9). Here we show that cationic lipid-mediated *in vivo* delivery of Cas9:guide RNA complexes can ameliorate hearing loss in a mouse model of human genetic deafness. We designed and validated *in vitro* and in cultured cells genome editing agents that preferentially disrupt the dominant deafness-associated allele in the *Tmc1* (transmembrane channel-like 1) Beethoven (Bth) mouse model, despite the fact that the mutant *Tmc1* (*Bth*) allele differs from the wild-type allele at only a single base pair. Injection of cationic lipid-encapsulated Cas9:guide RNA complexes that target the *Bth* allele into the cochlea of neonatal *Bth/+* mice substantially reduced progressive hearing loss. We observed higher hair cell survival rates, lower auditory brainstem response (ABR) thresholds, and enhanced acoustic reflex responses for injected ears compared with uninjected contralateral control ears, or compared with ears injected with Cas9:guide RNA complexes that target an unrelated gene. Collectively, these findings suggest protein:RNA complex delivery of genome editing agents in somatic cells *in vivo* as a potential strategy for the treatment of some autosomal dominant hearing loss diseases.

490107 The Mammalian Linc Complexregulates Genome Transcriptionalresponses to Substrate Rigidity

Abstract:

Mechanical integration of the nucleus with the extracellular matrix (ECM) is established by linkage between the cytoskeleton and the nucleus. This integration is hypothesized to mediate sensing of ECM rigidity, but parsing the function of nucleus-cytoskeleton linkage from other mechanisms has remained a central challenge. Here we took advantage of the fact that the LINC (linker of nucleoskeleton and cytoskeleton) complex is a known molecular linker of the nucleus to the cytoskeleton, and asked how it regulates the sensitivity of genome-wide transcription to substratum rigidity. We show that gene mechanosensitivity is preserved after LINC disruption, but reversed in direction. Combined with myosin inhibition studies, we identify genes that depend on nuclear tension for their regulation. We also show that LINC disruption does not attenuate nuclear shape sensitivity to substrate rigidity. Our results show for the first time that the LINC complex facilitates mechano-regulation of expression across the genome.

491903 Electricity from Methane By Reversing Methanogenesis Using an Engineered Consortium

Abstract:

Given our vast methane reserves and the difficulty in transporting methane without substantial leaks, the conversion of methane directly into electricity would be beneficial. Microbial fuel cells harness electrical power from a wide variety of substrates through biological means; however, the greenhouse gas methane has not been used with much success previously as a substrate in microbial fuel cells to generate electrical current. Here we construct a synthetic consortium consisting of (i) an engineered archaeal strain to produce methyl-coenzyme M reductase from unculturable anaerobic methanotrophs for capturing methane and secreting acetate, (ii) microorganisms from methane-acclimated sludge (with the active population identified as *Paracoccus denitrificans* through genomic sequencing) to facilitate electron transfer by providing electron shuttles (confirmed by replacing the sludge with humic acids), and (iii) *Geobacter sulfurreducens* to produce electrons from acetate, to create the first microbial fuel cell that converts methane directly into significant electrical current (*Nature Communications* in press, 2017). Notably, this methane MFC operates at high Coulombic efficiency.

496993 Reverse Engineering of Short-Chain Fatty Acid-Tolerance and Production in *Escherichia coli*

Abstract:

Product toxicity is a common problem in the microbial production of biorenewable fuels and chemicals. According to Orgelâ€™s Second Rule that â€œevolution is cleverer than you areâ€, directed evolution is a promising approach to address this issue. Additionally, reverse engineering of the evolved strains can aid in the identification and development of rational design strategies for future strain improvement.

Directed evolution was previously used to improve the tolerance of *E. coli* to exogenously supplied octanoic acid at neutral pH in minimal media. The evolved strain had increased tolerance not only to short-chain fatty acids (C6, C8, C10), but also to n-butanol and isobutanol. This evolved strain was able to attain fatty acid titers five-fold higher than the parent strain when expressing the *Anaerococcus tetradius* thioesterase. In this work, we use reverse engineering to identify and understand the mechanisms of beneficial phenotypes of the evolved strain. Firstly, four interesting mutations were identified and investigated through gene replacement. With cell characterization study, we determined if each mutation effected octanoic acid tolerance and altered the cell membrane (membrane leakage, membrane fluidity, cell surface hydrophobicity). Two of the mutant genes are regulators, so we also subjected these strains to transcriptome analysis. According to the transcriptome analysis results, many of the genes with differential expression are related to cell membrane. We plan to use the TRMR method to target the genes which contribute to the beneficial phenotypes of the evolved strain. In short, our work provides design strategies which are able to directly or indirectly alter the cell membrane and to increase short-chain fatty acids tolerance and production in *E. coli.*

500162 Kinetics of Silver Cation Diffusion across an Algal Cell Wall during Silver Nanoparticle Biosynthesis

Abstract:

In this study, the freshwater alga *Chlamydomonas reinhardtii* was used as a living biocatalyst to synthesize silver nanoparticles (AgNPs) and the effect of the Ag^+ substrate concentration on diffusion into the cell, algal health and NP production rates were examined. Inorganic nanoparticles have attracted plenty of attention since their unique physicochemical properties, which can be utilized in wide range of fields including drug delivery, antibacterial applications, redox catalysis, biosensors, etc. Besides chemical synthesis, biosynthesis, a process where the desired product is converted within or with the assistance of living organisms, has been regarded as a potential process for inorganic NP synthesis which requires no chemical solvents, generates limited wastes and consumes comparably less energy. Among the platform of biosynthesized inorganic nanoparticles, the role of algae has been put in the spotlight and attracted increasing focus from scientists worldwide because of the sustainable nature thanks to requiring little more than metallic cation precursor, photons and CO_2, However, the mechanisms and methods to optimize these processes remain undetermined. This research focuses on two primary process parameters. The first is the influence of the cell wall that per preliminary diffusion-reaction models indicates a bottleneck should occur. The second focus is on the Ag^+/cell culture density ratio. Cultivations of *C. reinhardtii* (CC-1690) and its cell wall deficient strain (CC-4425), were carried out in a cultivation platform holding 36 mini photobioreactors. Silver Nitrate was the Ag salt added to the cell culture to produce the Ag^+ ions for conversion to AgNPs by the algae. PAM fluorometry, UV-Vis spectroscopy and inductively coupled plasma atomic emission spectroscopy (ICP-AES) were applied to determine photosynthetic activity and silver composition in the liquid and solid phases of the cell culture. The current result has shown that high cell density to Ag^+ ratios, even with high concentration of Ag^+, can sustain the presence of silver nitrate and is potential to establish the sustainable photobioreactor system for the continuous production. Thus, this study can improve the development of low pollution, low cost, energy-saving processes for nanoparticles production by demonstrating the first steps toward a scalable photobioreactor-based production system for NP biosynthesis.

494757 The Global Regulator Irre from *Deinococcus Radiodurans* Enhanced *Saccharomyces Cerevisiae* Tolerances Toward Furfural

Abstract:

High tolerance to inhibitors is crucial for microorganisms producing ethanol from lignocellulosic hydrolysates. However, strain tolerance involves multiple genes and several complicated regulation mechanisms, so is difficult to achieve by manipulating single genes. It has been reported that the general regulator IrrE from the radioresistant bacterium *Deinococcus radiodurans* R1 plays an important role in radiation, oxidation, and desiccation resistance. In this study, several irrE gene libraries were constructed by three rounds of error-prone polymerase chain reaction (PCR) and transformed into *Saccharomyces cerevisiae*. The libraries were then screened for strains exhibiting higher furfural tolerance. The so-called FR mutant grew more effectively than the control group in furfural and glucose/furfural resistance tests, and it showed better ROS eliminating ability than the control strains. According to the furfural stress-related gene expression, the increased furfural tolerance in the FR mutant was associated with NADPH-related pathways. Moreover, the FR mutant was more tolerant to other main inhibitors, such as 5-HMF, formic and acetic acid, than the wild-type. The result demonstrates that the insertion of IrrE from *D. radiodurans* R1 promises to improve the resistance of *Saccharomyces cerevisiae* to multiple stresses.

497384 Extremophilic Biopolymer-Based Films: Production, Characterization, and Application

Abstract:

Exopolysaccharides (EPSs) are high molecular weight carbohydrate biopolymers that are composed of sugar residues linked together through glycosidic bonds, and secreted by microorganisms into the surrounding environment with certain properties and functions. The EPSs are one of the adaptations developed by various kinds of extremophiles in order to survive in the environments with harsh conditions, such as extreme temperature, salinity, or pH. Extremophiles are considered as producers of EPSs with novel and unusual characteristics which may hardly be found in more traditional polymers. Recently some extremophilic EPSs have already been characterized, and the properties and functions of EPSs from related extremophiles were also partly elucidated. However cumulative knowledge of the structural and functional properties of extremophilic EPSs is still limited. In this research, a thermophilic strain, *Geobacillus* sp. (WSUCF1) which produced unique type of EPSs has been studied. WSUCF1 could utilize glucose as optimum carbon and energy source to produce significant amount of EPSs (382 mg/L). Two purified EPSs were obtained through ion-exchange and gel filtration columns, and both of them demonstrated higher degradation temperature (319.20 and 314.85 °C) than the thermophilic EPSs recently reported. Monosaccharide composition analysis showed that these two EPSs were composed of mannose and glucose with various relative molar ratios (1/0.2141 and 1/0.0270). FTIR and NMR spectrum analysis also showed several unique features of the EPSs. The EPS-based films were developed through a solvent casting technique using glycerol as plasticizer, and their characteristics, such as surface morphology, thermal stability, mechanical properties and biocompatibility are currently under investigation. EPSs from the relatively unexplored WSUCF-1 strain will likely provide a valuable resource for exploitation in novel biotechnological processes and opportunities in various industrial fields including bio-based polymeric materials and antitumor-drug carrier. Meanwhile, the research on EPSs from strain WSUCF1 will also expand the scope for microorganisms isolated from hostile life environments.

A Metabolic Platform for Biosynthesis of D-amino-acids

Qiuge Zhang, Kechun Zhang
Department of Chemical Engineering and Materials Science
University of Minnesota, Twin Cities

AIChE The Global Home of Chemical Engineers

CEMS Chemical Engineering & Materials Science

Overview of D-amino acids

- **What are D-amino acids(DAAs)?**

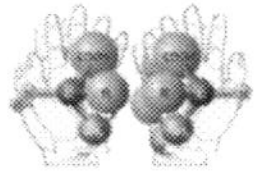

  - Based on α chiral carbon conformation, amino acids are classified into L-amino acids(**LAAs**) and D-amino acid(**DAAs**)
  - **LAAs**: more abundant, unit blocks to form proteins
  - **DAAs**: less common, but indispensable in certain metabolism

- **Where do DAAs exist in nature?**
  - Peptidoglycan in bacteria cell walls (D-alanine, D-glutamate)

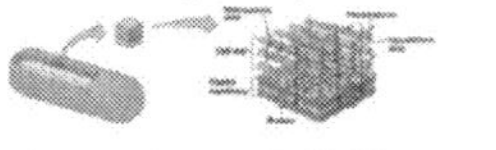

  - Neurotransmitter in mammalian brains (D-serine)

  - Some certain D-peptides (inhibitors of HIV entry)

  - D-aminoacyl-tRNA (D-tyrosine, D-aspartate, D-tryptophan, D- histidine, and D-lysine)

- **What are applications of DAAs?**

  - Pharmaceutical industry

  - Food flavorants

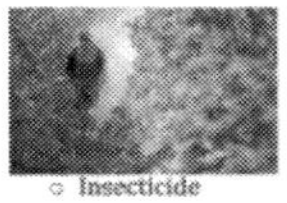

  - Insecticide

- **Natural enzymes for synthesis of DAAs?**
  - There are metabolic pathways in nature which involve DAAs

Traditional Production Methods

- **Asymmetric Chemical Synthesis**

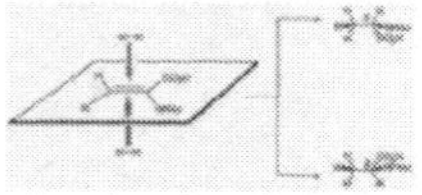

Asymmetric catalysts:

  - Expensive and complicated asymmetric catalysts
  - Unsustainable substrates
  - Low carbon economy

- **Enzymatic Synthesis**
  - L/D-stereospecific oxidoreductase

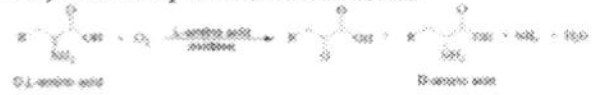

  - L/D-stereospecific amidohydrolase

➢ These strategies are basically L/D amino acids resolution with maximum theoretical yield: ***50%***

➢ High cost of enzyme purification in industrial production

DAAs Biosynthetic Pathway

- **Mimic nature?**

DAAs production by fermentation:

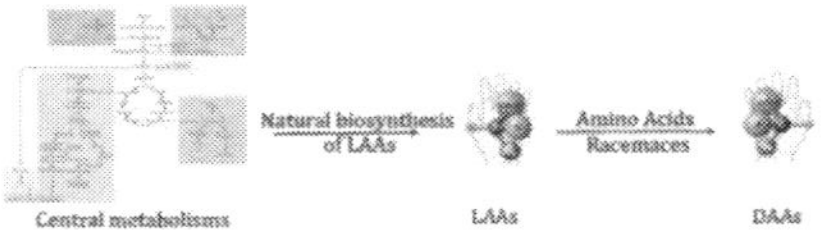

- **Major challenges associated with this pathway?**
  - DAAs are toxic to cell growth.
  - Amino acid racemases efficiency is generally low.
  - Product is tightly regulated by endogenous metabolism.

DAAs' Toxicity

- **Why toxic?**
  - DAAs can replace LAAs in protein synthesis and resulted proteins lose activities.
- **How to protect cells from DAAs' toxicity?**
  - Simple chemical modification on DAAs to help nature distinguish LAAs/DAAs

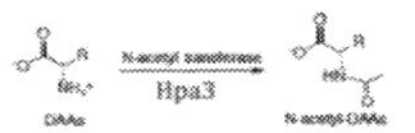

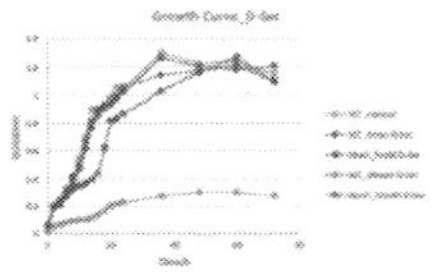

- **Why Hpa3?**

  - Hpa3 is a DAAs N-acetyltransferase found in *Saccharomyces cerevisiae*, acting specifically on DAAs.
  - Broad substrate range and high efficiency.
  - The acetyl group is easy to remove by hydrolysis after fermentation.

LAAs Overproduction

- **LAAs fermentation from glucose is industrially applicable**

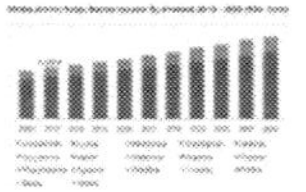

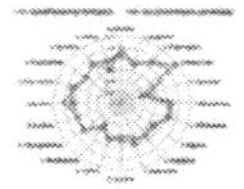

- **Metabolic engineering to overproduce L-serine (an example)**

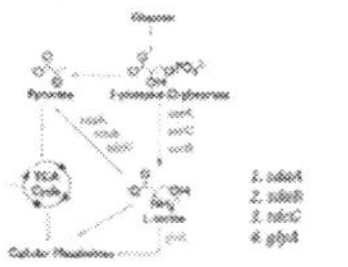

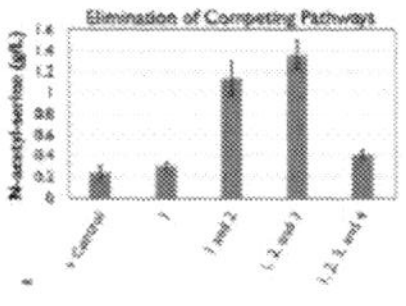

LAAs→DAAs Racemization

- **How to find racemases with high efficiency?**
  - Sequences Alignment (BLAST)
  - Screen racemases from different organisms

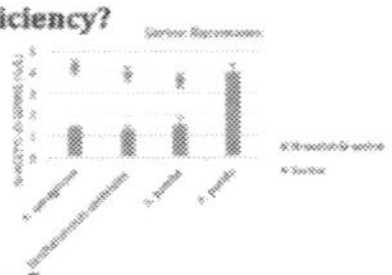

- **How to improve racemases' efficiency?**
  - Potential natural enzymes with high activity on D-amino acids
  - Abundant D-alanine in cell walls: high efficient alanine racemase
  - Engineer Alanine racemase to expand substrate range

AlaR crystal structure

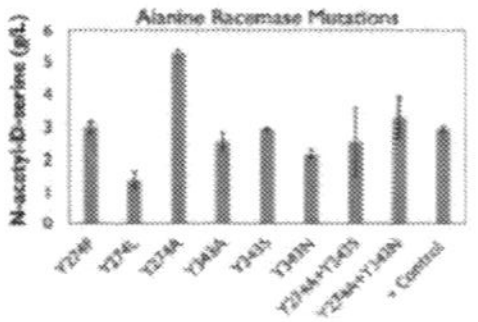

Future Work

- **Overall designed platform method to produce DAAs**

  - Economic methods using glucose as substrate
  - High conversion to DAAs from LAAs, driven by downstream acetylation
  - Less growth inhibition by DAAs

- **Future direction?**
  - Optimization of amino acid racemase for high efficiency and broad substrate specificity
  - Strains improvement to overproduce L-amino acids
  - Pathway expansion to other D-amino acids

References

- Gao, Xiuzhen, Qinyuan Ma, and Hailiang Zhu. "Distribution, industrial applications, and enzymatic synthesis of D-amino acids." Applied microbiology and biotechnology 99.8 (2015): 3341-3349
- Wolosker H, Dumin E, Balan L, Foltyn VN. "D-amino acids in the brain: D-serine in neurotransmission and neurodegeneration". The FEBS journal. 275 (2014): 3514–26.
- Welch, Brett D., et al. "Potent D-peptide inhibitors of HIV-1 entry." Proceedings of the National Academy of Sciences 104.43 (2007): 16828-16833.
- Yow, GY et al. "Physiological role of d-amino acid-*N*-acetyltransferase of *Saccharomyces cerevisiae*: detoxification of d-amino acids" Arch Microbiol (2006): 185: 39–46
- Balagurunathan et al. "Reconstruction and analysis of a genome-scale metabolic model for *Scheffersomyces stipitis*" Microbial Cell Factories (2012): 11-2.

Acknowledgement

- This work was supported by research grants from the McKnight Land Grant Professorship Program

492569 Homologous Constitutive Secretory Expression of Halo, Acid and Thermo-Tolerant β-Glucosidase in Marine *Aspergillus niger*

Abstract:

β-glucosidase, a key component of cellulase, plays an important role in the degradation of cellulose. Besides, it is widely used in food and beverage industry for hydrolyzing the flavor precursors of fruit, vegetable and tea to rich full flavors. However, the application of β- glucosidase is often limited by different environmental factors, such as high salt, high temperature and extreme pH, which lead to reduced enzyme activity.

A halo-tolerant marine *Aspergillus niger*, isolated from the East China Sea, can produce β-glucosidases with high enzyme activity. It was found that the β-glucosidases had good properties of halo, acid and thermo-tolerance. In order to further study the characteristics of the enzyme accurately, a part of one β-glucosidase gene was amplified from the genome of marine *A. niger* by using degenerate primers, which were designed according to the conserved sequences of β-glucosidases within *A. niger*. And then the complete gene, which includes introns, was obtained by hiTAIL-PCR. The target gene contains 7 exons and 6 introns by comparing with its mRNA, and the length of complete gene and coding region gene are 2603 bp and 2298 bp, respectively. Subsequently, the complete gene was placed under the constitutive promoter P*gpd*A and then transformed into its own host with the help of *Agrobacterium tumefaciens*. Finally, a recombinant which constitutively expresses β-glucosidase was obtained, and the expression quantity was about 1 $g\cdot L^{-1}$ in the supernatant determined by SEC-HPLC after 7 days of fermentation in the medium containing 0.5% yeast extract, 1% peptone, 3% NaCl, 2% glucose and 0.4% KH_2PO_4. Moreover, the enzyme accounted for more than 90% of the total proteins in the supernatant and its apparent molecular weight was 110 kDa. At the end, the characteristics of the enzyme were studied after removing the pigments, salt ions and other small molecular weight impurities by ultrafiltration. It was found that the optimum salinity, pH and temperature were 0.3-0.5 $mol\cdot L^{-1}$ NaCl, 2.2-3.8 and 60-70 °C, respectively. Significantly, the enzyme activity remained 60% at the condition of 4 $mol\cdot L^{-1}$NaCl, pH 1.0 or 80 °C.

Filamentous fungi have a strong ability of protein secretion. In this study, marine *A. niger* was used as the expression host for constitutive secretory expression of its own β-glucosidase. The expression level in the recombinant strain fully demonstrates the superiority of filamentous fungi in secretory expression of macromolecular protein which is more than 100 kDa. What's more, β-glucosidase is the dominant component in the supernatant at the non-induced condition since most of the self-secreted proteins are inducible proteins in *A. niger*. Thus, its enzymatic characteristics can be determined after a simple ultrafiltration of fermentation supernatant. Furthermore, the β-glucosidase, which has strong halo, acid and thermo-tolerance, shows a great application prospect in the chemical and food industry.

Keywords: β-glucosidase; marine *Aspergillus niger*; enzymatic characteristics; tolerance

Supported by the National Natural Science Foundation of China and Scientific Program of Zhejiang Province (2015C33219); Corresponding author : Shan-Jing Yao, E-mail: yaosj@zju.edu.cn

499403 Spray Delivery of Organoids to Reconstitute Intestinal Epithelium on Decellularized Native Extracellular Matrix

Abstract:

The native extracellular matrix (ECM) serves as a unique platform for tissue engineering because it provides an organ-specific scaffold in terms of both matrix composition and tissue architecture. However, efficacious cell-seeding techniques that would allow for recellularization of these ECM constructs with appropriate cell types to enable biologic function remain under development. In this study, the impact of spraying as a seeding technique for repopulation of decellularized small intestine was investigated. The effects of spray parameters on CaCo-2 cell viability and distribution were analyzed via fluorescent microscope. High cell viability and homogeneous cell distribution were obtained when cell suspensions were sprayed via airbrush at low pressure. The effect of spray pressure on the length and number of organoid clusters were evaluated by image analysis and monolayer formation was explored by H&E staining after 1 week of culture. While organoids seeded via pipetting onto the decellularized scaffold demonstrated clumping, organoids airbrush spray-seeded at 0.35 bar onto decellularized scaffold demonstrated monolayer formation. The results suggest that seeding cells via airbrush spraying holds great potential, especially for the recellularization of tubular organs.

501198 Establishing a Toxicity Threshold for Polymeric Nanoparticles in Pulmonary Cells

Abstract:

Lung disease is a major burden on the United States and global community. Diseases such as asthma, chronic obstructive pulmonary disease, pneumonia, and lung cancer are major causes of death and economic loss. In the US alone, 39.5 million people suffer from asthma and 12.7 million from COPD. Respiratory diseases have direct costs of $97.7 billion in hospital stays, medications, and primary doctor visits. Therapeutics by pulmonary administration offer targeted treatment for these illnesses with advantages not present in other routes of administration. Pulmonary administration avoids first-pass metabolism, eliminating distribution to unnecessary tissues and consequent side effects. Certified caregivers and needles are not required to administer the therapeutics, resulting in increasing patient compliance and reducing costs of treatment.

However, current methods of studying particle interactions with the respiratory system utilize *in vitro* and *in vivo* animal models, which are difficult to extrapolate into predictions of response in the human respiratory system. Typical preclinical models involve rats, mice, guinea pigs, and nonhuman primates, which have different modes of breathing and species-specific airway geometry, resulting in toxicity thresholds that differ from that of the human respiratory system. There is a need to establish toxicity thresholds in humans for materials commonly used in pulmonary therapeutics. CALU-3 and A549 cell lines were used to mimic the conducting airways and pulmonary airspaces of the human respiratory system, respectively. These human cell lines were used to study the effects of particle exposure and establish toxicity thresholds for polystyrene and poly(lactic-co-glycolic acid) nanoparticles in the main regions of the respiratory system. Culturing conditions of the airway and alveolar epithelial cells were optimized to obtain confluent polarized monolayers of cells. Characterization of cell lines included formation of functional tight junctions, by both florescence microscopy and transepithelial electrical resistance (TEER), and the secretion of mucus and surface proteins.

After exposure, cells were evaluated for effects of particles including viability assays observing mitochondrial activity and integrity of cell membranes, observation of resulting cell morphology, and RNA studies to quantify expression of various proteins including mucus secretion proteins, junctional adhesion molecules, and tight junction proteins. These studies allow us to conclude ideal particle sizes and dose concentrations for effective therapeutics.

495882 Construction of Human Bronchial Epithelium Culture Platform for Inhalation Drug Development

Abstract:

The culture of human bronchial epithelium is widely used to enable the *in vitro* research to understand the development of lung diseases and to facilitate the development of suitable treatment. Although many standard procedures for the culture of a bronchial epithelial cell line are readily available using transwell inserts, the culture steps are considered to be labor intensive and the cell conditions are varied with the handling (e.g. the remained medium in the apical side, the interval of mucus washing). In this work, we demonstrate a semi-automated platform for long-term mucociliary epithelium culture allowing the users to maintain the cell culture with minimum intervention. This highly integrated microfluidic-based platform was fabricated using a rapid prototyping technique based on laser patterned polymeric membranes/films/sheets and lamination. Microfluidic channels were patterned using laser cut acrylic, polycarbonate and PET sheets. Each layer was laminated using biocompatible silicone based adhesive transfer tapes to form the highly integrated bioreactor.

The platform mainly combines three parts: a cell culture device, a multi-functional reservoir, and a flow management system including a customized peristaltic pump, pinch valves, and fluid network. The cell culture steps can be simply controlled by switching the operation modes allowing: (1) cell seeding; (2) cell proliferation under liquid-liquid interface; (3) cell differentiation under air-liquid interface; (4) repeatedly mucus washing. The results demonstrate that the human airway epithelial cell (Calu-3) can be cultured on the platform with a high viability. The dynamic cell culture condition can simulate a physiologic environment that is capable of maintaining cell differentiation and tissue-specific functions. Finally, by introducing the aerosolized drug in the bronchial epithelium platform using a nebulizer, the adsorption ability of drug can be verified by measuring the concentration of dissolved drug in basolateral side of the tissue. In addition to bronchial tissue culture, the versatile combination of culture steps makes the tissue culture platform suitable to adapt to other *in vitro* human tissue culture models.

486412 New Approaches in Engineering Somatic Embryogenesis in Loblolly Pine Suspension Cultures

Abstract:

Many industries, including agriculture and healthcare, require efficient methods for replication of plants with optimal traits. The loblolly pine (*Pinus taeda*) is a valuable crop in the timber industry, occupying 30 million acres of U.S. land, and breeding efforts aim to produce a crop with ideal phenotypic traits, including superior growth and wood quality. One method to large-scale clonal crop propagation is somatic embryogenesis (SE), the process through which aggregated plant cells undergo differentiation *in vitro*, resulting in a germination-competent embryo. We seek engineering strategies in the earliest stage of growth to improve yield and more efficiently use resources. There are three stages of growth and development that lead to the production of embryos: 1) aggregated cells that form embryonic suspensor masses (ESMs) are grown and scaled up in maintenance cultures; 2) ESMs are plated on solid media to initiate SE; and 3) embryos are separated from the ESM and germinated to generate plants. However, the process is not fundamentally understood, leading to large, unpredictable variability in embryo yield, a number only determined 8-12 weeks downstream of SE initiation. Here, we present on a variety of process manipulations (both biochemical and biophysical) we have evaluated and their effects on SE process outcomes and performance. We have demonstrated that induction of stress-related pathways through exogenous addition of plant stress hormones prior to moving a culture into development improves the rate of SE, and have used this knowledge to explore a means for more efficient embryo production. We have separated cellular communities at different stages of development within maintenance cultures and determined how the different populations progress to form embryos. We will also comment on recent advances in culture protein modification without genetic transformation, by inhibiting or supplementing endogenous arabinogalactan proteins (AGPs), which we have found to significantly influence SE in loblolly pine. We will present the influence of these treatments on embryo yield as well as the molecular response of the cultures, including culture stress (determined by phenolic content), growth (using the Coulter Counter), as well as total and specific protein biomarkers. These are the first studies to consider process engineering as a simple and cost effective means to improve the overall feasibility of SE on both an academic and industrial scale.

492285 Using Ultrasound Standing Wave-Incorporated Dynamic Photobioreactor System to Enhance Medium Replacement Efficiency for Concentrated Microalgae Cultivation in Continuous Mode

Abstract:

Introduction: Microalgae have long been recognized as promising biosources for live feed, pharmaceuticals, and alternative fuel since they can provide abundance of essential chemicals such as vitamins, polysaccharides, and fatty acids. To minimize the potential contamination and provide quality cell products, microalgae culture in closed setting (i.e., photobioreactor) is generally considered as a favorable approach since all the microalgae growth-related factors, such as light, pH, temperature, and microorganism can be strictly controlled. However, decrease of nutrient and/or accumulation of cellular metabolites in the culture medium including excretions and dissolved oxygen generated from the microalgal photosynthesis may severely inhibit cell growth. Therefore, renewing the growth medium to maintain the balance and sufficiency of micro- and macronutrients in the living environment is crucial for massive microalgae cultivation. Although quite a few harvesting techniques including flotation, centrifugation, sedimentation, flocculation, filtration, and/or combinations of above methods have been achieved in the past decades, several drawbacks, such as increase of contamination risk, complexity of the process, labor consuming, and cost-ineffectiveness still remain obstacles for their practical use. To circumvent these issues, a synthetic ultrasound standing wave-incorporated dynamic photobioreactor system (USW-DPBRS) for high-efficiency microalgae culture was explored in this study.

Materials and Methods: The USW-DPBRS is consisting of 1) a culture chamber, 2) an acoustic chamber, 3) two medium reservoirs, and 4) three ultrasonic devices. The microalgae in the USW-DPBRS were harvested/separated from the culture medium driven by the USW-induced the primary and secondary acoustic radiation forces. The 1-MHz sinusoidal wave with 50% duty cycle and defined output intensity was served as the ultrasonic setting. To verify the efficacy of USW-DPBRS on cell culture, microalgae *Nannochloropsis oculata*were cultivated with and without medium change using USW-DPBRS with optimal settings including USW intensity, exposure time, and flow rate for 12 days, followed by analyses of cell growth and productions of microalgae including biomass and lipid.

Results and Discussion: Both mobile cells with and without USW exhibited similar profiles of cell growth within 7 days (growth rate = 0.802 ± 0.014 day^{-1}) and that was comparable to the group with neither circulation nor USW treatment (growth rate = 0.807 ± 0.037 day^{-1}). These results clearly show that the developed USW is harmless to microalgae. Based on the collection efficiency of microalgae, the USW operation was optimized by 1 MHz and output intensity of 8 W/cm^2 with circulating velocity of 2 mL/min whereby 93% of *N. oculata* in 30 mL can be collected within 2-h operation. The cells culture under USW-DPBRS in which the medium was changed every three days by optimal settings (growth rate = 0.033 day^{-1}) exhibited increased microalgae productions that the yields of biomass, total lipid, and eicosapentaenoic acid (EPA) of the *N. oculata* after 12-days cultivation significantly enhanced 2.6, 2.1, and 2.5 folds ($P < 0.05$ for each), respectively, as compared to the group without medium replacement (growth rate = 0.01 day^{-1}), and were all similar to the values obtained from the group with centrifugation (P= NS for each). These

results clearly demonstrated the applicability of the USW-DPBRS for use in the practice.

Conclusions: In this study, we have successfully developed a synthetic photobioreactor system, named USW-DPBRS for microalgae culture in which the medium replacement can be semi-automatically carried out through USW-mediated separation technique and gravity-driven perfusion. In comparison to conventional separation methods, the USF-DPBRS offers a non-fouling, labor-efficient, and cost-effective means for high-density microalgae culture in continuous mode.

494465 A Simple One-Step Deposition of Zwitterionic Polymer for Providing Biomaterials' Antifouling Ability Via Aminomalononitrile Polymerization

Abstract:

Non-fouling biomaterials surfaces that resist protein adsorption and cell adhesion are highly desirable for many biomedical applications such as blood-contacting devices and biosensors. A common feature for non-fouling materials is the ability to bind water molecules tightly. In comparison with a widely used anti-fouling polymer, poly(ethylene glycol), that achieves hydration via hydrogen bonding, zwitterionic materials, such as phosphobetaine, sulfobetaine and carboxybetaine, bind water molecules more strongly via electrostatically induced hydration. Decoration of biomaterials surfaces with zwitterionic polymers has shown to greatly reduce non-specific protein adsorption and cell adhesion. Techniques for surface conjugation of zwitterionic molecules are in great demand.

The surface-immobilization strategies can be categorized into "graft-to" and "graft-from" methods. Nevertheless, most of the methods could be only applied to specific types of substrates. A universal surface modification method is needed for the anti-fouling applications of zwitterionic polymers. In this study, we applied a facile and universal surface modification method based on prebiotic chemistry. The hydrogen cyanide trimer, aminomalononitrile (AMN) could polymerize to give a brown complex nitrogenous polymer on a wide variety of substrates. A copolymer of sulfobetaine methacrylate and 2-aminoethylmethacrylate (poly(SBMA-co-AEMA)) was synthesized, and then immobilized on biomaterials along with the deposition of polyAMN. The immobilization of poly(SBMA-co-AEMA) was demonstrated via the examination of X-ray photoelectron spectroscopy, water contact angle measurement, and atomic force microscope. We found that the zwitterionic polymer could be conjugated on biomaterials and display its anti-fouling capacity. Cell adhesion and protein adsorption were greatly reduced on the poly(SBMA-co-AEMA)-modified substrates. In conclusion, the one-step modification method provides a facile and effective approach to construct antifouling surfaces.

494157 N-Terminal Hypothesis for Alzheimer's Disease: Arguments for and Against

Abstract:

One possible reason why the amyloid hypothesis for Alzheimer's Disease (AD) pathogenesis is still controversial[1] is a lack of a mechanistic path from the cleavage products of the amyloid precursor protein (APP) to the deleterious effects on synaptic form and function[2-3]. From a review of our own published work and the recent literature including aggregation kinetics[4] and structural morphology[4], Aβ clearance[1, 4], molecular simulations[5], long term potentiation measurements with inhibition binding[6], and the binding of a commercial monoclonal antibody, aducanumab[7], we hypothesize that the N-terminal domains of neurotoxic Aβ oligomers are implicated in causing the disease. We call this the "N-Terminal Hypothesis for AD"[8]. In this talk, we present arguments for and against the N-terminal hypothesis.

References

1. Selkoe, D. J.; Hardy, J., The amyloid hypothesis of Alzheimer's disease at 25 years. *EMBO Molecular Medicine* **2016**, e201606210.

2. Welzel, A. T.; Maggio, J. E.; Shankar, G. M.; Walker, D. E.; Ostaszewski, B. L.; Li, S.; Klyubin, I.; Rowan, M. J.; Seubert, P.; Walsh, D. M.; Selkoe, D., J., Secreted amyloid beta-proteins in a cell culture model include N-terminally extended peptides that impair synaptic plasticity. *Biochemistry* **2014,** *53* (24), 3908-3921.

3. Willem, M.; Tahirovic, S.; Busche, M. A.; Ovsepian, S. V.; Chafai, M.; Kootar, S.; Hornburg, D.; Evans, L. D.; Moore, S.; Daria, A.; Hampel, H.; Muller, V.; Giudici, C.; Nuscher, B.; Wenninger-Weinzierl, A.; Kremmer, E.; Heneka, M. T.; Thal, D. R.; Giedraitis, V.; Lannfelt, L.; Muller, U.; Livesey, F. J.; Meissner, F.; Herms, J.; Konnerth, A.; Marie, H.; Haass, C., η-Secretase processing of APP inhibits neuronal activity in the hippocampus. *Nature* **2015,** *526* (7573), 443-7.

4. Murray, B.; Sorci, M.; Rosenthal, J.; Lippens, J.; Isaacson, D.; Das, P.; Fabris, D.; Li, S.; Belfort, G., A2T and A2V Aβ Peptides Exhibit Different Aggregation Kinetics, Morphology, Structure and LTP Inhibition. *Proteins* **2016,** *84* (4), 488-500.

5. Das, P.; Murray, B.; Belfort, G., Alzheimer's Protective A2T Mutation Changes the Conformational Landscape of the Aβ 1–42 Monomer Differently Than Does the A2V Mutation. *Biophys J* **2015,** *108* (3), 738-747.

6. Shankar, G. M.; Li, S.; Mehta, T. H.; Garcia-Munoz, A.; Shepardson, N. E.; Smith, I.; Brett, F. M.; Farrell, M. A.; Rowan, M. J.; Lemere, C. A.; Regan, C. M.; Walsh, D. M.; Sabatini, B. L.; Selkoe, D. J., Amyloid-beta protein dimers isolated directly from Alzheimer's brains impair synaptic plasticity and memory. *Nature*

medicine **2008,** *14* (8), 837-42.

7. Bussiere, T.; Weinreb, P.; Engber, T.; Rhodes, K.; Arndt, J.; Qian, F.; Dunstan, R.; Patel, S.; Grimm, J.; Maier, M. A Method of Reducing Brain Amyloid Plaques Using Anti-Ab Antibodies. US 20150315267 A1, 2015 Nov 5.

8. Murray, B.; Sharma, B.; Belfort, G., N-Terminal Hypothesis for Alzheimer's Disease. *ACS Chem Neurosci* **2017,** *8* (3), 432-434.

491547 Multiscale Analysis of Autotroph-Heterotroph Interactions in a High-Temperature Microbial Community

Abstract:

Microorganisms constitute a major portion of the biosphere, and drive planetary biogeochemical cycles through their metabolic activity. Iron-oxide mats in acidic (pH 2 â€“ 4), high-temperature (> 65 °C) springs of Yellowstone National Park (YNP) contain relatively simple microbial communities and are well-characterized geochemically. Consequently, these communities are excellent model systems for studying the metabolic activity of individual populations and key microbial interactions. The primary goal of the current study was to integrate *in situ* and *in silico* data across biological process-scales encompassing enzymatic activity, cellular metabolism, community interactions, and ecosystem geochemistry to predict and quantify the functional limits of autotroph-heterotroph interactions. Metagenomic and transcriptomic data were used to reconstruct carbon and energy metabolisms of an important autotroph (*Metallosphaera yellowstonensis*) and heterotroph (*Geoarchaeum* sp. OSPB) from Fe(III)-oxide mat communities. Standard and hybrid elementary flux mode and flux balance analyses of metabolic models predicted cellular- and community-level metabolic acclimations to simulated environmental stresses. *In situ* geochemical analyses, including oxygen depth-profiles, Fe(III)-oxide deposition rates, stable carbon isotope fractionation and mat biomass concentrations, were combined with cellular models to interpret autotroph-heterotroph interactions important to community structure-function. Integration of metabolic modeling with *in situ* measurements, including the relative population abundance of autotrophs to heterotrophs, demonstrated that Fe(III)-oxide mat communities maximize total community growth rate as predicted from the maximum power principle. Integration of multiscale data with practical ecological theory provide a basis for predicting autotroph-heterotroph interactions and community-level cellular organization.

495255 A Parallel Framework for Systematic Development of Multiscale Models Bridging Subcellular Biochemistry to Cell Population Dynamics

Abstract:

Cellular evolution takes place at much longer time scales compared to the subcellular molecular events in signaling and gene transcription. Bridging these processes at distinct time scales is imperative for the understanding of the evolutionary mechanisms of cells. Nevertheless, existing simulation techniques are inadequate to link such multiscale phenomena in a single model. Here, we introduce a new framework for the systematic development of multiscale cell population models. In an innovative approach, the framework expands a single-cell biochemical network model into a cell population model. By applying message passing interface (MPI) parallelism, it launches parallel simulations on a single-cell model and then treats each parallel thread as an individual cell object in a population. The parallel simulation threads (cell objects) evolve through death and division based on the intracellular network dynamics and temporal evolution of the extracellular environment. The framework simulates cell death by terminating a parallel thread. It simulates cell division by creating a new thread (daughter cell) from an existing thread (mother cell). The MPI mediates cell-to-cell and cell-environment communications. We demonstrate these capabilities by developing two cell population models using the framework. In one model, we treat the cells independent so that their death and division are not affected by other cells in the population. In the second model, we incorporate cellular interdependency from a quorum-sensing communication. In this latter model, each cell can affect other cells by modifying a shared growth environment. Using the models, we evaluate the computational performance of the framework. Our analysis indicates high model scalability and computation speed gain while a little loss in the accuracy of the model predictions.

485593 Development of Mathematical Approach to Studying Cholesterol Deposition in the Artery for Different Fluid Models

Abstract:

Atherosclerosis is an inflammatory disease that affects major arteries of the vasculature. It is a disease in which plaque builds up inside the arteries, narrowing of the arteries which are called stenosis. Atherosclerosis can lead to serious problems, including heart attack, stroke, or even death. A contributing factor in the growth and development of stenosis is the hydrodynamic point of view of blood where the development of atherosclerosis is often associated with local hemodynamics including shear stress and recirculation zones. There are many computational studies that have been addressed the blood flow behavior in such systems, but the modeling is not straightforward. This is due to the non-Newtonian nature of the blood at shear rates less than 100 s^{-1}, with the shear-thinning property of viscosity being considered to be the most significant non-Newtonian characteristic of blood.

In this work, the use of a mathematical model to analyze the effect of stenosis development on velocity profile and cholesterol deposition in the artery for different fluid models was explored. The results of this analysis indicated that in addition to the common parameters such as stenosis size, blood flow rate, stenosis height, blood viscosity is the most significant factor which has a substantial effect on the stenosis region. In summary, this work is a step forward towards the development of novel therapeutic and diagnostic approaches for atherosclerosis. An analytical mathematical approach is explained here that allows one to derive formulas for velocity profiles. Two main cases are considered in this research:

- Newtonian flow

· Non- Newtonian flow based on the Power Law Model, Casson Model, Carreau Model, Walburn-Schneck model.

The analysis involves the solution of the equation of the Navier-Stokes which is coupled with continuity equation. The research proposed here will have a highly beneficial impact on the area of clinical diagnostics and medicine.

Noninvasive Diagnostics for the Early Detection of Lower Respiratory Diseases : an *In-Silico* Study

Jianan Zhao, School of Chemical Engineering, Oklahoma State University, Stillwater, OK, USA
Yu Feng, School of Chemical Engineering, Oklahoma State University, Stillwater, OK,USA
Jun Wang, The University of Oklahoma Health Sciences Center, Oklahoma City, OK, USA
Xiaole Chen, School of Energy and Energy, Southeast University, Nanjing, China

According to the National Vital Statistic Report in 2012, chronic lower respiratory diseases is the 3rd leading cause of death in the US. Specifically, firefighters and soldiers are regularly exposed to lung irritants through ambient particulate matter present in their operational environment. The occupational exposure results in the inflammation and contraction of the airways and the obstruction of airflow in the deeper lung, which significantly increase the risks of developing respiratory diseases, such as chronic obstructive pulmonary disease (COPD), asthma, and constrictive bronchiolitis. To achieve better treatment outcomes, the diagnosis is necessary to detect the disease at earlier stages. However, there is strong evidence to suggest that the majority of patients are not aware of their condition, limited by the imaging resolution or the invasive nature of conventional diagnostic methods. Specifically, conventional methods to diagnose pulmonary diseases involve costly and invasive procures such as X-ray screening and bronchoscope. In this study, we developed a new noninvasive diagnostic methodology, i.e., "the exhaled aerosol-based analysis," to identify pulmonary diseases in the deeper lung by the measurable exhaled aerosol concentration pattern changes in human upper airways, combining the state-of-art technologies in both medical imaging analysis and computational fluid dynamics (CFD) simulations. The central hypothesis of the proposed "exhaled aerosol-based diagnosis" is that the changing of deeper lung structure will cause pattern shift of exhaled aerosols at upper airways. Mono-dispersed fludeoxyglucose (FDG) were employed as the biomarker aerosols in this study. CFD intersubject variability studies ascertained the validity, sensitivity, and robustness of the proposed diagnostic method. We simulated the inhalation and exhalation of FDG in three representative human respiratory system configurations (from mouth to Generation 17). Numerical results confirmed that the FDG concentration and size distribution exhaled is sensitive to the structural abnormality in lower airways caused by pulmonary diseases. The framework of the novel noninvasive pulmonary diagnostics is also developed to potentially improve the aerosol delivery protocol for future clinical trials. To achieve the detection at a subject-specific level, the individual normal lung morphology will be scanned for all firefighters and soldiers before the start of their services. CFD simulations will be performed to obtain the biomarker vapor concentration and size distribution of the healthy lung and artificially diseased lung as the subject-specific baseline as a *priori*. Using the baseline to compare with the vapor concentration measurement taken annually, clinical doctors will be able to detect any regional specific deeper lung diseases at their early stages.

495801 Fabrication of *in vitro* Human Breathing Lung Model for Inhalation Drug Development

Abstract:

Drug discovery through animal testing has been proven inefficient in many instances. Even some pharmaceuticals that successfully pass clinical trials are later found to have serious side effects that can lead to unwanted suffering, costly lawsuits, or even worse, the death of patients. The current technology developed to emulate organ-level functions in miniaturized tissue-engineered models is known as "micro-physiological systems" or "organs-on-a-chip". These models have been used to study the adsorption, distribution, metabolism, elimination, and toxicity (ADMET) of drugs in vitro. However, until now there are only a few such systems that can integrate both structures and flow mechanical features to recapitulate a complex lung with a physiological similar breathing motion. In this work, we developed a stepwise approach to construct a multilayered microfluidic platform that integrates both branched bronchiolar and deformed alveolar features to become a full lung model.

The device is composed of layers of acrylic, polyester (PET) and inflatable polydimethylsiloxane (PDMS) membranes that were pre-cut with a laser cutter followed by laminating with adhesive tapes. A water chamber was also integrated into the device to enable inflation of the flexible membrane that mimics the breathing motion by controlling the pressure difference in the water chamber. This non-pneumatic microfluidic aspiration mechanic that can stretch synthesized alveolar membranes and generate airflow in the airway. To realize the transport mechanism of inhalation drug in the breathing lung model, various sizes and charge properties of aerosolized drugs were generated and transported to the lung model through the breathing mechanism. The excessive aerosolized drugs were exhaled back to the aerosolization chamber while others remained in the *in vitro* lung model. To further improve the visualization ability, the aerosols were produced from fluorescein solution and observed using a fluorescent microscope. The distribution profile of the inhaled aerosols located in the different generation of bronchi was plotted by evaluating the remained aerosols. The *in vitro* lung model that mimics complex lung organ breathing mechanism is suitable to understand the transportation mechanism of drugs. These capabilities may be particularly useful in developing and accelerating clinical translation of inhalation drugs as well.

499983 Screening of Natural Osmolytes for Inhibiting Cancer Causing p53 Hot Spot Mutant Peptides Aggregation

Abstract:

Abstract

The p53 protein mutation is thought to be responsible for about fifty percent of all the cancers. When the p53 gene is mutated, the protein misfolds and aggregates, resulting in an inactive form of the protein, leading to abnormal cells to grow in an exponential manner [1]. Several studies show that the mutant p53 aggregation is observed in tumor tissue samples, biopsies, as well as in several cancer cell lines indicating the correlation between aggregation and tumor growth [2]. The studies further report that the aggregated mutant p53 proteins, loses their functionality. Hence, development of novel drug candidates that target the p53 mutant aggregations could aid in new directions in treating cancer. Here in this study we focus on the inhibitory effects of natural osmolytes on an aggregation prone 8 amino acid p53 conserved peptide PHITIITL, and hot sopt mutant peptides QPPHITIITL, WPPHITIITL *in vitro*. We are screening cationic, anionic, zwitterionic, and neutral osmolyte molecules to identify potential inhibitory candidates.

Experimental section:

The aggregation kinetics is studied by, Thioflavin T (ThT) binding assay, dynamic light scattering (DLS), turbidity, circular dichroism (CD), transmission electron microscopy (TEM), and atomic force microscopy (AFM).

Preliminary results:

Initial experiments show that osmolytes in micromolar to few millimolar concentrations significantly inhibits p53 conserved, and mutant peptides aggregation *in vitro*, and could be promising candidates for p53 mutant/misfolded protein aggregation associated cancer.

References:

1. Silva, J. L., De Moura Gallo, C. V., Costa, D. C. & Rangel, L. P. (2014) Prion-like aggregation of mutant p53 in cancer, *Trends in biochemical sciences.* **39**, 260-7.

2. Soragni, A., Janzen, D. M., Johnson, L. M., Lindgren, A. G., Thai-Quynh Nguyen, A., Tiourin, E., Soriaga, A. B., Lu, J., Jiang, L., Faull, K. F., Pellegrini, M., Memarzadeh, S. & Eisenberg, D. S. (2016) A Designed Inhibitor of p53 Aggregation Rescues p53 Tumor Suppression in Ovarian Carcinomas, *Cancer Cell.* **29**, 90-103.

498707 Novel Role of Transmembrane Domain of IRE1α Protein during Activation and its Implications in Progression of Cancer

Abstract:

The endoplasmic reticulum(ER) is the site for protein synthesis, folding and maturation. The unfolded protein response (UPR) is activated in response to a disruption of ER homeostasis. IRE1α is a transmembrane serine-threonine protein kinase/endoribonuclease and one of the ER stress sensor proteins that is activated during the UPR. Its failure to correct ER stress is implicated in the pathology of many diseases. The established method of activation of IRE1α is through dimerization mediated through its luminal domain.

Palmitate (PA) activates the UPR in various cell types. Long-chain fatty acids like palmitate (C16) are also known to induce epithelial-to-mesenchymal transition (EMT). An increase in free fatty acid levels is observed in various cancers (e.g. hepatocellular carcinoma) and the altered fatty acid metabolism is required to sustain tumor cells. However, the mechanism by which PA activates IRE1α to lead to EMT is unclear.

We hypothesize that unlike other ER stressors, PA activates IRE1α through a non-canonical luminal-domain independent mechanism. Specifically, we identified a W457 residue in the transmembrane domain of IRE1α that forms a part of the dimer interface on IRE1α. Mouse embryonic fibroblasts (MEF) IRE1α-/- knockout cells expressing wild-type (WT) IRE1α were more invasive than cells expressing IRE1α carrying a W457A mutation as demonstrated by wound healing studies. A bimolecular fluorescence (BiFC) assay was used to probe the importance of W457 in dimerization and activation of IRE1α. This assay uses the association of fragments of the Venus fluorescent protein driven by protein-protein interactions to demonstrate assembly of the IRE1α dimer. We confirmed that palmitate induces the dimerization of the transmembrane domain of IRE1α to activate IRE1α.

Studying the effect of fatty acids on the invasiveness and metastatic ability of cells through ER stress sensors like IRE1α provides insights into the biochemical changes induced by fatty acids that contribute to cancer progression.

499891 Validation of a Population Balance Model for Tumor Growth Using Zebrafish Melanoma Experiments

Abstract:

Validation of a Population Balance Model for Tumor Size Evolution Using Zebrafish Melanoma Experiments

Adeyinka Lesi, Silja Heilmann, Richard M. White and David S. Rumschitzki

Melanoma is a metastatic cancer that exhibits tremendous heterogeneity with typically ~70,000 mutations and is refractory to chemotherapy; it progresses by vertical invasion across the basement membrane. Advanced cases have very poor prognoses, yet recent discoveries of characteristic mutations of the BRAF gene and drugs that target them as well as dramatic advances in immunotherapy, specifically checkpoint inhibitors, are revolutionizing its prognosis. On the other hand, it turns out that the zebrafish naturally contains melanocytes that are susceptible to melanoma with the same key mutations as its human form. The White labâ€™s development of a stripeless, clear zebrafish (called casper) and a virulent, transplantable fluorescent melanoma cell line that can induce melanoma tumors in these zebrafish has created an excellent laboratory model with which to study tumor progression, shrinkage and metastasis since one can observe the tumor state of the casper fish without the need to sacrifice it. As such, this model is an ideal system to study the behavior of tumor populations in general.

In a previous talk we presented a new mathematical population balance (partial differential equation) model for tumor growth, regression and metastasis, whose results predict the time evolution of the expected numbers of tumors of each size in an ensemble of tumors from one, or more likely, many individuals. Mathematical rearrangement yields both a traveling (in tumor size space) component (growth or shrinkage of tumors of all sizes) as well as a surprising new term that corresponds to a diffusion in tumor size space. The interaction of this diffusive term with the experimental observation that larger tumors are less susceptible to immunity and treatment than smaller tumors predicts, when certain parameters are nearly in balance, the occurrence of tumor dormancy and recurrence. That is, patients with various cancers can undergo treatment and appear cancer-free for years, only to find the (genetically identical) tumors recurring years later. The mystery of this dormancy and recurrence is the focus of much research, mostly biological and genetic; our theory presents a possible explanation for it.

Naturally, to take such a model seriously, one must validate it experimentally. That is, one must first see if the model fits experimental data for some values of its parameters and then examine if those experimental parameters have the appropriate relations among them to predict dormancy and recurrence. In this talk we present detailed data for zebrafish whose immunity has been compromised by irradiation, thereby leaving only tumor growth and metastasis. We examine in detail the resulting histograms as a function of time and show that one can easily find parameters for which the model explains the data very well. This complements our excellent analogous fits to literature data for hepatocellular carcinoma. We also present data for immune-competent fish and use them to obtain the remaining tumor regression parameter. We examine the relative sizes of these parameters and predict what changes in these parameters would likely lead to dormancy and recurrence according to our theory. One could conceivably realize such adjustments

by, e.g., the use of chemo- and/or immuno-therapy drugs to raise, or immunosuppressive or moderate radiation, to lower the immunity parameter. We shall couple theory and experiments to make further predictions that we will be testing.

483938 Modeling of Fluid Flow and Oxygen Distribution in Solid Tumors

Abstract:

Most types of human cancer develop into solid tumors whereby cancer cells multiply uncontrollably to form neoplastic lesions embedded in tissues. In tumors, the interactions between cancer and normal cells causes abnormal physiological and biological changes to the tissue that leads to the development of the tumor microenvironment which inhibits the effectiveness of therapy and promotes tumor growth and heterogeneity. Some features of the tumor microenvironment include heterogeneous vascular structure and distribution, high interstitial fluid pressures, dense extracellular matrix and the presence of hypoxic regions. These properties are quite dynamic, vary from patient to patient and can influence each other. Currently, most forms of treatment used for tumors involve of the use of drugs that pass through systemic circulation whereby the drug must transport through the blood vessels of the tissue, across the vessel walls and make their way through the interstitial fluid and extracellular matrix in the interstitium to reach cancer cells in concentrations significant enough to induce cell death. Alternative treatments include the use of radiotherapy however this requires oxygen to be present in the target region to be effective. This highlights the need for a deeper understanding of the tumor microenvironment and its influence on transport so that the performance of treatments may be enhanced.

In this work, we employ computational mathematical modelling tools to allow us to understand the influence of various properties of the tumor microenvironment on fluid flow and oxygen distribution. Previous studies have attempted this however the tumor models considered were usually oversimplified where the heterogeneous vascular properties were not explicitly represented. In this study, we address this issue by building a model that can describe fluid flow and oxygen distribution in vasculature formed by tumor induced angiogenesis. Anderson and Chaplain's mathematical model for tumor induced angiogenesis is used as a basis to construct 3D tumor vasculature and geometry [1]. The vessels are segmented into straight cylindrical pipes and Poiseuilles law is used to describe flow through the vessels. Transvascular and interstitial flows are described by starlings and Darcy's equation respectively and are together coupled with vascular flow. A greens functions formulation developed by Pozrikidis (2010) and Secomb et al. (2004) is used to describe fluid and oxygen flux as a set of source points along a vessel [2][3]. Doing this allows interstitial fluid pressure, flow and oxygen distribution to be represented as function of vascular structure and density. To validate the model, morphological and hemodynamic analyses were performed and the data was found to correspond quite well with experimental values found in literature [4]. Fluid flow results from the model predicted elevated interstitial fluid pressure values between 1000-2000 pa which were corroborated by values determined experimentally. So far, our results show how interstitial fluid flow and pressure are affected by changes in properties such as vascular structure, heterogeneity, vessel permeability and ECM density. Spatial distributions of interstitial fluid pressure indicate that the discrete nature of the tumor vasculature can affect interstitial fluid pressure in some regions of the tumor which can in turn influence the penetration of drugs into these regions and allow cancer cells to survive treatment cycles. We present further results describing the effect of different vascular properties on the spatial distribution of oxygen in the tumor model.

[1] A. R. A. Anderson and M. A. J. Chaplain, "Continuous and Discrete Mathematical Models of Tumor-induced Angiogenesis," *Bulletin of Mathematical Biology*, vol. 60, pp. 857-900, 1998.

[2] C. Pozrikidis, "Numerical simulation of blood and interstitial flow through a solid tumor," *Journal of Mathematical Biology*, vol. 60, pp. 75-94, 2010.

[3] T. W. Secomb, R. Hsu, E. Y. H. Park and M. W. Dewhirst, "Green's Function Methods for Analysis of Oxygen Delivery to Tissue Microvascular Networks," *Annals of Biomedical Engineering*, vol. 32, pp. 1519-1529, 2004.

[4] S. K. Stamatelos, E. Kim, A. P. Pathak and A. S. Popel, "A bioimage informatics based reconstruction of breast tumor microvasculature with computational blood flow predictions," Microvascular Research, vol. 91, pp. 8-21, 2014.

498342 Optimization of Vincristine Infusion Time

Abstract:

Vincristine (VCR) is a core chemotherapeutic drug administered to pediatric Acute Lymphoblastic Leukemia (ALL) patients. It is administered via an IV bolus injection or an infusion minibag for approximately 10 minutes. In a subgroup of population, VCR administration leads to Vincristine Induced Peripheral Neuropathy (VIPN), which is the dose-limiting toxicity. For few patients, VIPN is severe with long-term effects. Even though VCR has been used as a chemotherapeutic drug for more than 50 years now, predictors and mechanism of VIPN induction are unclear. VIPN incidence is associated with VCR cumulative dosage. Hence, we are interested in finding an optimal infusion time that would lead to a reduction in VCR dosage.

A population balance model was developed to describe the mechanism of VCR in cells in different phases. The model was a function of time and cell age. This mechanistic model was combined with pharmacokinetics of VCR. Infusion time was optimized by maximizing number of leukemia cancer cells being killed by the end of induction phase of the treatment, with VCR being administered weekly. A preliminary analysis with a simplistic model showed that a schedule of 30 minute infusion in week 1, 12 minute infusion in week 2, 30 minute infusion in week 3, and 12 minute infusion in week 4, produced a better efficacy than the usual 10 minute infusion every week. Lesser amount of drug with same efficacy as before may reduce the chances of VIPN.

Currently, rigorous analysis is being done with a more complex model. In this model, cell phase transition rates are assumed to be functions following a lognormal distribution with cell age as a variable. Detailed pharmacokinetics is incorporated as well. Initially, the model consisted of G1 phase, and S, G2, M phase lumped together, followed by mitotic arrest. A quiescence phase is added to it. Further, bone marrow toxicity is incorporated to account for killing of normal cells. In the future, we intend to eventually estimate a personalized infusion time for every patient.

507748 Quantitative *in vivo* & *ex vivo* Multimodality Cell Imaging of Antigen-Specific T-cells in Murine Metastatic Ovarian Cancer

Abstract:

Cellular immunotherapies are an emerging treatment for many cancer types, including epithelial ovarian cancer. Adoptive transfer of antigen-specific T-cells has shown promise in pre-clinical and clinical studies, and has demonstrated efficacy in settings of low tumor burden. We are interested in the quantitative aspects of homing and cell-mediated destruction by antigen-specific T-cells. Specifically, how T-cell number influences the ability to detect and destroy individual metastatic lesions as a function of metastatic lesion growth rate, absolute size, and location, is poorly understood. First, we comprehensively characterized tumor kinetics and distribution in a metastatic ovarian cancer model. We used serial MRI imaging and optical *in vivo* imaging, *ex vivo* imaging, and *ex vivo* luminometry. Our *in vivo* MRI and *in vivo* bioluminescent data indicated that after intraperitoneal injection, ID8 ovarian tumor cells primarily home to the abdomen within four hours of transplantation (n = 10). Further, our data indicates that from day 0 (4-hour timepoint), abdominal signal decreased by over 90% by week 3, followed by linear growth until week 7 (n = 2). This data suggests that week 3 and week 5 tumor differ by a 15-fold increase in signal, indicating a large difference in tumor mass, confirmed by *ex vivo* bioluminescence and luminometry. Using this information, we developed a non-invasive, quantitative platform for imaging T-cells *in vivo* and *in vitro* in this well-defined model of metastatic ovarian cancer. First, OVA+ IE9-MP1 ovarian tumor cells were cultured and administered intraperitoneal in female C57BL6/J mice at a dose of 10 million each. Primary T-cells were then cultured from OT-1 TCR transgenic mice, in which the CD8+ T-cells express a TCR specific to ovalbumin (OVA), and employed to target the OVA+ IE9-MP1 cells. CD8+ T-cells were isolated, cultured, transduced with retrovirus (Fluc+ GFP+), and retroorbitally injected into lymph node-depleted mice. We hypothesized that following adoptive transfer, T-cells would traffic to major tumor locations (omentum), metastatic sites, and lymphatic organs. After 3 weeks of tumor growth, we administered luciferase-positive OVA-specific T-cells at doses of 3×10^6, 1×10^6 and 3.33×10^5 cells. Next we performed serial *in vivo* and end point *ex vivo* bioluminescence. *In vivo* imaging of abdominal signal indicates increase in T-cell proliferation between days 2-6, and days 6-8 demonstrate at least a 10-fold signal increase compared to day 2 signals for all three doses (n = 3 per dose). Interestingly, *in vivo* imaging demonstrated that approximately 25-50% of cells home to primary tumor site (omentum), while the remaining T-cells home to other abdominal locations. Normalized data reveals a >50% shift in T-cell location from primary to metastatic tumor site between days 1-5 for all doses (n = 9). T-cell homing analysis of tumor-bearing mice after 5 weeks are currently underway. Our data indicates that tumor-bearing mice can be separated into 3 week (rare mets) and 5 week (early mets) models, and that T-cells distribute between primary and metastatic sites. Future studies will determine T-cell homing in the 5-week model and decipher mechanisms of T-cell distribution at metastatic sites.

507749 Engineering the Liver Diverticulum from Human Pluripotent Stem Cells

Abstract:

Liver tissue demand is escalating due to the greatly increasing incidence of liver disease.

We hypothesize that the structures and cues that initiate 3D liver formation can be mimicked with human pluripotent stem cells (hPSC). Thus we aimed to engineer an in vitro model of the liver diverticulum (LD), a key structure that: 1) arises in mouse development (E9.5) and human development (d26) and 2) forms the 3D liver. From inside the gut to outside, the LD is composed of a single layer of hepatic endoderm (HE), encased by a single layer of endothelial cells, and surrounded by the septum transversum mesenchyme (STM). 3D liver formation starts when the hepatic endoderm (HE) delaminates, joins with the endothelial cells, and migrates into the STM. We first determined the appropriate LD dimensions with an online mouse database. Next, to model the LD, we used a commercially available cell migration system, which enabled selective cell seeding and cell adhesion of two migrating cell types, on 0.22 cm^2 square islands separated by 500 μm. HepG2 liver cancer cells, a model of the HE, and human mesenchymal stem cells (hMSC), a model of the STM, were cultivated and analyzed using time lapse microscopy. At 8 hours we observed that 60% of MSC and 30% of HepG2 cells on the opposing edges of each cell island, had migrated to the opposing side, and at 24 hours, the cells had intermixed (n=2). In a second LD model, we wished to specifically evaluate HepG2 motility towards MSC, but not the reverse. Using a Transwell cell invasion assay, HepG2 were seeded on a matrigel-coated insert (8 μm pore) and inserted above cultured MSC. After 24 hours, there was a 70% increase in liver cell invasion through the insert compared to the "no MSC" condition (n=2). Overall, both 2D assays showcase migration of HepG2 in proximity of MSC's. We then explored further 3D systems to self-assemble the LD with HepG2 (HE), MSC (STM), and human umbilical cord vein endothelial cells (HUVEC). Dispase-treated HepG2 cell fragments (100 μm x 100 μm) were initially seeded within hydrogels composed of Matrigel (MG). We added HUVEC 48 hours later, and performed microscopy. After 12 hours, HUVEC cell migration resulted in 80% of the HepG2 fragments having interconnected tubular, cord-like HUVEC structures, but not in no MG controls. In our final model of LD, we further refined this system. Singular HepG2 cell fragment was seeded into a single well of a 384-well plate. HUVEC were added to coat the fragment surface, and the system was submerged in a hydrogel (MG) containing MSC. We observed that average HUVEC/HEPG2 fragment area increased by 22%. This was due to 3-4 multicellular protrusions characteristic of collective cell migration (n=2) towards MSC, as occurs in the LD. Ultimately, our data suggests that HepG2 do migrate towards MSC in 2D systems, and that HepG2 and HUVEC together migrate towards MSC in 3D systems. Future studies will encompass PSC-Heps in a micro-engineered liver tissue, for an in-vivo liver cell therapy model.

508751 Controlling Endodermal Cell State by Understanding and Re-engineering Developmental Master Regulatory Gene Circuits

Abstract:

Human pluripotent stem cell-derived (hPSC)-derived endoderm, which gives rise to liver, pancreas, and lung, is heterogeneous, and the further investigation into the factors that regulate its differentiation are warranted. The overall objective of our research is to understand how Foxa2 regulates developmental regulatory gene circuits (DRGC) composed of master regulatory transcription factors (TF) and can eventually be used to control hPSC-derived endoderm induction, maintenance, and differentiation. Foxa1 and Foxa2 (Foxa1/2), together with other endodermal transcription factors (TF), compose the poorly understood DRGC. Foxa2 was identified as the first known pioneer TF, which binds to silent differentiation genes, like Albumin, to prime them for activation. The binding of Foxa2 appears to open normally inaccessible heterochromatin and provides access for other TF of DRGC, like Hnf1, HNF4a, and HNF6 to bind. This enables thousands of genes to be activated, within endoderm derivatives, in a relatively short amount of time during development. Consistent with this, mouse genetic studies of Foxa1/Foxa2 double knockdown (Foxa1/2-/-) within hepatic endoderm, demonstrates a complete absence of the liver bud and a failure to initiate the hepatic differentiation (Lee et al., Nature 2005). Similarly, Foxa1/2-/- phenotype in pancreatic endoderm, severely blocks pancreatic growth and differentiation. The developmental role of Foxa1/2 is in contrast to its role in mature liver tissue, in which Foxa1/2 knockdown in adult hepatocytes has no effect on transcription factors or liver-specific genes. To understand how Foxa1/2 contribute to the endodermal DRGC and orchestrate endoderm/hepatic differentiation, we first analyzed the role of Foxa1/2 in human hepatocellular carcinoma (HepG2) cells. We hypothesized that Foxa1/2 regulates hepatic endoderm transcription factors expressed within the HepG2 cell, thereby serving as a model for endoderm/hepatic endoderm. We used an RNA interference approach for the Foxa1/2 (-/-) phenotype followed by qRT-PCR/western blot to measure mRNA/protein level in components of the DRGC and other master transcription factors involved in differentiation. HepG2 cells were plated in multiple wells of a 24 well plate at 50,000 cells/well density and transfected with 50 nM of siFoxa1 (4 pools), siFoxa2 (4 pools) and both siFoxa1 and siFoxa2 molecules. qRT-PCR was performed after 48 hours of transfection and all the data was normalized to control sample with siPPIB as negative control. Our results indicate that the Foxa1/2(-/-) phenotype, but not the Foxa1(-) or the Foxa2(-) phenotype, is characterized by downregulation of key hepatic endoderm markers like AFP (alphafetoprotein), Alb (albumin) and Hhex (Hematopoietically expressed homeobox) (n=2). We observed heterogeneity of the effects of on Foxa1/2(-/-) phenotype gene expression of DRGC components like Gsc (mesendoderm TF), Hnf4a and Hnf6, as well as the endoderm cell surface marker Cxcr4 (n=2) . Surprisingly, we observed the upregulation of pancreatic markers such as Pdx1 and Pax6 (n=2). We confirmed Foxa1/2(-/-) phenotype by immunofluorescence, which demonstrated a ~80% reduction in Foxa2 levels, and also demonstrated that Foxa1/2(-/-) are intact morphologically and viable. Finally, we performed a western blot of Alb, demonstrating an ~80% knockdown in the Foxa1/2(-/-) phenotype. Overall, our data suggests the Foxa1/2(-/-) phenotype in HepG2 cells result in downregulation of liver differentiation markers, upregulation of pancreatic markers, and heterogeneous changes in DRGC components. Future work includes improving knockdown, testing the Foxa1/2(-/-)
phenotype on hPSC-derived endoderm, and performing analysis of chromatin remodeling to
assess epigenetic effects.

486586 Modeling Alzheimer's Disease Using Cortical Organoids Derived from Human Induced Pluriptotent Stem Cells

Abstract:

Alzheimer's disease (AD) is one of the most common neurodegenerative disorders and causes cognitive impairment and memory deficits of the patients. The mechanism of AD is not well known, due to lack of human brain models. Recently, the novel neural tissue patterning technology has been developed using human induced pluripotent stem cells (hiPSCs) to derive mini-brain tissues called organoids for modeling human brain development and neurological diseases. Thus, the objective of this research is to establish a novel in vitro AD model using 3-D forebrain cortical organoids derived from hiPSCs to study the response to the drug treatment. It is hypothesized that the 3-D forebrain organoids derived from hiPSCs with AD-associated genetic background may partially recapitulate the age-related disease pathology. To test this hypothesis, AD-patient derived hiPSCs with presenilin-1 mutation were used for cortical organoid generation. AD-induced inflammatory responses, matrix remodeling and drug testing were investigated. The results indicate that the cortical organoids derived from AD-associated hiPSCs exhibit a high level of Aβ42 comparing with healthy control. In addition, the AD-derived organoids result in an elevated expression of pro-inflammatory cytokines including IL-6 and TNF-α and alter matrix remodeling protein expression. Our study demonstrates the capacity of hiPSC-derived organoids for modeling AD using neural patterning technology and provides a potential approach for AD-related drug screening.

501564 pH-Dependent PDGF-BB-Induced Chemokinesis and Chemotaxis of NIH 3T3 Fibroblasts and Rat Bone Marrow-Derived Mesenchymal Stem Cells

Abstract:

Cell migration plays an essential feature in many physiological and pathological processes as well as in wound healing and scaffold-based tissue engineering. In the microenvironment experienced by the pH of the cell is a key factor that influences their behavior, such as motility. For example, in wound healing, a low pH is known to be beneficial, while a high pH is disadvantageous. However, surprisingly few studies are available on the effect of pH on cell migration, and its influence on cellular directional decision-making processes. Moreover, there is no consensus on how pH affects the chemokinesis and chemotaxis of different cell lines, or whether it can be used to control cell migration. We hypothesize that the modulating the pH can lead to increases or decreases in cell speed, as well as can provide the ability to control cell directional decisions. Therefore, we aim to examine the effect of pH on the migration of mouse embryo fibroblast cells (NIH 3T3) and mesenchymal stem cells (MSCs) - two important cell types for tissue engineering. To that end, fibroblasts and MSCs were exposed to different pH levels ranging from 4.0 to 12.0, either in the presence or absence of a chemotactic factor (i.e. Platelet-derived growth factor â€" BB); and their migration was tracked over time using an automated Olympus microscope. Migration speed, directionality, directional persistence, morphology, and viability of the cells were quantitatively determined and descriptive statistics are reported. It was found that the motility of both cell types were inhibited at acidic and alkaline conditions even in the presence of 25 ng/ml of PDGF-BB. Cellsâ€™ migration was completely restricted at pH 4.5 even though their viability was not affected, as confirmed by a viability assay. The fibroblast and mesenchymal stem cells migration was greatest at pH 7.4. The presented approach is expected to yield an enhanced fundamental understanding of cell migration for both cells lines NIH 3T3 and MSCs. Moreover, it was found that the change in pH conditions can be used to control the various modes of cell motility. Our results highlight a prominent role of pH in chemotaxis and provide potential tools for controlling tissue development, wound healing in scaffold-based tissue engineering. For example, pH could be used to obtained a desired cell speed and/or induce a specific type to migrate into a scaffold (while excluding other cell types). Furthermore, the data provided by our study can be beneficial to further understanding the mechanism of cancer metastasis, where pH can also be a significant factor.

496349 Investigation of the Variation in Exosome Release By Human Pluripotent Stem Cells in Static and Stirred Suspension Cultures

Abstract:

Human pluripotent stem cells (hPSCs) have the capacity to differentiate into multiple cell types of great significant to regenerative medicine. Exosomes, which are nanovesicles released from cells and with size ranging between 50-130 nm, carry proteins, mRNAs, and microRNAs (miRNAs) [1]. As vehicles for information transfer among cells, exosomes can be used as gene/protein delivery therapeutics with advantages over conventional technologies. We hypothesize that exosomes released from hPSCs and their differentiated products influence the process of stem cell fate specification.

The isolation of exosomes from hPSCs has not been reported to date. To that end, we successfully obtained exosomes from hPSCs and identified morphological attributes and relevant protein markers. Messenger RNAs and miRNAs linked to the hPSC pluripotent state were detected in these exosomes. Specifically, we discovered miR302A and miR302C, which induce and maintain pluripotency in hPSCs in a POU5F1 (OCT4)-dependent manner [2], in the hPSC-derived exosome cargo. Exposure of human embryonic kidney 293 (HEK293) cells to the exosomes from hPSCs led to upregulation of the aforementioned miRNAs thereby confirming exosomal capacity to transfer information between cells.

While investigation of the role of exosomes on hPSC differentiation is ongoing, we also noted that hPSCs secrete minute quantities of exosomes in dishes when compared to stirred-suspension cultures. Human PSCs cultured in stirred suspension release as much as 10 times higher amounts of exosomes than the same number of hPSCs cultured in dishes. Based on our findings, this difference in exosome production could not be attributed to potentially increased apoptosis or aberrant differentiation. We have investigated the TP53 (p53) downstream products STEAP3 and CHMP4C, which are known to regulate vesicle secretion [3], as causal agents for increased exosome secretion. TP53 may be activated due to increased shear stress experienced by cells in stirred suspension. Based on qPCR and Western blot analyses, the expression of STEAP3 is higher in stirred-suspension hPSC cultures compared to dishes, based on both mRNA (three-fold; qPCR) and protein levels (Western blot).

Our study has ramifications for the use of stirred-suspension culture platforms to obtain large yields of exosomes from self-renewing and differentiating hPSCs and for identifying markers which may facilitate hPSC specification. Moreover, the modulation of STEAP3 expression can conceivably provide a handle on the amount of exosomes released from hPSCs in culture.

References:

1. Kourembanas, S., Exosomes: Vehicles of intercellular signaling, biomarkers, and vectors of cell therapy. *Ann. Rev. Physiol.* **77**(7): 7.1-7.15,2014.
2. Li, H. L., et al., miR-302 regulates pluripotency, teratoma formation and differentiation in stem cells via an AKT1/OCT4-dependent manner. *Cell Death Dis.* **7**: e2078,2016.
3. Yu, X., et al., The regulation of the endosomal compartment by p53 the tumor suppressor gene. *FEBS J.* **276**(8): 2201-2212,2009.

498560 Differential Expression of Neuron-Glial Antigen 2 (NG2) and Melanoma Cell Adhesion Molecule (CD146) in Mesenchymal Stem Cells

Abstract:

Cellular heterogeneity of mesenchymal stem cells (MSCs) impedes their use in regenerative medicine. The objective of this research is to identify potential biomarkers for the enrichment of progenitors from heterogeneous MSC cultures. To this end, the present study examines variation in expression of neuron-glial antigen 2 (NG2) and melanoma cell adhesion molecule (CD146) on the surface of MSCs derived from human bone marrow in response to culture conditions and among cell populations. Multipotent cells isolated from heterogeneous MSC cultures exhibit a greater than three-fold increase in surface expression for NG2 and greater than two-fold increase for CD146 as compared with parental and lineage-committed MSCs. For both antigens, surface expression is downregulated by greater than or equal to six-fold when MSCs become confluent. During serial passage, maximum surface expression of NG2 and CD146 is associated with minimum doubling time. Upregulation of NG2 and CD146 during loss of adipogenic potential at early passage suggests some limits to their utility as potency markers. The relationship between proliferation and antigen expression suggests that NG2 and CD146 may have utility in isolating progenitors with high proliferation potential from heterogeneous MSC cultures. This research was funded in part with a grant from the National Science Foundation.

486911 Heterotypic Cell-Cell Interactions of Human Induced Pluripotent Stem Cells and Human Mesenchymal Stem Cells for Neural Differentiation

Abstract:

Organ buds, the condensed 3-D tissues emerged at the early stage of organogenesis, are a promising transplantation entity to regenerate functional and vascularized organs. However, complex heterotypic interactions of different cell types, such as human mesenchymal stem cells (hMSCs) and human induced pluripotent stem cell (hiPSC) derived neural progenitors, and their secretory activities during neurogenesis have not been well understood. The objective of this study is to investigate the impact of the composition and structure of 3-D hybrid spheroids of hiPSCs and hMSCs on neural differentiation and the secretion of extracellular matrices and trophic factors in vitro. The hybrid spheroids were formed at different hiPSC:hMSC ratios (100:0, 75:25, 50:50, 25:75, 0:100) using direct mixing and pre-hiPSC aggregation method, which generated dynamic spheroid structure. The cellular organization, proliferation, neural marker expression, the secretion of extracellular matrix proteins and the cytokines were characterized. The incorporation of MSCs upregulated the neural marker expression, matrix remodeling proteins and the secretion of transforming growth factor (TGF)-Î²1 and prostaglandin E2. This study indicates that the composition and structure of hiPSC-MSC spheroids differentially influence neural differentiation and trophic factor and matrix secretion due to the modulation of heterotypic cell-cell interactions.

497500 Development of HER2-Postive Breast Tumor Spheroids As a Better Approach to Study the Effectiveness of Novel Anticancer Therapies

Abstract:

Abstract

Until these days, one of the first steps in the development of novel anticancer therapies has been the *in vitro* validation of their pharmaceutical activity in 2D cell cultures. However, the limitations of such kind of cultures are now well known: cells are force to polarize for being anchored to a rigid and a flat surface and they are super-nourished and super-oxygenated. As consequence, modifications in their orientation, in the expression of membrane receptors and in the production of extracellular matrix (ECM) take place. Moreover, it has been demonstrated that 2D cultures promote the immortalization of cancer cells, favoring the survival of those who divide faster but are less resistant to chemotherapy. Thus, taking into account all these disadvantages, understanding why most successful 2D studies do not have proper *in vivo* results is easier and the need of the implementation of novel validation methods is clear.

In this context, numerous research groups have already started to develop tumor spheroids, i.e., 3D cancer cell cultures that mimic natural tumor conditions, growth in special gels or matrix [1]. They are usually characterized by fluorescence microscopy and it can be found in literature that this type of cultures has helped to study hypoxia conditions inside tumors, the angiogenesis process [2], the importance of cell plasticity in tumor malignancy progression [3] and the pharmaceutical action of different anticancer drugs [4].

The work here presented is related with this last application. Spheroids of HER-2 positive metastasic breast cancer cells (BT474 cell line) have been developed in both, DMEM/F12, supplemented with B27, hEGF, insulin and hydrocortisone, and in a Matrigel® matrix. Their growth has been checked by optic microscopy and they have been characterized by confocal fluorescence microscopy. Finally, the efficacy of a novel paclitaxel and Trastuzumab vehicle, consisting on nanoparticles of alginate and piperazine [5] with the mentioned drug and antibody attached to their surface, has been validated with these breast cancer spheroids and results have been compared to nanoparticles´ antitumor effect in conventional 2D cultures. In this manner, the establishment of a more accurate *in vitro* validation method is pretended, useful to validate different therapeutic agents with results closer to that got with *in vivo* studies.

References

[1] Alemany-Ribes M., Semino C.E. *Adv. Drug Deliv. Rev., **2014,*** 79-80, 40-49.

[2] Benton G., Arnaoutova I., George J., Kleinman H.K., Koblinski J. *Adv. Drug Deliv. Rev.*, **2014**, 79-80, 3-18.

[3] Liang Y., Jeong J., DeVolder R.J., Cha C., Wang F., Tong Y.W., Kong H. *Biomaterials*, **2011**, 32, 9308-9315.

[4] Gurski L.A., Jha A.K., Zhang C., Jia X, Farach-Carson M.C. *Biomaterials,* **2009**, 30, 6076-6085.

[5] Román J.V., Rodríguez- Rodríguez J.A., Martín Del Valle E.M., Galán M.A. *Polymer Adv. Tech.*, **2016**, 27(5), 623-629.

496494 Construction of a Multi-Culture Human Lung Platform for Tumor Metastasis Study

Abstract:

Lung cancer is one of the most common malignancies and the leading causes of cancer-related death in the world. Many clinical data has shown cancer metastasis is the major reason that causes the death of patients. To investigate the mechanism of metastasis, reconstruction of tumor microenvironment *in vitro* is considered as an easily accessible approach that can mimic the complex and multiple interactions between lung tumor cells and other surrounding tissue. However, the existing devices usually involve two or three types of cell co-culture together which are not able to represent the complex tumor microenvironment. Here we developed a microfluidic-based platform that can culture lung cancer cells, epithelial cells, fibroblasts and muscle cells in the same format.

Studies are aimed at understanding if the invasion mechanism of carcinoma cells can be regulated by the surrounded stromal cells. We have developed a technique to fabricate a microfluidic device that contains five of cell culture chambers. The cells in each chamber can grow in their own medium and share the same flow system that simulates the microvasculature. The culture chamber for each cell line was spatially separated by the embedded -hydrogel in the microfluidic channels so that the migration results of cancer cells can be dynamically and quantitatively obtained. We found that when the cancer cells were cultured surrounded by other stromal cells, the migration of cancer cells can be regulated. For example, epithelial cells started to lose their characteristic proteins which indicated that the epitheliaÌâ €"mesenchymal transition (EMT) might occur in the tumor microenvironment. Furthermore, the fibroblasts were activated into myofibroblast by cancer cells. These changes also increased the stiffness of hydrogel surrounding cancer cells and improved the migration ability of cancer cells. Taken together our results showed that with the existing stromal cells, the multiple interactions within the microenvironment could further enhance the migration of cancer cells compared to the culture with a single type of cells. By cultivating other cell types, this dynamic tumor culture platform could be served as an *in vitro* model to study initiation, progression, and remediation of cancers.

497779 Enhanced Cancer Immunotherapy By Chimeric Antigen Receptor-Modified T Cells Engineered to Secrete Checkpoint Inhibitors

Abstract:

Despite favorable responses of chimeric antigen receptor (CAR)-engineered T cell therapy in patients with hematologic malignancies, the outcome has been far from satisfactory in the treatment of solid tumors, partially owing to the development of an immunosuppressive tumor microenvironment. To overcome this limitation, we engineered CAR-T cells secreting checkpoint inhibitors (CPIs) targeting PD-1 (CAR.αPD1-T) and evaluated their efficacy in a human lung carcinoma xenograft mouse model. In an effort to evaluate the effector function and expansion capacity of CAR.αPD1-T cells *in vitro*, we measured the production of IFN-γ and T cell proliferation following antigen-specific stimulation. Furthermore, the antitumor efficacy of CAR.αPD1-T cells, CAR-T cells, and CAR-T cells combined with anti-PD-1 antibody was determined using a xenograft mouse model. Finally, the underlying mechanism of enhanced tumor eradication of CAR.αPD1-T cells was investigated by analyzing the expansion and functional capacity of tumor-infiltrating lymphocytes. Human anti-PD-1 CPIs secreted by CAR.αPD1-T cells efficiently bound to PD-1 and reversed the inhibitory effect of PD-1/PD-L1 interaction on T cell function. PD-1 blockade by continuously secreted anti-PD-1 prevented T cell exhaustion and significantly enhanced T cell expansion and effector function both *in vitro* and *in vivo.* In the xenograft mouse model, we demonstrated that the secretion of anti-PD-1 enhanced the antitumor activity of CAR-T cells and prolonged overall survival.

484816 Combination Cancer Therapy Using Chimeric Antigen Receptor Engineered Natural Killer Cells As Drug Carriers

Abstract:

Cancer treatments typically include surgical removal, chemotherapy, radiation, or some combination of these therapies. While these methods can remove the majority of the tumor cells, tumor recurrence and metastasis remain a major obstacle in cancer treatment. Numerous nanoparticle-based active targeting approaches have emerged to enhance the intracellular concentration of drugs in tumor cells; however, efficient delivery of these systems to the tumor site while sparing healthy tissue remains elusive. Cancer immunotherapy has garnered much attention within the last decade, and more recently, the concept of using immune cells as vehicles to transport anticancer agents directly to the tumor site has gained traction. Natural killer (NK) cells are part of the innate immune system and mediate rapid, short-lived responses by releasing cytokines that directly lyse infected or abnormal cells including tumor cells. The NK92 cell line is identical to the parental NK line isolated from a leukemia patient. NK92 cells are well-documented and have antitumor effects against various types of cancer including melanoma, leukemia and breast cancer in both preclinical and clinical settings. NK cells are not specifically cytotoxic to certain antigen-expressing target cells, but their specificity and efficacy can be enhanced with CARs. CARs are antigen-specific engineered receptors that can be expressed on human immune cells. They are composed of an extracellular antigen-binding domain derived from an antibody fragment. This allows the CAR-expressing immune cells to bind to specific surface antigens overexpressed on cancer cells. Intracellular signaling domains within the CAR provide signals to the immune cell to attack the antigen-expressing cell. Typically, CARs are expressed in T cells, but recent studies show CARs are effective tumor-targeting domains in NK cells as well. The inclusion of CARs in NK92 cells increases homing, specificity, and efficacy of tumor killing.

We modified our CAR.NK cells with crosslinked multilamellar liposomal vesicles (cMLVs) containing a chemotherapeutic drug, paclitaxel (PTX). cMLVs are liposomes functionalized with thiol-reactive maleimide headgroups, which allows them to be stably conjugated to the thiol-rich NK cell surface. Previously, we have demonstrated that cMLVs can act as a novel agent for combinatory drug delivery by co-localizing two drugs with distinct physicochemical properties to a single site, inducing a synergistic anti-tumor effect *in vitro* and *in vivo*. Here, our work presents the combination of immunotherapy and chemotherapeutic drug delivery by utilizing CAR.NK cells as carriers for PTX-loaded crosslinked multilamellar liposomal vesicles (cMLV (PTX)) to enhance antitumor efficacy in Her2 and CD19 overexpressing cancer models.

We ensured that the conjugation of cMLVs to the CAR.NK cell surface does not affect the functionality of the CAR.NK cell itself. After confirming the functionality of our cMLV(PTX)-conjugated CAR.NK cells *in vitro*, we established a mouse xenograft model to observe the effects of the anti-CD19 CAR.NK cells *in vivo*. Mice treated with conjugated cMLV(PTX)-CAR.NK had the most dramatically slowed tumor growth of all groups. We performed *ex vivo* analysis of our mouse xenograft tumor model to support our hypothesis that CAR.NK cells facilitate PTX delivery into the tumor site. The conjugated group had significantly higher PTX concentrations within the tumor tissue compared to all other groups. Overall, our work demonstrates that the efficacy of chemotherapy can be enhanced *in vitro* and *in vivo* while reducing off-target toxicity by using CAR-engineered NK92 (CAR.NK) cells as carriers to guide drug-loaded nanoparticles to the target site.

495006 Investigating the Drug Delivery Effect for Anti-Cancer Compounds Using Graphene Oxide Nanoparticles

Abstract:

The application of nanoparticles in the drug delivery industry has always been of substantial importance for enhancing the delivery efficiency of drugs. As one of the most widely used materials among all the applied nanoparticles, superparamagnetic iron oxide (SPION) has shown a great range of medical applications such as magnetic resonance imaging (MRI) contrast agents and also drug delivery along with applications in tumor therapy and cardiovascular diseases. Furthermore, having remarkable interacting capabilities with blood proteins and cancer cells, SPION has proved to be one of the most promising drug carriers for cancer treatment. To be applicable, SPION should be attached to a biocompatible and water-soluble polymer compound, such as PEG/PLA in combination with graphene oxide. This delivery vehicle has been investigated using SEM, FTIR, and Raman spectroscopies. In addition, according to the calculated rate constants by the pseudo-first order kinetic model and pseudo-second order kinetic model, the drug loading time can decrease with an increase in drug concentration. In this project study, appropriate models for the drug absorption and desorption parameters have been investigated to better determine the range of application for this drug delivery method. The results demonstrate a drug delivery vehicle that is biocompatible with a high degree of water dispersity, owing to a superior delivery vector.

Furthermore, UV-Vis spectroscopy was used to determine adsorption and desorption isotherms for naphthol blue black dye, a drug surrogate, as well as bovine serum albumin (BSA) and doxorubicin hydrochloride (DOX) onto a graphene oxide layer. This study compares the relationship of drug size, adsorption/desorption isotherms, and kinetic models to drug delivery, particularly the average percentage cumulative drug release rate when subjected to various pH and buffered environments. The use of nanoparticles, both passive and active targeting strategies, to enhance intracellular concentrations of drugs in cancer cells while avoiding toxicity to normal cells will be detailed in this talk.

496229 Genetic Polymorphisms in Inflammasome-Dependent Innate Immunity Among Pediatric Patients with Severe Renal Parenchymal Infections

Abstract:

Inflammasome innate immune response activation has been demonstrated in various inflammatory diseases and microbial infections. However, to our knowledge, no study has examined the inflammasome-dependent pathways in patients with urinary tract infection. Defective or variant genes associated with innate immunity are believed to alter the host's susceptibility to microbial infection. This study investigated genetic polymorphisms in genes encoding inflammasomes and the subsequent released cytokines in pediatric patients with severe renal parenchymal infections. This study included patients diagnosed with acute pyelonephritis (APN) and acute lobar nephronia (ALN) who had no underlying disease or structural anomalies other than vesicoureteral reflux (VUR). Single nucleotide polymorphism (SNP) genotyping was performed in the genes associated with inflammasome formation and activation (*NLRP3*, *CARD8*) and subsequent IL-1β cytokine generation (*IL-1β*). A total of 40 SNPs were selected for initial genotyping. Analysis of samples from 48 patients and 96 controls revealed that only nine SNPs (five SNPs in *NLRP3*; three SNPs in *CARD8*; one SNP in *IL-1β*) had heterozygosity rates >0.01. Hardy–Weinberg equilibrium was satisfied for the observed genotype frequencies of these SNPs. Analysis excluding patients with VUR, a well-known risk factor for severe UTIs, revealed a lower frequency of the CC genotype in *NLRP3* (rs4612666) in patients with APN and ALN than in controls. Correction for multiple-SNP testing showed that the non-VUR subgroup of the APN+ALN combined patient groups remained significantly different from the control group ($P < 0.0055$). This study is the first to suggest that the inflammasome-dependent innate immunity pathway is associated with the pathogenesis of pediatric severe renal parenchymal infections. Further investigation is warranted to clarify its pathogenic mechanism.

500769 Therapeutic Effect of Inhaled Tacrolimus-Loaded Nanocomposite Microparticles (nCmP) in a Pulmonary Hypertension Induced Rat Model

Abstract:

Abstract:

Pulmonary arterial hypertension (PAH) is a progressive cardiovascular disease, which hemo-dynamically is defined as an increased mean pulmonary arterial pressure higher than 25 mmHg at rest and 30 mmHg with exercise. Pulmonary drug delivery options have attracted increasing attention recently as they are capable of delivering therapeutics directly and efficiently to the lungs. As the result, pulmonary delivery systems exhibit advantages including increased local drug concentration, reduced side effects, rapid onset of action due to the enormous surface area and plentiful capillary vessels in the lung, and avoidance of the first-pass metabolism of the liver. Tacrolimus (TAC) is an immunosuppressant recently found to be useful in treating PAH as it reverses the BMPR-2 mutation associated with certain types of PAH. This study aimed to formulate TAC-loaded nanocomposite microparticles (nCmP) and investigate the therapeutic effect of inhaled TAC for the treatment of pulmonary hypertension in a rat model.

Acetalated dextran (Ac-Dex) is an acid-sensitive, biodegradable, and biocompatible polymer prepared in a one-step reaction by reversibly modifying hydrophilic dextran to acetalated dextran with hydrophobic groups. As compared to other commonly used polymers such as poly (lactic-co-glycolic acid) (PLGA), polylactic acid (PLA), and poly-Ïµ-caprolactone (PCL), Ac-Dex exhibits attractive properties suitable for the controlled release of therapeutic payloads. By controlling the reaction time during the formation of Ac-Dex, the ratio of cyclic acetal groups with a slower degradation rate to acyclic acetal groups with a faster degradation rate can be tuned. Moreover, the acid-sensitivity of Ac-Dex enables it to degrade faster in lower pH environments such as the lysosomes of macrophages or tumor cells, resulting in faster release of drug in such environments, while controlling the release before reaching them. Furthermore, Ac-Dex degrades into neutral by-products, which avoids undesirable changes of the micro-environmental pH in the body. Therefore, Ac-Dex was used to form nanoparticles (NP) loaded with TAC via a emulsion/solvent evaporation method followed by spray drying to form dry powder nCmP.

To induce PAH, adult Sprague-Dawley rats weighing 150â€"175 g were placed into a normobaric hypoxia chamber (10% FiO_2; Biospherix Ltd., Parish, NY) or room air for three weeks. A DP-4M insufflator (Penn-Century Inc., Philadelphia, PA) was used to administer the TAC-loaded nCmP into PAH-induced rat lungs. The pharmacokinetics and pharmacodynamics properties were assessed after the TAC nCmP were administered to the rats. In addition, the efficacy of the nCmP formulation towards the treatment of PAH was evaluated.

Keywords: Pulmonary hypertension; Tacrolimus; Acetalated dextran; Nanoparticles; Microparticle

496471 Engineering the Endothelial Glycocalyx to Restore Its Structure and Function

Abstract:

The glycocalyx (GCX) coats and protects the vascular endothelial cell (EC) layer and mediates EC function [1, 2, 3]. However, the specific pathways by which GCX regulates EC function remain to be clarified. Studies suggest that the EC GCX transmits signals, via the cytoskeleton, to cell-to-cell junctions where biological responses occur [1, 2]. Of the cell-to-cell junctions, gap junctions are potentially an important downstream player in GCX-mediated signaling, because gap junctions transport ions and signaling molecules which regulate numerous vessel functions [4, 5, 6] that rely on the GCX. The goal of our current research is to test the hypotheses that (i) GCX controls gap junction protein expression and gap junction coupling of neighboring ECs, and (ii) repair of damaged GCX can reverse dysfunctional EC communication. We studied EC with intact GCX, heparinase III enzyme degraded GCX (to disrupt GCX HS, the most abundant GCX component), or regenerated GCX (by replacing and stabilizing GCX HS). Inter-endothelial gap junction communication was evaluated using immunofluorescence confocal microscopy to examine gap junction protein (connexin43) expression and localization and by quantifying Lucifer yellow dye spread via gap junctions in EC monolayers. The results demonstrated gap junction communication dependence on GCX function. In addition, GCX repair by treating cells with exogenous HS and a co-factor restored gap junction protein placement which translated to the reactivation of gap junction channel activity. The results of this study are encouraging and will lead to future development of new therapies targeted at the GCX to address endothelial-dependent aspects of vascular disease.
References: [1] Tarbell and Ebong, Sci Signal, 2008; [2] Thi et al., P Natl Acad Sci USA, 2004; [3] Yao et al., Am J Physiol Heart Circ Physiol, 2007; [4] Segal and Beny, Am J Physiol, 1992; [5] Figueroa and Duling, Antioxid Redox Signal, 2009; and [6] Ebong et al., Pflugers Arch, 2013.

489396 Single-Cell Profiling of Dynamic Cytokine Secretion and the Phenotype of Immune Cells

Abstract:

Quantifying the heterogeneity of immune cells at the single-cell level in a high-throughput manner across multiple biological dimensions — from the genome and transcriptome, to intracellular and extracellular signaling, and to interaction with other kinds of cells — can have a direct impact in improving therapeutic discovery in biotechnology, in diagnosis of diseases, and in facilitating immunotherapy. It is desirable to have a modular method that can quantify and screen for cellular functionality such as motility, interaction with other cells, protein secretion to enable in-depth characterization of individual cells and that is able to integrate these parameters to provide multiplexed molecular profiles of single cells. We have developed and validated an integrated methodology that combines nanowell arrays and bead-based molecular sensors for detecting cytokine secretion dynamically without the need for encapsulation of single natural killer (NK) cells or T cells.

Although NK cells were classically defined as pre-activated effector lymphocytes empowered with innate cytolytic functionality, more recent data suggest that NK cells are endowed with complex functionalities including cytokine secretion and activation of antigen-presenting cells, and can thus act as a bridge between innate and adaptive immunity. Human NK cells are identified as $CD3^{-}CD56^{+}$ cells and are typically classified into different subsets based on the relative expression of the cell surface markers CD56 (adhesion marker) and CD16 (FcγRIIIA, low-affinity Fc receptor). The majority of NK cells in peripheral blood (>90%) are the $CD56^{dim}CD16^{+}$ phenotype, which is primarily believed to be responsible for cytolytic functionality including antibody-dependent cell-mediated cytotoxicity (ADCC) mediated by CD16. By contrast, the $CD56^{bright}CD16^{-}$ phenotype is the minor population in peripheral blood and is described as primarily responsible for secretion of cytokines including interferon gamma (IFN-γ). We utilized our Timelapse Imaging Microscopy In Nanowell Grids (TIMING) platform, which is based on incubation of individual cells within polydimethylsiloxane (PDMS) nanowell arrays, to directly examine the relationship between phenotype and dynamic IFN-γ secretion profile. Despite the fact that PDMS is widely adopted for the fabrication of microfluidic devices, PDMS tends to display a high level of non-specific protein adsorption. We demonstrated the robust passivation of the PDMS-based nanowell arrays with polyethylene glycol (PEG) and validated our assay by comparison to enzyme-linked immunospot (ELISPOT) assays. Numerical simulations were used to optimize the molecular density of capture antibodies on the surface of the beads to improve the capture efficiency of cytokines by the beads within an open-well system. Analysis of hundreds of individual human peripheral blood NK cells revealed that $CD56^{dim}CD16^{+}$ NK cells are immediate secretors of IFN-γ upon activation by phorbol 12-myristate 13-acetate (PMA) and ionomycin (< 3 h); there was no evidence of cooperation between NK cells leading to either synergistic activation or faster IFN-γ secretion. Within secretor NK cells population, early-secretor NK cells tended to have a higher level of CD16 surface expression, suggesting the existence of an elite population of $CD16^{+}$ NK cells capable of both lytic functionality and rapid cytokine secretion.

In a second application, we have quantified the dynamic secretion of IFN-γ from T cells using the same platform. Chimeric antigen receptor T cells (CAR^{+} T cells), which are genetically engineered to retarget and program anti-tumor functionality of T cells, are under intense clinical investigation as personalized cancer therapeutics. By utilizing CD19-specific third-generation CAR^{+} T cells, we have explored the relationship between basal motility of CAR^{+} T cells and their ability to secrete IFN-γ.

Collectively, these results establish our methodology as an investigational tool for combining phenotyping, cellular functionality and real-time protein secretion of individual cells in a high-throughput manner.

499392 Influence of Hepatic Function Due to Co-Culturing withÂ Endothelial Cell from Different Tissue Origins

Abstract:

Development of *in vitro* environments capable of mimicking physiological function has long been an aspiration of tissue engineers. The use of liver mimicking environments would be extremely beneficial in evaluating drug safety and efficacy, increasing the efficiency of the drug development process. In an effort to mimic *in vivo* architecture, three dimensional (3D) structures including cell aggregates and porous scaffolds have been investigated and reportedly exhibit more organitypic behavior. Also, the introduction of other cell types native to the organ of interest presents another way to mimic the physiological environment. We wanted to evaluate the effect of 2D and 3D, monoculture and co-cultures to determine which combination resulted in more closely mimicking *in vivo* behavior. We initially investigated the effect of culturing hepatocytes in 2D versus 3D on hepatic function by evaluating protein secretion, enzyme activity, cell viability, and metabolic profiles. Evaluation of enzymatic activity and metabolic profiles required challenging of cell cultures with a drug/substrate. We chose acetaminophen, due to readily available clinical data in literature. Metabolic profiles of the percentage of APAP consumption and the produced concentration of APAP metabolites (APAP-sulfate and APAP-glucuronide) were compared. The distribution of cells in the porous scaffold was also examined to ensure the culture environment was truly in 3D. Two hepatocyte cell lines, HepG2 and HepaRG, were investigated to determine the effect of cell line on function. We next investigated the addition of endothelial cells to hepatocyte cultures (2D and 3D) to determine the effect of co-culture on hepatic function using the same analysis techniques. Two endothelial cell lines, human umbilical vein (HUVEC) and liver sinusoidal (LSEC), were explored to determine effect on hepatocyte function. Protein secretion was higher in 2D compared with 3D in monocultures. The addition of endothelial cells closed the gap between protein secretion observed in monocultures. Co-cultures led to increased protein secretion in 2D and 3D compared with monocultures. Enzyme activity increased in 2D co-cultures compared with 2D monocultures. Metabolic profiles showed increased function in co-culture environments compared with monocultures.

501787 An All-in-One High Throughput Microfluidic Platform for Cell Culture and Migration Control

Abstract:

High-throughput and automated cell culture, conditioning, and screening platforms are becoming increasingly attractive with the advent of lab-on-a-chip technology which synergically couples the understanding of microfluidics with integrated sensing and controlling automation. Even though earlier works have demonstrated various platforms of either generating well-defined flow rates or controlling temperature at high consistency, the combination of these two ends in one single unit has not been established. Herein, we demonstrate an automated system that enables the on-demand regulation of both flow rate and temperature of a microfluidic system using a proportional-integral-derivative (PID) mechanism. As a proof-of-concept, we provided temporally and spatially high-resolution controls over flow and temperature in a microfluidic device which have two parallel flow channels (50 μm high (H) and 200 μm wide (W)) connected by an array of 50 diffusion channels (20 (W) x 15 (H) μm). As measurement is concerned, flow rate is calculated from the pressure drop inside these flow channels, which is measured via differential pressure sensors while temperature is measured by RTD sensors. These sensors are connected to a computer through appropriate IO modules. Fluctuations in these parameters are handled by Matlab code which could both communicate with high-precision microfluidic valves in regulating the flow rate and a custom-made ITO-glass heating unit in compensating for a temperature difference. With this platform, we were able to maintain flow rates as small as 2 μL min^{-1} in both flow channels. Temperature in the chip could be kept constant at 37 ± 0.1 °C. Following that, fibroblast cell chemotaxis experiment was performed in the chip. Platelet-derived growth factor-BB (PDGF-BB) was introduced in one of the flow channels and cells were seeded in another. A stable well-defined gradient of the chemoattractant was established inside diffusion channels. Fibroblasts were found to migrate inside the diffusion channels following the gradient of PDGF-BB over the course of 24 hours.

500997 Sustained Delivery of Phosphates from Crosslinked Peg Hydrogel Nanoparticles Suppress Collagenase Activity of Intestinal Pathogens

Abstract:

Introduction & Purpose:

Iatrogenic injury to the intestinal tract, such as that which occurs when surgeons remove tumors via endoscopy or major surgical resection, can disrupt the normal microbiota and leave a large wound to heal in the presence of highly pathogenic bacteria. Previous studies indicate that certain bacteria (i.e. *E. faecalis, S. marcescens* & *P. aeruginosa*) are capable of secreting high levels of collagenase, a key phenotype involved in healing impairment. Our previous studies have shown that normal inhabitants of the intestinal microbiota (*E. faecalis*) can be provoked to express enhanced collagenase in the gut during surgical injury leading to post-operative complications such as obstruction and anastomotic leak (a surgical problem that occurs when the resected and reconnected intestine fails to heal)[1]. A common approach to address this issue is the administration of oral and intravenous antibiotics to eliminate pathogens in the gut. However, clinical studies indicate that patients remain colonized by strains of *E. faecalis* and *P. aeruginosa* for as long as one week after surgery[2]. We have previously shown that depletion of extracellular phosphate occurs in the intestinal tract following surgical injury and that this depletion is a major cue that triggers bacterial virulence. As a result of extracellular phosphate depletion, pathogens must scavenge phosphate from host tissues, leading to the disruption of the epithelium and impaired healing. Earlier *in-vivo* and *in-vitro* studies have shown that the maintenance of monophosphate (Pi) in the epithelium via oral supplementation prevents virulence expression while maintaining bacterial survival[3]. Our recent studies indicate that a polyphosphate hexamer, sodium hexametaphosphate (PPi), suppresses virulence[4] as well as collagenase production *in-vitro* and prevents anastomotic leak when administered orally in mice subjected to intestinal injury and intestinally inoculated with collagenolytic bacteria[5]. Therapeutic drug delivery approaches focused on localized and sustained release of phosphates from nanoparticles to suppress collagenase production, while allowing commensal bacteria to proliferate normally offer great promise for the promotion of healing following surgical interventions in the intestinal tract.

We have developed Pi- as well as PPi- loaded poly(ethylene) glycol (PEG) hydrogel nanoparticles, NP-Pi or NP-PPi, respectively, using inverse phase miniemulsion polymerization, which result in sustained release of phosphates. Our previous studies have shown that NP-PPi suppress *P. aeruginosa* virulence in a dose-dependent manner *in-vitro* while maintaining bacterial survival. The goal of the present study is to determine the effects of NP-Pi and NP-PPi on the suppression of clinically relevant gram-negative and gram-positive collagenolytic bacterial strains.

Methods:

Nanoparticles were created using inverse miniemulsion polymerization[4]. This process involved free-radical initiated co-polymerization of PEG diacrylate (M_n= 575 Da) and N-vinylpyrrolidone (NVP). These reactants, as well as the water-soluble surfactant TWEEN 20 constituted the aqueous phase. The organic phase consisted of cyclohexane and Span 80 (an organic-soluble surfactant). The total concentration of reactive double bonds in the aqueous precursor was fixed at 0.685 M. A thermal initiator, potassium persulfate, was used to begin polymerization and was added to the precursor at a concentration based on a 1.5% molar ratio of initiator to total reactive double bonds. The reaction was allowed to proceed for 4.5 hours at 56°C under an ultra-high purity nitrogen blanket. Monophosphate (Pi) was physically encapsulated within the hydrogel nanoparticles during the inverse miniemulsion polymerization. After the reaction was complete, nanoparticles were precipitated in acetone and purified via three cycles of acetone washes with sonication and centrifugation. Particle size distribution was quantified using nanoparticle tracking analysis on a Nanosight LM10 system. Nanoparticle surface charge characteristics and swelling ratio were quantified using dynamic light scattering on a Malvern ZetaSizer NanoZS and gravimetric swelling measurements, respectively. Phosphate release kinetics from nanoparticles into bacterial tryptone yeast (TY) medium was determined using a molybdenum blue absorbance method. Blank nanoparticles were created by rinsing NP-Pi with excess deionized water until phosphate could no longer be detected by absorbance. NP-PPi were created by post-loading blank NPs with PPi. Blank NPs were exposed to a high concentration PPi solution for 14 days at 40°C ± 5°C. Post-loaded PPi-loaded NPs were then rinsed briefly with deionized water and characterized as described above. Clinically relevant collagenolytic pathogens were obtained from the Center for Surgical Infection Research at the University of Chicago Medical Center. Pathogens were prepared from frozen stock and were plated on tryptic soy broth agarized medium and incubated overnight at 37°C. A few colonies were transferred to 2mL of TY medium and grown in suspension for 24 hours at 37°C under shaking, after which experiments were performed to quantify bacterial virulence and survival. In order to quantify bacterial growth in solutions supplemented with NP treatment, bio-luminescent strains of *S. marcescens* and *P. aeruginosa* were obtained. Bio-luminescent *P. aeruginosa* with collagenase activity analogous to the isolated strain was available for purchase through Perkin Elmer *in-vivo* imaging reagents. Bio-luminescent *S. marcescens* was constructed by transformation with pAKlux2 plasmid containing *luxCDABE* luciferase cassette. To determine the ability of NP-Pi or NP-PPi to suppress collagenase, *S. marcescens* and *P. aeruginosa* were cultured in phosphate depleted media with varying concentrations of the following groups, free phosphate (Pi or PPi), NP-Pi or NP-PPi and blank (unloaded) nanoparticles. Bacterial growth was measured by bio-luminescence measurements. Collagenase activity was measured using a DQ™ gelatin and fluorescein conjugate assay kit with fluorescence excitation at 495nm and emission at 515nm.

Results:

NP-Pi and NP-PPi were found to have an average particle diameter of 181.7 nm ± 52.5 nm and 179.6 nm ± 71.8 nm respectively, a zeta potential of -17.92 mV ± 1.05 mV and -18.34 mV ± 1.44 mV respectively, a swelling ratio of 2.8 ± 0.08 and 4.0 ± 0.24 respectively, and allowed for sustained release of phosphate for over three days.

Our *in-vitro* studies indicate that for the gram-negative pathogen *S. marcescens*, NP-PPi are effective at suppressing collagenase activity and have no significant effect on bacterial survival when compared to control. Secondly, the blank NP control elicits no significant response in bacterial growth or collagenase levels for *S marcescens*. For the gram-negative pathogen *P. aeruginosa*, NP-PPi demonstrate a significant suppression of collagenase activity and a slight promotion of bacterial growth. The blank NP control unexpectedly shows mild attenuation of collagenase activity as well as significant promotion of bacterial growth. In the case of both gram-negative pathogens, our data indicate that PPi is more effective than Pi at suppressing collagenase activity. PPi (either delivered free or from NP-PPi) suppresses collagenase activity of the gram-positive pathogen *E. faecalis*, however, this effect was not found to be as significant as that observed in the case of the gram-negative strains.

Conclusions & Future Directions:

PEG hydrogel nanoparticles allow for sustained release of phosphates (Pi and PPi). NP-PPi allow for significant attenuation in collagenase production of gram-negative pathogens as a result of PPi release from these nanoparticles.

Current studies are being performed to optimize the effectiveness of NP formulations in suppressing collagenase production across gram-positive and gram-negative pathogens. Future work includes (1) determining the resistance of NP formulations to washout in a parallel plate flow chamber, (2) *in-vitro* and *in-vivo* studies to determine the mucoadhesiveness of NP polymer formulations on cultured epithelial cells and on intestinal mucosa, and (3) determining the functional durability and efficacy of NP formulations in animal models of intestinal injury.

References:

1. (Shogan BD, et al. Sci Transl Med. 2015:7(286):286ra68.)

2. (Ohigashi, S, et al. J Gastrointest Surg. 2013:17(9):1657-64.)

3. (Zaborin A, et al. Proc Natl Acad Sci. 2009:106(5):6327-32.)

4. (Yin Y, et al. Ann Biomed Eng. 2016:45(4):1058-1068.)

5. (Hyoju, SK, et al. Ann Surg. 2017 [Accepted].)

501611 Coarse-Grained Molecular Simulations Reveal Regulatory Insights into Immature HIV-1 Assembly Dynamics

Abstract:

The packaging and release of immature human immunodeficiency virus-1 (HIV-1) particles from host cells is a critical step in the viral replication cycle. Disrupting this highly-regulated process is a potential target for anti-retroviral treatments. It is well-known that Gag (group-specific antigen) polyprotein must multimerize into an immature lattice, which is also bound to a dimer of the viral ribonucleic acid (RNA) and is enveloped by the cell membrane. However, the molecular details of Gag assembly, especially as influenced by the presence of RNA and the cell membrane, remain unclear. To explore these dynamics, we developed a coarse-grained (CG) computational model that is derived from sub-nanometer resolution structural data and investigated the mechanisms that work in concert between Gag polyproteins, RNA, and the membrane to promote immature lattice growth. Our analysis, which is further supported by single particle tracking photoactivated localization experiments, demonstrates that viral RNA and the membrane are critical constituents that expedite Gag multimerization through scaffolding while short competitor RNA suppresses assembly. We also discuss the importance of weak anisotropic interactions at the helical spacer peptide 1 junction to mitigate defects in the otherwise continuous and hexameric Gag lattice. Finally, we reveal that lattice growth kinetics appears to be coupled to membrane deformation dynamics owing to intrinsic curvature that develops in the protein lattice. Overall, these findings elucidate a simple network of interactions that regulate the early stages of HIV-1 assembly and budding.

501404 Multidrug-Resistant *Escherichia coli*: Measurement of Membrane Mechanical Properties, Nanoscale Adhesion, and Biofilm Formation

Abstract:

A report from the United States' Center for Disease Control and Prevention (CDC) in 2013 estimated that at least two million people get antibiotic-resistant infections yearly and more than 23,000 die because of such infections. Rapid detection of multidrug resistance (MDR) in bacteria is a clinical challenge and pose a serious task to effective clinical management and control of MDR-related infections. To combat resistance, fundamental understanding of how microbial cell morphology and surface properties (elasticity, charge, and adhesion) differ among bacterial strains with variable drug resistances and abilities to form biofilms in response to antibiotics is needed. Here, we employed atomic force microscope (AFM) to characterize cell morphology and surface properties of *wild-type* multidrug-resistant *Escherichia coli* (MDR-*E.coli*) strains with various biofilm capacities before and after exposure to ampicillin at different minimum inhibitory concentrations (MICs). Forces between the hydrophobic or hydrophilic AFM cantilever and at least 15-25 locations on the bacterial surface were quantified. The MIC was determined using the broth dilution method.

Our results showed that individual *E. coli* strains with variable biofilm abilities responded differently to ampicillin as observed from morphological changes of their surfaces as well as quantified mechanical and adherence properties of their cells. For example, ampicillin at a MIC of 45 µg/ml and above the Clinical and Laboratory Standards Institute (CLSI) breakpoint concentration of susceptibility (8 µg/ml) induced morphological changes to the weak and moderate biofilm former strains in forms of cellular elongation as well as surface roughness and membrane disorientation within 1-8 hours of exposure. On the contrary, no morphological changes were observed in strong biofilm or no biofilm former *E. coli* strains.The elongated cells recovered after some time and maintained their normal morphology thereafter. When quantified, Young's moduli of elasticity of the moderate and the strong biofilm former *E. coli*strains were not changed from those of the untreated cells upon exposure to ampicillin at MIC of 50 µg/ml and 45 µg/ml, respectively. It is, therefore possible that ampicillin did not disrupt cell wall synthesis in these strains. However, ampicillin at MIC of 50 µg/ml significantly reduced the adhesion strength of the strong biofilm former strain while increased the adhesion force for the moderate biofilm former strain.

When all results are combined, weak biofilm former *E. coli* strains to elongate their cells and make their surfaces rougher to adapt to antibiotics. On the other hand, strong biofilm former strains rely on their biofilms to help them resist antibiotics more than on induced changes to their mechanical or morphological properties. This investigation provides a proof-of-concept that AFM can be used on the basis of detection of ampicillin resistance.

Keywords: AFM, multidrug-resistance, *E. coli*, surface properties, adhesion force and biofilm.

499492 Mechanical Force-Based Regulation of Protein Assemblies

Abstract:

Mechanical force-based regulation of protein assemblies

Ravi Chawla, Katie M. Ford, Pushkar P. Lele*

Artie McFerrin Department of Chemical Engineering

Texas A&M University, College Station, TX-77843

*Corresponding author: plele@tamu.edu

The bacterial flagellar motor undergoes structural changes in order to adapt to long-lived perturbations in the cell's thermal, chemical, electrical and mechanical environments. Remodeling of motor-assemblies improves the chances of survival. For example, the remodeling of the motor-switch extends the range of signal detection and enhances the ability to respond to chemical signals during chemotaxis.

The stator-complex that rotates the motor is responsible for surface sensing and remodels in response to perturbations in viscous loads [1]. The stator builds itself by recruiting additional stator units under high loads; however the mechanism for such load-dependent self-assembly is currently unknown. Here, we tested a hypothesis that the amount of force generated by each stator-unit modulates its association with the rotor. To do this, we measured stator-binding in mutants strains in which the motors reportedly develop lower torque compared to wild-type motors. Our measurements indicated that the stator retains at best one stator-unit when delivering a low force to the rotor. On the other hand, when delivering a higher force the stator retains as many as eleven stator-units. An analytical model was developed that incorporated an exponential dependence of the off-rate for individual units on the force delivered to the rotor. The model provided accurate fits to measurements of stator-rotor binding over a range of loads. It is possible that the tensile forces that develop when a stator-unit delivers a high torque uncover cryptic binding sites that stabilize the stator-rotor association. Our experimental and modeling results represent the first steps towards establishing a plausible mechanism for mechanical force-based regulation of the flagellar stator.

[1] Lele et al., Proc Natl Acad Sci USA, 110 (29), 11839-11844, 2013.

[2] Chawla et al., Sci Rep, under review.

498138 Modeling Drug Delivery By Electrokinetic-Based Methods in Cancer Tumor Treatment

Abstract:

According to the National Institute of Health, cancer is one of the leading causes of death worldwide. In 2012 alone, there were 14 million new cases and 8.2 million cancer-related deaths. Because of the prevalence of this disease, there has been a national push to develop novel new therapies and drug delivery systems for cancer treatment. The work reported herein is aimed at modeling drug delivery under the influence of an external electric field, with the goal of creating more of a "targeted" cancer tumor treatment. This is very important because of the complex, irregular vasculature of cancerous tumors that makes drug delivery very difficult. Alternative approaches, such as those based on electrokinetics, to those currently used in cancer tumor treatment could become very helpful if additional understanding of the role of the electrical field in the convective-diffusive transport of the drug can be gained. To further understand how to harness the promising technique mentioned above about the "targeted" drug delivery to the tumor site, the role that electrokinetic transport phenomena plays must be understood. In this presentation, we selected a cylindrical geometry to model the biofluid motion and the transport of drugs driven by the electrical field. In order to accomplish this, an expression for the electrohydrodynamics must be found by studying the classical hydrodynamics coupled with the electroosmotic flow to determine their influences on the velocity profile. From there, the transport of solute (drug) moved by the electrokinetic phenomena will be analyzed.

500187 Decoding Icy Metabolism: Flux Topology of a Psychrophilic Extremophile

Abstract:

Recent advancements in the mass spectrometry and system biology tools such as RNA-Seq have allowed for the precise characterization of microorganisms' metabolic functions. Representative species from many extreme environments have been characterized, from mesophilic to thermophilic and acidophilic organisms. Even though the volume of cold-temperature environments dwarfs that of other habitats, characterization of cold-adapted microorganism has remained relatively unexplored.

Here, we present the characterization of the marine psychrophilic extremophile, *Colwellia psychrerythraea* 34H, which grows at 0°C. We experimentally determined the biomass composition as well as the proteinogenic and lipid profiles of the bacteria. We used steady state ^{13}C-fingerprinting and ^{13}C-metbaolic flux analysis in conjugation with ^{13}C-dynamic labeling experiments to elucidate the central metabolic pathways of 34H grown on glucose and lactate. In tandem, we performed RNA-Seq analysis at icy temperature and room temperature to determine the global rearrangement of 34H's metabolism. We also used flux balance analysis to understand the cellular 'objective function' of 34H by comparing the computationally predict fluxes to the experimental determined fluxes. Finally, we used thermodynamic comparisons to explore the driving forces that may have led to the prevalence of the Entner–Doudoroff (ED) pathway among marine bacteria.

497112 CRISPRâ€"Mediated Genome Editing and Gene Repression inÂ *Scheffersomyces stipitis*

Abstract:

Scheffersomyces stipitis is an important nonconventional yeast known for its strong native xylose-utilizing capacity and has been recently shown as a well-suited host to produce compounds derived from shikimate pathway. However, its significant industrial application is hampered by the lack of genetic engineering tools. Here we developed CRISPR-Cas9 system in *S. stipitis* by using a codon-optimized Cas9 from *Streptococcus pyogenes* and single guide RNA (SgRNA), which improved targeted indel knockout efficiency by at least 80-fold. We further modified the CRISPR system *via* shortening *SNR52* promoter for easy-cloning and improving electroporation voltage for high transformation efficiency. Then, we constructed *ku* genes deficient strain and succeeded in increasing homologous recombination (HR)-based genome editing efficiency by at least 2-fold. Furthermore, we developed a transcription-modulating tool using CRISPR-dCas9 and validated that the repressor Mxi1 could promote the function of dCas9 to repress gene expression by 33%. Finally, we implemented our CRISPR system in the shikimate pathway, tuned gene expression, and improved shikimate and its derivatives production. Our methods enable delicately designed genetic/metabolic manipulations in *S. stipitis*, and provided good references for developing efficient CRISPR-Cas9/dCas9 tools in other nonconventional yeasts.

497963 Application of the Genome-Scale Modeling Approach to Exoelectrogenic Microorganisms in Microbial Fuel Cells

Abstract:

Abstract:

Microbial fuel cells (MFCs) use microogranisms to oxidize a substrate and produce electrical current. They are of growing interest as sources of renewable energy and for wastewater remediation. However, the application of MFCs is limited due to low power output and high material and designs costs. Research efforts to address these limitations have been primarily focused on experimental work, but there has been increasing interest in modeling and simulation [1]. MFCs have been modeled using kinetic models and genome-scale metabolic models. Kinetic models are used to model the dynamic changes and are based on the reaction kinetics and mass balance equations of the system. The resulting ordinary or partial differential equations are solved using numerical methods with boundary conditions based on the inputs to the system and the history [2]. There are numerous examples of kinetic models in the literature and a comprehensive review of kinetic models has been recently performed [1].

Genome-scale metabolic models are used to model the fluxes through the metabolic reactions in a system. A metabolic model is based on the genome annotation of the microorganism and the construction of the metabolic network. Metabolic models are analyzed using flux balance analysis based on the reaction stoichiometry [3]. In this work, a comprehensive review of the current state of genome-scale metabolic models is performed. Several different exoelectrogenic microorganisms have been identified. In particular, *Geobacteraceae* are most commonly studied and are well-suited for use in MFCs [4]. Two recent works have developed genome-scale models based on the microorganisms *Geobacter sulfurreducens* and *Geobacter metallireducens* [5][6]. Both works have further advanced previously developed models based on updates to the genome annotations for the microorganisms and the authors present results. The 2013 *G. sulfurreducens* model predicts the maximum theoretical power output to be 2.716 W/gDW [5]. The 2014 *G. metallireducens* model does not directly compute power output, instead it calculates electron donor and acceptor uptake rates, which are related to power output. In addition, the 2014 model identified a new culturing condition with formate as the electron donor, which was subsequently verified experimentally.

This new study has two objectives. First, a comprehensive review was performed of metabolic models from earlier works to the present with a focus on the two recent *G. sulfurreducens* and *G. metallireducens* models. The general process by which genome-scale metabolic models are developed is summarized and the processes by which the two recent models were developed based on new genome annotations for *G. sulfurreducens* and *G. metallireducens* are reviewed. Second, this study further analyzed the two recent models by running additional simulations in MATLAB to further evaluate two key questions: 1) what are the culturing conditions to maximize current output in MFCs based on the two microorganisms and 2) what are the best metabolic strategies to maximize current output. To evaluate the first question, the critical nutrients were identified for each microorganism based on the models. Phase diagrams have been generated for growth rates and fluxes through reactions related to current output as a function of critical nutrients. For example, for the *G. sulfurreducens* model, growth rate and fluxes through key reactions are shown to be linearly proportional to acetate uptake rate. To evaluate the second question, gene knockout analysis has been performed. For example, for the *G. metallireducens model*, 390 essential genes were identified and strategies for perturbing the metabolism, such as targeting genes for overexpression, were evaluated to optimize growth rate and fluxes through key reactions. Finally, a detailed comparison of the 2013 *G. sulfurreducens* model and the 2014 *G. metallireducens* model was conducted.

References:

[1] V.M. Ortiz-Martinez, M.J. Salar-Garcia, A.P. de los Rios, F.J. Hernandez-Fernandez, J.A. Egea, L.J. Lozano, "Developments in Microbial Fuel Cell Modeling," *Chemical Engineering Journal*, vol. 271, pp. 50-60, 2015. [2] J. Almquist, M. Cvijovic, V. Hatzimanikatis, J. Nielsen, M. Jirstrand, "Kinetic Models in Industrial Biotechnology - Improving Cell Factory Performance," *Metabolic Engineering*, vol. 24, pp. 38-60, 2014.

[3] J. Orth, I. Thiele, B. Palsson, "What is Flux Balance Analysis," *Nature Biotechnology*, vol. 28, no. 3, pp. 245-248, March 2010.

[4] D. Lovley, "The Microbe Electric: Conversion of Organic Matter to Electricity," *Current Opinion in Biotechnology*, vol. 19, pp. 1-8, 2008.

[5] L. Mao, W. Verwoerd, "Model-Driven Elucidation of the Inherent Capacity of the *Geobacter Sulfurreducens* for Electricity Generation," *Journal of Biological Engineering*, vol. 7, no. 14, pp. 1-21, 2013.

[6] A. Feist, H. Nagarajan, A. Rotaru, P. Tremblay, T. Zhang, K. Nevin, D. Lovley, K. Zenger, "Constraint-Based Modeling of Carbon Fixation of Electron Transfer in *Geobacter Metallireducens*," *PLOS Computational Biology*, vol. 10, issue 4, pp. 1-10, April 2014

492855 Multi-Paradigm Multi-Scale Metabolic Modeling of a Nitrogen Fixing Cyanobacterium with Two Distinct Metabolic Modes

Abstract:

Trichodesmium erythraeum is a marine dwelling nitrogen fixing cyanobacterium which is responsible for up to 42% of the annual biological nitrogen fixation. Nitrogenase, the enzyme which converts ambient nitrogen to ammonia, is poisoned by oxygen; cyanobacteria, which produce oxygen as a product of the water splitting reaction in photosynthesis, must segregate the two reactions. One approach is to create specialized cells (called heterocysts) that keep the microenvironment anaerobic and another is to fix nitrogen at night, when metabolism is respiratory. *T. erythraeum* is unique in how it handles this challenge: cells of similar structure to the rest of the filament carry out nitrogen fixation with a specialized metabolism (no photosystem II, carbon source is glycogen). Thus, the filament grows with two unique metabolic modes simultaneously. In order to model this organism more accurately the traditional flux balance analysis, we have developed a multi-scale multi-paradigm metabolic model which accounts for the diffusion of metabolites, spatial distribution and cell cycle.

The basis of this model is a genome-scale metabolic reconstruction of *T. erythraeum* which includes 971 reactions, 986 metabolites, and 647 unique genes and a biomass formation equation was based on experimental evidence collected in our laboratory. Constraints were developed for each specific cell type based on the differences in metabolism reported in literature. This general model was then used to predict fluxes for exponential growth and also used in dFBA to predict equilibrium compositions. Without any additional constraints, the dFBA model predicts an equilibrium composition of 15.5% diazotrophs which agrees well with published *in situ* data which report filaments are 10-20% diazotrophs. Moreover, the model predicts that nitrogen leakage is an essential condition of optimality for *T. erythraeum*; cells leak approximately 29.4% total fixed nitrogen when growing at the optimal growth rate, which agrees with values observed *in situ*. These results will be compared to predictions from the multi-scale multi-paradigm metabolic model we have created in an agent based framework. We will describe how placement in the filament and distance from the nitrogen fixing cells affects the metabolism of each cell. How this modeling framework can then be applied to more complex consortia will also be discussed.

496978 Uncovering and Correcting the Effect of Biomass Molecular Weight Discrepancies in FBA Calculations

Abstract:

In genome-scale metabolic models, the biomass has by definition a molecular weight of 1 g mmol-1. This is critical for correctly predicting growth yields, contrasting multiple models and more importantly modeling microbial communities. Nevertheless, we identified significant departures after examining 64 published models. The biomass weights of 34 models differed by 5 â€“ 50%, while 8 models have discrepancies >50%. 20 models were manually curated. By maximizing the original vs. corrected biomass reactions, flux balance analysis revealed >10% differences in growth yields for 12 of the curated models. Biomass molecular weight discrepancies are accentuated in microbial community simulations as they can cause significant and systematic errors in the community composition. Microbes with underestimated biomass molecular weights are overpredicted in the community whereas microbes with overestimated biomass weights are underpredicted. The observed departures in community composition are disproportionately larger than the discrepancies in the biomass weight estimate. Here we introduce a systematic procedure for checking and correcting the biomass weight, which we propose as a standard practice for metabolic reconstructions.

494710 Anaerobic Digestion of Kitchen Waste to Produce Biogas

Abstract:

The present work on kitchen waste was conducted to estimate the biogas production potential in local scenario with high alkalinity of water, and overall high mean temperature in the Sultanate of Oman. A 32 L anaerobic digester was designed and the studies were conducted in two combinations; namely biogas only from Cow dung manure (CD) and, biogas from Kitchen Waste and Cow dung mixture. Cow dung was used as a source of methanogens and was obtained from a local cattle farm in Sur, Oman. Temperature, pH, TSS, loading rate, retention time, and C/N ratio are among the key parameters studied in this biogas production process. All the experiments were run in triplicate. The digestion was carried out in a 32 liters anaerobic digester at a temperature of 31-32°C and under controlled pH in the range of 6.5-7 for a period of one month. The pH of the batch-fed digester was adjusted using NaOH as and when required during the entire cycle. A maximum of 5.75 L of biogas was produced on the 20th day of the study.

495625 Elucidation of Carbon Flux Topology Representing Photoautotrophic Growth in *Synechocystis* PCC 6803 Using Genome-Scale Isotopic Instationary Metabolic Flux Analysis

Abstract:

The reliability of flux estimation in photoautotrophic systems depends on the completeness and accuracy of carbon path coverage of the metabolic model. In this study, metabolic fluxes under photoautotrophic growth conditions in *Synechocystis* PCC 6803 are quantified by re-analyzing an existing dataset using genome-scale isotopic instationary ^{13}C-Metabolic Flux Analysis (INST-MFA) with the carbon mapping model *i*mSyn617. The mapping model traverses 18 novel carbon paths spanning Calvin cycle, photorespiration, an expanded glyoxylate metabolism, and corrinoid biosynthetic pathways and 190 additional metabolites absent in core models currently used for MFA. Flux elucidation reveals that 88% of the assimilated bicarbonate is fixed by RuBisCO and 12% off-gassed as CO_2. This sub-optimal carbon fixation is compensated by conversion of fixed CO_2 to biomass with near 100% efficiency while minimizing the yield of organic acids and glycogen. A newly discovered modality is the bifurcated topology of glycine metabolism using parts of photorespiration and the phosphoserine pathways to avoid carbon losses associated with glycine oxidation. The TCA cycle is shown to be incomplete with a bifurcated topology, fueled by anaplerotic metabolism and pyruvate oxidation. Inactivity of futile cycles and alternate routes results in pathway usage and (in)dispensability predictions consistent with experimental findings. The resolved flux map confirms maximization of biomass yield as the cellular objective with carbon fixation as the major metabolic bottleneck. Flux prediction departures from the ones obtained with the core model demonstrate the importance of constructing mapping models with global coverage to reliably glean new biological insights using labeled substrates.

507082 N-acetylchitohexaose Producing Chitinase Identified from Chitinase Profiles of *Aeromonas schubertii* by Enzymomics, a Novel Technique

Abstract:

Chitin, the abundant polysaccharide and main solid waste in the seafood processing industry, is mainly obtained from crustaceans now. Derivatives of chitin, inclusive of polysaccharides, oligosaccharides and monosaccharides, are biodegradable, histocmpatible and nontoxic. It attracts lots of attention since the chitin-oligosaccharides have been found to enhance the functions of immunological system and they might display as tumoricidal agents. While, the biological activity of chitin-oligosaccharides depends on the solubility and polymerization. The N-acetylchitohexaose with strong biological medical function has been demonstrated.

Chitinases are the enzyme cleaving chitin and distributed wildly. The chitinases in plants are part of defence as pathogenesis-related (PR) proteins. Few chitinases or chitinase-like proteins are identified in mammalian. Most functions of them are still unclear. In insects, crustaceans, and arachnids, chitin are the exoskeletons. Their chitinases play the roles in developing. In microorganisms with chitinases, those enzymes are essential for nutrients. For facilitate the chitin assimilation, many bacteria produce multiple types of chitinase with different biochemistry characteristics, such as molecule weights and pI. This multiplicity is presumed from different genes or modification of post-transcription and post-translation. Until now, the whole chitinases from one livestock have never profiled.

Recently, a technique, enzymomics, consolidating proteomics and zymography, in gel-activity assay, was developed. Enzymomics is the investigation of enzymome, the set of specific enzymes encoded by the genome. The approach was applied on a micoorganis, *Aeromonas schubertii* from the soil of mangroves marsh of south Taiwan. A bunch of the chitinases of *A. schubertii* were screened and profiled by enzymomics. Among those enzymes, one secretory chitinases with potent and SDS-resistant chitinolytic acititiy was selected and then isolated. This chitinsase was characterized with the molecular weight of 61,202 and produces hexameric oligochitinosaccharide from chitin degradation. It shows this enzyme has great potentialities for mass production of N-acetylchitohexaose in industry.

511416 Development of a Microfluidic Protocol for Fabricating PLGA Microparticles for Controlled Release Drug Delivery

Abstract:

As compared to traditional emulsion-based methods, microfluidics allows for precise size control of polymeric microparticles that encapsulate drug. Microparticle size is a key factor impacting *in vivo* biodistribution and potential for cellular phagocytosis. Optimal, localized drug delivery reduces the required drug load, enhances therapeutic effects, and improves overall safety for patients. For these reasons, a microfluidic method to fabricate poly-lactic-co-glycolic acid (PLGA) microparticles was explored. Two commercial microfluidic chips (Micronit, The Netherlands) with different channel widths (500mm and 100mm) were tested. Larger channels allowed for better stabilization of flow rates providing sufficient control to make microparticles and were better cleared of residual material allowing for reuse. The capacity to encapsulate drug with this approach was verified using a corticosteroid desired for local delivery to the sinuses. Comparable drug loading to emulsion microparticles demonstrated feasibility of creating monodispersed microparticles with this microfluidic system for controlled release drug administration.

489275 *Ex vivo* Study of Nanowires in Miniguts

Abstract:

Introduction: Telluride nanowires have many potential applications owing to their unique electronical, optical, and thermal properties. However, their use in medical applications is seldom considered due to the unclearness of their nanotoxicity. In this study, we demonstrated a case study on Bi2Te3 nanowires. Through live/dead cell viability testing and surface area analysis, we showed that 50 µg/mL Bi2Te3 exerts minimum influence on shrinking crypts. Figure A is a scheme of the experiment.

Methods: Primary small intestinal crypts were isolated from mouse small intestine. A 10mm small intestine was taken out and cut in pieces of 1 mm to 2 mm in length and washed five times using ice-cold PBS. Crypts were isolated, implanted within Matrigel on a pre-incubated 24-well plastic plate. Nanowires were added on the seventh day of growth.

Results: Figure B shows the live/dead cell images, where green indicates live cells and red indicates dead cells. Upon comparing the merged images of each concentration with those of the control, 200 µg/mL and 400 µg/mL showed more cells stained red and less cells stained green, which indicates that most of the cells were unhealthy. 50 µg/mL showed no significant difference compared with the control, which further indicates nanowires at 50 µg/mL pose the least harm to organoids and can be used as the optimal concentration.

Conclusions: We have demonstrated that Bi2Te3 nanowires at 50 µg/mL possess minimal cytotoxicity for organoids. Due to the toxicity of nanowires, concentration intake is limited. Because of their small size, nanowires can be constructed into any desired shape. They could be engineered into a microvillus-like shape for nutrition supply. Certain types of nanowires with unique thermoelectric properties could be incorporated into organoids and provide electrical/thermo potential, which can be utilized to manipulate the functionality of organoids.

Figure:

498678 Comprehensive Proteomic Analysis of High Productivity CHO Cells

Abstract:

Chinese hamster ovary (CHO) cells have been widely used as host cells for the production of recombinant therapeutic protein. However, the limited understanding of the regulation mechanism of exogenous gene expression in CHO cells has hampered the further improvements of protein production. In this study, a comprehensive proteomics approach was taken to better understand intracellular regulation of recombinant protein expression in CHO cells. Specifically, the global proteomic maps of a high antibody producing cell line (CHO DG44/IgG, volumetric titer >1g/L in shaker flasks) and its host (CHO DG44) were analyzed and compared. The intracellular proteins associated with transcription, translation, post translation modification, as well as carbon and energy metabolism were investigated, and the potential regulators were identified. In addition, proteomic analysis of the antibody producing CHO cells cultivated in batch process and different fed-batch processes were carried out, which was integrated with glucose, glutamine, and amino acid metabolism. This proteomic study can improve comprehension of cellular mechanisms in CHO cells and direct future cell engineering and process optimization.

499491 Labeling Neural Stem Cells Using Trackable Ultrasmall Iron Oxide Nanoparticle for Cell Transplantation Therapy

Abstract:

Neural stem cell (NSC) transplantation is starting to be used in clinical trials for regenerative rehabilitation in spinal cord injury, Parkinsonâ€™s disease, amyotrophic lateral sclerosis, and other neurodegenerative diseases. However, proper tracking of the transplanted NSCs is challenging. Currently, magnetic resonance imaging (MRI) is broadly used after NSC transplantation. Gadolinium (Gd) has been used as a MRI contrast agent because of its high resolution imaging. However, Gd contrast agents cause several adverse reactions such as nephrogenic systemic fibrosis, allergy-like reactions, and Gd retention in tissues. Another disadvantage is that Gd is not able to be track the NSCs long term while they are migrating to lesions. To circumvent these disadvantages of Gd, we labeled NSCs using ultrasmall iron oxide nanoparticles (Fe-USNPs). We were able to confirm the tracking of Fe-USNPs in NSCs over a week after their uptake by Prussian blue staining and MRI. We were also able to grow NSCs as neurospheres followed by the addition of the Fe-USNPs and verified their viability and stemness using stem cell markers and differentiation markers. These early results show promise to adapt the Fe-USNP labeling strategy for improved tracking of transplanted NSCs in the clinic in the future.

495043 Using Rnaseq to Investigate Population Heterogeneity Among Human Pluripotent Stem Cells Cultured on 2D or in 3D Biomaterial Scaffolds

Abstract:

Human pluripotent stem cells, with potential for indefinite self-renewal, may be an unlimited source of cells for a variety of biomedical applications including cell replacement therapy, disease modeling or drug screening. Often, success in these applications rely on our ability to control the stem-cell fate during expansion or differentiation. However, culture platform and stochasticity are two of the major factors that influence stem cell fate and affect the desired output. Single-cell RNAseq is a powerful, high-throughput, rapidly-evolving technique that may help us understand the effect of these factors on stem cell fate. Here, we use single-cell analysis to investigate the population heterogeneity within human induced pluripotent stem cells (hiPSCs) cultured either on 2D or in 3D scaffolds. We simultaneously analyzed ~10,000 different cells and identified ~26000 different genes. Subsequently, we developed a customized data analysis pipeline and observed distinct transcriptomic differences (specifically, ~8000 differentially expressed genes) between cells cultured on 2D and in 3D, and biological pathway analysis revealed phenotypes that correlated with our experimental observations. Importantly, single-cell analysis identified several subpopulations within the cells cultured using each platform, with unique biological properties. Further investigation of the population heterogeneity may elucidate the heterogeneous stem cell response to homogenous inputs.

493554 Screening and Characterization of Rabbit Scfv Antibodies for Sensitive Detection of C-Reactive Protein in Clinical Diagnosis

Abstract:

Screening technology of recombinant antibody fragments such as VHHs, scFvs and Fabs is critical and significantly important for drug discovery, affinity chromatography and diagnosis. Phage display system has been utilized for this purpose, while those with insufficient affinity were sometimes major components and therefore, identification of high-affinity recombinant antibodies were sometimes laborious process after biopanning. Here, we developed and proposed a promising methodology for identification of high-affinity antibodies from phage library. In a new biopanning procedure, negatively-charged liposomes were used as a solid support to immobilize antigen, CRP. By use of antigen-coupled liposomes, a variety of scFv clones with antigen-binding activities were successfully isolated. On the basis of aminoacid sequences of CDR H3, scFv clones with different epitopes were distinguished. All of scFv clones were ranked according to dissociation rate constant, koff which was calculated from the data obtained by SPR sensor. Production levels and dissociation constants Kd of scFv clones ranked in top 10 were further evaluated. Antigen-binding activities of scFvs adsorbed or coupled on the solid supports were also compared with those of the conventional mouse monoclonal antibodies. Consequently, the best variants of rabbit scFv clones successfully were immobilized on the surface of PS latex beads, and quantitatively detected antigen by latex turbidimetric assay. Thus, the methodology of scFv identification which was developed in the present study will be significantly useful for screening of diagnosis-grade rabbit scFv clones with high antigen-binding activity and specificity.

496826 Optical Nanosensors for Monitoring 3D Oxygen Gradients and Oscillations in Biofilms

Abstract:

Bacterial bioï¬lms can form persistent infections on wounds, on implanted medical devices, and are associated with many chronic diseases, such as cystic fibrosis. These infections are medically difficult to treat due to slow antibiotic penetration through the biofilm matrix making biofilms more resistant to antibiotic attack compared to their planktonic counterparts. The development of nanosensors to spatially and temporally monitor the behavior of biofilms via target analytes is a critical component towards improved understanding of biofilms and treatments. To this end, we developed oxygen-sensitive optical nanosensors to measure 3D oxygen gradients and growth oscillations within *Pseudomonas aeruginosa* biofilms.

Using this approach, we improved on traditional electrode-based 1D methods of measuring oxygen profiles by investigating the spatial and temporal variation of oxygen concentration during bioï¬lm growth and the changes in oxygen consumption rates when bioï¬lms are challenged with antibiotic attack. We observed spatial oxygen oscillations as the biofilm grew due to the dichotomy between the protection of interior cells and the nutrient consumption that occurs during growth. Further, we studied 3D oxygen gradients where oxygen was present at greater biofilm depths under antibiotic attack compared to the untreated controls.

500145 Rapid and Facile Fabrication of Thermoplastic Organs-on-Chips

Abstract:

Introduction: Micro physiological cell culture systems or "organs-on-chips" have garnered interest from both academia and industry. The scientific community has predicted that organs-on-chips will be used to study disease pathology and facilitate drug discovery. Furthermore, it has been postulated that seeding patient derived cells within organs-on-chips will enable personalized medicine. However, the currently prevalent microfabrication of organs-on-chips via poly(dimethylsiloxane) (PDMS) soft lithography may limit widespread access to microphysiological systems and their use toward drug discovery. PDMS soft lithography requires specialized microfabrication training and facilities which aren't widely accessible. Furthermore, PDMS absorbs hydrophobic molecules which is problematic for screening hydrophobic drug candidates. Finally, PDMS is highly gas permeable which is problematic for modeling hypoxic tissues like the intestine.

Methods: Here we present the facile, rapid, and scalable fabrication of a human intestine-on-a-chip using a commercial laser engraver systems, acrylic sheets, double sided adhesives, and a polycarbonate track etch membrane. The human intestine was recapitulated by culturing human Caco-2 cells in the apical fluidic compartment on a 1.0μm membrane while a syringe pump introduced cell culture medium into both the apical and basal fluid compartments. Caco-2 cells on chip were compared to control cultures of Caco-2 cells on commercial Transwell inserts via immunofluorescent staining for tight junction protein, F-actin, and cell nuclei. Mucus production on chip was compared to control Transwell cultures via immunofluorescent staining for mucin protein MUC2 and alcian blue staining.

Results: The fabrication methodology presented here boasts several key improvements compared to PDMS soft lithography: chip fabrication throughput was increased from days to hours while keeping material cost < $2 per chip, the technique doesn't require any specialized microfabrication, the thermoplastic chip construction enables oxygen tension control, and the double sided adhesives simultaneously act as fluidic compartments and provide a leak free bond without additional surface treatments like oxygen plasma or silanization. Immunofluorescent staining revealed the chip construction was biocompatible with human Caco-2 cells that similarly expressed tight junctions and F-actin compared to Transwell control cultures. Immunofluorescent staining and alcian blue staining showed that Caco-2 cells on chip produced more mucus compared to Transwell control cultures. We envision that the low cost, rapid, facile fabrication technique presented here will enable widespread access to micro physiological systems for researchers without microfabrication infrastructure or training.

487727 Byproduct Cross Feeding and Community Stability in an *in silico* Biofilm Model of the Gut Microbiome

Abstract:

The Human Microbiome Project was launched in 2008 with the goal of developing improved understanding of the human microbiome and its role in human health and disease. To date, the most widely studied system is the gut microbiome due to its critical role in food metabolism, profound influence on the immune system and suspected role in a wide variety of diseases including gut infections, inflammatory bowel and Crohns diseases, obesity, diabetes, cardiovascular disease, rheumatoid arthritis, colorectal cancer and even depression. The human gut microbiome is a highly complex multispecies system thought to consist of at approximately 1,800 genera and 15,000-36,000 species of microbes. The two dominant phyla in healthy humans are Firmicutes and Bacteroidetes, which comprise more than 90% of the community. Other important but much less prevalent phyla are Proteobacteria, Actinobacteria, Euryarchaeota and Verrucomicrobia as well as Eukaryota such as fungi. A critical metabolic function of the gut microbiota is to convert dietary fiber into short-chain fatty acids (SCFAs) that can be absorbed by the host intestine as an energy source.

Gut microbiome function is usually robust to dietary changes and other perturbations that alter species composition and SCFA synthesis. Correspondingly, metagenomic studies have shown wide variations in microbiota diversity and composition within healthy human populations. However, large perturbations in susceptible individuals can result in long-term microbiome alteration. Many diseases are associated with the gut flora being perturbed from their normal state through a poorly understood process known as dysbiosis. The role of species interactions in maintaining healthy community function or promoting dysbiosis are poorly understood. Despite some recent progress, the types of species interactions required for stable community dynamics in biofilm communities where nutrient gradients provide niches for different metabolic lifestyles have not been elucidated.

We developed a biofilm metabolic model of a very simple gut microbiome community consisting of a representative bacteroidete (*Bacteroides thetaiotaomicron*), firmicute (*Faecalibacterium prausnitzii*) and proteobacterium (*Escherichia coli*) to investigate the putative role of metabolic byproduct cross feeding between species on community stability, robustness and flexibility. The model predicted co-existence of the three species only if four essential cross-feeding relationships were present. We found that cross feeding allowed co-existence to be robustly maintained for large variations in biofilm thickness and nutrient levels. Our model predictions provide new insights into the impact of byproduct cross feeding and yield experimentally testable hypotheses about gut microbiome community stability.

487277 Mathematical Modelling of Salt Transport in Dry Salted Cheeses

Abstract:

Sodium chloride is vital to the production of safe and consistent cheese products through its effects on preservation, texture and functionality, and flavour. Understanding the mechanisms and modelling the movement of salt through fresh cheese curd, especially for dry-salted cheeses, is essential to optimizing production and improving consistency to meet growing global market demand. Current models for salt transport in cheeses have primarily used an unsteady-state Fickian diffusion approach to describe salt mass transfer in cheeses. While simple, these models fail to account for the underlying mechanisms and interactions between the transported salt ions, moisture, and proteinaceous curd matrix that ultimately control the rate of salt uptake during the dry salting process, especially the effects of local moisture content and moisture expulsion. A fundamentally-derived, one-dimensional mathematical model was developed to model the uptake and transport of salt and moisture through fresh, dry-salted cheese curds by defining the curd as a proteinaceous matrix with whey-filled, interconnected pores and accounting for the effects of osmotic pressure induced moisture expulsion. The model provides an essential first step to fundamentally describing the mechanisms that control the uptake and transport of salt and moisture during the salting process of cheese making.

501530 Effect of Drying Temperature, Humidity and Time on the Physico-Chemical Properties of Sugar Kelp (*Saccharina latissima*)

Abstract:

Sugar kelp (*Saccharina latissima*) is marine macro-algae and is a rich source of fibers, vitamins, minerals and antioxidants. Due to high moisture (~92%), it is highly susceptible to microbial attack and enzymatic deterioration and is either conventionally sun dried or hot air dried for extending its shelf life. Sun drying is one of the oldest techniques for food preservation, it is very slow, requires clear weather conditions which makes it an unreliable technique for Maine seaweed farmers. Hot air drying induces faster drying rate, but also leads to deterioration of texture and reduction in heat sensitive nutrients including vitamin C, antioxidants, phytochemicals, total flavonoid content and total phenolic content. Drying also creates void space and stress at the cellular level, leading to shrinkage that affect the quality of the dried seaweed. Since, shrinkage is governed by glass transition temperature (T_g). The aim of this research is to optimize the drying parameters by studying the effect of glass transition phenomenon on shrinkage of sugar kelp and hot air drying temperature, humidity and time on the physico-chemical properties (water activity, moisture content, pH, color, water holding capacity (WHC), oil holding capacity (OHC), ash content, fat content, vitamin C, antioxidant capacity and total phenolic content) of sugar kelp.

Freeze-dried sugar kelp was placed in air tight jars with different water activities (0.1–0.9). Moisture sorption isotherm was fitted non-linearly to BET and GAB model. Equilibrated samples were analyzed using a differential scanning calorimeter for T_g and freezing point (Tf). Samples were scanned from -90 to 100°C at the rate of 2°C/min. State diagram was developed based on Gordon-Taylor and Chen's model. Fresh sugar kelp samples of approximately 450g were dried at an air temperature of 30°C, 40°C, 50°C, 60°C and 70°C with relative humidity levels of 25% and 50% and air velocity of 10.0 m/s in the convective dryer. Dried sample were packed in air tight zip lock bags and stored at -80°C for further analyses of physicochemical properties.

The temperature gradient between air temperature and glass transition temperature had a clear impact on the shrinkage rate of sugar kelp. Drying temperatures below T_g resulted in lower shrinkage rates as compared to higher drying temperatures. At higher drying temperatures, the amorphous form of water continued to stay in the rubbery state, exhibiting a more flexible solid matrix. As the drying was entirely in the rubbery state region, the diffusion mechanism did not deviate from ideal Fick's law, which can thus be applicable for modeling thin layer diffusion for the falling rate period. The ash content of the samples were found to be in the range of 23.32% - 33.05% (w.b) and are inversely correlated to the water holding capacity ($r = -0.84$) and oil holding capacity ($r = -0.84$), which indicate the textural properties are highly dependent on the ash content irrespective of the drying temperature and humidity conditions. Heat sensitive nutrients such as vitamin C showed a positive correlation with respect to the drying temperature (30°C to 70°C) indicating drying time have significant effect ($p<0.05$) and it increases as the drying temperature increases corresponding to drying humidity of 25% and 50%, respectively.

This is the first study to report the effect of T g on shrinkage during drying of sugar kelp. This study also demonstrated the potential application of dehumidified drying in food industry as it might preserve the heat-sensitive nutrients. Results from this study will also help in estimating storage and processing conditions for the seaweed industry. This important information is essential for developing an innovative technology focused on clean, energy efficient and closed drying system for producing top-notch and local finished products for American consumers.

493508 Stability of Virucidal Effects of Green Tea Extracts

Abstract:

Viral infections often pose tremendous public health concerns as well as economic burdens. Despite the availability of vaccines or antivirals, personal hygiene is often considered as effective means as the first-hand measure against viral infections. The green tea polyphenols, in particular, epigallocatechin-3-gallate (EGCG), are known to exert potent antiviral activity. Here we evaluate the stability of virucidal effects of green tea extract (GTE). The temporal stability of antiviral effects of GTE was determined by measuring its virucidal activity against human influenza virus (A/H1N1) using the plague assay. The virucidal activity was observed in a dose dependent manner, and about 4–6 $\log_{10}$ reduction of infectivity was achieved in 0.01–0.1% GT solution. >99.99% inactivation of the influenza virus was achieved within five minutes exposure to GT solution. When formulated with 0.2% ascorbic acid, the potency of virucidal activity was compromised, and the exhaustion of antiviral activity was delayed, prolonging the duration of the virucidal function over extended period. A potent antiviral activity with about 4 $\log_{10}$ reduction of viral titers was maintained with wide range of temperatures (4–37 ºC) for two months as solution type, and for prolonged period as power. These data suggest that green tea extracts could be formulated as a safe and environmentally friendly personal hygiene against viral infections.

495499 Exploring the Potential Applications of Molecular Simulations to the Food Industry

Abstract:

Molecular dynamics simulations (MD) is the tool of choice for many researchers in their effort to investigate in atomistic (or near-atomistic) detail the structure, dynamic, or thermodynamic properties of biological and chemical systems. Employing MD in the food industry can produce a deep understanding of the structure-function relationship of the molecules of interest. Most of the application of MD to food-related problems revolves around the behavior of simple carbohydrates molecules in water and their interaction with proteins or other components. In this paper, we will discuss the impact that MD simulations could have in food research and, particularly, in Cargill. We will focus on two examples that show great potential for application to that area.

First, we studied the hydration properties of Scleroglucan (Sclg), which is a polysaccharide of D-glucose that is produced by fermentation of fungi. It consists of Î²-1,3 beta-glucans with Î²-1,6 branching. Sclg is quite an interesting molecule, as it has been shown to form a rod-like, triple helical structure, stabilized by both backbone hydrogen bonds at the center of the helix and side chain water bridges. Moreover, it has been suggested that multiple triple helices approach each other, creating sub-nanometer water channels between them that allow the permeation of small molecules only. We performed MD simulations to investigate the behavior of the helices and of the water surrounding them. We try to answer the questions regarding the stability of the triple helices and the impact they have on their environment.

Next, we discuss our work on elucidating the interaction between steviol glycosides with the bitter taste receptor that they bind to. There is an increasing demand for natural, zero-calorie sweeteners. Stevia is such a sweetener, extracted from leaves of the plant species *Stevia rebaudiana,* and it can be used as a sugar substitute. Stevia is a mixture of different components, with the most active being RebA, which is up to 150 times sweeter than sugar. Experimental studies have revealed that RebA binds to a specific bitter taste receptor that belongs to the GPCR group, which is a well-studied group of transmembrane proteins. We have employed structure-prediction methods and MD simulations to study the interactions that take place between the receptor and RebA. Our goal is to create a reliable model that can be used to explore the structure-activity relationship of RebA, and to potentially be able to predict the taste profiles of RebA analogues.

Assessing Environmental Impact from Acid Whey to Value-Added Products

Jasmina Burek[1], Daesoo Kim[1], Peggy Tomasula[2], Winnie Yee[2], Greg Thoma[1]

[1] University of Arkansas, Fayetteville, AR, [2] USDA ARS Eastern Regional Research Center, Wyndmoor, PA

Greek yogurt, America's rising star (38% market share), has 40% and 60% higher climate change impact and water depletion, respectively compared to set or stirred yogurt due to a straining process to remove excess acid whey[1] (Figure 1).

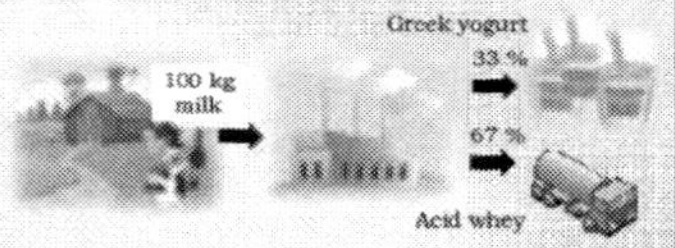

Figure 1. Greek yogurt and acid whey produced from 100kg of milk.

Previous research showed that anaerobic digestion can potentially reduce impact of Greek yogurt plant up to 11%[1] (Figure 3).

A problematic by-product

- production of Greek yogurt is expected to increase
- existing dairy wastewater treatment (WWT) plants cannot treat acid whey due to pH, BOD, and COD limits
- options such as whey-to-fertilizer and whey-to-feed have a low profitability
- installing anaerobic digester is costly
- concentrated acid whey products are not approved for use in food preparations.

A second life of acid whey (Figure 2)

(1) Whey-to-concentrate by reverse osmosis (RO) and drying – for use in animal feed and food preparations, if approved as such

(2) Whey-to-concentrate and lactose by ultrafiltration (UF), RO, and drying - for use in animal feed preparations

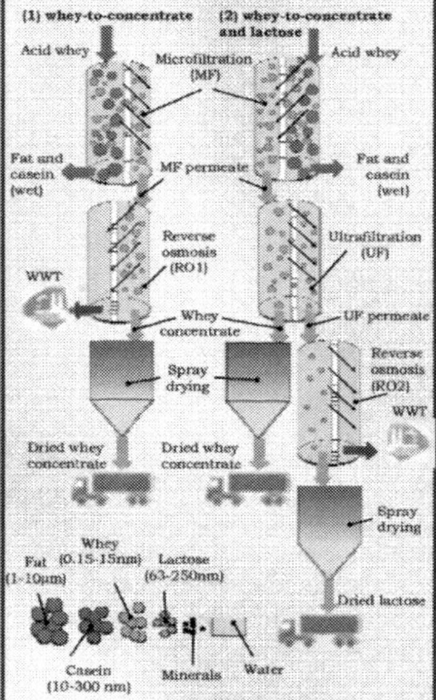

Figure 2. System boundary: from acid whey to value added dried whey concentrate and lactose.

Goal and scope

(A) cradle-to-gate life cycle assessment (LCA) of the yogurt processing plant including acid whey membrane scenarios and comparison to baseline whey-to-feed scenario

(B) gate-to-gate hot spot LCA of (1) and (2).

Allocation of environmental impacts between Greek yogurt and by-products was omitted in both (A) and (B).

We used SuperPro Designer© software to obtain the life cycle inventory (LCI) of membrane scenarios (1) and (2), Table 1[1,5].

Table 1: The LCI for acid membrane whey treatments including MF, RO, UF, and spray drying (per kg of wet or dried by-product).

	(1)&(2) MF	(1) RO+spray drying	(2) UF+spray drying	(2) RO+spray drying
Electricity (kWh)	2E-02	0.393	0.771	0.423
Natural gas (m^3)	5E-03	1.12	2.05	0.396
Membrane (kg)	3E-05	3.9E-04	2.9E-04	5.0E-04
Water (L)	0.376	1.37	5.23	4.07
Chemicals (kg)	0.002	0.012	0.049	0.036
WWT (L)	0.376	6.37	5.23	19.8
Transport (tkm)	0.108	1.20	0.304	0.278

MF- microfiltration, RO – reverse osmosis, UF – ultrafiltration.

Life cycle impact assessment (LCIA) included climate change (CC), non-renewable fossil energy (NonRF), water depletion (WD), marine eutrophication (MEU), photochemical ozone formation (POF), freshwater eutrophication (FEU), human toxicity (Htox), and ecotoxicity (Etox)[2,3,4]. Baseline Greek yogurt whey-to-feed cradle-to-gate LCIA results are shown in Table 2. Membrane scenarios show maximum 7% increase in cradle-to-gate environmental impacts. Primary hotspots are the same for both membrane scenarios (Figure 4).

Table 2. Baseline (whey-to-feed) Greek yogurt cradle-to-gate LCIA results.

	CC kg CO_2e	NonRF MJ	WD m^3	MEU kg N eq	POF kg NMVOC	FEU kg P eq	Htox CTUh	EcoTox CTUe
Greek yogurt	4.6	40.3	0.5	8E-03	2E-02	2E-03	4E-07	29.5

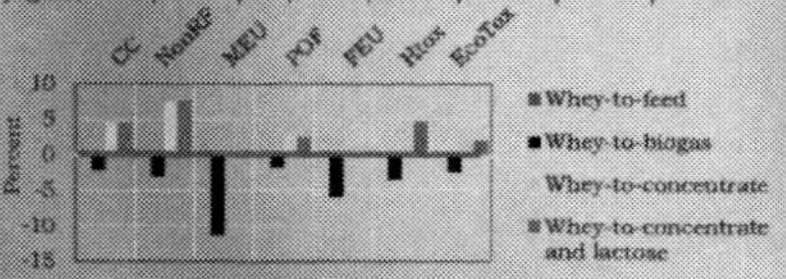

Figure 3. Cradle-to-gate relative increase/decrease of acid whey scenarios compared to baseline scenario.

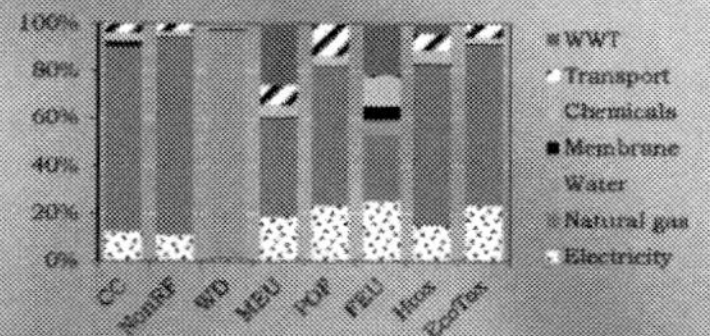

Figure 4. Hot spot gate-to-gate LCA of membrane scenarios.

Production of dried whey concentrate and lactose increased the Greek yogurt plant CC (5%) and NonRF (7%), Figure 3. Natural gas for spray drying was primary impact driver for most categories (Figure 4). While, anaerobic digestion reduced FEU and MEU, the additional WWT of RO permeates, which had BOD and COD above discharge limits, did not affect FEU and MEU (Figure 3). Dried products required half the size trucks and 50% less trucks compared to acid whey.

Anaerobic digestion shows the highest potential to reduce environmental impact of the Greek yogurt plant. Primary benefits of whey concentrates and lactose production include reducing transport cost, removing incentive for farmers to use acid whey, and potential use in food preparations, if approved by the FDA.

References

[1] Thoma et al 2016. Life Cycle Assessment of Yogurt Production and Consumption in the United States.
[2] Goedkoop et al 2009. ReCiPe 2008.
[3] Henderson et al 2011. USEtox fate and ecotoxicity factors.
[4] Solomon et al 2007. Fourth assessment report of the Intergovernmental Panel on Climate Change.
[5] Tomasula, P., Yee, W., 2016. Greek yogurt processing plant model.

Acknowledgments

DMI DAIRY MANAGEMENT INC

Whey to go

- acid whey enzyme treatment and nanofiltration
- Greek yogurt and by-products allocation
- consequential LCA

495470 VALUE Added Products from AGRO-Industrial Wastes. Colombian Andes Berry (*Rubus glaucus benth*) Residues

Abstract:

VALUE ADDED PRODUCTS FROM AGRO-INDUSTRIAL WASTES. COLOMBIAN ANDES BERRY (*Rubus glaucus benth*) RESIDUES

Javier D‡vila, Department of Engineering, Universidad Jorge Tadeo Lozano, Bogot‡ Colombia, Gonzalo Taborda, Chemistry Department, Universidad de Caldas, Manizales Colombia, Moshe Rosenberg, University of California, Davis, USA, Carlos Cardona, Department of Chemical Engineering, Universidad Nacional de Colombia sede Manizales.

Email of presenting author: javiera.davilar@utadeo.edu.co

Colombia has one of the most diverse varieties of fruits in the world, due to its thermal floors and its variety of soils that allow cultivating a broad spectrum of fruit species. Colombia leads in biodiversity per square kilometer and enjoys 433 species of native fruits and has the second larger number of plant species with 51,220, some of which are exotic fruits containing important compounds. In this sense, the productive chain of Andes Berry (*Rubus glaucus benth*) in Colombia is well established however, it lacking of research to find additional benefits of the residues after their processing. Thus, this research investigated Andes berry residues from its chemical characterization for suggesting potential uses and possible value added products derived from this residue. Moisture, extractives, cellulose, hemicellulose, lignin and ash contents were determined as well as other important characteristics such as total phenolic compounds, total antioxidant activity, anthocyanins and volatiles constituents. Results revealed that blackberry residues (BR) had 61.54% of holocellulose (cellulose and hemicellulose); a total phenolic compounds content of 2,700 mg of GA/100 g BR and an anthocyanins content of 126.41 mg/kg of BR. The total antioxidant activity was 174.8 μmol TE/g of BR that was due in 93.58% to the contribution of total phenolic compounds present in BR. Volatiles profile revealed that 2-heptanone, 2-heptanol and benzoic acid were, after ethanol (3,226.09 μg/kg of BR), the most significant compounds present in BR with 1,064, 707 and 542 μg/kg of BR respectively. Finally, a biorefinery approach is suggested (See figure 1) to take advantage of all valuable compounds present in this residue.

Figure 1. Biorefinery based on Andes berry residues

497633 Fe(III)-Induced Rapid Deposition and Polymerization of Dopamine on Microfiltration (MF) Membranes

Abstract:

Fe(III)-induced Rapid Deposition and Polymerization of Dopamine on Microfiltration (MF) membranes

Xuehua Ruan, Xuhang Liao, Yan Dai, Xiaobin Jiang, Gaohong He*

School of Petroleum and Chemical Engineering, State Key Laboratory of Fine Chemicals, Dalian University of Technology, Panjin, Liaoning, P.R. China, 124221

Corresponding Author Email: hgaohong@dlut.edu.cn

Polydopamine deposition, a promising strategy for material surface modification, is widely attempted to manufacture high-flux hydrophilic MF membranes. Nevertheless, the prospect is subjected to the time-consuming and uneven deposition arisen in the general coating routes. In this work, Fe(III) was employed to accelerate deposition and avoid agglomeration. According to the coordination theory, one Fe(III) ion and three dopamine molecules can be complexing together in the solution, which accelerates the deposition effectually. Besides, the multivalent Fe(III)-Fe(II) redox system, helpful to catalyze dopamine oxidization under dissolved oxygen, could promote the polymerization. The mechanism is illustrated in Figure 1a. In this instance, polydopamine coating layer can be yielded homogeneously and rapidly on the MF membrane surface. As a result, the growth rate of PDA with Fe(III) is accelerated by about 5 times, from 11.5 to 62.0 nm/h. The microstructure comparison between modified PVDF MF membranes without Fe(III) for 12 h and with Fe(III) for 2 h is shown in Figure 1b. PDA thickness coated on the former is about 139.5 nm, and that on the latter is comparable to be 124.8 nm. Much more, the samples modified in the PDA-Fe(III) solution possess the surface pore distribution similar to the pristine membranes. In comparison with the membranes modified by a general coating route, the PDA-Fe(III)-coated membranes behave excellently with the water flux three times higher. On the whole, the Fe(III)-induced PDA surface modification is a potential way to manufacture high-flux hydrophilic MF membranes facilely and efficiently.

Figure 1 Fe(III)-induced dopamine reaction mechanism and membrane microstructure.

Acknowledgement:

The authors are grateful to the financial support from National Natural Science Foundation of China (No. 21606035, U1663223, 21527812), Education Department of the Liaoning Province of China (LT2015007), and Changjiang Scholars Program (T2012049).

496038 Crystal Morphology Modifid and Solution Recovery Improved Membrane Crystallization

Abstract:

Crystal morphology modifid and solution recovery improved membrane crystallization

Xiaobin Jiang, Dapeng Lu, Guannan Li, Gaohong He*

State Key Laboratory of Fine Chemicals, School of Chemical Engineering, Petrochemical Energy-efficient Separation Technology Engineering Lab of Liaoning Province, Dalian University of Technology, Dalian, P. R. China

*Corresponding author: Email: hgaohong@dlut.edu.cn

Abstract

For the multiple requirements of environmental protection and biochemical resource reuse, the treatment of industrial wastewater and recovery of the high value-added products in the food and bioprocess with high efficiency is a pivotal topic.[1,2] Traditional treatment approach of multi-component wastewater with high salinity and organic solvent (which is commonly found in petrochemical, biochemical, food and pharmaceutical engineering) include vacuum evaporation crystallization (VEC), membrane based-hybrid separation technology, etc.[3-5] Among these approaches, membrane distillation crystallization (MDC) that combined membrane distillation with traditional crystallization process was attractive and significant by simultaneously recycling valuable salt crystals with high purity and specified molecular structural properties, pure volatile solvents with with high recovery.[6-8]

In this work, the comprehensive investigation is focused on the crystal nucleation and growth kinetics by considering both the membrane property and operational conditions of MDC. One of typical multi-component high sality wastewater, ethylene glycol (EG)-NaCl aqueous solution system, is introduced for the evaluation. The nucleation barrier, diffusion controlled crystal grwoth and interface reaction processes present a significant difference under various feed conditions and operation temperatures, which have profound effect on the crystal morphology (Fig. 1) and size distribution (Fig. 2). As a consequence, the resulting crystal with specified morphology can furtherly improve the separation efficiency. The advantage operating range of MDC for crystal morphology modification and solution recovery improvement on the researched system can be then outlined and optimized.

Acknowledgment

We acknowledge fnancial contribution from from National Natural Science Foundation of China (Grant No. 21527812, 21676043, U1663223), Changjiang Scholars Program (T2012049), the Fundamental Research Funds for the Central Universities (DUT16TD19) and Education Department of the Liaoning Province of China (No. LT2015007).

70- 0 to 80 % EG

Fig. 1. NaCl crystal morphologies under different feed solution composition $M_{EG}/(M_{EG}+M_{H2O})$: (a), 0 %; (b), 20 %; (c), 40 %; (d), 60%; (e), 80% (P_v=7 kPa£¬T=70 ¡æ).

Fig. 2. Crystal particle properties comparison between with various feed composition and operation temperature. (P_v=7 kPa, M_{NaCl}: M_{H2O}=6.6: 40; Number in the figure: average crystal size and C.V. (underlined).

References

1. Uribe IO, Mosquera-Corral A, Rodicio JL, Esplugas S. Advanced technologies for water treatment and reuse. *AIChE Journal.* 2015;61(10):3146-3158.

2. Chekli L, Phuntsho S, Kim JE, et al. A comprehensive review of hybrid forward osmosis systems: Performance, applications and future prospects. *Journal of Membrane Science.* 2016;497:430-449.

3. Ye C, Zheng H, Wu T, Fan M, Feng J, Li W. Optimization of solvent crystallization process in obtaining high purity anthracene and carbazole from crude anthracene. *Aiche J.* 2014;60(1):275-281.

4. Li W, Van der Bruggen B, Luis P. Integration of reverse osmosis and membrane crystallization for sodium sulphate recovery. *Chemical Engineering and Processing: Process Intensification.* 2014;85:57-68.

5. Yong A, Obanijesu EO. Influence of natural gas production chemicals on scale production in MEG regeneration systems. *Chemical Engineering Science.* 2015;130:172¨C182.

6. Quist-Jensen CA, Ali A, Mondal S, Macedonio F, Drioli E. A study of membrane distillation and crystallization for lithium recovery from high concentrated aqueous solutions. *Journal of Membrane Science.* 2016;505:167¨C173.

7. Xu S, Wang J, Zhang K, et al. Nucleation behavior of eszopiclone-butyl acetate solutions from metastable zone widths. *Chemical Engineering Science.* 2016;155:248-257.

8. Anisi F, Thomas KM, Kramer HJM. Membrane-assisted crystallization: Membrane characterization, modelling and experiments. *Chemical Engineering Science.* 2017;158:277-286.

502779 Engineer Bacteria Consortia to Execute Concerted Enzymatic Reactions

Abstract:

A microbial consortium is two or more microbial groups living symbiotically. It is a group of different types of bacteria that summatively act as a unit, with each bacteria contributing to the function of the consortium. The study and engineer of microbial consortium are critical to understand microbiome, environmental bioremediation and complex biochemical processes to produce biofuels or other important chemicals. In this presentation, we are going to highlight our recent progress to create and design a novel function multi-species microbial consortium. Exogenous enzymatic reactions will be engineered into the consortium. Efficiency, stability and evolvement of the consortium will be characterized. The ability to engineer robust multi-species communities would enable modulation of microbe consortium for human health, industrial and environmental applications.

492101 Effect of Heat Stress on Rice Seed Development: Discovering Global Regulatory Players and Modeling of Rice Metabolism

Abstract:

Increased global temperature is detrimental to plant growth and development and significantly reduce crop yields. In rice, heat stress during early seed development adversely affects the seed size and quality at maturity. Plants use a suite of strategies to respond to abiotic stresses, including changing the abundance of stress responsive genes/proteins that ultimately lead to the large-scale changes in gene expression levels, protein abundances, and metabolites. Complex gene-protein-reaction associations as well as regulatory mechanisms constitute a challenge to elucidate stress response mechanisms in plants. This research aims to understand the stress response mechanisms in developing rice seed using temporal transcriptomic analyses in control and stress conditions and using *in silico* metabolic analyses.

Rice plants were grown in a greenhouse under control conditions prior to flowering. A few days prior to flowering, plants were transferred to reach-in growth chambers with a 16h-light/8h-dark cycle at 28°C/ 25°C (control) . Fertilized seeds were marked on two consecutive days and plants were maintained under control conditions. Plants were exposed to heat stress (35°C) at 12 and 36 hours after fertilization (HAF) on day 1 and day 2, respectively, and young developing seeds were collected from control plants and stressed plants having two biological replicates. Total RNA isolated from developing seeds was used for differential gene expression analysis, which yielded ~7000 significantly stress-responsive genes. Clustering analysis based on Pearson's correlation was used to develop minimal regulatory network and sub-networks, and identifying global regulators. The highly connected "hub" genes in the co-regulatory network included previously identified MADS-box genes as well as a large number of novel regulatory genes. A comprehensive, genome-scale metabolic model of rice seed is under development using draft genome reconstruction in ModelSEED and KBase databases and previous models. The coexpression information obtained from our experiments will be incorporated as regulatory constraints in the model.

Work on other tissue types of rice and modeling the interactions between them is underway. The important interactions of the rhizobiome with the plant root not only affects root metabolism, but significantly affects the plant metabolism in all tissue types. We are using multi-level and multi-objective modeling framework to integrate different tissue-specific models into a robust plant-scale model. This systems level study will identify bottlenecks in the metabolic pathways and subsequently propose genetic intervention strategies improving crop yield. Our predictive mathematical model will come up with biologically important and non-intuitive solutions to problems related to stress response mechanisms, plant-microbiome interactions and developing tolerant plant species in an efficient and accurate fashion.

500479 Characterization and Heterologous Expression of Iron Hydrogenase Ethha_0031 of *Ethanoligenense harbinense* in *E. coli* Blr(DE3)

Abstract:

Hydrogenase is a critical enzyme in hydrogen evolution. Finding hydrogenases with high ability to evolve hydrogen will make big significance for the improvement of fermentative hydrogen production. *Ethanoligenens harbinense* is a new hydrogen-producing bacteria isolated with high hydrogen production rate of 2.81 mole hydrogen per mole glucose. Therefore, the hydrogenases of *Ethanoligenens harbinense* has attracted wide attention.

In this study, iron hydrogenase Ethha_0031 (hyd1) was successfully cloned from the genome of *Ethanoligenens harbinense* Yuan-3^T. Hyd1 encodes 418 amino acids and is predicted to own a molecular weight of 45.91 kilodaltons. The amino acid sequence of hyd1 presents 63% identity and 58% identity with the iron hydrogenases of *Clostridium cellulosi* and *Desulfosporosinus* sp. OT, respectively. Hyd1 is a monomer consisting of four [4Fe-4S] clusters, one [2Fe-2S] cluster and one [2Fe-2S-5O-2H2O] cluster. It has a conserved domain named NuoI super family which is subunit of formate hydrogenlyase /NADH: ubiquinone oxidoreductase.

Hyd 1 gene fragment was inserted after T7 promoter to construct a recombinant plasmid pBENT-hyd1. The plasmid was then transferred into *E. coli* BLR(DE3), confirmed by resistance screening and PCR. SDS-PAGE confirmed the high-level expression of hyd 1. The mutants acquired the ability to evolve hydrogen as compared to the wild BLR(DE3). Furthermore, hydrogenase activity assay was tested and confirmed that hyd1 showed a high ability to evolve hydrogen in mutant. This research is a new addition to the existing knowledge for the construction of hydrogen-producing engineered strains.

500347 Metabolic Engineering of Clostridium Aceticum for Acetone, Butanol and Ethanol Fermentation

Abstract:

Clostridium aceticum is a homoacetogenic bacteria which converts CO2 and H2 to acetate without losing any carbon, thus resulting in a higher potential yield than fermentation from heteroacetogenic bacteria such as Clostridium acetobutylicum for ABE (acetone-butanol-ethanol) fermentation. However, wild-type C. aceticum only produces acetate, and no paper has been published on the metabolic engineering of C. aceticum.

In this study, we established methods for metabolic engineering of Clostridium aceticum DSM 1496 (ATCC 35044). Acetone and solvent production by C. aceticum will be studied individually and together. For acetone production, butyrate-acetoacetate CoA-transferase gene (ctfA/B) and acetoacetate decarboxylase (adc) gene from Clostridium acetobutylicum ATCC 824 will be introduced into C. aceticum. Also, ethanol and butanol production pathway was introduced by overexpressing aldehyde/alcohol dehydrogenase 2 (adhE2), thiolase (thl), butanol dehydrogenase (bdh) and BCS cluster (including crotonase (crt), putative butyryl-CoA dehydrogenase (BCD), electron-transfer flavoprotein (etf), and 3-hydroxybutyryl-CoA dehydrogenase (hbd) genes) from Clostridium acetobutylicum ATCC 824. Fermentation kinetics will be studied from sugar and CO2, as well as from CO2 and H2. This study aims at producing chemicals from CO2, H2 and sugar, which could help alleviate CO2 emission problem as well as increase fermentation yield.

भारतीय प्रौद्योगिकी संस्थान हैदराबाद
Indian Institute of Technology Hyderabad

A Technique for the Usage of Agricultural Solid Substrate Extract for the Production of L-Asparaginase by Immobilized Ca-alginate beads

Anup Ashok, **Vaibhav Lendekar and Devarai Santhosh Kumar***

Industrial Bioprocess and BioProspecting Laboratory (IBBL), Department of Chemical Engineering,
Indian Institute of Technology Hyderabad, Kandi-502285, Sangareddy, Telangana State, India.

17AIChE
Annual Meeting, Minneapolis, MN

Abstract

*Immobilization is a technique used to prevent the contamination of active enzymes, increase substrate mobility and to prevent cell washout. Solid state fermentation (SSF) has been always observed as an alternative to Submerged fermentation (SmF). Optimization of the parameters will help in further improvement. The primary enzyme being produced is L-asparaginase that is used as an anti-neoplastic and anti-carcinogenic material. Different solid substrate extracts were tested and the best out of it was selected as wheat bran extract, the activity of crude enzyme is expected to be around **11.05 U/ml** in case of the extract media as opposed to **4-5 U/gds** in the solid media. **In a first of its kind study, an extract of the solid substrate was obtained using buffer solution and this extract is treated as liquid media, SmF process is run using calcium alginate beads as the immobilized material.***

Introduction

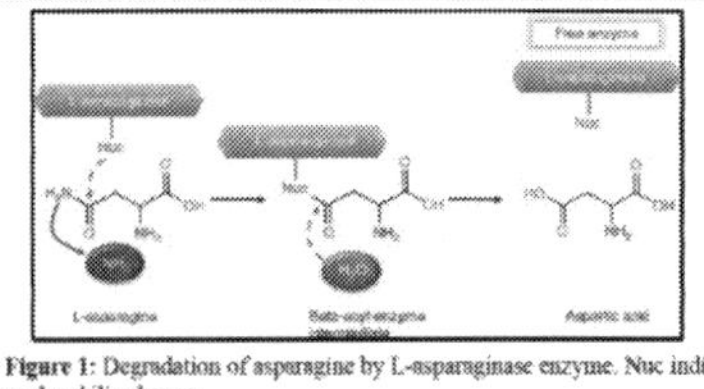

Figure 1: Degradation of asparagine by L-asparaginase enzyme. Nuc indicates nucleophilic change.

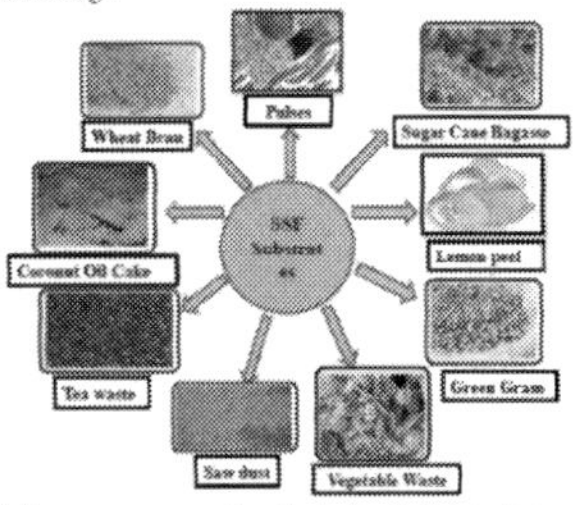

Figure 2: Some common examples of agricultural substrates that can be used in the process

Micro-organism	Nutritional requirement	L-Asparaginase Activity (U/ml)	References
Streptomyces griseus NIOT-VKMA29	Asparagine dextrose salt medium	56.78	Sarquis M.I.M (2004) *MEM I OSWALDO CRUZ.*, 99: 489-492.
Erwinia Chrysanthemi 3937	Peptone (1.0 %); Yeast extract (0.5 %); NaCl (1.0 %)	2.25	Babu & Satyanarayana (1996). *J SCI IND RES INDIA*, 55: 464-467
Fusarium equiseti	Glucose (0.5 %); Ammonium sulphate (0.5 %); Yeast extract (0.5%)	8.51	Hosmani and Kaliwal (2011). *INT J DRUG DISCOV*, 3: 88-99.
Aspergillus terreus MTCC 1782	L-Asparagine (1.0%); Corn flour (1.5%); Glucose (0.2%).	33.25	Bhaskar *et al* (2010). *CHEM BIOCHEM ENG Q*. **24**: 467–472.
Aspergillus terreus MTCC 1782	**Agricultural Substrate (Wheat Bran) Extract**	**11.05**	**This study**

Table 1: Reported enzyme activity values for conventional SmF process and the value reported in using the solid substrate extract

Methods

Solid Substrate Extract and Fermentation:

- 100 ml of extract of desired solid substrate was obtained by mixing Phosphate buffer maintained at pH 7 with 10 gram solid substrate,
- Sterilized in an autoclave.
- Filtered samples were collected.
- Clear liquid extract was obtained by centrifuge at 10000 rpm for 5 minutes.
- Immobilization of the cells was done using Ca-Alginate beads.
- Media required for the production was prepared.
- Required quantity (100 ml) was added to a flask into which the cells of *Aspergillus terreus* MTCC 1782 were added.
- Parameters studied was pH, temperature, number of cells, percentage of the alginate beads and the rotational speed (rpm).
- Experiment was run for 96 hrs (4 days) and samples were collected every 24 hrs (1 day).

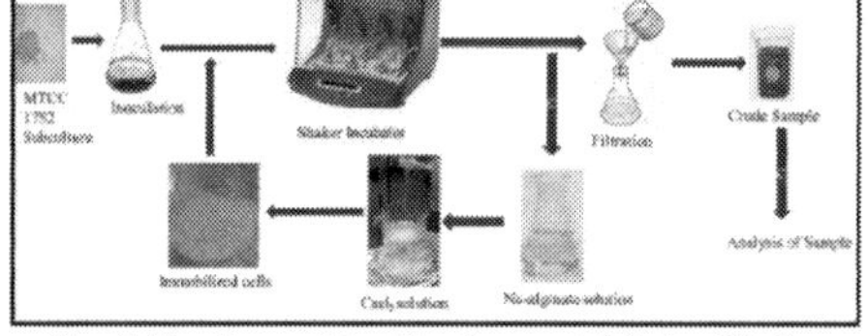

Figure 3: Schematic representation of the entire process including immobilization

Results

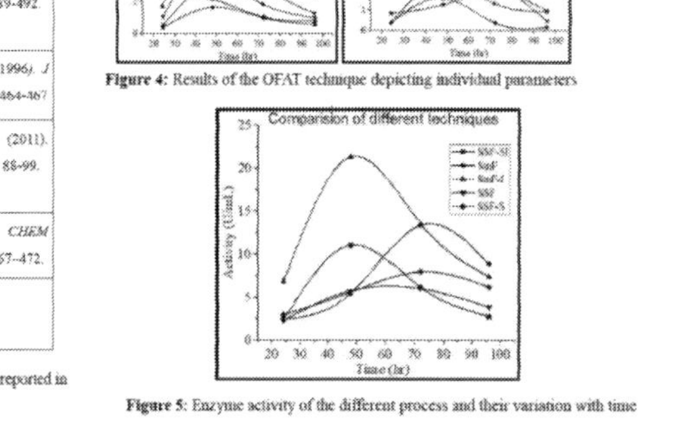

Figure 4: Results of the OFAT technique depicting individual parameters

Figure 5: Enzyme activity of the different process and their variation with time

Discussions

- Substrate Extract of wheat bran has shown maximum activity.
- Parameters:- pH: 4-8, temperatures: 15°C-50°C; initial cell conc: $1*10^6$-$10*10^6$ cell/mL, mixing speed: 120-250 rpm, and bead conc: 1-5% were used.
- One Factor At a Time (OFAT) optimization technique for the determination of individual parameters.
- Immobilization was done by adding cells to Na-alginate solution which was later added to $Cacl_2$ solution dropwise.
- Comparative study was done among SmF, SSF, SmF with immobilization (SmF-I), SmF of substrate extract (SmF-S) and SmF of Immobilized substrate extract (SmF-SI).
- The results in decreasing order of activity– SmF-I> SmF> SSF-SI> SSF-S> SSF, for max activity at optimized conditions.
- The economic analysis showed that SSF-SI technique has lower costs than both SmF and SmF-I .
- The SSF-S and SSF-SI can be used as substitute for the conventional techniques.
- The wastes produced have lower effect on the environment.

Conclusions

- The new methodology has shown increased enzyme activity against the conventional SSF technique .
- Different substrates can be used to see the increase/decrease in production of the enzyme.
- The optimum parameters were obtained after repeated experimentation, which are pH: 7, temperature: 30°C, using $4x10^6$ cells entrapped in 3% of alginate bead at 180 rpm shaker speed for wheat bran substrate.
- These techniques have a positive economic and environment advantage over the conventional techniques

Future Work

- Purification of the produced crude enzyme
- Scale-up the process to pilot plant and further to industrial scale.

Acknowledgements

- IIT Hyderabad, SERB-DST (SB-EMEQ-048/2014), IBB Lab - IITH

References

- Doriya K, Jose N, Gowda M, Kumar DS (2016). Solid-State Fermentation vs Submerged Fermentation for the Production of L-Asparaginase. *Advances in Food and Nutrition Research*, **78**:115-135.
- **Ashok A**, Doriya K, Rao DRM, Kumar DS (2017). Design of solid state bioreactor for industrial applications: An overview to conventional bioreactors. *Biocatalysis and Agricultural Biotechnology*, **9**:11-18
- **Ashok, A.** & Kumar, D.S (2017). Different methodologies for sustainability of optimization techniques used in submerged and solid state fermentation. *3 Biotech* **7**: 301

***corresponding author: devarai@iith.ac.in**

487084 The Enhanced Butanol Production and High-Efficient Product Recovery with Reduced Wastewater Generation

Abstract:

Abstract: Butanol as an important chemical and potential fuel could be produced via ABE fermentation from lignocellulosic biomass. Biobutanol production from Jerusalem artichoke stalk (JAS) was investigated in order to improve cellulose digestibility and efficient biobutanol fermentation. Compared with 9.0 g/L butanol (14.7 g/L ABE) production by 2% NaOH pretreatment of JAS, 11.8 g/L butanol (17.6 g/L ABE) was produced in the best scenario conditions of NaOH-H_2O_2 pretreatment, washing times and citrate buffer strengths etc. Furthermore, more than >64% water in washing pretreated JAS process could be saved, with improving butanol production by >25.0%. To mimic *in situ* product recovery for ABE fermentation, the vapor strippingâ€"vapor permeation (VSVP) process steadily produced 323.4â€"348.7 g/L butanol (542.7â€"594.0 g/L ABE) in condensate, which showed more potentials than pervaporation for butanol recovery. Therefore, the present study demonstrated an effective strategy on efficient biobutanol production with less wastewater generation using lignocellulosic biomass.

500940 Suite of Bacteria from the *Enterobacter* Genus Suitable for Lignin Degradation

Abstract:

Lignin is an amorphous three dimensional biopolymer composed of phenyl propanoid units linked by the recalcitrant C-C and C-O bonds. It constitutes between 15-30% of the lignocellulosic biomass. Biofuel industries which are mostly focused on conversion of the celluloses to biofuels are expected to generate large quantities of lignin waste. From the point of waste remediation and carbon efficiency, it is necessary to identify routes to covert lignin into valuable products. Here we explore a few species from the Enterobacter genus suitable for lignin bioconversion.The adaptive plasticity of the Enterobacter species to switch between aerobic, anaerobic respiration and fermentation has enabled survival in lignin rich ecological niches such as the soil, subsurface, sewage and also as epiphytes and plant pathogens.Recently, several novel lignolytic Enterobacter species were isolated from soil, paper and pulp waste and bamboo slips. We have isolated the lignin degrading Enterobacter RC202 from 4850 feet deep subsurface soil surrounding decomposing wood. The Enterobacter RC202 and other lignolytic bacterial species from the Enterobacter genus can utilize complex carbon substrates such as the kraft lignin and lignin model compounds as sole sources of carbon. The bacteria use mechanisms such as demethoxylation, dehydroxylation and aromatic ring cleavage to partial degrade or mineralize these compounds. We have compared the diverse approaches used by Enterobacter cloacae subsp. Cloacae ATCC® 35929™ , 35930™, Enterobacter cloacae subsp. dissolvens ATCC® 23373™ and the Enterobacter RC202 species to utilize selected aromatic compounds as sole carbon sources under aerobic and anaerobic conditions. A temporal product profile was developed from the HPLC and GC-MS analysis of spent culture media. Utilization of kraft lignin and lignin derived aromatic compounds by the lignin degrading Enterobacter species and application of these species in lignin valorization will be discussed.

504263 Large-Scale Synthesis of Dihydrostreptomycin via Hydrogenation of Streptomycin in a Membrane Dispersion Microreactor

Abstract:

In this work, a membrane dispersion microreactor was applied for the hydrogenation of streptomycin (STR) to produce dihydrostreptomycin (DHS). The reductant was dispersed into the continuous phase containing STR, leading to a significant increase in the mass-transfer specific surface area, and the conversion rate of STR reached 99.81 wt% within 3 min. In contrast, in stirred tanks, the conversion rate of STR only reached 99.5% in 30 min. Moreover, the minimun amount of KBH_4 used in the membrane dispersion microreactor was about 35.71% lower than that in traditional tanks, when getting the same conversion rate at about 99.5%.

487921 Direct Comparison Between Batch Bulk Mixing and Continuous Millifluidics in the Synthesis of Amorphous Drug Nanoparticles

Abstract:

Amorphous drug nanoparticles in the form of drug-polysaccharide nanoparticle complex (or drug nanoplex) had emerged as an ideal supersaturating delivery system of poorly-soluble drugs attributed to its many attractive characteristics. Herein we presented for the first time direct comparison between two nanoplex synthesis platforms, i.e. millifluidics and bulk mixing, representing continuous and batch production modes, respectively. They were compared by the resultant nanoplexâ€™s (1) physical characteristics (size, zeta potential, payload), (2) preparation efficiency, (3) storage stability, (4) dissolution rate/supersaturation generation, and (5) production consistency. For this purpose, the effects of key variables in drug-polysaccharide complexation (pH, charge ratio) were investigated in both platforms. Perphenazine and dextran sulfate were used as the drug and polysaccharide models, respectively.The results showed that both platforms shared similar dependences on pH and charge ratio with similar optimal preparation conditions. The pH was found as the governing variable through its influence on size and zeta potential, while the other variables (e.g. charge ratio, millifluidicsâ€™ residence time, tube diameter) were insignificant. The other aspects of nanoplex were minimally affected by the process variables. Nanoplexes having mostly similar characteristics were produced by both platforms, except for payload and supersaturation generation. Nevertheless, millifluidics was favored owed to its superior production consistency and well-demonstrated scalability.

Novel Method of Evaluating Liquid Absorption with Intra-Particle Pore of Pharmaceutical Porous Materials using a Capillary Rise Method

17AIChE

Takuma Oba [a),c)], Yasutomi Kato [b)], Ryoichi Sonoda [a)], Kohei Tahara [c)], Yoshiaki Kawashima [d)], Hirofumi Takeuchi [c)]

[a)]Formulation Department, CMC Center, Kaken Pharmaceutical Co. Ltd., [b)]R&D Planning & Project Management Department, Kaken Pharmaceutical Co. Ltd., [c)]Gifu Pharmaceutical University, [d)]Aichi Gakuin University

Summary

Introduction

Porous materials are attractive stuff for pharmaceutical particulate designing, such as for powdering oily or semi-solid drugs and improving the solubility of poorly soluble drugs. Understanding the mechanism and behavior of liquid absorption into intra pores and interstices of porous carrier particles should be important to effectively load active pharmaceutical ingredients (APIs) with the carriers. We established a simple and practical method to measure the amount of liquid or API solution absorbed intra-particle pores and trapped in interstices of particles separately.

Purposes

- To understand the absorption mechanism and behavior of liquid with the porous material
- To establish a simple and practical method to evaluate the liquid absorption of porous carriers
- To predict the amount of model drug loaded with particles by using the established method.

Experimental

Materials

- Porous Carrier: Porous calcium silicate ($CaSiO_3$) with petal-like pores (Florite®-RE, FR, from Tomita Pharmaceutical Co., Ltd.)
- Model API as a model of poorly water soluble drug: Itraconazole (ITCZ from Hanmi Pharm. Co., Ltd.)
- Solvent: Water, Ethanol, Dichloromethane and Glycerin

Physicochemical Properties Evaluated

Measurement of Contact angle (Sessile drop method)
— Contact angle meter (DSA30, KRÜSS GmbH)

- Absorption experiment (Capillary Rise Method)
— Force tensiometer (K12, KRÜSS GmbH)

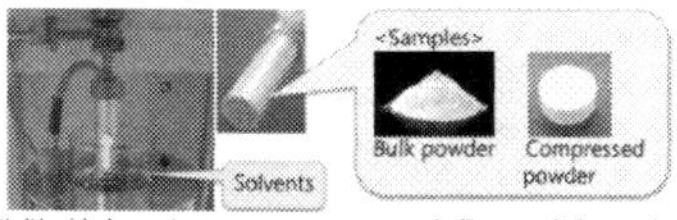

1) Liquid absorption amount was automatically recorded over time until the amount became constant
2) K was obtained from the slope of absorption kinetics
3) Penetration rate (cm^2/sec) was calculated from modified Washburn equation below

$$\frac{h^2}{t} = K \cdot \left(\frac{V_{void}\,\rho}{l_{powder}}\right)^2$$

- h: height reached by the liquid front at time t
- l_{powder}: the apparent height of packed powder sample
- V_{void}: the volume of intra-particle pores and interstices of particles
- ρ: the density of liquid

- Mercury intrusion porosimetry
— Mercury Porosimeter (9305 pore sizer, Shimadzu Co.)
- Preparation of ITCZ loaded FR (ITCZ-FR)

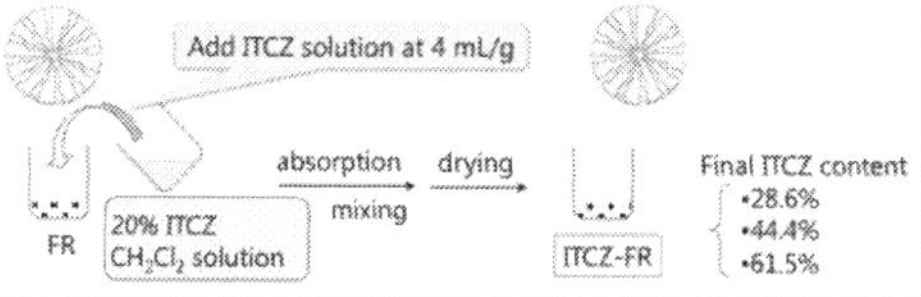

- Evaluation of ITCZ loaded FR (ITCZ-FR)
— Morphology by Scanning Electron Microscopy (JSM-6390LA, JEOL Ltd.,)
— Specific surface area by gas absorption method (Gemini2375, Shimadzu Co.)
— Crystal states by X-ray diffraction (RINT2200Ultima II/PC, Rigaku Co.,Ltd.,)

Conclusions

- ✓ **A liquid absorption experiment using a compressed sample of porous carriers was possible to predict the available intra-particle pore volume of the porous carriers.**
- ✓ **Prepared API loaded particle based on the findings from this experiment was found to adsorb API efficiently in the intra-particle pores in amorphous state.**
- ✓ **This novel method was found to be useful for understanding the liquid absorption mechanism to the porous carriers and for designing the API loaded particles for medicinal drug products.**

Results & Disccussion

Absorption experiment using Bulk FR powder

Table 1 Physicochemical properties of solvents employed

Sample	Water	Ethanol	Dichloro-methane	Glycerin
Viscosity (mPa·s)	0.89	1.10	0.44	924.7
Surface tension (mN/m)	72.3	22.5	28.6	65.2
Density(g/cm^3)	0.99	0.79	1.33	1.26
Contact Angle(°)*	24.4 ($\cos\theta$ = 0.91)	18.0 ($\cos\theta$ =0.95)	17.7 ($\cos\theta$ =0.95)	69.6 ($\cos\theta$ =0.35)

* Contact angle to Calcium Silicate surface

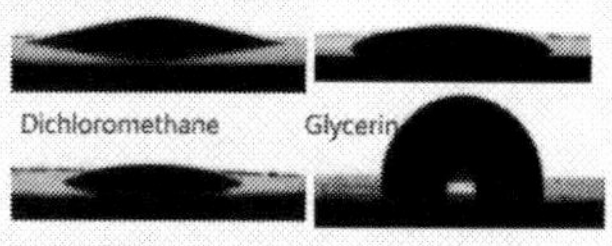

Fig. 1 Images of solvents drop placed on the compressed calcium silicate surface

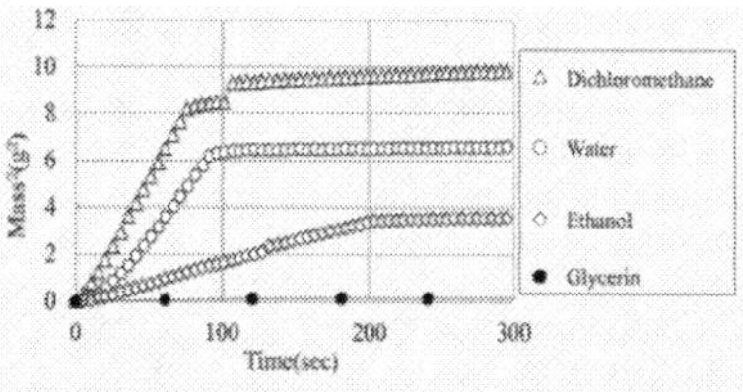

Fig. 2 Kinetics of capillary rise of solvents in FR powder bed

Table 2 Results of Absorption experiment of FR with various solvents

Sample	Water	Ethanol	Dichloro-methane	Glycerin
Absorption amount	7.96 mL	7.47 mL	7.19 mL	7.39 mL
Absorption time	2.1 min	4.7 min	1.8 min	430.6 min
K from slope part (g^2/sec)	0.0857	0.0160	0.1149	0.0004
Penetration rate (normalize value*)	0.094 (1.00)	0.011 (0.12)	0.223 (2.37)	0.001 (0.01)

*normalized by the result of Water

- The wettability of FR with various solvents was clarified by the Penetration rate.
- Penetration rate was strongly affected by the viscosity of the solvent.
- The absorption amount of liquid from this method was apparent amount including both penetrated intra particle pores and trapped in interstices voids of particles.

⇒ Liquid absorption test using a compressed porous carriers was established for precise evaluation

Absorption experiment using compressed FR

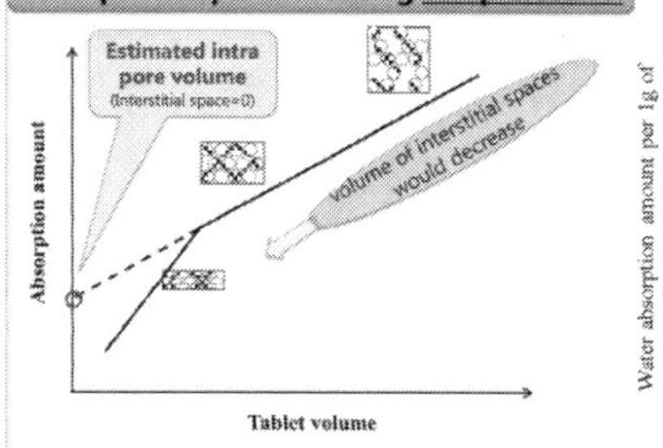

Fig. 3 Expected diagram of the relationship among tablet volume, absorption amount and inner states of tablet

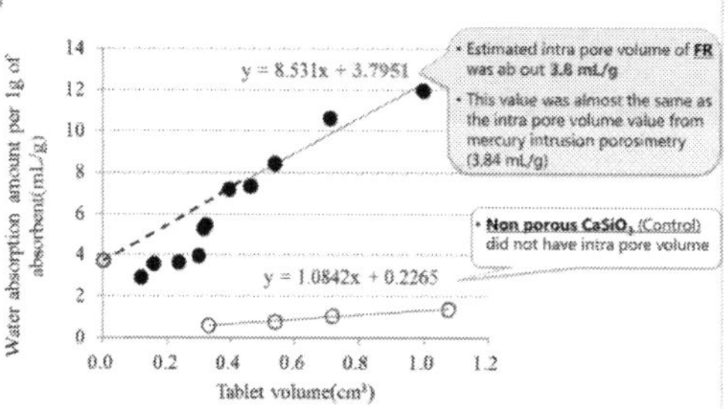

Fig. 4 Diagram of the relationship between FR(●) & non porous $CaSiO_3$(○) tablet volume and water absorption amount.

- This method can evaluate the absorption of intra particle pores and interstices voids of particles separately.
- FR was found to be able to absorb about 3.8 mL liquid into its intra-particle pore per 1 g FR.

⇒ It was possible to predict available intra-particle pore volume of the porous carriers by using this method

Preparation and evaluation of ITCZ loaded FR

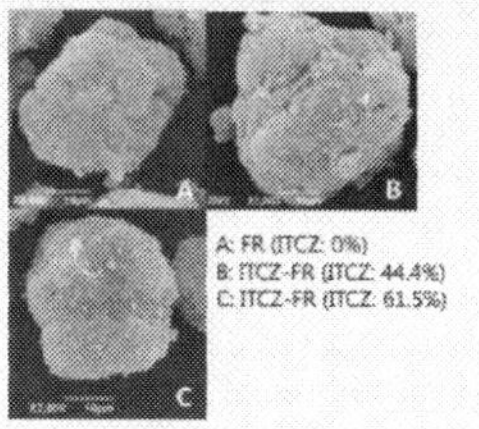

Fig. 5 SEM images of ITCZ-FR

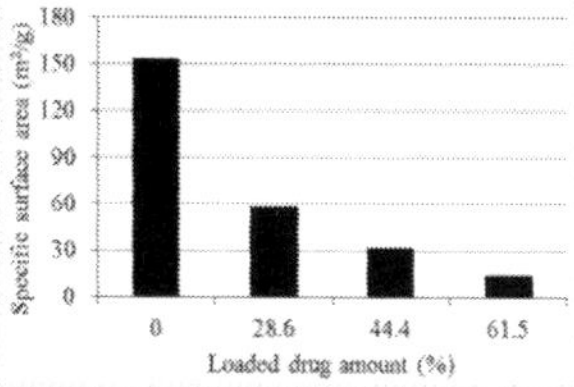

Fig. 6 Effect of loaded ITCZ amount on specific surface area of FR

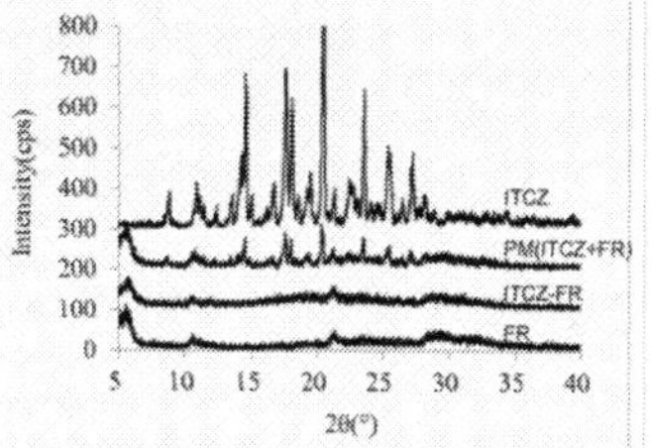

Fig. 7 Powder XRD Patterns of ITCZ-FR

- Absorption test using compressed FR can predict the amount of ITCZ solution to be absorbed inside the pore of FR particles.
- The ITCZ absorbed in the pore of FR was found to be amorphous state.

⇒ Our novel method to measure the intra-pore volume was found to be practically useful for formulation study

489919 Evaluation of Experimental Methods for Solubility Determination of Carbamazepine in Ethanol for Cooling Crystallization Process DesignÂ

Abstract:

Solubility data is important for pharmaceutical formulation and process design. For example, solubility data is essential for estimating the supersaturation and understanding the relative stability of crystal forms. The determination of an accurate solubility curve is critical for efficient design of the crystallization process. Inaccurate determination can lead to use of a higher or lower supersaturation level that would result in undesired extra nucleation or crystal growth. However, there are various solubility measurement techniques. Here we will evaluate several popular ones and discuss how to obtain accurate solubility data.

In this study, carbamazepine (CBZ) is chosen as the model drug due to its various polymorphic forms. The goal was to evaluate the effect of (a) dissolving methods, (b) sampling separation, and (c) quantification technique on the accuracy of solubility data. Also, the solubility curve was developed for form III and II at various temperatures for future crystallization applications. The two dissolving methods examined for Form III CBZ are: (1) the shake-flask, (2) mechanical stirring. The shake-flask method was performed as the base scenario to examine solubility differences by two off-line sampling separation techniques: (1) centrifuging or (2) filtering. The concentration was determined by three quantification techniques: (1) off-line gravimetrically, or (2) off-line high liquid chromatography method (HPLC) or (3) on-line Raman method. In addition, mechanical stirring, filtering and gravimetrical options were chosen to measure the solubility data of form II for a comprehensive understanding of solubility curves of CBZ form II and III.

The mechanic stirring method demonstrated similar mean values but with less variance compared to the shake-flask method. Through analysis of means, there were no significant differences between using centrifugation or filtration as the sampling separation method. But it is noted that a minimum of 30 minutes is required for settling time to avoid significant effect on the equilibrium concentration. Between the two off-line concentration quantification methods (gravimetric and HPLC), the HPLC method always showed lower mean values, but their difference from the gravimetric values are < 4.6%; the standard deviation for gravimetric methods is ~ 1.5% while the standard deviation for HPLC is ~ 3%. The reason for the standard deviation difference is probably due to the multiple-step sample dilution procedure for HPLC. On-line measurements in mechanic stirring with Raman spectroscopy was done to determine the solubility curve of form III as a function temperature (5 - 35 °C). The solubility curve was compared to off-line concentration quantification methods, and the differences for the mean values are normally < 5%. This on-line Raman method allows for quick determination of solubility curve from a single experimental study, and could be used as a quick solubility method if a low accuracy is acceptable. In addition, the dissolution profile of the metastable form (form II) was analyzed between 5 °C and 35 °C in order to develop determine whether the solubility curve for CBZ form II and form III are mono-tropic or enantio-tropic. The results suggest that the system shows a mono-tropic system between 5 °C and 35 °C, while the solubility ratio decreases from 1.20 at 35 °C to 1.12 at 5 °C. This work presents a comprehensive study on the solubility data of CBZ and the solubility determination methods that are important during the process development and design of the crystallization process.

490293 Continuous Solvent Exchange Washing of Pharmaceutical Intermediates

Abstract:

Continuous Solvent Exchange Washing of Pharmaceutical Intermediates

M. Kreimer[1], I. Aigner[1], S. Sacher[1], M. Krumme[2], T. Mannschott[2], P. van der Wel[3], A. Kaptein[3], J.G. Khinast[1,4]

[1]Research Center Pharmaceutical Engineering (RCPE) GmbH, Inffeldgasse 13, 8010 Graz, Austria

[2]Novartis Pharma AG, Novartis Campus, 4056 Basel, Switzerland

[3]Hosokawa Micron B.V., Gildenstraat 26, 7005 BL Doetinchem, Netherlands

[4]Graz University of Technology, Institute for Process and Particle Engineering, Inffeldgasse 13, 8010 Graz, Austria

In recent years, a lot of effort has been invested by the pharmaceutical industry to move away from batch processing to more agile and flexible continuous manufacturing processes. Going continuous has been discussed and described intensively by academia, industry and regulators, as cost-efficient manufacturing methods and easy up-scaling is strongly demanded. An entire continuous manufacturing plant includes several process steps from API production to the final oral dosage form such as synthesis, crystallization, washing, filtration, drying, milling, blending, granulation and tableting/capsule filling.

After API synthesis the input material for solid dosage form production can be prepared by crystallization through different methods like cooling, evaporation or sublimation. The produced suspension is usually a mixture of several liquids and solids, including the crystallized API. Before drying of the suspension mixture, various washing and filtration steps are involved to purify the drug product. The amount of impurities has to be decreased below a certain threshold to ensure patient safety and drug efficacy. Furthermore, tailored particle properties such as particle size distribution must not be influenced by washing to ensure consistent product quality and process ability within downstream process steps.

A typical washing and filtration process is arranged in cascades to guarantee a highly uniform and pure crystalline material. Therefore, anti-solvent is added in a precisely defined mixing ratio to the mother liquor. An optimal process design is necessary to decrease the number of multiple washing/filtration steps to a minimum to avoid changes of the solid product by mechanical stresses. In addition, considering environmental aspects, energy consumption and production of hazardous waste can be diminished as well.

In this work a newly developed, continuous solvent exchange washing process was established suitable for low

throughputs in the range of 2-12 l/h. A robust process design can be shown by means of model substances with a particle size below 100 µm. Variations in the composition of the mother liquid suspension and the washing medium revealed the influence of the solvent exchange washing process on the product particle size distribution. Within the given design space, suitable configurations were achieved for consistent product quality. In the specific case, changes of the particle size distribution were diminished below 15%.

493431 Design and Characterization of Fast-Release Clofazimine Nanoparticles to Improve Bioavailability

Abstract:

Clofazimine is a lipophilic riminophenazine antibiotic recommended by World Health Organization (WHO) as multidrug therapy of leprosy. To improve the drug dissolution rate, clofazimine was encapsulated in nanoparticles prepared with three affordable surface stabilizers, hypromellose acetate succinate (HPMCAS), lecithin, and zein, through the Flash Nanoprecipitation (FNP) process. Lyophilization and spray drying were applied and optimized to solidify nanoparticles. Dissolution tests were carried out to evaluate release profile of clofazimine nanoparticles prepared by different surface coatings in simulated gastric and intestinal fluids, comparing to raw clofazimine powder and commercial product Lamprene®. Remarkably improved dissolution rate and supersaturation levels were observed with all clofazimine nanoparticles despite different release kinetics. In addition, differential scanning calorimetry (DSC) was explored to study the physical state of clofazimine in nanoparticles. The results strongly suggest that the nanoparticles prepared with affordable coatings can be used as viable cryptosporidiosis therapeutics.

494873 Design of Efficient Metal Nanocatalysts for Continuous Synthesis of Drug Substances Via Cross Coupling Reactions

Abstract:

Design of Efficient Metal Nanocatalysts for Continuous Synthesis of Drug Substances via Cross Coupling Reactions

Andishaeh Dadgar, Farshid Mohammadparast, Marimuthu Andiappan, School of Chemical Engineering, Oklahoma State University, Stillwater, OK, USA.

Cross coupling reactions are widely used reactions in pharmaceutical industry. Heterogeneous metal nanocatalysts can be used as a platform to switch cross-couplings from traditionally-used batch mode to continuous flow mode. As the use of metal nanocatalysts in cross coupling reactions continues to grow, there is becoming a greater need for fast and efficient characterization methods. There is a group of metal nanocatalyst that exhibit a phenomena called surface plasmon resonance (SPR). SPR is a unique trait to each catalyst based on composition, size, and shape of the particles. SPR then shows up in the UV-Vis region of light easily measureable using UV-Vis spectroscopy. This gives a powerful tool to use in characterizing plasmonic metal nanocatalyst.

In this contribution, we show that surface plasmon resonance (SPR) spectroscopy can be used as platform to characterize nanometal catalyst quickly and efficiently. Using finite-difference time-domain optical simulations, we show the use of UV-Vis extinction spectrums to monitor nanoparticle geometry to predict the spectroscopic pattern we will see in experiments. We demonstrate that for a variety of nanocatalysts there is a measurable SPR peak shift and/or change in SPR intensity in the UV-Vis extinction spectrum that can be observed and predicted when the catalyst change size or leach. This novel and simple UV-Vis surface plasmon spectroscopic technique will help to (a) fundamentally understand the catalytic pathway and (b) develop efficient catalysts for a variety of reaction sets.

496128 Downstream Processing of Extrudates: Polymer Platform Development for Hot-Melt Extrusion/Tableting Via in-Line Monitoring of Compaction Properties

Abstract:

INTRODUCTION

As the number of applications for polymers in pharmaceutical development is increasing, there is need for fundamental understanding on how such compounds behave during processing. This research is focussed on the tableting behaviour of polymers used as carriers for solid dispersions manufactured via hot-melt extrusion (HME). Despite the fact that HME has been widely used for solid dispersion technologies there is still limited knowledge on how this processing technique might influence the downstream processing such as the tableting behaviour of pharmaceutical polymers and their formulations (Iyer et al., 2013). Therefore, a novel approach was developed to monitor and analyse 'in-die' compaction properties directly on a rotary tablet press during downstream processing of milled extrudates. This is advantageous due to the possibility of performing in-line measurements during the tableting process. During the first part of this research, a polymer platform was developed from which researchers can select the adequate polymer for both HME and tableting purpose by using formulation properties and in-line measured compaction properties (Grymonpré et al., 2017). Afterwards, a promising polymer was selected from this platform and subjected to a quality-by-design (QbD) approach with different active pharmaceutical ingredients (API's) to understand the influence of process and formulation parameters on the critical quality attributes (CQA) of extrudates and tablets of a hot-melt extruded solid dispersion formulation.

MATERIALS AND METHODS Soluplus (SOL), Kollidon VA 64 (VA 64) and Eudragit EPO (EPO) were selected as amorphous polymers, PVA 4-88 as semi-crystalline polymer, while Celecoxib (CEL) was chosen as BCS class II model drug for development of the polymer platform. Neat polymers and physical mixtures (35% drug load) were processed via HME (Prism Eurolab 16, Thermo Fisher, Germany) using co-rotating twin-screws containing 3 mixing zones.

The extrusion temperature was determined based on the extrudable range of each material as defined via a rheological screening, i.e. the temperature range corresponding to a complex viscosity between 1000 and 10000 Pa.s (Gupta et al., 2015). After cooling the extrudates were milled and sieved to obtain powders with comparable particle sizes as the neat polymer. For each polymer, 3 formulations (neat polymer, neat polymer extrudates and polymer-CEL extrudates) were compressed to tablets (270 ± 10 mg) on a rotary tablet press (MODUL^{TM} P, GEA Pharma Systems) equipped with cylindrical flat-faced Euro B punches of 10 mm diameter, using 6 main compaction pressures without a pre-compression step. All tablets were analysed for 'out-of-die' properties (tabletability, compressibility, compactibility) immediately after ejection. Punch deformation at each compaction pressure was calculated and corrected for during this study. An experimental approach was developed for in-line analysis of the compaction properties (plasticity factor (PF), in-die elastic recovery (IER) and Heckel analysis) on a rotary tablet press using linear variable displacement transducers (LVDT) incorporated inside the turret and clamped onto one pair of punches and a complementary data acquisition analysis system (CDAAS, GEA Pharma Systems). This approach enabled the monitoring of punch stroke movements during a compression cycle directly on a rotary tablet press. Principal component analysis (PCA) was executed on the relevant characterization and compaction data in order to classify the different materials according to their flowability and compaction behaviour using the multivariate data analysing software SIMCA 13.0.3 (Umetrics, Umeå, Sweden).

Furthermore, a statistical Design of Experiments (DoE) was used to evaluate the influence of three HME process parameters: barrel temperature (160-200 °C), screw speed (50-200 rpm), throughput (0.2-0.5 kg/h) and one formulations parameter: drug load (0-20 %) on the CQA's of extrudates and tablets of the selected SOL-CEL solid dispersion formulation.

RESULTS AND DISCUSSION

Solid state characterization showed that HME yielded stable glassy solutions for all amorphous polymer formulations containing 35 % CEL as these were stable for at least 6 months under stress conditions (40 °C and 75% relative humidity). In general, higher tablet tensile strengths were observed for glassy solutions formulated with all polymers due to higher interparticulate bonding strengths per unit bonding area, while no changes were detected in the interparticulate bonding areas between formulations. There was need for analysing the 'in-die' tablet compaction properties of these formulations to understand the mechanisms which modified the mechanical properties such as tablet tensile strength. At higher compaction pressures, glassy solutions of SOL and EPO underwent more plastic deformation which was correlated to a higher compactibility and tabletability of these formulations. Additionally, the formation of stable glassy solutions altered the formulations towards more fragmentary behaviour under compression which was beneficial for the tabletability. No differences in plasticity factor were detected when tableting the neat polymers compared to their extrudates, indicating that HME did not alter the volume reduction mechanism of the amorphous polymers without API. However, this was not the case when using the semi-crystalline polymer (PVA 4-88) as a significant decrease in tabletability was detected after HME of the neat polymer which was linked with a change in crystalline fraction (32 to 21.4%). With the inclusion of amorphous CEL, the tablet tensile strength (TS) was slightly recovered due to similar transition towards a more fragmentary behavior. Principal component analysis (PCA) was applied to summarize the behaviour during HME, milling, feeding and compaction of all formulations, enabling the selection of polymers which have the highest potential as carrier during HME for solid dispersions with CEL in respect of further downstreaming by tableting. By application of PCA on this polymer platform, SOL and VA 64 were selected as more favourable polymers compared to EPO and PVA 4-88 for HME/tableting purpose. The QbD-study on SOL-CEL enabled to evaluate the HME-process as well as the impact of the tableting behaviour on this formulation. Torque was mainly influenced by barrel temperature and drug load since both factors impacted the melt viscosity. In general, drug load had the most significant impact on the extrudate properties compared to the process parameters indicating that for this specific formulation the HME process was very robust. Tabletability was clearly affected by the drug load, while changing the process parameters had no impact on this property. Tablets manufactured from the glassy solutions yielded a significantly higher TS. Drug load also altered the 'shape' of the curves (i.e. inflection point), indicating changes in the mechanical properties of the formulations. Conclusion

Monitoring the punch movement using the described instrumentation and CDAAS-software was an effective tool for in-line measurement of compaction properties on a rotary tablet press. By combining 'in-die' and 'out-of-die' techniques, it was possible to investigate in a comprehensive way the impact of HME on the tableting behaviour of pharmaceutical polymers and their formulations. While HME had only a limited influence on the compaction properties of the amorphous polymers when no drug was included, HME changed the compaction properties of glassy solutions towards a more fragmentary behaviour, independent of the polymer type. The semi-crystalline PVA 4-88 encountered more impact of HME on tabletability of the neat polymer due to a change in crystalline fraction. In respect to the milling efficiency of the extrudates, the used amorphous polymers were significantly beneficial over the semi-crystalline polymer since they were more brittle. A polymer platform was successfully established using PCA on the large dataset, enabling selection of the most suitable polymers for a specific formulation in respect to HME, milling and further downstream processing towards tablets. Afterwards, a promising polymer was selected from this platform and successfully subjected to a quality-by-design (QbD) approach which emphasized that HME was a robust process for the combination SOL-CEL. However, the quality of extrudates and tablets can be optimized by adjusting specific formulation parameters (e.g. drug load).

REFERENCES

1. **1.** Iyer, R., et al. The impact of hot melt extrusion and spray drying on mechanical properties and tableting indices of materials used in pharmaceutical development. J. Pharm. Sci. 102, 3604–3613 (2013).
2. **2.** Grymonpré, W., et al. In-line monitoring of compaction properties on a rotary tablet press during tablet manufacturing of hot-melt extruded amorphous solid dispersions. Int. J. Pharm. 517, 348–358 (2017).
3. **3.** Gupta, S.S.., et al. Effect of carbamazepine on viscoelastic properties and hot melt extrudability of Soluplus. Int. J. Pharm. 478, 232-239 (2015).

498661 Virtual High-Throughput Screening Pipeline: Size and Classification Distribution Effects on Experimentally Validated Hit-Rates

Abstract:

Drug leads are a small fraction of small organic compounds, which itself is a small fraction of conceivable organic molecules. High-Throughput screening (HTS) is used to screen multitudes of compounds in parallel (test more compounds at the same time) and in experimental working volumes (reduce the amount of reagents and biological substances used). However, HTS has a nominal hit-rate of a few percent at best (Dobson, 2004) and thus most resources are used to test inactive compounds. In efforts to increase efficiency, computational methods are used to pre-screen compound libraries: likely active molecules are focused on and likely inactive or negative molecules are removed.

Previously, we developed a virtual HTS (vHTS) pipeline: using available experimental data, models predicting class and activity are trained and applied to compound databases to identify compounds likely to be active and to focus experimental efforts on (Chen and Visco, 2016), with the option of retraining the models to improve performance (hit-rate) or to identify more leads. The pipeline was applied several NCBIâ€™s PubChem Bioassay datasets: AID 825 (target: Cathepsin L, 1st iteration hit-rate: 19%, 2nd iteration hit-rate: 75%) (Chen and Visco, 2016), AID 728 (target: Factor XIIa, 1st iteration hit-rate: 43%, 2nd iteration hit-rate: 100%) (Chen and Visco), and AID 846 (target: Factor Xia, 1st hit-rate: 27%, 2nditeration hit-rate: 62%) (Chen and Visco).

To determine the hit-rate enrichment ability of the pipeline, we have applied our pipeline on more datasets, specifically targeted to examine how the pipeline responds to datasets of different sizes and classification distributions. Determining the effect of these two dataset parameters will indicate what datasets the pipeline has the most enrichment value on and/or what the expected enrichment ability is for a given dataset.

In this poster, we present work characterizing the effects of size and classification distribution. While controlling for one parameter (size or classification distribution), the other is varied. Experimental validation is conducted based on the protocol specified by the original dataset to determine vHTS hit-rates. The hit-rates determined using the same experimental protocol will allow for direct comparison of hit-rates and to identify the hit-rate enrichment ability of the pipeline for a given dataset. Based on the results, there will be a clearer idea what datasets will have the most impact on and what an expected hit-rate for a given dataset would be. We also aim to show, indirectly, the pipelineâ€™s hit-rate enrichment ability is repeatable and the pipeline is robust enough to handle many different kinds of datasets.

Chen J.J.F., Visco D.P. Jr., Developing an in silico pipeline for faster drug candidate discovery: Virtual high throughput screening with the Signature molecular descriptor using support vector machine models, *Chemical Engineering Science*, 2 March 2016. http://dx.doi.org/10.1016/j.ces.2016.02.037.

Dobson, C.M., 2004. Chemical space and biology. Nature 432, 824-828.

Chen, J.J.F, Visco, D.P. Jr. Identifying Novel Factor XIIa Inhibitors With PCA-GA-SVM Developed vHTS Models. *Manuscript in preparation.*

Chen, J.J.F, Visco, D.P. Jr. Identifying Novel Factor XIa Inhibitors With PCA-GA-SVM Developed vHTS Models. *Manuscript in preparation.*

499892 PAT on Oscillatory Systems: Monitor and Control Continuous Crystallization with Fourier Transform Infrared (FTIR) Spectrometer

Abstract:

Most pharmaceutical manufacturing processes include a crystallization step to purify active pharmaceutical ingredients (APIs). Despite its long history, difficulties associated with crystallization control prevail due to its complex kinetics.[1] However, process analytical technologies (PAT) have improved swiftly allowing for crystallization monitoring and control.[2] Traditionally, crystallization is carried out in batch which often results in batch-to-batch variation.[3] Continuous crystallization, on the other hand, is promising for its enhanced reproducibility and narrower property distributions amongst other improvements.[4] Continuous crystallization has become a key step in the current trend to reform the pharmaceutical industry from batch to continuous manufacturing. A well-studied system for continuous crystallization is the Mixed-Suspension-Mixed-Product-Removal (MSMPR) crystallizer in which the hydrodynamics are controlled with an agitator. There are many challenges associated with stirred tank systems such as poor local mixing, low heat and mass transfer, and varying shear rate.[5] To address these challenges, a novel commercial oscillatory baffled reactor (OBR) has been studied. It consists of a reactor tank and oscillating â€˜donutâ€™ shaped baffles to provide more uniform mixing.[6] Such systems have been well studied for API synthesis and are gaining interest for crystallization given their enhanced mass and heat transfer characteristics while imposing less shear rate, ideal for crystallization.[7] However, control strategies for continuous crystallization in such oscillatory systems have yet to be studied in the literature. The difficulty controlling continuous crystallization in an OBR stems from inadequate measurement of key factors. Vibrations caused by the oscillating motion often interfere with the signals of the PAT tools which consequently result in inconsistent readings.

A model-free control strategy for continuous cooling crystallization of paracetamol in the OBR is implemented in this study. Supersaturation control (SSC) is a model-free technique where concentration is monitored and supersaturation level is controlled by adjusting operating conditions such as temperature for the case of cooling crystallization. SSC has been well studied in MSMPR systems.[8,9] An SSC implementation was carried out for continuous crystallization in the OBR via CryMOCO, a commercially available integrated control software, a Huber thermo-regulator with a temperature sensor and a FTIR spectrometer. The FTIR is equipped with the Interferogram correcting the signal interference from oscillations. It provides online measurements of solute concentration in solution used by CryMOCO to control supersaturation. This study serves as a proof-of-concept for continuous crystallization SSC control in oscillatory systems.

1. Braatz RD. Advanced control of crystallization processes. *Annu Rev Control*. 2002;26 I:87-99. doi:10.1016/S1367-5788(02)80016-5.

2. FDA. Guidance for Industry PAT: A Framework for Innovative Pharmaceutical Development, Manufacuring, and Quality Assurance. *FDA Off Doc*. 2004;(September):16. doi:http://www.fda.gov/CDER/guidance/6419fnl.pdf.

3. Quon JL, Zhang H, Alvarez A, Evans J, Myerson AS, Trout BL. Continuous crystallization of Aliskiren hemifumarate. *Cryst Growth Des*. 2012;12(6):3036-3044. doi:10.1021/cg300253a.

4. Alvarez AJ, Singh A, Myerson AS. Crystallization of Cyclosporine in a Multistage Continuous MSMPR Crystallizer. *Cryst Growth Des*. 2011;11(10):4392-4400. doi:10.1021/cg200546g.

5. Kacker R, Regensburg SI, Kramer HJM. Residence Time Distribution of Dispersed Liquid and Solid Phase in a Continuous Oscillatory Flow Baffled Crystallizer. *Chem Eng J*. 2017;317:413-423. doi:10.1016/j.cej.2017.02.007.

6. Hewgill MR, Mackley MR, Pandit AB, Pannu SS. Enhancement of gas-liquid mass transfer using oscillatory flow in a baffled tube. *Chem Eng Sci*. 1993;48(4):799-809. doi:10.1016/0009-2509(93)80145-G.

7. Ni X, Mackley MR, Harvey AP, Stonestreet P, Baird MHI, Rama Rao NV. Mixing Through Oscillations and Pulsationsâ€"A Guide to Achieving Process Enhancements in the Chemical and Process Industries. *Chem Eng Res Des*. 2003;81(3):373-383. doi:10.1205/02638760360596928.

8. Yang Y, Nagy ZK. Combined Cooling and Antisolvent Crystallization in Continuous Mixed Suspension, Mixed Product Removal Cascade Crystallizers: Steady-State and Startup Optimization. *Ind Eng Chem Res*. 2015;54(21):5673-5682. doi:10.1021/ie5034254.

9. Yang Y, Nagy ZK. Advanced control approaches for combined cooling/antisolvent crystallization in continuous mixed suspension mixed product removal cascade crystallizers. *Chem Eng Sci*. 2015;127:362-373. doi:10.1016/j.ces.2015.01.060.

500265 Continuous Manufacturing of Oral Disintegrating Films: A Quality By Design Approach

Abstract:

The ICH guidelines and FDA guidances have encouraged pharmaceutical industries to implement risk based (ICH Q9), systematic and science–based approaches (ICH Q8 (R2) and ICH Q11), a robust pharmaceutical quality system (ICH Q10) and introduce Quality by Design (QbD) concepts, emphasizing science and risk-based approaches to assure product quality. Innovative continuous manufacturing (CM) technologies have a great potential to improve control over quality, reduce costs, enhance process safety, and significantly reduce the timelines currently involved across the medicines' supply chain. When compared to traditional batch manufacturing, continuous manufacturing also has enhanced ability to control the process and product quality through knowledge of residence times and the implementation of process analytical technology (PAT) tools. The objective of this work is to understand the relationships between product and process variables and critical quality attributes (CQAs) of ondansetron (a model drug compound) to support development and review of control strategies for continuous manufacturing of ODFs. This can aid in the understanding of common failure modes that provide the foundation for systematic risk assessment of continuous film processing.

The ODFs of ondansetron were prepared in-house using the solvent casting method in a continuous tape caster. These films were characterized in-line, at-line or off-line for drug assay, content uniformity, disintegration, dissolution, morphology, surface roughness, wettability, thickness, % moisture uptake, % moisture content, mechanical properties.

Principles of QbD along with appropriate design of experiments (DoE) were used to obtain a comprehensive understanding about the effect of critical material attributes (CMA's) such as % w/w of plasticizer, surfactant and critical process parameters (CPP's) such as mixing speed, coat thickness, air flowrate and drying temperature on the quality and performance of ondansetron ODF. Suitable Design of Experiments (DoE) was used to screen and optimize critical parameters that impact the product quality of ODFs. Data analysis using ANOVA and multifactor analysis was performed in elucidating interactions between different variables, rank order the various CMA's and CPP's, and help provide a predictive model for the process. A design space was established and verified by performing a failure mode analysis.

500607 Controlled Delivery of Fluorophores from 3D Two-Photon Photolithographic Printed Poly(Ethylene Glycol) Methacrylate Scaffolds

Abstract:

Controlled drug delivery systems have been utilized to enhance the therapeutic effects of many current drugs by effectively delivering drugs in a time-dependent and repeatable manner. With the aid of 3D printing technology, a novel drug delivery device was fabricated and tested to appraise functionality. Poly(ethylene glycol) methacrylate(PEGMA) scaffolds were fabricated using a Photonic Professional GT 3D printer. These scaffolds were printed with varying degrees of hatching, slicing, and pore sizes in order to observe the effects of these parameters on drug released. To test for controlled drug release, rhodamine B was used as a model drug and was homogenously mixed into the PEGMA to be trapped inside the polymer matrix after photopolymerization. Scaffolds were analyzed using a fluorescence microscope and quantified through an imaging software. Scaffolds were also imaged under the scanning electron microscope(SEM) to visualize the ultra-morphology and fine resolution of the micron size printed scaffolds. PEGMA scaffolds were then investigated for toxicity, where MTS assays were performed on induced pluripotent stem cells (IPSCs). The variability of the hatching, slicing, and pore size of the fabricated PEGMA scaffolds provided controlled release of rhodamine. When incubated with the iPSC, the scaffolds were observed to have no cytotoxic effects. The SEM images reaffirms the ability to fabricate scaffolds at the micron and nano- scale with high precision. Using the two-photon photolithographic printer, complex and miniature scaffolds can be printed with fine details, while providing control drug release.

500678 Controlled and Sequential Delivery of Fluorophores from 3D Printed Alginate-PLGA Tubes

Abstract:

Controlled drug delivery systems, that include sequential and/or sustained drug delivery, have been utilized to enhance the therapeutic effects of many current drugs by effectively delivering drugs in a time-dependent and repeatable manner. In this study, with the aid of 3D printing technology, a novel drug delivery device was fabricated and tested to evaluate sequential delivery functionality. With an alginate shell and a poly(lactic-co-glycolic acid) (PLGA) core, the fabricated tubes displayed sequential release of distinct fluorescent dyes and showed no cytotoxicity when incubated with the human embryonic kidney (HEK293) cell line or bone marrow stromal stem cells (BMSC). The controlled differential release of drugs or proteins through such a delivery system has the potential to be used in a wide variety of biomedical applications from treating cancer to regenerative medicine.

Journal of chemical & engineering data

Article

pubs.acs.org/jced

Pressurized-Synthetic Methodology for Solubility Determination at Elevated Temperatures with Application to Paracetamol in Pure Solvents

Brian de Souza,* Leila Keshavarz, Giuseppe Cogoni, and Patrick J. Frawley

Solid State Pharmaceutical Cluster (SSPC), Materials & Surface Science Institute (MSSI), Lonsdale Building, Department of Mechanical, Aeronautical and Biomedical Engineering, University of Limerick, Castletroy, Co. Limerick, Ireland

ABSTRACT: This paper describes a new nonintrusive method for the determination of high-temperature solubility data. Accurate high-temperature solubility data is vital to many industrial manufacturing processes such as cooling crystallization with direct implications for yield, throughput, and solvent usage. However, the provision of such data is notably absent from published literature for many active pharmaceutical ingredients. Pressurized-synthetic methodology is presented as a new technique for determining high-temperature solubility data. Paracetamol (acetaminophen) is used as a reference active pharmaceutical ingredient to validate the methodology. Solubility data determined using the pressurized-synthetic approach is reported for several pure solvents across a significantly extended temperature range. In the case of methanol, solubility data is obtained up to 354.15 K, above the atmospheric boiling point of the solvent, 337.65 K, and far in excess of the temperature range for which data exists in the literature, 268.15–303.15 K. The data obtained using the pressurized-synthetic method is validated against an extended gravimetric data set at temperatures up to the atmospheric boiling point for each solvent. Sensitivity studies were conducted to determine the influence of factors such as temperature gradient on the ultimate solubility determination. A temperature-based standard deviation of 0.1 K was established for paracetamol in 2-propanol at 303.15 K, comparing favorably with the temperature-based equivalent standard deviation of 0.2 K for the gravimetric approach. Binary interaction parameters for the pressurized-synthetic solubility data are derived and estimated for four different activity coefficient models, namely Margules, Van-Laar, Wilson, and non-random two-liquid (NRTL), along with the empirical solubility equation of Apelblat. For each solvent, the quality of fit of each of the activity coefficient models is analyzed. The NRTL model was found to best fit the experimental data for methanol, ethanol, 2-propanol, and acetone with mean square errors of 5.73×10^{-5}, 3.00×10^{-4}, 1.70×10^{-4}, and 7.35×10^{-5}, respectively. The pressurized-synthetic approach provides a nonintrusive, validated, and readily automated approach for the provision of valuable high-temperature solubility data that can be readily extended to binary and ternary systems.

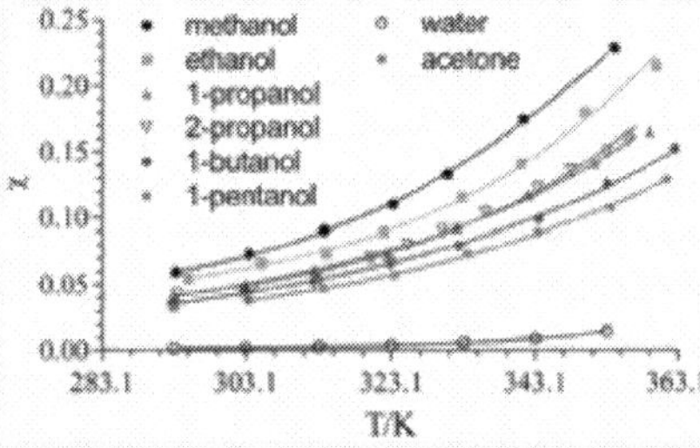

1. INTRODUCTION

The solubility of bioactive compounds is of critical importance to the pharmaceutical industry. Not only is accurate solubility data vital in terms of quantifying drug absorption and bioavailability, it is also essential for the development of manufacturing operations such as separation, liquid extraction, and crystallization.

In the case of the pharmaceutical cooling crystallization process, the difference in solubility as measured between upper and lower temperature bounds is by definition implicitly linked with the critical crystallization design metrics of yield, throughput, and solvent usage. For maximizing yield and throughput of cooling crystallization while minimizing solvent usage, a wide temperature span is normally desired. The provision of high- and low-temperature solubility data is therefore of upmost importance for industrial crystallization design. In general, it can be seen that API solubility curves show the most pronounced gradients at elevated temperatures. Thus, small increases in the upper temperature can potentially lead to significant process improvements. Accurate high-temperature solubility data is therefore imperative for designing crystallization processes; however, remarkably, the provision of high-temperature solubility data is notably absent from published literature to date.

There are several widely used methods for thermodynamic solubility determination, the most common of which include the gravimetric and shake flask methods.[1] In the gravimetric approach, the temperature of equilibration is fixed, and a saturated solution is achieved by contacting excess solute with solvent. Equilibrium is approached asymptotically, whereupon a sample of the equilibrated solution is filtered and weighed. The solvent is then gently evaporated, thus allowing for the evaluation of residual crystalline material mass and solubility. In the case of the shake flask method, again, an excess of drug solute is added to the solubility medium and the phases are again separated using filtration or centrifugation. Quantification

Received: January 31, 2017
Accepted: April 11, 2017
Published: April 19, 2017

of drug concentration in the saturated solution is measured by techniques such as ultraviolet (UV) spectrophotometric analysis or high-pressure liquid chromatography (HPLC).

Multiple drawbacks are associated with both gravimetric and shake flask methods. At elevated temperatures, and with volatile solvents, solubility measurements using either of these methods can be compromised by virtue of evaporation of the solvent. Unavoidable cooling occurs despite efforts to minimize the time between sample collection, filtering, and gravimetric weighing. Filtration of fine particles may not be entirely successful with some particles passing to the sample of equilibrated solution. Neither approach lends favorably to high-throughput screening studies. The maximum temperature for which solubility data can be obtained using both approaches is bound by the atmospheric boiling point of the solvent. In the case of UV or HPLC, samples may need to be diluted further for analysis, introducing the potential for measurement error. Regulation and maintaince of equilibrium temperature during analysis is a further potential concern. For viscous solvents in particular, settling of particles for analysis of clear solution may not occur.

This paper introduces a newly developed methodology for solubility determination, namely the pressurized-synthetic method. This nonintrusive method is based on the dissolution and consequent disappearance of the solute as monitored using laser instrumentation. The use of reactor pressurization extends the capability of the synthetic method, providing an innovative way to determine solubility even beyond the atmospheric boiling point of the solvent. The approach overcomes all the drawbacks previously listed for the gravimetric and shake flask methods. It does not require equilibrated sample collection or filtration. The approach can be readily automated, and the methodology does not occupy resources of specialized instrumentation.

For evaluating the pressurized-synthetic methodology, a reference active pharmaceutical ingredient (API) with well-established solubility data was required. Paracetamol (also known as acetaminophen) was chosen for this purpose. Paracetamol is an important analgesic and antipyretic drug. As a benchmark compound, it has, and continues to be, widely applied in crystallization research.[2−4] A range of pure solvents are employed in the manufacturing of paracetamol; in this paper, solubilities of paracetamol in eight of the most commonly used pure solvents are considered (methanol, ethanol, 1-pentanol, 1-propanol, 2-propanol, 1-butanol, water, and acetone).

There are several examples of solubility studies regarding paracetamol at low temperatures. However, an extensive literature search revealed no experimental data at elevated temperatures. Perhaps the most relevant of these studies is that of Granberg and Rasmuson,[5] who determined the solubility of paracetamol in a range of pure solvents using the gravimetric approach. The range of temperatures in that study was limited to 268.15−303.15 K. Likewise, Fernandez[6] gravimetrically determined the effect of solubility of paracetamol in ethanol for solution temperatures of 283.15−308.15 K, returning a slightly lower solubility when compared to the Granberg and Rasmuson data. A further study by Granberg and Rasmuson[7] examined the solubility of paracetamol of binary and ternary mixtures of water + acetone + toluene, again with a temperature range of 268.15−303.15 K.

Shakeel et al.[8] did present data covering a higher temperature range of 298.15−333.15 K for paracetamol using the shake flask method. However, the elevated temperature in that study was facilitated by a transcutol-water mixture, where transcutol has a high boiling point under atmospheric conditions of 475.15 K.

No previous literature was sourced in which the solubility of paracetamol was experimentally established at a temperature in excess of the atmospheric boiling point of the solvent.

The aims set out for the present work were to (1) establish solubility data for paracetamol up to and beyond the boiling point in pure solvents using the new pressurized-synthetic methodology; (2) validate the approach with comparisons with earlier and new gravimetric data, if necessary; (3) provide thermodynamic modeling of the extended solubility data; and finally (4) perform a comparative evaluation of activity coefficient models in terms of quality of fit.

2. MATERIALS AND METHODS

2.1. Materials. Paracetamol (BioXtra ≥99% NMR) was obtained directly from Sigma-Aldrich [CAS 103-90-2]. Paracetamol is known to exist in three polymorphic forms: monoclinic, commercially available, and room temperature relatively thermodynamically stable form I, orthorhombic (form II, metastable), and the highly unstable form III, which is rarely observed outside of fusion reactions. The crystal structures for both monoclinic and orthorhombic polymorphs have previously been reported in the literature.[9] Powder X-ray diffraction was utilized as per Figure 1 to confirm the crystal structure of the as-received paracetamol as being that of the commercially used form I with characteristic peaks (2Θ) including 12.1 ± 0.2, 13.6 ± 0.2, 15.5 ± 0.2, 18.2 ± 0.2, 23.5 ± 0.2, 24.4 ± 0.2, and 26.6 ± 0.2.

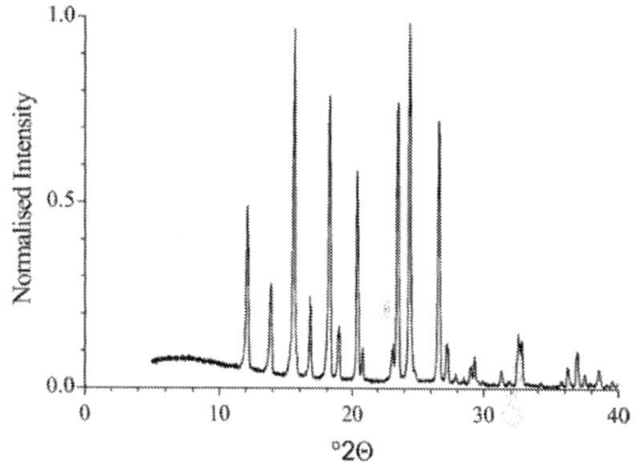

Figure 1. Experimental powder X-ray diffraction pattern for monoclinic paracetamol (form I) as received.

X-ray diffraction was utilized to eliminate the possibility of a polymorphic transformation during the solubility experiments across the full range of solvents considered. High-temperature samples of solute were prepared through rapid filtration followed immediately by X-ray diffraction analysis to mitigate the risk of solvent-mediated polymorphic transformation during storage. Powder diffraction data were collected on a Philips X'Pert-MPD PRO diffractometer (PW3064 sample spinner) with nickel-filtered Cu Kα radiation (λ = 1.5418 Å) run at 40 kV and 35 mA, 2θ = 5−40°, with a step size of 0.02° 2θ and a scan speed of 0.02° s^{-1}. Samples were prepared by light grinding prior to placement in a sample holder and flattening with a glass slide. The possibility of a polymorph transformation occurring during grinding was eliminated by analyzing a sample before and after grinding. Analysis of the

Table 1. Source, Mass Fraction Purity, and Boiling Points[a] of the Materials Used in This Work

chemical name	source	mass fraction purity	analysis method	additional purification	boiling point (1 atm)/(1.25 bar(g)) (K)
methanol	Sigma-Aldrich	≥0.999	HPLC[b]	none	337.65/359.35
ethanol	Sigma-Aldrich	≥0.999	HPLC	none	351.48/373.01
1-propanol	Sigma-Aldrich	≥0.999	HPLC	none	370.35/393.25
2-propanol	Sigma-Aldrich	≥0.999	HPLC	none	355.38/376.77
1-butanol	Sigma-Aldrich	≥0.995	HPLC	none	390.65/414.82
1-pentanol	Sigma-Aldrich	≥0.99	GC[c]	none	410.65/437.67
deionized water	distilled				373.15/396.99
acetone	Sigma-Aldrich	≥0.998	HPLC	none	329.44/354.83
monoclinic paracetamol	Sigma-Aldrich	≥0.99	NMR[d]	none	

[a]All normal boiling point data listed were obtained from Sigma-Aldrich and estimated for elevated pressure using the Clausius−Clapeyron relation. [b]High-performance liquid chromatography. [c]Gas chromatography. [d]Nuclear magnetic resonance spectroscopy.

spectra indicated no polymorphic transformation with consistent peaks before and after equilibration, indicative of form I.

The eight pure solvents and one solute utilized in the present work are listed in Table 1 alongside supplier and purity data and boiling points at atmospheric pressure and under pressurization (1.25 bar(g)). The water utilized was distilled, deionized, and filtered (0.2 μm).

2.2. Pressurized-Synthetic Solubility Determination.

Suitable reactors were required to perform the pressurized solubility measurements. Three glass-lined stainless steel pressure vessels were custom designed with a nominal capacity of 300 mL. These were fitted with flush mounted 15 bar pressure-rated sight glasses to facilitate the synthetic laser measurement. Vessel lids were mechanically fastened securely in place with Viton seals except in the case of acetone where EDPM seals were used for solvent chemical compatibility. The pressure vessel lids incorporated PTFE-sealed compression fittings for a PT-100 temperature gauge, chemically resistant 316 stainless steel pressure gauge, and pressure supply fittings. A pressure regulator was installed in addition to a pressure relief valve for safety reasons. For each solubility data point, predetermined masses of solvent and solute were added to, and weighed in, each reactor using an analytical balance (Mettler Toledo MS4002S, weighing capacity up to 4200 g, sensitivity ± 0.01 g).

To suspend the crystals in the solvent and further to minimize temperature and concentration gradients during the experiments, each vessel was agitated by means of a PTFE-coated magnetic stirrer bar driven by a serial magnetic stirrer plate (2mag MIXdrive 1) with 40 W stirring power and a maximum stirring volume of 3,000 mL per stir point. A maximum stir rate of 500 rpm was determined due to the potential for the formation of vortex and consequent air entrapment. A conservative 450 rpm was adopted for the solubility experiments, which is more than sufficient to ensure full suspension of paracetamol crystals.

The pressure vessels were sealed, purged, and slowly pressurized with N_2 to a pressure of 1.25 bar(g), which is sufficiently high to ensure minimal evaporation of any of the solvents used during the experiments. The pressure vessels were positioned within a thermostatic stainless steel bath (Grant GR150; 38L; stability ±0.005 K; uniformity ±0.02 K). For experiments up to 343.15 K, the bath was filled with deionized water, and the surface was covered with polypropylene spheres to avoid evaporation and minimize thermal gradients. For higher-temperature experiments, silicone oil was used to ensure appropriate thermal conduction while avoiding evaporation.

For the determination of solubility, a continuous-wave 635 nm red laser (MRL-III-635L) was utilized as is commonly adopted in laser diffraction measurements. The laser power was adjustable from 0 to 30 mW. A low power setting of 4 mW was selected to avoid excessive heating of the solution.

Silicon photodiode-based laser detectors (Coherent UV/vis Quantum Power Sensor, 325−1065 nm, calibration uncertainty ±2%, 5 μW to 100 mW wavelength dependent) were used to evaluate laser power and obscuration. The complete assembly of the apparatus used for pressurized-synthetic solubility measurements is illustrated in Figure 2.

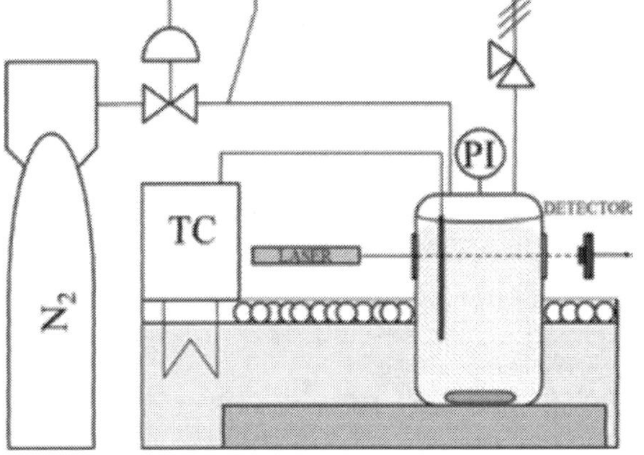

Figure 2. Schematic diagram of the experimental rig including temperature and pressure sensor, submersed magnetic stirring plate, thermostatic bath with temperature controller, 635 nm red laser, detector, magnetic stirring bar, and polypropylene spheres.

The experimental procedure adopted was broadly based on the laser monitoring approach.[10−12] This approach is also referred to as the synthetic or last crystal disappearance method.[13] The temperature of the reactors were slowly but progressively increased while monitoring obscuration of the laser beam by means of particle scattering. For each concentration of solute in solvent, a solubility temperature consistent with a sharp change in obscuration was detected as the particles dissolved. A representative sample result illustrating the sharp change in obscuration commensurate with the progressive, slow, and linear temperature increase of the reactor is shown in Figure 3 for the 0.05 molar fraction of paracetamol in 2-propanol.

Temperature profiles were implemented using the Labwise control software based on the feedback of the external

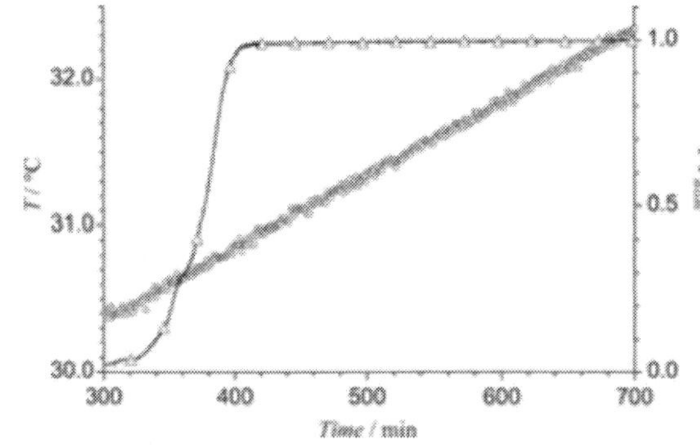

Figure 3. Pressurized-synthetic result demonstrating step change in laser obscuration as recorded for the 0.05 molar fraction of paracetamol in 2-propanol with a temperature gradient of 0.3 K/h. (O) Temperature data points and (△) normalized laser power detected.

temperature probe positioned within the reactor. A sensitivity study was performed to evaluate the effect of varying the temperature gradient, the results from which are shown in Figure 4.

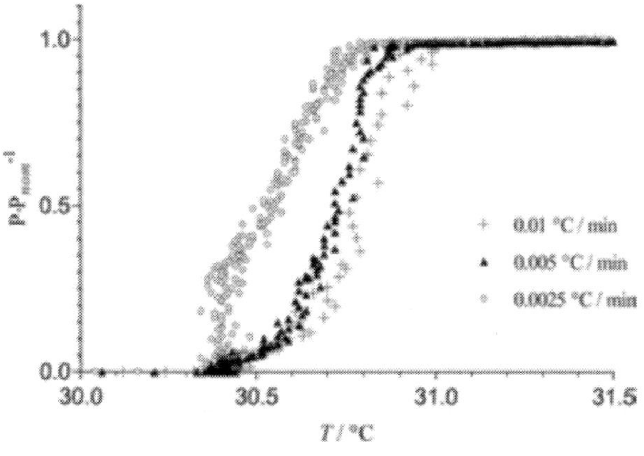

Figure 4. Sensitivity study investigating the influence of vessel temperature change gradient on the observed solubility point temperature measurements. The laser power measurement is normalized by the maximum laser power.

The results of the sensitivity study indicated a high level of consistency in the solubility temperature measurements with a standard deviation of 0.1 K. The uncertainty in the measurement is comparable with the earlier gravimetric work of Granberg and Rasmuson[5] who reported a mass-based standard deviation approximately equivalent to a temperature-based standard deviation of 0.2 at 303.15 K for paracetamol in 2-propanol.

For maximizing accuracy and ensuring sufficient time for equilibration, a conservative 0.3 K/h temperature gradient was adopted, which is significantly slower than the maximum suggested 2 K/h value[10] for synthetic measurements.

To monitor laser obscuration, data points were recorded every ten seconds with the average and standard deviation of laser power over 100 measurements returned. All experiments were conducted in the presence of a fume hood.

2.3. Gravimetric Solubility Measurements. Equilibrium solubility measurements were conducted for paracetamol in each of the eight pure solvents. The focus of the gravimetric measurements was to extend the earlier data set of Granberg and Rasmuson[5] such as to allow comparison with synthetic solubility data at more elevated temperatures. As with the pressurized-synthetic solubility determination, a Grant GR150 thermostatic stainless steel bath was used with a magnetic stainless steel stirrer plate placed on the base. Verification of temperature was obtained using a calibrated PT-100 resistance thermometer.

For experiments up to 343.15 K, the bath was filled using deionized water insulated by a layer of polypropylene balls. For higher-temperature gravimetric samples, high-temperature silicone bath oil was used. The data sets of low boiling point temperature solvents, namely acetone and methanol, were extended to 313.15 K; 2-propanol and ethanol were extended to 343.15 K, and higher boiling point temperature solvents were extended to 343.15 K in 10 K increments.

Excess solute was added to each of the solvent mixture tubes, which were then sealed to prevent solvent loss by evaporation. The solution was stirred at 500 rpm using Teflon-coated stirrer bars for a minimum of 72 h at constant temperature. Following this time, the solid residue was allowed to settle for at least 12 h, without stirring, until the saturated solution was observed to be visibly clear. This was briefly further verified using a laser light of 532 nm. The clear solution was then sampled (~10 mL) using a preheated syringe and filtered into preweighed dry glass vials using a 0.2 μm PTFE membrane (15 mm diameter) syringe filter.

The use of a cap incorporating te PTFE septum reduced the risk of solvent evaporation. The combined mass of vials and sample was measured without undue delay. The caps were then removed, and the solvents were allowed to evaporate slowly until only the solid residue of paracetamol was found to remain in the vials. In the case of volatile solvents with a boiling point of <393.15 K, drying was performed using a vacuum oven at 313.15 K for approximately 1 week with a further verification measurement performed several days later to ensure the sample was completely dry. In the case of less volatile solvents, the temperature of the vacuum oven was increased from 313.15 to 343.15 K after 3 days and sustained for approximately 1 week before the first mass measurement. A verification measurement was again performed several days later to ensure the sample was fully dry. All masses were weighed using an analytical balance (Mettler-Toledo AX054, weighing capacity up to 520 g, sensitivity ±0.1 mg).

Triplicate measurements were performed under the same conditions, and the arithmetic average value of the solubility was calculated. The estimated relative standard uncertainty based on repeat experiments (calculated as 100% standard deviation/average value) of the solubility measurement was less than 2% in each case. Negligible mass changes were noted between initial and verification measurements, thus indicating that the samples were completely dry.

3. THEORY

As noted earlier, the synthetic solubility studies conducted in this paper were carried out in pressurized vessels. The purpose of this pressurization was to control evaporation and furthermore to raise the boiling point of the solvent. It was not the intention of the study to directly affect the solubility of paracetamol in any of the pure solvents by means of the pressurization alone. Indeed, a theoretical demonstration was necessary to verify the appropriateness of negating the influence of pressure, as detailed below.

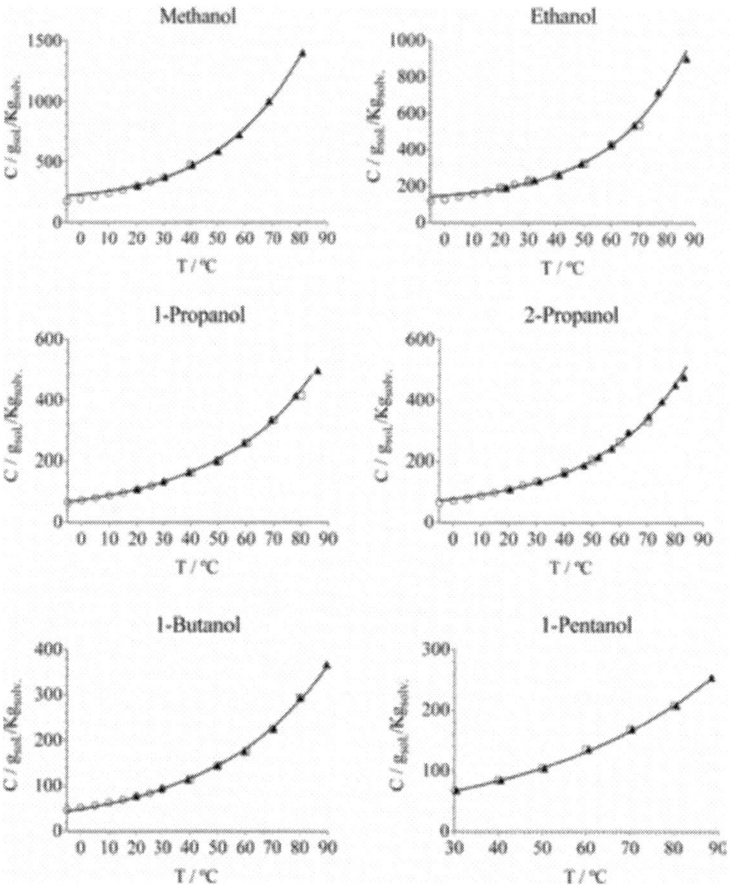

Figure 5. Solubility of paracetamol *C* versus temperature *T* in each of the alcohols as determined using (○) published literature values,[5] (▲) pressurized-synthetic data, and (□) the extended gravimetric data set. Pressurized-synthetic data are shown fitted using the NRTL model.

Early studies by Gibson[14] recognized that only in exceptional cases is the solubility of solids in liquids influenced by pressure. These exceptional cases include solutions of the carbonates, sulfates, sulfides, fluorides, and hydroxides of some alkalis, the alkaline earths, and the heavy metals in water, especially where the saturated solutions are very dilute. Pressure on the order of 1000s of atmospheres can significantly increase the solubility of these substances.

The variation of solubility with pressure for solids in pure liquids at equilibrium follows the following relationship[14]

$$\frac{dx_2}{dP} = -\frac{v_2 - v_2^S}{\left.\frac{\partial\mu_2}{\partial x_2}\right|_{P,T}} \qquad (1)$$

where x_2 is the molar fraction referring to the solute, P is the pressure in pascals, v_2 is the partial volume of the solute in the solution, and v_2^S is its specific volume in the solid phase, both expressed in m^3/mol, and $(\partial\mu_2/\partial x_2)_{P,T}$ is the rate of change of the chemical potential of the solute in the solution with a change of composition at the temperature and pressure of the experiment.

Considering the chemical potential change between solid and liquid solution and in solid phase[7]

$$\mu_2^L = \mu_2^S + kT\ln\frac{a_2}{a_{2,eq}} = \mu_2^S + kT\ln\frac{x_2\gamma_2}{x_{2,eq}\gamma_{2,eq}} \qquad (2)$$

where k represents the Boltzmann constant equal to 1.38×10^{-23} J/K, γ_2 is the activity coefficient of the solute in the liquid phase, T is the temperature in kelvin, and the subscript "eq" denotes the equilibrium conditions. Assuming the activity coefficients are approximately equal to 1 for the sake of simplicity, it is possible to evaluate the change in chemical potential with a change in solution composition

$$\left.\frac{d\mu_2}{dx_2}\right|_{P,T} = \frac{kT}{x_2} \qquad (3)$$

Substituting eq 3 into eq 1 and integrating leads to the expression

$$\ln\frac{x_2}{x_2^*} = -\frac{v_2 - v_2^S}{kT}(P - P^*) \qquad (4)$$

$$x_2 = \exp\left[-\frac{v_2 - v_2^S}{kT}(P - P^*)\right]x_2^* \qquad (5)$$

Incorporating the pressure applied for the experiments (1.25 bar(g)) and a volume variation for polymorphic form I of paracetamol of 1.75×10^{-6} m^3/mol at the melting temperature (442.2 K),[15] it is apparent that the term inside the exponential, relative to the atmospheric conditions in eq 5, must have a magnitude order of $O(10^{-19})$ for the range of temperatures under consideration. This implies that the RHS of eq 5 must

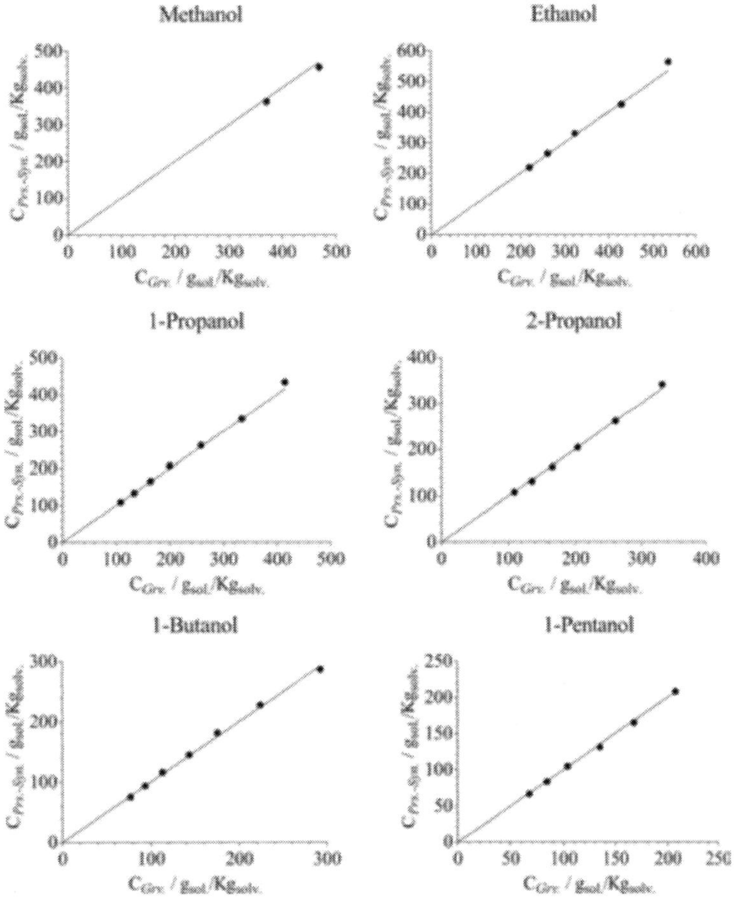

Figure 6. Comparison of gravimetric and pressurized-synthetic solubility results at temperatures up to the boiling point of each alcohol. The straight line represents an ideal correlation with gravimetric results, and the point *x* and y axis coordinates are obtained directly from gravimetric data and best-fit NRTL interpolation of synthetic solubility results, respectively.

tend to zero to balance the equation. Therefore, it can be concluded that the direct effect of pressure on solubility is negligible in the case of the present work.

4. RESULTS AND DISCUSSION

The pressurized-synthetic, as well as the extended gravimetric data from the present work and Granberg and Rasmuson[5] solubility data for paracetamol as measured in each of the different pure alcohols, is presented in Figure 5. No discernible discontinuity was observed between data obtained through the gravimetric and synthetic approaches. Each of the experimental data points represents a single concentration measurement. Given that the synthetic solubility curves were not known in advance and were estimated by extrapolation of existing data, some nonuniformity in the distribution of pressurized-synthetic data points is present. The pressurized-synthetic solubility is shown correlated using the non-random two-liquid (NRTL) model, which was found to have the best fit of the four activity coefficient models used for the thermodynamic modeling of solubility.

Using the pressurized-synthetic approach, solubility data for paracetamol in methanol was obtained up to 354.15 K, well beyond the atmospheric boiling point of methanol (337.65 K). The availability of such high-temperature solubility data overcomes the need for extrapolation with implicit error. For instance, in the case of applying the modified Apelblat correlation (eq 8) to the published data for paracetamol in methanol, an 11.2% error was calculated at the highest temperature measured (354.15 K) when compared to pressurized-synthetic measurements.

As previously noted, in the case of cooling crystallization, operations beyond the boiling up can enhance yield, throughput, and solvent usage. Consider a cooling crystallization of paracetamol in methanol; cooling from the atmospheric boiling point of methanol 337.65 K to the lowest temperature solubility point available (268.15 K) results in a theoretical yield of 80.2%. Increasing the upper temperature limit to 354.15 K facilitates an improvement in yield, which is increased to 87.5%. Indeed, further yield enhancements may be gained through extension of this method to temperatures in excess of those considered in the present work. Crystallization is a late-stage process in pharmaceutical manufacturing wherein the costs of the precursors have largely been absorbed. Therefore, the economic impact of process intensification can be very significant on an industrial scale. These benefits illustrate the clear need for high-temperature solubility data in excess of solvent atmospheric boiling points.

For validating the pressurized-synthetic method, a comparison between gravimetric- and pressurized-synthetic-derived solubility data was required. A direct comparison of the gravimetric-derived solubility data with the NRTL-modeled pressurized synthetic data is shown in Figure 6 for the alcohols.

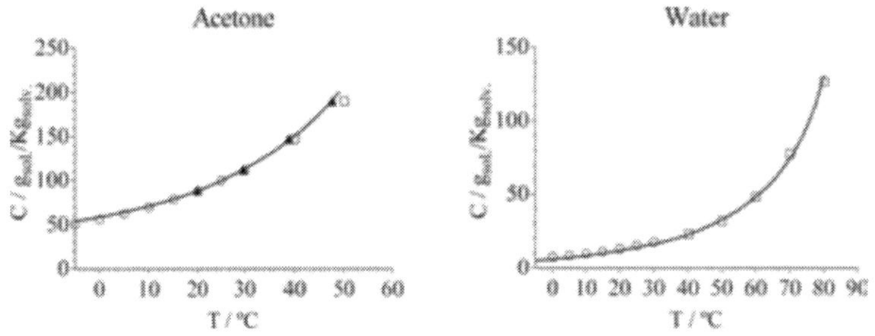

Figure 7. Solubility of paracetamol C versus temperature T in each of the nonalcohols as determined using (○) published literature values,[5] (▲) pressurized-synthetic data, and (□) the extended gravimetric data set. Pressurized-synthetic data are shown fitted using the NRTL model.

The data shown in Figure 6, in general, show excellent agreement between the solubility data obtained using the gravimetric and pressurized-synthetic methods. This is particularly true in the case of higher boiling point alcohols such as 1-butanol and 1-pentanol. Lower boiling point temperature alcohols such as ethanol, 1-propanol, and 2-propanol returned a marginally higher solubility using the pressurized-synthetic method at temperatures close to the atmospheric boiling point. This was anticipated in advance given the potential for error of the gravimetric approach at elevated temperatures.

In the case of the nonalcohols, solubility data for acetone was obtained up to a maximum of 323.15 K. The aggressive nature of hot acetone necessitated the provision of ethylene propylene diene monomer (EPDM) rather than Viton seals.

One of the drawbacks of the synthetic, and consequently the pressurized-synthetic method, is that the solute must be freely suspended within the solvent and physically obscure the laser path. The solubility of paracetamol in water is significantly less than that of all other solvents tested. Despite vigorous agitation, a significant mass of paracetamol was found to be suspended as a froth on the surface of the water. As such, it was not possible to obtain accurate solubility data for paracetamol in water using the pressurized-synthetic approach. In the case of gravimetric testing, even with limited mixing, equilibrium is ultimately reached with the opportunity to sample the clear solution between the paracetamol froth and settled crystals below. Additional gravimetric test results were taken to extend the solubility curve for paracetamol in water to 353.15 K as shown in Figure 7.

Table 2 presents the full data set of pressurized-synthetic results in both mole fraction x_2 and as $g_{solute}/kg_{solvent}$ (solute-free basis) at the measured saturation temperature under a pressure of 1.25 bar(g). Also presented is the calculated mole fraction solubility of paracetamol in each of the solvents using four activity coefficient models and the empirically modified Apelblat equation. Table 3 provides the extended set of gravimetric data for each solvent. For some data points, the mole fraction solubilities in Tables 2 and 3 are identical. In such cases, the concentration of paracetamol in pure solvent utilized for the pressurized-synthetic measurements was informed by the preceding gravimetric results.

The conversion between molar solubility x_2 and mass solubility C and vice versa was performed using the relationship

$$C = \frac{1000 x_2 MW_{SOLUTE}}{MW_{SOLVENT}(1 - x_2)} \quad (6)$$

$$x_2 = \frac{\frac{C}{MW_{SOLUTE}}}{\frac{C}{MW_{SOLUTE}} + \frac{1000}{MW_{SOLVENT}}} \quad (7)$$

5. SOLUBILITY MODELING

Many methods have been proposed to describe the solubility of a solute in different pure solvents; for instance, the perturbed-chain statistical-associating fluid theory (PC-SAFT) modeling approach has been widely applied for the correlation and prediction of solubility even at elevated temperatures.[16,17] In this work, the solubility of paracetamol in the selected solvents at different temperatures is correlated with the modified Apelblat equation,[18] and Margules, Van-Laar, NRTL and Wilson activity coefficient-based models.

The modified Apelblat equation is expressed in eq 8.

$$\ln C = a_A + \frac{b_A}{T} + c_A \ln T \quad (8)$$

where C represents the equilibrium concentration in grams of solute per kilograms of solvent at temperature T expressed in kelvin; the three adjustable parameters of the empirical solubility model are indicated as a_A, b_A, and c_A. The values of these parameters were found for each solvent through regression and are summarized in Table 4. The empirical model shows good agreement with the underlying experimental data in the case of all solvents tested.

Although the correlation of Apelblat shows good agreement with the underlining experimentally derived solubility data, the correlation is empirical in nature rather than thermodynamically derived. Activity coefficient models, such as Margules, Van-Laar, Wilson and NRTL, offer a more thermodynamically rigorous approach, accounting for deviation from the ideal solubility. For calculating the binary interaction parameters utilized by these models, it was first necessary to calculate the experimental activity coefficients based on the solubility data.

The experimental activity coefficients were calculated using a nonideal solid–liquid equilibrium equation defined as[18]

$$\ln(x_2\gamma_2) = \frac{\Delta H_f|_{T_m}}{R}\left(\frac{1}{T_m} - \frac{1}{T}\right) - \frac{\Delta C_p|_{T_m}}{R}\left[\ln\left(\frac{T_m}{T}\right) - \frac{T_m}{T} + 1\right] \quad (9)$$

in which the exchange of enthalpy of fusion is represented by ΔH_f, and ΔC_p signifies the difference in heat capacity of solute between the solid and liquid states, both evaluated at the normal melting temperature T_m. Using different scanning calorimetric analyses, Hojjati and Rohani[19] measured the

Table 2. Pressurized-Synthetic and Calculated Mole Fraction Solubility of Paracetamol as Calculated Using Margules, Van-Laar, Wilson, NRTL, and Modified Apelblat Equations across the Entire Range of Solvents at Saturation Temperature T and Pressure P = 1.25 bar(g)[a]

T (K)	C (g/kg)	x^{exp}	x^{apel}	x^{Marg}	x^{VL}	x^{W}	x^{NRTL}
			Methanol				
293.45	297.81	0.0594	0.0613	0.0633	0.0587	0.0575	0.0597
303.55	371.61	0.0730	0.0658	0.0766	0.0740	0.0729	0.0725
313.95	469.93	0.0906	0.0853	0.0933	0.0925	0.0927	0.0902
323.35	586.13	0.1105	0.1099	0.1115	0.1122	0.1147	0.1111
330.95	722.80	0.1328	0.1338	0.1289	0.1308	0.1358	0.1336
341.55	1000.00	0.1749	0.1719	0.1578	0.1609	0.1705	0.1739
354.15	1400.00	0.2288	0.2228	0.2004	0.2045	0.2192	0.2291
			Ethanol				
295.15	190.61	0.0549	0.0577	0.0545	0.0525	0.0520	0.0551
305.35	232.75	0.0662	0.0596	0.0667	0.0650	0.0649	0.0654
314.15	260.93	0.0737	0.0726	0.0794	0.0777	0.0781	0.0750
322.15	323.59	0.0898	0.0894	0.0933	0.0927	0.0932	0.0896
332.95	429.13	0.1157	0.1182	0.1161	0.1177	0.1181	0.1150
341.25	536.38	0.1405	0.1445	0.1375	0.1412	0.1416	0.1402
350.05	718.80	0.1797	0.1759	0.1649	0.1719	0.1720	0.1769
360.15	902.08	0.2156	0.2154	0.2021	0.2119	0.2123	0.2205
			1-Propanol				
293.05	108.09	0.0412	0.0432	0.0406	0.0411	0.0409	0.0412
302.95	132.77	0.0501	0.0464	0.0500	0.0502	0.0501	0.0501
312.55	163.47	0.0610	0.0581	0.0613	0.0612	0.0612	0.0609
322.55	199.45	0.0735	0.0749	0.0757	0.0754	0.0755	0.0748
332.25	257.84	0.0930	0.0947	0.0934	0.0932	0.0930	0.0925
342.15	334.10	0.1173	0.1180	0.1158	0.1162	0.1153	0.1153
351.25	415.29	0.1417	0.1418	0.1411	0.1422	0.1405	0.1410
359.15	498.00	0.1653	0.1639	0.1673	0.1694	0.1667	0.1680
			2-Propanol				
293.65	108.78	0.0415	0.0434	0.0411	0.0407	0.0403	0.0415
304.05	135.00	0.0509	0.0468	0.0512	0.0505	0.0504	0.0506
313.05	161.63	0.0604	0.0579	0.0619	0.0610	0.0613	0.0605
320.25	188.25	0.0696	0.0700	0.0721	0.0712	0.0717	0.0702
325.35	214.88	0.0787	0.0799	0.0805	0.0798	0.0803	0.0788
330.25	241.50	0.0876	0.0904	0.0894	0.0889	0.0895	0.0879
336.15	294.75	0.1049	0.1042	0.1020	0.1024	0.1026	0.1022
343.45	348.00	0.1215	0.1229	0.1193	0.1207	0.1207	0.1208
353.05	450.00	0.1518	0.1501	0.1473	0.1509	0.1500	0.1524
348.15	397.21	0.1364	0.1359	0.1323	0.1348	0.1343	0.1356
356.15	476.47	0.1593	0.1594	0.1573	0.1614	0.1604	0.1631
			1-Butanol				
293.05	77.07	0.0364	0.0379	0.0341	0.0361	0.0360	0.0355
302.95	93.64	0.0439	0.0452	0.0423	0.0438	0.0438	0.0436
312.55	113.13	0.0526	0.0542	0.0522	0.0530	0.0531	0.0532
323.05	143.81	0.0659	0.0669	0.0658	0.0656	0.0658	0.0662
332.55	175.38	0.0792	0.0817	0.0811	0.0797	0.0799	0.0806
343.65	224.08	0.0990	0.1038	0.1037	0.1005	0.1006	0.1015
353.15	292.76	0.1255	0.1276	0.1289	0.1238	0.1235	0.1236
362.65	366.49	0.1523	0.1569	0.1596	0.1522	0.1514	0.1505
			1-Pentanol				
303.55	67.82	0.0380	0.0386	0.0363	0.0383	0.0381	0.0378
313.85	84.20	0.0468	0.0459	0.0458	0.0471	0.0471	0.0471
323.65	103.77	0.0571	0.0571	0.0571	0.0575	0.0577	0.0580
333.65	135.07	0.0730	0.0719	0.0720	0.0711	0.0713	0.0718
343.55	168.19	0.0893	0.0894	0.0903	0.0878	0.0880	0.0885
353.75	207.54	0.1080	0.1100	0.1139	0.1092	0.1094	0.1094
361.55	254.14	0.1291	0.1275	0.1368	0.1302	0.1298	0.1287
			Acetone				
292.95	88.09	0.0327	0.0327	0.0343	0.0325	0.0324	0.0327
302.85	111.65	0.0411	0.0410	0.0425	0.0416	0.0417	0.0413
311.95	146.94	0.0534	0.0534	0.0521	0.0533	0.0534	0.0532

Table 2. continued

T (K)	C (g/kg)	x^{exp}	x^{Apel}	x^{Marg}	x^{VL}	x^{W}	x^{NRTL}
			Acetone				
320.65	189.83	0.0680	0.0678	0.0634	0.0678	0.0677	0.0682

[a]Standard uncertainties u are $u(T)$ = 1 K, $u(P)$ = 0.05 bar, and $u_r(x_2)$ = 0.02.

Table 3. Extended Gravimetric Data Set of the Solubility of Paracetamol in Pure Solvents Expressed in Both Mole Fraction x_2 and as $g_{solute}/kg_{solvent}$ (Solute-Free Basis) at Temperature T and Pressure P = 0.1 MPa[a]

T (K)	C (g/kg)	x^{exp}	C (g/kg)	x^{exp}
	Methanol		Ethanol	
293.15	297.81	0.0594	188.00	0.0542
303.15	371.61	0.0730	220.00	0.0628
313.15	469.99	0.0906	260.93	0.0737
323.15			323.59	0.0898
333.15			429.13	0.1157
343.15			536.38	0.1405
	1-Propanol		2-Propanol	
293.15	108.09	0.0412	108.78	0.0415
303.15	132.77	0.0501	135.01	0.0509
313.15	163.47	0.0610	165.54	0.0618
323.15	199.45	0.0735	203.70	0.0749
333.15	257.84	0.0930	262.13	0.0944
343.15	334.10	0.1173	331.50	0.1165
353.15	415.29	0.1417		
	1-Butanol		1-Pentanol	
293.15	77.07	0.0364	61.79	0.0348
303.15	93.64	0.0439	67.82	0.0380
313.15	113.13	0.0526	84.20	0.0468
323.15	143.81	0.0659	103.77	0.0571
333.15	175.38	0.0792	135.07	0.0730
343.15	224.08	0.0990	168.19	0.0893
353.15	292.76	0.1255	207.54	0.1080
	Acetone		Water	
293.15	88.09	0.0327	12.78	0.0015
303.15	111.65	0.0411	17.39	0.0021
313.15	146.94	0.0534	22.94	0.0027
323.15	189.83	0.0680	31.74	0.0038
333.15			48.03	0.0057
343.15			77.67	0.0092
353.15			126.48	0.0149

[a]Standard uncertainties u are $u(T)$ = 1 K, $u(P)$ = 0.05 bar, and $u_r(x_2)$ = 0.02.

Table 4. Apelblat Parameters and Root Mean Square Error (RMSE) Calculated for Each Solvent

solvent	a_A	b_A	c_A	RMSE
methanol	−193.9	7236	30.79	0.0020
ethanol	−193.3	7303	30.55	0.0034
1-propanol	−129.2	4322	20.97	0.0021
2-propanol	−156.6	5545	25.06	0.0014
1-butanol	−170.4	6393	26.93	0.0029
1-pentanol	−105	3207	17.26	0.0011
acetone	−270.4	10300	42.2	0.0128
water	−341.6	13280	52.64	3.7×10^{-5}

thermal properties of paracetamol to be T_m = 442.2 K, ΔH_f = 28.1 kJ/mol, and ΔC_p = 99.6 J mol^{-1} K^{-1} (at the fusion temperature).

Through insertion of the enthalpy, melting temperature, and heat capacity values into eq 9, and given that the experimental solubility values x_2 and temperatures are known, it is possible to evaluate the experimental activity coefficients by rewriting eq 9 for γ_2. The ideal solubility of paracetamol may also be estimated (setting γ_2 = 1).

Plots of the logarithm of experimental activity coefficients of paracetamol in alcohol- and nonalcohol-based solvents are given in Figure 8(a) and (b), respectively. Referring to eq 9, it is apparent that, at any given temperature, the activity coefficient is inversely proportional to the measured solubility. The ideal solubility is represented by the value 0 on the logarithmic activity coefficient scale with higher values relating to lower solubility. Increasing chain length in alcohols, as is shown in Figure 8(a), results in greater activity coefficient values and consequently reduced solubility. Activity coefficient trends for all solvents are shown to remain relatively flat or decrease slightly with increasing temperature.

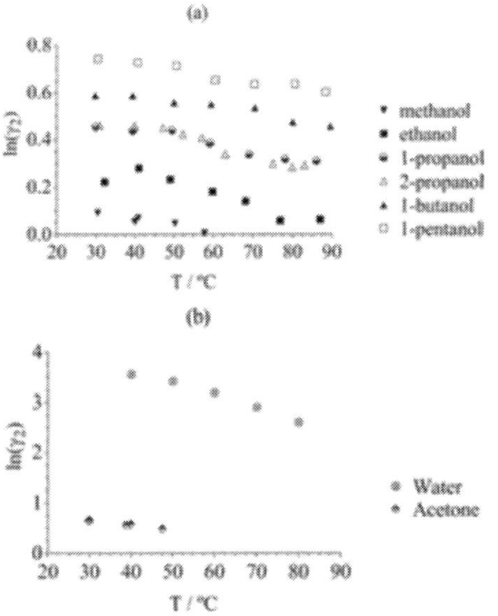

Figure 8. Logarithm of experimental activity coefficients of paracetamol at selected temperatures in (a) alcohols, (b) water, and acetone.

The highest activity coefficient values are observed with water, a highly polar solvent. Physically, this can be explained by the significant influence of the aromatic ring and methyl group in paracetamol on surrounding water molecules despite the fact that the alcohol and amide groupings in the paracetamol molecule are also polar, facilitating the formation of hydrogen bonds.

Having established the experimental activity coefficients, activity coefficient models could now be fitted to the experimental data. The four models selected are described in Table 5. Utilization of each of the models is commonly found in the literature. They can be differentiated by such factors as

Table 5. Activity Coefficient Models Used in the Present Work Complete with Their Respective Number of Parameters and Temperature Dependency Information

model	equation[a]	number of parameters	temperature dependency
Margules	$\ln \gamma_2 = \frac{A}{RT}(1 - x_2)^2$	1 [A]	explicit
Van-Laar	$\ln \gamma_2 = \frac{B_{VL}}{\left(1 + \frac{B_{VL}x_2}{A_{VL}x_1}\right)^2}$	2 [A_{vl}, B_{vl}]	implicit
Wilson	$\ln \gamma_2 = -x_2\left(\frac{\Lambda_{12}}{x_1 + \Lambda_{12}x_2} - \frac{\Lambda_{21}}{x_2 + \Lambda_{21}x_1}\right) - \ln(x_2 + \Lambda_{21}x_1)$ where $\Lambda_{ij} = \frac{v_j}{v_i}\exp\left(-\frac{\lambda_{ij}}{RT}\right)$ with $i \neq j$ and $i, j = 1, 2$	2 [λ_{12}, λ_{21}]	explicit
NRTL	$\ln \gamma_2 = x_1^2\left[\tau_{12}\left(\frac{G_{12}}{x_2 + x_1G_{12}}\right)^2 + \frac{\tau_{21}G_{21}}{(x_1 + x_2G_{21})^2}\right]$ where $G_{ij} = \exp(-\alpha\tau_{ij})$ and $\tau_{ij} = \frac{g_{ij}}{RT}$ with $i \neq j$ and $i, j = 1, 2$	3 [g_{12}, g_{21}, α]	explicit

[a] R is the ideal gas constant expressed in J mol^{-1} K^{-1} (8.314 J mol^{-1} K^{-1}) and v_i or v_j are the molar volumes of the ith or jth components in cm^3/mol.

Table 6. Estimated Binary Coefficients for Paracetamol in Different Pure Solvent Using Margules, Van-Laar, Wilson, and NRTL Models

	Margules	Van-Laar		Wilson		NRTL		
solvent	A	A_{VL}	B_{VL}	λ_{12}	λ_{21}	g_{12}	g_{12}	α
methanol	127.62	17.695	−0.085	549.958	1.46×10^3	1.50×10^4	-7.77×10^3	0.1627
ethanol	624.49	0.4312	0.0903	587.877	614.984	1.844×10^4	-1.07×10^4	0.1019
1-propanol	1.28×10^3	0.5542	0.2805	1.897×10^3	−531.1754	8.86×10^3	-5.83×10^3	0.1033
2-propanol	1278.31	0.6022	0.2429	1.00015×10^3	377.5298	12597.57	−7409.39	0.1257
1-butanol	1.72×10^3	0.6497	0.6108	3.35×10^3	-7.48×10^2	1.56×10^3	−45.657	−2.94
1-pentanol	2.16×10^3	0.802	0.733	3.67×10^3	-1.61×10^2	1.67×10^3	1.49×10^2	−9.468
acetone	1.69×10^3	1.037	0.1632	−558.174	3740.595	11752.49	−5241.13	0.2406
water	8.84×10^3	3.799	0.2622	4.990×10^3	4.871×10^3	11912.37	−1477.12	0.5141

Table 7. Mean Squared Error (MSE) of the Fitted Models for Each Solvent

solvent	Margules	Van-Laar	Wilson	NRTL	MSE_{max}/MSE_{min}
methanol	7.60×10^{-3}	5.10×10^{-3}	1.50×10^{-3}	5.73×10^{-5}	132.63
ethanol	3.63×10^{-3}	1.80×10^{-3}	2.21×10^{-3}	3.00×10^{-4}	12.10
1-propanol	2.30×10^{-4}	8.98×10^{-5}	2.33×10^{-4}	2.27×10^{-4}	2.59
2-propanol	7.11×10^{-4}	3.29×10^{-4}	5.19×10^{-4}	1.70×10^{-4}	4.18
1-butanol	1.93×10^{-3}	1.28×10^{-4}	1.80×10^{-4}	5.45×10^{-4}	15.07
1-pentanol	2.21×10^{-3}	3.30×10^{-4}	2.63×10^{-4}	2.94×10^{-4}	7.58
acetone	3.76×10^{-3}	1.02×10^{-4}	1.92×10^{-4}	7.35×10^{-5}	51.15
water	5.50×10^{-2}	3.80×10^{-3}	4.21×10^{-4}	6.47×10^{-4}	130.95

complexity and the inherent nature of the temperature dependency.

The binary interaction parameters were estimated using the following mathematical optimization problem solved using the MATLAB nonlinear least-squares algorithm *nlinfit*

$$\min_{\theta \in \mathcal{R}} \sum_{i=1}^{N_p} [\ln\gamma_{2,i}|_{mod}(T, \theta) - \ln\gamma_{2,i}|_{exp}(T)]^2 \tag{10}$$

where θ is the set of binary interaction parameters to be estimated for each model and each solvent used, N_p is the number of experimental points carried out for each solvent used, and the subscripts mod and exp refer to the model and experimental activity coefficients, respectively.

Insertion of the binary coefficients from Table 6 into the activity model equations of Table 5 allows for the prediction of molar solubility, which can then be converted to mass solubility as per eq 6. Returning to Table 2, the solubility derived from each of the activity coefficient model is directly compared to the experimental data.

The selection of activity coefficient model is influenced by the quality of fit of the model to the underlying data as well as the complexity of the model. Although the quality of fit of each activity coefficient model is solvent-dependent, across all solvents it is evident that the greatest deviation between experimental and model results occurs in all plots at the extremes of the temperature measurements. Further points in this temperature space could potentially reduce these discrepancies. Indeed, the methodology adopted in the current work facilitates further extension of the temperature range, thus potentially increasing the robustness of prediction within the current temperature range.

In seven of the eight solvents examined, the Margules model is shown to have the largest mean square error (Table 7). The NRTL model has the smallest mean square error for methanol, ethanol, 2-propanol, and acetone; Wilson is smallest for 1-pentanol and water, Van-Laar is smallest for 1-propanol and 1-butanol.

6. CONCLUSIONS

Solubility measurements are essential for pharmaceutical process development. This paper utilizes the pressurized-synthetic approach to facilitate measurement of solubility data for the reference compound paracetamol at elevated temperatures, in some instances exceeding the atmospheric boiling points of the solvent. Such high-temperature data is particularly important in the case of cooling crystallization, where the difference in solubility between upper and lower temperature limits is implicitly associated with yield and industrial efficiency. The error associated with extrapolation of solubility data was demonstrated to be potentially significant (11.2% error) using the modified Apelblat model for paracetamol in methanol and extrapolating from 303.15 to 354.15 K.

The data obtained from the pressurized-synthetic measurements were validated for all solvents with the exception of water by comparison with the gravimetric data. No discernible discontinuity was observed between data obtained through the gravimetric and synthetic approaches. Excellent agreement was seen for higher boiling point solvents at elevated temperatures, whereas for more volatile solvents, the pressurized-synthetic approach indicated a marginally higher solubility when compared to that of the gravimetric approach as was expected.

The solubility data obtained for all pure solvents was modeled using the modified Apelblat empirical equation, as well as Margules, Van-Laar, NRTL, and Wilson activity coefficient models. The NRTL model was shown to have the smallest mean square error for methanol, ethanol, 2-propanol, and acetone.

The methodology presented is this paper is an important extension of the synthetic method, which allows nonintrusive access to valuable high-temperature solubility measurements. The method can be readily automated. Care must be taken when applying the method to ensure appropriate suspension of solute in solvent. When applying the methodology, one should be aware of the potential for polymorphism at elevated temperatures, and if applying to crystallization, one should carefully consider the maximum solid loading.

■ AUTHOR INFORMATION

Corresponding Author

*Tel: +353 (0)61 213134. Fax: +353 (0)61 202944. E-mail: Brian.de.souza@ul.ie. URL: http://www.ul.ie/sspc.

ORCID

Brian de Souza: 0000-0002-4469-7802

Funding

This research has been conducted as part of the Synthesis and Solid State Pharmaceutical Centre (SSPC) and funded by Science Foundation Ireland (SFI) under grant 12/RC/2275.

Notes

The authors declare no competing financial interest.

■ NOMENCLATURE

Acronyms

ACE acetone
ACN acetonitrile
API active pharmaceutical ingredient
DIW deionized water
IPA isopropyl alcohol
MSE mean square error
MSZW metastable zone width
MW molecular weight [g/mol]
PBE population balance equation
PSD particle size distribution

Variables with Units and Symbols

a_A first Apelblat parameter
a_2 Solute activity
A Margules binary interaction parameter
A_{VL} first Van-Laar interaction parameter
b_A second Apelblat parameter
B_{VL} second Van-Laar interaction parameter
c_A third Apelblat parameter
C_{eq} equilibrium concentration [g_{sol}/kg_{solv}]
ΔC_p exchange of heat capacity at constant pressure evaluated at melting point [J mol^{-1} K^{-1}]
g_{ij} NRTL binary interaction parameter
G_{ij} corrected temperature-dependent NRTL binary interaction parameter
ΔH_f exchange of enthalpy of fusion evaluated at melting point [kJ/mol]
k Boltzmann constant [J/K]
N_p number of experimental points for each solvent
P pressure [Pa]
P^* reference pressure [Pa] (atmospheric condition)
R ideal gas constant [J mol^{-1} K^{-1}]
T temperature [K]
T_m melting temperature of paracetamol [K]
v_2^S molar volume of solute in the solid state [cm^3/mol]
v_i or v_j molar volume of the ith or jth component [cm^3/mol]
x_2 solute molar fraction
x_2^* reference solute molar fraction (atmospheric conditions)

Greek Letters

α NRTL nonrandomness parameter
γ_2 activity coefficient referring to the solute
λ_{ij} Wilson binary interaction parameter
Λ_{ij} temperature-dependent Wilson binary interaction parameter
τ_{ij} temperature-dependent NRTL binary interaction parameter
θ set of binary interaction parameters for each activity equation model

■ REFERENCES

(1) Higuchi, T.; Connors, K. A. Phase solubility techniques. *Anal. Chem. Instrum* **1965**, *4*, 117–212.

(2) Worlitschek, J.; Mazzotti, M. Model-Based Optimization of Particle Size Distribution in Batch-Cooling Crystallization of Paracetamol. *Cryst. Growth Des.* **2004**, *4*, 891−903.
(3) Nagy, Z. K.; Fujiwara, M.; Woo, X. Y.; Braatz, R. D. Determination of the Kinetic Parameters for the Crystallization of Paracetamol from Water Using Metastable Zone Width Experiments. *Ind. Eng. Chem. Res.* **2008**, *47*, 1245−1252.
(4) Mitchell, N. A.; Ó'Ciardhá, C. T.; Frawley, P. J. Estimation of the growth kinetics for the cooling crystallisation of paracetamol and ethanol solutions. *J. Cryst. Growth* **2011**, *328*, 39−49.
(5) Granberg, R. A.; Rasmuson, Å. C. Solubility of Paracetamol in Pure Solvents. *J. Chem. Eng. Data* **1999**, *44*, 1391−1395.
(6) Fernandez, C. Effect of the nature of the solvent on the crystallization of paracetamol. In *Proceedings of the 14th International Symposium on Industrial Crystallisation*; Cambridge, UK, 1999.
(7) Granberg, R. A.; Ducreux, C.; Gracin, S.; Rasmuson, Å. C. Primary nucleation of paracetamol in acetone−water mixtures. *Chem. Eng. Sci.* **2001**, *56*, 2305−2313.
(8) Shakeel, F.; Alanazi, F. K.; Alsarra, I. A.; Haq, N. Solubilization Behavior of Paracetamol in Transcutol−Water Mixtures at (298.15 to 333.15) K. *J. Chem. Eng. Data* **2013**, *58*, 3551−3556.
(9) Nichols, G.; Frampton, C. S. Physicochemical characterization of the orthorhombic polymorph of paracetamol crystallized from solution. *J. Pharm. Sci.* **1998**, *87*, 684−693.
(10) Jouyban, A.; Fakhree, M. A. A. *Toxicity and Drug Testing*; Acree, B., Ed.; InTech, 2012.
(11) Sheng, X.; Wang, Q.; Chen, L.; Shen, Z.; Pei, Y. Determination and Correlation for the Solubility of Glutaric Acid in Cyclohexane + Cyclohexanol + Cyclohexanone Solvent Mixtures. *J. Chem. Eng. Data* **2016**, *61*, 3761−3769.
(12) Luo, W.; Ruan, D.; Liu, D.; Xie, K.; Li, X.; Deng, W.; Guo, C. Measurement and Correlation for Solubilities of Adipic Acid in Acetic Acid + ε-Caprolactone Mixtures and Cyclohexanone + ε-Caprolactone Mixtures. *J. Chem. Eng. Data* **2016**, *61*, 2474−2480.
(13) Hao, H.; Hou, B.; Wang, J.; Zhang, M. Solubility of Erythritol in Different Solvents. *J. Chem. Eng. Data* **2005**, *50*, 1454−1456.
(14) Gibson, R. E. General considerations of the effect of pressure on solubility. *Trans., Am. Geophys. Union* **1938**, *19*, 273.
(15) Espeau, P.; Céolin, R.; Tamarit, J.-L.; Perrin, M.-A.; Gauchi, J.-P.; Leveiller, F. Polymorphism of paracetamol: relative stabilities of the monoclinic and orthorhombic phases inferred from topological pressure-temperature and temperature-volume phase diagrams. *J. Pharm. Sci.* **2005**, *94*, 524−539.
(16) Ruether, F.; Sadowski, G. Modeling the Solubility of Pharmaceuticals in Pure Solvents and Solvent Mixtures for Drug Process Design. *J. Pharm. Sci.* **2009**, *98*, 4205−4215.
(17) Reschke, T.; Zherikova, K. V.; Verevkin, S. P.; Held, C. Benzoic Acid and Chlorobenzoic Acids: Thermodynamic Study of the Pure Compounds and Binary Mixtures With Water. *J. Pharm. Sci.* **2016**, *105*, 1050−1058.
(18) Prausnitz, J. M.; Lichtenthaler, R. N.; Azevedo, E. G. de. *Molecular Thermodynamics of Fluid-Phase Equilibria*; Pearson Education: Upper Saddle River, NJ, 1998.
(19) Hojjati, H.; Rohani, S. Measurement and Prediction of Solubility of Paracetamol in Water−Isopropanol Solution. Part 2. Prediction. *Org. Process Res. Dev.* **2006**, *10*, 1110−1118.

494685 Evaluating Combined Heat and Power Deionization Systems for Efficient Water Reuse at Thermoelectric Power Plants

Abstract:

Efforts to increase both the United States and global water supply has primarily taken place through seawater and brackish water desalination. According to the International Desalination Association, the United States capacity for desalination increased in 2013 by 50% from 4 million to 6 million m^3/day. Unlike most areas of the world, nearly 77% of all desalination in the United States comes from the treatment of brackish water. Brackish water has lower salt concentrations (~<10 g/l) than seawater (35 g/l) and thus the energy consumed to remove salt ions in theory is much less than seawater desalination. Current technologies utilized for seawater desalination, reverse osmosis (RO) and Membrane distillation (MD), require a significant amount of energy regardless of the type of water (brackish or seawater). Furthermore, these technologies have approached their thermodynamic energy limits (1-3 kWh m$^{â€“3}$), indicating that further energy savings cannot be achieved. RO and MD are energy intensive technologies because they operated through extracting the main component water (H_2O) from the salt (NaCl). Alternatively, capacitive deionization (CDI), an emerging separation-based technology, operates by adsorbing the minority component (NaCl) from the feeding brackish water. Traditional CDI cells operate in a cyclic process where the capacitive electrodes are brought in contact with a concentrated salt solution (e.g. brackish water) and charged. Once the charged electrodes are fully saturated with salt ions, the electrodes are regenerated through discharging the electrodes in a brine solution. Analogous to a P-V diagram for traditional heat engines, the enclosed area of the charge versus voltage represents the energy that is consumed during this charging-discharging process. Until recently the CDI three step intermittent process has been relatively untouched in terms of literature and system operation. In this study, we explore the effect of temperature on salt removal with CDI. Experimentally, we use traditional electrochemical characterization methods, cyclic voltammetry (CV) and electrochemical impedance spectroscopy (EIS), to evaluate the capacitance and resistance changes with temperature. Here it was found that through increasing the operating temperature from 25°C to 45°C, the specific capacitance of carbon electrodes increased from 34.60±2.15 to 42.56±3.57 F/g, and the cell resistance decreased from 20.20±2.97 to 15.57±2.52 ohm. Additionally, the salt adsorption capacity decreased from 13.82±1.13, to 6.58±1.01 milligram NaCl/mg carbon. This indicates that although the electrochemical performance of CDI can be increased with temperature, the ion storage within the electric double layer is hindered at higher temperatures.

493557 Biochar Amendments for Increased Crop Yields: How Can Biochars Improve Crop Nutrient Availability?

Abstract:

Background:

Maintaining a sustainable food supply for the foreseeable future without excessive environmental degradation will require novel engineering approaches for remediating the human disruption of the nitrogen cycle. Over the past several decades, food production has been able to keep pace with human population growth only thanks to the development of new high-yielding crop varieties grown with the help of fertilizers. Thus, vast amounts of energy are consumed every year to produce the approximately 200 Mt of ammonia needed to make the nitrogen (N) fertilizers used by farmers around the globe.

Unfortunately, more than half of the N fertilizer applied to a field is not available for plant growth due to losses caused by surface runoff, leaching into surface and ground water, or volatilization. Because of these losses, the increased use of nitrogen fertilizers has been linked to a variety of water pollution problems ranging from the expansion of the hypoxic zone in the Gulf of Mexico caused by eutrophication to the contamination of wells and groundwater with nitrate N. Moreover, denitrification reactions convert nitrates to nitrous oxide, a major greenhouse gas and nitrogen oxide, a major atmospheric pollutant and significant contributor to the depletion of atmospheric ozone.

The amendment of soils with biochar has been heralded as a sustainable method for increasing crop yield and preventing pollution problems caused by fertilizer runoff. However, the available experimental results are highly variable [1-3]. While some studies reported significant beneficial effects with up to 60% higher yield after biochar addition, other studies reported that biochar amendments either had no effect or even resulted in up to 30% lower agricultural productivity [3]. One of the most important reasons for this variability is that biochar is not a single entity. Dozens of biomass feedstocks are used to produce biochars in multiple types of reactors under varying temperature and oxygen conditions, leading to thousands of biochars with widely varying properties. To complicate things even more, soil properties, climate conditions (like rainfall and temperature), plant requirements and many other parameters vary from application to application. As a result, such studies and meta-analyses have not provided much insight into the fundamental mechanisms that modulate the environmental performance of biochar amendments.

Our Previous Work on this Problem:

To begin bridging this knowledge gap, we focused our recent research on the mechanisms that control the ability of biochars to improve crop nutrient availability, one of the main mechanisms proposed to explain crop yield increases in biochar-amended soils [2].

As a first step in this direction, we have recently developed and tested a mathematical model that uses first principles to describe nutrient transport in soils amended with biochar [4]. Using a simple Langmuir isotherm to describe nutrient adsorption, our simulation results revealed that nutrient transport in biochar-amended soils is modulated by a complex interplay between the physical mechanisms governing flow in porous media and nutrient transport, and the chemical mechanisms modulating nutrient adsorption of the pore surface of biochar solids. Nutrient transport rates through the topsoil decreased when biochars with the proper adsorption capacity and affinity were mixed with the soil. This was due to the differential dynamics of the adsorption and desorption processes. Nutrient adsorption was significantly faster than desorption, with biochar essentially acting as a slow (or controlled) release medium for the adsorbed nutrient [4].

Current Work and Summary of Results:

The present study extends our previous work with a systematic study of the adsorption and desorption dynamics on a much broader class of biochars. We report the development of a mathematical model that describes the response of biochar particles to a pulse of nutrient that simulates a single fertilizer application. Nutrient adsorbs on the biochar particle when the fertilizer solution flows past the particle and starts desorbing when the fertilizer solution is replaced by pure water. Two nutrient transport mechanisms are considered: (a) diffusion through liquid-filled pores and (b) surface diffusion. We assume that nutrient adsorption follows a Sips isotherm whose parameters can be varied to describe adsorption on both homogeneous and heterogeneous solids. The Freundlich and Langmuir isotherms commonly used to describe adsorption of ammonium or nitrates on biochars are special cases of the Sips isotherm. The highly nonlinear partial differential equations describing this transient adsorption/desorption problem are discretized using high-order finite differences and the resulting system of first order differential equations are integrated with a variable-step, variable-order integrator based on numerical differentiation formulas.

As noted previously, a "good" biochar should be able to quickly adsorb the nutrient when it is present in the liquid flowing around it, but it should desorb it much more slowly in order to act as slow (or controlled) release medium. This is the mechanism through which a "good" biochar will improve crop nutrient availability [2, 4].

Our simulations clearly show that the ability of a biochar to slowly release the adsorbed nutrient depends on a complex interaction of (a) external mass transfer, (b) intraparticle diffusion (both pore and surface diffusion), and (c) adsorption dynamics (affinity constant and adsorption capacity).

The interaction between external mass transfer and intraparticle pore diffusion is a key factor for modulating the release of nutrient that has adsorbed on the biochar. Our system is characterized by low values of the Biot number, the dimensionless number that represents the ratio of external mass transport rate over the intraparticle diffusion rate. This is because of the low interstitial water velocities (10-90 cm/day) in biochar-amended soils. Low Biot numbers are one of the key factors responsible for the long desorption tails that characterize the slow release of nutrients from "good" biochars.

Our simulations reveal that all the processes mentioned above can significantly affect the nutrient release rates. For example, surface diffusion accelerates nutrient desorption from small (2 mm in diameter) and highly porous biochar particles (0.6 total porosity). But, surface diffusion will slow down desorption from larger (4-8 mm in diameter) and less porous particles, even though the latter biochar particles may have exactly the same adsorption isotherm parameters. Adsorption affinity and site heterogeneity can also interact with porosity and particle size to modulate nutrient desorption. Finally, surface diffusion again improves the ability of a biochar to act a slow release medium at high interstitial velocities.

By revealing that the response of biochar particle to a pulse of nutrient is governed by complex interactions among external mass transfer, intraparticle diffusion and adsorption, this study provides new insights into the fundamental mechanisms that influence the environmental performance of biochar amendments. It may also explain why biochars that share some key properties (e.g. carbon content) or produced at the same pyrolysis temperature can potentially have widely different impacts on crop yields, as has been repeatedly reported in the literature [2,3]. Much better characterization of biochar properties (porosity, adsorption isotherm, intraparticle transport mechanism) and more precise control of amendment or soil parameters (like particle size, biochar aging, soil permeability) are necessary before we can fully elucidate the mechanisms of biochar action.

References:

[1] Spokas, K. A.; Cantrell, K. B.; Novak, J. M.; et al. "Biochar: A Synthesis of Its Agronomic Impact beyond Carbon Sequestration." *J. Environmental Quality* **41**(2012) 973-989.

[2] Jeffery, S; F.G.A. Verheijena; M. van der Velde; A.C. Bastos. "A quantitative review of the effects of biochar application to soils on crop productivity using meta-analysis." *Agriculture, Ecosystems and Environment* **144**(2011) 175–187.

[3] Crane-Droesch, A; S. Abiven; S. Jeffery; M.S. Torn. "Heterogeneous global crop yield response to biochar: a meta-regression analysis." *Environmental Research Letters* **8**(2013) 044049.

[4] Sun, S; C. E. Brewer; C. A. Masiello; K. Zygourakis. "Nutrient Transport in Soils Amended with Biochar: A Transient Model with Two Stationary Phases and Intraparticle Diffusion" *Industrial and Engineering Chemistry Research*, **54** (2015) 4123–4135.

501354 Moisture Retention in Emulated Soil Micromodels: Development and Performance of Sustainable Agriculture Biotechnology

Abstract:

Soil bacteria living near plant roots secrete exopolysaccharides (EPS), a polymeric substance which can help modulate soil moisture and promotes plant growth. EPS functions in several different ways: (i) EPS behave as hydrogel, swelling during wet soil conditions and remaining hydrated during dry soil conditions; (ii) EPS promotes aggregation of soil particles and thereby promotes water retention in dry soil conditions through stronger capillary forces; and (iii) EPS alters surface properties of soil, creating more water repellent surfaces to inhibit rapid re-wetting in wet soil conditions. EPS functions in concert with the physical microstructure of real soil; therefore, it is necessary to study EPS performance in physically realistic micro-structures. Here we show pore-scale changes in moisture distribution within emulated soil micromodels that have systematically varied physical microstructures and surface water repellencies, and can be loaded with well-defined pore solutions. Experimental results showed dilute EPS solutions took twice as long to dry compared with pure water in identical micromodels. Microstructure geometry was found to influence drying rate over a range of pore saturation values as well as control the spatial distribution of liquid and gas phases within unsaturated soil micromodels. Deionized water took four times longer to completely dry from micromodels with higher water repellency surfaces compared with micromodels with lower water repellency surfaces. These results illustrate the various mechanisms whereby EPS impacts the spatial distribution and evaporation resistance of soil moisture. We provide a systematic method to evaluate performance of natural EPS solutions in modulating soil water and promoting resiliency of terrestrial ecosystem; we anticipate our approach may speed the development of sustainable agriculture biotechnology to enable food production using less water.

498287 Molecular Modeling and Simulation Studies of the Structural and Energetic Evolution during Dehydration of Food Systems

Abstract:

From a molecular point of view, foodstuffs are mostly comprised of biopolymeric molecules that form highly porous structures in which solvents, solutes, additives, and other natural or artificial ingredients are accommodated. Because of the complexity of the food structures and compositions and because of the sensitivity of the pore structures and properties of food to physical and chemical changes arising externally or internally, foodstuffs are one of the most complex types of soft materials and are inclined to exhibit significant structure-property relationships in the physicochemical, functional, processing, and nutritional aspects of food science and engineering. In this respect, water (moisture) is a major factor in determining the properties of foods in various stages as it is not only a major component of food materials but also acts as space filler for stabilizing food structures, solvent for dissolving additive molecules, medium for heat and mass transfer, and participant in the chemical and/or biological reactions in food systems. .

One of the important ways for preparing and preserving foods is dehydration (water removal), which is a process that requires energy to be supplied to enable water molecules to break loose from their interactions and escape from the porous structures of the foods. The process of food dehydration and the preservation of dehydrated foods can thus be directly related to the interactions of water molecules with food macromolecules, solute ions, and additive molecules, because these interactions dictate the energy requirement for water removal during food dehydration as well as the structures of foods before, during, and after dehydration. Moreover, the magnitudes of these interactions depend on the densities and intermolecular distances of mutually interacting molecules, rendering the water molecules in foods to exist in different energetic states and making single system-wide thermodynamic parameters (e.g, water activity) inadequate in representing the diverse states of water molecules in food porous structures and in elucidating the microscopic and macroscopic structural evolution of food materials caused by dehydration. Capable of tackling these challenges is the molecular dynamics (MD) method which employs molecular-based interaction potentials to construct a model system that provides a satisfactory physicochemical representation of the real system and allows various structural and dynamic properties to be studied at a molecular level. Recently, we have employed MD modeling and simulations to construct an amylose-based food material and study the dehydration process in a continuous dynamic manner by periodically examining the energetics of all water molecules at the water-vapor interface and removing the one with the highest energy from the model food system. It was found that the strong interactions of water molecules with the food polymer stabilize the porous structure of the food material and makes the dehydration energy requirement to be greater than that involved in pure water vaporization. They also cause the water interaction energetics to be highly nonuniform in all three spatial dimensions. As dehydration proceeds, the surface tension of the descending water interface coupled with strong water-polymer interactions exert downward stresses on the food polymer to induce polymer conformational compaction and food structural reduction.

We further added two functional food ingredients to the model food systems in our MD simulation studies. The first one is potassium sorbate, which is a commonly used food preservative against the growth of bacteria, fungi, molds, and yeast in foods. The second one is fructose-1,6-diphosphate (FDP), which is a bioprotective agent and a promptly available high energy substrate and can promote growth and improve health. The stability of the food porous structures was found to be enhanced by the presence of the food additive molecules which function as additional fillers in the food pore space and interact strongly with the polymer chains and water molecules of the food material.Â During dehydration, the descending water level tends to accumulate the additive molecules, in particular sorbate with a nonpolar hydrocarbon moiety, at the vapor-liquid interface, which helps to form dense clusters of food polymer chains at the interface, reduces the level of structural reduction, but increases the energy requirement for water removal.Â The additive molecules in the interior regions of the food structures have strong tendencies to stay close to food polymer and hence exhibit nonuniform density distributions as well. Extensively dehydrated food porous structures were found to have substantially improved stability due to the water molecules remaining in the pore structure that effectively â €œbindâ€ the food polymer and additive molecules together. Which makes it very difficult to recover the original pore structures of the food systems that existed before the commencement of dehydration. The findings of these MD studies suggest that the original pore structures could be reduced less during dehydration and could be recovered more effectively by rehydration, if (i) the food material has a higher polymer density, (ii) solute or additive molecules have elongated shapes to act as cross-linkers, and (iii) the effects of water surface tension can be minimized or eliminated using special dehydration methods such as freeze drying. The MD modeling and simulation approach developed in this work could form a basis for systematic evaluation of food porous structures and their structural and energetic evolution and mechanisms in molecular detail and for providing scientific explanations to the macroscopic trends and empirically observed phenomena during food dehydration.

493109 Metallic Membranes for N_2 Separation and NH_3 Production

Abstract:

The availability of nitrogen, phosphorus, and water are the three main factors that limit our ability to produce enough food to feed the growing population of the planet. The nitrogen cycle is one of the most significant biogeochemical cycles on Earth, as nitrogen is an essential nutrient for all forms of life. Although freely available in the atmosphere as dinitrogen (N_2), access to fixed forms of nitrogen constitutes, in many cases, the most limiting factor for plant growth. The industrial production of ammonia for fertilizers via the current Haber-Bosch process is an energy-intensive process that consumes 1-2% of the worldâ€™s annual energy supply. For these reasons, the need for advanced catalytic methods for the reduction of N_2 to ammonia remains a requirement for sustainability in the food, energy and water systems cycle.

The aim of this work is to explore the potential of metallic membranes for N_2 separation with the final intent to produce NH_3. Based on a preliminary theoretical investigation using density functional theory, the Group V transition metals (e.g., vanadium (V), niobium (Nb) and tantalum (Ta)) show strong affinity toward N_2. Moreover, from solubility and diffusivity values taken from the literature, iron (Fe) is a suitable fit for this application. Therefore, V, Nb, Ta and Fe foils with a thickness of 40 Î¼m have been chosen as membrane materials in this study.

Permeation tests with pure gases (He, N_2, CO_2 and CH_4) have been performed to characterize the membranes in terms of the N_2 permeating flux and ideal selectivity at different temperature and pressure conditions, varying from 400 to 600 Â°C and from 2.0 to 60 bar, respectively. Moreover, scanning electron microscopy (SEM) and an electron microprobe analyzer (EMPA) have been used to investigate the effect of the different operating conditions on the membrane surface. From the experimental results, all of the membranes show infinite selectivity towards N_2 permeation at each operating condition investigated and V showed better performance with respect to Nb, Ta, and Fe.

The V membrane is housed in the membrane reactor where a H_2 stream is used as the sweep gas to promote the NH_3 reaction. Lower pressure than the conventional Haber-Bosch process is used. Specifically, an operating pressure of 60 bar, instead of 200 bar (Haber-Bosch process), and a temperature of 500 °C are used as operating conditions to produce NH_3. The performance of the membrane reactor in terms of NH_3 conversion is, hence, investigated.

485442 Capacitive Deionization of Brackish Water for Irrigation and Energy Storage

Abstract:

Irrigated agriculture water represents the bulk of demand for water in most of the countries suffering from acute water scarcity in the Near East and North Africa, as well as Mexico, Pakistan, South Africa, and large parts of China and India. Greenhouses are emerging solution. Current testing in Mozambique and Sierra Leone has demonstrated significantly reduced water use. Yet even a few gallons per day per greenhouse is challenging because many local water sources are brackish. Capacitive deionization (CDI) is a sustainable, energy efficient, and cost effective technology for removal of low or moderate salinity typical of brackish water as compared to technologies that remove the water instead. Indigenous production of the carbon for the electrodes holds the key to such technology being adopted as a source of irrigation water, simultaneously facilitating the use of such using renewable energy sources. Biochars have recently been explored as electrode materials in CD, and can be easily produced from local plant-based agricultural waste materials.

The goal of this study is thus to identify biomass process conditions to enable the best performance as electrode material in CDI. Studies on CDI so far have been targeted towards achieving the best adsorption capacity, which directs towards the use of highly processed and tailored carbons. This is mostly achieved through chemical activation processes using toxic or corrosive and expensive materials which require deionized water for subsequent washing of the carbon, for achieving better CDI performance. Hence, with the effort directed towards making the process viable for use locally in regions with acute water scarcity, a change in the approach to tailor the electrode material for use in CDI is proposed. This approach is to determine the best possible material utilization by ensuring maximum possible contact between the liquid and the biochar surface (wettability) and by generating the least Ohmic loss thus resulting in highest electrical efficiency.

485402 Optimization of Hyaluronic Acid Production By *Streptococcus zooepidemicus* for Biomedical Applications

Abstract:

Hyaluronic acid is one of the polymeric materials that has been widely used in biomedical applications. Those include the applications in tissue engineering, dermal filler, and osteoarthritis treatment. Hyaluronic acid is found as the active ingredient in cosmetics and skin protection products. As aforementioned, the market demand of hyaluronic acid tends to increase as the consequence of consumer cares. Nonetheless, the production capacity of hyaluronic acid is still limited since the production from animal tissue extraction usually experiences high risk of contamination while bacterial production often yields the low productivity. To increase the market size of hyaluronic acid, one needs to develop the production process that can provide sufficient amount with the low cost. In this study, medium optimization was conducted for enhancing hyaluronic acid production by *Streptococcus zooepidemicus* UN-7. Among 4 media studied, the medium containing sucrose and lactic casein gave the highest hyaluronic acid production while the glucose based media gave a better biomass production. Further optimizing the carbon to nitrogen ratio gave improved final product titer. In addition, the agitation exhibited the great effect on hyaluronic acid production. Enhancing the hyaluronic acid production was conducted by the addition of the chemical precursors in the synthesis pathway. It was believed that under the optimized conditions, improved hyaluronic acid could be achieved and the process could be applied as the production platform for a large-scale operation.

498070 Mathematical Study for Prediction of Shelf Life of Curly Hydroponic Lettuce

Abstract:

Refrigerated vegetable preservation is a widely discussed topic in the world in order to define ideal conditions for fresh food storing from the harvest to consumer's table. Leafy vegetables are very consumed. However their physicochemical characteristics, during storage, undergo changes such as weight loss, moisture loss and degradation of chlorophyll, which reflect changes in sensory aspects such as color and wilting. These characteristics refer to the vegetables quality and the changes have a direct impact on consumer perception. The objective of this study was to evaluate the physicochemical and sensory stability of leaves of hydroponic curly lettuce for 14 days stored at 2 °C, 6 ° C and 10 °C (± 1 °C) in household refrigerator refrigeration, packed in expanded polystyrene (styrofoam) packaging wrapped with polyvinyl chloride film (PVC). Through physicochemical analyzes of mass loss, total chlorophyll, sensorial analyzes of color change, wilted appearance, edge darkening and central vein darkening, the results were evaluated by Tukey test at 5% of significance and for shelf life estimation was used Principal Component Analysis (PCA). These allowed to verify the influence of the cooling parameters of this study, on the quality factors of lettuce leaves. The PCA data were regressed a nonlinear mathematical model, Weibull model, to describe the degradation kinetics. According to Tukey's test, the best refrigerated storage isothermal conditions to maintain the desired physicochemical and sensory characteristics over the storage period were 2 °C and 6 °C. When estimating the shelf life using the Weibullian model was obtained 12.43 shelf life days for refrigerated storage at 2 °C; 8.76 days at 6 °C and 9.87 days at 10 °C. After comparing the results of analysis of the physicochemical, sensorial and shelf life estimation, was observed a good fit of the model to experimental data to predict the hydroponic curly lettuce degradation during the 14 days of refrigerated, resulting in determination coefficients of 0.94 for the storage temperature of 2 °C, 0.98 for 6 °C and 0.99 for 10 °C.

502649 Production of Polymalic Acid (PMA) and Malic Acid from Food Processing Byproducts By Aureobasidium Pullulans

Abstract:

PMA is a biodegradable and biocompatible water-soluble polymer with potential application as prodrugs or drug carriers in biomedical industry. PMA can be readily hydrolyzed to malic acid, which is widely used as an acidulant in foods and pharmaceuticals. In this study, PMA is produced by *Aureobasidium pullulans* from agricultural wastes such as soybean hull, soybean molasses, and sugarcane juice. PMA production from soybean hull hydrolysate supplemented with corn steep liquor (CSL) gave a malic acid yield of ~0.4 g/g at a productivity of ~0.5 g/L·h. ZX-10 can also ferment soy molasses without enzyme hydrolysis, converting all carbohydrates including the raffinose family oligosaccharides to PMA, giving a high titer (71.9 g/L) and yield (0.69 g/g) at a productivity of 0.29 g/L·h in fed-batch fermentation under nitrogen limitation, and a higher productivity of 0.64 g/L·h was obtained in repeated batch fermentation with cell recycle and CSL supplementation. Sugarcane juice can be used as an economical substrate without any pretreatment or nutrients supplementation. For PMA production from sugarcane juice, a high PMA titer of 116.3 g/L and yield of 0.41 g/g were achieved in fed-batch fermentation, and a high productivity of 0.66 g/L·h was achieved in repeated-batch fermentation with cell recycle. Cost analysis shows that malic acid can be produced at $1.10/kg from soy molasses, $1.33/kg from sugarcane juice, $1.37/kg from corn, and $1.74/kg from soybean hull. At the market price of $1.75/kg, malic acid production from soy molasses and sugarcane juice via PMA fermentation offers an economically competitive process for industrial production of bio-based malic acid.

494734 Growth Kinetics and Viability Studies of Common Probiotic Bacteria on Date Syrup

Abstract:

Date syrup is a proved inexpensive functional food with increasing demand in the consumer market. It contains many functional components such as polyphenols, carotenoids, and phytosterols along with high amount of total sugars. Date syrup, therefore, could be a good culture for probiotic bacteria to develop a functional fermented beverage. The objective of the present work is to study the growth and survival of two probiotic bacteria, *Lactobacillus* sp. and *Bifidobacetrium* sp. in unbleached date syrup. The growth kinetics, pH, total sugars, total organic acids, and total phenolic content were investigated over a 5-day fermentation cycle at 37°C, and cell viability was determined for 8 weeks at 2-4°C. Both species were found growing in date syrup without any additional nutrient requirements and increased to more than 10^6cfu/mL after 24 h. *Lactobacillus* sp. shown a more net decrease in pH with high sugar consumption and much better growth rate than *Bifidobacterium* sp. An insignificant difference in values of total phenolics was observed. This study proves that date syrup can be used in the development of a probiotic beverage and the common probiotic bacteria shows a promising growth on this plant product which is abundent in Middle East and North Africa (MENA) region.

495686 Shape-Changing Food Transforms from 2D to 3D By Water Interaction through Cooking

Abstract:

We developed a new method to make 3D-shaped food by cooking 2D edible film in water. The transformation process is triggered by anisotropic swelling, and it can substantially reduce shipping costs and storage space of 3D food. With the guidance of material-based simulation, we explored a variety of food materials and characterized their performance. Together with a chef, we created interactive dish, and demonstrated 2D-to-3D folding, hydration-induced wrapping, and temperature-induced self-fragmentation, which indicated the change of food shape, texture, as well as demonstrated the interaction with food materials. These shape changing food as a result of translational research presented the endeavor towards healthy and entertaining food, through effort from chemical engineering, material science, design, and food industry. The 2D-to-3D technology provides a new way for making functional food, in the area of futuristic nutrient and drug supply.

Optimization of growth media and functionality characterization of new potential probiotic *L. salivarius*, isolated from human milk

Malek, R.A[1*], Salehhuddin Hamdan[2], Ramli,S[1]., El Enshasy, H.A[1,3]

[1]Institute of Bioproduct Development (IBD), UniversitiTeknologi Malaysia (UTM), 81310 Skudai, Johor, Malaysia

[2]Faculty Biosciences and Bioengineering, Universiti Teknologi Malaysia (UTM), 81310, Skudai, Johor, Malaysia

[3] Bioprocess Development Department, City for Scientific Research and Technology Applications (CSAT), New Burg Al Arab, Alexandria, Egypt

Introduction

Breast milk is known as the best nutrient for newborn. It contain mixtures of protein, carbohydrate, minerals, and vitamins for infant development. The bioactive compound in it is beneficial in promoting the immune system maturation and protecting against infection. Breast milk also has immunomodulatory action which causes a better antibody production response after vaccination against poliomyelitis, tetanus, and diphtheria. Neonates who were given breast milk will have a more favourable intestinal microbiota compared to formula-fed infants. This may be caused by the presence of probiotic bacteria and other bifidogenic compounds in breast milk.

Probiotics are lived microorganisms that can exert health benefit to host when ingested in the right amount. The word probiotics are derived from the Greek words which means pro-life. Thus, probiotic concept as been defined by Fuller (1989) means "a live microbial feed supplement which beneficially affects the host animal by improving its intestinal microbial balance". Salminen et al.(1998) subjected that probiotic are microbial cells preparation or compound or microbial cells that have a beneficial effect on the health and wellbeing of the host. A good probiotic requires the following characteristics to function as bacteriotherapheutic agents (Baick and Kim, 2015: Dunne *et al.*, 2001). The characteristics include strains must come from genera of human origin and isolated from healthy human gastrointestinal tract, good stability and able to adhere to the intestinal mucosa, easily colonized in the human gastrointestinal tract, free from pathogenic history, able to produce antimicrobial substances, not associated with any disease, do not carry transmissible antibiotic resistance, and can demonstrate efficacy and safety (Sarao and Arora, 2017; Sornplang and Piyadeatsoontorn 2016; Roca and Lucia 2014; Huys*et al.*, 2013;Ashraf and Shah, 2011). Probiotics also must have good longevity, viability of the strain, and stability in commercial product for at least 10^6 to 10^8 live microorganism per g or mL (Sarao and Arora, 2017: Mitropoulou*et al.*, 2013) before it lost its effect during production, storage, handling, and passage through the human gastrointestinal tract.

The growth of *Lactobacillus salivarius* depends much on the culture medium compositions such as carbon source, nitrogen source, micronutrients and macronutrients. This also represents a large part of the bioprocess costs in large scale manufacturing (. One of the raw materials used are yeast extract, which is commonly used as a costly nitrogen source and growth factors for lactic acid bacteria. Screening and selection of the media components is important to achieve a suitable medium for optimum growth. Therefore, in this present work, the studies of cell growth of *L. salivarius* in different media and cultivation conditions in shake flask and bioreactor have been conducted. The medium which yielded the highest cell mass and the lowest lactic acid in shake flask was optimized using different meat peptone, glucose, and yeast extract concentrations. This is purposely done to accelerate the optimum cultivations performance of *L. salivarius* in shake flask levels before subjected to large scale production using pilot scale bioreactor (16-L) in batch fermentations under controlled and uncontrolled pH conditions.

Despite medium optimization, previous studies had shown that acidity and bile as other important factors that affect the probiotics to remain and exert their potential functionalities in a host. This leads the present work to study on the tolerance of *L. salivarius* towards acidity and bile environment. Studies were carried out by subjecting the cell to grow in synthetic gastric juice at pH ranges from 1 to 4 and in MRS broth containing ox gall at concentrations of 0.5% to 4% bile for 0 min, 30 min, 60 min, 90 min, and 120 min.

Apart of being tolerated to acid and bile salts, probiotics should also have the susceptibility to antibiotic so that they will not create any potential to transmit antibiotic resistance genes to other bacteria in gastrointestinal tract, especially to the pathogenic microorganism. They should possess a susceptibility to antibiotics such as streptomycin, gentamycin, erythromycin, rifampicin, ampicillin, and tetracylinesin this study. In the present work, cultivation of *L. Salivarius* was carried out to study the optimum conditions of important growth factors that affect the cultivations performance of *L. salivarius* and the viability of *L. salivarius* when exposed to low pH, gastric juice, bile, and susceptibility to different types of antibiotics

Materials and method

Microorganism

The probiotic strain *Lactobacillus salivarius WICC-B08* originally isolated from human breast milk obtained from Wellness Industry Culture collection (WICC) was used in this study. This strain was cultivated in MRS agar for cell growth.

Master cell bank

MRS agar was used for cell propagation. This medium composed of (g L^{-1}): glucose, 20; yeast extract, 5; peptone, 10; meat extract, 10; polysorbate 80,1; ammonium citrate,2.0; sodium acetate,5; magnesium sulfate,0.1; dipotassium phosphate,2; and agar, 20. The pH was adjusted to 7.0 before sterilization.

Cultivation conditions

Shake flask cultivation and inoculums preparation

For shake flask study levels, Erlenmeyer flasks (250 mL) with 50 mL working volume were used. After inoculation, flasks were incubated on rotary shaker (Innova 4080, New Brunswick, NJ, USA) at 200 rpm and 37 °C for 24 h for biomass production. For bioreactor cultures, cultivations were run in pilot scale bioreactors of 16 L with working volume of 8 L (BioEngineering, Wald, Switzerland). Cultivation was conducted at 400 rpm, 37 °C and pH was controlled at 7.0 by addition of 2.5 M $(NH_4)OH$.

Bioreactor cultivation

For bioreactor experiments, cultivations were carried out in a 16 L stirred tank bioreactor, (BioEngineering, Wald, Switzerland), with a working volume of 6.0 L. Agitation was performed using stirrer of two-four bladed Rushton turbines and the bioreactor was equipped with pH probe, oxygen probe, and foam sensor. Dissolved oxygen concentrations were analyzed by polarographic electrode (Ingold, Switzerland). Foam was suppressed, when necessary, by the addition of silicon antifoam reagent (Silicone antifoam, Sigma, USA).

Analytical Method

Sample Preparation

For both shake flasks and bioreactor, samples were taken at different time intervals. Every sample was prepared in a proper method in order to get accurate answers.

Cell dry weight determination

For both shake flasks and bioreactors, samples (20 mL) were taken at different time intervals and the optical density was measured spectrophotometrically against un-inoculated medium at 600 nm using single beam spectrophotometer (HACH, Loveland, CO.). The OD of culture was converted to dry cell mass through pre-determined standard curve between the OD value and cell dry weight.

Glucose and lactic acid determination

Both glucose and lactic acid were determined by using HPLC method. The falcon tube filled with 10.0 mL sample was being centrifuged using centrifuge machine (Hettich, ZENTRIFUGEN) at 3000 rpm for 10 minute. After the centrifugation, the clear supernatant was kept in a new universal bottle for ethanol determination.

Results and Discussion

Medium optimization for high biomass production by *L. salivarius*

Set of experiments were conducted to improve the medium components to increase biomass production. One approach was applied to achieve this target using one factor at time method (OFAT). The result of medium optimization is summarized in Table 1.

Table 1. Medium composition obtained after different optimization methods to increase biomass production by *L. salivarius*

Components	**Un-optimized**	**One factor at time (OFAT)**	**Biomass ($g L^{-1}$)**
Glucose ($g L^{-1}$)	25	20	3.4
Yeast extract ($g L^{-1}$)	25	20	4.4
Peptone ($g L^{-1}$)	25	35	5.75

Batch cultivation of *L. Salivarius* for high biomass production in a 16-L stirred tank bioreactor

Table 2 illustrates the data comparison between un-optimized and optimized in shake flasks and data performed in bioreactor. In order to develop sustainable probiotic production in industrial scale, high biomasses with a low cost and effective process are important. Cultivations were carried out in bioreactor with controlled and un-controlled pH. Results showed that higher dry cell mass can be achieved in controlled pH cultivations and harvested only at 10 hours of cultivation with only the production of lactic acid about 4.32 g L^{-1}. The cultivations time could be reduced compared with un-optimized medium 1 (16 h), to get higher biomass. The new formulated and modified SPY medium able to achieve high biomass; however, it only used three main components as a substrate.

Table 2. Comparison between cell growth and lactic acid production when cultivating *L. salivarius* in shake flasks level and bioreactor

	Shake flask		Bioreactor	
	Un-optimized	Optimized	Controlled pH	Uncontrolled pH
Dry cell mass, g L^{-1}	2.685	5.715	7.574	5.376
Lactic acid production, g L^{-1}	5.41	4.7	4.32	6.34
Specific growth rate, h^{-1}	0.0836	0.1794	0.2492	0.2332
Yield, $Y_{p/x}$	2.0149	0.8224	0.57037	1.17932
The highest growth time, hr	16	11	10	9

Simulated gastrointestinal tract conditions

The present study evaluated the viability of *L. Salivarius* under acidic conditions in artificial simulated gastric juice solutions (pH 1.0, 2.0, 3.0, 4.0) during 2 hours were incubated for 0 min, 30 min, 90 min, and 120 min at 37 °C, and assayed by viable cells counting. Survival rates of the probiotic *L. Salivarius* are varied during incubation. The pH of the stomach varies with time. The gastric juice in the stomach may be very acidic and may go to as low as 1 in several cases. When the food were consumed and get in the stomach, the pH may rise from 3.0 to 4.0 due to the content of saliva in the food and the buffering of the proteins. Small intestine and colons of humans and animals contain relatively high concentration of bile acid, which can inhibit the growth of probiotic bacteria. For probiotic bacteria, it is considered effective if it can survive at such harsh environment. *L. salivarius* showed higher resistant to SGJ at pH 1, 2, 3, and 4; cell viability was 22.9%, 38.8%, 63%, and 65%; and proved that this strains have higher resistance to SGJ under all pH tested.

Viability of *L. salivarius* in bile salts conditions

'*L. salivarius* of the tested organism showed survival rates of >95% in MRS that contains 0.25%, 1%, 2%, and 4% bile salt. The cells are tolerance to bile salts even the concentrations were increased. However, the reduction number still happen but did not cause the loss of *L. salivarius* potential as probiotics. The number of viable cells was reduced from 12.4 to 12.21 (1.84% inhibition), 12.33 to 12.29 (0.32% inhibitions), 10.51 to 10.16 (3.3%)10.72 to 10.23 (4.57%), when incubated in 0.5%, 1%, 2%, and 4% bile salt solution after 2 hours exposure. As a conclusion from this study, *L. salivarius* is tolerant to bile salt from concentration 0.5% to 4%.

Antibiotic resistance

L. salivarius is susceptible to antibiotics like erythromycin, rifampicin, tetracycline, ampicillin, and resistance to streptomycin and gentamycin.

Conclusion

In the present work, new industrial medium formula was developed for high cell mass cultivation of biotherapeutic/probiotic *L. salivarius*. Furthermore, the newly isolated *L. salivarius* from human milk has a big potential uses as a probiotic bacteria for human consumption.

Acknowledgement

The authors would like to thanks the MOHE and UniversitiTeknologi Malaysia (UTM) for Contract research grant no R.J130000.7344.4B200 and R.J130000.7646.4C110 that have supported this research.

Reference

1. Ashraf, R and Shah,N.P (2011). Review Article Antibiotic resistance of probiotic organisms and safety of probiotic dairy products. *International Food Research Journal* 18(3): 837-853
2. Baick, S-C and Kim, C-H (2015). Assessment of Characteristics and Functional Properties of *Lactobacillus* Species Isolated from Kimchi for Dairy Use. *Korean J Food Sci Anim Resour.* 35(3): 339–349.
3. Dunne,C., O'Mahony,L., Murphy,L.,Thornton,G., Morrissey,D., O'Halloran,S., Feeney,M.,Flynn,S., Fitzgerald,G., Daly,C., Kiely,B., O'Sullivan,G.C., Shanahan,F., Collins,K (2001). In vitro selection criteria for probiotic bacteria of human origin: correlation with in vivo findings. *Am J ClinNutr ;73(2)386-392*
4. Fuller R. Probiotics in man and animals. J. Appl. Bacteriol. 1989;66:365–378.
5. Huys,G., Botteldoorn,N., Delvigne,F., Vuyst,L.D., Heyndrickx, M., Pot,B., Dubois,J-J., Daube,G (2013). Microbial characterization of probiotics–Advisory report of the Working

Group “8651 Probiotics” of the Belgian Superior Health Council (SHC). *MolNutr Food Res*. 57(8): 1479–1504.

6. Mitropoulou,G., Nedovic,V., Goyal,A., Kourkoutas,Y (2013). Immobilization Technologies in Probiotic Food Production. *J NutrmMetab*. 716861.
7. Roca, C and Lucia,C (2014). Characterization of Commercial Probiotics: Antibiotic Resistance, Acid and Bile Resistance, and Prebiotic Utilization .Dissertations, Theses, & Student Research in Food Science and Technology. 46.
8. Sarao,L.K and Arora,M (2017). Probiotics, prebiotics, and microencapsulation: A review. Critical Reviews in Food Science and Nutrition 57 (. 2).
9. *Sornplang,P and Piyadeatsoontorn,S (2016).* Probiotic isolates from unconventional sources: a review. *Journal of Animal Science and Technology,* 201658:26

Biotechnology Journal
www.biotechnology-journal.com

DOI 10.1002/biot.201500068

Research Article

Dual utilization of NADPH and NADH cofactors enhances xylitol production in engineered *Saccharomyces cerevisiae*

Jung-Hyun Jo[1,*], *Sun-Young Oh*[1,*], *Hyeun-Soo Lee*[2], *Yong-Cheol Park*[2] *and Jin-Ho Seo*[1]

[1] Department of Agricultural Biotechnology and Center for Food and Bioconvergence, Seoul National University, Seoul, Republic of Korea
[2] Department of Bio and Fermentation Convergence Technology, Kookmin University, Seoul, Republic of Korea

Xylitol, a natural sweetener, can be produced by hydrogenation of xylose in hemicelluloses. In microbial processes, utilization of only NADPH cofactor limited commercialization of xylitol biosynthesis. To overcome this drawback, *Saccharomyces cerevisiae* D452-2 was engineered to express two types of xylose reductase (XR) with either NADPH-dependence or NADH-preference. Engineered *S. cerevisiae* DWM expressing both the XRs exhibited higher xylitol productivity than the yeast strain expressing NADPH-dependent XR only (DWW) in both batch and glucose-limited fed-batch cultures. Furthermore, the coexpression of *S. cerevisiae ZWF1* and *ACS1* genes in the DWM strain increased intracellular concentrations of NADPH and NADH and improved maximum xylitol productivity by 17%, relative to that for the DWM strain. Finally, the optimized fed-batch fermentation of *S. cerevisiae* DWM-ZWF1-ACS1 resulted in 196.2 g/L xylitol concentration, 4.27 g/L h productivity and almost the theoretical yield. Expression of the two types of XR utilizing both NADPH and NADH is a promising strategy to meet the industrial demands for microbial xylitol production.

Received 13 FEB 2015
Revised 07 SEP 2015
Accepted 07 OCT 2015
Accepted article online 16 OCT 2015

Supporting information available online

Keywords: Cofactor regeneration · Glucose-limited fed-batch fermentation · NADH-preferring xylose reductase · *Saccharomyces cerevisiae* · Xylitol

1 Introduction

Xylitol is a five-carbon sugar alcohol with a number of advantageous properties. As xylitol is as sweet as sucrose and has a 33% lower calorie, and extreme cooling and caries-preventative effects, xylitol has been commercially used as a sugar substitute in various products such as chewing gums, candies and toothpastes [1]. Xylitol was also chosen as one of the top 12 value-added chemicals from biomass and could be used as a platform chemical for polymer synthesis such as xylaric acid and glycols, directly polymerized for production of unsaturated polyester resin [2]. Ethylene glycol, an important chemical intermediate, can be produced by ruthenium-mediated hydrogenolysis of xylitol [3]. Due to these properties, demand of xylitol has increased in food and chemical industries. Currently, the annual xylitol market was estimated to be 125 000 tons of demand and priced at 4.5–5.5 US dollars per kg [4].

Xylitol is chemically synthesized by hydrogenation of xylose, one of major sugars in cellulosic biomass, in the presence of a metal catalyst and hydrogen gas at over ambient conditions such as 120°C of temperature and up to 5.5 MPa of pressure [5, 6]. Due to the sustain-

Correspondence: Prof. Jin-Ho Seo, Department of Agricultural Biotechnology and Center for Food and Bioconvergence, Seoul National University, 1 Gwanak-ro, 08826 Seoul, Republic of Korea
E-mail: jhseo94@snu.ac.kr

Additional correspondence: Prof. Yong-Cheol Park, Department of Bio and Fermentation Convergence Technology, Kookmin University, 77 Jeongneung-ro, 02707 Seoul, Republic of Korea
E-mail: ycpark@kookmin.ac.kr

Abbreviations: ACS1, acetyl-CoA synthetase; **NAD**, nicotinamide adenine dinucleotide; **NADP**, nicotinamide adenine dinucleotide phosphate; **XR**, xylose reductase; **ZWF1**, glucose-6-phosphate dehydrogenase

* These authors contributed equally to this work.

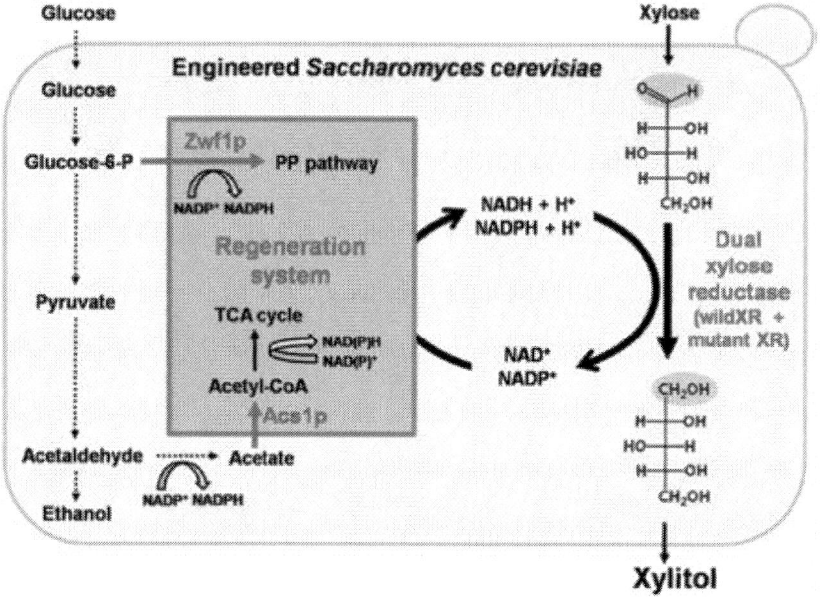

Figure 1. Metabolic pathways for xylitol biosynthesis and regeneration of NADPH and NADH cofactors in engineered *S. cerevisiae*. Zwf1p, glucose-6-phosphate dehydrogenase; Acs1p, acetyl-CoA synthetase; PP pathway, pentose phosphate pathway; TCA cycle, tricarboxylic acid cycle.

ability issue, especially biological production of xylitol has gained an interest from academia and industries [7]. Instead of metal catalysts, a biocatalyst of xylose reductase can undertake the hydrogenation reaction of xylose with aid of NADPH and NADH cofactors (Fig. 1). Because of instability and high cost of xylose reductase and the cofactors, the biological production of xylose should be devised by microbial systems. To date, robust microorganisms have been developed to express xylose reductase highly and to supply the cofactors sufficiently. Among several microorganisms, *Candida* species is well established to produce high concentration of xylitol over 237 g/L with high productivity [8]. *Candida* species is recognized as a good xylitol producer, however, it consumes xylose as an energy source so the xylitol yield does not reach a commercially applicable value over 90% [8, 9]. *Saccharomyces cerevisiae* is a GRAS (generally recognized as safe) microorganism not only used in food and alcoholic beverage industries but also applicable for production of biochemicals, biofuels and biopharmaceuticals. In spite of its outstanding properties, *S. cerevisiae* is unable to produce xylitol from xylose because it does not have an enzyme, xylose reductase (XR) endogenously. For xylitol biosynthesis in metabolically-engineered *S. cerevisiae*, various research efforts have been made including genetic introduction of the XR gene, supply of the cofactors, control of the pentose phosphate pathway [10–13]. Because engineered *S. cerevisiae* expressing XR cannot metabolize xylitol into the central carbon metabolism, it converts xylose to xylitol at almost the theoretical yield (~100%) [9].

Xylose reductase catalyzes the one-step hydrogenation of xylose to xylitol in the presence of NADPH or NADH cofactor (Fig. 1). Most XRs in nature depend on NADPH, of which the oxidized form should be reduced again inside cells. Since the NADPH regeneration is done by a few metabolic steps, insufficient supply of NADPH limits rapid production of xylitol. Meanwhile, a NADH-preferring XR mutant was developed by protein engineering and applied to conversion of xylose to ethanol in engineered *S. cerevisiae* [14, 15]. An engineered *S. cerevisiae* expressing the mutant XR showed much higher xylose consumption rate than the same yeast harboring the wild-type XR dependent on NADPH. Considering these results, we hypothesized that utilization of both NADPH and NADH as cofactors would improve xylitol biosynthesis in engineered *S. cerevisiae*. In this study, we constructed engineered *S. cerevisiae* strains expressing the NADPH-dependent wild XR (wXR) and the NADH-preferring mutant XR (mXR), which were originated from *Scheffersomyces stipitis*. Additionally, glucose-6-phosphate dehydrogenase for NADPH regeneration and acetyl-CoA synthetase for NADPH and NADH regeneration were overexpressed to improve the xylitol production performance further. Finally, fed-batch fermentation strategies were optimized to produce xylitol with a high titer and productivity.

2 Materials and methods

2.1 Strains and plasmids

Escherichia coli TOP 10 (Invitrogen, Carlsbad, CA, USA) was used for gene manipulation. LB medium (10 g/L tryptone, 5 g/L yeast extract and 10 g/L NaCl) was used for *E. coli* cultivation. *S. cerevisiae* D452-2 was used as a host strain for xylitol production and a source of the yeast genomic DNA. Plasmids p426GPD and p425GPD with the *GPD* promoter and *CYC1* terminator were used for the expression of the wild (Genbank accession No. 4839234) and mutant genes of *XYL1* encoding *S. stipitis* XRs, and the *S. cerevisiae ZWF1* (No. 855480) and *ACS1* (No. 851245) genes coding for glucose-6-phosphate dehydrogenase (Zwf1p) and acetyl-CoA synthetase (Acs1p), respectively. Plasmids pRS403, pRS405 and pITy-3 were used for chromosomal integration of the *XYL1* genes. Yeast strains and plasmids used in this study are listed in Supporting information, Table S1.

2.2 Genetic manipulation

To construct the expression cassettes of the wild XR (wXR) and mutant XR (mXR), the two XR genes were amplified by polymerase chain reaction (PCR) from previously constructed vectors of YEpM4-XR^{WT}and YEpM4-XR^{MUT} with the corresponding PCR primers: F1 and R1 (Supporting information, Table S2) [14]. Each amplified XR gene was treated with DNA restriction enzymes of *Bam*HI and *Xho*I, and inserted between the *GPD* promoter and *CYC1* terminator in plasmid p426GPD, resulting in the construction of plasmids p426GPD_wXR and p426GPD_mXR. The XR expression cassette of the *GPD* promoter, each XR gene and *CYC1* terminator in a row, was PCR-amplified with DNA primers of F2 and R2. After digestion with *Sac*I, the cassette was introduced into chromosomal integration vectors, resulting in the construction of plasmids pRS406_wXR, pRS403_mXR, pITy-3_wXR and pITy-3_mXR. Plasmids pRS406_wXR and pRS403_mXR were linearized by *Xcm*I and *Nhe*I treatment, respectively, transformed and integrated into the chromosome of *S. cerevisiae* D452-2 to construct the DWM strain. Engineered *S. cerevisiae* strains expressing multiple copies of the wXR gene (DWW) and mXR gene (DMM) were constructed by δ-integration of plasmids pITy-3_wXR and pITy-3_mXR in *S. cerevisiae* D452-2, respectively [16]. The *ZWF1* and *ACS1* genes were amplified from the *S. cerevisiae* D452-2 genomic DNA, digested with *Xma*I and *Xho*I for *ZWF1*, *Eco*RI and *Xho*I for *ACS1* and cloned into p425GPD vector to generate plasmids p425GPD_ZWF1 and p425GPD_ACS1. After amplifying the *ZWF1* expression cassette with primers of F2 and R2 and digesting with *Sac*I, the cassette was introduced into plasmid p425GPD_ACS1 to construct plasmid p425GPD_ZWF1_ACS1. Each resulting plasmid was transformed into the DWM strain, of which transformants were named DWM-ZWF1, DWM-ACS1 and DWM-ZWF1-ACS1. The schemes for plasmid construction are displayed in Supporting information, Fig. S1.

A yeast EZ-Transformation kit (BIO 101, Vista, CA, USA) was used for plasmid transformation into *S. cerevisiae*, of which transformants were selected on YNB medium containing 20 g/L glucose, appropriate amino acids and nucleotide (histidine, leucine or uracil). Yeast strains transformed with delta-integration plasmids were propagated on YP medium supplemented with 20 g/L glucose and an appropriate concentration of G418 antibiotic (Calbiochem Co., USA).

2.3 Fermentation conditions

S. cerevisiae was grown in 5 mL of YP (10 g/L yeast extract and 20 g/L peptone) or YNB leu$^-$ medium (6.7 g/L yeast nitrogen base without amino acids and 1.6 g/L yeast synthetic drop-out medium supplement w/o leucine) containing 20 g/L glucose at 30°C and 250 rpm for 20 h. The cells were harvested and inoculated into a 500 mL baffled flask with 100 mL YP medium containing glucose and xylose for batch fermentation at 30°C and 250 rpm. For glucose-limited fed-batch fermentation, the pre-grown cells were transferred into 100 mL of YNB leu$^-$ medium with 20 g/L glucose. After 20 h of culture at 30°C and 250 rpm, the cells were harvested and suspended in sterilized water. The suspended cells were inoculated to a 3.7 L fermentor (Bioengineering AG, Wald, Switzerland) with 1 L working volume of YP medium containing 19 g/L glucose and 76–98 g/L xylose at 30°C. Acidity was controlled at pH 5.0 by addition of 2 N HCl and 2 N NaOH. An initial optical density of the cultures was adjusted at about 0.5. An agitation speed and an aeration rate was set at 500 rpm and 1 vvm, respectively. A concentrated glucose solution (600 g/L) was fed constantly at 1.8 g/L h of feed rate after depletion of glucose initially added. In addition, 600 g/L xylose solution was supplemented intermittently or continuously into the culture broth to control xylose concentration between 35 and 80 g/L or around 80 g/L.

2.4 Determination of intracellular NAD(P)H and NAD(P)$^+$ concentrations

Concentrations of NADPH, NADP$^+$, NADH and NAD$^+$ were measured using the NADP$^+$/NADPH and NAD$^+$/NADH assay kits (BioAssay Systems, Hayward, CA, USA). *S. cerevisiae* cells grown in YNB leu$^-$ medium with glucose were harvested at the mid-exponential growth phase and suspended in 125 µl of the NAD(P)$^+$ extraction buffer. The concentration measurement followed the manufacturer's instruction.

2.5 Determination of cell and metabolite concentrations

Optical density (OD) of the yeast cells was measured using a spectrometer (Ultrospec 2000, Amersham Pharmacia Biotech, Uppsala, Sweden) at 600 nm, which was converted to dry cell weight by a pre-determined conversion factor, 0.3 dry cell weight (g/L)/OD. Concentrations of metabolites in the culture broth were determined by a high performance liquid chromatography (HPLC, Agilent 1100LC, Santa Clara, CA, USA) equipped with a refractive index detector at 35°C and a Rezex ROA-Organic Acid H^+ (8%) column (Phenomenex, Torrance, CA, USA). The column was eluted with 0.01 N of H_2SO_4 at a flow rate of 0.6 mL/min and heated at 60°C.

3 Results

3.1 Dual expression of wild XR and mutant XR improved xylitol productivity

In order to extend the availability of the two cofactors (NADPH and NADH), and examine the influence of the extended availability of reducing agents on xylitol production, engineered *S. cerevisiae* strains were constructed to express the wild XR (DWW) or mutant XR (DMM) only, and both the XRs (DWM) (Supporting information, Table S1). In vitro activity assay of the wild and mutant XRs were carried out to assess their expression and preference of cofactors in the engineered *S. cerevisiae* strains (Table 1). The DWW and DMM strains were highly specific for NADPH and NADH, respectively, whereas the DWM strain having dual XRs showed similar activities toward NADPH and NADH as expected. The DWM strain had a similar total XR activity to the DWW strain or 1.4 times higher total activity than the DMM strain.

To analyze the xylitol-producing performance, batch cultures of the DWW, DMM, and DWM strains were performed using YP medium with 21 g/L xylose and 21 g/L glucose (Supporting information, Fig. S2). All the engineered *S. cerevisiae* strains consumed glucose at the same consumption rate. After glucose added initially was utilized, the three strains started to consume xylose with different consumption rates. *S. cerevisiae* is a Crabtree positive strain able to produce ethanol from glucose even in aerobic condition [17]. Ethanol produced from glucose was consumed after glucose depletion and used for maintenance of cell viability because xylose cannot flow into the central energy metabolism in all strains. And ethanol and xylose were consumed simultaneously since ethanol was used for the regeneration of the cofactors for xylitol production. Xylitol production was linearly correlated to the consumption of xylose and ethanol (Supporting information, Fig. S2). Fermentation parameters of these batch fermentations were summarized in Table 1. The DWM strain expressing two types of XR exhibited the highest capability of xylitol production. The batch fermentation of the DWM strain resulted in improved xylitol concentration by 6.6% and 47% compared to the corresponding values for the DWW and DMM strains expressing each type of XR, respectively. As expected, the xylitol yield in all cases reached the theoretical xylitol yield (1 g-xylitol/1 g-xylose). Interestingly, the DMM and DWM strains showed 1.3 times higher final dry cell mass than the DWW strain, indicating that the partial or complete change of the wild XR to the mutant XR probably reduced a metabolic burden by expression of NADPH-consuming XR and hence elevated final cell mass.

3.2 High xylitol productivity was obtained in glucose-limited fed-batch fermentation

In batch fermentations, the DWM strain gave the best performance for xylitol production. To confirm its ability as xylitol producer, glucose-limited fed-batch fermentations of the DWW-C and DWM-C strains were performed in YP medium with 100 g/L xylose and 21 g/L glucose (Supporting information, Fig. S3). The DWW-C and DWM-C strains were constructed by introduction of plasmid p425GPD into the DWW and DWM strains, respectively. After depletion of glucose initially added, a concentrated glucose solution was fed continuously to maintain its basal level in the medium, which is called a glucose-limited feeding strategy. To relieve the glucose inhibition on xylose transport [18, 19] and to supply the cofactors and cellular energy, the glucose-limited feeding strategy was chosen in this study. By this glucose feeding strategy, xylose was linearly consumed to produce xylitol with different xylitol productivities. The glucose-limited fed-batch fermentation of the DWM strain exhibited 72 g/L xylitol concentration and 2.00 g/L h xylitol productivity which were 13% and 14% higher than those for the DWW strain. Based on the XR activity analysis and the xylitol-producing ability in batch and glucose-limited fed-batch fermentations, the balanced expression of the wild and mutant XRs under high XR expression levels could be a plausible reason for the enhanced production of xylitol by the DWM strain.

Table 1. Summarized results of batch cultures of engineered *S. cerevisiae* DWW, DMM and DWM strains using 21 g/L xylose and 21 g/L glucose

Strain	Final dry cell mass[a] (g/L)	Xylitol concentration[a] (g/L-h)	Xylitol productivity[a] (g/L)	Xylitol yield (g xylitol /g xylose)	Specific XR activity[b] (U/mgprotein)			XR activity ratio[c]
					NADPH	NADH	Total	
DWW	5.97	16.5	0.827	~ 1	2.11 ± 0.02	1.53 ± 0.03	3.64	1.39
DMM	7.41	12.0	0.599	~ 1	0.19 ± 0.01	2.47 ± 0.03	2.66	0.08
DWM	8.14	17.6	0.881	~ 1	1.89 ± 0.10	1.77 ± 0.03	3.66	1.07

a) These values are the mean of triplicate experiments.
b) Specific XR activity was measured three times independently.
c) This value was calculated by the division of XR activities on NADPH to NADH.

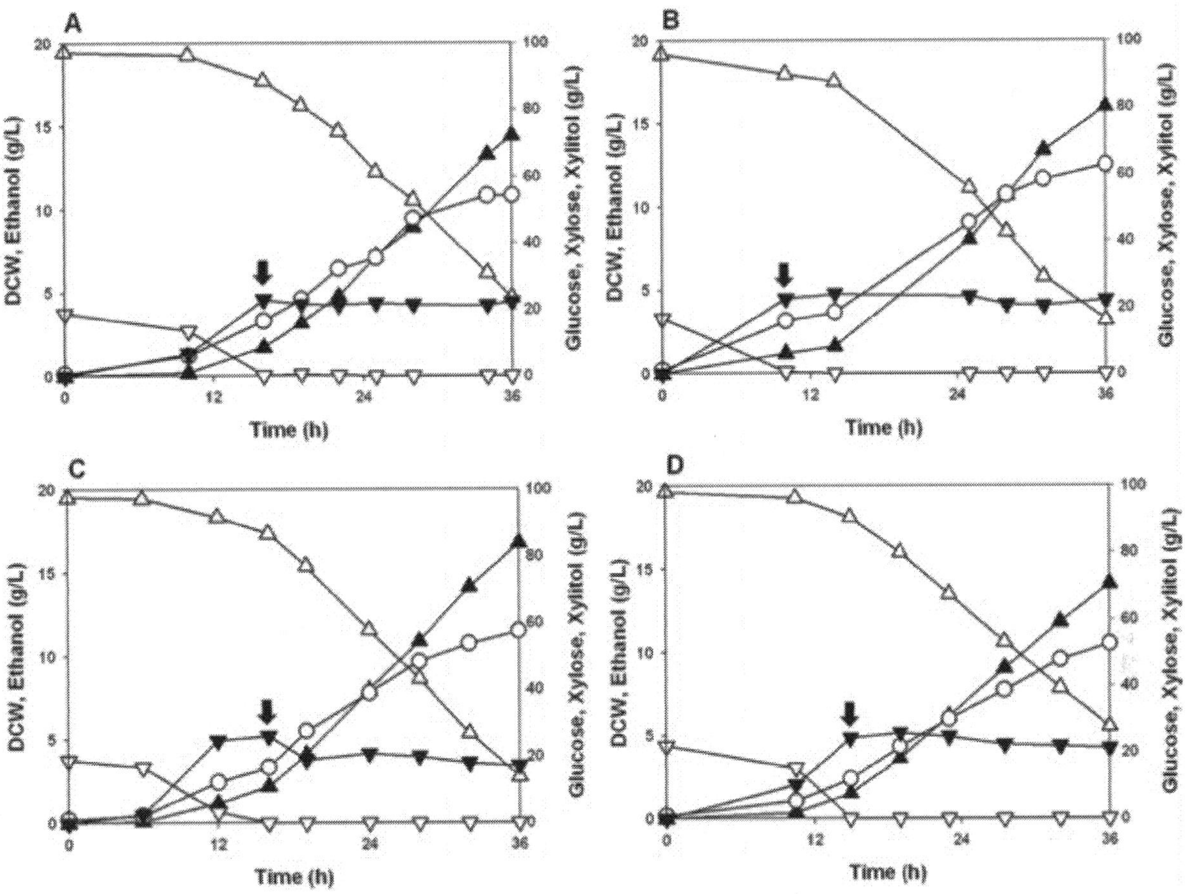

Figure 2. Effect of Zwf1p and Acs1p overexpression on xylitol production in glucose-limited fed-batch fermentations of engineered *S. cerevisiae* DWM-ZWF1 (**A**), DWM-ACS1 (**B**), DWM-ZWF1-ACS1 (**C**) and DWW-ZWF1-ACS1 (**D**) at 30°C, 500 rpm, 1 vvm and pH 5.0. A mixture of 18 g/L glucose and 98 g/L xylose was added initially. The arrow indicates the start point of the glucose feeding at a rate of 1.8 g/L h. Symbols are denoted as follows: ○, dry cell weight (DCW); ▽, glucose; △, xylose; ▲, xylitol; ▼, ethanol.

3.3 Overexpression of *ZWF1* and *ACS1* increased intracellular pools of NADPH and NADH

As shown in Fig. 1, biological production of xylitol is controlled by the activity and cofactor dependency of XR, and the supply of NADPH and NADH. As harmonized expression of the wild and mutant XRs significantly improved the xylitol-producing performance as presented above, sufficient supply of NADPH and NADH is expected to enhance the xylitol production further. In microbial systems, NADPH and NADH are regenerated by many metabolic enzymes. Among them, glucose-6-phosphate dehydrogenase encoded by the *ZWF1* gene and acetyl-CoA synthetase by the *ACS1* gene were selected as the NADPH and NADH regenerating enzymes, because of their high efficiency.

To investigate the effects of the intracellular levels of NADPH and NADH on xylitol production, Zwf1p and Acs1p were overexpressed in the DWM strain. Plasmids p425GPD_ZWF1 and p425GPD_ACS1 were constructed to harbor each coding gene and transformed into the DWM strain to make the DWM-ZWF1 and DWM-ACS1 strains, respectively. When intracellular concentrations of $NADP^+$, NADPH, NAD^+ and NADH in the DWM-C, DWM-ZWF1 and DWM-ACS1 strains were measured (Supporting information, Table S3), overexpression of Zwf1p and Acs1p changed the pools of the hydrogen donors and increased the intracellular concentrations of both NADPH and NADH with different extents. The *ZWF1* overexpression increased the ratios of NADPH to $NADP^+$ and of NADH to NAD^+ by 19%, compared to the DWM-C control strain. The *ACS1* overexpression led to 2.26- and 1.27-fold improvements in NADPH and NADH concentrations, respectively, relative to those of the DWM-C strain. In addition, the DWM-ACS1 strain showed 4.15 and 1.25 times higher $NADPH/NADP^+$ and $NADH/NAD^+$ ratios, respectively, than those of the DWM-C strain. As a result, intracellular pools of NADPH and NADH in the engineered strain were increased by overexpression of the NADPH and NADH-regenerating enzymes in the DWM strain.

To investigate the effects of the overexpression of *ZWF1* or *ACS1* on xylitol production, glucose-limited fed-batch fermentations of DWM-ZWF1 and DWM-ACS1 were performed using 18 g/L glucose and 98 g/L xylose (Fig. 2). The DWM-ZWF1 and DWM-ACS1 strains showed 2.04 and 2.15 g/L h maximum xylitol productivity, which were 4–15% higher than that of the DWM-C strain (Supporting information, Fig. S3B). It indicated that either *ZWF1* or *ACS1* overexpression enhanced the xylitol concentration and productivity. As expected, the increments of NADPH and NADH concentrations, and the NADPH/$NADP^+$and NADH/NAD^+ ratios elevated the xylitol producing performances of the DWM-ZWF1 and DWM-ACS1 strains. Irrespective of the *ZWF1* and *ACS1* overexpression, xylitol yields in each case were close to the theoretical value.

3.4 Coexpression of *ZWF1* (for NADPH regeneration) and *ACS1* (for NADH regeneration) enhanced xylitol productivity synergistically

Individual expression of the *ZWF1* and *ACS1* genes conferred the improved ability of xylitol production to the DWM strains. To maximize the merit of the DWM strain with the broad specificity of the cofactors, plasmid p425GPD_ZWF1_ACS1 was constructed to express both the *ZWF1* and *ACS1* genes, and transformed into the DWM strain. In the analysis of the intracellular cofactor contents (Supporting information, Table S3), overexpression of both Zwf1p and Acs1p in the resulting strain of DWM-ZWF1-ACS1 highly elevated the pools of both NADPH and NADH, and the ratios of the reduced cofactors (NADPH and NADH) to the oxidized cofactors ($NADP^+$ and NAD^+) by a four-fold maximally, relative to those for the DWM-C strain. The coexpression also led to higher ratios of NAD(P)H to $NAD(P)^+$ than the single expression of Zwf1p and Acs1p.

To evaluate the xylitol-producing performance, the DWM-ZWF1-ACS1 strain was cultured using 18 g/L glucose and 98 g/L xylose, and by the same fermentation strategy as the two fed-batch cultivations performed above (Supporting information, Fig. S3). As a control, the DWW-ZWF1-ACS1 strain was cultured by the same fermentation strategy. As shown in Fig. 2C and 2D, the patterns of xylose consumption and xylitol production were similar to other glucose-limited fed-batch fermentations in this study. The DWM-ZWF1-ACS1 strain produced 83.7 g/L xylitol with the final xylitol productivity of 2.33 g/L h, while the DWW-ZWF1-ACS1 strain did 70.6 g/L xylitol with 1.96 g/L h productivity. Especially, the DWM-ZWF1-ACS1 strain showed 17% higher xylitol productivity than that of the DWM-C strain (Fig. 2C). In comparison to the sole expression of either *ZWF1* or *ACS1*, the DWM-ZWF1-ACS1 strain exhibited 8–14% improvement in xylitol productivity. Considering these fermentation results, it was certain that the balanced expression of the wild and mutant XRs with an aid of the NADPH and NADH regeneration systems exhibited a synergistic effect on xylitol production. To ensure functional expression of Zwf1p and Acs1p in the DWM-ZWF1-ACS1 strain, crude extracts of the yeast cells were collected in the fed-batch culture and their enzyme activities were measured. As shown in Supporting information, Fig. S4, the DWM-ZWF1-ACS1 strain had at least five-fold higher specific activity of glucose-6-phosphate dehydrogenase than the control DWM-C strainat all times. Specific acetyl-CoA synthetase activity of the DWM-ZWF1-ACS1 strain was also five times higher than that of the control strain.

3.5 Control of xylose concentration in fed-batch fermentation improved xylitol productivity

The final strain (DWM-ZWF1-ACS1) showed the highest productivity than other *S. cerevisiae* strains engineered for xylitol production. To extent the period for the highest xylitol productivity and hence elevate final xylitol titer, fed-batch cultivation strategies were designed using intermittent addition or continuous feeding of a concentrated xylose solution to maintain an optimized concentration of xylose in the culture medium [9]. In the glucose-limited fed-batch fermentation of the DWM-ZWF1-ACS1 strain (Fig. 2C), the highest xylitol productivity was achieved when xylose concentration in the medium ranged from 40 g/L to 80 g/L. So, this range of xylose concentration was adopted to keep the highest xylitol-producing ability. As shown in Fig. 3A and 3B, glucose-limited fed-batch cultures of the DWM-C and DWM-ZWF1-ACS1 strains were carried out by intermittent addition of 600 g/L xylose solution to a bioreactor when xylose concentration in the medium dropped below 40 g/L. The concentrated xylose solution was added four times to control the xylose concentration range between 40 and 80 g/L. Finally, 185.3 g/L xylitol concentration and 3.71 g/L hr productivity were obtained in 50 h fed-batch culture of the DWM-ZWF1-ACS1 strain. On the other hand, the DWM-C strain cultured by the same fermentation strategy produced 150.2 g/L xylitol. Generally, sugar uptake rate depends on extracellular sugar concentration [20]. Xylose consumption rate decreased gradually whenever xylose concentration fluctuated. For maximizing xylitol production, a glucose-limited fed-batch fermentation was performed by continuous feeding of 600 g/L xylose solution at a constant rate of 10 mL/h (Fig. 3C). The feed rate was decided from the stage showing the highest xylose consumption rate in Fig. 2C. When initial glucose was depleted, the continuous xylose feeding started at 6 h culture time. During 24 h of the feeding, xylose concentration in the medium was maintained well at 80 g/L as

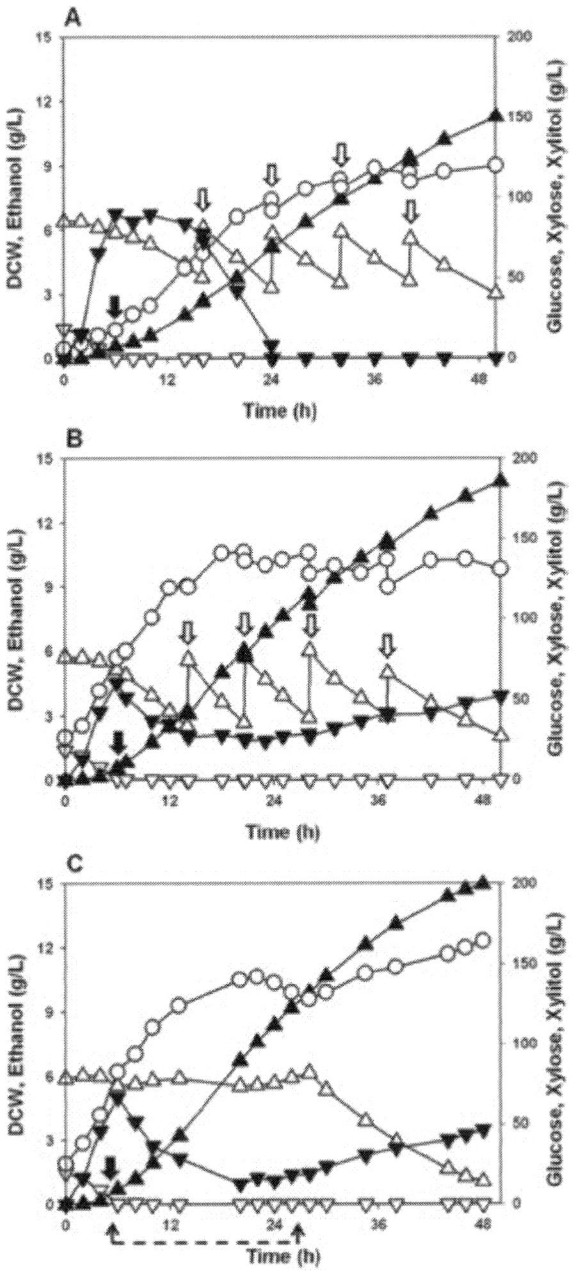

Figure 3. Fed-batch fermentations of engineered *S. cerevisiae* DWM-C (**A**), DWM-ZWF1-ACS1 (**B**, **C**). The closed arrow indicates the initiation of the glucose feeding at a rate of 1.8 g/L h. In the panels (**A**) and (**B**), a concentrated xylose solution of 600 g/L was added intermittently (open arrow). In the panel (**C**), the concentrated xylose solution was fed continuously at a rate of 10 mL/h and the dotted line indicates the period of the continuous feeding. Symbols are denoted as follows: ○, dry cell weight (DCW); ▽, glucose; △, xylose; ▲, xylitol; ▼, ethanol.

planned and most xylose added was converted into xylitol with the theoretical yield (~1 g/g). As a result, the final xylitol concentration of 196.2 g/L and xylitol productivity of 4.26 g/L h were obtained in the fed-batch culture by the continuous xylose-feeding strategy.

4 Discussion

Xylose is biologically reduced to xylitol by xylose reductase (XR). The wild XR from *S. stipitis* used in this study is known to be able to utilize both NADPH and NADH as cofactors. But the activity of wild XR on NADPH was 30% higher than that on NADH [21], so most engineered *S. cerevisiae* strains expressing the wild XR gene as well as conventional xylitol-producing yeasts convert xylose to xylitol using NADPH primarily. Because the intracellular NADH level is much higher than the NADPH level [22], balanced utilization of both NADPH and NADH is suggested to avoid the depletion of cofactors and hence improve xylitol production in engineered *S. cerevisiae*. Previously, a mutant XR preferring NADH was developed by protein engineering [15], and its expression reduced a metabolic disorder of redox-imbalance in the xylose metabolism of engineered *S. cerevisiae* [14]. As shown in the batch and glucose-limited fed-batch cultures (Supporting information, Fig. S2, Fig. S3; Table 1), the DWM strain expressing both the wild and mutant XRs showed higher xylitol producing performances than the DWW and DMM strains, indicating that the harmonized expression of two XRs with different cofactor specificities is a certain reason for the enhanced xylitol production by the DWM strain.

Meanwhile, supply of NADPH or NADH cofactor is a rate-limiting factor for xylitol production as well as the XR expression. Among several metabolic enzymes regenerating NADPH or NADH in the substrate level, glucose-6-phosphate dehydrogenase (Zwf1p) catalyzes the main metabolic pathway for NADPH regeneration in *S. cerevisiae* [11, 23]. Acetyl-CoA synthetase (Acs1p) converts acetate to acetyl-CoA, which is metabolized further into the tricarboxylic acid cycle with the NADH and NADPH regeneration metabolism. Since acetate is produced from acetaldehyde by NADPH regeneration, overexpression of acetyl-CoA synthetase is expected to modulate both NADH and NADPH levels simultaneously and indirectly [24]. Overexpression of Zwf1p and Acs1p enhanced the intracellular pools of NADPH and NADH (Supporting information, Table S3), resulting in maximum 17% increases in xylitol productivity than the control strain (Supporting information, Fig. S3B; Fig. 2C and Table 2). The similar effects of improved production of target molecules by the regeneration of the cofactors could be found in some reports. The overexpression of Zwf1p triggered the production of xylitol, GDP-L-fucose, ε-caprolactone by enhanced NADPH levels [12, 25, 26].

Table 2. Summarized results of glucose-limited fed-batch fermentations using 76–98 g/L initial xylose in this study

S. cerevisiae strain	Xylose feeding strategy	Final dry cell mass (g/L)	Final xylitol concentration (g/L)	Final xylitol Productivity (g/L-h)	Xylitol yield (g/g xylose)
DWW-C	No xylose feeding	8.8	63.5	1.76	~1
DWW-ZWF1-ACS1	No xylose feeding	10.5	70.6	1.96	~1
DWM-C	No xylose feeding	10.9	72.0	2.00	~1
DWM-ZWF1-ACS1	No xylose feeding	11.5	83.7	2.33	~1
DWM-C	Intermittent xylose addition[a]	9.1	150.2	3.00	~1
DWM-ZWF1-ACS1	Intermittent xylose addition[a]	10.4	185.3	3.71	~1
DWM-ZWF1-ACS1	Continuous xylose feeding[b]	12.2	196.2	4.27	~1
BJ3505/δXR[c]	Intermittent xylose addition	22.0	116	2.34	~1
C. tropicalis ATCC 13803[d]	Continuous xylose feeding	10.4	187	3.90	0.75

[a] A concentrated xylose solution of 600 g/L was added four times to control 40–80 g/L xylose concentration in the medium.
[b] A concentrated xylose solution of 600 g/L was fed continuously to maintain 80 g/L xylose concentration in the medium.
[c] These data were obtained from cell-recycle fermentation [10].
[d] These data were obtained by optimizing initial xylose concentration and feeding rate [9].

The Acs1p overexpression influenced the production of many biochemicals such as hydrogen, isopropanol, butanol by increasing the carbon flux into a desired product and decreasing acetate accumulation, which was toxic to cell growth [27, 28]. When cellobiose and xylose were consumed simultaneously, overexpression of *S. cerevisiae* Idp2p (isocitrate dehydrogenase) increased the NADPH availability and allowed a 63% improvement in xylitol productivity [29]. As a result, the modulation of the pools and ratios of NADPH and NADH was one of the effective ways to improve xylitol production.

Dual expression of NAD(P)H regenerating enzymes would be an easy way to increase target molecules by changing the pools of NADPH and NADH cofactors. In case of microbial xylitol production, however, this strategy was not always effective. Coexpression of glucose-6-phosphate dehydrogenase and 6-phosphogluconate dehydrogenase, two main NADPH-regenerating enzymes, increased xylitol productivity by engineered *Candida tropicalis* PP [30]. Coexpression of Acs1p and $NADP^+$-dependent aldehyde dehydrogenase 6 (Ald6p) in engineered *S. cerevisiae* BJ3505:δXR exerted a negative effect on xylitol production, compared to their single expression [24]. As shown in Fig. 2C and 2D, simultaneous expression of Zwf1p and Acs1p in the DWM strain exhibited 19% higher enhancement of xylitol productivity compared to the coexpression of Zwf1p and Acs1p in the DWW strain. It was certain that the coexpression of Zwf1p and Acs1p with different mechanisms of cofactor regeneration was beneficial for xylitol production mediated by both the wild and mutant XRs in the DWM-ZWF1-ACS1 strain.

The results of the glucose-limited fed-batch fermentations in this study are summarized in Table 2. The control of xylose concentration at an optimized range improved both xylitol concentration and productivity considerably. The continuous xylose-feeding strategy gave the best performances of xylitol production which were 2.3- and 1.8-fold increases in xylitol concentration and xylitol productivity, respectively, relative to the culture of the same strain without controlling xylose concentration. In comparison to the previous results done by other microorganisms, the DWM-ZWF1-ACS1 strain showed 82% and 9% higher xylitol productivity than a engineered *S. cerevisiae* BJ3505 strain expressing the NADPH-dependent XR from *S. stipitis* [10] and a wild type of *C. tropicalis* ATCC13803 [9], respectively, which were cultured using a similar fed-batch fermentation strategy. All cultures of the *S. cerevisiae* strains expressing foreign XRs in this study resulted in almost the theoretical yield (~1 g/g), whereas a traditional xylitol producer of *C. tropicalis* produced xylitol below the theoretical yield [9, 31]. Expression of two XRs highly specific for each cofactor along with two metabolic enzymes for increasing the pools of the reduced cofactors could clearly overcome both the process limitation of engineered *S. cerevisiae* (low productivity) and *C. tropicalis* (low yield) reported previously.

In conclusion, insufficient supply of NADPH cofactor and xylose utilization to cellular energy prevent the commercialization of microbial xylitol production. To overcome these hurdles, *S. cerevisiae* unable to metabolize xylose was engineered to express two types of XR using both NADPH and NADH cofactors. Additionally, Zwf1p and Acs1p were coexpressed to facilitate the NADPH and NADH regeneration. Finally, the best engineered *S. cerevisiae* DWM-ZWF1-ACS1 coexpressing NADPH-dependent and NADH-preferring XRs, and Zwf1p and Acs1p produced almost 196.2 g/L xylitol with over 4.27 g/L-h productivity and almost theoretical yield, which might meet the industrial demands for commercialization of microbial xylitol production.

This research was supported by The Advanced Biomass R&D Center (ABC) of Korea Grant (2011-0031359) and The National Research Foundation of Korea Grant (2013M1A2A2072600 and 2013R1A2A2A01004806) funded by the Ministry of Science, ICT & Future Planning and Technology.

The authors declare no financial or commercial conflict of interest.

5 References

[1] Makinen, K. K., Sugar alcohol sweeteners as alternatives to sugar with special consideration of xylitol. *Med. Princ. Pract.* 2011, *20*, 303–320.

[2] Clark, J. H., Deswarte, F., *Introduction to Chemicals from Biomass*, John Wiley & Sons 2011.

[3] Yue, H., Zhao, Y., Ma, X., Gong, J., Ethylene glycol: Properties, synthesis, and applications. *Chem. Soc. Rev.* 2012, *41*, 4218–4244.

[4] de Albuquerque, T. L., da Silva Junior, I. J., Macedo, G. R. d., Rocha, M. V. P., Biotechnological production of xylitol from lignocellulosic wastes: A review. *Process Biochem.* 2014, *49*, 1779–1789.

[5] Yadav, M., Mishra, D. K., Hwang, J.-S., Catalytic hydrogenation of xylose to xylitol using ruthenium catalyst on NiO modified TiO_2 support. *Appl. Catal., A* 2012, *425*, 110–116.

[6] Park, E. H., Genome-wide screening of *Saccharomyces cerevisiae* genes regulated by vanillin. *J. Microbiol. Biotechnol.* 2015, *25*, 50–56.

[7] Pérez-Bibbins, B., Torrado-Agrasar, A., Salgado, J. M., Mussatto, S. I., Domínguez, J. M., Xylitol production in immobilized cultures: A recent review. *Crit. Rev. Biotechnol.* 2015, *15*, 1–14.

[8] Kim, T.-B., Oh, D.-K., Xylitol production by *Candida tropicalis* in a chemically defined medium. *Biotechnol. Lett.* 2003, *25*, 2085–2088.

[9] Kim, J., Han, K., Koh, Y., Ryu, Y., Seo, J., Optimization of fed-batch fermentation for xylitol production by *Candida tropicalis*. *J. Ind. Microbiol. Biotechnol.* 2002, *29*, 16–19.

[10] Bae, S.-M., Park, Y.-C., Lee, T.-H., Kweon, D.-H. et al., Production of xylitol by recombinant *Saccharomyces cerevisiae* containing xylose reductase gene in repeated fed-batch and cell-recycle fermentations. *Enzyme Microb. Tech.* 2004, *35*, 545–549.

[11] Oh, Y.-J., Lee, T.-H., Lee, S.-H., Oh, E.-J. et al., Dual modulation of glucose 6-phosphate metabolism to increase NADPH-dependent xylitol production in recombinant *Saccharomyces cerevisiae*. *J. Mol. Catal. B: Enzym.* 2007, *47*, 37–42.

[12] Kwon, D.-H., Kim, M.-D., Lee, T.-H., Oh, Y.-J. et al., Elevation of glucose 6-phosphate dehydrogenase activity increases xylitol production in recombinant *Saccharomyces cerevisiae*. *J. Mol. Catal. B: Enzym.* 2006, *43*, 86–89.

[13] Chung, Y.-S., Kim, M.-D., Lee, W.-J., Ryu, Y.-W. et al., Stable expression of xylose reductase gene enhances xylitol production in recombinant *Saccharomyces cerevisiae*. *Enzyme Microb. Tech.* 2002, *30*, 809–816.

[14] Lee, S.-H., Kodaki, T., Park, Y.-C., Seo, J.-H., Effects of NADH-preferring xylose reductase expression on ethanol production from xylose in xylose-metabolizing recombinant *Saccharomyces cerevisiae*. *J. Biotechnol.* 2012, *158*, 184–191.

[15] Watanabe, S., Saleh, A. A., Pack, S. P., Annaluru, N. et al., Ethanol production from xylose by recombinant *Saccharomyces cerevisiae* expressing protein-engineered NADH-preferring xylose reductase from *Pichia stipitis*. *Microbiology* 2007, *153*, 3044–3054.

[16] Parekh, R. N., Shaw, M. R., Wittrup, K. D., An integrating vector for tunable, high copy, stable integration into the dispersed Ty δ sites of *Saccharomyces cerevisiae*. *Biotechnol. Prog.* 1996, *12*, 16–21.

[17] Piškur, J., Rozpędowska, E., Polakova, S., Merico, A., Compagno, C., How did *Saccharomyces* evolve to become a good brewer? *Trends Genet.* 2006, *22*, 183–186.

[18] Lee, W.-J., Ryu, Y.-W., Seo, J.-H., Characterization of two-substrate fermentation processes for xylitol production using recombinant *Saccharomyces cerevisiae* containing xylose reductase gene. *Process Biochem.* 2000, *35*, 1199–1203.

[19] Lee, W.-J., Kim, M.-D., Ryu, Y.-W., Bisson, L., Seo, J.-H., Kinetic studies on glucose and xylose transport in *Saccharomyces cerevisiae*. *Appl. Microbiol. Biotechnol.* 2002, *60*, 186–191.

[20] Shuler, M. L., Kargi, F., *Bioprocess Engineering*, Prentice Hall, Englewood Cliffs, NJ 2002.

[21] Verduyn, C., Van Kleef, R., Frank, J., Schreuder, H. et al., Properties of the NAD(P)H-dependent xylose reductase from the xylose-fermenting yeast *Pichia stipitis*. *Biochem. J.* 1985, *226*, 669–677.

[22] Vemuri, G., Eiteman, M., McEwen, J., Olsson, L., Nielsen, J., Increasing NADH oxidation reduces overflow metabolism in *Saccharomyces cerevisiae*. *Proc. Natl. Acad. Sci. U.S.A.* 2007, *104*, 2402–2407.

[23] Lee, W. H., Kim, M. D., Enhanced production of ε-caprolactone by coexpression of bacterial hemoglobin gene in recombinant *Escherichia coli* expressing cyclohexanone monooxygenase Gene. *J. Microbiol. Biotechnol.* 2014, *24*, 1685–1689.

[24] Oh, E.-J., Bae, Y.-H., Kim, K.-H., Park, Y.-C., Seo, J.-H., Effects of overexpression of acetaldehyde dehydrogenase 6 and acetyl-CoA synthetase 1 on xylitol production in recombinant *Saccharomyces cerevisiae*. *Biocatal. Agric. Biotechnol.* 2012, *1*, 15–19.

[25] Lee, W.-H., Chin, Y.-W., Han, N. S., Kim, M.-D., Seo, J.-H., Enhanced production of GDP-L-fucose by overexpression of NADPH regenerator in recombinant *Escherichia coli*. *Appl. Microbiol. Biotechnol.* 2011, *91*, 967–976.

[26] Lee, W.-H., Park, J.-B., Park, K., Kim, M.-D., Seo, J.-H., Enhanced production of ε-caprolactone by overexpression of NADPH-regenerating glucose 6-phosphate dehydrogenase in recombinant *Escherichia coli* harboring cyclohexanone monooxygenase gene. *Appl. Microbiol. Biotechnol.* 2007, *76*, 329–338.

[27] Krivoruchko, A., Serrano-Amatriain, C., Chen, Y., Siewers, V., Nielsen, J., Improving biobutanol production in engineered *Saccharomyces cerevisiae* by manipulation of acetyl-CoA metabolism. *J. Ind. Microbiol. Biotechnol.* 2013, *40*, 1051–1056.

[28] Wang, R.-Y., Shi, Z.-Y., Chen, J.-C., Wu, Q., Chen, G.-Q., Enhanced co-production of hydrogen and poly-(*R*)-3-hydroxybutyrate by recombinant PHB producing *E. coli* over-expressing hydrogenase 3 and acetyl-CoA synthetase. *Metab. Eng.* 2012, *14*, 496–503.

[29] Oh, E. J., Ha, S.-J., Rin Kim, S., Lee, W.-H. et al., Enhanced xylitol production through simultaneous co-utilization of cellobiose and xylose by engineered *Saccharomyces cerevisiae*. *Metab. Eng.* 2013, *15*, 226–234.

[30] Ahmad, I., Shim, W. Y., Jeon, W. Y., Yoon, B. H., Kim, J.-H., Enhancement of xylitol production in *Candida tropicalis* by co-expression of two genes involved in pentose phosphate pathway. *Bioprocess Biosyst. Eng.* 2012, *35*, 199–204.

[31] Bisaria, V. S., Kondō, A., *Bioprocessing of Renewable resources to Commodity Bioproducts*, Wiley Online Library 2014.

Cover illustration

This special issue, in collaboration with the Asian Federation of Biotechnology and edited by Professors Yoon-Mo Koo and Penjit Srinophakun , covers advances in bioprocess engineering. Several articles in this issue discuss different aspects of lignocellulose-based biorefining. Miscanthus (shown on the cover) is one example of a plant whose lignocellulosic biomass can be used for bioethanol production.

Biotechnology Journal – list of articles published in the December 2015 issue.

489395 Focal Adhesion Architecture and Composition Regulate Cell Adhesion and Traction Generation Independent of Integrin Class

Abstract:

Integrins are a class of heterodimeric transmembrane proteins that mediate attachment to the extracellular matrix (ECM) and serve as templates for focal adhesions (FAs), sophisticated protein complexes that regulate mechanical and chemical signals. At present, the mechanisms by which specific focal adhesion architectures affect traction generation and cellular adhesion remain poorly understood. Here, we use Förster resonance energy transfer (FRET)-based molecular tension sensors (MTSs) to measure the distribution of mechanical forces exerted by individual integrin heterodimers in living cells at the single-molecule level. We engineered two MTS variants: MTS_{low}, to probe a low force range (0-8 pN), and MTS_{int}, for examining intermediate forces (8-12 pN). We find the majority of bound integrins bear forces <12 pN and that integrins within defined adhesion complexes exert tractions that fall into distinct subpopulations. Using integrin knock-out lines rescued with specific integrin classes, we find that the single-molecule force distributions remain remarkably similar, indicating that these subpopulations are not primarily determined by the integrin class. Furthermore, we find that although integrin linkages bearing more than 6 pN are actin mediated, they are not required to support cell adhesion. In ongoing work, we investigate how perturbations to FA architecture and composition alter traction on the molecular and cellular levels.

490325 Quantifying Tissue-Induced Collagen Fiber Alignment in 3D Microfabricated Tissues

Abstract:

Introduction: The extracellular matrix (ECM) is a complex mixture of macromolecules that provides both biophysical and biochemical cues to cells and tissues. Cell-ECM interactions, which are known to drive the remodeling of the ECM, influence cell- and tissue-level behavior and are important in biological processes such as tissue morphogenesis and cancer progression [1, 2]. In particular, cell-induced collagen fiber alignment is widely observed, but the dynamics and mechanisms by which this alignment occurs remain unclear. Here, we quantified the dynamics of tissue-induced type-I collagen fiber alignment and evaluated the roles of contractility and proteolysis in the alignment process.

Materials and Methods: Three-dimensional (3D) microfabricated tissues were embedded in a matrix consisting of acid-extracted type-I bovine collagen and Matrigel using a previously developed soft-lithography approach [3]. 3D tissues were fabricated using functionally normal mouse mammary epithelial cells or breast cancer cells. Cytoskeletal contractility was manipulated using a Rho activator, a Rho kinase inhibitor, and a myosin II ATPase inhibitor. Proteolysis was inhibited using a broad-spectrum matrix metalloproteinase (MMP) inhibitor. Collagen fibers were visualized using confocal reflection microscopy, and fiber alignment was quantified from changes in the orientation of pixel intensity gradients over a period of 24 hours.

Results and Discussion: Tissue-induced collagen fiber alignment occurred rapidly and reached a plateau within 6 hours when ~50% of collagen fibers were aligned perpendicular to the tissue surface. During fiber alignment, a decrease in the tissue radius was observed, which indicated that tissue-exerted strain occurred coincident with the alignment process. Collagen alignment was not observed when tissues were treated with a Rho kinase or myosin II ATPase inhibitor. Tissues treated with a Rho activator showed an increase in both the rate and extent of collagen alignment. These results suggested that Rho-mediated cytoskeletal contractility has an important role in collagen alignment. We observed that inhibiting MMP activity did not affect the rate or extent of collagen fiber alignment, which further supported the idea that Rho-mediated cytoskeletal contractility is the dominant mechanism for collagen alignment. When comparing alignment in normal and cancer cells, we unexpectedly found that breast cancer cells, which have higher contractility but weaker cell-cell adhesions as compared to mammary epithelial cells, took longer to align the collagen fiber matrix. These results demonstrated that collective contractility is important for controlling the rate of collagen fiber alignment.

Conclusions: We quantified the dynamics of type-I collagen fiber alignment using 3D microfabricated tissues. Our results suggest that collective cellular contractility is the dominant mechanism that drives tissue-induced collagen fiber alignment. We found that tissues exerted strain on the surrounding collagen matrix and primarily used a mechanical mechanism to align collagen fibers.

References:

1. Provenzano PP et al. (2006) *BMC Med* 4(1):38.
2. Gjorevski N et al. (2015) Sci. Rep. 5:11458.
3. Piotrowski-Daspit AS et al. (2016) *JoVE* 113: e54283.

491521 Evaluation of Preliminary Adhesion Processes of Cyanobacteria (blue-green algae) on Photobioreactor Materials

Abstract:

Evaluation of preliminary adhesion processes of cyanobacteria (blue-green algae) on photobioreactor materials

Suvarna N L Talluri[1], Robb M Winter[1] and David R Salem[1,2]

[1]Department of Chemical and Biological Engineering

[2]Composites and Polymer Engineering (CAPE) Laboratory

South Dakota School of Mines & Technology, Rapid City, SD

Abstract

Due to increasing interest in replacing fossil fuels with biomass based fuels, cyanobacteria are considered as an attractive third generation renewable feed stock for biofuel production for their advantages such as CO_2 fixation, use of sunlight, waste water treatment and ease of genetic engineering. In spite of growing demand and valuable biofuels being already produced by cyanobacteria, their cultivation in photobioreactors (PBRs) poses a number of problems one of which is their tendency to cause rapid biofouling of PBR materials. Cyanobacteria fouling on PBR materials can be a significant operating challenge that can decrease the overall biofuel productivity of the system by interfering with the light penetration to cyanobacteria cells. Despite the fact that cyanobacteria are a common group of microorganisms that cause fouling in photobioreactors, studies focused on prevention of cyanobacteria are scarce. A detailed understanding of the adhesion strategies of cyanobacteria on PBR materials is needed in order to successfully develop antifouling materials. Therefore, in this study, we investigated the preliminary adhesion processes of a selected cyanobacterium (genetically engineered to enhance biofuel production) on a range of existing or potential PBR material surfaces. This was done by evaluating the physicochemical changes occurring during the initial adhesion and biofilm development of the cyanobacteria. Experimental techniques such as surface contact angle measurements, bright field microscopy, attenuated total reflectance-Fourier transform infrared (ATR-FTIR) spectroscopy and atomic force microscopy (AFM) were used to characterize the adhesion processes of cyanobacteria on the PBR materials. Contact angle measurements revealed that surface hydrophobicity/hydrophilicity of PBR materials was altered within 12h due to the formation of conditioning films. Atomic force microscopy revealed that changes in surface roughness, surface height and thickness of the conditioning films were substrate specific. ATR-FTIR spectroscopy showed that proteins and polysaccharides were the dominant groups involved in adhesion of the genetically engineered cyanobacteria to the PBR materials.

Keywords: Biofouling, photobioreactors, cyanobacteria, adhesion mechanisms.

493485 Cdc42 Regulates Formation of Branches during Chemotactic Invasion of 3D Angiogenic Sprouting

Abstract:

Angiogenesis is a process where new blood vessels form from existing vasculature. Endothelial cells from capillaries respond to chemotactic angiogenic cues by extending filopodial protrusions, digesting the vascular basement membrane, and invading into the interstitial matrix as tip cells with following stalk cells. These multicellular sprouts eventually develop into mature vessels with branched architectures. During the process of forming branched vascular network, the cytoskeleton of the endothelial cells undergoes dynamic changes. Rho GTPase proteins are known to regulate actin cytoskeletal dynamics in cell migration during organ development and tissue morphogenesis. Among Rho GTPase proteins, Cdc42 has a conserved role in regulating actin cytoskeleton dynamics, filopodial extensions, cell polarity, and migration. However, its role in angiogenic sprouting, especially in the context of chemotactic migration in angiogenesis, remains elusive due to the lack of existing in vitro platforms to truly capture angiogenic invasion in 3D microenvironment.

Recently, we have developed an AngioChip, an *in vitro* model wherein endothelial cells lining a perfusable lumen can be induced with angiogenic factors to sprout and invade into the surrounding extracellular matrix [1]. In this model, many of the morphogenic features of sprouting angiogenesis are recapitulated, including tip cells, stalk cells, and multicellular branched networks. Here, using this system, we sought to investigate the effects of Cdc42 on the morphogenic processes of angiogenesis by using a small molecule inhibitor of Cdc42. Unlike other studies of tubulogenesis and Matrigelâ„¢ assays, where endothelial cell network formation occurred in a uniform distribution of biochemical stimuli, our system employed a biochemical gradient not only to stimulate the formation of multicellular sprout structures, but also to unveil the effects of Cdc42 in the context of chemotactic migration in angiogenic sprouting. We observed that Cdc42 mediates several aspects of the morphogenetic processes of angiogenesis. Partial abrogation of Cdc42 activity reduces formation of new branches but not on the length of branches. Disruption of Cdc42 activity also reduces the extent of collective cell migration and unexpectedly results in many more filopodial protrusions [2]. Together our 3D biomimetic model provides a unique opportunity to study the role of Cdc42 in branching angiogenesis. A concrete understanding of vessel branching and how it is regulated by Cdc42 will provide valuable insight into designing and engineering vascular networks to support functional tissue implants.

[1] **Nguyen, D.H.,** Stapleton, S.C., Yang, M.T., Cha, S.S., Choi, C.K., Galie, P.A., Chen, C.S. (2013). *Biomimetic model to reconstitute angiogenic sprouting morphogenesis in vitro*. **Proceedings of the National Academy of Sciences**. 110(17): 6712-6717.

[2] **Nguyen, D.H.,** Gao, L., Wong, A., Chen, C.S. (2017). *Cdc42 regulates branching in angiogenic sprouting in vitro*. **Microcirculation** (Epub ahead of print).

494221 Solid Surface Tension of Biomaterials Direct Cellular Behaviors through Integrins

Abstract:

Cells physically interrogate their extracellular environment to make decisions related to cell proliferation, differentiation, migration and other critical processes. In addition to biochemical properties, physical properties of the extracellular matrix, including its stiffness, are key regulators for cell behaviors. Typically, on stiff substrates, cells display large spreading areas, assemble robust integrin-based adhesion complexes, whereas on soft substrates these functions are suppressed. However, cell behaviors that defy expectations based on substrate rigidity alone have been observed.

Recent studies have illustrated that solid surface tension can have a dominant role in the mechanical behaviors of soft materials with vanishingly small elasticity. A central hypothesis underlying our proposed work is that surface tension can override directives from elasticity in controlling cell response. We have now demonstrated that cells interacting with soft materials with appreciable surface tension primarily sense and respond to surface tension and not the bulk elastic moduli of the materials. Our results are consistent with theory that predicts that solid surface tension can dominate over elasticity at cellular length scales. Even on very soft substrates (E=100 Pa), if the surface tension is appreciable, cells assemble robust adhesion complexes and cytoskeleton stress fibers, spread over large areas, upregulate canonical integrin-based signal transduction, proliferate and migrate efficiently. Remarkably, we find that cells can even spread and migrate on liquid biomaterials that have high surface tension. On these materials, cells migrate through a highly collective, swarming-type motility that is significantly faster than migration on solid biomaterials. Together, our results indicate that material surface tension is important criteria for the design of biomaterials for cell growth and morphogenesis in tissue engineering.

497708 Migration Against the Direction of Flow Is LFA-1 Dependent in Human Hematopoietic Stem and Progenitor Cells

Abstract:

The recruitment of immune cells during inflammation is regulated by a multi-step, leukocyte adhesion cascade of cell rolling, activation, adhesion, and transmigration through the endothelial barrier. Similarly, Hematopoietic Stem and Progenitor Cells (HSPCs) use this pathway to migrate and home to the bone marrow in order to restart hematopoiesis post-transplantation. After selectin-mediated braking and slow rolling, HSPCs migrate along the vascular endothelium on a variety of adhesion ligands including Intracellular Adhesion Molecule-1 (ICAM-1), Vascular Cell Adhesion Molecule-1 (VCAM-1) or Mucosal Addressin Cellular Adhesion Molecule-1 (MAdCAM-1). Previous work by ourselves (1) and others (2) has demonstrated an interesting phenomena where both murine and human T lymphocytes, which have a remarkably similar integrin profile as HSPCs, can migrate efficiently upstream (against the direction of shear-flow) on ICAM-1 surfaces while migrating downstream (with the direction of shear flow) on VCAM-1. Other immune cell types, such as neutrophils, do not show this behavior.

Here we describe for the first time how both the KG1a immortalized stem cell line and primary bone marrow $CD34^+$ HSPCs can migrate against the direction of fluid flow on surfaces coated with cell adhesion molecules (CAMs). Using a laminar flow chamber, we demonstrate that KG1a cells and primary HSPCs migrate upstream on ICAM-1, downstream on VCAM-1, and both upstream and downstream on MAdCAM-1. Also, we demonstrate that KG1a cells and HSPCs display upstream migration on mixed surfaces of multiple CAMs and surfaces containing both E-selectin and CAMs, much like the activated endothelial surface. Furthermore, both KG1a and Primary HSPCs migrate upstream on activated HUVEC monolayers at shear rates seen in the vasculature. Blocking with monoclonal antibodies, we show that Lymphocyte Function-associated Antigen-1 (LFA-1) is the sole cell-borne integrin responsible for upstream migration along the endothelium. Signaling events downstream of LFA-1 (Î±$_L$Î²$_2$) engagement are critical to this mode of motility and furthermore, the localization of activated LFA-1 and the F-actin network are fundamentally different during upstream as compared to downstream migration. In all, this upstream motility has direct implications in understanding of how HSPCs traffic during bone marrow transplants and holds potential in having HSPCs better home to the marrow.

REFERENCES:

1. Dominguez GA, Anderson NR, & Hammer DA (2015) The direction of migration of T-lymphocytes under flow depends upon which adhesion receptors are engaged(). *Integrative biology : quantitative biosciences from nano to macro* 7(3):345-355.

2. Valignat M-P, Theodoly O, Gucciardi A, Hogg N, & Lellouch AnnemarieÂ C (2013) T Lymphocytes Orient against the Direction of Fluid Flow during LFA-1-Mediated Migration. *Biophysical Journal* 104(2):322-331.

498089 Patterned Cell Alignment in Response to Macroscale Curvature and Rho Activation

Abstract:

The formation of spatial behavior patterns in tissues is a long-standing problem in biology. Decades of research have focused on understanding how biochemical signaling and morphogen gradients establish cell patterns during development and tissue morphogenesis. Here, we show that geometric cues on the order of a cell length scale can drive organization and pattern formation. We find that mouse embryonic fibroblasts and human vascular smooth muscle cells sense curvature differently when in monolayers than when isolated on surfaces with various amounts of Gaussian curvature. While the long, apical stress fibers within these cells align in the direction of minimum curvature on cylindrical substrates, a subpopulation of stress fibers beneath the nucleus aligns in the circumferential direction. We find that the actin cytoskeleton is dramatically reorganized on curved substrates upon activation of Rho, which is associated with increased contractility of the stress fibers. Thus, stress fiber alignment is likely a result of a complex balance between energy penalties associated with stress fiber bending, contractility, and the dynamics of F-actin assembly. With a deeper understanding of how cells sense and respond to macroscale curvature, we may be able to use these geometric cues to guide cell behaviors such as migration and contraction.

501976 How Raw Material Attributes and Process Parameters Affect the Global Residence Time Distribution Behavior in Continuous Direct Compression

Abstract:

Title: *How Raw Material Attributes and Process Parameters Affect the Global Residence Time Distribution Behavior in Continuous Direct Compression*

Author: Samantha Hurley

The pursuit of drug product continuous manufacturing will revolutionize the pharmaceutical industry by improving product flow in our supply chains, such as leveraging fast cycle times and flexible batch sizes, and by reducing lead times and inventory levels. A successful batch to continuous manufacturing conversion requires a very robust control scheme. One commonly described control scheme involves the use of NIR spectroscopy for blend composition monitoring. This study evaluates the use of an alternate method, a Residence Time Distribution (RTD) based process model for control of the continuous direct compression (CDC) process without direct measurement of blend uniformity. In some cases, this method is preferred over traditional spectroscopic techniques as the primary control across a wide range of throughputs, blender speeds, and material properties.

The CDC system studied consists of two continuous convection blenders in series followed by a tablet press. Each of the blenders and the tablet press are modeled as a Plug Flow Reactor (PFR) in series with two Continuously Stirred Tank Reactors (CSTRs). Since each PFR and CSTR is characterized by a single time constant, t_{Lag} and t respectively, each blending unit can be described by threeparameters. Tracer experiments were conducted by spiking API into either the inlet of blender 1 or blender 2. The effluent API concentration was measured either by blend NIR spectroscopy at the exit of blender 2 or tablet HPLC at the exit of the tablet press. The RTD model parameters were optimized by using a least squares algorithm between the blend NIR/tablet HPLC and the RTD model, previously described. The model parameters were developed across various processing conditions and material properties (e.g., bulk densities and flow indices).

The results from the tracer experiments performed show that a ratio of $t{:}t_{Lag}$ is constant within each blending unit across all the various conditions. This universal behavior thus allows the development of a simple empirical relationship which describes RTD model parameters across the entirety of the design space. The results indicate that the relationship between the blender 1 and blender 2 model parameters versus throughput (TP) and blender speed (BS) can be described as

where A, B, C, and D are constants. The results also indicate the relationship between the tablet press model parameters versus throughput can be described as

where A, B, and C are constants. Knowledge of this universal behavior allows interpolation of model parameters anywhere within the design space, enables simpler model validation, and significantly simplifies process model development for future products.

502628 Development of a Continuous Blending System

Abstract:

Continuous processing equipment for solid dosage manufacture at a pilot scale can be costly and consume more space than is typical for pilot plants. Merck developed its own custom continuous blending system to fit within limited space within its clinical pilot plant. This discussion will walk through the design process, key learnings, challenges and development experiences as the pilot plant team designed a system to fit inside an existing pilot plant space.

In this discussion, the team will use experiences, system design and early data to show how a clinical continuous blending system can be designed to mimic the commercial process while meeting vertical and footprint limitations.

500878 Investigating Continuous Powder Blending at Different Scales Using Residence Time Distribution Studies

Abstract:

Powder mixing or blending is an essential unit operation required to achieve the homogenization of excipients (fillers, disintegrates, lubricants, etc.) and active pharmaceutical ingredient(s). Pioneering developments in continuous drug product manufacturing have facilitated the adoption of continuous powder blending platforms.[1-3] The residence time distribution (RTD) is a probability function that describes the amount of time that a fraction of material spends within any continuous processing system. The RTD is a critical tool in continuous processing equipment and rig design, in pharmaceutical process understand and is also used to ensure the quality of the manufactured product. [4]

In this work, continuous powder lubrication performance was compared using continuous dry powder blenders at both lab scale and production scale by varying total mass flow rate and blender blade speed with a view to translating the lubrication performance between blender systems. At steady state, the RTD and residence mass hold-up in the blender were measured. The tensile strength of tablets compacted from the blend at steady state were also measured in order to characterize the lubrication response.[5] Three different regimes of blending were observed with increasing blade speed – 'mixing', 'transitioning' followed by 'conveying'. Peclet number (Pe# - ratio of axial transport to axial dispersion) was determined by fitting RTD curves using the axial dispersion model. A slower line rate was required at the lab scale in order to achieve the same Pe# at production scale. The mean residence time, τ, was found to be the key predictor of lubrication performance, independent of blender scale. For both blender systems, τ decreases with increasing mass flow rate and increasing blade speed and, where blender speed was increased, changes in τ were minimized by selecting higher mass flow rates. Longer τ produced greater variation and skewness of the RTD across each blending system which has implications for material tracking and product yield within any continuous manufacturing process. Significantly, we show that the mean residence time, τ can be calculated directly from knowledge of the residence mass hold-up and mass flow rate for both systems.

References

1. Byrn, S. *et al.* Achieving Continuous Manufacturing for Final Dosage Formation: Challenges and How to Meet Them May 20–21 2014 Continuous Manufacturing Symposium. *Journal of Pharmaceutical Sciences* **104**, 792-802, doi:http://dx.doi.org/10.1002/jps.24247 (2015).

2. Teżyk, M., Milanowski, B., Ernst, A. & Lulek, J. Recent progress in continuous and semi-continuous processing of solid oral dosage forms: a review. *Drug Development and Industrial Pharmacy* **42**, 1195-1214, doi:10.3109/03639045.2015.1122607 (2016).

3. Gao, Y., Vanarase, A., Muzzio, F. & Ierapetritou, M. Characterizing continuous powder mixing using residence time distribution. *Chemical Engineering Science* **66**, 417-425,

doi:http://dx.doi.org/10.1016/j.ces.2010.10.045 (2011).

4. Engisch, W. & Muzzio, F. Using Residence Time Distributions (RTDs) to Address the Traceability of Raw Materials in Continuous Pharmaceutical Manufacturing. *Journal of Pharmaceutical Innovation* **11**, 64-81, doi:10.1007/s12247-015-9238-1 (2016).

5. Kushner Iv, J. & Moore, F. Scale-up model describing the impact of lubrication on tablet tensile strength. *International Journal of Pharmaceutics* **399**, 19-30, doi:http://dx.doi.org/10.1016/j.ijpharm.2010.07.033 (2010).

491220 Leveraging Raw Material and Blend Properties with the Die Filling Step of the Tableting Process

Abstract:

Leveraging raw material and blend properties with the die filling step of the tableting process

V. Vanhoorne[a*], B. Van Snick[a*], W. Grymonpré[a], J. Dhondt[c], C. Segers[a], K. Van Vooren[a], T. Eerdekens[a], G. Di Pretoro[b], T. De Beer[c], C. Vervaet[a]

[a]*Laboratory of Pharmaceutical Technology, Ghent University, Ottergemsesteenweg 460, B-9000 Ghent, Belgium*

[b]*Pharmaceutical Research and Development, Division of Janssen Pharmaceutica, Johnson & Johnson, Turnhoutseweg 30, B-2340 Beerse, Belgium*

[c]*Laboratory of Pharmaceutical Process Analytical Technology, Ghent University, Ottergemsesteenweg 460, B-9000 Ghent, Belgium*

**Both authors contributed equally*

Tablets are the most widely used dosage forms for oral drug delivery, mainly due to their ease of manufacturing, accurate dosing and high patient compliance. Today, tablets still account for more than 80 % of all pharmaceutical preparations. Despite the high quantities of tablets produced daily, a lot of the established processes and formulations are the results of trial-and-error experiments. Knowledge and understanding of the influence of process parameters and formulation properties on the final product quality is of great importance for the industry, offering economic efficiency and high product quality.

Therefore, the aim of this study is to investigate the influence of a broad range of formulation properties and process parameters related to the feed frame on the tablet properties and tablet weight variation. Based on characterization of a wide range of fillers for direct compression (13 materials) and active pharmaceutical ingredients (API) (14), thirty divergent blends were composed and evaluated on the tablet press. Characterization of the individual materials included size and shape analysis (laser diffraction, static image analysis, scanning electron microscopy), specific surface analysis, tapped, bulk and true density, flowability (drained angle of repose, dynamic angle of repose, flowability index, basic flowability energy, specific energy, aeration index, permeability), compressibility, charge density, moisture content (loss-on-drying) and dynamic vapor sorption. Based on PCA analysis of the fillers and API's, 5 fillers (Pearlitol 100SD, Emcompress, Avicel PH101, Tablettose 80, Avicel PH200) and 6 API's (Caffeine, Metoprolol tartrate, Intelence, micronized, powdered and dense powder grades of Paracetamol) were selected to be combined

into 30 blends (API/filler/magnesium stearate 9.93/89.33/0.75) in order to cover a broad range of formulations. Additionally, these blends were characterized. An experimental design with 6 runs (including 2 centerpoints) varying paddle speed (paddle 1: 15 – 100 rpm, paddle 2: 18 – 120 rpm) and turret speed (10 – 100 rpm) was performed on an instrumented rotary tablet press (Modul P, GEA Pharma systems). Experiments were performed at a fixed main compression force (15 kN), with a fixed pre-compression displacement (0.5 mm) and overfill (2 mm). Steady state of the tablet press was determined by evaluation of the tablet mass flow signal on a catch scale. After reaching steady state, the press was run for 3, 6 or 30 min at turret speeds of 10, 55 and 100 rpm, respectively. Samples were taken during 8, 2 and 2 s at intervals of 100, 20 and 10 s at turret speeds of 10, 55 and 100 rpm, respectively. Tablets of the individual samples were weighed and the pooled mean and pooled relative standard deviation was calculated. Additionally, the residence mass in the feed frame (RMF) was determined after each run and the bottom punch position to obtain a main compression force of 15 kN (BPMC) and ejection force were recorded. Multivariate data analysis was applied to investigate the relation between the process parameters, blend characteristics, product and process responses.

Blends with metoprolol tartrate as API and Pearlitol 100 SD or Emcompress as filler tended to stick on the punches. This behavior could be linked to the high wall friction value of metoprolol. Cellulose-based fillers did not exhibited this behavior because of their self-lubricating behavior. Additionally, high ejection forces were observed using Pearlitol SD100 as filler, resulting in picking and capping in combination with paracetamol.

Overall, the density of the blend impacted the mean tablet weight and residence mass in the feed frame. In contrast, the tablet weight variability could be related to the flowability of the blend expressed by the flow function coefficient obtained with the ring shear test. For the formulations with a well flowing, high density filler (e.g. Emcompress) no effect or a minor effect of the paddle speed and turret speed was detected on the tablet weight, tablet weight variability, RMF and BPMC. In contrast when a poorly flowing, low density filler (e.g. Avicel PH101) was used in the formulation, the tablet weight, RMF and BPMC were negatively and positively affected using a higher turret speed and paddle speed, respectively. These effects were most pronounced when the poorly flowing, low density filler was combined with an API with similar properties (e.g. Metoprolol tartrate, micronized Paracetamol). This signifies that die filling is incomplete and that the mass in the feed frame is low when poorly flowing, low density formulations are tableted. However, interestingly, the variability of the tablet weight of these formulations was affected by the turret speed or paddle speed. Higher paddle and turret speeds resulted in a reduction and increase of the tablet weight variability, respectively. The tablet weight variability of a very cohesive formulation, e.g. micronized paracetamol combined with Avicel PH101 (ffc=2.55), varied from 0.7 – 7.9% depending on the turret and paddle speed. Operating the tablet press at the highest paddle speed and lowest turret speed resulted in the lowest tablet weight variability, whereas operating the tablet press at the lowest paddle speed and highest turret speed resulted in the highest tablet weight variability. Intermediate settings of turret and paddle speed (center point conditions) resulted in a tablet weight variability of 3.6%. Therefore, it could be concluded that even poorly flowing formulations with a low density (e.g. consisting of a filler intended for wet granulation and a micronized API) could be tableted at high speed on a rotary tablet press, yielding acceptable tablet weight variability when the paddle speed is optimized.

The developed characterization database will be further expanded with more materials and additional material characteristics (e.g. compression properties as elastic recovery, plasticity constant, fragmentation) and a PLS model will be developed to link the material characteristics to process settings and responses of the tableting process. This way the model will be a valuable tool to compose formulations with new APIs during drug product development.

498979 Modelling of Pharmaceutical HME Processes

Abstract:

Modelling of Pharmaceutical HME Processes

Josip Matić*,**, Milica Stanković-Brandl**, Amrit Paudel**,Christophe Herkenne***, Jessica Lovey Martinetti***, Sophie Martel***, Johannes Khinast*,**

*Institute for Process and Particle Engineering, Graz University of Technology, Inffeldgasse 13/III, 8010 Graz, Austria

**Research Centre Pharmaceutical Engineering GmbH, Inffeldgasse 13/III, 8010 Graz, Austria

***Debiopharm Research&Manufacturing S.A., Campus "après-demain", Rue Levant 146, CP 368, 1920 Martigny, Switzerland

Email for correspondence: khinast@tugraz.at

Introduction

Hot melt extrusion (HME) is a continuous manufacturing process primarily using co-rotating intermeshing twin-screw extruders (TSE). The continuous manufacturing nature of the process allows for a steady product quality and a decrease in production costs. Furthermore, the process has the potential of increasing the bioavailability of poorly soluble drugs by enabling manufacturing of amorphous solid dispersions, making it interesting to the pharmaceutical industry. Although the process has been used in the polymer and rubber industry for years it is still not well enough understood and fairly new in the pharmaceutical industry. The HME setup for new API-polymer system and the technical transfer between different extruder scales (scale-up) are still open question.

Methodology

HME process simulations are done using fully resolved 3D simulations, or by simplified 1D simulation approaches [1-4]. The 1D mechanistic model developer in-house is able to simulate the HME process as a whole, giving information about the filling degree, pressure and temperature distribution, as well as the specific mechanical energy input (SMEC) across the screw configuration[1]. In order to setup the 1D model, the individual screw elements have to be characterized by determining the inherent conveying capacity and the pressure build up capacity[2]. The characterization is done using the mesh-free SPH simulation

approach, giving additional insight into the mixing and sheer rate distribution for every screw element, allowing for a comparison between triple- and double-flighted extruder screw element geometries [3].

Results

An HME process on a triple-flighted 16mm extruder, with two different materials, was investigated, providing recommendations for the process setup and scale-up. The validation was done by comparing the experimentally obtained torque; melt temperature and residence time distribution to the *in-silico* predictions.

Literature

[1] A. Eitzlmayr, G. Koscher, G. Reynolds, Z. Huang, J. Booth, P. Shering, and J. Khinast, "Mechanistic Modeling of Modular Co-Rotating Twin-Screw Extruders," *Int. J. Pharm.*, vol. 474, no. 1–2, pp. 157–176, 2014.

[2] K. Kohlgrüber, *Co-Rotating Twin-Screw Extruders*. Munich: Carl Hanser Publishers, 2008.

[3] A. Eitzlmayr, J. Matić, and J. Khinast, "Analysis of Flow and Mixing in Screw Elements of Co-Rotating Twin-Screw Extruders via SPH," *AIChe J.*

[4] R. Baumgartner, J. Matić, S. Schrank, S. Laske, J. G. Khinast, and E. Roblegg, "NANEX: Process Design and Optimization," *Internaitonal J. Pharm.*, vol. 506, pp. 35–45, 2016.

492270 In-Depth Experimental Analysis of Pharmaceutical Twin-Screw Wet Granulation in View of Detailed Process Understanding

Abstract:

Twin-screw wet granulation is gaining increasing interest within the pharmaceutical industry for the continuous manufacturing of solid oral dosage forms. However, limited prior fundamental physical understanding has been generated relating to the granule formation mechanisms and kinetics along the internal compartmental length of a twin-screw granulator barrel, and about how process settings, barrel screw configuration and formulation properties such as particle size, density and surface properties influence these mechanisms. One of the main reasons for this limited understanding is that experimental data is generally only collected at the exit of the twin-screw granulator barrel although the granule formation occurs spatially along the internal length of the barrel. The purpose of this study is to analyze the twin-screw wet granulation process using both hydrophilic and hydrophobic formulations, manufactured under different process settings such as liquid-to-solid ratio, mass throughput and screw speed, in such a way that the mechanisms occurring in the individual granulator barrel compartments (i.e., the wetting and different conveying and kneading compartments) and their impact upon granule formation are understood. To achieve this, a unique experimental setup was developed allowing granule characteristic data-collection such as size, shape, liquid and porosity distribution at the different compartments along the length of the granulator barrel. Moreover, granule characteristic information per granule size class was determined. The experimental results indicated that liquid-to-solid ratio is the most important factor dictating the formation of the granules and their corresponding properties, by regulating the degree of aggregation and breakage in the different compartments along the internal length of the twin-screw granulator barrel. Collecting appropriate and detailed experimental data about granule formation along the internal length of the granulator barrel is thus crucial for gaining fundamental physical understanding of the twin-screw wet granulation process.

496103 Understanding the Impact of Feedframe Design and Process Parameters on the Die-Filling Step of the Tableting Process

Abstract:

Powder compression on rotary tablet presses is widespread in various industries and for pharmaceuticals over 80 % of all oral dosage formulations are manufactured as tablets. The ease of tablet manufacturing combined with a high economic efficiency and patient compliance all contribute to the prominent role of tablets as pharmaceutical oral dosage form. With the concept of continuous processing gaining momentum in pharmaceutical manufacturing, a thorough understanding of how process and formulation parameters impact the critical quality attributes (CQA) of the end product is more than ever required. Regarding tableting processes, consistent die-filling is a crucial control variable since this will impact the final product quality (e.g. drug content and tablet weight). Although a wide range of feedframe designs is available in terms of paddle-type, chamber height, powder-flow 'enhancers' and feedframe aeration, there is limited knowledge on which specific feedframe design is the most appropriate for a specific formulation in order to obtain consistent die filling during tabletting.

The aim of this study is to investigate the impact of feedframe design and tableting process parameters on the die-filling step and other critical quality attributes (e.g. tablet tensile strength). The knowledge gained via this study will contribute to a rational selection of the feedframe design and tabletting settings in function of the formulation characteristics. Different paddle-types were selected based on their prevalence in the pharmaceutical industry and these were implemented in a standard feedframe consisting of 2 paddle wheels, whereby the feeding wheel had 8 paddles and the recovery wheel 12 paddles. The following 4 feedframe designs were evaluated: curved paddles, straight paddles, thin cylindrical paddles with varying height in the paddle wheel and curved paddle-wheels with lower thickness and a 'scraper', together with the possibility of aeration (present or not). An experimental design with 48 runs including 4x2 centerpoints for each paddle-type was conducted on a fully instrumented rotary tablet press (Modul P, GEA Pharma systems) with an additional fill-level sensor mounted on the feeding-tube which ensured a constant powder volume above the feedframe throughout all experiments. Paddle speed (15 - 100 rpm), the ratio of paddle speed 1 (i.e. feeding wheel) over paddle speed 2 (i.e. recovery wheel)(1-1.4), turret speed (10-100 rpm) and overfill-level (0-4 mm) were varied in the DOE as process parameters.

Experiments were performed at a bottom punch position generating a main compression force (15 kN), with a fixed pre-compression displacement (0.5 mm). Steady state of the tablet press was determined by monitoring the tablet mass flow signal on a catch scale. Each experimental run was performed on an empty and cleaned feedframe. After reaching steady state, tablets were sampled at different time points. The sampling duration (required to sample 20 tablets per time point) and sampling interval depended on the turret speed: a 20s sampling duration with an 80s interval at 10 rpm compared to a sampling duration of 2s with an 8s interval at 100 rpm. Tablets were analysed on their weight, tablet dimensions and hardness using a combitester (Pharmatron Smarttest 50), and the pooled mean and pooled relative standard deviation were calculated for tablet weight, tablet tensile strength and tablet porosity as responses. Additionally, the time needed to reach steady state conditions for each experimental run was calculated and the displacement signals on the pre-compression were monitored to evaluate fluctuations in the tablet weight variation. With respect to the latter, main compression forces were also recorded. Based on a PCA analysis of fillers and API's, microcrystalline cellulose (Vivapur 101) was initially selected as poorly flowing formulation for performing the experiments.

Overall, feedframe design affected the die filling and consequently the mean tablet weight: significantly higher tablet weights were obtained using straight paddles, while the feedframe with 'scraper' and lowered curved paddles resulted in significantly lower tablet weights. At some process settings with the latter paddle design, feedframe obstruction occurred due to intensive compaction inside the feedframe. However, no significant influence on the tablet weight variability was detected, indicating that the consistency of die filling seemed guaranteed for each paddle type, specifically since no interaction was observed between paddle type and overfill-level for RSD on tablet weight. The process parameters strongly impacted the tablet weight and weight variability: (a) higher turret speeds resulted in significantly lower mean tablet weight (i.e. less die-filling) and a higher tablet weight variability, (b) more overfill resulted in a higher tablet weight without an effect on tablet weight variability, (c) paddle speed 1 had a similar impact as the overfill-level on the mean tablet weight at higher paddle speeds, yielding a higher tablet weights and lower variability indicative of a more consistent die-filling at higher paddle speed 1 (i.e. feeding wheel) for this formulation. Similar trends were observed when investigating the tablet tensile strengths at each experimental run for this formulation.

The developed experimental design will be repeated with other powder formulations with varying density and flow behavior in order to investigate the robustness of this study towards other formulations. As a result, this research study will contribute to an improved approach for the set-up of future experimental tableting studies.

495257 Constructing Predictive Kinetic Models of Metabolism with Transcriptional Regulation

Abstract:

In this study, we describe the development of metabolic models using the Ensemble Modeling (EM) paradigm coupled with a description of transcriptional regulation for two organisms: 1) *C. thermocellum*, a cellulolytic microbe and 2) *E. coli*. For *C. thermocellum*, we constructed a core kinetic model k-ctherm118 [1], parameterized by simultaneously imposing fermentation yield data in lactate, malate, acetate and hydrogen production pathways for 19 measured metabolites spanning a library of 19 distinct single and multiple gene knockout mutants along with 18 intracellular metabolite concentration data for a *Î"gldh* mutant and ten experimentally measured Michaelis-Menten kinetic parameters. k-ctherm118 captures significant metabolic changes caused by (i) nitrogen limitation leading to increased yields for lactate, pyruvate and amino acids, and (ii) ethanol stress causing an increase in intracellular ammonia and sugar phosphate concentrations due to upregulation of cofactor pools. Robustness analysis of k-ctherm118 alludes to the presence of a secondary activity of ketol-acid reductoisomerase and possible regulation by valine and/or leucine pool levels. Overall, the *C. thermocellum* case study demonstrated that the developed kinetic model k-ctherm118 predicts phenotypes under genetic perturbations with a higher degree of accuracy than stoichiometric model as well as provides insight into missing metabolic pathways and regulations. Kinetic models learn metabolic redirections through careful parameterization that aims to recapitulate metabolic responses seen in multiple flux datasets with the aid of ^{13}C-MFA data [2]. A recent study [3] provides a comprehensive metabolic flux characterization of wild-type *E. coli* and 22 knockouts of enzymes in the upper part of central carbon metabolism, including glycolysis, pentose phosphate pathway and ED pathway. We apply these flux datasets and the EM paradigm to construct a genome-scale kinetic model of *E. coli* containing 787 model reactions, 674 metabolites and 295 substrate-level regulatory interactions.

Kinetic models simulate genetic perturbations by modifying specific enzyme levels *a priori* based on the mutant genotype information, which in turn modulates the concentrations of metabolites that the enzyme acts upon. However, the enzyme levels of an organism are also regulated by the available transcriptional machinery (global regulation) as well as by the changes in concentration levels of regulatory metabolites (specific regulation) under any given genetic or environmental perturbation. Thus, we expand the scope of the developed kinetic models by incorporating transcriptional regulatory layer which refines enzyme levels based on a linear combination of log-normalized changes in growth rate (global) and select intracellular metabolite pool (specific) levels. We use transcriptomic datasets that provide genome-wide collections of mRNA levels under genetic and environmental perturbations to estimate the transcriptional regulatory model parameters and validate their statistical significance. The developed transcriptional regulatory model captures the impact of perturbations in key metabolite pools on transcript levels of 85 enzymes spanning glycolysis, amino acid and nucleotide metabolism pathways. Ultimately, the transcriptional regulation integrated kinetic model captures the impact of enzyme level changes on metabolite concentrations as well as the regulation of enzyme levels by metabolite concentration changes increasing prediction fidelity.

1. Dash S, Khodayari A, Zhou J, Holwerda EK, Olson DG, Lynd LR, Maranas CD: **Development of a core Clostridium thermocellum kinetic metabolic model consistent with multiple genetic perturbations.** *Biotechnol Biofuels* accepted.

2. Khodayari A, Zomorrodi AR, Liao JC, Maranas CD: **A kinetic model of Escherichia coli core metabolism satisfying multiple sets of mutant flux data.** *Metabolic Engineering* 2014, **25:**50-62.

3. Long CP, Gonzalez JE, Sandoval NR, Antoniewicz MR: **Characterization of physiological responses to 22 gene knockouts in Escherichia coli central carbon metabolism.** *Metab Eng* 2016, **37:**102-113.

499252 Scalable and Efficient Bayesian Metabolic Modeling with Linear-Logarithmic Kinetics

Abstract:

Mathematical modeling of metabolic networks has shown significant promise in determining computational strain designs and improving bioprocess economics. Strain design methods based on flux balance analysis (FBA) quantify how metabolic networks can feasibly operate, and can determine optimal genetic interventions that seek to constrain organisms to a desirable metabolic state. These methods are most effective at predicting which reactions to knock out, since predicting the effects of enzyme overexpressions typically requires a kinetic description of the involved reactions. Metabolic ensemble modeling (1) has emerged in recent years as an effective tool for estimating confidence intervals in these kinetic descriptions from observable steady-state flux and concentration measurements. The method operates by sampling feasible kinetic parameters and filtering to sets of parameters that match experimental observations. There are therefore two computational bottlenecks that hinder an accurate determination of parameter identifiability given observed experimental results:

- Repeatedly solving the set of coupled ordinary differential equations for expected steady-state fluxes and concentrations following a change in enzyme expression, and
- the efficient sampling of parameter values that give rise to behavior consistent with experimental observations.

Recently, Saa & Nielsen (2) demonstrated that metabolic ensemble modeling could be understood through a Bayesian context, which allows Monte Carlo Markov chain samplers to reduce the number of parameter draws required to find those consistent with experimental observations. However, the slow integration of ODEs and lack of derivative information for the likelihood function continues to limit the scalability of the method to large data sets.

Here, we demonstrate how coupling linear-logarithmic kinetics for reaction flux to a metabolic ensemble modeling framework solves the two computational bottlenecks described above. The resulting method therefore represents a scalable, flexible framework for the integration of multiple '-omics' datasets to predict flux control coefficients. Specifically, linear-logarithmic kinetics enable steady-state fluxes to be predicted linearly from kinetic parameters (3), removing the computational burden associated with solving for steady-state flux. Additionally, since linear solutions permit the easy determination of likelihood gradients, advanced Bayesian inference techniques such as Hamiltonian Monte Carlo and automatic differentiation variational inference can be applied to reliably estimate posterior parameter distributions even for high-dimensional models. We demonstrate the method through a number of case studies, including medium-sized metabolic models (with hundreds of metabolites and reactions) to simple *in vitro* reaction systems with novel experimental data.

1. L. M. Tran, M. L. Rizk, J. C. Liao, *Biophysical Journal*. **95**, 5606–5617 (2008).
2. P. A. Saa, L. K. Nielsen, *Sci. Rep.*, 1–13 (2016). (DOI: 10.1038/srep29635)

3. L. Wu, W. Wang, W. A. van Winden, W. M. van Gulik, J. J. Heijnen, *European Journal of Biochemistry*. **271**, 3348–3359 (2004).

500382 The *E. coli* Proteome and Metallome Under Oxidative Stress

Abstract:

How tolerant a cell is to oxidative stress is a function of many complex and interacting mechanisms. We reconstructed a multiscale constraint-based model (stressME) of metabolic, proteome, and metallome responses to oxidative stress in *Escherichia coli*. Oxidative stress induced a global metabolic shift from respiration toward respiro-fermentation. The global shifts were mediated by the sigma factor RpoS, which was correctly predicted to be up-regulated. Oxidative stress decreased the overall translational efficiency due to the mismetallation of peptide deformylase (Def). The model predicted a compensatory up-regulation of Def, which was confirmed by RNA-Seq. Oxidative stress damaged iron-sulfur clusters in enzymes catalyzing amino acid biosynthesis. Auxotrophies were correctly predicted for aromatic, branched-chain, and sulfur-containing amino acids under oxidative stress. We further identified 32 stress-associated genes, which were more conserved across bacteria and archaea. We then simulated macrophage-mediated killing of *E. coli* through the interaction of Zn(II) overloading and reactive oxygen species (ROS). We found that the combination of ROS and a constraint on translational efficiency sensitized *E. coli* to Zn(II) overloading. We expect stressME to be useful for investigating the interplay between reactive oxygen species, metal homeostasis, and cell physiology in the context of disease and metabolic engineering.

497686 Modeling Stochasticity in the Cell Cycle

Abstract:

Stochasticity is endemic to biology. Biological systems are intrinsically noisy and heterogeneous at all levels, from the molecular to the population scale. On the cell-to-cell level, differences typically arise due to a randomness in reaction events [1]. The intrinsic noise in biological systems present an element of challenge for models attempting to emulate them appropriately. Although deterministic models are often the preferred method due to the ease of integration, they sometimes fail to represent biological processes which are susceptible to noise in a realistic manner [2]. Erratic influences present in nonlinear dynamic systems are sometimes so significant that they dictate the state and stability of that system [3]. This phenomenon, termed stochastic bifurcation, occurs when the level of noise drives the system from one steady state to another [4]. The meaning of these states can be polar opposite; in the context of inflammation for example, Reynolds *et al.* report that the level of noise can drive the system from a healthy state to that of asceptic death [5]. As the mathematical representation of biological models grows in complexity, the need to extend the capabilities of such models to capture stochastic behavior and to better understand noise-induced phenotypes is becoming increasingly apparent.

As with all biological responses, stochasticity occurs naturally in the cell cycle, the biological process regulating the rate and extent to which a population of cells proliferate. The cell cycle is generally considered to have four distinctive phases: the S phase, when DNA replication occurs, the mitotic (M) phase, and the G1 and G2 growth phases. Non-dividing cells exist in a quiescent state (G0) [6]. As a consequence of the outstanding work by Tyson, Novak, and Goldbeter, models have been developed that describe the key mechanisms regulating the cell cycle [7-9]. These models, however, remain deterministic in nature. While some simulations have incorporated stochastic components into the modeling network [10], a more detailed exploration of the impact of stochasticity on cell cycle progression is required.

This work aims to further examine the effects of noise on modulating the cell cycle by considering three main hypotheses. The first consideration explores the source of stochasticity. Which portion(s) of the cell cycle exhibit stochasticity? Does the noise originate from the microenvironment, such as variability in a growth factor, or does it arise out of the complex interactions among the cyclin/cyclin-dependent kinase (CDKs) network? Does noise propagate from one phase of the cycle to the next? Stemming from these questions, the sensitivity of the network must also be considered: are certain variables or parameters more robust to stochasticity? This analysis could provide information about whether cells have more rigorous control and regulation of certain aspects of the cell cycle. If the focus of the cell cycle is to ensure proper replication of genetic material [7], then perhaps one can hypothesize that phases involved in DNA replication are less tolerant to stochastic influences. Lastly, this work examines the time-dependence of noise on the cell cycle. The time at which noise is introduced has been shown to influence the state of dynamic systems if the system has multiple stable (or unstable) states.

Understanding the core regulatory mechanisms of the cell cycle and the role played by intrinsic stochasticity provides interesting therapeutic opportunities. The deregulation of the cell cycle has been affiliated with numerous pathologies,

such as cancer, where aberrant cell division occurs [11]. Guided by a more detailed insight of the underlying influences of stochasticity, drugs could be more appropriately designed to either restore a disrupted cell cycle or hinder the proliferation of malignant cells.

1. Kim, K.H. and H.M. Sauro, *In search of noise-induced bimodality*. BMC Biol, 2012. **10**: p. 89.

2. Szekely, T., Jr. and K. Burrage, *Stochastic simulation in systems biology*. Comput Struct Biotechnol J, 2014. **12**(20-21): p. 14-25.

3. Ditlevsen, S. and A. Samson, *Stochastic Biomathematical Models*, ed. M. Bachar. 2013.

4. Namachchivaya, N.S., *Stochastic bifurcation*. Applied Mathematics and Computation, 1990. **38**(2): p. 101-159.

5. Reynolds, A., et al., *A reduced mathematical model of the acute inflammatory response: I. Derivation of model and analysis of anti-inflammation*. J Theor Biol, 2006. **242**(1): p. 220-36.

6. Feillet, C., et al., *Coupling between the Circadian Clock and Cell Cycle Oscillators: Implication for Healthy Cells and Malignant Growth*. Frontiers in Neurology, 2015. **6**: p. 96.

7. Kar, S., et al., *Exploring the roles of noise in the eukaryotic cell cycle*. Proc Natl Acad Sci U S A, 2009. **106**(16): p. 6471-6.

8. Gerard, C. and A. Goldbeter, *Dynamics of the mammalian cell cycle in physiological and pathological conditions*. Wiley Interdiscip Rev Syst Biol Med, 2016. **8**(2): p. 140-56.

9. Novak, B. and J.J. Tyson, *Numerical analysis of a comprehensive model of M-phase control in Xenopus oocyte extracts and intact embryos*. J Cell Sci, 1993. **106 (Pt 4)**: p. 1153-68.

10. Barik, D., et al., *A Stochastic Model of the Yeast Cell Cycle Reveals Roles for Feedback Regulation in Limiting Cellular Variability*. PLoS Comput Biol, 2016. **12**(12): p. e1005230.

11. Chow, A.Y., *Cell cycle control by oncogenes and tumor suppressors: driving the transformaion of normal cells into cancerous cells*. Nature Education, 2010. **3**(9).

496784 Application of Machine Learning and Active Learning to Enhance Chemical Yields in Microbes

Abstract:

Microbes can be used to produce a wide variety of important chemicals from low cost substrates. The biochemical production capabilities of microbes can be improved through metabolic engineering, where metabolic and regulatory processes are adjusted using genetic manipulations. Efforts have been made to develop computational methods to study metabolic and regulatory networks of microbes and identify the genetic interventions needed to enhance the production of the desired metabolite(s). While kinetic models can be used to suggest strategies based on changes in enzyme levels and/or kinetic properties to improve flux through a metabolic pathway, these kinetic models need a lot of experimental data (e.g., proteomic, metabolomic, and fluxomic data) to parameterize them. Hence, there is a need for computational methods that can predict expression levels needed to achieve metabolic engineering goals with limited amounts of experimental data and no kinetic details about the biochemical pathways.

We developed an active learning framework called **ActiveOpt** to design expression constructs for a metabolic pathway of interest. ActiveOpt does not need a detailed kinetic model and instead uses a Support Vector Machine (SVM) classifier to predict product yields or productivities (either high or low) from ribosome binding site strengths estimated by the RBS Calculator [1]. ActiveOpt initially trains a SVM classifier from a few experiments, where RBSs in gene expression constructs are varied and product yields are measured and labeled (as high/low yield), and then proposes subsequent experiments to be conducted. ActiveOpt, with relatively little experimental data and no mechanistic or kinetic details of the pathway, can be used to design experiments to achieve high biochemical yields in a small number of experiments.

ActiveOpt was tested on two separate datasets: (i) a newly generated valine yield dataset and (ii) a published neurosporene productivity dataset [2]. The valine dataset included 91 experiments, in which plasmids expression valine biosynthesis and exporter genes (*ilvBNIHCDE* and *ygaZH*) with varying RBS strengths were transformed into *Escherichia coli* and valine yields measured. A leave-one-out cross validation showed that SVM classifiers built from this dataset have high precision (75%) and recall (87%). Starting with just a few of the 91 possible experiments, ActiveOpt could identify expression constructs resulting in at least 95% of the highest measured valine yield (across all conducted 91 experiments) in a small number of experiments (typically <7) and identify the genes whose RBS strengths significantly affect valine yield. ActiveOpt was also tested on a previously published neurosporene dataset, and the algorithm could again identify the expression constructs with high productivity in less than 10 experiments as compared to the 101 experiments conducted in the original study.

These results show that ActiveOpt can efficiently design gene expression constructs that lead to high chemical yield in organisms in very small numbers of experiments. It can also identify the genes whose expression (as predicted by RBS strengths) significantly influence biochemical production.

References

1. Salis, H.M., E.A. Mirsky, and C.A. Voigt, *Automated design of synthetic ribosome binding sites to control protein expression.* Nat Biotechnol, 2009. **27**(10): p. 946-50.

2. Farasat, I., et al., *Efficient search, mapping, and optimization of multi-protein genetic systems in diverse bacteria.* Mol Syst Biol, 2014. **10**: p. 731.

491510 *In silico* and *in vitro* Analysis of Energy Conservation and Bifurcating Enzymes in *Clostridium Thermocellum*

Abstract:

Clostridium thermocellum str. DSM1313 is a fast growing (poly)saccharide fermenter that produces H_2, ethanol, acetate, formate, and lactate as major byproducts, making this organism of interest to consolidated bioprocessing. The metabolic capabilities of this organism have been extensively studied, particularly to understand and optimize the bioconversion of sugars to biomass and/or byproducts. Less is known about the energy conserving pathways, specifically the role enzymatically catalyzed electron bifurcating (BF-) transhydrogenase and BF-hydrogenase reactions play in electron flux and energy generation. The objectives of this study were to, 1) characterize growth, byproduct production, and redox poise of *C. thermocellum* cultured under low or high H_2 partial pressures (pH_2); 2) quantify the thermodynamic limits of reactions catalyzed by enzymes implicated in electron flux; and 3) identify molecular components and design principles which are necessary for enhanced electron-mediated energy conservation. Growth characterization, byproduct production, and redox poise (i.e., [NAD(P)H]/[NAD(P)$^+$]) were determined by culturing *C. thermocellum* on cellobiose and in the presence of either Ar or H_2 headspace (1 bar). Classic thermodynamic modeling of redox reactions associated with electron flow were constrained by *in vivo* measurements to predict catalytic bias and reaction direction under defined pH_2 conditions. Stoichiometric metabolic networking, specifically Elementary Flux Mode Analysis (EFMA) and Flux Balance Analysis (FBA), was used to integrate growth data, genomics, and thermodynamic analysis to model energy and byproduct yields under simulated conditions. *C. thermocellum* exhibited extensive metabolic redundancy (e.g., pyruvate metabolism, hydrogenases, transhydrogenases), which enabled multiple pathways to accomplish redox balance. Redox tuning of ferredoxin (Fd) midpoint potentials was found to be the most sensitive parameter for determining the equilibria of Fd-dependent reactions, and although the *in vivo* ranges of [NAD(P)H]/[NAD(P)+] did not vary enough to significantly effect reaction equilibria, these measured ratios enabled the prediction of Fd midpoint potentials associated with the BF-NADH dependent NADPH:Fd transhydrogenase (NfnI) and the energy conserving Rnf complex. Finally, the electron bifurcating [FeFe] hydrogenase reactions were predicted to improve byproduct and ATP yields per mol of substrate oxidized with the highest ATP yields requiring concomitant evolution of H_2. Production of H_2 was also required for metabolic strategies associated with acclimation to low pH stress. The results of this study revealed the importance of electron bifurcating reactions in electron-mediated energy conservation of *C. thermocellum* and identified important control points that can be targeted for metabolic engineering and optimization.

505056 Model-Guided Engineering of Microbial Biocatalysts

Abstract:

Microbes have been engineered to produce a variety of chemicals, including biofuels, commodity chemicals, specialty chemicals, and therapeutics. Chemical production can be enhanced by connecting synthesis pathways to host metabolism, re-wiring regulatory networks, improving precursor production, and optimizing gene expression. A number of computational systems biology approaches have been developed to facilitate metabolic engineering efforts by suggesting which combination of genetic changes would improve chemical production. Network analysis methods can be used to identify paths from inexpensive substrates (e.g., sugars) to high-value chemical products and to identify central metabolic precursors that can be converted into a variety of chemical products. Genome-scale metabolic models can be used to predict how gene deletions, gene additions, and gene expression changes would impact chemical product yields, growth rates, and/or productivities. Additionally, machine learning and active learning algorithms can be used to optimize gene expression constructs to efficiently convert metabolic precursors into desired products. Case studies will be presented that show how computational tools can guide development of strains with enhanced sugar utilization, precursor production, and chemical production. Together this work illustrates how integrating computational and experimental efforts can lead to the rapid development of microbial biocatalysts.

505295 Professor Wei-Shou Hu: Dedicated to Service

Abstract:

I have enjoyed the privilege of having Professor Hu as a friend and colleague for several decades. Throughout all that time, I have always been impressed by Wei-Shou's selfless approach to his professional life. This talk will begin by describing the early days of CHO-based mammalian cell culture process development at Genentech to place in perspective his many contributions to that very important technology. I will also reminisce about some of Wei-Shou's pioneering use of computer simulations and early artificial intelligence advances. Did you know that he also used these tools to advance forestry practices? Finally, it will be a pleasure to relate some of Wei-Shou's many contributions to our professional societies including his incredible organizational abilities. These became especially apparent when I co-chaired a Biochemical Engineering meeting with him in 1997. It will be a pleasure and an honor to be part of this special day.

505296 Leveraging Bioprocessing Fundamentals for the Development of Autologous Cell-Based Gene Therapies

Abstract:

In many ways, the current state of process development for cell-based gene therapies mirrors the early years of protein therapeutic bioprocessing. Similar to challenges faced when the first biologics products were emerging, gene therapy companies are tasked with understanding complex products and developing robust, scalable manufacturing processes. Many of the critical advances that have been made over the past 30 years to understand more traditional cell culture processes, as well as the structured approach of Quality by Design, can be applied to gene therapy products in development. The challenges of autologous cell-based gene therapy additionally include patient-to-patient (raw material) variability, complex supply chains, and a requirement to develop novel analytical methods. The demonstration of biochemical, biophysical and cellular comparability across multiple facilities can be challenging. This presentation will summarize the advances made in the process and product development of autologous cell-based gene therapies.

505297 Opportunities and Challenges in the Commercialization of Bioproducts from Biomass and Algae

Abstract:

Thanks to advances in bioprocessing during the last 30 years, biomass and algae have become promising sources of sustainable fuels, chemicals, and biologicals. Fuels include ethanol, biodiesel, and jet fuel for the transportation sector. Chemicals include organic acids that serve as building blocks for manufacturing plastics, resins, films, and other products. Biologicals include a wide variety of nutraceuticals, cosmetics, and enzymes. With the use of industrial microbiology and biotechnology, biomass is enzymatically converted to sugars that are converted to the desirable end product through fermentation by natural or genetically engineered microorganisms, such as bacteria, yeasts, and fungi. In the case of microalgae, photosynthetic species are cultivated in open ponds and photobioreactors to sequester carbon dioxide and synthesize nonpolar lipids for jet fuel and biodiesel, polar lipids for cosmetics, various antioxidants and pigments, as well as protein for animal and fish feed. Like every new technology, sustainable bioproducts face challenges on their way to cost competitiveness, most importantly feedstock production and consistency, process scalability, water and energy utilization, and life-cycle sustainability. The presenter will review those issues and share his experience from partnering with industry and venture capital to close technology gaps, generate and protect intellectual property, and stimulate entrepreneurship and commercial deployment.

505300 Wei-Shou Hu and the Cho 'Omics Revolution

Abstract:

Wei-Shou Hu has been a pioneer and innovator in mammalian cell technology for more than three decades. He continues to act as a trail blazer in applying new technologies to mammalian bioprocessing; perhaps none more than in the area of 'omics analyses. His approaches have been inspirational and challenged the rest of the community to "up our game for the better." Indeed, it is upon this foundation that many researchers in mammalian biotechnology have established an improved framework for mammalian cell culture systems characterization. In Chinese hamster ovary (CHO) cells, Professor Hu was the first to apply a systems biology paradigm to enhance our understanding of this critically important cell factory. He continues to this day to implement inventive methodologies and techniques to help better describe CHO cells at the molecular level and how molecular events impact physiology and the production of complex biologics. In this presentation, some of the many important and informative 'omics contributions undertaken by the Hu lab will be described and how they have helped the mammalian biotechnology community better understand CHO cells and apply this cell factory for biomanufacturing applications.

505303 Viral Vaccines Manufacture in the Wei-Shou Hu Era

Abstract:

At the dawn of the 1980â€™s, viral vaccine manufacturing was still primitive, with a very few advanced exceptions. These newer technologies, many of which were supported by the research of Professor Wei-Shou Hu, paved the way for modern recombinant biologics. The past 30 years has seen the modernization and significant scale-up of vaccine production, as a large number of new products has arisen and demand has soared. At the same time, increases in understanding of cell and virus cultivation, enabled by Prof. Huâ€™s research, has enabled scale-out and scale-down strategies to be applied to economize the manufacture of viral vaccines. This talk will review the evolution of viral vaccine technologies and directions for their future.

505304 *An Industrial Perspective on the Impact of Cell Culture Technology Improvements on Biologics Drug Development*

Abstract:

Over the last 30 years, a series of technology advances have improved our ability to use mammalian cell culture to produce important human therapeutics. These include cell line, media & processing improvements that have led to dramatic increases in the efficiency & speed of developing production processes as well as enabling cost-effective production of antibody and protein products at the metric ton scale. These production improvements have provided the opportunity to pursue new approaches and opportunities for the treatment of important diseases resulting in significant benefit to patients and opening new doors for further future benefits. In this presentation, several examples of these technological advances will be reviewed and connected with resulting benefits as well as future opportunities for further impact.

Intensification of a Continuous Process Guided by Mechanistic Modeling

Rudi Oliveira, Ana Cruz, Filipe Ataide, Nuno Matos
October 2017

Hovione

Hovione has extensive manufacturing capacity spread over three continents

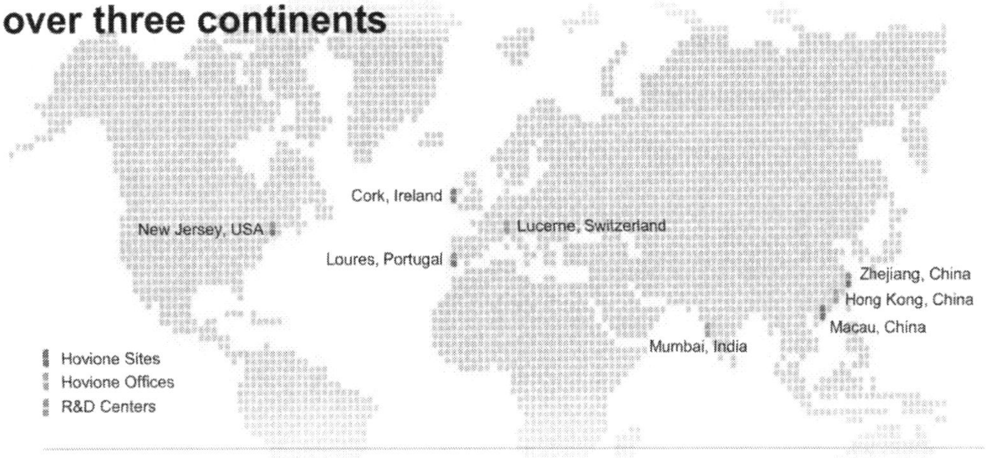

New Jersey, USA
Manufacturing facilities small volume
R&D Labs, Kilo and pilot plant
Sales and marketing for North America

Loures, Portugal
Manufacturing facilities including
R&D Labs, kilo and pilot plants

Macau, China
Manufacturing facilities

Cork, Ireland
Manufacturing facilities

Taizhou, China
Manufacturing facilities

Hovione on the Pharmaceutical Supply Chain

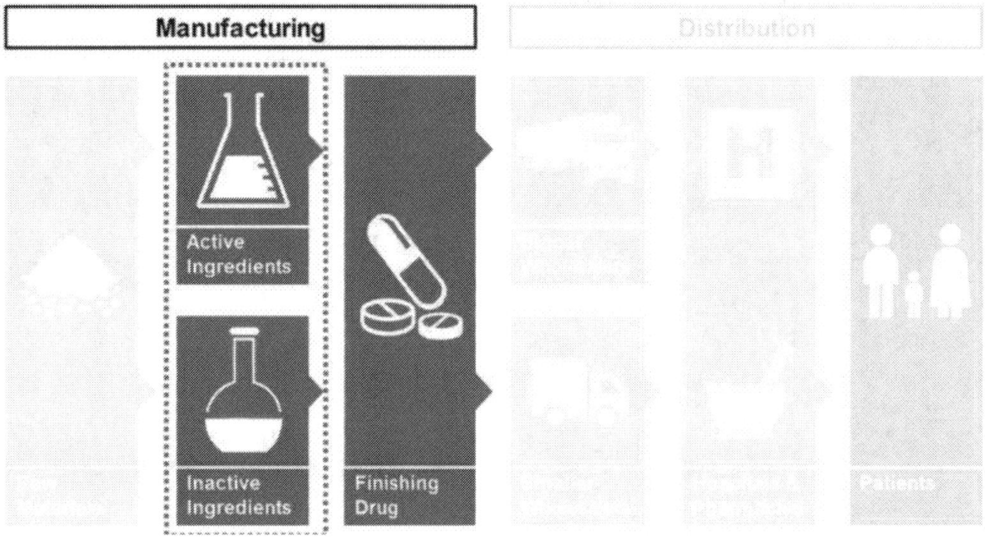

Drug Development

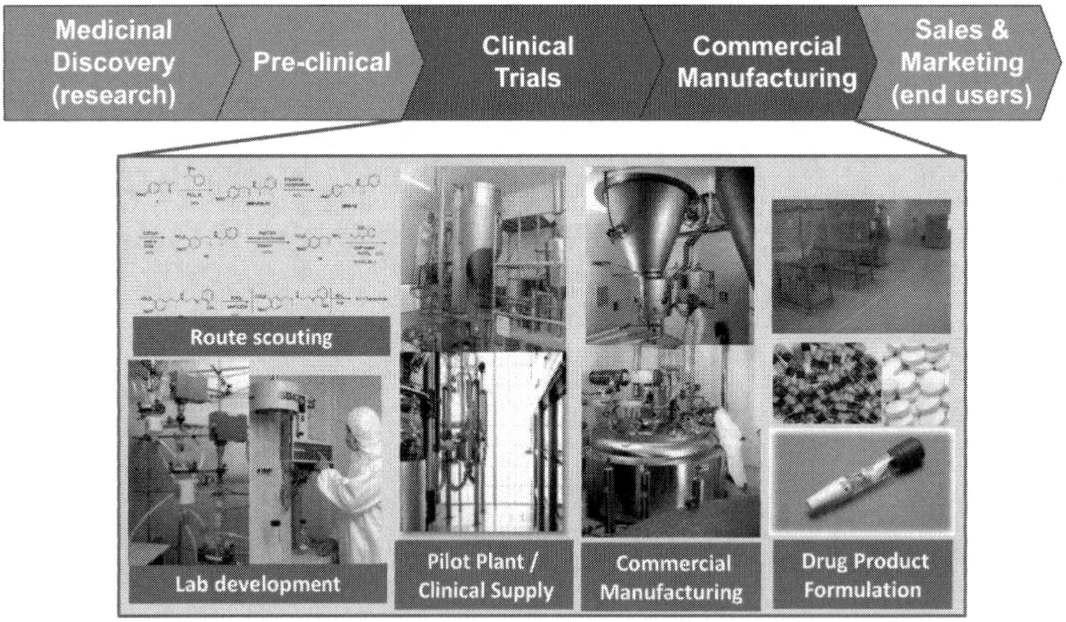

Agenda

Flow Chemistry: Features & Capabilities

Approach to Development: QbD

Feasibility Study: Batch-to-Continuous

Conclusions

Agenda

Flow Chemistry: Features & Capabilities

Approach to Development: QbD

Feasibility Study: Batch-to-Continuous

Conclusions

Batch vs Flow
Understand the Drivers

Main Features

Process Drivers

Safety	Efficiency	Quality	Environmental Impact	Time to market	Batch Size Flexibility	Costs

Flow Capabilities

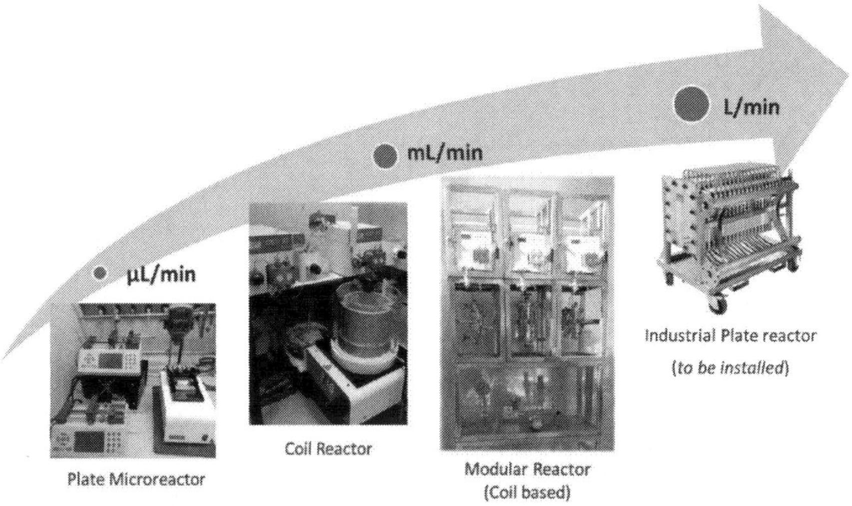

Agenda

Flow Chemistry: Features & Capabilities

Approach to Development: QbD

Feasibility Study: Batch-to-Continuous

Conclusions

Process Development at Hovione

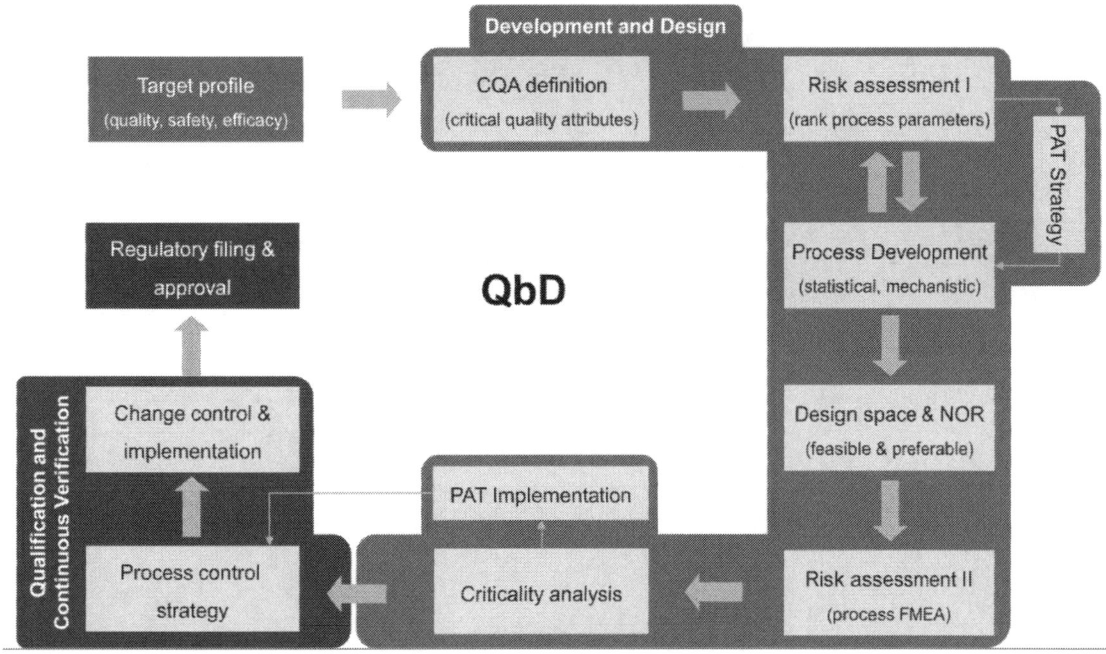

Approach to Development

Systematic workflow with gate reviews

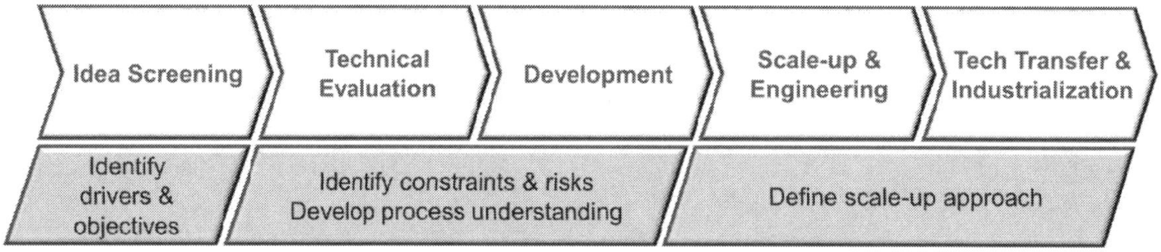

Agenda

Flow Chemistry: Features & Capabilities

Approach to Development: QbD

Feasibility Study: Batch-to-Continuous

Conclusions

Idea Screening | Technical Evaluation | Development

Project X

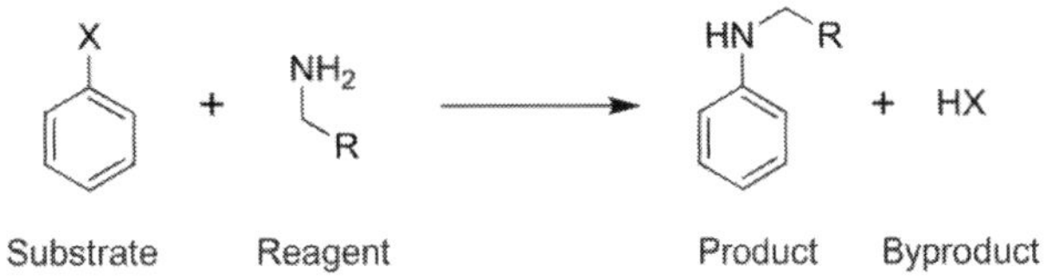

Significant exotherm at > 150°C

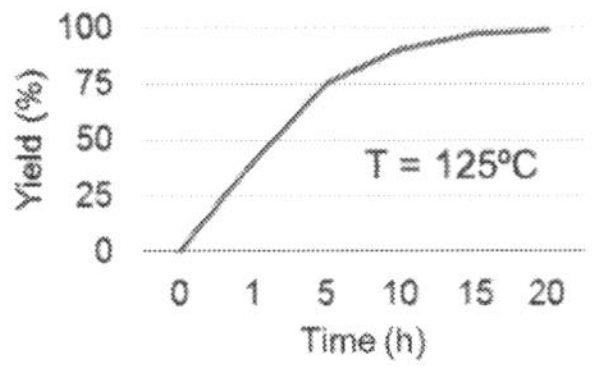

Low b.p. of solvent

Idea Screening | Technical Evaluation | Development

Challenge
Batch-to-continuous

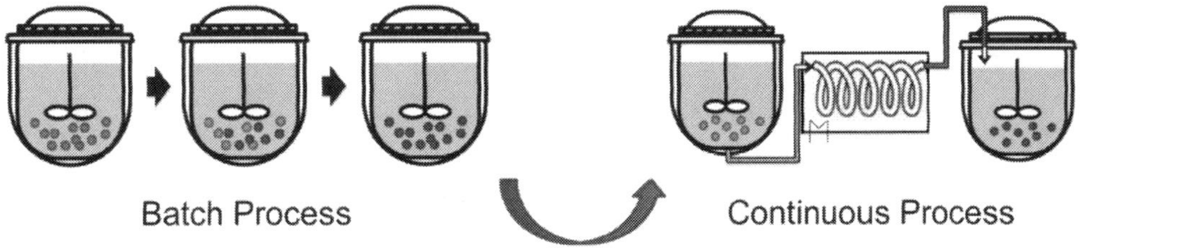

Where do you start?

Modeling

Idea Screening | Technical Evaluation | Development

Batch data

DynoChem

- Design space
- Predict the relation between process parameters and relevant quality attributes

Batch Model

Idea Screening | Technical Evaluation | Development

Reaction Scheme
Substrate + Reagent ⟶ Product + Byproduct

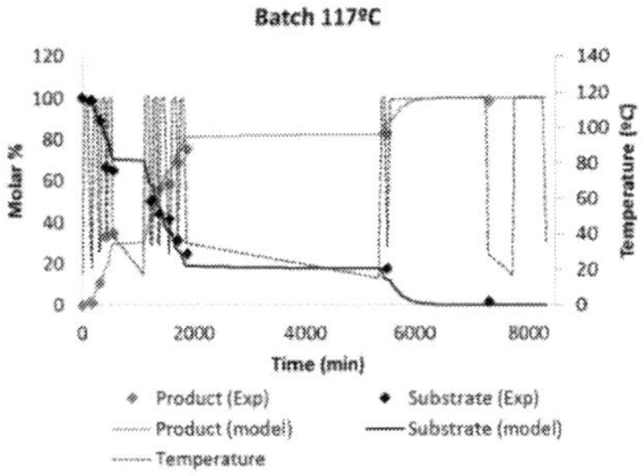

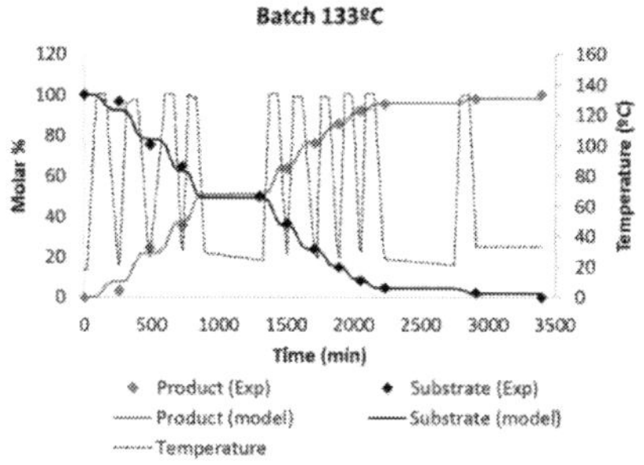

		Confidence Interval
Kinetic constant (*k* @ 117ºC)	2.72 x 10^{-4} L/mol.s	11.8%
Activation energy (E_a)	58.66 kJ/mol	11.8%

Performance Estimation
under Continuous Mode

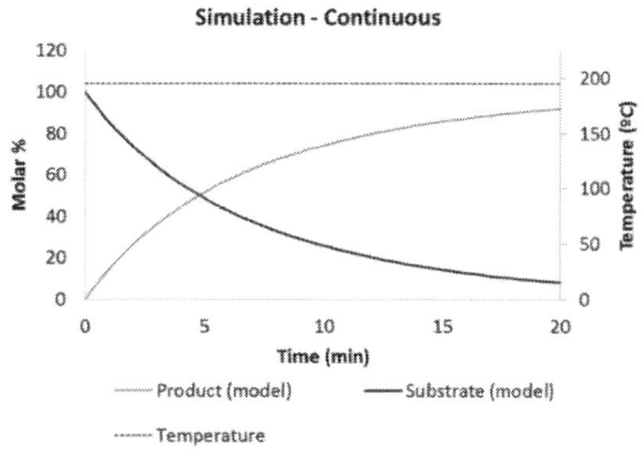

Temperature	Residence Time	Yield
195ºC	20 min	92 %

Lab Experiments
under Continuous Mode

Considerations:

- Selection of the equipment:
  - Material compatibility (MoC)
  - Ranges of operation (flowrate, temperature, pressure)

Temperature	Residence Time	Yield
195ºC	20 min	92 %

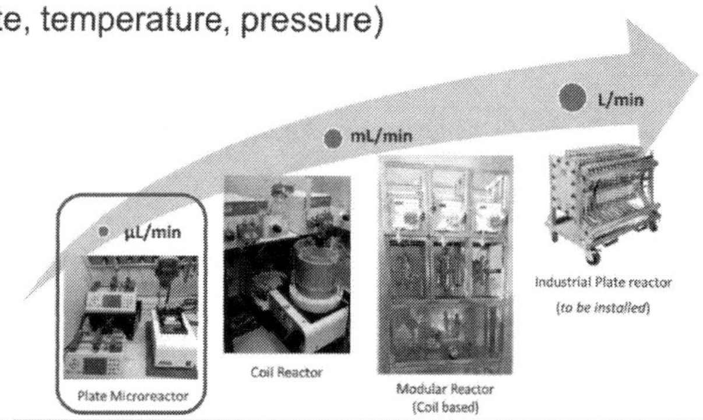

Idea Screening | Technical Evaluation | Development

Lab Experiments
under Continuous Mode

Considerations:

- Selection of the equipment;
  - Material compatibility (MoC)
  - Ranges of operation (flowrate, temperature, pressure)
- Solubility limits.

Idea Screening | Technical Evaluation | Development

Continuous Model

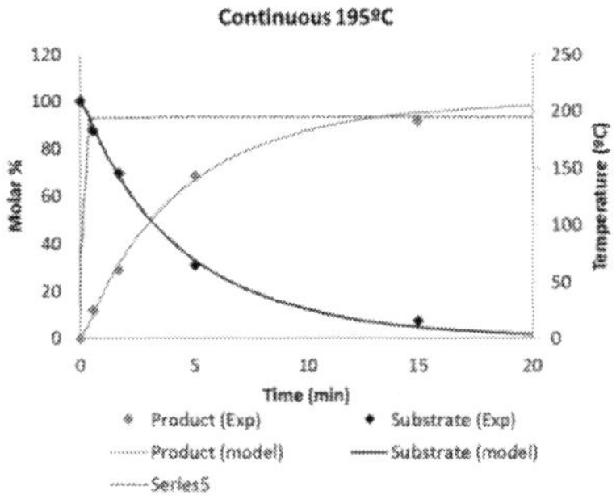

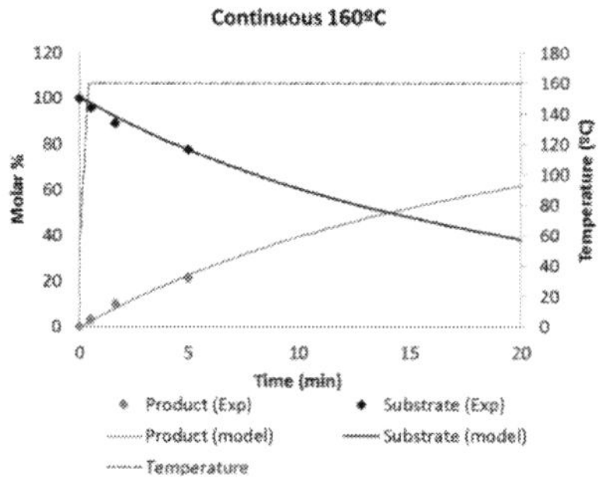

			Confidence Interval
Flow	**Kinetic constant** (*k* @ 195°C)	8.92×10^{-4} L/mol.s	6.9%
	Kinetic constant (*k* @ 117°C)	1.06×10^{-6} L/mol.s	*calculated*
	Activation energy (E_a)	76.66 kJ/mol	5.9%

			Confidence Interval
Batch	**Kinetic constant** (*k* @ 117°C)	2.72×10^{-4} L/mol.s	11.8%
	Activation energy (E_a)	58.66 kJ/mol	11.8%

Continuous Model

Idea Screening | Technical Evaluation | Development

Test Trials

Considerations:

- Different scale
- Different equipment

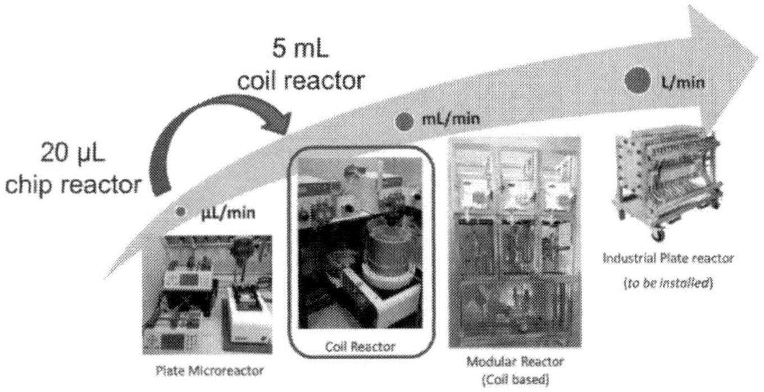

Continuous Model

Idea Screening | Technical Evaluation | Development

TestTrials

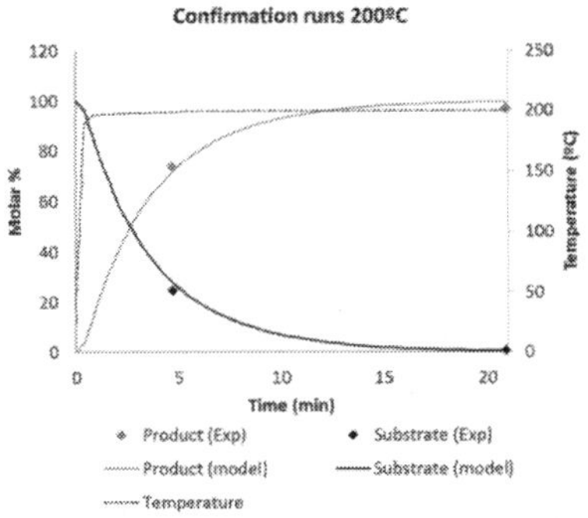

Batch performance
> 97 molar%

Temperature (°C)	Residence time (min)	Product - model (molar%)	Product - experimental (molar%)
200	20	99.4	97.2
200	5	74.8	73.9
210	10	97.2	94.4
210	5	83.4	85.6

Stage Gate Review

Idea Screening | Technical Evaluation | Development

Process Feasible

Similar Performance to Batch

Safer scale-up

Solubility Concern

Improve Mechanistic Understanding (explore other CPPs)

Study PAT application

Agenda

Flow Chemistry: Features & Capabilities

Approach to Development: QbD

Feasibility Study: Batch-to-Continuous

Conclusions

Agenda

Flow Chemistry: Features & Capabilities

Approach to Development: QbD

Feasibility Study: Batch-to-Continuous

Conclusions

- **Development Workflow**
  - ✓ **Organize** prior knowledge
  - ✓ Plan for **faster** development
- **Modeling**
  - ✓ **Application** of prior knowledge
  - ✓ **Save material**
  - ✓ Predict process performance in a **different**: **manufacturing mode**, **scale**, **equipment**

Hovione

Thank you for your attention.

Rudi Oliveira
rsoliveira@hovione.com
hovione.com

489646 Simultaneous Estimation of Kinetic Parameters and Curve Resolution of Spectral Data -Â Applications and Extensions

Abstract:

In the context of studying reaction kinetics, nowadays a common practice is to use spectroscopic instrumentation to indirectly monitor the concentration of the species in the reaction. Recently, Chen et al[1] proposed an approach based on maximum likelihood principles for simultaneous estimation of reaction kinetics and curve resolution. This work presents further learnings and developments in the application of this technique.

Three cases studies will be presented highlighting first the criticality of pre-estimation steps, specifically: *i*) the importance of the pre-processing algorithm applied to the spectral data prior to the estimation, and *ii*) wavelength selection where a nonlinear programming (NLP) formulation is proposed as a means to efficient wavelength selection. We also present an application of mixed integer NLP (MINLP) approaches to identify which active species can be observable in the spectra (since this information is not always available). Finally, we present an Estimability Analysis (EA) that identifies the appropriate subset of model parameters to be estimated from the available spectral data. Overall, the simultaneous technique presented by Chen et al. is shown to be a reliable and flexible method to deal with the complexities of spectral data and system dynamics in a reaction system.

References

1. Chen W, Biegler LT, Muñoz SG. An approach for simultaneous estimation of reaction kinetics and curve resolution from process and spectral data. *Journal of Chemometrics.* 2016;30(9):506-522.

489834 Modeling and Monitoring of a Continuous Silylation Reaction

Abstract:

A multidisciplinary approach was utilized to develop a mechanistic kinetic model of a continuous silylation reaction within a drug substance chemical synthesis. Analysis by mid-infrared spectroscopy and ^{1}H and ^{29}Si-NMR were both valuable monitoring tools toward obtaining insight into the reaction kinetics.

A multi-step kinetic mechanism was proposed based upon the experimentation and orthogonal simulation results. The impact of reaction temperature, solvent concentration, and recipe were investigated. The model has been applied to both batch and continuous processes. Discussion will include sharing key learning from the model development and its application toward risk management and understanding the feasible operating region.

490063 Estimability Analysis for Improved Parameter Estimation in Deterministic Models: Pharmaceutical Case Studies

Abstract:

Often in the early stages of model development, one needs to estimate the model parameters using data with insufficient information. This information insufficiency in the experimental data can arise from either a poor design, or lack of measurements from the system.

This work illustrates the application of *Estimability Analysis* to aid the parameter estimation of a deterministic mathematical model. This technique has proven useful when the information content of the data available is insufficient to obtain a good estimate of all the unknown model parameters. *Estimability Analysis* ranks the parameters from the most to least estimable; taking into account the uncertainties in both the initial guesses of the parameters and the data [1]. This ranking enables the modeler to determine which parameters to estimate, and which to fix. The obtained model parameters are a good initial guess to a formal model-based experimental design exercise to determine the necessary experiments to optimally parametrize the model.

Case studies are presented from drug substance process development where a kinetic model of a reactive system is being developed; and a third case study from the modeling for a continuous direct compression line for drug product.

References

[1] Wu et al., 2011, *International Journal of Advanced Mechatronic Systems,* 3, 188-197.

490870 Solubility Data Mining and Predictive Modeling: AI+ChE

Abstract:

In the synthesis of pharmaceutical drug substances, selecting the optimal solvent system as early as possible is critical to develop efficient, commercially viable processes. Process safety, environmental impact, and yield are driven in large part by the choice of solvent system. Yet early in process development, material availability limits the amount of experimental data that can be collected. Models to predict solubility enable more focused experimentation, but with current approaches there are trade-offs associated with accuracy, solute characterization, and accessibility for process development engineers.

Performing a historical analysis of experimental data using the tools of data scientists has the potential to unlock insights and streamline future development work. Applying these tools to solubility prediction, data was mined from an internal catalog of automated solubility screening reports across dozens of projects. In all, over 64,000 solubility measurements for >700 pharmaceutically relevant organic compounds in pure and mixed solvents at various temperatures have been aggregated from these reports and analyzed. Such a data set allows for the rapid testing of hypotheses related to solvent selection, such as correlations and synergies between solvent pairs and temperature effects.

Access to this large data set enables machine learning approaches to create quantitative structure property relationship models for solubility prediction that show an improvement over the current standard approaches. Deployed web applications can allow any researcher to rapidly calculate solubility predictions by providing only a molecular structure and a single benchmark solubility measurement. This presentation aims to show how the recently available combination of algorithms (random forests, support vector machines, neural networks, etc.), data-oriented programming languages (R, Python), and cloud computing capabilities can enable chemical engineers to better utilize their data for pharmaceutical process development.

492564 Toward Efficient Development and Reliable Scale-up of Agitated Filter Drying Protocol through DEM Modeling and Simulation

Abstract:

Drying of an active pharmaceutical ingredient (API) is an important unit operation in the pharmaceutical industry as it is used to control residual solvent content, in addition to controlling appropriate physical properties. Drying operation is often time consuming, and may turn out to be a bottleneck in the overall process cycle time. Many APIs are potent compounds that require filtration and drying to be done in a single contained vessel, known as an agitated filter dryer (AFD), to minimize operator exposure.

Gradual evolution of powder bed during the drying results in continuously changing physical properties. This poses significant challenges in experimentally driven understanding of the impact of drying process parameters on final physical properties. Agglomeration and attrition, which are governed by the shear, are the major issues encountered during design and scale-up of filter drying operation. The extent of shear and its distribution changes significantly across various scales and is not amenable to experimental measurements. With advancements in computer algorithms and cheaper computational power, more fundamental insights into particle dynamics including the shear stress distribution can be assessed using Discrete Element Methods (DEM).

In this talk, we will present some of the key attributes that govern the particle dynamics calculated using DEM in the context of AFD operation. We will also provide a decision tree to guide efficient design and reliable scale-up of agitated filter drying.

497704 A Novel Microfluidic Platform for Screening of Pharmaceutical Polymorphs Under Hydrodynamically-Controlled Crystallization Conditions

Abstract:

The majority of oral drugs are delivered in solid forms such as capsules, tablets, powders or suspensions. The physicochemical properties and quality of such drugs strongly depend on their crystalline structure or polymorph, which is controlled during the crystallization process. Various factors such as solvent, temperature, co-solvent, supersaturation, and additives can affect the formation of polymorph. Screening of polymorphs with respect to these factors can be combinatorially tedious and time-consuming. High-density microtiter plates with as many as 1586 wells are commonly employed to screen ~10,000 factors per day. However, the application of microtiter plates for polymorph screening has some practical limitations as the environmental conditions in stagnant wells of the microtiter plate is much different than those in the batch crystallizers. Alternatively, the microfluidic platforms have been used to screen polymorphs using very small quantities (~100 μg) of the drugs in the early stages of developments. Such microfluidic devices locally produce a mixture of crystalline material which is not suitable for the study of nucleation and growth of polymorphs. Here we propose a novel microfluidic platform which mitigates the issues of the previous designs by facilitating nucleation and growth of single crystals. This microfluidic device can effectively screen the aforementioned factors to generate single crystals of polymorphs at the stagnation point while implementing the hydrodynamic conditions which are not considered in the microtiter plate method. In this work, we have screened polymorphs of Carbamazepine using various mole fractions of chloroform:ethanol mixtures and temperatures in the microfluidic device. Further analysis for the measurement of growth rate and morphology are also discussed. The high throughput screening system is also proposed to evaluate multiple crystallization conditions within one single device.

1

POLYMORPHISM IN DRUG COMPOUNDS BY CONTROLLING FLUID DYNAMICS DURING CRYSTALLIZATION

Stephanie Guthrie, Gaurav Giri

AICHE 2017: Solid Form Selection: Cocrystals, Salts, Solvates, Polymorphs, and Beyond

University of Virginia
Department of Chemical Engineering

2

Current drug development

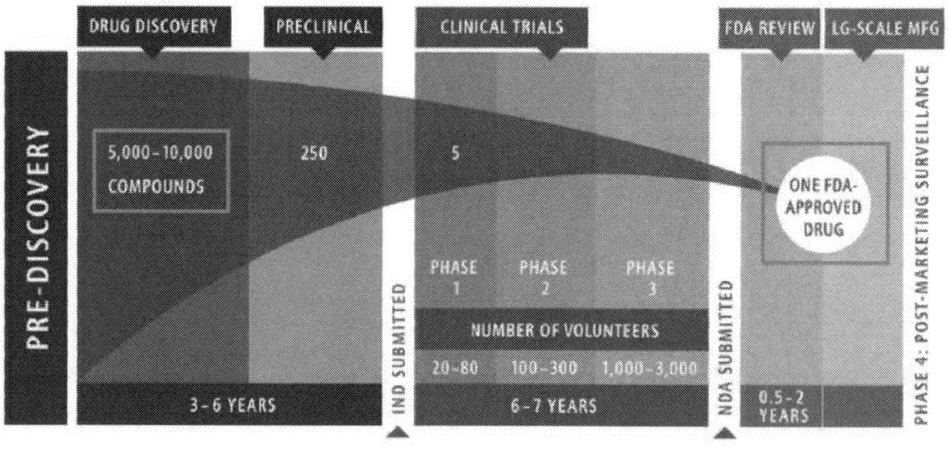

Source: Pharmaceutical Research and Manufacturers of America

Need to optimize polymorph detection

- Need ability to detect and isolate polymorphs
  - Polymorphs have unique physical and chemical properties
  - Color, solubility, bioavailability

ROY: 5-Methyl-2-[(2-nitrophenyl)amino]-3-thiophenecarbonitrile

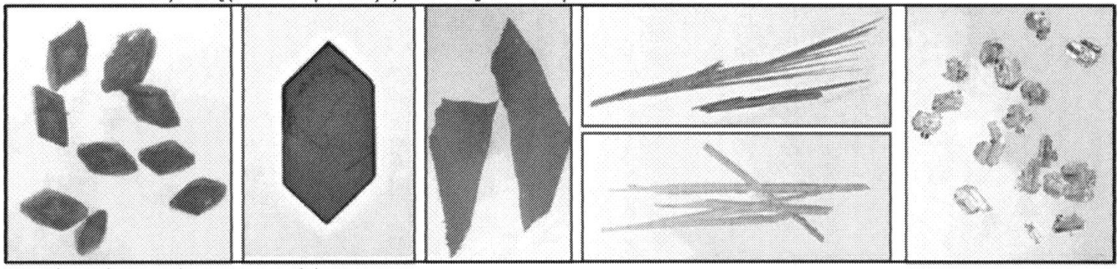

Yu, L. (2010). *Acc. Chem. Res.*, 43(9), 1257-1266.

Solid state changes solubility

Increased solubility

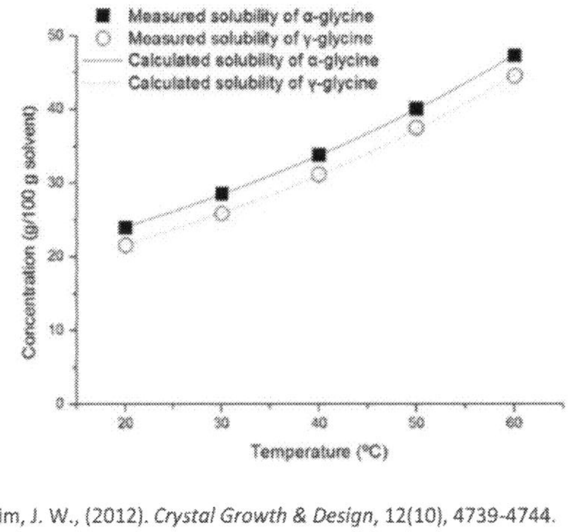

Kim, J. W., (2012). *Crystal Growth & Design*, 12(10), 4739-4744.

Enhanced dissolution rate

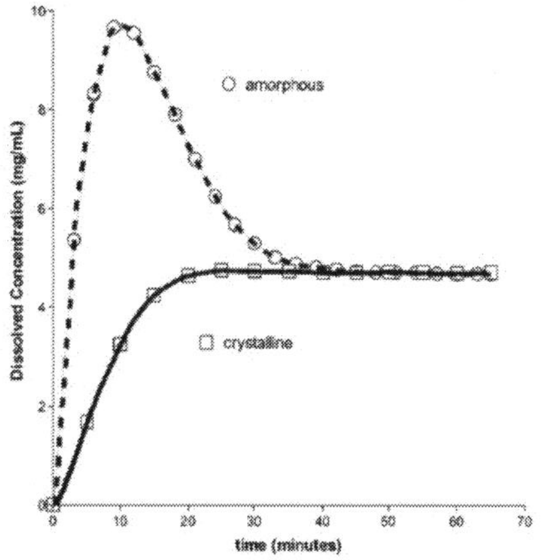

Brittain, H. (2014). *Am Pharmaceut Rev.*

5

Fluid dynamics can impact polymorphism

- Increased charge carrier mobility based on polymorphism in organic semiconductors

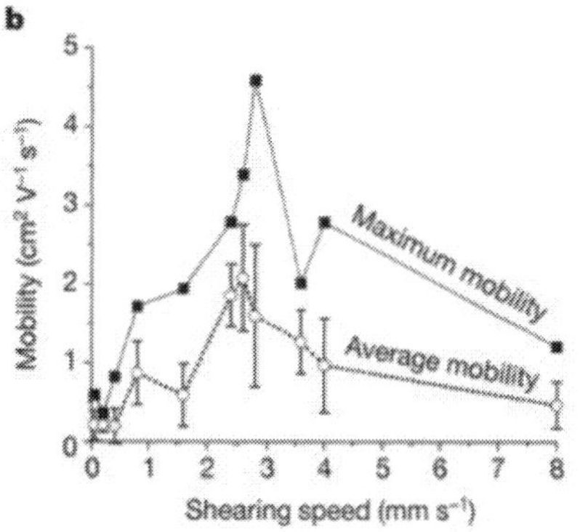

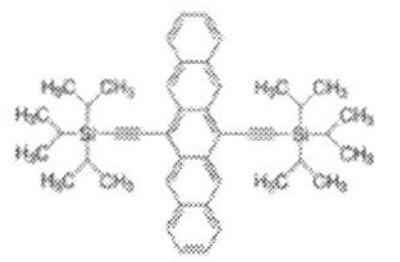

TIPS-pentacene

Giri, G. et al. (2011). *Nature* 480(7378), 504–508.

6

Solution shearing technique for pharmaceuticals

- Two pharmaceutically relevant molecules
  - Glycine, acetaminophen
- 10 mg/mL drug in deionized water
- Change processing parameters
  - Shearing speed: 0.01 – 3 mm s^{-1}
  - Temperature: 70-100 °C

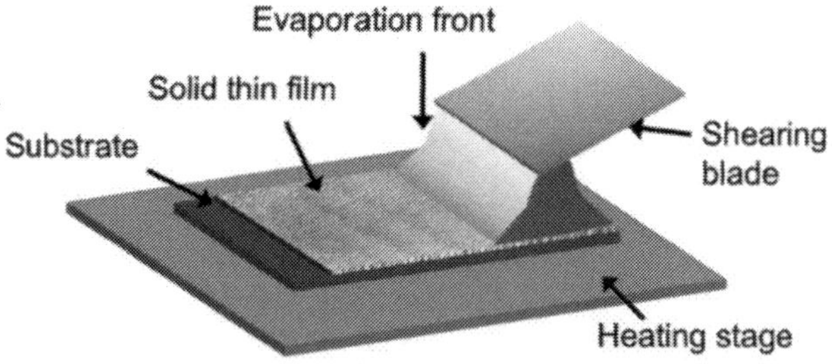

Shearing regimes

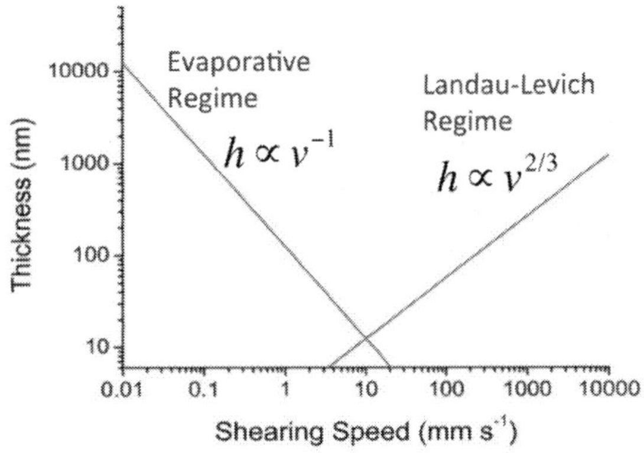

Le Berre, et al. (2009). *Langmuir*, *25*(5), 2554-2557.

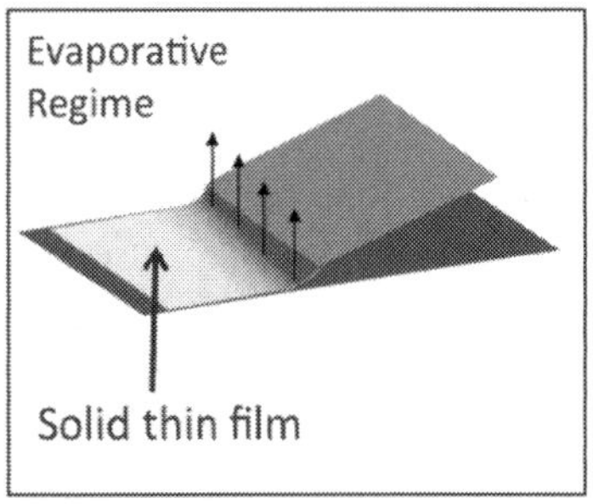

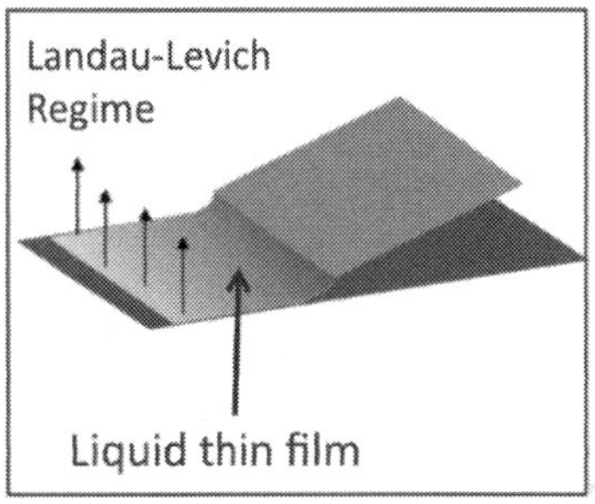

Film thickness characterization

70°C

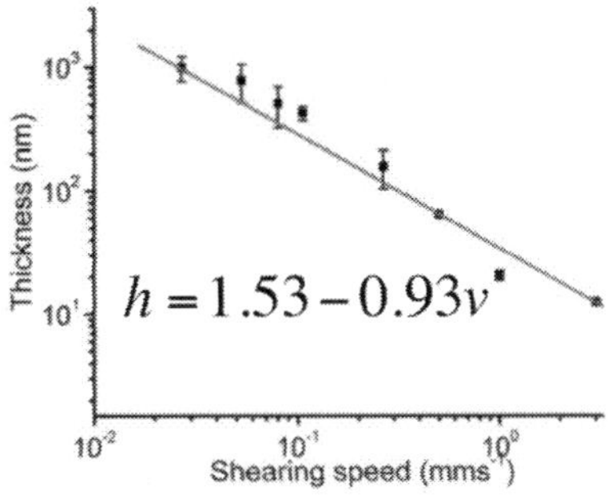

100°C

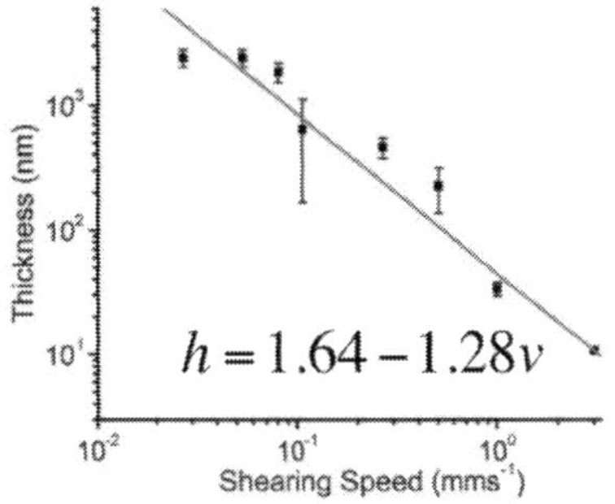

Glycine morphology

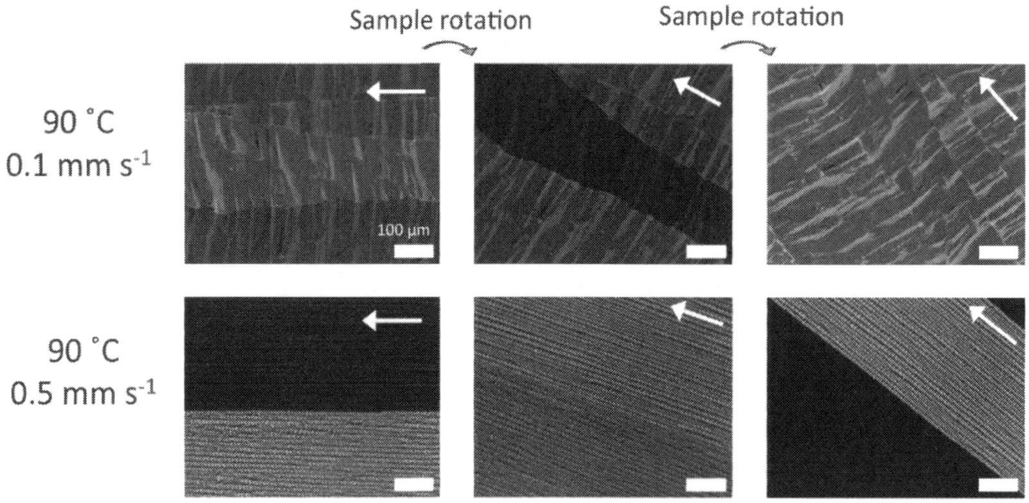

Acetaminophen morphology

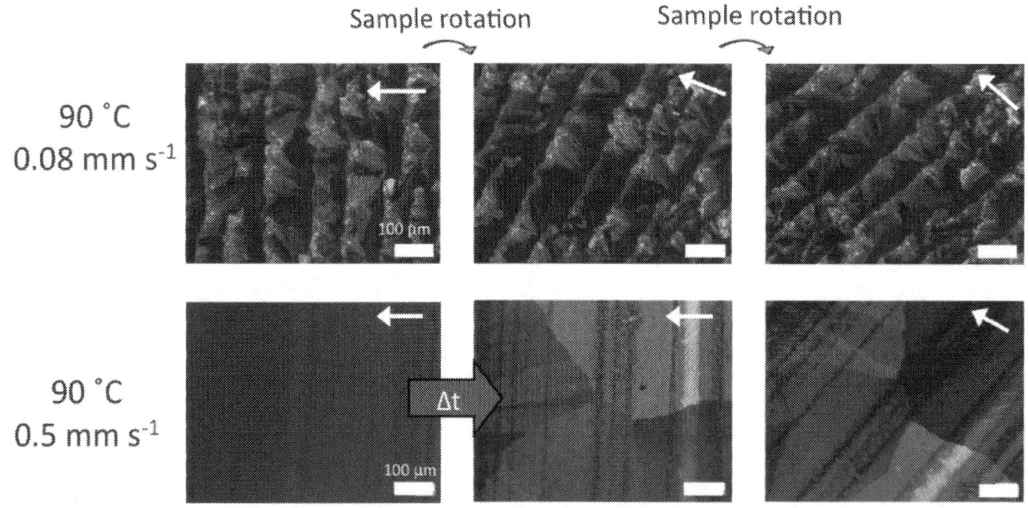

Grazing incidence X-ray diffraction (GIXD)

1. Incident X-ray beam is directed towards sample

2. X-rays are diffracted on film crystallites

3. Diffracted beam lands on area detector

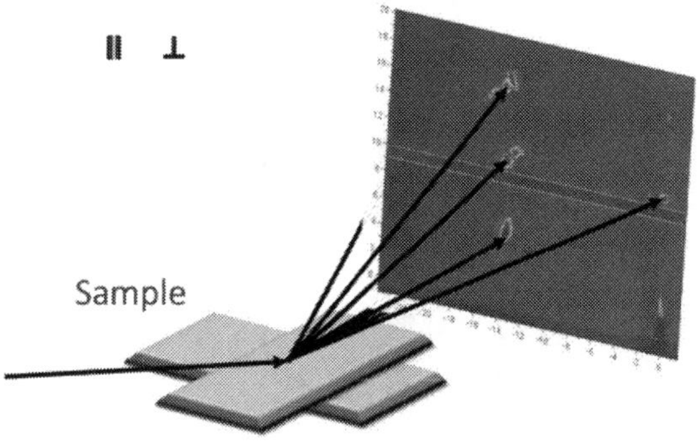

Samples are rotated parallel and perpendicular to shearing direction

- Orients new planes in the line of X-rays

Increasing order ↓

Sample type	Film Appearance	Characteristic GIXD	Description
Isotropic powder		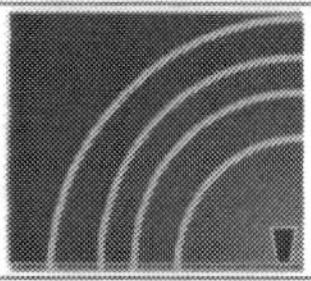	Annular rings
Anisotropic textured film		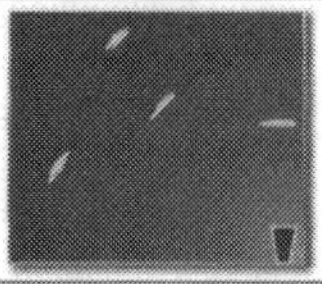	Peak smearing
Oriented film		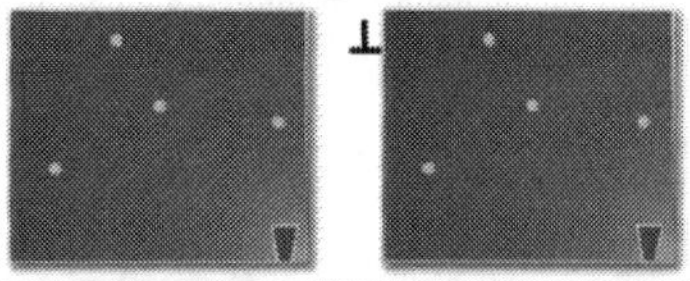 ∥ ⊥	Identical peaks upon rotation
Aligned film	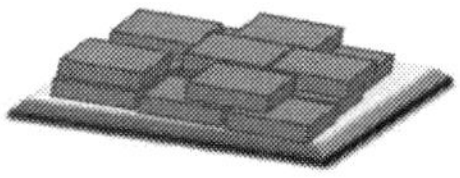	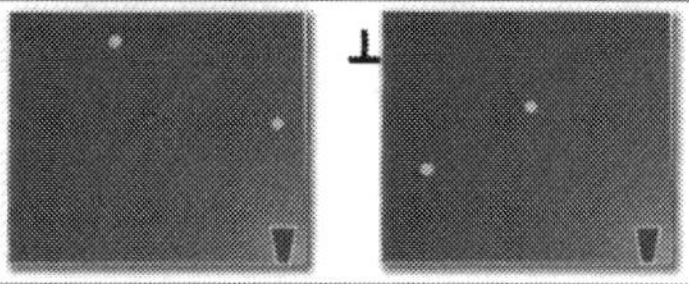 ∥ ⊥	Unique peaks upon rotation

13

Morphological characterization for glycine

Sample type 1

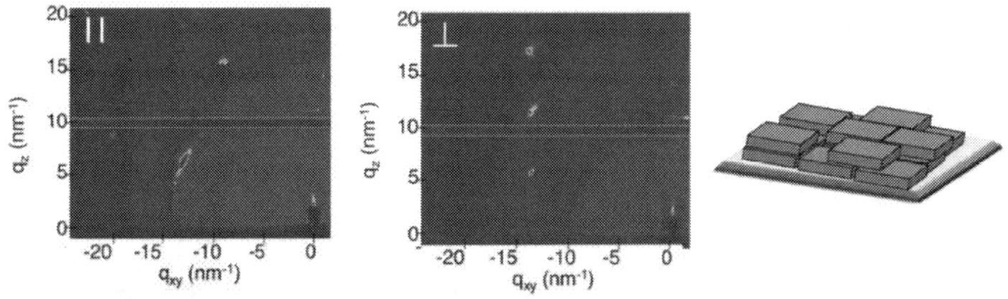

Sample type 2

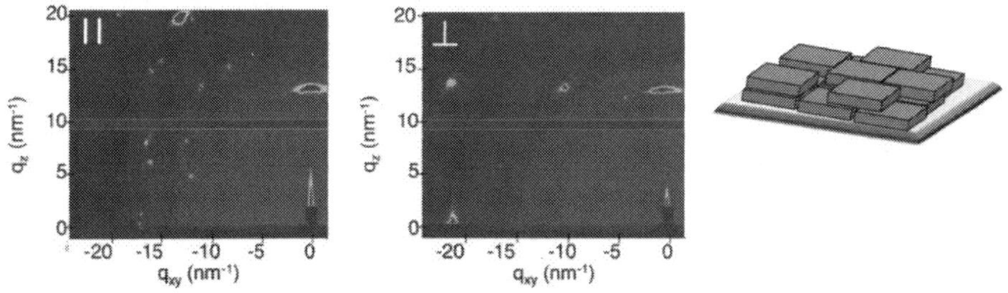

14

Morphology characterization for acetaminophen

Sample type 1

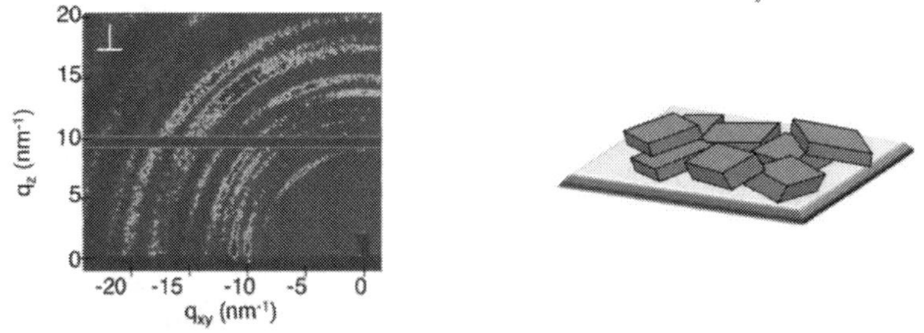

Sample type 2

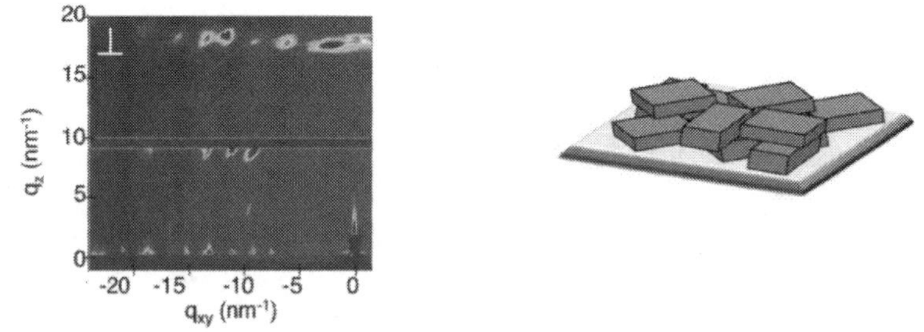

Integrated GIXD for glycine

- Confirmation of polymorphism

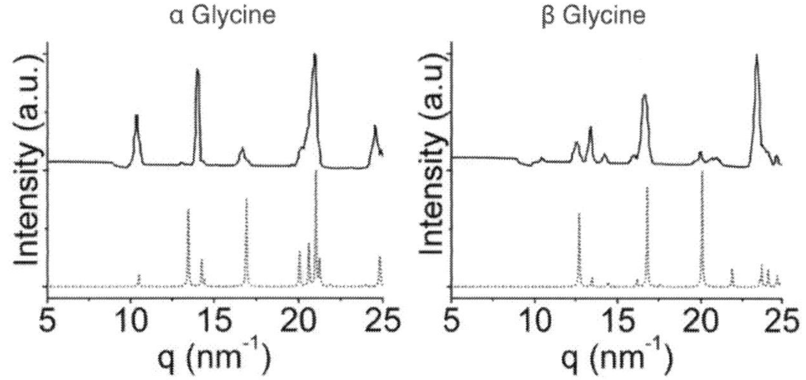

(—): Experimental   (- -): Integrated from crystal structure

Integrated GIXD for acetaminophen

- Confirmation of polymorphism

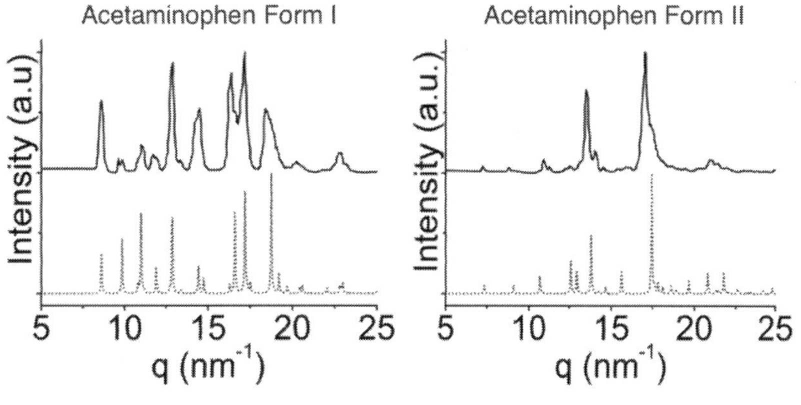

(—): Experimental   (- -): Integrated from crystal structure

Data reduction

Acetaminophen, 0.27 mm s^{-1}, 90˚C	
Sample	Polymorph assignment
1	Form II
2	Form II
3	Form I
4	Both
5	Form I

3/5 samples contain Form I
3/5 samples contain Form II

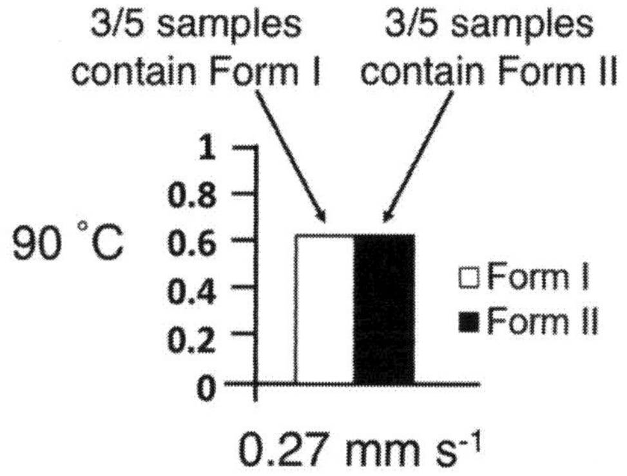

Tracking glycine polymorph formation

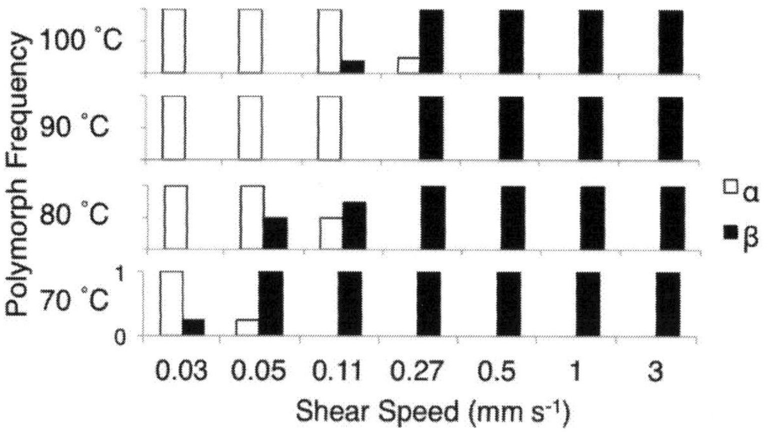

19

Tracking glycine polymorph formation

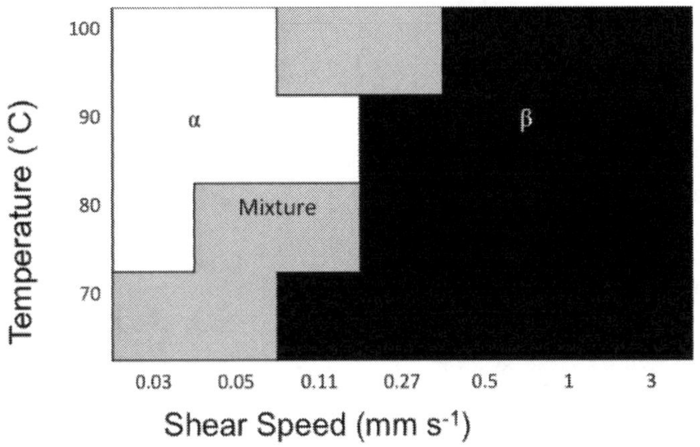

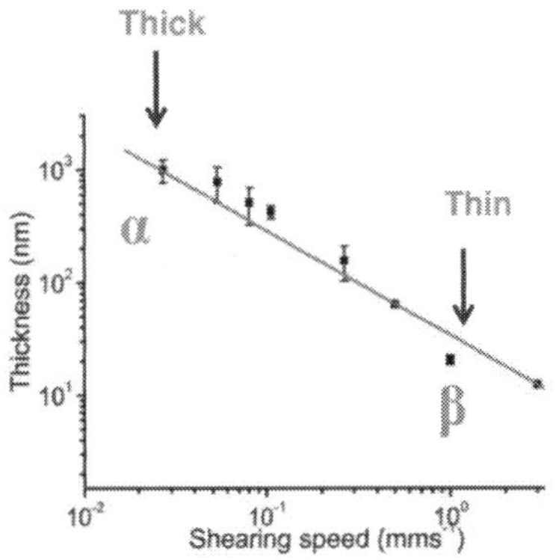

20

Tracking acetaminophen polymorph formation:

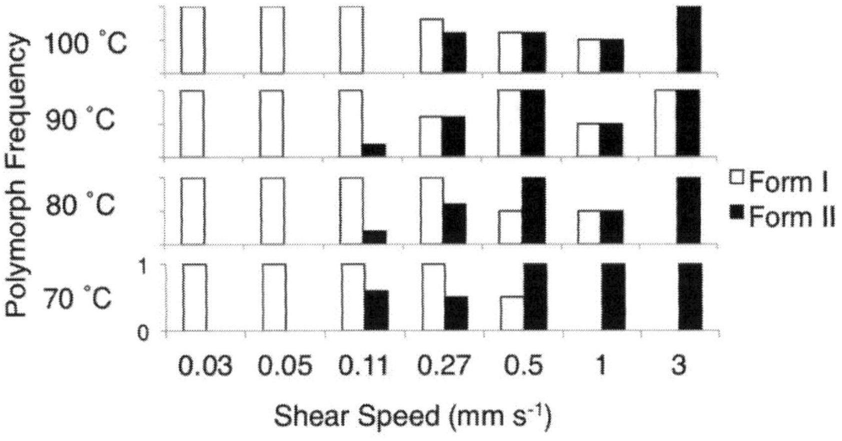

Tracking acetaminophen polymorph formation

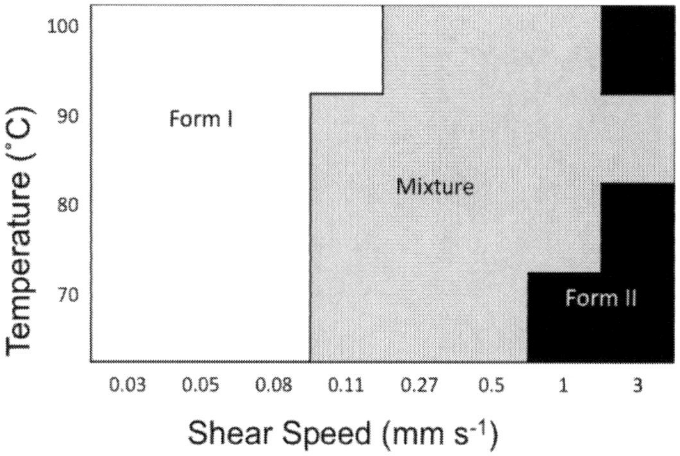

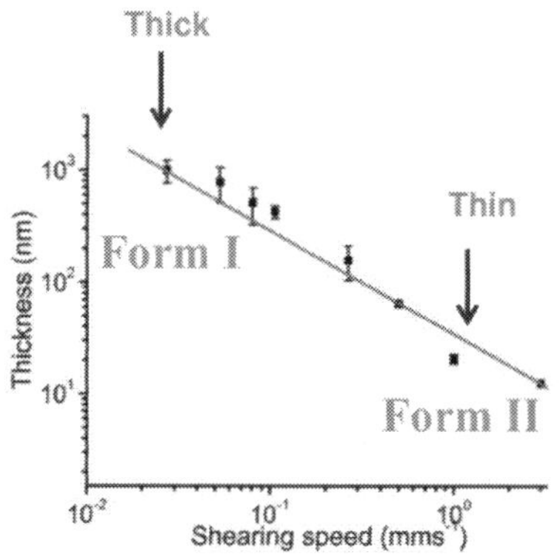

Rationale for polymorphism

- Thick films and thin films display polymorphic changes
- More stable form in thicker films
- Thin films provide a confined environment for isolation of metastable polymorphs

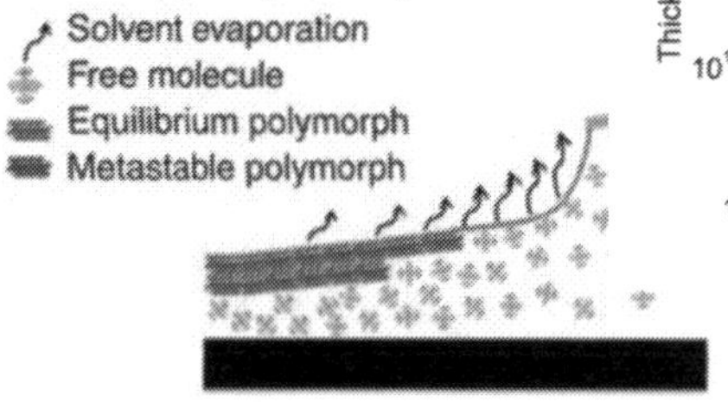

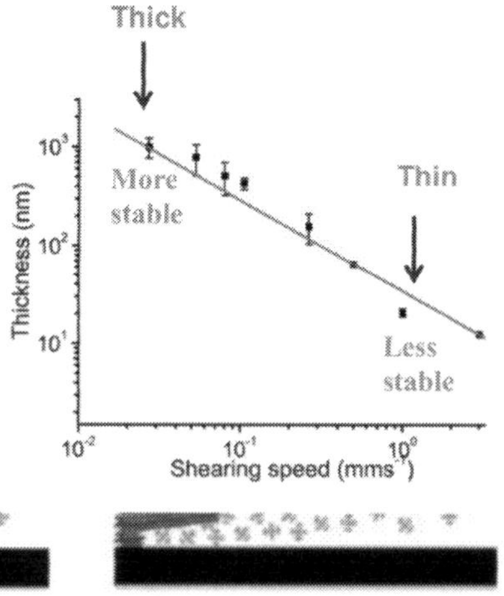

Giri, G. (2014). *Nature communications.* ***5**, 3573*

From II from amorphous transformation

- Amorphous phase isolated first
- Crystallizes over time

Acetaminophen

t =30 seconds t = 90 seconds t = 300 seconds

Carbamazepine

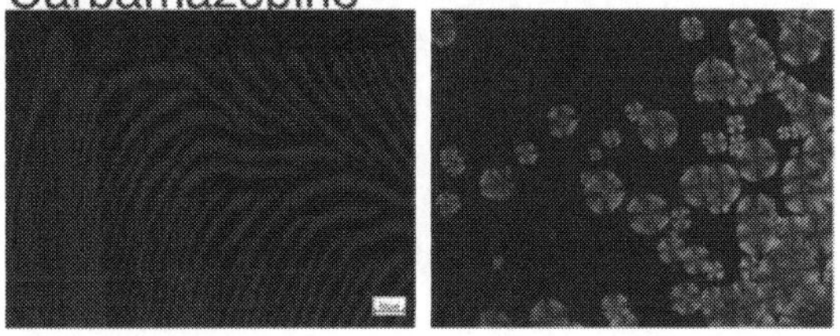

t = 0 seconds t = 10 minutes

Time temperature transformation

- Processing temperature impacts final phase formation
- Different microstructures
- Balance between nucleation and growth creates 'C' shape

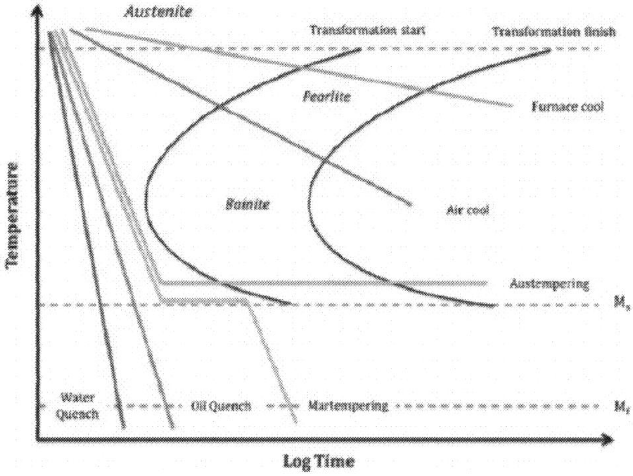

Brog, J. P (2013). *Rsc Advances*, 3(38), 16905-16931.

Crystallization kinetic diagram

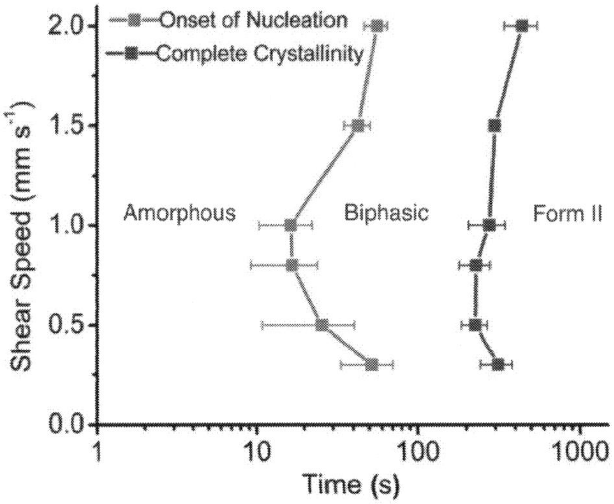

Summary

Processing techniques | Polymorphism | Processing diagram

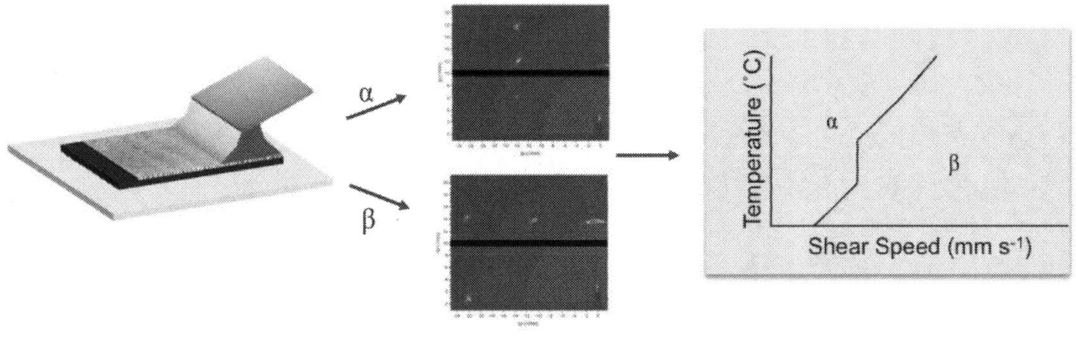

Acknowledgement

- Detlef Smilgies (CHESS)
- Luke Huelsenbeck
- Arian Ghorbanpour
- Giri group members

Virginia Space Grant Consortium
Aerospace Partnerships in Education, Research and Industry

Health Resources in Action®
Advancing Public Health and Medical Research

Cornell University
Cornell High Energy Synchrotron Source

Questions

488669 Mechanism of Selective Co-Crystallization of Cresol Isomers with Urea

Abstract:

The separation of *m*-cresol from mixtures of *m*-cresol and *p*-cresol has been a great challenge in industry since the physicochemical properties of *m*-cresol and *p*-cresol are pretty similar. In this work, co-crystallization method was used to separate *m*-cresol and *p*-cresol and the co-crystal preparation was attempted by cooling crystallization. The selective co-crystallization of *m*-cresol and *p*-cresol with urea was evaluated, based on the principle of supramolecular chemistry and crystal engineering strategy. It was found that the co-crystal former, urea, can only form co-crystal with m-cresol. One supramolecular synthon was formed between *m*-cresol and urea with the help of intermolecular mutual recognition and self-assembly process among them. Consequently, *m*-cresol can be effectively separated from *p*-cresol by the way of selective co-crystallization. The *m*-cresol _urea co-crystal (MC_U) was prepared on 1:1 molar ratio, which exhibits a regular hexagonal shape and belongs to orthorhombic system with space group P b c a and cell parameters a = 11.2502(5)Å, b = 7.1880(4) Å, c = 21.7868(15) Å. The hydroxyl group of *m*-cresol forms a H-bond with the carbonyl group in amide moiety of urea. The MC_U molecules adopt a chained layer structure. The solid-state properties of MC_U were characterized using X-ray Power Diffractometry (PXRD), Differential Scanning Calorimetry (DSC), ^{1}H nuclear magnetic resonance (^{1}H NMR) and Fourier Transformed Infrared Spectrometer (FTIR). Meantime, the selective co-crystallization process was on-line monitored by process analytical technologies/tools (PAT), such as Raman spectroscopy (Raman), Attenuated Total Reflection Fourier Transformed Infrared Spectroscopy (ATR-FTIR) and Focused Beam Reflectance Measurement (FBRM). Combined with the on-line monitoring data, ^{1}H NMR and the single crystal structure data, the mechanism of molecular recognition and self-assembly between m-cresol and urea is revealed. It was found that the whole co-crystal formation process can be divided into three steps: (1) *m*-cresol-urea dimers formation process, (2) MC_U nucleation process, and (3) MC_U growth process. And the reason for the selective co-crystallization is elucidated from the molecular level by using the corresponding calculation method. The mechanisms of selective co-crystallization as well as molecular recongnition and self-assembly between *m*-cresol and urea were proposed by analyzing the structure of *m*- and *p*-cresol, the single crystal structure data of cocrystal and the on-line monitoring data of Raman and ATR-FTIR. It was found that a chain supramolecular synthon was formed by the carbonyl oxygen (C=O) of the urea and the hydroxyl hydrogen (O-H) of the *m*-cresol relying on stronge hydrogen bonds. However, no obvious non-covalent bonds between *p*-cresol and urea molecules were observed. This failure is probably due to the position of the phenolic hydroxyl in toluene, which is affected by the methyl group. The electron cloud density of hydroxyl groups on *p*-cresol is reduced because of the presence of methyl groups, which could lead to the decrease of the electronegativity of oxygen, and thus will reduce the degree of exposure to the hydrogen atom attached to it. Obviously, *m*-cresol _urea co-crystal could be formed because of the contribution of the more exposed hydrogen atoms.

Continuous co-crystallization of benzoic acid and isonicotinamide by mixing-induced supersaturation: Exploring opportunities between reactive and antisolvent crystallization concepts

Vaclav Svoboda[1], Pól MacFhionnghaile[2], John McGinty[1], Lauren E. Connor[3], Iain D.H. Oswald[4], Jan Sefcik[2,*]

1. EPSRC Doctoral Training Centre in Continuous Manufacturing and Crystallisation, Department of Chemical and Process Engineering, University of Strathclyde, James Weir Building, 75 Montrose Street, Glasgow, G1 1XJ, United Kingdom

2. EPSRC Centre for Innovative Manufacturing in Continuous Manufacturing and Crystallisation, Department of Chemical and Process Engineering, University of Strathclyde, James Weir Building, 75 Montrose Street, Glasgow, G1 1XJ, United Kingdom

3. Collaborative International Research Programme, Strathclyde Institute of Pharmacy and Biomedical Sciences, University of Strathclyde, Technology and Innovation Centre, 99 George Street, Glasgow, G1 1RD, United Kingdom

4. Strathclyde Institute of Pharmacy and Biomedical Sciences, University of Strathclyde, 161 Cathedral Street, Glasgow, G4 0RE, United Kingdom

Corresponding Author:
Prof. Jan Sefcik
Department of Chemical and Process Engineering,
University of Strathclyde,
James Weir Building,
75 Montrose Street,
Glasgow,
G1 1XJ,
United Kingdom
jan.sefcik@strath.ac.uk

ABSTRACT

This study combines reactive and antisolvent crystallization concepts via mixing-induced supersaturation to demonstrate a wider range of options for solvent system selection in multicomponent crystallization. This approach was applied to investigate continuous crystallization of 1:1 and 2:1 co-crystals of benzoic acid and isonicotinamide. Design of Experiments was used to identify conditions where pure co-crystal phases are obtained and a continuous mixing-induced co-crystallization process was implemented to selectively produce either 1:1 or 2:1 co-crystals.

INTRODUCTION

Multicomponent crystalline materials are composed of more than one molecule in its crystal lattice. This includes co-crystals, solid solutions, solvates, and salts. The interest in multicomponent systems is a growing trend within the pharmaceutical industry[1]. Co-crystals are of benefit due to their ability to tailor physical and pharmaceutical properties of active pharmaceutical ingredients (APIs) such as solubility, bioavailability, stability, and the processability of the solid powder within industrial manufacturing processes[2]. They can also be utilized to expand IP portfolios[3]. However, co-crystallization is inherently more complex than single component crystallization as it involves an additional component with new potential solid phases. Having additional components and related process variables makes navigating the phase diagram and crystallization process more challenging.

Several methods for manufacturing co-crystals have been previously reported including both dry powder processes and solvent based processes, typically in batch processing mode. Dry powder processes are mechanochemistry based and utilize grinding at room temperature[4] or cryo-temperatures[5], polymer assisted grinding[6], high shear granulation[7] or hot melt extrusion[8]. Co-crystals from solution-based processes have been reported from evaporative and cooling[9], antisolvent[10] or reactive crystallization[11]. Cooling co-crystallization has previously been implemented in continuous platforms[12,13] and with co-crystal stoichiometry control[14]. Co-crystallization using antisolvent has been previously studied in batch processes[10], but not using continuous methods. Antisolvent crystallization of single component materials has been shown adaptable to continuous manufacturing and scale up in a number of processes[15–17]. Reactive co-crystallization mentioned in the work of Rodríguez-Hornedo et al.[11] does not involve a chemical reaction as such but a solution-based formation of a multicomponent solid phase from a mixture of two solutions undersaturated with respect to individual components using the same solvent.

Processes which rely on mixing to induce supersaturation such as reactive and antisolvent crystallizations, are well amenable to scale-up under continuous conditions[18]. Continuous manufacturing of pharmaceuticals can provide benefits such as decreased plant footprint, easier scale-up, shorter lead times and better unit performance through process intensification[19]. Since mixing can have strong impact on generation of supersaturation profile and subsequent nucleation, especially under kinetically controlled conditions[20], well-controlled mixing is essential for a control over final crystal properties, such as solid form and particle size distribution.

Antisolvent and reactive crystallization approaches can be combined in order to design a process where supersaturation is induced by mixing and is determined by the position in a quaternary phase diagram involving two crystal co-formers and two solvents. Depending on the shape of the phase diagram, which often shows a highly unsymmetrical nature[21], the co-crystal would be the thermodynamically most stable phase under some conditions. However, sometimes it may not be readily crystallized due to kinetic limitations (slow nucleation or growth) even if thermodynamically favored. Munshi and co-workers pointed out opportunities to use mixed solvents in controlling solid phase outcomes in cooling co-crystallization[22]. Solvent selection, a key design choice in crystallization, becomes more challenging in multicomponent systems. The key parameter of co-crystallization process design is the supersaturation with respect to the co-crystal phase, rather than the supersaturations of the individual co-formers. While having four components increases the complexity of mapping the phase diagram, it also allows for more options how to access solid phase regions which might not be easily accessible at a fixed solvent composition. For example, it may be possible to start with a solution of both coformers undersaturated in one solvent and add a second solvent to induce supersaturation, as in antisolvent crystallization. Alternatively, one can start with one co-former undersaturated in a given solvent mixture, and the other co-former undersaturated in the same solvent mixture and generate supersaturation by mixing these two solutions together, as in reactive crystallization. These decisions will be driven by the shape of the phase diagram and the nature of the target solid phase.

This study combines antisolvent and reactive crystallization concepts to develop a continuous co-crystallization process to produce benzoic acid (BZA) – isonicotinamide (INA) co-crystals. These molecules form two different co-crystals in 2:1 and 1:1 stoichiometric ratios which have been previously isolated and characterized from small scale cooling crystallization[23,24]. In this work, benchtop screening crystallization experiments were scaled to run in a continuous process. A Design of Experiment was carried out for the benchtop screening to better understand the effects of different process conditions on the crystallization of the two co-crystal phases. DoE screening was done in order to cut down the time and number of experiments to find suitable continuous crystallization conditions to selectively produce either 1:1 or 2:1 co-crystals by continuous crystallization.

EXPERIMENTAL SECTION

Materials. Benzoic acid (≥99.5%), isonicotinamide (99%), and ethanol (≥99.8%) were supplied by Sigma-Aldrich (Gillingham, UK). Deionised water was produced using the in-house Millipore Milli-Q system.

Batch Screening. Two sets of screening experiments were carried out. The first was an initial screening and the second was a systematic Design of Experiments (DoE) approach used for mapping of a limited region of the quaternary phase diagram (BZA, INA, Water, Ethanol) in order to identify suitable operating conditions for continuous crystallization. For both sets of screening experiments, an aqueous solution of isonicotinamide was added to an ethanolic solution of benzoic acid in a 20 mL vial, illustrated in Figure 1. Solutions were mixed at various ratios to obtain the total of 10g of solution and agitated using magnetic stirrer bar for 10 minutes past observed nucleation. All experiments were carried out at 25°C. For the initial screen, solutions of isonicotinamide in water and benzoic acid in ethanol were prepared at concentrations of 68.1 g/kg water and 213 g/kg ethanol, respectively, and mixed in ratios of 2:1, 1:1, 1:2, 1:4 and 1:9 by mass. The solid product obtained was filtered using 0.2 μm PTFE filter without washing, dried at 40°C for 24 hours and analyzed by X-ray powder diffraction (see solid characterization section). The design space for DoE was selected based on information from the initial screen. Experimental plan was created using MODDE in a 2 level full factorial design. The experimental worksheet, model fitting and results analysis was carried out in MODDE by Umetrics. Data were fitted using Partial Least Squares (PLS). The three variable parameters chosen for the DoE were: BZA-Ethanol solution concentration, INA-Water solution concentration, and the mass ratio of the two solutions mixed. Values of these three parameters combined determine the final composition of the mixture and thus the position on the phase diagram. The responses measured were induction time, solid yield, solid phases present and resulting slurry flow properties. Induction time was estimated as time after mixing until first particles were visually observed. Solid yield was taken as a percentage of total solute that was recovered as solids by filtration 10 minutes after the estimated induction time while agitating. Slurry flow was assessed by a rating system where a number was assigned on a scale from 1 to 5: 1 for an easily flowing mixture and 5 for a slurry too thick to flow with gravity. Solid phase was determined by XRPD (see solid characterization section). Based on this DoE, conditions for the production of 2:1 co-crystal were selected. Although the initial DoE was aimed to find suitable conditions for crystallization of both co-crystal forms, conditions leading to crystallization of 1:1 co-crystal resulted in solid loadings which would be too high for a continuous operation. Based on results from the initial DoE, a set of further experiments was carried out in an expanded design space to find suitable conditions for crystallization of 2:1 co-crystal. These experiments were carried out at lower solution concentrations and resulting solid phases, induction times and slurry flow ratings were determined in the same way as in the initial DoE runs.

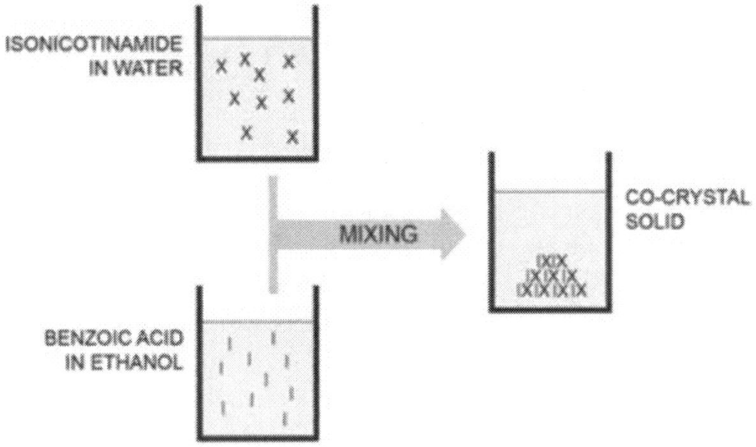

Figure 1. Solvent system used for mixing-induced co-crystallization

Solubility measurement. Solubility of the 2:1 co-crystal has been determined by a solvent addition method in the Crystalline Reactor System (Technobis), similar to a previously published procedure[25]. As compared to temperature variation or gravimetric methods, solvent addition relies on slow dilution of a suspension with a solvent (mixture) of a given composition under isothermal conditions until complete dissolution occurs, when a clear point can be detected. Solvent was added at a constant rate using PHD Ultra syringe pumps (Harvard Apparatus) to the Crystalline vials. The solubility measurement setup is shown in Figure 2. Multiple addition rates were tested (0.5 ml/hr and 0.75 ml/hr) to verify that dissolution kinetics have a negligible effect on the result at the addition rates used. Clear point was determined using both transmissivity measurement and images from the Crystalline camera. Image analysis produced more consistent results, which is in line with previous findings from Reus et al[25]. A study was also carried out to monitor the solid state transformation of the co-crystal in order to check phase stability in the same setup. Suspensions with concentrations as used for the start of solubility measurements were held for varying amount of time from 1 to 5 hours, after which the remaining solid was filtered, dried and analyzed by XRPD.

Figure 2. Technobis Crystalline and syringe pumps for solubility determination using the solvent addition method

Continuous crystallization. Continuous crystallization runs were performed using a concentric capillary mixer (Figure 3) as well as the Ehrfeld modular micro-reactor system equipped with Valve Mixer 30 module (Figure 4). The BZA-Ethanol solution was supplied to the concentric capillary mixer through the inner PEEK capillary, while the INA-Water solution was fed to the outer glass tube as shown in Figure 3. The capillary has internal and outer diameters of 0.51mm and 1.59mm, respectively. The outer glass tube has an internal diameter of 3mm. The capillary stream entered into the outer tube stream 6 cm from the T-junction. The feed solutions were pumped using Bronkhorst Mini CORI-FLOW system coupled with gear pumps allowing for accurate control of mass flow rates. The resulting slurry from the mixer was sampled 0.7m from the mixing point at the start and end of experiments. Collected slurry was filtered after a holding period in a Buchner funnel with a 0.45µm filter paper and dried overnight at 40°C. The slurry holding time without agitation was 10 minutes for 2:1 co-crystal runs, and 18 minutes for 1:1 co-crystal run. A camera was used for visual analysis of any fouling in the glass tube. Temperature of both streams was measured throughout experiments and it varied less than 2°C from 25°C. The total mass flow rates through the capillary mixer were 20, 40 and 60 g/min in 50:50 (w/w) ratio for producing the 2:1 co-crystal and 115 g/min at a 15:100 (w/w) ratio of benzoic acid solution flow rate to isonicotinamide solution flow rate for the 1:1 co-crystal. In order to demonstrate the transferability of the process to a commercial platform, the Ehrfeld modular micro-reaction system, shown in Figure 4, was used at the total mass flow rate of 20 g/min for producing the 2:1 co-crystal.

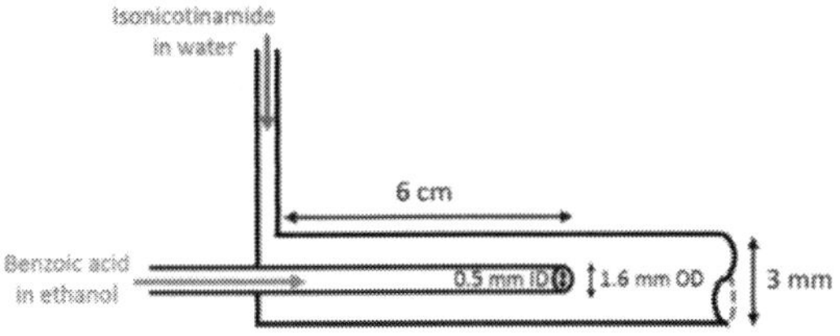

Figure 3. Concentric capillary mixer diagram with dimensions indicated

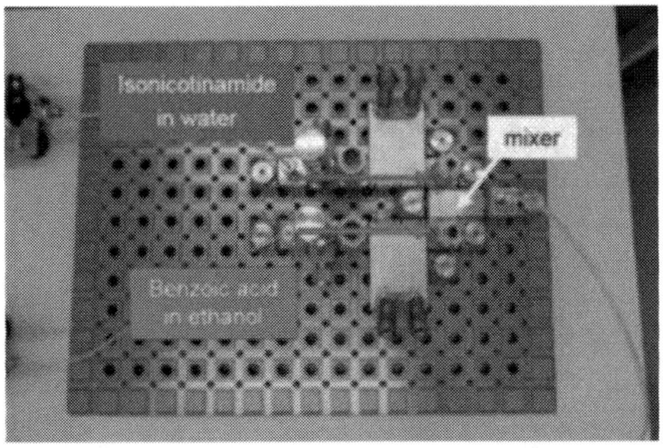

Figure 4. Ehrfeld modular micro-reaction system fitted with valve assited mixer 30

Solid characterization. Filtered and dried crystalline powder from the screening and continuous experiments was analyzed by X-ray Powder Diffraction (XRPD) and IR spectroscopy. Samples from continuous runs were further analyzed by Differential Scanning Calorimetry (DSC) paired with Thermogravimetric Analysis (TGA), and microscope image analysis for particle size measurement. Solids from continuous runs targeted to produce the 2:1 co-crystal were also analyzed by NMR. XRPD fingerprinting was performed on a sample placed in a 28 well plate, supported by Kapton film (7.5 µm thickness). Data were collected on a Bruker AXS D8 Advance transmission diffractometer equipped with θ/θ geometry, primary monochromatic radiation (Cu Kα_1 λ = 1.54056 Å), a Braun 1D position sensitive detector, and an automated multiposition x–y sample stage. Data were collected from 4 to 35° 2θ with a 0.015° 2θ step size and 1 s step^{-1} count time. FT-IR measurements were taken using Bruker Tensor II, using 32 scans with 4 cm-1 resolution from 450 to 4000 cm^{-1} with diamond tip ATR sampling plate. DSC/TGA data was obtained from Netszch STA 449 F1 Jupiter. Measurements for DSC/TGA were taken from 20 to 180 °C with a ramp of 10°C/min. For Particle Size Distribution (PSD), image analysis was carried out on dry powder using Malvern Morphologi G3 using low pressure dispersion and 2.5x optics. ^{1}H and ^{13}C NMR analysis was carried out using Bruker Advance 3 at 400 MHz by dissolving the solid product in deuterated DMSO in 5mm NMR vials.

RESULTS and DISCUSSION

Mixing-induced supersaturation. A case study on benzoic acid – isonicotinamide co-crystallization presented here explores the space between reactive and antisolvent crystallization concepts in generating mixing-induced supersaturation to target specific solid phases of co-crystals. The illustrative phase diagram in Figure 5 is similar to the phase diagrams presented previously for the system investigated here[23]. We use it to demonstrate different modes of inducing supersaturation through mixing. The blue lines illustrate examples of reactive and antisolvent crystallization. In reactive crystallization, two undersaturated solutions, each containing a single coformer in the same solvent mixture, are mixed to supersaturate with respect to a desired co-crystal solid. In antisolvent crystallization, an undersaturated solution containing both coformers in one solvent is mixed with antisolvent to generate supersaturation. The red line illustrates a new approach used in this study, where antisolvent and reactive crystallization concepts were combined to target a specific region of the phase diagram, in this case 2:1 and 1:1 co-crystal regions. In this combined approach, two undersaturated solutions, each containing a single co-former in a single (pure) solvent, are mixed to generate supersaturated solution with respect to desired co-crystal phase in the mixed solvent.

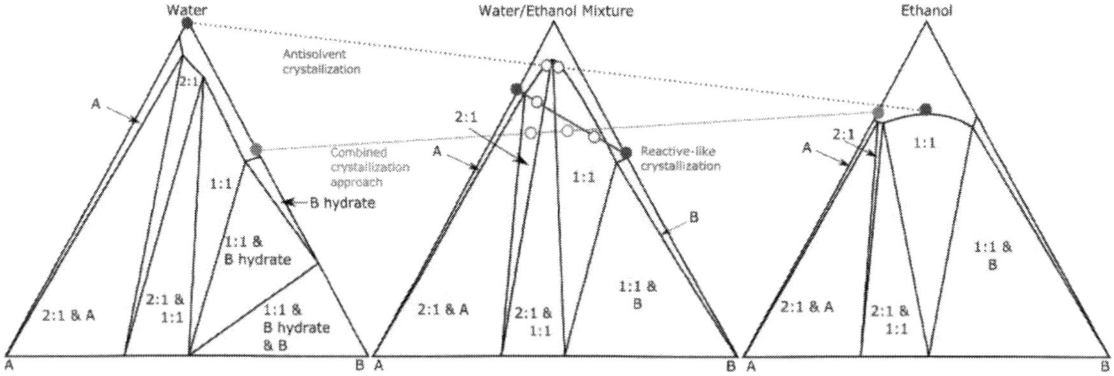

Figure 5. Illustration of three modes of inducing supersaturation through mixing: Reactive, antisolvent and combined approach. Targeted solid phases are 1:1 and 2:1 co-crystals of A and B and solvent is a mixture of water and ethanol. Dotted lines do not correspond to lever rule.

Screening. An initial screen was used to determine design space boundaries and was followed by a detailed DoE driven screen to map a design space corresponding to a region of interest in the phase diagram. XRPD from the initial screen has shown that the method of mixing solutions of isonicotinamide in water and benzoic acid in ethanol can produce both 2:1 and 1:1 co-crystals. DoE responses were chosen as key outcomes relevant for development of a continuous crystallization process. High solid loadings at the end of experiments limited solution concentrations from above, as too high solid loadings could lead to downstream blockages. On the other hand, long induction times limited solution concentrations from below, as relatively rapid nucleation is desirable. Due to the probabilistic nature of primary nucleation, induction time estimates from batch experiments are only intended to give an approximate guidance, nevertheless there was a good consistency among repeated experiments.

Despite not necessarily providing direct information on phase stability under thermodynamic equilibrium conditions, the DoE mapping provided a rapid way to find potentially suitable crystallization conditions while also giving indication about kinetics. Contour plots displayed in Figure 6 show responses at a fixed solution ratio where at higher concentrations of starting solutions the yield is higher, slurry flow rating diminishes due to higher solid loading and induction time decreases due to higher supersaturation. The co-crystal phase formed is primarily dependent on the solution flow ratio rather than the solution concentrations within the design space investigated. Contour plots for other solution ratios are located in supplementary information (ESI Figures 3-5). After targeting the desired phase by adjusting the solution flow ratio, slurry flow ratings of 2-3 were deemed as acceptable for a continuous crystallizer. The induction time model prediction is the least reliable as can be expected for a probabilistic process. The design space selected was large enough to find suitable conditions for the crystallization of the 2:1 co-crystal, while the conditions to crystallize the 1:1 co-crystal phase had to be extrapolated and subsequently validated. The extrapolation was done by choosing various concentrations of initial solutions outside of the original design space and analysing the same response parameters. The conditions from the DoE chosen as most suitable to produce the 2:1 co-crystal are 50/50 w/w solution flow ratio, with the solution concentrations of 109.4 g BZA/kg ethanol and 48.5 g INA/kg water. For the 1:1 co-crystal the optimal conditions were the solution flow ratio of 13/84 w/w (benzoic acid solution to isonicotinamide solution) with the solution concentrations of 30g BZA/kg ethanol and 40g INA/kg Water.

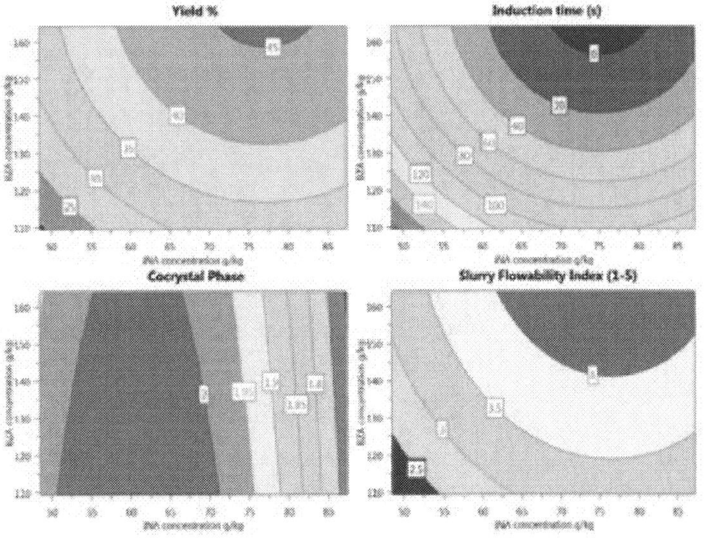

Figure 6. Contour plot showing responses from the Design of Experiments at a fixed solution ratio of 50:50 (w/w)

Solubility. During the phase stability test, the 2:1 co-crystal began to transform to the 1:1 co-crystal over time at the crystallizer outlet conditions. Transformation commenced between 158 to 264 minutes in an isothermal stirred environment as detected by XRPD, see Figure 7. This makes solubility determination more difficult as the 2:1 solid form is not thermodynamically stable at the chosen process conditions. However, the solid-liquid equilibrium of the metastable solid phase was determined for the purpose of thermodynamic yield calculations. The solubility test was timed to avoid any significant transformation. Solubility measurements were carried out in 48.8% (w/w) ethanol/water mixture, which is the same as the solvent mixture at the continuous mixer outlet during crystallization of 2:1 co-crystal. The results from six experiments suggest the solubility to be 0.0446 g/ml solvent with a high level of reproducibility based on clear points from image analysis. The spread of various measurements is illustrated in Figure 8. The high spread of clear point measurements based on transmission may be attributed to the limited ability to detect solids by transmission measurements at low concentrations, especially for thin needles. The different solvent addition rates show comparable results, thus the result is not dependent on the kinetics of crystal dissolution. The findings about the 2:1 co-crystal metastability point out that the outlet composition of the crystallizer slurry is either in the 1:1 co-crystal or a mixture of the 1:1 and 2:1 co-crystal regions on the phase diagram. The solubility of the 2:1 co-crystal has been previously reported between 0.0367 g/ml and 0.0482 g/ml for 50% water/ethanol mixture by[26] using two methods. However, the measurements reported therein have a high variability and the gravimetry method implemented did not seem to account for a possibility of a phase transformation. Therefore, it is possible that in some cases the solubility of the 1:1 co-crystal or a mixture of both co-crystal phases could have been measured. A recent study by Munshi et al.[22] shows the ternary phase diagram at 50% ethanol/water, although it was not reported whether this was 50% by mass or by volume[22]. As we have shown, solvent mediated transformation may occur in this system under conditions studied in this and previous works, and this may perhaps also help to explain the unusual shape of the phase diagram reported in the work of Munshi et al.

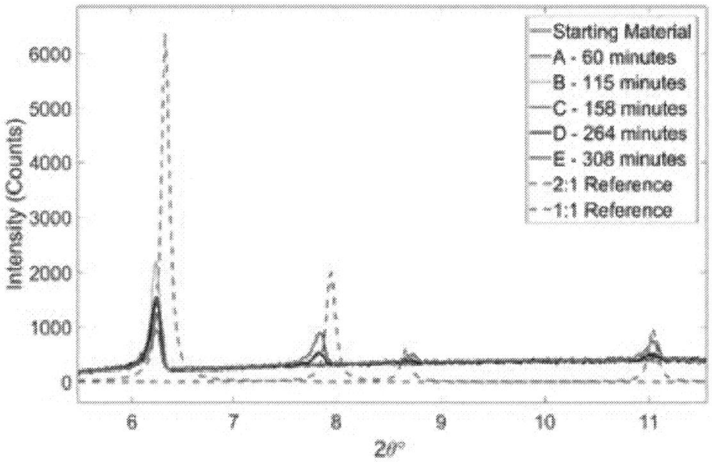

Figure 7. XRPD from suspension stability study of 2:1 co-crystal in 48.8% ethanol – water (w/w)

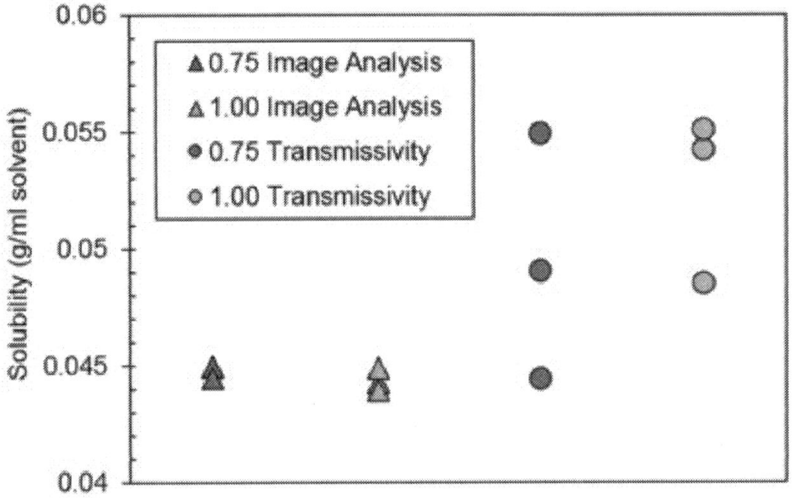

Figure 8. Solubility of 2:1 co-crystal in 48.8% ethanol – water (w/w) at 25 °C

Continuous crystallization. Moving from benchtop vial-based screening to a continuous platform introduces new process parameters and challenges. The overall flow rate was found to have a strong impact on the mixing regime and thus final particle properties. Three different overall flow rates were examined with the aim of producing the 2:1 co-crystal: 60 g/min, 40 g/min and 20 g/min at 50:50 ratio (w/w) of solution streams. The continuous runs revealed fouling and subsequent blockages to be a challenge, however blockage issues decreased with lower flowrates. Antifouling measures such as low power sonication could be used to mitigate this issue[27]. The solid output was 0.507 g/min, 0.342 g/min and 0.0259 g/min for overall flowrates of 60 g/min, 40 g/min and 20 g/min, respectively. The L50 of particles obtained at overall flow rates of 60 g/min, 40 g/min and 20 g/min was 361μm, 317μm and 238μm, respectively. Despite producing the desired solid form (Figure 10) from all conditions, mixing performance drastically reduced at the overall flow rate of 20 g/min resulting in a lower solid yield. The low velocity difference between the capillary jet and the outer flow did not provide sufficient mixing to rapidly achieve the desired supersaturation. The same process at the overall flow rate of 20 g/min resulted in 0.0484 g/min solid output in the Ehrfeld platform. Clearly, mixing conditions influence nucleation kinetics of co-crystal as the solid yield and resulting particle size distribution (see ESI Figure 6) varies with overall flowrate although the overall solution composition is the same in all cases. In order to achieve a higher yield in the crystallization process, increasing of residence time downstream from the mixing point combined with additional cooling stages should be considered. Further process optimization would benefit from more detailed knowledge of the relevant phase diagram and additional experiments in an expanded design space.

A continuous crystallization was also carried out to produce the 1:1 co-crystal, using flow rates of 15 g/min of benzoic acid solution through the capillary and 100 g/min of isonicotinamide solution through the outer tube. For this experiment the solutions were less concentrated to avoid blocking due to potentially high solid loading of the resulting slurry. The induction time at these concentrations was relatively long compared to the 2:1 co-crystal experiments and particles were not immediately visible in the mixer outlet. The solid output after holding the slurry for 18 minutes was 0.0726 g/min.

Solid characterization. Solid state characterization techniques were selected in order to establish solid phase identity and purity. Crystal unit cell parameters were previously documented for both co-crystal phases which makes powder X-ray diffraction suitable for phase verification. The co-crystal phases also have unique melting points verifiable by DSC. NMR was also used to independently verify the stoichiometry of co-formers in the crystalline sample. The morphological differences in crystals for the 2:1 and 1:1 forms were minor. The needle shaped morphology of both co-crystal forms can be seen in Figure 9. Results from Malvern Morphologi G3 are shown in supplementary information and similar aspect ratio distributions were found for both co-crystal forms. Analysis by XRPD on the dry samples from the capillary mixer is shown in Figure 10 (XRPD on samples from the Ehrfeld platform are in supplementary information ESI Figure 9). Lines A, B and C are samples from continuous runs designed to produce the 2:1 co-crystal. Simulated patterns are generated from unit cell data available from the Cambridge Structural Database under structure labels: BUDWEC[28] and MOVTOH[23] for the 1:1 and 2:1 co-crystals respectively. Comparison with D – the simulated pattern suggests that a pure 2:1 co-crystal phase has been produced by all runs. Similarly, line E shows pure 1:1 co-crystal by comparison to the 1:1 co-crystal simulated pattern – line F. This is further supported by results from DSC shown in Figure 11. DSC for the 2:1 co-crystal samples shows a sharp thermal event previously reported at 142 – 143°C for the 2:1 co-crystal[29]. The TGA curves also show a decrease in mass from 148-150°C onwards indicating sample decomposition. Similarly, DSC for 1:1 sample shows a unique endotherm above 160 °C, at the expected melting point of the 1:1 co-crystal[29] with subsequent decomposition.

For additional confirmation, the samples were analyzed using ATR-IR. The absorbance spectra are in Figure 12. The figures show a spectrum for each starting material as well as samples from continuous experiments. It is clear that the co-crystal phases have a well-defined unique spectra and it is possible to distinguish the 2:1 and 1:1 co-crystals using IR spectrometry.

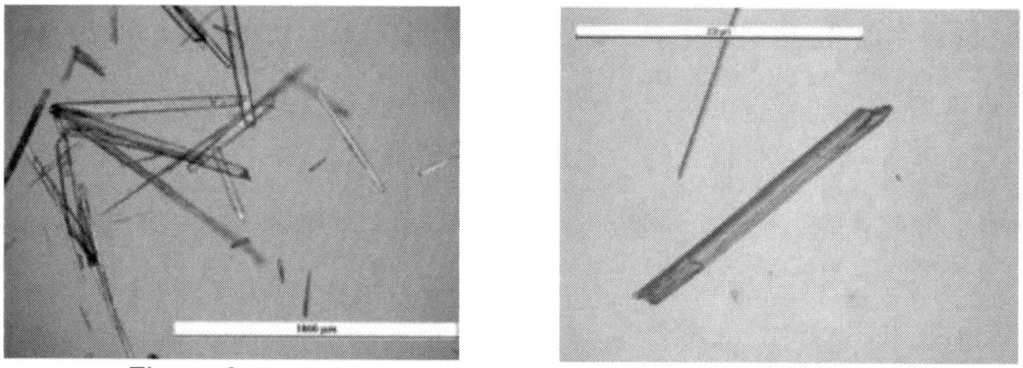

Figure 9. Optical microscope images: 2:1 co-crystal (Left) and 1:1 co-crystal (Right)

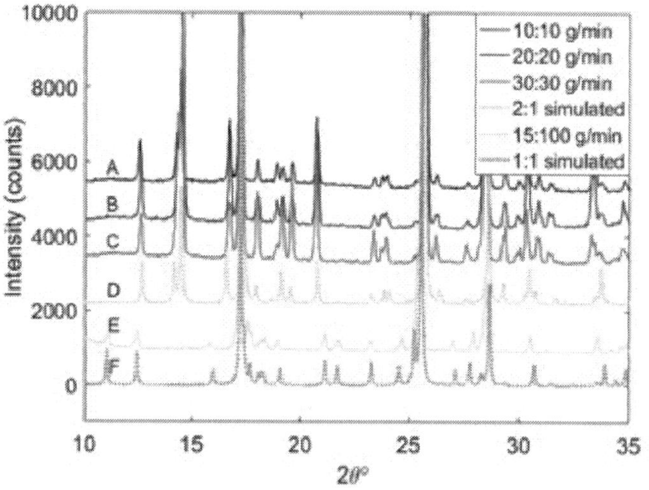

Figure 10. XRPD for samples from all continuous runs together with calculated patterns

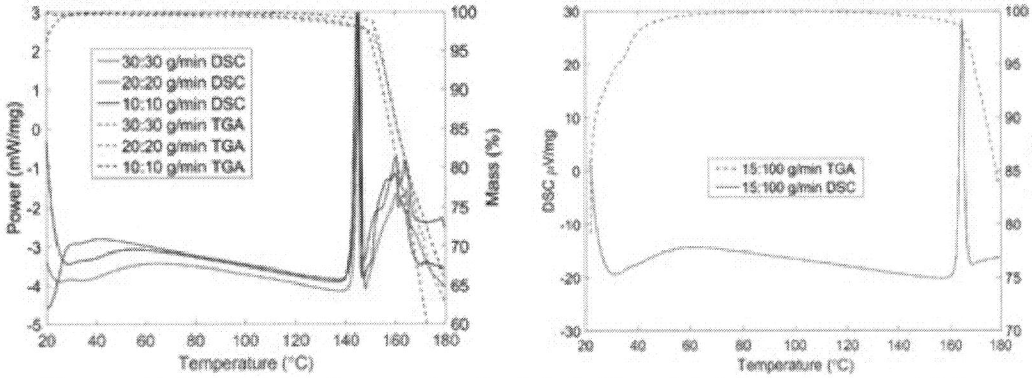

Figure 11. DSC/TGA for samples for 2:1 (left) and 1:1 (right) co-crystal from continuous runs

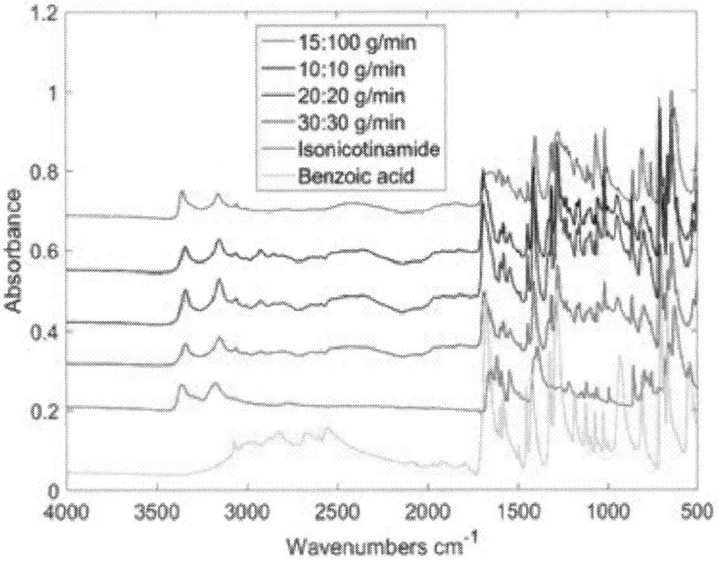

Figure 12. ATR-IR spectra for starting material and samples from continuous runs

NMR analyses on solid samples dissolved in deuterated DMSO from each continuous run confirmed (within 2% error) a 2:1 ratio for each sample from the continuous runs aimed to produce 2:1 co-crystal (ESI Figure 8). Continuous experiments using 40 g/min and 60 g/min flow rates produced particles of very similar particle size and shape characteristics. Results from 20 g/min continuous experiment show significantly smaller particles at much lower solid yield (ESI Figure 6). The aspect ratios for the particles from 20 g/min run are more needle-like with a narrower distribution.

We showed that both 2:1 and 1:1 pure forms of the co-crystal can be produced using the combined approach method based on mixing-induced supersaturation in a continuous crystallization process. The combined antisolvent and reactive crystallization approach allows for additional flexibility in reaching specific phase regions in multicomponent systems using multiple solvents.

CONCLUSIONS

We have used Design of Experiments to find conditions suitable for formation of 2:1 and 1:1 co-crystals of benzoic acid and isonicotinamide by mixing ethanolic solutions of benzoic acid and aqueous solutions of isonicotinamide in isothermal conditions. The input parameters were the two solution concentrations and the mass ratio of the two solutions, while the output variables were the solid form produced, solid yield, induction time and slurry flow rating. Optimal conditions were then identified for both 2:1 and 1:1 co-crystal formation and continuous crystallization experiments were performed at several total flow rates. Both solid forms were selectively produced in continuous crystallization as confirmed by XPRD, IR, DSC and NMR. Mixing conditions were found to influence the solid yield and particle size distribution of crystal produced although not the solid form.

The combined antisolvent and reactive crystallization approach can be a useful tool for reaching solid phases that are otherwise difficult to access through mixing-induced supersaturation. Under conditions of thermodynamic control, it may be possible to reach certain regions of phase diagrams that would be not accessible by other approaches due to non-ideal shapes of some phase diagrams. Under conditions of kinetic control, it may be possible to influence nucleation kinetics through the initial mixing of two solutions which would depend on compositions of both solutions as well as mixing conditions applied.

ACKNOWLEDGMENT

We would like to thank Maria Briuglia for assistance with Technobis Crystalline. The authors would like to thank EPSRC and the Doctoral Training Centre in Continuous Manufacturing and Crystallisation (Grant Ref: EP/K503289/1) for funding Vaclav Svoboda and John McGinty. The authors would like to thank EPSRC and Centre for Innovative Manufacturing in Continuous Manufacturing and Crystallisation (Grant Ref EP/I033459/1) for funding Pól MacFhionnghaile. The authors would like to thank the International Strategic Partnership between University of Strathclyde and Nanyang Technological University Singapore for funding Lauren E. Connor. The authors would like to thank EPSRC for Fellowship (Grant Ref EP/N015401/1) for funding Iain D.H. Oswald. The authors would like to acknowledge that this work was carried out in the CMAC National Facility supported by UKRPIF (UK Research Partnership Fund) award from the Higher Education Funding Council for England (HEFCE) (Grant ref HH13054).

REFERENCES
(1) Domingos, S.; André, V.; Quaresma, S.; Martins, I. C. B.; Minas da Piedade, M. F.; Duarte, M. T. J. Pharm. Pharmacol. 2015, 67, 830–846.
(2) Elder, D. P.; Holm, R.; Diego, H. L. de. Int. J. Pharm. 2013, 453, 88–100.
(3) Trask, A. V. Mol. Pharm. 2007, 4, 301–309.
(4) Friščić, T.; Jones, W. Cryst. Growth Des. 2009, 9, 1621–1637.
(5) Macfhionnghaile, P.; Hu, Y.; Gniado, K.; Curran, S.; Mcardle, P.; Erxleben, A. J. Pharm. Sci. 2014, 103, 1766–1778.
(6) Hasa, D.; Carlino, E.; Jones, W. Cryst. Growth Des. 2016, 16, 1772–1779.
(7) Rehder, S.; Christensen, N. P. A.; Rantanen, J.; Rades, T.; Leopold, C. S. Eur. J. Pharm. Biopharm. 2013, 85, 1019–1030.
(8) Moradiya, H. G.; Islam, M. T.; Halsey, S.; Maniruzzaman, M.; Chowdhry, B. Z.; Snowden, M. J.; Douroumis, D. CrystEngComm 2014, 16, 3573.
(9) Wittering, K.; King, J.; Thomas, L.; Wilson, C. Crystals 2014, 4, 123–140.
(10) Lee, M.-J.; Chun, N.-H.; Wang, I.-C.; Liu, J. J.; Jeong, M.-Y.; Choi, G. J. Cryst. Growth Des. 2013, 13, 2067–2074.
(11) Rodríguez-Hornedo, N.; Nehm, S. J.; Seefeldt, K. F.; Pagán-Torres, Y.; Falkiewicz, C. J. Mol. Pharm. 2006, 3, 362–367.
(12) Zhao, L.; Raval, V.; Briggs, N. E. B.; Bhardwaj, R. M.; McGlone, T.; Oswald, I. D. H.; Florence, A. J. CrystEngComm 2014, 16, 5769.
(13) Powell, K. A.; Bartolini, G.; Wittering, K. E.; Saleemi, A. N.; Wilson, C. C.; Rielly, C. D.; Nagy, Z. K. Cryst. Growth Des. 2015, 15, 4821–4836.
(14) Powell, K. A.; Croker, D. M.; Rielly, C. D.; Nagy, Z. K. Chem. Eng. Sci. 2016, 152, 95–108.
(15) Alvarez, A. J.; Myerson, A. S. Cryst. Growth Des. 2010, 10, 2219–2228.
(16) Brown, C. J.; Adelakun, J. A.; Ni, X. Chem. Eng. Process. Process Intensif. 2015, 97, 180–186.
(17) Ferguson, S.; Morris, G.; Hao, H.; Barrett, M.; Glennon, B. Chem. Eng. Sci. 2013, 104, 44–54.
(18) Lindenberg, C.; Schöll, J.; Vicum, L.; Mazzotti, M.; Brozio, J. Chem. Eng. Sci. 2008, 63, 4135–4149.
(19) Baxendale, I. R.; Braatz, R. D.; Hodnett, B. K.; Jensen, K. F.; Johnson, M. D.; Sharratt, P.; Sherlock, J.-P.; Florence, A. J. J. Pharm. Sci. 2015, 104, 781–791.
(20) Bałdyga, J.; Makowski, Ł.; Orciuch, W. Chem. Eng. Res. Des. 2007, 85, 745–752.
(21) Leyssens, T.; Springuel, G.; Montis, R.; Candoni, N.; Veesler, S. Cryst. Growth Des. 2012, 12, 1520–1530.
(22) Munshi, T.; Redha, B.; Feeder, N.; Meenan, P.; Blagden, N. Cryst. Growth Des. 2016, 16, 1817–1823.
(23) Seaton, C. C.; Parkin, A.; Wilson, C. C.; Blagden, N. Cryst. Growth Des. 2008, 9 (1), 47–56.
(24) Boyd, S.; Back, K.; Chadwick, K.; Davey, R. J.; Seaton, C. C. J. Pharm. Sci. 2010, 99, 3779–3786.
(25) Reus, M. A.; van der Heijden, A. E. D. M.; ter Horst, J. H. Org. Process Res. Dev. 2015, 19, 1004–1011.
(26) Redha, B. H. Impact of mixed solvent on co-crystal solubility, ternary diagrams and crystallisation scale-up. Crystallisations of Isonicotinamide- Benzoic Acid Co-crystals from Ethanol-Water Co-solvent System., University of Bradford, 2013.
(27) Al Nasser, W. N.; Pitt, K.; Hounslow, M. J.; Salman, A. D. Powder Technol. 2013, 238, 151–160.
(28) Aakeröy, C. B.; Beatty, A. M.; Helfrich, B. A. Angew. Chem. Int. Ed. 2001, 40, 3240–3242.
(29) Buanz, A.; Prior, T. J.; Burley, J. C.; Raimi-Abraham, B. T.; Telford, R.; Hart, M.; Seaton, C. C.; Davies, P. J.; Scowen, I. J.; Gaisford, S.; Williams, G. R. Cryst. Growth Des. 2015, 15, 3249–3256.

497864 A Multiscale Computational Method for Prediction of Polymorphs

Abstract:

Physical properties of solid pharmaceutical drugs stem from their crystalline structure or polymorph. Pharmaceutical molecules can undergo polymorphism to attain more than one crystalline structure depending on the crystallization conditions. Experimental screening of polymorphs is expensive and time-consuming. Therefore, the accurate methods to predict polymorphs have become essential to build in quality into the pharmaceutical products. The most prominent method to predict crystalline polymorph is the lattice energy minimization technique, which uses appropriate sampling and optimization algorithms to identify space groups and lattice energy. This approach has limitations in treating the solute-solvent interactions and capturing the effect of additives, solvents, co-solvents and hydrodynamic conditions of the crystallizer. Therefore, there is a need to develop a model which considers these effects involved in multiple length and time scales during the crystallization process. Here we propose a consistent multiscale approach to predict dynamics of polymorph formation, which includes hierarchical simulation of sequential events â€“ i) lattice formation via self-assembly of molecules, ii) growth of lattice and formation of molecular clusters, and iii) development of faces and evolution of faceted crystal.Â The effect of the solute-solvent interactions, additives, solvents, co-solvents and hydrodynamic conditions are embedded into the calculation of free energy barriers for molecular attachment/detachment in these three events. An example of polymorphism in Glutamic acid crystals using the multiscale method will be shown.

499540 Polymorph Selection By Continuous Precipitation

Abstract:

Due to polymorphic variation in mechanical, electrical, stability, bioavailability and physicochemical properties, there always exists an optimal polymorph for a given application. Sometimes, a thermodynamically metastable form is preferred. In these cases, intelligent process design must be utilized to produce the desired polymorph. By judiciously selecting a solvent, temperature, inlet supersaturation, and residence time, one can typically produce a pure, preferred polymorph in a continuous crystallization process regardless of the polymorphâ€™s thermodynamic stability.Â Design rules exist to guide the process engineer for systems in which heterogeneous nucleation and size independent linear crystal growth are the dominant mechanisms of solid deposition from solution. [1] Here, we hypothesize that these findings remain qualitatively true in precipitative systems in which homogeneous nucleation and/or agglomeration are also important crystal growth mechanisms. This hypothesis is supported by our recent experiments in which we continuously produce metastable forms of agglomerative $CaCO_3$. We have also extended the population balance analysis to include systems in which agglomeration is significant. Here we report these experiments and calculations and explain the implications for general polymorphic systems that exhibit significant agglomeration.

[1] Farmer, T. C., Carpenter, C. L., & Doherty, M. F. (2016). Polymorph selection by continuous crystallization. *AIChE Journal*, *62*(9), 3505-3514.

501674 Metabolic Engineering and Synthetic Biology with Human Milk Oligosaccharides

Abstract:

Human milk oligosaccharides (HMOs) are prebiotic compounds found in breast milk that favor the growth of beneficial bacteria in the infant gut. Often these compounds are derivatives of lactose or n-acetyllactosamine that contain either a fucose or sialic acid moiety. While there are many efforts underway to synthesize these beneficial complex carbohydrates, the current standard for their detection is low-throughput HPLC-MS. We present a high-throughput, fluorescence-based detection system for fucosylated and sialylated human milk oligosaccharides in vivo in E. coli. Such high-throughput systems facilitate directed evolution for increased in vivo HMO production and we report the detection of HMOs in metabolically engineered strains. Additionally, the reporter enables the use of HMOs as orthogonal inducers of cellular circuits in synthetic biological applications. Finally, we demonstrate HMO catabolism for the production of important gut metabolites, mimicking the effect of important probiotic gut bacteria.

495215 Engineered Probiotics for the Treatment of Multidrug Resistant E.coli

Abstract:

A single extraintestinal E.coli strain, ST131, has recently emerged as the most dominant strain in clinical laboratories across the globe. ST131 causes a variety of extraintestinal infections but is notorious for those in the urinary tract (UTIs), causing up to 95% of UTIs worldwide. Since its identification in 2008, ST131 has rapidly risen to â €˜superbugâ€™ status, now being identified as the single most common multidrug resistant E.coli strain. The resistance phenotype of ST131 is primarily attributed to the H30-Rx subclone which is fluoroquinolone and extended-spectrum cephalosphorin resistant, making it increasingly difficult to treat in the clinic. There is pressing need to develop novel antimicrobial strategies to mitigate the dissemination and spread of ST131 and to more effectively treat infected patients.

Probiotics are safe-to-consume organisms that have been proposed as a potential treatment option for ST131.In collaboration with the Veteranâ€™s Affairs Medical Center, we have engineered the probiotic *E.coli* strain, Nissle 1917 (EcN), to elicit strong antagonistic activity against the H30-Rx subclone, JJ1886. We have designed an EcN strain that is capable of producing high-titers of three diverse antimicrobial peptides (AMPs), Microcin C7, Microcin V, and Microcin N, using a combination of different peptide secretion mechanisms. AMPs have been researched for years for their antibiotic properties but have failed in translational success due to their high synthesis costs and rapid degradation rates in the body. Using a probiotic host, we can overcome these hurdles enabling localized production at the site of infection. In this work, we have screened dozens of AMPs for activity against JJ1886. We have chosen three peptides that act by orthogonal mechanisms on JJ1886 which we demonstrate decreases resistance development. We also take an in-depth look at the remaining resistant subpopulation, examining growth characteristics and screening for antibiotics and human-produced peptides that have increased susceptibility on this persistent subset.

497723 Synthetic RNA-Inhibitor Antibiotics in Non-Traditional Antibiotic Pathways to Treat Multi-Drug Resistant Bacteria

Abstract:

Multidrug-resistant (MDR) infections are a pressing concern to global health which is made worse by the lack of new antibiotics. The overuse and misuse of current antibiotics has created pandrug-resistant bacteria for which there are no longer effective antibiotics treatments. In particular, an increasing class of gram-negative MDR pathogens like Enterobacteriaceae including carbapenem-resistant *Escherichia coli*Â and extended spectrum Î²-lactamase producingÂ *Klebsiella pneumoniae*Â are severely antibiotic resistant and were recently designated priority 1 critical class bacterial pathogens in urgent need of effective antibiotics by the World Health Organization. Here we show that developing synthetic sequence-specific antimicrobials enables us to target non-traditional antibiotic pathways to discover new antibiotic targets and investigate potential combination therapies. Using *E. coli* as a model organism we created peptide nucleic acid based antisense molecules which target essential genes in non-traditional antibiotic pathways including metabolism, cell signaling, and stress response and have predicted homology in varied species of Enterobacteriaceae. Although the antisense molecules were designed against essential genes in non-pathogenic, drug-sensitive *E. coli*, these molecules demonstrate therapeutic potential against pathogenic, MDR clinical isolates of *E. coli*, *K. pneumoniae*, and *S. enterica*, thus highlighting the potential to create broad-spectrum yet gene-specific antibiotics using predictive homology. The clinical isolates used in this study are highly resistant to most classes of antibiotics, yet when the antisense molecules are used in combination with traditional antibiotics we observe a strong synergistic effect that significantly inhibits cell growth greater than either mono-therapy. Our findings highlight the potential utility of applying antisense technology to discover novel targets in non-traditional antibiotic pathways.

501408 Optogenetic Regulation of Insulin Secretion in Pancreatic Beta-Cells

Abstract:

Proper insulin production by pancreatic beta-cells, which is central to blood glucose homeostasis, is orchestrated by a complex circuitry involving intracellular molecules including cyclic AMP (cAMP). Tackling aberrations in glucose-stimulated insulin release such as in diabetes with pharmacological agents, which boost the secretory capacity of beta-cells, is linked to adverse side effects. We hypothesized that a photoactivatable adenylyl cyclase (PAC) can be employed to modulate cAMP in beta-cells with light thereby enhancing insulin secretion. To that end, the PAC gene from the bacterium *Beggiatoa* (bPAC) was delivered to beta-cells. Upon illumination a cAMP increase was noted within 5 minutes and a significant drop at 12 minutes post-irradiation. The concomitant augmented insulin secretion was comparable to that from beta-cells treated with secretagogues. Greater insulin release was also observed over repeated cycles of photoinduction without adverse effects on viability and proliferation. PAC-expressing beta-cells cultured as pseudoislets displayed a more pronounced light-triggered increase in insulin secretion versus that from monolayers. The enhanced hormone release was curtailed upon Ca^{2+} channel blocking. This optogenetic system for modulating cAMP and insulin release can be employed for the study of beta-cell function and for enabling new therapeutic modalities for diabetes.

498123 An Oncogene Activity-Dependent "Suicide Gene" Vector System for Selective Targeting of Cancer Cells

Abstract:

The aim of suicide gene therapy is to introduce to mammalian cells exogenous genes that drive the selective conversion of non-toxic molecules to lethal substances. The HSV-tk/GCV (Herpes simplex virus thymidine kinase/ganciclovir) prodrug system is one common suicide gene strategy. Integration of the HSV-tk gene into the host cell genome allows for phosphorylation of GCV, an acyclic analog of the 2'-deoxyguanosine nucleotide. Phosphorylated GCV competes with guanine for DNA integration during synthesis. Due to its acyclic structure, GCV is a poor substrate for DNA chain elongation and causes DNA damage and cell death[1]. Thus, insertion of the HSV-tk gene in host cancer cell genomes, typically through viral vectors, and subsequent administration of GCV, enables selective cell killing. While this system has been studied extensively as a therapeutic strategy in cancer, discovering techniques to localize gene delivery to cancerous tissues, rather than healthy cells, has been challenging[2]. This work outlines a new strategy in which HSV-tk expression is directly regulated by the activity of the extracellular regulated kinase (ERK), an oncogenic kinase that is commonly hyper-activated in cancer cells with mutation or upregulation of oncogenes such as RAF, RAS, and EGFR, among others. Consequently, while both cancerous and non-cancerous cells may be transduced by a viral delivery system, this system can be tuned such that only cancer cells displaying elevated ERK activity will have sufficient HSV-tk expression to induce cell death. By modifying an existing fluorescence-based reporter for ERK activity[3], we engineered a viral vector that expresses a fusion of HSV-tk with a protein domain from the transcription factor fos-related antigen 1 (FRA1). ERK phosphorylation of specific sites within the FRA1 fragment slows the turnover of the HSV-tk fusion protein. Therefore, cells with high ERK activity express more of the suicide gene product and selectively die upon GCV administration. We have begun to explore the efficacy of this new system in glioblastoma multiforme cell lines to demonstrate its selectivity in killing cells with high ERK activity. Additional ongoing work is focused on demonstrating the vector's usefulness in a heterogenous population of cancerous and healthy cells, with an ultimate goal of applying the strategy *in vivo*. The outlined suicide gene strategy is novel example of targeting an aberrant signaling process without perturbing the signal directly. This approach could prove beneficial because strategies that target ERK directly eventually result in resistant subpopulations that are highly aggressive and lead to recurrence.

1 - Fillat, C. et al. *Current Gene Therapy*, 3: 13-26, 2003.
2 – Karjoo, Z. et al. *Advanced Drug Delivery Reviews*, 99: 113-128, 2016.
3 – Albeck, J. et al. *Mol Cell*, 49: 249-61, 2013.

500894 Genetically Encoded, Synthetic Glycopolymers for Tunable Control of Plasma Membrane Shapes and Organelle Biogenesis

Abstract:

Highly curved, plasma-membrane organelles, such as microvilli, cilia, filapodia, axons, and cytonemes, are ubiquitous and essential in multicellular life. The understanding of the underlying mechanisms for formation of these essential organelles remains limited, as do modular strategies and tools for precision control over them. Here, we find that the plasma membrane bending is driven entropically by flexible glycoproteins and glycopolymers anchored to the plasma membrane. Based on these observations, we developed a library of natural, semi-synthetic, and fully synthetic cDNAs that encode glycoprotein products of varying configurational entropy for controlling membrane shapes. Specifically, we developed a modular library of repetitive mucin domains of varying length and extent of glycosylation, and fused these domains to membrane anchors for cell surface expression. The mucin-type proteins with their pendant O-glycans are structurally analogous to semi-flexible bottlebrush polymers. We show that our constructs can program membrane shapes from nearly flat to fully tubulated. Furthermore, we developed an entropic polymer brush model, which predicts the structure of the experimentally observed cell-membrane tubes and informs the design of additional glycoprotein modules. In addition to its application in synthetic biology, our work reveals a new potential function and mechanism by which flexible glycans and glycoproteins regulates cell membrane shape and organelle biogenesis.

504008 Enabling Design-Driven Medicine with Synthetic Biology: Engineering Programmable Cell-Based Therapies

Abstract:

The complementary disciplines of systems biology and synthetic biology are enabling a transformation in the development and implementation of effective treatments for a broad range of patients and diseases. These technologies allow us to consider health and disease at a systems level, and increasingly, to translate this understanding into the design of novel therapies that overcome existing clinical barriers. Here, I will describe our efforts to harness this new paradigm to enable cancer immunotherapy. In part, we are building understanding of the local immune dysfunction that currently limits therapeutic efficacy, and our systems-level investigations have revealed novel insights into how these multicellular networks develop and persist. In order to translate these insights into therapies, we are also developing technologies that enable us to engineer programmable mammalian cell-based therapies. The emerging field of synthetic biology provides an approach for building novel cellular functions from the bottom up, and the â €œtoolboxâ€ of biological parts that operate in mammalian cells is rapidly expanding. To date, however, we lack the ability to construct synthetic cell-based biosensors that detect and respond to many physiologically relevant environmental cues. Here, I will discuss our work developing novel synthetic biology tools that address this need, enabling the constructs of living devices that interface robustly with host physiology. I will also discuss our application of these technologies to probe and modulate interactions between cancer and the immune system, illustrating the potential of this emerging paradigm of design-driven medicine.

499582 Direct Quantification of Deubiquitinating Enzyme Activity in Intact Cells Using a Protease Resistant, Cell Permeable, Peptide-Based Reporter

Abstract:

Tremendous attention has recently shifted towards the implementation of molecularly-targeted therapeutics in treatment of different cancers. In the case of multiple myeloma, proteasome inhibitors Bortezomib and Carfilzomib have gained clinical success and FDA approval. However, one major challenge facing cancer patients is the presence of drug-resistant tumor cells that is further complicated by the high patient-to-patient variability due to the heterogeneous, personalized nature of cancer. Thus, there is an essential need for sensitive, high throughput biochemical assays to determine whether patients would benefit from such molecularly-targeted treatments. Deubiquitinating enzymes (DUBs) are an essential part the ubiquitin-proteasome system (UPS), the primary cellular pathway for protein degradation. Recently, DUBs have been indicated to play an important role in multiple myeloma patients developing a resistance to proteasome-based therapeutics. The goal of this project was to synthesize a novel peptide-based reporter of DUB activity that is cell permeable, long-lived, fluorescent, and specific to DUBs. The reporter (termed RW-LRGG-AFC) consisted of a Î²-hairpin motif [RWVRVpGOWIRQ] that acts as both of cell penetrating peptide (CPP) and â€˜protectideâ€™. The Î²-hairpin was conjugated onto the N-terminus of a DUB substrate consisting of the final four amino acids on ubiquitin (LRGG) with a C-terminal fluorophore (AFC, 7-amino-4-trifluoromethylcoumarin) to provide a detectable metric of DUB activity. The RW-LRGG-AFC reporter was initially characterized in HeLa and OPM2 cell lysates and shown to exhibit superior kinetics when compared to the unprotected four amino acid peptide (Z-LRGG-AMC). After validating the performance in cell lysates, the RW-LRGG-AFC reporter was demonstrated to be internalized by intact HeLa and OPM2 cells allowing for direct visualization of DUB activity in both population-based studies by fluorometry and single cell-based studies by microscopy. Reporter performance was determined to be both time- and concentration dependent. Inhibitor experiments with PR-619 (a commercially available DUB inhibitor) were performed to determine the specificity of the reporter to DUBs as well as demonstrate its ability to detect DUB inhibition at the single cell level. Intracellular stability of the peptide was assessed using an established degradation assay in cell lysates followed by RP-HPLC analysis. Ultimately, the novel peptide-based reporter was observed to rapidly penetrate the cells and remain stable under cytosolic conditions during the course of several hours. This new biosensor allows for the dynamic measurement of DUB activity in intact cells and has the potential to be incorporated into high-throughput screening (HTS) platforms to allow for personalized diagnostics.

497442 Engineering a Bioluminescence-Based Protein Kinase Reporter for *in vivo*, Longitudinal Studies

Abstract:

A hallmark of many cancers is the aberrant activation of protein kinases that promote tumorigenic phenotypes. To measure the activity of these kinases, cancerous tissues must typically be manipulated *ex vivo* and then processed using immunological techniques such as western blotting or immunofluorescence microscopy. Newer techniques such as fluorescence resonance energy transfer (FRET) and fluorescence-lifetime imaging microscopy (FLIM) have been developed to monitor the activity of kinases in live cells, but these methods are often challenging to utilize for live-animal studies. These challenges stem from inherent limitations of utilizing light microscopy for *in vivo* tumor studies, mainly low light penetrance and high tissue autofluorescence. As a result, most in vivo studies assess the activity of a given kinase at the end-point of an experiment without any knowledge of signal dynamics. Here, we outline our work adapting an extracellular signal-regulated kinase (ERK) activity reporter for use in longitudinal animal experiments. This reporter expands upon a previously developed construct that exploits ERKâ€™s ability to phosphorylate and stabilize a protein domain in the transcription factor fos-related antigen 1 (FRA1)[1]. First, we validated that a fluorescent version of the FRA1-based Integrative Reporter (FIRE) faithfully reported ERK activity in a cancer cell line *in vitro*. We then used cloning techniques to modify the FIRE reporter plasmid, replacing fluorescent proteins with click beetle luciferase enzymes. This cloned vector was used to engineer a glioblastoma multiforme cell line to express the bioluminescence-based FIRE reporter. Finally, we used this cell line for subcutaneous xenograft mouse studies to evaluate the ability to measure ERK activity quantitatively *in vivo*. Ultimately, the ability to monitor the activity of oncogenic kinases in real-time during longitudinal, *in vivo* tumor studies can considerably reduce the number of animals needed for experiments that require activity measurements throughout an experimental timeline. Furthermore, this method eliminates concerns regarding how *ex vivo* manipulation and processing might perturb kinase activity. This work serves as a proof of concept in the development of future reporters for other oncogenic kinases, demonstrating significant technical improvements for kinase activity measurement *in vivo*.

1 - Albeck, J. et al. *Mol Cell*, 49: 249-61, 2013.

486716 Engineering a DNA Polymerase to Become a Calcium-Sensitive Biosensor/Recording Device

Abstract:

DNA is an excellent information medium. This is proven not only by biological history, but recent advances in which modern DNA reading and writing technologies have successfully utilized the four-base code as a substitute for the 1's and 0's of binary. The next frontier of this technology is to engineer a device that harnesses the information capacity of DNA to record data in real-time. Application in biological contexts would appear both obvious and advantageous, with the DNA polymerase an ideal signal transducer. Accordingly, if a DNA polymerase were engineered to have a have a functional change in response to an environmental cue, it could function as a "ticker-tape" recorder. Here we apply a rational mutagenesis approach to graft a calmodulin-like calcium binding motif to the exonuclease domain of a DNA polymerase to deliver a step change in fidelity (loss in proof-reading capacity) upon calcium binding. Our hope is for this device to contribute to brain mapping efforts, currently lacking technologies with sufficient space-time measurement resolution, by recording the calcium spikes associated with neuronal firing into DNA.

500473 Development of Single-Virion Fusion Tool for Assessing Influenza Virus Pandemic Risk

Abstract:

Pandemic influenza remains a formidable threat to human health, but it is challenging, if not impossible, to predict which influenza virus subtype will cause the next pandemic. An alternative way is to assess the pandemic risk of all animal influenza viruses that have infected humans, identify those that show elements of pandemic potential, and work to alleviate that potential risk. To achieve these goals, a risk assessment framework was created to evaluate the risk elements including: genomic variation, receptor-binding properties, transmissibility in animal models, and antiviral treatment susceptibility[1]. In recent years, our group has shown that the membrane fusion process is an essential entry step of influenza virus that varies significantly with subtype and may also be an important factor that determines its pathogenicity [2]. Single-virion fusion assays combine total internal reflective fluorescence microscopic (TIRFM) and a microfluidic device coated with supported lipid bilayers that mimic host cell surfaces, to allow the direct observation of fluorescence-labeled viral particles binding to, and fusing with, the target cell membranes. This assay quantitatively characterizes: 1) the highest pH that initiates fusion, 2) the fusion kinetics across pH values, from early to mature endosomes, 3) the extent of fusion of the population, and 4) acid stability of hemagglutinin (HA). These fusion data characterize the release of viral genomes along the endosomal pathway, and we have shown that there are clear differences among various strains, e.g., clinical strains vs. attenuated strains. Hence, it may be necessary to include virus fusion evaluation into the influenza risk assessing framework.

Here, we performed single-virion fusion assay to assess the pandemic potential of influenza virus H10N8, a strain that transmitted from chickens to humans in China in 2013. This strain was lethal to three out of four infected human cases. Considering its high fatality, we used it as a model for studying a newly transmitted influenza virus. In our work, we employed pseudoviruses that contain the H10 and N8 proteins on a benign backbone for the purpose of classifying the fusion properties of this HA. In this presentation, we describe the particle construction, the fusion assay, and the analysis of the data to build a picture of this virusâ€™s pandemic potential as it depends on its fusion properties. We believe fusion could be a key factor, that when taken with the other risk elements, may create a more complete framework to help scientists identify high-risk strains.

1. Trock, S.C., S.A. Burke, and N.J. Cox, *Development of framework for assessing influenza virus pandemic risk.* Emerging infectious diseases, 2015. **21**(8): p. 1372.

2. Costello, D.A., G.R. Whittaker, and S. Daniel, *Variations in pH Sensitivity, Acid Stability, and Fusogenicity of Three Influenza Virus H3 Subtypes.* Journal of virology, 2015. **89**(1): p. 350-360.

500748 Sort-Seq Approach to Engineering an *E. coli* Formaldehyde-Inducible Promoter

Abstract:

Complex synthetic regulatory networks require orthogonal transcription factors with preferably modular DNA-binding and sensing domains. When controlling the expression of measurable reporters, such as fluorescent proteins or antibiotic resistance markers, these transcription factor-based biosensors have shown promise by increasing the efficiency of high-throughput screens and selections, and in dynamic pathway regulation by allowing real-time monitoring of intracellular metabolite concentrations. Here we expand the synthetic biology toolbox through the characterization of an *E. coli* transcription factor-based formaldehyde biosensor and engineer its corresponding operator sequence guided by information obtained through a robust and rapid sort-seq approach to achieve tight and tunable gene expression. The resulting analysis identifies two 19-bp repressor binding sites, one upstream of the -35 RNA polymerase (RNAP) binding site and one overlapping with the -10 site, and assesses the relative importance of each position and base therein. Key mutations were identified for tuning expression levels and were used to engineer formaldehyde-inducible promoters with predictable activities. Engineered variants demonstrated up to 14-fold lower basal expression, 13-fold higher induced expression, and a 3.6-fold stronger response as indicated by relative dynamic range.

491301 Medium-Throughput Detection of Microbially Produced Serotonin Via a GPCR-Based Sensor

Abstract:

Improving the biosynthetic performance of chemical-producing microbes is limited due to reliance on low-throughput screening technologies (~100 samples per day) to assess the microbial production of chemicals. Chemical biosensors, where chemical detection triggers a fluorescent signal, has the potential to accelerate the screening process. Recently we developed G-protein coupled receptor (GPCR)-based chemical sensors by coupling human GPCRs to the yeast cell machinery to trigger cell fluorescence upon chemical binding. Here, we engineer a GPCR-based sensor for serotonin and apply it to the detection of microbially produced serotonin in a medium-throughput fashion (96-well plates). Validation of the serotonin sensor for medium-throughput screening applications using 96-well plates sets the stage for the rapid engineering of microbes for the production of serotonin using sensor-guided engineering.

504476 The Role of Nanosensors for Biodiagnostics and Bioprocess Monitoring

Abstract:

Our lab at MIT has been interested in how the 1D and 2D electronic structures of certain nano-materials can be utilized to advance new concepts in molecular detection, specifically for the real time monitoring of human physiology, feedback control of drug release, as well as bioprocessing monitoring to aid manufacturing. Our work demonstrates that nanosensors have a key advantage in their ability to achieve the ultimate detection limit of single molecules. Reducing a sensor transducer to nanometer scale dimensions lowers the detection limit until the stochastic fluctuations of single molecule can be detected. This level of sensitivity can be best capitalized on with concomitant advances in molecular recognition. To this end, our laboratory at MIT has introduced CoPhMoRe or **Co**rona **Ph**ase **Mo**lecular **Re**cognition[1] as a method of discovering synthetic antibodies, or nanotube-templated recognition sites from a heteropolymer library. We show that certain synthetic heteropolymers, once constrained onto a single-walled carbon nanotube by chemical adsorption, also form a new corona phase that exhibits highly selective recognition for speciï¬c molecules. To prove the generality of this phenomenon, we report three examples of heteropolymersâ€"nanotube recognition complexes for riboï¬,avin, L-thyroxine and estradiol. The platform opens new opportunities to create synthetic recognition sites for molecular detection. We have also extended this molecular recognition technique to neurotransmitters, producing the first fluorescent sensor for dopamine. Another area of advancement in biosensor development is the use of near infrared fluorescent carbon nanotube sensors for in-vivo detection[2]. Our vision is the realization of multiplexed sensor arrays that can provide real time biochemical detection of physiological markers in humans. In our recent work, we show that PEG-ligated $d(AAAT)_7$ DNA wrapped SWNT are selective for nitric oxide, a vasodilator of blood vessels, and can be tail vein injected into mice and localized within the viable mouse liver. We use an SJL mouse model to study liver inflammation in vivo using the spatially and spectrally resolved nIR signature of the localized SWNT sensors. Lastly, we discuss new concepts for real time, bioprocess monitoring to aid pharmaceutical production and other applications.

1. Zhang, JQ et. al. Molecular recognition using corona phase complexes made of synthetic polymers adsorbed on carbon nanotubes. Nature Nanotechnology, 8, 12, 2013, 959-968

2. Iverson, NM, et. al. In vivo biosensing via tissue-localizable near-infrared-fluorescent single-walled carbon nanotubes. Nature Nanotechnology, 8, 11, 2013, 873-880

October 31, 2017

Fibroblasts Affect Each Other's Directional Decision Making Process during Chemotaxis in Microfluidic Mazes

Quang Long Pham, David Chege, Timothy Dijamco, Anh Tong, Sagnik Basuray and Roman Voronov

New Jersey Institute of Technology
Department of Chemical, Biological and Pharmaceutical Engineering

1

SUMMARY

- ❑ Introduction & Background
- ❑ Hypothesis
- ❑ Method
- ❑ Results and Discussion
- ❑ Conclusion
- ❑ Future work

2

Introduction: Chemotaxis

EMBRYOGENESIS 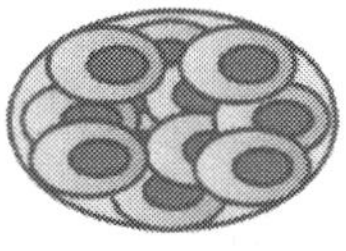

WOUND HEALING 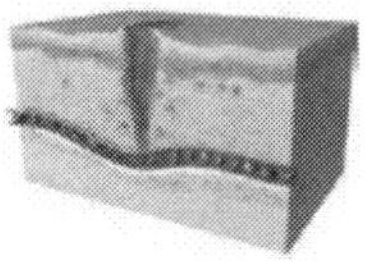

IMMUNE RESPONSE 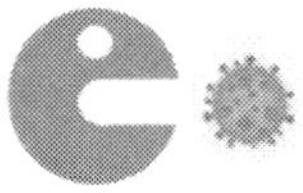

CANCER INVASION 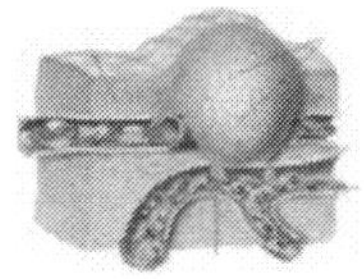

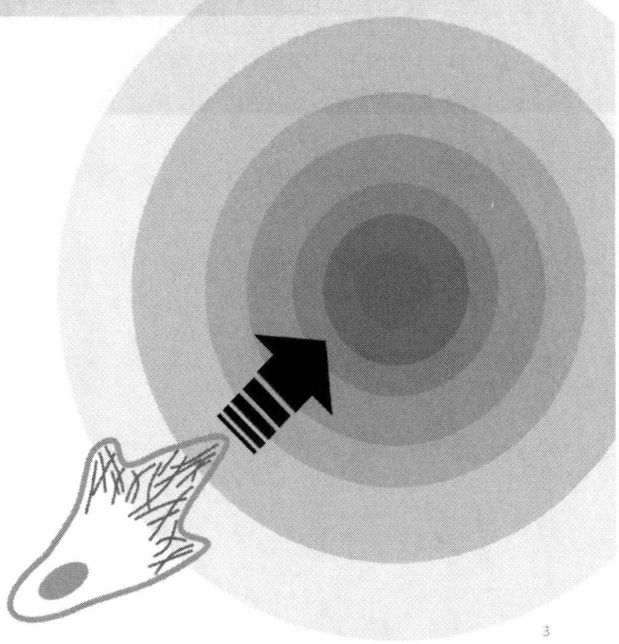

3

Directional Decision Making in confined environment:

No cell sequence! No fibroblast!

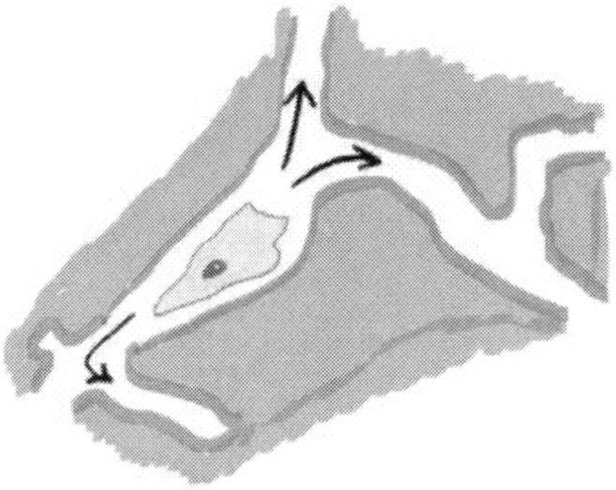

Neutrophil

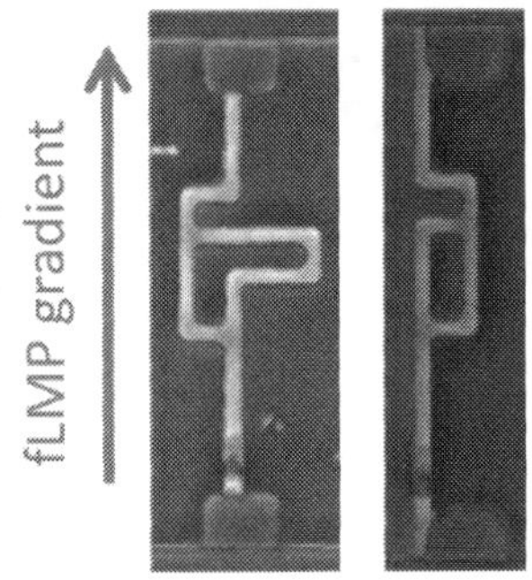

Ambravaneswaran et al, 2010

Epithelial cancer cell (no initial gradient!)

Scherber et al, 2012

* fLMP: Formyl-methionyl-leucyl-phenylalanine

4

Why fibroblasts?

Fibroblast: critical to extracellular matrix formation:
wound healing & tissue engineering

- ❑ Remodeling
- ❑ Deposition
- ❑ Synthesis

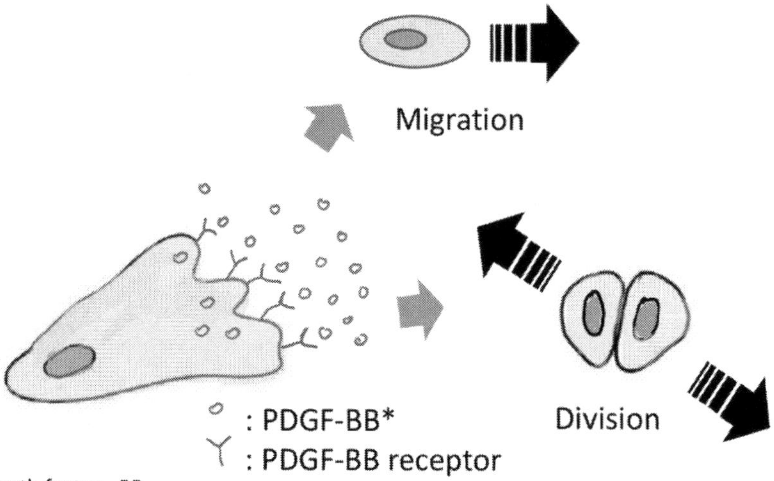

* PDGF-BB: platelet derived growth factor - BB

5

Hypothesis

- ❑ Cell sequence influences directional choices
- ❑ Cell division influences directional choices

6

Device Concept

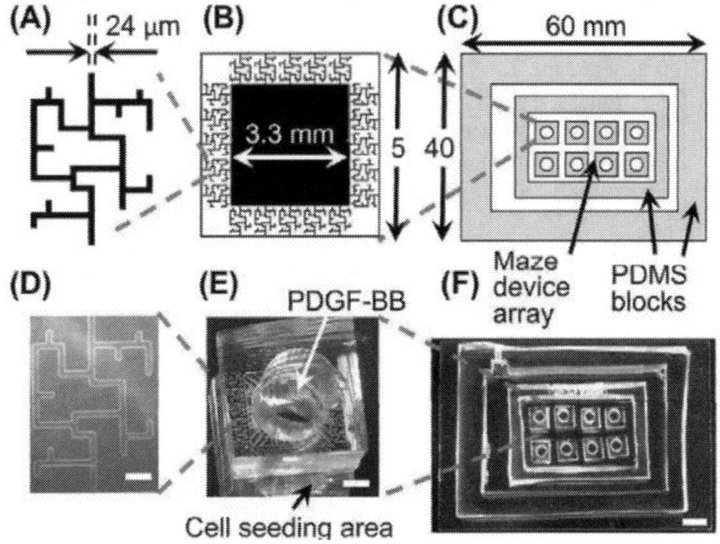

* PDMS: poly (dimethyl siloxane) elastomer

7

Device Fabrication

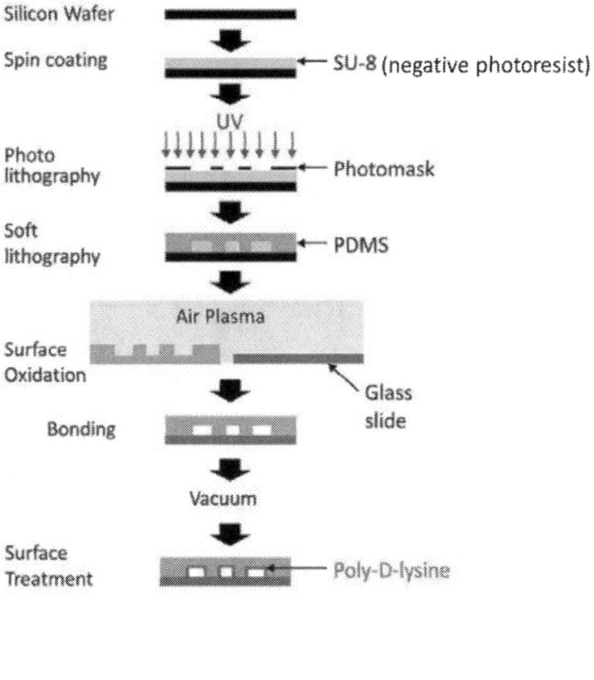

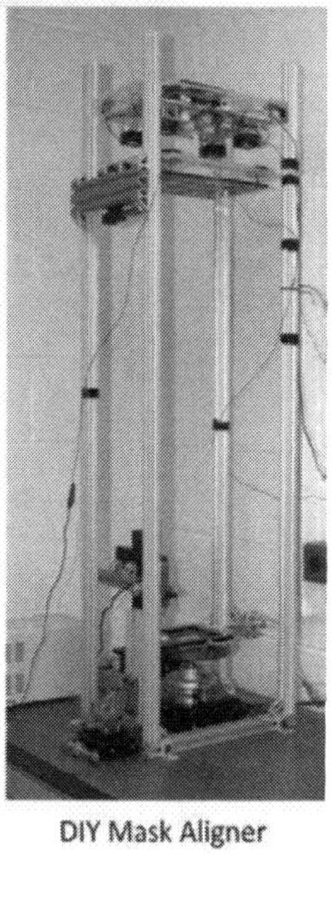

DIY Mask Aligner

8

Experimental setup & data acquisition

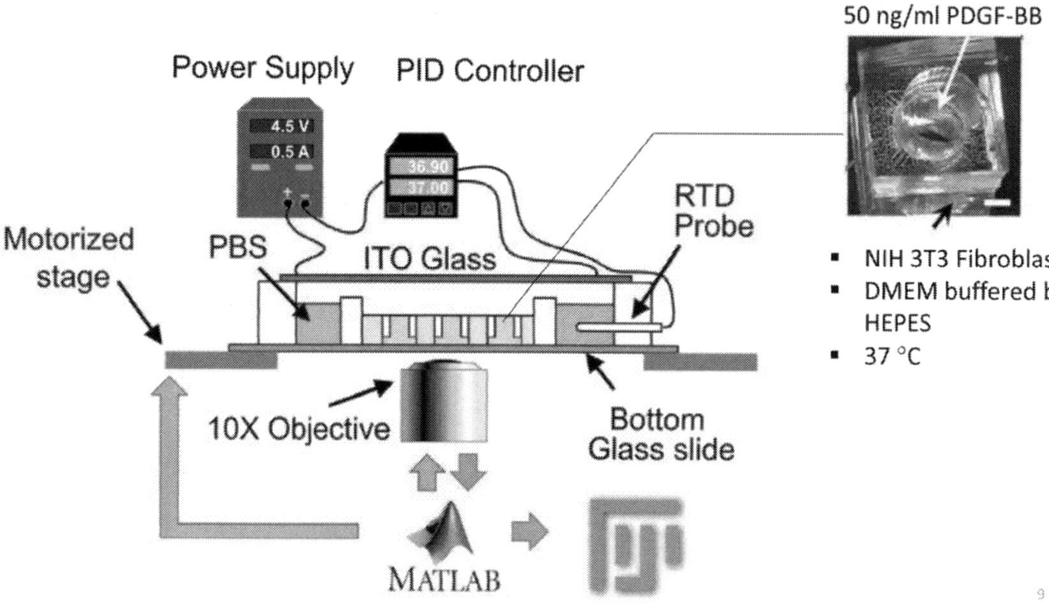

- NIH 3T3 Fibroblast
- DMEM buffered by HEPES
- 37 °C

9

Results and Discussion

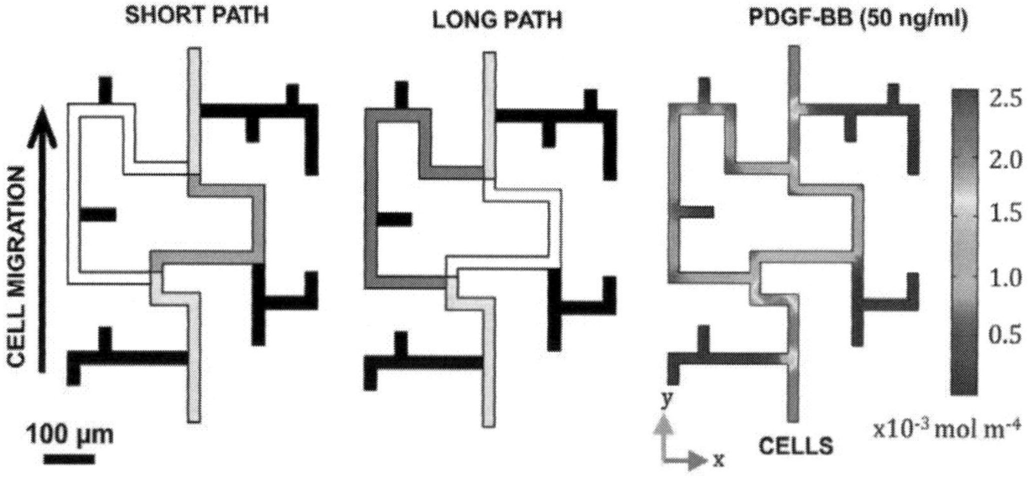

10

Cells influence each other in decision making

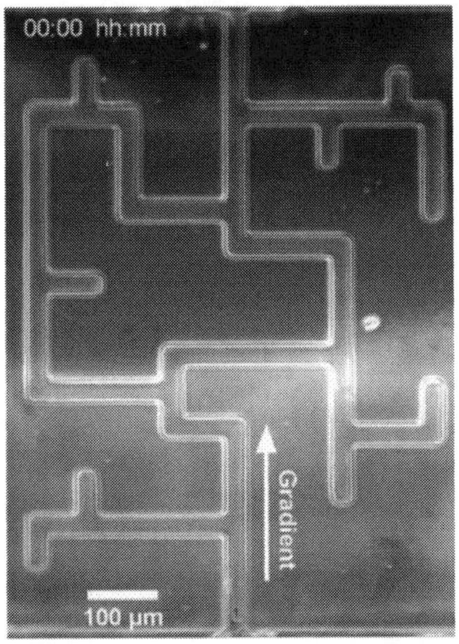

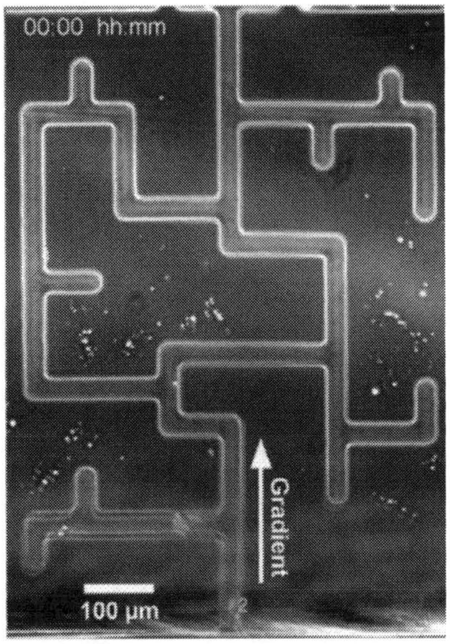

11

First cells prefer the short path over the long path

Cell population tend to equalize the directional bias

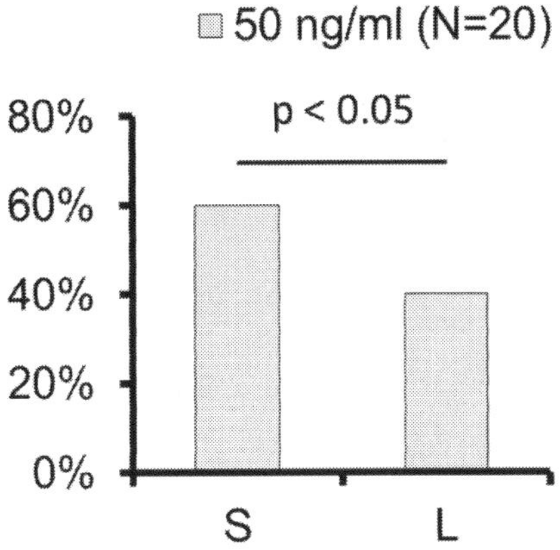

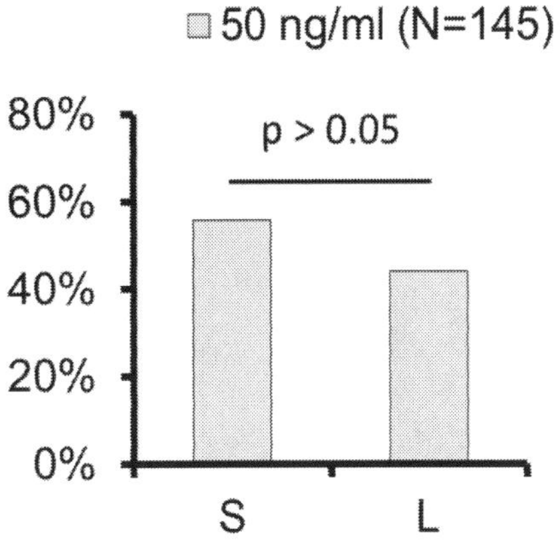

$p \geq 0.05$, no statistically significant difference

$p < 0.05$, statistically significant difference

12

Alternate sequence is preferred by <u>the first two cells</u>

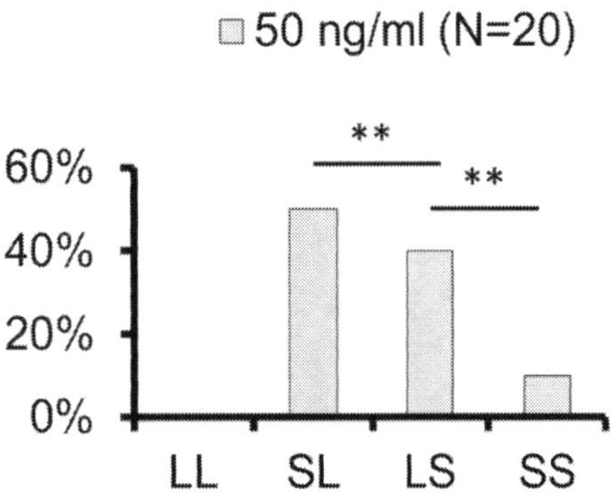

Alternate sequence is also preferred by <u>any two consecutive cells</u>

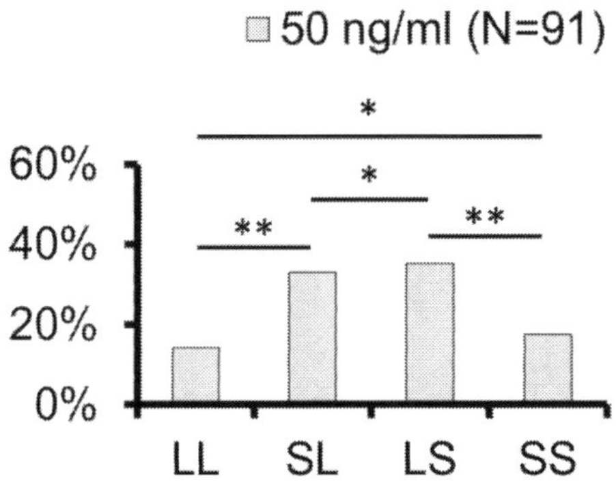

∗ : $p \geq 0.05$, no statistically significant difference

∗∗ : $p < 0.05$, statistically significant difference

13

Cell division influence decision making

One daughter migrate <u>against gradient</u>

Both daughters migrate <u>along gradient</u>

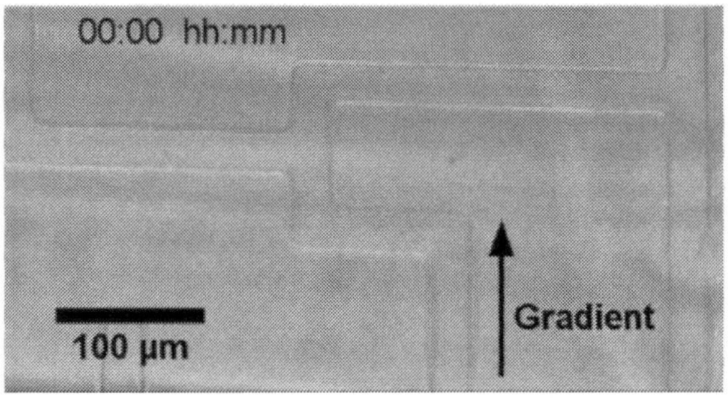

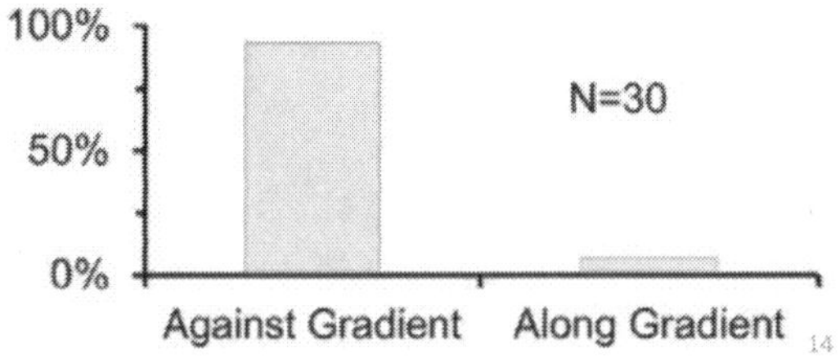

14

Division occurs more frequently at low PDGF-BB concentration

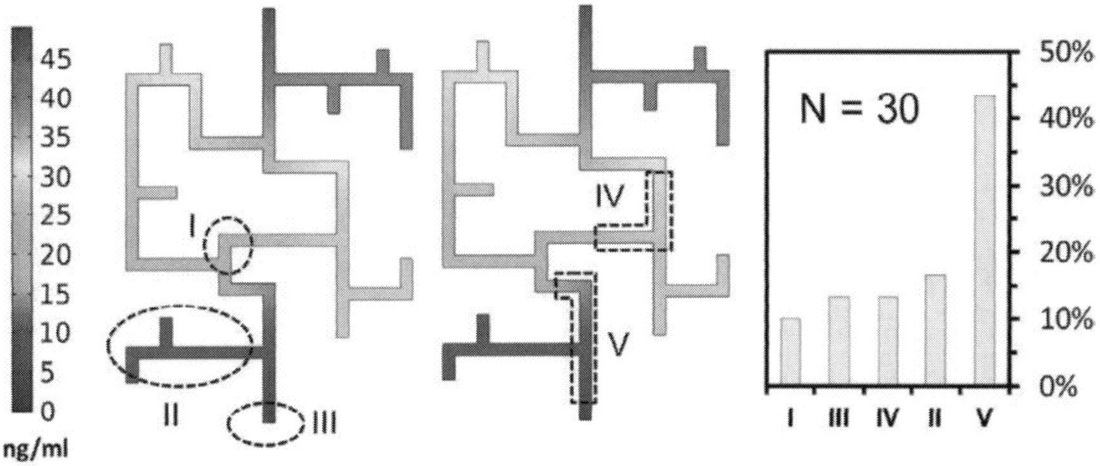

15

Proposed cell division mechanism: highly asymmetric receptor distribution → daughter cells of distinctive sensing capacities

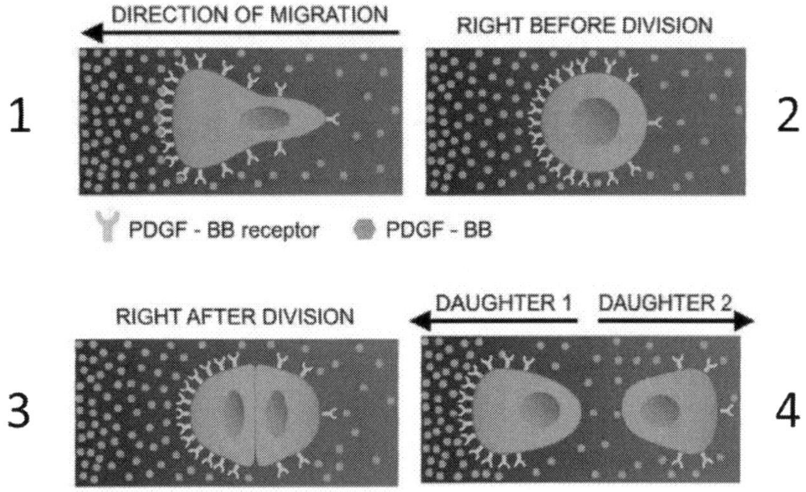

16

Conclusion

We test influences of cell sequence, PDGF-BB concentration, and cell division on fibroblast chemotaxis using a microfluidic maze

- ❑ Cells alternate between long and short paths
- ❑ Mitotic cells go in opposite directions, one can go against gradient and common traffic of oncoming cells

17

Future Work

- ❑ Quantify PDGF-B receptors expressed by divided cells
- ❑ Characterize consumption rate of PDGF-BB by fibroblasts
- ❑ Build prediction model on fibroblasts' decision making in sophisticated constructs (more representative of tissue architecture *in vivo*)

18

Acknowledgement

Guidance and advice:

- Prof. **Roman Voronov**: PhD advisor
- Prof. Sagnik Basuray: PhD co-advisor

Research Fund:

- Gustavus and Louise Pfeiffer Foundation Research Funding
- New Jersey Institute of Technology Start-up Grant

My co-authors

- David Chege
- Timothy Dijamco
- Elisabeth Stein
- Anh Tong

19

20

Migration speed

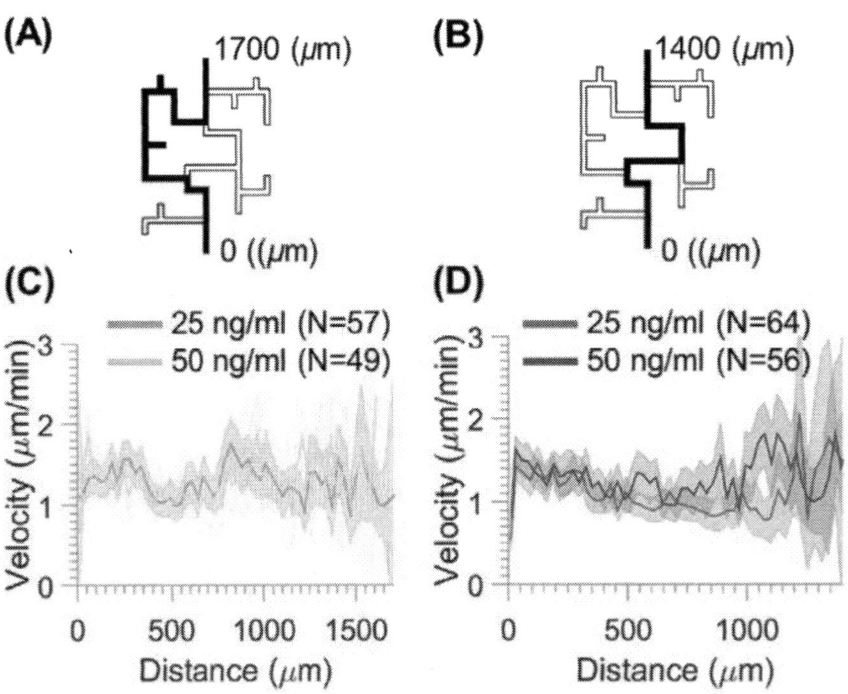

21

498142 Nuclear Cytoskeletal Linkages Are Required for Normal Cell Function

Abstract:

Introduction: The linker of nucleoskeleton and cytoskeleton (LINC) complex, which is composed of outer nuclear membrane proteins Klarsicht, ANC-1, and Syne homology (KASH) and inner nuclear membrane proteins, Sad1 and UNC-84, transmits mechanical forces from the cytoskeleton to the nucleus. Here we asked if an intact LINC complex is required for normal cell function.

Result: The LINC complex has been proposed to transmit forces required for nuclear shaping. However, we did not find differences in nuclear shapes between control cells and LINC-disrupted cells. Although LINC complex disruption did not affect the nuclear shaping, we found that LINC disrupted MCF10A cells failed to form normal acini with clear lumens. Unlike WT cells, LINC disrupted cells failed to heal wounds in a wound healing assay. LINC disrupted fibroblasts also were unable to migrate. The lack of migration however did not correlate with changes in ROCK-Rho-CDC42 activities in fibroblasts.

Conclusion: An intact LINC complex is required for normal cell migration and tissue development, but not for nuclear shaping in MCF10A cells. We propose that mechanical force transmission by the LINC complex to the nucleus allows cells to migrate efficiently, which may explain the observed effects of LINC disruption on migration and tissue development.

499199 Sprouty2 Regulation of Glioblastoma Adhesion and Invasion

Abstract:

Glioblastoma multiforme (GBM) is a devastating form of brain cancer where patients typically succumb within two years of diagnosis. Most tumors will recur even after surgical resection due to the presence of residual tumor cells beyond the margin of the resected tumor mass, highlighting the importance of identifying the key regulators of GBM cell invasion. Previous work from our group identified the protein Sprouty2 (SPRY2) as a possible tumor promoter in GBM. Specifically, SPRY2 knockdown reduced proliferation of human glioblastoma cells and antagonized the growth of subcutaneous tumor xenografts in mice. New observations from our lab suggest that SPRY2 may also regulate GBM cell adhesion and migration phenotypes. In 3D cultures of tumor cell spheroids, GBM cells demonstrated extensive invasion into the matrix as revealed by live-cell microscopy studies. This phenotype was largely abrogated by SPRY2 knockdown. In 2D cultures, SPRY2-replete cells tended to grow as monolayers, with most cells clearly adhered to tissue culture plastic. In contrast, similarly dense SPRY2-deficient cells displayed a tendency to grow on top of each other, suggesting a preference for cell-cell adhesions over spreading on tissue culture plastic. Based on these findings, we hypothesized that SPRY2 may play a role in the invasion of GBM cells away from the main tumor body. Further, this role of SPRY2 may contribute to the ineffectiveness of initial surgical resections in resolving GBM. Quantitative immunofluorescence measurements of phosphorylated paxillin (p-paxillin, Y118) indicate that SPRY2 knockdown antagonizes formation of p-paxillin containing focal adhesions. Based on this finding, we hypothesized that SPRY2 plays a role in the regulation of p-paxillin containing focal adhesions in a manner that allows GBM cells to more efficiently invade their surrounding matrix. Focal adhesion kinase (FAK) is reported to be an upstream regulator of paxillin phosphorylation, and thus we hypothesized that FAK may drive cell spreading. Surprisingly, treatment with a FAK inhibitor increased cell spread area and antagonized the formation of cell-cell adhesions. Ongoing studies will aim to elucidate a mechanistic model that unites these apparently conflicting findings. Ultimately, these studies are relevant to the identification of targeted therapeutics that may limit putative SPRY2-mediated GBM invasion, and thus has relevance to the improvement of patient outcomes.

501406 The Mechanobiology of Cancer Cell Motility Under Vertical and Lateral Confinement

Abstract:

The mechanobiology of cancer cell motility under vertical and lateral confinement

Cell migration is a fundamental cellular phenomenon that plays a pivotal role during cancer metastasis. Our current understanding of cell migration relies mainly on studies using two-dimensional (2D) substrates or 3D gels. However, in vivo, cancer cells migrate not only within 3D extracellular matrices but also through longitudinal tissue microtracks, which impose varying degrees of confinement on cells. Evidence suggests that physical confinement alters intracellular signaling pathways, thereby affecting the mechanisms and phenotypes of cell migration. Recent work reveals that compression on the top and bottom of cells (vertical confinement), as opposed to compression on the sides (lateral confinement), induces significantly higher levels of nuclear rupture. This suggests a fundamentally different mechano-sensing mechanism which depends on the direction of confinement. Thus, we hypothesized that cells employ distinct migration mechanisms under vertical vs lateral confinement. To address this hypothesis, we developed PDMS- (stiffness of ~1MPa) and hydrogel- (10-50 kPa) based microchannel devices of prescribed cross-sectional areas: *width(W)*=3Î¼m and h*eight(H)*=10Î¼m for lateral confinement and *W*= 10Î¼m and *H*=3Î¼m for vertical confinement. Using HT-1080 fibrosarcoma cells as a model, we demonstrate that vertically confined cells migrate more slowly than laterally confined cells. This observation holds true for a number of different normal-like (fibroblasts and smooth muscle cells) and cancer cells (HOS osteosarcoma cells). We also report that the vertically confined cancer cells employ a slower bleb-based mode of migration, whereas they switch to the faster protrusion-based mode of migration in lateral confinement. In addition, our work explores the molecular mechanisms responsible for the differential cell migration speed in vertical vs lateral confinement. We employ a multi-disciplinary approach involving state-of-the-art bioengineering (Micropipette Aspiration, Atomic Force Microscopy), imaging (confocal, FRET) and molecular biology techniques, to study the role of perinuclear actin (ARP2/3, FMN2), nuclear stiffness (LMNA, EMD), the nucleo-cytoskeleton linker (SUN1, SUN2) and small GTPases during vertically and laterally confined migration. Collectively, our work enhances our understanding of the complex process of confined cell migration and provides a novel perspective on how cells sense and respond to different types of confinement.

492999 Investigating Changes in Cellular Based Forces in Monolayers By Tracking Sub-Nuclear Sensors

Abstract:

In mechanobiological processes, cells generate forces to modulate stiffness and are responsive to mechanical perturbations. Measuring mechanical forces within monolayers of cells is particularly challenging where there are unknown balances of forces among cells and between the cells and the substrate. This is further complicated when considering heterogeneities within a monolayer with additional spatial variations. However, these mechanical heterogeneous variations in the monolayer are of interest since they lead to cellular transformations in development, wound healing, and disease. Additionally, the extent that forces propagate through a given cell to its nucleus can alter the expression of genes through the process of mechanotransduction. Current techniques for cellular force measurement often require an extracellular probe, which the cells perturb. Forces are then calculated from these perturbations. The use of external probes, or specific extracellular environments, for force measurement limits current techniques to particular experimental set ups, often limiting their biological relevance. Here we present a novel technique, termed **S**ensors from **I**ntra**N**uclear **K**inetics (SINK), which uses the cell nucleus as a relative force sensor. SINK utilizes fluorescently tagged genome bound proteins, whose motion is observed in live cells and tracked over time. SINK takes advantage of the interconnectedness of the cytoskeleton to the nucleus where we show the extent to which molecular motors as well as nuclear to cytoskeletal connections alter chromatin movement. We further show the utility of SINK by investigating how structural changes affect force propagation through cells as they transition from isolated cells to confluent monolayers. Additionally, we investigate the extent that substrate stiffness affects force propagation in monolayers, demonstrating that cells sense their underlying substrate, which ultimately influences motion within the nucleus. Finally, we begin to show how point defects in a relatively homogeneous monolayer affect surrounding cells, with future applications as a disease model for cancer or cardiovascular diseases, as these often stem from point defects in tissues. Because SINK relies on the cell nucleus, it could be extended to *in situ* and *in vivo* systems, or may be used in conjunction with current force measurement techniques to get a more complete understanding of how cells propagate force from focal adhesions, through the cytoskeleton, and ultimately into the nucleus where the chromatin fibers are affected. Thus, the future utility of SINK lies in the ability to investigate the extent to which physiologically based forces affect genome expression through physical perturbation of the chromatin fiber.

492006 Compression Affects Short-Time Subdiffusion of Loci and Inclusion Bodies in E. coli Cells

Abstract:

We have tracked the loci and inclusion bodies (Î¼NS particles) motions in *E. coli* cells under compression on a microfabricated device systematically in real time. We observed that both fluorescently labeled loci and GFP-Î¼NS particles performed subdiffusion in *E. coli* cells. This subdiffusion may arise from an intrinsic viscoelasticity of bacterial cytoplasm, as extensive research has demonstrated that proteins, inclusion bodies, and loci all exhibit subdiffusion in a crowded cytoplasm. However, our experiments show that compression affects the subdiffusion of loci and inclusion bodies in very different ways. As the main external compression force might be exerted on the cell membranes of high modulus, the structure of nucleoid contributing to the loci subdiffusion can remain unchanged. Thus, whilst the cytoplasmic diffusion is slowed down significantly under compression, as evidenced by notable reductions in both the diffusion coefficient and travel distance of Î¼NS particles, the diffusion coefficient of DNA loci remains almost the same under compression. These results suggest that the dynamics of bacterial chromosome is decoupled from the viscoelastic environment of the cytoplasm, in contrast to some previous ideas proposed in literature. It would seem that DNA elasticity and organization (which are probably weakly affected by the deformation) play a more important role in loci subdiffusion. In addition, since the compression on a cell has similar cytoplasmic effects as suppressing metabolic activities in a cell, i.e. the onset of glass-like behavior, these particle tracking experiments in compressed cells provide new insights into the physical properties of the cytoplasm.

504535 Investigations of Mechanisms of Force Transduction in Tissues

Abstract:

This talk focuses on our investigations of mechanisms of force transduction in tissues. Specifically, I will describe our recent identification of a mechano-transduction cascade that integrates intercellular adhesion proteins, cadherins, growth factor receptor signaling, and integrins to regulate tissue functions in response to mechanical stimuli. Integrins are well-known force-transducing adhesion proteins that regulate cell spreading and contractility on compliant matrices. However, cadherinsâ€"crucial adhesion proteins at cell-cell contactsâ€"also form mechanically responsive adhesion complexes. Cadherin mediated force-transduction impinges on a range of tissue functions, including the maintenance of tissue barrier integrity, the shear alignment of vascular endothelial cells, tissue morphogenesis, and cell proliferation. One identified cadherin-mediated force-transduction mechanism involves an actin binding protein, alpha-catenin, that associates with cadherin complexes at the plasma membrane. We and others showed that alpha-catenin undergoes a conformational change in response to force on cadherin receptors, to facilitate cytoskeletal remodelling, which mechanically reinforces perturbed cadherin adhesions. However, in more recent studies using mechanical probes and live cell imaging, we identified a second, cadherin-mediated, force transduction mechanism that activates epidermal growth factor receptor signalling and integrins in a global, mechano-sensitive signalling cascade. These findings reveal an integrated signalling network in cells that coordinately regulates cell and tissue mechanics. Furthermore, matrix rigidity modulates both integrin and cadherin-mediated force transduction through this integrated network. These results suggest a mechanism underlying the growth factor sensitivity of epithelial cells cultured on rigid versus soft matrices.

489642 Assessing the Adequacy of a Sampling Frequency in the Application of PAT to a Continuous Manufacturing Process for Drug Product

Abstract:

The implementation of spectroscopic instruments to determine the concentration of active pharmaceutical ingredient (API) in powder samples is vast and well documented in the scientific literature. The introduction of continuous manufacturing lines for solid oral drug product has triggered increased interest in the application of such tools to determine the concentration of API in moving streams of powder. A key question in the development of such methods is the determination of an adequate frequency of sampling for the PAT devices, as it pertains the overall dynamics of the system

Our contribution will discuss a hybrid approach that uses information from a deterministic model, and the variability in the NIR spectra, along with time-series analysis techniques (such as ARX models, partial auto correlation coefficients, and cross-correlation coefficients) and the Shannon theorem of sampling to assess the adequacy of proposed frequencies of sampling and averaging in the spectrometer located in the feed-frame of the tablet press.

492244 Continuous Production Technology for Pharmaceutical Semi-Solid and Liquid Formulations: Processability and Influence of Process Parameters on the Product Quality

Abstract:

N. Bostijn[1,*], T. De Beer[1], W. Dhondt[2], J. Van Renterghem[1], C. Vervaet[3]

[1]Laboratory of Pharmaceutical Process Analytical Technology, Ghent University, Ottergemsesteenweg 460, 9000 Ghent, Belgium

[2]Janssen Global Technical Operations, Pharmaceutical Mfg (PM) Platform, Turnhoutseweg 30, 2340 Beerse, Belgium

[3]Laboratory of Pharmaceutical Technology, Ghent University, Ottergemsesteenweg 460, 9000 Ghent, Belgium

INTRODUCTION

Pharmaceutical production today mainly consists of batch processes. These are time-consuming, expensive and sub-optimal processes. Continuous processes on the other hand, allow for more flexibility, an easier scale-up and better quality control. The aim of this study is to investigate whether the Contimix system, a modular continuous production system, initially developed for the continuous production of semi-solid and liquid cosmetics, can be employed for pharmaceutical applications as well. A commercially available pharmaceutical cream, which is originally produced via a batch process, was manufactured with the Contimix system. In a first step, the processability of the cream with the Contimix system was investigated. Furthermore, the influence of the process parameters on the product quality was studied through statistical Design of Experiments (DoE), where the different process parameters were varied simultaneously.

MATERIALS AND METHODS

Materials

A pharmaceutical cream (Janssen Pharmaceutica N.V., Beerse, Belgium) was selected as model formulation.

Methods

All creams were produced with a continuous production system (Contimix, Hebold systems, Cuxhaven, Germany). A two level fractional factorial design was applied to study the influence of five process parameters on the product quality: process temperature, throughput, mixing speed in mixing vessel 1, mixing speed in mixing vessel 2 and water phase temperature. The studied quality attributes of the cream (DoE responses) were Active Pharmaceutical Ingredient (API) concentration (at the beginning, middle and end of a DoE run), yield point, viscosity of the yield point, viscosity (at low, intermediate and high shear rate) and pH. The data was analyzed using the MODDE Pro 11.0 software (MKS Umetrics, Umeå, Sweden). The API concentration of the creams was measured using a Raman Rxn2 spectrometer (Kaiser Optical Systems, Ann Arbor, MI, USA), equipped with a fiber-optic PhAT probe. The laser wavelength was 785 nm. Spectra were collected off-line with an exposure time of 20 s and a resolution of 5 cm^{-1}. The spectral range of the system was from 150-1890 cm^{-1}. Rheological measurements were conducted with a Haake Mars III rheometer (Thermo Fisher Scientific, Karlsruhe, Germany). The experiments were carried out at a fixed temperature (25°C), using a 60 mm parallel plate geometry. A stress ramp and viscosity curve were obtained from each produced cream. The pH of the creams was measured with a glass HI 3220 probe (Hanna instruments, Temse, Belgium).

RESULTS AND DISCUSSION

Prior to the experiments, the raw materials of the cream were divided into three pre-phases, which were batch mixed before the actual continuous production. This additional step was necessary since the Contimix system does not feature as many pumps as there are components in the cream. Furthermore, there is no possibility of introducing powders directly into the process. The results of the DoE show that the influence of the process parameters on the pH was negligible. The API concentration at the three time points between the experimental runs was not influenced by the process parameters. However, when the API concentration within the runs was evaluated, a downward trend in API concentration was detected for runs with a high throughput. This was caused by an inhomogeneity of pre-phase 3, which contained the API. Even for runs produced at a lower throughput, a similar trend was observed, suggesting that the homogenization of pre-phase 3 was not optimal, leading to an inhomogeneity of the API in the final product. The stress ramp experiments showed an influence of the process temperature. A higher process temperature was linked with a lower viscosity. The viscosity curve experiments showed an influence of the process temperature and throughput, but both influences were minimal. However, when the viscosity curves were investigated, a difference between the runs at low and high process temperature was visible. A possible explanation for the influence of the process temperature on the yield point and viscosity can be found in the short time between manufacturing the creams and conducting the rheological experiments. To draw conclusions about this effect, a number of runs at low and high process temperature should be executed and the viscosity of these creams should be measured directly (1 day) and a longer period (4 weeks) after production. Despite achieving processability, further improvements are necessary to transfer this continuous production technology to the pharmaceutical industry. The batch pre-step should be replaced by a continuous step. In order to achieve this, a way of feeding powder directly into the system is necessary. This would make the production system completely continuous, with all the advantages this entails. The original batch produced cream was measured with the same techniques as the creams of the DoE runs and compared with a center point run (run 18). The pH and API concentration were similar for both creams. From the stress ramp experiment, it can be concluded that the yield point of run 18 was slightly higher than that of the original product, indicating that both creams have similar rheological properties. This was confirmed by the viscosity curves, where the viscosity decrease with increasing shear rate was similar.

CONCLUSION

Processability of a pharmaceutical cream with the Contimix, a continuous production system, was achieved. From the screening design, it can be concluded that the process temperature had a significant influence on the viscosity of the creams, i.e. higher process temperature resulted in creams with a lower viscosity. Changing the other process parameters had no significant impact on the measured quality attributes. Furthermore, the Contimix was capable of producing a commercially available pharmaceutical cream in a continuous way with a similar quality as the original formulation.

494641 Modeling of Residence Time Distribution of a Continuous Dry Granulation Tableting Line

Abstract:

In recent time academia and industry have shown the economic and process related benefits of continuous pharmaceutical manufacturing. Regulatory authorities support strongly this approach of processing as well. This trend leads to a need for advanced control strategies, which enable automated operation of integrated production lines. With the first continuous plants being implemented for continuous production of solid oral dosage forms, out-of-spec material discharge strategies and process control with the aim to ensure robust product quality and to save resources have to be developed. Malfunctions of process equipment, unexpected process events and start-up or shut-down phases can lead to out-of-spec material. Due to material holdup and back mixing effects in specific unit operations, the out-of-spec material distribution can spread over process time. Therefore, knowledge of the residence time distribution (RTD) is essential not only for batch definition and material tracking aspects. For the development of cost-effective discharge strategies a fundamental tracking of the material through the process based on RTD is required.

In this work a dry granulation line, as one of the common ways of solid dosage form manufacturing, was investigated. The line consists of a feeding, blending, dry granulation, pneumatic transport and tableting step. For the determination of the residence time distribution in the considered unit operations as well as in the entire line a step change of API was performed at different total throughputs. As a system response the concentration of API was measured via in-line NIR spectroscopy at the outlet of blender, granulation and tablet press. Additionally, samples of powder, granules and tablets were drawn and analyzed off-line by means of UV/Vis as a reference.

Based on the experimental RTD data a transfer function was calculated and the transfer behavior of single unit operations modeled. The RTDs for single unit operations were interlinked and the entire process simulated with customized models in the process simulation software gSolids. The selected approach allows the prediction of out-of-spec material amount and the discharge time for different throughputs. This work shows promising results for both the inter- and extrapolation of RTDs for various throughputs and comparison with validation runs of the entire line.

494709 Robust State Estimation of Feeder and Blender Systems in Continuous Pharmaceutical Manufacturing Systems

Abstract:

Abstract

The development and application of advanced monitoring, operational- and design-making strategies for continuous pharmaceutical manufacturing require good understanding of the current state of the system. Moving horizon estimation (MHE) can estimate the current system state by solving a well-defined optimization problem where process complexities are explicitly taken into consideration as constraints. Traditional MHE techniques assume random measurement noise governed by some normal distributions. However, this assumption is not always valid since the actual distributions are often unknown or exhibiting departures from normal form. In addition, conventional state estimators fail to consider gross or systematic errors, such as bias, outliers, and drifts. The presence of gross errors can contaminate the state estimates and the reconciled measurements, causing undesirable â€œsmearingâ€ effects [1]. As a result, research on state estimation is closely related to topics of data reconciliation (DR) and gross error detection (GED).

Robust estimators, also known as the maximum likelihood estimators or M-estimators, are commonly used in reconciling measurements contaminated with gross errors. Robust estimators have been implemented in many research areas, including state estimation, data reconciliation, and parameter estimation [2, 3, 4]. Systematical comparisons between various robust estimators can be found in literature [1, 5]. To improve the accuracy and robustness of state estimation, we incorporate robust estimators within the standard MHE skeleton, leading to an extended MHE for dynamic state estimation with gross errors.

In continuous tablet or capsule production lines, the optimal operation of the feeder-blender system (FBS) is crucial in order to meet the critical quality attributes of the final product. Advanced monitoring and control strategies of FBS require fast and accurate estimates of the current process state. In this work, the proposed MHE framework is implemented on a simple FBS configuration which includes two loss-in-weight (LIW) feeders and one continuous blender. Based on our experiment results, we propose a first-principle dynamic model for screw feeders and a reduced-order, compartment-based model for continuous blenders. Both dynamic models are used in our MHE framework, to result in a dynamic optimization (DO) problem constrained by differential algebraic equations (DAEs).

The proposed MHE strategy is first tested on simulation runs of the FBS where measurements are contaminated with gross errors such as outliers and drifts. Numerical results show significant benefits in applying the extended MHE technique with robust estimators. In addition, we implement our MHE framework in a pilot plant FBS, which is part of a continuous dry granulation line, and demonstrate on-line dynamic state estimation and data reconciliation. Experimental results confirm that the proposed MHE framework is robust to gross errors and can provide accuracy state estimates. Moreover, the efficient solution of the MHE realized in this work, suggests feasible application of on-line state estimation on more complex continuous pharmaceutical processes.

Bibliography

[1] Özyurt, Derya B., and Ralph W. Pike. "Theory and practice of simultaneous data reconciliation and gross error detection for chemical processes." Computers & chemical engineering 28.3 (2004): 381-402.

[2] Tjoa, I. B., and L. T. Biegler. "Simultaneous strategies for data reconciliation and gross error detection of nonlinear systems." Computers & chemical engineering 15.10 (1991): 679-690.

[3] Arora, Nikhil, and Lorenz T. Biegler. "Redescending estimators for data reconciliation and parameter estimation." Computers & Chemical Engineering 25.11 (2001): 1585-1599.

[4] Nicholson, Bethany, Rodrigo López-Negrete, and Lorenz T. Biegler. "On-line state estimation of nonlinear dynamic systems with gross errors." Computers & Chemical Engineering 70 (2014): 149-159.

[5] Prata, Diego Martinez, José Carlos Pinto, and Enrique Luis Lima. "Comparative analysis of robust estimators on nonlinear dynamic data reconciliation." Computer Aided Chemical Engineering 25 (2008): 501-506.

496016 Large-Scale Experimental Comparison of Batch and Continuous Technologies for Pharmaceutical Tablet Manufacturing

Abstract:

This work presents experimental comparison of batch and continuous technologies for pharmaceutical tablet manufacturing at a scale of 100 kg. The investigation covered the units of mixing, granulation, drying, blending, and compression, for which three processes were established based on batch and continuous technologies. For the batch process, conventional facilities were used, with employing fluidized bed granulation and high shear granulation as the granulation unit. For the continuous process, a recently developed facility was used, which interconnects all the units and is capable of producing tablets at the speed of 25 kg/hr. Common to these three processes, a 100 kg of raw materials were provided that contained ethenzamide as an active pharmaceutical ingredient (API), and mannitol and crystalline cellulose as excipients. As a preparation, experiments at a small scale were performed in order to determine process parameters in the experiments at a scale of 100 kg. During or after the production, samples were taken for investigating the following three items. First was the quality of granules as intermediate products, e.g., particle size distribution, mean diameter, and loose bulk density. Second was the quality of the tablets, e.g., tablet hardness, content of API, and dissolution profiles. Third was the yield of the overall process considering various causes of losses, such as stabilization of the continuous process, sticking to the equipment, or condition setting of compression. For the continuous process, stability of product quality during the process operation was also investigated because the total run time can be up to hours.

The experimental result indicated that the product quality of all processes fulfilled the standard values which are often used in the industry. Granules made by fluidized bed granulation were porous, while granules made by high shear granulation and the continuous process were spherical. Although hardness of tablets made by the high shear granulation process was lower than that by the other two processes, there are no significant differences of dissolution profiles among the three processes. In the continuous process, product quality measured by mean particle diameter of granules was stable for the entire run time except for the initial nine minutes. Because of this reason, yield of the continuous process was lower than those of the other two processes, which indicates that stabilization would be one of the important challenges in continuous technology. In conclusion, the three investigated processes using continuous, fluidized bed, and high sheer granulation showed similarity in terms of product quality. However, intermediate products showed different quality characteristics such as size/shape, and especially in continuous technology, yield was found to be an improvement opportunity. The actual decisions should be based on various aspects such as quality, economy or safety, and thus, development of multiobjective design methodology is desired in future.

496075 A Cost-Benefit Analysis of Continuous Manufacturing Using Discrete Event Simulation

Abstract:

In the pharmaceutical industry, the demand for small volume, market specific products has increased in recent years. However, current batch processes do not allow the flexibility to meet smaller demands without a high cost of inventory and risk of discards. Continuous processing may offer advantages to traditional batch processing including shorter lead times, reductions in inventory, decreases in management overhead, and fewer discards or loss of tenders. However, the specific conditions which make continuous manufacturing beneficial are currently not well understood. To quantify the benefit and determine the conditions where continuous manufacturing is an attractive alternative for products, a discrete event simulation model was developed. The model compares continuous to batch tablet formulation processes under numerous different scenarios including; various market conditions, production planning strategies, capacity constraints, and production volumes. The model uses a product currently produced at a Merck production facility as a base case. The outputs from the model define the product, market, and production conditions best suited for a continuous manufacturing platform. This ultimately enables a mechanism for identifying products that are a good fit for a continuous manufacturing platform.

496960 Real-Time Monitoring and Control of API Concentration in a Tablet Press for Continuous Manufacturing of Tablets

Abstract:

Active pharmaceutical ingredient (API) concentration in powder blends is a critical material attribute (CMA) that is considered one of the most important product critical quality attributes (CQAs) for their final product (tablets). To determine the API concentration before tableting, real-time monitoring of API concentration before and after compaction has been performed for a variety of batch and continuous processes via Near-Infrared (NIR) spectroscopy [1, 2]. This analytical technology, which has been highly encouraged within the Quality by Design (QbD) and Process Analytical Technology (PAT) frameworks, has shown to be a valuable for detecting issues in CMAs during the process and has become a critical in-process sensor in the pharmaceutical industry [3].

This work presents a feed-forward controller for continuous manufacturing by using a NIR-based real-time monitoring technology at a tablet press feed frame. The basic idea behind feed forward control is to measure important process variables and take corrective action before the disturbances upset the process. The feed forward controller therefore needs to receive information about the known and predictable effects of process disturbances before those occur [4]. The proposed methodology involves measuring the inherent variability in the API concentration of material coming from the process and adjusting tablet weight to maintain overall API potency in tablets. We measured the residence time distribution (RTD) information from the rotary tablet press together with the NIR from the feed frame to develop the proposed framework as follows.

We evaluated the RTD of our Fette 1200i tablet press unit. A step change of API concentration of powder blend was given before the feed frame and the API concentration by the feed-frame NIR and compressed tablets were measured [5]. Based on these profiles, RTD of the feed frame was shown to have a similar RTD profile as a continuous stirred tank reactor (CSTR). RTD of the turret was found to be negligibly short and no difference was observed between the profiles of the feed-frame and tablets. The estimated RTDs are useful for prediction of the CQA and its controls. The formulation for our case study consisted of micronized acetaminophen, microcrystalline cellulose (Avicel® PH102), and magnesium stearate at a ratio of 15:84:1 by weight, respectively. A portable JDSU NIR probe was attached in the feed frame, sampling the powder spectra through one of the unit's glass windows. Five (5) different blends ranging from 10 – 20% API concentration were made as calibrations for the NIR model. NIR spectra were measured in-line using the 5 powders blends to build the calibration model *in situ*. A calibration model with small calibration error was constructed by partial least squares (PLS) analysis. Validation experiments with four different acetaminophen concentrations demonstrated that the constructed model has high accuracy during long-term operation.

Overall, based on the feed-frame's RTD profile and the information from the NIR sensor's model, a feed-forward controller was developed to account for changes in powder blend API concentration and adjust tablet weight to decrease variability of API potency in tablets. It was verified by both software simulations

and experiments that API potency could be maintained at the target value by the feed-forward control compared to the open-loop process.

References

1. Vanarase, A.U., et al., *Real-time monitoring of drug concentration in a continuous powder mixing process using NIR spectroscopy.* Chemical Engineering Science, 2010. **65**(21): p. 5728-5733.

2. Jarvinen, K., et al., *In-line monitoring of the drug content of powder mixtures and tablets by near-infrared spectroscopy during the continuous direct compression tableting process.* Eur J Pharm Sci, 2013. **48**(4-5): p. 680-8.

3. USFDA, *Guidance for Industry: PAT — A Framework for Innovative Pharmaceutical Development, Manufacturing, and Quality Assurance*, U.D.o.H.a.H. Services, Editor. 2004: Rockville, MD.

4. Myerson, A.S., et al., *Control Systems Engineering in Continuous Pharmaceutical Manufacturing. J. Pharm. Sci.*, 2015. **104**: p. 832-839.

5. Ward, H.W., et al., *Monitoring blend potency in a tablet press feed frame using near infrared spectroscopy.* Journal of Pharmaceutical and Biomedical Analysis, 2013. **80**: p. 18-23.

498201 Causal Network Modeling of Calcium Wave Propagation in Rat Liver Lobules

Abstract:

The liver performs a wide variety of physiological functions such as regulation of intermediary metabolism, lipid synthesis, bile production and xenobiotic detoxification. Normal liver function requires tight regulation of intracellular processes as well as intercellular coordination. Regulation of intracellular localization and levels of Ca^{2+} is critical to maintain hepatocyte metabolism and bile secretion, and regulate proliferation, autophagy and apoptosis. Consequently, disruption of Ca^{2+}dynamics can potentially lead to pathological conditions.

Evidence from whole tissue Ca^{2+} imaging studies points towards wave-like spatial patterns of cytosolic Ca^{2+} in response to circulating hormonal stimuli in liver lobules. Experimental studies have indicated that gap-junction mediated transfer of inositol 1,4,5-trisphosphate (IP_3) between adjacent hepatocytes is required for sequential induction of Ca^{2+} oscillations in neighboring cells, resulting in a spatial wave that spreads across a liver lobule. Alternatively, it has been proposed that paracrine signaling, spatial gradients of intracellular Ca^{2+} signaling components as well as mechanical stress induced by blood flow can lead to wave-like propagation of Ca^{2+} signals. In previous work, we developed an ODE based computational model that accounts for propagation of Ca^{2+} signals along a hepatocyte cord within a lobule. Based on simulation and analysis of the model, we predicted that in addition to gap junction mediated IP_3 exchange, spatial gradients in intracellular Ca^{2+} signaling components were required to yield a wave-like propagation of Ca^{2+}signals.

In the present study, we evaluated the above hypotheses pertaining to calcium wave propagation by obtaining and analyzing a unique data set on lobular scale Ca^{2+} signal dynamics measured at single cell resolution. We adopted a data driven, bottom-up approach to characterize relative contributions of cell autonomous, IP_3 mediated and paracrine signaling mediated Ca^{2+} wave dynamics starting at the single cell level. We imaged cytosolic Ca^{2+} levels in response to hormonal stimuli in perfused rat liver, to obtain a large-scale data set on Ca^{2+} transients in 1300 hepatocytes measured at 4 second intervals for 1600 seconds. Consistent with previous results, we observed wave-like propagation of Ca^{2+} signals throughout liver lobules. We analyzed the data set in three distinct ways including deterministic modeling, correlation networks, and causal network modeling, to uncover the contribution of proposed alternative mechanisms to spatial patterns of Ca^{2+}signals. Note that our interpretation and analysis was based on a simplification of cell-cell interactions and lobular organization in two dimensions, and hence would miss the interactions that occur in the actual three-dimensional liver tissue.

We initially utilized our ODE model and evaluated whether cell autonomous responses without intercellular molecular interaction/exchange can yield the observed experimental results. We employed a rejection sampling approach to

identify parameters for 1300 individual cellular Ca^{2+} time traces. Starting with priors of uniformly distributed parameter values centered on the nominal values from previous ODE model, we fitted the Ca^{2+} time traces of each hepatocyte to obtain cell-specific posterior parameter distributions corresponding to good data fits. Spatially correlated shifts in parameter posterior distributions would have suggested the absence of entrainment of Ca^{2+} responses of physically adjacent hepatocytes, and therefore of intercellular communication. However, our simulations yielded large overlaps in the posterior distributions of model parameters for cells lying within hepatocyte cords. The absence of shifts in posterior parameter distributions suggested that spatial gradients alone do not account for the observed Ca^{2+} wave propagation.

We next computed pair-wise correlations of Ca^{2+} traces to develop an undirected correlation network that was analyzed for alignment with the spatial patterns of potential interactions between hepatocytes. Our analysis suggested that Ca^{2+} signal correlation was spatially restricted to clusters of hepatocytes that were not limited to cells lying within the hepatocyte cords. These results suggest additional mechanisms of Ca^{2+} response synchronization besides gap junction mediated molecular exchange, such as release of paracrine signals by hepatocytes in their immediate microenvironment.

In order to assess whether the observed spatially clustered correlations in Ca^{2+} signals could be interpreted for directed interactions, we performed data-driven causal network modeling based on transfer entropy estimates of pair-wise information transfer between cells. Note that these formal measures of information flow are based on Shannonâ€™s information theory, and do not directly translate to the colloquial meaning of the term. Based on the findings from undirected correlation networks, all hepatocytes lying within 50Âµm of each other were considered as candidates for information transfer. Causal network analysis identified information transfer edges between co-localized hepatocytes lying within as well as across hepatocyte cords, further indicating that synchronization of hepatocyte Ca^{2+}responses does not arise exclusively due to gap junction mediated molecular exchange. Additionally, we found that information transfer between adjacent hepatocytes lying along a hepatocyte cord was not unidirectionally aligned.

The results of our study indicate that gap junction mediated IP_3 transfer as well as paracrine signaling contribute simultaneously towards shaping spatio-temporal dynamics of calcium wave propagation in liver lobules . The two mechanisms possibly contribute towards islands of highly synchronized hepatocytes. Additionally, spatial gradients of intracellular Ca^{2+} signaling components between the extremities of liver lobules might lead to an apparent Ca^{2+} wave propagation originating from multiple foci. These novel aspects must be included in our computational model to accurately capture Ca^{2+} signal propagation in liver lobules.

493122 Characterizing Host Immune Cell and Local Transcriptional Dynamics during Influenza Infection

Abstract:

Motivation

Influenza virus is a threat to human health, which results in millions of cases, hundreds of thousands of hospitalizations, and tens of thousands of deaths every year in the United States [1]. Several aspects of the immune response are believed to play a role in influenza infection pathology. Host targeted therapies are promising therapeutics but their effectiveness is limited due to the lack of a systematic characterization of the immune system during infection.

Background

Transcriptional profiling is a promising approach to multi-scale characterization of the intracellular signaling and immune cell population changes that occur during disease. While traditional statistical tests work well for identifying genes and signaling pathways differentially expressed in diseased tissue, they are not well suited for evaluating the effect of immune cell migration on gene expression.

Deconvolution algorithms seek to exploit gene expression to infer cellular demographics. Basic deconvolution algorithms postulate that expression profiles of cell mixtures are linear combinations of the expression profiles of pure cells, and use linear regression tools to enumerate cell proportions. These algorithms have been utilized to de-convolute mixtures of different kinds of tissues or cells [2], subdivisions of immune cells [3], or immune cells at different states (resting or activated) [3]. Most of these algorithms are only tested by artificial data of easily distinguishable mixtures, such as mixtures of lung, liver and brain [2], or mixtures of T cells, monocytes, B lymphoblasts, and B-cell lymphomas [3]. Furthermore, the cell proportions can only be accurately determined if all cell types of the sample are known *a priori*. Other approaches, such as CTen, use dynamic clustering analysis of time-course transcriptomic data to identify groups of genes sharing common expression patterns that can be associated with specific immune cells [4]. This method shows high accuracy but can only determine relative fold changes in immune cells (as opposed to fraction of cell population or cell counts). Several tools, including CIBERSORT by Newman et al [5] and DCQ by Altboum et al [6], are designed for complex tissues and temporal samples. However the reported estimations by these tools may not be satisfactory, and the error rates of deconvolution algorithms in cell quantification of complex and time course samples are still not understood.

Methods and results

Here, we evaluate the ability of several computational tools to infer immune cell counts using time-course, gene expression data. We constructed a library of microarray data from multiple published datasets of pure immune cells to obtain their transcriptomic signatures. The temporal microarray data of infected lung tissue [7] are used to evaluate predictions of the deconvolution tools. Pearson correlation coefficients are calculated by comparing estimated abundances at different time points to cell count data determined by fluorescence-activated cell sorting. We find that current approaches result in discontinuities and dramatic oscillations of immune cell proportions in infected tissue over time.

Discussion

Current deconvolution models demonstrate that transcriptomic signatures of pure cells can be used to estimate the composition of cell mixtures. However, predictions of cell populationsâ€™ dynamics during the immune response are not continuous and oscillate heavily due to neglecting dependencies between each time point. Our future algorithm will focus on dynamic models of cell infiltration to better predict immune cell demographics from transcriptional data.

References

[1] Centers for Disease Control and Prevention, 2016. [Online]. Available: https://www.cdc.gov/flu/about/qa/disease.htm. [Accessed 3 2017].

[2] Y. Zhong, Y.-W. Wan, K. Pang, L. M. Chow and Z. Liu, "Digital sorting of complex tissues for cell," BMC Bioinformatics, vol. 14, no. 89, 2013.

[3] A. Abbas, K. Wolslegel, D. Seshasayee, Z. Modrusan and H. Clark, "Deconvolution of Blood Microarray Data Identifies Cellular Activation Patterns in Systemic Lupus Erythematosus," Plos One, 2009.

[4] J. E. Shoemaker, T. J. Lopes, S. Ghosh, Y. Matsuoka, Y. Kawaoka and H. Kitano, "CTen: a web-based platform for identifying enriched cell types from heterogeneous microarray data," BMC Genomics, 2012.

[5] A. M. Newman, C. L. Liu, M. R. Green, A. J. Gentles, W. Feng, Y. Xu and C. D. Hoang, "Robust enumeration of cell subsets from tissue expression profiles," Nature Methods, p. 453â€"457, 2015.

[6] Z. Altboum, Y. Steuerman, E. David, Z. Barnett-Itzhaki and L. Valadarsky, "Digital cell quantification identifies global immune cell dynamics during influenza infection," Molecular Systems Biology, vol. 10, no. 2, 2014.

[7] J. E. Shoemaker, S. Fukuyama, A. J. Eisfeld, D. Zhao, E. Kawakami and S. Sakabe, "An Ultrasensitive Mechanism Regulates Influenza Virus-Induced Inflammation," PLoS Pathogens, 2015.

496560 A Two-State Model-Based Cell Clustering and Network Inference for Single-Cell Gene Expression Data

Abstract:

Recently, the advent of new single-cell profiling technologies have provided a promising means to elucidate the mechanisms regulating dynamic cellular processes, including cell development, maturation, and differentiation (Trapnell, 2015; Paul *et al.*, 2015; Richard *et al.*, 2016). In particular, conventional techniques for the generation of transcriptomics data such RNA-sequencing and real-time quantitative PCR have been successfully adapted for the quantification of gene expression levels in individual cells (Tang *et al.*, 2009; White *et al.*, 2011). Single cell analysis allows us to study the significance of cell-to-cell heterogeneity, which would be impossible using population-averaged measurements (Sandberg, 2013). Single cell profiling technologies are now able to provide high-throughput (many cells) high-dimensional (many genes) expression data in a single experiment, pioneering a whole new series of computational methods for the data analysis. In this context, the identification and characterization of different cell types within a heterogeneous population is an important step in the analysis of single-cell data (Liu and Trapnell, 2016), which may lead to new insights on the underlying biological process. Standard unsupervised clustering strategies such as *k*-means (Kanungo *et al.*) and hierarchical clustering (Johnson, 1967) have previously been applied. However, the high variability in gene expression patterns and the high dimensionality of the data make the clustering problem not trivial. In recent years, different clustering strategies have been developed specifically for single-cell expression data by modifying traditional clustering algorithms and/or by using dimensional reduction techniques (Lin *et al.*, 2017; Xu and Su, 2015; Buettner *et al.*, 2015; žurauskiene and Yau, 2016; Kiselev *et al.*, 2017). In addition, time-variant cell clustering methods have been implemented to elucidate the appearance of cell types in cell differentiation stages (Marco *et al.*, 2014; Huang *et al.*, 2014). Until recently (Ezer *et al.*, 2016), the bursty and stochastic dynamics of gene expression in single cells, which has been suggested as a major source of the variability (K¾rn *et al.*, 2005), were not directly addressed in the clustering of single cell transcriptomics data.

In this work, we implemented a clustering algorithm that explicitly takes into account the stochastic dynamics of gene transcriptional process using the two-state model (Kim and Marioni, 2013). The two-state model describes the gene expression processes involving (1) the promoter switching between ON and OFF state, (2) in the OFF state (a closed chromatin state), the gene is not accessible for the transcription and (3) in the ON state (an open chromatin state), the gene transcription could occur, producing the mRNA molecules in bursts (Munsky *et al.*, 2012). A total of four kinetic parameters fully describe the two-state model, namely Kon (rate of activation), Koff (rate of inactivation), Kt (rate of transcription) and Kd (rate of degradation). Our clustering approach is based on the idea that cells belonging to the same cluster should share the same parameters of the two-state model. This concept has been recently illustrated by Ezer et al. in the implementation of their single cell clustering method called SABEC (Simulated Annealing for Bursty Expression Clustering) (Ezer *et al.*, 2016).

Our clustering method starts by randomly assigning cells into clusters, and iteratively reassigns cells to clusters until convergence or until a maximum number of iterations, as follow: (1) for each cluster and each gene, determine the parameters for the two-state model that best fit the distribution of expression among cells in the cluster, (2) calculate the probability (likelihood) for each cell to be in each of the clusters given the cellÕs gene expression data and the cluster parameters from step (1), and (3) re-assign cells to the cluster that gives the maximum likelihood for the cellÕs gene expression data. To improve both the accuracy and stability of the solutions, we performed the iterations above multiple times with different initial cell assignments, and combine the outcomes in a consensus matrix that summarize how often two cells are clustered together (Fig. 1). Finally, we implemented k-medoids (Bhat, 2014) using the consensus matrix to obtain the cell clustering. The number of clusters can be user-defined or automatically evaluated based on heuristic approaches such as gap-statistics or silhouette scores. In comparison to SABEC (Ezer *et al.*, 2016), the greedy optimization in our algorithm converged much faster (seconds vs. minutes) while producing comparable clustering results. For example, in an application to single-cell expression data from hematopoietic stem cell (HSC) differentiation (Moignard *et al.*, 2013), each repeat of SABEC took 45 minutes to complete on a standard workstation, while our greedy algorithm converged to the final solution in less than one minute.

In the next step, we employed the cell clustering outcome for two different tasks: (1) to define the trajectory of gene expression that describes the cell differentiation progression, and (2) to infer the gene regulatory networks that govern the cell differentiation process. In these two tasks, we viewed each of the clusters as a ÒstateÓ in the cell differentiation. In the first task, we first define the cell state of each cluster based on prior information of either the division stages or the time stamps of the cells in the cluster. For each cell, we used the two-state model to compute the cell pseudo-stage or pseudo-time, specifically by taking a weighted average of the cluster states with the log-likelihood values of the cell to be in any given cluster as the weights. The cell trajectory was finally constructed by ordering cells according to their pseudo-stage or pseudo-time.

In the final task, we employed the cluster and cell trajectory to portray the cell differentiation progression, and inferred the gene regulatory network that governs this process. More specifically, we employed our previous algorithm, SINCERITIES (Papili Gao *et al.*, 2016), on the ordered cells. Briefly, SINCERITIES is able to reconstruct the (causal) gene-gene regulations by employing regularized linear regression, based on changes in the distributions of gene expressions over the differentiation trajectory. Meanwhile, the signs/modes of the gene regulations (activation and repression) are inferred by computing partial correlation analysis between pairs of genes. . We demonstrated the efficacy of our algorithms for single cell clustering, cell trajectory construction and gene regulatory network inference on the differentiation of chicken erythrocytic cells (Richard *et al.*, 2016). The results further confirmed the importance of sterol pathways in initiating the cell differentiation process in this particular cell line.

References

Bhat,A. (2014) K-MEDOIDS CLUSTERING USING PARTITIONING AROUND MEDOIDS FOR PERFORMING FACE RECOGNITION. *Int. J. Soft Comput. Math. Control*, **3**.

Buettner,F. *et al.* (2015) Computational analysis of cell-to-cell heterogeneity in single-cell RNA-sequencing data reveals hidden subpopulations of cells. *Nat. Biotechnol.*, **33**, 155–160.

Ezer,D. *et al.* (2016) Determining Physical Mechanisms of Gene Expression Regulation from Single Cell Gene Expression Data. *PLOS Comput. Biol.*, **12**, e1005072.

Huang,W. *et al.* (2014) Time-variant clustering model for understanding cell fate decisions. *Proc. Natl. Acad. Sci. U. S. A.*, **111**, E4797-806.

Johnson,S.C. (1967) HIERARCHICAL CLUSTERING SCHEMES. *Psychometrika*, **32**.

K¾rn,M. *et al.* (2005) Stochasticity in gene expression: from theories to phenotypes. *Nat. Rev. Genet.*, **6**, 451–464.

Kanungo,T. *et al.* An Efficient k-Means Clustering Algorithm: Analysis and Implementation.

Kim,J. and Marioni,J.C. (2013) Inferring the kinetics of stochastic gene expression from single-cell RNA-sequencing data. *Genome Biol.*, **14**, R7.

Kiselev,V.Y. *et al.* (2017) SC3: consensus clustering of single-cell RNA-seq data. *Nat. Methods*.

Lin,P. *et al.* (2017) CIDR: Ultrafast and accurate clustering through imputation for single-cell RNA-seq data. *Genome Biol.*, **18**, 59.

Liu,S. and Trapnell,C. (2016) Single-cell transcriptome sequencing: recent advances and remaining challenges. *F1000Research*, **5**.

Marco,E. *et al.* (2014) Bifurcation analysis of single-cell gene expression data reveals epigenetic landscape. *Proc. Natl. Acad. Sci. U. S. A.*, **111**, 5643–50.

Moignard,V. *et al.* (2013) Characterization of transcriptional networks in blood stem and progenitor cells using high-throughput single-cell gene expression analysis. *Nat. Cell Biol.*, **15**, 363–72.

Munsky,B. *et al.* (2012) Using Gene Expression Noise to Understand Gene Regulation. *Science (80-.).*, **336**.

Papili Gao,N. *et al.* (2016) SINCERITIES: Inferring gene regulatory networks from time-stamped single cell transcriptional expression profiles. *bioRxiv*.

Paul,F. *et al.* (2015) Transcriptional Heterogeneity and Lineage Commitment in Myeloid Progenitors. *Cell*, **163**, 1663–1677.

Richard,A. *et al.* (2016) Single-Cell-Based Analysis Highlights a Surge in Cell-to-Cell Molecular Variability Preceding Irreversible Commitment in a Differentiation Process. *PLOS Biol.*, **14**, e1002585.

Sandberg,R. (2013) Entering the era of single-cell transcriptomics in biology and medicine. *Nat. Methods*, **11**, 22–24.

Tang,F. *et al.* (2009) mRNA-Seq whole-transcriptome analysis of a single cell. *Nat. Methods*, **6**, 377–382.

Trapnell,C. (2015) Defining cell types and states with single-cell genomics. *Genome Res.*, **25**, 1491–8.

White,A.K. *et al.* (2011) High-throughput microfluidic single-cell RT-qPCR. *Proc. Natl. Acad. Sci. U. S. A.*, **108**, 13999–4004.

Xu,C. and Su,Z. (2015) Identification of cell types from single-cell transcriptomes using a novel clustering method. *Bioinformatics*, **31**, 1974–80.

žurauskiene,J. and Yau,C. (2016) pcaReduce: hierarchical clustering of single cell transcriptional profiles. *BMC Bioinformatics*, **17**, 140.

Figure 1: Cell clustering in HSC differentiation using single cell transcriptional expression data. (a) The single cell expression data came from five populations of cells during hematopoiesis: LMPP (lymphoid-primed multipotential progenitor), PreM (premegakaryocytes), GMP (granulocyte-macrophage progenitor) and CLP (common lymphoid progenitors). (b) Maximum-likelihood based clustering of cells with our greedy algorithm.

485318 Maximum Entropy Approach for Parameter Estimation in Signaling Networks

Abstract:

Predictive models of signaling networks are key to our understanding of cellular function and in designing rational interventions in disease. However, using network models to predict signaling network behavior is challenging due to inherent cell-to-cell variability of network parameters, such as reaction rates and species concentrations that govern network dynamics. In this work, we present an inference framework based on the principle of maximum entropy (ME) to estimate the joint probability distribution over network parameters that is consistent with experimentally observed cell-to-cell variability in concentrations of network species. We apply the framework to study the signaling cascade activated by the epidermal growth factor (EGF) resulting in phosphorylation of protein kinase B (Akt); a central signaling hub in mammalian cells. Notably, the inferred parameter distribution allows us to computationally predict the distributions of phosphorylated Akt (pAkt) levels at early and late times after EGF stimulation as well as the distribution of cell surface EGF receptors (sEGFRs) after prolonged stimulation with EGF. We discuss how the developed framework can be generalized and applied to other types of data as well as problems beyond signaling networks.

491313 Computation-Driven Mechanistic Understanding of the Cellular Cost and Regulation of Melanin Production

Abstract:

Computation-Driven Mechanistic Understanding of the Cellular Cost and Regulation of Melanin Production

Wheaton Schroeder, Jyothi Kumar, Rajib Saha and Steve Harris

The University of Nebraska – Lincoln, Lincoln, NE

Exophiala dermatitidis, a melanized extremophilic black yeast, is of interest primarily for its ability to synthesize melanin as a stress protectant. In addition, as an emerging fungal pathogen, melanin deposition in the cell wall hides this organism from the host immune system, often leading to high mortality rates, and without which its pathogenic efficacy is significantly decreased. Thus, with a small genome *E. dermatitidis* may serve for dual purposes: first, as a model system for melanocytes to explore treatments for melanin-related human skin disorders such as albinism, and second, as a model pathogenic organism to study how to combat infections by disrupting melanin production, not only from *E. dermatitidis*, but also for other melaninized fungal pathogens such as *Cryptococcus neoformans*. At present, relatively little is known about the regulation and kinetics of melanin production in E. dermatitidis, and in particular whether synthesis of melanin imposes a severe metabolic cost on the fungus. Our chosen tool for filling this knowledge gap is computational modeling, which is a promising approach that has not yet been systematically applied to extremophilic fungi or to human fungal pathogens.

Computational modeling of metabolism is now an indispensable tool to drive the processes of understanding, discovering, and redesigning of biological systems. Although Flux Balance Analysis (FBA) is the primary tool used for this purpose, it has significant limitations due to the lack of reaction kinetics, chemical species concentration, and metabolic regulation. Hence, the development a kinetic model framework will allow for a greater understanding of *E. dermatitidis* as a pathogen and as a model organism. At present, one of the greatest challenges to this approach is that this organism's metabolism is not well studied. Specific challenges to overcome include that the genome is less than 50% annotated, there is no metabolic model available in either literature or in public databases such as ModelSeed, and there are no genes or genome entries for this organism in the Kyoto Encyclopedia of Genes and Genomes (KEGG). Hence, the necessary first step in computational modeling is to build a genome-scale model by usingthe available annotations of *E. dermatitidis*, published models, and homology information of related species (such as *Aspergillus*-genus).

Once the genome-scale model is constructed, a novel approach to building a kinetic model framework using FBA, transcriptomic, metabolomics, experimental, and genomic data is used. The completed kinetic model framework currently has about 1700 reactions and 2100 metabolites. This modeling framework will next be used for exploring important questions of biosynthesis, regulation, and metabolic cost related to melanin production, as a freestanding organism, as a pathogenic organism, and as a model system for human melanocytes. In addition, the computational model will be useful for *in silico* discovery and exploration of possible treatment options for infections cause by *E. dermatitidis* and related species by inhibiting or downregulating melanin synthesis, presenting significant cost savings in developing viable treatment options. Further, this computational model can be used as an *in silico* model of human melanocytes, to study and hopefully suggest strategies to cure melanin-related skin disorders such as albinism.

487716 Metabolic Modeling of Interactions Between *Pseudomonas aeruginosa* and *Staphylococcus aureus* in Cystic Fibrosis Biofilm Infections

Abstract:

Pseudomonas aeruginosa and *Staphylococcus aureus* are two bacteria most commonly isolated from the lung infections of cystic fibrosis (CF) patients. Individuals co-infected with these two bacteria are more prone to exacerbations and hospitalization than patients infected with either bacterium alone; thus, such coinfections contribute to worse clinical outcomes in this patient population. CF infections are often characterized by the formation of biofilms in which bacteria reside in a protective matrix of self-produced polymers. More effective treatment strategies are needed for these polybacterial biofilm infections to improve the quality of life and reduce mortality in CF patients. We hypothesize that interactions between these two bacteria are attributable to metabolites that are secreted by each species and diffuse over long distances to shape spatial biofilm structure. Our *in vitro* studies show that *S. aureus* shifts from respiratory to fermentative metabolism in the presence of *P. aeruginosa* supernatant, resulting from inhibition of the *S. aureus* electron transport chain (ETC) by siderophores and 2-heptyl-4-hydroxy quinoline-N-oxide (HQNO). Moreover, we observed that *S. aureus* is more tolerant to the antibiotics when grown in the presence of *P. aeruginosa* supernatant.

In this study, we developed a multispecies biofilm metabolic model to further investigate the hypothesis that *P. aeruginosa* can exert its influence on *S. aureus* at a distance via secreted factors. Simulations were performed by supplying glucose, amino acids and oxygen at the top of the biofilm to mimic the *in vivo* CF environment. *P. aeruginosa* was forced to secrete HQNO, which diffused through the biofilm and inhibited *S. aureus* ETC through cytochrome b. Lactate synthesized by *S. aureus* was modeled to be the preferred carbon source for *P. aeruginosa*, as has been shown experimentally. The multispecies biofilm model was shown to reproduce experimentally observed behavior, including dominance of *P. aeruginosa*, upregulation of *S. aureus* fermentative metabolism and enhanced *P. aeruginosa* growth due to lactate cross feeding. These results represent the first step towards integrating laboratory experiments with clinically isolated bacterial strains and *in silico* modeling of biofilm metabolism to unravel the enhanced pathogenicity of this CF coinfection.

505520 Hybrid Models Explain Emergent Dynamics in Complex Cell Populations

Abstract:

Computational modeling is an essential tool for integrating and understanding complex biological systems. With increasingly high-resolution, high-throughput, and dynamic experimental data, the *in silico* systems biology community is able to develop better informed models to interrogate the complex, heterogeneous, and multiscale nature of cellular microenvironments. My lab employs machine learning with dynamical systems and agent-based models to study emergence in heterogeneous cell populations. Specifically, we investigate the impact of intracellular network dynamics on intercellular signaling in context of both cancer and immune systems. Our simulations interrogate how the whole is greater than the sum of its parts by predicting cell population dynamics from the composition of simpler signaling, or biological modules. Elucidating the compositionality problem is fundamental to advancing our understanding of basic science; to promoting the impact of synthetic biology; and to designing precise dynamic therapeutic strategies.

505305 Process Analytical Utility of Raman Microspectroscopy for Monitoring Cellular Therapy Manufacturing

Abstract:

A major wave of promising cellular therapies is progressing through clinical trials, with the potential to one day improve treatments for cancer, diabetes, cardiac and other degenerative diseases. Given the complexity of mammalian cells, the success of these therapies will to a great extent depend on developing robust methods to reliably monitor both the production processes and assess the quality of the manufactured cell products. Raman spectroscopy offers a non-invasive and label-free approach to distinguish different cell types and physiological states by analyzing macromolecular composition changes in living cells. Spectral markers, such as the ratio of nucleic acid to protein-associated band intensities, have been used to discriminate between stem cells and their differentiated progeny. Both the types of cell death and apoptotic stages can be distinguished. Thus, Raman spectral indicators have a great potential to provide a highly informative, non-destructive means to monitor cellular changes in manufacturing processes for cell-based therapy products.

AUTHOR INDEX

AUTHOR INDEX

AUTHOR INDEX

AUTHOR INDEX

AUTHOR INDEX

AUTHOR INDEX

AUTHOR INDEX

AUTHOR INDEX

AUTHOR INDEX

AUTHOR INDEX

AUTHOR INDEX

AUTHOR INDEX

AUTHOR INDEX

AUTHOR INDEX

AUTHOR INDEX

AUTHOR INDEX

AUTHOR INDEX

AUTHOR INDEX

AUTHOR INDEX

AUTHOR INDEX

AUTHOR INDEX

AIChE
120 Wall Street, FL 23
New York, NY 10005-4020

ISBN 978-1-5108-5802-2

9 781510 858022

Food, Pharmaceutical and Bioengineering Division 2017

Core Programming Area at the 2017 AIChE Annual Meeting

Minneapolis, Minnesota, USA
29 October - 3 November 2017

Volume 2 of 2

ISBN: 978-1-5108-5802-2

Printed from e-media with permission by:

Curran Associates, Inc.
57 Morehouse Lane
Red Hook, NY 12571

Printed by Curran Associates, Inc. (2018)

For permission requests, please contact AIChE
at the address below.

AIChE
120 Wall Street, FL 23
New York, NY 10005-4020

Phone: (800) 242-4363
Fax: (203) 775-5177

www.aiche.org

Additional copies of this publication are available from:

Curran Associates, Inc.
57 Morehouse Lane
Red Hook, NY 12571 USA
Phone: 845-758-0400
Fax: 845-758-2633
Email: curran@proceedings.com
Web: www.proceedings.com

TABLE OF CONTENTS

VOLUME 1

VOLUME 2

Some abstracts in this book refer to images that were not made available by the publisher

505306 Engineering Metabolic Reaction NetworksÂ

Abstract:

The improved productivity of a bioreactor depends to a large extent on the development of stable cell lines with optimized pathways that convert nutrients into desired products at the maximum possible yield. We are using elementary mode analysis to theoretically evaluate metabolic networks for the best pathway structure. Such pathways can be practically implemented using knockout mutations that reduce complex metabolic networks to reaction networks of minimal but optimized functionality. Moreover, the identification of the complete set of elementary modes in a metabolic network represents an objective set of discrete states of the metabolism. The set of possible reaction trajectories can be analyzed using tools of statistical thermodynamics which permit the prediction of the most probable and stable pathway structure in a metabolic network. While this approach has been validated in several prokaryotic systems it is expected to be applicable and useful also in more complex, eukaryotic cell systems.

505307 Stochastic Behavior of Reaction Diffusion Systems. Direct Evaluation of Average Behavior

Abstract:

Understanding transfer of drug resistance among bacterial species in biofilms is essential for combating the phenomenon towards fighting infection. Biological signaling processes occurring in biofilms present stochastic reaction diffusion systems. Computation of average behavior of stochastic systems in which nonlinearity of chemical reactions interact with diffusion involves time consuming sample path evaluations. This talk will discuss recent methods for circumventing the calculation of sample paths for direct evaluation of averages.

505308 From Synthetic Biology to Nano Biotechnology: Rational Antimicrobial Engineering Approaches Towards Combating Drug-Resistant Pathogens

Abstract:

The rapid rise of drug-resistant superbugs and the declining antibiotic pipeline are serious challenges to global health. In this talk, I will describe multi-pronged systems, synthetic biology and nanobiotechnology approaches being devised in our lab to engineer antimicrobials that can overcome antimicrobial resistance. Transcriptome studies in our lab have shown that, when exposed to antimicrobials, bacteria enter an â€œadaptive resistanceâ€ state by exploring multiple pathways sampling a dynamic gene regulatory space. We investigate these adaptive pathways by controllably up-regulating and down-regulating the expression of genes known to be involved in bacterial tolerance. We employ emerging synthetic biology techniques to investigate gene regulatory networks involved in controlling adaptive resistance. Using CRISPR based technology, we rationally engineer library of synthetic genetic devices to activate and inhibit native gene expression of key essential gene and stress-response networks. Here we show that in the presence of these synthetic perturbations, significant control over bacterial fitness can be achieved during adaptation to sub-minimal inhibitory concentrations of a range of toxins, including disinfectants and antibiotics. Moreover, we employ this approach to perturb expression of multiple genes simultaneously, and observe predominant negative epistasis with severe loss in fitness. To translate our findings in to the clinical setting, we engineer antisense therapeutics that can block translation of any desired gene in a pathogen-specific manner for targeted inhibition. Using this approach I will demonstrate a platform for accelerated development of novel antibiotics against multi-drug resistant (MDR) bacterial clinical isolates as well as any emergent bacterial threats. Finally, I will also present an engineered nanoparticle-based therapeutic strategy using quantum dots which, when activated by stimuli, release Reactive Oxygen species to eliminate a broad range of MDR bacterial clinical isolates including methicillin-resistant *Staphylococcus aureus*, extended-spectrum Î²-lactamase producing *Klebsiella pneumoniae* and *Salmonella typhimurium*, and carbapenem-resistant *Escherichia coli*. The techniques presented in this talk offer a novel approach for both impeding the intrinsic adaptive pathways leading to antibiotic resistance, and re-sensitizing antibiotic resistant pathogens to traditional therapies employed at the clinical setting.

505309 Cell Culture Engineersâ€™ Influence in Cell Therapies

Abstract:

Over the past 30 years, bioengineering has enhanced, by several orders of magnitude, the productivity and robustness of industrial-scale, stirred-tank therapeutic protein production processes. Progress continues with creative methods of process intensification and time-honored chemical engineering methods for continuous processing. It is expected that over the next decades, cell culture engineering should provide the same enhancements to industrial production of cell-based therapeutics. Cell therapies are a fast-growing and rapidly evolving and improving area of biotechnology which have the potential to provide unprecedented efficacy for devastating diseases globally. Among the many challenges in this area that cell culture engineers will tackle are high costs of goods, rare and ill-characterized raw materials (often patientsâ€™ own cells), complex supply chains, and unpredictable process performance. This presentation will highlight the influence cell culture engineers will have in addressing these challenges with parallels to the challenges of therapeutic protein production.

505310 Glycosylation: Trace Metals and Design for Robustness

Abstract:

Glycosylation profile may affect the efficacy of therapeutic monoclonal antibodies, and is impacted by macro and micro nutrient levels. Trace metals are known to be cofactors of glycosylation enzymes as well as inhibitors. Variability in trace metal concentrations can thus impact glycoform consistency. This may result either from a metabolic effect causing changed availability of intracellular nucleotide-sugars or a more direct effect on enzyme activity. Knowledge of these effects and their probable cause will help design robust processes targeting a specific glycoform.

In this work, we have evaluated the impact of change in concentration of several trace metals on glycosylation of IgG in CHO cells to identify trace elements which affect glycosylation in a dose dependent manner. Medium supplementation strategies are used to understand the probable causes of the observed effects. A mathematical model is used to help evaluate the hypotheses. Based on these, a medium formulation strategy to reduce the effect of variability in the concentration of such trace elements on the glycoform of recombinant IgG will be discussed. Tools developed in our group to facilitate evaluation of some feeding strategies will be discussed.

489666 Use of Model Discrimination Method in Drug Substance Process Development

Abstract:

Drug substance process development often includes several reaction steps to produce the Active Pharmaceutical Ingredient (API), where each step leads to a desired API intermediate. Mathematical models with different structures can be built to represent the proposed reaction mechanisms. It is common for multiple model constructs to equally explain the data gathered at this early stage. At this point it is important to select a model structure among the multiple candidates proposed. This work highlights the application of a model discrimination method to achieve this goal.

The model discrimination method was used to identify a discriminative experimental recipe for a reaction system of interest [1]. The experiment was then conducted to generate data that was qualitatively compared against the model predictions. The model that better explains the trends of the experimental data is deemed to be the best representation of the system among the candidates. To illustrate the use of this method, case studies for different reactions in drug substance process development will be given.

Once a model is discriminated, further investigation can be performed to optimally parameterize the model by following a model based experimental design method. A representative model can help the process development team determine safer and more efficient operating conditions.

Reference

[1] Espie, D. and Macchietto, S. (1989), The optimal design of dynamic experiments. AIChE J., 35: 223â€“229. doi:10.1002/aic.690350206.

489846 Model-Aided Development for a Mixed Anhydride Formation in a Drug Substance Manufacturing Process

Abstract:

Mathematical models are valuable tools in process development to determine acceptable parameter ranges and control strategies for manufacturing. Some of the benefits of modeling in pharmaceutical development are its ability to reduce material costs and resource needs, while simultaneously increasing process understanding and quality control.

A case study is presented here that demonstrates the use of automated experimentation to gather high-value reaction data for kinetic parameter estimation of a six reaction step mechanism with nine components. The model was used to simulate kinetic and thermal profiles for varying addition strategies of a key reagent. The resulting model was used to define parameter targets and ensure process robustness in preparation for a successful registration stability campaign.

494892 Design of Efficient Metal Nanocatalysts for Cross Coupling Reactions Using Finite Difference Time-Domain Simulations

Abstract:

Design of Efficient Metal Nanocatalysts for Cross Coupling Reactions using Finite Difference Time-Domain Simulations

Andishaeh Dadgar, Farshid Mohammadparast, Marimuthu Andiappan, School of Chemical Engineering, Oklahoma State University, Stillwater, OK, USA.

Cross coupling reactions (e.g., Suzuki and Heck couplings) have been traditionally carried out using homogeneous palladium (Pd) based catalysts, and in recent years, the use of heterogeneous metal nanocatalysts has been of great interest. However, the metal nanocatalysts (e.g., Pd, Au) require more drastic reaction conditions (e.g., high temperatures) because of their low catalytic activity. The operation of cross couplings under these drastic reaction conditions compromises long-term stability of the nanocatalysts due to leaching of metal atoms. Therefore, there remains a critical need to develop a heterogeneous metal nanocatalyst with high activity and long-term stability for cross couplings.

In this contribution, we show that finite-difference time-domain (FDTD) simulations can be used as platform to design efficient heterogeneous metal nanocatalysts. Our FDTD simulations predict that the UV-Vis surface plasmon extinction spectra of the metal nanoparticles are very sensitive to the local environment, and therefore, changes in even a small fraction of the surface atoms (e.g., leaching of the surface atoms) can be identified from the changes in the extinction spectra. We will show that a combination of FDTD simulations and a simple UV-Vis surface plasmon spectroscopic technique can be used to develop efficient heterogeneous metal nanocatalysts for a variety of cross coupling reactions.

495908 Nonlinear Optimisation of Continuous Artemisinin Crystallisation with Explicit NRTL Model-Based Solubility Prediction

Abstract:

Continuous processing is studied for use in pharmaceutical production to overcome the inherent drawbacks of batch processing, which is the current paradigm in the pharmaceutical and fine chemicals industry. Developments in small-scale flow technologies (and in microreactors in particular) have accelerated this effort; flow chemistry has not only allowed higher efficiencies but also enabled access to reaction pathways that have not been previously available for batch production (Jensen, 2017). This sustained interest in Continuous Pharmaceutical Manufacturing (CPM), along with the development of Process Analytical Technology (PAT) explains the strong interest towards Quality by Design (QbD) for pharmaceutical production processes, which enjoys significant regulatory body encouragement and support (Lee et al., 2015).

One crucial aspect of continuous processing is the continuous separation of Active Pharmaceutical Ingredient (API) products, often at low scales of process throughput and strict purity requirements. Research in product separation, including downstream drug product formulation, is a vibrant area (Ierapetritou et al., 2016; Vural Gürsel et al., 2017; Yabuta et al., 2017). The recent technological development of plug flow crystallisers and Continuous Oscillatory Baffled Crystallisers (COBC, which operate in an oscillatory plug flow regime) offers production advantages, including enhanced heat and mass transfer and improved profile temperature control (Brown and Ni, 2011; McGlone et al., 2015).

Our recent publications explore process modelling, simulation and optimisation of the continuous synthesis and separation of ibuprofen (Jolliffe and Gerogiorgis, 2017a) and artemisinin (Jolliffe and Gerogiorgis, 2017b). By using systematic kinetic parameter estimation, solubility prediction via the UNIFAC method, and comprehensive economic evaluation, optimal total costs have been determined.

Artemisinin is a key antimalarial, and its continuous synthesis has recently been demonstrated (Kopetzki et al., 2013; Gilmore et al., 2014); product recovery and purity enhancement via crystallisation is a viable technique (Malwade et al., 2016). The present paper investigates further the modelling and systematic optimisation of continuous artemisinin crystallisation incorporating explicit solubility predictions via the Non-Random Two-Liquid (NRTL) model. Modelling a plug-flow COBC helps ensure high heat and mass transfer rates, with minimal impact of temperature and concentration gradients on the API product. Toluene, used in a promising CPM process for artemisinin (Horváth et al., 2015), is assumed to be the process solvent; the upstream portion of the flowsheet (reactor) performance is analysed on the basis of previous work (Jolliffe and Gerogiorgis, 2016). The antisolvents considered encompass three different binary mixtures (ethanol and ethyl acetate, ethanol and acetone, or acetone and ethyl acetate) chosen due to to their pharmaceutical processing suitability and the availability of solubility data and NRTL parameters (Nti-Gyabaah et al., 2010). The ratio of antisolvents in the binary mixture ranges continuously between 1:9 to 9:1 by weight. Similarly, the ratio of process solvent (toluene) to binary antisolvent mixture ranges between 1:1 to 1:5 by weight. A continuous crystalliser operating temperature range (between 40 °C and 5 °C) has been considered. Crystal growth and productivity have been modelled on the basis of similar systems (Horváth et al., 2015), and impurities have been assumed to not co-crystallise with the API product.

The nonlinear optimisation (NLP) formulation encompasses the optimal choice of antisolvent, rate of antisolvent addition and the operating temperature which is preferable for continuous artemisinin crystallization. Our results include objective function (minimal cost) surfaces for a range of key variables, such as API recovery; insightful three-dimensional illustrations are obtained for process cost as a function of decision variables, e.g. crystallisation operating temperature and ratio of binary antisolvent mixture addition over incoming reactor effluent. The systematic optimisation of technoeconomic indices under explicit process equilibria and thermodynamic solubility constraints offers unprecedented insight into fine-tuning Continuous Pharmaceutical Manufacturing (CPM) process design and operation.

LITERATURE REFERENCES

Brown, C.J., Ni, X., 2011. Online evaluation of paracetamol antisolvent crystallization growth rate with video imaging in an oscillatory baffled crystallizer. Cryst. Growth Des. 11, 719–725.

Gilmore, K., Kopetzki, D., Lee, J.W., Horváth, Z., McQuade, D.T., Seidel-Morgenstern, A., Seeberger, P.H., 2014. Continuous synthesis of artemisinin-derived medicines. Chem. Commun. 50, 12652–12655.

Horváth, Z., Horosanskaia, E., Lee, J.W., Lorenz, H., Gilmore, K., Seeberger, P.H., Seidel-Morgenstern, A., 2015. Recovery of artemisinin from a complex reaction mixture using continuous chromatography and crystallization. Org. Process Res. Dev. 19, 624–634.

Ierapetritou, M., Muzzio, F., Reklaitis, G., 2016. Perspectives on the continuous manufacturing of powder-based pharmaceutical processes. AIChE J. 62, 1846–1862.

Jensen, K.F., 2017. Flow chemistry—Microreaction technology comes of age. AIChE J. 63, 858–869.

Jolliffe, H.G., Gerogiorgis, D.I., 2017a. Technoeconomic optimisation of a conceptual flowsheet for continuous separation of an analgaesic Active Pharmaceutical Ingredient (API). Ind. Eng. Chem. Res., in press.

Jolliffe, H.G., Gerogiorgis, D.I., 2017b. Technoeconomic optimisation and comparative environmental evaluation of continuous crystallisation and antisolvent selection for artemisinin recovery. Comput. Chem. Eng., in press.

Jolliffe, H.G., Gerogiorgis, D.I., 2016. Process modelling and simulation for continuous pharmaceutical manufacturing of artemisinin. Chem. Eng. Res. Des. 112, 310–325.

Kopetzki, D., Lévesque, F., Seeberger, P.H., 2013. A continuous-flow process for the synthesis of artemisinin. Chem.–Eur. J. 19, 5450–5456.

Lee, S.L., O'Connor, T.F., Yang, X., Cruz, C.N., Chatterjee, S., Madurawe, R.D., Moore, C.M.V., Yu, L.X., Woodcock, J., 2015. Modernizing pharmaceutical manufacturing: from batch to continuous production. J. Pharm. Innov. 1–9.

Malwade, C.R., Buchholz, H., Rong, B.-G., Qu, H., Christensen, L.P., Lorenz, H., Seidel-Morgenstern, A., 2016. Crystallization of artemisinin from chromatography fractions of Artemisia annua extract. Org. Process Res. Dev. 20, 646–652.

McGlone, T., Briggs, N.E.B., Clark, C.A., Brown, C.J., Sefcik, J., Florence, A.J., 2015. Oscillatory Flow Reactors (OFRs) for continuous manufacturing and crystallization. Org. Process Res. Dev. 19, 1186–1202.

Nti-Gyabaah, J., Gbewonyo, K., Chiew, Y.C., 2010. Solubility of artemisinin in different single and binary solvent mixtures between (284.15 and 323.15) K and NRTL Interaction Parameters. J. Chem. Eng. Data 55,

3356–3363.

Vural Gürsel, I., Kockmann, N., Hessel, V., 2017. Fluidic separation in microstructured devices – Concepts and their Integration into process flow networks. Chem. Eng. Sci.

Yabuta, K., Hirao, M., Sugiyama, H., 2017. Process model for enhancing yield in sterile drug product manufacturing. J. Pharm. Innov. 1–12.

499096 Multi-Layered Modelling Techniques for the Development of Continuous Manufacturing Processes

Abstract:

Traditionally the majority of active pharmaceutical compounds have been manufactured at large plant scales using batch methodologies. Continuous manufacturing is becoming an increasing attractive alternative however due to apparent benefits such as smaller manufacturing footprints, increased process safety, lower manufacturing costs and increased supply-chain flexibility. Realising all of the above benefits using conventional continuous equipment is however quite challenging and sometimes not achievable. In this project, a new approach to the design of continuous manufacturing platforms is being presented which is not restricted to using such conventional equipment but which is based around the bespoke design and manufacture of all unit operations, optimised to fit the specific process being examined. Processes are quantified in detail using first principle models which in turn are used to inform computational fluid dynamics (CFD) simulations leading to tailored equipment designs which can be manufactured with high precision using metallic 3D printing.

This presentation will focus on a number of case studies which demonstrate the linking of these various modelling techniques for several different unit operations required for the continuous manufacturing of pharmaceutical compounds including liquid-liquid separations, biphasic reactors and homogenous reactors. In particular however, the talk will focus on the development of a novel distillation column methodology, employing an inert gas to enable the concentration of solvent streams at low temperatures (required by temperature sensitive APIs). First principles theoretical models were used in order model the behaviour of such gas-liquid systems in bubble columns, enabling verification with experimental data. However as a bubble column does not represent the optimum design of such a gas-liquid contacting system, CFD was then employed to examine possible designs concepts suitable for 3D printing and to quantify evaporation rates in such possible configurations.

Following the modelling processes, these unit operations are 3D printed from stainless steel to enable their use as part of manufacturing processes. This approach leads to significant process intensification as all unit operations are designed specifically for the process being examined and are hence manufactured to ensure the processâ€™s optimum performance under the required conditions. Utilising 3D printing informed by process models to manufacture the process train also leads to significant cost reductions as unit operations become essentially disposable, reducing both the upfront capital costs and also essentially eliminating ongoing cleaning costs.

Several results demonstrating this approach are presented, including a biphasic reactor which is able to reduce the required reaction residence time by approximately 100 fold compared to an agitated batch reactor and a continuous homogenous reactor with dimensions of 15cm x 7cm x 3cm which is capable of achieving a throughput of 580,000 kg/L/day, representative of manufacturing output levels for many modern pharmaceutical compounds.

495661 A Refined Non-Random Two-Liquid Segment Activity Coefficient Model for Solubility Modeling

Abstract:

Accurate correlation and robust prediction for complex molecule solubility or solid-liquid equilibrium (SLE) in solvents and solvent mixtures have drawn much interest recently[1]. Solubility models such as the Non-Random Two Liquid Segment Activity Coefficient (NRTL-SAC) model[2] and its variations such as UNIQUAC-SAC[3] and COSMO-SAC with apparent sigma profiles[4] are now routinely used in the industry. This paper presents a refinement to the successful NRTL-SAC model in several perspectives. Specifically, the model is refined by modifying the Flory-Huggins expression in the entropy calculation. Additionally, a new conceptual segment has been proposed to better characterize polar components. At last, temperature dependency of activity coefficient is addressed in the refined model. We present the rationale for the refined model, along with the updated conceptual segment numbers for common solvents. Both the original and the refined NRTL-SAC models are employed to calculate vapor-liquid and liquid-liquid equilibrium for binary solvent systems and solubility of several drug molecules in solvents and solvent mixtures. The refined model provides improved prediction performance when compared with the original model, especially for systems containing larger size molecules. The refined NRTL-SAC model will be further extended to provide predictive capability for complex molecule solubility in excipients and polymers.

1. Tanveer, S.; Hao, Y.; Chen, C.-C., Introduction to solid-fluid equilibrium modeling. *Chemical Engineering Progress* **2014,** *110* (9), 37-47.

2. Chen, C.-C.; Song, Y., Solubility Modeling with a nonrandom two-liquid segment activity coefficient model. *Industrial & Engineering Chemistry Research* **2004,** *43* (26), 8354-8362.

3. Haghtalab, A.; Yousefi Seyf, J., Vaporâ€"liquid and solidâ€"liquid modeling with a universal quasichemical segment-based activity coefficient model. *Industrial & Engineering Chemistry Research* **2015,** *54* (34), 8611-8623.

4. Islam, M. R.; Chen, C.-C., COSMO-SAC sigma profile generation with conceptual segment concept. *Industrial & Engineering Chemistry Research* **2014,** *54* (16), 4441-4454.

502902 Modcell: A Prototype for Modular Cell Engineering

Abstract:

To produce a target chemical at high yields, titers, and productivities, various combinations of available genetic parts for expression system can result in a large number of microbial cell factories generated for characterization. This engineering approach will become increasingly laborious and expensive when seeking to develop optimal strains for production of a large space of biochemicals due to extensive screening. Our recent theoretical development of modular cell (MODCELL) design principle can offer a promising solution for rapid generation of optimal microbial cell factories in a systematic manner using a modular cell and exchangeable production modules. In this study, we experimentally validated the MODCELL design and developed a prototype of modular cell engineering by demonstrating: i) a modular (chassis) cell is required to tightly couple with a production module (e.g., alcohols and esters), ii) degree of coupling between the modular cell and production modules can be modulated to enhance product synthesis, iii) a modular cell can be used as a host to select and discover functions of hypothetical proteins, and iv) metabolic pathway evolution based on growth selection could enhance growth and product rates of the weakly coupled cells. We envision that the MODCELL design provides a powerful platform for rapid development of optimal microbial cell factories for combinatorial synthesis of biochemicals.

498171 A Synthetic Biosensor to Determine Peroxisomal Acetyl-CoA Concentration for Compartmentalized Metabolic Engineering

Abstract:

Sub-cellular compartmentalization is used by all eukaryotes and some prokaryotes as a means to create favorable microenvironments for various metabolic reactions. These compartments can concentrate enzymes, separate competing metabolic reactions, and isolate toxic intermediates in metabolic pathways. Such advantages have been recently harnessed by metabolic engineers to improve the production of various high-value chemicals via compartmentalized metabolic engineering. However, one challenge in compartmentalized metabolic engineering is to determine key metabolite level in these compartments. Conventional techniques such as metabolomics analysis and transcription-based biosensors could only reflect cytosolic metabolite concentration instead of compartmental metabolite concentration. To this end, we developed a synthetic biosensor to determine a key metabolite, i.e., acetyl-CoA, in a representative compartment of yeast, i.e., peroxisome. This synthetic biosensor used highly efficient enzyme re-localization via PTS1 signal peptides to construct a metabolic pathway in peroxisome, which solely converted peroxisomal acetyl-CoA to polyhydroxybutyrate (PHB) via three enzymes, *phaA phaB* and *phaC*. By quantifying the PHB level in yeast, we successfully determined peroxisomal acetyl-CoA level under various culture conditions. We next demonstrated our biosensor by screening a library of single knockout yeast mutants and identified one yeast mutant, Î”RPD3, which had elevated level of peroxisomal acetyl-CoA compared to wild type yeast. We expect our synthetic biosensors can be widely used to deepen our understanding of sub-cellular compartmental metabolism and facilitate the â€œdesign-build-testâ€ cycle of compartmentalized metabolic engineering.

500579 Tailoring Pathways for Balanced Expression by Short Mufflers

Abstract:

Production of biosynthesized compounds requires the construction of complex metabolic pathways and optimization of these pathways is usually sought to enable high-yield production. Interesting approaches have been demonstrated in the last decade to fine-tune the expression level of individual genes to achieve a balanced pathway expression. Examples include engineering of regulatory elements, such as promoters, terminators, ribosome binding sites and enhancers.

In this work, we develop a novel strategy to rapidly modulate gene expression in individual cassettes by using short library-generated muffler sequences. A study using GFP as a reporter showed that significant variation in gene expression was easily obtained when randomizing 30 bp sequences flanking the GFP expression cassette. This approach was subsequently applied to carotenoid biosynthetic pathway and cellobiose utilizing pathway, leading to identification of variants with balanced expression in a very effective manner. This success supports that coupling of library creation using expression mufflers to high-throughput screening methods is a useful approach in the construction of tailored pathways.

501461 Development of a Formaldehyde Biosensor and its Application to Engineering of Methanol Metabolism inÂ *E. coli*

Abstract:

Formaldehyde is a prevalent environmental toxin, and a key intermediate in single carbon metabolism. The ability to monitor formaldehyde concentration is therefore of interest for both environmental monitoring and for metabolic engineering of native and synthetic methylotrophs, but current methods suffer from low sensitivity, complex workflows, or the requirement for expensive analytical equipment. Here we develop a formaldehyde biosensor based on the FrmR repressor protein and cognate promoter of *E. coli*, coupled to either luciferase or GFP. Optimization of the native repressor binding site and regulatory architecture enabled detection at levels as low as 3 Î¼M. We then used the sensor to benchmark the *in vivo* activity of several NAD-dependent methanol dehydrogenase (MDH) variants, the rate-limiting enzyme that catalyzes the first step of methanol assimilation. In support of MDH directed evolution using this assay, we examined the potential for cheaters in a mixed population, and developed a novel strategy to prevent cross-talk by using glutathione as a formaldehyde sink to minimize intercellular formaldehyde diffusion. We then applied phage-assisted continuous evolution (PACE) using the sensor architecture developed here to evolve high-activity variants of MDH. Finally, we show the utility of the reporter in balancing expression of MDH and the formaldehyde assimilation enzymes HPS and PHI in an engineered *E. coli* strain to minimize formaldehyde build-up while also reducing the burden of heterologous expression, and evaluating strains designed for increased RuMP pathway flux to enable high rates of formaldehyde assimilation. This biosensor offers a quick and simple method for sensitively detecting formaldehyde, and has the potential to be used as the basis for directed evolution of MDH and dynamic formaldehyde control strategies for establishing synthetic methylotrophy.

501565 Synthesis of Isobutanol and n-Butanol Using Engineered Coenzyme a-Dependent Pathways in *Ralstonia Eutropha* H16

Abstract:

The use of Coenzyme A-dependent pathways has been well established at a primary method for the biosynthesis of varying chain-length alcohols for use in fuels and chemicals. However, only linear chain alcohols have been able to be produced though this pathway. Recently a new enzyme, isobutryl-CoA mutase, has been theorized to convert linear acyl-CoA molecules to branched molecules. In this work, we characterize this enzyme, and demonstrate the ability to generate isobutanol in *Ralstonia eutropha* via a renewable feedstock. This is the first example of isobutanol production via a renewable feedstock. The *Ralstonia eutropha* strain was initially engineered via heterologous gene expression, codon optimization, and promoter strength manipulation of the engineered production pathway to achieve a strain capable of producing ~200mg/L n-butanol via fructose or ~30mg/L via formate. After strain optimization, the isobutryl-CoA mutaste was added to generate isobutanol at ~30mg/L via fructose. This work generated a novel pathway for the production of isobutanol. Further exploration of this enzyme class could expand the biological toolkit available for the bioproduction of other previously unattainable chemicals from a renewable source.

495440 Layered Regulation Strategies to Improve D-Glucaric Acid Production

Abstract:

Microbial synthesis can be a sustainable method of chemical and material production. In order for microbial production to become economically viable, the titer, yield, and selectivity frequently must be improved through strain engineering. One strain engineering technique for pathway improvement is dynamic regulation, where cells are designed to sense extracellular or intracellular changes and alter their behavior accordingly. Dynamic regulation has been used to redirect flux from byproducts to products, balance the usage of key intermediates, and minimize production of toxic intermediates. We applied dynamic regulation to the production of D-glucaric acid, a precursor to nylons and detergents. Two orthogonal, autonomous, and tunable dynamic regulation systems were engineered into the D-glucaric acid pathway, resulting in the highest glucaric acid titer reported thus far in the *Escherichia coli* K-12 strains. The first regulation scheme is a pathway-independent quorum sensing system that knocks down the glycolytic enzyme PfkA at a critical cell density, switching the culture from growth to production mode. The second regulation system controls expression of a downstream enzyme on an as-needed basis â€" if sufficient levels of a key intermediate are detected via a biosensor, the enzyme is transcribed. While the two regulation strategies independently increase titers 2-to-3-fold, the combination of the two systems achieves a 10-fold increase in titer over the un-regulated control. This work demonstrates the use of synthetic biology tools to implement layered dynamic regulation, a promising strain engineering strategy for improving microbial production.

504104 Strategies for Exploiting Non-Growth Metabolism in Biosynthesis

Abstract:

Exploiting non-growing cells for biosynthesis has recently been recognized as a promising approach for economical biosynthesis. At the same time, dynamic metabolic engineering strategies have enabled engineers the ability to use non-static expression programs, which account for the unique phases of a bioreactor. By inoculating cells with the production program 'off,' and turning the program 'on' in late exponential phase has many advantages. The cells (i.e. catalyst) growth can be stopped, so there is no additional loss in carbon yield due to growing more cells. As well, engineering strategies that are toxic or detrimental to cell growth can be readily used if no more cell growth is required. Examples include: knockout of essential enzymes, overexpression of toxic genes, and draining a metabolic pool that is otherwise needed for growth. There are also advantages to having the biosynthesis program 'off' during growth. Cell growth is shortened (and therefore batch time), because there is not a burden for making product. As well, contaminating microbes have a diminished opportunity to overtake the culture, because uninduced strains grow quickly.

Our lab has recently addressed two challenges with exploiting non-growth metabolism: (a) late exponential phase induction must be done on the industrial scale making typical inducers prohibitively costly, and (b) metabolic activity is typically suppressed when cells stop growing, which is detrimental to productivity. To address the need for late exponential phase induction, we have developed a glucose starvation toggle switch. As with the original toggle switch, after a cell has experienced an induction signal, the response remains active, even after the induction signal is removed. We have altered this switch to use glucose starvation as an induction signal. From a bioprocessing standpoint, this enables growth on glucose to a defined density (based on the initial glucose concentration). After glucose is exhausted, the switch is toggled and the biosynthesis program is activated. Once the switch has toggled, glucose can be reintroduced and consumed by the cells to produce the desired product.

To address low metabolic activity in non-growing cells, we have identified gene knockouts that maintain high metabolic activity for different parts of metabolism when growth is stopped due to their essentiality. The strategy would be to allow expression of these essential enzymes to grow cells, but repress them once an adequate cell density is reached. Using metabolic modeling, we have identified thirty gene knockouts that are essential on minimal media and characterized their metabolic capabilities. Several mutants maintained high glucose uptake rates and secreted a variety of byproducts once limited by the essentiality. I will discuss our characterization of these strains for both uptake/secretion fluxes, as well as intracellular metabolomics. In total, we have developed tools to enable facile switching to production phase at industrial scale and identified metabolic operating conditions that should enable high productivity in non-growing cells. This work should be enabling for a number of industrial biotechnology processes.

500617 Inter-Individual Variability in Physiological Response to Lipid Infusions Predicts Considerable Heterogeneity in Outcomes for Lipid Resuscitation: A Physiologically Based Pharmacokinetic-Pharmacodynamic Study in a Virtual Population

Abstract:

Local anesthetic systemic toxicity (LAST) is a life-threatening event in which the introduction of anesthetic to the bloodstream induces cardiac arrest. Anesthetic-induced cardiac arrest is refractory to standard life support measures and is thus often fatal. A bid to understand the etiology of hypersensitivity to local anesthetics led to the discovery of a potential antidote for LAST: intravenous lipid emulsions. These suspensions, composed primarily of vegetable oil, egg phospholipids, and water, are freely available in clinical settings. Their off-label use as antidotes to anesthetic toxicity first yielded a successful clinical resuscitation in 2006. Two years later, a lipid suspension was used to revive a patient suffering from >70 minutes of cardiac arrest that was unresponsive to all other resuscitation efforts. Intriguingly, the offending toxin in this case was not an anesthetic, but a combination of 2 other pharmaceutical compounds. An ensuing series of small animal studies and anecdotal case reports hints at the possibility of lipid emulsions being broad-spectrum antidotes to drug toxicity.

Unfortunately, our understanding of when and how lipid interventions prove beneficial is still limited to clinical anecdotes and small animal experiments. Both sources of information are plagued by heterogeneous outcomes and limited exploration of the multivariate parameter space characterizing cardiac life support. Hence, even the underlying mechanism of therapeutic efficacy has long been the subject of speculation and heuristic reasoning with limited possibility for experimental validation. What we now know about the multifunctional mechanism of action has been obtained through our groupâ€™s integrative studies incorporating mechanistic mathematical modeling.

In our most recent work, we sought to further demonstrate how mathematical analysis can be used to gain insight into the likely success of lipid-mediated resuscitation. Importantly, we embraced the inherent variability between pre-clinical subjects and established how estimation of relevant physiological parameters on an individual basis can inform treatment protocols by permitting the analysis of outcome heterogeneity over simulated virtual populations. Using data previously collected in a study involving rats (N=7), we employed uncertainty quantification approaches to identify a subset of 2 key parameters that could be estimated for each subject. One parameter was determined to be correlated with the magnitude of cardiac response to lipid infusion, while the other played a role in determining the smoothness of the response over time. To account for the dynamic nature of these underlying physiological processes, we adjusted our existing pharmacokinetic-pharmacodynamic model to allow time variation of the 2 selected parameters; these were permitted to vary according to four linear splines of equal length. Model comparison tests were employed to ascertain that the choice of 4 splines was most appropriate for our data set.

From our individualized estimates, we constructed population distributions for each parameter. We then sampled from

these distributions to create a virtual population (N=10,000) undergoing emulsion therapy after a simulated overdose of the anesthetic bupivacaine. The virtual population exhibited a wide range of expected outcomes, with only a subset of the population expected to recover baseline cardiac function upon treatment. We thereby demonstrate the limitations of using mean responses from small preclinical studies to heuristically posit guidelines for lipid-mediated cardiac resuscitation.

496390 Effect of Circadian Disruption on Hepatic Gluconeogenesis

Abstract:

The circadian machinery is biological time-keeping mechanisms which orchestrate the biological changes that follow a 24-hour cycle. In mammals, the superchiasmatic nuclei (SCN) in the hypothalamus takes the light/dark cycle as the entraining signal and acts as the central pace maker that synchronizes the clock genes in the peripheral organs, such as muscles, liver, heart, and kidneys. The rhythmicity of hormonal and metabolic signals play an important role in immune [1] and metabolic functions [2] by maintaining and controlling the phase relations among the various clocks. The robustness of circadian rhythms is associated with a homeostatic state of the host while disruption in the circadian rhythms is observed in a diseased state of the host [3].

Aside from the light/dark cycle, food intake is a strong *zeitgeber*, a cue given by the environment on host to reset the internal body clock. Feeding rhythms are a more potent *zeitgeber* compared to the light/dark cycle eventually synchronizing the peripheral clocks away from the SCN, especially in organs governing the energy homeostasis of the host such as the liver. Therefore, an interconnection between the circadian clocks and metabolism is evident, and it is supported by a multitude of studies. At a clinical level, several studies correlate the disruption of circadian rhythms to increased rate of metabolic syndrome. Shift-work and sleep deprivation dampen the circadian rhythms, and are linked with obesity and high level of triglycerides [4]. At a molecular level, Clock mutant mice showed decreased metabolic rate [5] while liver-specific *Bmal1* knockout disrupted glucose homeostasis in mice [6]. These studies suggest that a bidirectional relationship exists between the circadian clocks and metabolism, probably as parts of a network of complex signaling cascades that intricately regulate one another.

Moreover, one can realize from these observations that a broader environmental or behavioral cues (such as feeding patterns) can influence specific, low-level targets (such as liver enzymes). To explore this hypothesis, we would like to demonstrate by a mathematical model how the feeding/fasting cycle influences a specific metabolic function in relation to the circadian rhythms entrained to the light/dark cycle. As a first step toward answering this question, we have previously demonstrated via mathematical modeling how the rhythms of the peripheral clock genes (PCGs) in the liver can be completely entrained to the feeding rhythms, synchronizing away from the light/dark cycle. We further developed this model to explore the effects of light-feeding phase relations on hepatic metabolism. We model the transcription of two genes, *G6pc* and *Pck1*, as a surrogate for glucose metabolism in the human liver.

In our model, PPARγ coactivator-1a (PGC1-a) is a key molecule that incorporates the effects of light and feeding to the transcription of the gluconeogenic genes. Transcription of PGC-1a is activated by forkhead box protein O1 (FOXO1) binding upstream of *Pck1* and *G6pc* [7]. When in the nucleus, PGC-1a is deacetylated by SIRT1, which results in an increased activity [8]. The glucocorticoid receptor is also required along with HNF-4a for full activation of *Pck1* promoter [9]. PGC-1 has shown to be induced synergistically in primary liver cultures by cAMP and glucocorticoids. Additionally, *Pck1* promotor requires co-activation of the glucocorticoid receptor and the liver-enriched transcription factor HNF-4a by PGC-1 for full transcriptional activation. HNF-4a is necessary for the action of PGC1, as hepatocytes from mice lacking HNF-4a loses the ability of PGC-1 to activate key genes of gluconeogenesis such as *Pck1* and *G6Pc* [10]. The role of HNF-4a and PGC-1a were lumped for simplicity in our model. FOXO1 also plays an important role in gluconeogenesis. PGC-1a binds and co-activates FOXO1 in a manner inhibited by Akt-mediated phosphorylation [11]. FOXO1 is known to be required for robust activation of gluconeogenic gene expression in hepatic cells and in mouse liver. Since FOXO1 binds to a DNA motif upstream of the gluconeogenic genes, we used it to model the signal transmission from PGC-1a to transcription of gluconeogenic genes [11]. Finally, PGC-1a also activates the transcription of Bmal1 (Equation 7) [12].

Our model builds upon earlier works [13] of a semi-mechanical model for light entrainment on the peripheral clock genes through the HPA axis. The main driver for oscillations in the HPA axis are the negative feedback between glucocorticoid and corticotropin-releasing hormone (CRH)/adrenocorticotropic hormone (ACTH) using a modified Goodwin oscillator. The glucocorticoid released from the central compartment regulates the peripheral clock genes (*Per/Cry*, *Bmal1*, and *Clock*). The feeding/fasting cycle influences the periphery by modifying the redox reactions between NAD^+ and NADH [14]. The feeding-entrained NAD^+ activates human sirtuin 1 (SIRT1), a class II histone deacetylase. SIRT1 then acts in coordination with CLOCK/BMAL1 complex, interacting with the PCGs in several ways. It facilitates the transcription of NAMPT, which is the rate-limiting enzyme for the NAD^+ salvage cycle, regulated by the CLOCK/BMAL1 complex. NAD^+, in turn, activates SIRT1, as well as entraining glucocorticoids through a transit compartment that represents neural input through the ventromedial arcuate nucleus [15]. SIRT1 also facilitates the degradation of PER/CRY proteins in the nucleus, exerting more influence on the peripheral clock genes. Finally, the transcription of *G6pc/Pck1* (gluconeogenic gene) is controlled by binding of glucocorticoid-receptor complex to the glucocorticoid response element on the promotor region, as well as FOXO1 binding also to the promotor region.

Current model successfully predicts some observations from experimental time-restricted feeding studies. Some basic predictions include that restricting feeding during the active phase results in phase shift of total gluconeogenic enzyme amount and the increase of amplitude in PEPCK total enzyme amount [16]. Another interesting prediction is that when the system is subjected to constant lighting, the oscillation in G6Pase activity becomes less robust [17]. Furthermore, overexpression of CRY protein results in decreased transcription of gluconeogenic genes [18]. Disruption of circadian clocks by the means of Clock gene knockout resulted in over-expression of *G6pc* and *Pck1*, while the nominal expression level could be partially recovered by altering the light-feeding phase relations. To analyze this complex network, sensitivity analysis of the model was performed in three parts: 1) local sensitivity analysis of all parameters; 2) selection of important elementary effects by Morris method; and 3) global uncertainty analysis employing the high dimensional model representations (HDMR). Interpreting and integrating the results from all three methods revealed that transcription and activation of PGC-1a and FOXO1 along with *Per/Cry* transcription had the greatest influence over the expression level of gluconeogenic genes, demonstrating the close relationship between circadian rhythms and hepatic metabolic activities.

References

1. Silver, A.C., et al., *The circadian clock controls toll-like receptor 9-mediated innate and adaptive immunity.* Immunity, 2012. **36**(2): p. 251-61.

2. Feillet, C.A., U. Albrecht, and E. Challet, *"Feeding time" for the brain: a matter of clocks.* J Physiol Paris, 2006. **100**(5-6): p. 252-60.

3. Fang, M., et al., *Enhancement of NAD(+)-dependent SIRT1 deacetylase activity by methylselenocysteine resets the circadian clock in carcinogen-treated mammary epithelial cells.* Oncotarget, 2015. **6**(40): p. 42879-91.

4. Spiegel, K., et al., *Effects of poor and short sleep on glucose metabolism and obesity risk.* Nat Rev Endocrinol, 2009. **5**(5): p. 253-61.

5. Turek, F.W., et al., *Obesity and metabolic syndrome in circadian Clock mutant mice.* Science, 2005. **308**(5724): p. 1043-5.

6. Lamia, K.A., K.F. Storch, and C.J. Weitz, *Physiological significance of a peripheral tissue circadian clock.*

Proc Natl Acad Sci U S A, 2008. **105**(39): p. 15172-7.

7. Daitoku, H., et al., *Regulation of PGC-1 promoter activity by protein kinase B and the forkhead transcription factor FKHR.* Diabetes, 2003. **52**(3): p. 642-9.

8. Schwer, B. and E. Verdin, *Conserved metabolic regulatory functions of sirtuins.* Cell Metab, 2008. **7**(2): p. 104-12.

9. Yoon, J.C., et al., *Control of hepatic gluconeogenesis through the transcriptional coactivator PGC-1.* Nature, 2001. **413**(6852): p. 131-8.

10. Rhee, J., et al., *Regulation of hepatic fasting response by PPARgamma coactivator-1alpha (PGC-1): requirement for hepatocyte nuclear factor 4alpha in gluconeogenesis.* Proc Natl Acad Sci U S A, 2003. **100**(7): p. 4012-7.

11. Puigserver, P., et al., *Insulin-regulated hepatic gluconeogenesis through FOXO1-PGC-1alpha interaction.* Nature, 2003. **423**(6939): p. 550-5.

12. Green, C.B., J.S. Takahashi, and J. Bass, *The meter of metabolism.* Cell, 2008. **134**(5): p. 728-42.

13. Mavroudis, P.D., et al., *Mathematical modeling of light-mediated HPA axis activity and downstream implications on the entrainment of peripheral clock genes.* Physiol Genomics, 2014. **46**(20): p. 766-78.

14. Diaz-Munoz, M., et al., *Anticipatory changes in liver metabolism and entrainment of insulin, glucagon, and corticosterone in food-restricted rats.* Am J Physiol Regul Integr Comp Physiol, 2000. **279**(6): p. R2048-56.

15. Yi, C.X., et al., *Ventromedial arcuate nucleus communicates peripheral metabolic information to the suprachiasmatic nucleus.* Endocrinology, 2006. **147**(1): p. 283-94.

16. Perez-Mendoza, M., J.B. Rivera-Zavala, and M. Diaz-Munoz, *Daytime restricted feeding modifies the daily variations of liver gluconeogenesis: adaptations in biochemical and endocrine regulators.* Chronobiol Int, 2014. **31**(7): p. 815-28.

17. Ashman, P.U. and J.R. Seed, *Biochemical studies in the vole, Microtus montanus. II. The effects of a Trypanosoma brucei gambiense infection on the diurnal variation of hepatic glucose-6-phosphatase and liver glycogen.* Comp Biochem Physiol B, 1973. **45**(2): p. 379-92.

18. Lamia, K.A., et al., *Cryptochromes mediate rhythmic repression of the glucocorticoid receptor.* Nature, 2011. **480**(7378): p. 552-6.

498747 A PK/PD Model of ACE Inhibition in Kidney Cells for Treatment of Diabetic Tissue Damage

Abstract:

Diabetes is the primary cause of chronic kidney disease (CKD). High glucose conditions have been known to initiate and exacerbate the pathophysiology of CKD particularly through damage to podocytes. Podocytes are terminally differentiated renal cells that play a key role in maintaining the structure and function of the kidneys. Podocytes express a local renin-angiotensin system (RAS) that is altered in diabetes. Studies have shown that Angiotensin II (ANG II), a blood pressure regulating hormone, produced by the RAS is upregulated in diabetic conditions and triggers podocyte injury and apoptosis. Angiotensin converting enzymes (ACE) inhibitor drugs inhibit the production of ANG II. However, these drugs are not well characterized for use in CKD. The progression of CKD could be slowed by controlling the ANG II levels to prevent irreversible podocyte loss. Due to sparse experimental evidence for glucose-dose-dependency of ANG II, especially for podocytes, we propose and parameterize a mathematical model to describe the glucose-stimulated RAS in podocytes that produces ANG II. The model can connect drug concentration to the inhibition of the ANG II in RAS in podocytes for the ACE inhibitor drugs in the cases of normal and impaired renal function.

We define a model for the pharmacokinetic RAS network that triggers the synthesis of ANG II and the pharmacodynamic physiological response to ACE inhibitors in a time and dose-dependent manner for normal and impaired kidney. The model takes glucose dynamics and drug dosage as the input. The glucose-sensitive podocyte RAS pathway is represented by a system of ordinary differential equations to track ANG II and other RAS peptides. Glucose-dependence is added through different enzyme activity parameters. A one-compartment model with first-order absorption from the site of administration and first-order elimination is used here for the oral administration of ACE inhibitor drugs. Previous experimental studies are used to estimate the unknown parameters and kinetic constants for the model. The parameters are analyzed by local and global sensitivity analyses along with other validation cases. MATLAB is used to solve the system of ordinary differential equations, fit parameters to data, and to develop a graphical user interface (GUI).

The model fits a mathematical function to experimental data for glucose dynamics and uses it to predict the corresponding change in ANG II concentrations with the pharmacokinetic RAS network model. The pharmacodynamic part of the model shows that ANG II concentration is lowered by the drugs. The model is used to explore the effects of different drug dosage on the levels of the diacid form of the drug and ANG II in diabetic or normal patients with normal or impaired kidney function over time.

We have developed a physiologically relevant PK/PD model for ACE inhibition parameterized for local RAS in podocytes for drugs in normal and impaired renal conditions. The model can use experimental data for glucose dynamics from normal and diabetic subjects and predict the dosage for each drug in normal or impaired kidney to control the ANG II levels. The model is packaged as a MATLAB app to facilitate reuse for research and educational purposes. The result of this work will enable further studies on the impacts of ACE inhibition in chronic kidney disease or other diseases.

495632 Dynamic Analysis of the DNA Sensing Pathway Predicts Host Immune Response

Abstract:

Introduction

To effectively respond to a pathogen, the host intracellular immunity must dynamically assess and change with the environment. Dysregulation of the immune response can increase disease pathology and impair viral clearance. This can lead to chronic, systemic infections which—left untreated—can be fatal [1]. It is therefore imperative to develop a predictive model describing how pathogens invade the host and abrogate the immune response.

Significant research has gone into identifying key host proteins that perpetuate the signaling cascades responsible for innate immunity. The recently identified DNA sensing molecule cGAS recognizes viral DNA and invokes a type I interferon response. This leads to paracrine signaling of the JAK-STAT pathway and subsequent transcription of interferon stimulated genes (ISGs). ISGs effect various elements of the virus life cycle, many of which have specific mechanisms that are not yet known [2]. While this is an excellent description, it fails to accurately characterize the dynamics of the viral DNA response. Understanding these dynamics can help explain observations that are not addressed by static, qualitative descriptions. For example, it has been well established that viruses with similar genotypes exhibit distinct kinetics which alter the course of infection [3]. Modeling will elucidate the discrepancies in experimental observation like this and motivate new hypotheses.

Methods

To characterize the DNA sensing pathway's behavior in response to viral DNA, ordinary differential equations (ODEs) are employed to simulate how host proteins coordinate and activate ISGs. The model is comprised of 16 differential equations that represent the forward and reverse rates for each species (proteins and mRNAs) in the pathway. Typical formulations for biochemical reactions are used to define these rates including mass action, Michaelis Menten, and Hill kinetics. The latter is useful for modeling highly cooperative binding events such as the interactions between cGAS, viral DNA, ATP, and GTP to form the signaling molecule cGAMP. The model contains 24 rate constants, some of which have estimates from literature while others are unknown.

Traditional parameter estimation techniques that rely on gradient descent or maximum likelihood estimation which can be insufficient for large nonlinear ODE systems [4]. To compensate, a Bayesian approach is implemented to survey the parameter landscape and return probabilistic distributions as estimates. Adaptive Markov chain Monte Carlo (MCMC) simulation provides a computationally sufficient method to estimate the posterior distributions for all the unknown parameters. Uniform priors were used to avoid biasing parameter estimation [5]. To perform simulations for MCMC, the R package Flexible Modelling Environment (FME) was utilized [6]. Data for parameter fitting comes from time-course gene expression experiments. A transcriptionally deficient variant of Herpes Simplex Virus (HSV) was used to infect human fibroblasts. This type of experiment is informative because it controls for the negative regulation from viral proteins and allows for the observation of the unabated DNA sensing pathway.

In addition to parameter fitting, global and local sensitivity analyses were performed to determine the robustness of the model. Relatively large sensitivities, while also revealing stability properties, can be effective for motivating new experimental hypotheses. A parameter sensitive in the model is a likely candidate for viral manipulation.

Results

Model simulations capture the relative timing and shape of the dynamics seen in experiment and literature [7]. Typical upregulation of interferon beta occurs within four to six hours given an input of viral DNA and follows a saturation curve. Precursor protein complexes exhibit low accumulation due to the high input and output rates. It is possible that this is an accurate description of the biology, but also could be an indication that model reduction is needed. The production of ISGs show a delayed upregulation starting at 24 hours and downregulation after 48 hours. The eventual decline is controlled by the negative regulators (SOCS1, SOC2, USP18) that overtake the interferon signal and shut down the pathway. In many instances biologically, chronic inflammation occurs from aberrant ISG production that was not controlled by the negative regulators [8]. The model predicts this behavior with continual DNA stimulation.

We show how perturbations in highly sensitive kinetic parameters abrogate immune responses making their associated proteins likely targets for viral inhibition. Feedback mechanisms are also excellent targets due to their ability to amplify signals and stabilize systems.

Future work aims to expand this model to incorporate the transcription of interferon stimulated genes (ISGs) that are positively autoregulated, as well as ISGs that are not directly activated by the interferon response. These subsets of ISGs can be determined by different knockout experiments along the DNA sensing pathway. This analysis is vital for modeling to determine the timeframes of individual ISG activation.

Additionally, the model can be implemented to simulate different types of viral infections and autoimmune diseases caused by the detection of self-DNA. Understanding the different contexts in which this signaling pathway is stimulated can provide additional information into the structure of the pathway and varying kinetics observed.

Bibliography

[1] D. Kobasa, S. M. Jones, K. Shinya, J. C. Kash, J. Copps, H. Ebihara, Y. Hatta, J. Hyun Kim, P. Halfmann, M. Hatta, F. Feldmann, J. B. Alimonti, L. Fernando, Y. Li, M. G. Katze, H. Feldmann, and Y. Kawaoka, "Aberrant innate immune response in lethal infection of macaques with the 1918 influenza virus," *Nature*, vol. 445, no. 7125, pp. 319–323, Jan. 2007.

[2] J. W. Schoggins, "Interferon-stimulated genes and their antiviral effector functions," *Curr. Opin. Virol.*, vol. 1, no. 6, pp. 519–525, 2011.

[3] J. E. Medeiros-Filho, "Differences in viral kinetics between genotypes 1 and 3 of hepatitis C virus and between cirrhotic and non-cirrhotic patients during antiviral therapy," *World J. Gastroenterol.*, vol. 12, no. 45, p. 7271, 2006.

[4] M. Ashyraliyev, Y. Fomekong-Nanfack, J. A. Kaandorp, and J. G. Blom, "Systems biology: parameter estimation for biochemical models," *FEBS J.*, vol. 276, no. 4, pp. 886–902, Feb. 2009.

[5] K. S. Brown and J. P. Sethna, "Statistical mechanical approaches to models with many poorly known parameters," *Phys. Rev. E*, vol. 68, no. 2, p. 21904, Aug. 2003.

[6] K. Soetaert and T. Petzoldt, "Inverse Modelling, Sensitivity and Monte Carlo Analysis in *R* Using Package

FME," *J. Stat. Softw.*, vol. 33, no. 3, pp. 1–28, 2010.

[7] M. Fribourg, B. Hartmann, M. Schmolke, N. Marjanovic, R. A. Albrecht, A. García-Sastre, S. C. Sealfon, C. Jayaprakash, and F. Hayot, "Model of influenza A virus infection: Dynamics of viral antagonism and innate immune response," *J. Theor. Biol.*, vol. 351, pp. 47–57, 2014.

[8] G. Trinchieri, "Type I interferon: friend or foe?," *J. Exp. Med.*, vol. 207, no. 10, 2010.

500733 Computational Modeling of Tuberculosis Granuloma Activation

Abstract:

Tuberculosis (TB) is one of the most common infectious diseases and deadliest diseases worldwide. It is estimated that one-third of the world's population is infected with TB, and 1.5 million TB-related deaths were reported in 2014. TB is spread by aerosol droplets containing Mycobacterium tuberculosis (Mtb). The Mtb bacteria enter through the respiratory system and are attacked by the immune system in the lungs. The bacteria are clustered and contained by aveolar macrophages into cellular aggregates called granulomas. These granulomas can hold the bacteria dormant for long periods of time in a condition called latent TB. The bacteria can be activated when the granulomas are compromised by other immune response events in a host, such as cancer, HIV, or aging. MMP-1 dysregulation has been recently implicated in TB activation through experimental studies, but the mechanism is not well understood. Animal and human studies currently cannot probe the dynamics of activation, so a computational approach is proposed to fill this gap. The overall objective of the study is to predict TB cavity formation (a hallmark of activation) in response to the dynamics of MMP-1 dysregulation.

A previously developed mathematical model of the immune response to Mtb in the lungs [1] is used as the basis for this work. This previous model can track the levels of T cells, bacteria, cytokines and chemokines and can simulate the possible outcomes of the infection: clearance, latency, and reactivation. The infection outcome of interest in this work is latency. The previous model is a system of 16 ordinary differential equations (ODEs). To model the effect of MMP-1 dysregulation, 2 additional ODEs for the concentration of MMP-1 and collagen were developed, and a term was added to a preexisting ODE to represent the leakage of extracellular bacteria. The preexisting model and new ODEs were incorporated into a Python class, and an ode solver included in the Scipy package was used to solve the system of ODEs. A global sensitivity analysis using SALibâ€™s implementation of Saltelli sampling and Sobol analysis was done on the model.

With the new equations to the model, a long term progression towards a leaking case is possible. Using a set of parameters that generates a latent case in the previous model, the new model at first starts to tend towards the latent case. After the bacterial population establishes itself, production of MMP-1 is increased. This drives the degradation of collagen, which allows bacteria to escape the granuloma. The case ultimately tends towards a steady state with a constant rate of bacterial leakage.

The model proposed here is intended to provide a possible mechanism for the link between MMP-1 dysregulation and TB activation. The new equations added into the model serve to add a leaking case that represents the formation of a necrotic cavity, which is characteristic of activation after a period of latency. Further work includes validating individual model parameters, sensitivity analysis, and model results under other conditions.

References: [1] Sud, Bigbee, Flynn & Kirschner. 2006. *The Journal of Immunology*, 176 (7), 4296â€“4314.

Acknowledgments: This work is supported in part by the Oklahoma Center for Respiratory Infectious Diseases,

supported by the National Institute of General Medical Sciences of the National Institutes of Health under Award Number P20GM103648.

498646 Investigating Cholera Toxin Binding Mechanism with Gangliosides Via Kinetic Modeling and Experimental Measurements

Abstract:

Cholera toxin (CTx), which is secreted by *Vibrio cholera,* is a toxin protein that is the main culprit of cholera. CTX is an AB_5 toxin, containing an enzymatic A-subunit and five identical cholera toxin B-subunits (CTB) [1]. CTB binds the gangliosides such as GM1 on host cell membranes leading to the endocytosis of CTx. In cytoplasm, the toxic A-subunit adversely increases the intracellular cyclic AMP level in the host cell [2]. The elevated cyclic AMP concentration results in severe fluid loss, which can be life-threatening if persists. In order to better cope with CTx invasion, the detailed mechanism of CTB-GM1 interactions, which is the prerequisite for CTx cell entry, needs to be elucidated. It is worth noting that a pentavalent CTB can simultaneously binds to multiple copies of GM1 in host cell membranes. The positive cooperativity between each binding subunit can further enhance the overall binding avidity [1, 3]. The conventional ligand-receptor analysis (monovalent binding model) is not sufficient to describe the complex multivalent CTx binding mechanisms. In this study, both *in silico* and *in vitro* approaches are employed to investigate complex interactions between CTB and GM1 in a cell-membrane-mimicking environment.

Lauer et al. developed a five-step kinetic model for CTB binding to GM1 on lipid bilayer [3]. However, the model did not account for the aforementioned positive cooperativity among CTB subunits. Here, we integrated the five-step model with other binding mechanism proposed by Turnbull et al. [1]. The new kinetic model incorporated association/dissociation constants with cooperativity factors. Based on the number of bound GM1 per CTB and the local configuration, seven CTB-GM1 complexes were considered in our simulation. The accuracy of kinetic model were verified by a novel protein binding assay. We used a quantitative nanocube-based biosensor to collect CTB-GM1 binding kinetics in a cell membrane mimicking environment [5]. The high-throughput and easy-to-use features of nanocube sensor allow collect massive data at various experimental conditions. In addition, we applied the same methodology to examine other host cell receptors such as GM2 and fucosyl-GM1, which exhibit distinct CTB binding kinetics. In summary, we integrated theoretical modeling and experimental measurements to explore cholera pathogenesis. Our work offers a systematic tool for the biomedical community to reveal the pathogenesis of other diseases.

Reference

1. Turnbull, W.B.; Precious, B.L.; Homans, S.W. Disseting the cholera toxin-ganglioside GM1 interaction by isothermal titration calorimetry. *J. Am. Chem. Soc.* **2004**, *126*, 1047-1054.
2. de Haan, L.; Hirst , T.R. Cholera toxin: A paradigm for multi-functional engagement of cellular mechanisms. *Molecular Membrane Biology* **2004**, *21*, 77-92.
3. Lauer, S.; Goldstein, B.; Nolan, R.L.; Nolan, J.P. Analysis of cholera toxin â€" ganglioside interactions by flow cytometry. *Biochemistry* **2002**, *41*, 1742-1751.
4. Lin, H.; Kitova, E.N.; Klassen, J.S. Measuring positive cooperativity using the direct ESI-MS Assay. Cholera toxin B subunit homopentamer binding to GM1 pentasaccharide. *J. Am. Soc. Mass. Spectrom.* **2014**, *25*, 104-110.
5. Worstell, N. C.; Krishnan, P.; Weatherston, J.D.; Wu, H. Binding cooperativity matters: A GM1-like ganglioside cholera toxin B subunit binding study using a nanocube-based lipid bilayer array. *Plos One* **2016**, *11 (4)*, e0153265.

498782 Biomarker Identification in Autism Spectrum Disorder: Common Pitfalls and Emerging Strategies

Abstract:

Biomarkers are measurements obtained from a patient that are predictive of some phenomenon, such as a disease [1] or environmental exposure [2]. Biomarkers link complex observations to a quantitative measurement (e.g., loss of consciousness in diabetes with hypoglycemia) and can be used for a variety of purposes, such as assigning diagnostic status, stratifying patients, or measuring clinical outcomes [3]. Consequently, biomarkers are integral to the understanding and treating disease. However, biomarkers are frequently assessed with inappropriate analytics methods [4] and restricted to single measurements [5], which contributes to the underwhelming numbers of proposed biomarkers that lead to clinical translation [5]. These issues with current biomarker development/reporting are further discussed below.

Distributions of biomarkers in the diseased and healthy cohorts are typically described by measures of central tendency and some measure of the dispersion or range. Furthermore, the most commonly reported difference in the biomarker distributions for diseased and healthy cohorts is a difference in the measure of central tendency [3]. However, this "statistical significance" is usually misinterpreted as being synonymous with "clinical significance". Instead, diagnostic biomarkers should be assessed on the separability of the diseased and healthy distributions (e.g. sensitivity/specificity, receiver-operating-characteristic curve, etc.) in order to achieve clinically-relevant biomarkers.

Potential biomarkers usually arise from discussions of or mathematical modeling of a biological network and how this network connects to the phenomenon of interest. Traditionally, only single measurements or metrics are used to construct a potential biomarker and the remaining portions of the underlying network is ignored [3]. While this approach can work well for simpler diseases, more complex diseases lacking a completely understood disease mechanism (e.g., neurological diseases) can benefit from incorporating multiple measurements into the potential biomarker without specific relationships between multiple measurements imposed *a priori* [3], [6].

These issues with current biomarker reporting/development are highlighted through case studies of biomarker identification from Autism Spectrum Disorder (ASD). It is shown that effective biomarkers can be developed from multiple measurements and that p-Values from Student's t-tests can lead to incorrect conclusions regarding biomarker quality. In comparison to this, multivariate statistical analysis and modeling techniques can result in the identification of biomarkers that can be used for disease classification and severity prediction even when univariate approaches fail [6].

[1] T. J. Lyons and A. Basu, "Biomarkers in Diabetes: Hemoglobin A1c, Vascular and Tissue Markers," *Transl. Res.*, vol. 159, no. 4, pp. 303–312, Apr. 2012.

[2] N. L. Benowitz, "Cotinine as a biomarker of environmental tobacco smoke exposure," *Epidemiol. Rev.*, vol. 18, no. 2, pp. 188–204, 1996.

[3] J. C. McPartland, "Considerations in biomarker development for neurodevelopmental disorders," *Curr. Opin. Neurol.*, vol. 29, no. 2, pp. 118–122, Apr. 2016.

[4] A. D. Barker, C. C. Compton, and G. Poste, "The National Biomarker Development Alliance accelerating the translation of biomarkers to the clinic," *Biomark. Med.*, vol. 8, no. 6, pp. 873–876, Jul. 2014.

[5] G. Poste, "Bring on the biomarkers," *Nature*, vol. 469, no. 7329, pp. 156–157, Jan. 2011.

[6] D. P. Howsmon, U. Kruger, S. Melnyk, S. J. James, and J. Hahn, "Classification and adaptive behavior prediction of children with autism spectrum disorder based upon multivariate data analysis of markers of oxidative stress and DNA methylation," *PLOS Comput. Biol.*, vol. 13, no. 3, p. e1005385, Mar. 2017.

491442 Stability Analysis of Stochastic SchlöGl Model

Abstract:

The behavior of chemical reaction networks is influenced by the number of molecules and the size of the system. When the number of molecules is asymptotically reaching large values, the system is in the macroscopic (thermodynamic) limit. In such cases, a deterministic modeling formalism is more suitable to describe the system. Away from it, at the microscopic limit, a stochastic approach is more suitable since the system's size is small and thermal molecular fluctuations become significant [1].

We have previously discussed the lack of knowledge of the behavior of networks in the mesoscopic area, where the system is sizable yet still under the influence of thermal noise [2]. This lack is largely due to numerical and computational difficulties in stochastically modeling mesoscopic system [2]. We have also provided insight into how these issues can be resolved with the help of ZI-closure scheme method [3]. This is a numerical method to solve chemical master equations by maximizing the entropy of the reacting system.

We employ the Schlögl model reaction network, as a general example of bistable systems. A wide variety of system sizes is studied with ZI-closure. In the cases that the system has one stable attractor, the stochastic system recovers the deterministic behavior for large system sizes. However, the stochastic mesoscopic behavior only partially matches the deterministic solution in the region of kinetic parameter values that result in two stable deterministic attractors. In such cases, the stochastic system recovers only one of the deterministic attractors in the mesoscopic limit.

Stability analysis of the stochastic system provides a better insight into the system's behavior. Eigenvalues of the stochastic system are calculated through the Jacobian matrix of the linearized system around the steady state [4]. The correlation between the eigenvalues of the stochastic system at the mesoscopic limit and the ones of the deterministic formalism reveals the underlying stability dynamics of the system. The stochastic formalism can distinguish between attractors and thus retrieves the more stable one. In contrast, the deterministic formalism finds all the stable attractors indistinguishably.

1. Constantino PH, Vlysidis M, Smadbeck P, Kaznessis YN, "Modeling stochasticity in biochemical reaction networks", J Phys D: Appl Phys 2016 Feb 1; 49(9): 093001 10.1088/0022-3727/49/9/093001
2. Vlysidis M, Kaznessis YN "The Mesoscopic Behavior of Stochastic Schlögl Model", 2016 AIChE Annual Meeting, San Francisco, CA
3. Smadbeck P, Kaznessis YN, "A Closure Scheme for Chemical Master Equations", Proc Natl Acad Sci U S A. 2013 Aug 27; 110(35): 14261-5 10.1073/pnas.1306481110
4. Smadbeck P, Kaznessis YN, "On a theory of stability for nonlinear stochastic chemical reaction networks", J Chem Phys 2015 May 8; 142: 184101 10.1063/1.4919834

490465 Overcoming Challenges for Scale-up of Continuous Pharmaceutical Drug Substance Processes

Abstract:

The area of continuous technology and reactor design applied to pharmaceutical drug substance processes has advanced significantly over the last decade. Primary drivers of flow versus batch operations are enhanced safety, greater process control, increased productivity, and better operational stability. However, several obstacles need to be overcome prior to successfully scaling-up flow processes from laboratory to manufacturing. Some examples are solid fouling, stability process hold, and mapping feasible operational space for process perturbations. This talk will focus on two case studies related to a: 1) Grignard reaction carried out in a continuous stirred tank reactor (CSTR) and, 2) hydrogenation reaction run in a pipes-in-series plug flow reactor (PFR). The investigated case studies will illustrate methodology and use of experimental tools in conjunction with mathematical models to overcome the above challenges. Specifically, the Grignard case study involved issues with instable process holds and solid fouling issues (affecting downstream performance) that were overcome by mechanistically probing the system with careful experimental design and analytical techniques. On the other hand, mathematical models were used for the hydrogenation case study to demonstrate their power in mapping out the feasible operational space. These efforts enabled scaling up these reactions to successfully run in various pivotal and validation campaigns.

485306 Preparation of Hazardous "on-Demand" Reagents Using Continuous Processing

Abstract:

Hazardous and unstable reagents are often forgone for use during manufacturing due to storage and safety concerns. These reagent are often replaced with more expensive alternatives. Continuous processing allows for the preparation of hazardous and/or unstable reagents that can be used directly in the production of pharmaceuticals. This talk will focus on using the Corning ® Advanced Flow Reactor™ (AFR) for the continuous preparation and use of several hazardous and unstable reagents.

489921 Model-Aided Development for a Continuous Amide Bond Formation in a Drug Substance Manufacturing Process

Abstract:

Mathematical models are valuable tools in development to determine acceptable parameter ranges and control strategies for continuous manufacturing. Some of the benefits of modeling in pharmaceutical development are its ability to reduce material costs and resource needs, while simultaneously increasing process understanding and quality control.

A case study is presented here that demonstrates the use of NMR reaction monitoring (ReactNMR), to gather high-value reaction data in batch mode for kinetic parameter estimation. The model describes the amide bond formation between an aniline and a mixed anhydride. The kinetic model was then employed to describe a continuous process that includes a plug flow reactor (PFR), a delay coil, two mixed-suspension, mixed-product-removal (MSMPR) crystallizers in series, and a filter/dissolution vessel. The resulting model was used to enable risk assessment, test failure modes, and build decision trees for use in the event of a process upset at manufacturing scale.

492008 Effects of Catalyst and Metals Pretreatment for the Continuous Hydrogenation of a Halogenated Aromatic Nitro Compound

Abstract:

Continuous processing for the synthesis of active pharmaceutical ingredients (APIs) continues to be a growing area of interest and investment. Continuous heterogeneous catalysis offers both safety and capital cost advantages over a traditional batch autoclave.

This investigation explored impurity control for the continuous hydrogenation of a halogenated aromatic nitro compound. The hydrogenation utilized a palladium catalyst for the reduction of an aromatic nitro compound to an aniline. In particular, the work involved development of a packed bed reactor for the continuous heterogeneous catalysis. Several challenges appeared in transitioning the chemistry from batch to continuous processing. These challenges were characterized using advanced monitoring techniques (flow-NMR, online HPLC, etc.) and investigated using a combination of batch and flow processing. As part of that effort, a catalyst pretreatment was investigated which had been demonstrated as effective in batch mode for suppressing an impurity. Parameters investigated included duration and temperature of pretreatment, as well as liquid flow profile. Also investigated was the effect of residual metals in the feed solution entering from a previous synthetic step. Feed pretreatment and its impact on reaction impurity profile and bed life were evaluated.

492261 Large Scale Production ofÂ Pharmaceutical Intermediates in the Flow Mode

Abstract:

In recent years, there have been significant efforts in the Pharmaceutical Industry to adopt continuous processing technologies to support the chemical/unit operation steps. Major drivers for developing flow processes and continuous manufacturing for specific steps include process safety issues, poor heat transfer, poor mass transfer in multiple phase reactions, presence of unstable intermediates, extreme reaction conditions requirements, and the need for short residence time. Since 2015, STA Pharmaceutical, a WuXi AppTec company, has established a leading continuous processing technology platform to support process R&D as well as plant scale production of pharmaceutical intermediates. The reaction types supported to date include cryogenic and room temperature lithiations, deprotection, photochemistry, ozonolysis, oxidations, dehydrations, coupling, cyclization, amination, and azide chemistries among other reactions. Flow steps included the above chemical steps as well as unit operations including phase separation, crystallization, filtration, and dissolution. The scale of the operations ranged from a few hundred grams to a few hundred kilograms under GMP and non-GMP modes. This presentation gives an overview of the continuous processing platform in STA followed by case studies illustrating the key benefits of this technology platform.

494952 Plasmonic Metal Nanocatalysts As Platform for Continuous Synthesis of Drug Substances

Abstract:

Plasmonic Metal Nanocatalysts as Platform for Continuous Synthesis of Drug Substances

Farshid Mohammadparast, Andishaeh Dadgar, Marimuthu Andiappan, School of Chemical Engineering, Oklahoma State University, Stillwater, OK, USA

Approximately 25% of all the reactions performed in the pharmaceutical industry consist of cross-coupling reactions. The cross-couplings have been traditionally carried out by homogeneous palladium (Pd) catalyzed batch processes. The National Science and Technology Council (NSTC) has recently identified the continuous manufacturing of pharmaceuticals as one of five manufacturing areas of emerging priority in the USA, revealing that continuous manufacturing of pharmaceuticals has the potential to reduce manufacturing cost by up to 40-50%. Given these advantages, in recent years, metal nanoparticle catalyzed cross-couplings have been of great interest since they can be used as a platform to switch cross-couplings from traditionally used batch mode to continuous flow mode. Additionally, heterogeneous metal nanocatalysts can be immobilized on a rigid support in the form of a packed-bed reactor, and therefore, can avoid the need for subsequent difficult steps to separate the catalyst and product.

In this contribution, we show that heterogeneous plasmonic metal nanocatalysts can be used as efficient catalysts to drive cross couplings. Specifically, we demonstrate that visible light can be used as a powerful tool to make plasmonic metal nanocatalysts as efficient photo-catalysts with activity and stability superior to that of traditionally used homogeneous Pd based catalysts. These plasmonic metal nanocatalysts can be used to design efficient cross-coupling flow processes to produce drug substances.

Abstract, AIChE 2017

Developing Scale-Up Approach for Fast Chemical Reactions in Continuous Flow

Because of potential high API demand for a particular Merck product currently under development, a new chemical synthesis was developed with improved yield. One of the chemical transformations in the synthesis was developed as a continuous flow step, improving reaction conversion by over 15% compared from batch mode. The reaction exhibits very fast kinetics (in the order of ms) requiring very good mixing. Successful scaling up of such fast reactions requires two important pieces of information:

1. Understanding of how reaction performance is affected by mixing.

2. A reliable way of measuring/scaling up mixing across different scales/geometries of static mixers.

In order to gain understanding on the effect of mixing, the continuous reaction was ran at various flow rates, static mixer geometries/sizes and temperatures. The collected data for reaction conversion was analyzed statistically in order to obtain a mathematical model for its dependence on the above process parameters. The model had acceptable predictive capabilities and confirmed an expected trend: for each mixer, the conversion was increased with increasing flow rate (improved mixing) and decreasing temperature (slower kinetics, allowing for better mixing). The drawback of this approach was that all these models were static mixer specific. If a different static mixer was to be used, a new study was needed in order link the flow rates to reaction conversion. A different, more generalized method was needed to measure/characterize mixing, independent of the mixer type.

Many research attempts to characterize and scale-up continuous mixing in static mixers usually involve correlations using scale-up parameters such as Reynolds number (Re), residence time (τ) or turbulent energy rate (ε). To link all these scale up parameters to the mixing intensity, a method known as the Fourth Bourne Reaction is usually employed. Some of these parameters usually work well, however they may fail to capture any differences in the internal geometry of the static mixer. Instead of using the above scale-up parameters, the conversion from the Fourth Bourne Reaction was used directly as a surrogate process parameter. This direct method of measuring mixing is more accurate as it will capture any internal changes to the mixing pattern (such as back mixing), which may not be captured if the other scale-up approaches are used. It can be seen as a snapshot of the mixing pattern inside the static mixer.

In order to show the applicability of this method, the Fourth Bourne Reaction tests were performed on the same mixers used in the continuous reaction, within the same flow rate ranges. For each mixer type/flow rate combination, the Bourne reaction conversion (BRC) number was assigned and used as a surrogate process parameter. Then the BRC and temperature only were used in the same statistical analysis. Both BRC and temperature were found to be statistically significant and the overall model had the same predictive capabilities. The advantage of this method was the ability to substitute the mixer type/flow rate combinations with the single BRC parameter. This allowed the three model plots produced previously for each of the mixers to be reduced to only one: BRC and temperature only. This methodology also makes scale-up much more straight-forward: one need to show only that the BRC is the same across scales in order to ensure identical mixing (provided the BRC test is executed identically).

506740 Food, Pharmaceutical and Bioengineering Division Award in Chemical Engineering: Directed Evolution of New Viruses for Therapeutic Gene Delivery

Abstract:

Gene therapy – the delivery of genetic material to a patient for therapeutic benefit – has experienced an increasing number of successful human clinical trials, particularly ones using delivery vehicles or vectors based on adeno-associated viruses (AAV). These include trials for hemophilia B, Leber's congenital amaurosis (LCA2), spinal muscular atrophy, and lipoprotein lipase deficiency (LPLD). As a result, AAV is the basis for a clinically approved gene therapy product to treat LPLD in the European Union, and an AAV-based gene therapy will likely receive FDA approval this year for LCA2. AAV is thus capable of safe and therapeutic gene delivery to some cell targets. However, vectors in general face a number of challenges that limit their efficacy, not surprisingly since the parent viruses upon which these vectors are based were not evolved by nature for our convenience to use as human therapeutics. For most applications, there is unfortunately insufficient mechanistic knowledge of underlying virus structure-function relationships to empower rational design to improve such vectors.

As an alternative approach, we were the first to develop and have since been implementing directed evolution – involving the iterative genetic diversification of a viral genome and functional selection for desired properties – to address a number of problems with AAV. Genetic diversification has included the random diversification of peptide sequences at defined locations in the capsid, random point mutagenesis of the cap gene, and recombination of cap genes from a number of parental serotypes to create random chimeras. Using a range of in vitro and in vivo strategies to select the resulting large (~108) libraries for improved function, we have evolved AAVs for evasion of neutralizing antibodies, enhanced biodistribution and spread within a target tissue, greatly improved delivery efficiency, and targeted delivery in vitro and in vivo, thereby substantially improving the vectors' capacity to meet human therapeutic needs.

506741 Area 15A Food Plenary Award - Efficient Biosynthesis of Microalgal DHA-Rich Oil: From Lab to Factory Scale

Abstract:

The Nutrient Chemicals are a number of chemicals with multiple functions which are important for good health, such as antioxidant, anti-radiation, improving memory, regulating blood lipid etc [1]. Our work is concentrated on one of lipid nutrient chemicals, **docosahexaenoic acid (DHA)**. Docosahexaenoic acid is an important w-3 fatty acid, and contributes greatly to infant mental development and cardiovascular diseases prevention [2]. It has been widely used in food additives and pharmaceutical industries. The research work in our group mainly focused on how to realize the efficient DHA production by marine microorganism and improve the product quality. The detailed research work include strain improvement, nutrition and oxygen regulation, improvement of lipid quality, pathway modification, omics study and bioreactor design and scale up. We believe our work will be interested not only for those who focused on DHA-rich oil production, but also for the other nutrient chemicals related researchers.

Reference:

[1] REC Wildman. Handbook of nutraceuticals and functional foods, 2016, CRC press

[2] Agostoni C, Riva E, Trojan S, Bellu R, Giovannini M. Docosahexaenoic acid status and development quotient of healthy term infants. Lancet. 1995, 346(8975):638

506742 Area 15B/PD2M Plenary Award: Biorelevant Phase Transformations in Supersaturated Solutions of Poorly Water Soluble Drugs

Abstract:

For oral delivery, a drug must first dissolve in the intestinal fluids, and then be absorbed across the gastrointestinal membrane. Unfortunately, many new drugs are poorly soluble in water and are consequently difficult to deliver to the patient.To improve this situation, drugs can be formulated as supersaturating delivery systems such as an amorphous drug-polymer blends or other metastable solid forms. Such systems generate a supersaturated solution in vivo, improving membrane transport rate and ultimately absorption relative to stable crystal forms. However, supersaturated solutions are meta- or unstable and phase transformation to a lower energy state(s) is thermodynamically favored. Such phase transformations include liquid liquid phase separation and crystallization. The nature of the phase transformation and the characteristics of any precipitate formed is important since these can impact the supersaturation profile and the redissolution kinetics of the phase separated material. Herein, different phase transformations and their potential implications for the oral delivery of poorly water soluble drugs are considered.

506743 Area 15C Bioengineering Plenary Award: Orthogonal Metabolism for Industrial Biomanufacturing

Abstract:

The bow-tie architecture of metabolism affords biological systems with inherent robustness and plasticity, which also appears to be the case for engineered systems like the power grid, manufacturing, and the Internet (Trends Biotechnol. 2004 22:446). This architecture allows microbial metabolic networks to catabolize a myriad of substrates to a small number of precursor metabolites, ATP and reducing equivalents, which in turn serve as the basis to biosynthesize the building blocks and macromolecules required to sustain life. Efforts to metabolically engineer microbial systems to produce small molecules have been almost exclusively based on harnessing the aforementioned bow-tie architecture and corresponding pathways (Cell 2016, 164:1185). However, a number of inherent drawbacks associated with the use of such an approach have motivated the development of pathways orthogonal to the host metabolism (Nat Commun. 2017, 8:15188). For example, despite the high efficiency of pathways involved in catabolic and central metabolic functions, native biosynthetic pathways typically suffer from energy, carbon, and catalytic inefficiencies (Biochemistry, 2011, 50:4402), which in turn limit product synthesis. In addition, exploitation of native metabolic architectures requires extensive understanding of the host metabolism and can cause problematic cross-talk between product-forming and growth-sustaining functions.

In recent years, our laboratory has focused on designing and engineering novel biosynthetic pathways with varying degrees of orthogonality. For example, we have repurposed a catabolic pathway, the b-oxidation cycle, to operate in a biosynthetic manner (Nature 2011, 476:355) and extended this concept to develop a biosynthetic platform that is orthogonal to the host anabolic network (Nat. Biotechnol. 2016, 34:556). More recently, we have developed a synthetic metabolic pathway that allows conversion of single-carbon (C1) substrates to multi-carbon products in a manner that is orthogonal to the universally accepted bow-tie architecture of metabolism. In this talk, I will highlight our recent work on engineering these pathways and its implication for industrial biomanufacturing.

506744 Area 15D/E Life Sciences Plenary Award: "Physiome on a Chip": How Integration of Systems Biology with "Organs-on-Chips" May Humanize Therapeutic Development

Abstract:

"Mice are not little people" – a refrain becoming louder as the strengths and weaknesses of animal models of human disease become more apparent. At the same time, three emerging approaches are headed toward integration: powerful systems biology analysis of cell-cell and intracellular signaling networks in patient-derived samples; 3D tissue engineered models of human organ systems, often made from stem cells; and micro-fluidic and meso-fluidic devices that enable living systems to be sustained, perturbed and analyzed for weeks in culture. This talk will highlight the integration of these rapidly moving fields to understand difficult clinical problems, with an emphasis on translating academic discoveries into practical use. Applications in cancer and chronic inflammatory diseases are highlighted. In particular, technical challenges in modeling complex diseases with "organs on chips" approaches include the need for relatively large tissue masses and organ-organ cross talk to capture systemic effects, as well as new ways of thinking about scaling to capture multiple different functionalities from drug clearance to cytokine signaling crosstalk. An example of how gut-liver interactions can be parsed at these levels will be featured, along with new approaches for culturing complex 3D tissues with synthetic extracellular matrix.

501505 A Method to Reduce DimensionalityÂ of Powder Flow Characterization

Abstract:

Background: Material flow properties often have substantial impact on the performance of the manufacturing processes. For solids materials, the flow behavior is a multi-dimensional characteristic. Currently, a practical way is to measure powder properties using several standardized lab scale tests so that the flow behavior of a material can be described using a large group of indices. However, this approach is often time-consuming, and some of the indices obtained may be redundant. **Purpose:** The purpose of this study is to examine the intrinsic relations between different flow indices in a multivariate approach, and to develop a methodology that identifies an optimal set of measurements depending on the availability of time, materials, and instrument. **Method:** The proposed methodology includes techniques to characterize material flow properties and multivariate analysis. Pharmaceutical materials with varying flow properties were firstly characterized. The flow properties of each material was represented by multiple flow indices. To reduce the dimensionality of the database, the mathematical correlations between flow indices within each test were firstly examined. Computational iterations of principal component analysis was then performed. Depending on availability of instruments, the amount of materials, and time needed for each test, each flow index can be associated with an assigned weight. The criteria of selecting the optimal set was based on correlation coefficient of the Weighted Euclidean distance. **Results:** Results suggested that by assigning different weights to each flow index depending on the availability of time, materials, and instrument, different optimal sets of characterization measurements can be identified. The optimal set significantly reduced the cost of measurements. In addition, compared to the original dataset, the optimal set was able to capture maximal information of the material flow behavior. The optimal set of measurements can also be used prior to establishing the predictive correlations between material flow properties and process performance. **Conclusion:** The work presented here has shown an efficient approach to reduce dimensionality of material flow testing. This approach is especially powerful when the amount of time, instrument, or a material is limited, or when latent variables representing material flow properties are needed to develop process control strategy.

Using Mechanistic Modeling of Chromatography to Increase Process Understanding

Tim Fattor, Steve Hunt*, Jonathan Rocher*, and Bob Todd**
**KBI Biopharma*

Although the past decade has seen a rise in the use of mechanistic modeling of chromatography unit operations to increase process understanding, the biopharmaceutical industry's preferred approach to characterization of these processes is still the resource-intensive and experimentally-based design of experiments (DoE). Although this traditional approach does increase process understanding to a limited degree, resource pressures invariably limit the size (number of experiments) and scope (number of parameters investigated) of these studies and therefore result in a diminished level of process understanding. Due to inherent limitations, DoE based process characterization studies often do not adequately address important sources of process variability, including the impact of: feed stream variability, difficult to control parameters (e.g. column bed height, resin-lot), and variability in buffer preparation. Further, DoE studies are not well suited to identify the edges-of-failure for a process. As mechanistic modeling of chromatography unit operations is uniquely suited to address all of these limitations, we have modeled an intermediate cation exchange purification step using a targeted set of calibration experiments. Using this model, we demonstrate the increased process understanding that can be achieved through mechanistic modeling of chromatography unit operations.

OUTLINE

1. INTRODUCTION
   - Continuous processes: set up, monitoring and control
   - Multivariate monitoring
2. METHODOLOGY
   - Construction of the multivariate model
   - Informatics set up
   - Monitoring
3. RESULTS
   - R&D site: real time monitoring and retrospective analysis
   - Manufacturing site: real time monitoring
4. IMPLEMENTATION
5. FUTURE PERSPECTIVES
6. CONCLUSIONS

1. INTRODUCTION

Continuous processing in pharmaceutical industry

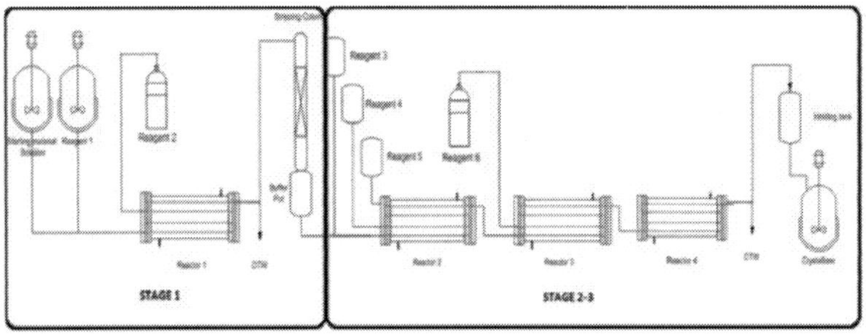

- Smaller footprint
- No inventory
- Flexible batch size
- Lower solvent and energy consumption
- Consistent quality

- Sensitive to solid formation
- Little regulatory precedents

1. INTRODUCTION

Set up of continuous API synthesis: process and monitoring

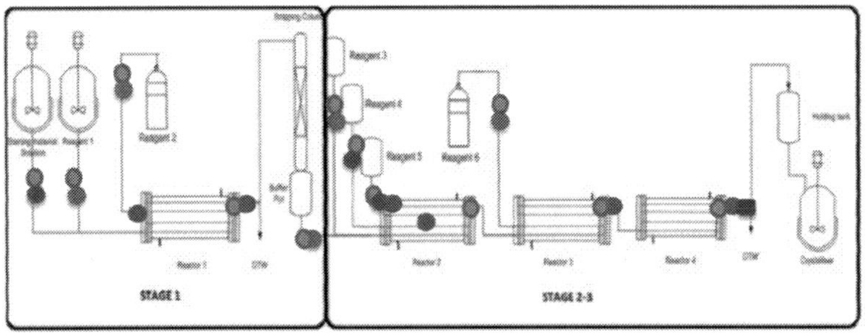

- Conductivity sensors
- Pressure sensors (each pump, reagent and reactor)
- Flow rate sensor (each pump and reagent)
- Pump speed sensor (each pump: not shown)
- Temperature sensors across reactors and heat transfer fluid (not shown)
- Online HPLC

R&D: 30 sensors
Manufacturing: 29 sensors

R&D: 40 sensors
Manufacturing: 52 sensors

1. INTRODUCTION

Set up of continuous synthesis: monitoring and control

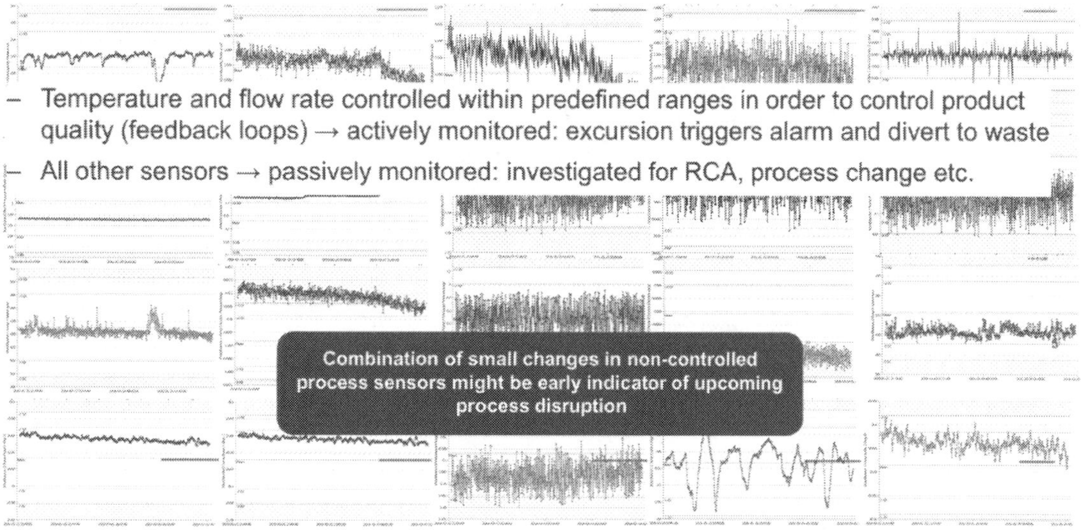

- Temperature and flow rate controlled within predefined ranges in order to control product quality (feedback loops) → actively monitored: excursion triggers alarm and divert to waste
- All other sensors → passively monitored: investigated for RCA, process change etc.

1. INTRODUCTION

Multivariate monitoring

- MSPC model can summarize variability of multiple process sensors readings over time
- Summarizes main trends and correlation patterns in a new trend and calculates a limit based on the 95% confidence interval of normal distributed data.
  - Simplified visualization of process performance
  - Understand natural versus impactful process variation
  - Enhanced process control resulting in more consistent quality
  - Early drift detection can trigger correction actions and prevent forced shut down
  - Facilitates root cause analysis

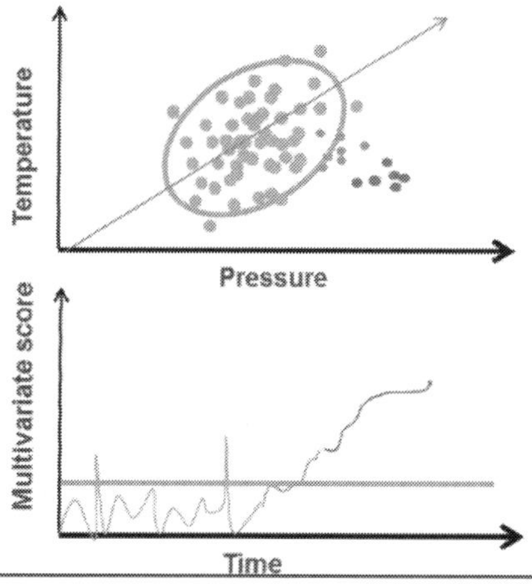

2. METHODOLOGY

1. Construction of the multivariate model

- Data selection
  - State-of-control measurements only (↑sensitivity)
  - Fixed process settings (↑sensitivity)
  - Sufficient process length to ensure representative for future processes (↑robustness), e.g. include full feed bottle cycle

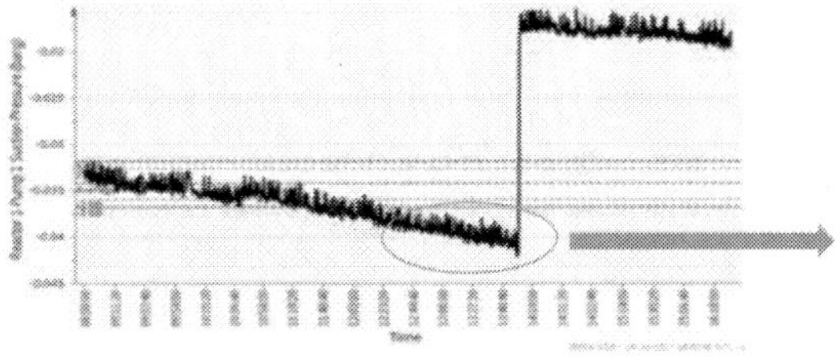

2. METHODOLOGY

1. Construction of multivariate model

- Pre-processing
  - Process sensors different scales and ranges → scale to unit variance (divide by SD)
  - Controlled sensors: observed standard deviation <<< accepted standard deviation → down weight
- Model selection
  - Cross validation prone to over fit model due to collinearity of data (i.e. consecutive time points not independent) → check validity of selected model components (e.g. >5%)

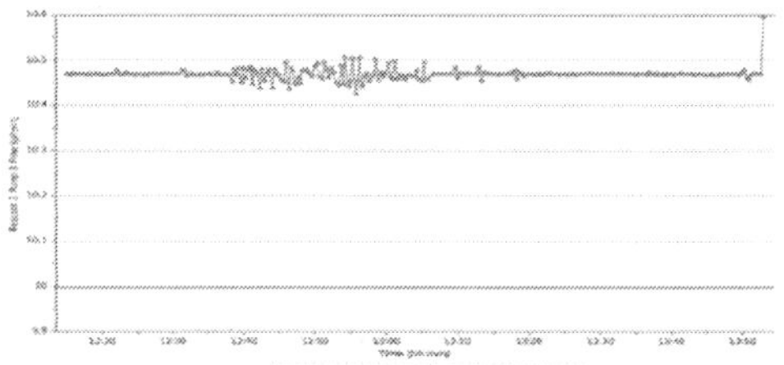

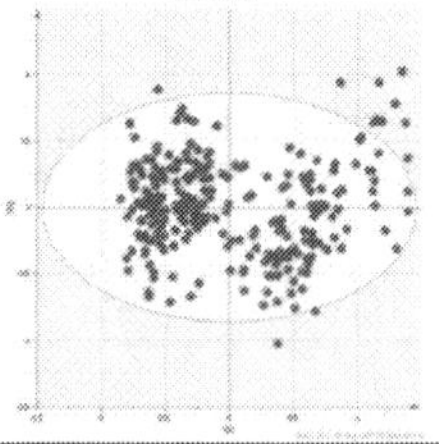

2. METHODOLOGY
Informatics set up

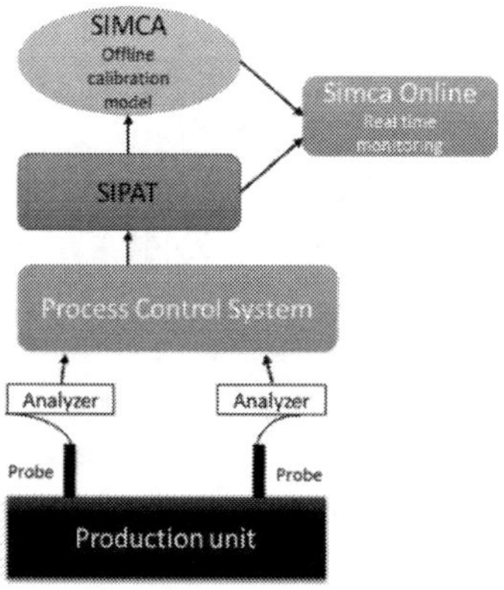

- Sensor readings from probe every second
- Sensor readings collected by SIPAT every 20 seconds
  - Statistical analysis confirmed no information lost by decreased frequency
- Historical data extracted via SIPAT and imported to Simca to build model defining operational space off line
- Multivariate model imported to Simca Online and set up sensor and alarm configuration
- Monitoring in real time started via Simca Online

2. METHODOLOGY
3. Monitoring

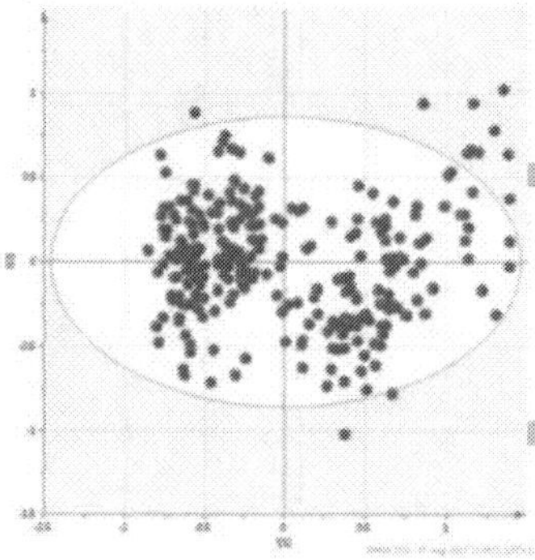

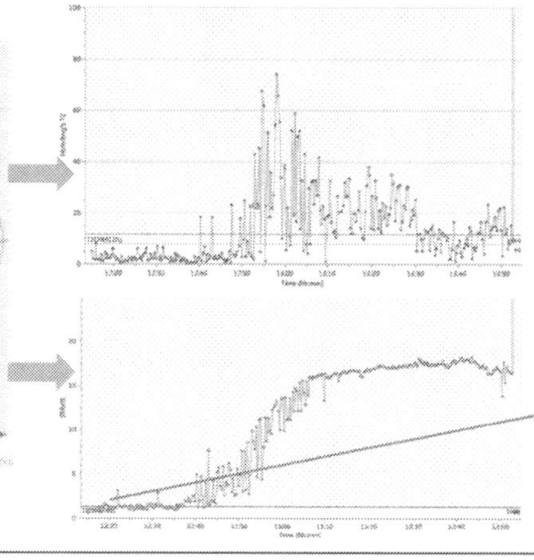

By default multivariate alarm flagged if:

- Three consecutive measurements exceed 99% statistical limit
- 10 out of 100 consecutive measurements exceed 99% statistical limit

Needs optimizing to reduce false positive alarms, whilst maintaining sensitivity for relevant process changes

3.RESULTS

1. R&D site monitoring: evaluation during development campaign

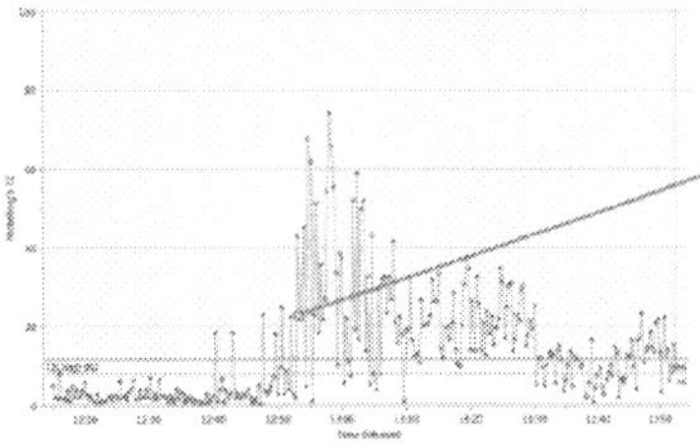
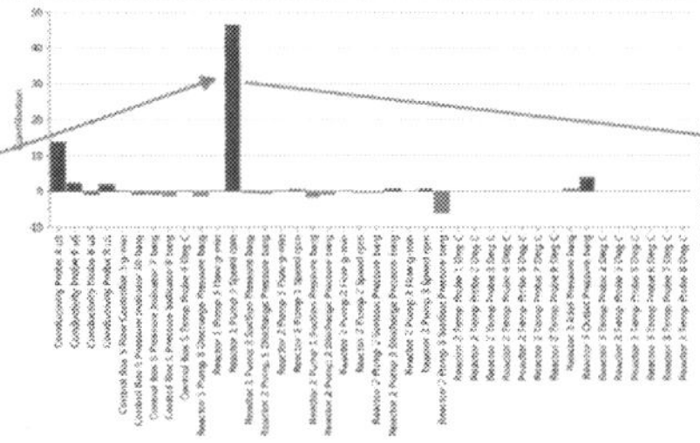
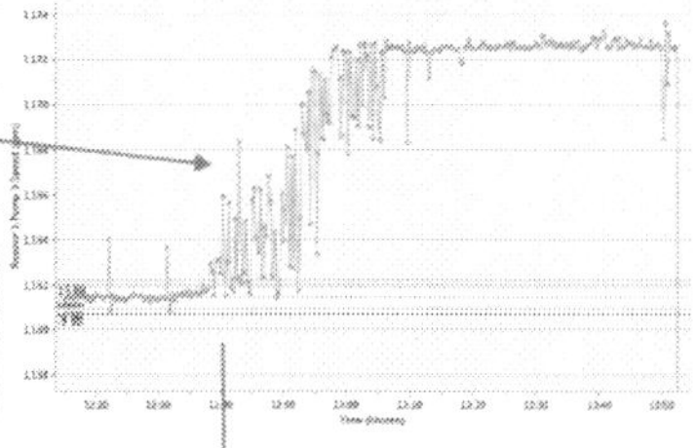
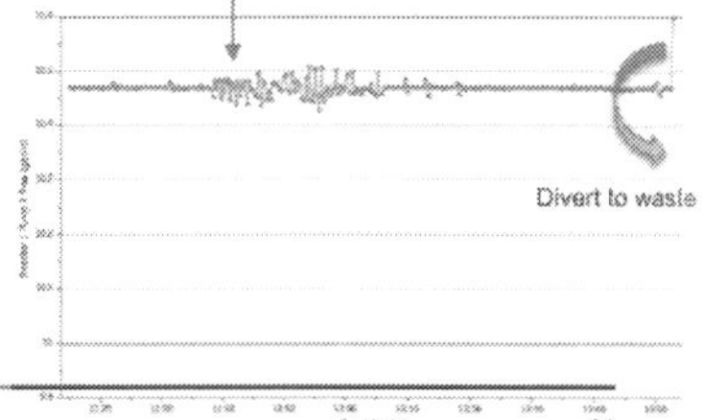

- Multivariate alarm: 12:48
- Root cause multivariate alarm: 15 rpm change in pump speed
- Univariate alarm: 13:52
- Action: pump switch at 13:52
- Increase in pump speed detected one hour before the univariate flow rate alarm → pump switch based on multivariate alarm could prevent flow rate excursion and divert to waste

3. RESULTS

1. R&D site monitoring: evaluation during development campaign

Observed process failure	Multivariate detection	Process change	Univariate Alarm	Action
Stage 2 feed pump failure	12:48	15 rpm increase pump speed	13:52	Switch pump
Reagent 2 flow failure	13:18	0.6 barg increase upstream p	15:28	Increase pressure on reagent + divert to waste
7 more similar flow failures for reagent 2 flow				
Reagent 3 pump failure	02:19	12 rpm increase in pump speed	06:38	None (naturally restored)

3. RESULTS

1. R&D site: retrospective RCA for recurrent reagent stream blockage

Largest process changes before blockage:

- Controller differential pressure
- R3 downstream pressure
- Column pressure

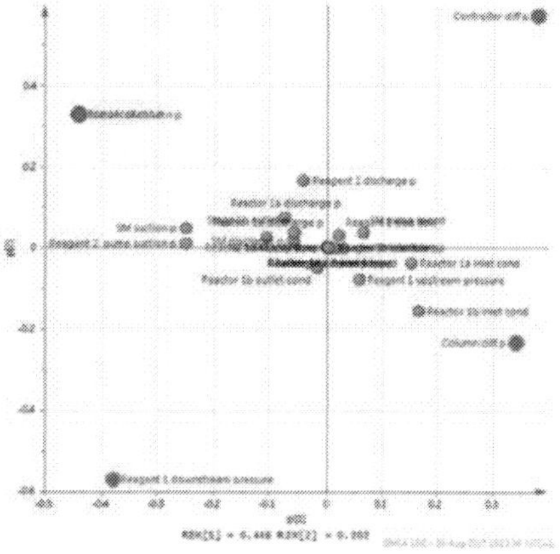

- MVA confirmed root cause blockage
  - Originates from reagent stream
  - Does not start in the reactor

Wealth of real time process data facilitates retrospective root cause analysis

3. RESULTS

2. Manufacturing site monitoring: evaluation during development campaign

Observed process failure	Multivariate detection	Process change	Univariate Alarm	Action
Reagent 2 flow failure	14:53	0.35 barg increase p after flow controller	14:05 (24 h later)	Divert to waste
2 more reagent 2 flow failures as described above				
Reagent 3 pump failure	12:16	Spikes (2 g/min increase) flow rate	12:36	Switch pump
Reagent 1 pump failure	06:06	0.015 bar increase supply filter differential p	06:23	Switch pump
Reagent 6 depletion	23:14	3 % increase flow control output	23:27	Divert to waste and top up reagent 6

4. IMPLEMENTATION

Plan: routine implementation for predictive maintenance by operators

- Procedural flow and decision tree
- List with specific actions in case of detected drift, e.g. switch pump in case of 5 consecutive pump pressure excursions

Aim: detect early onset process failures enabling actions to maintain consistent process performance and avoid divert to waste

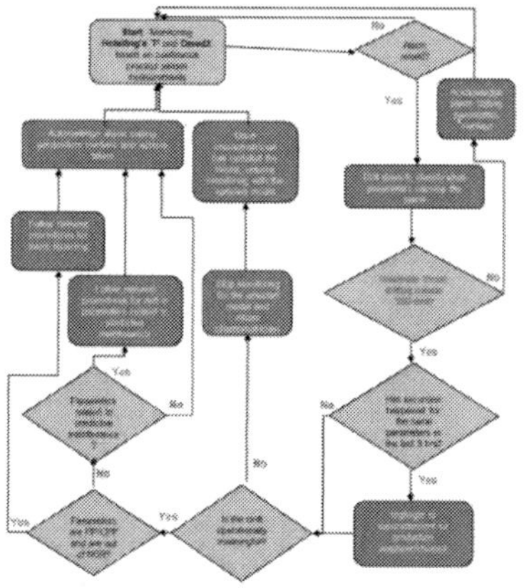

5. FUTURE PERSPECTIVES

Learning

- Increased understanding model sensitivity/robustness and improve implementation strategy
- Combine process sensor data with quality data (e.g. online HPLC): enable detailed quality information at high frequency
- Combine system models and multivariate process monitoring to investigate opportunities for process control

Implementation

- Implement for all high volume processes (API and drug product) containing multiple continuous synthesis stages or unit operations as part of the control strategy (quality assurance and batch release)
- Develop model validation and maintenance plan compliant with regulatory requirements and accommodating high flexibility (e.g. frequent model updates to enable set point changes within the design space)

Advanced Control

- Real-time control of product quality using feedback and feedforward loops

6. CONCLUSIONS

- Multivariate monitoring enabled early detection onset process failure
  - Prolonged time window from detection till process excursion
  - Enables preventative actions to ensure process performance and avoid divert to waste
- Right balance between model sensitivity and model robustness (minimized false alarms) is critical:
  - Optimize data selection to define operational space
  - Optimize monitoring diagnostics limits and alarm settings based on manufacturing experience
- Routine implementation for predictive maintenance by operators
  - Standardized procedures (prevent subjective decisions, actions changing product quality, e.g. set point change)
  - Rigorous yet agile plan for model validation and maintenance accommodating frequent updates
- Future: embed more quality aspects or achieve real time optimization and control

6. ACKNOWLEGEMENTS

- Malcolm Berry: project lead
- Hannah Robinson, Peter Clements for engineering advice
- Peter Shapland for chemistry advice
- Peter Hamilton and Christian Airiau: platforms support and vision
- Ian Barylski: SIPAT and SOL support

500198 Global System Analysis of Interconnected Flow Sheet Models for Drug Product Manufacturing to Performance

Abstract:

Flowsheet models developed using gPROMS platform presents a holistic modeling and simulation approach to design drug product manufacturing processes. Often, these models involve estimation of several parameters using experimental data and measurements. Global Systems Analysis (GSA) is a powerful tool which can be used to identify the cardinal factors of the different parts of the process and their effect on critical quality attributes. In this presentation, we will share our recent experience using GSA through two case studies. First case study will focus on the development of API dry milling model and application of GSA in identifying inputs which are most influential for the model predictions. In the second case study, a comprehensive flowsheet that combines API crystallization, downstream solids operations and drug performance is thoroughly analysed using GSA.

499041 Closed Loop Dynamics of Ribbon Density in a Dry Granulation Process

Abstract:

End to end continuous processing is currently a key focus area in pharmaceutical manufacturing. The increase in research and development is evident from recent FDA approvals of drug products manufactured using continuous downstream processes [1]. Dry granulation is one of the unit operations added for improving material handling and blend uniformity in the continuous tableting line.

Dry granulation is an inherently continuous process. The particle size distribution and bulk density of the granules are critical quality attributes of the process, which depend primarily on the ribbon relative density. The ribbon density is hence a crucial intermediate quality attribute of the unit operation. It is thereby valuable to monitor and control the ribbon density for ensuring desired granule qualities.

Assurance of product quality in real-time requires satisfactory process models with a robust sensor network and control architecture. A dynamic, simplified process model along with real-time monitoring capability is important to the implementation of supervisory control of the process to achieve Quality by Control.

The Johanson model [2] describes the fundamental solid mechanics of roller compaction as a standalone steady state process. FEM analyses have suggested modifications to the model for improved prediction of ribbon density [3]. A process scale model, however, needs to consider the production rate and requires integration of previous unit operations that feed into the roller compactor, in addition to the solid mechanics of the roller compactor.

The process model by Reynolds et al [4] integrates the screw conveyor with the roller compaction system, describing the effect of screw speed on the ribbon density and production rate at steady state. A dynamic process model is, however, essential for the implementation of a control system, as proposed by Hsu et al [5]. The model did not include the flow rates through the system and lacked experimental validation.

This work focuses on a modified process model incorporating the solid mechanics, process dynamics and throughput. The model combines the dynamics of material flow from Hsu et al (2010a) and Reynolds et al (2010) and solid mechanics suggested by Liu and Wassgren (2016). Demonstration of real-time ribbon density monitoring using NIR and microwave sensors [6,7] enable the validation of such a dynamic model. Additionally, the model evaluates the integrated system of a hopper and screw conveyor feeding into the roller compactor.

Blends comprising of acetaminophen and microcrystalline cellulose at inlet flow rates of 8-12 kg/h and varying roll pressures are used for real-time monitoring of ribbon density. The solid mechanics material parameters are estimated using ribbons. The process variables and ribbon density are recorded and monitored using DeltaV DCS.

This work progresses to demonstrate the implementation of ribbon density as proposed by Hsu et al. The integrated

process model and real time ribbon density monitoring enable the development of control strategies for the desired ribbon density and eventually the granule properties.

References:

1. https://blogs.fda.gov/fdavoice/index.php/2016/04/continuous-manufacturing-has-a-strong-impact-on-drug-quality/
2. Johanson JR. A Rolling Theory for Granular Solids. *J Appl Mech*. 1965;32(4):842-848
3. Liu Y, Wassgren C. Modifications to Johansonâ€™s roll compaction model for improved relative density predictions. *Powder Technol*. 2016; 297:294-302.
4. Reynolds G, Ingale R, Roberts R, Kothari S, Gururajan B. Practical application of roller compaction process modeling. *Comput Chem Eng*. 2010;34(7):1049-1057
5. Hsu S-H, Reklaitis G V., Venkatasubramanian V. Modeling and Control of Roller Compaction for Pharmaceutical Manufacturing. Part I & II: Process Dynamics and Control Framework. *J Pharm Innov*. 2010;5(1-2):14-36
6. Gupta A, Austin J, Davis S, Harris M, Reklaitis G. A Novel Microwave Sensor for Real-Time Online Monitoring of Roll Compacts of Pharmaceutical Powders Online-A Comparative Case Study with NIR. *J Pharm Sci*. 2015;104(5):1787-1794.
7. McAuliffe MAP, Oâ€™Mahony GE, Blackshields CA, et al. The Use of PAT and Off-line Methods for Monitoring of Roller Compacted Ribbon and Granule Properties with a View to Continuous Processing. *Org Process Res Dev*. 2015;19(1):158-166.Â

493080 Integrated Control and Data Management System for Continuous Pharmaceutical Manufacturing Process

Abstract:

Continuous manufacturing (CM) is evolving as a preferred platform for pharmaceutical products involving solid dosages forms. One advantages of CM among others is that the product quality can be controlled in real time and thereby its open the possibilities of achieving Quality by Control (QbC) and real time release (RTR) paradigm. However, there are different levels of challenges associated with design and implementation of an efficient control system into pharmaceutical manufacturing plant. Identification of right control strategies such as feedforward (FF), feedback (FB), or combined FF/FB[1] as well as right control algorithms such as PID, MPC, or hybrid MPC-PID is often a tough decision that need to be made prior to implementation of integrated control system into the manufacturing plant[2]. Selection of a right control strategy and algorithm is mandatory task both for efficient manufacturing as well as regulatory point of view. Therefore, the different control options need to be evaluated and the best integrated control system needs to be implemented into the pilot-plant. The integrated control system is then need to be coupled with material traceability framework for real time identification of batches so that the product can be recall in case if needed. The continuous manufacturing is a data rich process and therefore a systematic data management framework is also desired to be implemented to collect, store and manage all the relevant data in real time.

In this work, different control strategies for direct compaction continuous tablet manufacturing process has been developed and implemented into our pilot-plant facility using industry standard control platform. The critical control variables are drug concertation, powder level, main compression force, tablet weight and hardness. Both, feedback (FB) as well as feedforward (FF) control strategies have been implemented into the pilot-plant. Feedback (FB) control strategy has been compared with a combined feedforward/feedback (FF/FB) control strategy. A novel method for real time monitoring of powder density has been developed and integrated with feedforward control loop. The feedforward control loop takes the powder bulk density signal as an input and mitigate the effect of variations in powder bulk density on CQAâ€™s proactively and thereby reduces the rejections significantly in compare to only feedback control strategy. Three control algorithms (PID, MPC, hybrid MPC-PID) have been also implemented and compared for feedback loops. The flexible control system is designed in such a way that any control scheme can be easily selected. The control system is also modular in nature that can be easily transferred to any other control platform and manufacturing plant. The continuous tablet manufacturing pilot-plant runs under each control scheme and their performance are evaluated and thereby a best control system has been identified. A systematic methodology for material traceability has been also developed and coupled with integrated control system. The methodology starts with material and process specifications followed by scanning the powder drums and transmitting the material and time information to the control platform. The control platform then utilizes the information stored in the database, generated through residence time distribution model and information collected through real time monitoring to track the material during continuous manufacturing. All the relevant data generated during continuous manufacturing has been systematically collected, stored and organized in a data hub (OSI PI). The PAT data has been collected through PAT data management tool (synTQ) and then transferred to the PI system while the other operational data has been collected in control platform (DeltaV) and send to PI system via OPC communication. Finally, the data has been transferred to the cloud system for enterprise wise efficient sharing and applications.

The objective of this presentation is two-fold: first to evaluate the performance of integrated control system of continuous pharmaceutical tablet manufacturing process and then to highlight the material traceability and data management framework.

References

1. Singh, R., Muzzio, F., Ierapetritou, M., Ramachandran, R. (2015). A combined feed-forward/feed-back control system for a QbD based continuous tablet manufacturing process. PROCESSES Journal, 3, 339-356.
2. Haas, N. T., Ierapetritou, M., Singh, R. (2017). Advanced model predictive feedforward/feedback control of a tablet press. Journal of Pharmaceutical Innovation, 1-14, DOI 10.1007/s12247-017-9276-y.

497291 *In silico* Process Characterization of Drug Product Filling Operations: Predicting Quality Attributes from First Principles

Abstract:

Process characterization of drug product filling operations is typically a time- and labor-intensive process requiring substantial manufacturing line down-time due to the large number of process inputs and lack of mechanistic understanding of the fluid dynamics and filler systemsâ€™ sensitivities to drug product properties (e.g. viscosity, density, surface tension), or variations in site, equipment type, settings, and ambient conditions.

As an alternative to empirical process characterization, we have developed a computational model which predicts process quality attributes â€“ for instance flow rate, fill volume/weight, cavitation, droplet formation, and product concentration â€“ from first principles of product dynamic properties, fluid dynamics, structural mechanics and control theory. Formulation of the model couples finite element and finite volume solutions to structural deformation and fluid dynamics in pinch valves, and leverages analytical solutions to ordinary differential and algebraic equations accounting for orifice and pipe flow as well as equipment control laws.

We present an application of the model to a Time/Pressure filling operation, including formulation, deployment as a software application, and development of a workflow for in-silico process characterization utilizing design space response studies, local sensitivity studies, and uncertainty analysis.

489865 Use of Site Specific Recombinases to Engineer New Cell Lines for Therapeutic Protein Biomanufacturing

Abstract:

Traditional cell line development for therapeutic protein biomanufacturing involves random integration and amplification of a product gene to endow cells with high transcript level of the product gene and high production capabilities. Subsequently, single cell cloning is performed and a clone with high productivity that retains its productivity over a long cultivation period is selected as the production cell line. This process is time and resource intensive, and could be shortened by improving the stability and homogeneity of cells for production. We have established a streamlined cell line development process by generating a template cell line with a â€œlanding padâ€ that contains an antibody gene surrounded by recombination sites. Through recombinase mediated cassette exchange (RMCE), a new product gene can be quickly swapped in to take its place in the landing pad and generate a new producing cell line. The template cell line was created using single copy integration of lentivirus and screened for high productivity. Additionally, the cell line has a high transcript level of the product gene and has been shown to be stable in its productivity over long term culture, demonstrating the feasibility of single copy integration for protein production.

By RMCE, new cell lines producing TNFR:Fc fusion protein or Erythropoietin were generated from the template cell line. First, we optimized the use of RMCE by performing recombination transiently between plasmids to determine the optimal donor DNA to recombinase ratio. Then, using this information, we generated pools of cells with new product genes. The pools were then subcloned to characterize the variation between cells that have had targeted integration of a new product gene. We compared the productivity, transcript levels, and growth of different clones to identify the potential for RMCE to replace traditional cell line development. If a high producing cell line can be changed to produce a new product gene and maintain its productivity, we could vastly improve the cell line development process by shortening the time to obtain a high producing, stable cell line.

500859 Implications of Protein Sequence Variants Derived from Systematic Starvation

Abstract:

Many metabolic and biosynthetic pathways can influence protein product attributes during translation and post-translational modification (i.e. sequence variation, glycosylation, oxidation, proteolysis, etc.). For both recombinant and native protein biosynthesis, the resulting protein product attributes are analogous to a historical record of cellular events. Of particular interest are sequence variants caused by amino acid misincorporation. Believed to be universal, amino acid misincorporation is the inherent errant replacement of an amino acid by another. In biological systems, high-level amino acid misincorporation can cause loss of proteostasis and disease through the compounded accumulation of errors. And in biopharmaceutical applications, since protein product function is derived from the productâ€™s attributes, elevated sequence variant frequency has been shown to result in adverse effects such as product instability, structure change, aggregation, and the potential for immunogenicity. Amino acid starvation, fast cell growth, overproduction, and oxidative stress, which are physiological conditions that can be observed in disease states and during biopharmaceuticals production, can all significantly exacerbate mistranslation by orders of magnitude. In an effort to elucidate the major sequence variant mechanisms, to our best knowledge, our work is the first investigation of amino acid misincorporation through systematic amino acid starvation. Our results reveal that one-base mismatches during codon recognition and tRNA misacylation are both common and major mechanisms. Also, different cell lines share common misincorporation patterns suggesting that a conserved set of mechanisms and error propensities exists.

498957 Anti-Adhesive Glycosylated Proteins As a Technology for Improved Suspension Cell Culture in Bioreactors

Abstract:

Anti-Adhesive Glycosylated Proteins as a Technology for Improved Suspension Cell Culture in Bioreactors

Mucins are brushlike biopolymers found in biointerfaces, including joints and on the surface of secretory glands (Figure 1A). Mucin biopolymers have anti-adhesive properties that mediate lubrication, resist cell adhesion, and protect cell surfaces, amongst other functions. Although mucin glycoproteins have broad potential in medicine and biomaterials applications, their use has been limited due to the challenge of biomanufacturing sufficiently large quantities. For example, prokaryotic host production, despite its efficiency, cannot properly reproduce the defining O-linked glycosylation of mucins. To address these issues, we have created a protocol to efficiently produce and purify structurally correct recombinant mucins from eukaryotic hosts, allowing us to better understand the molecular basis of mucin structure and investigate downstream applications, including their use as (1) biocompatible lubricants and (2) anti-clumping agents in eukaryotic-cell bioreactors. In another application, we engineered a membrane-anchored mucin, and observed that cells expressing the engineered mucins spontaneously transition from adherent to floating in the media, indicating decreased cell-cell and cell-matrix interactions (Figure 1C, D). Remarkably, these cells displayed an increased ability to survive in suspension compared to non-mucin expressing control cells (Figure 1E). These results suggest that cell surface engineering with genetically encoded mucins can improve the viability of eukaryotic cells, possibly streamlining suspension adaptation and bioreactor performance. Furthermore, the decreased cell-cell adhesivity of engineered cells should minimize aggregation, a common problem that has limited the use of human cell hosts in biomanufacturing. Overall, our engineered cells have potential to become a new production technology for biomanufacturing of complex biologics and recombinant proteins with human-specific glycosylation.

Figure 1

488935 Multi-Omics Approaches to Unravel Cellular Metabolism Towards Enhancing Process Robustness

Abstract:

Progress in process development, from cell line screening to online monitoring technologies have enabled the generation of highly productive processes, to meet the growing clinical demand for biological therapeutics. This has shifted the focus of development activities to guarantee process robustness and consistency in product quality attributes, as these processes are still susceptible to variability in raw materials or operational errors such as overfeeding. Advances in various omicâ€™s methodologies enable a comprehensive systems biology approach into understanding cellular behavior in response to such process deviations and enable formulation of mitigation strategies.

In this work, a multi-omics approach was applied to a chemically defined CHO fed-batch process resulting in high peak cell densities and titers up to 10 g/L. Upon dosing a single supplement to this process, the cellular metabolism changed resulting in even higher peak cell densities, decreased ammonia generation and further increased productivity. However, at different doses corresponding product quality attributes such as glycosylation and protein aggregation were affected. A multi-omics approach encompassing transcriptomics, proteomics and metabolomics methods was used to investigate the underlying cause for differences in cellular phenotype from a global perspective. A total of 16000 transcripts, 7000 proteins, and 700 metabolites were evaluated for each condition. Differences in glycosylation were driven by a multitude of changes in metabolic pathways as confirmed by the complementary omics data. Among the key pathways affected were those involved in central carbon metabolism, amino acid metabolism and cellular redox homeostasis. Protein aggregation on the other hand was driven by an extracellular mechanism involving the supplement discussed above and was confirmed through independent cell free studies. This comprehensive analysis resulted in further media optimization and in establishing potential cell line engineering targets to improve process efficiency and robustness.

501767 Epigenomic Analysis on Chinese Hamster Ovary (CHO) Cells for Enhanced Production Cell Line Stability

Abstract:

For a biomanufacturing of protein therapeutics, the producing CHO cell line must have high expression level of the product gene. This is mostly achieved through a gene amplification process that increases the copy number of transgene in the genome. The increased mRNA expression of the transgene leads to high protein production, but often results in unstable cell lines that lose the productivity over long-term culture. Many have pursued the construction of robust production cell lines in which the transgene is integrated in a single, active and stable site in the genome. Ideally, such integration sites are rationally selected. The stability of the expression of transgene is affected by the structural integrity and vulnerability of the locus of integration as well as many factors affecting gene expression, among them epigenetic regulation. We surveyed the epigenetic landscape in multiple CHO cell lines to better understand the relationship between epigenetic status and cell line stability.

Chromatin accessibility is a strong indicator of regulatory activity in the genome. Local chromatin can become exposed by RNA polymerase occupancy at promoters and transcription factor occupancy at enhancers. To assess chromatin accessibility we employed Assay for Transposase-Accessible Chromatin followed by sequencing (ATAC-seq) that uses a hyperactive Tn5 transposase to insert sequencing adapters preferentially into open chromatin. We found that the presence of ATAC-seq peaks near transcription start sites correlated with gene expression data. We also identified clusters of open chromatin peaks with very high ATAC-seq signal, commonly known as super-enhancers. These features are thought to mediate high transcription of cell-type specific genes. We hypothesize that integrating transgenes into loci with high ATAC-seq signal will confer increased expression stability due to reduced likelihood of epigenetic silencing.

This study provides a global view of transcriptional regulatory signatures and their potential for mediating strong transgene expression. Epigenomic analysis can benefit cell line development processes by identifying loci that maintain transgene stability without compromising transcriptional activity.

498977 CRISPR/Cas9-Mediated Knock-in of an Optimized TetO Repeat for Live Cell Imaging of Endogenous Loci

Abstract:

CRISPR-mediated gene targeting has been used for various applications including epitope or GFP tagging of endogenous proteins. One application that has not been explored yet is the tagging of the DNA itself for visualizing the position and dynamics of specific chromosomal regions. Nuclear organization has an important role in determining genome function; however, it is not clear how spatiotemporal organization of the genome relate to the functionality. Thus, in order to relate genome function to the 4D nuclear organization, a direct, microscopy-based method for live cell tracking of the dynamics of any specific endogenous locus of interest is necessary. A commonly used method for imaging a specific region of interest within the chromosome is DNA fluorescence *in situ* hybridization (DNA-FISH); however, the procedure is technically challenging and expensive, making it a low throughput method. Moreover, DNA-FISH is performed in fixed cells, which cannot be used for tracking dynamics of DNA. Live cell imaging of DNA in mammalian cells was previously done by using fluorescent repressor-operator system (FROS). The idea was based on enrichment of a fluorescent protein (FP) at a specific site on the DNA to obtain enough fluorescent signal that is above the background noise. For this purpose, fluorescent proteins such as GFP were fused to DNA binding proteins that can recognize a repeating DNA sequence. If the repetitive DNA is positioned close enough, it can serve as a tag for the region of interest. In the two commonly used systems, repeating sequences of lac operators (LacO) or tet operators (TetO) were used as a DNA tag and FP fused Lac repressors (LacI) or Tet repressors (TetR) were used for visualization of the tag. In previous examples, tagging was mostly done either within a plasmid or BAC containing the gene of interest, which might not fully represent the endogenous locus. An attempt for tagging an endogenous locus yielded very low efficiency. Discovery of novel modular proteins such as transcription activator-like effectors (TALEs) and the CRISPR/Cas9 system, whose DNA recognition specificity can be easily tailored, lead to the development of alternative approaches for visualizing endogenous loci. Coupled with fluorescent tags, both TALEs and a catalytically inactive version of Cas9 (dCas9) were used for visualization of naturally occurring repetitive sequences within mammalian genomes. However, repetitive sequences are not equally distributed throughout the genome, necessitating other strategies for visualizing such non-repetitive regions. Visualization of non-repetitive sites by TALEs or dCas9 requires introduction of a set of many TALEs or sgRNAs recognizing multiple different sites, which could be challenging and inefficient. Thus, an easy and efficient method for visualization of non-repetitive endogenous loci is highly desirable.

In this work we developed a new strategy for live cell imaging of specific chromosomal regions. Our approach was based on knock-in of a TetO repeat into endogenous loci via HR. HR is normally inefficient in mammalian cells. Thus, in our strategy we took advantage of the CRISPR/Cas9 system to increase the efficiency of knock-in of a TetO repeat via HR in HCT116 colorectal carcinoma cell line. One bottleneck in the traditional HR method is the use of long homology arms, which requires the arms to be PCR-amplified from the genomic DNA first, followed by their cloning into a donor vector. Construction of such a donor vector could be challenging, especially for higher throughput applications requiring creation of many such donor vectors. Based on a previous work using limited homology arms of down to 50 base pairs, we utilized a one-step, PCR-based method to create a linear donor DNA containing a TetO repeat. In this strategy, the TetO repeat and a selection marker are PCR-amplified using primers with 50 nt 5â€™ extension sequences, which are homologous to the target site sequence. The resulting linear donor DNA contains 50 bp homology arms flanking the insert, and the donor can then be gel or PCR-purified to be used directly for transfection.

Direct repeats of TetO could be unstable during plasmid preparation from bacteria, making the protocol tedious. Moreover, it is also difficult to PCR-amplify direct repeats. Thus, to improve the TetO repeats and add more beneficial features to it, we created a multimer of TetO (19 nt) repeats separated by random 10 nt sequences that do not contain CpG sites. Optimized repetitive sequences enabled easy PCR of both 48mer and 96mer TetO donors with a clear band. With the help of the improved donor DNA design and an optimized TetO repeat, it was possible to achieve knock-in efficiency of up to 20% using the 48mer TetO repeat. 48 repeats of TetO were also efficient at visualizing endogenous loci when TetR-EGFP was expressed. The observed number of spots was consistent with the genotyping results; thus, our suggested technique for visualization of endogenous sites was highly specific. Moreover, a comparison with wild-type cells showed that our method could resemble natural compartmentalization of the loci.

We also demonstrated applicability of our method in various different regions, representing both gene-rich, transcriptionally active and nuclear speckle-associated sites as well as gene-poor, transcriptionally inactive and nuclear lamina-associated sites. Knock-in efficiencies were 20% and 18% in two different gene-rich loci and the efficiency was 2% and 13% in two different gene-poor loci. As an interesting finding, we observed that gene-poor and heterochromatin regions exclusively preferred non-homologous end joining (NHEJ) for integration of the knock-in cassette, instead of HR. A direct ligation of the donor DNA after the expected cut site was observed in all the knock-in clones analyzed. Our results show that an NHEJ-based approach can be more successful at knock-in into heterochromatin sites. An encouraging finding was the fact that we could also visualize heterochromatin/gene-poor regions by expressing TetR-EGFP. Our method was also successful at live cell tracking of the dynamics of different loci.

The ease and the potential of scalability of the suggested method can help analysis of various loci in parallel to increase our understanding of relationships between gene compartmentalization and function. Given the general applicability of our method, we anticipate that it can be utilized to understand nuclear localization of various gene-rich and poor regions and enhance future causative analysis between gene function and compartmentalization.

504099 Continuous Processing Strategies for Biologics Manufacturing

Abstract:

There has been a growing interest in developing and implementing continuous processing for biologics manufacturing. Compared to traditional batch technologies, continuous processing offers more flexibility and can improve product quality and lower the costs of production. The US FDA has been supportive of the implementation of continuous processing using a science- and risk-based approach and there are no regulatory hurdles for its implementation. This presentation will discuss the drivers behind the rising utilization of continuous processing technologies and the challenges in development and scale up of continuous processing platforms. Case studies will be shown on developing high cell concentration perfusion cell cultures and direct product capture unit operations for manufacturing of recombinant proteins and monoclonal antibodies.

497879 Pinducer Analysis for the Design of a Nucleation Subsystem for Continuous Slug-Flow Crystallization

Abstract:

The reliable and convenient control of crystallization processes has received increased academic and industrial interest. Ultrasonication has been used to aid crystallization, such as accelerating nucleation, reducing crystal size, reducing aggregation, or changing polymorphic ratios [1]–[3]. Most bench-scale continuous sonocrystallization studies are based on configurations involving an ultrasonic probe directly inserted into the solution/slurry or a tube placed within a bath. An alternative multiphase millifluidic tubular crystallizer design combined the advantages of using a probe (focused energy) and a bath (indirect ultrasonication) for continuous nucleation, with the nuclei size tuning by changing the ultrasonic amplitude [4].

To better understand and improve the design of a continuous crystallization process, we employ a needle-tip pressure transducer ("pinducer") to measure and analyze the "primary effect" of ultrasonication, which is the generation of pressure waves (the term "primary effect" is used as in Ref. [5]). This analysis is used to gain understanding on the crystallization outcome (the "secondary effect"). This separation of the analysis of the primary effect from the secondary effect provides greater insight and understanding than past studies that only reported the crystallization results of applying ultrasonication. The pressure waves of ultrasonication are measured for different equipment configurations, and different positions and orientations of the pinducer relative to the ultrasonication probe. The quantification of the pressure waves as a function of frequency provides insights into bubble dynamics and is used to guide the design of the nucleation subsystem and its positioning relative to other subsystems of the continuous crystallization apparatus. The analysis of the spatial localization of energy dissipation that occurs during ultrasonication informs the design of the focused indirect ultrasonication zone to enhance reproducibility and the control of nuclei size distribution [6]. Although the focus of this study was on a particular continuous crystallizer design, the same procedures for the analysis of physical phenomena via pinducers employed in this study could be applied to provide insights and guidance into other ultrasonication-facilitated crystallization designs.

References

[1] J. R. G. Sander, B. W. Zeiger, and K. S. Suslick, "Sonocrystallization and sonofragmentation," *Ultrason. Sonochem.*, vol. 21, no. 6, pp. 1908–1915, 2014.

[2] O. Narducci, A. G. Jones, and E. Kougoulos, "Continuous crystallization of adipic acid with ultrasound," *Chem. Eng. Sci.*, vol. 66, no. 6, pp. 1069–1076, 2011.

[3] R. J. P. Eder, S. Schrank, M. O. Besenhard, E. Roblegg, H. Gruber-Woelfler, and J. G. Khinast, "Continuous sonocrystallization of acetylsalicylic acid (ASA): Control of crystal size," *Cryst. Growth Des.*, vol. 12, no. 10, pp. 4733–4738, 2012.

[4] M. Jiang, C. D. Papageorgiou, J. Waetzig, A. Hardy, M. Langston, and R. D. Braatz, "Indirect ultrasonication in continuous slug-flow crystallization," *Cryst. Growth Des.*, vol. 15, no. 5, pp. 2486–2492, 2015.

[5] V. S. Sutkar and P. R. Gogate, "Design aspects of sonochemical reactors: Techniques for understanding cavitational activity distribution and effect of operating parameters," *Chem. Eng. J.*, vol. 155, no. 1–2, pp. 26–36, 2009.

[6] M. Jiang, C. Gu, and R. D. Braatz., "Analysis of focused indirect ultrasound via high-speed spatially localized pressure sensing and its consequences on nucleation," submitted.

499788 Understanding Metastable Zone Width without Nucleation Rate

Abstract:

Traditionally, the metastable zone width (MZW) in solution crystallization is thought to originate from the path-dependence of the steady-state nucleation rate of a solute. However, new experimental findings appear to go against this popular notion. In this work, we propose an alternate viewpoint to understand the MZW, and show that MZW may not necessarily be a function of the nucleation rate as conventionally understood.

Our work presents experimental data on the MZWs of various solutes and show that these data can be explained without invoking the concept of a nucleation rate. We use data from our own work as well as that from the recent literature. The applicability of the new concept of MZW to both evaporative and cooling crystallizations is shown. The stochastic view point of nucleation is contrasted against a more â€˜deterministicâ€™ view point of the classical nucleation theory. The pitfalls of obtaining nucleation kinetics from MZW experiments are discussed. Both induction time and MZW measurements are recommended as complementary experimental techniques to characterize the kinetics of nucleation in solution crystallization.

499208 Molecular Interactions between Nanocellulose and Crystallizing Pharmaceuticals

Abstract:

Cellulose is a highly abundant natural polymer that, being a plant-based material, exhibits renewability and biodegradability. It is frequently used in the form of microcrystalline cellulose as a filler and binder in pharmaceutical tablets. A nanoscale form of cellulose can be prepared by removing much of the amorphous cellulose material and leaving small crystalline particles with a large concentration of hydrogen bonding groups on the surface. Due to its strong hydrogen bonding tendency and the presence of charges on the nanocellulose surface, it has the potential to bind with drug molecules and act as a release inhibitor, a stabilizer or a site for heterogeneous crystallization of the drug itself. Beyond hydrogen bonding and charge, however, the nanocellulose surface can be readily altered to display a variety of functional groups, leading to flexibility for rational design of nanocellulose-drug composites. We use nanocellulose functionalization, such as carboxylation and sulfonation, to design composites containing nanocellulose and crystallized pharmaceutical products, with an aim towards using the molecular interactions between nanocellulose and drugs to direct formation of desired polymorphs and crystal sizes. The high surface area of the nanocellulose provides a high concentration of interaction sites and the small size of the nanocellulose excipient particles has interesting performance in promoting assembly and packing of the composite in drug products such as compressed tablets and dense free standing films.

500635 A New Model for Solubility Prediction to Guide Solvent Selection for Process Development

Abstract:

Solvent selection to obtain adequate yield and impurity rejection is a critical activity in the development of crystallization processes for pharmaceutical molecules. This activity requires the determination of solubilities for a solute molecule (moreover a polymorph of a solute molecule) in a variety of different solvents and solvent combinations. This is referred to as solvent screening. There are roughly fifty different solvents commonly used in synthetic chemical manufacturing. Although experimentally determining the solubility of a single solute in each of these solvents is an approachable (be it impractical) problem, experimental screening becomes unapproachable for mixtures of two or more solvents. Conventionally, intuition has driven the reduction of the complete set of binary blends to a more tractable set of mixed solvent systems on a case-by-case basis. As such the majority of solvent combinations remaining unexplored for a given molecule. A predictive model was developed to better facilitate the exploration of single solvent systems and blends.

This model was developed starting from a mechanistic perspective for crystal growth and dissolution, which is most applicable at temperatures far from the melt. The model can be implemented based on either the regression of a limited set of solubility data and the subsequent prediction of solute characteristics (magnitude of polar, dispersive and hydrogen bonding interactions, size) or using molecular modelling to approximate these characteristics directly. The regression approach has proven to be efficient (requiring minutes) and accurate for single solvent systems. This model has been extended to mixtures of two solvents, and coupled with crystallization process design calculations to predict yield, and provide a rough process outline to the end user. The intention being that once a â€œgoodâ€ system has been found (determined using the model) it can be verified experimentally. This model has been applied to several systems (~20 solutes and 200 solute/solvent combinations) including the modelling of a solubility data set for paracetamol and caffeine collected within the Enabling Technologies Consortium.

Heterogeneous Nucleation of Small Molecule Crystals in the Presence of Nanoparticles

Lucrèce H. Nicoud, Leia M. Dwyer, Allan S. Myerson*

Chemical Engineering Department, Massachusetts Institute of Technology, Cambridge, MA, U. S. A.
*Corresponding author: myerson@mit.edu

Abstract
In this study, we evaluated the possibility to use nanoparticles as a means to obtain fundamental insights on the heterogeneous nucleation of organic molecules. To do so, we investigated the crystallization of diphenhydramine hydrochloride in ethanol in the presence of silica coated magnetic nanoparticles. We performed crystallization experiments at two different scales, namely in 1 mL vials and in 100 mL vessels. At both scales, crystallization was induced by two different ways, either by cooling or by anti-solvent (heptane) addition. It was found that the presence of nanoparticles promotes crystallization, in particular at low supersaturation. However, the high variability of induction times and the aggregation of the nanoparticles makes it difficult to extract quantitative information regarding the impact of the nanoparticles on the crystallization kinetics.

Introduction

Surfaces are ubiquitous in crystallizers, in particular due to the walls of the vessel, the impeller or the presence of dust. Surfaces generally lower the energy barrier to nucleation, so that heterogeneous nucleation commonly dominates over primary homogeneous nucleation in most laboratory or plant settings. However, the physical mechanisms underlying heterogeneous nucleation remain poorly understood. Because the early stages of solution crystallization play a decisive role in determining the final crystal properties, such as the crystal size, shape and polymorphism, a fundamental understanding of nucleation is a key requisite to improve control over crystallization processes [1, 2].

Experimental studies aiming at investigating crystallization in the presence of surfaces employed various types of templates including crystals [3], polymers [4] and self-assembled monolayers (SAMs) [5, 6]. The latter are highly ordered and oriented molecular assemblies that form spontaneously by the adsorption of a surfactant on a solid surface [7]. SAMs have been reported to influence crystal size [8], enantiopurity [9], morphology [10], and polymorphism [11].

In this work, we evaluate the possibility to use nanoparticles to investigate the mechanisms of heterogeneous nucleation. Nanoparticles seem particularly suitable for this purpose due to their large surface area and tunable properties. Nanoparticles are indeed available in a wide range of sizes (from few nanometers to few micrometers), shapes (most often spheres, but also cubes, rods and nanostars) and functionalities (mainly thanks to the large portfolio of thiols and silanes, which allow the surface functionalization of gold and silica nanoparticles, respectively). One may therefore envision performing studies with various types of well characterized nanoparticles in an attempt to understand how the uncontrolled presence of foreign objects such as dust may impact nucleation in regular crystallizers. Nanoparticles have been shown to promote the formation of protein crystals at the microliter scale [12, 13, 14], but their quantitative impact on the crystallization kinetics still needs to be elucidated.

In this context, we investigate the impact of silica coated magnetic nanoparticles on the crystallization behaviour of diphenhydramine hydrochloride (DPH), which is a commercial pharmaceutical ingredient. Silica coated magnetic nanoparticles were selected as they combine two main advantages: *(i)* the magnetic core allows performing magnetic decantation, which is very practical for several preparation steps (e.g. to wash, concentrate) and for a potential reuse after crystallization, *(ii)* the silica shell allows functionalizing the surface with a wide variety of silanes. We start our study with experiments performed at the milliliter scale and then increase the crystallization volume to 100 mL. At both scales we induce crystallization by two different ways, namely anti-solvent (heptane) addition and cooling.

1. Materials and Methods

1.1 Materials

1.1.1 Suppliers

DPH was purchased from BeanTown Chemical at a purity $\geq 99\%$. Anhydrous heptane 99% was purchased from Sigma-Aldrich and 200 proof ethanol from Deacon Labs. Silica coated magnetic nanoparticles with a total diameter of 200 nm were purchased from Spherotech Inc (catalog number SIM-025-10H). N-decyltriethoxysilane was purchased from Gelest (catalog number SID2665.0) to perform functionalization of the silica surface of the nanoparticles with decane.

1.1.2 DPH solution preparation

DPH solutions were prepared by dissolving DPH in ethanol at 35 °C under magnetic stirring. The solutions were then filtered with a PTFE syringe filter prior to performing crystallization experiments. Filters with a 0.2 μm and 0.45 μm cut-off were used for crystallization studies at the small scale (1 mL) and large scale (100 mL), respectively. DPH solutions were prepared at a concentration of 315.8 g/L^{EtOH} so as to reach 300 g/L^{EtOH} after addition of the nanoparticle solution (95 % DPH solution and 5 % nanoparticle solution in volume). For control experiments performed in the absence of nanoparticles, the corresponding quantity of pure ethanol was added to reach the same final DPH concentration. In the following, the nanoparticle concentrations are given in grams of nanoparticles per liter of DPH solution (i.e. DPH dissolved in ethanol before a potential addition of heptane). The units are abbreviated as g/L^{DPH} for the sake of simplicity.

1.1.3 Nanoparticles solution preparation

Before usage, the nanoparticles (Nps) were washed, first with water and then with ethanol. To do so, a given volume of nanoparticle solution was placed in a tared vial on top of a magnet. After magnetic decantation, the solvent was removed and a volume of water equal to twice the initial volume of the nanoparticle solution was placed on top of the nanoparticles. The nanoparticles were then redispersed by sonicating the solution for 2 min (Branson 3510 ultrasonicator bath). This wash procedure was then repeated with ethanol. After magnetic decantation and solvent removal, the last traces of ethanol were eliminated by drying the nanoparticles in a vacuum oven at around 60 °C. The vial was then weighted to determine the mass of nanoparticles and an appropriate volume of ethanol was added to reach a target nanoparticle concentration. The nanoparticles were redispered by placing the vial in the bath sonicator for 10 min.

It was checked with dynamic light scattering (DLS) that the nanoparticles are properly dispersed after sonication. Measurements were performed with a DynaPro NanoStar instrument (Wyatt Technology) at a 90° angle using a laser whose wavelength was 658 nm.

1.2 Nanoparticles functionalization

Before functionalization, the silica coated magnetic nanoparticles were washed as described in the previous section. A 5 g/L solution of nanoparticles in a 95/5 % v/v ethanol/water mixture was prepared. The water was acidified with acetic acid at a concentration of 10^{-2} M. The silane was added to the solution to reach a 2 % v/v concentration. The solution was sonicated in the bath sonicator during 2 min. The functionalization reaction was then performed at 25 °C under shaking at 500 rpm during 24 h. The temperature was subsequently increased to 70 °C during 2 h. The solvent was removed by magnetic decantation and the nanoparticles were dried in the vacuum oven at 90 °C during 4 h. Finally, the nanoparticles were washed twice with ethanol using sonication to redisperse the nanoparticles at each wash step.

The extend of functionalization was assessed by Fourier transform infrared (FTIR) spectroscopy (IdentifyIR Smiths). To do so, few microliters of a concentrated nanoparticle suspension were deposited on the sample holder and 256 scans were recorded after ethanol evaporation.

1.3 Solubility

Saturated solutions of DPH were prepared by magnetically stirring an excess of DPH crystals in contact with a selected solvent (either pure ethanol or a mixture of ethanol and heptane) during 24 h. The temperature was maintained at a set value thanks to a water bath. After 24 h, the solution was filtered with a 0.2 μm PTFE syringe filter and the filtrate was placed in tared vials. The vials were weighted immediately after introducing the filtrate and after evaporation of the solvent. The solvent was first evaporated under the hood and the last traces were removed by drying the crystals in a vacuum oven at around 60 °C. Once the mass of the samples was constant over time, the solubility was computed as the mass of crystals divided by the mass of solvent that had evaporated.

1.4 Crystallization likelihood (1 mL)

The tendency of a solution to crystallize was assessed by preparing at least 5 samples per investigated condition and visually inspecting after 24 h whether crystals formed. To do so, 1 mL samples were placed in glass vials (ThermoScientific, C4000-1W) incubated in a thermoshaker (MultiTherm, heat/cool, Benchmark Scientific) at a selected temperature and under 500 rpm shaking. For anti-solvent crystallization, the temperature was set to 25 °C. The formed crystals were observed with an optical microscope (Nikon Eclipse ME 600).

1.5 Crystallization kinetics (100 mL)

Batch crystallization kinetic experiments were performed using an EasyMax 102 worksation (Mettler Toledo). The solution was stirred at 250 rpm with a pitched-blade impeller (Alloy-C22, 38 mm diameter, downwards). The total solution volume was 100 mL. For anti-solvent crystallization, the reactor temperature was maintained at 25 °C and heptane was poured at once on top of the DPH solution to reach a heptane concentration of 33 % v/v^{tot}. For cooling crystallization, the reactor temperature was decreased from 25 °C to 15 °C at a rate of 1 °C/min.

The crystallization kinetics was monitored by focused beam reflectance measurements (FBRM) using a lasentec probe (S400Q).

In few experiments, particle vision measurements (PVM) were performed to visually observe the nanoparticle aggregates and the crystals (ParticleView V19, Mettler Toledo).

2. Results and Discussion

2.1 Choice of nanoparticles

The core-shell nanoparticles selected for this study combine the advantages of a practical handling and of a possible functionalization thanks to the magnetic core and to the silica surface, respectively. Indeed, the magnetic behaviour of the nanoparticles is very convenient for all the preparation steps such as washing, concentration and surface functionalization. The silica shell instead prevents the nanoparticles from extensive aggregation and allows the grafting of a wide variety of silanes, which then confer various surface properties to the nanoparticles.

A transmission electron microscopy picture of the nanoparticles is shown in Figure 1, where it is seen that the magnetic core is around 160 nm and the silica shell approximately 20 nm thick. The total diameter of 200 nm was confirmed by DLS analysis.

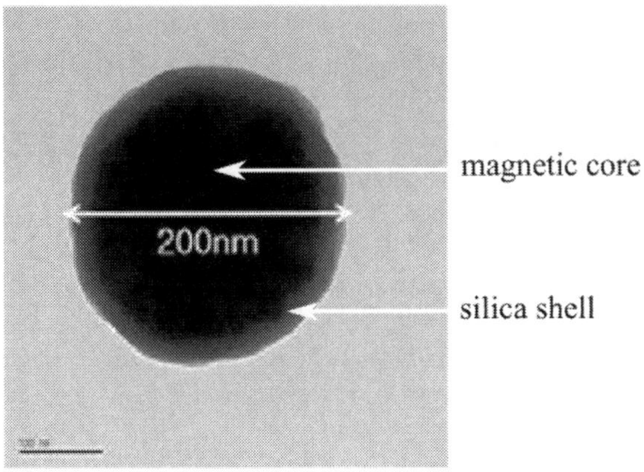

Figure 1. Transmission electron microscopy of the silica coated magnetic nanoparticles. Provided by Spherotech Inc.

To give an order of magnitude, a 1 g/L solution of such nanoparticles corresponds to a concentration of 6×10^{13} nanoparticles per liter and to a volume fraction of 0.03 %. Therefore, even a quite dilute solution of nanoparticles provides a large number of foreign objects, which are as many potential nucleation sites.

2.2 Nanoparticle functionalization

The silica shell of the nanoparticles allows functionalizing the surface with various silanes. As an example, the FTIR spectrum of nanoparticles functionalized with n-decyltriethoxysilane, which provides a functionalization with decane molecules, is shown in Figure 2. As a comparison, the spectra of the non-functionalized (bare) nanoparticles and of the silane are also shown. In contrast to the bare nanoparticles, the functionalized nanoparticles exhibit three small additional peaks in the range from 2800 to 3000 cm^{-1}. These peaks coincide with those of the silane, thus suggesting that the functionalization reaction was successful. The low intensity of the peaks corresponding to the grafted decane as compared to the pure silane can be attributed to the small mass of the surface groups with respect to the bulk of the nanoparticles.

2.3 Solubility

Figures 3(a) and (b) show experimental data of DPH solubility as a function of the heptane volumetric fraction and temperature, respectively. The solid line in Figure 3(a) represents the dilution line upon heptane addition starting from a DPH solution at 300 $mg/mL^{EtOH} = 382.2\ mg/g^{EtOH}$. It is observed that crystallization is expected to occur from a thermodynamic point of view at heptane concentrations above approximately 10 % in volume. In the case of cooling, the DPH concentration is kept constant, as indicated by the flat solid line in Figure 3(b). It is observed that crystallization is expected to occur at temperatures roughly below 23 °C.

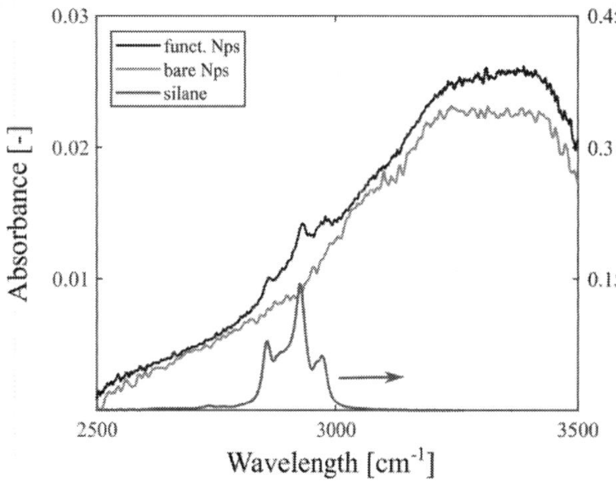

Figure 2. FTIR spectrum of nanoparticles functionalized with decane as compared to the spectra of the bare nanoparticles and of the corresponding silane (whose intensity is read on the right axis).

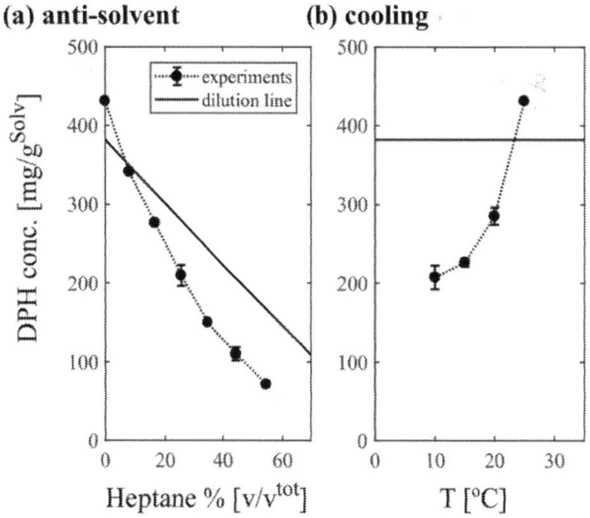

Figure 3. DPH solubility as a function of (a) heptane volumetric fraction and (b) temperature. Symbols correspond to experimental data obtained by gravimetry, whereas solid lines indicate the operating conditions, i.e. a DPH solution at 300 $mg/mL^{EtOH} = 382.2\ mg/g^{EtOH}$ diluted with heptane or cooled down.

2.4 Crystallization likelihood (1 mL)

2.4.1 Non-functionalized nanoparticles

In a first set of experiments, the impact of the nanoparticles (Nps) on the crystallization behaviour of DPH was investigated at the milliliter scale in glass vials. For each condition under investigation, at least five samples were prepared and visually inspected after 24 h of shaking at a target temperature to determine whether crystals formed during the experiment time frame. The results for anti-solvent crystallization and cooling crystallization are shown in Figure 4(a) and (b) respectively.

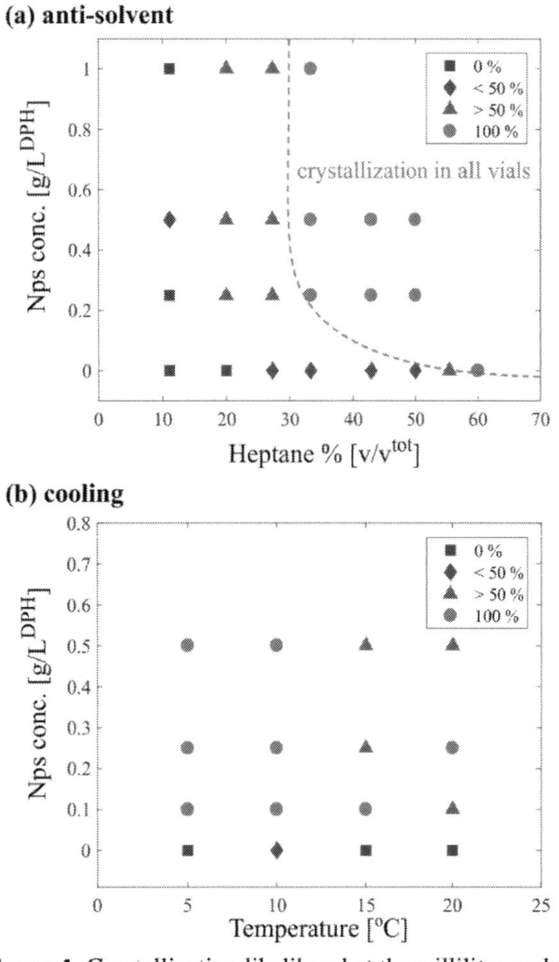

Figure 4. Crystallization likelihood at the milliliter scale in (a) anti-solvent crystallization and (b) cooling crystallization. For each condition, at least five samples were prepared and visually inspected after 24 h of shaking to determine whether crystals formed. The percentages in the caption indicate the fraction of samples that crystallized. In the case of anti-solvent crystallization, the temperature was set to 25 °C. Non-functionalized nanoparticles were used.

Let us first consider the samples without the nanoparticles, i.e. the row of symbols at the bottom of both diagrams. According to the solubility data shown in Figure 3, the selected conditions of heptane concentration and temperature were expected to induce crystallization in all samples. However, it is observed in Figure 4 that most of the samples did not crystallize within 24 h. For example, none of the samples whose heptane concentration was below 20 % crystallized (blue squares). At heptane concentrations between 25 and 55 %, only some of the samples crystallized (purple symbols) and it was necessary to reach 60 % heptane so as to observe crystallization in all the vials (red circles). To verify that there was no mistake in sample preparation, a seed was added in the samples that did not crystallize within 24 h. This resulted in rapid crystallization, thus confirming that all the solutions were supersaturated. This means that, although all the samples were expected to crystallize from a thermodynamic point of view, a significant kinetic barrier prevented crystallization to occur during the time frame of the experiment.

Interestingly, it is observed in Figure 4 that the addition of non-functionalized nanoparticles resulted in the crystallization of many more samples as compared to the situation without nanoparticles. For instance, all the samples at 33 % heptane crystallized in the presence of nanoparticles, whereas less than half crystallized in the absence of nanoparticles. Similarly, none of the samples at 15 °C crystallized in the absence of nanoparticles whereas the vast majority of the samples with the nanoparticles crystallized.

Overall, these data show that nanoparticles promote crystal formation, in particular at low supersaturation values, i.e. in the metastable zone.

2.4.2 Functionalized nanoparticles

Following the idea of a previous study [15], crystallization experiments were performed with nanoparticles functionalized with decane, which is an anti-solvent for DPH. As a reference, the addition of 50 % decane in volume reduces the solubility of DPH from 432 to 137 mg/g^{Solv}. The functionalization of the silica coated magnetic nanoparticles with decane was discussed in Section 2.2.

According to the original work, the addition of nanoparticles functionalized with anti-solvent molecules was able to induce the crystallization of slightly under-saturated solutions of sodium chloride, D-mannitol and fenofibrate. Here, we investigated whether the addition of decane functionalized nanoparticles could induce the crystallization of DPH solutions at 25 °C with heptane concentrations varying from 0 to 10 % in volume. According to the solubility data shown in Figure 3(a), such conditions lead from slightly under-saturated up to saturated solutions. Several nanoparticle concentrations ranging from 0.25 g/L^{DPH} to 1 g/L^{DPH} were tested and several repetitions were performed for each condition. However, none of the samples crystallized within 24 h.

After observing that decane functionalized nanoparticles were not able to induce the crystallization of DPH solutions in the range 0-10 % heptane, larger heptane volume fractions were investigated. As in the case of the bare nanoparticles, crystallization was observed in all the samples whose heptane concentration was above 33 %. Nevertheless, one noteworthy difference was observed in terms of crystal color. Figures 5(a) and (b) show pictures of the crystals formed with non-functionalized nanoparticles and with decane functionalized nanoparticles, respectively. The top pictures were taken with a regular camera, whereas the bottom pictures were taken with an optical microscope. In both samples, the heptane concentration was 43 % v/v^{tot} and the nanoparticle concentration 0.25 g/L^{DPH}. Due to the red/brown colour of the nanoparticles, it is possible to visually assess the entrapment of nanoparticles in the crystal lattice. It is seen in Figure 5 that the crystals formed in the presence of the bare nanoparticles are darker than those formed in the presence of the nanoparticles functionalized with decane. This might suggest that the surface functionalization with decane reduced nanoparticle entrapment.

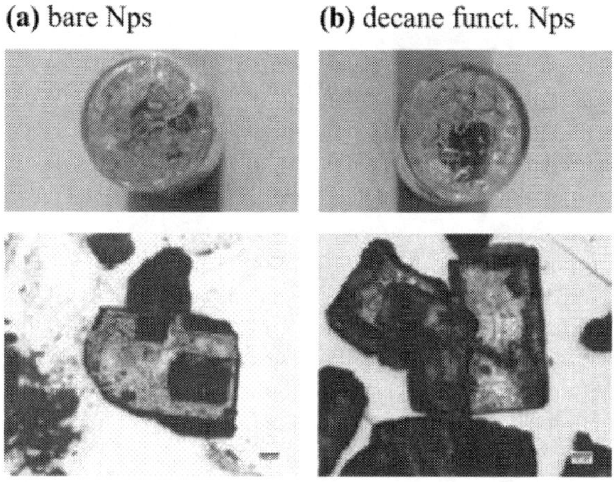

Figure 5. Pictures of DPH crystals formed in the presence of (a) bare and (b) decane functionalized nanoparticles. The heptane concentration was 43 % v/v^{tot} and the nanoparticle concentration 0.25 g/L^{DPH}. After crystallization, the crystals were washed with heptane. The top pictures were taken with a regular camera, whereas the bottom pictures were taken with an optical microscope.

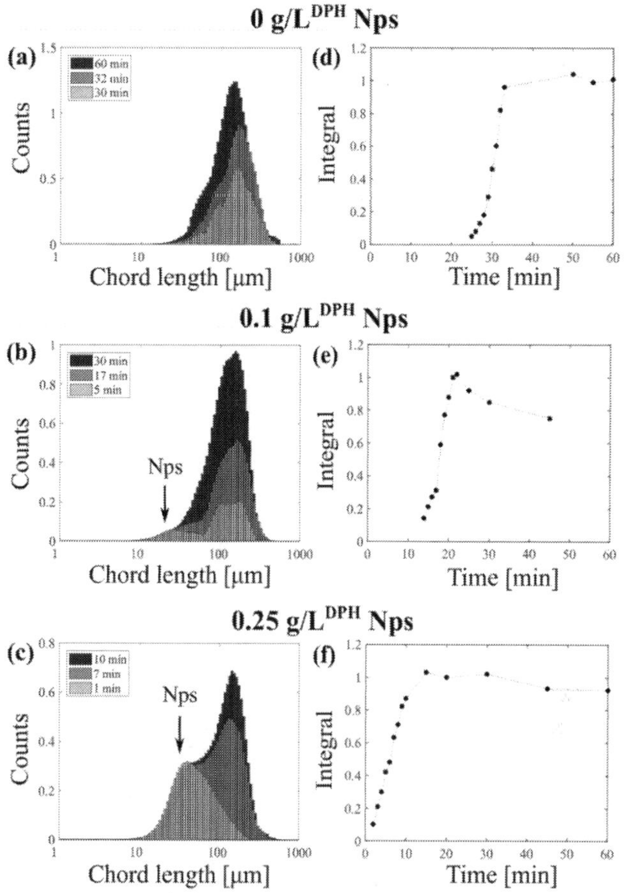

Figure 6. (a-c) Cube weighted chord length distributions monitored by FBRM at different time points and nanoparticles concentrations, as indicated in the legend. (d-f) Integral of the chord length distribution as a function of time for the same three nanoparticle concentrations. The peak corresponding to the nanoparticles (centered at around 20 μm) was substracted before computing the integral and the curves were normalized between 0 and 1. Bare nanoparticles were used.

2.5 Crystallization kinetics (100 mL)

After investigating the crystallization likelihood at the milliliter scale, batch crystallization kinetics were performed in a 100 mL stirred vessel equipped with a FBRM probe. For anti-solvent crystallization, a heptane concentration of 33 % v/v^{tot} was selected, whereas for cooling crystallization a temperature of 15 °C was chosen.

Figure 6(a) shows cube weighted chord length distributions recorded in the absence of nanoparticles at three different time points during anti-solvent crystallization. It is observed that the crystal size is in the order of 100 μm. The number of crystals increases with time, but their size remains relatively constant. This indicates that crystal growth is significantly faster than crystal nucleation.

Figures 6(b) and (c) show the chord length distributions at three different time points in the presence of 0.1 g/L^{DPH} and 0.25 g/L^{DPH} bare silica coated magnetic nanoparticles, respectively. As compared to the situation without the nanoparticles, an additional peak centered at around 20 μm is observed. This peak does not correspond to crystals but is due to the presence of nanoparticles. This was evidenced by an experiment performed below the saturation (25 °C, no heptane), during which a similar peak was observed, as further discussed in Section 2.6.

In order to evaluate the mass of crystals formed as a function of time, the peak corresponding to the nanoparticles was substracted and the integral of the cube weighted chord length distribution was computed. The results are presented in Figures 6(d), (e) and (f) for 0, 0.1 and 0.25 g/L^{DPH} nanoparticles, respectively. In the absence of nanoparticles, a lag time of approximately 25 min is observed before crystallization starts. Then, the mass of crystals increases sharply until it reaches a plateau. In few experiments, a slight decrease in the number of counts was observed at large reaction times (see Figures 6(e)). This might be due to crystals intermittently sticked to the FBRM probe falling off the probe window.

Several repetitions of these anti-solvent experiments were performed and the results are reported in Figure 7(a) in terms of lag times. It is seen that the average lag time decreases with the concentration of nanoparticles.

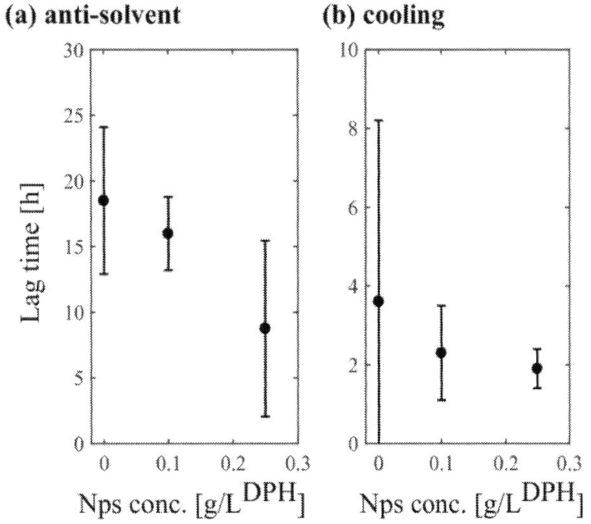

Figure 7. Lag time as a function of the concentration of nanoparticles in the case of (a) anti-solvent and (b) cooling crystallization. Non-functionalized nanoparticles were used.

Similar studies were carried out in the case of cooling crystallization. Chord length distributions comparable to those obtained during anti-solvent crystallization were obtained. The lag time is shown as a function of the nanoparticle concentration in Figure 7. Once more, it is seen that the average lag time decreases upon nanoparticle addition. However, the poor reproducibility makes it difficult to properly quantify the impact of the nanoparticles.

2.6 Nanoparticle aggregation

In this section, we investigate in more details the aggregation behaviour of the nanoparticles. Figure 8(a) shows the total number of counts (circles) and the average size (squares) of the cube weighted distribution measured with FBRM in an experiment performed in the presence of 0.25 g/L^{DPH} of bare nanoparticles below the supersaturation, i.e. where no crystals can form. In this experiment, the temperature was maintained at 25 °C and no heptane was added. It is seen that both the number of counts and the average size increase during the first two minutes and then become almost constant with time. Note that an average size close to 45 µm is reached, which is more than two order of magnitude larger than the initial nanoparticle diameter (200 nm as measured by DLS).

The nanoparticle aggregates were visualized by PVM and a representative picture recorded after 8 min is shown in Figure 8(c). It is seen that particulates in the order of few tens of microns are observed, in agreement with the FBRM data shown in Figure 8(a).

After performing particle vision measurements, the temperature was decreased to 15 °C so as to induce crystal formation. Figure 8(d) shows a representative picture recorded after 3 h of cooling, where both nanoparticle aggregates and crystals are observed. The crystals appear much larger than the nanoparticle aggregates, in line with the bimodal distribution shown by the FBRM data in Figure 6.

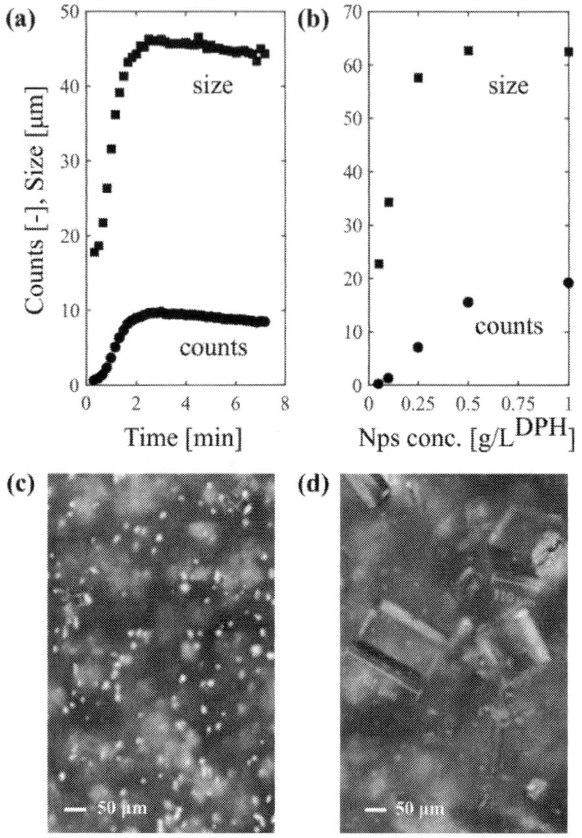

Figure 8. (a) Total number of counts (circles) and average size (squares) of the cube weighted distribution measured with FBRM as a function of time in a 300 g/L^{EtOH} DPH solution at 25 °C in the presence of 0.25 g/L^{DPH} of bare nanoparticles. (b) Total number of counts (circles) and average size (squares) of the cube weighted distribution measured with FBRM at the early stages of an anti-solvent crystallization experiment. (c) PVM images recorded after 8 min of the experiment reported in panel (a). (d) PVM images recorded after 3 h of cooling at 15 °C following the experiment reported in panel (a).

The impact of the concentration of nanoparticles on their aggregation behaviour was studied under supersaturated conditions in the case of anti-solvent crystallization. It was already observed in the context of Figure 6 that an increase in the nanoparticle concentration from 0.1 to 0.25 g/L^{DPH} leads to an increase in the height of the peak corresponding to the nanoparticles. Figure 8(b) shows how the total number of counts and the average size related to the peak of the nanoparticles increase as a function of the nanoparticle concentration. A sharp rise in the size of the nanoparticle aggreates is observed at nanoparticle concentrations below

0.25 g/L^{DPH}. Above that value, the nanoparticle aggregates reach an average size of around 60 μm.

Nanoparticle aggregation represents a major issue in this study. Indeed, aggregation reduces both the the surface area and the number of particles in solution. Assuming that the aggregates are compact spheres with a uniform size distribution, the surface area and the number of particles are inversely proportional to the aggregate diameter and to the cube of their diameter, respectively. Considering that in the present case aggregates are approximately 100 times larger than the primary particles, aggregation roughly reduces the surface area by a factor 100 and the number of particles by 10^6.

Besides reducing the surface area and the number of particles, aggregation also complicates the interpretation of experimental data. Indeed, aggregates are non regular structures, which are difficult to characterize with accuracy in terms of shape and size distribution. In addition, aggregation is affected by a number of parameters, such as time and nanoparticle concentration as shown in Figure 8, but also potentially temperature, solvent type, solute concentration, nanoparticle functionalization, etc. It is therefore difficult to rationalize the quantitative impact of nanoparticles on the crystallization kinetics monitored under various experimental conditions. For instance, the observation that the aggregate size increases with nanoparticle concentration prevents a detailed analysis of the impact of the nanoparticles concentration on the kinetics of crystallization. Similarly, studies aiming at investigating different surface functionalization would require the decoupling between the impact of surface functionalization on the aggregation of nanoparticles and on the kinetics of crystallization.

3. Conclusion

In this study, we assessed the possibility to use nanoparticles to gain fundamental insights into heterogeneous nucleation mechanisms. This was motivated by the fact that nanoparticles offer a large surface area and are tunable in terms of various properties, such as size, shape and surface chemistry.

At the milliliter scale, it was shown that silica coated magnetic nanoparticles promote the crystallization of diphenhydramine, which was induced either by heptane addition or cooling. Then, the crystallization kinetics was monitored with FBRM in 100 mL stirred reaction vessels and it was found that the presence of nanoparticles reduces the lag time before crystallization occurs. However, the effect of the nanoparticles seems less pronounced than at the milliliter scale and the large variability in the lag time measurements makes it difficult to draw solid conclusions.

Finally, it was shown that the nanoparticles significantly aggregate under the considered experimental conditions, which reduces the surface area and the number of particles in solution. This complicates the interpretation of experimental results and hinders the quantitative analysis of the impact of the nanoparticles on the crystallization kinetics, at least for the system under investigation.

Acknowledgments

Financial support from the Swiss National Foundation is gratefully acknowledged (grant P2EZP2_168909). The support of DARPA for a portion of this project is gratefully acknowledged.

References

[1] A. S. Myerson. *Hanbook of industrial crystallization, 2nd Edition*. Butterworth-Heinemann, 2002.

[2] P. G. Vekilov. Nucleation. *Cryst Growth Des*, 10(12):5007–5019, 2010.

[3] M. D. Ward. Bulk crystals to surfaces: Combining x-ray diffraction and atomic force microscopy to probe the structure and formation of crystal interfaces. *Chem Rev*, 101(6):1697–1725, 2001.

[4] C. P. Price, Ad. L. Grzesiak, and A. J. Matzger. Crystalline polymorph selection and discovery with polymer heteronuclei. *J Am Chem Soc*, 127(15):5512–5517, 2005.

[5] A. Singh, I. S. Lee, K. Kim, and A. S. Myerson. Crystal growth on self-assembled monolayers. *CrystEngComm*, 13(1):24–32, 2011.

[6] J. Aizenberg, A. J. Black, and G. M. Whitesides. Control of crystal nucleation by patterned self-assembled monolayers. *Nature*, 398(6727):495–498, 1999.

[7] J. C. Love, L. A. Estroff, J. K. Kriebel, R. G. Nuzzo, and G. M. Whitesides. Self-assembled monolayers of thiolates on metals as a form of nanotechnology. *Chem Rev*, 105(4):1103–1170, 2005.

[8] A. Y. Lee, I. S. Lee, S. S. Dette, J. Boerner, and A. S. Myerson. Crystallization on confined engineered surfaces: A method to control crystal size and generate different polymorphs. *J Am Chem Soc*, 127(43):14982–14983, 2005.

[9] J. Chen and A. S. Myerson. Pasteur revisited: Chiral separation by crystallization on self-assembled monolayers. *CrystEngComm*, 14(24):8326–8329, 2012.

[10] A. Y. Lee, A. Ulman, and A. S. Myerson. Crystallization of amino acids on self-assembled monolayers of rigid thiols on gold. *Langmuir*, 18(15):5886–5898, 2002.

[11] R. Hiremath, J. A. Basile, S. W. Varney, and J. A. Swift. Controlling molecular crystal polymorphism with self-assembled monolayer templates. *J Am Chem Soc*, 127(51):18321–18327, 2005.

[12] Y. Takeda and F. Mafuné. Induction of protein crystallization by platinum nanoparticles. *Chem Phys Lett*, 647(Supplement C):181–184, 2016.

[13] S. Ko, H. Y. Kim, I. Choi, and J. Choe. Gold nanoparticles as nucleation-inducing reagents for protein crystallization. *Cryst Growth Des*, 17(2):497–503, 2017.

[14] F. Hodzhaoglu, F. Kurniawan, V. Mirsky, and C. Nanev. Gold nanoparticles induce protein crystallization. *Cryst Res Technol*, 43(6):588–593, 2008.

[15] S. A. Kulkarni and A. S. Myerson. Reversible control of solubility using functionalized nanoparticles. *Chem Comm*, 53(8):1429–1432, 2017.

487832 The Influence of Solution Conditions on the Self-Assembly of Pre-Nucleation Clusters

Abstract:

The concept of pre-nucleation clusters has been established to, successfully, explain phenomena, in nucleation kinetics, that they were inconsistent with the classical nucleation theory.[1] The key idea is that pre-nucleation clusters are formed providing a comfortable niche for nuclei to be formed.[2] Phenomenological theories exist describing this two-step nucleation process.[3] However, there is a lack of qualitative understanding of the fundamental phenomena underpinning the formation of pre-nucleation clusters both in a supersaturated and an understaturated environment.

Using the models proposed in more recent works is being shown that bigger pre-nucleation clusters provides an environment giving more chances of survival to the nucleus inside it. These findings were tested experimentally using a number of industrially relevant molecules (arginine, glycine, glutamic acid). These molecules are particularly attractive as they have relatively large nucleation times. In addition the formation of pre-nucleation clusters as nucleation precursors for them has been reported in literature. Using DLS it was shown that in aqueous systems, the size of the pre-nucleation clusters scales with the supersaturation. These findings are in accordance with the findings proposed from previous works. Upon filtration, both in a supersaturated and an undersaturated regime, the pre-nucleation clusters recover their initial size in a few hours, in the absence of any agitation.

Experiments were conducted in systems containing water and ethanol to investigate the influence of solutionâ€™s chemical potential on the size of the pre-nucleation clusters. It is shown that, in both the undersaturated and the supersaturated regime, the formation of the pre-nucleation clusters follows the fundamentals of self-assembly in solution.[4] The findings suggest that the influence of the solvent can overcome that of supersaturation, for example an aqueous system containing arginine at Sâ‰ˆ0.5 (0.8 g of L-Arginine) gives bigger pre-nucleation clusters than a solution containing 30 % v/v ethanol at Sâ‰ˆ0.9.

The findings of this work verify the majority of the findings reported in literature. However, it reports different kinetics of recovery of the pre-nucleation clusters size upon filtration. It highlights the applicability of the fundamentals of self-assembly in solution. The findings suggest that in depth understanding of the self-assembly phenomena are required for the development of efficient strategies for the control of crystallization kinetics via stirring[5] and for the formulation of amorphous solid dispersions.

References

1. P.G. Vekilov. Nucleation. *Crystal Growth & Design*. **2010**, 10, 5007.
2. B. Peters. On the coupling between slow diffusion transport and barrier crossing in nucleation. *The Journal of Chemical Physics*. **2011**, 135, 044107.
3. W. Pan, A.B. Kolomeisky, P.G. Vekilov. Nucleation of ordered solid phases of proteins via a disordered high-density state: Phenomenological approach. *The Journal of Chemical Physics*. **2005**, 122, 174905.
4. J.N. Israelachvili. Intermolecular and Surface Forces. 3rd ed. Oxford, Academic Press; **2011**.
5. A. Jawor-Baczynska, B.D. Moore, H.S. Lee, A.V. McCormick, J. Sefcick. Population and size distribution of solute-rich mesospecies within mesostructured aqueous amino acid solutions. *Faraday Discussions*, **2013**, 167, 425.]
6. M. Diwan, H. Li, F. Nordstrom. Primary Nucleation Under High Shear Conditions. *AICHE Conference Proceedings*, **2015**

490516 Simulations and Experiments Validation on the Membrane Based Response Technology for Nucleation Detection and Crystallization Control

Abstract:

Simulations and Experiments validation on the Membrane based Response Technology for Nucleation Detection and Crystallization Control

Xiaobin Jiang, Wu Xiao, Dapeng Lu, Rui Zhao, Gaohong He*

State Key Laboratory of Fine Chemicals, School of Chemical Engineering, Petrochemical Energy-efficient Separation Technology Engineering Lab of Liaoning Province, Dalian University of Technology, Dalian, P. R. China

*Corresponding author: Email: hgaohong@dlut.edu.cn

Abstract

The mass transfer flux decline and the crystallization on the membrane interface is often considered as a fouling behavior [1,2]. For now, non-attention had been paid on applying this sensitive nucleation response phenomenon for the crystal nucleus detection. While, the inevitable nucleation on the membrane interface under certain operational conditions inspired us that the membrane interface could perform as the response medium of the nucleation detection. The responding mechanism is constructed base on the chain effect of ¡®nucleation on the pores of the membrane interface-transmembrane flux sharply decrease¡¯. The inflection point of the transmembrane flux confirms the nucleation timing on the membrane interface. The accurate recording of the transmembrane flux data can be utilized to calculate the solution concentration to determine the metastable zone width (MSZW), which is key data to determine the crystallizaiton operative window [3,4].

Obviously, without relying on the optical signal, membrane based nucleation response method would be easy and more meaningful in the case of opaque solutions that laser intensity response method is invalid [5]. This advantage will improve the controlling accuracy and flexibility during MSZW measurement. Moreover, with the proposed method, it is easy to enhance control over the process. The real operative conditions during crystallizaiton can be achieved in this membrane based technology. Detailed simulations and experiments on the typical binary aqueous solution systems were implemented to represent the role of T_f, T_p and A/V_0 on the supersaturation degree adjustment.

In addition, it also reported that the polymer interface with nanometer scale pores had significant impact of on inducing crystallization and polymorphism control [6,7]. The foundational work can promote the theoretical research and application of the MDC based nucleation response and crystallization control by providing the qualified microporous membrane as desired nucleation detection and response interface.

Acknowledgment

We acknowledge fnancial contribution from from National Natural Science Foundation of China (Grant No. 21527812, 21676043, U1663223), Changjiang Scholars Program (T2012049), the Fundamental Research Funds for the Central Universities (DUT16TD19) and Education Department of the Liaoning Province of China (No. LT2015007).

References

[1] Meng, S.; Ye, Y.; Mansouri, J.; Chen, V., Fouling and crystallisation behaviour of superhydrophobic nano-composite PVDF membranes in direct contact membrane distillation. *Journal of Membrane Science* **2014,** 463, 102-112.

[2] Chen, G.; Lu, Y.; Krantz, W. B.; Wang, R.; Fane, A. G., Optimization of operating conditions for a continuous membrane distillation crystallization process with zero salty water discharge. *Journal of Membrane Science* **2014,** 450, 1-11.

[3] Ulrich, J.; Jones, M. J., Industrial Crystallization. *Chemical Engineering Research and Design* **2004,** 82, (12), 1567-1570.

[4] Ulrich, J.; Strege, C., Some aspects of the importance of metastable zone width and nucleation in industrial crystallizers. *Journal of Crystal Growth* **2002,** 237, 2130-2135.

[5] Jiang, X.; Ruan, X.; Xiao, W.; Lu, D.; He, G., A novel membrane distillation response technology for nucleation detection, metastable zone width measurement and analysis. *Chemical Engineering Science* **2015,** 134, 671-680.

[6] Curcio, E.; L¨®pez-Mej¨ªas, V.; Di Profio, G.; Fontananova, E.; Drioli, E.; Trout, B. L.; Myerson, A. S., Regulating Nucleation Kinetics through Molecular Interactions at the Polymer¨CSolute Interface. *Crystal Growth & Design* **2014,** 14, (2), 678-686.

[7] Diao, Y.; Helgeson, M. E.; Siam, Z. A.; Doyle, P. S.; Myerson, A. S.; Hatton, T. A.; Trout, B. L., Nucleation under Soft Confinement: Role of Polymer¨CSolute Interactions. *Crystal Growth & Design* **2012,** 12, (1), 508-517.

500706 Immunomodulatory Peptide Amphiphile Micelles for Prophylactic Vaccination

Abstract:

Peptide amphilphiles (PAs) are a class of di-block materials comprised of hydrophilic peptide head(s) and hydrophobic tail(s) which self-assemble into peptide amphiphile micelles (PAMs) in water. PAMs have shown their great ability in biomedical applications. More recently, PAMs are studied for the application of prophylactic vaccines.

This study demonstrates that micelle vaccinesâ€™ physical properties such as size, shape, and surface charge are controllable by chemical modification. These different physical properties have shown different abilities to interact with biological systems. For instance, these interactions include different cell uptake ability by Antigen Presenting Cells(APCs), immunogenicity, and cytotoxicity. These discoveries will help the future design of prophylactic vaccines.

Vaccine adjuvants are vaccine agonists that can help the body produce protection. Molecular adjuvants are a type of adjuvants that can stimulate Toll Like Receptors (TLRs) on APCs. Studies have shown that co-delivery of antigen and adjuvants can produce desirable immune responses, so numerous research has been done to associate antigen and molecular adjuvant utilizing engineering methods. This study compares two co-delivery strategies of the model ovalbumin antigen peptide and the TLR-2 agonist Pam_2C, a lipid based adjuvant, either chemically linking with Pam_2C and antigen to form micelles (strong association force), or entrap Pam_2C in antigen micelles (weaker association force). Surprisingly, the stronger chemical association between antigen and Pam_2C yielded lower immunogenicity both *in vitro* and *in vivo*. This may lead to more extensive studies on the topic of the co-delivery of antigen and adjuvant for vaccine research.

501180 Aptamer-Peptide-Drug Conjugates: Delivery of Precise Synergistic Drug Ratios for Enhanced Cancer Selectivity

Abstract:

Multi-drug regimens, a widely used treatment method for several cancers, are synergistic only at certain ratios and tumor drug accumulation at these precise ratios is necessary for dose reduction. More importantly, it has been shown that, drug ratios can be selected by empirical testing to induce synergy specifically on the cancer cell line of interest while having an antagonistic effect on non-cancerous cell lines. By employing this strategy, current limitation with combination chemotherapy, of administering highly toxic doses, can potentially be remedied. However, distinct transport and distribution properties of chemically diverse small molecule drugs makes it a non-trivial challenge to deliver them at the desired ratios to tumors. We have developed a novel Aptamer-Peptide Drug conjugate (APDC) platform, in order to deliver drug combinations in precise ratios specifically to cancer cells while avoiding healthy cells. This platform enables the unification of the biodistribution and pharmacokinetic properties of drug combinations and simultaneously allows delivery of an exact synergistic molar drug ratio in a highly preferential fashion to cancer cells via aptamer targeting. Aptamers offer a wide range of clinical advantages over targeting antibodies such as non-immunogenicity, higher tissue penetration and potential to target a wider range of cancer markers. Additionally, superior thermal and chemical stability over antibodies and faster synthesis time makes their industrial scale production quite cost-effective.

To promote cancer specific uptake of our APDCs, we employ a well known aptamer AS1411 that targets overexpressed cell surface nucleolin on several types of cancer. Specifically, we chose a triple negative human metastatic breast cancer cell line, MDA-MB-231, that overexpresses cell surface nucleolin and has an intermediate response to chemotherapy. To make APDCs, chemotherapeutic drugs doxorubicin (DOX) and camptothecin (CPT), a previously employed combination for aggressive metastatic breast cancer, were loaded in a controlled covalent fashion to the aptamer using peptide scaffolds. Subsequent testing revealed that our APDCs are highly selective to nucleolin overexpressing breast cancer cells compared to a control epithelial cell line, MCF 10A. Systematic invitro evaluation of several drug combinations revealed a molar ratio dependent synergy, and drug ratios, 0.25 â‰¤ R â‰¤1 (R - CPT/DOX) were found to be significantly synergistic over others. Futhermore, certain molar ratios amongst these were found to be more effective at killing MDA-MB-231 cells compared to MCF-10A cells. Finally, we demonstrated that by delivering these synergistic molar ratios in a targeted fashion via APDCs, significant enhancement in toxicity exclusive to MDA-MB-231 cells can be obtained. These results reinforce the importance of delivering precise drug combination ratios in a targeted fashion to cancer cells so as to exploit the full potential of synergistic interactions of specific drug ratios in the tissue of interest.

488673 Heterotrifunctional Scaffolds for Measuring Intracellular Bond Cleavage Kinetics

Abstract:

Cleavable bonds have long been an attractive intracellular switch for nanoparticle and conjugate-based therapeutics. A variety of cleavable bonds have been designed to respond to intracellular stimuli such as pH gradients, reducing environments, and enzymatic activity. However, the rationale behind choosing an optimal cleavable chemistry has remained elusive. To establish design rules, we have sought to quantify the kinetics of sub-cellular bond cleavage in a cell type and endocytic pathway dependent manner. Towards this end, we have developed a novel, cleavable heterotrifunctional chemical cross-linker capable of incorporating targeting ligands and environmentally responsive bonds in a modular fashion. This sequence-defined, oligothioetheramide-based platform addresses many of the synthetic and functional limitations of traditional cleavable, heteromultifunctional chemical cross-linkers.

In a proof-of-principle study, we synthesized four heterotrifunctional oligoTEA cross-linkers with up to two orthogonal internal cleavage sites[1]. To demonstrate their utility as multifunctional probes, these linkers were conjugated to a pair of fluorophores that undergo Föster resonance energy transfer (FRET) and a model protein - human transferrin. Confocal microscopy of these transferrin probes showed differential fluorescence emission intensities, indicating disulfide bond reduction within the endocytic transferrin pathway. To quantify the kinetics of this cleavage event, a quenched fluorescence probe was analyzed via quantitative live-cell imaging in three different cell lines. The data presented herein will inform the design of a variety of drug delivery platforms fine-tuned for cell-specific and pathway-specific release of active drug cargo.

1. Sorkin, M. R., Walker, J. A., Brown, J. S. & Alabi, C. A. Versatile Platform for the Synthesis of Orthogonally Cleavable Heteromultifunctional Cross-Linkers. *Bioconjugate Chemistry* (2017). doi:10.1021/acs.bioconjchem.7b00033

500896 Evaluation of Enzyme-Loaded Nano-Polymersome Treatment on Neurodegenerative Disease

Abstract:

Delivery of therapeutics to the brain through non-invasive administration is a difficult task due to the blood-brain barrier (BBB), which prevents the transport of 98% of therapeutics. In GM1 gangliosidosis, patients are missing β-galactosidase (βgal), an enzyme necessary for cellular digestion, with major central nervous system (CNS) manifestation. GM1 gangliosidosis is fatal in infancy with no clinically available treatment. We are designing and characterizing the first nanoparticle-mediated treatment of GM1 gangliosidosis using self-assembled polymersomes due to their high physiological stability and tunable release for IV enzyme delivery. When coupled with apolipoprotein, delivery through the BBB and to the lysosome of neural cells will occur, treating patients without invasive surgery

Polymersomes formed using poly(ethylene glycol)-b- poly(lactic acid) (PEGPLA) can encapsulate, protect, and deliver the lysosomal enzyme, βgal, under low-pH conditions. PEGPLA polymersomes form via solvent injection with an average diameter of 145 ± 21 nm. PEGPLA polymersomes encapsulate βgal at 72.0 ± 12.2% efficiency and demonstrate simultaneous encapsulation and ligand attachment at 86.7 ± 11.6% efficiency. Amine-reactive PEG facilitated the attachment of CF 350 Amine, a blue fluorescent ligand, for fluorescent imaging, and apolipoprotein E (ApoE), a target to the LDLR family of receptors, for BBB delivery, to the polymersome surface. In vitro, PEGPLA polymersomes demonstrate limited release in physiologic environment, pH 7.4, with a burst release upon membrane poration in lysosomal environment, pH 4.8, which is as desired for delivery of βgal to the lysosome. Cellular studies, using GM1 gangliosidosis-diseased fibroblasts, confirm that βgal-loaded polymersomes increase enzyme activity to normal levels with doses as low as 0.7 mg/cm^2. The addition of apolipoprotein E as a targeting ligand decreases the effective dose to 0.175 mg/cm^2, indicative that receptor mediated endocytosis is effective.

Results demonstrate strict control over carrier size formed, therapeutic payload, release location, and ligand attachment possible when utilizing polymersomes as an enzyme delivery vehicle. This novel carrier provides a brain delivery platform for currently untreatable diseases and has the potential to cause a paradigm shift in the way we treat the central nervous system. Initial animal studies are promising towards the goal of creating the first clinical treatment for GM1 gangliosidosis, using a combination of enzyme replacement therapy and nanotechnology methods to cross the BBB.

501367 Enhanced Fibrinolysis Using Magnetically Powered Colloidal Microwheels

Abstract:

Background: The mechanical stability of blood clots is derived from the biopolymer fibrin. Blood clots can be biochemically decomposed via systemic or local delivery of tissue plasminogen activator (tPA) or urokinase plasminogen activator (uPA). Such drugs activate plasminogen to plasmin, which binds to and lyses fibrin. Fibrinolytic treatment is limited by two processes: the transport of plasminogen activator (PA) to a clot, and the dissolution of the clot itself. Delivery of fibrinolytics depends on method of administration. Blood flow is minimal in vessels that are occluded, and intravenously administered PA must diffuse to clot interface. Both the rate of diffusion and the rate of clot dissolution can be increased by using a higher systemic concentration of PA, but high concentrations of PA can induce hemorrhaging in the patient. In large arteries, catheterization can be used to deliver a high local concentration of PA while keeping systemic concentration relatively low; however, catheters provide no solution to thrombi that occur in smaller arteries.

Clot dissolution rates depend on several factors, including the pressure gradient across the thrombus, the composition of the thrombus, and PA concentration; however, lysis rates are typically limited by transport of PA to and throughout the thrombus. Microparticles functionalized with tPA offer an alternative method for delivery of PA to a thrombus that can generate high localized concentrations and accelerate transport of PA to and through a thrombus. Such a method overcomes the transport restrictions associated with PA delivery and accelerates fibrin degradation. Previously, we have demonstrated assembly of microparticles into "microwheels" which translate under a magnetic field with rapid velocities. Such microwheels are well-suited to carry PA to a thrombus in an occluded blood vessel.

Methods: One micrometer superparamagnetic beads subjected to a 9 mT rotating magnetic field (f= 100 Hz) assemble spontaneously into microwheels because of net induced dipole interactions. In this work, the magnetic field is generated externally by five solenoids. A sinusoidal function of alternating current is passed through two pairs of coils on the x-y plane. A $\pi/2$ phase lag between the coils on the x-axis and those on the y-axis cause the field to rotate at the center of the coils. A fifth solenoid positioned on the z-axis cants the magnetic field such that it rotates at an angle with respect to the horizontal. Microwheels align with the angle of the field and roll because of wet friction between the wheel and the surface. The direction of microwheel translation is readily controllable in real time, or can be pre-programmed to direct the wheels along user-defined paths.

Fibrin gels were formed inside a microfluidic channel from recalcified normal pooled plasma (NPP) or platelet rich plasma (PRP) (20 mM $CaCl_2$) with 4.5 nM thrombin. Fibrinolysis rates achieved by $1.5*10^6$ tPA-beads/µL, corresponding to an effective tPA concentration of 3.6 ug/mL, were compared to lysis rates induced by 1 and 10 µg/mL soluble tPA in a microfluidic device. The beads were guided by the magnetic field to translate with two different modalities: direct and corkscrew motions. Under the direct modality, wheels translate unidirectionally towards the fibrin front. In contrast, the corkscrew modality causes wheels to approach the fibrin front in spiral pattern with a directional bias.

Results: We observe fibrinolysis rates of 0.81 ± 0.25 and 5.0 ± 1.5 um/min for 1 ug/mL and 10 ug/mL soluble tPA, respectively. Lysis rates achieved by functionalized microwheels driven by direct and corkscrew modalities were 4.8 ± 0.3 and 9.6 ± 1.5 um/min. This data supports the approach of a magnetic field-powered microwheel as a viable drug delivery mechanism for tPA. Similar lysis rates for microwheel experiments were observed in clots formed from both NPP and PRP, indicating that microwheels can bust fibrin networks despite the presence of platelets. It should be noted that natural tPA inhibitors such as plasminogen activator inhibitor one (PAI-1) are present in physiologic concentrations since the experiments are performed in plasma.

Both the unidirectional and corkscrew translation modalities offer improved performance over soluble tPA. Translational velocities of beads approaching the fibrin front were 4.5 ± 3.8 um/s and 2.1 ± 1.8 um/s for unidirectional and corkscrew motion. Since these velocities are high relative to fibrinolysis rates, microwheels accumulated at the gel interface, resulting in a local concentration of tPA of up to 50-fold higher than the initial bulk concentration. Moreover, field-driven microwheels penetrated the fibrin gel, transporting drug within the fibrin network itself. The direct motion modality drove small wheels consisting of 1-4 microparticles approximately 100-200 um into the gel, whereas the corkscrew modality drove much larger microwheels (consisting of >4 colloids) from 20-60 um into the fibrin network. Roughly twice as many beads penetrated the gel under the corkscrew modality than the unidirectional one, corresponding to an increase in lysis velocity. The combination of biochemical and mechanical action contributed to enhanced fibrinolysis.

Conclusions: Fibrinolysis using soluble tPA is a transport-limited process that relies heavily on diffusion. Colloidal microparticles functionalized with tPA offer a lysis strategy that overcome the transport limitations inherent in soluble tPA fibrinolysis. Colloids are dispersed into a microfluidic system, and assembled into microwheels via subsequent application of a magnetic field. The low-density functionalized microwheels are driven to the fibrin interface and rapidly accumulate near the gel surface. Uniquely, microwheels driven by the magnetic field also penetrate into the fibrin gel, transporting tPA within the fibrin network and enhancing dissolution rates. Microwheels have potential as a drug delivery method against thrombi occurring in small vessels where catheterization and systemic soluble PA delivery are ineffective.

491418 Intracellular and Extracellular Delivery of Proteins and Small Biomolecules Via Near-Infrared Light

Abstract:

Small biomolecule delivery is done via "caged" compounds. However, each bioactive requires the chemical synthesis of its own "cage". As yet, no NIR triggered caged compounds are effective under typical experimental conditions. We have developed liposomes tethered to plasmon-resonant hollow gold nanoparticles (HGN) that can be triggered to release their contents by picosecond pulses of physiologically friendly, deeply penetrating near infrared (NIR) light. Liposomes tethered to HGN can encapsulate almost any water-soluble biologically active molecule by confining high concentrations in liposomes tethered to HGN. We could delivery of multiple agent at different times and locations, which is impossible with current liposome or caged compound technologies. Chemically disparate calcium, and ATP are all released at near 100 % efficiency from liposomes within msec. We could also control release rates and windows of each biomolecule in a mixture independently, by delivering two species or even changing the order of release.

505033 Repurposing Mechanosensitive Channels to Study Confined 3D Cell Migration

Abstract:

Biological membranes are involved in a large number of cellular processes including cell migration, membrane trafficking, and cell signaling. My lab is interested in studying the mechano-chemical responses of biological systems. we have reconstituted the function of a bacterial mechanosensitive channel MscL in mammalian cells. Using this system, we investigated the role of actin cytoskeleton in mediating local membrane tension that activates MscL. Cell migration in 3D confined environment has recently been shown to involve an osmotic engine mechanism where there is polarized water and ion influx at the leading edge and efflux at the trailing edge. We hypothesize that MscL expression could potentially disrupt the osmotic engine during confined migration. I will describe how we are repurposing MscL for studying cell migration in in vitro microfluidic and in vivo environments. We hope MscL can be used for drug delivery and for novel mechanotransduction.

MAPÚA UNIVERSITY

Department of Agriculture PHILRICE PHILIPPINE RICE RESEARCH INSTITUTE

Quality Improvement of PhilRice "*Tapuy*" (Rice Wine) by Ultrafiltration Hollow Fiber Membrane

Michelle Almendrala, Ph.D.
School of Chemical Engineering and Chemistry,
Evelyn H. Bandonill, Marissa V. Romero, PhD,
Philippine Rice Research Institute

17AIChE
Annual Meeting, Minneapolis, MN

Tapuy is made by:

- inoculating steamed glutinous rice with a starter *'bubod'* to undergo
  - ✓ saccharification and
  - ✓ fermentation

BUBOD – an oven dried dough (made of rice-flour) with

- pure cultures of the mold called *Rhizopus oryzae* and
- the yeast *Saccharomyces cerevisiae*

Tapuy

Tanimura et al. (1977) reported that traditional rice wine made from glutinous rice has

- 13.5%-19.1% alcohol,
- 2.78%-5.18% reducing sugars,
- 3%-5.44% total sugars,
- 9.3°Bx - 15°Bx, and
- pH of 3.3-4.9.

Rice wine

Chinese rice wine /yellow wine – have beneficial health effects on the prevention of cancer and cardiovascular disease

➢ **antioxidant activity** (nonvolatile)

➡ **total phenolic content** : *Syringic acid and (+)-catechin*

F. Que, L.C. Mao, X. Pan, Antioxidant activities of five Chinese rice wines and the involvement of phenolic compounds, Food Res. Int. 39 (2006) 581–587.

F. Que, L.C. Mao, C.G. Zhu, G.F. Xie, Antioxidant properties of Chinese rice wine, its concentrate and volatiles, LWT-Food Sci. and Techn. 39 (2006) 111–117.

- Export of rice wine was only reported in 1991 until 2001, - country has exported a total of 7.22 million liters of this product with a corresponding value of FOBUS$4,753,964 (Loria, 2006).

- Philippines has imported a total volume of 0.4 million liters of rice wine from 1991 to 2001 according to the National Statistical Coordination Board (NSCB), which amounted to FOBUS$556,222 (Loria, 2006).

- With the improvement of tapuy wine by ultrafiltration, it is expected that the export of Philippine rice wine will increase in the near future.

Significance of the Study

Current Production Process of PhilRice *Tapuy*

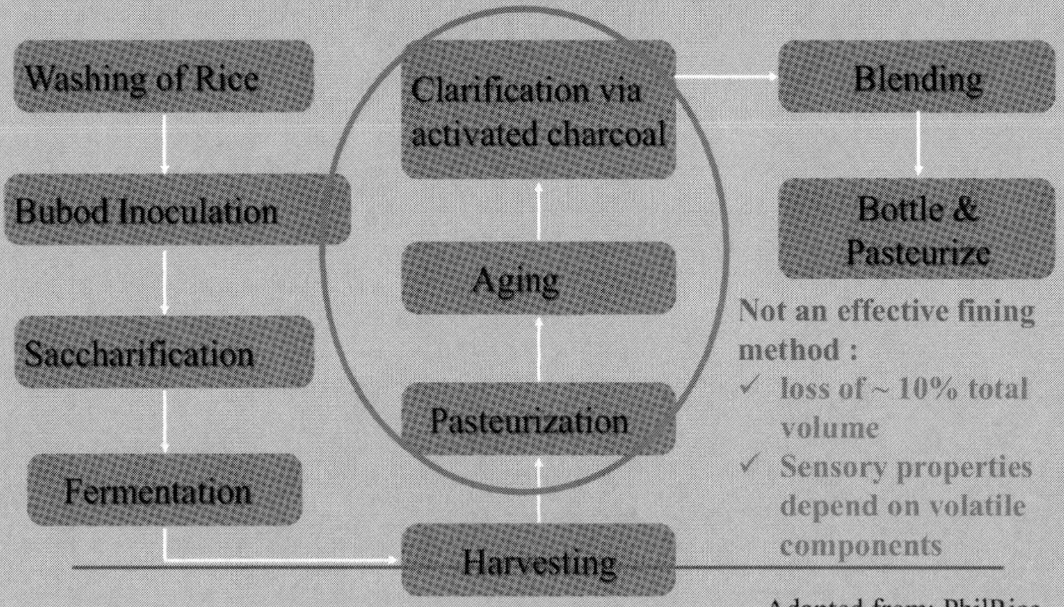

Adapted from: PhilRice

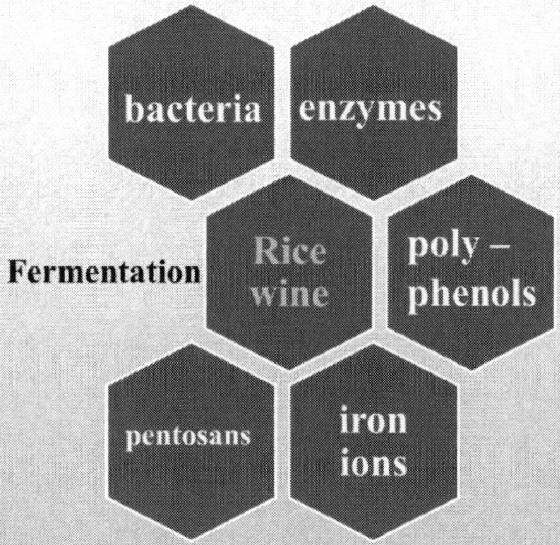

➡ **Non-biological stability and characteristics of rice wine**

- Ethyl carbamate (EC, urethane, $C_2H_5OC(O)NH_2$)
  - ✓ known carcinogen classified in group 2B by the International Agency for Research on Cancer (IARC)
  - ✓ generated during the processes of water boiling, jar-filling and storage,

❖ Remedy: strict control of time and temperature

Note: EC limit - 30 µg/L for wine and 400 µg/L for fruit brandies

EC – a carcinogen

Wu, H. *et al.* Study on the changing concentration of ethyl carbamate in yellow rice wine during production and storage by gas chromatography/ mass spectrometry. Eur Food Res Technol (2012) 235:779–782

• **Amines – Biogenic amines**

✓ present in rice wines: Seratonin, tyramine, histamine, cadaverine, and putrescine

✓ in high amount, biogenic amines may cause not only organoleptic defects but also adverse effects in sensitive human individuals, due to the toxicity

Zhong et al., 2012

Darkening

- Proteins and sugar undergoing Maillard-like reactions
- Polymerization of polyphenols

Off-flavor and odor

- Polymerization of polyphenols

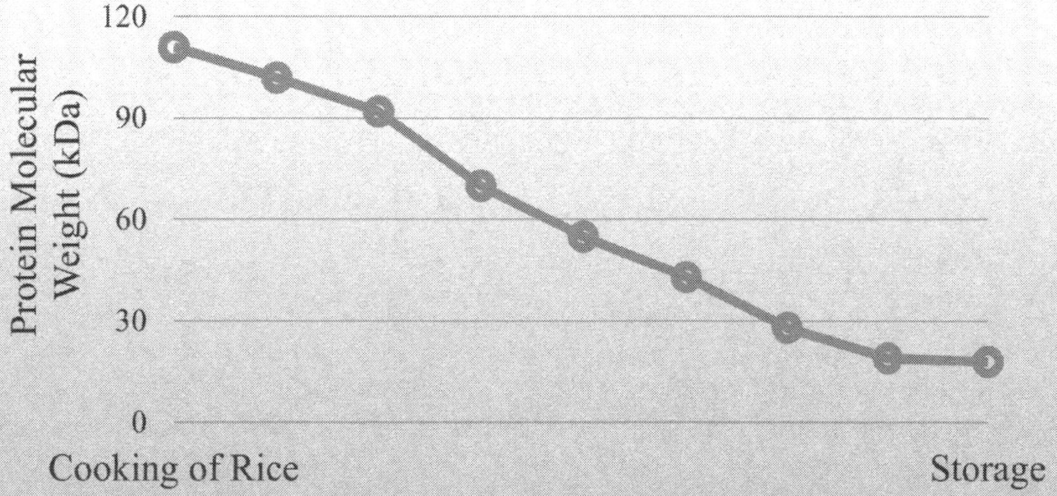

Protein Instability of Tapuy

Adapted from: Bergonio *et. al.* (2012)

- Predominance of 85, 76, 48, and 29 kDa in the liquid fractions and 85 and 28 kDa proteins in the solid fractions were also observed by Bergonio et al (2012).

- In **pasteurized** samples:
  - 1-month aged samples, only the 76 kDa protein was noted while only 67 kDa remained in the final product and

  - 6-month-old sample, which suggests that there is a gradual degradation of a high molecular weight protein to an 85, 76, then a 68 kDa protein.

Causes of wine degradation

- Significantly reduced the concentration of bacteria
  (Shu-Sen et al. ,1989, Shirikhande, 1993)
- Retained flavor and aroma
- Improved protein stability
- Less costly

Membrane Ultrafiltration for clarification and as a fining method for wine stabilization

- To clarify raw rice wine using hollow-fiber (HF) membrane module with 500 kDa MWCO
- To apply ultrafiltration using hollow fiber membranes with MWCO of 10, 50 and 100 kDa as a fining technique.

General Objectives

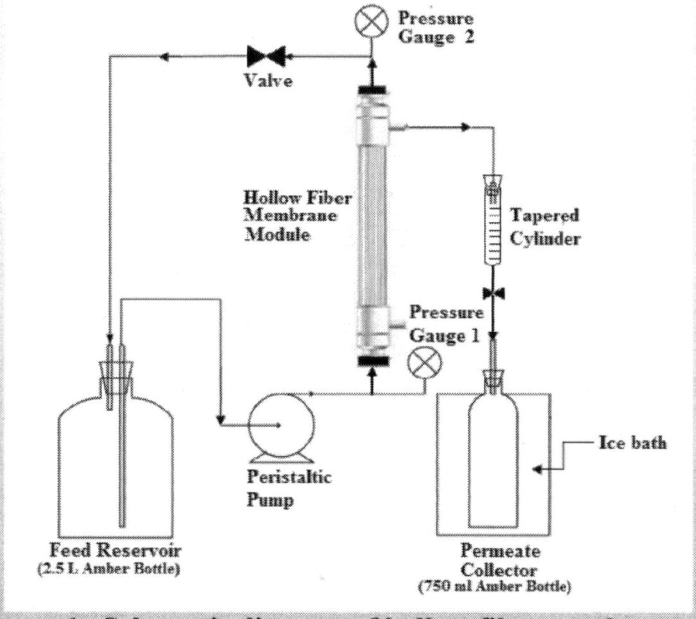

Figure 1. Schematic diagram of hollow-fiber membrane ultrafiltration system.

METHODOLOGY

T1 - PM500 + PM10, PM50, PM100, bottled and pasteurized
T2 - PM500, bottled and pasteurized
T3 - Pasteurized, PM10, PM50, PM100, bottled and pasteurize
T4 (Control)- Pasteurized, conventional filtration, bottled and pasteurized

Test Samples

T3: Raw rice wine pasteurized and aged for one month (not clarified by UF PM500) , then stabilized with PM10, 50 and 100 kDa.

T1 Sample: Ultrafiltered raw rice wine and aged for one month and passed thru PM10, PM50, PM100
T2 Sample: Ultrafiltered raw rice wine, aged, bottled wand pasteurized

[redacted]

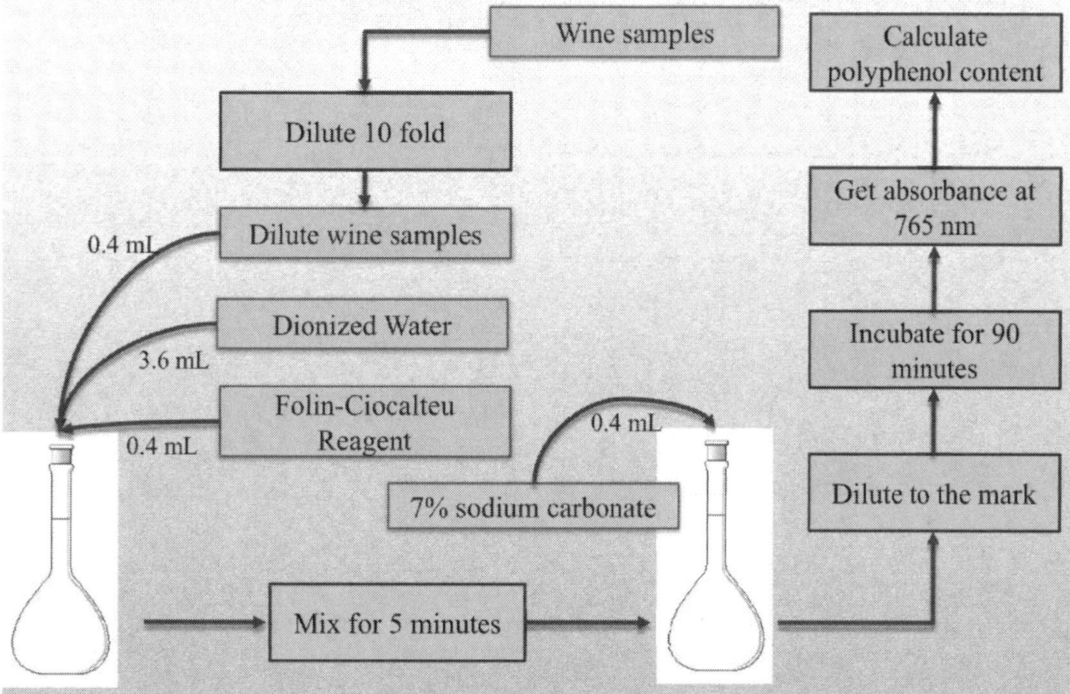

Analysis of Polyphenol: Folin-Ciocalteu Method for Wine

[redacted]

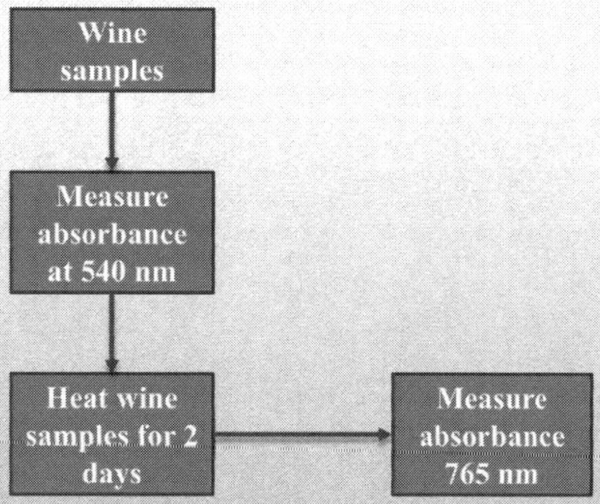

Analysis of Protein Stability: Accelerated degradation by heating

Data and Results

Table 1. Aged rice wine properties for T1 and T4

Sample Treatments	Physico-chemical Properties				COLOR			Microbial Count		
	°Brix	pH	% Alcohol	Crude Protein	L	a	b	TPC (cfu/ml)	Yeast (cfu/ml)	E.coli and coliform (cfu/ml)
Untreated raw wine	21.40	4.07	16.89	1.26	85.00	0.30	11.72	$3.40x10^4$	$3.05x10^5$	ND
T1 after PM500	21.40	4.07	16.89	1.26	93.86	-1.11	5.42	ND	ND	ND
T1 after 2 weeks	20.70	4.02	15.92	1.13	94.21	-1.19	6.10	ND	ND	ND
T1 after 4 weeks	20.85	4.03	16.10	1.08	94.36	-1.36	6.96	ND	ND	ND
T4 after Pasteurization	20.75	4.07	16.77	1.28	71.13	0.31	11.42	65.00	ND	ND
T4 after 2 weeks	21.50	4.08	16.77	1.28	94.26	-1.18	6.85	70.00	ND	ND
T4 after 4 weeks	21.25	4.05	16.78	1.24	94.76	-1.41	7.23	55.00	ND	ND

L = a, b values **a = +a (red), -a (green)**

l = 100 (white), 0 (black) **b = +b (yellow), -b (blue)**

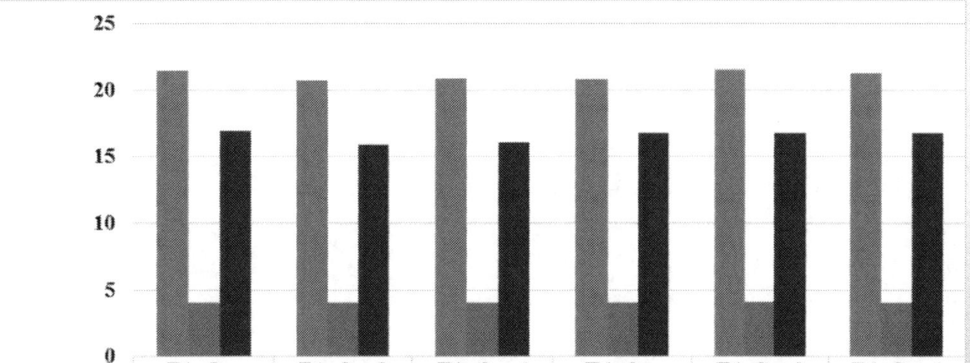

	T1 after PM500	T1 after 2 weeks	T1 after 4 weeks	T4 after Pasteurization	T4 after 2 weeks	T4 after 4 weeks
°Brix	21.40	20.70	20.85	20.75	21.50	21.25
pH	4.07	4.02	4.03	4.07	4.08	4.05
% Alcohol	16.89	15.92	16.10	16.77	16.77	16.78

Figure 3. Physico-chemical properties of rice wine after one month processing.

Sample Treatments	Physico-chemical Properties				COLOR			Microbial Count		
	°Brix	pH	% Alcohol	Crude Protein	L	a	b	TPC (cfu/ml)	Yeast (cfu/ml)	E.coli and coliform (cfu/ml)
Untreated (Raw Wine)	21.40	4.07	16.89	1.26	85.00	0.30	11.72	3.40×10^4	3.05×10^5	ND
T1 after PM500	20.70	4.02	15.92	1.13	93.86	-1.11	5.42	ND	ND	ND
T1 PM100 (Unpasteurized)	19.70	3.82	14.87	1.07	94.63	-1.84	8.45	ND	ND	ND
T1 PM100 (Pasteurized)	20.90	3.78	16.18	1.11	94.41	-1.97	9.60	ND	ND	ND
T1 PM50 (Unpasteurized)	19.50	3.91	14.26	1.42	95.26	-1.86	7.50	ND	ND	ND
T1 PM50 (Pasteurized)	19.90	3.88	15.16	1.46	94.59	-2.41	10.58	ND	ND	ND
T1 PM10 (Pasteurized)	20.45	3.93	14.93	1.41	94.94	-1.97	8.44	ND	ND	ND
T1 PM10 (Unpasteurized)	19.35	3.84	14.24	1.34	95.04	-1.89	7.88	ND	ND	ND
T2 after PM500	20.70	4.02	15.92	1.13	93.86	-1.11	5.42	ND	ND	ND
T2 after bottling and past.	20.90	3.86	16.02	0.98	94.44	-1.92	9.31	ND	ND	ND
T3 after Pasteurization	20.75	4.07	16.77	1.28	71.13	0.31	11.42	65.00	ND	ND
T3 PM100 (Unpasteurized)	19.75	3.84	14.81	1.10	94.61	-2.09	9.54	ND	ND	ND
T3 PM100 (Pasteurized)	21.05	3.87	15.95	1.16	94.76	-2.49	10.48	ND	ND	ND
T3 PM50 (Unpasteurized)	12.50	3.72	15.27	0.79	95.00	-2.48	10.83	ND	ND	ND
T3 PM50 (Pasteurized)	20.95	3.88	15.95	1.17	94.43	-2.18	10.66	ND	ND	ND
T3 PM10 (Pasteurized)	19.30	3.87	14.99	1.34	94.97	-1.68	7.46	ND	ND	ND
T3 PM10 (Unpasteurized)	20.75	3.85	14.21	1.36	94.77	-1.88	8.53	ND	ND	ND
T4 after Pasteurization	20.75	4.07	16.77	1.28	71.13	0.31	11.42	65.00	ND	ND
T4 after filtration, bottling and pasteurization	20.15	3.84	15.53	0.95	95.81	-0.47	1.69	No Analysis		

Data after one month processing, from harvesting to bottling.

Table 4. Sensory attributes of rice wine using ultrafiltration.

Rice Wine Samples	Sample No.	Mean Sensory Scores											
		Aroma (pleasant)[1]	Off-odor[1]	Clarity[2]	Color[3]	Alcohol content[1]	Sweetness[1]	Sourness[1]	Bitterness[1]	Astringency[1]	Smoothness[4]	Off-flavor[1]	Overall acceptability[5]
Treatment 1													
PM100 (unpasteurized)	1	8.80^{a}	0.28^{a}	14.23^{a}	8.63^{a}	9.68^{a}	8.20^{a}	1.98^{a}	6.92^{a}	5.74^{a}	9.69^{a}	0.93^{a}	10.34^{a}
PM100 (pasteurized)	2	**9.39^{a}**	0.19^{a}	14.47^{a}	10.10^{a}	10.93^{a}	9.04^{a}	1.27^{a}	6.60^{a}	5.72^{a}	**10.08^{a}**	0.53^{a}	10.72^{a}
PM50 (pasteurized)	3	8.58^{a}	0.64^{a}	13.90^{a}	**11.39^{a}**	10.23^{a}	6.83^{a}	1.90^{a}	**6.99^{a}**	5.52^{a}	10.00^{a}	0.50^{a}	8.70^{a}
PM10 (pasteurized)	4	8.56^{a}	0.28^{a}	14.43^{a}	9.26^{a}	10.46^{a}	6.99^{a}	1.86^{a}	6.08^{a}	6.10^{a}	8.40^{a}	0.50^{a}	9.03^{a}
PM10 unpasteurized)	5	9.36^{a}	0.19^{a}	14.53^{a}	8.49^{a}	9.78^{a}	7.23^{a}	1.41^{a}	5.47^{a}	4.70^{a}	9.46^{a}	0.20^{a}	8.97^{a}
Treatment 2 (after bottling and pasteurization)	6	7.57^{b}	**0.99^{a}**	14.14^{a}	9.96^{a}	8.47^{b}	7.42^{a}	2.43^{a}	5.80^{a}	5.48^{a}	9.01^{a}	**1.63^{a}**	7.81^{a}
Treatment 3													
PM100 (unpasteurized)	7	8.09^{a}	0.17^{a}	14.32^{a}	9.78^{a}	9.64^{a}	7.57^{a}	2.01^{a}	6.66^{a}	5.82^{a}	9.32^{a}	1.19^{a}	8.92^{a}
PM100 (pasteurized)	8	8.62^{a}	0.39^{a}	14.29^{a}	9.66^{a}	9.88^{a}	7.61^{a}	2.05^{a}	6.44^{a}	5.58^{a}	9.42^{a}	0.78^{a}	9.21^{a}
PM50 (pasteurized)	9	8.74^{a}	0.28^{a}	14.46^{a}	11.29^{a}	9.67^{a}	**8.44^{a}**	1.64^{a}	6.17^{a}	5.80^{a}	9.68^{a}	0.16^{a}	10.09^{a}
PM10 (pasteurized)	10	8.68^{a}	0.10^{a}	14.44^{a}	8.76^{a}	10.18^{a}	7.10^{a}	**2.97^{a}**	6.70^{a}	5.60^{a}	7.66^{a}	0.44^{a}	8.67^{a}
PM10 (unpasteurized)	11	9.01^{a}	0.19^{a}	14.21^{a}	10.41^{a}	**11.23^{a}**	7.63^{a}	1.97^{a}	6.62^{a}	6.80^{a}	9.19^{a}	0.20^{a}	9.87^{a}
Treatment 4 (control)	12	8.82^{a}	0.44^{a}	**14.58^{a}**	1.44^{b}	11.07^{a}	8.29^{a}	2.24^{a}	6.22^{a}	**7.11^{a}**	8.72^{a}	0.41^{a}	**10.76^{a}**

[1]Mean values with the same letter within a column are not significantly different at p=0.05 (n=9).
[1]1= none; 15= intense [2]1=turbid; 15=clear [3]1= none; 15= yellow [4]1= harsh; 15= smooth [5]1=dislike extremely; 15= like extremely

Permeate Flux vs Time
(Filtered August & October*)

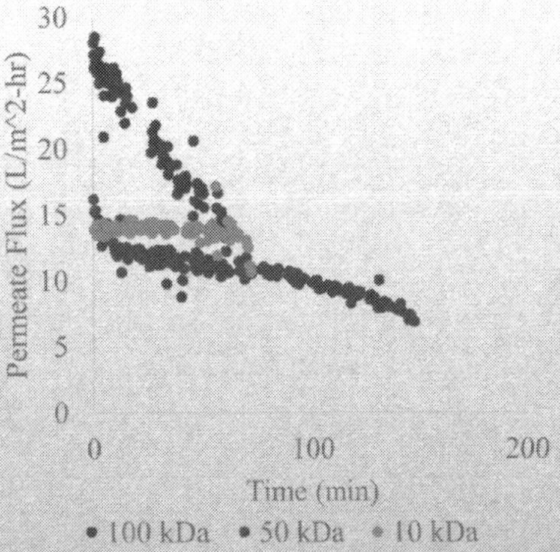

Permeate Flux vs Time
(Filtered December)

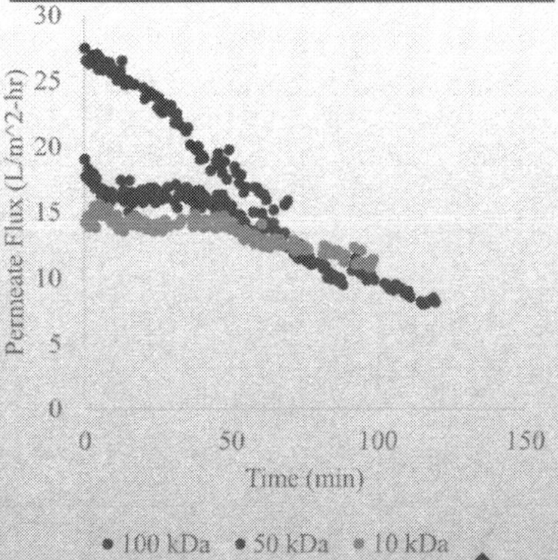

Effect of storage time on permeate flux: for T3 Sample

*10 kDa filtration was performed on October 2015

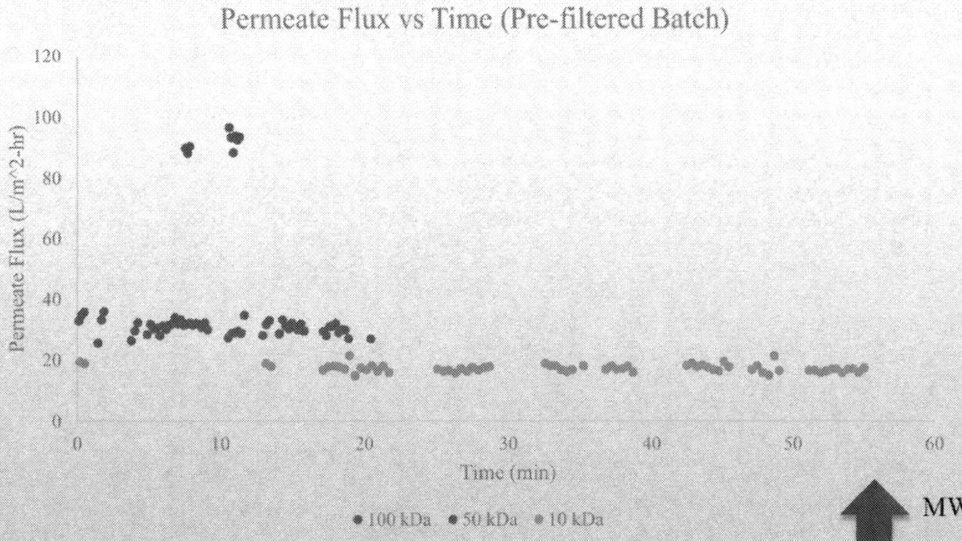

Effect of MWCO on Permeate Flux (T1:Pre-filtered with PM500)

100 and 50 kDa filtrations were performed on August 2015
10 kDa filtration was performed on October 2015

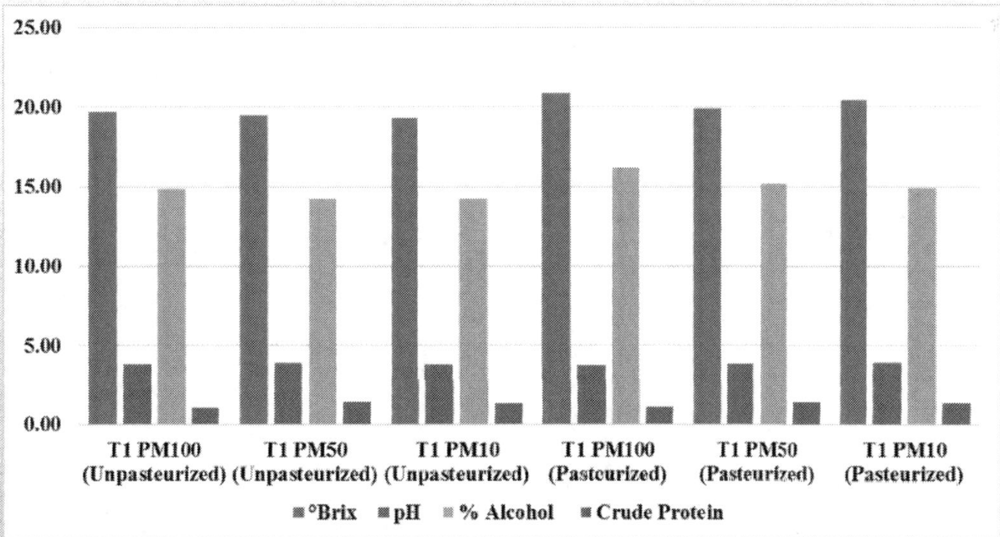

Figure 5. Effect of membrane cut-offs on wine physico-chemical properties.

No.	Wine Sample
1	T2
2	T3 PM100 Dec batch
3	T3 PM50 Dec batch
4	T3 PM10 Dec batch
5	T4 Control
6	New Bottle
7	T1 PM100
8	T1 PM50
9	T1 PM10
10	T3 PM100 Aug batch
11	T3 PM50 Aug batch
12	T3 PM10 Aug batch

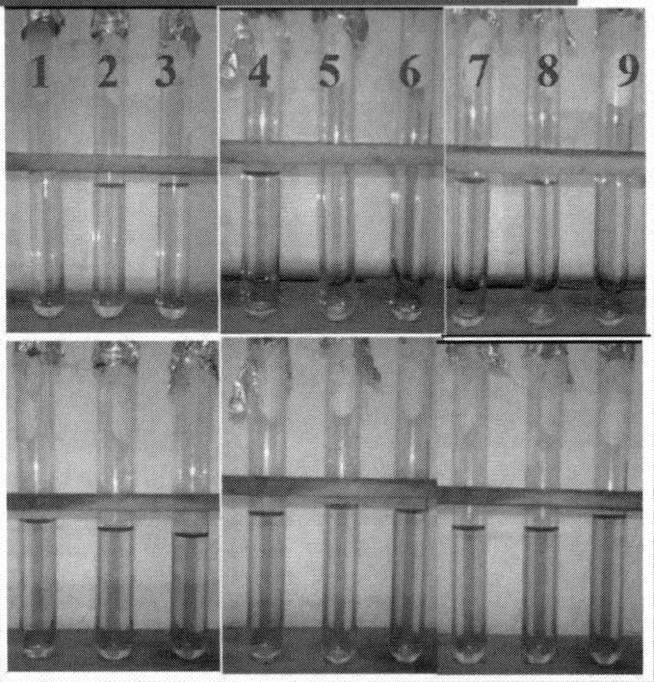

Heat Stability of Protein (2 days at 60°C)

No.	Wine Sample	pH	Polyphenol, mg/L	Brix	pH	Protein	Alcohol	Sensory	Stability
1	T2 (PM500)	3.53	614.30	22.35	3.49	0.98	16.02	7.81	0.0487
2	T3 PM100 (Dec. 2015)	**3.61**	**691.72**	23.09	3.54	1.16	15.95	9.21	0.0470
3	T3 PM50 (Dec. 2015)	3.635	716.45	21.76	3.57	1.17	15.95	10.09	0.0910
4	T3 PM10 (Dec. 2015)	**3.715**	**787.42**	21.82	3.55	1.34	14.99	8.67	0.0300
5	T4 (raw wine Pasteurized)	3.535	504.62	22.18	3.5	1.28	16.77	-	0.0003
6	T4 (Filtered by AC)	3.54	490.65	22.18	3.54	0.95	15.53	10.76	0.0050
7	T1 PM100 (Aug. 2015)	3.585	627.20	22.59	3.52	1.07	14.87	10.34	0.0480
8	T1 PM50 (Aug. 2015)	**3.74**	**741.18**	23.12	3.71	1.42	14.26	8.7	0.1517
9	T1 PM10 (Oct. 2015)	3.66	680.97	-	3.66	1.34	14.24	8.97	0.0693
10	T3 PM100 (Aug. 2015)	3.615	697.10	22.80	3.63	1.1	14.81	8.92	0.0097
11	T3 PM50 (Aug. 2015)	3.61	688.49	22.44	3.62	0.79	15.27	-	0.0357
12	T3 PM10 (Oct. 2015)	3.64	701.40	22.68	3.6	1.36	14.21	9.87	0.0580

Polyphenol content (GAE) and stability of protein

Rice wine sensory evaluation: clarity and color stability

1. After **5 months** of storage at room and refrigerated temperature
(February to July 2016)

2. After **15 months** storage Sampling done on May 16, 2017
Start of storage: February 16, 2016

Sample	Description
1	T1 PM100 (Unpasteurized)
2	T1 PM100 (Pasteurized)
3	T1 PM50 (Pasteurized)
4	T1 PM10 (Pasteurized)
5	T1 PM10 (Unpasteurized)
6	T2 after bottling and past.
7	T3 PM100 (Unpasteurized)
8	T3 PM100 (Pasteurized)
9	T3 PM50 (Pasteurized)
10	T3 PM10 (Pasteurized)
11	T3 PM10 (Unpasteurized)
12	T4 after filtration,bottling and past.
13	PM50 permeate
14	Pressed sample, pasteurized, Trial 1
15	Pressed sample, Ref upon UF & aging

Mean values with the same letter within a column are not significantly different at $p = 0.05$
*clarity= 0 turbid; 15 clear
**color = 0 none; 2.15 moon mist; 4.30 lemon; 6.40 yellow; 8.60 gold; 10.70 amber; …………12.80 brown; 15 dark brown
n= 9 in 5 mos.; 10 in 15 mos.

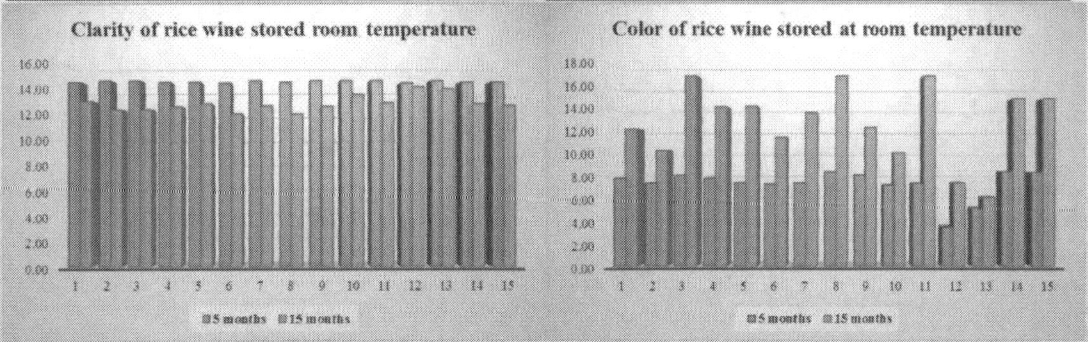

Room Temperature Sensory Evaluation

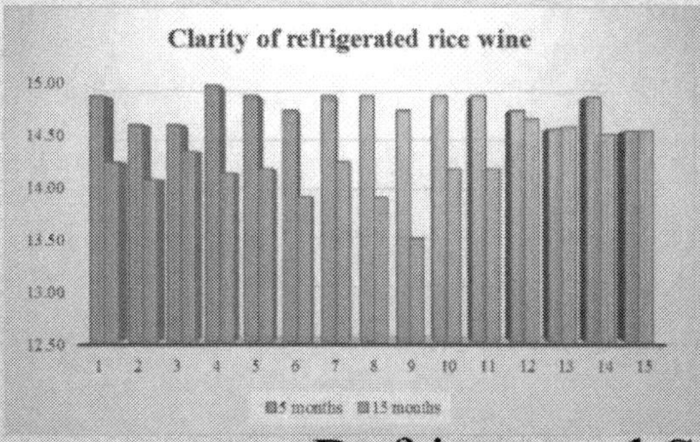

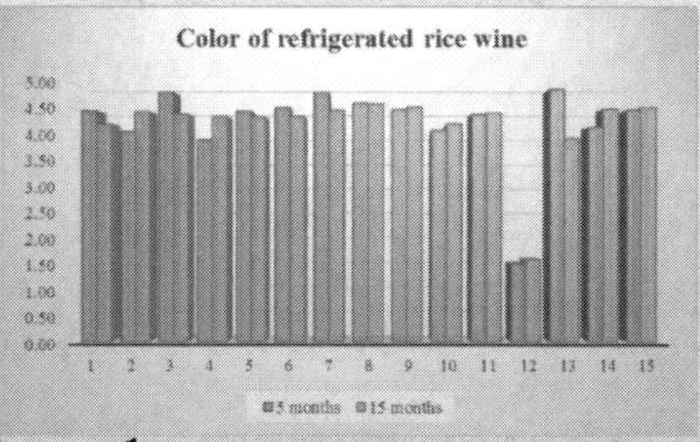

Refrigerated Sample

Sensory Evaluation

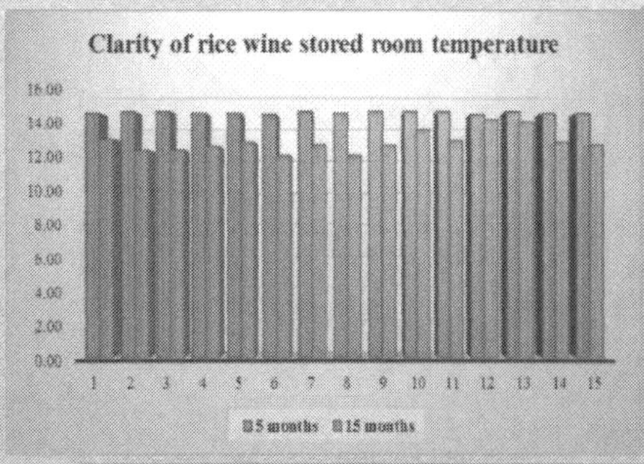

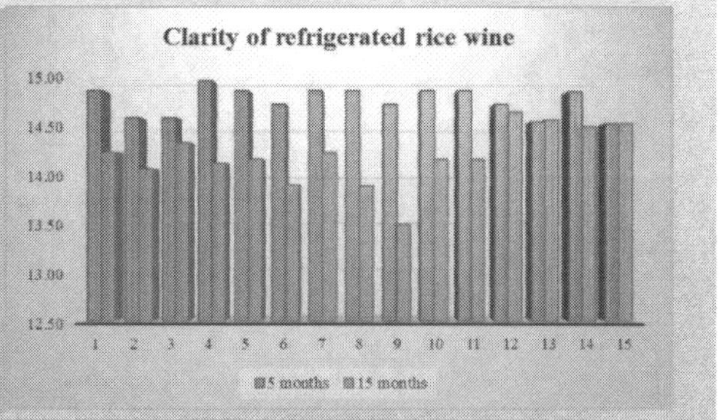

CLARITY 5 vs. 15 Months

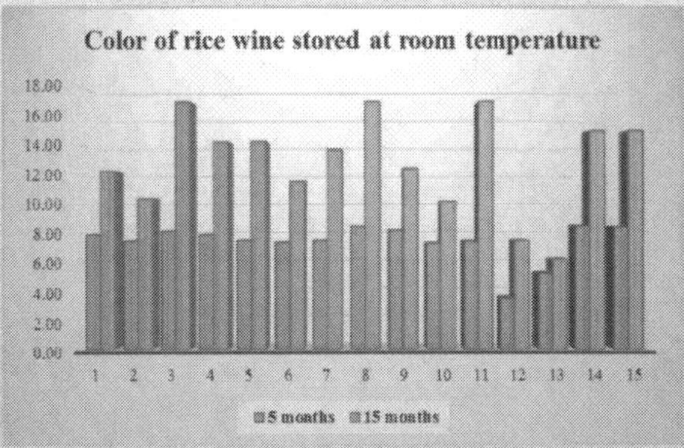

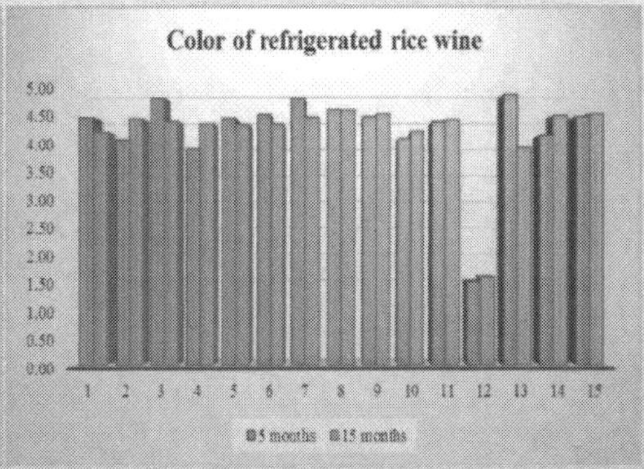

COLOR 5 vs. 15 Months

- Induced the formation of

✓ *sub-micronic metastable aggregates of protein*

→ stabilized by:

✓ electrostatic repulsions, or

✓ rapid formation of micrometer-sized aggregates

⇩ sedimentation

Cooling Stabilization

Protein Aggregation in White Wines: Influence of the Temperature on Aggregation Kinetics and Mechanisms. J. Agric. Food Chem., 2010, 58 (18), pp 10209–10218

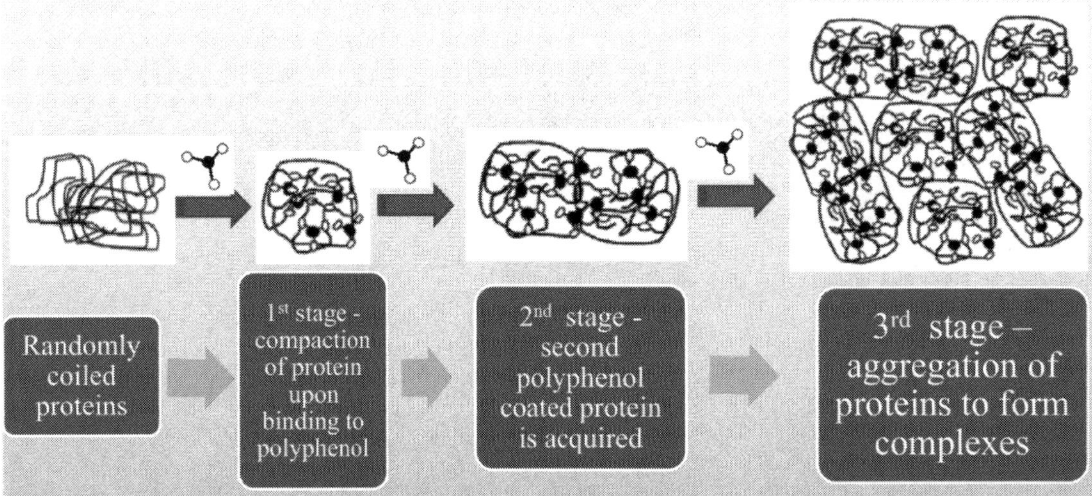

Polyphenol-protein 3-stage binding model

Jo¨bstl, E. et al. Molecular Model for Astringency Produced by Polyphenol/ Protein Interactions, Biomacromolecules, 2004, 5, 942-949,

- intermediate size polyphenols bind to salivary proline-rich proteins (PRPs), precipitate as tannins, lead to the sensation of astringency

"feeling of puckeriness and dryness in the palate"

Astringency

Jo¨bstl, E. et al. Molecular Model for Astringency Produced by Polyphenol/ Protein Interactions, Biomacromolecules, 2004, 5, 942-949,

[redacted]

- Primary source of color, flavor, aromatic characteristics of wine

- Involve in wine hazing and oxidative degradation (browning)

  → Polymerization of phenolic compounds

Flavanoids phenols – have more tendency to brown more intensely

- SO_2 – prevents polymerization and oxidative browning
- Ultrafiltration – removes unstable protein, phenols, and polyphenol oxidase

Phenolic compounds

[redacted]

- PM500 HF membrane can be used to clarify raw rice wine but is not enough to produce good quality wine with longer shelf-life.
- Fining technique using ultrafiltration (with PM10, 50 and 100) did not prevent the browning of rice wine after five months.
- The conventional method of fining using activated charcoal is more effective in removing the small molecules of proteins and polyphenols causing oxidative browning and maillard-like reaction in rice wine

- It is recommended that activated charcoal treatment followed by ultrafiltration as a better clarification option to produce good quality rice wine.

Conclusion

- Angelo King Institute, School of Economics
De La Salle University, Manila Philippines

- Matten Technologies Incorporated, Laguna, Philippinies

Acknowledgement

- Bergonio, K.B., Duldulao, J.B.A., Nonog, K.A.R., and Hipolito, M.C.(2012). *The proteins of PhilRicetapuy: From pre- to post-production.* 25th National Rice R&D Conference Book of Abstracts. pp 92-93.
- Tanimura, W., P. Sanchez, & M. Kozaki. (1977). The Fermented Foods in the Philippines I: Tapuy (Rice Wine). *J. Agr. Sci. of Tokyo Univ. of Agr. 22* (1): 135-41.
- Zhong, J., Ye, X., Fang, Z., Xie, G., Liao, N., Shu, J., et al. (2012). Determination of Biogenic Amines in Semi-dry and Semi-sweet Chinese rice Wines from the Shaoxing Region. *Food Control, 28*(1), 151-156.

References

- Goncalves, F., Fernandes, C., & de Pinho, M. (2001). White wine clarification by micro:ultrafiltration: effect of removed colloids in tartaric stability. *Separation and Purification Technology*, 423-429.
- Li, M., Zhao, Y., & Xing, W. (2010). Clarification of raw rice wine by ceramic microfiltration membranes and membrane fouling analysis. *Desalination, 256*(1-3), 166-173.
- Loria, IFB. 2006. Draft Businessplan for rice wine commercialization. RCFS, PhilRice.
  Science City of Muñoz, Nueva Ecija.

References

- Shrikhande, A.J., & Kupine, S. A. (1993). Ultrafiltration A New Approach for Quality Improvement of Pressed Wine. In B. Gump (Ed.), ***Beer and Wine Production Analysis, Characterization and Technological Advances*** (pp. 197-218). California, CA: American Chemical Society.
- Shu-Sen, W., Ling-Ling, W., & Al-Lian, L. (1989). Ultrafiltration of Rice Wine for Removal of Bacteria. *Journal of Membrane Science, 44*(2-3), 245-252.

References

499141 Novel Applications of Walnut Husk Extracts

Abstract:

During the processing of nuts, a considerable amount of material such as husks and shells are produced and often discarded as compost or landfill. In producing walnuts for retail sale, the husks are often discarded. Walnut husks contain tannins and polyphenolic compounds and the husk may be used for treatment of intestinal parasites and fungal infections. Green walnut husks harvested from Missouri and Michigan were extracted in ethanol using ultrasonic assisted extraction and extracted using supercritical carbon dioxide with an ethanol modifier. Ultrasonic assisted extraction was performed at room temperature for an hour, while supercritical fluid extraction was performed at 60°C with an ethanol mole fraction of 15%. The effects of pretreatment drying of walnut husks upon antioxidant potential as measured by the total phenolic content (TPC) assay were evaluated. A variety of diverse applications were evaluated including antioxidant and corrosion inhibition of extracts blended in paints, oxygen scavenging of extracts infused in plastics, and antimicrobial effects.

498598 Nutritional Value Improvement of Corn Ethanol Co-Product By Yeast Engineering

Abstract:

With the common use of corn ethanol as a transportation fuel additive in the United States, the major co-product Distillers Dried Grains with Solubles (DDGS) is massively produced and reached 36 million metric tons in 2016. The high energy, mid-protein, and high digestible phosphorus content of DDGS make it a very attractive, partial replacement for some of the more expensive, traditional energy (corn), protein (soybean meal), and phosphorus (mono- or dicalcium phosphate) used in animal feeds. However, variability in amino acids content and digestibility has been observed and remains a challenge when DDGS is included in animal diets. Excessive nitrogen will be excreted if more crude protein is included to meet the animal requirement for amino acids. Thus, when formulating diets with DDGS, some key amino acids is often incorporated as supplement. The key amino acids that is lacking here are lysine, tryptophan and arginine. Lysine is the first limiting amino acids. Tryptophan and arginine are too costly for external addition. The present study is focused on improving nutritional value of co-products in corn ethanol biorefinery by increasing the level of the key amino acids. This was accomplished by genetic engineering of *Saccharomyces cerevisiae* to accumulate more key amino acids in the cell biomass, thereby increasing its level in the final co-products. Our previous results on lysine engineering showed that the free lysine content could be increased by nearly 4 time-fold in engineered strains compared with wild-type strain. The lysine in the biomass increased by around 1 time in the engineered strains. The possibility of ethanol fermentation could not be significantly affected by modifying lysine synthesis pathway was also confirmed. The engineering work on arginine and tryptophan is being in process. The polypeptides rich in arginine and tryptophan is expressed in yeast and it is expected the yeast biomass with higher ratio of arginine and tryptophan could be obtained. The better nutritionally balanced co-products would minimize the cost associated with nutrient utilization and decrease the discharge of nitrogen to the environment, which will benefit corn ethanol biorefinery, animal feed industry as well as ecosystem.

494228 Effects of Artificial Electron Carriers on Increasing NADH Availability, Acetate Re-Assimilation and Butyric Acid Production By *Clostridium Tyrobutyricum*

Abstract:

Clostridium tyrobutyricum produces butyric and acetic acids from glucose. The maximum theoretical yield and selectivity of butyric acid in the fermentation depends on the availability of NADH, which favors the more reduced product. In this study, three kinds of artificial electron carriers (benzyl viologen (BV), methyl viologen (MV) and neutral red (NR)) that inhibit hydrogen production, was used to increase NADH availability and butyric acid production while simultaneously eliminate the accumulation of by-product, acetic acid by facilitating its re-assimilation. In order to understand the mechanism and find the optimum condition contributing to artificial electron carriers function efficiently, the effects of BV, MV and NR at various concentrations, and their addion time on butyric acid production were studied. Compared with the control (C. tyrobutyricum mutant strain HydEm), the addition of BV increased butyric acid final concentration by 10%, yield by 32% and productivity by 6% in the feb-batch fermentation using bioreactor with accurate pH control. Most importantly, acetate was completely re-assimilated and 99% purity of butyric acid was obtained, which would greatly decrease the recovery cost and increase economic feasibility of bio-based butyric acid production. The result demonstrates that redox equilibrium can be exploited to drive carbon flux redistribution for desirable metabolic product.

492266 Zinc-Mediated Pleiotropic Effects and Regulatory Mechanism on Acetone-Butanol-Ethanol (ABE) Fermentation by *Clostridium acetobutylicum*

Abstract:

Zinc-mediated pleiotropic effects and regulatory mechanism on acetone-butanol-ethanol (ABE) fermentation by *Clostridium acetobutylicum*

You-Duo Wu[1*], Chuang Xue[1] and Li-Jie Chen[1]

[1]School of Life Science and Biotechnology, Dalian University of Technology, Dalian, China

Recently, micronutrient zinc as a key cofactor has been demonstrated to play pleiotropic effects on cellular metabolism associated with differential carbohydrate utilization, cell growth, stress response as well as metabolic transition of acidogenesis and solventogenesis. Therefore, uncovering the zinc-mediated regulatory mechanism on preferred cellular traits was essential for exploring beneficial interactions of responsive genes and expanding regulatory network at system levels. The potential regulatory network involving in multigenic interactions and multiple cellular subsystems will contribute to the sustainable improvement of genetic and metabolic engineered solventogenic clostridia.

LC-ESI/MS analysis for intracellular metabolites revealed that, despite of rapid glucose utilization, glycolysis-derived metabolites such as G6P, F6P, FBP and PEP were dramatically decreased along with increased levels of energy supply ATP and reducing power NADH at the early stage of glucose-based ABE fermentation, which could well account for the enhanced exponential growth and earlier initiation of solventogenesis due to excessive NADH-derived oxido-reduction re-balance. In addition, branch node intermediates such as AcCoA, AcAcCoA and BuCoA involved in carbon flux distribution of acidogenesis/solventogenesis were also decreased coupled with extracellular products re-distribution, suggesting that micronutrient zinc could facilitate glycolysis and central carbon metabolism.

Global transcriptional analysis demonstrated that the expression levels of mutiple genes responsible for carbohydrate transport/metabolism were differentially upregulated with zinc supplementation, including glucose-PTS genes *glcG*, CAC1353 and CAC1354, fructose operon *fru* and 14 genes involved in xylose and arabinose transport/metabolism. Among the glycolytic genes, *glcK*, *pfkA*, *pgK* and *pykA* expression levels were slightly upregulated while H_2 biosynthetic gene *hydA* downregulated, which was consistent with the enhanced glycolysis and inhibited hydrogenesis. Furthermore, several key acidogenic and solventogenic genes *thlA*, *etfAB*, *bcd*, *crt*, *ctfAB*, *adc* and *bdhB* were differentially upregulated, which accelerate cascading conversion of central carbon metabolism, indicating the zinc-mediated regulatory mechanism involved in metabolic and genetic network of *C. acetobutylicum* at global levels.

Key Words: *Clostridium acetobutylicum*; Micronutrient zinc; ABE fermentation; Intracellular metabolites analysis; Transcriptional analysis.

497144 A Computational Modeling to Integrate Multi-Omics in *Clostridium cellulovorans* to Guide Metabolic Engineering

Abstract:

The cellulolytic *Clostridium cellulovorans* has been engineered to produce n-butanol from low-value lignocellulosic biomass in consolidated bioprocessing (CBP). The objective of this study was to integrate fermentation data and the multi-Omics experimental data by developing a mathematic model, targeting to predict the key regulators for metabolic engineering. In the process engineering, the corn corb cellulosic fermentation produced butanol with >3 g/L, yield >0.14 g/g, and selectivity >3 g/g by *C. cellulovorans-adhE2*. Our metabolomics study identified a total of 474 intracellular metabolites, including amino acid, carbohydrate, lipid, cofactor, nucleotide, peptide, and secondary metabolism. Meanwhile, about 1000 proteins were identified in the proteomics study. The primary static flux analysis indicated that cellulose favored butyrate production and the metabolic flux distribution from C2 to C4. The first-generation enzyme catalyzed reactions-based dynamic model was built by integrating the fermentation results, metabolomics data and proteomics data that involved in the metabolic pathways from pyruvate to butanol formation. This dynamic model simulated the intracellular metabolite fluctuation, included the effect of enzyme expression, and predicted the limiting steps at a systematic level. For example, we found that the carbon flux distribution from C2 to C4 was reduced in the mutant using cellulose as substrate, which was caused by the low expression of thiolase (thl) that catalyzes the reaction of acetyl-CoA to acetoacetyl-CoA and crotonase (crt) that catalyzes the reaction of 3-hydroxybutyryl-CoA to crotonyl-CoA. It was predicted that the butanol production can be improved to titer of >5 g/L and selectivity of >0.6 g/g from cellulose under optimized conditions.

504049 (Keynote) Increasing Tolerance to High Salt Concentrations in Acid Hydralysate of Cassava Pulp By Cell Immobilization in a Static Bed Fermentor

Abstract:

Utilizing inexpensive cellulosic feedstocks or agricultural residues as the sole carbon substrates is one key approach in fermentation cost reduction. Pretreatment and hydrolysis are common techniques employed in sugar recovery from such feedstocks. Nevertheless, some impurities remaining after pretreatment and hydrolysis such as salts, acetic acid, phenolic compounds, and furfurals are claimed to be responsible to lower fermentation efficiency due to their inhibitory effects on microbial metabolism. To eliminate such problems, post-treatment after hydrolysis is required and this undoubtedly results in cost contribution. An acceptable threshold of impurities remaining in the hydrolysate also depends on the robustness of the microbes to be used in fermentation. In this work, the performance of immobilized *Rhizopus oryzae* in a static bed fermentor was tested against the salt impurities remaining in the cassava pulp hydrolysates for its metabolic responses and lactate production. Similar to that observed in the typical glucose fermentation, after inoculation, spore germination and immobilization on the cotton sheet occurred, generating complete immobilization on the cotton cloth and cell-free fermentation broth. Cell growth and end product formation kinetics were observed during the fermentation using the cassava pulp hydrolysates. The fermentation performance somehow indicated the evidence of stress and impaired metabolic rate as the responses of the remaining impurities in the hydrolysates. The scanning electron micrographs show the salt crystals attached to the growing mycelia resulting in cell impairment. Nevertheless, by cell immobilization in the correct morphological stage, *R. oryzae* could withstand and subsequently produce lactate and ethanol as the end products under this drastic condition. As a consequence, we can claim the effectiveness of the static bed fermentor configuration.

510408 Using Metabolomics and 16S rRNA Sequencing to Investigate the Impact of Environmental Chemical Perturbations on Gut Microbiota Community Composition and Function

Abstract:

Recent findings suggest that a significant alteration of the gut microbiota, or dysbiosis, contributes to various diseases and disorders, including developmental disorders of the brain. Dysbiosis may result from infection, diet, or other environmental perturbations. One example of an environmental perturbation is exposure to biologically active synthetic chemicals present in household and commercial products. In recent years, exposure to these chemicals, labeled as endocrine disrupting compounds, have emerged as a significant public health concern. Interestingly, several of these chemicals, e.g. bisphenol A and phthalate esters, have been linked to neurodevelopmental disorders, including autism spectrum disorder (ASD).

This presentation describes a study on the effects of di-ethylhexyl phthalate (DEHP), a ubiquitous plasticizer. The effects of this chemical were studied in an *in vitro* culture model of the gut microbiota. Using 16S rRNA sequencing and untargeted LC-MS based metabolomics, we found that DEHP significantly modifies both the microbiota community structure as well as metabolic profile. Applying a novel metabolomics data annotation tool developed in our laboratory, we confidently identified more than 100 products of microbiota metabolism. Co-analyzing the 16S and metabolomics data using a metabolic model of the microbiota revealed that the chemically induced increases and decreases in specific metabolites can be attributed to the depletion or enrichment of particular groups of bacteria. Notably, DEHP exposure significantly increased the level of *p*-cresol, while expanding the abundance of *Clostridium bolteae*, both of which have been identified recently as potential biomarkers of ASD. Our results suggest that environmental chemicals could cause significant dysbiosis of the gut microbiota leading to an altered milieu of bioactive metabolites in the intestine, consistent with other studies linking environmental chemical exposure to developmental disorders involving gastrointestinal conditions.

510409 Genome-Scale Engineering of Lactococcus Lactis for Probiotic Outcomes

Abstract:

Humans have long reaped the benefits of probiotics via food fermentation with lactic acid bacteria. In this talk, I will discuss efforts in our research group to engineer the lactic acid bacterium Lactococcus lactis for improved probiotic outcomes. For example, besides its use in fermentation and in industry as a lactic acid producer, L. lactis has also been engineered for delivery of therapeutic proteins in vivo in the gut environment. We seek to use the toolbox of synthetic biology to expand the capabilities of this organism as an engineered probiotic. Our approach takes two routes: (1) Increasing the efficiency of genome engineering using the versatile CRISPR-Cas9 technology in this organism and (2) Discovery of genes and pathways that contribute to complex phenotypes by creation of genome-scale libraries and selection for desired phenotypes such as chemical tolerance and protein secretion. We also discuss the use of prebiotics to synergistically improve outcomes by encouraging lactococcal growth in complex environments. Increased throughput of genome engineering and identification of genomic targets will broaden the range of uses of this already food-safe organism for both industrial and human health purposes.

510410 Harnessing the Power of the Human Gut Microbiome to Treat Debilitating Disease

Abstract:

The human gut microbiome has become a hot topic in the search for new therapeutics to solve unmet medical needs. As the world's knowledge has expanded, the gut microbiome has been implicated in many conditions as far ranging as infectious disease to autism. In 2011, Rebiotix was founded to develop a microbiome-based drug to treat recurrent *Clostridium difficile* infection – a highly contagious disease of the gastrointestinal track – where standard of care therapeutics fall short. Join Lee Jones, UMN ChE '79, as she shares the journey and the science behind what is expected to be the first commercially available microbiome product in the world.

493110 Probiotic *Escherichia coli* Inhibits Biofilm Formation of Pathogens

Abstract:

Numerous chronic bacterial infections are involved in biofilm state and it is difficult to eradicate them by antibiotics. Biofilm formation is an inter-species phenomenon that involves dynamic interactions between bacteria in the biofilms. However, little is known to understand molecular interactions between biofilm bacteria especially in multi-species biofilms while majority biofilm studies has been done in mono-species. *Escherichia coli* strain Nissle 1917 (EcN) is a probiotic bacterium that has antagonistic effects on adherence, growth, and biofilm formation of other *E. coli* strains. In this study, we investigated the ability of EcN to outcompete with the biofilm formation of diverse pathogens such as enterohemorrhagic *E. coli* (EHEC), *Enterococcus faecalis*, *Pseudomonas aeruginosa*, and *Serratia marcescens*. When dual-species biofilms are formed, EcN reduced EHEC biofilm population up to 100-fold compared to EHEC mono-species biofilm formation, *E. faecalis* 10-fold, and *P. aeruginosa* 10-fold. EcN was dominant in the dual-species biofilms by occupying up to 90% of the total biofilm population. To see whether the molecules secreted from EcN affect EHEC biofilm formation, we grew EHEC biofilms under EcN cell-free supernant. EcN-supernatant inhibited EHEC biofilms up to 100-fold which is similar level of reduction to the result of EcN-EHEC biofilms. As a control, non-probiotic and non-pathogenic *E. coli* BW25113 did not reduce the biofilm formation of pathogenic bacteria. These results imply that probiotic *E. coli* inhibits the biofilm population of pathogenic bacteria, and provide insights that probiotics can be applied as an effective treatment for the biofilm-related infectious diseases.

498115 Understanding and Engineering the Metabolic Environment of a Healthy Skin Microbiome

Abstract:

The skin is the largest organ and the first line of protection against external factors. The skin environment is generally cool, acidic, and dry with inter- and intra-personal variation in pH, temperature, moisture content, elasticity and metabolite profiles. Different skin sites are colonized by diverse bacteria from *Actinobacteria*, *Firmicutes*, *Proteobacteria* and *Bacteroidetes* and fungi from *Ascomycota* and *Basidiomycota*. Some of these organisms are beneficial and contribute to protective properties of the skin. Skin commensal bacteria influence wound healing and infection by pathogens through direct antimicrobial activity and signaling host cells to boost innate and adaptive immune responses. To study these interactions and harness the protective power of commensal skin organisms, we are developing *in vitro* systems to monitor metabolic and immune responses of microbial skin isolates and skin cell cultures. Metabolic predictions of microbial growth states indicate that the skin pathogens are highly complementary with commensal isolates, suggesting an ecological mechanism for colonization. *Staphylococcus aureus*, a pathogenic skin organism, produces a small molecule which kills cultured skin models, while *Staphylococcus epidermidis,* a commensal isolate, protects cultured skin cells from pathogen growth. Further, in one-on-one interaction assays *S. epidermidis* inhibits the growth of *S. aureus* and other pathogens. Towards understanding the physical and chemical interactions important for the skin microbiome, we have developed reporter strains for skin isolates and have imaged infection dynamics using confocal and time-lapse microscopy. Collectively, our experiments suggest that commensal strains can have direct antimicrobial activity against pathogens and can activate host immune responses to further inhibit growth of pathogens. Improved understanding of these systems will allow us to examine interactions between resident microorganisms of the human skin and ultimately enable engineering the metabolic environment of the skin to encourage growth of beneficial microorganisms and enhance human health.

499557 Micro-Droplet Co-Cultivation and Characterization of Vaginal Bacteria

Abstract:

The vaginal microbiome (VMB) plays a fundamental role in womenâ€™s health and their susceptibility to diseases. For instance, Bacterial vaginosis (BV), the most common genital tract disorder in women of reproductive-age in the United States, is associated with preterm birth, an increased acquisition of HIV, and a higher risk of developing pelvic inflammatory disease. BV is characterized by the replacement of *Lactobacillus* species by other obligate and facultative anaerobic bacteria, such as *Gardnerella vaginalis*. Despite its significance, the fundamental questions of how diverse microorganisms in this microbial community interact with one another and with their host still remain largely unanswered. Current approaches for investigating these questions include conventional *in vitro* microbial cultivation and animal models, of which each has severe limitations such as low throughput and high resource requirement. More recently, microfluidic droplet co-cultivation has emerged as a new technology for dissecting microbial communities in a high-throughput fashion that renders substantial reduction of required reagents, time, and labor. In this work, we employ microfluidic droplet co-cultivation and several downstream analysis methods to recapitulate an antagonistic relationship between *G. vaginalis,* associated with BV, and *Lactobacillus jensenii,* a favored lactic acid producing vaginal bacterium. Pools of droplets can be characterized readily by fluorescence *in situ* hybridization (FISH) or quantitative polymerase chain reaction (qPCR). Furthermore, we show isolation of single droplets and colony formation on agar plates for quantitative analysis based on cell viability. Additionally, multiple displacement amplification (MDA) amplifies DNA content within a single droplet enabling metagenomic analysis. This work demonstrates the foundation of our microdroplet technology and represents the first step of our efforts aiming to elucidate microbial interactions in the human vagina and subsequently to intervene rationally the vaginal microbiome through engineering of probiotics and prebiotics.

501327 Targeted Species Removal in a Model Microbiome Via Phage Lytic Enzymes

Abstract:

Surveying microbiome composition (genomics, metagenomics) and assessing function through metatranscriptomic and metabolomic approaches have transformed our understanding of biodiversity and the roles microbes play in human health. However, new approaches and technologies are needed to develop mechanistic and causative insights into why and how communities vary over space and time, how communities adapt to change, and to develop new strategies for predictably modulating microbial community composition and function. Here, we focus on the targeted removal of a single species in a community. While adding an organism to an established model microbiome is relatively easy, the targeted removal of one species is more challenging. To address this challenge, we have developed a targeted bacteriophage lytic enzyme-based approach to selectively kill one organism in a model microbial community. Unlike conventional antibiotics, phage lytic enzymes have exquisite species specificity that target a single bacterial species or even a strain within a species. These lytic enzymes possess one of several cell wall peptidoglycan hydrolase activities with highly selective cell wall-binding specificities, which combine to confer activity and specificity toward a specific bacterial species. We have observed that *Staphylococcus aureus* can be selectively removed from co-cultures using the phage lytic enzyme lysostaphin (LST). Similarly, a second cell lytic enzyme, PlyPH, can be used to selectively remove *Bacillus cereus*. We have observed that the effectiveness of the selective killing depends on population density, lytic enzyme concentration and growth conditions. We will also present our work towards using Lst and PlyPH to selectively remove these organisms from a model human skin microbiome. *S. aureus* is a skin commensal in more than 30% of humans, while *B. cereus* is an opportunistic human skin pathogen. Such efforts are key to our long-term goal of being able to predictably modulate a target microbiome, based on causative or mechanistic insights, in order to alter microbiome function, or improve human health. The proposed project will also lay the groundwork for using targeted antimicrobial strategies for exploring microbiome dynamics and function in broad range of communities and environments.

495067 Niimbl: The National Institute for Innovation in Manufacturing Biopharmaceuticals

Abstract:

Technology innovation, workforce development, and regulatory advancement have been a hallmark of the industry for the past several decades. As the industry continues to expand, mature and evolve, there are new opportunities for shared risk in technology innovation, for partnerships to facilitate training, and for new approaches to engage with health authorities. It can be argued that collaborative efforts in this space, done appropriately, can reduce the burden and investment that would otherwise be required by individual organizations to advance their technology needs and address issues with speed, quality, and cost. At the same time, there is significant diversity in the industry in product types as well as projected needs which may mitigate some of the benefits for collaborative efforts moving forward. This presentation is intended to catalyze a discussion regarding the opportunities and challenges associated with collaborative technology development and will include a discussion about the ways that health authorities and federal agencies can foster collaboration among biopharmaceutical manufacturers, suppliers, government, and academic partners through a Manufacturing USA institute.

498214 The Transcelerate Comparator Network: Transforming the Supply Chain for Clinical Comparator Medicines

Abstract:

The procurement of comparator medicines required for use in clinical trials for the development of new medicines, poses a formidable challenge for supply chain professionals within biopharmaceutical research and development. This case study describes the synthesis and evolution of the TransCelerate Comparator Network, a biopharmaceutical industry forum that facilitates the procurement of comparator medicines and exchange of information between participating member companies. The goal of this network is to enable member companies to conduct clinical trials more efficiently and with enhanced integrity.

The TransCelerate Comparator Network marks a transformative shift in the manner in which biopharmaceutical companies have historically secured comparator medicines to conduct clinical trials. It is a game changer in that it has led to the creation of a platform that provides participating member companies access to clinical medicines and related documentation which facilitates their ability to conduct clinical trials more efficiently and mitigates risk to the continuity of clinical trials. Additionally this network has fostered an environment where member companies can collaborate and develop community based knowledge and learn from the experiences of others.

The TransCelerate Comparator Network is a great example of how multiple companies across an industry can come together and collaborate in the precompetitive space to develop creative solutions that address long standing challenges and in the process simplify the supply chain to increase efficiency, boost resilience, enhance integrity and reduce operating costs.

490317 Accelerating Pharmaceutical Development through Precompetitive Collaboration: The Enabling Technologies Consortium

Abstract:

The Enabling Technologies Consortium (ETC), which originated as a Working Group within the International Consortium for Innovation and Quality in Pharmaceutical Development (IQ Consortium), is a consortium of pharmaceutical and biotech companies focused on the optimization of technologies that enable pharmaceutical development. The goal of ETC is to simplify and accelerate the process for establishing collaborations by standardizing the procedures and agreements under which they operate; to proactively identify high-value opportunities to deliver innovative technologies where the business case is compelling and collaboration with the broader external community (vendor, academic, government) is required. The current scope of the ETC is in technology within the Chemistry, Manufacturing and Controls (CMC) of drug substance development. This presentation provides an update on the progress of the overall â€˜Enabling Technologiesâ€™ effort since its inception in August 2015 using the drug substance drying working group as a case study

502398 Cmac: A Partnership Approach to Precompetitive Collaboration for Pharmaceutical Manufacturing Research

Abstract:

Modern research increasingly demands multidisciplinary approaches with many national funders increasingly recognizing the potential benefits for supporting the development of close relationships between industry and academia. CMAC is a large collaborative UK based manufacturing research centre that is focused on advanced crystallization and accelerating the adoption of continuous manufacturing to enable future medicines supply chains. Led from the University of Strathclyde, CMAC brings together a multidisciplinary academic team from across several UK universities (Bath, Cambridge, Imperial College, Loughborough, Leeds and Sheffield) as well as a range of active international academic partnerships. Now in its second phase and starting a new 7 year, £50M programme, CMAC was co-created with industry partners to ensure activities and outputs align with areas of existing and emerging industrial need. Our members include GSK, AZ, Novartis, Bayer, Eli Lilly, Takeda and Roche as well as over 20 technology providers and SMEs. This scale and breadth of expertise brings excellent opportunities to inform research direction and inspire researchers however it also demands that all aspects of the collaborations are well co-ordinated and supported. The collaborative approach to establishing this critical mass activity has also attracted significant investments from UK research councils & agencies (EPSRC, SFC/HEFCE) as well as charities and pharmaceutical industry and has stimulated a number of subsequent initiatives aimed at transforming the UK research and innovation system. In addition to the manufacturing research programme, CMAC has active programmes in other key areas, namely: skills and training, facilities development and industrial translation, aimed at maximisingthe benefits to all of our partners.

This presentation will review the model and different approaches used across CMACâ€™s precompetitive programme and will highlight challenges and lessons from such large collaborative activities involving academia and industry.

488365 A Perspective on a Pre-Competitive NSF Engineering Research Center Focused on Pharmaceutical Manufacturing: C-Sops

Abstract:

Throughout the last ten years, The Center for Structured Organic Particulate Systems (C-SOPS) has existed as a National Science Foundation ERC focused on the pre-competitive development of pharmaceutical manufacturing products and processes. During peak levels of activity, six universities and over fifty corporations were involved in the program. A major focus of the program throughout and even today is continuous solid dose manufacturing. Through this program, much of the groundwork resulting in todayâ€™s commercialized continuous solid dose manufacturing approaches by member organizations such as Vertex, Janssen, Eli Lilly and others was developed pre-competitively.Â Today, C-SOPS continues to work closely with top pharmaceutical companies adopting advanced manufacturing technology such as continuous.Â In this talk, a brief history of C-SOPS will be presented along with lessons learned on pre-competitive efforts and where the program is going today.

496512 The Development of an Interactive Knowlege Epicenter for the Advancement of Continuous Manufacturing

Abstract:

Continuous manufacturing has increased in popularity amongst pharmaceutical organisations in recent times, and much research is currently being conducted both in academic and industrial settings. However a large portion of the research which is being conducted may be classified as precompetitive and thus a significant opportunity exists for the industry as a whole to augment knowledge accumulation rates by working together and developing one common, central space where information can be easily presented, located and interacted with. The Continuous Manufacturing Repository is being developed in order to achieve this objective. It is an easily accessible online environment in which participating companies can selectively share their publically available presentations and/or publications, creating the 'one-stop-shop' for continuous manufacturing knowledge. This presentation will focus on how the CM repository works, how companies can get involved and will demonstrate the long term vision for the repository.

498905 Engineering and Profiling the Substrate Specificity of Human Proteases for Therapeutic Applications

Abstract:

Autoimmune diseases such as inflammatory bowel disease (IBD) are characterized in part by an overproduction of regulatory proteins, such as tumor necrosis factor α (TNF-α). Antibodies that target TNF-α have been at the forefront of treatments for IBD. While antibody therapies can be effective, they are often limited by a loss of efficacy over time and can lead to an increased risk of opportunistic infections. To address these issues, an alternative approach is to degrade target proteins using proteases, some of which have been used as therapeutic agents for several decades. However, in order to expand the pharmacopeia of many human proteases, their substrate specificity needs to be fine-tuned to enhance their efficacy and avoid adverse side effects, as well as to exhibit resistance to *in vivo* inhibitors.

Recently, we reported a high-throughput yeast platform named Yeast Endoplasmic Reticulum (ER) Sequestration Screening (YESS). In YESS, a protease substrate display cassette, which is composed of selection and counter-selection substrates flanked by epitope tags, is co-expressed in the yeast ER along with a protease. Peptide display is detected by fluorescent labeling and cleavage is screened by fluorescence-activated cell sorting (FACS). Human neutrophil elastase (HNE), an extracellular serine protease with broad specificity, may modulate immune inflammatory responses through the degradation of both membrane-bound and soluble TNF-α. In order to develop a protease-based drug to treat IBD, we challenged a mutagenized library of HNE through YESS for increased specificity towards TNF-α. After four rounds of FACS, we isolated HNE variants that specifically degrade peptide substrates found in soluble TNF-α with minimal activity towards membrane-shedding TNF-α sequence preferred by the WT enzyme. Fortuitously, the HNE variants no longer cleaved peptides derived from serpin B1, a potent HNE *in vivo* inhibitor. Similarly, in an effort to develop a treatment for SLE, we repurposed the specificity of human matrilysin (hMMP7), a metalloprotease involved in the breakdown of extracellular matrix, to specifically cleave the IgG hinge region which leads to IgG deactivation. To characterize these protease variants biochemically, we are concurrently developing methods for human protease expression and purification. In the future, we will characterize these variants *in vivo* using mice models of IBD and SLE.

In addition, we coupled YESS with Next-Gen sequencing (YESS-NGS) to profile the substrate specificity of wild-type and our engineered HNEs. In this respect, YESS offers several advantages over exiting protease profiling methods. YESS circumvents the need to have purified proteins, enables access a large sample of sequences (10^7) to determine fine patterns of reactivity not previously discovered. We have applied YESS-NGS to profile several other proteases, including TEV proteases and their variants, human kallikreins, neutrophil elastase and their variants, neprilysin, cathepsin E, among others.

486551 A Novel Transthyretin Peptide Mimic As a Therapeutic for Alzheimer's Disease

Abstract:

Alzheimer's disease (AD) is the most common form of dementia and one of many diseases caused by amyloidogenic proteins. While amyloidogenic proteins differ in many ways, they all undergo aggregation in which monomeric protein spontaneously self-assembles into soluble oligomers and insoluble fibrils. In the case of AD, it is the amyloid-β (Aβ) protein that aggregates, resulting in the formation of extracellular senile plaques in the brain. While Aβ monomer occurs naturally in blood, plasma, and cerebrospinal fluid (CSF), Aβ aggregates, particularly oligomers, induce cellular toxicity. Several CSF proteins may naturally sequester Aβ aggregation, with one of the most promising being transthyretin (TTR). TTR is a homotetrameric protein that functions as a carrier for thyroxine and retinol binding protein. A double mutation (F87M/L110M) produces an engineered protein that is stable as a monomer (mTTR). Both TTR and mTTR bind Aβ, inhibit Aβ aggregation, and reduce Aβ toxicity. This has led our lab to synthesize a series of cyclic conformationally constrained peptide mimics of TTR that can bind to Aβ and inhibit its toxicity in vitro.

In this study, we evaluated the therapeutic potential of one of these cyclic peptides, cG8, based on its selectivity, specificity, and stability. We explored its interaction with specific Aβ aggregates (monomer, oligomer, fibril), evaluated its stability and binding to Aβ under physiologically relevant conditions, and determined its ability to inhibit the aggregation of other amyloidogenic proteins. We also compared cG8 to mTTR, to evaluate the relative advantages of a peptide-based versus protein therapeutic.

To measure selectivity for binding to Aβ monomer, oligomers, or fibrils, cG8 or mTTR was biotinylated and adsorbed to pre-blocked NeutrAvidin ELISA plates. Aβ monomer (fresh), oligomers (24h, RT), or fibrils (72h, 37°C) were added, and binding was assessed with an anti-Aβ 6E10 antibody. cG8 and mTTR both bound significantly more Aβ oligomers and fibrils than Aβ monomer, indicating selectivity for the toxic forms of Aβ. To further evaluate the selectivity under biologically relevant conditions, binding of Aβ to cG8 in the presence of fetal bovine serum (FBS), brain lipids, and mouse brain tissue lysate was evaluated using the same assay. Although binding of Aβ decreased for both cG8 and mTTR in the presence of tissue lysate, cG8 retained Aβ binding capacity significantly better than did mTTR. Repeating the assay but with FBS or lipids indicated that it is largely the proteins in tissue lysate that cause the decreased Aβ binding. Immunoprecipitation and mass spectrometry studies are currently underway to identify the co-adsorbing proteins responsible for decreased Aβ-cG8 or Aβ-mTTR binding. These results indicate that the peptide cG8 suffers less from interference from other proteins than mTTR because it was designed to contain a minimal Aβ-binding domain, whereas the protein mTTR also contains other domains that will bind nonspecifically to serum or CSF proteins.

We used Thioflavin T (ThT) fluorescence to assess the ability of cG8 and mTTR to prevent Aβ aggregation or to disaggregate pre-formed fibrils. When co-incubated with freshly prepared Aβ, both cG8 and mTTR were very effective at inhibiting Aβ aggregation even at sub-stoichiometric concentrations. However, neither cG8 nor mTTR were effective at disaggregating preformed Aβ fibrils (48 h, 37°C) even at equimolar concentrations. Thus, although both cG8 and mTTR are highly effective at preventing the maturation of Aβ monomers into fibrils, neither disaggregate pre-formed fibrils into the more toxic oligomeric species.

Besides Aβ, there are many other proteins that aggregate into amyloid fibrils. Two well-known examples are amylin (islet amyloid polypeptide), associated with type II diabetes, and α-synuclein, associated with Parkinson's disease. To evaluate whether cG8 and mTTR are selective for Aβ or are general anti-amyloid inhibitors, we incubated amylin or α-synuclein with cG8 or mTTR. Both cG8 and mTTR proved ineffective at inhibiting α-synuclein aggregation, as measured by ThT fluorescence. However, both significantly reduced amylin aggregation even at 20-fold excess amylin. We explain this outcome by hypothesizing that the aggregation mechanism of amylin and Aβ may be similar to each other, and distinct from that of α-synuclein.

The stability of cG8 under biologically relevant conditions was evaluated by incubation in either simulated gastric fluid (SGF, pH 1.2) or simulated intestinal fluid (SIF, pH 6.8) for 24 h at 37°C. After incubation, degradation of cG8 and its ability to inhibit Aβ aggregation was assessed via mass spectrometry (MS) analysis and ThT fluorescence, respectively. No chemical degradation of cG8 was observed after incubation in SGF and SIF. Additionally, cG8 exposed to SGF and SIF not only retained its ability to significantly inhibit fibril formation but did so at a level comparable to mTTR.

This study demonstrates that cG8 binding to Aβ is affected significantly less by the presence of biological materials than is Aβ-mTTR binding. Additionally, cG8 is stable and retains its inhibitory capability when exposed to biological fluids. Although cG8 does bind to Aβ fibrils, it does not cause fibril disaggregation which could potentially lead to a toxic response. cG8 was also able to prevent the aggregation of amylin, which is a common occurrence in type II diabetes. Combined, these results support the use of cG8 as an AD therapeutic and indicate that cG8 may be beneficial in the treatment of other amyloidogenic diseases. *In vivo* experiments to evaluate cG8 distribution and efficacy in transgenic mice are currently underway.

485797 Directing Reversible Cell-Cell Interactions with Evolved Fibronectin Domains

Abstract:

Targeted molecular recognition continues to be a cornerstone of contemporary therapeutics and diagnostics. Though antibodies are commonly employed as targeting scaffolds, their large size, long plasma half life, and lack of facile bioconjugation make antibody-based ligands less than ideal for many basic and clinical applications. As the number of clinically relevant biomarkers continues to grow, there is an increasing need to develop new agents capable of selectively identifying and targeting these markers.

To meet this need, we have engineered a targeted protein scaffold based upon the human tenth type III fibronectin domain that binds to the highly overexpressed carcinoma antigen epithelial cell adhesion molecule (EpCAM). Using yeast surface display, mammalian cell panning, and a novel titratable avidity-reduction selection technique, we successfully evolved fibronectin clones exhibiting high affinity (11 nM) and robust selectivity for cellular EpCAM. These new selection methodologies can be applied to other cellular biomarkers and should increase the success rate of isolating scaffolds whose binding efficacy translates to *in vivo* applications.

Furthermore, we incorporated these EpCAM-targeting ligands into a multivalent chemically self-assembled nanoring (CSAN) for use as a cell-directing scaffold. The CSAN is formed when bivalent dihydrofolate reductase ($DHFR^2$) fusion proteins are spontaneously oligomerized by a chemical dimerizer, bis-methotrexate. When the CSAN is functionalized with our anti-EpCAM fibronectins and monovalent streptavidin, heterobifunctional CSANs are formed. These bispecific nanorings are capable of interfacing between EpCAM-expressing tumor cells and biotinylated surfaces â€" including other cells. Importantly, the CSAN scaffold can be disassembled via exposure to the FDA-approved antibiotic trimethoprim, providing a pharmacologic mechanism for reversing the interactions. Using this system, we have successfully directed and reversed targeted cell-cell interactions *in vitro*.

In conclusion, we have engineered novel, EpCAM-targeted fibronectin domains and incorporated them into a modular cell-directing scaffold. This CSAN scaffold affords a versatile, non-genetic, and pharmacologically reversible platform for modifying mammalian cell membranes with specific targeting elements. Thus, our CSAN platform is directly applicable to therapeutic regimes such as cell-based immunotherapy and regenerative medicine and may offer significant advantages over established cell-directing methods.

497254 Design of a Cholesterol-Binding Peptide to Inhibit Bacterial Toxin Activity

Abstract:

Membrane cholesterol is a common target of pathogens, including bacteria and viruses, as well as bacterial toxins. Many of these pathogens exploit a cholesterol recognition amino acid consensus (CRAC) motif to facilitate cholesterol recognition. We have previously identified and characterized the CRAC motif used by a member of the repeats-in-toxin (RTX) family of toxins, the leukotoxin (LtxA) produced by *Aggregatibacter actinomycetemcomitans*, to bind to target cells. We hypothesized that a peptide consisting of the LtxA CRAC motif would bind strongly to cholesterol to inhibit the activity of LtxA. Using solid-phase peptide synthesis, we created a series of peptides based on the CRAC motif of LtxA and used a variety of techniques, including isothermal titration calorimetry, surface plasmon resonance and confocal microscopy, to measure the peptidesâ€™ affinity for cholesterol and ability to inhibit LtxA activity. Our wildtype CRAC peptide binds cholesterol with an almost equal affinity as the toxin from which the sequence was derived (LtxA). This peptide specifically inhibits binding of LtxA to cholesterol and thus inhibits internalization and cytotoxicity of LtxA. A control peptide, in which the CRAC sequence has been scrambled, has a reduced affinity for cholesterol and does not inhibit the activity of the toxin, demonstrating the specificity of the interaction. A panel of peptides containing conservative residue substitutions demonstrated that the charge of the peptide significantly influences affinity for cholesterol and subsequent inhibition activity; highly charged peptides had no affinity for cholesterol, but surprisingly, a decreased charge did not enhance affinity. Our experimental and simulation results indicate that the peptide resides on the surface of the membrane, and the peptide must maintain the proper balance of hydrophobic and hydrophilic character to retain activity. Importantly, the peptide does not affect cell viability over extended time periods, supporting its possible therapeutic use to inhibit the activity of cholesterol-binding pathogens. Although recognition of membrane cholesterol is a vital step in the mechanisms used by numerous pathogens, a method to inhibit this process without disrupting cellular function has not yet been uncovered. Here, we have designed a cholesterol-binding peptide for this purpose and demonstrated that it is effective in blocking bacterial toxin activity and is non-toxic to host cells.

497052 Enhancing the Chemical Versatility of Yeast Display to Target the Tumor Microenvironment

Abstract:

Although cancer-promoting activities of extracellular proteases and peptidases in the tumor microenvironment are well known, the disruption of individual enzymes in this environment remains a fundamental challenge. The development of highly specific enzyme inhibitors would enable characterization of enzyme functions in the microenvironment and could lead to identification of potential leads for therapeutic applications. We are exploring the hypothesis that integrating additional chemical functionality into binding proteins will provide opportunities for discovering potent inhibitors that simultaneously exploit the best features of proteins and small molecules. In order to construct, evaluate, and screen protein-small molecule "hybrids," we have established a platform enabling yeast display of proteins containing noncanonical amino acids (ncAAs). This platform will enable us to leverage the quantitative and high throughput capabilities of yeast display to determine how best to assemble these structures and, once assembled, identify potent, selective inhibitors. Here, we present advances in the technological capabilities of our ncAA-compatible yeast display platform.

In order to broaden the range of chemical functionality that we can use with our platform, we are pursuing quantitative approaches to evaluate the site-specific addition of a 21st amino acid to proteins in yeast. The use of display-based reporter constructs with tags before and after the intended site for ncAA insertion (via stop codon suppression) enables us to evaluate the efficiency and fidelity of ncAA incorporation. Using a series of orthogonal translation system components that facilitate insertion of aromatic ncAAs into proteins, we observe stop codon readthrough efficiencies of approximately 10 percent and misincorporation of canonical amino acids at levels of 20 percent or lower using flow cytometry-based readouts. This level of performance is sufficient for some applications but will need to be improved in order to realize its full potential. Toward this end, we have begun to investigate several potential strategies for enhancing both stop codon readthrough and ncAA incorporation fidelity. Our findings suggest that changing orthogonal translation system component expression levels or manipulating the yeast genome both appear to be promising strategies for enhancing genetic code manipulation. However, flow cytometry-based readouts also reveal that high cell-to-cell variability and promiscuity of "orthogonal" translation system components can confound uniformly efficient, high fidelity ncAA incorporation. We anticipate that these findings, further quantitative measurements, and screening approaches will reveal important principles of genetic code manipulation and provide opportunities for adding a range of functionalities to proteins displayed on the yeast surface. Expanding the range of genetically encodable chemical groups will facilitate a range of chemical transformation on the yeast surface and the direct encoding of functional groups that can disrupt protease and peptidase activities in ways that the canonical amino acids cannot.

To begin to exploit the existing capabilities of our platform, we are investigating strategies for incorporating ncAAs into antibody fragments at positions located close to antibody complementarity determining regions. We have identified several conserved antibody positions that appear to tolerate aromatic ncAA incorporation and facilitate chemical modification. The insertion of ncAAs at one of several conserved positions results in antibody variants with similar binding properties to those of the parent clones, confirming that the sites we have identified tolerate ncAA substitutions. The introduction of azide or alkyne functionalities at these positions facilitates chemical modification on the yeast surface. Experiments with five antibody variants have uncovered general reactivity trends: 1) conserved sites appear to react similarly in different antibodies; 2) subtle changes in amino acid side chain properties affect chemical reactivity; 3) changes in the structures of small molecules used to derivatize antibody fragments affect apparent reactivity. These findings suggest that chemical modifications of antibodies on the yeast surface are generalizable, provided that empirical tests are used to identify suitable reaction conditions for the individual ncAA-small molecule pair to be employed. We have exploited our understanding of these trends to robustly and consistently construct "hybrid" structures containing functional groups suitable for disrupting major families and superfamilies of enzymes including matrix metalloproteinases and serine hydrolases. Efforts to identify protein-small molecule hybrids that disrupt the activities of members of these groups of enzymes are ongoing. Our approach to incorporating more chemical functionality into proteins displayed on the yeast surface will enable us to explore structures not accessible with current inhibitor discovery technologies in high throughput. These efforts may lead to inhibitors with enhanced specificity and potency with potential applications for evaluating the biology of the tumor microenvironment and could further lead to the identification of new therapeutic candidates.

nature chemical biology

ARTICLE

PUBLISHED ONLINE: 14 NOVEMBER 2016 | DOI: 10.1038/NCHEMBIO.2233

Tunable thermal bioswitches for *in vivo* control of microbial therapeutics

Dan I Piraner[1-3], Mohamad H Abedi[1-3], Brittany A Moser[1], Audrey Lee-Gosselin[1] & Mikhail G Shapiro[1*]

Temperature is a unique input signal that could be used by engineered microbial therapeutics to sense and respond to host conditions or spatially targeted external triggers such as focused ultrasound. To enable these possibilities, we present two families of tunable, orthogonal, temperature-dependent transcriptional repressors providing switch-like control of bacterial gene expression at thresholds spanning the biomedically relevant range of 32–46 °C. We integrate these molecular bioswitches into thermal logic circuits and demonstrate their utility in three *in vivo* microbial therapy scenarios, including spatially precise activation using focused ultrasound, modulation of activity in response to a host fever, and self-destruction after fecal elimination to prevent environmental escape. This technology provides a critical capability for coupling endogenous or applied thermal signals to cellular function in basic research, biomedical and industrial applications.

Rapid advances in synthetic biology[1,2] are driving the development of genetically engineered microbes as therapeutic[3–6] and diagnostic[7–10] agents for a multitude of human diseases. A critical capability for many envisioned applications is the ability to control the function of engineered microbes *in situ* to enable spatially and temporally specific activation at anatomical and disease sites such as the gastrointestinal tract or tumors[2]. However, among existing control methods, systemic chemical administration typically lacks the spatial precision needed to modulate activity at specific anatomical locations, while optical approaches suffer from poor light penetration into tissues[11]. On the other hand, temperature can be controlled both globally and locally—at depth—using technologies such as focused ultrasound[12], infrared light[13] and magnetic particle hyperthermia[14]. In addition, body temperature can serve as an indicator of microbial entry into and exit from the host organism and of the host's condition (e.g., fever).

Notwithstanding this potential, remarkably few high-performance thermal bioswitches are available to control gene expression in engineered microbes. The ideal bioswitch should have a sharp thermal transition resulting in a large change in activity (>100-fold over a few degrees), and its switching temperature should be tunable to enable a broad range of applications. In addition, the bioswitch should be orthogonal to endogenous cellular machinery and compatible with other thermo-responsive components to allow multiplexed thermal logic. Existing temperature-dependent regulators of gene expression—including microbial heat shock factors, membrane-associated proteins, RNA thermometers and transcriptional repressors—fail to fulfill these criteria. Microbial heat shock promoters undergo a relatively low level of thermal induction (~10-fold)[15], have crosstalk with other stimuli such as chemical stress[16] and may be difficult to tune without deleterious effects on the cell. Likewise, the switching temperature of membrane-associated proteins depends on bilayer composition and occupies second-messenger pathways[17]. Meanwhile, RNA thermometers, while orthogonal to the host and amenable to tuning, generally have small dynamic ranges (<10-fold) and broad transitions (>10 °C)[18–23]. Of the available molecular machinery, several natural and mutant transcriptional repressors have the most robust thermal switching and potential for orthogonality[24–30]. However, their relative performance has not been characterized, they have not been systematically engineered to operate at specific transition temperatures, and their potential utility for *in vivo* microbial therapy applications has not been demonstrated.

To address the need for robust, tunable, orthogonal thermal control of engineered microbial systems, we systematically screened candidate transcriptional regulators, developed two orthogonal families of high-performance thermal bioswitches with tunable thresholds within the biomedically relevant range of 32–46 °C, and demonstrated the potential utility of these devices in three distinct *in vivo* scenarios relevant to mammalian microbial therapy. These scenarios include spatially selective activation within a mammalian host using focused ultrasound, sensing and response to a fever, and self-destruction at ambient temperatures to prevent environmental contamination after leaving the intended host.

RESULTS

High-performance thermal bioswitches

In order to engineer new families of robust, tunable, orthogonal thermal bioswitches, we began by characterizing the performance of six temperature-dependent transcriptional repressors and six heat shock promoters. Our panel included TlpA, a transcriptional autorepressor from the virulence plasmid of *Salmonella typhimurium*. This protein contains an approximately 300-residue C-terminal coiled-coil domain that undergoes sharp, temperature-dependent uncoiling between 37 °C and 45 °C, and an N-terminal DNA-binding domain that, in its low-temperature dimeric state, blocks transcription from the 52-bp TlpA operator–promoter[24,30]. In addition, we tested a well-known temperature-sensitive variant of the bacteriophage λ repressor cI (mutant cI[857], here referred to as TcI) acting on a tandem pR–pL operator–promoter[25]. In most previous applications, TcI repression has been modulated via large changes in temperature (e.g., steps from 30 °C to 42 °C)[25]. However, its original description as a virulence factor suggested that much sharper switching may be possible[26]. Alongside TlpA and TcI, we tested four reported temperature-sensitive mutants of the *Escherichia coli* repressors TetR (A89D and I193N)[27] and LacI (A241T and G265D)[28,29], together with a panel of *E. coli* heat shock promoters, including GrpE, HtpG, Lon, RpoH, Clp and DnaK (**Fig. 1a**).

The performance of these constructs is summarized in **Figure 1b**. TlpA and TcI had by far the largest dynamic ranges (355 ± 45 and >1,432, respectively), reflecting a combination of tighter

[1]Division of Chemistry and Chemical Engineering, California Institute of Technology, Pasadena, California, USA. [2]Division of Biology and Biological Engineering, California Institute of Technology, Pasadena, California, USA. [3]These authors contributed equally to this work. *e-mail: mikhail@caltech.edu

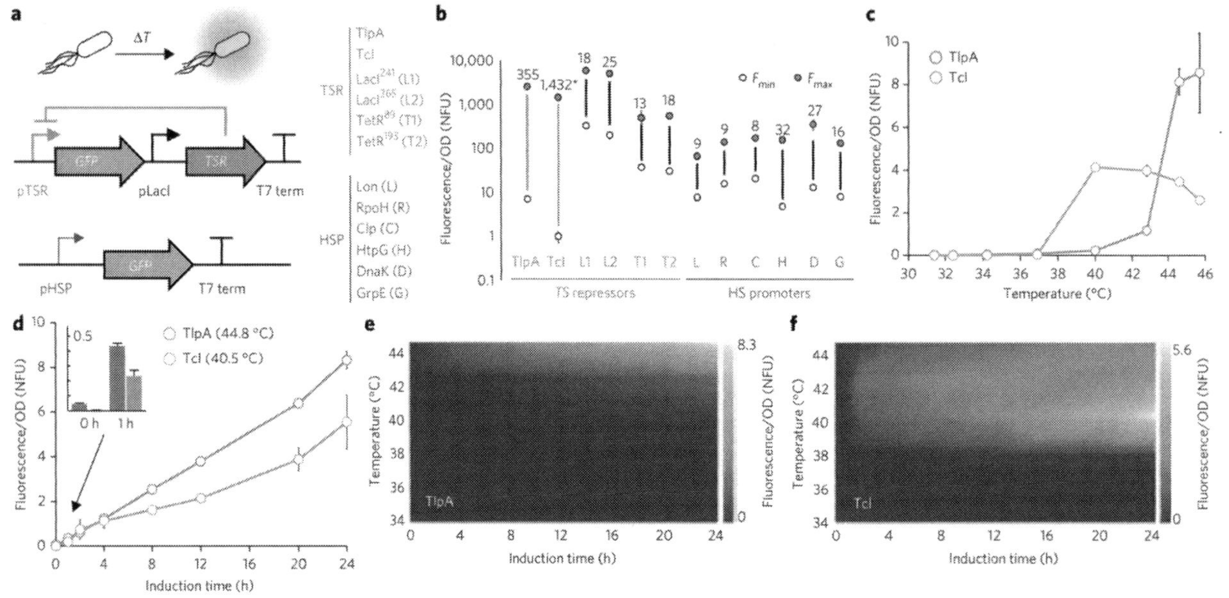

Figure 1 | High-performance thermal bioswitches. (**a**) Constructs used to assay the performance of temperature-sensitive repressors (TSR, top) and heat shock promoters (HSP, bottom). The specific repressors and promoters assayed are listed in blue and gray, respectively. (**b**) Optical density (OD)-normalized fluorescence after 12 h of thermal induction for the constructs shown in **a**. NFU, normalized fluorescence units. F_{min} represents expression at 31.4 °C; F_{max} is the maximum fluorescence intensity measured for each construct, measured up to 45.7 °C. The fold changes between F_{min} and F_{max} are listed above each sample. Where not seen, error bars are smaller than the symbol. $N = 4$ for TSRs and $N = 3$ for HSPs. The F_{min} for TcI (indicated by *) is reported from measurement at 34.2 °C because expression at lower temperatures was below the detection limit of the assay. (**c**) OD-normalized fluorescence from the TlpA- and TcI-regulated constructs as a function of induction temperature for a fixed duration of 12 h. $N = 4$. (**d**) OD-normalized fluorescence as a function of thermal induction duration at the maximal induction temperature for the TlpA and TcI constructs. $N = 4$. (**e**,**f**) OD-normalized fluorescence landscapes for TlpA- and TcI-gated constructs, respectively, as a function of both incubation temperature and duration. Data shown interpolated from an 8 × 8 sampling matrix. All samples in **d**–**f** were maintained at 30 °C after the indicated period of thermal induction for a total experimental duration of 24 h before measurement. Error bars represent ± s.e.m.

repression at subthreshold temperatures and stronger promoter activity above threshold. Both of these repressors show sharp thermal transitions, with more than 30-fold induction over ranges of 5 °C and 3 °C centered at 43.5 °C and 39.5 °C for TlpA and TcI, respectively (**Fig. 1c**). Furthermore, both systems are capable of rapid induction, with greater than ten-fold changes in expression observed after a 1-h thermal stimulus (**Fig. 1d**). Complete time-temperature induction profiles for TlpA and TcI are shown in **Figure 1e,f**. In addition to their high performance, TlpA and TcI are expected to be more orthogonal to cellular machinery than both the native heat shock promoters and the engineered TetR and LacI repressors, the latter of which are used in multiple endogenous and engineered gene circuits[31–33]. A homology search[34] showed that TlpA and TcI repressors are present in far fewer bacterial species than either TetR or LacI (**Supplementary Results**, **Supplementary Fig. 1**). On the basis of these factors, we chose TlpA and TcI as our starting points for further bioswitch engineering.

Since the behavior of the TlpA operator–promoter has not been studied in *E. coli*, we characterized its molecular mechanisms to inform its utilization in genetic circuits. As shown in **Supplementary Figure 2**, the TlpA operator is a strong promoter (88-fold stronger than $LacI^{Q}$) driven by the transcription factor σ^{70}. Interestingly, this promoter has bidirectional activity, with identical thermal regulation in both orientations but approximately 200-fold lower maximal expression in the reverse direction (**Supplementary Fig. 2c,d**). This property will enable convenient adjustment of TlpA-regulated expression according to circuit requirements.

Tuning bioswitch activation temperatures

Applications in microbial therapy require thermal bioswitches that activate at different transition temperatures. For example, a host colonization sensor should be activated at 37 °C, whereas a fever detector may work best with a thermal threshold of 39 °C, and a focused ultrasound-activated switch may require a transition point of 41 °C to avoid nonspecific actuation. Synthetic biology applications outside biomedicine may likewise have a variety of thermal requirements. It is thus highly desirable to be able to tune thermal bioswitches to activate at new temperatures while retaining sharp, robust switching performance. To enable such tuning of TlpA and TcI, we devised a simple and effective high-throughput assay based on colony fluorescence. We grew *E. coli* expressing GFP under the control of mutant repressors (generated by error-prone PCR) on solid medium and replica-plated the colonies onto separate plates for simultaneous incubation at desired 'off' and 'on' temperatures (**Supplementary Fig. 3a**). We then imaged the plates through wide-field fluorescence, as shown in **Figure 2a**. As expected, many colonies showed constitutive expression (ostensibly due to loss of repressor function) while others failed to derepress (most likely retaining their original high transition point). However, several colonies showed thermal induction in the desired regime. Within each screen, we selected several such colonies to undergo liquid-phase characterization of induction temperature, switching sharpness and expression levels (**Fig. 2b**). From these variants, we selected mutants that retained the desirable performance characteristics of the wild-type repressor, but with shifted transition temperatures.

Screening of TlpA mutants at off–on temperatures of 30–37 °C and 37–40 °C produced high-performance bioswitches centered at 36 °C and 39 °C, respectively, which we named $TlpA_{36}$ and $TlpA_{39}$ (**Fig. 2c**). For TcI, we selected both downshifted (TcI_{38}, $T_m = 38$ °C) and upshifted (TcI_{42}, $T_m = 42$ °C) variants relative to the original protein (**Fig. 2d**). Together, the engineered TlpA and TcI repressor

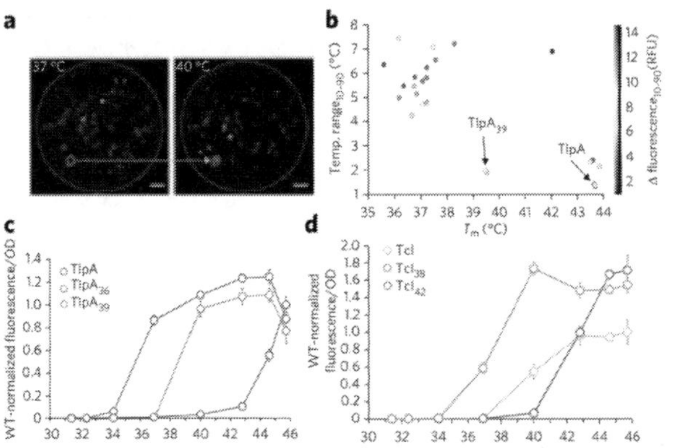

Figure 2 | Tuning the transition temperature of thermal bioswitches. (**a**) Fluorescence image of replica plates used to screen for TlpA variants turning on between 37 °C and 40 °C. One colony selected for further assay is indicated by the orange circle. Scale bars, 1 cm. (**b**) TlpA variants plotted by their measured midpoint transition temperatures (T_{50}) and 10–90% transition range (T_{10-90}), estimated by linear interpolation. The color of each data point maps to the change in fluorescence over the T_{10-90} span. (**c**) OD-normalized fluorescence of the novel TlpA variants normalized to wild type (WT). (**d**) OD-normalized fluorescence of the novel TcI variants normalized to wild type. $N = 4$ for **c** and **d**. RFU, relative fluorescence units. Error bars represent ± s.e.m.

families cover the biomedically relevant range of 32 °C to 46 °C (**Supplementary Fig. 3b**) while demonstrating a dynamic range similar to that of the wild-type protein (**Supplementary Table 1**). The observed decrease in fluorescence at the highest temperatures tested may be due to thermal instability of the cell's transcriptional and translational machinery. The amino acid substitutions identified in these bioswitch variants are shown in **Supplementary Figure 4** and listed in **Supplementary Table 3**. Remarkably, a single round of mutagenesis was sufficient in all cases to obtain at least one variant with the desired switching behavior, suggesting that both TlpA and TcI are highly tunable for a broad range of applications.

Thermal logic circuits using orthogonal bioswitches

To enable microbial therapy applications, it is useful to develop thermal logic circuits capable of controlling multiple functions at different temperatures or confining activity to within a narrow thermal range. This would enable cells to, for example, initiate one therapeutic function upon host colonization and switch to a different function during a host fever response or local activation with focused ultrasound. We hypothesized that since TlpA and TcI act on orthogonal target sequences, we could combine them in circuits designed for multiplexed thermal control or bandpass activation of microbial function. To assess the first possibility (**Fig. 3a**), we made a construct encoding a GFP modulated by $TlpA_{36}$ and an RFP regulated by TcI (**Fig. 3b**). As predicted, upon exposure to a range of temperatures, the two reporter genes were activated independently at their expected thresholds, with no apparent crosstalk in their induction (**Fig. 3c**). Independent thermal control of the coexpressed circuits is illustrated by spatially patterned bacterial variants incubated at 37 °C and 42 °C (**Fig. 3d**). Next, to develop a thermal bandpass filter (**Fig. 3e**), we engineered a circuit placing the expression of RFP under the control of the lambda promoter, gated by both TcI (turning on above 36 °C) and the temperature-independent wild-type cI repressor, which was itself placed under the control of TlpA (activating above 43 °C) as shown in **Figure 3f**. The cI open reading frame was preceded by a synthetic terminator (BBa_B1002) terminator and a weak ribosome-binding site to reduce buildup of this repressor at 40–43 °C due to leakage of the upstream TlpA operon. This resulted in RFP expression confined between 36 °C and 44 °C, while simultaneously turning on GFP above RFP's turn-off temperature (**Fig. 3g,h**).

Spatially targeted control using focused ultrasound

After developing TlpA and TcI-based thermal bioswitches, we demonstrated their utility in three prototypical microbial therapy scenarios. First, we tested the ability of thermal bioswitches to mediate spatially selective control of microbial therapies using focused ultrasound, a modality that is well established in its ability to elevate temperatures in deep tissues with millimeter spatial precision[12] and is used clinically to treat diseases such as cancer[35] and essential tremor[36]. Focused ultrasound has been used to activate gene expression in mammalian cells[37], but it has not, to our

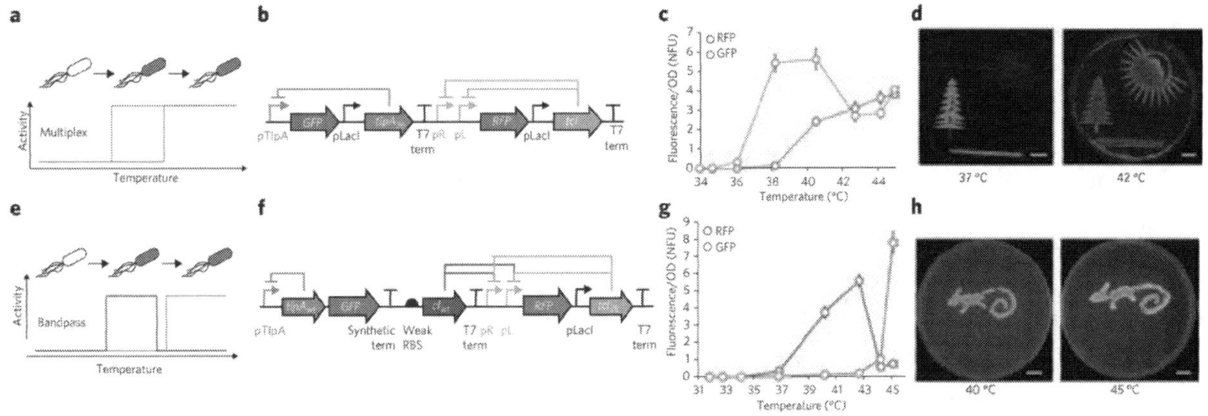

Figure 3 | Thermal logic circuits. (**a**) Diagram illustrating multiplexed thermal activation. (**b**) Circuit diagram of the pCali2 plasmid, which contains GFP gated by $TlpA_{36}$ and RFP gated by TcI. (**c**) Expression of GFP and RFP from pCali2-containing *E. coli* over the indicated range of temperatures (12-h incubation). (**d**) Plate images of overlaid GFP and RFP fluorescence from the pCali2 plasmid (grass) and plasmids expressing only the green (tree) and red (sun) components. Note that at 42 °C the grass shows both green and red fluorescence. (**e**) Diagram illustrating a thermal bandpass filter. (**f**) Circuit diagram of the pThermeleon plasmid, in which RFP is gated by TcI_{38} and also by the wild-type cI repressor. GFP is gated by $TlpA_{wt}$ on the same plasmid, which also weakly drives the expression of cI_{wt} through a T7 terminator and weak ribosome-binding site. (**g**) Thermal expression profile of RFP and GFP from pThermeleon-containing *E. coli* (12 h incubation). (**h**) Overlaid GFP and RFP fluorescence images of plated bacteria containing pThermeleon cultured at 40 °C and 45 °C. Scale bars, 1 cm. $N = 4$ for **c** and **g**. Error bars represent ± s.e.m.

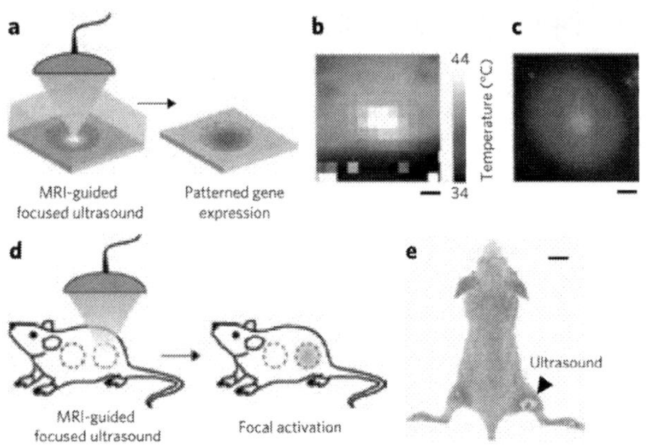

Figure 4 | Remote control of bacterial agents using focused ultrasound. (**a**) Illustration of the *in vitro* focused ultrasound experiment: focused ultrasound is used to heat a target area of a bacterial culture lawn through a tofu phantom (depicted as translucent) under MRI guidance, followed by fluorescence imaging. (**b**) MRI-based temperature map of the bacterial specimen during steady-state ultrasound application, overlaid on a raw grayscale MRI image of the phantom. (**c**) Fluorescence image of the region targeted by ultrasound, showing activation consistent with a bacterial construct expressing GFP under the control of $TlpA_{36}$ and RFP regulated by TcI. (**d**) Illustration of the *in vivo* experiment, in which focused ultrasound is used to activate subcutaneously injected bacterial agents at a specific anatomical site. (**e**) Representative thresholded fluorescence map of a mouse injected subcutaneously in both left and right hindlimbs with *E. coli* expressing GFP under the control of $TlpA_{36}$, following ultrasound activation directed at only the right hindlimb. Scale bars, 2 mm (**b**,**c**) and 1 cm (**e**).

knowledge, been employed to control the activity of microbes *in vivo*. Such control could be highly advantageous in applications in which the activity of a systemically administered microbial therapy needs to be localized to a specific anatomical site, such as a deep-seated tumor or section of the gastrointestinal tract, that would be difficult to reach with optogenetic triggers. To test this concept, we first activated gene expression using focused ultrasound in tissue-mimicking phantoms under the guidance of magnetic resonance imaging (MRI)[38] (**Fig. 4a**). This guidance enabled precise spatial targeting of the ultrasound focus and real-time monitoring and adjustment of local temperature. We applied this technique to a flat lawn of *E. coli* containing the multiplexed expression circuit shown in **Figure 3b**. This specimen was assembled with a tissue-mimicking tofu phantom, and steady-state focal heating over 45 min resulted in a radial thermal gradient with an average focal temperature of 42 °C, as observed by real-time MRI thermometry (**Fig. 4b**). A corresponding pattern of spatially localized fluorescence is seen in **Figure 4c**.

To establish the feasibility of this approach *in vivo*, we injected *E. coli* expressing GFP under the control of $TlpA_{36}$ subcutaneously into both hindlimbs of a nude mouse and applied MRI-guided focused ultrasound to one location (**Fig. 4d**) to produce a local steady-state temperature of 41 °C for 45 min to 1 h. This thermal dose is below the damage thresholds for mammalian tissues such as muscle and brain[39,40]. *In vivo* fluorescence imaging 4 h after ultrasound treatment showed robust expression of GFP specifically at the ultrasound-targeted anatomical site (**Fig. 4e**). Two additional animals undergoing the same procedure are shown in **Supplementary Figure 5**. $TlpA_{36}$ was selected as the thermal bioswitch for these experiments because its activation threshold is approximately 4 °C above the typical murine cutaneous temperature[41], a sufficient difference for site-specific ultrasound activation.

Programmed responses to mammalian host temperature

Next, we sought to develop autonomous thermosensitive microbes responsive to endogenous changes in host temperature. First, we investigated whether bacteria can be engineered to sense and respond to a host fever (**Fig. 5a**). We subcutaneously injected one flank of a nude mouse with *E. coli* expressing GFP under the control of $TlpA_{36}$, and the other flank with *E. coli* expressing GFP controlled by wild-type TlpA as a high-threshold control for nonspecific activation. The mouse was then housed at 41 °C for 2 h in an established fever model paradigm[42]. *In vivo* fluorescence imaging 4 h after fever induction showed robust expression of GFP in the flank injected with $TlpA_{36}$-regulated bacteria (**Fig. 5b**). No significant activation was seen in the opposite flank or in a mouse housed at room temperature (**Fig. 5c**). Two additional replicates of this experiment are shown in **Supplementary Figure 6**.

Second, we tested whether a thermal bioswitch operating at 37 °C could be used to confine the activity of genetically engineered microbes to the *in vivo* environment of a mammalian host and thereby limit the potential for environmental contamination. Toward this end, we designed a genetic circuit in which $TlpA_{36}$ controls the expression of CcdA, a bacterial antitoxin, while constitutively expressing the toxin CcdB, thereby restricting growth to temperatures above 37 °C (**Fig. 5d**). A degradation tag was fused to CcdA to accelerate cell death at nonpermissive temperatures. Bacteria carrying this plasmid grew normally at 37 °C, while bacteria incubated at 25 °C had significantly reduced survival as demonstrated by their colony-forming unit (CFU) counts (**Fig. 5e**). We administered these bacteria to mice by oral gavage and collected fecal pellets after 5 h to allow transit through the gastrointestinal tract. The pellets were kept for 24 h at either 25 °C, corresponding to excretion into the ambient environment, or at 37 °C, equivalent to persistent residence in the gut, and subsequently assayed for colony formation. The survival of cells excreted into ambient temperature was reduced by 10,000-fold as compared to cells maintained under host conditions (**Fig. 5f**).

DISCUSSION

Our results establish two new families of high-performance, orthogonal thermal bioswitches with tunable activation thresholds to enable a variety of biotechnology applications. Both TlpA- and TcI-based switches respond to temperature with hundreds-fold changes in gene expression. If needed, this response could be further boosted using well-established strategies such as tandem operators and positive feedback amplification[43,44]. In addition, the temporal response of thermal bioswitch circuits could be made either more transient, by manipulating the lifetime of the resulting transcripts and proteins, or longer lasting, by using genetic toggle switches or recombinase-based architectures[28,45]. This may enable persistent functions to be controlled with thermal stimuli shorter than those used in this study.

Our strategy for tuning the thermal response of TlpA and TcI is rapid and simple to implement. The fact that we could identify high-performance variants with new transition temperatures by screening several hundred mutants suggests that many different sequences could satisfy a given thermal requirement. Here, we focused on bioswitches operating between 32 °C and 46 °C, in keeping with potential therapeutic applications and the thermal tolerance of our bacterial chassis. We expect that a similar selection strategy using thermophilic or cryophilic bacteria could be used to tune TlpA and TcI over a yet broader temperature range for industrial applications such as biofuel production. The thermal stability of the regulated gene product will need to be taken into account in these scenarios. Within the temperature range tested in this study, GFP, RFP and CcdA were functional.

The presented bioswitches have sequences orthogonal to bacterial host machinery and to each other, enabling multiplexed thermal

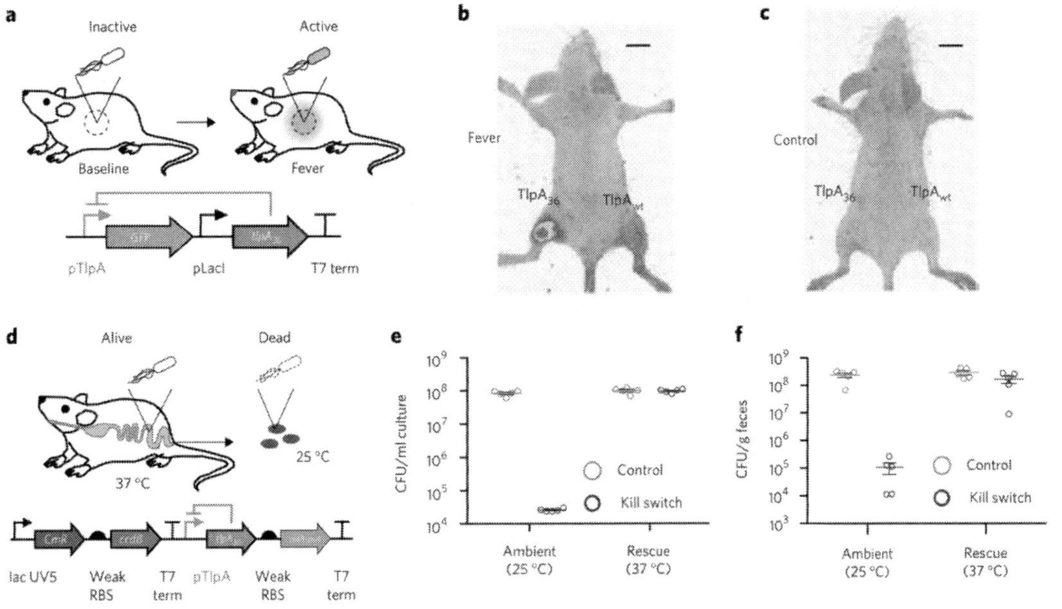

Figure 5 | Programmed responses to mammalian host temperature. (**a**) Illustration of the fever-induced activation experiment and circuit diagram of the corresponding *E. coli* construct. (**b**) Representative thresholded fluorescence map of a mouse that underwent fever induction after being injected subcutaneously with plasmids expressing $TlpA_{36}$- and TlpA-regulated GFP into the left and right hindlimbs, respectively. (**c**) Representative thresholded fluorescence map of a mouse that was prepared identically to the animal in **b** but maintained at room temperature. (**d**) Illustration of the temperature-based host confinement strategy, and circuit diagram of the thermal kill switch permitting bacterial survival only at temperatures above 36 °C, at which antitoxin expression is derepressed by $TlpA_{36}$. (**e**) Colony counts from liquid cultures of kill-switch-containing cells and controls (containing no toxin system) after 24 h of incubation at the indicated temperature. $P = 0.0002$ for kill switch versus control cells at 25 °C and $P < 0.0001$ for kill switch at 25 °C versus 37 °C. $N = 5$. (**f**) Colony counts in fecal samples freshly collected from $N = 5$ mice 5 h after oral gavage of kill-switch-containing *E. coli* or controls. The feces were incubated at a temperature representative of post-defecation conditions (25 °C) or were rescued at 37 °C. $P = 0.0067$ for kill switch versus control cells at 25 °C and $P = 0.0275$ for kill switch at 25 °C versus 37 °C. $N = 5$. Error bars represent ± s.e.m. Scale bars, 1 cm.

actuation. If desired, additional multiplexing could be accomplished by replacing the DNA-binding domains of TlpA or TcI with those of other dimeric repressors. Additional engineering may be needed to adapt this technology to other host organisms. Certain species of therapeutic microbes, such as *Lactobacillus* spp., are able to use some promoters transferred directly from *E. coli*[46]. Others may require incorporation of the relevant operator sequences into promoters native to the host[47]. Alternatively, fusions of TlpA and TcI with DNA-binding domains from other microbes or eukaryotes could adapt TlpA and TcI for use in these species.

The three *in vivo* scenarios demonstrated in this work will inform the use of thermal bioswitches in future microbial therapy applications. For example, the ability to detect a host's fever provides a potential safety mechanism with which to curtail effector activity in response to runaway inflammation, a major and sometimes lethal side-effect of antitumor cell therapy[48]. In addition, temperature-dependent kill switches can be used to restrict the survival of enterically administered microbes to host body temperature and thereby mitigate the risk of patients shedding genetically modified, pharmaceutically active organisms into the surrounding environment. Such kill switches can be incorporated into recently developed multilayered and multi-input containment circuits for greater efficiency in preventing mutational escape[49,50]. Furthermore, the ability to activate microbial function at specific anatomical sites using focused ultrasound opens new therapeutic avenues by potentially allowing a physician to locally target therapeutic effects that would be intolerable via systemic administration. Additionally, the ability to trigger gene expression *in vivo* can be combined with genetically encoded genomic or proteomic tools[51–53] to enable the study of cellular signaling within the context of mammalian hosts.

Received 12 May 2016; accepted 16 September 2016; published online 14 November 2016

METHODS

Methods and any associated references are available in the online version of the paper.

References

1. Ford, T.J. & Silver, P.A. Synthetic biology expands chemical control of microorganisms. *Curr. Opin. Chem. Biol.* **28**, 20–28 (2015).
2. Fischbach, M.A., Bluestone, J.A. & Lim, W.A. Cell-based therapeutics: the next pillar of medicine. *Sci. Transl. Med.* **5**, 179ps177 (2013).
3. Steidler, L. *et al.* Treatment of murine colitis by *Lactococcus lactis* secreting interleukin-10. *Science* **289**, 1352–1355 (2000).
4. Daniel, C., Roussel, Y., Kleerebezem, M. & Pot, B. Recombinant lactic acid bacteria as mucosal biotherapeutic agents. *Trends Biotechnol.* **29**, 499–508 (2011).
5. Claesen, J. & Fischbach, M.A. Synthetic microbes as drug delivery systems. *ACS Synth. Biol.* **4**, 358–364 (2015).
6. Wells, J.M. & Mercenier, A. Mucosal delivery of therapeutic and prophylactic molecules using lactic acid bacteria. *Nat. Rev. Microbiol.* **6**, 349–362 (2008).
7. Courbet, A., Endy, D., Renard, E., Molina, F. & Bonnet, J. Detection of pathological biomarkers in human clinical samples via amplifying genetic switches and logic gates. *Sci. Transl. Med.* **7**, 289ra283 (2015).
8. Danino, T. *et al.* Programmable probiotics for detection of cancer in urine. *Sci. Transl. Med.* **7**, 289ra28 (2015).
9. Kotula, J.W. *et al.* Programmable bacteria detect and record an environmental signal in the mammalian gut. *Proc. Natl. Acad. Sci. USA* **111**, 4838–4843 (2014).
10. Archer, E.J., Robinson, A.B. & Süel, G.M. Engineered *E. coli* that detect and respond to gut inflammation through nitric oxide sensing. *ACS Synth. Biol.* **1**, 451–457 (2012).
11. Ntziachristos, V. Going deeper than microscopy: the optical imaging frontier in biology. *Nat. Methods* **7**, 603–614 (2010).

12. Haar, G.T. & Coussios, C. High intensity focused ultrasound: physical principles and devices. *Int. J. Hyperthermia* **23**, 89–104 (2007).
13. Huang, X., El-Sayed, I.H., Qian, W. & El-Sayed, M.A. Cancer cell imaging and photothermal therapy in the near-infrared region by using gold nanorods. *J. Am. Chem. Soc.* **128**, 2115–2120 (2006).
14. Thiesen, B. & Jordan, A. Clinical applications of magnetic nanoparticles for hyperthermia. *Int. J. Hyperthermia* **24**, 467–474 (2008).
15. Zhao, K., Liu, M. & Burgess, R.R. The global transcriptional response of *Escherichia coli* to induced sigma 32 protein involves sigma 32 regulon activation followed by inactivation and degradation of sigma 32 in vivo. *J. Biol. Chem.* **280**, 17758–17768 (2005).
16. de Marco, A., Vigh, L., Diamant, S. & Goloubinoff, P. Native folding of aggregation-prone recombinant proteins in *Escherichia coli* by osmolytes, plasmid- or benzyl alcohol-overexpressed molecular chaperones. *Cell Stress Chaperones* **10**, 329–339 (2005).
17. Inda, M.E. *et al.* A lipid-mediated conformational switch modulates the thermosensing activity of DesK. *Proc. Natl. Acad. Sci. USA* **111**, 3579–3584 (2014).
18. Kortmann, J., Sczodrok, S., Rinnenthal, J., Schwalbe, H. & Narberhaus, F. Translation on demand by a simple RNA-based thermosensor. *Nucleic Acids Res.* **39**, 2855–2868 (2011).
19. Neupert, J., Karcher, D. & Bock, R. Design of simple synthetic RNA thermometers for temperature-controlled gene expression in *Escherichia coli*. *Nucleic Acids Res.* **36**, e124 (2008).
20. Waldminghaus, T., Kortmann, J., Gesing, S. & Narberhaus, F. Generation of synthetic RNA-based thermosensors. *Biol. Chem.* **389**, 1319–1326 (2008).
21. Hoynes-O'Connor, A., Hinman, K., Kirchner, L. & Moon, T.S. De novo design of heat-repressible RNA thermosensors in *E. coli*. *Nucleic Acids Res.* **43**, 6166–6179 (2015).
22. Satija, R., Sen, S., Siegal-Gaskins, D. & Murray, R.M. Design of a toolbox of RNA thermometers. Preprint at *bioRxiv* http://dx.doi.org/10.1101/017269 (2015).
23. Wieland, M. & Hartig, J.S. RNA quadruplex-based modulation of gene expression. *Chem. Biol.* **14**, 757–763 (2007).
24. Hurme, R., Berndt, K.D., Namork, E. & Rhen, M. DNA binding exerted by a bacterial gene regulator with an extensive coiled-coil domain. *J. Biol. Chem.* **271**, 12626–12631 (1996).
25. Valdez-Cruz, N.A., Caspeta, L., Pérez, N.O., Ramírez, O.T. & Trujillo-Roldán, M.A. Production of recombinant proteins in *E. coli* by the heat inducible expression system based on the phage lambda pL and/or pR promoters. *Microb. Cell Fact.* **9**, 18 (2010).
26. Sussman, R. & Jacob, F. Sur un systeme de repression thermosensible chez le bacteriophage lambda d'*Escherichia coli*. *C. R. Hebd. Seances Acad. Sci.* **24**, 1517–1519 (1962).
27. Wissmann, A. *et al.* Selection for Tn10 tet repressor binding to tet operator in *Escherichia coli*: isolation of temperature-sensitive mutants and combinatorial mutagenesis in the DNA binding motif. *Genetics* **128**, 225–232 (1991).
28. Chao, Y.P., Chern, J.T., Wen, C.S. & Fu, H. Construction and characterization of thermo-inducible vectors derived from heat-sensitive lacI genes in combination with the T7 A1 promoter. *Biotechnol. Bioeng.* **79**, 1–8 (2002).
29. McCabe, K.M., Lacherndo, E.J., Albino-Flores, I., Sheehan, E. & Hernandez, M. LacI(Ts)-regulated expression as an in situ intracellular biomolecular thermometer. *Appl. Environ. Microbiol.* **77**, 2863–2868 (2011).
30. Hurme, R., Berndt, K.D., Normark, S.J. & Rhen, M. A proteinaceous gene regulatory thermometer in Salmonella. *Cell* **90**, 55–64 (1997).
31. Wilson, C.J., Zhan, H., Swint-Kruse, L. & Matthews, K.S. The lactose repressor system: paradigms for regulation, allosteric behavior and protein folding. *Cell. Mol. Life Sci.* **64**, 3–16 (2007).
32. Bertram, R. & Hillen, W. The application of Tet repressor in prokaryotic gene regulation and expression. *Microb. Biotechnol.* **1**, 2–16 (2008).
33. Jensen, P.R., Westerhoff, H.V. & Michelsen, O. The use of lac-type promoters in control analysis. *Eur. J. Biochem.* **211**, 181–191 (1993).
34. Altschul, S.F. *et al.* Gapped BLAST and PSI-BLAST: a new generation of protein database search programs. *Nucleic Acids Res.* **25**, 3389–3402 (1997).
35. Al-Bataineh, O., Jenne, J. & Huber, P. Clinical and future applications of high intensity focused ultrasound in cancer. *Cancer Treat. Rev.* **38**, 346–353 (2012).
36. Elias, W.J. *et al.* A pilot study of focused ultrasound thalamotomy for essential tremor. *N. Engl. J. Med.* **369**, 640–648 (2013).
37. Deckers, R. *et al.* Image-guided, noninvasive, spatiotemporal control of gene expression. *Proc. Natl. Acad. Sci. USA* **106**, 1175–1180 (2009).
38. Fite, B.Z. *et al.* Magnetic resonance thermometry at 7T for real-time monitoring and correction of ultrasound induced mild hyperthermia. *PLoS One* **7**, e35509 (2012).
39. McDannold, N.J., King, R.L., Jolesz, F.A. & Hynynen, K.H. Usefulness of MR imaging-derived thermometry and dosimetry in determining the threshold for tissue damage induced by thermal surgery in rabbits. *Radiology* **216**, 517–523 (2000).
40. McDannold, N., Vykhodtseva, N., Jolesz, F.A. & Hynynen, K. MRI investigation of the threshold for thermally induced blood-brain barrier disruption and brain tissue damage in the rabbit brain. *Magn. Reson. Med.* **51**, 913–923 (2004).
41. Rudaya, A.Y., Steiner, A.A., Robbins, J.R., Dragic, A.S. & Romanovsky, A.A. Thermoregulatory responses to lipopolysaccharide in the mouse: dependence on the dose and ambient temperature. *Am. J. Physiol. Regul. Integr. Comp. Physiol.* **289**, R1244–R1252 (2005).
42. Pritchard, M.T. *et al.* Protocols for simulating the thermal component of fever: preclinical and clinical experience. *Methods* **32**, 54–62 (2004).
43. Illing, A.C., Shawki, A., Cunningham, C.L. & Mackenzie, B. Substrate profile and metal-ion selectivity of human divalent metal-ion transporter-1. *J. Biol. Chem.* **287**, 30485–30496 (2012).
44. Nistala, G.J., Wu, K., Rao, C.V. & Bhalerao, K.D. A modular positive feedback-based gene amplifier. *J. Biol. Eng.* **4**, 4 (2010).
45. Andersen, J.B. *et al.* New unstable variants of green fluorescent protein for studies of transient gene expression in bacteria. *Appl. Environ. Microbiol.* **64**, 2240–2246 (1998).
46. Natori, Y., Kano, Y. & Imamoto, F. Characterization and promoter selectivity of Lactobacillus acidophilus RNA polymerase. *Biochimie* **70**, 1765–1774 (1988).
47. Mimee, M., Tucker, A.C., Voigt, C.A. & Lu, T.K. Programming a human commensal bacterium, *Bacteroides thetaiotaomicron*, to sense and respond to stimuli in the murine gut microbiota. *Cell Syst.* **1**, 62–71 (2015).
48. Tey, S.-K. Adoptive T-cell therapy: adverse events and safety switches. *Clin. Transl. Immunology* **3**, e17 (2014).
49. Chan, C.T., Lee, J.W., Cameron, D.E., Bashor, C.J. & Collins, J.J. 'Deadman' and 'Passcode' microbial kill switches for bacterial containment. *Nat. Chem. Biol.* **12**, 82–86 (2016).
50. Gallagher, R.R., Patel, J.R., Interiano, A.L., Rovner, A.J. & Isaacs, F.J. Multilayered genetic safeguards limit growth of microorganisms to defined environments. *Nucleic Acids Res.* **43**, 1945–1954 (2015).
51. Lang, K. & Chin, J.W. Cellular incorporation of unnatural amino acids and bioorthogonal labeling of proteins. *Chem. Rev.* **114**, 4764–4806 (2014).
52. Handley, A., Schauer, T., Ladurner, A.G. & Margulies, C.E. Designing cell-type-specific genome-wide experiments. *Mol. Cell* **58**, 621–631 (2015).
53. Grammel, M. & Hang, H.C. Chemical reporters for biological discovery. *Nat. Chem. Biol.* **9**, 475–484 (2013).

Acknowledgments

The authors thank J. Szablowski for assistance with focused ultrasound, A. Mukherjee, J. Bois and A. Gluhovsky for helpful discussions, and S. Zemsky, R. Rezvani, Y. Jiang and G. Ha for experimental assistance. D.I.P. was supported by the NIH fellowship for Predoctoral Training in Biology and Chemistry (T32GM007616). M.H.A. was supported by an NSF graduate research fellowship and the Paul and Daisy Soros Fellowship for New Americans. This work was supported by a DARPA Young Faculty Award (D14AP00050), the Weston Havens Foundation, a Burroughs Wellcome Career Award at the Scientific Interface and the Heritage Medical Research Institute (M.G.S.).

Author contributions

D.I.P. co-conceived and planned the study, generated genetic constructs, evaluated their performance *in vitro* and *in vivo*, and co-wrote the manuscript. M.H.A. co-conceived and planned the study, generated genetic constructs, evaluated their performance *in vitro* and *in vivo*, and co-wrote the manuscript. B.A.M. generated genetic constructs and evaluated their performance *in vitro*. A.L.-G. conducted *in vivo* experiments. M.G.S. co-conceived and supervised the study and co-wrote the manuscript. All authors provided input on the final manuscript.

Competing financial interests

The authors declare no competing financial interests.

Additional information

Any supplementary information, chemical compound information and source data are available in the online version of the paper. Reprints and permissions information is available online at http://www.nature.com/reprints/index.html. Correspondence and requests for materials should be addressed to M.G.S.

ONLINE METHODS

Plasmid construction and molecular biology. All constructs were made via restriction cloning, KLD mutagenesis, or Gibson Assembly using enzymes from New England BioLabs. All plasmids and their sources of genetic material are described in **Supplementary Table 2**. All constructs were cloned in Mach1 *E. coli* (Thermo Fisher) and the sequence-validated plasmids were assayed in NEB10β *E. coli* (NEB). Fluorescent reporters referred to in the text as GFP and RFP are mWasabi and mCherry, respectively[54,55].

Thermal regulation assay. 2 mL cultures of 2× YT medium with 100 μg/mL ampicillin were inoculated with a single colony per culture and grown at 30 °C, 250 rpm for 20 h. After dilution to OD_{600} = 0.1 in LB (Sigma) with 100 μg/mL ampicillin, the cells were propagated at 30 °C, 250 rpm for 1.5 h, after which OD_{600} was measured using a Nanodrop 2000c (Thermo Scientific) in cuvette mode every 10 min. At OD_{600} = 0.25, the cultures were dispensed in 25 μL aliquots into 8 well PCR strips with optically transparent caps (Bio-Rad) using a multichannel pipette and placed into a spatial temperature gradient formed by a Bio-Rad C1000 Touch thermocycler with the lid set to 50 °C. The temperature in each thermocycler well was verified using a TEF-30-T thermocouple (J-KEM Scientific) immersed in 25 μL of pure water within a PCR tube. After the prescribed thermal stimulus, PCR strips were removed, vortexed, and spun down on a tabletop centrifuge and the fluorescence was measured using a Stratagene MX3005p qPCR (Agilent). Immediately after measurement, the cultures were diluted with 75 μL LB/Amp and mixed, after which 90 μL of culture was transferred into 96-well plates (Costar black/clear bottom) for measurement of OD_{600} using a SpectraMax M5 plate reader (Molecular Devices). For studies of gene expression as a function of thermal induction time (**Fig. 1d–f**), samples were returned to incubation at 30 °C after their indicated thermal induction periods such that the total experimental duration was 24 h. Fluorescence measurements were made at the end of this period. Gene expression (E) was determined according to equation (1):

$$E = \frac{F_{\text{sample}} - F_{\text{blank}}}{OD_{\text{sample}} - OD_{\text{blank}}} - \frac{F_{\text{background}} - F_{\text{blank}}}{OD_{\text{background}} - OD_{\text{blank}}} \quad (1)$$

Here, F is the raw fluorescence of the given sample and OD is the OD of the given sample at 600 nm. Raw OD measurements for all experiments are provided in **Supplementary Figure 7**. As expected, bacterial growth is highest in the physiological range of 35 °C to 39 °C. The value of blank fluorescence was determined as the average of all 96 wells in a qPCR plate filled with 25 μL LB. Blank OD was taken as the y-intercept of a standard curve of 90 μL nonfluorescent *E. coli* cultures whose OD_{600} values were determined by cuvette measurements in a Nanodrop 2000c spectrophotometer (96 samples total). Background fluorescence was measured from a nonfluorescent construct derived by mutating the chromophore of mWasabi[56] in the pTlpA-Wasabi plasmid (pTlpA-Wasabi-NF). Fluorescence measurements for the thermal expression landscapes of TlpA and TcI were performed using the plate reader due to signal saturation of the qPCR at the 24 h time point (Sample $N = 3$; Background $N = 2$ for each time point and temperature). Errors from background measurements were propagated by addition in quadrature. Errors from blank measurements were negligible relative to sample-to-sample variation (relative s.d. <2%) and were omitted from the calculation.

Colony screening for TlpA tuning. Error-prone PCR was performed on the repressor gene sequences of pTlpA-Wasabi (Stratagene GeneMorph II kit) and pTcI-Wasabi (NEB Taq Polymerase/0.2 mM $MnCl_2$), and the PCR products were inserted into the parent constructs using Gibson Assembly. The resulting libraries were transformed into NEB10β *E. coli* and plated on LB Agar. Following overnight incubation at 30 °C and the appearance of colonies, a Replica-Plating Tool (VWR 25395-380) was used to replicate each seed plate into two receiver plates. One receiver plate was grown overnight at the desired repressed temperature and the other at the intended activation temperature. Upon the appearance of visible colonies, plates were imaged in a Bio-Rad ChemiDoc MP imager using blue epifluorescence illumination and the 530/28 nm emission filter. Images were examined manually for colonies that appeared dark or invisible on the 'off plate' but showed bright fluorescence on the 'on plate'. Approximately 10^3 colonies were screened per library. These colonies were picked and subjected to the liquid culture thermal activation assay described above, whereupon their thermal induction profile was compared to that of their parent plasmid. Variants that demonstrated sharp switching and large dynamic range between the desired new transition temperatures were sequenced, re-transformed, and assayed using a higher number of replicates.

***In vitro* toxin–antitoxin assays.** NEB10β cells were transformed with the thermally regulated toxin–antitoxin plasmid and allowed to grow at 37 °C overnight. Because reversion of plasmids carrying toxic genes such as CcdB is known to be a common phenomenon, we used a replica plate screen to isolate colonies that maintained a functional thermal kill switch after transformation. To this end, we replica plated the original transformation into two new plates, one incubated at 25 °C and the other maintained at 37 °C. Colonies that grew at the permissive temperature of 37 °C and not at 25 °C were used in downstream *in vitro* or *in vivo* experiments. For *in vitro* experiments, the selected colonies were grown in 2× YT medium with 100 μg/mL ampicillin at 37 °C with shaking until OD_{600} of 0.6 whereupon they were diluted and plated onto LB agar plates. The plates were incubated overnight at either 25 °C or 37 °C, after which colony forming units (CFU) were counted.

Focused ultrasound. MRI-guided focused ultrasound treatment was performed using a 16-channel ultrasound generator, motorized MRI-compatible transducer positioning system and an annular array transducer operating at 1.5 MHz (Image Guided Therapy, Pessac, France). Targeting and real-time imaging was performed using a Bruker Biospec/Avance 7T MRI system with RF excitation delivered by a 7.2 cm diameter volume coil and detection via a 3 cm diameter surface coil. Temperature monitoring was performed using a continuously applied Fast Low Angle Shot sequence with a T_R of 75 ms and T_E of 2.5 ms, matrix size of 32 × 32, and varying FOVs as listed below. Phase images were processed in real time using ThermoGuide software (Image Guided Therapy), and temperature was calculated from the per-pixel phase accumulation due to a decrease in proton precession frequency of 0.01 p.p.m./°C.

For *in vitro* heating, 100 μL of a saturated NEB10β culture expressing the temperature-inducible reporter circuit was plated overnight at 30 °C and incubated for approximately 12 h to form a lawn on a plate containing 0.24% w/v LB (Sigma) and 0.32% w/v Bacto Agar (BD). An approximately 3 cm × 3 cm square of agar was excised from the plate and placed, with the bacterial side facing up, onto a comparably sized pad of 1 cm thick extra firm tofu (O Organics) coated with SCAN ultrasound gel (Parker Laboratories) to exclude air at the interface. A 1 cm high plastic washer made by drilling through the lid of a VWR 35 mm plastic tissue culture dish was placed onto the bacteria, and the assembly was inverted and placed onto the surface coil such that the bacterial lawn, facing down, was supported by the washer. The ultrasound transducer was positioned above the assembly, in contact with the tofu through another thin layer of ultrasound gel. To provide a reference to compensate for global phase drift during the experiment, a second piece of tofu was placed within the field of view but spatially separated by a 1 cm air gap from the object under insonation. A fiber optic thermometer (Neoptix T1) was inserted into the reference tofu, and the difference between the MRI-derived reference temperature and thermometer-reported temperature was accounted for at the site of insonation when calculating the true focal heating.

Ultrasound was applied with the focus aimed at the tofu immediately adjacent to the agar layer with manual control of power level and duty cycle so as to maintain a temperature of 41.5–43 °C for 45 min. Imaging was performed as described above with a matrix size of 5.39 × 5.05 cm and a slice thickness of 2 mm. The plate was subsequently returned to 30 °C for 5 h and imaged using a Bio-Rad ChemiDoc MP imager with blue epi-illumination and a 530/28 nm emission filter (mWasabi) and also green epi-illumination and a 605/50 nm filter (mCherry).

Animal procedures. All animal procedures were performed under a protocol approved by the California Institute of Technology Institutional Animal Care and Use Committee (IACUC). 9-week-old BALB/c female mice and 4-week-old NU/J 2019 female mice were purchased from Jackson Laboratory (JAX); 4-week-old SCID/SHC female mice were purchased from Charles River.

For *in vivo* ultrasound actuation, *E. coli* expressing the pTlpA36-Wasabi plasmid were grown to OD 0.6, pelleted, and resuspended to OD 24. A 100 µL bolus was injected subcutaneously into both hindlimbs of a nude mouse (SCID or NU/J2019). Mice were anaesthetized using a 2% isoflurane-air mixture and placed on a dedicated animal bed with the surface coil positioned below the target limb of the mouse. Anesthesia was maintained over the course of the ultrasound procedure using 1–1.5% isoflurane. Respiration rate was maintained at 20–30 breaths per minute, and temperature and respiration rate were continuously monitored using a pressure pad (Biopac Systems) and a fiber optic rectal thermometer (Neoptix). The target limb was thermally activated by elevating the temperature to 41 °C and maintaining the elevated temperature for 45 min to 1 h. Temperature monitoring and adjustment was performed as described above for *in vitro* experiments. Following ultrasound treatment, the mouse was returned to its cage for 4 h, anesthetized, and imaged using a Bio-Rad ChemiDoc MP imager with blue epi-illumination and the 530/28 nm emission filter (mWasabi).

For host fever sensing experiments, SCID mice injected with bacteria as described above were housed in an incubator preset to 41 °C for 2 h and control mice were housed at room temperature. Following treatment, all mice were housed at room temperature for 4 h, anaesthetized, and imaged using a Bio-Rad ChemiDoc MP imager with blue epi-illumination and the 530/28 nm emission filter (mWasabi).

Mouse images are representative of three independent *in vivo* experiments. Fever-induced and control mice were littermates randomly selected for each experimental condition. Investigators were not blinded to group allocation because no subjective evaluations were performed.

For host confinement experiments, BALB/c mice were given drinking water containing 0.5 mg/mL of ampicillin for 24 h, and then starved for food overnight. *E. coli* were grown in 2× YT medium containing ampicillin at 37 °C with shaking until OD_{600} of 0.6. Cultures were pelleted and resuspended at 10^8 cells/mL in PBS containing 1.5% $NaHCO_3$. 200 µL of the suspension was administered orally using a gavage needle. Food was returned to the mice and the drinking water contained ampicillin throughout the entire experiment. Fresh fecal samples were collected from each mouse 5 h after gavage and incubated at 37 °C or 25 °C for 24 h, then weighed, homogenized in PBS at 0.1 g/mL, diluted and plated onto LB agar plates containing ampicillin. Plates were then incubated overnight at 25 °C and 37 °C. Bacterial colonies were counted as described above for *in vitro* toxin–antitoxin experiments. The sample size was $N = 5$ mice, which was chosen based on preliminary experiments indicating that it would be sufficient to detect significant differences in mean values.

Electrophoretic mobility shift assay. Interaction between TlpA, σ^{70}-RNAp holoenzyme and DNA was demonstrated using a gel shift assay. For this, 50 pmol of fluorescein-labeled double stranded DNA representing the TlpA operator with flanking padding sequences (70 base pairs in total) was incubated with either 50 pmol of TlpA protein or 5 Units (8.5 pmol) σ^{70}-RNAp holoenzyme (NEB M0551S) individually in 50 µL reaction buffer comprising 40 mM Tris-HCl, 150 mM KCl, 10 mM $MgCl_2$, 0.01% Triton X-100 and 1 mM DTT at a pH of 7.5. As a negative control, the wild-type TlpA operator was replaced with a scrambled version. Following incubation at 37 °C for 30 min, 10 µL of the reaction mixture was supplemented with glycerol to a final concentration of 5% and loaded in a nondenaturing 4% polyacrylamide resolving gel. The gel was run at 65 V for 90 min in buffer comprising 45 mM Tris-borate and 1 mM EDTA at a pH of 8.3. DNA was visualized using Bio-Rad ChemiDoc MP imager using blue epifluorescence illumination and the 530/28 nm emission filter.

Statistics and replicates. Data is plotted and reported in the text as the mean ± s.e.m. Sample size is $N = 4$ biological replicates in all *in vitro* experiments unless otherwise stated. This sample size was chosen based on preliminary experiments indicating that it would be sufficient to detect significant differences in mean values. *P* values were calculated using a two-tailed unpaired heteroscedastic *t*-test.

54. Ai, H.W., Olenych, S.G., Wong, P., Davidson, M.W. & Campbell, R.E. Hue-shifted monomeric variants of Clavularia cyan fluorescent protein: identification of the molecular determinants of color and applications in fluorescence imaging. *BMC Biol.* **6**, 13 (2008).
55. Shaner, N.C. *et al.* Improved monomeric red, orange and yellow fluorescent proteins derived from *Discosoma* sp. red fluorescent protein. *Nat. Biotechnol.* **22**, 1567–1572 (2004).
56. Kutrowska, B.W., Narczyk, M., Buszko, A., Bzowska, A. & Clark, P.L. Folding and unfolding of a non-fluorescent mutant of green fluorescent protein. *J. Phys. Condens. Matter* **19**, 285223 (2007).

doi:10.1038/nchembio.2233

505045 Now You See Me, Now You Don'T—Building Better Biologics through Immune Evasion

Abstract:

Biotherapeutics are helping to reshape modern medicine, but these powerful drugs bear unique risks and limitations compared to chemotherapeutics. For example, protein drugs are subject to immune surveillance in the body, and ensuing anti-drug immune responses can manifest clinical complications including altered pharmacokinetics, loss of efficacy, formation of toxic immune complexes, and allergic reactions of varying severity. Thus, understanding and mitigating biotherapeutic immunogenicity is becoming an integral part of biopharma's drug development process.

T cell epitopes are immunogenic peptides proteolytically processed from intact proteins and displayed on the surface of antigen presenting cells via class II MHC. Subsequent binding of peptide-MHC complexes by receptors on $CD4^+$ helper T cells initiates a series of signaling events that culminate with B cell maturation and production of high affinity antibodies able to bind the offending, intact biotherapeutic. These detrimental anti-drug immune reactions can be short-circuited by reengineering a biotherapeutic such that its constituent peptides evade molecular recognition by class II MHC, however "deimmunizing" mutations should not undermine the stability or activity of the folded and functional drug. Deimmunization via T cell epitope deletion must therefore balance two competing objectives: reduce immunogenicity yet maintain function.

To address this molecular engineering challenge, our collaborative groups have developed deimmunization algorithms that simultaneously optimize proteins for both low immunogenic potential and high molecular fitness. These computational tools map the dual-objective protein design space, identifying the Pareto optimal frontier, or the set of variants whose predicted immunogenicities and functionalities are not simultaneously dominated by any other design. A preemptive understanding of the tradeoffs between extent of deimmunization and maintenance of function can provide valuable guidance during the biotherapeutic engineering and optimization process, and we have validated the algorithms in extensive proof-of-concept studies.

Most recently, we have employed these algorithms to design next-generation antibiotics based on lysostaphin, a peptidoglycan hydrolase with potent bactericidal activity towards drug-resistant *Staphylococcus aureus*, commonly called MRSA. Lysostaphin's anti-MRSA activity is well documented, but the enzyme is highly immunogenic due to its own microbial origins. In this presentation we describe detailed characterization of our lead candidate's immunogenicity and antibacterial activity. Our experimental systems include ex vivo human cellular immunoassays and in vivo model studies in humanized HLA-transgenic mice. The data from these studies provide a unique opportunity to connect the dots between mutagenic deletion of putative T cell epitopes, silencing of T cell activation, suppression of the anti-drug antibody response, and improving in vivo efficacy. This deimmunized antibacterial drug candidate is now undergoing IND-enabling studies with an eye towards first-in-human trials. More broadly, our protein deimmunization algorithms represent a platform technology with the potential to accelerate and de-risk biotherapeutic development.

494957 Regulation of Multispanning Membrane Protein Topology Via the Post-Translational Flipping of Charged Protein Elements

Abstract:

Multispanning membrane proteins â€" i.e., proteins with multiple transmembrane domains that thread back-and-forth across the cell membrane - are essential for cellular functions that include signal transduction, material transport, and energy conversion. Performing these functions requires the proteins to integrate within the membrane in the correct topology, or the overall orientation of the protein relative to the membrane. Conventionally, it is believed that a single membrane protein topology is determined during translation by the orientation of the first transmembrane segment that integrates into the membrane. Challenging this view, the multispanning, dual-topology protein EmrE obtains two possible topologies with equal likelihood. Moreover, the topology of EmrE is highly sensitive to single charge mutations, including C-terminal mutations, indicating that topology is adjustable well after the onset of translation. Understanding the origin of this topological plasticity could yield new insight into membrane protein structure and function, as well as providing guidelines for engineering proteins for specific topological orientations.

In this work, we use coarse-grained simulations to investigate the integration of the multispanning membrane protein EmrE on realistic biological timescales. We employ a simulation model that enables access to a timescale of minutes while retaining sufficient chemical accuracy to capture the forces that drive membrane integration. We find that EmrE establishes its topology post-translationally by the stochastic â€œflippingâ€ of charged protein elements across the membrane over long time scales. This behavior is surprising because the transport of charged moieties across the membrane is thought to occur on timescales too long to be biologically relevant. We further use atomistic molecular dynamics simulations to demonstrate that such timescales can be dramatically reduced by cooperative interactions between amino-acid side chains that minimize the free energy barrier for flipping. Together, these results suggest a new mechanism for multispanning membrane protein topogenesis and indicate that post-translational topological rearrangements may play an important role in determining membrane protein structure and function.

497542 Characterization of Pro-Regions in Mammalian Growth Factors Using Deep Sequencing and Yeast Surface Display

Abstract:

Growth factors comprise a large family of proteins that play a critical role in proliferation and differentiation, function as a cellular growth stimulator and inhibitors, and promote survival of cells. These proteins are associated with several diseases including neurodegenerative disorders, among others. These growth factors represent important targets for the development of potential therapeutic candidates. As one example, Brain-derived Neurotrophic Factor (BDNF) is a candidate to promote the regeneration of the lost or damage cells in combination with progenitor cells after a spinal cord injury[1]. However, a major limitation is mammalian growth factor expression in active form in high titers. In their native context, certain growth factors are expressed as a pro-protein in which the pro-peptide may act as a molecular chaperone of the mature protein to guide correct folding and processing through the secretory pathway. In the attempt to achieve high titers, this pro-region has been engineered to enhance the folding and secretion of human BDNF for yeast expression[2]. In this talk, we will present the improvement of the active titers of a particular growth factor using comprehensive site saturation mutagenesis (SSM) libraries of the pro-region coupled to yeast surface display and deep mutational scanning. These SSM libraries were also sorted against two different anti-growth factor monoclonal antibodies (mAbs) that are conformationally selective. We evaluated the function of 3,296 mutants and found over a 10-fold improvement in binding and folding activities after two round of sorting for approximately 86% of all possible single nonsynonymous mutations. Mutations that increased proper growth factor folding largely overlapped between the two mAbs, showing the reproducibility of the deep mutational scanning method. Distinct features of the pro-region that enable high yeast secretion will be discussed, as will downstream applications like conformational epitope mapping that are enabled by the high active surface expression of the growth factor.

1. Sahni, V. & Kessler, J. A. Stem cell therapies for spinal cord injury. *Nat. Rev. Neurol.* **6,** 363â€"372 (2010).

2. Burns, M. L. *et al.* Pro-region engineering for improved yeast display and secretion of brain derived neurotrophic factor. *Biotechnol. J.* **11,** 425â€"436 (2016).

501401 Intrinsically Disordered Proteins As Sensors of Membrane Curvature

Abstract:

The ability of proteins to detect membrane curvature is essential for the initiation and growth of endocytic structures. To date, all known mechanisms of curvature sensing rely on specific protein structural features. Key examples include wedge-like amphipathic helices, which insert into defects on the surfaces of highly curved membranes and crescent-shaped BAR domains, which bind with greatest affinity to highly curved membranes that match their inherent curvature. In contrast, here we report a novel mechanism of membrane curvature sensing that arises from a lack of protein structure. Specifically, our results demonstrate that intrinsically disordered protein (IDP) domains, which lack a well-defined secondary structure, are highly sensitive to membrane curvature. In particular, quantitative measurements of protein-membrane binding as a function of membrane curvature reveal that the chain entropy of IDPs drives them to bind more strongly to highly curved membrane surfaces, which pose fewer steric restrictions in comparison to flat membranes. Using a tethered vesicle assay to quantify protein binding as a function of membrane curvature, the IDP C-terminal domains of the endocytic adaptor proteins AP180 and Epsin1 bound to highly curved lipid vesicles (20-50 nm in diameter) with approximately 5-fold greater affinity than to vesicles of lower curvature (100-200 nm in diameter). This level of curvature sensitivity is comparable to that observed using the same assay to examine the amphipathic helix-containing ENTH domain and the BAR domain of Amphiphysin. These findings are particularly important since IDP domains are found within many of the best studied endocytic proteins, including the adaptors of the clathrin pathway, the COPII pathway and others. In particular, this work suggests that entropic mechanisms of membrane curvature sensing - which arise from changes in protein conformational entropy rather than protein structure - may provide a highly potent, yet previously unknown mechanism of membrane curvature sensing. These findings elucidate the interplay between structured and unstructured domains in curvature sensing proteins, and paradoxically illustrate how a lack of protein structure can result in protein functionality.

501782 Effect of Secondary Structure of Cell-Penetrating Peptides on Their Interaction with Fungal Cells

Abstract:

Cell-penetrating peptides (CPPs) are a class of peptides that have the ability to cross cell membranes either alone or while carrying molecular cargo. Although their interactions with mammalian cells have been widely studied, much less is known about their interactions with fungal cells, particularly at the biophysical level. To improve this biophysical understanding, we analyzed the membrane interactions of eight CPPs with the fungal pathogen *Candida albicans* using experiments and simulations. Circular dichroism (CD) was used to evaluate the peptide structure in hydrophobic and hydrophilic environments and in the presence of cells. All peptides exhibited a structural transition from a random coil secondary structure (cecropin B, MAP, MPG, penetratin, pVEC, SynB, and TP-10) or a weak Î±-helix (Pep-1) in an aqueous buffer to an Î±-helical structure in a hydrophobic solvent. When cells were added to the aqueous buffer, a shift in the secondary structure of the peptides was observed for cecropin B, MAP, pVEC, SynB, and TP-10. With the exception of SynB, which exhibited Î²-sheet structure, these peptides were Î±-helical in the presence of cells. MPG, penetratin, and Pep-1 did not undergo structural changes when *C. albicans* cells were added. Monte Carlo simulations were performed to more mechanistically probe the interaction of the peptides with the cell membrane, and these simulations suggest that pVEC, MAP, TP-10 and cecropin B strongly penetrate into the hydrophobic domain of the membrane lipid bilayer, inducing the structural transition to an Î±-helical conformation observed for most of the peptides by CD in the presence of cells. In contrast, Pep-1 and MPG remained in the hydrophilic region without a shift in conformation. Combined with data on membrane depolarization and our previous work evaluating the translocation mechanisms of these peptides, our results suggest that strong insertion into fungal cell membranes generally leads to shifts in peptide secondary structure and depolarization of the cell membrane, while a lack of deep membrane interaction and a lack of conformational shift tend to be associated with translocation mechanisms involving endocytosis. However, results of simulations and various experimental approaches are not always consistent, suggesting additional phenomena may contribute to the interaction of CPPs with *C. albicans* cells and that multiple translocation mechanisms may be possible.

504093 Predicting and Engineering Protein Complex Structures, from Antibodies to Glycoproteins

Abstract:

Biological function often arises from the interactions of proteins with other biomolecules, and these interactions are grounded in the three-dimensional structures of the complexes. My lab develops computational tools to study the atomic structure of protein interfaces to explain not only how biological and disease processes work but also how one might alter these processes at the molecular level. Creating computational tools first requires solving problems of basic science including (1) how to sample the myriad conformations available to proteins and (2) how to accurately calculate the energy of each conformation. In this talk, I will discuss our work on protein docking (including capturing backbone conformational changes), antibody-antigen modeling (including the challenging CDR H3 loop and applications to immune repertoires) and carbohydrate modeling (including protein-oligosaccharide docking and engineering of glycotransferases).

499197 Continuous Flow Nitration: Research and Process Development

Abstract:

Bi-phasic nitration is commonly executed for making nitro intermediates in the pharmaceutical industry. Their scale-up, however, is generally problematic because of the following concerns: high exothermicity, need for proper mixing and liquid-liquid dispersion, and formation of energetic di-nitro by-products. A detailed understanding of these aspects is required to ensure proper scale-up of nitration.

In this talk we will present a case study of a scale-up of 1-methyl-4-(trifluoromethyl)benzene nitration using KNO_3/H_2SO_4 as the nitrating agent. To understand these aspects, targeted experiments were performed to prepare a mass transfer with chemical reaction model (using film theory) to quantify the relation between liquid-liquid dispersion with the conversion and di-nitro impurity levels. Model predictions showed that acceptable conversion required a Hatta number of at least 1.9, with 80% of reaction done in the film. To meet these requirements, a CSTR was chosen as the preferred contacting device, with proper heat transfer protocols in place to handle the exotherm.

The work presented here was sponsored by AbbVie. AbbVie contributed to the design, research, and interpretation of data, writing, reviewing, and approving the presentation. All authors are employees of AbbVie.

494380 On-Demand Medicinal Chemistry and Compound Synthesis in Oscillating Droplets

Abstract:

There is increasing pressure in drug discovery to deliver a steady stream of active compounds for physicochemical profiling and potency testing. In order to accelerate lead optimization, it is imperative to reduce the time required for each iteration of design, synthesis, and screening [1]. With recent advances in automation and high-throughput biological screening, the bottleneck has crept toward the synthesis step [2, 3, 4]. Medicinal chemistry laboratories have also benefitted from this trend in automation [5, 6, 7, 8], where one integrated platform has even been used for the discovery of novel Abl kinase inhibitors [9].

Here, we describe an automated chemistry platform that can efficiently screen a wide range of reactions, including single/multi-phase, single/multi-step, and photochemical reactions at the 20 microliter scale. Individual droplets are prepared by a liquid handler and moved through fluoropolymer tubing by a carrier gas at elevated pressures. Reactions occur in a heated reactor where the liquid droplet is oscillated back and forth to ensure thorough mixing â€“ even for biphasic liquid-liquid reactions or liquid-gas reactions â€“ without being limited to a finite residence time [10]. Optional inlet and/or outlet injections enable multistep chemistry. A portion of the crude product mixture is sampled and sent directly to an online HPLC/MS for analysis, purification, and product collection, typically of 10-500 Î¼g.

The system offers the enhanced heat and mass transfer characteristics, increased safety, and opportunity for automation associated with flow chemistry while enabling the following key advantages: (a) reduction of material consumption, preparing just 40 microliters for each reaction condition, (b) elimination of residence time dispersion, (c) elimination of the inverse relationship between residence time and mass transfer rate for a flow reactor of fixed length, (d) simplification of continuous and discrete variable screening through the use of a liquid handler to prepare reaction mixtures, and (e) elimination of the time- and material-waste associated with waiting for flow reactors to reach steady state.

Importantly, because the reaction droplet is equivalent to one segment in a continuous segmented flow, reaction conditions identified using the platform can be directly translated to a continuous synthesis. This helps bridge the connection between the early research stages of drug discovery and the later stages, where material demands grow from the Î¼g-mg scale to the g-kg scale.

As a demonstration of this platform, 36 examples of frequently used N-X bond forming reactions using typical small-molecule building blocks are screened. A representative biphasic Suzuki-Miyaura cross coupling reaction, a multi-step synthesis of the drug product diclofenac, and several visible-light catalyzed oxidations are also reported.

[1] K. C. Nicolaou, *Angew Chem Int Edit* **2014**, *53*, 9128-9140.

[2] A. G. Godfrey, T. Masquelin, H. Hemmerle, *Drug Discovery Today* **2013**, *18*, 795-802.

[3] X. Niu, F. Gielen, J. B. Edel, A. J. deMello, *Nat Chem* **2011**, *3*, 437-442.

[4] R. Ostafe, R. Prodanovic, W. L. Ung, D. A. Weitz, R. Fischer, *Biomicrofluidics* **2014**, *8*.

[5] M. Peplow, *Nature* **2014**, *512*, 20.

[6] B. Ahmed-Omer, E. Sliwinski, J. P. Cerroti, S. V. Ley, *Org Process Res Dev* **2016**.

[7] A. Buitrago Santanilla, E. L. Regalado, T. Pereira, M. Shevlin, K. Bateman, L.-C. Campeau, J. Schneeweis, S. Berritt, Z.-C. Shi, P. Nantermet, Y. Liu, R. Helmy, C. J. Welch, P. Vachal, I. W. Davies, T. Cernak, S. D. Dreher, *Science* **2015**, *347*, 49-53.

[8] B. J. Reizman, K. F. Jensen, *Accounts of Chemical Research* **2016**, *49*, 1786-1796.

[9] B. Desai, K. Dixon, E. Farrant, Q. Feng, K. R. Gibson, W. P. van Hoorn, J. Mills, T. Morgan, D. M. Parry, M. K. Ramjee, C. N. Selway, G. J. Tarver, G. Whitlock, A. G. Wright, *J. Med. Chem.* **2013**, *56*, 3033-3047.

[10] M. Abolhasani, N. C. Bruno, K. F. Jensen, *Chem. Commun.* **2015**, *51*, 8916-8919.

501379 Ultrahigh-Throughput Screening of Chemical Synthetic Pathways Using Desorption Electrospray Ionization Mass Spectrometry

Abstract:

We describe a state-of-the-art automated platform for ultrahigh-throughput screening of chemical reactions based on reaction acceleration in charged microdroplets and utilizing Desorption Electrospray Ionization Mass Spectrometry (DESI-MS). The goal is to deliver an extremely rapid yet insightful qualitative yes/no analysis for the discovery of new chemistry in the order of up to 10,000 reactions per hour. The positive results are then used to guide synthesis in a more scalable setting, including in batch and flow. Using this system, particular chemical routes are prepared in parallel using reagents which are robotically pipetted and mixed in chemically-inert well plate. The routes are tested by stamping the reaction mixtures onto an inert reaction substrate, e.g. PTFE membrane, followed by electrospraying of various solvents at an angle relative to the substrate to generate secondary ionized reaction microdroplets. These microdroplets are propelled into the mass spectrometer for mass analysis. The results are then analyzed using an automated MS data analysis system developed in-house, which provides chemical reaction monitoring and intelligence for optimization of future reaction screening conditions to increase the probability of pathway discovery.

494694 Heuristics for Implementing Photoredox Catalysis in Flow Reactors

Abstract:

Heuristics for Implementing Photoredox Catalysis in Flow Reactors Eric G. Moschetta,* Steven M. Richter, and Steven J. Wittenberger AbbVie, Inc. Process R&D, 1401 Sheridan Road North Chicago, IL 60064 *corresponding author: eric.moschetta@abbvie.com Photoredox catalysis is reemerging in organic synthesis as a powerful method for constructing small molecules under mild conditions. Such transformations often employ transition metal catalysts to absorb visible light, exciting the catalysts into energetic states that facilitate single electron transfer events between the catalysts and organic substrates.[1] These reactions frequently occur at mild temperatures, including room temperature, using visible light sources such as LEDs that provide intense light with narrow wavelength emissions. As such, materials for conducting photoredox chemistry at the lab scale are readily available for chemists and engineers. However, significant challenges remain in visible-light photoredox catalysis. The primary challenge in industrial applications is scaling up photoredox reactions from laboratory scale batch reactors to reactors that are capable of multi-kilogram scale production. Proper reactor design relies on rigorous understanding of the inherent kinetics of photoredox reactions, which are largely unexplored. The photon flux, the overall observed rate at which photons are received in the reactor, must be high enough to ensure that the inherent kinetics of the reaction are the operating conditions of the reactor. The lamp-reactor geometry, therefore, is a critical design factor in scalable photoredox flow chemistry. Maximizing the photon flux also likely depends on careful tuning of the catalyst concentration, reactant concentration, and light intensity. Continuous flow reactors can facilitate improved photon fluxes and increased throughput through significantly decreased reaction times compared to batch reactions, which are both necessary for industrial photoredox catalysis.[2] Flexible fluoropolymer tubing is chemically resistant to many materials and highly transparent to visible light, meaning that tubing can be wrapped around light sources such as LEDs for access to customizable tubular flow reactors for photoredox applications. The short tube diameters allow for shorter path lengths, meaning that photons are more evenly distributed throughout the reactor, increasing the apparent rate of reaction. However, general principles and heuristics for designing such flow reactors are not yet available to process chemists and engineers. This work establishes heuristics for employing flow reactors in photoredox chemistry. The effects of reactant and catalyst concentration, tube diameter, and LED intensity on the observed photon fluxes and rates of reaction will be discussed. The model reaction for establishing the heuristics will be the oxidation of 9,10-diphenylanthracene using molecular oxygen and a common ruthenium photoredox catalyst, $Ru(bpy)_3Cl_2$. The quantum yield for this reaction is known,[3] meaning that the photon flux under a given set of reaction conditions can be calculated directly once the rate of reaction is measured. These heuristics will then be examined against a pyrazole synthesis driven by photoredox catalysis, using the same ruthenium catalyst, to see if heuristics accurately predict how the observed rates of reaction change with changes in reaction conditions in flow. **References**

[1] C. K. Prier, D. A. Rankic, D. W. C. MacMillan, *Chem. Rev.* **2013**, *113*, 5322-5363.

[2] Y. Su, N. J. W. Straathof, V. Hessel, T. Noel, *Chem. Eur. J.* **2014**, *20*, 10562-10589.

[3] S. P. Pitre, C. D. McTiernan, W. Vine, R. DiPucchio, M. Grenier, J. C. Scaiano, *Sci. Rep.* **2015**, *5*, 16397.

This abstract presentation was sponsored by AbbVie. AbbVie contributed to the design, research, and interpretation of data, writing, reviewing, and approving the publication. All authors are AbbVie employees.

500424 Operando Raman of the Nitroldol Reaction over Pore-Expanded Fdu-12 â€“ Dendron Hybrids

Abstract:

Many labs have used the Nitroaldol reaction to probe the accessibility and reactivity of supported organic ligands.[1] Our lab has reported prior work on simple ligands and branched melamine-dendrons as catalysts for this reaction. The current work describes our effort on using pore-expanded FDU-12[2] as a novel support for growing melamine dendrons and their use as catalysts for this reaction. Kinetics measurements were obtained using a suite of dendrons with different diamine spacers and peripheral groups to determine how the diamine identity and density influences reactivity. To understand the chemistry at a more basic level operando Raman spectroscopy was used to monitor the various intermediates formed.

A library of samples of first through third generation dendrons supported on pore expanded FDU-12 were synthesized, characterized, and tested for the Nitroaldol reaction. Consistent with prior work the conversion/rate decreases with increasing dendron generation for all samples. We ascribe this finding to mass transfer limitations. We observe that diamines with secondary amines give higher conversions than primary amines. In the case of the G1 samples these samples all have within +/-5% the same number of functional groups.

On a per site basis the dendrons with bulky diamines such as trimethylenedi(piperidien) have a rate a factor 2-3 times higher than the dendrons made with 4-aminomethylpiperidine. Our work shows the effect of primary versus secondary amines, and the effect of the free volume on reactivity, wherein high free volumes are advantageous for reactivity. In all cases we see high (>80%) selectivity to the alcohol product except for the amino-silica.

We have also performed operando Raman[3] measurements to monitor the reaction. A band at 1715 cm^{-1} (C=O) can be seen to systematically decrease with conversion, and the appearance of the C=C double bond of the nitroalkene at 1650 cm^{-1} is clearly observed. Finally a feature at 1350 cm^{-1} assigned to C=N intermediates is observed to increase during reaction, illustrating the utility of Raman as a tool to probe this reaction.

In summary our results show that 1) while we can synthesize more uniform dendrons on pore expanded FDU-12 than SBA-15 the windows between the cages impose severe transport restrictions which are clearly discernible in our results, and 2) operando Raman can be used to easily monitor the formation of the alkene product (C=C bond) as well as to observe the imine (C=N) intermediate. The results show these hybrid materials are interesting catalytic materials for this reaction and that they are amenable to analysis using operando methodologies which have not been widely used prior in this sub-community.

References

1. Bass, J. D., Solovyov, A., Pascall, A. J., and Katz, A. *J. Am. Chem. Soc.* 128 373 (2006).

2. Fan, J, Yu, C., Gao, F., Lei, J., Tian, B., Wang, L., Luo, Q., Tu, B., Zhou, W. and Zhao, D. *Angew. Chem. Int. Ed.* 42 3146 (2003).

3. Grabow, K., and Bentrop, U. *ACS Catal.* 4 2153 (2014).

500195 Scale-up of a Reaction Step from Lab to Commercial Scale, Switching from Batch to Continuous Manufacturing Using a Mechanistic Modelling Approach

Abstract:

This work presents the journey of scaling-up a manufacturing process in a multi-purpose plant that is focused on the reaction step. The scale-up process is based on a mechanistic modelling approach and resorts to both batch and flow data that was generated at lab scale. The modelling applied in this process is used not only to increase process understanding and further optimize the process to highest performance possible, but also to guide the outcome in terms of process safety.

The case-study to be shown is composed of a multi reaction mechanism, comprising one very exothermic reaction, in a gas-liquid system that will be run at commercial scale in a continuous mode. The commercial continuous reactor is composed of several plates, where the first plate is intended to promote mixing of the gas and liquid streams, while the following plates will be use to carry out the reaction. In the last two plates, there will be a quench step, before proceeding to the downstream of the process.

The lab data was generated both in a batch reactor and also in a flow reactor. In the first reactor, it was possible to monitor not only the reaction itself by HPLC sampling of main compounds, but also following the reagent feeding profile, allowing to gathering of knowledge on reaction kinetics. Additional experiments were run at different temperatures in order to improve knowledge on the activation energy of the multiple reaction mechanism. In the continuous reactor, it was possible to gain experience on the reaction outcome in a continuous mode and to apply scale-up engineering rules. On top of that, it was possible to improve the kinetic model using screening experiments around typical process parameters (flowrate, temperature, reagent-to-starting material ratios).

With the accumulated knowledge built upon both reactors, it is possible to simulate the large scale reaction in a commercial scale continuous reactor, based on the kinetic, heat and mass transfer models derived from the lab experiments. The modelling approach developed using Dynochem will allow the optimization of the large scale conditions of the process but also a much safer scaling-up of the process to the commercial scale.

Microreactors driven by electric field for enzyme catalysis in ATPS

Alexandr Romanov, Lucie Vobecka,
Zdenek Slouka and **Michal Pribyl**

Department of Chemical Engineering,
University of Chemistry and Technology, Prague, Czech Republic

Content

- Motivation
- Microreactor concepts
  - With droplet flow
  - With parallel flow
- Aqueous two-phase systems (ATPS)
  - Basic properties
  - Droplet addressing by electric field
- Case study
  - Enzyme hydrolysis of penicillin G a cephalexin synthesis
- Conclusions

University of Chemistry and Technology, Prague

Motivation – Why micro?

- Large interfacial area – intensive mass and heat transport
- Short transport distances – reduction of transport resistances
- Laminar (Stokes) flow – predictable behavior
- Fast prototyping, fast debugging
- Possible integration of several unit operation into one chip – mixing, reaction, separation, detection, quantification

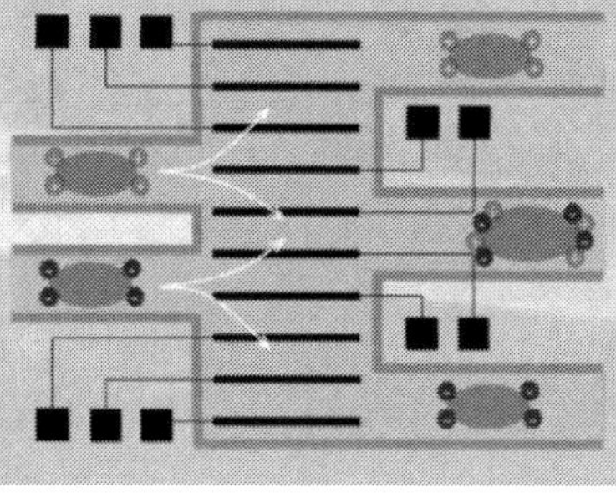

University of Chemistry and Technology, Prague

Motivation – Why electric field?

- Driving force – electric field instead of pressure gradient $\Delta p \approx \frac{32 l v \eta}{d^2}$
- No moving parts and complex mechanical devices are necessary
- Different electrokinetic phenomena can be exploited – electrophoresis, electroosmosis, dielectrophoresis, electrowetting, ...
- Pumping of solutions and multi-phase mixtures
- Differences in ion charge numbers and molecule sizes lead to electrophoretic separation
- Noncharged droplet and particles can be separated by dielectrophoresis
- Droplet addressing into particular parts of microfluidic structures
- Induced electrocoalescence, electrojetting, electrosplitting

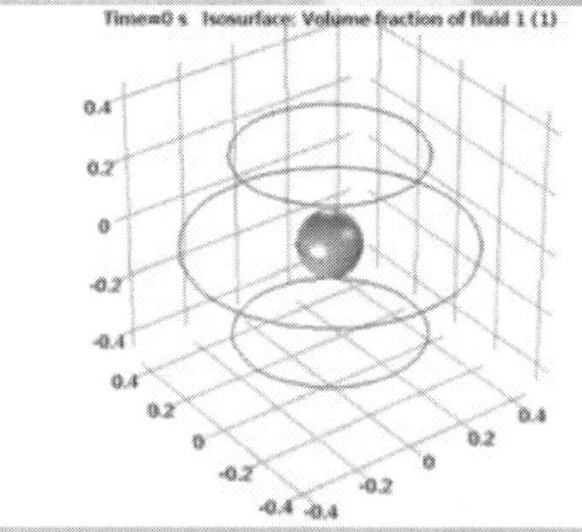

Flittner & Pribyl, J. Fluid Mechanics 822, 31-53, 2017.

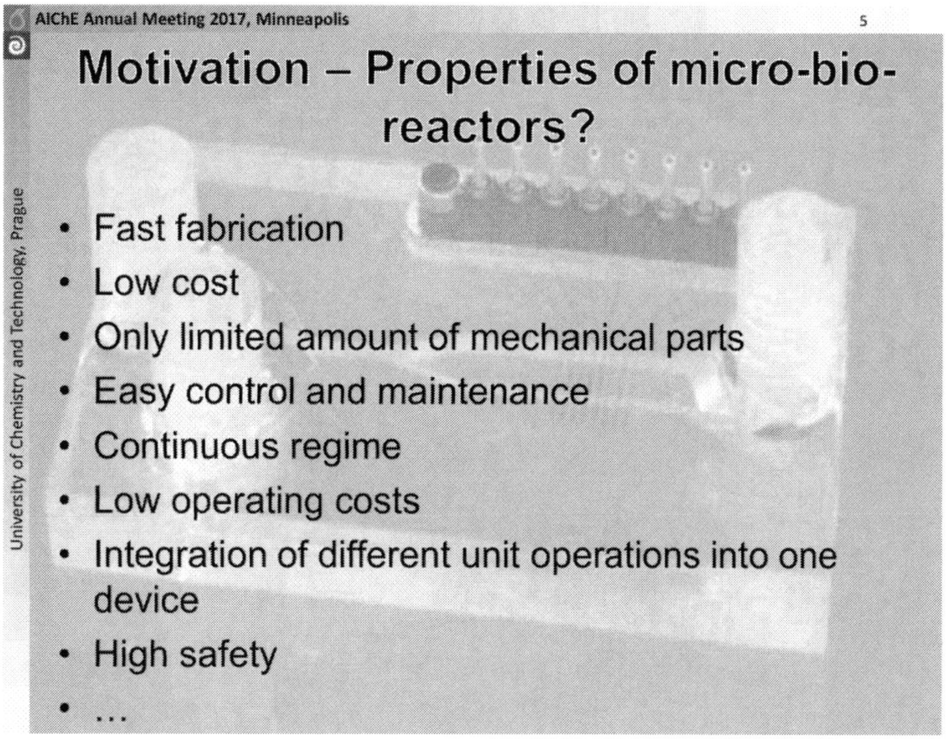
AIChE Annual Meeting 2017, Minneapolis
5
Motivation – Properties of micro-bio-reactors?
• Fast fabrication
• Low cost
• Only limited amount of mechanical parts
• Easy control and maintenance
• Continuous regime
• Low operating costs
• Integration of different unit operations into one device
• High safety
• ...
University of Chemistry and Technology, Prague

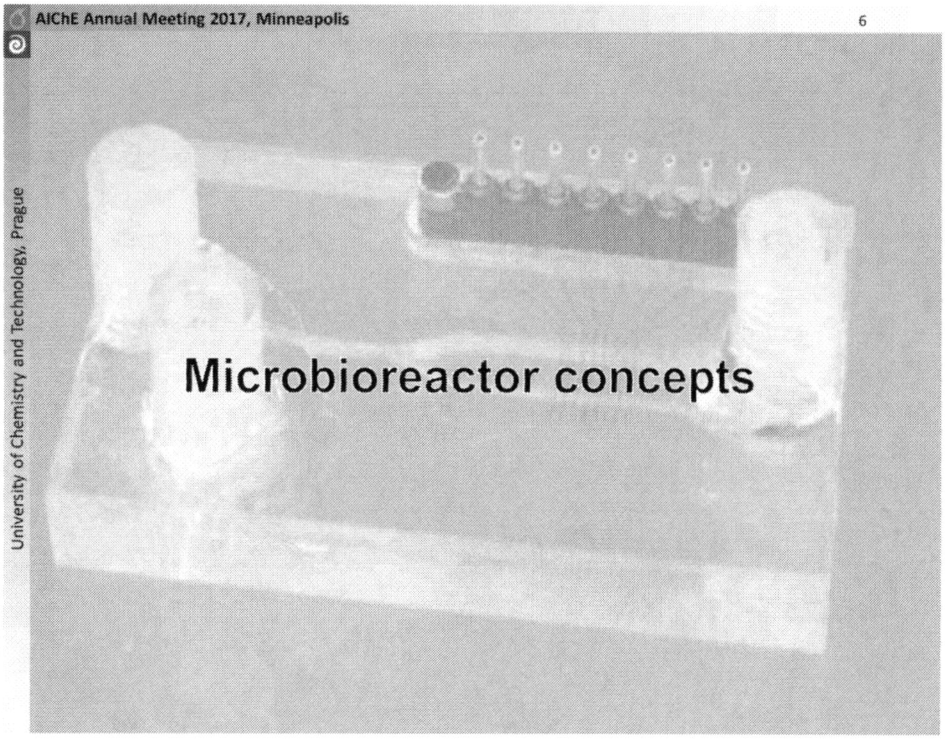
AIChE Annual Meeting 2017, Minneapolis
6
Microbioreactor concepts
University of Chemistry and Technology, Prague

University of Chemistry and Technology, Prague

Bioreactor with droplet flow

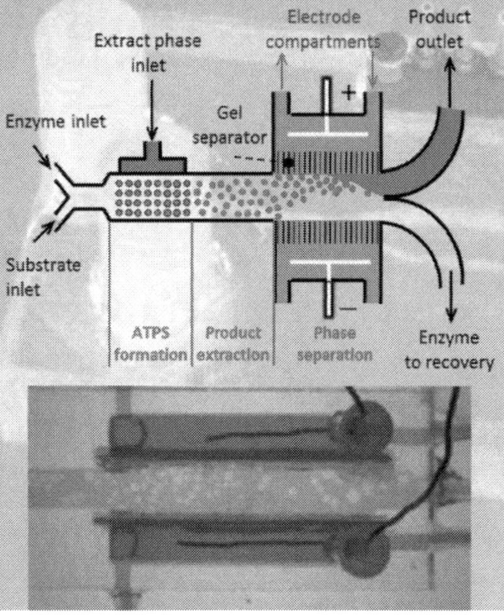

- Integrated reactor-separator
- Droplet flow of ATPS to extract reaction products
- Large interfacial area
- Electric field assisted separation of ATPS
- Possible recycle of enzyme solution or enzyme recovery

University of Chemistry and Technology, Prague

Bioreactor with parallel flow

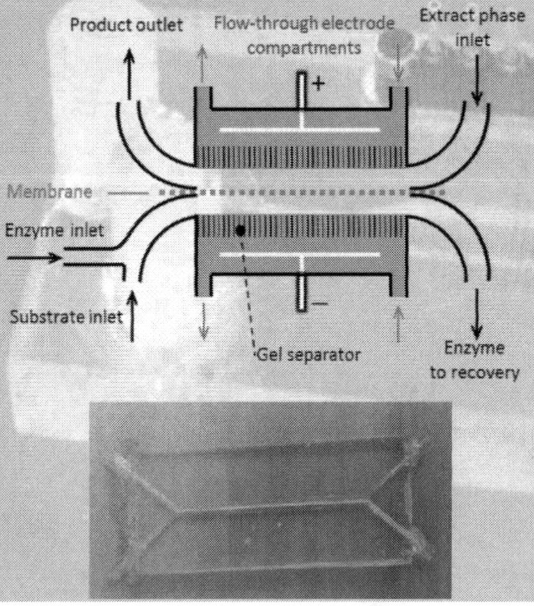

- Integrated reactor-separator
- Parallel flow of ATPS
- Thin membrane for flow stabilization
- Co-current or counter-current arrangement
- Enhancement of concentration of ionic species (products)
- Enzyme recovery

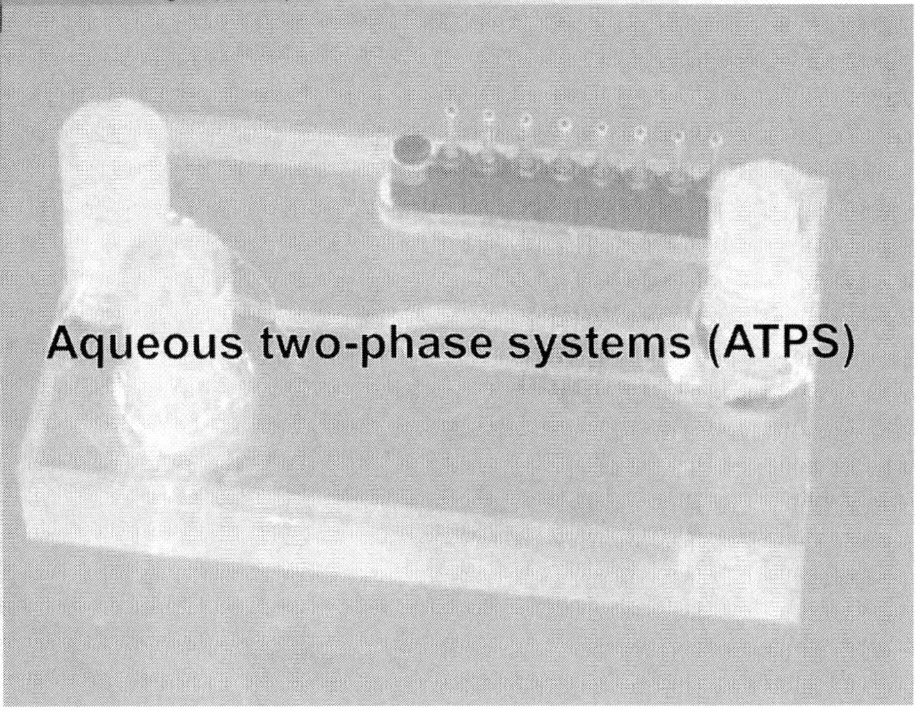

Aqueous two-phase systems (ATPS)

Aqueous two-phase systems (ATPS)

- Two immiscible aqueous phases
  - Polymer/polymer/water (polyethylene glycol, dextran, water)
  - Polymer/salt/water (polyethylene glycol, phosphates, water)
- Milder conditions than in the extractions by organic solvents
  - Protein, organelles, cells, DNAs separation
- Extractive fermentation
  - Biological reaction in one phase, extraction of products or reactant dosing to/from the other phase
  - Elimination of substrate/ product inhibitions
  - Shift of reaction equilibrium towards the product side

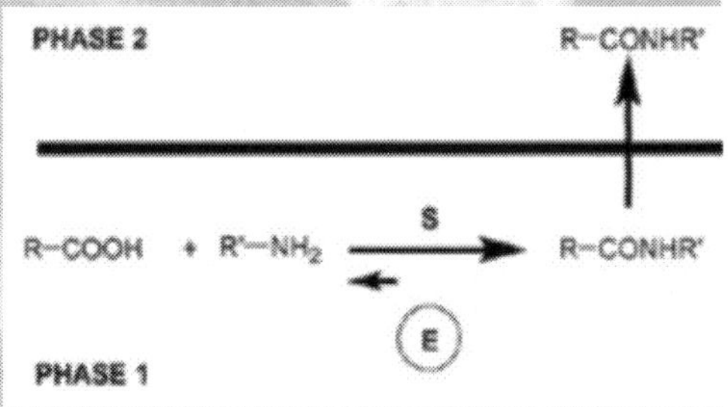

Terreni et al., Enzyme and Microbial Technology 36, 2005, 672–679.

University of Chemistry and Technology, Prague

Aqueous two-phase systems

- Mixture of polyethylene glycol (PEG) 4000, salt, water
- Two immiscible phases
  - Top PEG phase
  - Bottom salt phase
- Both phases conductive!

Low interfacial tension (σ ~ 1 mN/m)

Despite of it, generation of relatively stable slug flow is possible

Extraction in microscale

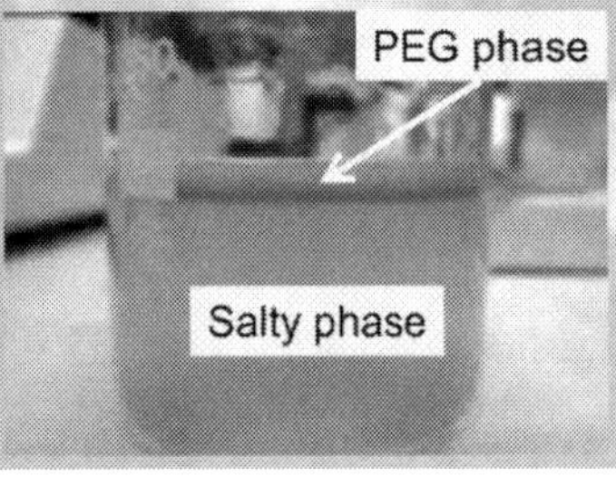

University of Chemistry and Technology, Prague

ATPS manipulation by electric field

Open chip

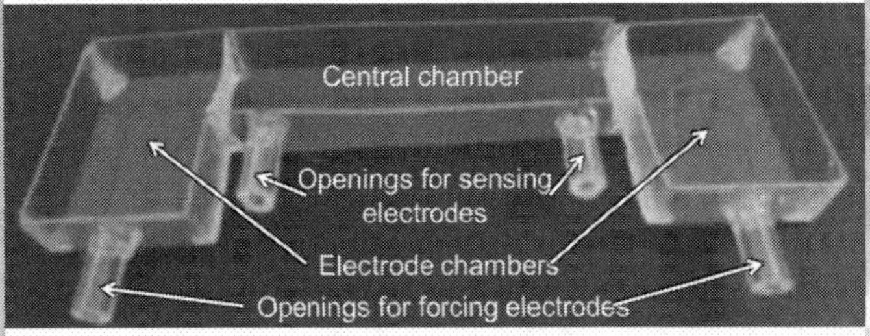

Closed chip

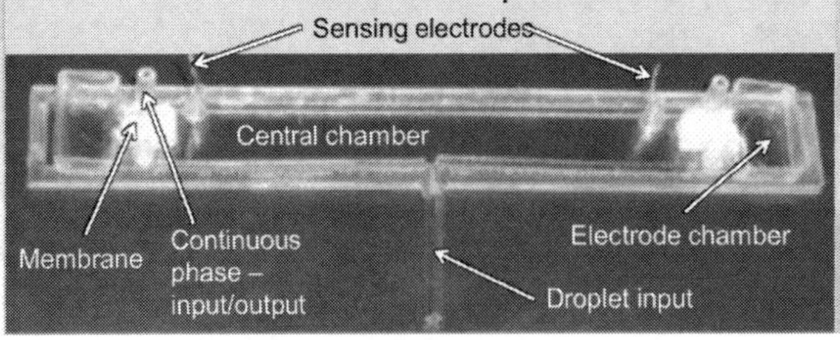

- Droplet of salty (bottom) phase dropped on PEG (top) phase
- Salty phase is heavier but small droplets are kept at the PEG surface
- Both phases are conductive
- Electrode reservoirs are separated membranes treated by agar gel

University of Chemistry and Technology, Prague

Time course of a typical experiment

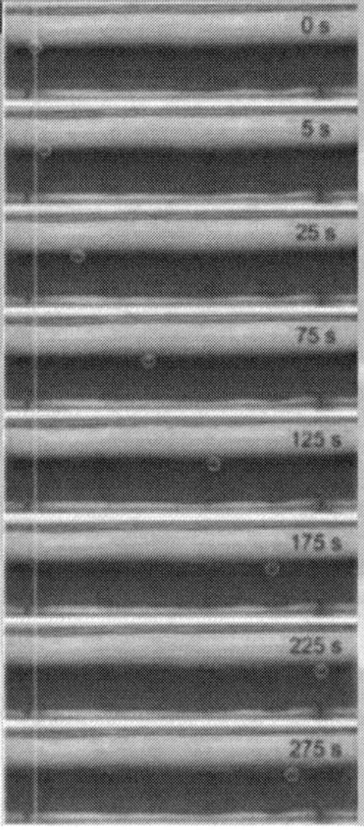

- PEG – continuous phase
- Droplets of the salty phase always move to negatively charged electrode
- Typical electric field strength 10 V cm^{-1}

Position-time plot

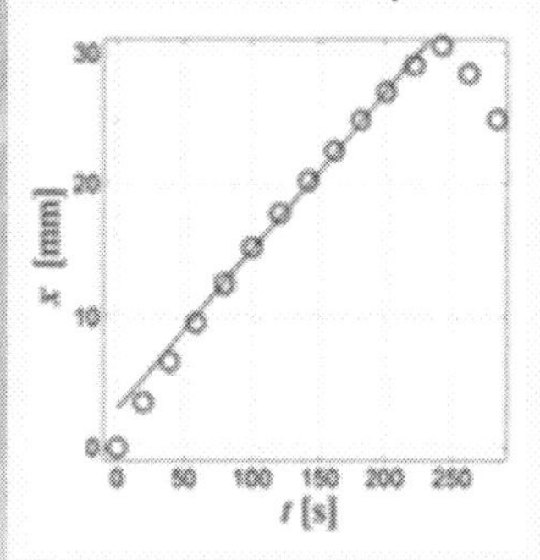

Vobecka et al., Microfluid Nanofluid (2017) 21:51

University of Chemistry and Technology, Prague

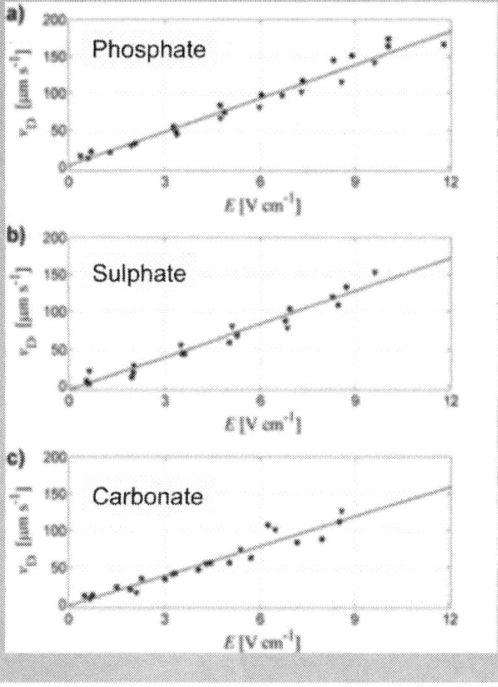

Open system – droplet mobilities

Mobility values

Phosphate>Sulphate>Carbonate

ATPS	$\mu_E/[m^2\ V^{-1}\ s^{-1}]$
PEG/phosphate	$(1.52 \pm 0.12) \times 10^{-7}$
PEG/sulphate	$(1.48 \pm 0.16) \times 10^{-7}$
PEG/carbonate	$(1.34 \pm 0.17) \times 10^{-7}$

- Linear dependences of droplet velocity on applied electric field strength
- Mobilities higher by at least factor 10 if compared to water-oil systems
- Low field is necessary for addressing

University of Chemistry and Technology, Prague

Closed system – droplet mobilities

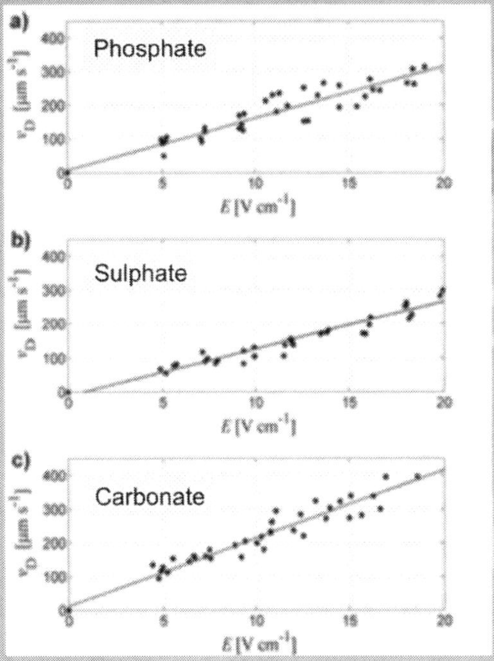

Mobility values

Carbonate>Phosphate>Sulphate

ATPS	$\mu_E/[m^2\ V^{-1}\ s^{-1}]$
PEG/phosphate	$(1.54 \pm 0.16) \times 10^{-7}$
PEG/sulphate	$(1.38 \pm 0.10) \times 10^{-7}$
PEG/carbonate	$(2.02 \pm 0.16) \times 10^{-7}$

- Linear dependences of droplet velocity on applied electric field strength
- In PEG/carbonate significantly higher mobility if compared to open system

University of Chemistry and Technology, Prague

Droplet addressing

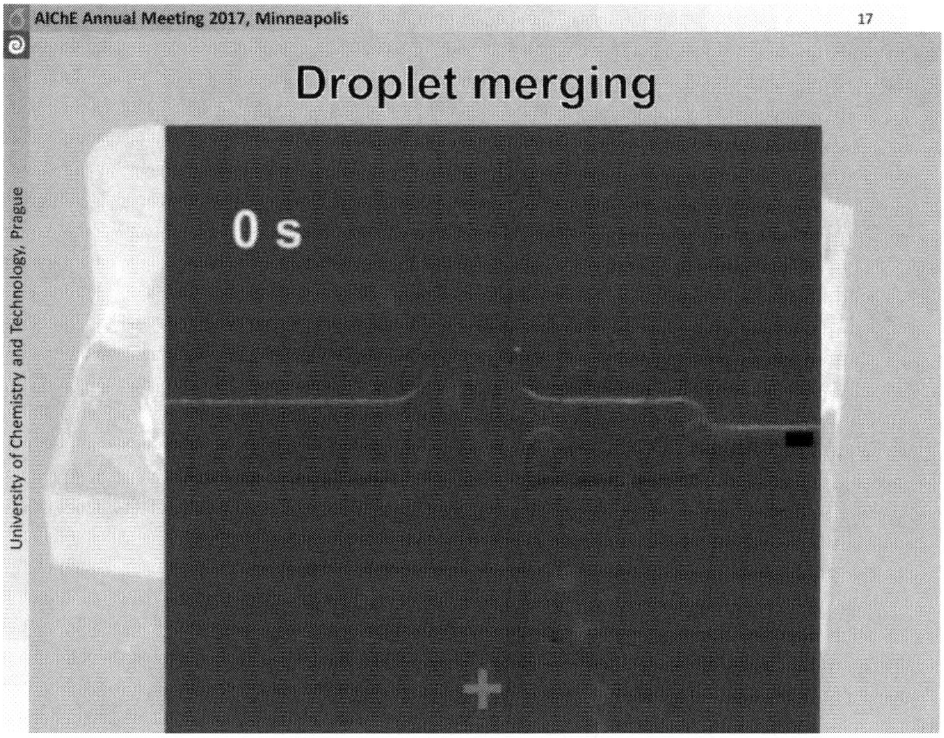
AIChE Annual Meeting 2017, Minneapolis
17
Droplet merging
0 s
University of Chemistry and Technology, Prague

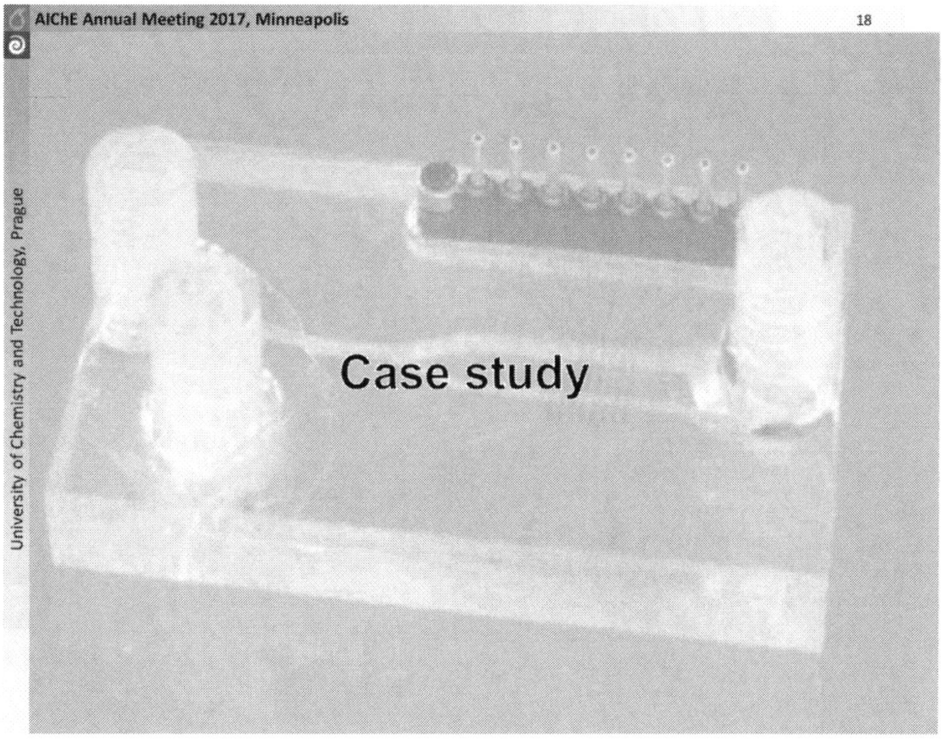
AIChE Annual Meeting 2017, Minneapolis
18
Case study
University of Chemistry and Technology, Prague

Optimization of ATPS composition

$$PenG^- + H_2O \underset{pH \approx 8}{\overset{\text{Penicillin acylase}}{\rightleftharpoons}} 6\text{-}APA^- + PAA^- + H^+$$

$$7\text{-}ADCA + PGME \underset{pH \approx 6.5}{\overset{\text{Penicillin acylase}}{\rightleftharpoons}} \text{Cephalexin} + MeOH$$

ATPS system appropriate for the continuous production and separation of 6-APA has to fulfil the following conditions:

(i) To provide partition coefficients significantly higher than one for the reaction products and significantly lower than one for PA or vice versa.

(ii) To buffer reaction mixture close to the pH optimum of the enzyme.

(iii) To allow for electrophoretic separation of the aqueous phases.

(iv) To avoid high expenses.

University of Chemistry and Technology, Prague

Initial ATPS screening

ATPS	Composition
A	PEG 400 - 20 wt %, phosphates - 15 wt%, water - 65 wt%
B	PEG 4000 - 10 wt %, citrates - 12 wt%, water - 78 wt%
C	PEG 4000 - 10 wt %, dextran 40 - 8 wt%, 0.05 M phosphate buffer - 82 wt%
D	PEG 4000 - 10 wt %, dextran 40 - 8 wt%, 0.25 M phosphate buffer - 82 wt%
E	**PEG 4000 - 15 wt %, phosphates - 10 wt%, water - 75 wt%**

Partitioning tests

ATPS	Experiments T2		Experiments T3	
	L_{6-APA}	L_{PAA}	M_{6-APA}	M_{PAA}
A	9.1	25.5	5.6	5.4
B	2.1	3.1	0.36	0.21
C	1.1	1.4	0.23	0.19
D	1.6	2.0	0.22	0.18
E	2.7	6.1	0.13	0.09

The partition coefficient L_i is defined as the ratio of the concentrations of i-th component in the top and bottom phases, respectively.

Enzyme partition coefficient M_i

Effect of dextran molecular weight

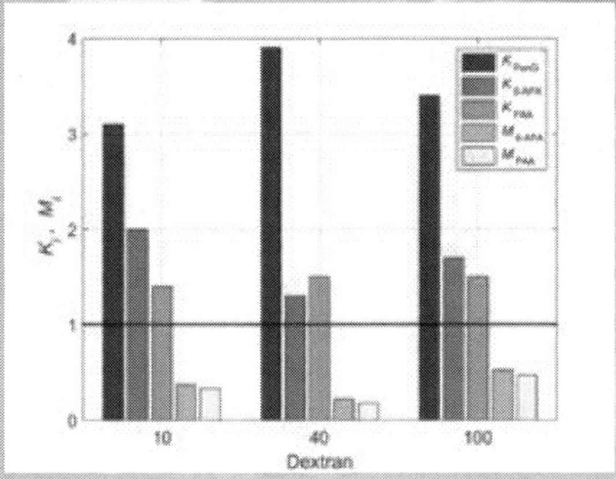

Problems:

- Relatively low values of the partition coefficients for 6-APA
- PEG/dextrans ATPS are more expensive compared to PEG/salt ATPS
- Dextran phase is much more viscous that any salty phase

Effects of PEG molecular weight and salt content

Enzyme partitioning
Effect of PEG MW

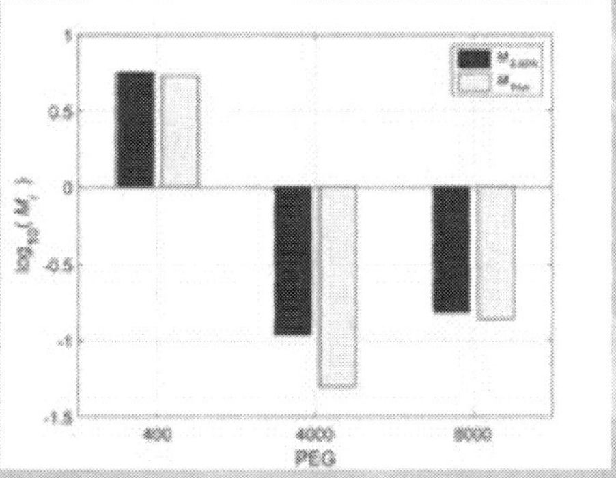

Enzyme and product partitioning
Effect of salt content

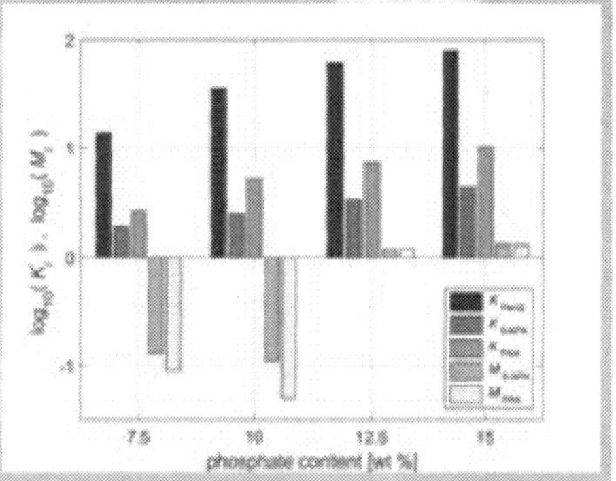

Extraction experiments with ATPS based on PEG 4000, phosphates, water

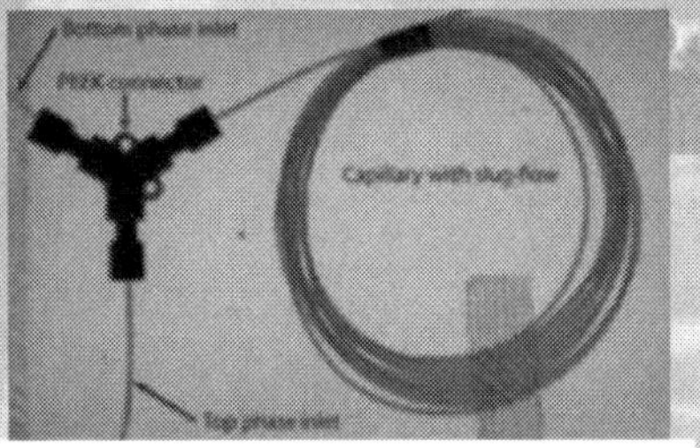

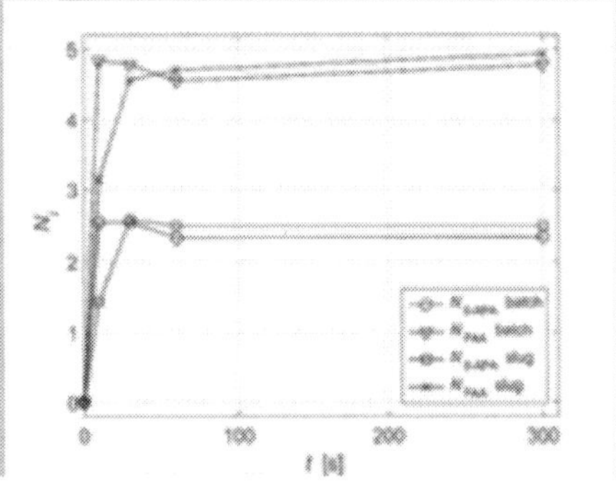

- Extraction of products from the reaction mixture in slug flow and reference batch arrangement
- Equilibrium is attained within 1 min in both arrangements
- Lower energy consumption in slug flow arrangement
- Possible numbering up

University of Chemistry and Technology, Prague

Concluding remarks

- The use of ATPS allows for the efficient separation of reaction products (antibiotics or their precursors) from homogeneous reaction mixture
- Enzyme recovery is possible
- Integrated microfluidic reactors-separators are currently under development
  - Droplet flow
  - Parallel flow
- DC electric field can improve their performance by two ways
  - Easy phase separation
  - Enhancement of concentrations of reaction products in outlet streams
- Apparent electrophoretic mobilities of salty droplets in ATPS are at least 10 times higher than those in typical water-oil systems particle-water systems

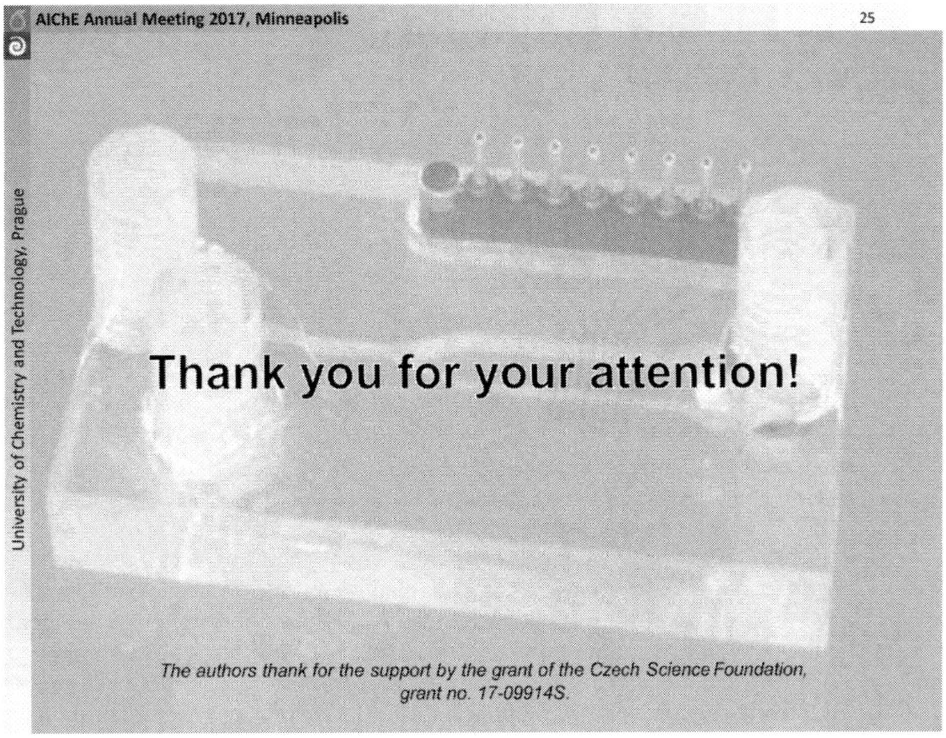
AIChE Annual Meeting 2017, Minneapolis
25
University of Chemistry and Technology, Prague
Thank you for your attention!
The authors thank for the support by the grant of the Czech Science Foundation, grant no. 17-09914S.

491414 Modulating Antibody/Antigen Affinity By Triggered Assembly and Disassembly of an Artificially Split Protein M

Abstract:

While antibodies are the fastest growing therapeutic drug class in the US, promising high efficacy for patients and profits from drug companies, many antibody therapies suffer from off-target activity. While typically occurring in a fraction of patients, promiscuous antibody interactions are associated with treatment-related complications and costly failures in clinical trials. Therefore, a generalizable means of restricting antibody therapies from functioning outside of their intended target areas is greatly needed.

Herein, we propose that Protein M (ProtM)â€"a recently discovered antibody binding domain found on the surface of *Mycoplasma genitalium*â€"might meet this demand. ProtM binds with high affinity to the variable light chain (V_L) of human IgG, and its C-terminus is able to interact loosely with the complimentary determining region (CDR) of those antibodies. By imposing a steric barrier between the CDR and its respective target, ProtM is able to reduce greatly antibody/antigen affinity. To demonstrate ProtMâ€™s ability to modulate this affinity, we cloned artificial C-terminal truncations to identify a ProtM variable capable of antibody binding without antigen-blocking activity. We then employed this variant to engineer a protease-responsive antibody binding switch. To achieve this, we ligated the ProtM truncation to the remainder of its C-terminus (pMCT) using Sortase A (SrtA) to activate ProtM activity, and we inserted a thrombin cut site to allow for C-terminus removal to deactivate the full-length ligation product. To test our switch, we used an MBP/anti-MBP dot blot as a model system: MBP was dotted onto a nitrocellulose membrane and then mixed with anti-MBP antibody pre-incubated with different ProtM combinations. The truncation product showed a vastly reduced ability to inhibit antibody binding, and pMCT showed no inhibition. However, the truncation ligated to pMCT returned to full-length activity. Upon cleavage with thrombin, ProtM once again lost its ability to inhibit antigen recognition. Because ProtM operates similarly on a wide vary of different antibodies, this work represents a first step towards a generalizable, universal approach to reduce off-target antibody activity in therapies.

497581 Systems Engineering N-Glycans of Recombinant Therapeutic Proteins Produced in Mammalian Cells

Abstract:

Glycan patterns on therapeutic proteins affect the stability, half-life, and biological activities of those medicines and are an important quality attribute. The glycosyltransferase reactions form a complex network and are distributed into compartmentalized Golgi apparatus. Furthermore, long pathways are involved in the synthesis of nucleotide sugar precursors requisite for glycosylation, and the transport rate of those precursors into Golgi apparatus is controlled by specific transporters.

We have taken a systems engineering approach to engineer the glycosylation pathway in CHO cells to modulate the N-Glycan patterns in recombinant immunoglobulin G (IgG). Using a network visualization tool developed in our lab (GlycoVis) and a reaction network model to analyze the N-Glycan profile of a recombinant IgG, the potential limiting step of galactosylation was identified and the components of the glycosylation network, including glycan synthesis, nucleotide sugar synthesis and transporter were compiled. Transcript levels of the candidate genes were evaluated for the merit of genetic manipulation. Candidate genes, individually as well as in the form of cassettes of multiple genes with potential synergistic effects, were introduced into CHO for overexpression. Resulting N-Glycan patterns of recombinant protein produced in glycoengineered CHO cells were analyzed by HPLC using 2-AB labeling. Several gene construct gave the desired the results. The results also serve to refine the model parameters and sharpen its prediction. Such a systems approach will enhance our capability to steer N-Glycan patterns and enhance the control of the quality of therapeutic proteins.

498354 A Framework for Development of Integrated and Computationally Feasible Models of Large-Scale Mammalian Cell Bioreactors

Abstract:

The growing demand for therapeutic proteins and increasing focus from regulatory bodies have confronted the biopharma industry with challenges of ensuring affordability, delivery, and consistent quality of products. In the absence of mature manufacturing technology platform, the industry relies on capacity expansion, operational agility and efficiency of supply chain to enhance the overall production economics. The capacity of manufacturing sites has increased up to 200,000L due to utilization of large-scale bioreactors (as large as 25,000L) while scale-up methodologies is still based on overall characterization of the system characteristics such as aeration and agitation mechanisms. Despite growing revenue, the return on innovation as the engine of the industry has been declining. Considering total cost of launching an asset and future net cash inflows an asset is forecast to deliver, the rate of return has been showing a declining trend from 10.1% in 2010 to 3.7% in 2016 [1]. Robust mechanistic bioreactor models facilitate improvement in equipment utilization, medium design, and feeding strategy, and shed light on some of the challenges of scale-up including productivity and byproduct formation.

Mammalian cell cultures have become responsible for half of the annual revenue earned by the biotechnology industry. Vulnerability of cells to physical environmental stimuli and spatial heterogeneity in industrial scale bioreactors demand a modeling framework which captures both biology and hydrodynamics of the system and their interactions. Existing cell culture models, in which hydrodynamics is often excluded, usually consider all causes of cell loss in one parameter, e.g. growth rate. The accuracy of these models deteriorates when process conditions change as their parameters are highly dependent on strain, cultivation medium and fermentation conditions [2]. In addition to narrow confidence intervals, often achieving a satisfactory fit of experimental data requires considering extra terms, which are theoretically difficult to explain. Overall, despite the practical and commercial applications of CHO cells there are only few reports on their kinetics of growth and production. More specifically there is no literature on the kinetic parameters of CHO cell batch culture related to mAb production [3].

In this work we propose a framework for modeling of bioreactor operation integrating biological and physical phenomena which maintains computational feasibility and permits linking the model to nonlinear solvers for optimization [4]. Biological processes are captured by modeling relevant portions of the cellular activities that control the product of interest. Unstructured modeling utilizes a reduced number of reactions to capture cellular metabolism macroscopically. Computational fluid dynamics (CFD) simulation is employed for prediction of flow characteristics inside the reactor. The main attributes of the flow are velocity fields, gas hold up, rate of dissipation of energy and gas-liquid mass transfer. Bubbles and turbulent eddies have been recognized to directly affect viable cell density [5]. To include their interactions with cells the parameters of damaging threshold of mechanical shear and critical distance of interaction with bubbles are introduced into the model. Computing the evolution of the system requires dynamic multi-phase CFD simulation, which demands many CPU hours to simulate processing time of few seconds. For computational tractability the metabolic model is solved dynamically while hydrodynamics switches between defined steady states. This is based upon the assumption that the time for the flow to reach a new steady state in response to change in operational conditions is negligible compared to the total processing time. Operating conditions are described by process parameters of impeller rotation speed, gas sparging flowrate, and liquid fill level. A set of admissible operating states is defined by discretization of process parameters. The steady state solutions are obtained in advance from CFD simulations. Further model reduction is achieved by time-space decomposition through dividing the reactor into compartments small enough to be assumed homogenous. It provides the necessary formulation and modeling environment for coupling the model with nonlinear programming solvers and optimize bioreactor operation. There are two basic categories of optimization algorithms that use reduced models. Those are differentiated based on their dependencies on the original detailed model for calculation of gradients [6]. For the proposed formulation most of the evaluations and construction of the reduced models are done before starting the optimization since minimal communication between the optimization algorithm and the original model is preferred.

Development of a surrogate model for hydrodynamics is investigated to exploit data obtained from CFD simulations and replace some of the discrete variables. To use large datasets generated by the simulations, first a sampling method based on Latin hypercube design is implemented. It accounts for spatial dependency of data and improves the uniformity of spread of samples. The surrogate model consists of a linear regression term and a stochastic error which follows a Gaussian process. Simultaneous estimation of coefficients in both terms improves the predictive power of the surrogate. This is achieved through maximizing a penalized likelihood function with spatial dependence incorporated into the penalty term. To reduce the computational burden on the solver an iterative method based on local linear approximation of the penalty term is used. With careful initial guess and choice of regularization parameter for the penalty term, one step of iteration has been shown to provide a statistically efficient approximation of the coefficients.

Quantification of the extent of effects of mechanical shear and bubble interactions on viability of cells is a valuable outcome which is achievable through integrated modeling, experimental design, and parameter estimation. Integration also gives better estimations of cellular growth and death rates as functions of metabolites concentrations. It addresses lack of universality of cell culture models, which is a major hindrance towards implementation of mechanistic models in biotechnology industry. Additionally a well-posed model which captures interactions of system components and is coupled with optimization solvers can recommend substantial improvements for bioreactor operations.

References

1. Terry, C. and N. Lesser, *Balancing the R&D equation*. 2017, Deloitte Center for Health Solutions.

2. Rocha, I., *Model-based strategies for computer-aided operation of recombinant E. coli fermentation*. 2003, Universidade do Minho.

3. Lopez-Meza, J., et al., *Using simple models to describe the kinetics of growth, glucose consumption, and monoclonal antibody formation in naive and infliximab producer CHO cells*. Cytotechnology, 2016. **68**(4): p. 1287-1300.

4. Farzan, P. and M.G. Ierapetritou, *A Framework for Development of Integrated and Computationally Feasible Models of Large-Scale Mammalian Cell Bioreactors*. Submitted to the Journal of Industrial & Engineering Chemistry Research, 2017

5. Chalmers, J., *Animal cell culture, effects of agitation and aeration on cell adaption*, in *Encyclopedia of Cell Technology*, R.E. Spier, Editor. 2000, Wiley-Interscience.

6. Biegler, L.T., Y.D. Lang, and W.J. Lin, *Multi-scale optimization for process systems engineering*. Computers & Chemical Engineering, 2014. **60**: p. 17-30.

501796 Multi-Enzyme Display on Bacterial Spore Surface for Plastic Degradation

Abstract:

Bacterial endospores released from cells as a survival mechanism are resistant to various harsh environments. Spores from *Bacillus subtilis* have been used to display diverse proteins including enzymes and vaccines. Typically the displayed protein is translationally fused to a surface anchor protein and the fusion protein is assembled onto the spore surface during the sporulation process. Thus, no further purification is needed and this significantly reduces the production costs. In addition, due to the robustness of spores, the displayed proteins often gain resistance to a wider range of temperatures and pHs as well as other unfavorable conditions. The spore surface display platform proves to be an ideal protein expression and display system either for easier long term storage or for enhanced robustness. However, so far only a single protein has been displayed on one spore surface. For multiple proteins involved in the same pathway, localizing them on the same spore surface with close proximity has obvious advantages to promote the overall efficiency. In this project, we take a two-enzyme pathway to break down plastic PET to its monomers as an example to show how to modulate their expression level on the spore surface. Regulation using synthetic inducible promoters at the transcription level and designer scaffold based regulation at the post-translational level are both used to vary the ratio of the two enzymes: PETase and MHETase. These are further compared with a mixture of spores carrying the two enzymes separately. Depending on the choice of anchor proteins, certain pairs do offer spatial advantages to improve the overall efficiency. This platform for display multiple proteins can be adapted to more complex pathways so that bacterial spores can be used as robust biocatalysts for *in vitro* biosynthesis or biodegradation.

Posttranslational modifications as regulators of activity in nucleus versus mitochondria

Lawrence I. Grossman, Siddhesh Aras, Neeraja Purandare, Stephanie Gladyck, Mallika Somayajulu, Maik Hüttemann

Center for Molecular Medicine and Genetics, Wayne State University School of Medicine, Detroit, MI

1

AIChE, Minneapolis, November 1, 2017

Outline

- MNRR1/CHCHD2
- CHCHD10
- Mitochondrial retrograde regulation and the effect of posttranslational modification of MNRR1 in nucleus and mitochondria and effect on function

2

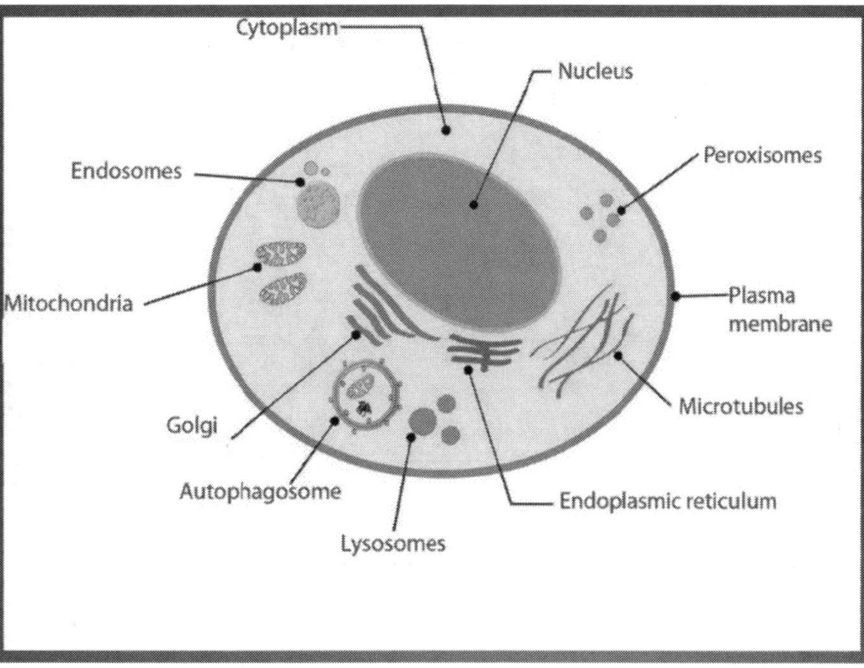

Outline

- MNRR1 /CHCHD2
- CHCHD10
- Mitochondrial retrograde regulation and the effect of posttranslational modification of MNRR1 in nucleus and mitochondria and effect on function

MNRR1 /CHCHD2 functions in both the nucleus and the mitochondria

- MNRR1 is a CX_9C protein.
  - CX_9C proteins are found in the mitochondrial intermembrane space and contain a coiled coil–helix–coiled coil–helix (CHCH) fold
- We have found several CX_9C proteins that are in more than 1 compartment and that help the cell sense and respond to mitochondrial stress
  - Orthologs MNRR1 (CHCHD2) and CHCHD10

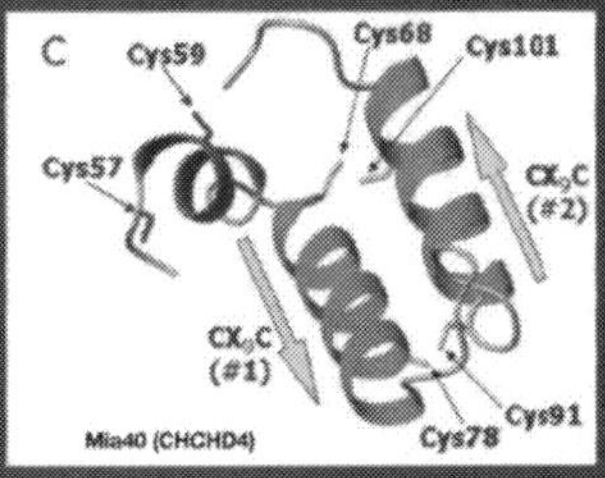

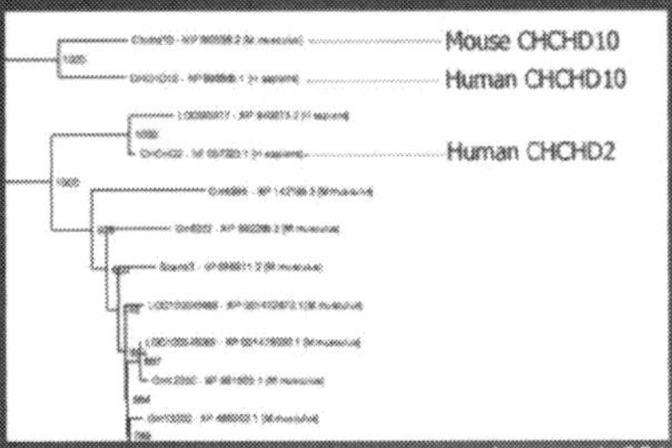

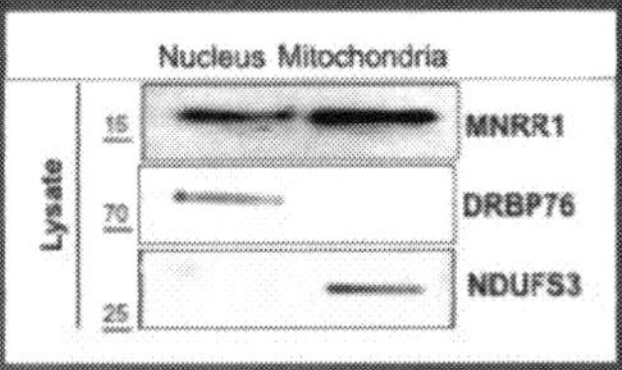

Aras et al, 2013; Aras et al, 2015; Aras et al, 2017; Grossman et al, 2017

Nuclear functions of MNRR1

6

MNRR1/CHCHD2 functions in the nucleus as a transcription factor

Example of unusual case where the same protein functions in more than 1 cellular compartment *and has a different function in each*

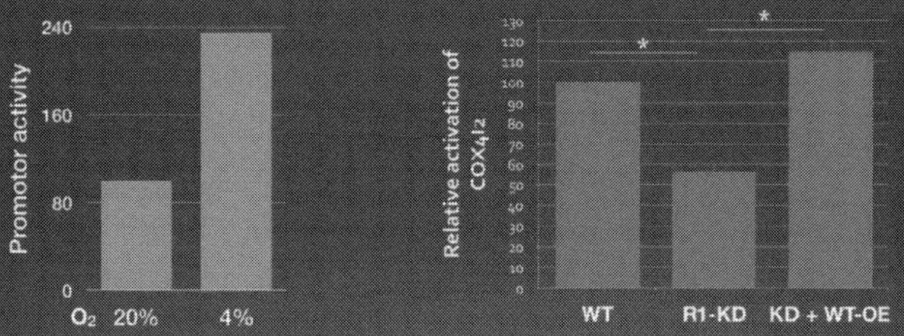

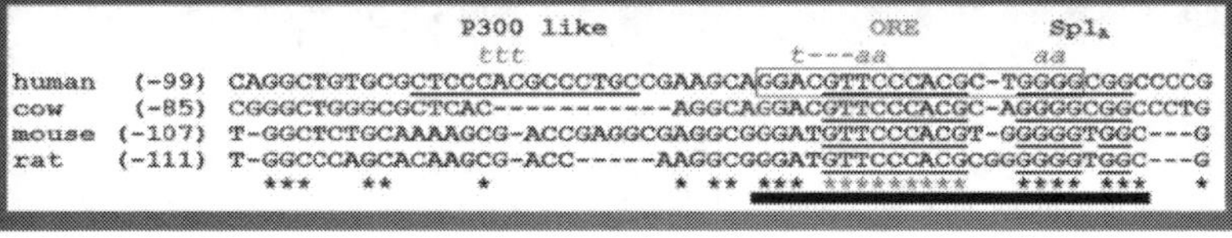

COX4I2

MNRR1 appears to regulate other genes

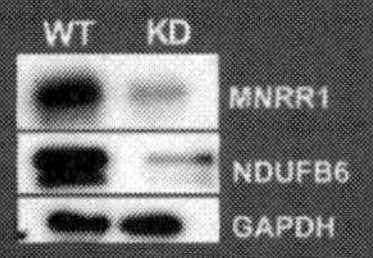

Gene	Function
ATG7	Autophagy
PGC-1α	Regulator of mitochondrial biogenesis
MFN-1	Mitochondrial fusion
NDUFB6	Complex I subunit
Mn-SOD	Converts superoxide to H_2O_2
Cu/Zn-SOD	Converts superoxide to H_2O_2
GPx	Converts H_2O_2 to H_2O
YME1L1	IMS protease
RICTOR	mTORC2 component
GLUT4	Insulin regulated glucose transporter
pAKT	Also PKB; kinase that activates cell survival pathways
Fox01	Transcription factor
Sirt1	Deacetylase

8

Nuclear function: MNRR1 binds to and stimulates the promoter oxygen responsive element (ORE)

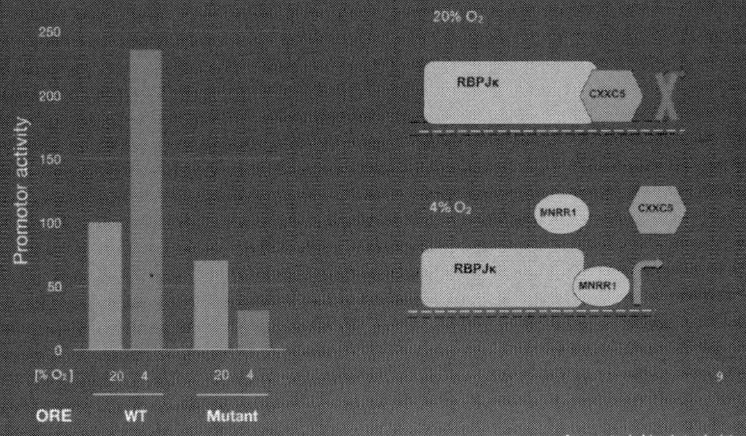

9

Aras et al, Nucleic Acids Res, 2013

Mitochondrial functions of MNRR1

10

MNRR1 functions in the mitochondria by regulating COX activity

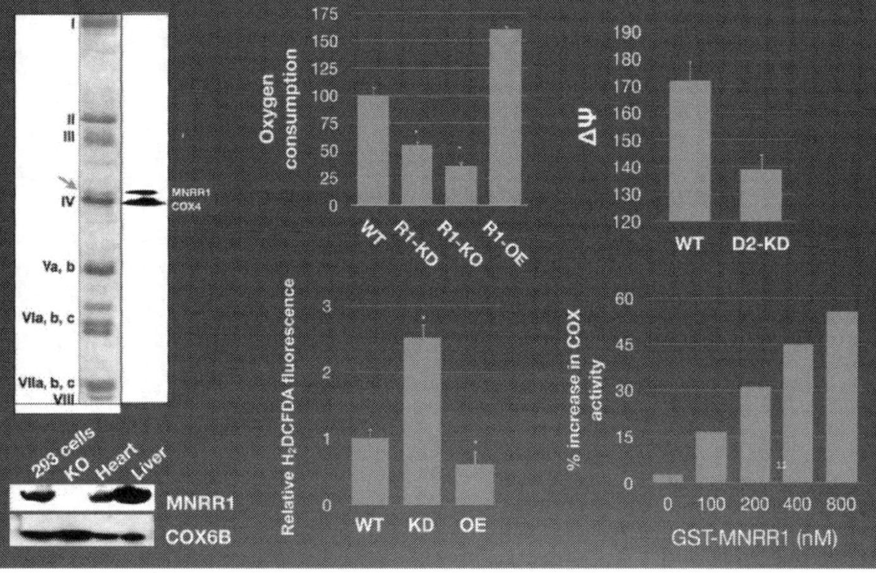

11

Outline

- Mitochondria and retrograde regulation
- MNRR1 (CHCHD2)
- CHCHD10
- Posttranslational modification of MNRR1 in nucleus and mitochondria and effect on function

12

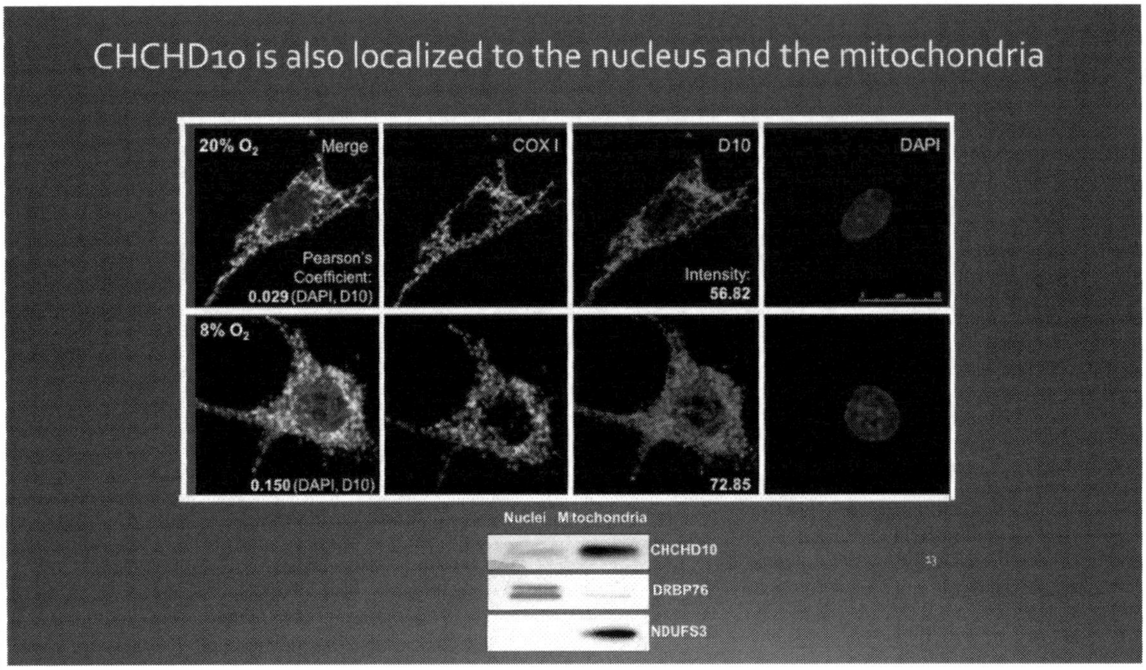
CHCHD10 is also localized to the nucleus and the mitochondria
20% O2
Merge
COX I
D10
DAPI
Pearson's
Coefficient:
0.029 (DAPI, D10)
Intensity:
56.82
8% O2
0.150 (DAPI, D10)
72.85
Nuclei
Mitochondria
CHCHD10
DRBP76
NDUFS3
33

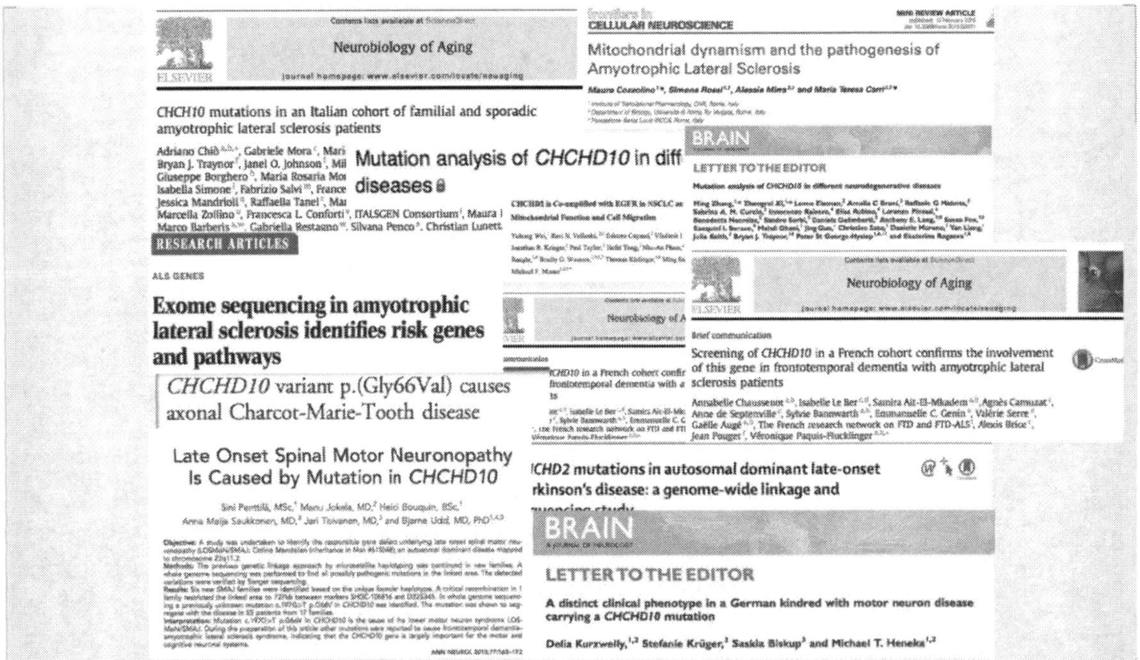
Neurobiology of Aging
CHCH10 mutations in an Italian cohort of familial and sporadic amyotrophic lateral sclerosis patients
frontiers in CELLULAR NEUROSCIENCE
Mitochondrial dynamism and the pathogenesis of Amyotrophic Lateral Sclerosis
Mutation analysis of CHCHD10 in diff
diseases
BRAIN
LETTER TO THE EDITOR
RESEARCH ARTICLES
ALS GENES
Exome sequencing in amyotrophic lateral sclerosis identifies risk genes and pathways
Neurobiology of Aging
Brief communication
Screening of CHCHD10 in a French cohort confirms the involvement of this gene in frontotemporal dementia with amyotrophic lateral sclerosis patients
CHCHD10 variant p.(Gly66Val) causes axonal Charcot-Marie-Tooth disease
Late Onset Spinal Motor Neuronopathy Is Caused by Mutation in CHCHD10
CHD2 mutations in autosomal dominant late-onset rkinson's disease: a genome-wide linkage and
BRAIN
LETTER TO THE EDITOR
A distinct clinical phenotype in a German kindred with motor neuron disease carrying a CHCHD10 mutation
Delia Kurzwelly, Stefanie Krüger, Saskia Biskup and Michael T. Heneka

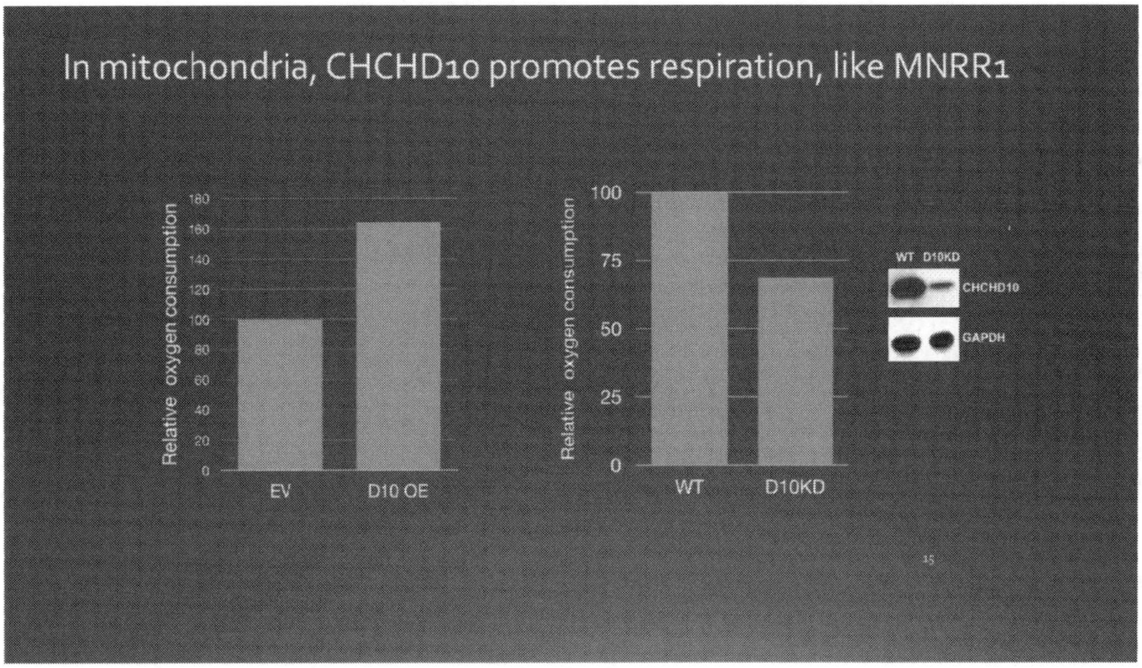
In mitochondria, CHCHD10 promotes respiration, like MNRR1
Relative oxygen consumption
EV
D10 OE
Relative oxygen consumption
WT
D10KD
WT D10KD
CHCHD10
GAPDH
15

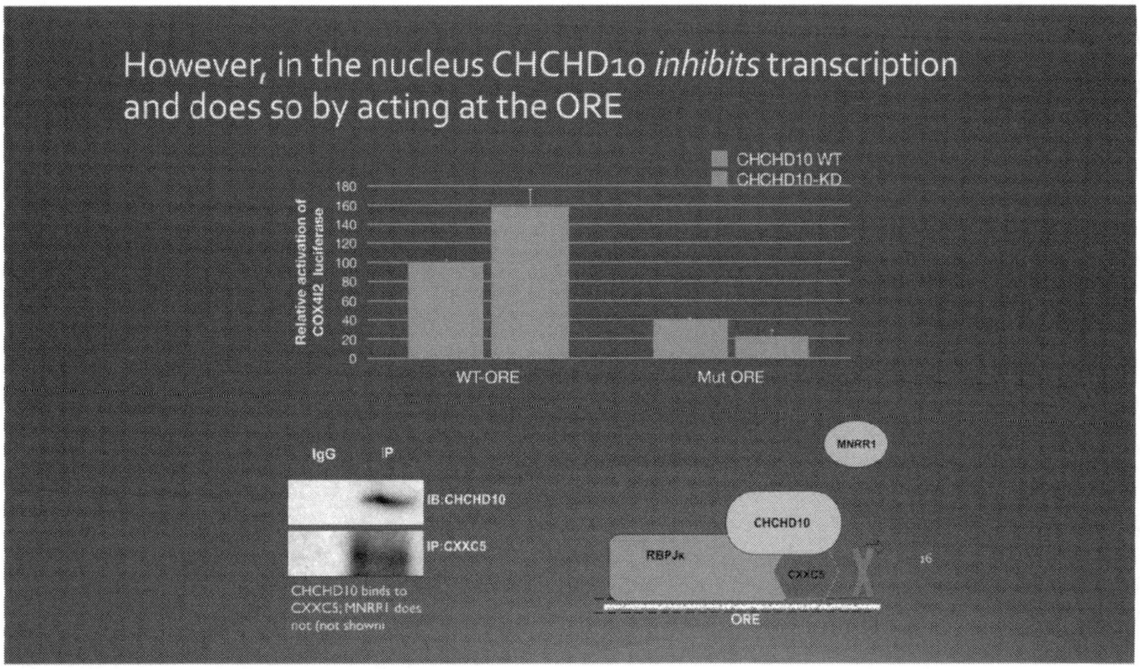
However, in the nucleus CHCHD10 *inhibits* transcription and does so by acting at the ORE
CHCHD10 WT
CHCHD10-KD
Relative activation of COX4I2 luciferase
WT-ORE
Mut ORE
IgG
IP
IB:CHCHD10
IP:CXXC5
CHCHD10 binds to CXXC5; MNRR1 does not (not shown)
MNRR1
CHCHD10
RBPJκ
CXXC5
ORE
16

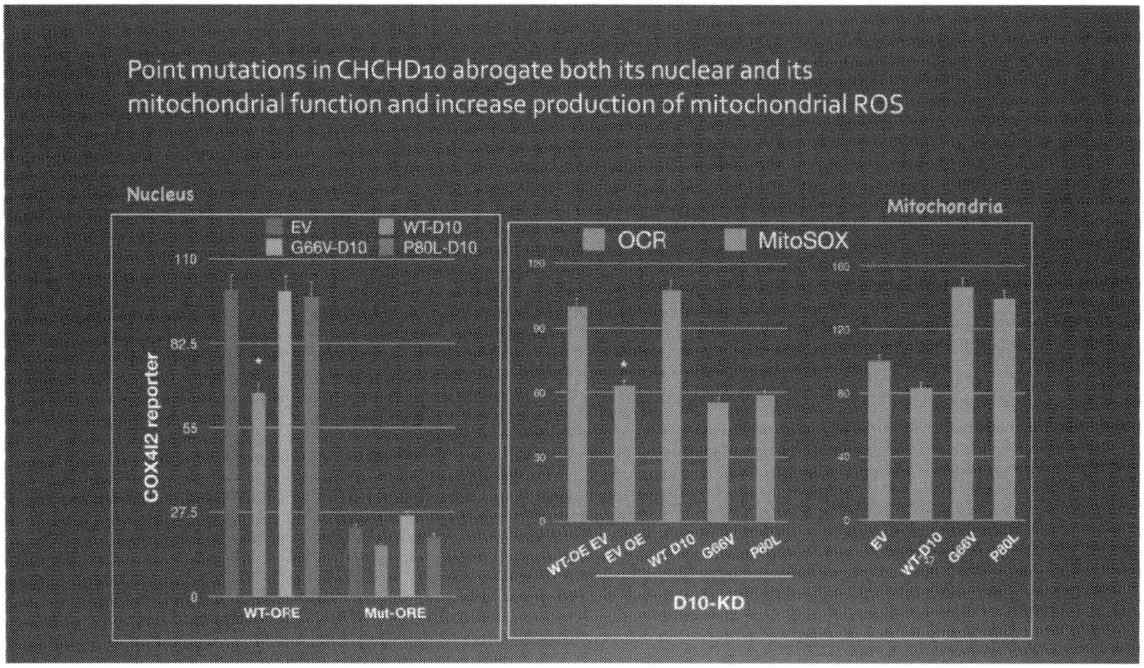
Point mutations in CHCHD10 abrogate both its nuclear and its mitochondrial function and increase production of mitochondrial ROS
Nucleus
Mitochondria
EV
WT-D10
G66V-D10
P80L-D10
COX4I2 reporter
110
82.5
55
27.5
0
WT-ORE
Mut-ORE
OCR
MitoSOX
WT-OE EV
EV OE
WT D10
G66V
P80L
D10-KD
EV
WT-D10
G66V
P80L

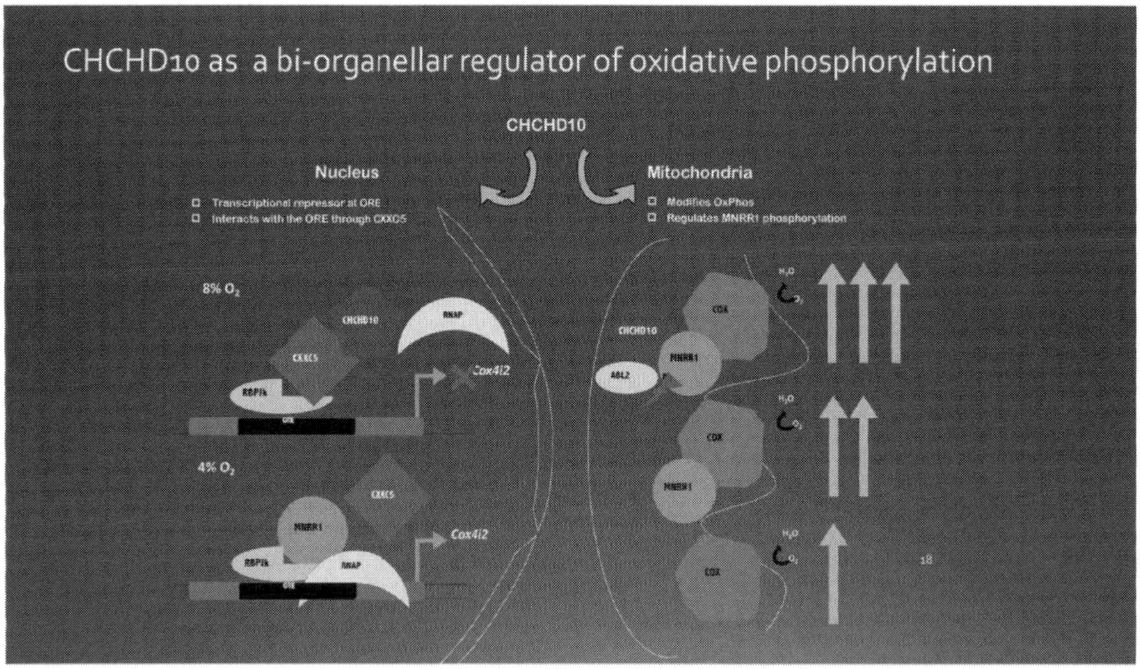
CHCHD10 as a bi-organellar regulator of oxidative phosphorylation
CHCHD10
Nucleus
Transcriptional repressor at ORE
Interacts with the ORE through CXXC5
Mitochondria
Modifies OxPhos
Regulates MNRR1 phosphorylation
8% O2
4% O2
CHCHD10
RNAP
CXXC5
Cox4i2
MNRR1
ABL2
COX
18

Outline

- Mitochondria and retrograde regulation
- MNRR1 (CHCHD2)
- CHCHD10
- Posttranslational modification of MNRR1 in nucleus and mitochondria and effect on function

19

Mitochondrial function is posttranslationally regulated

```
MPRGSRSRTS RMAPPASRAP QMRAAPRPAP VAQPPAAAPF SAVGSSAAAP    50
RQPGLMAQMA TTAAGVAVGS AVGHTLGHAI TGGFSGGSNA EPARPDITYQ   100
EPQGTQPAQQ QQPCLYEIKQ FLECAQNQGD IKLCEGFNEV LKQCRLANGL A 151
```

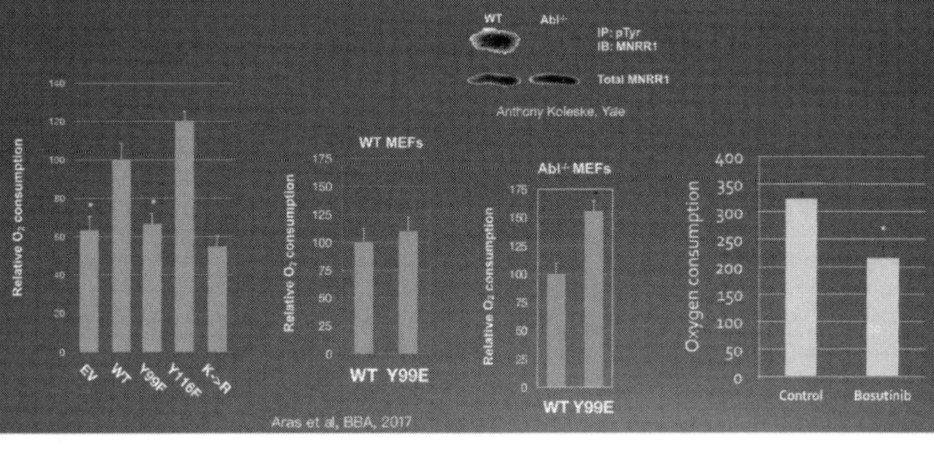

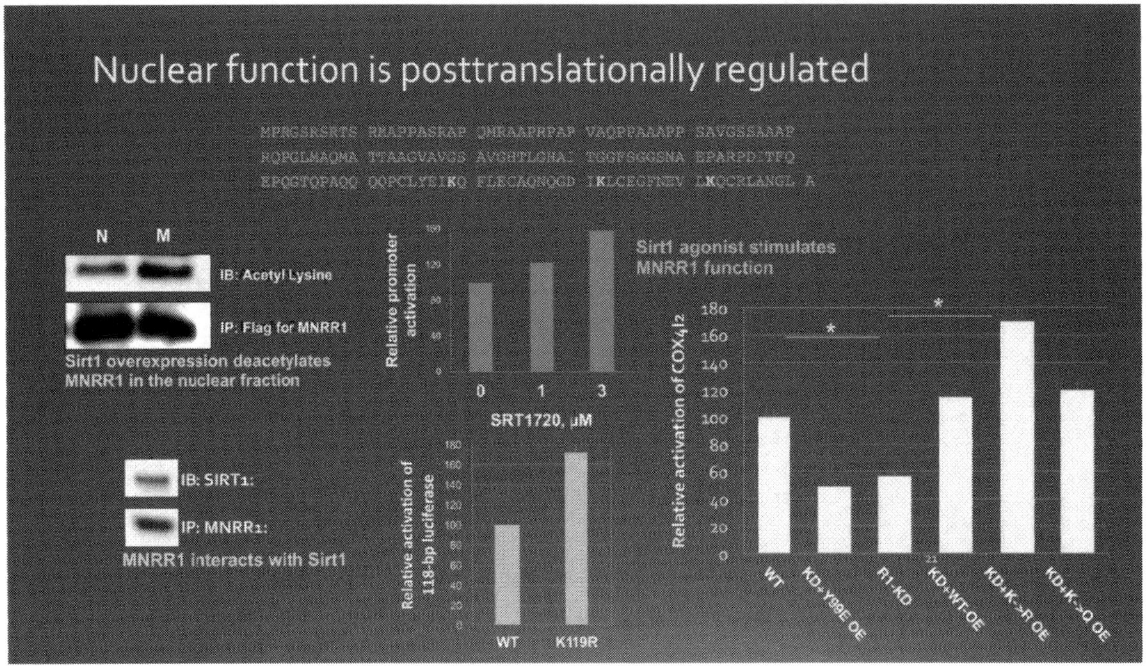
Nuclear function is posttranslationally regulated
MPRGSRSRTS RRAPPASRAP QMRAAPRPAP VAQPPAAAPP SAVGSSAAAP
RQPGLMAQMA TTAAGVAVGS AVGHTLGHAI TGGFSGGSNA EPARPDITFQ
EPQGTQPAQQ QQPCLYEIKQ FLECAQNQGD IKLCEGFNEV LKQCRLANGL A
N M
IB: Acetyl Lysine
IP: Flag for MNRR1
Sirt1 overexpression deacetylates MNRR1 in the nuclear fraction
IB: SIRT1:
IP: MNRR1:
MNRR1 interacts with Sirt1
Relative promoter activation
0 1 3
SRT1720, µM
Relative activation of 118-bp luciferase
WT K119R
Sirt1 agonist stimulates MNRR1 function
Relative activation of COX4I2
180 160 140 120 100 80 60 40 20 0
WT
KD+Y99E OE
R1-KD
KD+WT-OE
KD+K->R OE
KD+K->Q OE
21

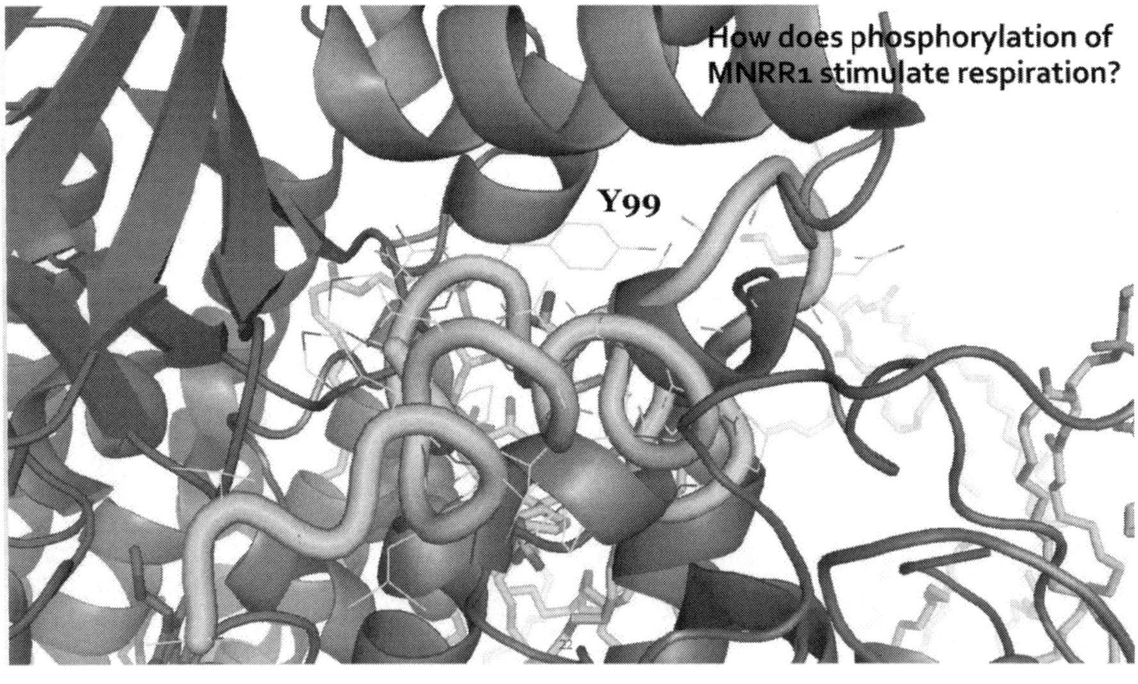
How does phosphorylation of MNRR1 stimulate respiration?
Y99
22

Summary

- MNRR1, a regulator of COX4I2 and other genes, functions in both the nucleus and the mitochondria
  - In the nucleus it transactivates genes containing an ORE promoter sequence
  - The mitochondrial function of MNRR1 is promoted by tyrosine phosphorylation carried out by Abl2 kinase (ARG)
- CHCHD10 also functions in both compartments, inhibiting transactivation at the ORE and promoting MNRR1 phosphorylation in the mitochondria
- Two pathogenic CHCHD10 mutations are dysfunctional in both the nucleus and the mitochondria

MNRR1 status

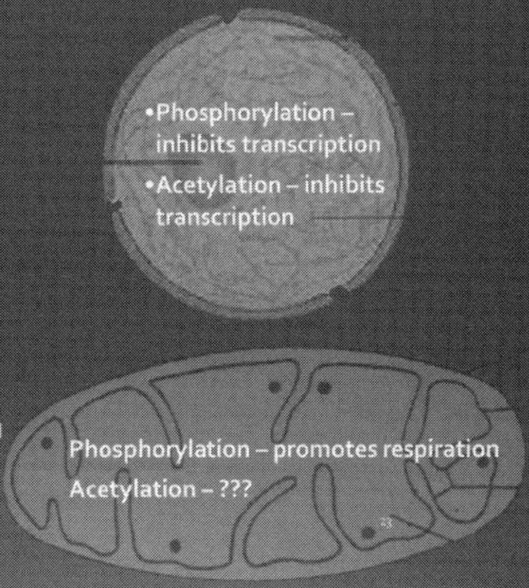

23

Acknowledgements

Grossman lab

Siddhesh Aras

Neeraja Purandare

Stephanie Gladyck

Mallika Somyajulu

Rooshan Arshad

Minbo Bai

Malati Vadapalli

Jordan Zhou

Hassan Arrabi

Justyna Resztak

Hüttemann lab

Maik Hüttemann

Icksoo Lee

Collaborators

Natascha Sommer
Norbert Weissmann
University of Giessen

Anthony Koleske
Yale University

Thomas Langer
University of Cologne

Andrey Kozlov
University of Vienna

David Lombard
University of Michigan

John Kamholz
University of Iowa

Deyu Fang
Richard Scarpulla
Northwestern University

Stephan Züchner
University of Miami

DoD (PRMRP)
Henry L. Brasza endowment

24

501392 Location-Dependent Effect of Post-Translational Pegylation on Pegylation Efficiency and the Activity and Protein Stability of T4 Lysozyme

Abstract:

Years of work conjugating the biocompatible polymer polyethylene glycol (PEG) to proteins have already resulted in the development of FDA-approved PEGylated protein therapeutics that are more stable and have longer functional lifetimes than their un-PEGylated counterparts. The post-translational covalent attachment of PEG groups, or PEGylation, increases a proteinsâ€™s molecular weight, thereby reducing clearance from the body. PEGylation also shields proteins from proteases and can thermodynamically stabilize a proteinâ€™s conformation. Thus PEGylation can also enhance protein biocatalysis stability and function. However, a central limitation to polymer-protein conjugation is polymer additions often interfere with protein folding and disrupt the protein or enzyme function. Thus, the attachment site of PEG on a protein must be optimized to achieve maximum stability and activity. Using cell-free protein synthesis and the incorporation of non-canonical amino acids, we developed a rapid system to compare PEGylation location on model protein T4-lysozymeâ€™s activity and stability. We also assess the impact of different PEG sizes, the accessibility of different locations for PEGylation, and the impact of PEGylation at multiple sites. The results of this study are compared to current design heuristics, simulations, and predictive tools for PEGylation. Overall, it is still challenging to predict a prior the optimal locations for PEGylation, however, additional structure-based guidelines to aid in the rational selection of optimal PEGylation sites are presented. Furthermore, the rapid screening system presented provides quick assessment of rationally selected PEGylation sites.

504081 Making Water-Soluble Integral Membrane Proteins Directly in Living Cells

Abstract:

Integral membrane proteins (IMPs) play crucial roles in all cells and represent attractive pharmacological targets. However, functional and structural studies of IMPs are hindered by their hydrophobic nature and the fact that they are generally unstable following extraction from their native membrane environment using detergents. As an alternative to extraction and *in vitro* solubilization, we devised a general strategy for *in vivo* solubilization of IMPs in structurally relevant conformations without the need for detergents or mutations to the IMP itself. This technique, called SIMPLEx (solubilization of IMPs with high levels of expression), allows direct expression of soluble products in living cells by simply fusing an IMP target with truncated apolipoprotein A-I, which serves as an amphipathic proteic â€œshieldâ€ that sequesters the IMP from water and promotes its solubilization. In one notable application, we converted *Escherichia coli* DsbB, a transmembrane enzyme that catalyzes the re-oxidation of the periplasmic oxidase DsbA by ubiquinone, into a water-soluble biocatalyst using SIMPLEx. When solubilized DsbB variants were co-expressed with an export-defective copy of DsbA in the cytoplasm of wild-type *E. coli* cells, artificial oxidation pathways were created that efficiently catalyzed *de novo* disulfide bond formation in a range of substrate proteins and in a manner that depended on both DsbA and quinone. Hence, DsbB solubilization was achieved with preservation of both catalytic activity and substrate specificity. Moreover, the generality of the solubilization technique should pave the way for unlocking the biocatalytic potential of other membrane-bound enzymes whose utility has been limited by poor stability of IMPs outside of their native lipid bilayer context.

500543 Regioselective Alcohol Ring Opening of Epoxides Using Lewis Acidic Catalysts

Abstract:

Epoxides are versatile chemical intermediates that can be conveniently synthesized from alkenes.[1] Opening of the epoxide ring can be achieved with various nucleophiles like water, alcohols, amines, and thiols yielding synthetically important di-substituted compounds.[2–5] Although heterogeneous acids and bases are known to catalyze this reaction, limitations like low selectivity and/or low stability can limit their scope.[6–8] In this work we have studied silica-based Lewis acidic catalysts with varying pore-sizes and Lewis acid-site strength. We report that these catalysts can selectively open the epoxide ring using alcohols under mild conditions for different systems including epichlorohydrin-methanol. Systematic investigations of the effect of acid site strength, temperature, and pore-confinement of these materials revealed that these factors affect the catalytic activity but not the regioselectivity. In case of straight chain mono-functional epoxides, the regioselectivity increases with an increase in nucleophile size. Overall, we have studied catalytic performance and reusability of different silica-based Lewis acidic catalysts for epoxide ring opening with alcohols.

References:

(1) van der Waal, J. C.; Rigutto, M. S.; van Bekkum, H. *Appl. Catal. A Gen.* **1998**, *167* (2), 331–342.

(2) Tang, B.; Dai, W.; Wu, G.; Guan, N.; Li, L.; Hunger, M. *ACS Catal.* **2014,** *4* (8), 2801-2810.

(3) Kahandal, S. S.; Kale, S. R.; Disale, S. T.; Jayaram, R. V. *Catal. Sci. Technol.* **2012**, *2*, 1493-1499.

(4) Dhakshinamoorthy, A.; Alvaro, M.; Garcia, H. *Chem. - A Eur. J.* **2010**, *16* (28), 8530–8536.

(5) Zhu, X.; Venkatasubbaiah, K.; Weck, M.; Jones, C. W. *J. Mol. Catal. A Chem.* **2010**, *329* (1–2), 1–6.

(6) Brunelli, N. A.; Long, W.; Venkatasubbaiah, K.; Jones, C. W. *Top. Catal.* **2012**, *55*, 432–438.

(7) Takeuchi, H.; Kitajima, K.; Yamamato, Y.; Mizuno, K. *J. Chem. Soc. Perkin Trans. 2* **1993**, 199–203.

(8) Feng, Y.; Lydon, M. E.; Jones, C. W. *ChemCatChem* **2013**, *5* (12), 3636–3643.

BEILSTEIN JOURNAL OF ORGANIC CHEMISTRY

Self-optimisation and model-based design of experiments for developing a C–H activation flow process

Alexander Echtermeyer[1,2], Yehia Amar[2], Jacek Zakrzewski[2] and Alexei Lapkin[*2,§]

Full Research Paper

Open Access

Address:
[1]Aachener Verfahrenstechnik – Process Systems Engineering, RWTH Aachen University, Aachen, Germany and [2]Department of Chemical Engineering and Biotechnology, University of Cambridge, Cambridge, United Kingdom

Email:
Alexei Lapkin[*] - aal35@cam.ac.uk

[*] Corresponding author
§ + 44 1223 334796

Keywords:
automated reaction system; C–H activation; design of experiments; flow chemistry; process modelling; self-optimisation

Beilstein J. Org. Chem. **2017**, *13*, 150–163.
doi:10.3762/bjoc.13.18

Received: 31 October 2016
Accepted: 05 January 2017
Published: 24 January 2017

This article is part of the Thematic Series "Automated chemical synthesis".

Guest Editor: I. R. Baxendale

Abstract

A recently described C(sp^3)–H activation reaction to synthesise aziridines was used as a model reaction to demonstrate the methodology of developing a process model using model-based design of experiments (MBDoE) and self-optimisation approaches in flow. The two approaches are compared in terms of experimental efficiency. The self-optimisation approach required the least number of experiments to reach the specified objectives of cost and product yield, whereas the MBDoE approach enabled a rapid generation of a process model.

Introduction

The development of manufacturing processes to produce functional molecules, such as pharmaceuticals or fine chemicals, often relies on experience and trial-and-error, rather than on mechanistic process models [1]. The only reason for this is the complexity of chemistry and the duration of time required for the development of good mechanistic models. A game changer in this area is the recently emerged field of automated continuous-flow experiments driven by algorithms for sequential design of experiments (DoE), which significantly reduce the effort in running routine reactions and generating data for optimisation of reaction conditions [2-7]. An illustration of the concept is shown in Figure 1.

Mainly, self-optimisation experimental platforms are used to rapidly obtain optimal reaction conditions using either flow [8-10] or batch experiments [11]. In these cases, the optimisation is driven by the global or target optimisation towards the selected performance criteria. This is rather different from the objectives of model development. In the case of model development, the key criterions are the ability of a model to describe

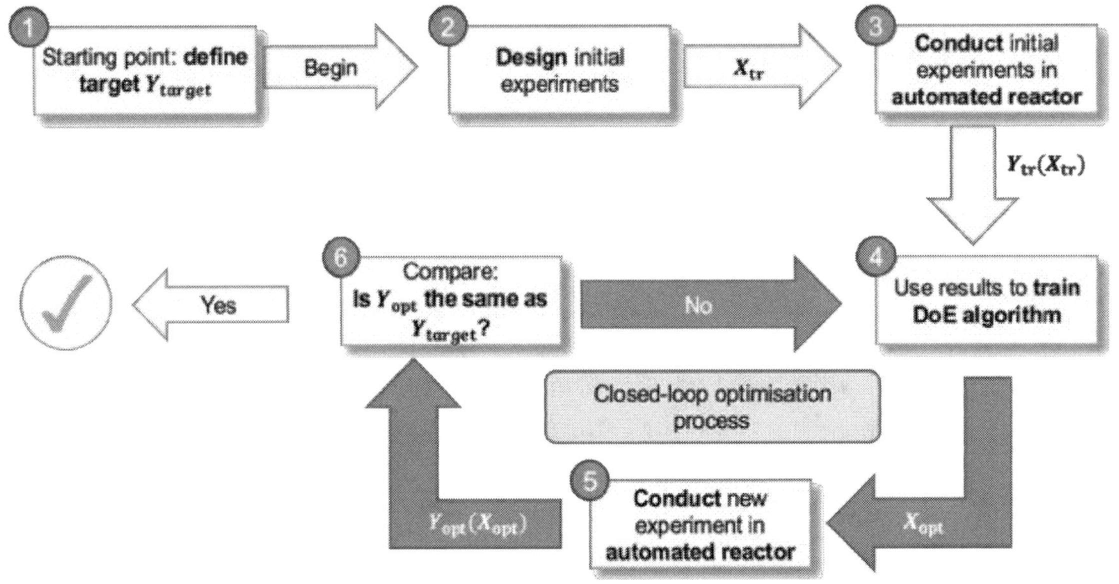

Figure 1: A framework of closed-loop or self-optimisation combining smart DoE algorithms, process analytics, chemoinformatics and automated reactor systems.

the observed experimental data and to predict process performance under unseen conditions. Thus, experiments required for model development are frequently what would be considered as 'bad' experiments in the case of optimisation.

A model-development framework has been demonstrated on the basis of an automated microreactor experimental system for several complex reactions [8,12,13]. The framework uses factorial design of experiments to obtain an initial data set for parameter estimation, followed by an iterative search with online model discrimination and parameter estimation, guided by D-optimal design. In a different approach, transient data from continuous-flow experiments were used to identify parameters of a known mechanistic scheme to discriminate between several alternative model structures and to identify model parameters, but no specific design of experiments method was used [14].

The framework proposed in the present publication is using a model-based design of experiments method (MBDoE) [15-17], which incorporates the model with its parameters, as well as details of the experimental setup, such as measurement accuracy and experimental limitations, to design the most informative experiments. This approach requires some model structures to be known a priori which may restrict the methodology to reactions with known mechanism, or to empirical parametric models. A discussion of how a priori knowledge of chemistry, i.e., reaction mechanisms, is included in self-optimisation and model-development frameworks is not well documented in the literature. Very recently we have shown that a priori knowledge in the form of density functional theory level (DFT) mechanistic calculations can be used to propose process models and to perform in silico design of novel flow processes [18]. In this publication, we present an extension of this methodology, in which an initial process model is developed through a MBDoE methodology coupled with an automated self-optimisation flow system.

This approach was tested on the Pd-catalysed C–H activation reaction of **1** resulting in the formation of an aziridine **2** (Scheme 1) [19]. The reaction was recently discovered [20] and its mechanism studied [21] and later proven [18]. A simplified mechanism is shown in Scheme 2. In the reaction of interest, the starting material **1**, an aliphatic secondary amine, is converted into an intermediate species **B** in a catalytic first step and

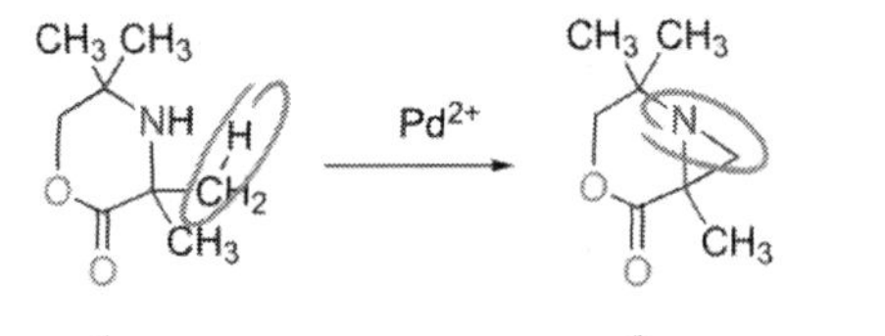

1 **2**

Scheme 1: Catalytic reaction scheme showing C–H activation of an aliphatic secondary amine **1** to form the aziridine product **2** [19,20]. Orange rings show C–H and C–N bonds in the substrate and the product, respectively, indicating the location of the C–H activation.

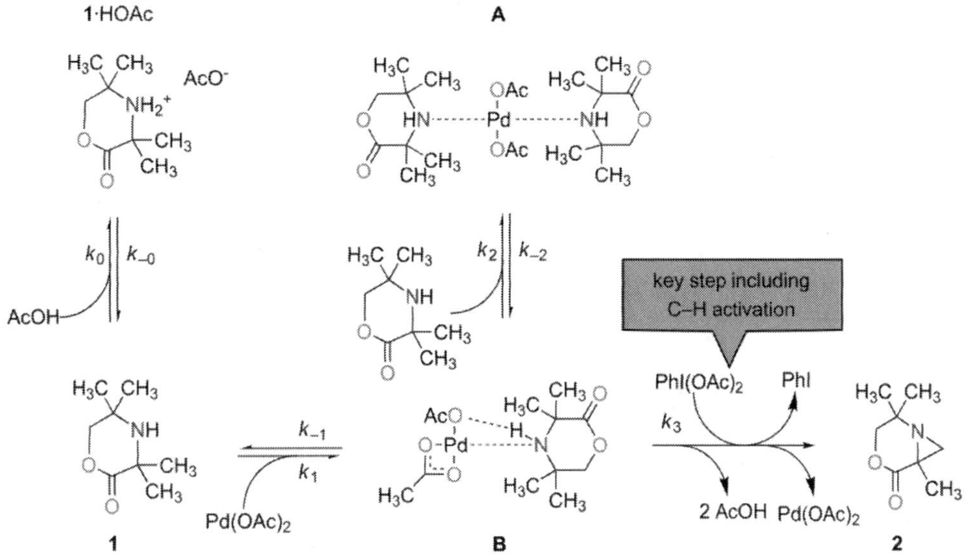

Scheme 2: A simplified reaction mechanism based on literature [21], showing intermediate **B** and the side reaction compounds **1**·HOAc and **A**. The key step includes the C–H activation. **1**: starting material, **1**·HOAc: coordinated starting material, $Pd(OAc)_2$: catalyst, **2**: product, $PhI(OAc)_2$: oxidant.

consecutively transformed to product **2** in the second step, which comprises the C–H activation. In addition to the main reaction pathway, **B** can form the relatively unreactive resting state complex **A**, and compound **1** can also form a coordinated species **1**·HOAc upon protonation with a molecule of acetic acid. This limits the formation of **A** due to reduced concentration of **1**.

Table 1 gives an overview of the a priori knowledge used in this study. Fast reaction steps were lumped into a single one, containing the critical C–H activation, and described by reaction rate constant k_3 in Scheme 2. Empirical information provided constraints of process conditions, such as temperature and concentration ranges, whereas initial values of kinetic parameters were estimated based on a DFT model. Further details can be found in Supporting Information File 1.

Here we demonstrate an MBDoE approach on the basis of the model structure and the initial model parameters from DFT calculations and using automated flow experiments. We then use the obtained process model to develop a surrogate model for optimisation, and compare the different methodologies: classical kinetic modeling approach, MBDoE with automated flow experiments and black-box optimisation in achieving the different objectives of the methods.

Results and Discussion

Experimental system for model development and optimisation in flow

Although a number of experimental systems for self-optimisation were reported in the literature, this number is fairly small and very few examples of using flow experiments for model development are reported [8,9]. In this study a commercial

Table 1: Details of information considered as a priori knowledge in this study and source of this knowledge.

A priori knowledge	Source
reaction mechanism, concentration constraints of species due to degradation of starting material and product.	[21]
Gibbs free energies of reaction, obtained from DFT study.	[18]
target values based on best results from previous experimental study.	[18]
physical constraints (maximum oxidant concentration to prevent crystallisation, maximum temperature to prevent excessive catalyst decomposition).	empirical
technical details of experimental set-up (e.g., variance of gas chromatography (GC) used in variance model for MBDoE, minimum and maximum flow rates).	empirical

Vapourtec R2+/R4 system was used with a standard 10 mL coiled reactor. To save on expensive reagents, reagents and catalyst were injected using 2 mL sample loops, with the solvent being continuously pumped between the reaction slugs. The two employed sample loops were filled with the same reaction mixture (further information on sample preparation is given in Supporting Information File 1) to avoid potential experimental errors due to inaccuracies of generating mixtures with specific concentrations by pumps. Laminar flow through long pipes will necessarily cause dispersion, which dictates the minimum reaction slug length that can be used. This was determined experimentally, which also allowed to develop the method of detection of the reaction slugs (by a flow UV cell) and the protocol for GC sampling. A schematic depiction of the experimental system is shown in Figure 2.

Physical model generation and refinement

The initial model structure and parameters were taken from the earlier published DFT study of the reaction [18]. Performing MBDoE in the process modelling software gPROMS [22] resulted in a design indicating the experimental conditions, the reaction times and the number of samples required in each experiment for the estimation of a particular parameter or the combination of parameters. Table 2 shows the different experiments conducted for estimation of the given parameters. Each experiment refers to a particular composition of the reaction mixture, but with various reaction times for each sample within the experiment. Neither in the MBDoE step for $k_{j,ref}$ nor in the step for $E_{a,j}$ could an experiment be designed for the estimation of all parameters simultaneously. This is likely due to correlations between the parameters, which is common for reaction networks and consecutive reactions. To overcome this problem, sophisticated decoupling techniques and special design criteria considering direct measures of correlation could be used [23,24]. However, as shown by Franceschini and Macchietto, a simple design-by-grouping method can also yield reasonable results [25]. Following this approach experiments were designed for either a single or groups of parameters. Parameters, which showed a maximum in their normalised local sensitivity curves in the same time interval were grouped together. This is reasonable, as a sample taken in this time interval likely yields sensible data for the estimation of the respective parameters. As can be seen from Figure S9 (Supporting Information File 1), all parameters of the same type showed maximum sensitivity in approximately the same time interval. Hence, all possible combinations of single and grouped parameters were tested in the two MBDoE steps and those with the lowest correlation, maximum number of included parameters and a t-value larger than a reference t-value were selected. It is worth noting, that this method overcomes problems with parameter correlations during the experimental design for parameter subgroups and the subsequent estimation. The effects of the neglected parameter correlation may reoccur during the overall parameter estimation, but can be reduced due to the refinement of the parameters in the subgroups.

The best possible design with minimum analytical effort was selected. It can be seen from Table 2 that the t-test is successful for the experiments 1–3. This is not surprising as possible correlations between the parameters are neglected by splitting them into subgroups or even singles. For the experiments 4, 5 and 6 (Table 2) not all parameters pass the t-test. The best possible ex-

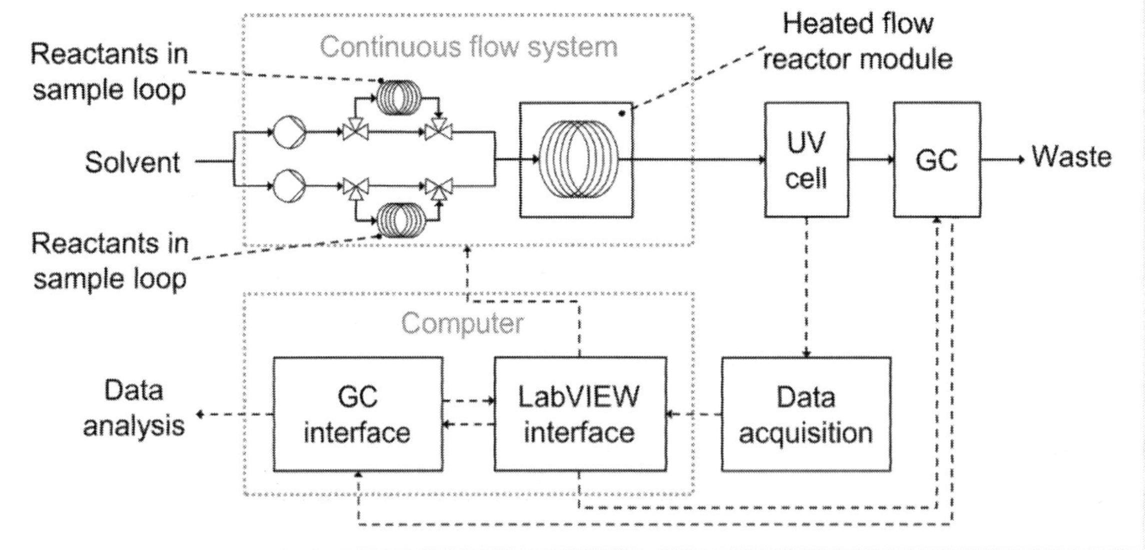

Figure 2: Schematics of the automated continuous-flow system used for model development and 'black-box' sequential optimisation.

Table 2: Results of the MBDoE for kinetic parameters, showing the number of samples needed in the experiment and the statistical t-test results.

Experiments	Parameter(s)	Number of samples	t-value	t_{ref}
1	$k_{0,ref}$	7	76.19	2.92
2	$k_{2,ref}$	6	23.36	2.92
3	$k_{3,ref}$	5	23.36	2.92
4	$k_{0,ref}$, $k_{2,ref}$, $k_{3,ref}$	11	5.34, 0.03, 6.42	1.94
5	$E_{a,0}$, $E_{a,2}$, $E_{a,3}$	11	0.05, 0.04, 2.88	1.94
6	$E_{a,0}$, $E_{a,2}$, $E_{a,3}$	11	0.63, 0.25, 2.33	1.94
7	$E_{a,0}$, $E_{a,2}$	10	2.79, 17.1	2.02
8	$E_{a,0}$, $E_{a,3}$	10	3.99, 46.8	1.94

perimental design was selected. Due to failed estimability analysis, no experiment design included parameters for the reaction $j = 1$. Experimental conditions associated with each experiment sequence are given in Supporting Information File 1, Table S4 and Table S5.

Parameter estimation and comparison of effort

For the investigated reaction in Scheme 2, Zakrzewski et al. generated and validated a kinetic model using a classical kinetic approach [18]. For this they used 38 batch experiments, each comprising approximately 10 sample points at different reaction times, which in total resulted in more than 400 sample points used for the estimation of kinetic model parameters. In contrast to that, we used MBDoE and flow experiments. As Table 2 shows, the MBDoE resulted in 8 experiments with a total of 71 samples required to determine the model parameters. These numbers highlight the benefit of MBDoE for parameter estimation, reducing the consumption of materials, cost and time associated with sample generation. Due to some failed experiments only 64 experimental sample points were used for the model development.

The parameter estimation was employed to obtain estimates of the kinetic parameters $k_{j,ref}$ and $E_{a,j}$, where $j \in \{0, 1, 2, 3\}$ in a two-step procedure using standard solver settings in gPROMS. By applying the initial guesses for the parameters, each experiment was first used to estimate only the parameter for which it was designed, while keeping the others fixed at their current values. Afterwards, all experiments were included in an overall estimation with the parameter values obtained from the previous estimations as new initial guesses to account for possible parameter correlations, which were neglected by grouping the parameters. Even though no experiment design comprised the parameters for reaction $j = 1$ specifically, they were still included in the overall estimation to refine their initial values as much as possible. To avoid stopping the estimation at undesired local optima, several such estimation runs were performed. The final results of the obtained parameter values are shown in Table 3. The final values of parameters $k_{1,ref}$ and $E_{a,1}$ do not differ much from the initial guesses, which is not surprising as the estimability analysis had already predicted a weak influence of $k_{1,ref}$ and $E_{a,1}$ on the model output, i.e., this cannot be estimated with precision. However, this was not necessary, as the parameters do not change the model prediction. Therefore, also the very large 95% confidence interval can be explained. For all other parameters, the difference between initial guess and final value is significant, which might be caused be the simplifications employed for computing the guesses and the uncertainty of the DFT calculation in the exponentially amplified van't Hoff equation. The 95% confidence intervals for the parameters of reaction $j = 0$ are both one magnitude smaller than the final parameter values indicating sufficiently low uncertainty and good significance. The confidence intervals for the parameters of reaction $j \in \{2, 3\}$ are larger than the final parameter value. However, as the values for $k_{j,ref}$ and $E_{a,j}$ can only be positive, this indicates still some level of uncertainty in the parameters estimates. This uncertainty is further revealed by a comparison of the t-values and the reference t-values. For the overall estimation of $k_{j,ref}$ it was impossible to attain t-values exceeding the reference t-value, even though $k_{0,ref}$ comes close. Furthermore, the predicted t-values from the MBDoE, shown in Table 2, could not be reached. In the overall estimation of $E_{a,j}$ three out of four parameters could not be estimated with high statistical significance. Only for $E_{a,0}$ the t-test was satisfied with a t-value close to the predicted one in Table 2. The problem of diminished statistical significance of the estimates is likely due to practical identifiability issues as the measurement data employed for the estimation was affected by experimental errors. Additionally, parameter correlation effects reappeared during the overall estimation making it more difficult to obtain useful results.

Table 3: Results of the parameter estimation showing the final values, initial guesses, 95% confidence intervals (CI) and t-values for each of the 8 parameters.

Parameter	Initial guess	Final value	Units	95% CI	t-values
$k_{0,ref}$	3.019	3.035	$L\ mol^{-1}\ s^{-1}$	±0.396	1.403^a
$k_{1,ref}$	2,551,604	2,728,600	$L\ mol^{-1}\ s^{-1}$	$\pm 4.817 \cdot 10^{11}$	$2.49 \cdot 10^{-6\ a}$
$k_{2,ref}$	8,591	16,997	$L\ mol^{-1}\ s^{-1}$	$\pm 6.282 \cdot 10^5$	0.012^a
$k_{3,ref}$	0.001756	0.140378	s^{-1}	±8.175	0.012^a
$E_{a,0}$	84,132	128,517	$J\ mol^{-1}$	$\pm 5.152 \cdot 10^4$	2.495^b
$E_{a,1}$	45,019	44,941	$J\ mol^{-1}$	$\pm 1.709 \cdot 10^{10}$	$2.63 \cdot 10^{-6\ b}$
$E_{a,2}$	59,508	20,995	$J\ mol^{-1}$	$\pm 3.525 \cdot 10^6$	0.006^b
$E_{a,3}$	98,831	144,942	$J\ mol^{-1}$	$\pm 3.517 \cdot 10^6$	0.041^b

[a]Refers to t_{ref} = 1.725 and [b]refers to t_{ref} =1.688.

Figure 3 shows a comparison of the simulated model response incorporating the final parameter values vs the experimentally observed product concentrations for the experiments 4 and 8 of the MBDoE in Table 2. These show a reasonably good model fit. Only experiments suggested by MBDoE were conducted to generate data for parameter estimation. Thus, as the method did not suggest samples to be taken at reaction times longer than 50 minutes in experiment 4, Figure 3a, or between zero and 24 minutes in experiment 8, Figure 3b, there was no data collected. In total, 8 such experiments were conducted, four for each of the two parameter types (shown in Supporting Information File 1).

Despite the remaining uncertainty in some of the parameters, indicated by the large 95% confidence intervals, the quality of model prediction was considered to be good-enough for the purpose of in silico training of the smart DoE algorithm for target optimisation. Thus, the final parameter values in Table 3 were accepted and used in the model employed for the subsequent in silico target optimisation steps.

Improvement of process conditions using an a priori model and in silico optimisation

Access to automated experimental systems allows to perform black-box sequential optimisation using sequential DoE algorithms. However, if a process model is available, there exist two more options for optimisation: optimisation using the available process model directly, or optimisation using a surrogate model. The latter is frequently used in expensive computer experiments, and in the case of large-scale process simulations, when evaluations of process models is computationally too expensive. In the case of our test reaction the MBDoE approach enabled us to develop a reasonably good process model in a small number of flow experiments. We can use this process model to perform

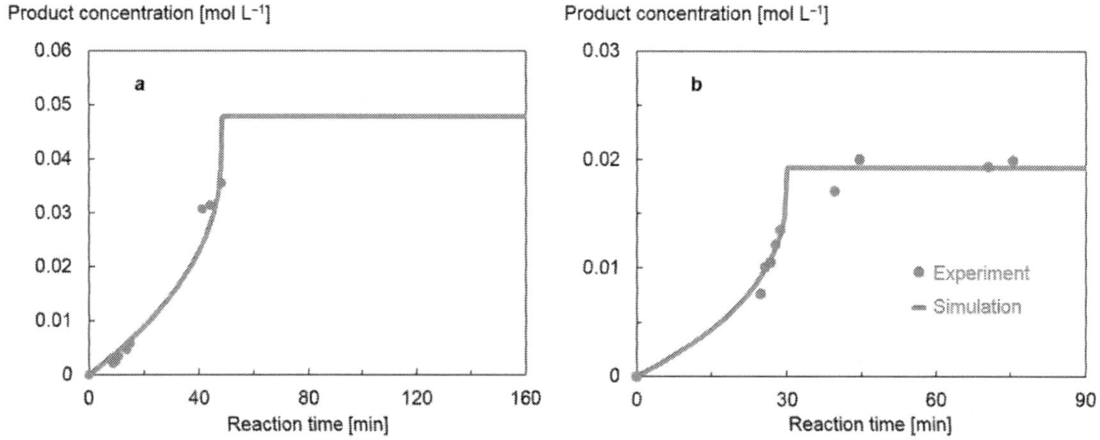

Figure 3: Results of experiments from the MBDoE in Table 2, conducted for parameter estimation, and their corresponding simulated model responses based on the estimated parameters. The only experiments conducted were those calculated by the MBDoE, defined by the sampling times and recipes suggested. (a) Experiment 4 conducted at a reference temperature of 70 °C to estimate rate constants. (b) Experiment 8 used for determining activation energies, at 75 °C.

optimisation. Although this model is not expensive to evaluate we resorted to building a surrogate model, which allowed us to use an efficient target optimisation algorithm we have demonstrated earlier [11,26]. Target optimisation is significantly easier compared to global optimisation as the optimiser is allowed to stop after finding only few conditions that satisfy a target, compared to the problem of finding a global optimal.

The target functions and their corresponding values in the optimisation presented below were the yield, y, of **2** defined in Equation 2, with a desired value of 100%, and a specific cost function given in Equation 3 with a target value of 2,108 £ h kg^{-1}. This cost function was selected to account for the material and energy consumption, and the reaction time with respect to the amount of product **2**. Thus, $cost_{el}$ and $cost_i$ represent the electricity and material costs, whereas W_{el} and $m_{i,0}$ denote the consumed electricity and materials, respectively. The product output $\dot{m}_2$ with

$$\dot{m}_2 = \frac{m_2}{t_{reaction}} \tag{1}$$

combines the amount of product **2** with the necessary reaction time. The cost target value was derived based on a reaction with the shortest reaction time and highest yield, identified from a series of prior experiments (see Table S2 and Figure S5 in Supporting Information File 1 for further details) [18].

$$y = \frac{m_2 - m_{2,0}}{m_{1,0}} \cdot 100\% = 100\% \tag{2}$$

$$cost = \frac{cost_{el} W_{el} + \sum cost_i m_{i,0}}{\dot{m}_2} = 2{,}108\ \text{£h kg}^{-1} \tag{3}$$

The surrogate model was trained on the 64 experimental points obtained for model parameter identification. In case of an unknown mechanism, the experiments for mechanism discovery could also be included in the training set, which leads to a data-efficient approach. However, they might not be the most informative for training the surrogate. The output of the surrogate model is the suggested next experiment to perform, which was used as an input to the process model. Upon reaching the targets in silico after a number of optimisation iterations, the successful input conditions were verified experimentally, to confirm the predictions.

The in silico results for the optimisation target cost and yields are shown in Figure 4 and Table 4. It can be seen that out of 174 iterations, several points were very close to the targets and two optimal sets of conditions satisfy both targets (these iterations are marked with stars). The simulation results of the two identified successful sets of conditions both predict a yield of 98.72%. The experimentally obtained yields in the validation of the two sets of conditions were determined to be >99%, which is caused by the uncertainty of the applied GC method including sample preparation, which lead to ±1% variance in the yield value. The algorithm is not expected to exhibit fast convergence, since it is exploring the reaction space to develop a better statistical model. The physical experiments performed thereafter confirmed prediction of the successfully attained target

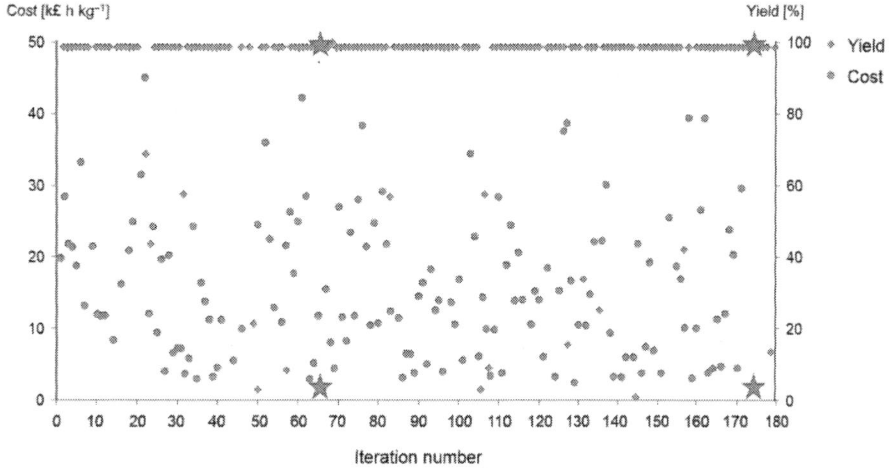

Figure 4: Results of in silico iterations of the multi-objective active learner (MOAL) algorithm [26]. Each iteration produces two resulting values, one for yield and one for the cost function. Targets were 100% for yield and 2,108 k£ h kg^{-1} for the cost function. Green stars signify experiments that satisfy the selected targets.

Table 4: Experimental conditions and results of the experimental validation of the two successful predictions that met the target specifications.[a]

Iteration	T [°C]	$t_{reaction}$ [min]	R_{acid-1}	R_{cat-1}	Yield [%]	Cost [k£ h kg^{-1}]
66	107	9	46.1	0.077	>99 (98.72)	1.79 (1.92)
174	101	10	41.4	0.077	>99 (98.72)	1.79 (1.93)

[a]R_{acid-1}: ratio of the concentrations of acetic acid and compound **1**, R_{cat-1}: ratio of the concentrations of catalyst and compound **1**. Values in brackets are the predicted values by the physical model. Further information regarding the experimental conditions is given in Supporting Information File 1, Table S6.

values. Hence, only the successful predicted experimental conditions X_{opt} were tested in real experiments, which saved time, cost and material, otherwise associated with testing false predicted reactions.

The cost target was more difficult to reach than the yield target, which was already fulfilled after the first iteration and later for most of the proposed experiments. This can be seen by the large fluctuation in the cost values for the proposed experiments over the 174 iterations. One possible reason might be the structure of the cost function with many input variables and strong sensitivity with regards to product amount and reaction time. The reaction conditions shown in Table 4 indicate relatively similar conditions with respect to temperature, reaction time as well as acid and catalyst loading, and do not at this stage demonstrate a case of multimodality.

We have also applied the same target optimisation algorithm for direct improvement of this chemical system as a 'black-box' sequential optimisation. For this approach five experiments were used as a training set, using Latin Hypercube space filling algorithm; the results are shown in Supporting Information File 1, Table S7 (Expt. 1–5). Figure 5 shows results of the initial set of experiments on the left side of the plot. It is noted that two of the five training experiments did incidentally meet the target value for yield at the conditions set. All outputs, regardless whether they reached the desired target values, were included into the training set and the algorithm was re-trained on the updated set once more. This iterative process was conducted six times. After a single iteration, the results of the first suggested set of conditions were already more promising than any of the training points. The target for yield has been met, as perhaps expected as it was optimal for two of the training exam-

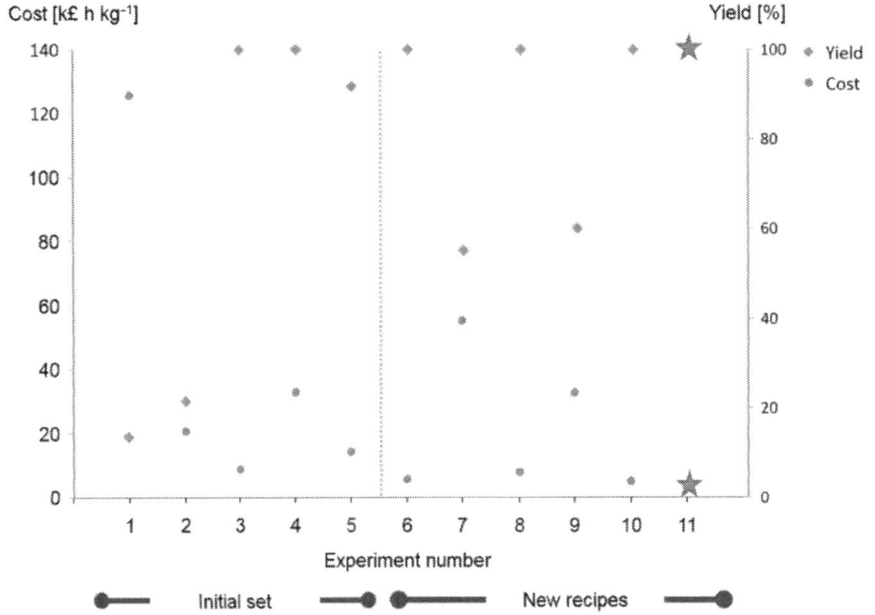

Figure 5: Results of the optimisation driven by a statistical algorithm and in the absence of a physical process model. Results of the training set and of sequential optimisation are shown. Information regarding the experimental conditions is given in Supporting Information File 1, Table S7.

ples as well, and the value for cost has been significantly reduced, getting closer to the pre-defined target. Whilst it is observed that experiments 7 and 9 have a large margin of error with regards to the targets, this is due to the exploration function of the algorithm.

Four of the new suggested experimental conditions achieved high yields with the accepted accuracy and had lower cost scores than even the lowest that was found in the initial set. The recipe at the 6th and final iteration following the training set corresponds to a temperature of 102 °C, reaction time of 15 minutes, acid–substrate ratio of 27.85 and catalyst–substrate ratio of 0.084. With these final conditions, the algorithm converged as both targets were met simultaneously for the first time.

Comparison of the two optimisation approaches

In this work we used the automated flow set-up combined with MBDoE approach to rapidly develop a good-enough process model, which was then used to train a surrogate model and perform a target optimisation. This resulted in two new sets of reaction conditions which both provided better results than the ones obtained previously. In our second approach we used the experimental flow system as a 'black-box' and employed the same statistical target optimisation algorithm to experimentally find the conditions that satisfy the set targets. In this specific case the 'black-box' target optimisation is extremely efficient and found suitable reaction conditions within a very small number of experiments. However, no knowledge about the system was generated. The approach of using automated flow experiments in combination with MBDoE allows to minimise experimental effort compared to classical kinetic studies, but results in a process model that can be directly used in optimisation. This approach is clearly preferred for the cases when a model structure could be identified. There would be many practical cases when due to complexity of chemistry it would be unrealistic to develop a physical model within a reasonable timescale. Then the 'black-box' approach is a viable option.

Conclusion

In conclusion, we present an approach of using model-based design of experiments, based on the first principles model structure, in automated flow experiments, and coupling of the process models with a statistical machine learning based target optimisation. We demonstrate that MBDoE offers a significant potential for efficient and rapid generation of process models in flow experiments. The developed process model enables in silico training of the optimisation surrogate model and cost effective determination of process conditions that satisfy the set performance targets. While this is certainly faster than physical experiments, we also show that the self-optimisation works well when trained on a space-filling method to avoid many necessary experiments for model generation. This results in a set of experiments that reach the pre-defined targets in six iterations, although it does not provide any process knowledge. Hence, a combined approach, leading to generation of a surrogate model and a physical model has unique advantages of rapid optimisation and simultaneous generation of process knowledge.

Experimental

Reaction system and analysis

All reactions were performed in continuous segmented flow using the $R2^+$/R4 system by Vapourtec, see Figure 2. The reaction mixture segments and the solvent were pumped through a 10 mL polytetrafluoroethylene (PTFE) tubular reactor and quenched in an ice bath at the reactor outlet. A minimum segment volume of 2 mL was found to be necessary to avoid dispersion effects in the centre of the segment. The segments were detected using the in-line UV cell, which allowed automatic triggering of the GC (Agilent 6850) to sample the segment at its centre. The flow GC vial was designed by Daniel Geier and Ralf Thelen from the Institut für Technische und Makromolekulare Chemie (ITMC) at RWTH Aachen University and manufactured in-house in Cambridge. GC analysis was performed for product **2** with an accuracy of ±0.0005 mol L^{-1}. Due to the decomposition of **1**·HOAc, **B** and **A** to **1** during sampling following a reaction, the reaction mixture was analysed for species **1** prior to beginning a reaction, with an accuracy of ±0.0003 mol L^{-1}. All communication between instruments was custom-coded in LabVIEW and communication with Vapourtec was via its proprietary Excel interface. Further details of the set-up and the on-line auto-sampling strategy, as well as a protocol for sample preparation and experiment execution are provided in Supporting Information File 1.

Materials

Toluene (Sigma-Aldrich, anhydrous, 99.8%), acetic acid (Sigma-Aldrich, ReagentPlus, ≥99.0%), acetic anhydride (Sigma-Aldrich, ReagentPlus, ≥99.0%), 1,1,2,2-tetrachloroethane (Sigma-Aldrich, reagent grade, ≥98.0%), palladium(II) acetate (Sigma-Aldrich, reagent grade, 98.0%, no further purification steps were applied, the same batch was used for all experiments, stored according to manufacturer's suggestions), (diacetoxyiodo)benzene (Sigma-Aldrich, ≥98.0%) were all used as received. 3,3,5,5-Tetramethylmorpholin-2-one was synthesised as described elsewhere [18].

Model development and analysis

A process model was developed on the basis of the previously discussed reaction mechanism and DFT estimates of the rate constants (see Table S1, Supporting Information File 1) [18].

The kinetic model was developed as a well-stirred tank reactor. The model includes kinetic equations, energy and material balances as well as constitutive equations. A lumped model was created as each reaction segment was assumed to be ideally mixed; thus no excess volume was considered for mixing. As the reaction takes place in a homogeneous liquid phase, and as the tube dimensions are small, there was no need to account for mass transfer effects. For simplicity, the slightly endothermic nature and hence the heat of reaction for the C–H activation was neglected.

The temperature effect in the reaction steps shown in Scheme 2 were expressed using the Arrhenius equation in its re-parametrised form, shown in Equation 4 and Equation 5 [27,28]. This facilitates subsequent parameter estimation by decoupling the kinetic parameters of each reaction. Assuming equilibrium for the three reversible reaction steps in Scheme 2, the 8 kinetic parameters of interest in this reaction sequence were the reference reaction rate constants ($k_{j,ref}$) and activation energies ($E_{a,j}$), where $j \in \{0, 1, 2, 3,\}$ given by Equation 4 and Equation 5).

$$k_j = k_{j,ref} \cdot e^{\frac{E_{a,j}}{R}\left(\frac{1}{T}-\frac{1}{T_{ref}}\right)} \tag{4}$$

$$k_{j,ref} = A \cdot e^{\frac{E_{a,j}}{RT_{ref}}} \tag{5}$$

The temperature-dependent volumetric reaction rates $R_{i,j}^V$ of compound i in the reaction j were modelled by Equation 6, in which $v_{i,j}$ are the stoichiometric coefficients of a compound i in the reaction j, c_i denotes the molar concentration of the compound i, k_j represents the reaction rate constant of the reaction j and $n_{i,j}$ gives the order of the reaction. All reaction steps in Scheme 2 were found to be first order with respect of the participating compounds, except for the oxidant $PhI(OAc)_2$ which is of zero-order dependency [21].

$$R_{i,j}^V = v_{i,j} k_j \prod c_{i,j}^{n_{i,j}} \tag{6}$$

In addition, the overall and the component mole balances, Equation 7 and Equation 8, were written for the process model, where V denotes the reaction volume.

$$\frac{dn}{dt} = V \sum_{i=1}^{n_{comp}} \sum_{j=1}^{n_{reac}} R_{i,j}^V \tag{7}$$

$$\frac{dn_i}{dt} = V \sum_{j=1}^{n_{reac}} R_{i,j}^V \tag{8}$$

The balances were constructed for a single reaction mixture segment, which was assumed to behave as a batch reactor, as samples were taken in the dispersion-free centre of the segment.

For the purpose of calculating the cost associated with heating the system, a steady state energy balance, Equation 9, was established.

$$W_{el} = \frac{Q_{heat}}{\eta_{heat}} = \frac{n_0 \overline{c_{p,0}}(T - T_0)}{\eta_{heat}} \tag{9}$$

were η_{heat} denotes an overall efficiency of conversion of electrical into thermal energy of the reaction mixture segment. This efficiency was determined experimentally for the employed reactor system by measuring the electrical power input to the Vapourtec heating system needed to increase the temperature of a reaction mixture stream with a set flowrate and of known composition, thus with known molar flow and heat capacity, from ambient temperature of approx. 20 °C to a reaction temperature of 70 °C. By inserting these values into Equation 10, the value for the energy-conversion efficiency was calculated and kept constant.

$$\eta_{heat} = \frac{\dot{n}_0 \overline{c_{p,0}}(T - T_0)}{P_{el,Vapourtec}} \tag{10}$$

To complete the process model, simple constitutive equations were applied and initial parameter values were computed. The latter were identified based on Gibbs free energies of reaction for the chemical system, which were obtained from a priori DFT calculations with an accuracy of ±7 kJ mol^{-1} [18]. These values were related to kinetic parameters through the exponential van't Hoff equation.

Subsequently, the model was investigated and tested for identifiability to ensure its structural soundness, i.e., that it can be used to uniquely determine its parameters. This was done using an established method, detailed elsewhere [29,30]. It is worth noting that structural identifiability is tested under the assumption of noise-free measurement data and no uncertainty of the model. Thus, it does not necessarily imply practical identifiability as the measurement data for the parameter estimation is usually superimposed by noise and errors [31]. Thereafter, an estimability analysis, based on visual inspection of the local dynamic sensitivity curves [32], confirmed that all parameters except those for reaction with $j = 1$ can be determined with sufficient accuracy. For this reaction step, low sensitivity of the measurable quantity, concentration of compound **2**, was found (see Supporting Information File 1, Figure S9 for further details). The local dynamic sensitivity curves of the remaining parameters were used to identify the time intervals with the maximum sensitivity of the parameters (see Supporting Information File 1, Figure S9), indicating reasonable sample points to obtain sensible measurement data for parameter estimation.

The physical model was implemented in gPROMS. The D-optimality criterion was selected to refine the model using the model-based design of experiments (MBDoE) and parameter estimation suite of gPROMS by employing standard solver settings. Further details on this procedure are given in Supporting Information File 1 together with constraints employed for the experiment control variables in the MBDoE optimisation problem (Equations S8–S21, Supporting Information File 1).

The experimental design and parameter estimation strategy included two steps. In the first step, experiments were designed at a reference temperature $T_{ref} = 70$ °C to eliminate $E_{a,j}$ as a parameter in each reaction j, see Equation 4 and Equation 5). A t-test was used as a statistical method for judging the increase in precision of the predicted parameter and, hence, the statistical significance of the estimates which is attained if the predicted t-value exceeds a reference value t_{ref}. The performed experiments would therefore generate data to enable estimation of the reference reaction rate constants $k_{j,ref}$ independent of $E_{a,j}$. After these parameters have been determined to sufficient accuracy, they were kept constant and experiments were designed at temperatures different from T_{ref} in the second step to obtain data for the subsequent estimation of $E_{a,j}$. The combination of MBDoE and subsequent parameter estimation was repeated twice in both steps to increase parameter precision, whilst keeping the experimental effort low. This was necessary to ensure good parameter improvement in the case of poor initial parameter guesses.

Algorithm for statistical optimisation

One key element of the proposed framework for self-optimisation of reaction conditions is the statistical multi-target optimisation method. For this purpose, the multi-objective active learner (MOAL) algorithm coded in the numerical computing environment MATLAB (v.2015b) was adopted, which combines Gaussian processes as a surrogate model with the concept of mutual information and a genetic algorithm [26]. To apply it to the chemistry under investigation, the algorithm was provided with specified targets Y_{target} for the optimisation and defined experiment design variables $X = [T, t_{reaction}, c_{1,0}, c_{AcOH,0}, c_{cat,0}]$ as the degrees of freedom. The latter were bounded by the corresponding constraints (see Supporting Information File 1 for details). A set of 2,000 randomly generated candidate solutions, uniformly distributed within the allowed design space was employed, because the algorithm works with discrete evaluation techniques for the optimisation. The initial training set $[X_{tr}, Y_{tr}(X_{tr})]$ contained the input variables X_{tr} and measurements of the corresponding target values $Y_{tr}(X_{tr})$, which were adopted from the MBDoE approach. It was updated continuously, so that at each iteration of the algorithm, a new training point was added. Binary Gaussian process classification was included into the algorithm to account for feasible and infeasible solutions in X, hence learning the promising regions of the design space and evolving some internal process knowledge stepwise with each new iteration. An infeasible solution could occur if an experiment fails in the laboratory. Thus, each point in the training was equipped with one more value, providing information on its feasibility (1) or infeasibility (−1). The current limitation of this approach is that it does not automatically distinguish if the experiment failure identifies the region of design space where the specific reaction is not working, or the failure was due to a random fluke and the same experiment, if repeated, would be successful. There is a way of dealing with this problem, which we will implement, when the algorithm will be published.

After classification of the candidate solutions and training samples, the Gaussian process was trained by fitting the so called hyperparameters of its covariance and likelihood functions by maximising the marginal likelihood with a conjugate gradient optimiser. In this way, a statistical surrogate model was created to provide an approximated response surface for the underlying problem of investigation. This response surface was used to evaluate the feasible candidate solutions and subsequently identify a best solution $\tilde{Y}_{opt}(\tilde{X}_{opt})$ with corresponding experimental conditions $\tilde{X}_{opt}$. As only discrete candidate solutions were evaluated by this method, the Non-dominated Sorting Genetic Algorithm-II (NSGA-II) was employed for perturbation of $\tilde{Y}_{opt}(\tilde{X}_{opt})$ to explore the neighbourhood for further refinement of the generated solution. The resulting combination of input and output conditions were subsequently assessed against the targets. If the targets were attained within acceptable tolerance, the results were accepted and the statistical algorithm converged. Otherwise, the results were fed back into the training set and a new iteration was started.

In silico optimisation

The in silico optimisation process was initiated by first training the MOAL algorithm (with the data generated for the purpose of parameter estimation of the physical model described above). This enables the algorithm to construct a statistical surrogate model and suggest a set of experimental conditions which might give results that are closer to the targets. This set is then fed into the physical model to predict what outputs are expected as though the experiment had been conducted. This process is repeated until the required tolerance is reached. A margin of tolerance was included, such that results within 10 and 1.5% of the target values for cost and yield, respectively, were taken as successful, due to the expected difficulty in achieving those targets. Subsequently, the successful reaction conditions found in silico were tested experimentally. In the case of failure, the experimental results were fed back into the algorithm and the target optimisation loop starts again. Otherwise, the algorithm converged and suitable experimental conditions were identified.

In principle, standard optimisation approaches employing the physical model directly to identify optimum operating conditions could be used, but would give poor results in case of uncertainty and restricted validity of the physical model. However, by applying the MOAL algorithm, technical difficulties regarding multi-objective global optimisation can be overcome. Furthermore, the proposed optimisation procedure can deal with potential uncertainties and restricted validity in the physical model. This is achieved by the machine learning functionalities of the MOAL algorithm, which retrain the algorithm not just on the physical model but also on unsuccessful experiments, erroneously predicted as suitable by the physical model. Thus, it obtains information beyond the capabilities of the physical model. An additional point is, that the MOAL algorithm proved to be especially suited for the detection of multiple possible solutions to indicate multimodality, which is challenging for standard optimisation methods, but can yield valuable information in the current case.

Statistical closed-loop optimisation

Statistical target optimisation was performed using the MOAL algorithm. Latin hypercube sampling (LHS) was used to discretise the experimental space initially. An overview of this sampling strategy is laid out for one variable in Figure S10 in Supporting Information File 1. A uniform distribution was taken for the input variables and hence the cumulative distribution function was linear. In this case k is five and N was the number of initial experiments to be conducted, which was decided to be five. This number of initial training experiments was selected as in the previous application of the MOAL algorithm for laboratory optimisation [11], the same number of training samples was applied for an optimisation of two targets, but with a 14 dimensional design space, instead of five dimensions in the current case. Still, the algorithm converged within 17 iterations (including the five training experiments). Hence, we assumed, that for the current work with less design space dimensions, the algorithm would learn the response surface as fast as it was the case for its previous application.

Table 5: Nomenclature.

Symbol	Definition	Units
η_{heat}	heat efficiency	–
$v_{i,j}$	stoichiometric coefficient of component *i* in reaction *j*	–
A	pre-exponential factor in Arrhenius equation	case dependent
$cost$	investigated target value	£ h kg^{-1}
$cost_{el}$, $cost_i$	cost of electricity, cost of material component *i*	case dependent
$c_{i,0}$	initial component concentration	mol L^{-1}
$\overline{c_{p,0}}$	average molar heat capacity	J mol^{-1} K^{-1}
$E_{a,j}$	activation energy of reaction *j*	J mol^{-1}
$k_{j,ref}$	reference rate of reaction in reaction *j*	case dependent
m_i, $m_{i,0}$	mass of component *i*, initial mass of component *i*	kg s^{-1}
n_i, n_0	number of moles of component *i*, total number of moles initially	mol
$P_{el,Vapoutec}$	electrical power uptake of Vapourtec flow system	W
Q_{heat}	heat	J
R	universal gas constant	J mol^{-1} K^{-1}
$R^V_{i,j}$	reaction rate of component *i* in reaction *j*	mol L^{-1} s^{-1}
t, $t_{reaction}$	time, reaction time	s, min
T, T_{ref}, T_0	temperature, reference temperature, initial temperature	°C
V	volume of system	L
W_{el}	electrical work	J
X	experiment design variables	–

Table 5: Nomenclature. (continued)

$\widetilde{X}_{opt}$	proposed suboptimal inputs	–
X_{opt_l}	proposed optimal inputs	–
X_{tr}	input training matrix	–
y	yield	%
Y_{target}	target outputs matrix	–
$\widetilde{Y}_{opt}(\widetilde{X}_{opt})$	proposed suboptimal outputs	–
Y_{opt_l}	proposed optimal outputs	–
Y_{tr}	output training matrix	–

Supporting Information

Supporting Information File 1

Details of experimental set-up and protocols, table of a priori data taken from our previous study, details of model development, MBDoE results, and LHS results. [http://www.beilstein-journals.org/bjoc/content/supplementary/1860-5397-13-18-S1.pdf]

Acknowledgements

Yehia Amar is grateful to UCB Pharma for his Ph.D. scholarship. Jacek Zakrzewski is grateful to the Department of Chemical Engineering and Biotechnology for funding his Ph.D. scholarship within Catalysis@Cambridge programme.

References

1. Dolas, R. T.; Siddheshwar, S. S.; Somwanshi, S. B.; Merekar, A. N.; Godge, R. K.; Pattan, S. R. *Int. J. Curr. Trend. Pharm. Res.* **2013,** *5,* 137–143.
2. Sans, V.; Cronin, L. *Chem. Soc. Rev.* **2016,** *45,* 2032–2043. doi:10.1039/C5CS00793C
3. Ley, S. V.; Fitzpatrick, D. E.; Myers, R. M.; Battilocchio, C.; Ingham, R. J. *Angew. Chem., Int. Ed.* **2015,** *54,* 2–17. doi:10.1002/anie.201410932
4. Houben, C.; Lapkin, A. A. *Curr. Opin. Chem. Eng.* **2015,** *9,* 1–7. doi:10.1016/j.coche.2015.07.001
5. McMullen, J. P.; Jensen, K. F. *Org. Process Res. Dev.* **2010,** *14,* 1169–1176. doi:10.1021/op100123e
6. McMullen, J. P.; Stone, M. T.; Buchwald, S. L.; Jensen, K. F. *Angew. Chem., Int. Ed.* **2010,** *49,* 7076–7080. doi:10.1002/anie.201002590
7. McMullen, J. P.; Jensen, K. F. *Annu. Rev. Anal. Chem.* **2010,** *3,* 19–42. doi:10.1146/annurev.anchem.111808.073718
8. Reizman, B. J.; Jensen, K. F. *Acc. Chem. Res.* **2016,** *49,* 1786–1796. doi:10.1021/acs.accounts.6b00261
9. McMullen, J. P.; Jensen, K. F. *Org. Process Res. Dev.* **2011,** *15,* 398–407. doi:10.1021/op100300p
10. Parrott, A. J.; Bourne, R. A.; Akien, G. R.; Irvine, D. J.; Poliakoff, M. *Angew. Chem., Int. Ed.* **2011,** *50,* 3788–3792. doi:10.1002/anie.201100412
11. Houben, C.; Peremezhney, N.; Zubov, A.; Kosek, J.; Lapkin, A. A. *Org. Process Res. Dev.* **2015,** *19,* 1049–1053. doi:10.1021/acs.oprd.5b00210
12. Reizman, B. J.; Jensen, K. F. *Org. Process Res. Dev.* **2012,** *16,* 1770–1782. doi:10.1021/op3001838
13. Moore, J. S.; Jensen, K. F. *Org. Process Res. Dev.* **2012,** *16,* 1409–1415. doi:10.1021/op300099x
14. Hone, C. A.; Holmes, N.; Akien, G. R.; Bourne, R. A.; Muller, F. L. *React. Chem. Eng.* **2017**. doi:10.1039/C6RE00109B
15. Walter, E.; Pronzato, L. *Automatica* **1990,** *26,* 195–213. doi:10.1016/0005-1098(90)90116-Y
16. Espie, D.; Macchietto, S. *AIChE J.* **1989,** *35,* 223–229. doi:10.1002/aic.690350206
17. Franceschini, G.; Macchietto, S. *Chem. Eng. Sci.* **2008,** *63,* 4846–4872. doi:10.1016/j.ces.2007.11.034
18. Zakrzewski, J.; Smalley, A. P.; Kabeshov, M. A.; Gaunt, M. J.; Lapkin, A. A. *Angew. Chem., Int. Ed.* **2016,** *55,* 8878–8883. doi:10.1002/anie.201602483
19. Hu, X. E. *Tetrahedron* **2004,** *60,* 2701–2743. doi:10.1016/j.tet.2004.01.042
20. McNally, A.; Haffemayer, B.; Collins, B. S. L.; Gaunt, M. J. *Nature* **2014,** *510,* 129–133. doi:10.1038/nature13389
21. Smalley, A. P.; Gaunt, M. J. *J. Am. Chem. Soc.* **2015,** *137,* 10632–10641. doi:10.1021/jacs.5b05529
22. Process Systems Enterprise. gPROMS. https://www.psenterprise.com/products/gproms. 1997-2005.
23. Franceschini, G.; Macchietto, S. *AIChE J.* **2008,** *54,* 1009–1024. doi:10.1002/aic.11429
24. Franceschini, G.; Macchietto, S. *Ind. Eng. Chem. Res.* **2008,** *47,* 2331–2348. doi:10.1021/ie071053t
25. Franceschini, G.; Macchietto, S. *Ind. Eng. Chem. Res.* **2007,** *46,* 220–232. doi:10.1021/ie060758c
26. Peremezhney, N.; Hines, E.; Lapkin, A.; Connaughton, C. *Eng. Optim.* **2014,** *46,* 1593–1607. doi:10.1080/0305215X.2014.881997
27. Asprey, S. P.; Naka, Y. *J. Chem. Eng. Jpn.* **1999,** *32,* 328–337. doi:10.1252/jcej.32.328
28. Agarwal, A. K.; Brisk, M. L. *Ind. Eng. Chem. Process Des. Dev.* **1985,** *24,* 203–207. doi:10.1021/i200028a034

29. Asprey, S. P.; Macchietto, S. *Comput. Chem. Eng.* **2000,** *24,* 1261–1267. doi:10.1016/S0098-1354(00)00328-8
30. Grewal, M.; Glover, K. *IEEE Trans. Autom. Control* **1976,** *21,* 833–837. doi:10.1109/TAC.1976.1101375
31. Galvanin, F.; Ballan, C. C.; Barolo, M.; Bezzo, F. *J. Pharmacokinet. Pharmacodyn.* **2013,** *40,* 451–467. doi:10.1007/s10928-013-9321-5
32. McLean, K. A. P.; McAuley, K. B. *Can. J. Chem. Eng.* **2012,** *90,* 351–366. doi:10.1002/cjce.20660

The definitive version of this article is the electronic one which can be found at:
doi:10.3762/bjoc.13.18

498655 Engineering Porous Polymer Hollow Fiber Microfluidic Reactors for SustainableÂ C-H Functionalization

Abstract:

The highly hydrophilic and solvent-stable porous polyamide-imide (PAI) hollow fibers were created by crosslinking of bare PAI hollow fibers with 3-aminopropyl trimethoxy silane (APS). The obtained composite hollow fibers were used as continuous-flow process for heterogeneously catalyzed the aldol and nitroaldol condensations of aromatic aldehydes with acetone and nitomethane at 50 and 40 °C, respectively. Furthermore, the APS-grafted PAI hollow fibers were also functionalized with salicylic aldehyde for binding catalytically active Pd(II) ions through covalent postmodification method. The catalytic activity of the composite hollow fiber microfluidic reactors (Pd(II) immobilized APS-grafted PAI hollow fibers) was tested via heterogeneous Heck coupling reaction of aryl halides under both batch and continuous flow reaction in polar aprotic solvents at high temperature (120 °C) and low operating pressure. XPS and ICP analyses of the startings and recycled composite hollow fibers indicated that the fibers contain very similar loading of Pd(II) implying no degree of catalyst leaching from hollow fibers during reaction. The composite hollow fiber microfluidic reactors showed long-term stability and strong control over the leaching of Pd species.

489997 Comparison of Pd Catalyst Activation Protocols in Suzuki Coupling and Associated Impacts on in-Process Control and Process Performance

Abstract:

Suzuki coupling of an aryl-bromide species to boronic acid forms a pharmaceutical process intermediate. The intermediate contains an atropisomeric stereo-center that affects downstream API. To set the atropisomeric center, a ligated-Pd pre-catalyst was generated in MeTHF or THF prior to being activated with aqueous base; the Suzuki coupling reaction occurs in a biphasic mixture.

Over the course of development, two different processes were developed and implemented on pilot-plant scale: In process [A], the pre-catalyst was activated with aqueous base and subsequently charged to the starting material solution. The resulting active catalyst is a Pd(0) species that is sensitive to oxidation and exhibited degradation over time. In process [B], a small amount of the aryl-bromide starting material (10%) is charged to the pre-catalyst solution. The resulting active catalyst is a Pd(II) species that is not sensitive to oxidation (given its 2+ oxidation state) at room temperature. The two processes generated catalyst species that are different but yet still part of the same catalytic cycle.

The catalyst stability afforded by process [B] had several impacts on in-process control, processing feasibility, and reaction performance. First and foremost, an in-process check was implemented to verify whether sufficient active catalyst was generated to complete the downstream Suzuki coupling; this improvement to the control strategy was not feasible with process [A] given the time instability of the Pd(0) species. Second, the number of processing vessels was reduced from three to two. Oxygen sensitivity during the Suzuki coupling required process [A] to use three separate vessels - one each for the catalyst, base, and starting materials. With process [B], the starting materials could be charged to the catalyst vessel after generating the stable Pd(II) species. Finally, process [B] exhibited faster reaction kinetics at equivalent Pd loading. The improved kinetics allowed for process [B] to be implemented on scale at a lower reaction temperature that resulted in a minor increase in product selectivity.

In summary, the two processes showed marked differences with respect to control, feasibility, and performance. The comparison demonstrates the effects and importance of catalyst stability in a pharmaceutical manufacturing process.

494865 Plasmonic Catalysts As Efficient Catalysts for Cross Coupling Reactions

Abstract:

Plasmonic Catalysts as Efficient catalysts for Cross Coupling Reactions

Andishaeh Dadgar, Farshid Mohammadparast, Mari Andiappan, School of Chemical Engineering, Oklahoma State University, Stillwater, OK, USA

Cross coupling reactions are a diverse set of reactions that are widely used throughout industry. For instance, cross coupling reactions make up 25% of all pharmaceutical reactions. They are also used for many advanced materials such as metal-organic frameworks (MOFs) and OLED materials. The main drawback that limits the reactions though is the wide use of homogenous palladium catalyst at high temperatures.

To overcome this limitation, we propose heterogeneous plasmonic catalysts to drive cross couplings under relatively mild conditions. With plasmonic nanocatalyst, we show improved rate of reaction and stability at a lower temperature than previously seen with conventional Pd catalysts. Furthermore, with in operando measurements we demonstrate the stability of the catalyst as a function of time, solvent, and temperature. The implications of this work are a.) find a suitable replacement for homogenous palladium b.) develop a heterogeneous catalyst capable of long term stability for scale up to a continuous flow system.

494177 Stereoconvergent Tandem Systems Combining Photocatalyzed Isomerization and Enzymatic Reduction for Chiral Building Blocks Synthesis

Abstract:

Abstract

The increasing demand for chiral building blocks, especially those used for the preparation of biologically active compounds, has motivated the development of novel strategies for enantioselective synthesis. Particularly, there is a growing interest in the development of one-pot tandem catalytic systems combining the reactivity of a chemical catalyst with the selectivity engendered by an enzyme. Such tandem systems can achieve levels of chemo-, regio- and stereo-selectivity that are unattainable with a small molecule catalyst. Here, we report the development of a conceptually distinct cooperative chemoenzymatic system by coupling a photosensitizer-catalyzed alkene isomerization in tandem with an enantioselective ene-reductase that selectively reduces one of the isomers for the synthesis of various chiral building blocks of drugs and biologically active compounds. With this system, we can stereoconvergently reduce the isomer mixtures of 2-substitued 1,4-dicarbonyl alkenes, cyano-substituted acrylate esters or ketones, cyano- or ester-substituted acrylamides to corresponding enantiopure reduced products at high yield (>80%) and excellent enantiomeric excess (>95%). These products can be used to easily synthesize γ-butyrolactones, γ-lactams, β-amino acids, γ-amino acids, amines and diamines, a structural motif found in many bioactive compounds, such as Phenibut and Baclofen. Importantly, this is the first example of combining photoisomerization with enzymes for enantiopure product synthesis rather than photocatalyst-directed co-factor regeneration in an oxidoreduction reaction. In addition, we show for the first time, that flavin mononucleotide (FMN) can catalyzes two different reactions, alkene isomerization and hydride transfer, in a single system, making our tandem reaction potentially metal-free. We anticipate our system to be a starting point for more synthetic applications when coupled with different enzymes.

A model-based approach for the identification of kinetic models from laboratory reactors

Lu Han, PhD

Pfizer Inc.

Eastern Point Road, Groton, CT

Session: Model Based Integrated Design of Pharmaceutical Drug Substance Processes

Mathematical models of complex chemical reactions in the form of ordinary differential equations provide the basis for systems understanding. In the early phase of modeling, there is a large pool of plausible mechanistic nonlinear models because the underlying mechanism is relatively unknown or uncertain. Thus, building high fidelity and validated kinetic models at limited experimental resource is an important activity. These models in turn can be used to guide process design and decision-making of pursuing batch or continuous manufacturing.

In this work, an application of model-based experimental design to understand the reaction kinetics of a pharmaceutical drug intermediate is presented. The motivation of this model-based guidance is to increase process understanding and control of a reaction impurity, for which purification can require additional post-processing steps downstream. Traditional batch-type experiments were leveraged early on to study the experimental conditions that favored this impurity. Model structures denoting different semi-mechanistic models were postulated and were equally successful at matching the experimental observations at normal operating conditions. Optimal experimental design for model discrimination was utilized to find an experimental condition for the purpose of identifying the most accurate model structure. Eq. 1 (below) outlines the optimization problem,

$$\boldsymbol{\phi}^{MD} = \arg\max_{\mathbf{u}(t)} \int_0^t \sum_{i=1;}^{N_M} \sum_{\substack{i=i+1;\\ i\neq j;}}^{N_M} \left(\mathbf{y}^i(\boldsymbol{\theta}^i, \mathbf{u}(t), t) - \mathbf{y}^j(\boldsymbol{\theta}^j, \mathbf{u}(t), t)\right)^2 dt \quad (1)$$

subject to:

$$\frac{d\mathbf{z}^i(t)}{dt} = \boldsymbol{f}^i(\mathbf{z}^i(t), \boldsymbol{\theta}^i, \mathbf{u}(t), t), \quad (2)$$

$$\mathbf{y}^i(t) = \boldsymbol{g}(\mathbf{z}^i(t), \boldsymbol{\theta}^i),\ i = 1, \dots, N_m. \quad (3)$$

which seeks to maximize the differences in response trajectories between the candidate models with any number of model structures N_M over the duration of the batch experiment. It's worth to note that the response matrix, **y**, is related to the concentration matrix **z** by function **g**. The model parameters, $\boldsymbol{\theta}$, contains the Arrhenius kinetics and absorbance matrix for the species. Parameter estimation was performed only on the kinetic parameters in this work.

At the optimal experimental conditions $\boldsymbol{\phi}^{MD}$, all except for one of the candidate models were rejected on the basis of high lack of fit. The best-fitting model was considered to have the most accurate model structure within the bounds of the design space. This model was subsequently used to design a continuous process in a homogeneous plug flow reactor. The flow process can offer some benefits over the batch process: better residence time control of the reaction, greater control of the process disturbances, real-time control to handle changing inputs, and smaller/more compact of equipment. A new solvent system was designed for this chemistry, in order to solubilize all species. An optimal experimental design procedure was performed to improve the estimation of kinetic parameters. An optimization problem was formulated to minimize a metric of the parameter uncertainty region predicted by the model:

$$\boldsymbol{\phi}^{PE} = \arg\min_{\mathbf{u}(t)}\{\psi[\boldsymbol{V}_\theta(\boldsymbol{\theta}, \mathbf{u}(t), t)]\} \quad (4)$$

where $\mathbf{V}_\theta$ is a function of the sensitivity matrix and variance-covariance matrix of the residuals and parameters, subject to the model equations (Eq. 2), response vector (Eq. 3), and experimental bounds. The quality of the optimized experiment design $\boldsymbol{\phi}^{PE}$ was compared to a factorial design. The effect of the different designs on the parameter estimation will be discussed. The applicability of the derived kinetic parameters will be demonstrated on the basis of predictions to a new experiment designed to maximize the discriminability between the two parameter sets.

Through these scenarios, we will show real-life examples of system engineering methodologies applied to pharmaceutical chemistries. Model-based methodologies offer a powerful tool to enhance process understanding of new chemical systems by exploiting the most informative regions of experimentation.

486415 Early Stress Activation Via Methyl Jasmonate Influences Loblolly Pine Somatic Embryo Yield

Abstract:

Somatic embryogenesis (SE) provides a means for large scale production of a clonal crop, thereby enabling the propagation of significant crops with ideal phenotypes. This is particularly important in crops with large time investments prior to crop maturity, including pine trees. The loblolly pine is an important timber crop that can be produced through SE, but variable yields of viable embryos are often reported and the process is not fundamentally understood. The process of SE involves a response to change in the culturing conditions of cells from maintenance conditions to conditions associated with embryo development, and some of the changes are associated with a stress on the cultures. Here, we explore the early activation of stress on the cultures prior to moving to embryo-producing media. We show that early addition of methyl jasmonate (MeJA), a largely conserved activator of specialized metabolism in numerous plant systems, initiates a response in the cultures that is different between cultures that produce a high yield and those that produce a low yield. In high yielding cultures, low MeJA concentrations improve SE yield and high concentrations impede SE. In low yielding cultures, little response to MeJA is observed, suggesting that the culture metabolism is not as responsive in these low yielding cultures, and providing a potential way to screen cultures prior to plating. We present a robust culture engineering strategy and characterization platform using the Coulter Counter to quantify biomass and cell types, an assay using the Folin-Ciocalteu reagent to measure total phenolic content, as well as methods to quantify culture growth using both weight and image processing techniques. Understanding the influence of stress on SE cultures can provide a simple process engineering strategy to influence yield without genetic manipulation.

Improving Biocompatibility of 3D Printed Stereolithography Resin

AIChE Annual Conference 2017

Cell Culture Engineering & Process Design II: Reactor Engineering

Christopher J. Hawxhurst, Andrea L. Kadilak, Charles M. Bridges, Daniel J. Gage, Leslie M. Shor

November 1st, 2017

UCONN
SCHOOL OF ENGINEERING

Example Microfluidic Devices

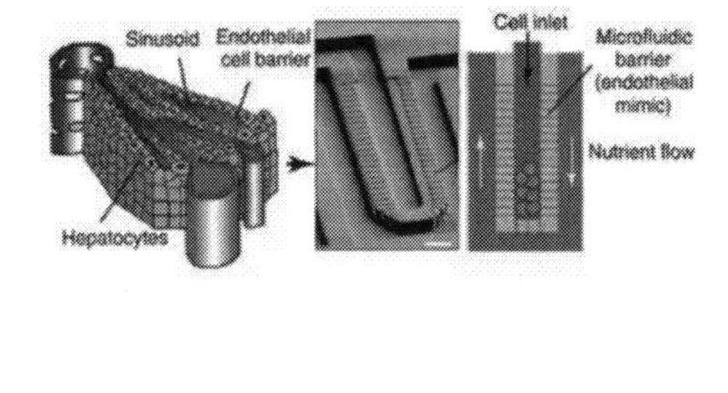

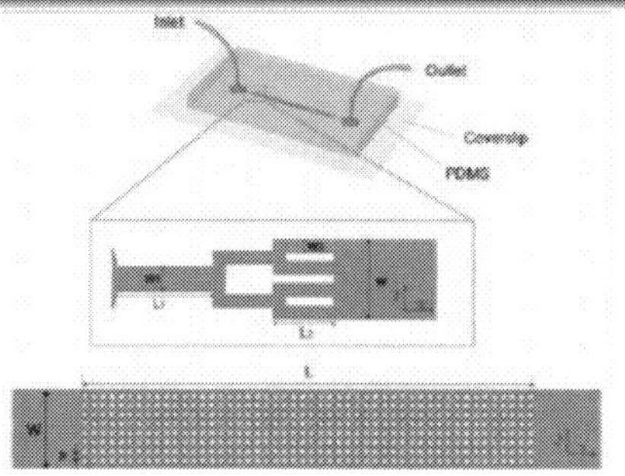

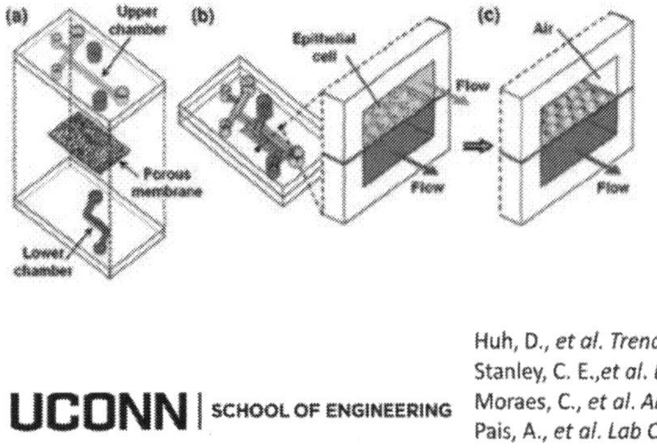

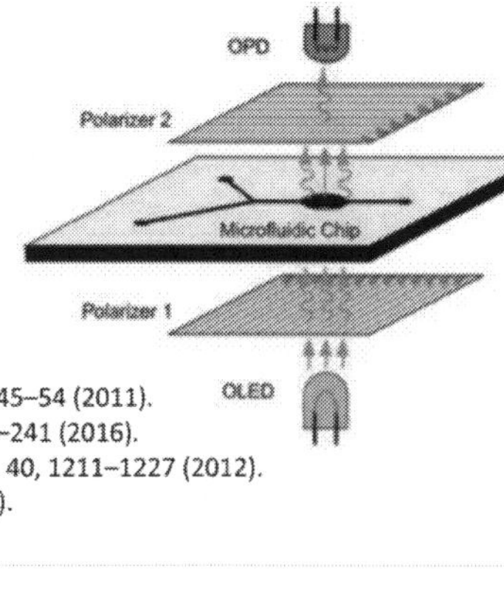

Huh, D., *et al. Trends Cell Biol.* 21, 745–54 (2011).
Stanley, C. E.,*et al. Lab Chip* 16, 228–241 (2016).
Moraes, C., *et al. Ann. Biomed. Eng.* 40, 1211–1227 (2012).
Pais, A., *et al. Lab Chip* 8, 794 (2008).

Why Microfluidics?

- *Microfluidics: allow precise control of fluid flow and behavior, at least one channel dimension will be in tens to hundreds of microns range*
- Benefits
  - Optically observable during experimentation
  - Environmental gradients possible
  - Small device and sample volume

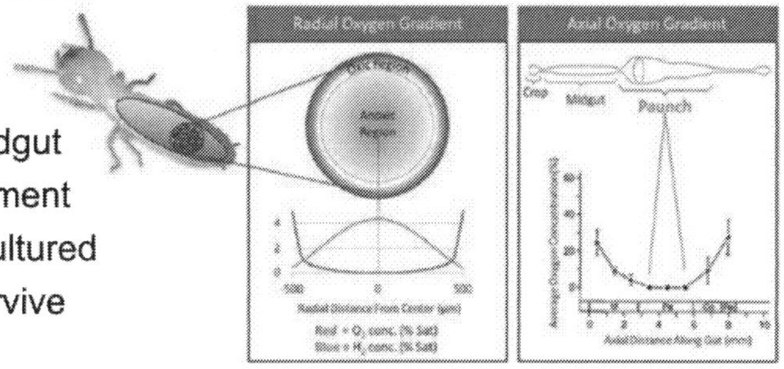

- Example: culturing termite hindgut
  - Complex gradient environment
  - Most species cannot be cultured
    - Need symbionts to survive

 Adapted from: Brune, A. *et al.* 61, 2681, (1995).

3D Printing Microfluidic Devices

- Advantages over soft photolithography
  - Cost efficiency
  - Faster prototyping
  - Fully 3D designs possible
  - More automatable
    - No clean room required
    - Less training necessary

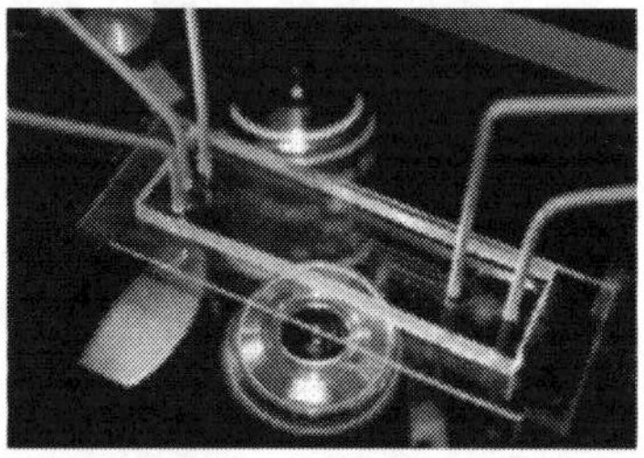

- Challenges
  - Resolution
  - Optical clarity/roughness
  - **Biocompatibility of resin**

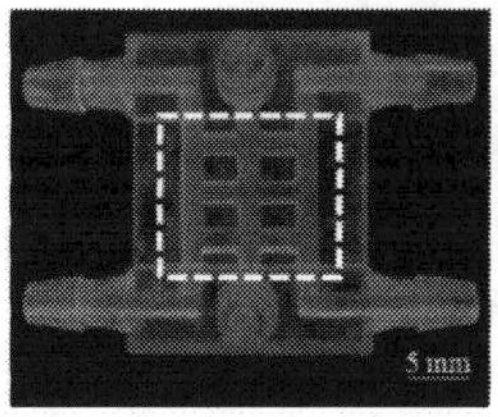

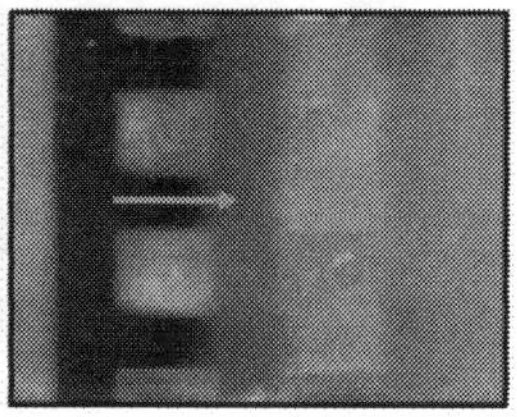

 http://www.ddw-online.com/
Kadilak, A. L., *et al.* Biomicrofluidics 11, 54109 (2017).

3D Printing Technology

- Fused Deposition Modeling (FDM)
  - Makerbot
  - Common personal home printer
  - Typically acrylonitrile butadiene styrene
    - Fully biocompatible
  - Limited resolution

- **Stereolithography** (SLA)
  - Formlabs Form 2
  - High resolution
  - Photo-curable resins
    - Cell toxicity
      - Labile photoinitiators
      - Uncross-linked monomers/oligomers

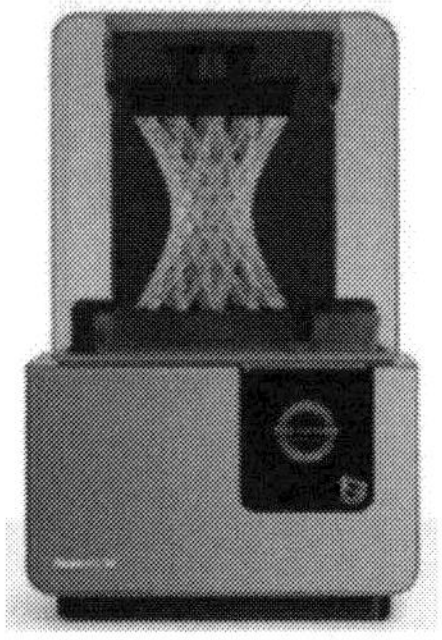

UCONN | SCHOOL OF ENGINEERING

https://makerbot.com/replicator/
https://formlabs.com/3d-printers/form-2/

5

Motivation

- 3D printers are gaining popularity for creating microfluidic devices
- Previous work on biocompatibility of 3D printed resins[1-3]
  - Resin treatments decreased adverse effects on zebrafish embryos
- Little is known about the effect on microbial organisms
  - Need to develop assays
    - Straightforward
    - High throughput
      - Well plate platform
    - Cover important parameters
      - Viability
      - Growth
    - Approach applicable dosages (SA:V ratios)
      - Channel 1x1 mm, 20 mm long: 40 cm^{-1}

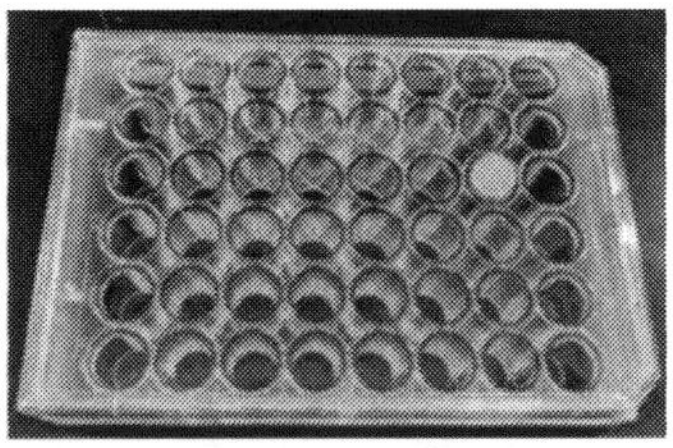

UCONN | SCHOOL OF ENGINEERING

1. Oskui, S.M. *et al. Environmental Science & Technology Letters*, **3**, 1-6 (2016).
2. Zhu, F. *et al. Biomicrofluidics*, **9**, 046502 (2015).
3. Macdonald, N. P. *et al. Lab Chip* **16,** 291–7 (2016).

6

Research Structure

- Cell viability assay
  - Determine number of viable cells after prolonged resin exposure
    - Avoid cell growth
  - Vary toxic compound dosage
    - Change SA:V ratio

- Cell growth assay
  - Determine effect of resin on microbial growth curve
  - Develop 3D model for use with 48 well plate
    - Easy to implement

Biocompatibility Treatments

- Manufacturer suggested treatment
  - 30 minute rinse in isopropanol (IPA)
    - Baseline treatment
    - Removes uncross-linked resin from surface
    - Possible not all uncross-linked resin removed
- Additional treatments after baseline
  - Thermal
    - May improve mechanical stability
  - Ultraviolet (UV) light
    - Crosslink underexposed resin
  - Solvent extraction
    - Remove toxic compounds
  - Combination

3D Printed Disk

- Disk
  - Extremely simple geometry
  - Low surface area
- Dimensions
  - 9.5 mm diameter
  - 0.5 mm width

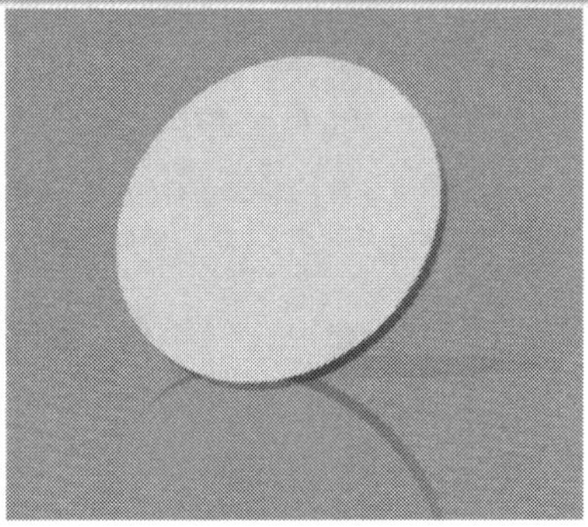

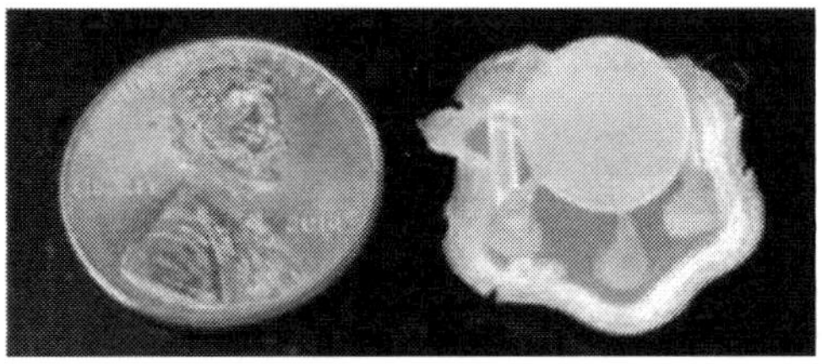

UCONN | SCHOOL OF ENGINEERING

Changes in Microbial Viability

- Bacteria incubated to stationary phase
  - At 30°C
  - Washed 3x
- 3D printed disks
  - 24 hours at 30°C
- Cell viable concentration
  - Plate counts of serial dilutions
  - 2 trials
    - 2 replicates

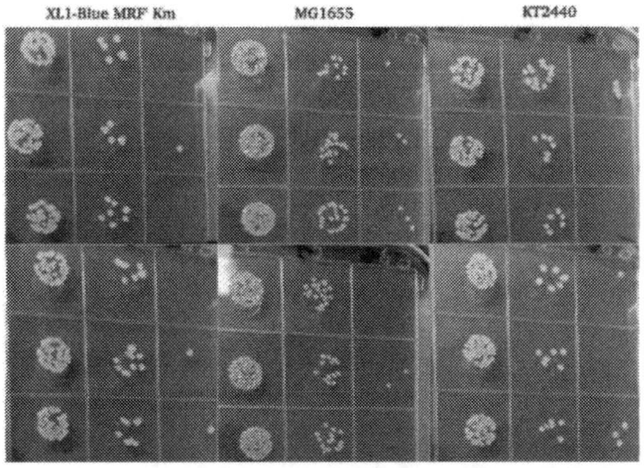

Top row: Control (0 disks)
Bottom row: 15 disks

UCONN | SCHOOL OF ENGINEERING

Cell Viability in Formlabs Resin

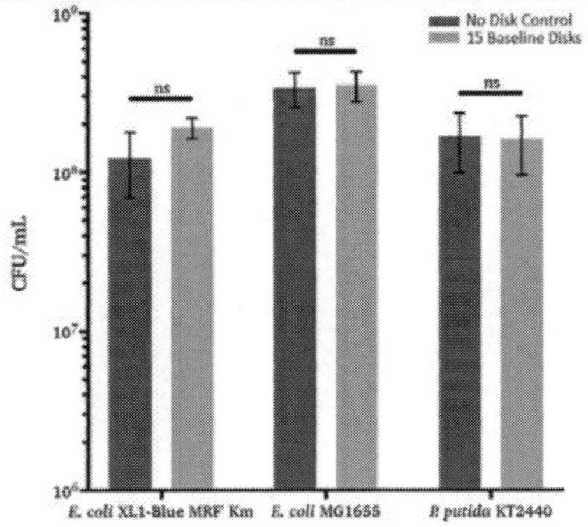

- Cells suspended in M9 minimal broth
- No significant difference between control and baseline treatment
  - Possible cell growth

UCONN | SCHOOL OF ENGINEERING

Cell Viability in Formlabs Resin

- Cells suspended in phosphate buffered saline (PBS)
  - Should prevent cell growth
- Decrease in cell viability for
  - Control vs 10, 15 baseline disks
  - 10, 15 baseline vs treated disks
- Treatment: 24 hours in DI followed by UV and thermal exposure
- Approx. 5 fold decrease in viable cells at SA:V 7.8 cm^{-1} with baseline treatment

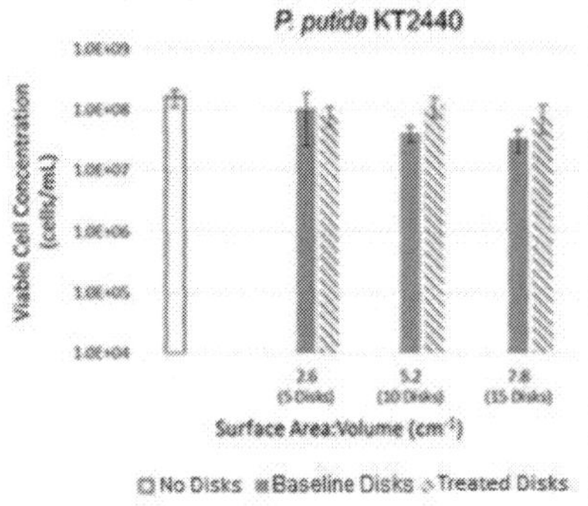

			Baseline		
		Control	5 disks	10 disks	15 disks
	Control		-	**	**
Treated	5 disks	*	-		
	10 disks	-		**	
	15 disks	*			*

* $p < 0.05$ ** $p < 0.01$

UCONN | SCHOOL OF ENGINEERING

Changes in Microbial Growth

- 3D printed material in 48 well plate
- Cultured *E. coli* overnight in lysogeny broth (LB)
  - Pelleted, resuspended in fresh LB
- Growth curve
  - 48 hours
  - 30°C
  - Measured OD_{600}, fluorescence
    - Every 10 minutes
- Disks mechanically polished for clarity

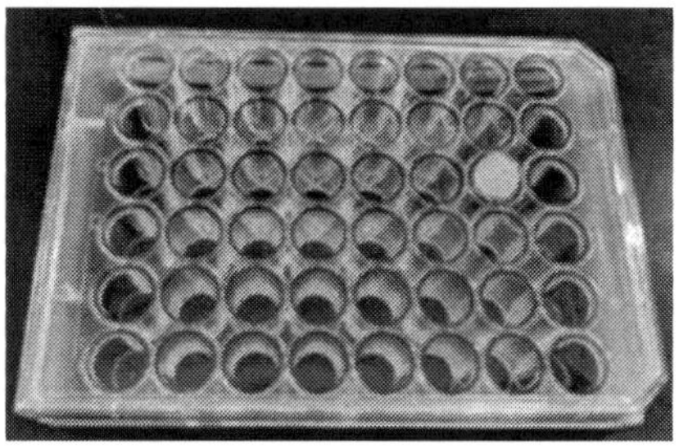

Growth Assay with Disks

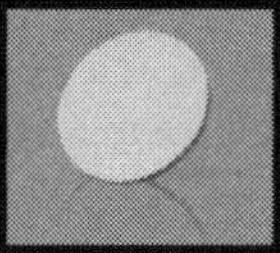

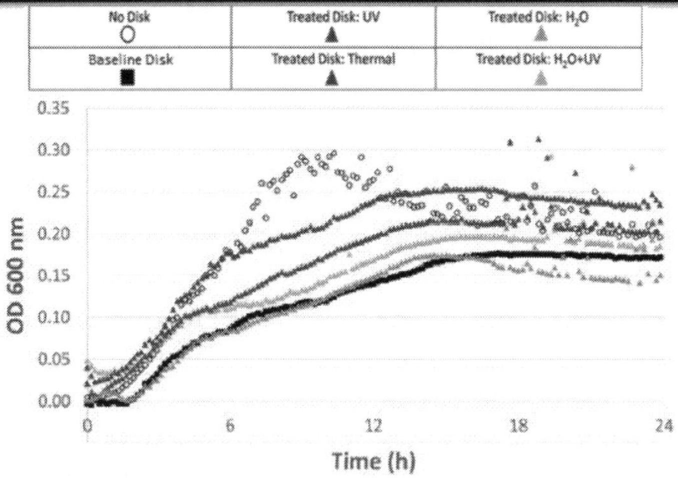

Growth Assay with Disks

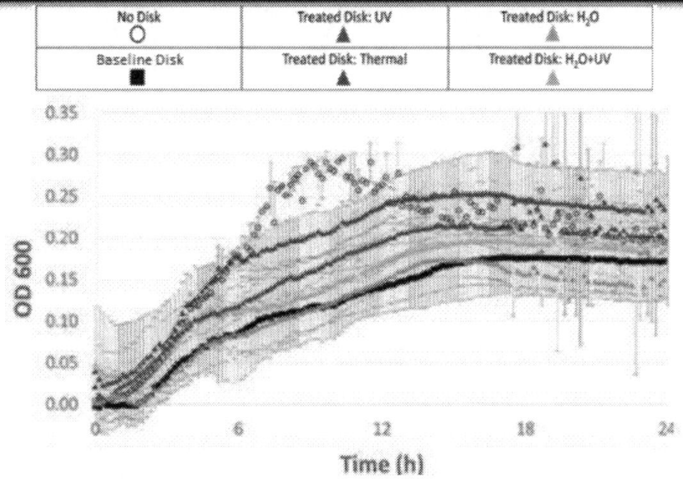

- Large standard deviations make any interpretation difficult
- Possibly due to:
  - Inconsistent refraction from physically polished disks
  - Small media volume

UCONN | SCHOOL OF ENGINEERING

15

3D Printed Cylinders

- Nested cylinders
  - High surface area
    - ~31 cm^{-1}
  - Large time investment
    - Fragile
  - Still impedes beam
    - Cylinders shift during shaking

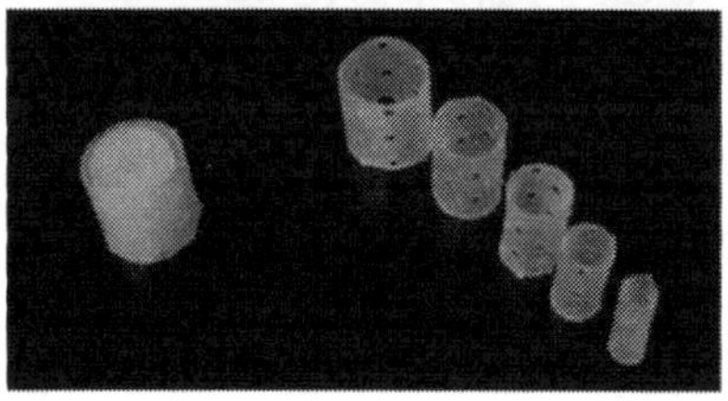

- Single cylinder
  - Robust, quick
  - Decent surface area
    - ~21 cm^{-1}
  - Does not impede beam

UCONN | SCHOOL OF ENGINEERING

16

Growth Assay with Cylinders

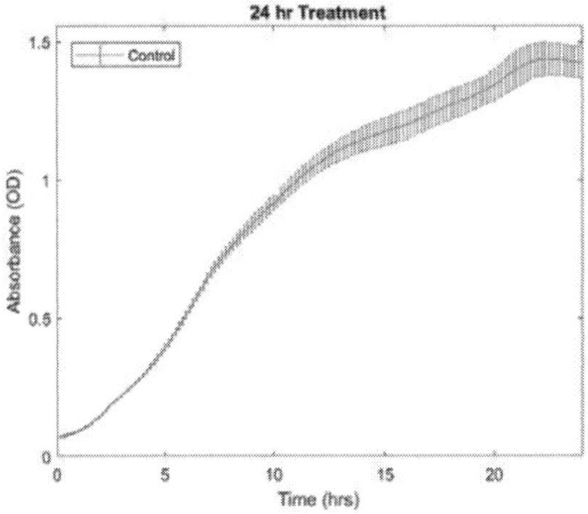

- Increased SA:V ratio versus disk causes marked decrease in growth curve for baseline treatment

Growth Assay with Cylinders

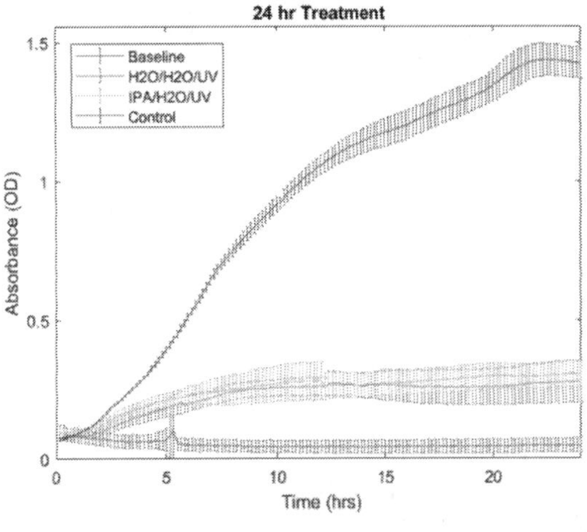

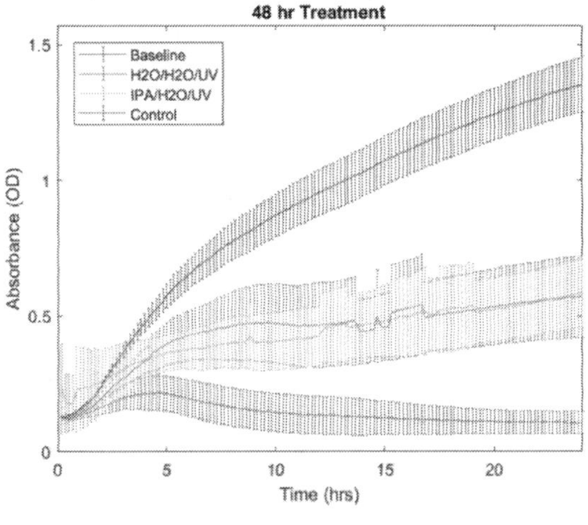

- Increase treatment time seems to decrease adverse effects of resin
- May require longer treatment times to reach full biocompatibility
  - Some geometry effects may be present as well

Conclusions

- SLA printers are popular due to resolution and ability to create enclosed channels in clear resin
- Assays developed for testing microbial biocompatibility
  - Can be easily adapted to a variety of systems

- Formlabs resin has a detrimental effect
  - Significant decrease in cell growth
    - Some effect on cell viability as well
  - Additional treatments reduced adverse resin effects
    - Need to treat past manufacture recommendations

- Toxic effects will likely be amplified by higher SA:V ratios found in milli- and microfluidic channels

Thanks To

- Advisor: Leslie Shor
- Andrea Kadilak
- Charles Bridges
- Daniel Gage
- Shor Group

- Funding: DOE
- Grant DE-SC0014522

U.S. DEPARTMENT OF ENERGY

Questions?

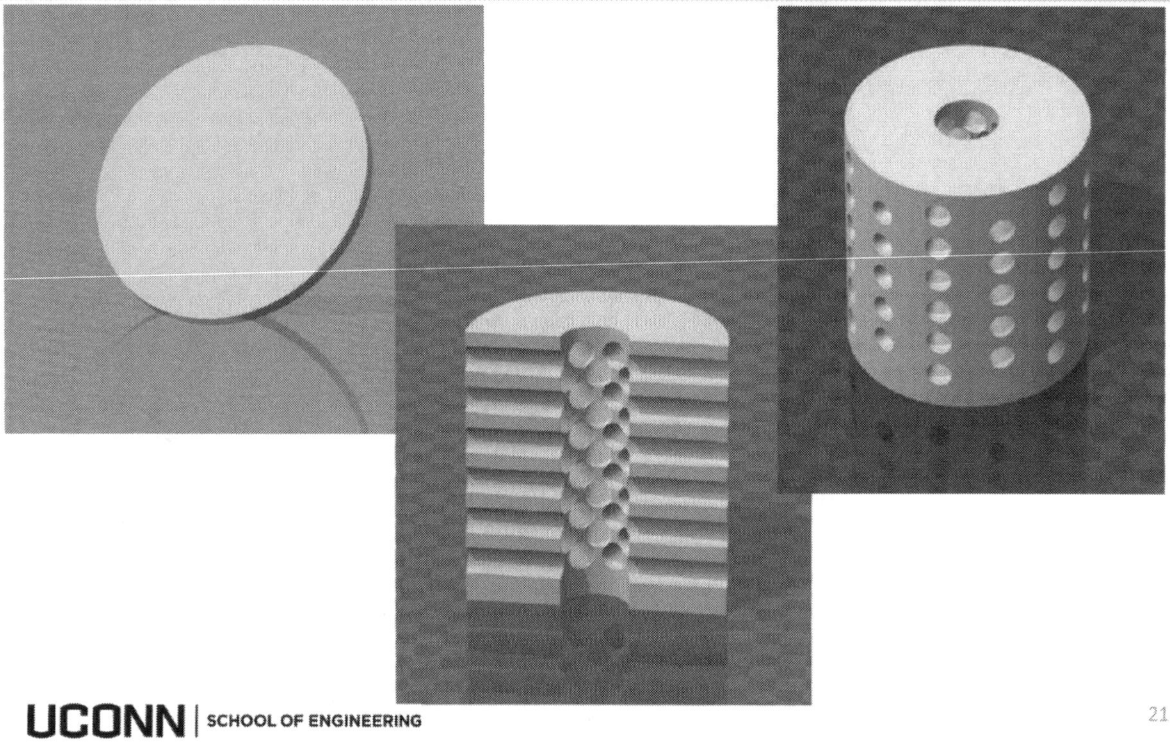

UCONN | SCHOOL OF ENGINEERING

21

500999 Monoclonal Antibody (mAb) Production in Continuous Microfluidic Systems

Abstract:

An unfulfilled need exists for cost-effective systems that can screen production conditions for biopharmaceutical compounds. At the lab scale, culture bottles and Erlenmeyer flasks continue to be the first and preferred reactor type for the development and selection of culture conditions (i.e., culture medium, addition protocols, temperatures, additives, among others) used to produce biopharmaceutical compounds (i.e., monoclonal antibodies and other recombinant proteins). These are batch systems that (a) do not provide a tight control of the cell microenvironment, and (b) fail to appropriately mimic commercial production scale conditions.

Here, we illustrate the use of continuous microfluidic systems to perform the operations of selection and screening of high-titer mAb producers. One system consists of a long continuous microfluidic channel (CMC) in which CHO cells remain anchored on the bottom surface while a stream of culture medium is fed continuously through the channel. We show that this simple system can be scaled up (by parallel scaling) to enable production in the order of tens to hundreds of mg day^{-1}.

A second system mimics the operation and dynamic behavior of a continuous stirred tank system (CSTR). We show that this system can achieve different steady state conditions at different flow rates. The low volume of this mini-CSTR (1,500 microliters) allows the use of extremely small quantities of culture medium to characterize the kinetic behavior of CHO cell mAb producers and to provide cost-effective screening of process conditions.

Both the CMC and the mini-CSTR are fully controlled by an Arduino system that regulates temperature and agitation, while monitoring cell concentration.

490759 Application of Mass Transfer Modeling to Improve Predictability of Scale-Down Industrial Fermentation Processes

Abstract:

Oxygen mass transfer is one of the largest constraints in scaling up aerobic fermentation processes. New fermentation conditions or strains that generate higher oxygen demands or broth viscosities often lead to oxygen limitations, thereby preventing implementation at industrial scale. Understanding which conditions will ultimately lead to oxygen limitation at industrial scale is of paramount importance for R&D programs that rely on scale-down models for process development. This presentation will describe a novel approach for using volumetric oxygen mass transfer coefficient (k_La) modeling to emulate large scale conditions in real-time. The construction of the k_La model will be reviewed and the utility will be demonstrated through a case study on a new strain for secondary metabolite production.

490719 Enhanced Oxygen Transfer and Cell Growth in a Continuous Rotating Bioreactor

Abstract:

A continuous bioreactor has been designed to increase cell growth rate and overall productivity by improving oxygen delivery. The operation of this bioreactor is similar to a roller bottle, but it can be operated either in batch or continuous mode, with inlet flows available to provide medium and oxygen to the bioreactor. The new design improved oxygen transfer in the bioreactor by at least two fold. The higher oxygen transfer performance was achieved by using a spiroid tube attached to the outer wall of the bioreactor, thereby increasing the gas-liquid interfacial surface.

Oxygen measurements were conducted to determine oxygen transfer in the bioreactor with and without the spiroid at different operating conditions. In addition, mathematical models were established to simulate the oxygen transfer in the system and to calculate the volumetric mass transfer coefficients (k_La) in both chambers and the spiroid tube at different operating conditions. Growth curves for E. coli with and without the spiroid were measured to confirm improvement in oxygen transfer.

Key words: continuous bioreactor, oxygen transfer, cell cultivation, spiroid

504100 Using Metabolic Modeling and Metabolomics to Improve Cell Culture-Based Bioprocesses

Abstract:

The number of new recombinant protein therapeutics and other biologics have quadrupled since 2008. Biologics now account for more than a quarter of all new drug products approved by the FDA. Most biologics are manufactured using eukaryotic cell cultures to enable post-translational modifications, with Chinese hamster ovary (CHO) cells as the dominant production platform. Many years of advances have achieved increased volumetric productivity through very high density fed-batch cultures. However, these high density cultures have also presented new challenges in maintaining cellular viability and productivity over the duration of the culture. A large body of published studies suggests that addressing metabolic inefficiencies could further improve CHO cell-based production of biologics.

This presentation describes recent efforts in our laboratory to better understand key metabolic events that occur over the course of *industrially relevant* CHO cell fed-batch cultures, and to leverage this understanding to devise strategies for process improvement. Our laboratory has pursued both modeling and metabolomics as complementary approaches to gain insights into metabolic inefficiencies present in high density CHO cell cultures.

The first part of this presentation will discuss a dynamic model of CHO cell metabolism that accurately predicts the effects of both process- and cell-level modifications. Using this model, we explored ~10^4 combinations of process and cell modifications, and found that knockdowns in glycolysis can raise cell viability across a range of process situations. However, depending on the process conditions, such knockdowns could reduce, rather than increase, the titer. Our results highlight the benefits of considering process and cell modifications together.

The second part of this presentation will discuss applications of targeted and untargeted metabolomics, respectively, to quantify key fluxes and identify potential growth inhibitory byproducts of CHO cell metabolism. Targeted LC-MS experiments of isotopically labeled central carbon intermediates identified a bottleneck in the oxidation of glucose and amino acids that is linked with cytosolic fatty acid synthesis, which could potentially explain the high lactate phenotype often observed in high density CHO cell cultures. Untargeted LC-MS experiments were utilized to compare the global profile of metabolic byproducts of genetically similar cells lines grown under identical conditions yet exhibiting significantly different growth rates. A major technical challenge in untargeted metabolomics is metabolite identification. A typical LC-MS experiment generates 10^3~10^4 unique data features. Despite progress, only a small fraction of these features can be reliably annotated with metabolite identities. To address this challenge, we assembled a novel computational workflow that utilizes the biological context of a sample to annotate and interpret the metabolomics data. The accuracy of our workflow is illustrated through comparisons with other currently available metabolite identification tools. Applied to LC-MS data collected on the CHO cell lines with different growth rates, our workflow identified several previously uncharacterized metabolic byproducts that are significantly elevated in low growth cell lines. Subsequent dose response studies confirmed that at least one of these byproducts potently inhibits CHO cell growth.

489947 Scientific Considerations on Continuous Crystallization and Its PAT Method Validation

Abstract:

Continuous manufacturing (CM) is an emerging technology in the pharmaceutical manufacturing sector. As the final purification/isolation step, crystallization can significantly affect the drug substance critical quality attributes (CQAs) and is one of the critical steps in an integrated continuous manufacturing (CM) process. Continuous crystallization has the potential to improve manufacturing efficiency and product quality, but both industry and FDA currently lack commercial experience with the technology. Here, we focus on the application of quality risk management to the development of a continuous crystallization system. Specifically, we designed and built an automated two-stage Mixed Suspension Mixed Product Removal (MSMPR) crystallization platform for a model compound (Carbamazepine, CBZ) that exhibits multiple polymorphs. The crystallization process includes integration of Process Analytical Technology (PAT) tools (on-line Raman microscopy and Focused Beam Reflectance Microscopy, FBRM) for real-time monitoring. Material transport, process control, and product collect approaches were investigated as the three key areas for risk identification and mitigation during continuous crystallization process development. Our proof of concept continuous crystallization system uses feedback controls to achieve constant levels in crystallizers, a centralized automation program coded in LabView, and PAT monitoring for polymorphs and particle size distribution (Raman and FBRM). We developed a PAT method utilizing Raman Spectroscopy to monitor solute concentration and polymorphs of CBZ. A calibration model was developed and validated; and the limit of detection for a metastable polymorphic form was quantified. The impact of multiple factors, such as solid content, and temperature, on the model prediction was studied. The model was validated following the principles described in USP <1225>. The results demonstrated that a linear model can predict the solute concentration with less than 10 percent error. The repeatability and intermediate precision was evaluated and the relative standard deviation was below 5 percent. The limit of detection and quantification were found to be 3.8 and 7.4 mg/ml, respectively. The limit of detection for the metastable form was determined by monitoring the ratio of characteristic peaks when increasing the percentage of the metastable form in relation to the total amount of crystals in the solution. A statistically significant change in the peak ratio was observed after the concentration of the metastable form reached 25 percent.

499363 An Engineering Approach to Concentration of Temperature-Sensitive Pharmaceutical Process Streams for Continuous Crystallization of an API

Abstract:

An Engineering Approach to Concentration of Temperature-Sensitive Pharmaceutical Process Streams for Continuous Crystallization of an API

P. Roche[1], G.K. Pascual[1], R. C. Jones*[1], P. Donnellan*[1], and B. Glennon[1]

[1] *Synthesis and Solid State Pharmaceutical Centre (SSPC), School of Chemical and Bioprocess Engineering, University College Dublin, Belfield, Dublin 4, Ireland*

Keywords: Solution Concentration, Continuous Processing, Bubble Column, Crystallization, MSMPR, PAT Technologies

Introduction

Traditionally, most active pharmaceutical compounds (API) have been manufactured at large plant scales using batch methodologies. Continuous manufacturing is becoming an increasing attractive alternative due to apparent benefits such as smaller manufacturing footprints, increased process safety, lower manufacturing costs and increased supply-chain flexibility. [1] A common problem faced in small molecule API production is the necessity for concentration steps between unit operations i.e. where the upstream process (e.g. reaction) is fully developed at a given concentration of solution; while the downstream process (e.g. crystallization) is operated at a different concentration range, driven by the metastable limits of the crystallisation process. To date, concentration steps have been carried out as evaporative crystallisation, i.e. boiling off a known amount of solvent until a desired concentration is reached. [2] However, some APIs have shown signs of degradation at high temperatures. [1]This study aims to provide a low temperature method under ambient pressure for concentration of API solvent solutions.

Section 1 â€“ Solution Concentration: A Novel Approach

This section of work focuses on employing a simple, lab-scale bubble column as a gas-liquid contacting system to achieve a desired removal rate of solvent. A model-based, mass transfer rate prediction is shown and compared with experiments performed for several common solvents used in the pharmaceutical industry. Thermodynamic predictions of Vapour-Liquid Equilibria were used in conjunction with mass transfer coefficients to accurately predict the rate of removal of solvent for the system and verified experimentally. [2]

For this work, a solution of Paracetamol in Methanol was chosen as the process, due to the well-researched saturation limits of paracetamol [3]and the reasonable results obtained from previously studied trials with methanol solutions in the column. The solution was reduced to an optimum concentration for the proceeding step from a dilute feed, while assuring that nucleation would not occur in any of the transfer lines by operating safely within the saturation limits.

Section 2 â€“ Continuous Crystallization Optimization

Following the establishment of model predictions, the column was employed immediately upstream of the feed of a continuously operated mixed suspension, mixed product removal (MSMPR) crystallizer with intermittent withdrawal via dip pipe [4, 5]using an automated pressure supply for the cooling crystallization of paracetamol in methanol. In-situ Focused Beam Reflectance Measurement (FBRM G400, Mettler Toledo) was used for real-time monitoring of chord length distribution and to accurately ascertain the point at which steady state operation had been reached; while in-situ PVM (ParticleView V819, Mettler Toledo) provides microscopy quality images in real time to monitor the crystal habit.

The overall objective of the work was to enable the adjustment of crystallization parameters, based on the optimum concentration provided by the gas-liquid contacting system, and to provide robust prediction of the crystallizer performance. Simple extraction of growth and nucleation kinetics from the MSMPR at steady state was accomplished using the population balance model.

References

1. Rawat, T. and I.P. Pandey, *Forced Degredation Studies for Drug Substances and Drug Products - Scientific and Regulatory Considerations*. Journal of Pharmaceutical Sciences and Research, 2015. **7**(5): p. 238-241.

2. Sandler, S., *Other Types of Phase Equilibria in Fluid Mixtures*, in *Chemical, Biochemical and Engineering Thermodynamics*. 2006, John Wiley & Sons, Inc. p. 575-593.

3. Granberg, R.A. and A.C. Rasmuson, *Solubility of Paracetamol in Pure Solvents*. Journal of Chemical Engineering Data, 1999. **44**: p. 1391-1395.

4. Power, G., et al., *Design and Optimization of a Multistage Continuous Cooling Mixed Suspension, Mixed Product Removal Crystallizer*. Chemical Engineering Science, 2015. **133**: p. 125-139.

5. Morris, G., et al., *Estimation of Nucleation and Growth Kinetics of Benzoic Acid by Population Balance Modeling of a Continuous Cooling Mixed Suspension, Mixed Product Removal Crystallizer*. Organic Process Research & Development, 2015. **19**(12): p. 1891-1902.

497499 The Effect of Axial Dispersion on Crystal Size Distribution in a Meso-Scale Continuous Oscillatory Baffled Crystallizer

Abstract:

The effect of axial dispersion on crystal size distribution in a meso-scale continuous oscillatory baffled crystallizer

Iyke Onyemelukwe[1], Qinglin Su[2], Zoltan K. Nagy[1,2], Chris D. Rielly[1]

[1] Department of Chemical Engineering, Loughborough University, Loughborough, LE11 3TU, UK

[2] Davidson School of Chemical Engineering, Purdue University, West Lafayette, IN 47906, USA

Abstract

Tubular crystallizers, specifically the continuous oscillatory baffled crystallizer (COBC), for continuous manufacturing have been an active research area in recent years, e.g. investigating how axial dispersion is affected by flow regime, operating conditions and the geometry of the periodic constrictions [1] and how the resulting residence time distribution (RTD) influences the supersaturation distribution and the resulting product crystal size distribution [2]. In this study, an experimental setup of a meso-scale COBC was developed to study the effects of axial dispersion on a cooling crystallization process of the glycine in water system. First, the axial dispersion effect, characterized by the Peclet number, $\mathbf{P_e}$, was measured experimentally by using an imperfect pulse method, tracking the concentration-time history of a tracer at two axial locations after the tracer input for both the liquid and particle phases. Dispersion models for the tracer concentration in each phase were then fitted to calculate the dispersion parameters. The method used a fast Fourier transformation (FFT) to convolute the measured input concentration profile with the parameterized model RTD to almost perfectly match the measured output profile; The effects of the oscillation frequency and amplitude on the dispersion coefficients were studied.

A seeded cooling crystallization of glycine and water system was run in the meso-COBC under the same range of through-flow and oscillatory operating conditions. Broad crystal size distributions of final crystals were observed. Furthermore, a steady-state mathematical model of COBC, consisting of population balance model of crystals, mass balance equation of solute, and energy balance equation for heat transfer, was developed to provide an in-depth understanding of the dispersion effect on crystal size distribution. It was found that the axial dispersion had a less significant effect on the crystal size distribution under the current operating conditions, whereas the fast cooling rate along the tube had a far greater impact, which results in extensive primary nucleation. An optimal trajectory of temperature along the tube was important in controlling the supersaturation and the crystal size distribution [3].

However, the nature of the counter-current heat transfer, which leads to a series of natural cooling curves, one for each change in jacket temperature, makes difficult the control of temperature trajectory in a limited number of tube segments. Perspectives on using the meso-scale COBC for crystallization kinetics screening for continuous manufacturing and for scaling up to larger COBC facilities will be given.

References

[1] L. N. Ejim, S. Yerdelen, T. McGlone, I. Onyemelukwe, B. Johnston, A. J. Florence and N. M. Reis, "A factorial approach to understanding the effect of inner geometry of baffled meso-scale tubes on solid suspension and axial dispersion in continuous, oscillatory liquid-solid plug flows," *Chemical Engineering Journal*, vol. 308, pp. 669-682, 2017.

[2] R. Kacker, S. I. Regensburg and H. J. Kramer, "Residence time distribution of dispersed liquid and solid phase in a continuous oscillatory flow baffled crystallizer," *Chemical Engineering Journal*, vol. 317, pp. 413-423, 2017.

[3] Q. Su, B. Benyahia, Z. K. Nagy and C. D. Rielly, "Mathematical modeling, design, and optimization of a multisegment multiaddition plug-flow crystallizer for antisolvent crystallizations," *Organic Process Research Development*, vol. 19, pp. 1859-1870, 2015.

495126 Continuous Heterogeneous Crystallization and the New Method of Making Tablets

Abstract:

A novel continuous heterogeneous crystallization process was developed, which the active pharmaceutical ingredient is crystallized directly on the surface of an excipient within the crystallizer. The product is then directly formed into tablets without the need for complex downstream processing steps such as milling, sieving, granulation, and blending. The innovative process has no particle processing and handling unit operation. The APIs and excipients systems selected by study of molecular interaction mechanisms, and surface functionality of substrate. The effect of various process parameters on the process stability and drug loading were studied.

Within the range of our analysis, proportional drug loading can be easily tailored for a wide range depending on required drug dosage per tablet. The fully continuous processes, and eliminating the downstream steps result in producing crystalline compounds and final form (tablets) in significantly faster, more efficient, with lower footprint, and more economical. This novel approach eliminates almost all the conventional unit operations in the drug product process, by connecting the API purification and crystallization step directly to the tablet making final stage.

KEYWORDS: Continuous manufacturing, Heterogeneous nucleation, Surface facilitated nucleation, Drug product

491965 A Comparative Study of Continuous Crystallization in an Oscillatory Baffled Crystallizer and a Mixed-Suspension-Mixed-Product-Removal Crystallizer

Abstract:

Crystallization has become a predominant technique in particle design technology and it is widely applied in the pharmaceutical industry. It directly dictates downstream processes and heavily influences overall drug properties such as bioavailability and dissolution rate. Traditionally crystallization has been carried out in batch mode where operating conditions vary with time, often resulting in inconsistent products.[1] Continuous crystallization, on the other hand, has been identified to improve reproducibility, reduce operating cost and simplify scale-up.[2] A well-studied system for continuous crystallization operation is Mixed-Suspension-Mixed-Product-Removal (MSMPR) crystallizer where an agitator is commonly used to provide mixing. There are challenges associated with stirred tank systems such as poor local mixing, low heat and mass transfer, and varying shear rate.[3] To address these challenges, a novel commercial oscillatory baffled reactor (OBR) has been studied. It consists of a reactor tank and oscillating â€˜donutâ€™ shaped baffles to provide more uniform mixing.[4] Oscillatory systems have been well studied for reactions and synthesis because of their improved mixing. Interest in the use of OBRs for crystallization processes has increased due to enhanced mass and heat transfer capabilities and reduced shear imposed on crystals.[5]

In the work proposed here, a comparison between continuous crystallization in an MSMPR versus an OBR is carried out to assess the differences in final product properties. Continuous operation in both vessels is achieved using peristaltic pumps and vacuum for slurry removal. The residence time distribution (RTD) is studied in each system using homogeneous and heterogenous tracers at different mixing conditions. Continuous cooling crystallizations of paracetamol is then evaluated in both crystallizers at the same feed concentration and mean residence time with varying power density levels. At each power density level, the start-up time and steady state crystal properties (i.e. mean size, CSD) are analyzed to determine the effect of hydrodynamics. A rigid and robust online FTIR probe, capable of correcting for signal interference caused by oscillations, is used to monitor the solute concentration during operation. This study will improve the understanding of oscillatory systems as an alternative crystallization vessel and the understanding of the impact of different mixing mechanisms on the final crystal size distribution.

1. Nagy ZK, Aamir E. Systematic design of supersaturation controlled crystallization processes for shaping the crystal size distribution using an analytical estimator. *Chem Eng Sci*. 2012;84:656-670. doi:10.1016/j.ces.2012.08.048.

2. Alvarez AJ, Singh A, Myerson AS. Crystallization of Cyclosporine in a Multistage Continuous MSMPR Crystallizer. *Cryst Growth Des*. 2011;11(10):4392-4400. doi:10.1021/cg200546g.

3. Kacker R, Regensburg SI, Kramer HJM. Residence Time Distribution of Dispersed Liquid and Solid Phase in a Continuous Oscillatory Flow Baffled Crystallizer. *Chem Eng J*. 2017;317:413-423. doi:10.1016/j.cej.2017.02.007.

4. Hewgill MR, Mackley MR, Pandit AB, Pannu SS. Enhancement of gas-liquid mass transfer using oscillatory flow in a baffled tube. *Chem Eng Sci*. 1993;48(4):799-809. doi:10.1016/0009-2509(93)80145-G.

5. Ni X, Mackley MR, Harvey AP, Stonestreet P, Baird MHI, Rama Rao NV. Mixing Through Oscillations and Pulsationsâ€"A Guide to Achieving Process Enhancements in the Chemical and Process Industries. *Chem Eng Res Des*. 2003;81(3):373-383. doi:10.1205/02638760360596928.

490309 Continuous Drying of Pharmaceuticals

Abstract:

Continuous Drying of Pharmaceuticals

M. Kreimer[1], I. Aigner[1], S. Sacher[1], M. Krumme[2], T. Mannschott[2], P. van der Wel[3], A. Kaptein[3], J.G. Khinast[1,4]

[1] Research Center Pharmaceutical Engineering (RCPE) GmbH, Inffeldgasse 13, 8010 Graz, Austria

[2] Novartis Pharma AG, Novartis Campus, 4056 Basel, Switzerland

[3] Hosokawa Micron B.V., Gildenstraat 26, 7005 BL Doetinchem, Netherlands

[4] Graz University of Technology, Institute for Process and Particle Engineering, Inffeldgasse 13, 8010 Graz, Austria

Traditional production of pharmaceutical products has been mostly conducted via batch processing during the last decades. This kind of production is inflexible and involves cost-intensive process development, especially during up-scaling within phases II and III of drug product development. A lot of effort has been made from academia, industry and regulators to overcome these challenges by development of fully continuous manufacturing processes for pharmaceutical goods. This paradigm shift from batch to continuous processing is demanded as more and more NME (new molecule entities) are entering the market and demanding flexible production processes. Furthermore, continuous pharmaceutical manufacturing facilities allow 100% real-time quality assurance and process control. A complete manufacturing line may include very different unit operations starting with API synthesis and ending up with compaction of final tablets. Although a range of manufacturing equipment is available for various unit operations, there is still a gap in linking primary and secondary manufacturing.

Continuous drying of APIâ€™s is one of those challenging process steps. Consistent quality of the dried solid product is crucial and has to be maintained after crystallization, washing and filtration. During these processes extensive effort is taken to optimize particle properties such as particle morphology, size or bulk flowabiltiy. Therefore, the particle properties of the intermediates defined and tuned in previous steps have to be at least unaltered during drying or even optimized. This raises several challenges during the drying process of wet powders, because agglomeration and attrition are present permanently as counteracting forces due to shear movement.

Besides matching of specified product quality, the key to successful continuous drying of powders is a robust process design and consistent process ability. These process design considerations include, amongst others, avoiding of dead zones and a narrow residence time distribution. The latter is favorable to avoid mixing of particles with different moisture content and to diminish the tendency of agglomeration or attrition. One approach to avoid those issues is the application of forced-feed.

In this work experiments were conducted on two different drying platform technologies with different model substances, including APIâ€™s and excipients with particle sizes below 100Â Âµm. The two investigated drying platforms are based on extrusion and a contact-convective drying technology. Different process settings were investigated and successful configurations achieved. Initial moisture content of the feed material was varied to analyze the drying rate and capacity of the dryers. Measurements were performed for residual moisture content, particle size distribution and flowability and results were compared between raw, wet and dried material. Residual moisture levels below 0.1% were reached. No agglomeration or attrition was observed during the drying procedure as particle size distribution was unchanged for the dried product. Furthermore, similar results were obtained for powder flowability prior and after drying.

496616 Continuous Fluidized Bed Drying of Pharmaceutical Materials

Abstract:

Recent years, the continuous manufacturing has become a trending technology in the pharmaceutical industries. Traditional batch manufacturing requires long processing times and consumes a tremendous amount of materials during process development. The US FDA is encouraging the companies to step forward into cleaner, flexible, more efficient continuous manufacturing. Although there are no regulatory hurdles for implementing continuous manufacturing, the lack of understanding of the process is limiting the integration of unit operations. Take the fluidized bed as an example; batch fluidized bed drying has been a traditional way of removing the moisture from the pharmaceutical ingredients due to its well-mixing and gentle drying behavior. The flow properties and the resulting tablet properties will be affected because of the change of moisture and particle size during the drying process. When a continuous process is adopted, the material property distribution can spread over process time. Therefore, understanding the residence time distribution is crucial to the process control.

In this study, experiments were conducted in a modified Glatt GPCG-2 continuous fluidized bed dryer/granulator. A twin-screw feeder was used to feed the wet materials. The residence time distributions for a pharmaceutical excipient (Fujicalin) and a placebo granule (Lactose/Avcil) were investigated using the colored tracer. Fujicalin is a free flowing porous material, and the Lactose/Avcil granule has very different flow properties before and after drying. The tracer concentration was calibrated and quantified with a color spectrophotometer that measures percent reflectance as a function of wavelength.[1] The residence time distribution curve was fitted by a dispersion model. The material properties were examined before and after the process. The effect of the operating conditions - material feed rate, air flow rate and initial loading were studied. The results showed that the Glatt continuous fluidized bed dryer behaved like a CSTR. The residence time could be reduced by applying a high material feed rate. The air flow rate and the initial loading did not have a significant effect on residence time for the conditions that we have investigated. As we previously found that the zero-order kinetic is dominant during the drying process, the outlet material moisture can be easily predicted by combining the RTD model and the drying kinetic.[2]

[1] Emady, H.N.; Wittman, M.; Koynov, S. *et al.* A simple color concentration measurement technique for powders. Powder Technology 2015, 286, 392-400.

[2] Chen, H.; Liu, X.; Bishop, C. *et al.* Fluidized bed drying of a pharmaceutical powder: A parametric investigation of drying of dibasic calcium phosphate. Drying Technology 2016, null-null.

494859 Transdermal Protein Delivery Via a Choline-Based Deep Eutectic Solvent

Abstract:

Transdermal Protein Delivery Via A Choline-based Deep Eutectic Solvent

Presenting Author: Kelly Ibsen[1]

Co-Authors: Amrita Banerjee[1], Yasunori Iwao[2], Michael Zakrewsky[1], Samir Mitragotri[1]

[1]University of California Santa Barbara, USA; [2]University of Shizuoka, Japan

The outer layer of the skin is a formidable barrier to drug delivery, and many of currently available methods to increase the permeability of skin, such as chemical penetration enhancers (CPE) ethanol and Transcutol, also cause irritation and damage. This research aims to develop low-irritation compounds that enhance transdermal delivery of proteins. We have shown that CAGE, a benign choline-geranate eutectic, can deliver small proteins.[1] Here, we investigate its ability to enable macromolecule penetration, using three proteins with a range of sizes and hydrophobicity: insulin bovine serum albumin (BSA) and ovalbumin (OVA). Transdermal delivery of macromolecules such as proteins could improve a myriad of treatments that currently require subcutaneous injections such as diabetes, where needle phobia results in poor patient compliance and treatment outcomes.[2]

To measure skin penetration compared to standard CPEs, fluorescently-labeled proteins in 1) CAGE, 2) ethanol, 3) Transcutol®, or 4) buffer were applied to porcine skin discs in a diffusion cell. Confocal laser imaging was conducted on skin cross-sections at 24 h. Insulin and BSA concentrations were quantitatively measured in the separate skin layers and the acceptor compartment after 24 and 48 h via fluorescence or isotope detection. Additionally, *in vivo* delivery of insulin was examined by measuring blood glucose levels in rats after topical application of insulin-CAGE compared to subcutaneous insulin injection.

Confocal laser imaging showed penetration of insulin, BSA and OVA with CAGE into the skin but not with 50% v/v ethanol, Transcutol or buffer after 24 h. The amount of insulin or BSA delivered through the skin section at 48 h, as measured by concentration in the acceptor chamber, was significantly higher with CAGE than buffer. Topical application of insulin-CAGE on rats resulted in a significant decrease of 40% in blood glucose compared to insulin in buffer after 4 hours that was sustained for the length of the 12-hour study, providing a more sustained effect on blood glucose than subcutaneous injection. These results demonstrate the effectiveness of CAGE as a potential transdermal delivery vehicle for therapeutic proteins.

Acknowledgements: The presenting author is supported by the National Science Foundation Graduate Research Fellowship under Grant No. DGE-1144085.

References

1. M. Zakrewsky, K. S. Lovejoy, T. L. Kern, T. E. Miller, V. Le, A. Nagy, A. M. Goumas, R. S. Iyer, R. E. Del Sesto, A. T. Koppisch, D. T. Fox, S. Mitragotri. *Proc. Natl. Acad. Sci.* 2014. *111*, 13313.

2. D. Nathan. Diabetes Care. 2014. 37(1): 9-16.

496510 Enhanced Delivery to Skin Using Novel Terbinafine Ionic Liquids

Abstract:

The delivery of drugs into and across the skin is non-invasive, painless, and well-accepted, and it represents an attractive alternative to oral and parenteral delivery. The skin is the largest organ of the human body, and it acts as a protective barrier against pathogens and other foreign substances. Skin is composed of several layers, including stratum corneum, viable epidermis and dermis. The outermost layer, i.e., stratum corneum, is usually the rate-limiting barrier for drug delivery to skin, in which solute (i.e., drug) transport across this layer is primarily via passive diffusion. Drug delivery to skin remains limited to a small number of drugs, because the stratum corneum has been shown to prevent skin penetration of most hydrophilic and large molecular weight drugs. In this study, we examined the enhancement of drug delivery to skin using novel room-temperature ionic liquids (RTILs) synthesized from an active pharmaceutical ingredient (API) with antifungal activity.

Ionic liquids (ILs) are organic salts that are composed entirely by ions. Many possible cation-anion combinations, where an active pharmaceutical ingredient (API) is one of the ions, can produce ILs with improved biological activity. Specifically, RTILs are very attractive candidates for drug delivery to skin, since there is no need for solvent to dissolve the API, and they can produce a liquid formulation containing 50% API on a molar basis at room temperature. In addition, the use of RTILs can take advantage of some of their enhanced physical and chemical properties that enable improved solubility, stability, bioavailability and provide an alternative route to deliver drugs to the skin.

Here, we demonstrate the enhanced transdermal delivery of terbinafine using RTIL formulations. We report the synthesis, characterization, and evaluation of novel RTILs with antifungal activity. These RTILs incorporated the API terbinafine in combination with inactive ingredients safely used in other pharmaceutical formulations as counterions. These RTILs are being developed to treat skin fungal infections, and to reduce the side effects often encountered with other delivery routes. They were synthesized through a salt metathesis reaction, and their physicochemical properties (i.e., MW, density, viscosity, log P, water solubility) were measured. *Ex-vivo* drug permeability experiments were performed using pig ear skin, as well as *in-vivo* drug permeability experiments using rodents. The drug permeability experimental results from the RTIL formulations showed a significant enhancement in skin permeability of up to two orders of magnitude when compared to the commercial topical formulation of terbinafine. In addition, our studies showed higher amounts of drug remaining in the skin layers when compared to the commercial formulation. *In-vitro* antifungal activity experiments in four different fungal strains of medical significance confirmed that the RTIL formulations remained active. We can conclude that terbinafine can be formulated as an RTIL with dramatically enhanced skin permeability and strong antifungal efficacy that has promise as a novel RTIL-formulated drug for transdermal delivery of terbinafine.

496350 Controlled Release of N-Acetylcysteine Positively Affects the Viability, Redox State and Morphology of Oligodendrocyte Progenitor Cells Under Oxidative Stress

Abstract:

Introduction: Overproduction of reactive oxygen species (ROS) is a major factor in the pathogenesis of multiple sclerosis. Localized antioxidant delivery is therapeutically relevant to counteract and neutralize ROS in such oxidant-stressed systems. Poly(lactic-co-glycolic acid) (PLGA) microparticles can provide a localized, controlled release of encapsulant, though this has been limited primarily to large proteins or small hydrophobic molecules. Here we investigate the therapeutic efficacy of controlled release of N-acetylcysteine (NAC) on oligodendrocyte progenitor cells (OPCs) under oxidative stress by examining viability, redox state and morphology. We aim to rescue OPCs from oxidative stress by using highly loaded PLGA microparticles encapsulating the hydrophilic, small molecule NAC.

Materials and Methods: NAC-loaded PLGA microparticles were prepared via the double emulsion, solvent evaporation technique including doping of all external water phases with NAC to maximize drug load (yielding between 40 and 133 ug NAC/mg particle). For all cell experiments, OPCs were grown in proliferation media for 24 hours. Experimental media was free of growth factors and antioxidants. For the 24-hour rescue experiment, OPCs were incubated for 30 minutes with 1mM H2O2. Concentrated free NAC, empty particles, or NAC-loaded particles were then added to the wells containing H_2O_2 to yield 1mM NAC. Particle dose was determined by cumulative NAC released within 24 hours. For the prolonged treatment experiment, OPCs were incubated for 30 minutes with 0.5 mM NAC or 48-hour equivalents of empty or NAC-loaded particles. Concentrated H_2O_2 was then added to the wells to yield 0.2mM in the well. After 24 hours of incubation, H_2O_2 was added again to yield 0.2mM in the well and incubated for an additional 24 hours. At the end of each experiment, wells were imaged with a Zeiss fluorescent microscope under brightfield and GFP channels. OPCs were lysed and ATP (CellGlo), DNA (PicoGreen) and glutathione (GSH-Glo) were measured.

Results and Discussion: Added 30 minutes after the onset of oxidative stress, controlled release of NAC for 24 hours rescues the viability of OPCs to 80% of healthy control, from 30% without treatment. There is an insignificant increase in viability relative to the free, soluble drug. Controlled released of NAC protects OPCs from multiple administrations of H_2O_2 over the course of 48 hours, yielding a 2-fold increase in reduced glutathione per cell which indicates a more reduced redox state. Controlled release of NAC also enhances OPC morphology, yielding 0.34 processes/cell with particle treatment compared to 0.29 processes/cell without treatment. Ongoing work aims to fabricate core-shell microparticles with a subdued initial burst release. Preliminary confocal images suggest that we have fabricated core-shell PLGA microparticles with two distinct polymer phases.

Conclusions: Highly loaded PLGA microparticles encapsulating the small-molecule antioxidant NAC rescued OPCs from oxidative stress, yielding a viability near control levels. NAC-loaded particles were shown to protect OPCs from multiple administrations of oxidants over 48 hours by increasing the levels of reduced glutathione above control levels. NAC-loaded particles also prevented degradation of OPCs. This work shows that efficiently loaded antioxidant microparticles capable of controlled and localized delivery could have therapeutic potential in oxidant-stressed systems by rescuing and protecting cells from harmful oxidants.

492195 Gold Nanocones: A Candidate for Cavitation-Enhanced Drug Delivery

Abstract:

Title: Gold Nanocones: A Candidate for Cavitation-Enhanced Drug Delivery

Authors: Xiaoqian Su and James J. Kwan

Ultrasound is a high frequency longitudinal pressure wave. With sufficient energy, ultrasound is capable of nucleating cavitation: the dynamic oscillations of a gas or vapour cavity. Cavitation promotes a multitude of mechanical effects, including, jetting, shockwaves, and microstreaming. These mechanical effects have been established to be key mechanisms in ultrasound-enhanced the delivery of therapeutics for a broad range of clinical applications such as cancer therapy. However, the nucleation of cavitation with ultrasound alone requires energies that are damaging to surrounding tissue. Therefore to remove off target damage, particles that generate cavitation, i.e., cavitation agents, are required.

Conventional cavitation agents, i.e., ultrasound contrast agents, have demonstrated the ability to enhance drug delivery. Yet the potential for enhanced therapy from these particles in certain therapies are limited because of their large size and rapid acoustic destruction short lifetime. We have previously demonstrated that the use of a single cavity polymeric nanoparticles that trap nanobubbles, provide sustained cavitation, and improved cancer drug delivery[1]. However, the acoustic energy requirements to nucleate sufficient cavitation was substantial. Here, we show an effective method to manufacture gold nanocone cavitation agents. Furthermore, the unique structure of the gold nanocones enables the formation of clustered bellflower super structures with multiple cavities, which will reduce the acoustic energy requirements for cavitation. To measure the acoustic responses of these novel cavitation agents, we expose these particles to increasing pressure amplitudes until cavitation is observed with a passive cavitation detector. The result here will aid in the design and manufacture of the next generation ultrasound-responsive nanoparticles for ultrasound-enhanced drug delivery.

1 Kwan, J. J. et al. Ultrasound©\Propelled Nanocups for Drug Delivery. *small* **11**, 5305-5314 (2015)

Keywords: ultrasound, drug delivery, cavitation, nanoparticles

493091 Crossing the Blood-Brain Barrier at the Nanoscale: Understanding the Influence of Physical Characteristics

Abstract:

As a variety of nanomaterials advance the treatment of various diseases including cancer, their utility to treat brain disorders remains an open question, owing to the exceptional barrier properties of the blood-brain barrier (BBB). Specifically, despite extensive studies on its biology, our understanding of the factors that influence nanoparticle translocation across the BBB is limited. To begin these investigations, we discuss our progress toward understanding the role of nanoparticlesâ€™ physical properties (*e.g.* size, shape, and flexibility) in their permeation across the BBB using *in vitro* models. Our microfluidic BBB platform recapitulates *in vivo* flow and architecture conditions while retaining sufficient simplicity to ensure a well-controlled environment. We demonstrate our ability to create and characterize these models, focusing on the device and protocol design. We also discuss the image analysis methods used to quantitatively measure particle transport in real time. The results provide important criteria to guide the rationale design of nanoparticles to facilitate their transport across the BBB.

496906 Targeted Drug Delivery to the Back of the Eye By Hydrogel Pushing in the Suprachoroidal Space

Abstract:

Drug delivery targeted to the choroid and retina of the eye and, in some cases the most posterior portion of these tissues at the macula is expected to increase drug efficacy and reduce side effects, because the site of action for ocular drugs treating diseases such as age-related macular degeneration, diabetic macular edema, and posterior uveitis is around macula. Previously, we have used microneedles to inject drug formulations into the suprachoroidal space (SCS) of the eye, which allows drug to flow circumferentially at the choroid-sclera interface. This targets drug to the choroid and adjacent retina, but does not specifically target the macula. This approach is currently the subject of multiple advanced clinical trials for treatment of ocular inflammation.

In this study, we sought to develop a method to better target drug delivery to the macula via the SCS. We hypothesize that injecting drug particles in a formulation that swells within the SCS can push the drug further toward the back of the eye and in that way flow to the macula. We accomplished this by injecting, from a single syringe, a first formulation containing drug particles followed by a second formulation containing a swellable hydrogel material without drug designed to push the drug particles in the first material posteriorly toward the macula within the SCS.

Albino New Zealand White rabbit eyes were used for ex vivo experiments. All SCS injections were accomplished using a 30-gauge hollow microneedle with 750 µm length. We filled two materials into each syringe. First, 30 µL of 4% (w/v) hyaluronic acid (HA, 2.6 MDa) hydrogel was filled into the syringe as a drug pushing material. Then 20 µL of a model drug formulation containing 0.5% (w/v) of red-fluorescent particles (2 µm diameter) in 1% (w/v) HA, was filled into the syringe without mixing with the first material. In this way, the model drug particles can come out of the syringe first during injection. The injection was performed 3 mm posterior to the limbus at supranasal location of the ocular globe. Intraocular pressure (IOP) of the eye ex vivo was measured using a tonometer and adjusted to 10-15 mmHg by adding HBSS buffer into the vitreous humor. To minimize backflow of the injected materials after injection, the syringe was maintained for 1 min at the injection site.

Particle distribution was analyzed right after the injection by freezing the eyes in chilled 100% isopropyl alcohol. Then, fully frozen eyes were dissected from posterior pole to limbus by making eight radial cuts like flower petals. To determine the particle distribution in the SCS, each petal was cut into 4 pieces depending on the distance from limbus â€“ 0-3, 3-6, 6-9, over 9 mm â€“ and the particles were extracted with 0.5X RIPA buffer. Particle content in each piece of tissue was measured by it signal by fluorometry.

To optimize the two formulations in the syringe, the HA concentration in the particle formulation and the pushing hydrogel formulation was varied. The objectives were (i) to minimize the hydrogel content in the particle formulation so that it would flow more easily to the back of the eye, (ii) to maximize the hydrogel content in the pushing formulation to increase hydrogel swelling and, therefore, pushing, (iii) to minimize viscosity of both formulations to reduce pressure during injection and (iv) to achieve an appropriate balance between the viscosities of the two formulations so that they would not mix during injection or pushing. Based on these considerations, we chose 1% (w/v) HA for a particle solution and 4% (w/v) HA as the pushing hydrogel.

Using this formulation, 14.1 ± 5.4% of particles was delivered toward the back of the eye (i.e., > 6 mm from the limbus) right after injection, 54.3 ± 10.2% of particles were delivered > 6 mm from the limbus after 6 h at 37 °C. This demonstrated the ability of the pushing formulation to move more than half of the particles to the back part of the SCS. This pushing was facilitated not only by the increase in hydrogel volume upon contact with water in the eye, but also because the viscosity of the HA pushing hydrogel was decreased at 37 °C (relative to room temperature at which it was injected).

In addition, we hypothesized that a high-salt (i.e., high osmotic pressure) HA hydrogel can increase osmotic flow of water into the hydrogel to increase pushing even more. To assess this hypothesis, we added 9 % sodium chloride to the HA hydrogel pushing material. We found that the high-salt HA hydrogel pushed 80.7 ± 6.1% of the particles > 6 mm from the limibus (and 38.0 ± 2.1% of particles were > 9 mm from the limbus) 6 h after injection at 37 °C.

We conclude that hydrogel pushing in the suprachoroidal space can be target delivery of particles to the back of the eye, which may be of use for drug delivery of ocular diseases.

504536 Polymerizable Prodrug Monomers for the Treatment of Serious Intracellular Bacterial Infections

Abstract:

Polymeric prodrugs consist of a therapeutic agent that has been covalently conjugated to a macromolecular scaffold via a hydrolytic or enzymatically degradable linkage. This strategy has been shown to substantially increase the solubility and stability of the parent drug while also enhancing drug circulation half-lives and reducing immunogenicity. An attractive route for the synthesis of polymeric prodrugs is the direct reversible deactivation radical polymerization (RDRP) of therapeutic agents that have been reversibly modified with suitable vinyl functionality. This versatile approach allows for one or more drug moieties to be incorporated into the final polymer at predetermined ratios without the need for additional conjugation and purification steps. In this talk we will discuss the synthesis and subsequent polymerization of prodrug monomers derived from the antibiotic ciprofloxacin with phenolic or aliphatic esters linking the drug to a polymerizable methacrylate group. In aerosolized and completely lethal F.t. novicida mouse challenge models, the fast-releasing phenyl ester linked ciprofloxacin macromolecular prodrug provided high cure efficiencies (75% survival rate under therapeutic treatment vs. 0 % survival for free ciprofloxacin), and the importance of release kinetics was demonstrated by the inactivity of the similar but slow-releasing prodrug system. Pharmacokinetics and biodistribution studies further demonstrated that the efficacious fast-releasing prodrug retained drug dosing in the lung above the MIC over a 48 h period with corresponding Cmax/MIC and AUC0â€“24h/MIC ratios being greater than 10 and 125, respectively; the thresholds for optimal bactericidal efficacy.

498490 Residence Time Distribution and Segregation Studies Trough Real Time Measurements By Near Infrared Spectroscopy

Abstract:

Currently, pharmaceutical companies are moving from batch to continuous manufacturing. Blender is a critical unit operation of continuous pharmaceutical manufacturing process that assures the blend uniformity and thereby drug content in final product [1]. The near infrared (NIR) spectroscopy is an important tool to measure the drug concentration at blender outlet and therefore to characterize the continuous blender [2-4]. The characterization of a continuous blender was achieved through residence time distribution (RTD) analysis and segregation studies by using a set of pharmaceutical excipients and active ingredients with material properties that defines a principal components analysis score plot.

RTD characterization was performed through in-line measurements by near infrared (NIR) spectroscopy of changes in composition adding a pulse of tracer into the system. Each material used was analyzed by using a continuous blending setup consisting in one feeder connected to a GEA continuous blender through a hopper. At the blender outlet, a chute was connected with a lateral window to control the level of powder with a rotary valve. Perpendicular to the powder flow, a near infrared probe with the light source was connected for spectral acquisition in diffuse reflectance mode. The reflected light is transferred to the interferometer for processing through fiber optic. Six experiments for each material has been performed at tow flow rates and 3 impeller speeds (revolution per minute (RPM)). A second feeder was necessary in cases where low density exceeds the gear capacity of a single feeder setup. Results of in-line RTD measurements demonstrated the selectivity of calibration models and low limit of detection required for analysis of tracers providing critical information for blender characterization.

Segregation studies were conducted for five formulations in continuous blending. For each configuration in blender RPM, samples were collected for further analysis in an ASTM D 6940-04 segregation tester. Near infrared prediction of ten samples collected for each experiment were performed by using 1-Dimentional sampling approach. Calibration models for each formulation were prepared and implemented. Results indicates that each blender configuration has a different impact in the segregation behavior of the API concentration and the NIR method provided a fast and reliable results.

Reference

[1]. Yijie Gao, Aditya Vanarase, Fernando Muzzio, Marianthi Ierapetritou, Characterizing continuous powder mixing using residence time distribution, Chemical Engineering Science, Volume 66, Issue 3, 1 February 2011, Pages 417-425, ISSN 0009-2509, http://doi.org/10.1016/j.ces.2010.10.045.

[2]. Singh, R.; Sahay, A.; Karry, K. M.; Muzzio, F.; Ierapetritou, M.; Ramachandran, R., Implementation of an advanced hybrid MPCâ€"PID control system using PAT tools into a direct compaction continuous pharmaceutical tablet manufacturing pilot plant. International journal of pharmaceutics 2014, 473 (1â€"2), 38-54.

[3]. ColÃ³n, Y. M.; Florian, M. A.; Acevedo, D.; Méndez, R.; Romañach, R. J., Near Infrared Method Development for a Continuous Manufacturing Blending Process. Journal of Pharmaceutical Innovation 2014, 9 (4), 291-301.

[4]. Cárdenas, V.; Cordobés, M.; Blanco, M.; Alcalà, M., Strategy for design NIR calibration sets based on process spectrum and model space: An innovative approach for process analytical technology. Journal of Pharmaceutical and Biomedical Analysis 2015, 114, 28-33.

499494 Use of Process Analytical Technology (PAT) in Advanced Manufacturing Process Monitoring and Control: Scientific Considerations and Regulatory Challenges

Abstract:

Significant scientific advancement made in PAT tool design and process analysis has resulted in the increasing use of PAT during process development. However, the actual implementation of PAT in a manufacturing setting has lagged despite its potential for advanced manufacturing process monitoring and control. In this presentation, we will discuss some of the scientific considerations and regulatory challenges when evaluating a PAT system for process monitoring and control for the manufacturing setting. Some aspects will be discussed are:

1) Identification of *critical process parameters or critical quality attributes* for monitoring and control purpose;

2) Risk-based strategy for identifying *measurement interface* between process analyzer and process stream and data acquisition frequency;

3) Model building framework, pros and cons of *each modeling methods* for process monitoring and control 4) Incorporating the PAT monitoring and control as part of *control strategy* for manufacturing.

Several examples will be included to highlight the use of PAT for advanced manufacturing process monitoring and control, in the spirit of facilitating broad adaptation of PAT in advanced manufacturing for robust manufacturing and product quality.

491877 3D Raman Imaging: A Method to Study the Effects of Lubrication on the Microstructure of Tablets

Abstract:

3D Raman Imaging: A Method to Study the Effects of Lubrication on the Microstructure of Tablets

Shashwat Gupta, Savitha Panikar, Fernando Muzzio

Rutgers- The State University of New Jersey

When it comes to pharmaceutical tablets, there are known knowledge gaps in the current manufacturing processes, particularly regarding the relationship between material properties of ingredients, processing conditions, and resulting tablet attributes. These three aspects can be linked through a common feature of understanding the microstructure of a sample. A better understanding of tablet microstructure has the potential to unlock the black box between process parameters and performance attributes. Product structure understanding of this nature can aid in prediction of product performance, process development as well as deviation investigations. Understanding the microstructure of a tablet requires knowing the spatial distribution of the different ingredients and void space, which can be enabled by Chemical Imaging (CI).

Raman imaging is a powerful tool for studying tablet structure, with superior chemical specificity than NIR and is also less affected by physical characteristics of the sample.

This study focuses on using 3D Raman Imaging system (H2Optx), as an off-line measurement system, to bridge this gap for tablets made using a Direct Compression (DC) continuous line, for the lubrication process. For pharmaceutical operations, lubrication is essential in order to reduce the friction between the surfaces of manufacturing equipment and that of organic solids as well as to ensure the continuation of an operation. The most common pharmaceutical lubricant used is Magnesium Stearate (MgSt). Till now, extensive structural studies have not been conducted to study the effects of lubricant(s) on tablet properties.

In this work, we investigate the particle size distribution, spatial variability (Variograms) of MgSt, using the 3D Raman Images. These parameters are then correlated with product performance (tablet hardness and dissolution profile) as well as formulation (amount of MgSt) and process parameters (Total exposure to shear and Residence Time), with the help of Principle Component Analysis.

Variographic data helped to quantify the distribution of MgSt in tablets of experimental conditions. On the other hand, to discern the variability in the distribution of MgSt within the layers of a tablet, we needed PCA score plots of the variographic data obtained from the 3D Raman images. They revealed a stark difference in the clustering of the layers for the tablets with different process and formulation parameters. The hardness of the tablets from the two experimental conditions agree with the expected trend. A wider distribution of MgSt resulted in tablets with a lower tablet hardness. However, there is not a clear trend in the dissolution profile with respect to the distribution of MgSt.

An example of a comparison between 3D Raman Images of tablets elucidating the effect of total exposure to shear is shown below:

(a)

(b)

Figure 1: 3D Raman Images (a) Lower exposure to shear: MgSt, green – particles <50 microns, red – particles from 50 to 250 microns and (b) Higher exposure to shear: MgSt, green – particles <50 microns, red – particles from 50 to 250 microns

The future research will involve studying how the distribution of MgSt changes in a tablet, as we move ahead in a continuous line i.e. comparison between post blending and post tableting samples. All the work will be used as a platform for an on-line measurement system.

493745 Development and Validation of an in-Line NIR Spectroscopic Method for Continuous Blend Potency Determination in the Feed Frame of a Tablet Press

Abstract:

INTRODUCTION

The desire within the pharmaceutical industry to shift towards continuous processing strengthens the need to invest in process analytical technology (PAT), because continuous real-time quality monitoring and control is essential for continuous production. Due to its fast, non-invasive and non-destructive character, near-infrared (NIR) spectroscopy is one of the most widely used PAT tools for the analysis of pharmaceutical tablets. The fact that NIR can be implemented into a process environment via fiber optics makes it an excellent in-line PAT tool.

Nowadays, NIR spectroscopy for potency analysis is one of the most widely used PAT applications for tablets [1–4]. This is often achieved by on-line or off-line NIR examination of tablet cores, resulting in a considerable lag time between the actual tablet sampling and the subsequent quality analysis.

To avoid this time lag, in this study, a measurement set
measurement of the circulating blend inside the feed frame of a tablet press (MODUL™P, GEA Process Engineering n.v., Halle, Belgium).

Secondly, an NIR calibration model was developed and validated, enabling in-line and real-time blend potency measurements within the optimised interfacing.

Next, the residence time distribution (RTD) inside the feed frame was studied to determine the potential lag time between the measured concentrations in the feed frame of the tablet press and the measured concentrations in the resulting tablets. Finally, influences of the process parameters on the in-line and off-line RTD were investigated.

MATERIALS

During the set-up optimisation and quantification model development study, a formulation was used consisting of sodium saccharine (JMC corporation, South Korea), microcrystalline cellulose (Avicel® PH 102, FMC biopolymer, Ireland), dibasic calcium phosphate (Emcompress™, JRS Pharma, Germany) and magnesium stearate (Mallinckrodt, Ireland)(0.5% w/w) as a lubricant for tableting.

For the RTD studies, a blend consisting of lactose (FastFlo 316, Foremost) and magnesium stearate (Mallinckrodt, Ireland)(0.5% w/w) was employed. Spiking was performed using a formulation consisting of lactose and sodium saccharine (in equal parts), and magnesium stearate (0.5% w/w) was added for lubrication.

NIR ACQUISITION

The in-line NIR spectra were collected in reflection mode using a diode array spectrometer: SentroPAT FO (Sentronic GmbH, Germany). A SentroProbe DL RS NIR (Sentronic GmbH, Germany) connected to the SentroPAT system, through a fiber-optic cable, was mounted inside the feed frame. Spectra were acquired at a speed of 15 spectra per second. A spectral range of 1150-2200 nm was covered with a spectral resolution of 2 nm.

The sampled tablets were inspected off-line in transmission mode using an FT-NIR analyser (Antaris II, Thermo Fisher Scientific, USA), furnished with a tablet analyser module. In transmission mode, a larger sample volume is measured, since the light penetrating through the sample is measured. This measurement mode is not suitable in-line because of the longer integration time and the need to be able to measure through the sample, which is impossible in the current design of the feed frame.

DEVELOPMENT AND OPTIMISATION OF SET-UP FOR IN-LINE MEASUREMENTS INSIDE THE FEED FRAME

An NIR probe was mounted on top of the feed frame in such a position that it measured the circulating blend just before compression. In order to ensure a reproducible and robust implementation of the NIR probe in the feed frame, a micrometer was installed in the tablet press, allowing a precise positioning of the probe. During tableting, NIR spectra were continuously acquired.

Initially, disturbances of the NIR signal, caused by the rotating paddle wheels of the feed frame were observed. The use of mathematical spectral filters to reduce spectral noise can be very complex and can result in deletion of important physical information about the powder blend within the feed frame [5]. Therefore small notches were created inside the paddle wheel fingers of the tablet feed frame, avoiding disturbances in the spectra and hence eliminating the need for mathematical or spectral filters.

The effect of paddle wheel type, probe height and paddle wheel speed on the stability of the NIR signal was investigated by means of a full factorial experimental design, developed in MODDE 10 (Umetrics, Umeå, Sweden). Optimal process settings, resulting in most stable NIR spectra, were used to make NIR calibration models.

DEVELOPMENT AND VALIDATION OF IN-LINE NIR CALIBRATION MODELS

Calibration models were developed by regressing the in-line NIR spectra against the true sodium saccharine concentration of the tablets (based on known weighed masses). In-line spectra were acquired during tableting of five blends with different concentrations. Models were calculated for the prediction of two different API target concentrations: 5% and 20% (w/w). Several calibration models for the prediction of this API content were developed, comparing the use of single spectra or averaged spectra, as well as using partial least squares (PLS) regression versus ratio models. The best models were selected and validated by means of accuracy profiles [6–8], by processing physical mixtures with four different API concentrations (validation set).

RTD DETERMINATION OF THE FEED FRAME

In order to determine the RTD of the feed frame, initially, the feed frame was filled with the lactose formulation. On top of this lactose blend, a pulse of 50 g of the spiking blend (saccharine/lactose) was introduced and on top of the spiking blend, 2500 g of lactose blend was added. During tableting, in-line NIR spectra were continuously acquired and tablets were sampled for off-line potency analysis. This experiment was repeated using different tableting parameters, according to a full factorial experimental design.

The mean RTD was calculated via the approach used by Kumar et al.[9]. The influence of the tableting parameters on the RTDs was evaluated using MODDE 10. In addition, the difference and link between in-line measured RTDs and off-line measured RTDs was investigated to gain a better understanding of the mechanisms inside the feed frame.

CONCLUSION

In this work, an in-line NIR method with high speed acquisition of spectral information to allow real-time release (RTR) was developed for use in the feed frame of a tablet press.

An in-line NIR spectroscopic model was developed and validated for the monitoring of API concentration of a powder blend in the feed frame during a tableting process.

Finally, RTD of the feed frame was determined in-line and off-line, helping to gain a better understanding of the flow/mixing processes going on inside the feed frame.

This study showed that in-line NIR measurements in the feed frame of a high speed rotary tablet press can be applied as a method for powder composition monitoring during the tableting process. When coupled with existing tablet press control systems, this method represents, together with the knowledge of the RTD of the feed frame, a big step towards real time release testing and real time release.

REFERENCES

[1] W. Li, A. Bashai-Woldu, J. Ballard, M. Johnson, M. Agresta, H. Rasmussen, et al., Applications of NIR in early stage formulation development. Part I. Semi-quantitative blend uniformity and content uniformity analyses by reflectance NIR without calibration models, Int. J. Pharm. 340 (2007) 97–103. doi:10.1016/j.ijpharm.2007.03.040.

[2] W. Li, L. Bagnol, M. Berman, R. a. Chiarella, M. Gerber, Applications of NIR in early stage formulation development. Part II. Content uniformity evaluation of low dose tablets by principal component analysis, Int. J. Pharm. 380 (2009) 49–54. doi:10.1016/j.ijpharm.2009.06.032.

[3] J. Gottfries, H. Depui, M. Fransson, M. Jongeneelen, M. Josefson, F.W. Langkilde, et al., Vibrational spectrometry for the assessment of active substance in metoprolol tablets: A comparison between transmission and diffuse reflectance near-infrared spectrometry, J. Pharm. Biomed. Anal. 14 (1996) 1495–1503. doi:10.1016/0731-7085(96)01800-6.

[4] E. Peeters, A.F. Tavares da Silva, M. Toiviainen, J. Van Renterghem, J. Vercruysse, M. Juuti, et al., Assessment and prediction of tablet properties using transmission and backscattering raman spectroscopy and transmission NIR spectroscopy, Asian J. Pharm. Sci. 11 (2016) 547–558. doi:10.1016/j.ajps.2016.04.004.

[5] A.J. O'Neil, R.D. Jee, A.C. Moffat, The application of multiple linear regression to the measurement of the median particle size of drugs and pharmaceutical excipients by near-infrared spectroscopy., Analyst. 123 (1998) 2297–2302. doi:10.1039/a806001k.

[6] P. Hubert, J.J. Nguyen-Huu, B. Boulanger, E. Chapuzet, P. Chiap, N. Cohen, et al., Harmonization of strategies for the validation of quantitative analytical procedures: A SFSTP proposal - Part I, J. Pharm. Biomed. Anal. 36 (2004) 579–586. doi:10.1016/j.jpba.2004.07.027.

[7] P. Hubert, J.J. Nguyen-Huu, B. Boulanger, E. Chapuzet, P. Chiap, N. Cohen, et al., Harmonization of strategies for the validation of quantitative analytical procedures A SFSTP proposal – Part II, J. Pharm. Biomed. Anal. 45 (2007) 70–81. doi:10.1016/j.jpba.2004.07.027.

[8] P. Hubert, J.J. Nguyen-Huu, B. Boulanger, E. Chapuzet, P. Chiap, N. Cohen, et al., Harmonization of strategies for the validation of quantitative analytical procedures: A SFSTP proposal–Part III, J. Pharm. Biomed. Anal. 45 (2007) 82–96. doi:10.1016/j.jpba.2007.06.032.

[9] A. Kumar, J. Vercruysse, M. Toiviainen, P.E. Panouillot, M. Juuti, V. Vanhoorne, et al., Mixing and transport during pharmaceutical twin-screw wet granulation: Experimental analysis via chemical imaging, Eur. J. Pharm. Biopharm. 87 (2014) 279–289. doi:10.1016/j.ejpb.2014.04.004.

500787 Identification of Key/Non-Key Process and Model Parameters of Polishing Chromatography, Primary Containers and Autoinjectors Using Mechanistic Models

Abstract:

Identifying key/non-key process parameters is an essential step in the Quality-by-Design methodology. It demonstrates true process understanding and it enables risk-based decision making. Typically, this is done by FMEA and DOE studies which have some important limitations. First-principles and mechanistic models are being used to advance the vision of rational drug substance development via a rigorous and quantitative Quality-by-Design approach. In this work, we use three mechanistic models to study the factor/response of drug/process/device systems of practical interest to industry. We employ well-established global sensitivity analysis techniques (Sobol indices) to address the so-called factor screening, factor prioritization and factor fixing setting [Saltelli et al., 2008]. This enables a rigorous analysis of key/non-key process and model parameters to accelerate design iterations, manage/reduce risk, and ultimately assure quality via a probabilistic in-silico QbD design space. This new methodology can work as a complement (or a replacement) of more traditional and experiential-based FMEA approaches. This is the first time such a rigorous analysis has been published for a polishing chromatography system. The application to the primary container case-study reveals that only half of the factors of this process are non-essential. For the autoinjector device two sub-components happen to be the most relevant to system variability.

大连理工大学
Dalian University Of Technology

In-Situ Agglomeration Measurement during KDP Crystallization Based on Double-View Image Analysis

Yan Huo, Tao Liu*, Cai Y. Ma, Xue Z. Wang

Institute of Advanced Control Technology
Dalian University of Technology, Dalian, China

Institute of Advanced Control Technology

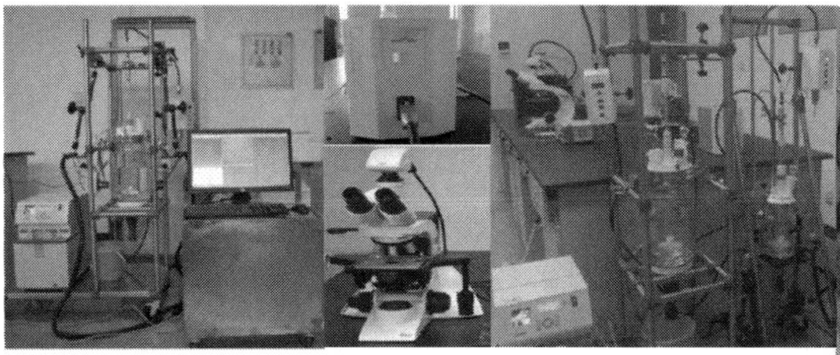

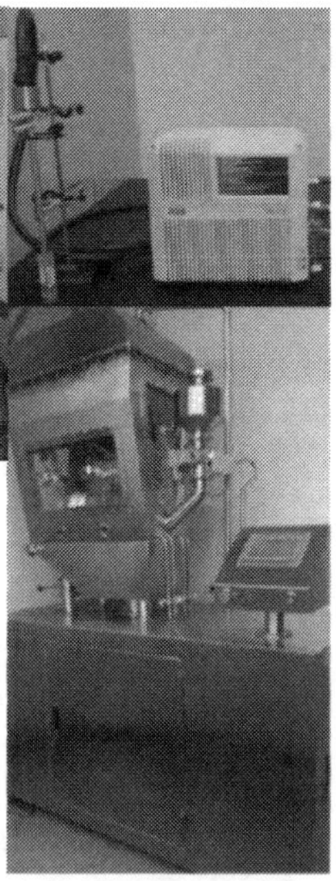

Research interests:

- In-situ crystallization & drying process monitoring using ATR-FTIR, FT-NIR, 2D/3D imaging systems;
- PBM and CFD modeling of crystallization processes;
- Online model predictive control and optimization of crystallization & drying processes;
- Design of in-situ measurement and control devices.

Outline

大连理工大学
Dalian University Of Technology

1. Introduction
2. Problem statement
3. Experimental setup
4. Agglomeration measurement
5. Conclusions & Outlook

1. Introduction

大连理工大学
Dalian University Of Technology

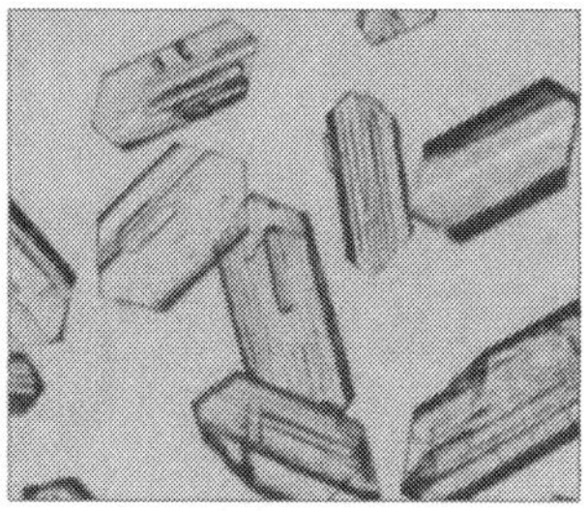

Particle shape

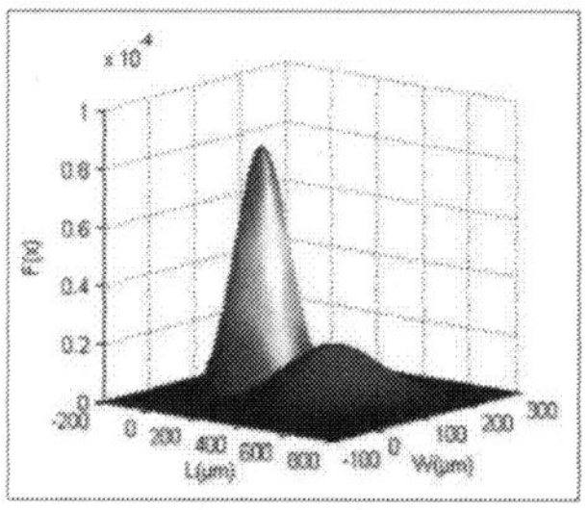

Particle size distribution

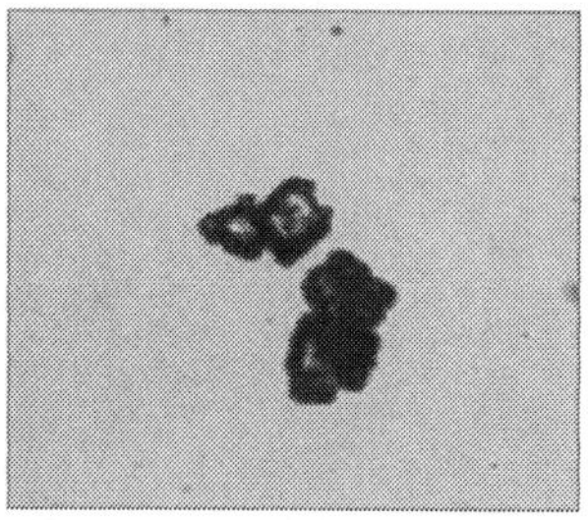

Particle agglomeration

On-line imaging systems have been increasingly applied to monitor industrial crystallization processes at the micro-scale in the past two decades, offering on-line observation of crystal shapes, the crystal size distribution, and particle agglomeration.

2. Problem statement

大连理工大学 Dalian University Of Technology

Pseudo agglomeration problem

Overlapping?

Agglomeration?

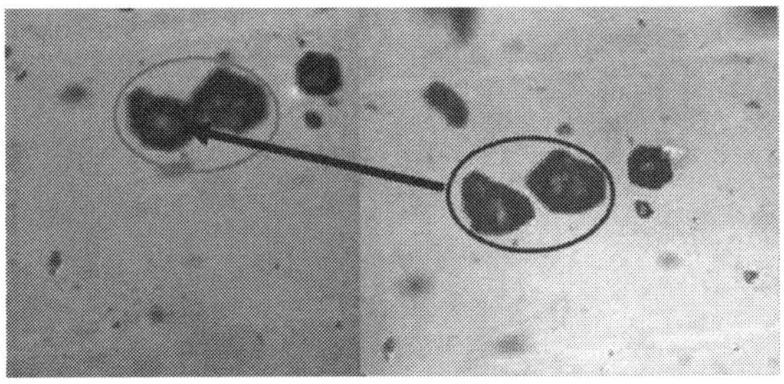

Left view   Right view

3. Experimental setup

大连理工大学 Dalian University Of Technology

Online monitoring of crystallization process

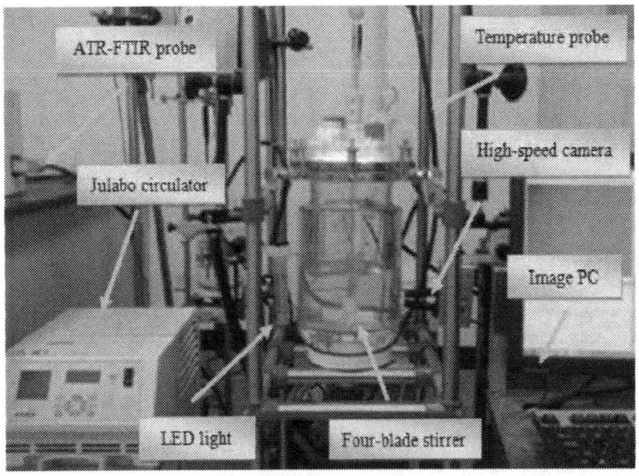

No	Device
1	4L glass vessel
2	Temperature probe
3	Julabo circulator
4	4-paddle agitator
5	Image PC
6	High-speed cameras
7	ATR-FTIR

Two cameras (UI-2280SE-C-HQ) with CCD sensors and USB Video Class standard were made by IDS Imaging Development Systems GmbH.

3. Experimental setup

大连理工大学
Dalian University Of Technology

Online monitoring of crystallization process

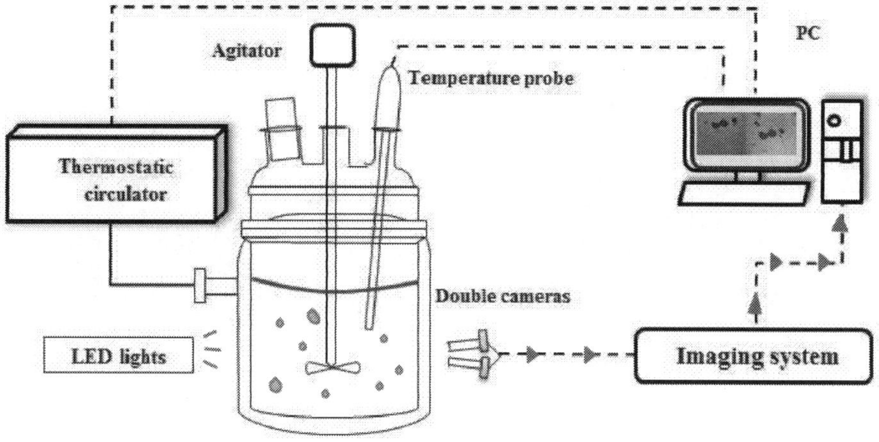

7

4. Agglomeration measurement

大连理工大学
Dalian University Of Technology

Image processing for agglomeration measurement

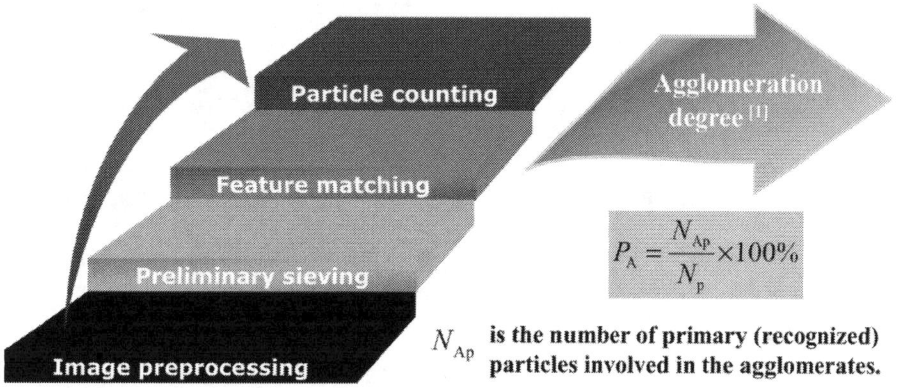

$$P_A = \frac{N_{Ap}}{N_p} \times 100\%$$

N_{Ap} **is the number of primary (recognized) particles involved in the agglomerates.**

N_p **is the total number of valid particles counted in the image.**

[1] Nývlt, J.; Karel, M., Crystal agglomeration. *Crystal Research and Technology*, 1985.

8

大连理工大学
Dalian University Of Technology

Step 1: Image preprocessing
(including image compression and segmentation)

Image compression

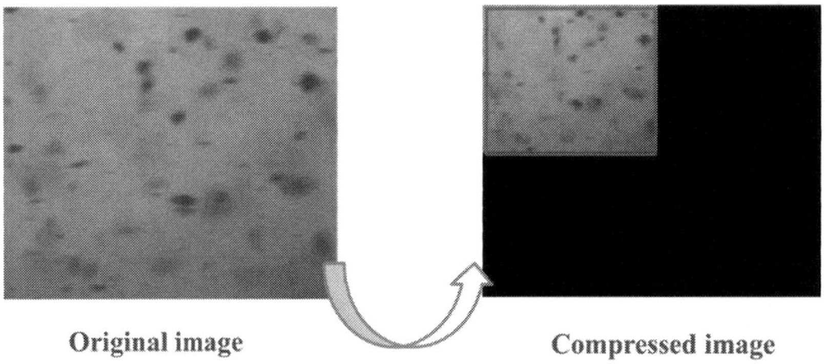

Original image **Compressed image**

Wavelet transform

Objective: reduce the raw image size to one quarter for saving computation effort.

大连理工大学
Dalian University Of Technology

Image segmentation

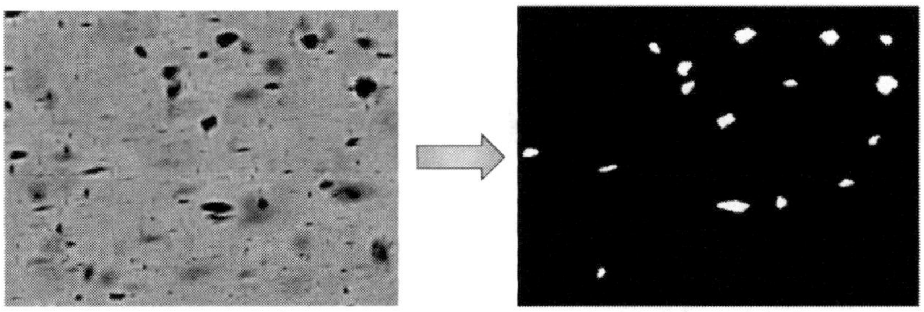

Reduced image **Segmented image**

Task: perform image sementation to extract the particle shapes from the image.

Solution: the uneven background is extracted by using an opening operation algorithm [2]; the segmentation is computed by the Otsu thresholding function[3]

[2] Chen, J.; Wang, X. Z., A wavelet method for analysis of droplet and particle images for monitoring heterogeneous processes. Chemical Engineering Communications 2005, 192, (4), 499-515.
[3] Gonzales, R. C.; Woods, R. E.; Eddins, S. L., Digital image processing using MATLAB. Pearson Prentice Hall: Upper Saddle River, NJ, 2004.

4. Agglomeration measurement

大连理工大学
Dalian University Of Technology

Step 2: Preliminary sieving

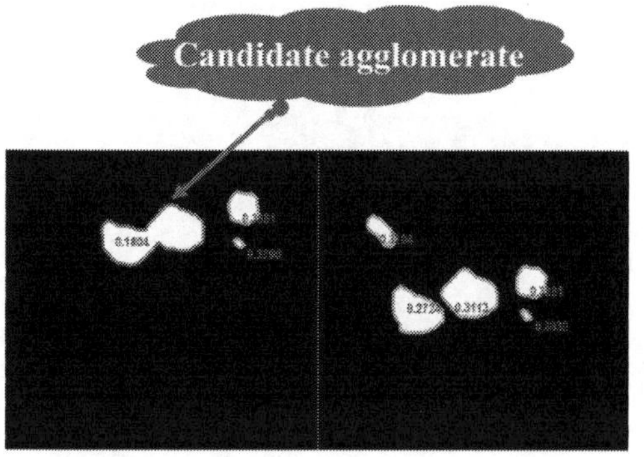

Key idea: sieve out candidate agglomerates by defining the particle shape feature.

Inner distance based curvature

$$\mathrm{IDC} = \frac{\sum_{n=1}^{N} c_n}{N}$$

$$x_c = \frac{1}{N}\sum_{n=0}^{N-1} x_n \qquad y_c = \frac{1}{N}\sum_{n=0}^{N-1} y_n$$

$$d_n = \sqrt{(x_c - x_n)^2 + (y_c - y_n)^2}$$

$$\rho_n = d_n - \bar{d} \qquad c_n = \frac{\rho_n''}{(1+\rho_n'^2)^{3/2}}$$

N is the number of particle boundary.

c_n s the curvature of the nth deviation distance.

4. Agglomeration measurement

大连理工大学
Dalian University Of Technology

Step 3: Feature matching

Interest point location

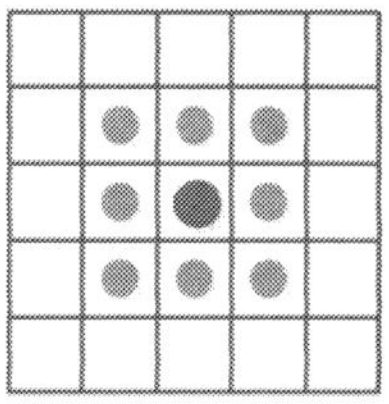

Objective: find the extremum value in a local region

The Hessian matrix for approximation is adopted to determine the extremum point

$$\mathrm{Det}(H) = D_{xx}D_{yy} - (0.9D_{xy})^2$$

D_{xx}, D_{yy}, D_{xy} are the approximations of the Gaussian second order derivatives for the image by using the box filters [4], respectively.

[4] Lowe, D. G., Distinctive image features from scale-invariant keypoints. *International Journal of Computer Vision*, 2004.

4. Agglomeration measurement

大连理工大学 Dalian University Of Technology

Step 3: Feature matching

Feature descriptor

Improved texture descriptors

(to cope with image noise and uneven illumination background)

DCLBP (deviation compensated LBP)

$$DCLBP_{P,R}(i,j)=\sum_{p=1}^{P}s(g_p-g_a)2^p$$

where g_p is the gray values, g_a is a deviation location parameter

$$s(g_p-g_a)=\begin{cases}1, & if\ g_p-g_a\geq T;\\ 0, & if\ g_p-g_a<T.\end{cases}$$

MDLBP (magnitude difference based LBP)

$$MDLBP_{P,R}(i,j)=\sum_{p=1}^{P}s(m_p-\overline{m_p})2^p$$

where $m_p=|g_c-g_p|$

$\overline{m_p}$ is the averaged value for $[m_0,m_1,...,m_{P-1}]$

$$s(m_p-\overline{m_p})=\begin{cases}1, & if\ m_p-\overline{m_p}\geq T\\ 0, & if\ m_p-\overline{m_p}<T\end{cases}$$

The local binary pattern (LBP)

[5] Ojala, T.; Pietikäinen, M.; Mäenpää, T., Multiresolution gray-scale and rotation invariant texture classification with local binary patterns. *IEEE Transactions on Pattern Analysis & Machine Intelligence*, 2002, 24, (7), 971-987.

4. Agglomeration measurement

大连理工大学 Dalian University Of Technology

Numerical example: three texture patterns with different features

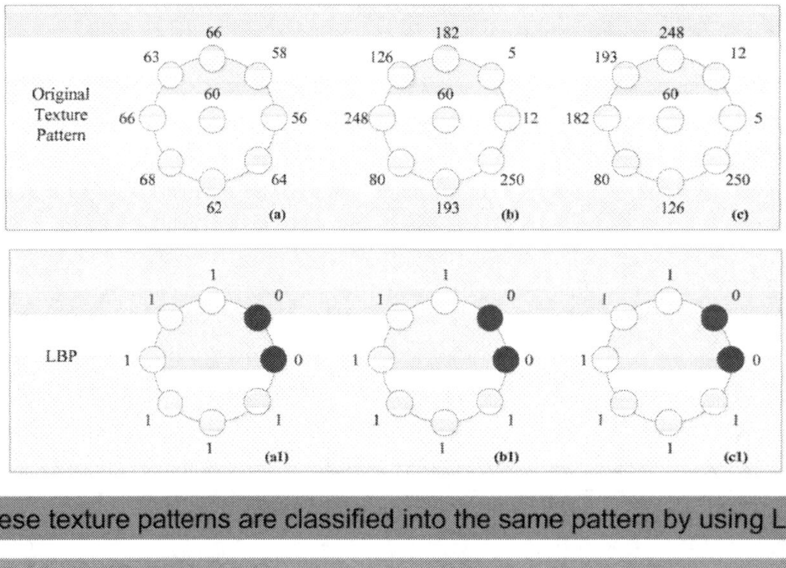

These texture patterns are classified into the same pattern by using LBP!

Effectively recognized by DCLBP and MDLBP

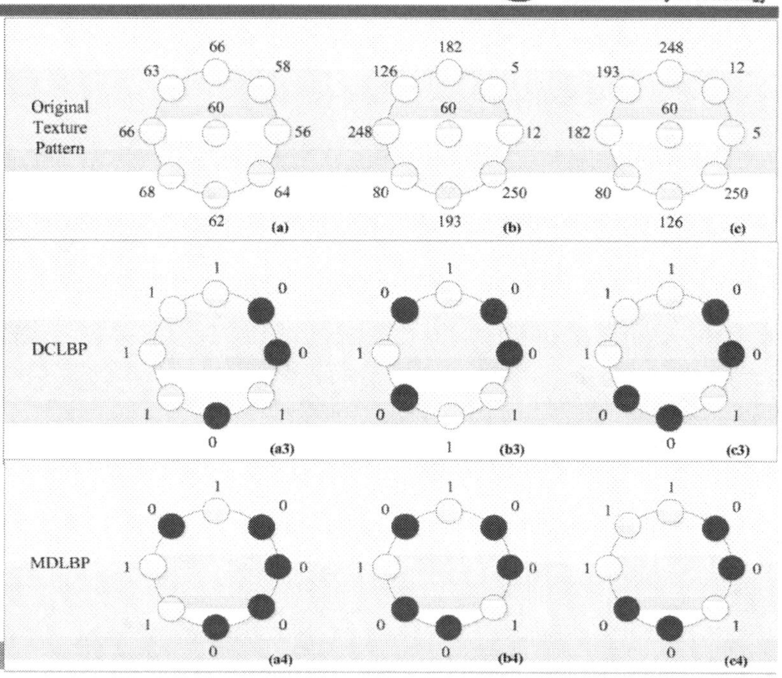

4. Agglomeration measurement

大连理工大学
Dalian University Of Technology

Descriptor matching

Exclude pseudo agglomerates due to overlapping

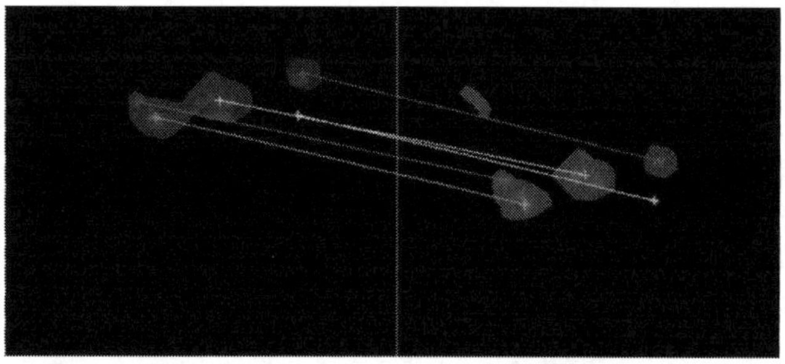

Left view Right view

Task: match the interest points between double-view images in real time.

Solution: the approximately nearest neighbor searching algorithm via K-dimension tree[6].

---

[6] Arya, S.; Mount, D. M.; Netanyahu, N. S.; Silverman, R.; Wu, A. An optimal algorithm for approximate nearest neighbor searching, *Acm-Siam Symposium on Discrete Algorithms*. 1994, Arlington, Virginia, 1994; 1994; pp 573-582.

Comparison of matching results by using different algorithms

$$P_{pr} = \frac{N_{pa}}{N_{pa} + N_{pm}} \qquad P_{pm} = \frac{N_a}{N_t}$$

Methods	Particle recall ratio (%)	Point matching accuracy (%)	Averaged computation time per frame (s)
Proposed	97.96	92.01	0.9347
SIFT [7]	94.90	91.83	1.3703
ASIFT [8]	93.88	91.65	1.8061
SURF [9]	95.92	88.47	0.8925
ORB [10]	90.82	83.95	0.4126

Higher   Higher   Faster

References:

[7] Lowe, D. G., Distinctive image features from scale-invariant keypoints. *International Journal of Computer Vision*, 2004, 60, (2), 91-110.

[8] Morel, J. M.; Yu, G., ASIFT: A new framework for fully affine invariant image comparison. *Siam Journal on Imaging Sciences*, 2009, 2, (2), 438-469.

[9] Bay, H.; Ess, A.; Tuytelaars, T.; Gool, L. V., Speeded-up robust features (SURF). *Computer Vision & Image Understanding*, 2008, 110, (3), 346-359.

[10] Rublee, E.; Rabaud, V.; Konolige, K.; Bradski, G. In ORB: An efficient alternative to SIFT or SURF, *International Conference on Computer Vision*, 2011, 2564-2571.

Step 4: Particle counting

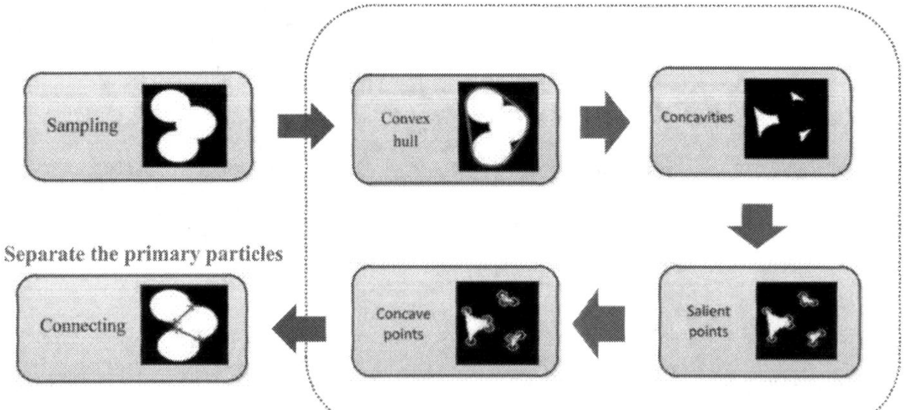

Solution:

A concave point on the agglomerate boundary is determined if all the connecting lines between the concave point and its adjacent salient points are inside the agglomerate.

Experiment: recognized results for two different cases

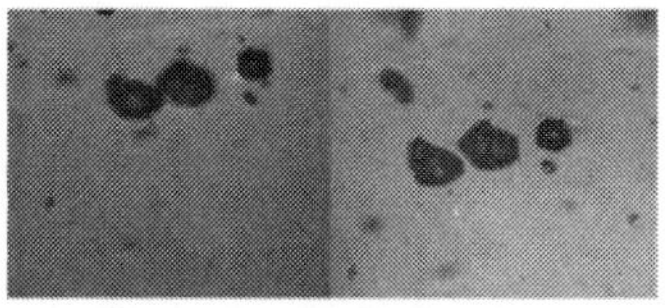

Case 1

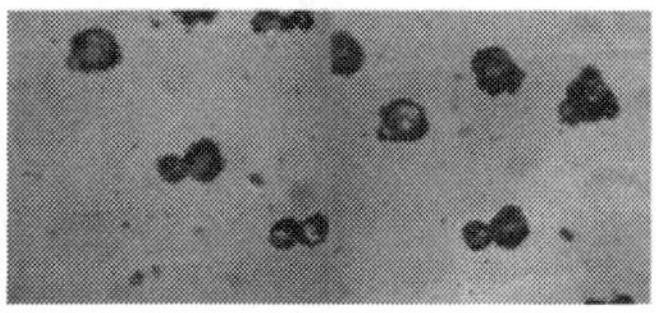

Case 2

Target parameters in the double-view images	Case 1	Case 2
The number of pseudo agglomerates	1	0
The number of true agglomerates	0	2
The number of unagglomerated particles	5	3
The number of primary particles involved in the agglomerates	0	4
The total number of all the primary particles	5	7
The agglomeration degree	0%	57.14%

Microscopic double-view images captured during the KDP (KH_2PO4) crystallization process

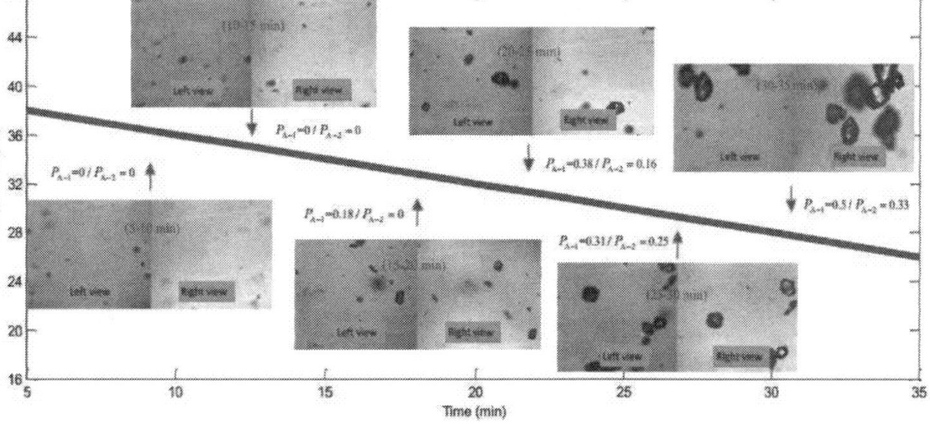

5. Conclusions & Outlook

大连理工大学 Dalian University Of Technology

- A synthetic image analysis method is proposed for monitoring particle agglomeration during a crystallization process.
- By defining an IDC index to detect possible agglomeration, a preliminary sieving algorithm is given to sieve out candidate agglomerates.
- Two texture descriptors, DCLBP and MDLBP, are proposed to recognize pseudo agglomerates with good accuracy.
- An efficient particle counting algorithm is established to assess the agglomeration degree.
- **It is interesting to develop a fast three-dimensional reconstruction method based on double-view images for CSD estimation in the future work.**

Yan Huo, Tao Liu*, Xue Z. Wang, Cai Y. Ma, Xiongwei Ni. On-line detection of particle agglomeration during solution crystallization by microscopic double-view image analysis. *Ind. Eng. Chem. Res.*, 2017, accepted, DOI: 10.1021/acs.iecr.7b02439.

Acknowledgement

Funding support

- **NSF China Grant (61633006)**
- **National Thousand Talents Program of China**

Institute of Advanced Control Technology

大连理工大学
Dalian University Of Technology

Thanks for your attention & comments!

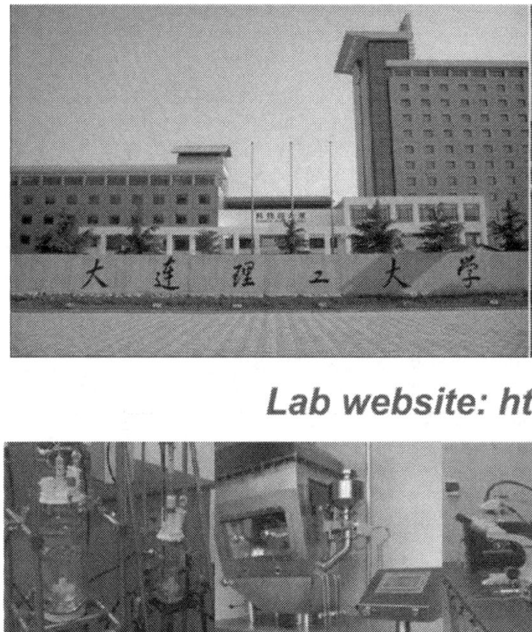

Lab website: http://act.dlut.edu.cn/

491432 Progress in Industrial Implementation of Optical Coherence Tomography (OCT) for Real-Time Evaluation of Pharmaceutical Coating Processes

Abstract:

The coating thickness of modified release coatings, enteric release coatings or coatings containing active pharmaceutical ingredients (APIs) is of biopharmaceutical importance. For modified release coating the coating thickness directly affects the dissolution profile. Thus, methods for in-line measurement of coating thickness can bring significant benefits with regards to end-point determination and monitoring of process trajectories.

Typical established methods to measure the coating thickness are weight and diameter gain of a limited number of drawn samples. In-line measurements mainly focus on spectroscopic techniques (NIR and Raman) or particle size distribution (PSD) analysis of beads. An emerging technology to measure the coating thickness is optical coherence tomography (OCT). OCT has several advantages compared to the state-of-the-art as it provides a direct measurement of coating thickness, only based on refractive index, not depending on chemometrics or on PSD statistical analysis. Additionally, OCT can deliver information of mean coating thickness and uniquely its variation within single samples and between the samples of a batch.

Optical coherence tomography has been presented as a proof-of-concept prototype at several conferences and papers are dealing with the data evaluation algorithms. This talk will focus on the application of an OCT system ready for industrial implementation in drum coating of tablets and fluid-bed coating of beads. Challenges in an industrial setting are for example the eccentricity of the drum, compensation of dispersion by thick process windows and the vast variation of coating materials, leading to different appearances of the OCT image.

The potential of OCT for monitoring and control of pharmaceutical coating processes is shown by in-line and at-line data for tablets and beads.

The authors would also like to acknowledge the contributions of Manuel Zettl, Philipp Jocham, Gregor Hauseder, Lukas Pichler and Michael Weissensteiner for experimental work and software development.

504512 Metabolic Modeling of Microbial Communities

Abstract:

Despite rapid progress, our ability to probe microbial communities for species-specific information of metabolite pools or metabolic fluxes remains limited. Most of the available information refers to species abundance and overall chemical exchanges with the environment. Computational modeling provides the missing link for explaining the metabolic interactions of microbial partners in communities in the context of conservation laws and optimality principles. While several modeling techniques have been developed for microbial communities, little emphasis has been placed on the need to impose a time-averaged constant growth rate across all members for a community to ensure co-existence and stability. To this end, we developed the SteadyCom optimization framework for predicting metabolic flux distributions consistent with the steady-state requirement. SteadyCom imposes restrictions on the allowable community species membership, composition and phenotypes. In contrast to the direct use of FBA, SteadyCom is able to reveal hidden interactions in a community as demonstrated in a community of four *E. coli* auxotrophic mutants and predict an abundance profile of a gut microbiota model with a good agreement to experimental data.

In separate efforts, we applied community metabolic modeling to analyze experimental observations in the gut microbiome. A working hypothesis regarding the production of a beneficial estrogen, (S)-equol by gut microbes was assessed in terms of metabolic consequences by simulating the interactions between three representative organisms (Bacteroidetes, Firmicutes and equol producer). The effect of diet and species abundance on the communityâ€™s ability to produce (S)-equol was explored and testable predictions were generated. In another case study, a ten-species gut microbiota metabolic model was developed to analyze the effect of bile salt hydrolase (BSH) activity and gut microbial productions under the treatment by an obesity-related drug, glycine-Î²-muricholic acid. Under the running hypothesis of the studies was that BSH activity expressed in *Lactobacillus* and Clostridia relieves the intestinal farnesoid X receptor (FXR) antagonism and in turn induces obesity, short-chain fatty acid productions and amino acid consumptions by the gut microbiota correlated to the experimental changes was predicted, supporting the hypothesis. These case studies allude to the promise of genome-scale metabolic modeling approach in microbiome research.

504513 Micro-Droplet Enabled Co-Cultivation, Isolation, Andinteraction Characterization of Microbial Communities

Abstract:

Microorganisms live in synergistic communities in nature and play key roles in numerous natural or engineered systems. For example, there are increasing evidences that microbes in our gut assist in digesting food, help prevent the invasion of pathogens, and stimulate the immune system. However, the basic physiological and ecological mechanisms governing these complex microbial communities remain largely unclear. We have been exploiting droplet microfluidics for co-cultivation and dissection of symbiotic microbial communities. By encapsulating subsets of a microbial community in nano-liter droplets, we provide highly parallel localized environments for co-growth and aim to identify microbial interactions. The conceptual and technological framework is being applied to the investigation of the human gut and vaginal microbiomes.

510413 A Synthetic Probiotic Engineered to Treat Phenylketonuria

Abstract:

Phenylketonuria (PKU) patients are unable to properly metabolize the amino acid phenylalanine (Phe), leading to neurotoxicity. Currently, few broadly-useful treatment options exist. To address this unmet need, we constructed a synthetic probiotic strain of the *Escherichia coli* Nissle, designated SYN-PKU. SYN-PKU is engineered to express Phe degradative genes that are transcriptionally activated in the anoxic conditions of the mammalian gut. In phenylketonuric mice, SYN-PKU catalyzed a significant decrease in blood Phe levels independent of dietary protein intake, suggesting active recirculation of systemic Phe to the intestinal tract. Phe enterorecirculation was also demonstrated for the first time in healthy non-human primates (NHPs), and recirculating Phe was converted by SYN-PKU to *trans*-cinnamate, which was hepatically metabolized to hippurate and excreted in the urine, allowing its use as a quantitative biomarker of probiotic activity. Importantly, SYN-PKU also inhibited serum Phe elevation in NHPs following an oral Phe dietary challenge. This work supports synthetic probiotics as a novel class of therapeutics for treating rare genetic metabolic disorders and defines a strategy for their progression.

498718 *In vitro* Fermentation to Understand Healthy and Stressed Gut Microbiome Metabolism

Abstract:

The abundant bacterial consortia within the gut microbiome plays a critical role in metabolism of dietary compounds that translate to beneficial influences on the host. From a military perspective, the gut microbiome serves as an ideal tool to not only enhance Soldier gut and immune health, but also improve survivability and performance. Our work employs *in vitro* fermentation as a means to elucidate the systematic processes of colonic bacterial metabolism of dietary inputs under both native and stressed conditions. The broad experimental capacity of *in vitro* fermentation enables the generation of multi-faceted data sets (i.e time- and dose-dependent, colonic domain specific, etc.) to gain a detailed understanding on bacterial community dynamics, inflammation and immunity. This talk will focus on the use of *in vitro* fermentation to investigate both the prebiotic potential of cranberry A-type proanthocyanidins (PAC) within a healthy microbiome, and the utilization of fermentable fiber and protein as a function of stress-induced dysbiosis on gut metabolism. Briefly, fecal inocula derived from three individuals was used to inoculate a nutrient-rich anaerobic media supplemented with purified PAC. Colonic fermentations were performed in parallel at 37°C under colonic domain-specific condtions using an automated bioreactor platform for precise control over a multitude of parameters. Total growth assessment through protein content analyses revealed domain-specific metabolism. After 5 hrs, PAC caused a 2.3-fold increase in total bacterial abundance under transverse colon conditions while ascending and distal colon showed 1.7 and 1.3 fold increase respectively. After 10 hrs, further PAC metabolism was not evident in ascending and distal colons; however, was still evident under transverse conditions until stationary phase was met. Bacterial population dynamics, determined through 16S rRNA sequencing, metabolic byproduct analysis via GC-FID and PAC metabolite discovery through LC/MS is underway. These analyses will enable identification of specific bacteria and the upregulation of certain SCFAs that collectively result in a beneficial health effect on the host. For stressed metabolism studies, fecal samples from volunteers exposed to an acute military-relevant stressor, a sudden change in diet for 21 consecutive days, were pooled (N=10) and inoculated in a nutrient-rich medium supplemented with resistant starch (RS II) or casein, at multiple dosages. Total growth, population and SCFA analysis for comparison to the pre-intervention microbiome state and to volunteers on a standard diet (N=10) will be discussed. Efforts to simulate the gut microbiome using continuous culture fermentation will also be presented. An understanding of gut microbiota metabolism dynamics under both stressed and unstressed conditions could direct future dietary supplementation strategies to build resiliency against military-relevant stressors and offset negative health and performance impacts of gut dysbiosis.

500987 Phylogenetic and Functional Profiling of the Microbiome Associated with a Phytoremediation System Targeting Air Quality in the Built Environment

Abstract:

We are spending majority of our time indoors, where we can be exposed to harmful chemicals and bacteria. The active modular phytoremediation system (AMPS) is a biowall where plants are grown aeroponically in a granular bulk media. Biowalls have strong potential for improving indoor air quality through bioremediation by both plants and their-associated microbes. Understanding whether the presence of soil affects the AMPS microbiome was targeted because the presence of soil in biowalls has been correlated to the presence of mold spores and increases in plant module replacement. We wanted to know if microbial community composition and function are maintained if soil is removed from the AMPS modules and plants are supported only by the granular bulk media. Microbiome samples were isolated from bulk media and roots from three modules without soil and bulk media, soil and roots from three modules containing both soil and bulk media. We determined the phylogenetic profile by sequencing the 16S rRNA gene. All microbiome samples were observed to contain diverse communities with many rare species, and were all dominated by members of the phyla *Proteobacteria* and *Planctomycetes*. These phyla are typical of plant-associated microbiomes. Small differences are observed in the number and diversity of species in roots from modules with and without soil. Samples were clustered based on abundant and rare community members by calculating pairwise Bray-Curtis and Canberra distances respectively. In both cases, root and soil samples from modules containing soil clustered separately from all other samples. Thus, the presence of soil does contribute to a minor increase in diversity and drives differences in community composition. We are exploring if genes capable of promoting plant growth or degrading pollutants such as volatile organics are present in the communities. These functions may be leveraged to enhance phytoremediation using AMPS-type systems. We are using a novel computational tool to query the presence of these target genes using the 16S rRNA data and available genomic and metabolic databases to make functional predictions. Predictions can be verified using PCR-based techniques. This study demonstrates that the microbiomes associated with plants grown with and without soil in the biowall are diverse. The results set the stage for forward engineering efforts to enhance the remediation capabilities of these microbiomes and improve indoor biodiversity.

488575 Localization of Matrix Production Reveals *B. subtilis* Biofilm Growth Mechanics

Abstract:

Bacterial biofilms are communities of microbial organisms that are encased in self-produced extracellular polymeric substances (EPS). The EPS matrix facilitates cellular signaling that leads to coordinated behavior at scales that are much larger than the single bacterium. This gives rise to sophisticated collective behavior that is absent in planktonic cells, as evidenced by the enhanced mechanical rigidity, advanced architecture for water retention and uptake of nutrients, and the elevated antibiotic resistance in biofilms. Existing studies have mainly focused on the regulatory networks involved in biofilm formation at the single-cell level. The relation between colony-scale physiological growth and behavior, with spatial and temporal patterns cell-di erentiation within the colony remains under-explored.

Here, using distinct fluorescent transcriptional reporters and *in situ* imaging of biofilm growth, I demonstrate that considering the spatial and temporal distribution of the matrix-producing, sporulating and motile cellular phenotypes is critical in resolving *B. subtilis* biofilm development. Our key result is that, beyond an initial transient phase, the production of the EPS matrix and the onset of sporulation is localized to a radially propagating front at the biofilm exterior. The width of the matrix producing front approaches a constant value at late times. By using a correlation analysis in the fluctuations of fluorescence reporter activity, we reveal that the propagating fronts correspond to a pair of travelling waves of *tapA* gene expression (matrix-related) and *sspB* gene expression (sporulation-related) of cells that are immobilized within the biofilm matrix. Importantly, we show that the spatiotemporal propagation of both traveling waves are coupled by a single length scale (~ 500 µm) and time scale (~1.4 h) throughout development. Furthermore, the spatial gene expression profiles of matrix production and sporulation exhibit a data collapse into a self-similar asymptotic shape, and the front displacement exhibits a square-root scaling - hallmarks of universal dynamics that governs matrix production and sporulation of cells within the front.

What are the implications of our discovery towards quantitatively characterizing physical mechanisms that give rise to the observed phenotypic diversity and physiological heterogeneity? In a single cell, it is well-known that nutritional stresses and starvation triggers a transition from matrix production to the transcription of genes involved in sporulation. Additionally, the physical expansion and spreading of the biofilm colony must result from the mechanical forces associated with EPS matrix production within the localized propagating front. Therefore, a full biomechanical model of biofilm shape and morphology requires coupling nutrient uptake and consumption that gives rise to localized patterns of matrix gene expression, with mechanical forces that govern colony expansion.

In the second part of the talk, I present a mathematical model that combines nutrient transport, a concentration dependent biomass production rate and a thin-film equation that balances biomass production and biofilm advection driven by mechanical forces due to osmotic pressure and cell growth. I show that the model reproduces both the shape and morphology of the biofilm colony, observed expansion rates, as well as the experimentally measured matrix-production and sporulation gene expression profiles. Finally, I discuss the broader implications of using this combined experimental and modeling approach to probe various aspects of understanding biofilm colony development, phenotypic diversity and growth strategies.

501420 Development and Characterization of a Promoter Library for Probiotic *E. coli* Nissle Using a High-Throughput Barcoding Approach

Abstract:

Knock-in of genes or pathways with therapeutic potential to the gut microbiota using engineered probiotic strains has the potential to revolutionize human health. However, the dosage of therapeutics delivered *in vivo* must be precisely controlled both spatially and temporally. This degree of tunability requires detailed knowledge of transcriptional regulator activity *in situ*. To date, engineered organisms have been largely developed in the aerobic, homogeneous, resource-rich, noncompetitive environment of the test tube. The gut environment is largely the opposite, so it is unclear how well existing databases of transcriptional regulators will function in gut-resident probiotics. However, accurate determination of gene expression in the gut context requires a vertebrate host, quickly making the traditional approach of testing each promoter in separate animals infeasible and unethical. We have optimized a high-throughput approach for *in situ* measurement of promoter activity using DNA barcodes, enabling expression assays for entire promoter libraries to be performed in single animals using high-throughput sequencing. To demonstrate this approach, we curated 32 candidate promoters through analysis of metatranscriptome datasets and promoter libraries, and transferred these synthetic constructs into *E. coli* Nissle. We confirmed that our barcoding strategy had an undetectable effect on gene expression, but interestingly found that each barcode exhibited a significant (up to 5-fold) and reproducible bias during amplification and sequencing. In response, we developed a model of this bias enabling accurate recapitulation of known ratios of barcode-containing DNA in standard curve experiments. Using this validated method, we then analyzed promoter activity in *E. coli* Nissle grown *in vitro* and *in situ* under a variety of environmental conditions (oxygen content, diet, and body site). Collectively, these experiments reveal exciting dynamics in synthetic constructs during passage through the mammalian gut and inform the design of probiotic strains engineered to improve human health. More broadly, these experiments also reveal PCR bias as a significant confounder of barcoded abundance measurements and suggest strategies to overcome its effects.

488055 Mutual Information to Inform Protein Library Design

Abstract:

Mutual Information to Inform Protein Library Design

Introduction:

Directed evolution has been widely used to develop new or improved proteins for research, diagnostic, and therapeutic applications. "You get what you select for" is the defining principle for success; however, this expression ignores the importance of the initial input library, which constrains the potential biomolecules that can be selected. Designed libraries must balance variability in positions that modify functionality against potentially disruptive mutations. Moreover, increasing variability at a single position is met with diminishing returns,[1] and combinatorial library diversity can easily dwarf practical limitations in sampling capabilities. We explore the use of information theory to inform library design and better optimize this balance. Specifically, we use mutual information (MI) as a tool to infer interdependence between sequence positions based on their entropies,[2] thus providing a means of objectively defining positions that provide specific functionality. In this study, we seek to determine if MI can provide insights for library design that a traditional design approach (e.g., one based on sequence alignment) would neglect, and we experimentally compare the results of these design approaches.

Materials and Methods:

For this study, libraries were designed based on the leucine-rich repeat (LRR) protein motif. These proteins mediate numerous protein-protein interactions in organisms throughout all orders of life, and thus are attractive as a class of therapeutics or nutraceuticals to improve human health. Additionally, their short, repetitive sequence provides a simple proof-of-principle model to test the capabilities of MI in library design. MI is calculated as

,

where represents residue at position and represents residue at position . MI benefits from large sample size, so sequence data of 4,930 LRR repeats were obtained from the SMART database.[3] Multiple sequence alignments were submitted to MISTIC,[4] a web tool for analyzing protein MI. These results were used to cluster amino acid positions based on their roles in defining protein structure or target binding.

Based on these results, three distinct LRR libraries were subsequently designed for selections with ribosome display,[5] a powerful cell-free method for directed evolution that can accommodate experimental libraries of up to 10^{13} sequences. Version HVP was fully sampled and it contained high amino acid variability in the binding paratope only.

Version MI was also fully sampled, but it had moderate amino acid variability in the binding paratope and light variability distributed in non-binding residues which were selected using the MI results. Finally, Version HVP+MI was a superset of Versions HVP and MI, with the same highly variable binding residues from HVP and the light, distributed variability from MI. This final library had increased theoretical diversity, but suffered from the practical limitation of not being fully sampled experimentally, underscoring the balance that must be struck in library design. The libraries were synthesized and validated for expected variability via Sanger sequencing of randomly selected clones.

The libraries were selected against a model target protein, chosen for its ease of handling and existing natural LRR binders, using ribosome display. Two successive rounds of selection against the target were performed. The output of each round is an RT-PCR product, so enrichment of binders against the target protein (relative to a negative control) is easily visualized via DNA gel electrophoresis. In this context, a low negative signal is expected when an enriched library is panned against an unrelated target.

Following the two selection rounds, the libraries were recovered using gel extraction and subsequently cloned into a sequencing vector. A set of clones from each enriched library was randomly chosen and sequenced. The sequences were analyzed using ClustalX2 and Microsoft Excel to study library enrichment. Interactive Tree of Life was used to create a phylogeny tree of the recovered sequences.[6] Selected individual clones were then subjected to one final round of ribosome display to check clone specificity. Selection stringency was increased in this analytical round, and competitive selection with unbound target was used as the background to demonstrate binder functionality.

Results and Discussion:

Figure 1 depicts a network visualization of MI among amino acids in the studied LRRs. As expected, highly conserved positions cluster into the core LRR motif, specifying the protein structure. One variable cluster represents the binding domain, since some of these positions are generally recognized as contributing to target recognition. A second variable cluster, termed the back-end, is likewise revealed. Also of note is residue 13, which retains high variability but does not share much MI with other residues, suggesting it evolves independently and should not be diversified in any library design. These results were used to design the three described 2-repeat LRR libraries. Naïve library sequencing confirmed that the desired variability in the three libraries was attained, so each was subjected to two rounds of ribosome display selection against the same target protein. The DNA gel in Figure 2 depicts RT-PCR products after second-round selection results for binding either the desired target protein or a negative control. The high signal-to-background ratio is suggestive of successful enrichment for binders in all three libraries.

Sequences of randomly selected clones from the enriched libraries were analyzed to understand the molecular nature of binder enrichment. Positions 8, 10, and 11 in the paratope showed heavy bias, indicating their importance in binding and consistent with the initial MI results. All three libraries showed bias towards G10/R10 and G11/R11, suggesting a similar pathway towards binder evolution. G10 and G11 enrichment are particularly interesting, as this residue appears to coevolve with a bulky amino acid at nearby positions to accommodate steric effects. The data set is too small to conduct further MI analysis on the enriched libraries, but this result highlights the key insights that coevolution analysis provides both before and after selections. Additionally, the dispersed variable positions in libraries MI and HVP+MI focused *away from* the consensus in library HVP. This enrichment would not be captured in a traditional library design that restricts diversity to only the paratope.

A phylogenetic tree of the selected sequences is shown in Figure 3. Only the Version MI library exhibited strong convergence, with two distinct sequences in particular appearing several times (sequence redundancy is quantified by the width of the horizontal line at the end of a branch). Since these two sequences were overrepresented, they were individually tested in one final round of single-clone ribosome display. With competitive selection as the background in this stringent round, the results in Figure 4 confirm that both sequences appear to be functional binders. Only the Version MI library, which was MI-informed and fully sampled, achieved this convergence under the selection conditions of this study.

Conclusions:

The use of MI in library design reinforces decisions from a traditional sequence information design approach, but it also provides new insights into residue positions that may or may not contribute to library function. Since initial library design is critical to successful biomolecule evolution but necessarily limited in its complexity, MI can serve as a useful quantitative tool to inform optimal design given these practical constraints. For a model LRR library, these results led to an alternative library design. Our study showed that a fully sampled library with MI-informed variability (Version MI) performed better than traditional approaches that either restricted variability to the paratope (Version HVP) or increased variability but prevented full sampling (Version HVP+MI). The fully sampled, MI-informed library converged most rapidly to functional binders in our experimental study.

Acknowledgements:

This work was supported by a 3M Fellowship, University of Minnesota Interdisciplinary Doctoral Fellowship, NIFA Predoctoral Fellowship (2017-67011-26204), and National Science Foundation grant (CBET1055231).

References:

[1]Shohei Koide and Sachdev S. Sidhu. *ACS Chem Biol.* 2009. 4(5) 325-334; [2]Louise C. Martin. *Bioinformatics*. 2005. 21 (22) 4116-4124; [3]Franco L. Simonetti *et al. Nucleic Acids Res*. 2013. 41 W8-W14; [4]Ivica Letunic *et al. Nucleic Acids Res*. 2012. 40 (D1): D302-D305; [5]Jozef Hanes and Adreas Plückthun. *Proc. Natl. Acad. Sci. USA* 1997. 94 4937-4942; [6]Ivica Letunic and Peer Bork. *Nucleic Acids Res*. 2016. 44 (W1): W242-245

Figure 1. MI network of LRRs. Nodes denote amino acid position in each repeat, and coloring represents position conservation. Edges depict high MI between positions. These data are used to cluster positions into the core LRR motif, binding residues, and back end positions. A cartoon representation of a single LRR repeat is provided for context.

Figure 2. Second-round ribosome display results (RT-PCR outputs from the round) for each library version binding either the target protein (+) or a dummy target (-).

Figure 3. Phylogenetic tree of sequences from all three enriched libraries. A convergent sequence cluster is highlighted, and two of these sequences are denoted for their dramatic over-representation in the enriched Version MI library.

Figure 4. Final ribosome display results for the two isolated Version MI sequences binding the target protein either with (+) or without (-) competitive selection in a stringent selection round.

491677 Blueprints for Diversifying Small Protein Scaffolds in the Context of Multiple Secondary Structures

Abstract:

Discovering proteins with new functionality gives rise to life saving therapeutics and is a necessity for expanding beyond current diagnostic and fundamental research applications. Our ability to quickly establish novel protein function is greatly hindered by the inefficiency at which combinatorial approaches explore potentially useful proteins. The protein sequence-function landscape is a sparsely populated, barren terrain of immense size. This unfortunate truth makes it difficult to randomly stumble upon a functional protein. Moreover, even when starting with a functional protein, making small random changes or mutations will tend to destabilize the protein. Improperly choosing one too many mutations will impede proper folding of the native structure and preclude function entirely. This concept has significant implications as it relates to protein library design. Libraries composed of broadly, uniformly diversified positions experience a large fraction of unstable variants. However, these effects can be mitigated by identifying and excluding detrimental amino acids from select positions as well as biasing inter- and intra-molecularly beneficial residues at library positions on a sitewise basis. Here, we will expand upon our recent work involving sitewise diversification of a three-helix bundle protein scaffold where we find biased amino acid diversity to be 13-fold more effective for discovering novel protein function compared to the traditional broad, uniform diversification scheme. We will then explore the common conclusions that exist between multiple distinct protein scaffold topologies. High-throughput sequencing, computational prediction of stability, and structural analyses guided the design of combinatorial libraries for affibody, Fn3HP, and Gp2. Thousands of high affinity ligands, having specific binding function to several targets, were isolated and deep sequenced. Amino acid preferences throughout binding hotspots, the effects of diversification on destabilization, and the importance of predictive ability of various input datasets will be discussed.

495092 A Structure-Based High-Throughput Screening Method for Multi-Step Enzymatic Reactions Using Optically-Guided Mass Spectrometry Profiling of Microbial Colonies

Abstract:

Matrix-assisted laser desorption/ionization (MALDI) mass spectrometry imaging (MSI) has been used for rapid phenotyping of enzymatic activities but mainly limited to single-step conversions. To engineer a multi-step reaction, however, modified intermediates must be accepted at each step of the catalytic sequence to obtain a final product. Engineering an individual enzyme in isolation ignores possible downstream effects, and MSI screening platforms primarily designed for single-step enzymatic reactions may be ill-suited for engineering multi-step pathways.

Herein we report a label-free, structure-based method for high-throughput engineering of multi-step biochemical reactions, based on optically-guided MALDI-MS analysis of bacterial colonies. The use of bacterial cells provides containment of multiple enzymes and access to substrates and cofactors via metabolism. Automatic MALDI-MS acquisition at a rate of 1~2.5 s per colony using machine vision for randomly distributed colonies allows simple procedures to prepare strain libraries without advanced liquid handling. To process the resulting large mass spectral datasets, computational tools were developed to survey structural variations and/or quantitative changes of analogs or congeners. Screening results are visualized in a manner similar to classical, colorimetric assays, allowing straightforward interpretation for researchers with limited MS experience.

The MALDI-MS profiling approach was utilized to screen both substrate and enzyme libraries for natural product biosynthesis. For analogs of a peptidic antibiotic, plantazolicin, multivariate analyses by t-distributed stochastic neighbor embedding successfully enumerated structural diversity and clustered colonies with isobaric residue substitutions. Follow-up analysis, including high-resolution MS and tandem MS, was readily performed from the same sample target following MALDI-MS screening. We identified both known and new plantazolicin analogs, and all assignment based on MS results were consistent with subsequent DNA sequencing analyses of the precursor peptide library. Separately, in the context of rhamnolipid synthesis, relative intensities of congeners with various lipid moieties were evaluated to engineer enzymatic specificity. The glycolipid profiles of each colony were overlaid with optical images to facilitate mutant recovery. Enzyme variants with altered specificities were successfully isolated to produce rhamnolipid mixtures with custom congeners compositions. In both cases, large populations of colonies were rapidly characterized at the molecular level, providing information-rich insights not easily accessible with traditional screening assays. Utilizing standard microbiological techniques with routine microscopy and MALDI-MS instruments, this simple yet effective workflow is applicable for a wide range of structure-based screening applications in microbial systems.

501159 Yeast Surface Display of Full Length Human Tau for Combinatorial Antibody Profiling

Abstract:

Microtubule-associated protein tau is expressed at high levels in neuronal cell types and is essential for stabilizing axonal microtubule assembly. In neurodegenerative disorders such as Alzheimerâ€™s disease, frontotemporal dementia, and Parkinsonâ€™s disease, mutations or post-translational modifications (PTMs) of tau cause its dissociation from microtubules and lead to formation of characteristic tau inclusions. In Alzheimerâ€™s disease, recent evidences show that tau modifications at specific sites cause toxicity in neurons and can be propagated through secretion and uptake of these species. Due to the large number of PTM sites that can potentially affect this process, many antibodies targeting these modified epitopes are being generated. However, methods to rapidly characterize the specificity of these antibodies are lacking. To address this challenge, we recently succeeded in displaying human tau on the surface of yeast cells. In this presentation, we show that physiologically important isoforms of the human tau, including the 441 amino acid full-length isoform can be displayed on yeast cells by genetic fusion to the cell wall protein Aga2. We demonstrate the use of yeast cells displaying human tau to determine the epitope of anti-tau monoclonal antibodies (mAbs) and quantify their specificity. By creating site-directed mutations, we confirmed or newly determined the epitope of several widely used anti-tau mAbs. Using known phospho-specific mAbs, we show that yeast displayed tau is largely non-phosphorylated at the target sites. These sites can be phosphorylated by treatment with kinases such as protein kinase A and glycogen synthase kinase 3 beta, which have been shown to phosphorylate human tau. We show that several antibodies previously thought to be phospho-specific have strong cross reactivity towards non-phosphorylated tau. We also demonstrate combinatorial approaches for de novo identification of anti-tau antibody epitopes using a high-throughput screening. These results show that yeast displayed human tau is a powerful platform for profiling and quality control of anti-tau antibodies.

500601 Saturation Mutagenesis Panning Libraries Enable Activity and Specificity Modulation of Microcin J25

Abstract:

Modern large-scale agricultural practices which incorporate high density farming with sub-therapeutic antibiotic dosing are a major contributor to the rise of antibiotic resistant bacterial infections. A leading foodborne pathogen, *Salmonella*, is present on as much as 20% of all packaged chicken products with 15% of those isolates found to be resistant to 4 or more antibiotics by the Centers for Disease Control. Alternatives to antibiotics, such as antimicrobial peptides (AMP) offer higher levels of specificity and activity towards targets of interest compared to traditional antibiotics. To this end, Microcin J25, a 21 amino acid AMP with a highly stable threaded lasso structure has been studied with single-site saturation mutagenesis to evaluate mutations contributing to the specificity and activity against multiple *Salmonella* and *Escherichia coli* strains.

A collection of 218 mutants (91% phenotypic coverage of the 240 possible variants) encompassing 12 positions in both the ring and loop of J25 was built and tested for activity versus *Salmonella* and *Escherichia coli* strains in a growth halo assay. Mutational tolerance of ring residues was lower than loop residues, with 24% and 50% of mutations retaining nonzero activity towards at least one target within the target organism test panel, respectively. Multiple mutations conferred differential response between the different targets. For example substitution of position 13 isoleucine with tyrosine retained wild-type activity level against *E. coli* JJ1887 (a urinary tract infector) while reducing activity against *E. coli* O157:H7 by 50%. In contrast, substitution with leucine improved activity by 10 and 80% against both *E. coli*, respectively, while reducing activity against *Salmonella enteritidis* and *Salmonella Tennessee by* 40%. In addition, substitution of position 3 alanine for cysteine displayed zero activity against both *E. coli* strains while retaining activity against both *Salmonella.* In total an improved specificity response was recorded for 78 of the 218 mutants ($p = 0.01$). A second generation library simultaneously mutating 6 positions between 2 to 5 phenotypic options was generated to assess the impact of multiple mutations on the specificity of J25. Of these multi-position mutants 53 of 1800 sampled retained some level of activity against the pathogen *Salmonella enteritidis*. The activity of these mutants will further be determined against a panel of commensal *E. coli* isolated from fecal and urine samples of healthcare patients, as well as pathogenic strains of enterobacteriaceae. This information will identify J25 variants with improved specificity for particular pathogenic targets.

These *in vitro* specificity improvements are likely to improve *in vivo* treatment efficacy by reducing clearance of commensal bacteria in the gastrointestinal tract of hosts. This both reduces selection of resistance within the microflora and reduces the likelihood of secondary infection from opportunistic pathogens, both problems with broad-spectrum antibiotic treatment. In addition, all variants generated were produced through the native microcin J25 synthesis pathway and could be adapted for use in previously described engineered probiotics utilizing J25 (Forkus et. al., 2017).

501464 Accessing the Location-Dependent Effects of Unnatural Amino Acids on Protein Expression and Activity with Cell-Free Protein Synthesis-Facilitated Rapid Screening

Abstract:

The incorporation of unnatural amino acids (uAA) can introduce novel functional groups into proteins with site-specific precision and control, with significant applications in basic science and protein engineering. However, uAA can decrease protein expression and functional activity depending on the location within the protein â€“ a process that is not yet completely understood and difficult to predict. Therefore, practical applications often necessitate a time-consuming optimization of uAA location by individual expressions, purification and evaluations for many variant proteins. Such optimization can be accelerated through an in vitro system. We utilize this technology to simultaneously screen 24 different T4-Lysozyme mutants with different uAA incorporation sites in a matter of hours. Screening data offered a mechanistic explanation to some effects of uAA incorporation on expression and activity. Such effects include the solvent accessible surface area, the protein secondary structure, the distance from the active site, and the mRNA sequence. These effects are evaluated and their relative importance compare. In the end, structure-based guidelines to aid in the rational selection of uAA locations are presented. However, it remains challenging to predict the overall impact of uAA incorporation a prior, justifying the need for the above described rapid screen.

Citation:

Schinn SM, Bradley W, Groesbeck A, Wu JC, Broadbent A, Bundy BC. Rapid In Vitro Screening for the Location-Dependent Effects of Unnatural Amino Acids on Protein Expression and Activity. Biotechnology and Bioengeering. 2017. Apr 11. doi: 10.1002/bit.26305. Accepted.

504142 RNA-Programmed DNA Methylation

Abstract:

Mammalian genomes exhibit complex patterns of gene expression regulated, in part, by DNA methylation. The study of the mechanisms of initiation, spreading and inheritance of DNA methylation is hindered without tools to efficiently and reproducibly direct methylation patterns to precise sites in model systems. Fusion of DNA methyltransferases (MTases) to DNA binding domains can bias methylation to CpG sites near the DNA binding domainâ€™s target sequence. However, since the MTase activity of the fusion does not require binding of the DNA binding domain to the chromosome, off-target methylation results. The most effective targeted MTases will require clear design rules to direct site-specific DNA methylation and minimize the unintended off-target DNA methylation. Here we report a targeted MTase composed of an artificially split CpG MTase (sMTase) with one fragment fused to a catalytically-inactive CRISPR/Cas9 protein (dCas9) that directs the functional assembly of sMTase fragments at the targeted CpG site. We precisely map RNA-programmed DNA methylation to targeted CpG sites as a function of distance and orientation from the protospacer adjacent motif (PAM). Expression of the dCas9-sMTase in mammalian cells led to predictable and efficient DNA methylation at targeted sites as directed by guide RNAs. Multiplexing guide RNAs enabled targeting methylation to multiple sites in a single promoter and to multiple sites in multiple promoters. This programmable de novo MTase tool might be used for studying mechanisms of initiation, spreading and inheritance of DNA methylation, and for therapeutic gene silencing.

498760 Engineering a Self-Assembling Peptide System Derived from Beta-Amyloid

Abstract:

The misfolding of beta-amyloid protein leads to neurotoxic aggregation in the brain. Amyloid fibrils and neurofibrillary tangles in the brain are hallmarks of Alzheimerâ€™s disease progression. The driving force of aggregation of the 40 or 42 residue peptide is an internal domain of seven residues known as the hydrophobic central domain. Our group has developed a novel, self-assembling, dual peptide system derived from this sequence. The novel peptides do not homo-assemble in solution, although rapid hetero-assembly occurs upon mixing. The resulting aggregates have been characterized extensively, including positive ThT fluorescence and beta-sheet secondary structure typical of amyloid aggregates. Beta-sheet composition and morphology can be altered with varying salt concentrations in the buffer and the addition of terminal capping groups. This system has useful applications in the development of novel functional biomaterials, as well as applications in identifying the mechanisms of amyloid toxicity in neurodegenerative diseases such as Alzheimerâ€™s disease.

492710 Competition between Ligands: How Retinol-Binding Protein and Beta-Amyloid Compete for Binding to Transthyretin

Abstract:

Transthyretin (TTR) is a soluble transport protein that circulates in blood and cerebrospinal fluid. One of its primary functions is the transport of retinol (vitamin A), via coordination with retinol-binding protein (RBP), to target cells. Once internalized, retinol and its oxidative products regulate transcription of a large number of genes. Retinol must also be delivered across the blood-brain barrier, where it plays essential roles in maintaining neuronal health. The mechanism by which retinol is taken up by cells or crosses the blood-brain barrier is still under investigation; both receptor-mediated and receptor-independent mechanisms have been proposed. TTR is a homotetrameric protein; each monomeric subunit contains two four-stranded β-sheets and a lone α-helix. RBP binds to multiple monomeric subunits in the tetramer, with contacts primarily involving TTR's α-helix. Although two RBP can bind per TTR, under biological conditions the binding stoichiometry is 1:1.

Recently it has been discovered that TTR may serve a second role, as an inhibitor of the Alzheimer-related beta-amyloid (Aβ). Neuronal dysfunction in Alzheimer's disease is generally believed to be caused by aggregation of Aβ. Multiple lines of evidence demonstrate that TTR binds to Aβ, thereby inhibiting Aβ aggregation and neurotoxicity. We have shown that TTR contains two Aβ binding domains: the solvent exposed α-helix and a β-strand that is in an interior hydrophobic cavity in the folded TTR tetramer. Our data indicate that Aβ binding to the α-helix triggers destabilization of TTR tetramers, thus opening up the interior binding sites on TTR for greater Aβ binding. This hypothesis is supported by strong evidence that TTR monomers (with exposed 'interior' sites) are significantly more effective than tetramers at binding Aβ and inhibiting its aggregation.

Importantly, the RBP binding site on TTR overlaps with the Aβ binding domain on TTR's α-helix. Thus, we wondered to what extent RBP and Aβ compete with each other for binding to TTR, or whether RBP interferes with TTR's inhibitory role against Aβ. We report that holo-RBP (hRBP, retinol complexed with RBP) significantly hindered TTR's protection against Aβ aggregation. This effect was characterized by thioflavin T fluorescence, light scattering, and nanoparticle tracking analysis. Next we determined that RBP does not directly compete with Aβ for binding to TTR. This is likely because the RBP-TTR binding stoichiometry is effectively 1:1, leaving unbound α-helix sites available for Aβ binding. The question remains, then, as to why RBP prevents TTR from inhibiting Aβ aggregation when it does not prevent Aβ binding. A plausible explanation arises from the recognition that RBP stabilizes TTR tetramer against dissociation into monomer subunits. Specifically, the tetramer stability enhancement by hRBP counteracts Aβ-induced tetramer dissociation, thus preventing access of Aβ to the interior binding sites on TTR. Fluorescence experiments, currently underway, are designed to provide direct evidence supporting this hypothesis. Our data demonstrate that RBP is competitive with Aβ in preventing TTR from carrying out its biological function of inhibiting Aβ aggregation. We are currently exploring whether the opposite is also true: does Aβ compete with RBP to prevent TTR from assisting with retinol delivery? Since there is evidence that Alzheimer's patients suffer retinol deficiencies, this is potentially a previously unrecognized mechanism

by which Aβ interferes with neuronal health.

496278 Understanding the Role of Glycine in Amyloid Protein Aggregation through Rationally Designed Protein Sequences

Abstract:

Aggregation of amyloid proteins is associated with a myriad of medical conditions including Alzheimer's disease (AD), Type II diabetes, and Parkinson's disease. While attributable to different amyloidogenic proteins, these proteins all share a common feature: a periodic glycine motif (GxxxG). This glycine motif, associated with increased backbone flexibility, is extended in a number of familial mutations associated with early onset AD (i.e. Flemish, Arctic) that significantly increase disease severity. A better understanding of the role that the glycine motif, and by extension chain flexibility, plays in protein aggregation will allow for new and give insight into therapeutic approaches.

In this study, the glycine motif in amyloid-β (Aβ), the amyloid protein associated with AD, is targeted via either extension, by introduction of additional periodic glycine, or contraction, by replacement of glycine with a bulkier amino acid. Modifications that extend periodic glycine include those that align with familial AD mutations (L17G/A21G) as well as off sequence mutations (V18G/E22G). Modifications that attenuate the glycine motif target G25 via replacement with either a bulky (G25I) or more conservative (G25A) mutation.

To examine how these mutations impact aggregation kinetics, aggregation was monitored via thioflavin-T (ThT), a dye that fluoresces in the presence of β-sheet structure, and aggregation curves fit using an adapted sigmoidal kinetic model that accounts for unique features observed at late stages of aggregation. Kinetically, extensions of the glycine motif (L17G/A21G, V18G/E22G) correlate with faster nucleation and a reduction in lag time. Interestingly, the on sequence extension (L17G/A21G) results in an increased equilibrium plateau while the off sequence extension (V18G/E22G) closely resembles the WT at plateau. Both L17G/A21G and V18G/E22G resulted in aggregates with a greatly diminished fibril morphology as visualized by TEM.

L17G/A21G aggregate were fractionated by size exclusion chromatography revealing an enhancement of small, soluble aggregates. Aggregate hydrodynamic radii (R_H), determined using dynamic light scattering, were smaller than WT; however, as observed with 8-anilino-1-naphthalenesulfonic acid (ANS), these aggregates have an unchanged surface hydrophobicity. Similarly, V18G/E22G soluble aggregates possessed a reduced R_H but were very similar in hydrophobicity to the WT, suggesting L17G/A21G and V18G/E22G, increases in glycine motif, corresponds with formation of smaller aggregates with a conformation very similar to that of WT Aβ.

Broadly, the reduction of the glycine motif has the opposite effect. Both G25A and G25I result in aggregates resembling traditional fibrils via TEM, however, differences are apparent in aggregation kinetics. The more conservative mutation, G25A, resembles the WT with a very similar kinetic lag but an enhanced plateau. However, G25I completely inhibits the formation of ThT-binding aggregates. As a result, circular dichroism was used to assess secondary structure characteristics revealing the propensity of G25I alone in creating β-sheet negative aggregates. Soluble aggregate fractions of G25A show a reduction in hydrophobicity

compared to the control, while G25I aggregates remain unchanged.

Taken together with well-studied observations that smaller aggregates are more physiologically active, these results support the hypothesis that an extended glycine motif, including that associated with familial mutations, may contribute to disease progression and early onset of the disease.

497749 N-Terminal Hypothesis for Alzheimer's Disease: Analyzing Dimers of a β peptide and its Protective and Causative Mutants

Abstract:

The amyloid (Aβ peptide) hypothesis dominates the pathogenesis of Alzheimer's Disease (AD), yet it is still controversial. This is due to a lack of a mechanistic path from the cleavage products of the amyloid precursor protein (APP) such as soluble Aβ monomer and soluble molecular fragments and to their deleterious effects on synaptic form and function. From a review of previously published work including aggregation kinetics and structural morphology, $A\beta_{1-42}$ clearance, molecular simulations, long-term potentiation measurements with inhibition binding, and the binding of a commercial monoclonal antibody, aducanumab, we hypothesize that the N-terminal domains of neurotoxic Aβ oligomers are implicated in causing the disease. We call this the "N-Terminal Hypothesis for AD". In order to understand the behavior of the N-terminus, we use brute force and replicate exchange molecular simulations guided by the recently reported NMR fibril structure of Aβ peptide and conduct kinetic aggregation and long-term potentiation measurements.

The structure of homozygous and heterozygous dimers of wild-type $A\beta_{1-42}$, and its causative (A2V) and protective (A2T) variants are analyzed. To study the role of N-terminus in the interactions between Aβ dimers, we constrain the central hydrophobic region and C-terminus to the known structure of Aβ fibers, while leaving the N-terminus free and examine the structural differences between these dimers. Furthermore, we performed unconstrained simulations of Aβ dimers to understand their association and the resulting structures. To analyze these dimers we examined the number of contacts and distances between the N-termini, and contact (2D heat) maps of their conformational landscape. Aggregation kinetics and long term potentiation of the three variants are also measured and compared in the presence and absence of monoclonal antibody, bapineuzumab, which targets N-terminal residues 1-5 of Aβ with high affinity.

499792 Computational Modeling of Amyloid Aggregation Kinetics to Gain Insights into the Effect of Ions on Amyloid Aggregation

Abstract:

The aggregation of amyloidogenic proteins through cross beta-sheet interactions depends upon two main factors: primary sequence of the protein and environmental conditions. The ordered fibrous aggregates formed as a result of this process are involved in a number of neurodegenerative diseases in mammals. The process of amyloid formation involves a two-step pattern of initial nucleation corresponding to a lag phase followed by a fiber elongation phase. This behavior has been modeled empirically using mathematical functions such as the logistic function, and other kinetic models, such as the Finke-Watzky (F-W) 2-step mechanism, which capture the near sigmoidal shape of the aggregation curve. However, these empirical models do not provide any information about the underlying mechanism of aggregation.

Previous work from our group on amyloid formation by Sup35NM protein containing the prion domain of yeast prion protein Sup35p has shown that strongly hydrated anions (kosmotropes) result in fast aggregation, whereas poorly hydrated anions (chaotropes) result in slower aggregation[1,2]. We have also previously shown that the ionic composition of the solution influences the formation of amyloid variants, or "strains"[2,3]. In this work, we have compared different population balance models for amyloid aggregation kinetics based on mechanisms comprising of nucleation, elongation, breakage, and secondary nucleation. Fitting kinetic data to population balance models has previously been used to arrive at plausible mechanisms for aggregation of amyloid beta peptides[4,5]. Here global fitting of aggregation data is performed to arrive at the most plausible model which is used to fit aggregation data obtained in the presence of different ions. Comparison of the fit parameters is then performed to understand how the presence of ions affects some of the above mentioned underlying processes.

[1] Yeh, V., Broering, J.M., Romanyuk, A., Chen, B., Chernoff, Y.O. and Bommarius, A.S., 2010. The Hofmeister effect on amyloid formation using yeast prion protein. Protein Science, 19(1), pp.47-56. doi: 10.1002/pro.281

[2] Rubin, J., Khosravi, H., Bruce, K. L., Lydon, M. E., Behrens, S. H., Chernoff, Y. O. and Bommarius, A. S., 2013. Ion-specific effects on prion nucleation and strain formation. Journal of Biological Chemistry, *288*(42), pp.30300-30308. doi: 10.1074/jbc.M113.467829

[3] Sharma, A., Bruce, K. L., Chen, B., Gyoneva, S., Behrens, S. H., Bommarius, A. S. and Chernoff, Y. O., 2016. Contributions of the Prion Protein Sequence, Strain, and Environment to the Species Barrier. Journal of Biological Chemistry, 291(3), pp.1277-1288. doi: 10.1074/jbc.M115.684100

[4] Cohen, S.I., Linse, S., Luheshi, L.M., Hellstrand, E., White, D.A., Rajah, L., Otzen, D.E., Vendruscolo, M., Dobson, C.M. and Knowles, T.P., 2013. Proliferation of amyloid-β42 aggregates occurs through a secondary nucleation mechanism. Proceedings of the National Academy of Sciences, 110(24), pp.9758-9763. doi: 10.1073/pnas.1218402110

[5] Meisl, G., Yang, X., Hellstrand, E., Frohm, B., Kirkegaard, J.B., Cohen, S.I., Dobson, C.M., Linse, S. and Knowles, T.P., 2014. Differences in nucleation behavior underlie the contrasting aggregation kinetics of the Aβ40 and Aβ42 peptides. Proceedings of the National Academy of Sciences, 111(26), pp.9384-9389. doi: 10.1073/pnas.1401564111

500654 Human Islet Amyloid Polypeptide: Identifying Early-Stage Aggregation Mechanisms through Molecular Simulation

Abstract:

Human islet amyloid polypeptide (hIAPP or human amylin) is implicated in the onset of type II diabetes, which afflicts millions of people worldwide. hIAPP is an amyloidogenic polypeptide which self-assembles into amyloid fibrils. However the prefibrillar species of hIAPP are of particular interest due to their cytotoxic properties. It is essential to investigate the mechanisms of early-stage hIAPP aggregation in order to better understand the role of hIAPP in disease onset, as well as to design effective therapeutics. Using atomistic molecular dynamics simulations with a host of advanced sampling techniques, we reveal mechanistic details of hIAPP dimer and trimer formation. Metadynamics calculations uncover a free energy landscape for the hIAPP dimer, which indicates multiple possible dimerization pathways. We then use finite temperature string method calculations to identify probable pathways for assembling the hIAPP dimer and trimer. We also examine the effects of physiological salt concentration on the aggregation process. Results shed light on the mechanisms and energetics of hIAPP aggregation.

504096 The Role of Cholesterol in Membrane Protein Activity

Abstract:

G-protein coupled receptors (GPCRs) represent the largest family of membrane proteins in the living world, and serve key roles in cellular signaling to diverse stimuli. For this reason, GPCRs are the largest target of therapeutics on the market. Despite this importance, our understanding of the relationship between structure and function, including downstream signaling, is rather limited. Since the first human GPCR, Î²$_2$-adrenergic receptor, was crystallized in 2007, a specific cholesterol binding site between helices I, II, III and IV â€“ the cholesterol consensus motif (CCM) â€“ was established, and as many as 25% of all class A GPCRs have been predicted to have a specific interaction with cholesterol. The role of the lipid environment including cholesterol may be very important in drug discovery and efficacy, and thus has been a focus of our laboratory.

For several years, we have been studying the role of cholesterol on the activity of the adenosine A_{2A} receptor ($A_{2A}R$), which is a class A GPCR whose natural ligands include adenosine and caffeine. In this talk, I will describe our studies of the role of bulk cholesterol on activity in human embryonic kidney (HEK-293) cells including ligand binding and downstream activity. I will also describe collaborative computational and experimental studies on mutagenesis of putative cholesterol interaction sites to test the specific interaction between cholesterol and $A_{2A}R$ both at the CCM and alternative proposed interaction sites predicted by both coarse grain and atomistic simulations.

497391 Engineering Selective Immunoglobulin E Inhibitors for Peanut Allergies

Abstract:

Food allergies are a growing concern in the US; drastic increases in patients who have severe allergies to such common foods as peanuts, tree nuts, milk, eggs, wheat and shellfish are a serious health concern that causes problems for patients, doctors and the food processing industry. This problem is further compounded by the lack of pre-reaction therapeutics; instead, patients can only manage symptoms after allergen exposure. We propose a new twostep, methodology for developing patient-specific therapeutics for food allergies by inhibiting allergen specific immunoglobulin E (IgE) molecules, the molecule primarily responsible for triggering severe allergic reactions. One of the challenges in allergen specific IgE inhibition is the clonal variability and the variety of potential IgE binding epitopes on food allergy proteins. Using our method of inhibitor development, each patientâ€™s serum is screened using nanoallergens, liposomes displaying potential IgE binding epitopes, in order to assess and determine the most immunodominant IgE binding epitopes for each patient. Next, we synthesize specific inhibitions, called covalent heterobivalent inhibitors (cHBIs) using the information gleaned from the nanoallergen assessment. The result is a personalized allergy treatment that can potently inhibit allergen specific IgEs with a combination of bivalent binding to improve specificity and covalent interactions to provide long term inhibition. We have demonstrate the utility of this technique by evaluating four peanut allergy patientsâ€™ sera using nanoallergens and subsequently crafting specific cHBIs for the most potent peanut protein epitopes for these patients. This cHBI cocktail showed specific binding only to allergy patient IgEs and *ex vivo* inhibition of degranulation using a standard cellular assay. This study provides the groundwork for the nanoallergen-cHBI method for developing targeted allergy therapeutics for any food allergy.

501274 Glycan-Specific Antibody Discovery through Designer Glycopeptide Immunization

Abstract:

Sugar groups, or glycans, specific to or enriched on particular cell types and strains, such as tumor-associated carbohydrate antigens and immunogenic lipopolysaccharide on pathogenic bacteria, serve as identification markers that can be leveraged for targeting with monoclonal antibodies. In addition to being a powerful laboratory tool for purification, screening, and imaging, monoclonal antibodies have in recent years become increasingly important in treatment of cancer, autoimmune disorders, and an expanding list of other indications. However, to date, therapeutic monoclonal antibodies have almost exclusively targeted proteins, and there continues to be a lack of anti-glycan monoclonal antibodies for use in the laboratory and clinic. The poor immunogenicity of glycans contributes to this limited availability. Unlike protein antigens, carbohydrates often elicit an immune response independent of T cells, characterized by a lack of class-switched high affinity antibodies relevant for discovering therapeutic and diagnostic monoclonal antibodies.

Historically, monoclonal antibodies have been produced using hybridoma and display technologies. However, the emergence of next-generation sequencing approaches offers a rapid, refined antibody discovery alternative compared to the labor-intensive, time-consuming cell screening or panning steps required with current methods. High-throughput sequencing of entire cell populations are now widely accessible methods for obtaining sequences of target-binding immunoglobulin genes. This has previously been used to identify high-affinity anti-protein antibodies, and we propose that this approach can be expanded to glycan targets through utilization of glycosylated peptide antigen that elicits a robust T cell-dependent immune response while mitigating off-target antibody production.

Here we demonstrate that *Escherichia coli* engineered to synthesize designer glycans serves as a tractable platform for the production of rationally designed glycopeptides. These peptides contain motifs permitting *in vivo* glycosylation linked to amino acid epitopes known to bind with cell surface MHC II molecules necessary for a T cell-dependent humoral immune response. Using the O-antigen polysaccharide (O-PS) from *Francisella tularensis* as a model pathogenic bacterial glycan, we show immunization of mice with purified *E. coli*-expressed glycopeptides results in class-switched high affinity antibodies specific to the O-PS antigen. Analysis of the repertoire of immunoglobulin genes, obtained via high-throughput sequencing of the antibody-producing splenocytes isolated from immunized mice, can then be used to elucidate and recapitulate monoclonal antibodies with high affinity towards O-PS. Due to the modularity of the glycopeptide platform, whereby new glycans can be conjugated, immunized, and sequenced, this rapid monoclonal antibody pipeline can be exploited for use beyond our model O-PS antigen to other carbohydrate targets of interest.

500816 Localized Multi-Component Delivery Platform Generates Local and Systemic Anti-Tumor Immunity

Abstract:

Introduction: Tumors employ complementary overlapping mechanisms to evade immune surveillance, many emerging cancer immunotherapies attempt to target multiple pathways to eradicate malignant cells. Though modulation of independent pathways by simultaneous administration of multiple immune has shown great promise, the clinical impact remains limited due to severe toxicity associated with high systemic dosages. Localized injections at tumor sites may reduce systemic toxicity, however such approaches still require frequent injections of large amounts of drug (soluble proteins are rapidly diluted by lymphatic drainage and systemic circulation). For example, intra-tumoral injections of anti-CTLA-4 and anti-OX40 plus CpG were successful in generating systemic immunity to tumors, but required daily intratumoral injections of 100 μg/day/mouse for 5 days. Therefore, novel platforms for efficient delivery of combination therapies at lower doses would be enabling. Here, we describe immunomodulatory molecule delivery system (iMods), which provides sustained extracellular delivery of checkpoint inhibitor (anti-PD-L1) and, targeted intracellular delivery of a tumor antigen (OVA) along with adjuvant (poly(I:C)), and the indoleamine deoxygenase (IDO) inhibitor 1-MT.

Methods: Contralateral melanoma B16F10 model was utilized. iMods containing OVA-poly(I:C), 1-MT, anti-PD-L1, in all possible combinations were injected in the periphery of the tumor (after the tumor was palpable). The size of the tumor was monitored till 30 days or till tumors were unbearable.

Results: In melanoma bearing mice, combinatorial delivery of these factors with iMods led to regression of both treated and untreated (contralateral) melanoma tumors and 100% survival. These promising therapeutic outcomes were attributed to significantly enhanced ratios of anti-tumor CD8 T-cell / tumor-protective regulatory T-cell (Treg) in tumors and tumor draining lymph nodes.

Figure. (A) Melanoma mouse model. **(B)** Survival curve for treatment groups and CD8/Treg ratios.

Conclusions: In summary, iMods develop potent systemic in vivo immune responses in the form of reduced numbers of Treg, and increased numbers of active CD8 T-cells in the tumor environment. Moreover, iMods formulations were able to generate systemic anti-tumor immunity via one-time local injection without lethal toxicity. iMods, as a platform, may be useful for the comparing a wide variety of combinatorial immunomodulating molecules in the context of cancer immunotherapy.

References: 1. *Acharya et.al. Advanced Functional Materials 2017,* **2.** *Acharya et.al. Journal of Drug Targeting 2016.*

496464 Microsphere-Assisted Peptide Screening (MAPs): High-Throughput Identification of Promiscuous MHCII-Binding Peptides for T-Cell Epitope Vaccine Designs

Abstract:

T cells recognize antigenic peptides as part of a peptide-major histocompatibility complex (pMHC) via T-cell receptors (TCRs). The cognate peptide is termed a T-cell epitope and can be used to develop a T-cell epitope vaccine. However, before T-cell epitopes can be identified and used as vaccines, candidate peptides must first be presented on MHC molecules. Therefore, peptide-MHC binding is the most discriminating step in the identification of T-cell epitopes and significant effort has been dedicated to identifying peptides that bind a specific MHC protein. This has proven to be a challenge in the field as MHC proteins are extremely polymorphic which results in different T-cell epitope repertoires in different persons. MHC allelic diversity is particularly challenging in the context of designing T-cell epitope vaccines, where in order to provide protection to a diverse population, the selected peptides should bind promiscuously to many different MHC alleles. Identifying peptide-MHC binders is even further complicated for class II MHC (MHCII), as MHCII have open-ended binding grooves that can accommodate long, variable-length peptides. Several computational and experimental systems have been developed to determine if a specific peptide will bind to an MHC allele. Current computational systems using complex matrix based systems and machine learning algorithms are incredibly high-throughput and can accurately predict peptide binding to many MHCI alleles. Despite the successful application of computational approaches to peptide-MHCI binding, even the most sophisticated *in silico* systems struggle to accurately predict peptides binding to MHCII. In contrast, most experimental systems are capable of providing incredibly accurate quantitative peptide-binding data for both MHCI and MHCII alleles. However, these systems tend to be labor-intensive, low-throughput, and are often applied to limited sets of MHC alleles and thus can miss promiscuously-binding peptides.

To address some of the limitations of current experimental systems used to screen peptide-MHC binding, we developed a novel high-throughput screening method to rapidly identify promiscuously binding peptides. Microsphere-assisted peptide screening (MAPs) coupled with our modular baculovirus-MHCII expression system allows one to screen large peptide libraries for binding to many MHCII alleles. The MAPs approach consists of performing peptide-exchange reactions in which an epitope-tagged peptide is loaded on to one of four common MHCII alleles. These exchanged-pMHC complexes are loaded to the surface of microspheres and analyzed using flow cytometry to determine the relative peptide binding. We validated the MAPs system by screening a 20mer peptide library with known binding data and analyzed the results using receiver operating characteristic (ROC) curves. This ROC analysis demonstrated the MAPs system is rapid and accurate with an area under curve greater than 0.90. We then applied the MAPs approach to successfully identify four promiscuously-binding peptides within a 26-member peptide-library of the rotavirus capsid protein VP7. Interestingly, structural analysis of the rotavirus VP7 protein revealed that these four promiscuously-binding peptides share a number of residues with known antibody neutralization sites on the native VP7 trimer. Finally, the complete peptide-binding dataset for the VP7 library screen was compared to peptide binding predictions by computational methods within the IEDB Analysis Resource including *SMM-align*, *NN-align*, and

NetMHCIIpan. The predictive-algorithms captured many of the peptide-binders observed using the MAPs system; however, the computational predictions had a false-negative rate of roughly 20% including the four promiscuously binding peptides discovered using MAPs. Therefore, in addition to serving as a robust experimental platform for identifying promiscuously binding peptides for potential T cell epitope vaccines, MAPs should also contribute to the improvement of computational systems that predict peptide-MHCII binding *in silico*.

494248 Kinetic Analysis of CD3Î¶ and CD28 Chimeric Antigen Receptor T Cell Activation

Abstract:

As T cell immunotherapy applications have expanded, it has become increasingly important to better understand the signaling events that lead to T cell activation. For example, T cells bearing chimeric antigen receptors (CARs) containing the CD3Î¶ T cell activating domain and the CD28 co-stimulatory domain have emerged as promising cancer therapies; but, to expand this therapy to a wider patient population, the CAR proteins need to be better engineered to control T cell activation. The six tyrosine sites on CD3Î¶ and four on CD28 are phosphorylated by lymphocyte-specific protein tyrosine kinase (LCK), resulting in downstream signaling that leads to T cell activation; however, the specific mechanisms that control LCK CAR binding and catalytic activity are not well defined. These ten phosphorylation sites on the CAR contribute to different signaling events in the T cells, so understanding the mechanisms through which these sites are activated can inform the best strategies for engineering the next generations of CARs. In this study, we have used quantitative phospho-proteomic mass spectroscopy and computational modeling to quantify the site-specific kinetics of CD3Î¶ and CD28 phosphorylation.

We constructed the model using BioNetGen, a rule-based formalism that allows us to account for the many species that arise due to the complexity of the multiple phosphorylation sites on the CARs. The model is implemented as a set of ordinary differential equations in MATLAB and fit to experimental data generated using an *in vitro* reconstituted membrane system that mimics the two-dimensional interactions that occur in T cells [1]. We used phospho-proteomic mass spectrometry to measure the level of phosphorylation at each individual tyrosine residue in the system over time. We have also tested how mutations in the CAR and changes in the lipid microenvironment could affect the model kinetic rates. The correct model predictions were validated experimentally. By combining this work with our existing model of LCK autoregulation [2], we can predict how changing CAR structure and microenvironment will affect the rate of T cell activation. Thus, we are establishing a quantitative framework that can be used to design optimal CAR activation of T cells for immunotherapy.

Citations

[1] Hui, E. and R. Vale. *Nat. Struct. Mol. Biol.*, 2014, 21(2), 133-142.

[2] Rohrs, J. A., Wang, P., and S. D. Finley. *Cell Mol. Bioeng.*, 2016, 9(3), 351-367.

491803 Catch Bonds at T Cell Interfaces

Abstract:

An open problem in immunology is how T cells, which orchestrate the adaptive immune response, reliably distinguish and respond to small numbers of antigenic ligands in a sea of non-stimulatory ligands. T cell activation is thought to depend largely upon the binding kinetics between the T cell receptor (TCR) and membrane-presented ligands (pMHC) on other cells, but binding kinetics *in situ* are influenced by physical properties of the cell-cell interface and are difficult to characterize experimentally. Recently, experiments have shown that TCR-pMHC bonds experience a variety of forces at the cell-cell interface and that stimulatory bonds behave as catch bonds, with an average bond lifetime that initially increases with an increasing tensile force. Because T cells are initially stimulated by small numbers of TCR-pMHC complexes, it is important to understand how bond formation drives dynamic changes in membrane organization, how these changes affect forces experienced by the bonds, and ultimately how these forces affect bond lifetimes.

Here, we use computational methods to characterize time-dependent forces on TCR-pMHC bonds in response to changes in the membrane shape and the organization of other surface molecules. We then determine the distributions of bond lifetimes using recent force-dependent lifetime data for T cell receptors bound to various ligands. Strong agonists, which exhibit catch bond behavior, are markedly more likely to remain intact than an antagonist whose average lifetime decreases with increasing force. Thermal fluctuations of the membrane shape enhance the decay of the average force on a bond, but also lead to fluctuations of the force. These fluctuations promote bond rupture, but the effect is buffered by catch bonds. When more than one bond is present, the bonds experience reduced average forces that depend on their relative positions, leading to changes in bond lifetimes. Our results highlight the importance of force-dependent binding kinetics when bonds experience time-dependent and fluctuating forces. This is suggestive from a mechanistic standpoint, as force-dependent regulation of TCR-pMHC binding times provides a physical mechanism that could help T cells discriminate between self and foreign ligands.

505035 Marrow-Derived & Surface-Engineered Macrophages Engorge, Accumulate, and Differentiate in Antibody-Targeted Regression of Solid Tumors

Abstract:

Marrow-derived macrophages are highly phagocytic, but whether they can also be made to traffic into solid tumors and engulf cancer cells is questionable, given well-known limitations of tumor-associated macrophages (TAMs). Here, SIRPa on macrophages from mouse and human marrow was inhibited to block recognition of its CD47 ligand, a "marker of self" on all other cells. These macrophages were then systemically injected into mice with fluorescent human tumors that had been antibody targeted. Within days, the tumors regressed, and fluorescence analyses showed that the more the SIRPa-inhibited macrophages engulfed, the more they accumulated within regressing tumors. Human-marrow-derived macrophages engorged on the human tumors, while TAMs were minimally phagocytic, even toward CD47-knockdown tumors. Past studies have opsonized tumors in situ with antibody and/or relied on mouse TAMs but have not injected SIRPa-inhibited cells; also, unlike past injections of anti-CD47, blood parameters remained normal here. Consistent with tumor-selective engorge-and-accumulate processes in vivo, phagocytosis in vitro inhibited macrophage migration through micropores that mimic features of dense 3D tissue. Accumulation of SIRPa-inhibited macrophages in tumors favored tumor regression for 1–2 weeks, but donor macrophages quickly differentiated toward non-phagocytic, high-SIRPa TAMs. Analyses of macrophages on soft (like marrow) or stiff (like solid tumors) collagenous gels demonstrated a stiffness-driven, retinoic acid-modulated upregulation of SIRPa and also a mechanosensitive nuclear marker, lamin-A. Mechanosensitive differentiation was similarly evident in vivo and likely limited the anti-tumor effects, as confirmed by re-initiation of tumor regression by fresh injections of SIRPa-inhibited macrophages. Macrophage motility, phagocytosis, and differentiation in vivo are thus coupled.

490572 Advanced Control Strategy Approaches for Continuous Pharmaceutical Manufacturing: The Regulatory Landscape

Abstract:

The FDA believes that strategic efforts to encourage and sustain improvements in pharmaceutical manufacturing quality are critical to the FDA mission of making safe, effective and high-quality drug products consistently available to the patients who need them. Over the past decade since the publication of the PAT guidance, there has been significant progress in modernizing the pharmaceutical manufacturing sector. There has been a general move away from fixed uni-variate parameter specifications towards a multivariate design space approach and the adoption of in-process measurement systems to aid process development. At the same time industry and regulators are being confronted with and increasing complex challenges including unprecedented drug recalls and shortages. The implementation of advanced control strategies can help address these challenges by increasing the efficiency, flexibility, and robustness of manufacturing processes. In FDAâ€™s experience to date, continuous pharmaceutical manufacturing is serving as a driving force for modernizing control strategy approaches including model based control, multivariate monitoring, analysis of large of data sets, and Real Time Release Testing (RTRT). Process models that can capture the dynamic relationships between process parameters, raw materials, and product attributes in a quantitative and predictive manner support the development and implementation of these advanced control strategy approaches. While there are no regulatory barriers for adopting advanced control strategy approaches, there still may be hurdles encountered when fitting these approaches into the existing regulatory framework. In this presentation, scientific and regulatory considerations for advanced control approaches designed to mitigate identified risk areas are discussed including performance criteria and lifecycle considerations. Continuous drug product and drug substance manufacturing case studies are utilized to highlight these considerations.

495150 Pharmacy-on-Demand: A Continuous Revolution in Pharmaceutical Manufacturing?

Abstract:

Unlike its counterparts in the petrochemical, food and drink industries; pharmaceutical manufacturing still applies a non-continuous or â€œbatch-wiseâ€ approach. Thus, current pharmaceutical processes are ill-equipped to react to market forces or surges in demand due to humanitarian outbreaks. Active pharmaceutical ingredients (API) are world travelers with complex supply chains and long batch syntheses, meaning it can take months to manufacture and deliver the final drug product to the patient. This enormous space/time demand can result in reduced profits and inflated capital and operating expenditure.

With the era of blockbuster drugs apparently over and Pharmaceutical companies racing to the bottom of the pricing market in recent times, a paradigm shift towards continuous manufacturing may provide the answer. The benefits of flow chemistry in the synthesis of API has now been successfully demonstrated on a number of occasions. Likewise, many Drug Product unit operations are currently conducted in a continuous manner. The present challenge is now the implementation of an end-to-end manufacturing strategy that can allow these methodologies to work in concert, in a streamlined manner that enables drug substance and drug product production in one manufacturing platform.

Pharmacy on Demand (POD) seeks to address these manufacturing inefficiencies through the use of a compact and reconfigurable platform, capable of multi-step syntheses for a catalogue of API. Substantial advances in complex reaction telescoping, real time formulation, reaction engineering and pumping technologies have enabled the design, construction and implementation of the platform. Tight control of process inputs and the implementation of in-line process analytical technologies enables high volumes of APIs (Atropine, Benadryl, Ciprofloxacin, Diazepam, Doxycycline, Lidocaine and Ibuprofen) to be manufactured over a three-week campaign period. This timeframe involved the synthesis, recrystallization and formulation of each API to their correct dosage form. These solid or liquid formulated Drug Products are produced to a standard that passes all USP specifications.

496111 Platforms for Integrated Continuous Drug Subtance Design and Manufacture

Abstract:

As pressure grows on pharmaceutical companies to deliver increasing levels of process performance, reduced capital and manufacturing costs the industry is looking more and more to continuous processing to help tackle some or all of these challenges.

Within this presentation case studies will be presented where unique continuous processing hybrid platforms have been utilised for the development and operation of continuous processes. These processes included heterogeneous and homogenous chemistries, work-up/distillations, crystallization and isolation. Specifically within the case studies presented details will be outlined, via PAT integration, engineering kinetic and statistical models, experimental and lab rig design via 3D printing, required to enable the design, optimisation and operation of these platforms . During the presentation the design considerations over CSTRs, PFRs, hybrid or novel 3D printed reactor design configurations will be presented and how such design selections can impact or tailor the design properties of the intermediates and importantly the final API solid state characteristics.

496380 Residence Time Distribution and Material Traceability for Continuous Drug Substance Processes

Abstract:

Title: Residence Time Distribution and Material Traceability for Continuous Drug Substance Processes

Authors: Christopher Polster[1], Venkata Ramana Reddy[3], Carla Luciani[1], Stephen Jeffery[2], Martin Johnson[1], Kevin Chinn[4], Hod Finkelstein[4]

[1] *Eli Lilly & Co., Indianapolis, IN-46221, USA*

[2] *Eli Lilly S.A. Irish Branch, Dunderrow, Kinsale, Co. Cork, Ireland, P17 NY71*

[3] *Lilly Capability Center India, Bangalore, India*

[4]*Trutag Technologies, Inc., 2045 Lauwiliwili St, Unit 301, Kapolei, HI 96744*

Abstract

The concept of residence time distribution (RTD) to analyze chemical reactor performances appeared earlier 1930â€™s.[1] Danckwerts materialized the field by defining the most common RTDs.[2] Currently, the use of RTDs to understand mixing characteristics of the units/trains as well as material traceability and batch definition is an established approach.

With the rapidly growing adoption of continuous manufacturing processes in the pharmaceutical industry, the understanding of RTD through modeling and characterization is imperative to support both R&D activities and long term commercial manufacturing of both drug substance and drug product.[3],[4],[5]

In this work, some of the techniques used to measure RTD are described. Also, libraries were developed to provide a tool for R&D and manufacturing scientists to develop and simulate plant flowsheets that can be used for material traceability, determining diversion boundaries, and guide decision trees for diversion scenarios. Specific case studies will be shared.

---

[1] MacMillin, R.B., Weber, M. The Theory of Short-Circuiting in Continuous-Flow Mixing Vessels in Series and the Kinetics of Chemical Reactions in such Systems. *Transactions of American Institute of Chemical Engineers* **1935**, 31,

409-458.

[2] Danckwerts, P.V. Continuous Flow Systems. Distribution of Residence Times. Chem Eng. Sci. 1953, 2, 1-13.

[3] Manley, L. Manufacturing Platformâ€™s Residence Time Distribution Using Major and Minor Unit Formula Excipient Step Changes and Soft Sensors. *2016 AIChE Annual Meeting* (342c).

[4] Garcia-Muñoz, S., Luciani, C., Shi, Z., Roth, W., Butterbaugh, A. Hanson, J., Manley, L., Barnes, L., Leavesley, I., Almaya, A. Model-Aided Development of a Continuous Drug Product Manufacturing Process.

[5] Engisch, W., Muzzio, F. Using Residence Time Distribution (RTDs) to Address the Traceability of Raw Materials in Continuous Pharmaceutical Manufacturing. *J. Pharm Innovation* **2016**, 11, 64-81.

494238 A Miniature CSTR Platform for Continuous Processing of Multiphase Systems

Abstract:

Continuous manufacturing has emerged as a powerful tool for producing fine chemicals (e.g. active pharmaceutical ingredients (APIs)) over the past decade. However, numerous important reactions in pharmaceutical industry involve multiphase interactions, which create challenges for flow chemistry to gain a wider acceptance. For example, the presence of stoichiometric amount of solids as reagents, by-products, or products leads to the issue of blocking a flow reactor, and the problem is more severe for the milli-scale or micro-scale flow systems. Besides, the mass transfer resistance at the phase boundaries sometimes becomes the rate-limiting step. Considering current widely used tubular reactors, only providing limited interfacial interactions imposed by the laminar flow pattern, a more universal and versatile reactor design is desired for handling solids in flow, generating well-controlled mixing performance, and at the same time, offer compatibility to withstand harsh chemical environment.

Here, we present a new miniaturized continuously stirred-tank reactor (CSTR) for continuous processing of multiphase systems. The design, with the feature of plug-and-play, takes efficient mass and heat transfer, chemical compatibility, and ease to fabricate and clean into account. Single-phase residence time distribution (RTD) measurements of the CSTR unit reveal nearly ideal CSTR mixing. It also has the capability to tune the RTD by connecting several units in series to achieve approximate plug flow reactor (PFR) RTD, which gives the flexibility to optimize the selectivity based on the reaction kinetics.

Two versions of miniature CSTRs were developed for: (i) solid handling in flow, and (ii) high-performance liquid-liquid mixing.

Solid handling CSTR: employing the non-stick polytetrafluoroethylene (PTFE) as the reactor body material with minimized inter-unit connection enables the CSTR to continuously handle high-concentrated slurries for hours without significant signs of clogging. Two solid-forming reactions, (i) glyoxal reacting with cyclohexylamine to form N,Nâ€²-dicyclohexylethylenediimine, and (ii) sulfonylation of 2-octanol with methanesulfonyl chloride, demonstrate its practical usefulness as a powerful for flow chemistry field.

High-performance liquid-liquid CSTR: incorporating the magnetic-coupling as the driving force to provide external agitation inside the confined reactor space gives the possibility to generate extraordinary two-phase mixing performance without losing the capability to seal the reactor. We demonstrate the reactor with a comparable liquid-liquid mass transfer coefficient to Corning Advance-flow Reactor, which is the state-of-the-art among commercial reactors. More importantly, it gets around the problem of coupling of flowrate and mixing, and is capable to accommodate reaction kinetics from seconds to hours. Phase transfer catalyzed reactions were demonstrated in this newly designed CSTR, with the corresponding kinetic model developed for Makosza phase-transfer mechanism. In addition, the success of carrying out biphasic Suzuki-Miyaura reactions further validates its suitability for the pharmaceutical applications.

Grams of API to Commercial-Scale Continuous GMP Drug Product Manufacture in Less Than a Year

Carl Hartmann

Phase I Clinical Supply Options

Phase I Clinical Demand: initial forecast **200,000+** doses

Drug-in-capsule
Output: 200 capsules / hr
Consider: RBA study, low throughput

Small-Scale Development
Req: 250 g API

Continuous Tablet Dev
Req: under 2 kg API
Output: 50k-100k /hr

CT Manufacture Decision

?

Time

Small-Scale Development Defined

- Dispense, Blend, Clean

Lubrication Scale Model

3+ hrs and 200-10,000 g API → 30 min and 1-20 g API

- Setup, Compress, Clean

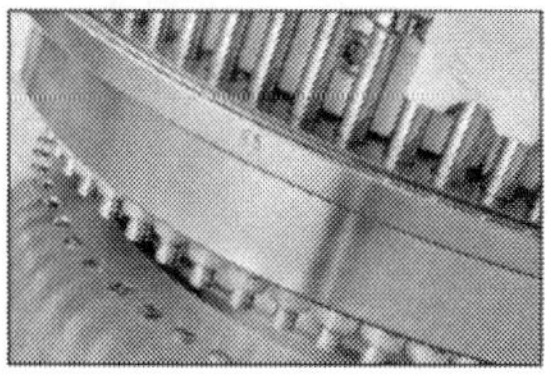

Intrinsic Material Properties Models

3+ days → 1 hr

3

Small-Scale, but Big Data

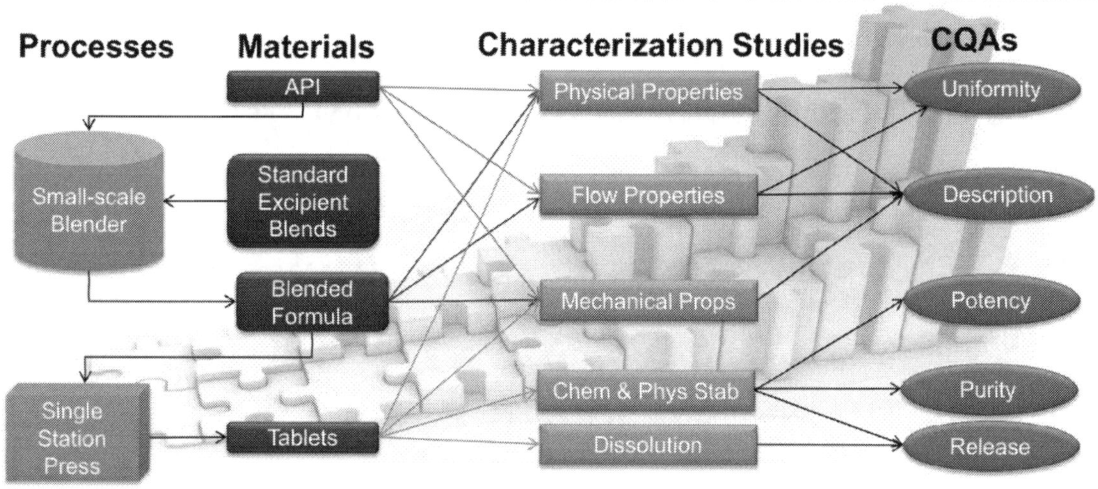

4

Small-Scale Development Blending

DS Lot	X90 (um)	Hausner Ratio	FFC (flow) values
A	192.4	1.41	4.10
B	179.8	1.46	9.18
C	129.7	1.41	4.86
D	135.7	1.34	4.84
E	72.2	1.56	2.85

Five drug substance (DS) lots of varying particle size distributions (PSD), each 50 gram quantity

Each DS lot blended into a common formulation for comparison, each blend targeting 50 gram quantity

Material Name	Common Formulation (%)
Drug Substance	28
Lactose Monohydrate	16
Microcrystalline Cellulose	48.5
Superdisintegrant	5
Lubricant	2.5

5

Small-Scale Development Tablets

- Standardized tablets:
  - 6 mm round, flat, 100 mg target weight

A preliminary blend at 42% drug load (DL). Observation of picking/sticking.

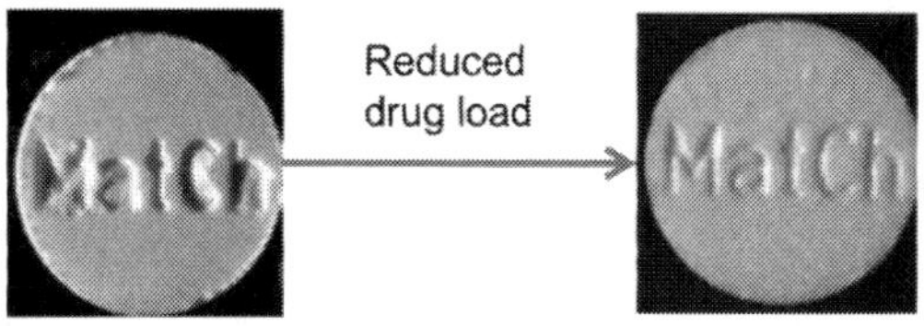

Comparison of five differing PSD DS lots conducted in blends at 28% DL. No visual picking/sticking.

- Single station instrumented press
- Compression profile and stability testing

6

Initial Tablet Results - Common Blend

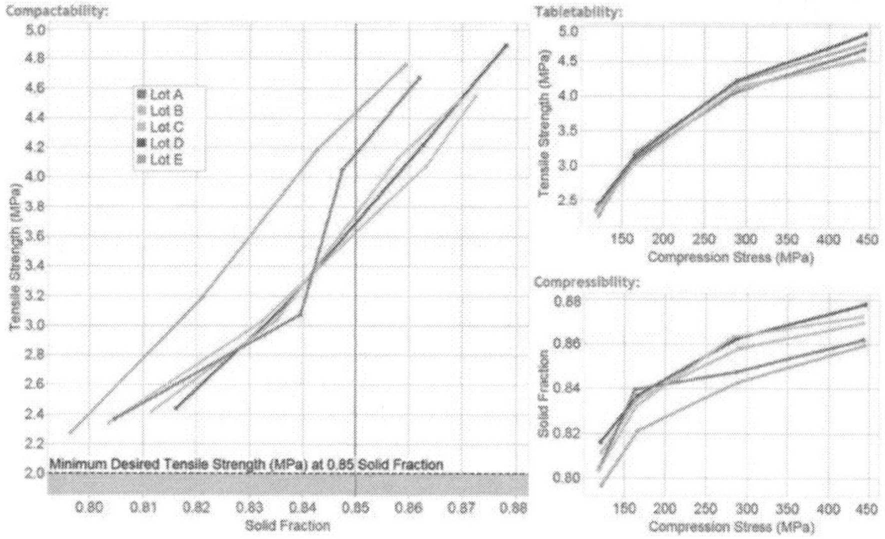

7

Adjustments and confirmation

- Initial risk assessment:
  - Smallest API, poor flow
  - No other red flags
- Formulation development:
  - Reduce drug load
  - Eliminate lactose

8

CM Direct Compression Flow Diagram

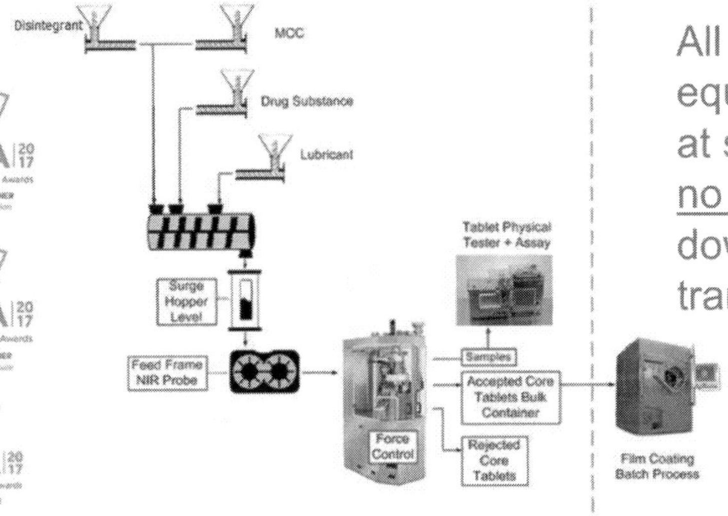

All CM equipment trains at same scale; <u>no scaling</u> up or down with tech transfer

9

CM Direct Compression Animation

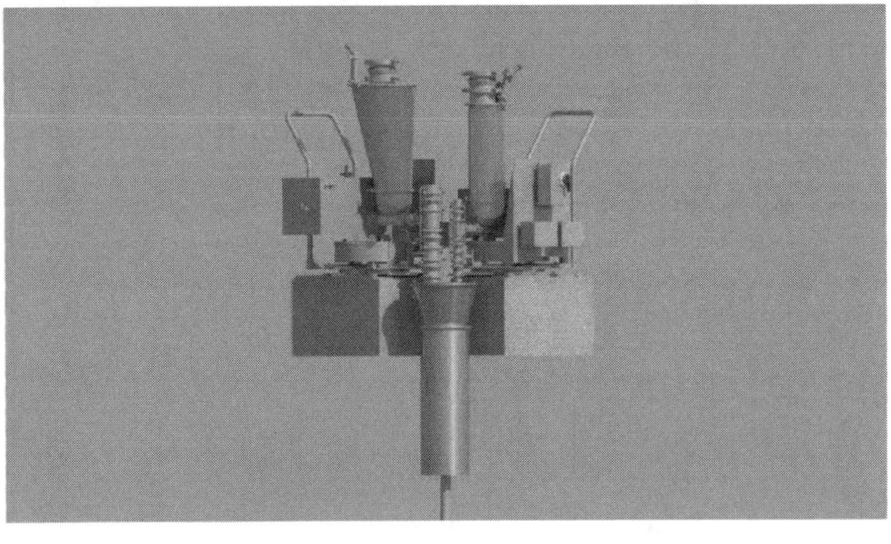

10

Leading into Demonstration Batches

- Off-line **feedability** (LIW feeder) for DS:
  - For twin-screw tooling selection
- Spectroscopy: **NIR** signal range of DS and DS blends
  - For tablet press feed frame NIR on-line monitoring
- Select **tablet tooling** shapes and sizes appropriate for CT
- Identify **coating** system and color appropriate for CT

5 mg strength
6 mm round
Pink coat

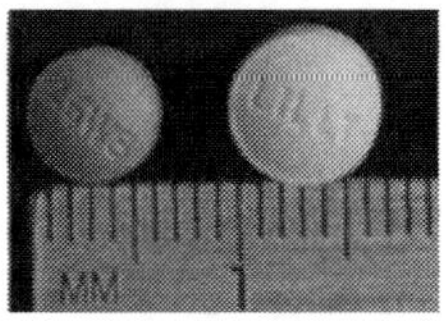

25 mg strength
7 mm round
Yellow coat

11

Development Demonstration Batches

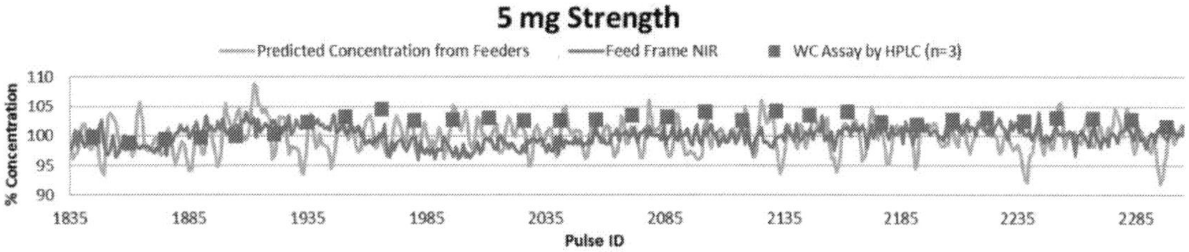

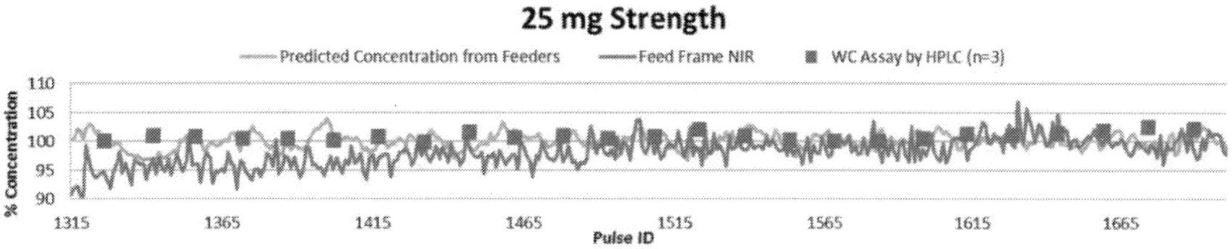

Less than 2 kg API required in total for both batches

12

Phase I Clinical Supply Decision

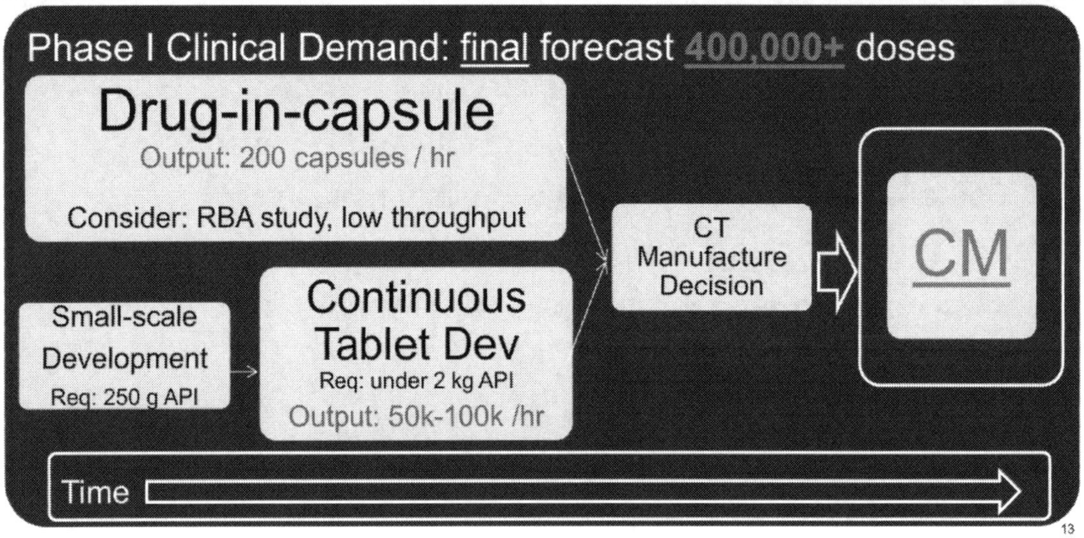

13

GMP Commercial Site Manufacture

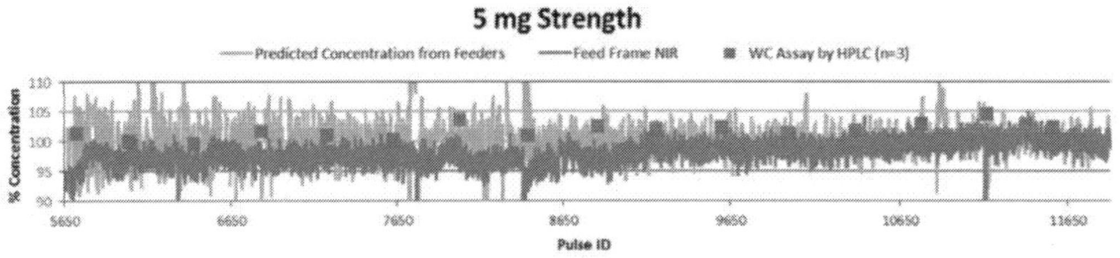

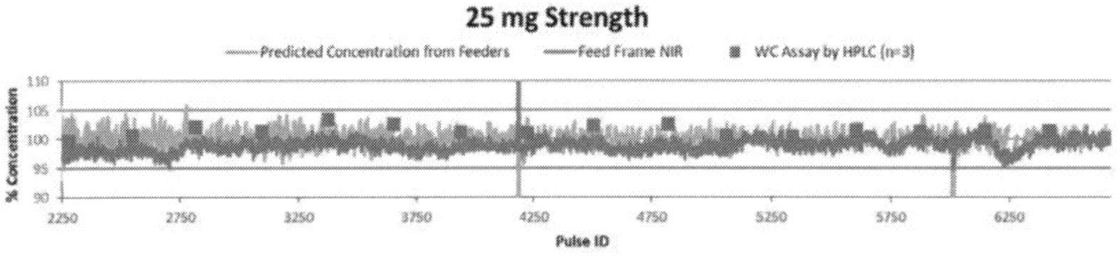

Both batches readily meet USP <905> and ASTM 2810 criteria

14

Timeline

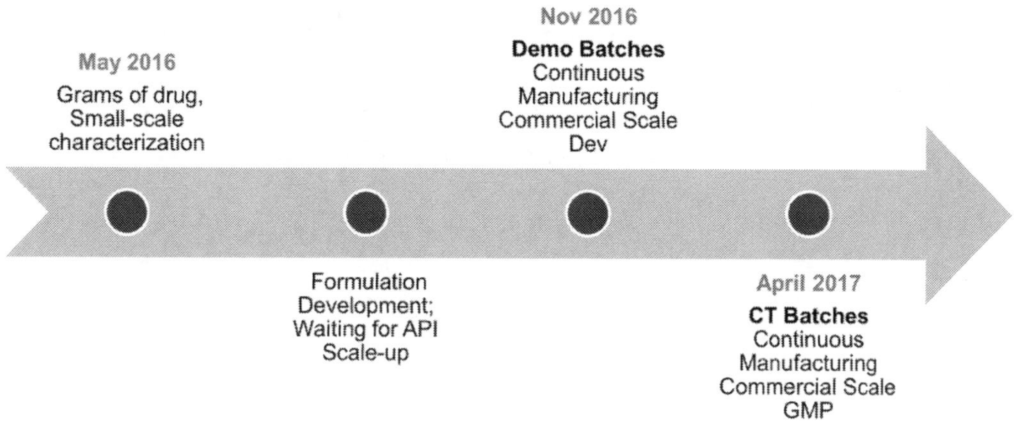

Grams to GMP commercial-scale in less than a year!

15

Conclusions

- Small-scale characterization and continuous manufacturing enables **rapid development** of a commercial-scale formulated tablet
- CM platform allows for **flexibility** in delayed decisions on supply requirement
- Early development work assures feasibility of CM platform to **buy down risk** with little API

16

Acknowledgements

Mark Polizzi
Dave Myers
Mike Miller
Pete Shi
Adam Butterbaugh
Josh Hanson
Melissa Keeney
Joe Martinelli
Adam McFarland
Moussa Boukerche
Dale Greenwood
Kirk VanderKamp
Tim Kramer
Sarah Schmitt

17

Questions?

18

487680 A Continuous and Controlled Pharmaceutical Freeze-Drying Technology for Unit Doses

Abstract:

Driven by growing needs in the biopharmaceutical market and regulatory pressure, a continuous and controlled freeze-drying technology for unit doses to preserve biopharmaceuticals has been developed. Such continuous process allows a more efficient, cheaper, greener and controllable manufacturing method compared to traditional batch production systems, offering competitive advantages and business opportunities.

Pharmaceutical freeze-drying (lyophilization) is a low-temperature drying process in which aqueous solutions of heat-labile biopharmaceuticals are converted into solids with sufficient stability for distribution and storage. Similar to all manufacturing processes of drug products (solids, semi-solids and liquids), conventional pharmaceutical freeze-drying is generally accomplished using batch processing that is considered time-consuming, costly, non-flexible and lacking robust quality control and real-time release.

Four major industrial drivers are demanding a more efficient and better controllable pharmaceutical freeze-drying technology for unit doses: cost-cutting, regulatory pressure, a fast growing biopharmaceutical market and an ageing population requiring more personalized medicines.

The continuous and controlled freeze-drying technology, developed following the principle of model based design, offers clear advantages over current batch production such as cost reduction (up to 50%), track-and-trace product quality control, and a significant reduction of processing time (> 40 times faster, e.g. 1 hour instead of 5 days at a vial level), reduced need for clean room and a substantial sustainability gain.

489155 Engineering Considerations for Dissolved Oxygen Sensor Response Across Silicone Membranes in 3-L Single Use Bioreactors

Abstract:

Single use bioreactors are becoming standard in biopharmaceutical development and manufacturing because they reduce capital expenditure, eliminate cleaning complications, and decrease down time between batches. Several successful bioreactor designs are available, but sensor integration poses different challenges between scales. For example, large-scale cultivations require a high degree of sensor fidelity, typically using costly disposable adaptors to sterilize and connect traditional reusable probes. In contrast, high throughput bioprocessing values cheaper sensor technology, often trading off robustness. With bench-scale bioreactors, a gap exists because their use in process characterization and scale-down studies still demands accuracy, precision, and longevity, but at a broad experimental scope that doesnâ€™t allow for high unit cost per run.

If industry leaders want the advantages of single-use vessels as well as the proven reliability of reusable sensors for 3L bioreactors, operators need to sterilize probes separately for aseptic insertion in a hood. To avoid the need for manipulation of dissolved oxygen (DO) sensors, we use a thin silicone membrane as a gas-permeable sterile-barrier. In doing so, we retain the measurement integrity in addition to the economic advantages of reusable probes.

This presentation discusses engineering considerations for DO sensor response when impeded by a secondary membrane. We applied first principles to separate the additional delay from overall probe response, quantified maximum lag times, and demonstrated a robust design window for acceptable performance of mammalian cell culture. We are demonstrating the long-term functionality of this solution by implementing across our routine process development cultivations.

496074 Process Development Tool to Rapidly Predict the Stability of Biopharmaceuticals

Abstract:

Process Development Tool to Rapidly Predict the Stability of Biopharmaceuticals

S. Hedberg, Y. Hutabarat, J. Heng, J. Haigh, D. Williams

ABSTRACT

Protein aggregation is a critical problem for the safety of biopharmaceuticals as they are linked to adverse immunologically related responses in patients. Much effort has been made to gain a better understanding of aggregation, however, the mechanisms leading to protein aggregation are still not fully understood. Protein-protein molecular interactions in solution are known to be involved in protein solution aggregation behaviour and are a common issue for the manufacturing of biopharmaceuticals such as monoclonal antibodies, mAbs.

Therefore, a major industrial and academic challenge is the development of fast and reliable methods, either theoretically or experimentally driven, which allow formulation scientists to identify optimal biopharmaceutical species and solution conditions which deliver stable solution products, thus avoiding aggregation. The osmotic second virial coefficient (B_{22}) is a fundamental physiochemical property that describes the thermodynamics of the protein-protein interactions in solution, which can be a useful tool for predicting the aggregation propensity of proteins.

One way of predicting aggregation propensity is self-interaction chromatography (SIC) that can be used for determining the osmotic second virial coefficient (B_{22}). This technique has recently demonstrated to be a tool that can successfully predict phase behaviour of proteins. However, due to the need to immobilise every protein in order to make B_{22} measurements makes this technique less ideal for use in process development. Another technique, cross-interaction chromatography (CIC), has shown to be an even more high-throughput technique than its predecessor with similar capabilities. In CIC only one protein needs to be immobilised in order to screen a range of other proteins.

In a case study a number of different monoclonal antibodies from both IgG1 and IgG4 subclasses, and a polyclonal antibody were immobilised and screened on each of the columns for a wide range of process development conditions. The aim was to see if the osmotic second virial cross coefficient (B_{23}) between two proteins can be used to predict B_{22}.

The results of the case study were very successful and proved to be a good prediction of the B_{22} value for all the proteins in all conditions. This will enable a rapid process development tool with only mg quantities of biopharmaceutical species that can be injected on a generic column e.g. a polyclonal IgG column.

Previously, the accuracy of the B_{22} values of several of the mAbs has been directly validated by the use of traditional gel filtration chromatography methods and dynamic light scattering, to investigate the size and quantity of aggregates over time. This has shown that the determined B_{22} held powerful information about the long term stability of biopharmaceuticals.

In summary, this work has developed a rapid and high-throughput tool, which manages to estimate the B_{22} values for different biopharmaceutical products using minimal amounts of protein. This will help scientists in the biopharmaceutical industry finding suitable conditions for process development as well as picking drug candidates with high long term stability.

NATIONAL INSTITUTE OF CHEMISTRY

Paper number ID# 489514

Model-Based Process Optimization for Upstream Bioreactor Production–Downstream Chromatographic Separation of CHO Cell Monoclonal Antibodies (mAb)

Uroš Novak, Miha Kastelic, Drejc Kopač, Miša Mojca Cajnko, Andrej Pohar, Matic Grom and Blaž Likozar

uros.novak@ki.si, blaz.likozar@ki.si

Minneapolis, 1.11.2017

next BioPharm DSP

BioPharm.Si

REPUBLIKA SLOVENIJA
MINISTRSTVO ZA IZOBRAŽEVANJE, ZNANOST IN ŠPORT

EVROPSKA UNIJA

Research areas of the department of Catalysis and Chemical Reaction Engineering

ADREM

ADREM (Spire, Horizon 2020)
Adaptable Reactors for Resource- and Energy-Efficient Methane Valorisation

Carbon dioxide activation

MefCO2 (SPIRE, Horizon 2020)
Methanol synthesis from captured CO2 with the use of surplus electricity from renewable resources

FReSMe

FReSMe (Horizon 2020)
From residual steel gases to methanol

Methane activation & conversion

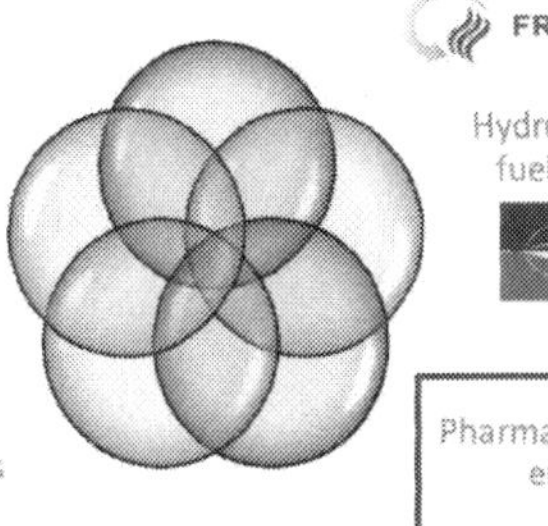

Hydrogen & fuel cells

NATO OTAN

NATO (SPS Programme)
Enhanced portable energetically self-sustained devices for military purposes

MarineBiotech

Biomass-derived building blocks

Interreg ITALIA-SLOVENIJA

CELKROG

Pharmaceutical process engineering

Biopharm.si :
New generation of biological drugs

nextBioPharmDSP (LEIT, Horizon 2020)
Next-generation Biopharmaceutical Downstream Process

https://www.ki.si/en/odseki/d13-department-of-catalysis-and-chemical-reaction-engineering/

Outline

- ❑ Introduction and aim of the work
- ❑ Upstream process:
  - mechanistical model describing the metabolism of CHO cells
  - metabolic flux analysis (MFA)
  - modeling of N-glycosylation
- ❑ Downstream process:
  - affinity chromatography model development
  - parametric sensitivity study
  - different options of operation (loading strategy)
- ❑ Conclusions

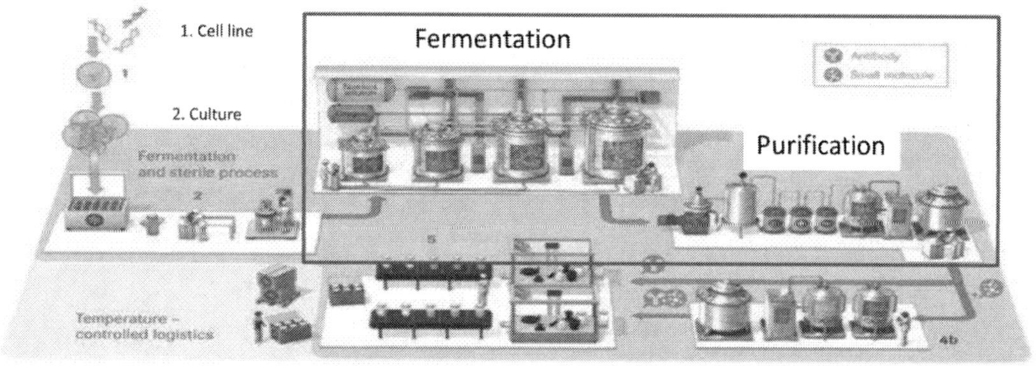

http://www.roche.com/dam/jcr:045db96e-6a0d-48ea-b37c-8ed42e5a59e7/en/supply_resp_2.jpg

Introduction

"The process is the product."

BioPharm.Si 2016-2020

The main goal of BioPharm.Si programme, involving high tech and biotech Slovenian companies and excellent research institutions, is development of continuous manufacturing processes and innovative devices and services that will facilitate transition from batch to fully continuous manufacturing.

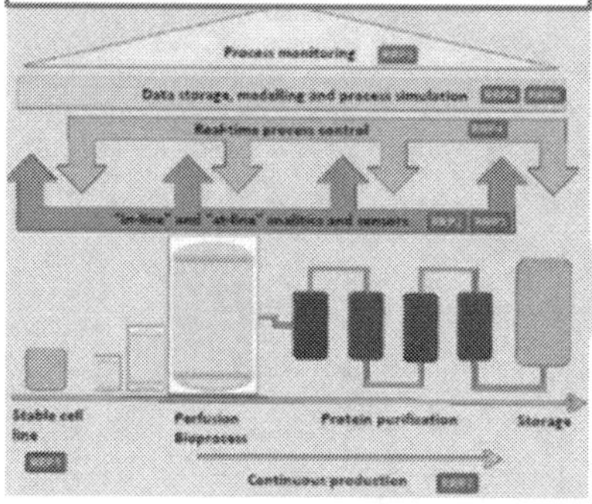

More info at www.biopharm.si/

Aim

Mechanistical mathematical modelling as a tool in biopharmaceutical development

Motivation

Systematic statistical evaluation of the data → Enhancing product quality control

Smart planning of experiments (analytics & formulation)

New process design and optimization → Reduce costs and increase productivity

nextBioPharmDSP 2015-2019

Implementation of a fully integrated manufacturing platform for biosimilar mAb based on continuous chromatography, in combination with single-use disposable technology for all unit operations of DSP on pilot/small production scale together with incorporation of advanced analytical tools.

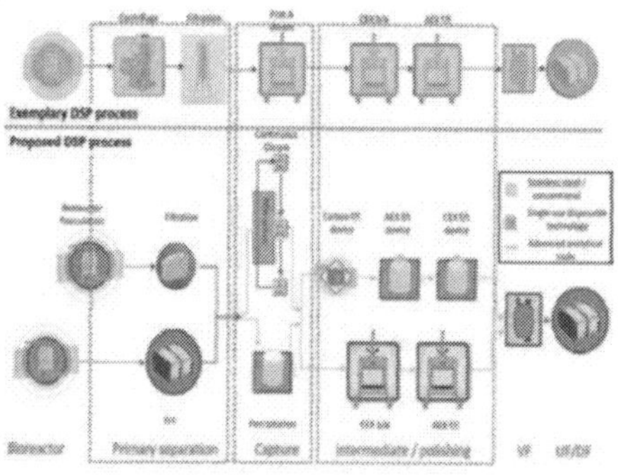

More info at http://nextbiopharmdsp.eu/

Upstream process - mechanistical model describing the metabolism of CHO cells

- Mathematical description of a series of bioreactions that have an impact on the dynamics of cell growth, metabolism and the production of intracellular products.
- Building a global model with a multi-model approach that takes into account the growth stages of the cell culture and could successfully predict the development of experimental data.
- Optimizing bioprocess *in-silico* through qualitative and quantitative insight into mechanisms through which conditions in bioprocess affect the glycosylation of therapeutic proteins.

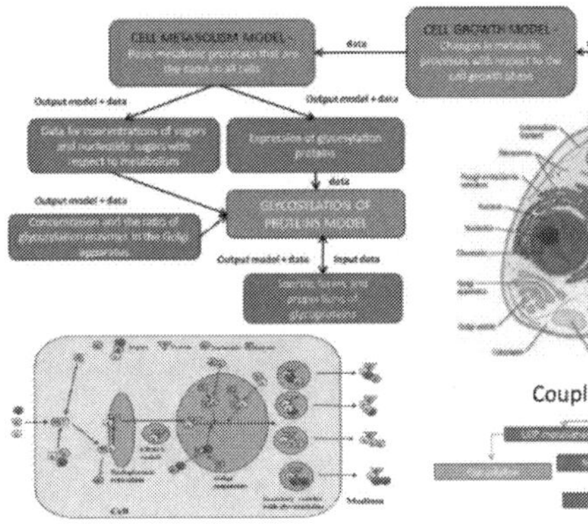

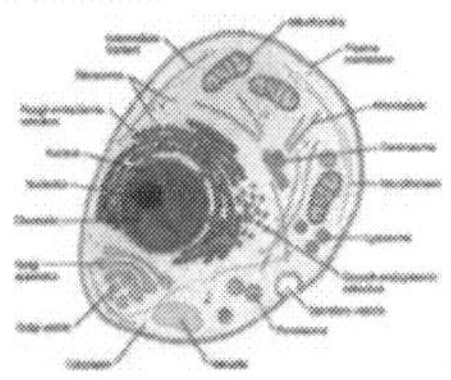

MFA- Methodology for quantification of metabolic pathways from limited number of experimental data

Scheme of metabolic flows in the CHO cell

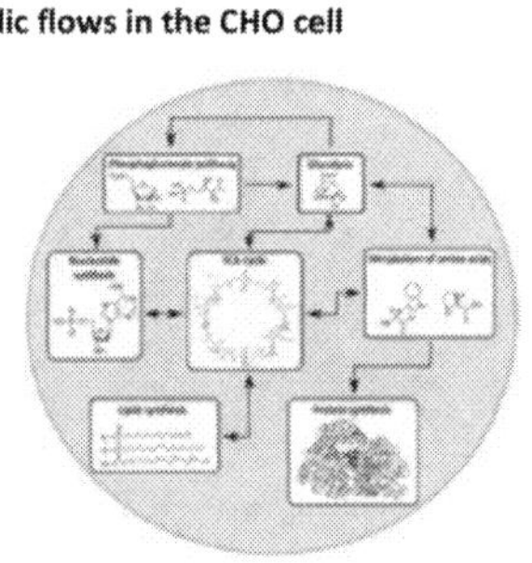

Coupling cell metabolism with N-glycosylation

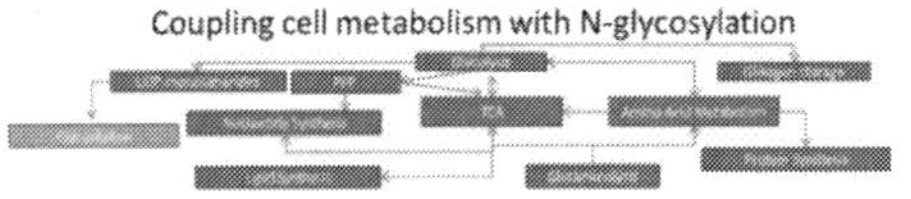

Metabolic flux anaylsis (MFA)

- The MFA allows you to calculate the value metabolic fluxes (v) for the selected metabolic scheme.
- Measured values of selected fluxes (typically: consumption and release of extracellular metabolites).
- The MFA allows you to shrink metabolic schemes to the several macro reactions.
- Easier modeling of reactions kinetics .

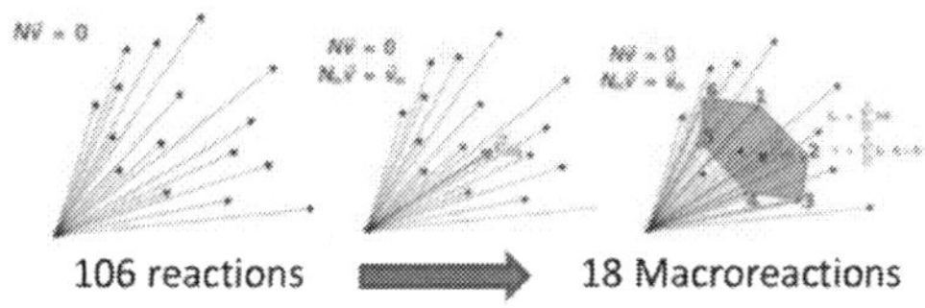

- Possibilities of addition of substrates and removal of products
- Dynamic profiles are available for each selected compound in metabolism

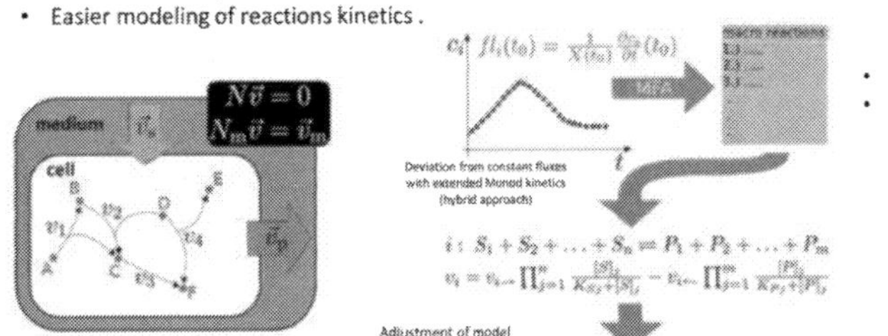

$\vec{N}$ - stoichiometric matrix

N_m - flux vector for reactions

$\vec{v}$ - matrix of measured metabolites

$\vec{v}_m$ - vector of measured fluxes

$\vec{v}_{p,s}$ - fluxes for membrane transport

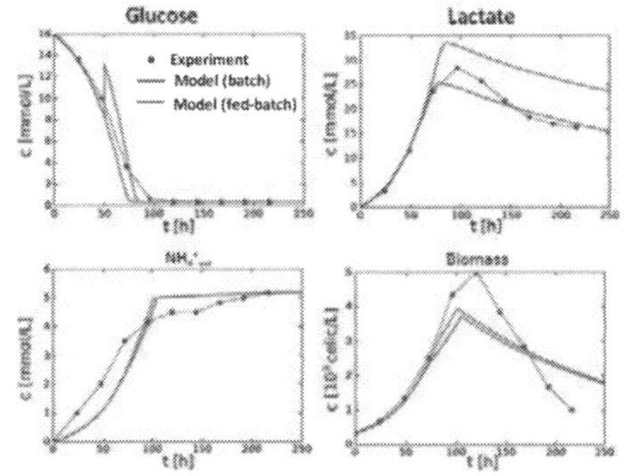

Experimental data; Zamorano F., Wouwer A.V., Jungers R.M., Bastin G., J. Biotechnol., 164, 409-422, 2013

Modeling of N-glycosylation

- Glycan binding rules determine the decision trees or N-glycosylation pathway
- calculate the rates (rj) at each step of the tree, and choose the most likely path in a balanced way
- model one transition of the initial glycan m9 through Golgi apparatus, repeat 100 times (statistics)

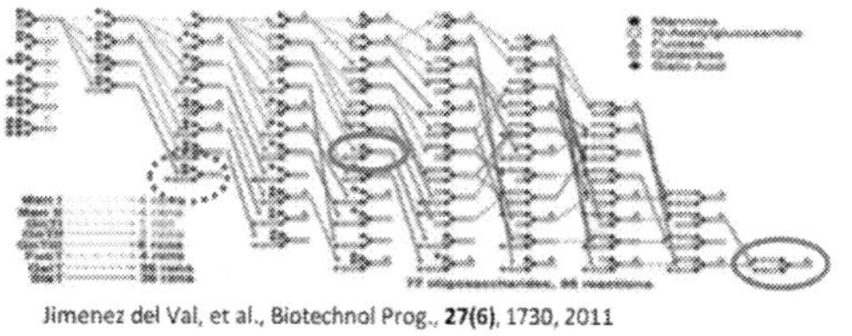

Jimenez del Val, et al., Biotechnol Prog., **27(6)**, 1730, 2011

I. Glycosylation depending on the position in Golgi, we take into account the enzyme concentrations vary according to the location in Golgi

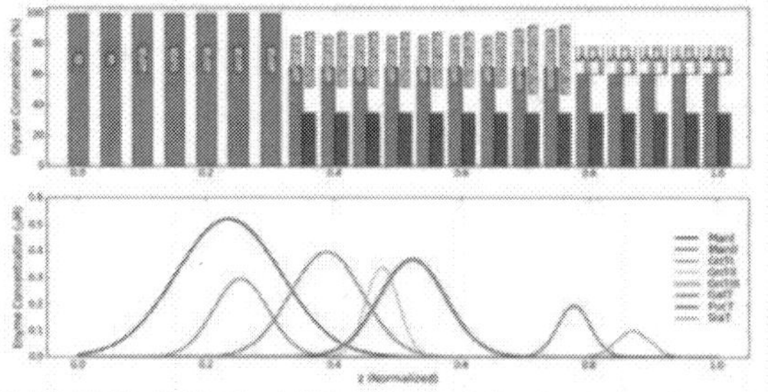

II. Finding parameters according to the final distributions of individual glycan

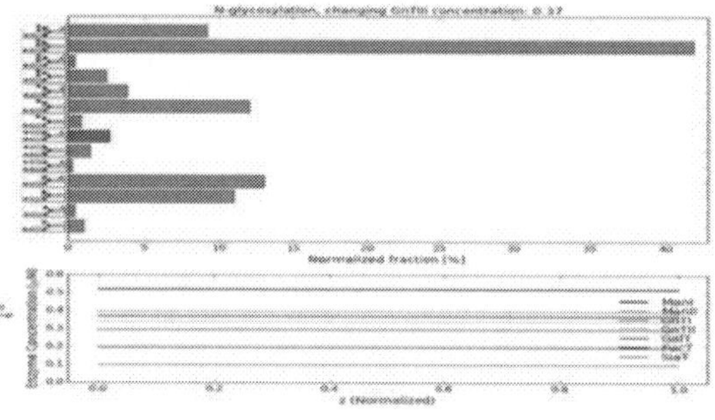

Coupling of cell metabolism with N-glycosilation

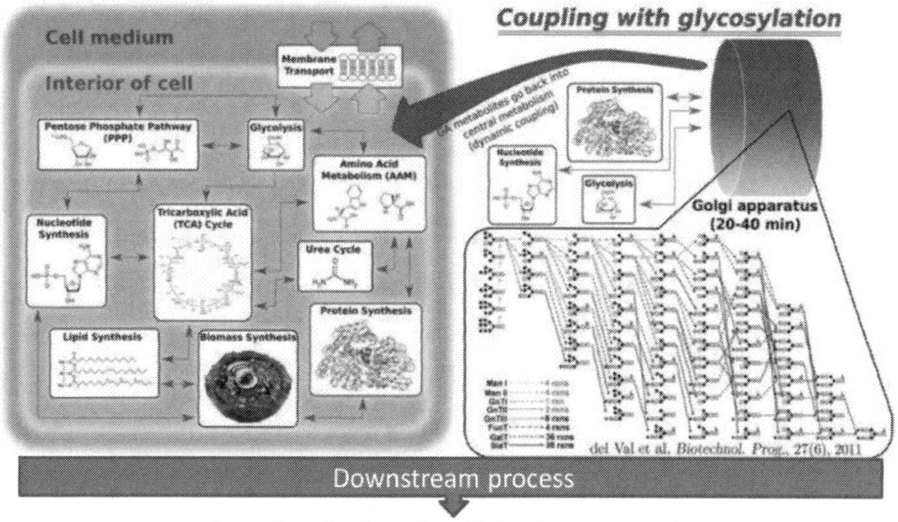

Capturing of gylcans by affinity chromatography

Downstream process – affinity chromatography

Batch affinity chromatography operates in cyclic mode.

1. loading of mAb on equilibrated affinity sorbent contained in chromatography column,
2. washing of column to remove impurities,
3. elution to desorb mAb bound on sorbent and
4. equilibration/regeneration step to prepare the column with affinity sorbent for next cycle.

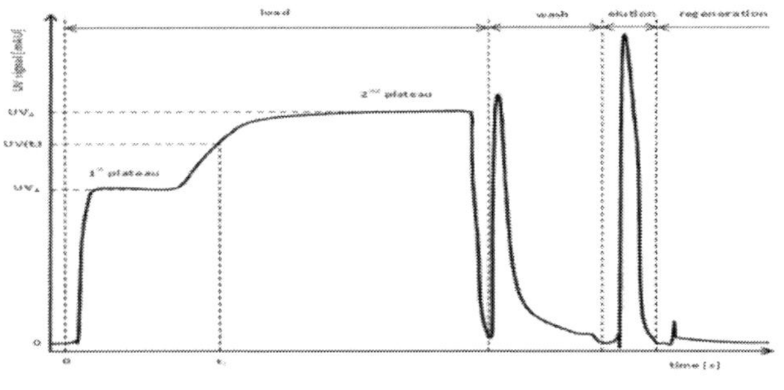

E. von Liers, J. Andresson, Computers and Chemical Engineering 34 (2010) 1180-1191

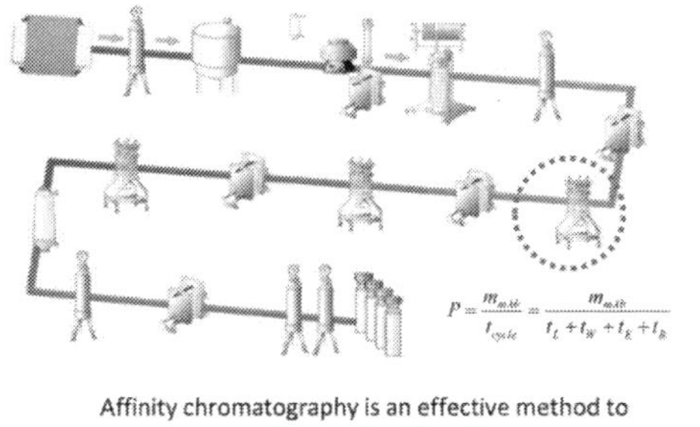

Affinity chromatography is an effective method to capture and purify mAb

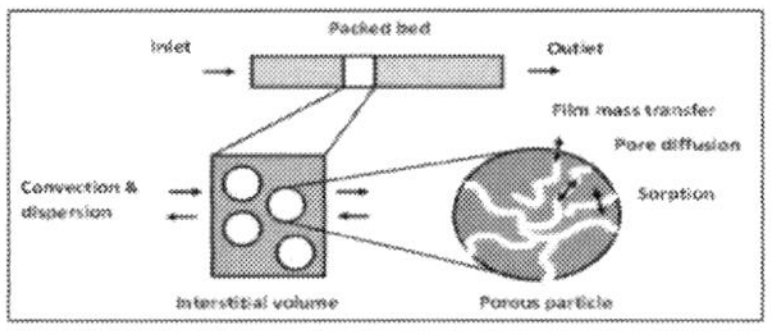

Mathematical model formation

Mass balance states that antibody accumulated in some (infinitesimally) point of bulk liquid is equal to sums of convection and exchange with affinity resin particles and liquid.

$$\varepsilon_c \frac{\partial c}{\partial t} = -u\varepsilon_c \frac{\partial c}{\partial z} + \varepsilon_c D_{ax} \frac{\partial^2 c}{\partial z^2} - (1-\varepsilon_c)\frac{3}{r_p} k(c - c_{p,s})$$

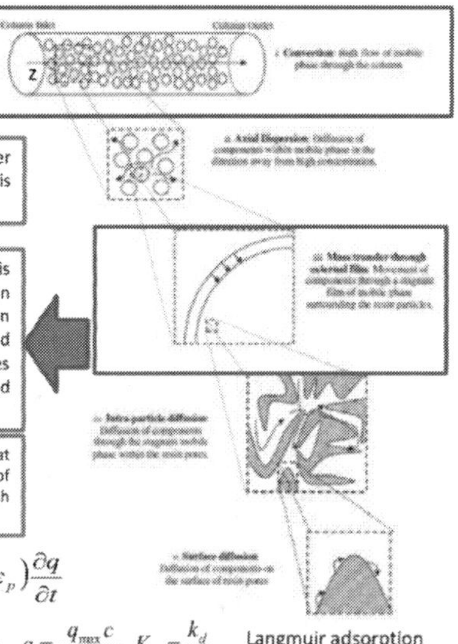

$$Sh = \frac{kd_p}{D_m} \qquad Re = \frac{ud_p\rho}{\eta} \qquad Sc = \frac{\eta}{D_m\rho}$$

$$Sh = \begin{cases} 1{,}85\left(\frac{1-\varepsilon_c}{\varepsilon_c}\right)^{0,33} Re^{0,33} Sc^{0,33}, Re < 40 \\ \frac{1{,}09}{\varepsilon_c} Re^{0,33} Sc^{0,33}, 40 \le Re < 55 \\ \frac{0{,}25}{\varepsilon_c} Re^{0,69} Sc^{0,33}, 55 \le Re < 1050 \end{cases}$$

If ratio of column length to particle diameter (L/dp) is above 350 axial dispersion is negligible.

When external film mass transfer resistance is negligible compared to internal diffusion resistance, particle surface concentration becomes equal to concentration in bulk liquid and diffusion at surface of particles determines mass transfer between bulk liquid and particles.

$$\varepsilon_c \frac{\partial c}{\partial t} = -u\varepsilon_c \frac{\partial c}{\partial z} + \varepsilon_c D_{ax} \frac{\partial^2 c}{\partial z^2} - (1-\varepsilon_c)\frac{3}{r_p} D_{eff} \left.\frac{\partial c_p}{\partial r}\right|_{r=r_p}$$

$$D_{ax} = \frac{d_p u \varepsilon_c}{0{,}339 + 0{,}033\,Re^{0,48}}$$

Mass balance for mAb in pores states that accumulation is equal to difference in transport of antibody by pore diffusion and reaction in which antibody binds to ligand within pores.

$$\varepsilon_p \frac{\partial c_p}{\partial t} = D_{eff}\left(\frac{\partial^2 c_p}{\partial r^2} + \frac{2}{r}\frac{\partial c_p}{\partial r}\right) - (1-\varepsilon_p)\frac{\partial q}{\partial t}$$

Boundary and initial conditions

$$c_p = c = c_{p,s}, r = r_p \quad \frac{\partial c_p}{\partial r} = 0, r = 0$$

$$c = c_0, 0 \le z \le L \quad c_p = c_{p0}, 0 \le r \le r_p, 0 \le z \le L$$

$$q = q_0, 0 \le r \le r_p, 0 \le z \le L$$

$$D_{eff} = K_p \frac{\varepsilon_p}{\tau} D_m \quad D_m = 2{,}74 * 10^{-9} M^{-1/3} \quad \frac{\partial q}{\partial t} = k_a(q_{max} - q)c_p - k_d q \quad q = \frac{q_{max} c}{c + K_d}, K_d = \frac{k_d}{k_a}$$

Langmuir adsorption isotherm

Parametric sensitivity study

Column and resin parameters (Mab Select SuRe and Mab Select SuRe LX) were taken from the literature (Hahn et al. 2005):

ε_c...column void fraction 0.36 [/]
ε_p...particle porosity 0.88 [/]
q_{max}...maximum binding capacity of affinity resin 5.47 [mol/m^3]
K_d...equilibrium desorption constant 7.7x10^{-4} [mol/m^3]
K_{fit}...parameter, taking resin particle pore diameter, antibody molecule diameter and pore tortuosity into account 0.083 [/]
c_{in}...concentration at piping system inlet 3.33 [mg/ml]
F... flow rate [m^3/s]
d_p...diameter of particles 85 [µm]
L...length of column 100 [mm]
D...column diameter 8 [mm]
D_{ax}...axial dispersion coefficient [m^2/s]
V_{PFR}...volume of plug flow elements [m^3]
V_{CSTR}...volume of perfectly mixed elements [m^3]

*R. Hahn, P. Bauerhansl, K. Shimara, C. Wizniewski, A. Tscheliessnig, A. Jungbauer, *J. Chromatogra. A*, 1093 (2005) 98-110

Changes at 80% of the initial value (%)		
Parameter	Mab Select SuRe LX	Mab Select SuRe
ε_c	14	3.3
ε_p	-13	-9.4
q_{max}	-19	-19
K_d	0.9	2.0
K_{fit}	-12	-8.7
d_p	24	12
L	-13	-9.6
D	-26	-21
k	/	3.9
D_{ax}	0.4	0.0
C_{in}	-1.5	-2.2
F	13	7.6
V_{CSTR}	0.0	-0.2
V_{PFR}	0.0	0.0

Dynamic binding capacity (DBC)
The amount of antibody that can bind to a chromatography media under flow conditions.
DBC_{10}...dynamic binding capacity at 10 % breakthrough [mg/mL]

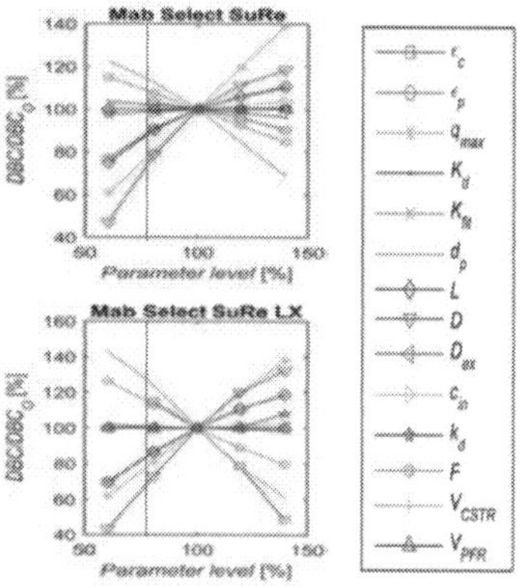

Different options of operation

Loading strategy	Constant flow rate	Flow ramp	Quadratic flow function	Reversed flow
Number	1	2	3	4
Description	Column is loaded with constant flow rate F until BT=0.6 % at which loading step ends.	Loading flow rate starts at value F_1 decreasing linearly to value of F_2 between times t_1 and t_2 until allowed BT=0.6 %.	Flow changes with quadratic function of time $F(t)=a_1+a_2t+a_3t^2$	Column is loaded with flow rate F_1 until time t_1 or BT=0.6 %. Then it is loaded from other side of column with flow rate F_2 until BT=0.6 %
Optimization parameters	Flow rate F	Flow rates F_1, F_2 and times t_1, t_2,	Quadratic function coefficients a_1, a_2, a_3	Flow rates F_1, F_2, time t_1
Reasoning	Optimum flow rate should correspond to compromise between amount of antibody captured per cycle (high amounts are favored by low flow rates) and time of load (short times are favored by high flow rates).	Column is first loaded quickly but less effectively with high flow rate, then with decreasing flow effectiveness loading increases while time needed also increases.	Similar as flow ramp only different flow function was tested.	When allowed breakthrough is reached, concentration of bound antibody falls from column inlet to column outlet. By reversing flow, fresh harvest first encounters less saturated part of column.
Productivity increase (%)	/	**13**	10	12

AC Column
Width 8 mm; Length 100 mm

$$P=\frac{m_{mAb}}{t_{cycle}}=\frac{m_{mAb}}{t_L+t_W+t_E+t_R}=\frac{m_{mAb}}{t_L+t_{WER}}$$

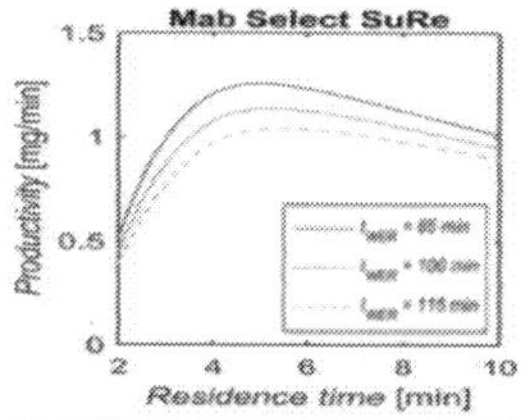

Affinity sorbent	Mab Select SuRe		
Parameter	RT	t_L	P
Unit	min	min	mg/min
t_{WER}=85 min	5.0	55	1.26
t_{WER}=100 min	5.3	60	1.14
t_{WER}=115 min	5.5	65	1.04

Optimum parameters of flow ramp loading strategy

Parameter	Unit	Mab Select SuRe
Start flow rate F_1	mL/min	3.18
Start of ramp t_1	min	0
End flow rate F_2	mL/min	0.77
End of ramp t_2	min	19
Loading time t_L	min	49
Productivity P	mg/min	1.42

Conclusions - Upstream

- An unique theoretical description that **continuously describes** all cell growth phases was developed.
- The **reversibility** of macro reactions is a key element that is needed for the correct description of the concentration profiles of measured components in the medium or cell.
- After determining model parameters any modifications in the medium could be simulated further enabling **smart design** of the experiments.
- The developed methodology can be **coupled** with the protein N-glycosylation and affinity chromatography (further work).

Conclusion - Downstream

- Developed model enables characterization of affinity resin properties with **minimum** experimental work requirements.
- Internal diffusion is **rate limiting step** of affinity chromatography process.
- Axial dispersion and film mass transfer resistance has **low influence**.
- Results of parametric sensitivity study showed, that volumes of external chromatography system have **minor effect**.
- Transport inside monoclonal affinity chromatography column is **controlled by pore diffusion**.
- Amongst loading strategies tested, **flow ramp loading** showed the best results compared to loading with constant flow rate.
- This strategy is the **simplest** to implement and required **no mechanical changes** to the existing equipment.

Acknowledgments

Assist. Prof. dr. Blaž Likozar
Dr. Miha Kastelic
Matic Grom
Dr. Andrej Pohar
Dr. Drejc Kopač
Dr. Miša Mojca Cajnko

BioPharm.Si

REPUBLIKA SLOVENIJA
MINISTRSTVO ZA IZOBRAŽEVANJE, ZNANOST IN ŠPORT

EVROPSKA UNIJA

next BioPharm DSP

Thank you for your attention

16

494003 Exploring the Multi-Minima Behavior of Organic Crystal Polymorphs at Finite-Temperature

Abstract:

Polymorph prediction is a critical aspect of pharmaceutical solid form design to ensure that the crystal with the highest stability and performance properties is selected for development. Computational crystal structure prediction (CSP) studies currently predict the experimental crystal forms by identifying minima on the lattice energy surface. One of the most pressing challenges in current CSP studies is that this minimized lattice energy approach typically suffers from a significant overprediction of stable crystal forms. Most polymorphic systems have on the order of ten stable solid forms. However, CSP studies can yield hundreds or thousands of unique lattice minima.

A common hypothesis put forward for why some lattice minima do not correspond with a unique experimental structure is that the thermal motions under working conditions will cause many lattice minima to rapidly restructure into a more stable form or cause multiple minima to coalesce into a single indistinguishable ensemble. Under this hypothesis, some potential energy basins explored at high temperature should be `rough' landscapes filled with multiple local energy minima rather than the classic picture of a smooth potential energy basin with a single global minimum-energy structure.

Here we demonstrate the existence of these rough basins with multiple lattice minima in a number of different polymorphic systems. The experimental crystal structures of twelve polymorphic systems were heated to a range of temperatures up to ambient conditions using molecular dynamics simulations. A set of configurations from each trajectory was then quenched to zero Kelvin using crystal energy minimization. We find that the high-temperature configurations of rigid-molecule crystals tend to all collapse to the same initial lattice minima after quenching. However, the configurations from more flexible molecules tend to minimize into a myriad of different lattice minima, suggesting that flexibility leads to multi-minima behavior. We also find that the number of minima found in the crystal basins increases with temperature, and the presense of multiple lattice minima within the free energy basin is polymorph-dependent in addition to being molecule-dependent. The existence of these multi-minima basins at ambient conditions lends credence to the theory that incorporating thermal motion can reduce the number of spuriously predicted structures in future CSP studies.

502617 Iterative Synthesis of Heteropolymers Using Organic Solvent Nanofiltration

Abstract:

Iterative Synthesis of Heteropolymers Using Organic Solvent Nanofiltration

Ruiyi Liu, Piers Gaffney, Patrizia Marchetti, Marc Schaepertoens, Ruijiao Dong, Andrew Livingston

Barrer Centre, Department of Chemical Engineering, Imperial College London

Sequence-defined heteropolymers underpin much of biology in the forms of proteins/ peptides and oligonucleotides (DNA and RNA). The pharmaceutical industry is investing heavily in the potential of these species to push through the bottleneck in the drug discovery pipeline. Such polymers feature a repeating backbone motif, but each monomer possesses a variable side-chain. It is the order and chemical nature of these side-chain residues that confers the function of the final heteropolymer. Beyond such familiar biomolecules, heteropolymers present opportunities in chemical bar-coding and security, and even for long-term information storage. Looking still further ahead, and taking a cue from Nature, engineered heteropolymers of non-biological monomers might provide uniquely tailored and functional engineering materials, if only they could be flexibly prepared at scale.

Oligonucleotides (oligos) and peptides are the dual test-beds upon which heteropolymer synthesis is founded. They are prepared by iterative synthesis, a process in which one monomer at a time is coupled to the terminus of a growing chain - like differently colored beads threaded onto a string. To ensure error-free sequence synthesis, other reactive groups are masked by protecting groups; permanent protecting groups block the side-chains throughout the synthesis, and a temporary protecting group is removed in the final stage of each chain extension cycle to expose a new reactive chain terminus. To simplify purification after chain extension, and to make the process easily generalizable, both oligos and peptides are widely synthesized on solid phase supports, where excess reagents are simply flushed from a packed bed or column of support beads.

Solid phase synthesis protocols are intrinsically difficult to up-scale. Furthermore, for the successful iterative synthesis of long sequences to succeed, the coupling efficiency must be as near to quantitative as possible, but mass transfer from the bulk liquid to the solid support is the rate-determining step for all but the very slowest coupling reactions. We overcome both these roadblocks to scalable iterative synthesis by performing coupling reactions in the bulk liquid phase and separating reaction debris by organic solvent nanofiltration (OSN).

OSN membranes separate solutes principally on the basis of size selectivity. During liquid phase iterative synthesis the growing heteropolymer is the largest solute present, whilst the monomer debris is the next largest. Therefore, the ability of the membrane to retain the growing heteropolymer, but to permeate the monomer debris, is the critical separation criterion. The molecular weight of the monomer debris from oligo or peptide synthesis falls in the range 400-700 Da. Indeed, it is hard to conceive of functional heteropolymers, either natural or man-made, with smaller monomer debris. Thus the development of OSN membranes with permeation pathways open enough to filter such large species, and yet maintain their physical and chemical integrity during repeated synthesis cycles, was an essential enabling technology for liquid phase heteropolymer synthesis.

We have previously shown that short peptides and oligos could be prepared by means of liquid phase synthesis and purification by single-stage OSN. We have now achieved pharmaceutical length oligo synthesis (20-mer) using a two-stage membrane purification apparatus that takes the cycle yield to near-quantitative recovery of the growing polymer. Despite harsh solvents (DMF and acetonitrile), and both acidic and basic treatments, the OSN membranes maintained constant performance over multiple cycles (> 19).

The same principles have been applied to the synthesis of poly-ethers. Uniform (or monodisperse) poly-ethylene glycols up to 60-mer were constructed by the stepwise concatenation of short units, purifying the growing chain by OSN after each cycle. We are now extending these principles to the synthesis of engineered heteropolymers with a chemically robust poly-ether backbone and functionalised chiral side-chains.

496577 Challenges in Process Development of Antibody Drug Conjugates (ADCs)

Abstract:

Antibody drug conjugates (ADCs) are a class of biopharmaceutical drugs in which a cytotoxic agent is chemically linked to an antibody that recognizes a tumor-associated antigen. The basic strategy is to combine target selectivity of antibodies with the potent cytotoxic molecules, with a goal of generating drugs that are highly efficacious and also safe.

Pfizer Biotherapeutics Pharmaceutical Sciences is responsible for developing scalable manufacturing processes for Pfizerâ€™s portfolio of ADCs. Process development group is engaged in initially receiving the technology from discovery groups/external collaborators and developing the technology further for preparation of these ADCs from small scale (several grams) to large scale (hundreds of grams to kilograms). Pfizerâ€™s early stage ADCs utilize several conjugation chemistries via cysteines, lysines or glutamines on the antibody as well as several different linkers and payloads. Each ADC molecule, even of the same type of conjugation chemistry, poses unique challenges for process and analytical development groups in the development and implementation of preparative and analytical technologies.

This presentation highlights conjugation and purification development of ADCs to illustrate Pfizerâ€™s development paradigm. In the process development arena the approach to maintain acceptable yields while maintaining quality attributes will be discussed. Our emphasis on thorough characterization of the final product will be discussed as well as consistency across multiple batches.

497845 Improved Manufacturing Process for Semi-Synthetic Calicheamicin Linker-Payload *En Route* to Antibody-Drug Conjugates

Abstract:

Calicheamicin is produced by fermentation of *Micromonospora echinospora* ssp. calichensis and isolated from the broth. The natural product is then synthetically modified to attach a linker, and the â€œlinker-payloadâ€ product is reacted with antibodies to produce antibody-dug conjugates (ADCs) used in the treatment of cancer.Â We have developed an improved manufacturing process for the production of calicheamicin linker-payload and gained a better understanding of the chemistry at each step. Process improvements include fewer synthetic steps, fewer unit operations, reduced solvent usage, improved yield, and increased robustness. The improved process is now approved for commercial production of calicheamicin linker-payload as a drug substance intermediate for ADCs. Highlights of the process, learnings, development strategy, and benefits of implementation will be presented, along with selected aspects of the conjugation process to produce calicheamicin ADCs.

498048 Intracellular Invasion of *Salmonella* Drives Tumor Colonization of the Bacteria *in vitro*

Abstract:

Salmonella colonize tumors at ratios greater than ten thousand to one over any other organ in the body [1]. Unlike small molecule drugs, which disperse into tumors through passive diffusion, *Salmonella* can actively penetrate into tumors using flagella. The bacteria can also be genetically modified to produce DNA, RNA and peptide therapies. These characteristics make the bacteria well suited to deliver targeted and patient specific cancer therapies.

However, the mechanisms governing tumor colonization of *Salmonella* are not completely understood. While motility and chemotaxis influence tumor colonization [2–5], the bacteria also have the ability to intracellularly invade cancer cells. The dependence of intracellular invasion on tumor colonization has not yet been elucidated. In addition, bacterial motility and intracellular invasion have cross regulatory mechanisms that make each phenomenon dependent on one another. This study, therefore, aimed to test two interrelated hypotheses: **(1) Overexpressing the master motility regulator, *flhDC*, in *Salmonella* increases dispersion and colonization as well as intracellular invasion in tumor masses and (2) Upregulating flagellar dependent motility increases intracellular invasion, which, is the main driver of *Salmonella* tumor colonization** *in vitro*. *Salmonella* utilize the transcription factor complex, *flhDC*, to regulate flagellar synthesis and thus, motility. Flagellar synthesis also influences type three secretion system (T3SS) synthesis, which, *Salmonella*use to intracellularly invade epithelial cells.

To determine whether flagellar-dependent motility could increase bacterial intratumoral dispersion and intracellular invasion, *flhDC* was overexpressed within *Salmonella* (*flhDC* overexpressing *Salmonella*, or, foS) using the arabinose inducible *pBAD* system. A microfluidic tumor-on-a-chip model was used to determine the dynamics of foS colonization and intracellular invasion of tumor tissue *in vitro*. This tumor-on-a-chip model recapitulates *in vivo* nutrient gradients and cancer cell heterogeneity present within tumors. In order to quantify tumor colonization and intracellular invasion dynamics of foS within the *in vitro* tumor masses, the arabinose inducible *flhDC* gene circuit was coupled with constitutively expressed red fluorescent protein, *DsRed*, as well as *pSSEJ* induced green fluorescent protein (GFP) gene circuits. The *plac* promoter drove constitutive *DsRed* expression while the intracellularly active *pSSEJ* promoter only drove *GFP* expression within intracellular *Salmonella*. Time lapse microscopy was used to determine the dynamics of tumor colonization and intracellular invasion of *Salmonella*. When foS was grown in the presence of the inducer, arabinose, aqueous motility increased 33% ($P<0.05$). In the tumor-on-a-chip model, *flhDC* overexpression in *Salmonella* increased distal tumor colonization and intracellular invasion of cancer cells ten-fold and eight-fold, respectively ($P<0.05$), compared to *Salmonella* only. These results illustrate that overexpressing *flhDC* in *Salmonella* increases both intratumoral dispersion and intracellular invasion of the bacteria.

In order to isolate which factor, either motility or intracellular invasion, was the main driver of tumor colonization, a *ΔSipB*, non-intracellularly invading, strain of *Salmonella* was created. This strain was transformed with the same *DsRed/GFP* reporter system to assess bacterial tumor colonization and intracellular invasion dynamics. Not surprisingly, the *ΔSipB* strain intracellularly invaded cancer cells in the tumor-on-a-chip model five-fold less than *Salmonella* alone ($P<0.05$). However, the *ΔSipB* colonized tumor tissue five-fold less than *Salmonella* alone ($P<0.05$). In order to isolate the dependence of motility on tumor colonization, *flhDC* was overexpressed in the *ΔSipB*, non-intracellularly invading *Salmonella*, and administered to the tumor-on-a-chip along with *ΔSipB Salmonella* as a control. Overexpressing *flhDC* did not alter tumor colonization in any way when compared to *ΔSipB Salmonella* alone. These results demonstrate that intracellular invasion into cancer cells, independent of bacterial motility, drives tumor colonization of *Salmonella*.

Since *Salmonella* are complex microbes with several colonization mechanisms, it is important to understand which of these most influences bacterial survival and growth within tumors. This study has demonstrated that bacterial motility by itself does not influence tumor colonization. However, motility strongly influences the ability of *Salmonella* to intracellularly invade tumor cells, which, greatly increases tumor colonization *in vitro*. This study has provided a provided a foundational framework to investigate how motility and intracellular invasion could influence bacterial tumor colonization *in vivo*.

1. Forbes NS, Munn LL, Fukumura D, Jain RK. Sparse initial entrapment of systemically injected Salmonella typhimurium leads to heterogeneous accumulation within tumors. Cancer Res. 2003;63(17):5188–93.

2. Toley BJ, Forbes NS. Motility is critical for effective distribution and accumulation of bacteria in tumor tissue. Integr Biol. 2012;4:165.

3. Zhang M, Forbes NS. Trg-deficient Salmonella colonize quiescent tumor regions by exclusively penetrating or proliferating. J Control Release. 2015;199:180–9.

4. Kasinskas RW, Forbes NS. Salmonella typhimurium lacking ribose chemoreceptors localize in tumor quiescence and induce apoptosis. Cancer Res. 2007;67(7):3201–9.

5. Kasinskas RW, Forbes NS. Salmonella typhimurium specifically chemotax and proliferate in heterogeneous tumor tissue in vitro. Biotechnol Bioeng. 2006;94(4):710–21.

499885 Engineering Lipid Nanoparticles for Targeting Inflammation Site in Atherosclerosis

Abstract:

Atherosclerosis is an arterial disease characterized by the formation of plaques composed of cholesterol and macrophages. Low density lipoprotein (LDL) particles get attached to the proteoglycan of endothelium inside artery and become highly susceptible to oxidation and other chemical modifications. Oxidized LDLs induce inflammation by increasing the leukocyte adhesiveness and permeability to the endothelium. This condition may lead to further complications such coronary heart disease and myocardial infarction which are leading causes of death in the developing world. Targeted delivery of drugs to the diseased tissue can be efficient method in treatment of atherosclerotic plaque. It has several advantages as compared to systemic drug delivery such as reduced dosage of drug, lesser side effects, decrease in cost of therapy and increased accumulation of drug within the target tissue. Vitamin E is known for its antioxidant properties and can play a vital role in treatment of atherosclerosis.

In this study we plan to develop vitamin E loaded lipid nanoparticles (LNPs) targeting the inflammation site in atherosclerosis. The fluorescently tagged lipid nanoparticles were engineered using several phospholipids and were crosslinked with Hyaluronic acid (HALNPs). The reactive oxidation species (ROS) studies was performed using stem cells where oxidative stress was created using peroxide (H2O2) and then the vitamin E loaded lipid nanoparticles were targeted to these cells. These nanoparticles were also delivered to 3D co-culture of smooth muscle cells (SMC) and endothelial cells (EC) to study the trafficking and specific targeting. We observed that the targeted delivery of HALNPs loaded with vitamin E significantly decreased the oxidative stress induced in stem cells due to peroxide. We further plan to perform ex-vivo HALNP delivery experiments using arteries extracted from pig.

500410 Using Ultrasound to Enhance Tumor Cell Killing By Anti-PD1 and Doxorubicin-Loaded Particles

Abstract:

Metastatic melanoma is currently an incurable disease for which alternative treatments to chemotherapy alone are sought. The use of ultrasound (US) with microbubbles is a promising regimen for tumor treatment. In this study, the addition of US provided a means for controlled drug delivery through inertial cavitation. The aim was to obtain a synergistic or additive effect with respect to tumor killing by combining US with chemotherapy and checkpoint blockades. In vitro experiments combining US with blank PLGA particles showed higher cytotoxicity towards B16.F10 melanoma cells compared to either treatment alone. Furthermore, in vitro experiments demonstrated an ability to control the release kinetics of doxorubicin (dox) from dox-loaded PLGA particles through the application of US. Utilizing a melanoma murine model, the application of US to dox-loaded particles and/or anti-programmed cell death protein 1 (anti-PD1) yielded higher tumor regressions and survival rates compared to any treatment alone.

COMPUTATIONAL STUDY OF DEFORMABLE FUNCTIONALIZED NANOCARRIER ADHESION TO THE CELL SURFACE

Dr. Samaneh Farokhirad, University of Pennsylvania, Philadelphia, PA
Dr. Natesan Ramakrishnan, University of Pennsylvania, Philadelphia, PA
Dr. Portonovo S. Ayyaswamy, University of Pennsylvania, Philadelphia, PA
Dr. David M. Eckmann, University of Pennsylvania, Philadelphia, PA
Dr. Ravi Radhakrishnan, University of Pennsylvania, Philadelphia, PA

A functionalized nanocarrier (NC) adheres to its target cell through the multivalent interactions of its surface ligands with their corresponding cell surface receptors. The binding efficacy of the NC is governed by a complex set of physico-chemical and physiological parameters, that include its shape, size, surface chemistry, the receptor expression levels, the chemical composition, the mechanical state of the target cell membrane, the flow conditions, and the thickness of the glycocalyx layer near the target site. Rational design and optimization of functionalized NCs both require a fundamental understanding of the interplay between these seemingly disparate factors. New classes of NCs possess unique and exquisite chemistries for which rational design principles are quite distinct from rigid and regular-shaped NCs. We have considered a new class of biocompatible core–shell polymer-based NCs consisting of a lysozyme core with a dextran corona, which has the capability to host small-molecule drugs. The polymer microstructure was modeled as a fixed number of arms attached to a core which was verified experimentally [1], and each arm is modeled as connected segments of freely jointed chain. Using the proposed model, the elasticity modulus of NC was shown to fall in the range of moderately soft biological materials, e.g., cell, and tissue [2].

We focus on the adhesion of deformable crosslinked polymer based NCs functionalized with antibodies to receptors on the cell membrane. We have developed a computational framework by accounting for the mechanical properties of the NC, its emergent shape, and the multivalent binding interactions with receptors on the host cell membrane. The conformational state of the functionalized NC interacting with the cell membrane is defined in terms of energy functional and interaction potentials and is evolved through a Monte Carlo (MC) approach. The binding interaction between the ligand molecule and the target receptor is modeled using the Bell potential. We model the ligand molecule as a spring that is attached to one of the beads of the NC, and the bead is connected to the core through number of freely jointed bonds. The conformational state of the membrane surface is governed by an elastic Hamiltonian representing the membrane bending energy, and the receptor molecules are defined as cylindrical rods, with flexural angles distributed according to beam theory, and in addition are allowed to diffuse on the membrane surface [3].

The complex interplay between various energetic and entropic contributions in the system results in the avidity of the functionalized NC. In particular, in a recent study for a rigid NC, the importance of the loss in translational entropy of receptor motion and the change in entropy of the cell membrane undulations were demonstrated as crucial factors that define NC avidity in a cell-type dependent fashion [3]. In this study, a third dimension is explored to this problem by considering the NC flexibility.

We systematically vary the contributions of the NC conformational entropy in the case of NC with different crosslinks, to show effect of crosslinking on the NC adhesion free energy landscape. The configurational entropy of NC is computed using a quasiharmonic analysis where the eigen modes of internal vibrations are computed using a principal component analysis of the covariance matrix of configurational fluctuations captured in the MC trajectories. The covariance matrix for bead position fluctuations is calculated and corrected for removal of the translational and rotational movement contributions. Similar procedure is followed for NC away from the wall. We found that the adhered NC due to increase in density of beads, show less vibration while NC in bulk show larger vibration.

We proposed two different methods in both real and Fourier spaces to compute the conformational entropy of the undulating membrane which yield similar entropy values for the undulating membrane. It was shown that the NC binding is sensitive to the degree of crosslinking from very soft, deformable NCs to rigid spheres, as well as to the membrane excess area that accommodates the additional area though its undulations. Moreover, we conclude that the interplay between the entropic terms of the deformable NC and the cell membrane undulations play a significant role on the enhancement of the NC avidity. It was also found that the entropic contributions of thermal undulations of membrane and crosslinking density of NC cause much higher multivalency of binding interaction for deformable NC than for a hard sphere. These results are directly applicable in the design of deformable NCs for targeted drug delivery and biomedical imaging applications.

We acknowledge funding from the National Institutes of Health through grants NIBIB 1R01EB006818 and 1U01EB016027.

References

1. M. C. C. Ferrer, V. V. Shuvaev, B. J. Zern, R. J. Composto, V. R. Muzykantov and D. M. Eckmann (2014), "ICAM-1 Targeted Nanogels Loaded with Dexamethasone Alleviate Pulmonary Inflammation" *PLoS One*, 9, e102329u.
2. A. Sarkar, D. M. Eckmann, P. S. Ayyaswamy, R. Radhakrishnan (2015),"Hydrodynamic Interactions of Deformable Polymeric Nanocarriers and the Effect of Crosslinking," *Soft Matter*, 11, 5955.
3. Ramakrishnan, N., R. W. Tourdot, D. M. Eckmann, P. S. Ayyaswamy, V. Muzykantov, and R. Radhakrishanan (2016), "Biophysically Inspired Model for Functionalized Nanocarrier Adhesion to Cell Surface: Roles of Protein Expression and Mechanical Factors," *J. Royal Society Open Science*, 3, 160260.

500435 Acetalated Dextran Nanoparticles for Rapid and Glucose Responsive Insulin Delivery

Abstract:

Acetalated Dextran Nanoparticles for Rapid and Glucose Responsive Insulin Delivery

Lisa R Volpatti, Robert S. Langer, Daniel G. Anderson.

Koch Institute of Integrative Cancer Research & Department of Chemical Engineering, MIT, Cambridge, MA 02139.

Statement of Purpose: Diabetes mellitus is a class of diseases in which the body either does not produce or is insensitive to insulin. This results in poorly controlled blood glucose levels, often leading to severe complications including heart disease, stroke, and kidney failure. An attractive diabetes therapy is self-regulated delivery of insulin to recapitulate deficient insulin production or signaling.[1] In order for a self-regulated system to be therapeutically relevant, it must rapidly respond to changes in glucose concentration. To achieve this goal, acetal-functionalized dextran (Ac-dex) was synthesized (Fig. 1a) with various degrees of modification and formulated into nanoparticles encapsulating insulin, glucose oxidase (GOx), and catalase. GOx converts glucose to gluconic acid and reduces the pH in the microenvironment of the nanoparticles. The acetal groups are subsequently cleaved from the polymer in an acid catalyzed reaction, solubilizing the native dextran and releasing the insulin (Fig. 1b). We demonstrate that nanoparticles synthesized from Ac-dex with 55% of residues containing a cyclic acetal modification release 100% of encapsulated insulin within the first 2 hours of exposure to hyperglycemic conditions. We further preconcentrate these nanoparticles by encapsulating them in alginate microgels to form an injectable subcutaneous depot for the modulation of glucose levels *in vivo*.

Fig. 1. Glucose responsive insulin release. a) Schematic diagram of the synthesis of acetalated dextran. b) Schematic of nanoparticle system containing enzymes and insulin. c) Percent dextran degraded over time for polymers with varying degrees of acetal modifications. d) Protein release from 55% modified acetalated dextran nanoparticles containing glucose oxidase and catalase, showing enhanced kinetics with increasing glucose concentrations.

Methods: Ac-dex was synthesized as previously described[2] with reaction times ranging from 10 min to 60 min, and a sonication/evaporation method was used to produce protein-encapsulated nanoparticles with an optimized weight ratio of approximately 6:1.5:92.5 insulin:enzymes:Ac-dex. *In vitro* release studies for Ac-dex degradation were performed in either pH 7.4 or pH 4.7 buffer to represent normoglycemic and hyperglycemic conditions, respectively. Phosphate buffered saline with varying concentration of glucose was used to show glucose-responsive insulin release. Samples

were incubated at 37 °C with shaking, and aliquots were taken and analyzed at designated time points.

Results: Ac-dex was synthesized (Fig. 1a) with varying reaction times to yield polymers containing between 55% and 83% of glucose residues modified with a cyclic acetal group, as confirmed by NMR analysis. Single emulsion Ac-dex nanoparticles were prepared, and the percent of soluble dextran in the supernatant was measured over the course of 24 hours to determine the degradation kinetics of the polymers (Fig. 1c). Over 80% of the 55% modified Ac-dex was degraded within the first half hour, while 80% of the 83% modified Ac-dex was degraded only after 18 hours of incubation. The polymers with the fastest response time were used to synthesize nanoparticles encapsulated with insulin, and their release correlated strongly to the degradation kinetics. Notably, there was no significant release or leakage of insulin under physiological pH over the course of 24 hours. GOx and catalase were then encapsulated into nanoparticles from the 55% modified polymer, and insulin release was monitored in response to glucose (Fig. 1d). Within 2 hours, upon reduction of the pH in solution to around 4.5, 100% of encapsulated insulin was released in response to physiologically relevant hyperglycemic conditions (400 mg glucose/dL). We next preconcentrate nanoparticles into alginate microgels with an average diameter of 400 μm and concentration of 20 mg nanoparticles/mL without altering the insulin release kinetics. We further show that by encapsulating a blend of nanoparticles of varying stabilities (55%, 71%, and 78% modification) into alginate microgels, we can form a rapidly responding self-regulated system that could be used as a once daily injectable therapy for diabetic patients.

Conclusions: Alginate microgels encapsulating nanoparticles synthesized from a range of modified dextran polymers show promise in creating a fast-acting glucose-responsive insulin delivery system that has the potential to modulate glucose levels on a therapeutically relevant timescale.

Ref: [1]Bawa P et. al. Biomed. Mater. 2009; (4:022001). [2]Bachelder EM et al. J. Am. Chem. Soc. 2008; (130:10494–10495.)

498077 Succinylated Polyethylenimine Derivatives Enhance Gene Expression and Serum Stability *in vitro*

Abstract:

Polyethylenimine (PEI) is considered the gold standard of polymeric, non-viral gene delivery vectors. The highly cationic polymer readily condenses DNA through electrostatic interactions and can subsequently be internalized and processed by cells to incorporate the genetic payload into the existing genome to produce a therapeutic effect. However, the standard properties of PEI may not be optimal when attempting to overcome common inter- and intracellular barriers. For example, the repeating amine structure of PEI results in a large positive charge density, which is conducive to condensing DNA and leads to favorable electrostatic interactions with the negatively charged cell membrane. Alternatively, the excess positive charge leads to nonspecific binding of negatively charged serum proteins and can produce a significant cytotoxicity. Intracellularly, the large number of amines in PEI results in a large buffering capacity, allowing for endosomal and lysosomal escape via the â€˜proton-spongeâ€™ mechanism, but the strong electrostatic interactions between PEI/DNA can prevent facile unpackaging. Thus, the degree of protonation can greatly affect the behavior of PEI polyplexes.

In an attempt to address these issues, we have previously shown that acetylating fractions of primary and secondary amines of PEI can lead to increases in gene delivery efficiency. It was shown that modifying up to 43% of the primary amines with acetic anhydride mediated a higher transgene expression, while further reaction led to lower levels of expression. Furthermore, with increasing degrees of acetylation, polymer/DNA interactions were weakened and resulted in a higher efficiency of polyplex unpackaging within the cell. However, the modified polymers all had significantly lower buffering capacities, suggesting that there may be an optimal balance between the strength of polymer/DNA interactions and buffering capacity. Despite the significant increases in gene delivery efficiency realized by acetylated PEI, transfections in the presence of serum resulted in gene delivery activity lower than that of unmodified PEI in the absence of serum, implying that the acetylated polymers still suffered from serum instability.

In this study we modified PEI by reacting varying fractions of primary and secondary amines with succinic anhydride, resulting in zwitterionic PEI derivatives. Derivatives with 5.8% ($zPEI_5$), 8.0% ($zPEI_8$), 15.4% ($zPEI_{15}$), 26.7% ($zPEI_{26}$), and 35.0% ($zPEI_{34}$) of the primary amines succinylated were successfully synthesized. *In vitro* gene delivery activity increased over that of unmodified PEI in $zPEI_5$, $zPEI_8$, and $zPEI_{15}$ in the absence of serum, with $zPEI_5$ resulting in an approximately 10-fold increase. More importantly, the succinylated polymers exhibited a higher stability in the presence of serum, with $zPEI_5$ gene delivery in serum surpassing gene delivery by unmodified PEI in the absence of serum. Cytotoxicity of the succinylated polymers decreased with increasing degrees of succinylation. Additionally, succinylation effectively reduced the Î¶-potential from 15.1 mV to 15.0-9.6 mV while increasing particle diameter from 61 nm to 174-669 nm. The succinylated PEIs also exhibited decreased polymer/DNA interaction strengths, most likely attributed to the presence of negatively charged carboxyl groups which electrostatically repel the DNA. Similarly, the enhanced transfection activities suggest the presence of negatively charged functional groups

shielded the succinylated polyplexes from the effects of the serum proteins. Much like the balance of polymer/DNA interactions to buffering capacity, our results suggest a similar balance between shielding particles from negatively charged serum proteins while maintaining favorable interactions with the negatively charged cell membrane. Future studies will explore and compare the differences between the acetylated and succinylated polymers to illuminate mechanisms and cellular pathways to identify design criteria for polymeric gene delivery vectors.

493285 Engineering CRISPR-Cas9 Plasmid Loaded PLGA Nanoparticles to Repair Mutations in the *Tlr4* Gene in Mice

Abstract:

Gene therapy involves the permanent alteration of DNA to fix genetic diseases. The Clustered Regularly Interspaced Short Palindromic Repeats (CRISPR) technique has received much attention recently due to its potential for revolutionizing targeted genome editing with unprecedented precision and control. However, one of the main setbacks to implementing this new process is the difficulty in effective delivery of high molecular weight DNA into the cell. This CRISPR technique typically uses a circular plasmid DNA, which is hard to transfect into the cell due to natural defense mechanisms that inhibit foreign DNA materials from entering.

Previous gene transfection approaches have used both chemical and physical stimuli, but these often destabilize cellular membranes resulting in unacceptable cytotoxicity especially in test animals.[1] We hypothesize that using polymer nanoparticles will protect the plasmid from degradation en route and increase delivery into the cell without the harmful side effects seen using other methods for transfection.

Using an FDA-approved polymer and a modified nanoprecipitation method[2], a poly(lactic-co-glycolic acid) (PLGA) nanoparticle carrier was engineered to deliver a CRISPR-Cas9 plasmid designed to correct a single point mutation in the *Tlr4* gene in mice that cause them to not react to endotoxins, such as lipopolysaccharides (LPS). The PLGA carrier is also fluorescently labeled by encapsulation of a commercially available fluorophore, 6,13-Bis(triisopropylsilylethynl) pentacene (TIPS pentacene), that had not been used before in the biomedical field until recently.[3]

However, before even getting to cell studies, the particles must first be appropriately designed, fabricated, and characterized to ensure that the requirements of the application are met. The focus of the research to be presented is on the particle processing and engineering that precedes the cell and animal studies. This work addresses the questions: What is the best solvent system to use for particle fabrication? What are the resulting sizes and zeta potentials? What is the composition of the polymeric carrier as well as what was successfully loaded? How can the particles be processed for long-term storage stability? Once these are first answered, then the method can be optimized to increase DNA loading. Initial studies showed a loading of 0.61 wt% of the ~8500 bp plasmid using an amine end capped PLGA at an encapsulation efficiency of 82%. These results are already comparable to the typical 1 wt% loadings seen for DNA with only 4500 bp.[2] From the promising framework set up, the next steps include increasing the target DNA loading, and studying the release kinetics of the biologic. This systematic design and characterization of the particles can be translated to a variety of different applications in the area of nanoparticle drug delivery.

References

1. Bauer, M.; Kristensen, B. W.; Meyer, M.; Gasser, T.; Widmer, H. R.; Zimmer, J.; Ueffing, M., Toxic effects of lipid-mediated gene transfer in ventral mesencephalic explant cultures. *Basic Clin Pharmacol* **2006,** *98* (4), 395-400.

2. Niu, X. M.; Zou, W. W.; Liu, C. X.; Zhang, N.; Fu, C. H., Modified nanoprecipitation method to fabricate DNA-loaded PLGA nanoparticles. *Drug Dev Ind Pharm* **2009,** *35* (11), 1375-1383.

3. McDaniel, D. K.; Jo, A.; Ringel-Scaia, V. M.; Coutermarsh-Ott, S.; Rothschild, D. E.; Powell, M. D.; Zhang, R.; Long, T. E.; Oestreich, K. J.; Riffle, J. S.; Davis, R. M.; Allen, I. C., TIPS pentacene loaded PEO-PDLLA core-shell nanoparticles have similar cellular uptake dynamics in M1 and M2 macrophages and in corresponding in vivo microenvironments. *Nanomedicine: Nanotechnology, Biology and Medicine* **2017,** *13* (3), 1255-1266.

489001 Achieving Long-Term Stability of Lipid siRNA Nanoparticles: Examining the Effect of pH, Temperature, and Lyophilization

Abstract:

The broadest clinical application of siRNA therapeutics will be facilitated by drug-loaded delivery systems that maintain stability and potency for long times under ambient conditions. In the present study, we seek to better understand the stability and effect of storage conditions on lipidoid nanoparticles (LNPs), which have been previously shown by our group and others to potently deliver RNA to various cell and organ targets both in vitro and in vivo. Specifically, this study evaluates the influence of pH, temperature, and lyophilization on LNP efficacy in HeLa cells. When stored under aqueous conditions, we found that refrigeration (2°C) kept LNPs the most stable over 150 days compared to storage in the -20°C freezer or at room temperature. Because the pH of the storage buffer was not found to influence stability, it is suggested that the LNPs be stored under physiologically appropriate conditions (pH 7) for ease of use. Although aggregation and loss of efficacy were observed when LNPs were subjected to freezeâ€"thaw cycles, their stability was retained with the use of the cryoprotectants, trehalose, and sucrose. Initially, lyophilization of the LNPs followed by reconstitution in aqueous buffer also led to reductions in efficacy, most likely due to aggregation upon reconstitution. Although the addition of ethanol to the reconstitution buffer restored efficacy, this approach is not ideal, as LNP solutions would require dialysis prior to use. Fortunately, we found that the addition of trehalose or sucrose to LNP solutions prior to lyophilization facilitated room temperature storage and reconstitution in aqueous buffer without diminishing delivery potency.

510417 Unlocking the Potential of the Animal Mycobiome for Bioenergy and Animal Health

Abstract:

Complex microbiomes within the digestive tract of herbivores are increasingly of interest for the discovery of novel lignocellulose degrading enzymes and biocatalysts for the production of biofuels. Central to this community are anaerobic fungi, which are prolific digesters of crude lignocellulose through a combination of powerful secreted enzymes and invasive growth. Despite their outsize contributions to digestion in large herbivores, the mycobiome has historically been overlooked due to its low abundance within the microbiome. Here, we enrich for these fungi and pursue a multi â€"omics approach integrated with biochemical characterization, growth phenotypes, and heterologous expression to probe the catalytic potential of these overlooked organisms for bioenergy and improved animal health. We have assembled a panel of novel fungal isolates from both zoo and farm animals representing several types of fungal genera. We demonstrate that these fungi express and secrete tightly regulated carbohydrate active enzymes (e.g. cellulases) that are active on a broad array of untreated agricultural residues, food wastes, and woody forestry products. More importantly, these enzymes are robust against increasing levels of syringyl lignin, which is a natural plant defense against microbial degradation. Thus, these anaerobic fungi have significant potential to pretreat diverse and difficult lignocellulosic feedstocks for bioenergy production without enzyme supplementation from other organisms. Also encoded within the genomes of these organisms are large reservoirs of natural product gene clusters that rival those of super-producing actinomycetes. These natural products are hypothesized to mediate microbiome function, and thus animal health, and offer potentially novel biocatalytic routes to advanced biofuels and medicines. We are currently validating these clusters and characterizing their products and regulation through metabolomics and co-culturing studies. Our ongoing efforts suggest that the animal mycobiome is a rich untapped reservoir for bioenergy, and provides a roadmap for its application in diverse engineering efforts.

510418 40,000+ Genomes and Counting: Computational Lessons from Building a Giant Culture Collection

Abstract:

Biological sequence data is as rich and complex as it is inexpensive to produce, and the pace of DNA sequencing technology development quickly renders current best practices and conventions for storing and organizing biological sequence information obsolete and/or irrelevant. This is a particular challenge for companies specializing in microbiome science and amassing large collections of microbial isolates -- how can we account for this ever-changing environment and maintain a stable data infrastructure at scale? At AgBiome we have selected a technology stack for genomic data science that enables rapid development, persistent relevance of analyses, and reproducibility -- we employ suite of opensource tools including NextFlow, Docker, Mash/Sourmash, and Jupyter. In this talk, I will contrast data analysis in academia versus industry and present an analysis case study involving one of AgBiome's products in development to demonstrate how we battle the head-spinning dynamism of genomics.

492809 Engineering Cross-Feeding Co-Cultures As a Platform for High-Throughput Screening of Microbial Strain Libraries for Enhanced Biomolecule Production

Abstract:

A variety of methods have been developed to explore vast phenotype landscapes in a controlled manner. The applicability of these methods to the development of bacterial production strains is, however, limited by the requirement of a screening method with high enough throughput to investigate all the variants generated. Products that require chromatography-based screens typically reach a throughput of only ~10^2 variants per day, substantially lower than those with colorimetric or growth-based assays. Here, we present a new method for screening *Escherichia coli* strains for improved production of amino acids and/or their intermediates. Amino acid intermediates can be diverted towards production of a variety of other useful compounds, including higher alcohol biofuels.

This method uses a cross-feeding auxotroph system to convert production levels of amino acid or amino acid intermediates into co-culture growth characteristics that are detectable in a high-throughput context. As proofs of concept we examine two target molecule cases: tryptophan and 2-ketoisovalerate (a precursor of valine, leucine, and the drop-in biofuel isobutanol). We demonstrate that suitable pairs can be selected so that improvements in production by the production strain lead to increased co-culture growth rate and changes in the final composition of the population. We explore implementation of this co-culture screening method on petri dishes and in microdroplets, demonstrating its ability to identify rare over-producers from a library. We investigate the dynamic ranges of our model systems and ways in which they may be expanded.

495657 Engineering Microbial Consortia for Bioelectrocatalysis Processes

Abstract:

Bioelectrochemical systems hold great promises for diverse bioenergy and environmental applications, including microbial fuel cells (MFC) for green and sustainable bioelectricity production from wastewater and biomass, autonomous power sources for sensors and beacons by marine sediment MFCs, sea waters desalination by microbial desalination cells, hydrogen production by microbial electrolysis cells, and microbial electrosynthesis for CO_2 reduction and chemicals production.

However, exoelectrogens, such as *Shewanella* and *Geobacter* being widely studied in MFCs, could only use limited spectrum of carbon sources. To expand the carbon source range being used in MFCs, we herein rationally designed a few microbial consortia to broaden the carbon sources that could be used in MFCs and bioelectrochemical systems. Firstly, we constructed a glucose-fed fungus-bacteria microbial consortium including a fermenter (*Saccharomyces cerevisiae*) in which the ethanol pathway was knocked out and the lactic acid biosynthesis pathway was introduced into *Shewanella cerevisiae*, and an exoelectrogen (*S. oneidensis* MR-1). We optimized the co-culturing conditions of the microbial consortium to achieve an optimal coordination between carbon source metabolism of the fermenter and extracellular electron transfer of the exoelectrogen, such that lactate, the metabolic product of glucose by the recombinant *S. cerevisiae*, was continuously supplied to *S. oneidensis* in a constant level until glucose exhaustion. This metabolic coordination between the fermenter and the exoelectrogen enabled bioelectricity production in a glucose-fed MFC. Secondly, we used the design principle of "division-of-labor" to construct a three-species microbial consortium for power generation, consisting of recombinant *Escherichia coli*, *Bacillus subtilis* and *Shewanella oneidensis*. In this consortium, *E. coli* digested glucose to produce lactate as a carbon source, *B. subtilis* produced riboflavin as an electron shuttle, and *S. oneidensis* served as the exoelectrogen to generate electricity. The three species formed a cross-feeding microbial consortium, performing "better together" for power generation.

Our study provided new insights into the rational design of more efficient, stable, and robust synthetic microbial consortia applicable in bioenergy and environments.

References:

1. Yue Liu, Mingzhu Ding, Wei Ling, Yun Yang, Xiao Zhou, Bingzhi Li, Tao Chen, Yong Nie, Miaoxiao Wang, Boxuan Zeng, Xia Liwi, Hong Liu, Baode Sun, Heming Xu, Jiamei zhang, Yi Jiao, Yanan Hou, Hui Yang, Sijia Xiao, Qucheng Lin, Xinzi He, Wenjie Liao, Zeqi Jin, Yufei Xie, Bofeng Zhang, Tianyu Li, Xi Lu, Jiabei Li, Fan Zhang, Hao Song*, Yingjin Yuan*, "A three-species microbial consortium for power generation", Energy & Environmental Science, DOI: 10.1039/C6EE03705D (in press, 2017)

2. Tong Lin, Xue Bai, Yun Yang, Jingyu Wang, Yidan Hu, Bingzhi Li, Ying-Jin Yuan, Hao Song*, "Synthetic Saccharomyces cerevisiae-Shewanella oneidensis consortium enables glucose-fed microbial fuel cell", AIChE Journal, (in press, 2017)

496525 Screening Rhizobacteria Interactions Using a High-Throughput Microwell Array Platform

Abstract:

Plant root-associated microbial communities consist of thousands of unique species that function together to offer beneficial properties to the host. These benefits include nutrient uptake, plant hormone regulation, and pathogen protection. The structure and function of these communities is shaped by microbe-microbe interactions, most of which are unknown or poorly characterized because traditional co-culture methods can only examine a few interactions in a single assay. In addition, traditional methods typically probe interactions between colonies or bulk cell populations, overlooking single-cell interactions that often play a pivotal role in community function. To improve these characterization capabilities, we are developing a unique screening tool termed the â€œmicrobe arrayâ€, designed to discover unknown inter-cellular interactions that exist between plant pathogens and root microbiome members. By implementing previously developed techniques in microfabrication, the platform uses micro-scale wells to form thousands of unique combinations between small numbers of fluorescently-labeled plant pathogens and isolates from a root microbiome. Each well serves as a unique and independent co-culture site, and the growth of the pathogen in each well of the array can be monitored in parallel. This allows for identification of wells where abnormal pathogen growth occurs. We are currently developing materials and methods that enable extraction of cells out of individual wells of interest, allowing for the sequencing and discovery of microbiome members that promote or inhibit pathogen growth. Current studies are focused on screening interactions between *Agrobacterium tumefaciens*, the agent of Crown Gall Disease, and microbiome members from *Heliantus annuus* root. Identification of the critical species that inhibit plant pathogen growth and colonization will inform the development of constructed microbial communities for use in new biocontrol strategies. Successful development will also result in a new and powerful screening tool for rapidly uncovering microbe-microbe interactions that occur in any microbiome of interest.

498112 Engineering Modular Microbial Communities for Cellulose Utilization and Bioproduct Synthesis

Abstract:

Recent environmental concerns have increased the need to develop systems which can utilize waste materials to make chemicals. Cellulosic biomass, an abundant waste material, can be degraded into sugar monomers to serve as an inexpensive carbon source for microbial systems. Current approaches for biomass utilization rely on pretreatment processes and expensive purification schemes. To address these challenges, microbial communities can be engineered to synergistically perform complex tasks, by dividing the demands among multiple specialized modules, or functional units, to maximize the efficiency of desired product formation. Therefore, we aim to build microbial systems with interchangeable modules, with the first module dedicated to cellulose degradation and the second module specialized for bioproduct synthesis. To engineer synthetic communities for the production of biofuels and pharmaceuticals, we are building and testing mixed cultures of *Bacillus megaterium* and *Escherichia coli* strains. We will describe our recent progress engineering these two modules and assembling them for activity.

Complete degradation of cellulose requires a combination of three cellulase activities â€" an endoglucanase, a cellobiohydrolase, and a beta-glucosidase. We have developed a cellulose degradation module consisting of *Bacillus megaterium* strains for the secretion of cellulases. A small library of signal peptides (SPs) was selected to tag each cellulase, and thereby used to trigger protein secretion in *B. megaterium*. Cellulase activity against amorphous cellulose was confirmed through a series of bioassays, and the most active signal peptide constructs were identified for each cellulase. Expression optimization for each strain focused on testing secretion and activity under different expression temperatures, induction time points, expression lengths, and media conditions. Activity of the optimized cellulase secretion strains was characterized individually and in tandem to assess synergistic cellulolytic activity. We have also demonstrated that the secreted cellulase activity can be used to degrade cellulose into glucose monomers to support growth of an *E. coli* production module. We will describe our progress investigating the effect of different inoculation ratios, expression conditions, and community compositions for the cellulose degradation module to determine optimized cellulase secretion and activity conditions for the system. Further system development has also targeted the second module, with the introduction of a violacein-producing *E. coli* strain. The cellulose degradation module can be integrated with many different bioproduct synthesis modules to build complex systems for the synthesis of high value products from cellulose.

498559 Symbiotic Microbial Communities for Cleaning Agricultural Waters and Bioenergy Production

Abstract:

With increase in intensive crop and livestock production, the excess application of P as fertilizer and manure contributes to the build-up of soil P levels causing eutrophication [1, 2]. Currently a curative solution to combat water pollution by recuperating N and P, recycling the P for food production by closing the nutrient loop to effectively manage dwindling rock P reserve are the major challenges. The most common phosphorus removal techniques for municipal and industrial wastewater are biological treatment and chemical precipitation. Bio-valorization offers an excellent alternative to chemical treatment technologies as the biomass produced can be a source of protein and other valuable bio-chemicals and bio-fuels in this low cost technology. In this present work, a novel symbiotic microbial community called mycoalgae biofilm [3] was developed using the isolated strains of polyphosphate accumulating fungi and microalgae for cleaning agricultural waters and bioenergy production. We want to apply this platform for the efficient recovery and recycling of nutrients in waste stream. The effect of various process conditions on the nutrient recovery efficiency and the changes in the biofilm biomass composition was studied. Since the biofilm will be used for nutrient recovery from an array of samples, a synthetic medium for the cell growth was used as the medium for the biofilm development to have different initial conditions. Depending on the nature of the wastewater and its composition, and the fungi-microalgae combination intended to use for bio-remediation, the developed mycoalgae biofilm can be a sustainable microbial process for food, feed or energy source. The composite biofilm shows promising results in wastewater treatment especially in removing the N, P, suspended solids and other organics from nutrient polluted water in a single-step reactor with easy biomass recovery and high efficiency, when compared to the individual pure cell cultures. The biofilm attaches to the polymer matrix, leaving behind the clean water for recirculation, and the composite biomass for nutrient recycle. The commercial expansion of this biofilm system could be a step ahead in having integrated and sustainable water treatment systems.

References

1. E.K. Wilson, Danger from microcystins in Toledo water unclear, Chem. Eng. News, 92 (32) (2014), p. 9
2. Ye, Y., Gan, J., & Hu, B. (2015). Screening of phosphorus-accumulating fungi and their potential for Phosphorus removal from waste streams. Applied Biochemistry and Biotechnology, 177(5), 1127-1136.
3. Aravindan Rajendran and Bo Hu, Mycoalgae biofilm: development of a novel platform technology using algae and fungal cultures, Biotechnology for Biofuels, 20169:112, DOI: 10.1186/s13068-016-0533-y

495978 Technoeconomic Evaluation of Cyclosporine Crystallisation Intensification Using a Cascade of M.S.M.P.R. Crystallisers

Abstract:

Continuous Pharmaceutical Manufacturing (CPM) has been recognised as a promising new production paradigm over the currently implemented batch methods, which imply high volumes of waste and inefficiencies [1]. Whilst the literature demonstrates many continuous flow syntheses of active pharmaceutical ingredients [2], technological innovation is required to realise the benefits of integrated end-to-end CPM processes, including purification and separation unit operations. Process systems engineering (PSE) methodologies allow the screening of candidate CPM processes for their feasibility and viability prior to committing to significant time and financial investments in experimentation and scaleup [3]. Recent efforts in modelling and optimisation of the upstream CPM of APIs have demonstrated the importance of conducting PSE studies in establishing economically viable separation processes [4], realising issues in modelling of particulate processes for solids handling in continuous operation [5] and establishing limitations attainable process improvements in pharmaceutical manufacturing [6].

Cyclosporine is an immunosuppressant drug for the prevention of organ transplant rejection. The continuous cooling crystallisation of cyclosporine has recently been investigated using continuous mixed suspension, mixed product removal (MSMPR) crystallisers, due to their broad applicability and ease of integration into existing pharmaceutical processes [7]. The effect of process configuration [8], mother liquor and solids recycle [9], number of stages [10] on crystallisation yields, purities and product size distributions have been studied to demonstrate the feasibility of continuous crystallisation of the API. Technoeconomic comparisons between batch and continuous crystallisation configurations for cyclosporine to validate the viability of continuous crystallisations are yet to be implemented.

The present paper work develops a steady-state process model for the continuous cooling crystallisation of 100 kg per annum of cyclosporine from acetone including a systematic technoeconomic comparison of various continuous process configurations with the batch method, with a focus on computationally exploring and demonstrating process intensification. The effect of the number of MSMPR stages and varying recycle considerations (mother liquor versus solids recycle) on attainable yields, process material efficiencies and costs are compared. Mother liquor recycle requires vacuum evaporation of solvent, whilst solids recycle uses gravity driven separation to allow recycle of concentrated slurry to MSMPR stages. The developed model simultaneously solves MSMPR mass balances, crystal population balances and crystallisation kinetics to calculated stage and overall crystallisation yields. Prediction of temperature-dependent cyclosporine solubilities in acetone uses a regressed surrogate polynomial established from published experimental data [11]. Temperature- and concentration-dependent growth and nucleation rates are computed on the basis of published kinetic parameters [10] for accurate calculation of attainable yields. Material efficiencies are quantified via popular green chemistry metrics, namely the process mass intensity (PMI) and the mass productivity (MP), which offer quantitative insight into process intensification potential, for each configuration considered. Unit volumes and material throughputs required to meet the plant API capacity are scaled to account for stage inefficiencies. Economic analyses are conducted via an established methodology for pharmaceutical manufacturing processes [12].

The continuous MSMPR crystallisation configurations investigated show potential for application in the CPM of cyclosporine from the calculated attainable material efficiencies and total costs. The sensitivity of process performance to various operating parameters has been explored as well, indicating the significance of rigorous design space investigation of CPM processes. Our work encourages subsequent technoeconomic studies and life cycle assessments (LCA) of the CPM of cyclosporine for further investigation of process feasibility and viability.

The present study exhibits the importance of conducting PSE studies of candidate continuous separation processes for CPM prior to further process development. We demonstrate the economic viability of CPM to facilitate the transition from batch methods, and quantify the potential intensification benefits of improved material efficiency of continuous separation processes in comparison to their batch counterparts. This study expedites further process modelling for the CPM of cyclosporine and other candidates, as well as optimisation of continuous separation processes to realise the full benefits of continuous operation in the pharmaceutical industry.

LITERATURE REFERENCES

[1] Poechlauer, P., Colberg, J., Fisher, E., Jansen, M., Johnson, M. D., Koenig, S. G., Lawler, M., Laporte, T., Manley, J., Martin, B., O'Kearney-McMullan, A., 2013. Pharmaceutical roundtable study demonstrates the value of continuous manufacturing in the design of greener processes, *Org. Process Res. Dev.* **17**(12): 1472–1478.

[2] Baumann, M., Baxendale, I. R., 2015. The synthesis of active pharmaceutical ingredients (APIs) using continuous flow chemistry, *Beilstein J. Org. Chem.* **11**: 1194–1219.

[3] Teoh, S. K., Rathi, C., Sharratt, P., 2015. Practical assessment methodology for converting fine chemicals processes from batch to continuous, *Org. Process Res. Dev.* **20**(2): 414–431.

[4] Diab, S., Gerogiorgis, D. I., 2017. Process modelling, simulation and technoeconomic evaluation of separation solvents for the Continuous Pharmaceutical Manufacturing (CPM) of diphenhydramine, *Org. Process Res. Dev.* DOI: 10.1021/acs.oprd.6b00386, in press.

[5] Rogers, A., Ierapetritou, M., 2014. Challenges and opportunities in pharmaceutical manufacturing modelling and optimization, *Comput. Aided Chem. Eng.* **34**: 144–149.

[6] Benyahia, B., Lakerveld, R., Barton, P. I., 2012. A plant-wide dynamic model of a continuous pharmaceutical process, *Ind. Eng. Chem. Res.* **51**(47): 15393–15412.

[7] Su, Q., Nagy, Z. K., Rielly, C. D., 2015. Pharmaceutical crystallisation processes from batch to continuous operation using MSMPR stages: Modelling, design, and control, *Chem. Eng. Process. Process Intensif.* **89**: 41–53.

[8] Alvarez, A. J., Singh, A., Myerson, A. S., 2011. Crystallization of cyclosporine in a multistage continuous MSMPR crystallizer, *Cryst. Growth Des.* **11**(10): 4392–4400.

[9] Li, J., Trout, B. L., Myerson, A. S., 2016. Multistage continuous mixed-suspension, mixed-product removal (MSMPR) crystallization with solids recycle, *Org. Process Res. Dev.* **20**(2): 510–516.

[10] Li, J., Lai, T. C., Trout, B. L., Myerson, A. S., 2017. Continuous crystallization of cyclosporine: the effect of operating conditions on yield and purity, *Cryst. Growth Des.* **17**(3): 1000–1007.

[11] Wong, S. Y., Tatusko, A. P., Trout, B. L., Myerson, A. S., 2012. Development of continuous crystallization processes using a single-stage mixed-suspension, mixed-product removal crystallizer with recycle, *Cryst. Growth Des.* **12**(11): 5701–5707.

[12] Schaber, S. D., Gerogiorgis, D. I., Ramachandran, R., Evans, J. M. B., Barton, P. I., Trout, B. L., 2011. Economic analysis of integrated continuous and batch pharmaceutical manufacturing: A case study, *Ind. Eng. Chem. Res.* **50**(17): 10083–10092.

489029 Application of Advanced Process Control Tools in Continuous Processing

Abstract:

Continuous manufacturing is now widely discussed in the pharmaceutical industry. The availability of turn-key lines offered by various suppliers allows the transfer of established and new products to the continuous manufacturing approach. At the moment the focus is mainly on solid oral dosage forms including the common routes via wet and dry granulation or direct compaction. From a processing point-of-view the shift to continuous mode is supported by the fact that most common unit operations are continuous machinery by design, such as feeding, tableting or granulation. One challenge is the integration of all units into a single line, which needs implementation of interfaces, transport units and buffers. The remaining challenge is the establishment of control concepts that enable a fully automated, yet still flexible process, capable of handling start-up and shut down, as well as of out-of-spec events. Tools for in-line monitoring of critical quality attributes and real-time acquisition of process data have been widely demonstrated. Process models, fast enough for design of controllers, have been developed recently. However, most continuous processes filed by pharmaceutical industry still remain on a level 2 control strategy based on in-process controls without control loops.

In this work different tools for advanced process control were developed and investigated and their implementation in processes or unit operations will be shown. This includes novel process analytical technologies enabling real-time data acquisition of statistical representative numbers or even each single dosage form, concepts for out of spec material handling based on modelling of residence time distribution or model predictive control. The latter approach was developed for a feeding and blending unit. The rotation speed of the blender is manipulated to control the mass hold-up and the outlet mass flow rate in order to mitigate fluctuations of homogeneity caused by feeder events. Another example deals with the discharge of out-of-spec material, in combination with an operating concept for a tablet press based on different pre-defined set points of the turret speed. Recent developments in the field of PAT will complete the presentation.

491286 Nonlinear First-Principle Model-Based Control of a Continuous Pharmaceutical Manufacturing Process

Abstract:

Model-based process control is one of the pillars of continuous pharmaceutical manufacturing process by the QbD approach. The first-principle approach, which comes from in-depth understanding of the governing transport phenomena and performance of unit operations, yields developing of predictive models with which the continuous processes can be manipulated, controlled, and optimized. Despite empirical modeling, where limited operation conditions and process status are being mapped within a DoE, other unseen nonlinearities in the process can be captured in a nonlinear first-principle model to optimally control the process.

In this work, a nonlinear first-principle distributed dynamic model was developed to control a drying process for continuous manufacturing of orally dissolving films loaded with a pharmaceutical compound. The drying model is the core of active control for the process, in which the process parameters (such as air temperature, or production rate), feed compositions (like solid content and concentration), or final productsâ€™ properties (for instance the required final moisture content) can be the manipulating variables. Simulation of transport phenomena, diffusion controlled mass transfer (mutual solvent-polymer vs. moisture sorption isotherm), the spatial distribution of solvent concentration in the film and the moving boundary on the interface due to the thin film shrinkage during drying are some of the complexities of the model development. The final quality attribute of the product is set as the control set point and other process variables were manipulated in the nonlinear model control (MIMO) to converge the process in the design space of variable inputs. The model funds the steps for future phases of PAT implementation, RTRt, and process optimization.

497129 A Fault-Tolerant Control Design for Real-Time Release in Continuous Manufacturing of Solid Dose Using Direct Compaction

Abstract:

Â A Fault-tolerant Control Design for Real-time Release in Continuous Manufacturing of Solid Dose using Direct Compaction

Qinglin Su, Mariana Moreno, Jianfeng Liu, Sudarshan Ganesh, Yasasvi Bommiready, Marcial Gonzalez, Gintaras V. Reklaitis, Zoltan K. Nagy

School of Chemical Engineering, Purdue University,
West Lafayette, IN 47907

Thomas Oâ€™Connor, Tian Geng

CDER, Office of Pharmaceutical Quality, U.S. Food and Drug Administration,
Silver Spring, MD 20993

Abstract

Advantages of continuous manufacturing in the pharmaceutical industry have been demonstrated through extensive research studies in the last decade [1], but mostly at the conceptual or theoretical level. Only few continuous manufacturing processes have been approved by the US Food and Drug Administration thus far, e.g., the Vertex Orkambi and Janssen Darunavir [2]. To move from a conceptual continuous design to a practically operating pilot plant or manufacturing process, the efforts of regulators, academics, and industry are now focused more on quality control strategies [3]. Specifically, the active process control design for real-time release and its risk management in continuous manufacturing [4, 5] have been considered as critical for successful implementation of continuous manufacturing processes.

In this study, a fault-tolerant control design for real-time release is implemented, according to our previous systematic framework for control design and analysis [5], to a pilot plant at Purdue University for continuous manufacturing of solid dosage pharmaceuticals using direct compaction. First, beside the fundamental control loops provided by the equipment vendors for unit operations of feeding, blending, tableting, etc., to manipulate the critical processing parameters (CPPs), supervisory active feedback and feedforward process control strategies are designed to maintain the critical quality attributes (CQAs) based on process analytical technology (PAT) tools and final product quality testing, e.g., Sotax Auto Test 4. In the active process control design, the major system dynamics are identified with state-space models using historical operating data in the pilot plant, and a series of rational control design and analysis metrics are employed to evaluate the control performance and screen the potential design risks [5]. For example, a Relative Gain Array (RGA) analysis is used to evaluate the system interactions and select the best-performing control pairings in the feeding-blending units, where the diagonal pairing is preferred under low system frequency domain, and is risky under higher frequency as is the case when the feeder nozzle is accidently attached to the blender hopper frame.

In the proposed fault-tolerant control design, a risk mapping, assessment and planning (Risk MAP) strategy is also considered to address the potential risks associated with the above active process control design. The active process control system is designed to mitigate or avoid quality risks. However, implementation of risk reduction measures can introduce new risks into the system or increase the significance of other existing risks (ICH Q9). In this study accepted residual risks and new potential risks to product quality after the implementation of the active process control system for the continuous manufacturing process are examined. . For example, when faults occurred in the feeding-blending units, the process performance indices, e.g., process capabilities index, are below their minimum targets. Hence, for real-time release purposes, the proposed fault-tolerant control design is evaluated experimentally and in simulation mode under these risk scenarios, e.g., process disturbance, probe fouling, instrument calibration errors, etc. It is as expected that the fault-tolerant control design with advanced process systems engineering tools, such as process monitoring, data reconciliation, state estimation, etc., can mitigate those potential risks that are common causes of variations in an otherwise stable process.

Steady-state operation of the pilot-plant is also executed over longer durations to validate the proposed fault-tolerant control design. Concluding remark will also be offered, suggesting potential guidelines for achieving real-time release in continuous solid dosage manufacturing.

References

[1] M. Ierapetritou, F. Muzzio and G. Reklaitis, "Perspectives on the continuous manufacturing of powder-based pharmaceutical processes," *AIChE Journal*, vol. 62, no. 6, pp. 1846-1862, 2016.

[2] L. Yu, "Continuous Manufacturing Has a Strong Impact on Drug Quality," US Food and Drug Administration, 12 April 2016. [Online]. Available: https://blogs.fda.gov/fdavoice/index.php/2016/04/continuous-manufacturing-has-a-strong-impact-on-drug-quality/. [Accessed 11 April 2017].

[3] S. L. Lee, T. F. O'Connor, X. Yang, C. N. Cruz, S. Chatterjee, R. D. Madurawe, C. M. V. Moore, L. X. Yu and J. Woodcock, "Modernizing pharmaceutical manufacturing: from batch to continuous production," *Journal of Pharmaceutical Innovation,* vol. 10, no. 3, pp. 191-199, 2015.

[4] R. Singh, M. Ierapetritou and R. Ramachandran, "An engineering study on the enhanced control and operation of continuous manufacturing of pharmaceutical tablets via roller compaction," *International Journal of Pharmaceutics*, vol. 438, pp. 307-326, 2012.

[5] Q. Su, M. Moreno, A. Giridhar, G. V. Reklaitis and Z. K. Nagy, "A systematic framework for process control design and risk analysis in continuous pharmaceutical solid-dosage manufacturing," *Journal of Pharmaceutical Innovation*, p. in review, 2017.

496819 Intensified Crystallization of Sulfathiazole and Sulfanilamide By Direct Co-Crystal Assembly

Abstract:

Most of the studies mainly focus on the co-crystallization step rather than on the synthesis and co-crystallization together. Based on the concept of direct co-crystal assembly, we have successfully produced 1:1 co-crystals of sulfathiazole-theophylline (STZ-THE) and 1:1 co-crystals of sulfathiazole-sulfanilamide (STZ-SNM) by integrating the chemical synthesis and co-crystallization. Interestingly, the yields of the integration method were higher than the ones of normal reactive crystallization methods, and the amount of purities were also reduced. For the chemical reaction that suffered from the low yields or had the impurity which was hard to be separated, the formation of co-crystals might be an alternative strategy to improve the efficiency of purification and the value of drugs as well. The modification of loading stoichiometry of reactants to co-formers could lead to the difference of product compositions. Both STZ-THE and STZ-SNM exhibited different pH solubility profiles. In other words, co-crystallization could change the pH effect active pharmaceutical ingredient.

497068 Direct Co-Crystal Assembly from Synthesis to Co-Crystallization

Abstract:

Pharmaceutical co-crystals were assembled directly via chemical synthesis by cooling under three new strategies. The screening method for direct co-crystal assembly by a chemical reaction was developed based on liquid-assisted grinding. The ternary phase diagram including a chemical reaction and co-crystallization was established for a 2 : 1 co-crystal of benzoic acidâ€“sodium benzoate.

502452 Continuous Preparation of 1:1 Haloperidol-Maleic Acid Salt By a Novel Solvent-Free Method Using a Twin Screw Melt Extruder

Abstract:

Salts are generally prepared by acid-base reaction in relatively large volumes of organic solvents, followed by crystallization. In this study, the potential for preparing a pharmaceutical salt between haloperidol and maleic acid by a novel solvent-free method using a twin-screw melt extruder was investigated. The pH-solubility relationship between haloperidol and maleic acid in aqueous medium was first determined, which demonstrated that 1:1 salt formation between them was feasible (pH_{max} 4.8; salt solubility 4.7 mg/mL). Extrusion of a 1:1 mixture of haloperidol and maleic acid at the extruder barrel temperature of 60 °C resulted in the formation of a highly crystalline salt. The effects of operating temperature and screw configuration on salt formation were also investigated, and those two were identified as key processing parameters. Salts were also prepared by solution crystallization from ethyl acetate, liquid-assisted grinding, and heat-assisted grinding and compared with those obtained by melt extrusion by using DSC, PXRD, TGA, and optical microscopy. While similar salts were obtained by all methods, both melt extrusion and solution crystallization yielded highly crystalline materials with identical enthalpies of melting. During the pH-solubility study, a salt hydrate form was also identified, which, upon heating, converted to anhydrate similar to that obtained by other methods. There were previous reports of the formation of cocrystals, but not salts, by melt extrusion. ^{1}H NMR and single-crystal X-ray diffraction confirmed that a salt was indeed formed in the present study. The haloperidol-maleic acid salt obtained was nonhygroscopic in the moisture sorption study and converted to the hydrate form only upon mixing with water. Thus, we are reporting for the first time a relatively simple and solvent-free twin-screw melt extrusion method for the preparation of a pharmaceutical salt that provides material comparable to that obtained by solution crystallization and is amenable to continuous manufacturing and easy scale up.

491840 Flexible-Backbone Protein Docking Using Motif Scoring and Large Conformational Ensembles

Abstract:

Computational prediction of protein-protein complex structures facilitates a fundamental understanding of biological mechanisms and enables drug design. Binding-induced conformational changes in proteins confound current protein-protein docking algorithms by greatly increasing the degrees of freedom to be sampled. While rotamer libraries have alleviated the sampling challenges for side-chains, backbone flexibility remains a principal challenge in docking. Previous studies have found limited success by varying the backbone along a restricted set of coordinates or residues or by docking a small number of backbones conformations of the two partners. In this study, we develop and benchmark the new EnsembleDock application in the Rosetta Molecular Modeling Suite. During the first, coarse-grained step, the method explores the relative orientation between the monomers while rapidly sampling from hundreds of backbone conformations of the unbound monomers, pre-generated using a variety of techniques to capture different backbone motions. It adapts to the backbone variance in the ensembles to ensure adequate backbone sampling for diverse ensembles; up to $O(10^6)$ combinations of backbone conformations and relative orientations are tested in 1-4 minutes. We have improved the near-native discrimination for these conformations by as much as 75% over the previous version by using *motif scoring*, a scheme based on a pre-tabulated set of implicit side-chain energies. In the second, all atom step, the interface is optimized by packing the side chains while performing small relative motions between the two monomers. To more aggressively simulate induced changes at the interface, we have developed an optional step where all atoms at the interface are moved along energy gradients in a reduced-coordinate framework. With these new methods, EnsembleDock can capture interface backbone changes greater than 1 Å between the unbound and bound states without any geometric restrictions. For targets with backbone changes greater than 2 Å, we verify the ability of the method to pick out near-native backbones by doping the existing ensembles with near-bound backbones.

500628 Systematic Redesigning of *E. coli* water Channel Porin, Ompf, for Desired Pore Size Using Iterative Protein Redesign and Optimization (IPRO) SuiteÂ

Abstract:

The increasing demand for freshwater for food production, processing and energy generation has made the efficient utilization of marginal water resources an imperative global environmental objective. Ultra-permeable membranes (dense solution, diffusion-based or channel-based) have emerged as key industrial workhorses for the same. Aquaporins (AQPs) have emerged as ideal candidates for channel-based membranes owing to their ~billion molecules/sec hydraulic permeabilities. Outer membrane porin type-F (OmpF) are beta-barrel proteins have also found relevance in channel-based membranes employed for aqueous-phase separations due to their ability to survive contacts with 100% ethanol. To this end, we systematically redesigned the pore constricting residues of OmpF using a modified version of Iterative Protein Redesign & Optimization (IPRO) suite (1) of programs to perturb the OmpF protein backbone, mutate and recursively repack the sidechains in order to obtain the selective internal structure of AQP (pore constriction diameter ~4 times smaller than OmpF) inside the stable beta-scaffold of OmpF (pore constriction diameter ~1.1nm). We first generated an initial set of designs and performed MD simulations and subsequent experiments using stopped-flow light scattering experiments to validate three designs, which are unique in their internal pore geometries. We subsequently generated 17 more designs with a modified objective of sampling pore sizes across the 0.2-1 nm range. The primary objective of this work is to redesign OmpF with any user-defined pore size without compromising on the stability of the mutated protein and thereafter assess its transport properties such as, hydraulic permeability and solute rejection.

Pore sizes between 0.3 nm and 0.5 nm are key to several industrial and environmental separations such as methane/CO2 and salt/water, which is accessible to the less stable AQP1. We classified the IPRO designs into three broad classes based on the final pore geometry. A cork-screw (CSD) class of IPRO-suggested design (pore constriction ~0.236 nm by ~0.194 nm) shows alternately stacked bulky groups along the pore periphery, which when expressed, purified and immobilized on a porous support, exhibits high hydraulic permeability and not only completely rejects glycine (75 Da) but also shows ~70% salt (58 Da) rejection. Another type of mutant (with 25 Trp mutations) shows a highly hydrophobic pore (pore constriction ~0.33 nm by ~0.35 nm) that exhibits several folds higher hydraulic permeabilities than AQP1. Herein we propose a platform for designing precisely tuned sub-nm membrane transporters which will play a pivotal role both in performing energy-efficient water treatment and vesicle mediated highly-selective transport of antibiotics. Subsequent efforts will be directed towards systematic mutation of membrane-facing residues for the OmpF mutant series of proteins to tune interactions with various biomimetic membrane materials.

1. Pantazes RJ, Grisewood MJ, Li T, Gifford NP, Maranas CD. The Iterative Protein Redesign and Optimization (IPRO) suite of programs. J Comput Chem. 2015;36(4):251-63.

493214 Rational Methods for Optimizing Antibody Specificity By Controlling the Net Charge of the Complementarity-Determining Regions

Abstract:

The intense interest in using antibodies as therapeutics is largely due to their attractive properties, including their high affinity, specificity, potency, stability, manufacturability and low toxicity. Nevertheless, the affinities of lead antibodies identified via *in vitro* or *in vivo* methods are typically not high enough for therapeutic applications and must be further enhanced via *in vitro* affinity maturation. A key challenge during *in vitro* affinity maturation is that gains in antibody affinity can come at the cost of reduced specificity. To understand the sequence determinants of antibody specificity, we selected a large number of antibody variants with common frameworks using multiple *in vitro* antibody libraries and different selection methods that enable identification of antibodies with low and high specificity. Sequence-based analysis of these variants reveals that one of the most important determinants of antibody specificity is the average net charge per residue in the complementarity-determining regions (CDRs). Increased negative charge in the CDRs – at least over moderate ranges of CDR net charge explored in this work – is linked to reduced non-specific binding. Conversely, increased positive charge – especially due to arginine mutations – is linked to increased non-specific binding. Interestingly, several other sequence-based properties (fraction of hydrophobic and aromatic residues) are weakly correlated with antibody specificity for the antibodies studied in this work. Moreover, we find that net charge per CDR residue is a useful metric for identifying clinical-stage antibodies with increased risk for non-specific interactions despite that these antibodies have significant sequence differences outside of the CDRs. We are currently using these insights to design antibody libraries in which the net charge of the CDRs is fixed within specific ranges in order to promote high specificity and reduce affinity/specificity trade-offs during affinity maturation. Our long-term goal is to use these and related findings to develop systematic and robust design methods for rapidly generating and optimizing antibodies for use in a range of diagnostic and therapeutic applications.

492380 Insertion of a Calcium-Binding Beta Roll Domain into a Thermostable Alcohol Dehydrogenase Enables Allosteric Control over Cofactor Specificity

Abstract:

The RTX domain from the adenylate cyclase toxin of *Bordetella pertussis* was cloned into thermostable alcohol dehydrogenase D (AdhD) from *Pyrococcus furiosus*. The insertion was made in loop A, between the 124th and 125th amino acid of AdhD, a member of the aldo-keto reductase (AKR) superfamily. The fusion protein, β-AdhD, retained the calcium-binding ability of the RTX domain. The formation of the β-roll in β-AdhD was confirmed via Förster resonance energy transfer, where terbium, a calcium analog, showed a higher fluorescence emission in the fusion protein compared to the wild type enzyme. Additionally, the fusion retained the thermostability of AdhD, as well as its ability to oxidize alcohols. Although β-AdhD is less active than AdhD, it gained the ability to modulate activity upon calcium addition and interestingly, this was only seen in the presence of NAD^+. Furthermore, calcium was determined to be a mixed inhibitor with β-AdhD with NAD^+ while no clear inhibition was determined in the presence of $NADP^+$. Therefore, not only does calcium modulate activity, but it also imparts cofactor preference in β-AdhD, where NAD^+ is preferred in the absence of calcium and $NADP^+$ is preferred in the presence of calcium. This work can be utilized in the development of multi-step catalytic cascades where activity can be controlled externally. Additionally, this approach is modular in that the insertion site, loop A, is found in all members of the AKR superfamily. Thus, we have demonstrated that by inserting a calcium-binding domain into a key catalytic loop in a cofactor-dependent enzyme, we have created an allosterically-regulated enzyme with tunable cofactor selectivity, reminiscent of a calcium-controlled cofactor selectivity rheostat switch.

Abdallah, W., Solanki, K., and Banta, S. "Insertion of a calcium-binding beta roll domain into a thermostable alcohol dehydrogenase enables allosteric control over cofactor specificity" *Journal of Molecular Biology* (Submitted).

491290 Engineering a Thioesterase for Improved Medium-Chain Fatty Acid Profile

Abstract:

Chemicals carrying a linear medium-chain (C8-C12) hydrocarbon have uses as fuels, and commodity and fine chemicals. Medium-chain chemicals are derived from medium-chain fatty acids (MCFAs) via derivatization with tailoring enzymes. Microbial production of MCFAs is hindered by the inefficient heterologous expression of medium-chain plant thioesterases and the broad fatty acid profile of bacterial thioesterases. Here, we use rational protein design to engineer a bacterial thioesterase for improved MCFA production in *Escherichia coli*. Further, using a previously developed G-protein coupled receptor (GPCR)-based MCFA sensor, we detect the different MCFA levels of produced by *E. coli* expressing the wild type and engineered thioesterases. This work sets the stage for sensor-guided engineering of MCFA producing microbes.

490627 Leveraging Disulfide Bonds to Stabilize Small Protein Scaffolds during Extensive Diversification

Abstract:

Cysteines play a unique role in natural and engineered proteins. Its high reactivity and susceptibility to oxidation gives cysteine the ability to drive molecular interactions as well as provide structural support. In protein design, cysteines are often avoided because of potential difficulties during production in bacteria as well as to enable site-specific conjugation via genetic introduction of a sole cysteine. While being mindful of these benefits afforded by cysteine-free proteins, cysteines also elicit stabilizing features upon formation of disulfides. This trade-off becomes especially important during protein library design. Heavily diversifying a protein in pursuit of novel function or improved biophysical properties can result in the accumulation of numerous destabilizing mutations within individual library variants. However, appropriate placement of cysteines can compensate for such destabilization by forming covalent disulfide bonds. In this study, we examined the impact of disulfides in the context of multiple small protein scaffolds with diverse secondary structures. We constructed combinatorial libraries within three protein scaffolds â€“ fibronectin beta sandwich, affibody three-helix bundle, and Gp2 alpha/beta topology â€“ and sorted them for specific, high affinity binders to a multitude of targets. Deep sequencing datasets corresponding to these high affinity binders were probed for cysteine content. Variants isolated from combinatorial protein libraries based on each of the three scaffolds gave rise to a varying prevalence of cysteine. Importantly, libraries composed of overly-diversified positions (i.e. first generation designs that failed to exclude detrimental amino acids at library positions) produced a strong increase (>7%) in cysteine at one-third of diversified sites. These less refined library designs demonstrated dramatic up-regulation, from 20% to 45%, in clones containing either two or four cysteines while showing a 15% decrease in the prevalence of clones containing either one or three cysteines. Characterization of cysteine-rich variants revealed disulfides contributing up to 21°C in thermal stability. Conversely, second generation library designs that acted on site-specific information to guide amino acid bias at each position, reducing the average Shannon entropy of library sites from 0.97 to 0.74, actually showed a down-regulation in cysteine content among high affinity variants. This depletion of cysteines was accompanied by a 15°C increase in thermal stability. Our results provide evidence for both the importance of site-wise amino acid bias as well as the potential for disulfides to compensate for overly-mutated, destabilized proteins.

505526 Next Generation Therapeutics to Combat Infectious Diseases

Abstract:

Of the 56 therapeutic monoclonal antibody products currently marketed in the US, four now target infectious disease indications. An additional 40 recombinant antibodies are in clinical trials for infectious indications, with 29 in phase II or III trials. The evolution of antibiotic-resistant bacteria, the emergence of new pathogens, and a growing population of immunocompromised individuals means that in many cases antibodies are an increasingly attractive therapeutic option. Next-generation antibody formats, including antibody-drug conjugates and single-domain antibodies as well as antibody mixtures and bispecific antibodies provide access to novel therapeutic mechanisms and allow for targeting a wider range of epitopes.

This talk will provide an overview of recent advances in the field and highlight two on-going projects in my lab. First, to address a resurgence in pertussis in high resource countries and continued high rates of morbidity and mortality in low resource countries, we have developed and antibody therapeutic neutralizing the toxin primarily responsible for symptoms. This antibody has been engineered for high affinity binding, reduced immunogenicity and extended serum half-life. We have also characterized its mechanism of action, using biochemical, structural and cellular assays. We have shown hu1B7 is protective against disease in mouse and adolescent baboon models of disease. A single dose can prevent disease symptoms in a neonatal baboon model when administered five weeks before experimental challenge. Second, to address issues with recurrent cyto-megalovirus infection in immuno-suppressed individuals, we envision a bispecific antibody able to redirect any passing T cells toward CMV suppression. In our first iteration of this therapeutic, we aim to target infected cells via a T cell receptor (TCR) binding the immuno-dominant peptide-HLA complex. Since TCRs are typically low affinity and express poorly as soluble molecules, we have used a novel eukaryotic-based cell display system that allowed us to rapidly identify variants with higher affinity and enhanced stability. We have generated TCR variants with up to 100-fold improved affinity that retain exquisite peptide selectivity. We will report our initial efforts to use this as a targeting molecule and as one arm of a bispecific molecule.

493416 Understanding the Role of Conformational Change in Product Inhibition of 2-(2' hydroxyphenyl)Benzenesulfinate Desulfinase (Dszb

Abstract:

Increasingly stringent regulation of sulfur oxide emissions and environmental stewardship necessitates more effective fossil fuel desulfurization technologies. While traditional catalytic means of desulfurization effectively remove simple sulfur compounds, more complex thiophenic molecules remain intact; these thiophenic molecules now account for the majority of sulfur emissions from liquid transportation fuels. Biodesulfurization via enzyme catalysis enables highly specific, rapid thiophenic desulfurization occurring at ambient temperature and pressure. The 4-step catabolic pathway converts dibenzothiophene (DBT), a common crude oil contaminant, into the sulfur-free molecule 2-hydroxybiphenyl (2-HBP) without the disruption of carbon-carbon bonds. 2-hydroxybiphenyl desulfinase (DszB), the rate-limiting enzyme in this biocatalytic process, is capable of selectively cleaving carbon-sulfur bonds. Accordingly, fundamental understanding of the molecular mechanisms of DszB must be developed. Based on crystallographic evidence, we hypothesize that DszB undergoes an active site conformational change associated with the catalytic mechanism. Moreover, we anticipate this conformational change is responsible, in part, for observed product inhibition. *Rhodococcus erythropolis* IGTS8 DszB was recombinantly produced and purified via Escherichia coli BW25113 to test these hypotheses. Activity and the conformational changes of DszB in the presence of 2-HBP were evaluated. The activity of recombinant DszB appears comparable to the natively expressed enzyme and is inhibited via competitive binding of the product, 2-HBP. Using circular dichroism, global changes in DszB conformation were monitored in response to 2-HBP; we found that both product and substrate generate similar structural changes. For more time sensitive characterizations, a fusion protein – DszB and a Small Ubiquitin-like Modifier (SUMO Tag) – was created to improve enzyme stability. The resulting fusion protein was then subjected to hydrogen deuterium exchange mass spectroscopy to monitor local changes in protein conformation in response to 2-HBP. Results linking conformational change to kinetic inhibition provide a basis for catalytic improvement of DszB via rational design through identification of groups of residues responsible for inhibitory effects.

502102 Structural, Thermodynamics and Kinetics Role of Novel Hot-Spot Mutations of BCR-ABL1 in Resistance Towards "Ibs" Inhibitors

Abstract:

Acquisition of mutations in the BCR-ABL1 kinase domain (KD) is frequently associated with tyrosine kinase inhibitor (TKI) failure in CML. Recently, we revealed a novel mutation "hot-spot" in the BCR-ABL1 KD region (residues 295 – 312), associated with high resistance and poor clinical outcomes [1].Some of these new BCR-ABL1 variants locate in a protein region, which might not be directly involved in TKI interaction. Therefore we hypothesized that some of these mutations might be linked to TKI resistance via a direct mechanism while others might exert an indirect effect on drug resistance.A question then arise: "Do the majority of the BCR-ABL1 KD mutations determined so far, studied and characterized by considering just the BCR-ABL KD domain, exert their resistance mechanism by directly interfering with TKI binding, as often claimed?"The present contribution integrates structural, computational, and molecular biology techniques to understand the eventual role of the newly reported hot-spot mutations of the BCR-ABL1 KD in TKI resistance observed in CML patients using a "two-tier level" investigation: isolated KDs and SH2-liker-SH3-KD BCR-ABL1 constructs.

Specifically, *in silico*experiments will disclose the effects exerted by the "hot-spot" BCR-ABL1 mutations on the protein structure, thermodynamic stability, and ability to interfere with BCR-ABL1 binding to specific TKIs (e.g., imatinib, dasatinib, and ponatinib).

Structural biology evidences of wild type and mutant isolated KDs and SH2-linker-SH3-KD (SSK) constructs (*per se*and in complex with TKIs) will be discussed to support in silico predictions.

Direct drug binding to both KD and SSK constructs will be measured using isothermal titration calorimetry and surface plasmon resonance.

*In vitro*kinase assays will monitor the ability of these mutant isoforms to auto-phosphorylate and to phosphorylate a substrate peptide.

"Prototypical" BCR-ABL1 KD mutations will be analyzed in parallel for comparison and to confirm/dispute the current, classical view of TKI resistance.

[1] Gibbons DL, Pricl S, Posocco P, et al., Molecular dynamics reveal BCR-ABL1 polymutants as a unique mechanism of resistance to PAN-BCR-ABL1 kinase inhibitor therapy. Proc Natl Acad Sci U S A. 2014 Mar 4;111(9):3550-5. doi: 10.1073/pnas.1321173111.

501471 Novel Computational Protocol for Small-Molecule : Protein Receptor Docking; Application to 1,4-Dhna and Tcdd Binding to AhR Mouse Protein

Abstract:

Docking algorithms are widely used to understand the relationship between the structure and function of interactions with small molecule : protein complexes that govern biological functions. Despite the advances made in computational chemistry, and the variety of docking frameworks to study small molecule : receptor binding, there is still significant room for improvement in obtaining highly accurate computational predictions of the structure of small-molecule : receptor complexes [1]. We have developed a CHARMM-based docking protocol that innovatively combines the following key features: (1) the use of multiple short molecular dynamics (MD) docking simulations, with residues within the protein receptor's binding pocket unconstrained, to nearly exhaustively search the binding modes of the ligand in the binding pocket of the receptor, (2) the use of both harmonic and quartic spherical potentials to constrain the ligand to the binding pocket during the multiple short MD docking simulations, and (3) the selection of the most probable binding modes using interaction energy calculations, as well as the use of explicit water, all-atom MD simulations and physical-chemistry based free energy calculations to computationally elucidate the most favorable binding mode of the ligand in complex with the protein receptor.

In this study, we applied our docking protocol to determine the binding of 1,4-dihydroxy-2-naphthoic acid (1,4-DHNA) and TCDD to the mouse AhR protein [2]. The most energetically favored binding mode of TCDD with the AhR based on our docking protocol is in high accordance with previous experimental and computational studies examining this interaction [2, 3, 4, 5, 6]. The most energetically favored binding mode of 1,4-DHNA shows that both 1,4-DHNA and TCDD predominantly form strong interactions with the same residues of AhR. More polar interactions occur in the AhR : 1,4-DHNA complex in comparison to the AhR : TCDD complex due, in part, to differences in overall charge and substituent interactions with various amino acid side-chains. Association free energy calculations suggest that 1,4-DHNA binds with a lower affinity. This result is consistent with experimental observations showing similar efficacies of 1,4-DHNA and TCDD as AhR agonists but differences in their potencies [2]. The strong agreement of our work in comparison with previous studies investigating TCDD in complex with AhR support the validity of our docking protocol and suggest that the computationally derived binding modes most likely represent the naturally occurring binding modes of TCDD and 1,4-DHNA with residues in the AhR ligand binding pocket.

[1] Wang Z, Sun H, Yao X, Li D, Xu L, Li Y, Tian S, Hou T. (2016). Comprehensive evaluation of ten docking programs on a diverse set of protein-ligand complexes: the prediction accuracy of sampling power and scoring power. *Phys Chem Chem Phys.* 18, 12964-12975.

[2] Cheng Y, Jin UH, Davidson LA, Chapkin RS, Jayaraman A, Tamamis P, Orr A, Allred C, Denison MS, Soshilov A, Weaver E, Safe S. (2017). Editor's Highlight: Microbial-Derived 1,4-Dihydroxy-2-naphthoic Acid and Related Compounds as Aryl Hydrocarbon Receptor Agonists/Antagonists: Structure-Activity Relationships and Receptor Modeling. *Toxicol Sci.* 155(2), 458-473.

[3] Motto I, Bordogna A, Soshilov AA, Denison MS, Bonati L. (2011). New aryl hydrocarbon receptor homology model targeted to improve docking reliability. *J Chem Inf Model.* 51, 2868–2881.

[4] Pandini A, Soshilov AA, Song Y, Zhao J, Bonati L, Denison MS. (2009). Detection of the TCDD binding-fingerprint within the Ah receptor ligand binding domain by structurally driven mutagenesis and functional analysis. *Biochemistry.* 48, 5972–5983.

[5] Xing Y, Nukaya M, Satyshur KA, Jiang L, Stanevich V, Korkmaz EN, Burdette L, Kennedy GD, Cui Q, Bradfield CA. (2012). Identification of the Ah-receptor structural determinants for ligand preferences. *Toxicol Sci.* 129, 86–97.

[6] Soshilov AA, Denison MS. (2014). Ligand promiscuity of aryl hydrocarbon receptor agonists and antagonists revealed by site-directed mutagenesis. *Mol Cell Biol.* 34, 1707–1719.

490851 *Vibrio fischeri* Aspartate 1-Decarboxylase Revealed By Model-Enabled Gene Search

Abstract:

Vibrio fischeri strain ES114 is a bioluminescent marine bacterium that forms a symbiotic relationship in the light-emitting organ of the Hawaiian bobtail squid, *Euprymna scolopes.* Its metabolic capabilities are representative of many other marine bacteria, including both beneficial and pathogenic members of the genus *Vibrio* and, thus, are of great interest to researchers. In this study, we developed model-enabled gene search (MEGS), which is an integrated computational and experimental approach, and identified that VF_0892 encodes the *V. fischeri* aspartate 1-decarboxylase (PanP). We also compared *in vitro* kinetic properties of PanP with *Escherichia coli* aspartate 1-decarboxlyase (PanD) and the *V. fischeri* glutamate decarboxylase.

Using the MEGS approach, PanP was first identified as missing from the genome annotation when discrepancies between metabolic model predictions and experimental culturing data arose. Then a selection strategy for PanP was designed, which consisted of a glucose minimal medium condition and an *E. coli* Δ*panD* strain whose growth in this medium was dependent on PanP. Finally, a genomic library of *V. fischeri* was transformed into *E. coli* Δ*panD*, and growth selection identified the VF_0892 as encoding the missing PanP enzyme.

VF_0892 is currently annotated in NCBI as a glutamate decarboxylase (E.C. 4.1.1.15). To compare the activities and substrate specificities of the VF_0892 and *E. coli* PanD enzymes towards the two potential substrates (aspartate and glutamate), decarboxylase activities were evaluated using the decarboxylase, phosphoenolpyruvate carboxylase, and malate dehydrogenase -linked assays. The VF_0892 enzyme was around 10-fold more active at 28°C compared to 37°C, which is consistent with the optimal growth of *V. fischeri*at 28°C and its intolerance to higher temperatures. Both PanD and the VF_0892 enzyme showed a much higher reaction rate when aspartate, rather than glutamate, was used as the substrate (PanD: 10-fold; VF_0892: 5-fold at 37°C, and 26-fold at 28°C). Using the same assays, we verified the VF_1064, annotated as a glutamate decarboxylase, truly encodes a decarboxylase with a strong substrate preference of glutamate over aspartate.

In addition to *V. fischeri* PanP, we have successfully used this integrated MEGS approach to determine the functions of ten metabolic genes in *V. fischeri*, *Gluconacetobacter xylinus*, and *Zymomonas mobilis*, which demonstrates the versatility of this method.

Mutating the ribosomal active site *in vitro*

Tasfia Azim[1], Anne d'Aquino[2] and Michael C. Jewett[1]
[1]Chemical and Biological Engineering, Northwestern University, Evanston, IL,
[2]Interdisciplinary Program in Biological Sciences (IBiS), Northwestern University, Evanston, IL

The ribosome, a 2.5-MDa molecular machine that polymerizes α-amino acids into proteins, is the catalytic workhorse of the translation apparatus. The catalytic capacity of the translation machinery has attracted extensive efforts to repurpose it for novel functions. One key idea is that the natural translation machinery can be harnessed to synthesize polymers containing non-natural building blocks. Expanding the repertoire of ribosomal substrates and functions is a difficult task, however, because the requirement of cell viability severely constrains the alterations that can be made to the ribosome, a catalyst that sustains the life of a cell. In practice, these constraints have made the natural ribosome nearly unevolvable and, so far, no generalizable approach for modifying the catalytic peptidyl transferase center (PTC) of the ribosome has been advanced. We propose to address this grand challenge by using cell-free systems that harness the biosynthetic potential of cellular machines without using intact cells, thus removing cell viability constraints.

Here, we use our *in vitro* ribosome synthesis, assembly, and translation system (termed iSAT) to generate variant ribosomes with mutations in the active site, and inquire how these modifications impact protein synthesis. The active site of the ribosome is composed solely of ribosomal RNA (rRNA). This rRNA is assembled into three key loops, which facilitate translation: the peptidyl transferase loop, the A-loop, and the P-loop. The peptidyl transferase loop plays a key role in positioning substrates, is involved with peptide stalling, and is a key target for antibiotic binding. The A- and P-loops flank the peptidyl transferase loop within the ribosome's active site. The A-loop is responsible for aminoacyl-tRNA interactions with the ribosome, and the P-loop is known to play a role in further facilitating peptide bond formation between amino acids. Because mutations to many of the nucleotides in these three active site loops confer lethal phenotypes, few to no studies exist in probing and fully characterizing them.

Understanding the effects of single-base mutations on these active site nucleotides will provide insight into the basic biochemistry of these bases in translation, but also provide the groundwork for engineering the catalytic center of the ribosome. Using iSAT, we have assembled mutant ribosomes possessing single-base substitutions of 23S rRNA nucleotides in these active site loops. By successfully quantifying full-length protein synthesis kinetics of iSAT-assembled wild type and mutant ribosomes, we unexpectedly found many key PTC mutations, which were expected to abolish ribosomal activity, still permitted full-length protein synthesis. We also assessed translation fidelity and ribosome assembly. Our work provides a comprehensive mapping of the impacts of multiple ribosomal mutations on protein synthesis. The understanding gained facilitates efforts to engineer and evolve ribosomes for synthetic biology.

496022 Mapping the Aggregation Behaviour of Biopharmaceuticals: A New Approach

Abstract:

Mapping the Aggregation Behaviour of Biopharmaceuticals: A New Approach

S. Hedberg, J. Heng, D. Williams

ABSTRACT

Protein aggregation is a critical problem for the safety of biopharmaceuticals as they are linked to adverse immunologically related responses in patients. Much effort has been made to gain a better understanding of aggregation, however, the mechanisms leading to protein aggregation are still not fully understood. Protein-protein molecular interactions in solution are known to be involved in protein solution aggregation behaviour and are a common issue for the manufacturing of biopharmaceuticals such as monoclonal antibodies, mAbs. Therefore, a major industrial and academic challenge is the development of fast and reliable methods, either theoretically or experimentally driven, which allow formulation scientists to identify optimal biopharmaceutical species and solution conditions which deliver stable solution products, thus avoiding aggregation.

In this work we will describe a SIC (self-interaction chromatography) method for determining the osmotic second virial coefficient (B_{22}), where B_{22} describes the thermodynamics of protein-protein interactions in solution. This experimental biophysical approach allows solution stability maps including solution pH and salt concentration effects to be obtained using mg quantities of species such as mAbs. The approach has been validated against traditional size-exclusion chromatographic methods on samples that were aged over many weeks at elevated temperature. Both methods gave similar stability condition maps, with the SIC method delivering results within days compared with traditional stability trials which would typically take many months. Over a wide range of test conditions good correlations were found between experimental B_{22} values and experimentally measured aggregation rates.

Similarly, the conformational and structural changes were investigated for a number of selected conditions over a longer time at continuously increasing temperature. The results proved that SIC easily could predict the conformational changes of mAbs too. From this study it was concluded that improvement in the SIC methods reported here allows accurate and robust mapping of solution stability conditions as well as a good prediction of conformational stability.

505540 Engineering Antimicrobial Peptides to Target Fungal Pathogens

Abstract:

Candida albicans and other *Candida* species are human commensal organisms but can cause disease when patients are immunocompromised. Increasing drug resistance and the limited number of available antifungal agents necessitate the search for new therapeutic strategies. To address current therapeutic challenges, we are engineering peptides for improved interactions with fungal cells. One of these peptides is histatin 5, is an antimicrobial peptide found in human saliva that has potent and specific activity against *C. albicans*. Although histatin 5 has promise as a therapeutic agent, the fungus produces secreted aspartic proteases that degrade the peptide and reduce its antifungal activity. We have designed analogs of histatin 5 that offer strong resistance to the fungal proteases without reducing antifungal activity. We are investigating the use of these analogs for treating and preventing disease, while also improving understanding of how the secreted aspartic proteases recognize and cleave peptide substrates. In addition to exploring peptides as potential therapeutics, we are also using peptides to target and deliver bioactive molecules to *Candida* pathogens. We are examining how properties of these cell-penetrating peptides affect their translocation across the cell wall and cell membrane of *Candida* cells and are defining the types of cargo that can be delivered into fungal cells. By applying protein engineering strategies to designing peptides for targeted interactions with fungal cells, we are gaining information on structure-function relationships that will enable more efficient design of biomolecules for specific interactions with biological systems.

502906 Engineering Islet-Specific Microvascular Network within hPSC Derived Pancreatic Islet Organoids

Abstract:

An emerging area in tissue engineering is the development of three dimensional engineered constructs, organoids, which are comprised of multiple organ-specific cell populations capable of recapitulating an *in-vivo* organ systemâ €™s structure and function in an *in-vitro* setting. Engineering tissue specific organoids from human pluripotent stem cells (hPSCs) has resulted in successful reproduction of similar organ functionality from renewable cell source for a variety of target organ systems. Such systems have included derivation of intestinal, brain, and renal organ models. Development of organoid systems requires organ-specific parenchyma cell source and a platform for self-organization and lineage specific induction of chosen cell types. Additionally, reproducing the organ-specific functional vasculature during reproduction of the complex organ structure is extremely vital for maintaining nutrient supply and appropriate organ function. This is particularly crucial while engineering pancreatic islet organoids, since a dense fenestrated intra islet vasculature is vital for supporting glucose delivery and insulin response.

We have investigated multiple 3D platforms which promote spontaneous aggregation of pre- differentiated hPSC derived pancreatic progenitor cells (hESC-PPs) into a 3D organoid which demonstrated functional insulin production *in vitro* as well as *in vivo* mouse model. The resulting spheroids are readily recoverable and amenable for size and cellular composition tuning. This hydrogel mediated aggregation allowed direct integration of isolated microvessel elements within the hESC-PP organoids. Continued *in vitro* culture of the multicellular islet organoids promoted microvascular expansion and formation of vascular networks, especially with inclusion of supporting stromal cell populations. Pancreatic phenotype of the vascularized organoids was strengthened, demonstrated by the gene expression of key pancreatic maturation markers (NKX6.1, PDX1, and INS). Moreover, the intra-organoid vasculature demonstrated an increase in islet endothelial specific API gene expression and PLVAP, an indicator of increased endothelial diaphragm and fenestration, key elements of islet vascular development. *In vivo* vascular development within organoids was further characterized in mouse implantation models. In conclusion, we believe the results present a major step in the *in-vitro* production of hPSC islet organoids with microvasculature, likely to enhance function and foster faster *in-vivo* integration, while also being conducive for organ-on-a-chip applications.

492087 Modeling Diseased BBB through Directed Differentiation of Patient-Derived iPSCs into Brain Microvascular Endothelial Cells

Abstract:

The blood-brain barrier (BBB) forms the interface between the microvascular system and the brain and is responsible for maintaining homeostasis in the central nervous system. As such, disruption to the BBB is implicated in many neurological diseases, but whether the loss in BBB integrity is an upstream or downstream effect of the disease can be difficult to ascertain. We aimed to investigate this question for the disease cerebral childhood adrenoleukodystrophy (ccALD), which is an X-linked disease that affects males ages 3-10 and results in death within 3-5 years of clinical onset. ccALD is caused by mutations in the ABCD1 gene which codes for a transporter responsible for bringing very long chain fatty acids (VLCFAs) into the peroxisome for subsequent beta oxidation. Buildup of VLCFAs in the brain leads to rapid neurological deterioration in a subset of patients. We hypothesized that the BBB integrity of ccALD patients is inherently decreased compared to wild-type (WT) individuals. To model the BBB, we differentiated induced pluripotent stem cells (iPSCs) that originated from WT individuals and from ccALD patients into brain microvascular endothelial cells (BMECs) to compare their molecular and functional characteristics. The differentiated BMECs were characterized via PCR analyses and immunocytochemistry to confirm that the BMECs possess requisite BBB markers. Functional characteristics were quantified by the transendothelial electrical resistance (TEER), sodium fluorescein permeability, and P-glycoprotein activity. A stark functional difference was observed in the TEER of the BMECs. BMECs from ccALD patient iPSCs (ccALD-BMECs) displayed significantly reduced TEER ($2590 \pm 330\ \Omega \cdot cm^2$; mean ± standard error) compared to that of the WT-BMECs ($5000 \pm 510\ \Omega \cdot cm^2$). This result signifies that the barrier function of the ccALD-BMECs is inherently worse than that of the WT-BMECs. With this experimental evidence, we aimed to improve the barrier function of the ccALD-BMECs utilizing block copolymers. Block copolymers of poly(ethylene oxide) (PEO) and poly(propylene oxide) (PPO) have been demonstrated to be able to protect damaged cell membranes. Treatment of the ccALD-BMECs with a diblock copolymer of PEO and PPO (PPO-*b*-PEO) resulted in improvement of barrier function. We observed an approximate 60% increase in the TEER of ccALD-BMECs treated with PPO-*b*-PEO during development compared to the TEER of the control. Furthermore, we observed an approximate 20% increase in TEER of ccALD-BMECs treated with PPO-*b*-PEO post-development compared to the TEER of the control. In summary, we have provided evidence for functional differences between BBB cells derived from iPSCs that originated from ccALD and WT individuals. Furthermore, we have demonstrated that the deficient functionalities of the ccALD-BMECs can be improved by treatment with block copolymers.

494948 hESC-Derived Striatal Cells Generated Using a Scalable 3D Hydrogel Promote Recovery in a Huntingtonâ€™s Disease Mouse Model

Abstract:

Huntingtonâ€™s disease (HD) is an inherited neurological disorder characterized by the progressive degeneration of striatal medium spiny neurons (MSNs), which eventually leads to fatal deficits in movement, cognition, and behavior. One promising approach for treating HD is cell replacement therapy, where lost cells are replaced by striatal MSNs or MSN progenitors derived from human pluripotent stem cells (hPSCs). While remarkable prior work paved the way to generate MSNs from hPSCs for application towards HD treatment, as with many stem cell differentiation processes the current state-of-the-art methods for producing striatal MSNs rely on 2D culture systems that typically include poorly defined components, limit scalability, and have yielded mixed therapeutic outcomes in animal models of HD. To take a next step towards clinical translation, here we develop an approach for the efficient generation of striatal progenitors from human embryonic stem cells (hESCs) within a fully defined and scalable PNIPAAm-PEG 3D hydrogel. Specifically, a 2-fold higher fraction of DARPP32^{+} striatal neurons were generated in the 3D biomaterial, which demonstrated functional electrophysiological maturity 30 days earlier, relative to cells differentiated on conventional 2D surfaces. Following their efficient derivation, transplantation of 3D-derived striatal progenitors into a transgenic mouse model of HD slowed disease progression, improved motor coordination, and significantly increased lifespan by 41%. Importantly, transplanted cells developed an MSN-like phenotype and formed synaptic connections with host cells. Our results illustrate the potential of scalable 3D biomaterials for generating striatal progenitors for HD cell therapy.

www.nature.com/scientificreports

SCIENTIFIC REPORTS

OPEN

Received: 07 September 2016
Accepted: 01 December 2016
Published: 06 January 2017

An Integrated Miniature Bioprocessing for Personalized Human Induced Pluripotent Stem Cell Expansion and Differentiation into Neural Stem Cells

Haishuang Lin[1], Qiang Li[1] & Yuguo Lei[1,2,3]

Human induced pluripotent stem cells (iPSCs) are ideal cell sources for personalized cell therapies since they can be expanded to generate large numbers of cells and differentiated into presumably all the cell types of the human body *in vitro*. In addition, patient specific iPSC-derived cells induce minimal or no immune response *in vivo*. However, with current cell culture technologies and bioprocessing, the cost for biomanufacturing clinical-grade patient specific iPSCs and their derivatives are very high and not affordable for majority of patients. In this paper, we explored the use of closed and miniature cell culture device for biomanufacturing patient specific neural stem cells (NSCs) from iPSCs. We demonstrated that, with the assist of a thermoreversible hydrogel scaffold, the bioprocessing including iPSC expansion, iPSC differentiation into NSCs, the subsequent depletion of undifferentiated iPSCs from the NSCs, and concentrating and transporting the purified NSCs to the surgery room, could be integrated and completed within two closed 15 ml conical tubes.

Cell therapies offer hope to treat many human diseases and injuries that cannot be treated or their progression cannot be altered by current treatments[1]. Autologous cells are ideal for cell therapies since they induce minimal or no immune rejection[2,3]. However, it is very challenging to isolate sufficient numbers of autologous cells from the patient for an effective treatment[4]. Furthermore, majority of human cell types exhibit limited growth and significant phenotype shifting when cultured, preventing generating the required numbers of autologous cells through *in vitro* expansion[1–5]. Human induced pluripotent stem cells (iPSCs) provide a solution for this challenge. With reprogramming factors, adult cells from the patient, such as fibroblasts, can be reprogrammed into iPSCs within about one month[6]. iPSCs can be cultured for long term and expanded into large numbers under complete defined conditions[7]. They can be differentiated into presumably all the cell types of the human body[2,8,9]. During the past decade, protocols for efficiently differentiating iPSCs into various human cell types, such as cortical neurons[10,11], gamma-aminobutyric acid (GABA)-ergic interneurons[12–14], midbrain dopaminergic (DA) neurons[15,16], endothelial cells[17–19], mesenchymal stem cells[20,21], cardiomyocytes[22–24], hepatocytes[25–27], beta cells[28,29] and other cells[2,8,9] have been developed. Many of these cells are being investigated for treating degenerative diseases and injuries[30], such as Parkinson's disease (PD)[15,16,31], Alzheimer's disease (AD)[32], stroke[33], spinal cord injury (SCI)[34–37], blindness[8,38,39], myocardial infarction (MI)[22,40], diabetes etc. The iPSC-derived retinal pigment epithelium has been tried in human[8]. In short, iPSCs are ideal cell sources for personalized cell therapies.

However, the advancement of iPSC-based personalized cell therapies is currently hindered by the high cost to biomanufacture the cells[1–5]. With the current bioprocessing[41], patient cells are collected and cultured for a few days[41]; then, reprogramming factors are delivered to these cells to reprogram them into iPSCs (which takes approximately one month). Next, high quality iPSC clones are selected, expanded and characterized for their pluripotency and genome integrity with a variety of assays (which takes approximately one to two months); then,

[1]Department of Chemical and Biomolecular Engineering, University of Nebraska, Lincoln, Nebraska, USA. [2]Mary and Dick Holland Regenerative Medicine Program, University of Nebraska Medical Center, Omaha, Nebraska, USA. [3]Fred & Pamela Buffett Cancer Center, University of Nebraska Medical Center, Omaha, Nebraska, USA. Correspondence and requests for materials should be addressed to Y.L. (email: ylei14@unl.edu)

iPSCs are expanded and differentiated into the desired cells. Finally, the produced cells are purified, characterized for their identities, purity, and potency and formulated for transplantation. The whole bioprocessing takes a few months and is mainly done using 2D, open culture systems (e.g., 2D cell culture flasks) through manual operations–a processing which leads to low reproducibility, high risk of contamination, and requirement for highly skilled technicians[42]. The whole bioprocessing is also required to comply with the current Good Manufacturing Practice (cGMP)[42]. In addition, 2D culture systems have low yield. For instance, only ~2×10^5 cells can be produced per cm^2 surface area, meaning that it will require ~85 six-well plates to produce the cells (~1×10^9 cells) sufficient for one patient[43,44]. Maintaining these plates requires large incubator and cGMP-compliant facility space, labor, and reagent. If large numbers of patients need iPSC-based personalized cell therapies, the cell production can only be done in large cell biomanufacturing centers (i.e. the centralized cellular biomanufacturing)[42]. Patient cells are sent to the center, and the produced cells are sent back to the point-of-care for transplantation. This centralized biomanufacturing has additional disadvantages[1,42,45], including: (i) patient cells may be cross-contaminated and (ii) there are high costs and risks associated with the transportation, logistics, tracking, and recording. In summary, the cost for biomanufacturing personalized iPSCs and their derivatives with current technologies is not affordable for the majority of patients[1–5].

One method to significantly reduce the biomanufacturing cost is to make cells in individualized, closed, computer controlled miniature cell culture device at the point-of-care (i.e. the cGMP-in-a-box production)[42]. Using closed culture devices avoids contamination risk and eliminates the requirement for cGMP processing. Automation of all key operations avoids output variations and reduces need for highly skilled operators. Biomanufacturing at the point-of-care reduces the cost and risk related to the logistics and transportation. Miniaturizing the culture system makes it possible to simultaneously biomanufacture cells for large numbers of patients at the point-of-care (i.e. high throughput biomanufacturing).

In this paper, we describe our effort to develop such a miniature bioprocessing for making NSCs from human iPSCs. The bioprocessing takes advantage of the discovery that human iPSCs could be expanded in 3 dimension (3D) thermoreversible Poly(N-isopropylacrylamide)-Poly(ethylene glycol) (PNIPAAm-PEG) hydrogels at high growth rate and yield[43,46]. In this paper, we first developed a protocol that could efficiently differentiate human iPSCs into NSCs in the PNIPAAm-PEG hydrogel. We then, with the assist of this hydrogel scaffold, integrated the bioprocessing including the iPSC expansion, iPSC differentiation into NSCs, the subsequent depletion of undifferentiated iPSCs from the product, and concentrating and transporting the produced cells to the surgery room into two closed, 15 ml conical tubes.

Methods

Culturing human pluripotent stem cells (hPSCs) in 2D. iPSCs (iPSCs reprogrammed from human mesenchymal stem cells) were obtained from George Q. Daley laboratory (Children's Hospital Boston, Boston)[47]. H9 hESCs were purchased from WiCell Research Institute. hPSCs (iPSCs and H9s) were maintained in 6-welll plate coated with Matrigel (BD Biosciences) in Essential 8™ medium (E8, Invitrogen)[7]. Cells were passaged every 4 days with 0.5 mM EDTA (Invitrogen). Medium was changed daily. Cells were routinely checked for the expression of pluripotency markers, OCT 4 and NANOG, their capability to form teratomas in immunodeficient mice, their karyotypes and bacterial contaminations.

Culturing hPSCs in 3D PNIPAAm-PEG hydrogels. To transfer the culture from 2D to 3D PNIPAAm-PEG hydrogels, hPSCs maintained in Matrigel-coated 6-welll plate were treated with Accutase (Life Technologies) at 37 °C for 5 minutes and dissociated into single cells[43,46]. Dissociated cells were mixed with 10% PNIPAAm-PEG (Cosmo Bio, USA) solution dissolved in E8 medium on ice and cast on tissue culture plate, then incubated at 37 °C for 10 minutes to form hydrogels before adding warm E8 medium containing 10 μM ROCK inhibitor (Y-27632, Selleckchem). Medium was changed daily. Cells were passaged every 5 days. To passage cells, medium was removed, and 2 ml ice-cold PBS was added to dissolve the hydrogel for 5 minutes. Cell spheroids were collected by spinning at 100 g for 3 minutes. Cells were incubated in Accutase (Invitrogen) at 37 °C for 10 minutes and dissociated into single cells.

hPSC differentiation in 2D. Single hPSCs (iPSCs and H9s) were plated in Matrigel-coated 6 well plates (2×10^6 cells/well) and cultured in E8 medium overnight to reach >90% confluency. E8 medium was removed and replaced with neural induction medium consisting of Essential 6™ medium (E6, Invitrogen) supplied with 100 nM LDN193189 (Selleckchem) and 10 μM SB431542 (Selleckchem) for 7 days. Note that E6 medium is equal to E8 medium minus the bFGF and TGFβ proteins. Medium was changed daily[48].

hPSC differentiation in 3D PNIPAAm-PEG hydrogels. Single hPSCs were encapsulated in the PNIPAAm-PEG hydrogels (1×10^6 cells/ml hydrogel) and cultured in E8 medium for 5 days. E8 medium was removed and replaced with neural induction medium for 7 days. Medium was changed daily. To differentiate NSCs into the cortical neurons, NSC spheroids were harvested on day 11 and plated to Matrigel-coated 6 well plates, and cultured in neural differentiation medium consisting of Neurobasal® Media (Life Technologies), B27 (50X, Life Technologies), BDNF (20 ng/ml, PeproTech), GDNF (10 ng/ml, PeproTech), L-ascorbic acid (200 μM, Sigma), DAPT (2.5 μM, Tocris), Dibutyryl-cAMP (0.5 mM, Santa Cruz Biotechnology) for another 19 days. Half medium was changed every two days.

To make ventral midbrain NSCs, hPSC spheroids were cultured in neural induction medium containing 50% DMEM/F12 + 50% Neurobasal medium, 1% N2, 2% B27 minus vitamin A, SB431542 (10 μM), LDN193189 (100 nM), CHIR99021 (0.7 μM, Selleckchem), purmorphamine (2 μM, Selleckchem) and sonic hedgehog (shh) C25 II (200 ng/ml, Home-made) in the 3D PNIPAAm-PEG hydrogels for 11 days. NSC spheroids were harvested on day 11 and plated to Matrigel-coated 6 well plates, and cultured in neural differentiation medium consisting of Neurobasal® Media (Life Technologies), B27 (50X, Life Technologies), BDNF (20 ng/ml, PeproTech), GDNF (10 ng/ml, PeproTech), L-ascorbic acid (200 μM, Sigma), DAPT (2.5 μM, Tocris), Dibutyryl-cAMP (0.5 mM, Santa Cruz Biotechnology) for another 19 days. Half medium was changed every two days.

Culturing hPSC-derived NSCs in 2D. NSCs made from hPSCs were incubated with Accutase at 37 °C for 5 minutes and dissociated into single cells. Cells were plated on Matrigel-coated 6 well plate at high density (2.0×10^6 cells/ml) and cultured in expansion medium consisting of KnockOut™ DMEM/F-12 (Life Technologies), StemPro Neural Supplement (50X, ThermoFisher Scientific), 20 ng/ml bFGFs (Peprotech), 10 ng/ml EGFs (Peprotech), Glutamax I (100X, Gibco) and 2 μg/ml heparin (Sigma). Medium was changed daily and cells were passaged every 3 days.

Culturing hPSC-derived NSCs in 3D PNIPAAm-PEG hydrogels. Dissociated single NSCs were encapsulated to 3D PNIPAAm-PEG hydrogels at 2×10^6 cells/ml and cultured in expansion medium containing KnockOut™ DMEM/F-12, StemPro Neural Supplement (50X, ThermoFisher Scientific), 20 ng/ml bFGFs (Peprotech), 10 ng/ml EGFs (Peprotech), Glutamax I (100X, Gibco) and 2 μg/ml heparin (Sigma). Medium was changed daily and cells were passaged every 7 days with Accutase.

Integrated miniature bioprocessing. With a syringe and needle, 4 °C PNIPAAm-PEG solution containing single iPSCs were injected into room temperature E8 medium in a 15 ml conical tube. Fibrous hydrogels were formed instantly. A long needle was inserted through the septum cap into the tube that its open reached the tube bottom. A Variable-Speed Peristaltic Tubing Pump (Control Company, USA) was used to continuously perfuse the culture medium into the tube through this long needle. A second short needle was placed in the septum cap so that the medium can flow out. Medium was stocked in a sealed and oxygen-permeable plastic bag. Medium in the bag was changed daily. The cell culture tube, pump and medium bag were placed in a cell culture incubator at 37 °C. E8 medium and neural induction medium was used for day 1 to 5, and day 6 to 12, respectively. On day 12, the cell culture tube was placed on ice for 5 minutes to liquefy the hydrogel and release the spheroids. Cells were collected by spinning the tube at 100 g for 5 minutes. The cell pellet was treated with Accutase at 37 °C for 10 minutes and dissociated into single cells. Single cells were collected by spinning at 300 g for 5 minutes. Cells were re-suspended with 80 μl PBS buffer and 20 μl of anti-SSEA-4 microbeads (Miltenyi Biotec) were added and incubated at 4 °C for 15 minutes. The SSEA4+ iPSCs were pulled down with a magnet and NSCs in the supernatant were transferred into a new tube. Cells were pelleted by spinning at 300 g for 5 minutes and transported to the surgery room for transplantation.

Transplanting iPSC-derived NSCs to rat brains. All animal protocols were approved by the Animal Care and Use Committee of the University of Nebraska, Lincoln. All experimental procedures involving animals were carried out in accordance with the guidelines of the Institutional Animal Care and Use Committee of the University of Nebraska, Lincoln. Sprague Dawley female rats were obtained from Charles River. Animals received intraperitoneal cyclosporine A (10 mg/kg, LC Laboratories) injection starting 1 day before transplantation. For transplantation, animals were anesthetized with 2–4% isoflurane. 2×10^5 cells suspended in 4 μl DMEM medium were injected into the striatum (AP + 0.5 mm; ML ± 3.0 mm; DV-6 mm) at 0.5 μl/minute using a 10 μl Hamilton syringe (Hamilton Company, USA) with a stereotaxic frame (RWD Life Science Inc). On day 7, rats were anesthetized with ketamine/xylazine and perfused with PBS followed by 4% paraformaldehyde. After fixation, the brain was serially sectioned (40 μm in thickness) with a Leica cryosection machine, and free-floating sections were stained with antibodies.

Immunostaining and imaging. Cells cultured on 2D surface were fixed with 4% paraformaldehyde (PFA) at room temperature for 15 minutes, permeabilized with 0.25% Triton X-100 for 10 minutes and blocked with 5% goat serum for 1 hour before incubating with primary antibodies at room temperature for 2 hours. After extensive wash, secondary antibodies in 2% BSA were added and incubated for another 1 hour.

To assess the pluripotency of hPSCs expanded in 3D PNIPAAm-PEG hydrogels, hPSCs were dissociated into single cells with accutase, stained in suspension. Cells were placed in 96 well plate and imaged with Zeiss Axio Observer Fluorescent Microscopy. The percentage of OCT4+ or NANOG+ nuclei was quantified with Image J software. This method was also used to quantify the PAX6+ or NESTIN+, or TUJ-1+ or TBR-1+ cells in Figs 1, 2, 3 and 5. At least 1000 cells were analyzed. To stain the brain sections, samples were incubated with PBS+ 0.25% Triton X-100 + 5% goat serum+ primary antibodies at 4 °C for 48 hours. After extensive wash, secondary antibodies in 2% BSA was added and incubated at 4 °C for 4 hours.

LIVE/DEAD® Cell Viability kit (Invitrogen) was used to stain the live and dead cells following the product manual. Antibodies were used: OCT4 (1:500, Santa Cruz Biotech), NANOG (1:200, Santa Cruz Biotech), PAX6 (1:300, Biolegend), NESTIN (1:1000, Biolegend), TUJ-1 (1:10,000, Sigma), TBR-1 (1:500, Abcam), TH (1:500, Pel-Freez), and GABA (1:1000, Sigma), HuNu (1:200, Millipore). Secondary antibody: Alexa Fluor 488-AffiniPure Donkey Anti-Mouse IgG, Cy3-AffiniPure Donkey Anti-Rabbit IgG.

Quantitative PCR (qPCR). Total RNA was extracted from the harvested cells using Trizol (Invitrogen, Carlsbad, CA, USA), according to the manufacturer's instructions. Reverse transcription is done with the Maxima First Strand cDNA Synthesis Kit. Quantitative real-time PCR was carried out in an Eppendorf MasterCycler RealPlex4 (ThermoFisher Scientific) using the Power SYBR Green PCR Master Mix (ThermoFisher), according to the manufacturer's instructions. The data were normalized to the endogenous GAPDH.

Teratoma formation *in vivo*. All animal protocols were approved by the Animal Care and Use Committee of the University of Nebraska, Lincoln. All experimental procedures involving animals were carried out in accordance

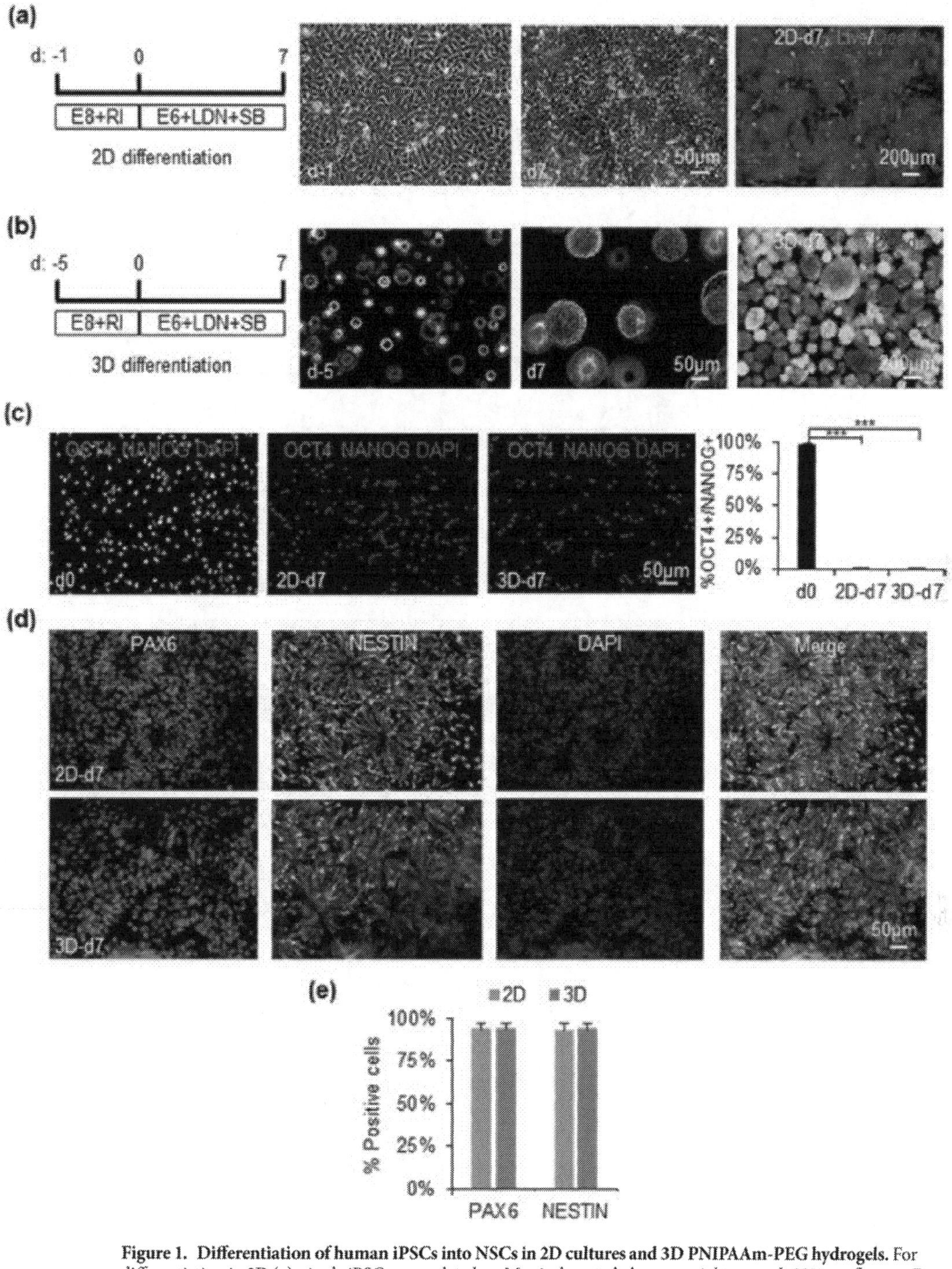

Figure 1. Differentiation of human iPSCs into NSCs in 2D cultures and 3D PNIPAAm-PEG hydrogels. For differentiation in 2D (**a**), single iPSCs were plated on Matrigel-coated plate overnight to reach 90% confluency. For differentiation in 3D (**b**), single iPSCs were cultured in the PNIPAAm-PEG hydrogel for 5 days to generate uniform spheroids. Cells were differentiated in neutral induction medium for 7 days. Phase images for cells on day 0 (d0) and 7 (d7), as well as live/dead cell staining for cells on day 7 are shown. (**c**) 98% of cells expressed the pluripotency markers, OCT4 and NANOG before the differentiation. Less than 2% of cells were OCT4+/NANOG+ after 7 days differentiation in 2D or 3D. Cells were dissociated into single cells and plated on Matrigel-coated plate for 6 hours before the fixation and staining. (**d,e**) On day 7, ~95% of cells were positive for the NSC markers, PAX6 and NESTIN. The 3D spheroids were plated on Matrigel-coated plate overnight to form monolayers before the fixation and staining. The triple asterisk (***) indicates statistical significance at a level of $p < 0.001$.

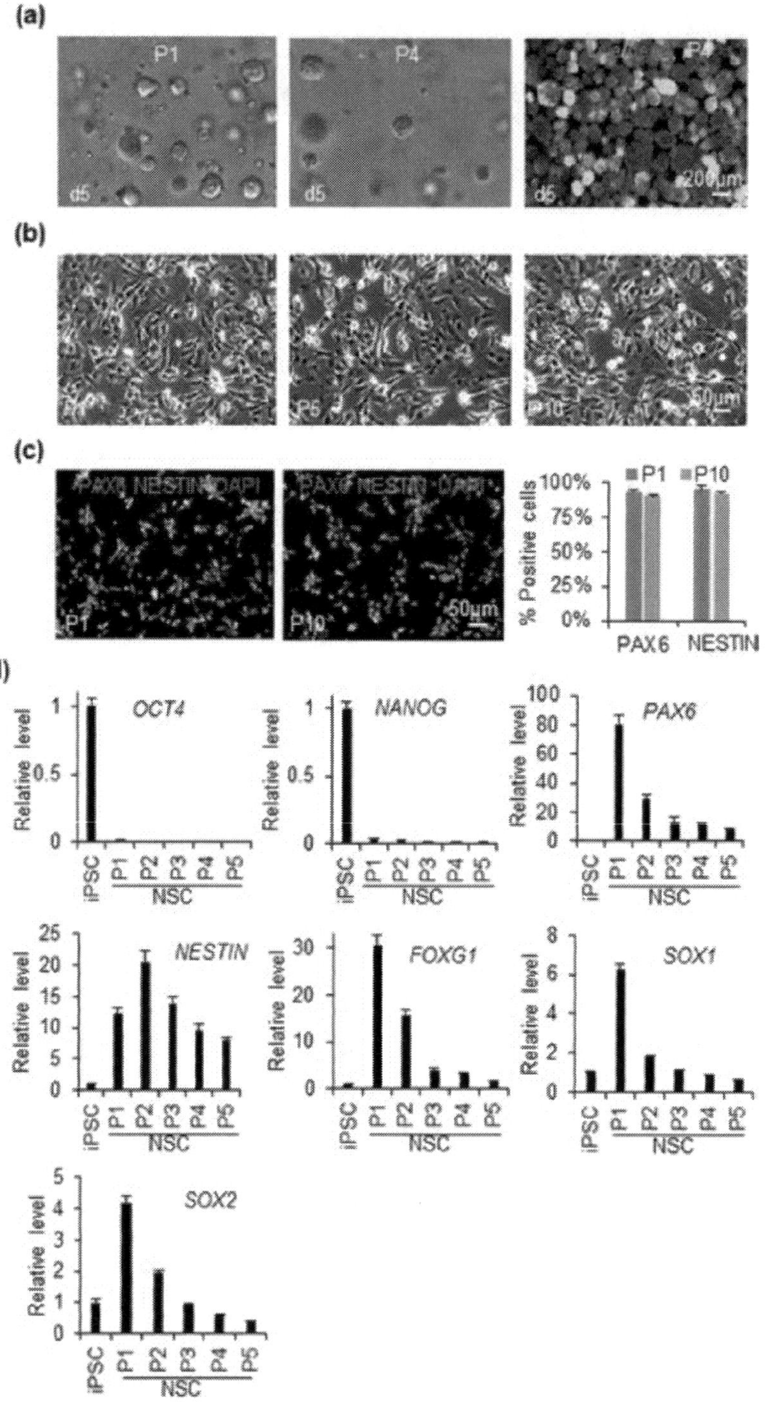

Figure 2. Culturing iPSC-derived NSCs in 3D PNIPAAm-PEG hydrogels and 2D cultures. NSCs derived from iPSCs in the 3D PNIPAAm-PEG hydrogel were dissociated into single cells and cultured in the 3D PNIPAAm-PEG hydrogel (**a**) or on Matrigel-coated 6 well plate (**b–f**). Cells were passaged every 5 days or 3 days for 3D and 2D cultures, respectively. (**a**) phase and live/dead staining images of day 5 spheroids at passage 1 (P1) and 4 (P4) in the 3D PNIPAAm-PEG hydrogels. (**b**) Phase images of NSCs on day 3 of passage 1, 5 and 10 (P1, P5, P10) in 2D cultures. (**c**) 90-95% of the NSCs at P1 and P10 in 2D cultures were positive for the NSC markers, PAX6 and NESTIN. (**d**) qPCR data on the expressions of pluripotency and NSC markers during P1 to P5 in 2D cultures.

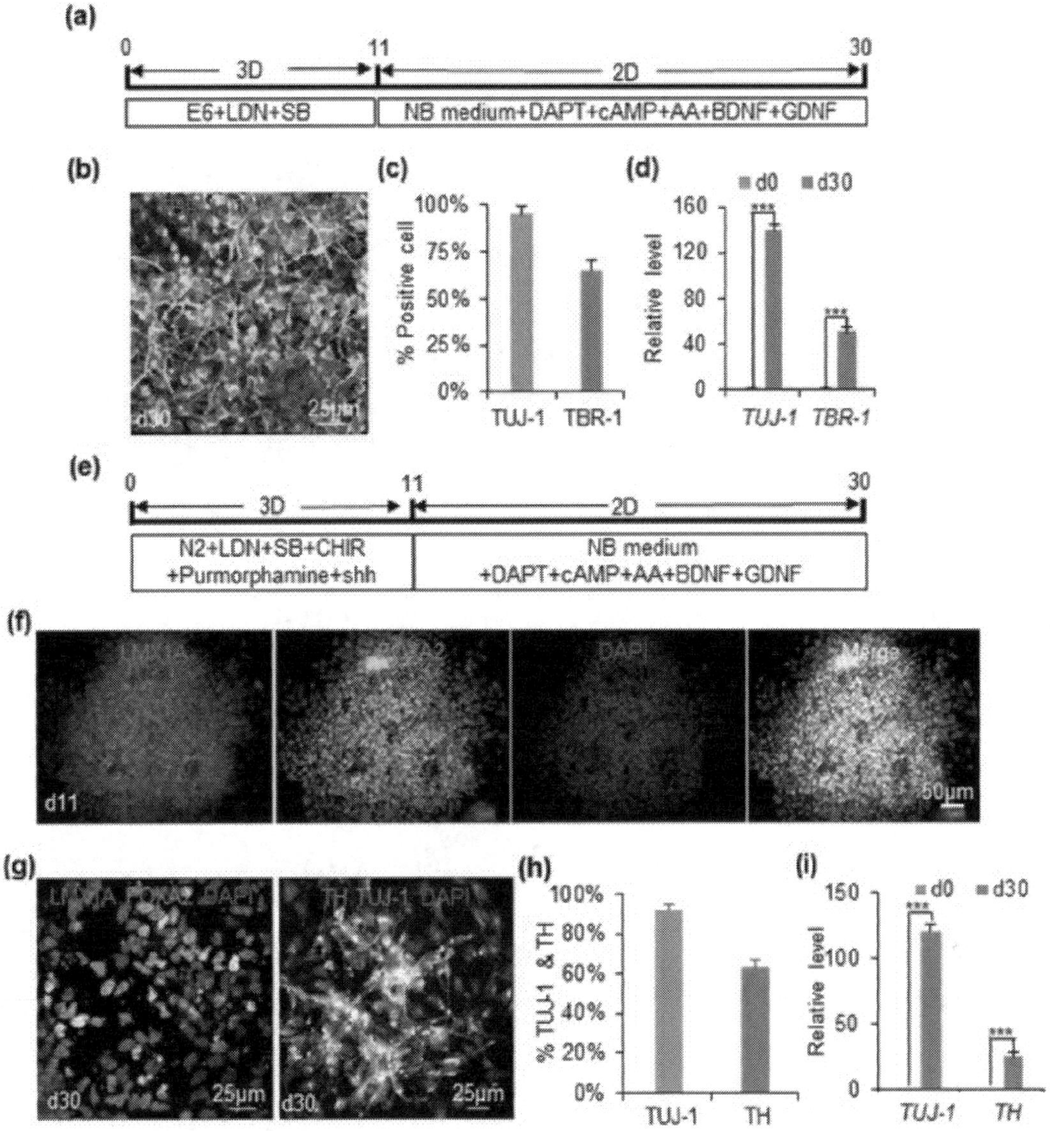

Figure 3. Differentiating iPSCs into cortical and ventral midbrain NSCs in 3D PNIPAAm-PEG hydrogels. (**a**) When iPSCs were differentiated in the E6 medium+ LDN193189+ SB431542 in the 3D PNIPAAm-PEG hydrogels for 11 days, and further differentiated in the neural differentiation medium in 2D for 19 days, cortical neurons could be made. 95% and 65% of the produced cells were positive for TUJ-1 and TBR-1, receptively (**b,c**). (**d**) qPCR showed the expressions of *TUJ-1* and *TBR-1* were increased 139 and 51-fold on day 30, respectively. (**e**) When iPSCs were cultured in the E6 medium + LDN193189 + SB431542 + CHIR99021 + shh + purmorphamine in the 3D PNIPAAm-PEG hydrogels for 11 days, they became FOXA2+ and LMX1A+ ventral midbrain progenitor cells (**f**). When they were further differentiated in the neural differentiation medium in 2D for 19 days, FOXA2+ and LMX1A+ ventral midbrain neurons could be made. 92% and 63% of the produced cells were positive for TUJ-1 and TBR-1, receptively (**g,h**). (i) qPCR showed the expressions of *TUJ-1* and *TH* were increased 120 and 25-fold on day 30, respectively. The triple asterisk (***) indicates statistical significance at a level of $p < 0.001$.

with the guidelines of the Institutional Animal Care and Use Committee of the University of Nebraska, Lincoln. 3×10^6 hPSCs (small clumps) were suspended in 25 µl PBS + 25 µl Matrigel (BD Biosciences) and injected subcutaneously at the back of the neck of the NOD-SCID mice (Charles River). Teratoma was harvested when its size reached 2 cm. Teratoma was fixed with 4% PFA for 48 hours, dehydrated with 70%, 95% and 100% ethanol sequentially, and de-fated with xylene for 2 hours before embedded in paraffin. 10 µm thick section and stained was cut and stained with hematoxylin and eosin. The structures from all 3 germ layers were identified by trained specialist.

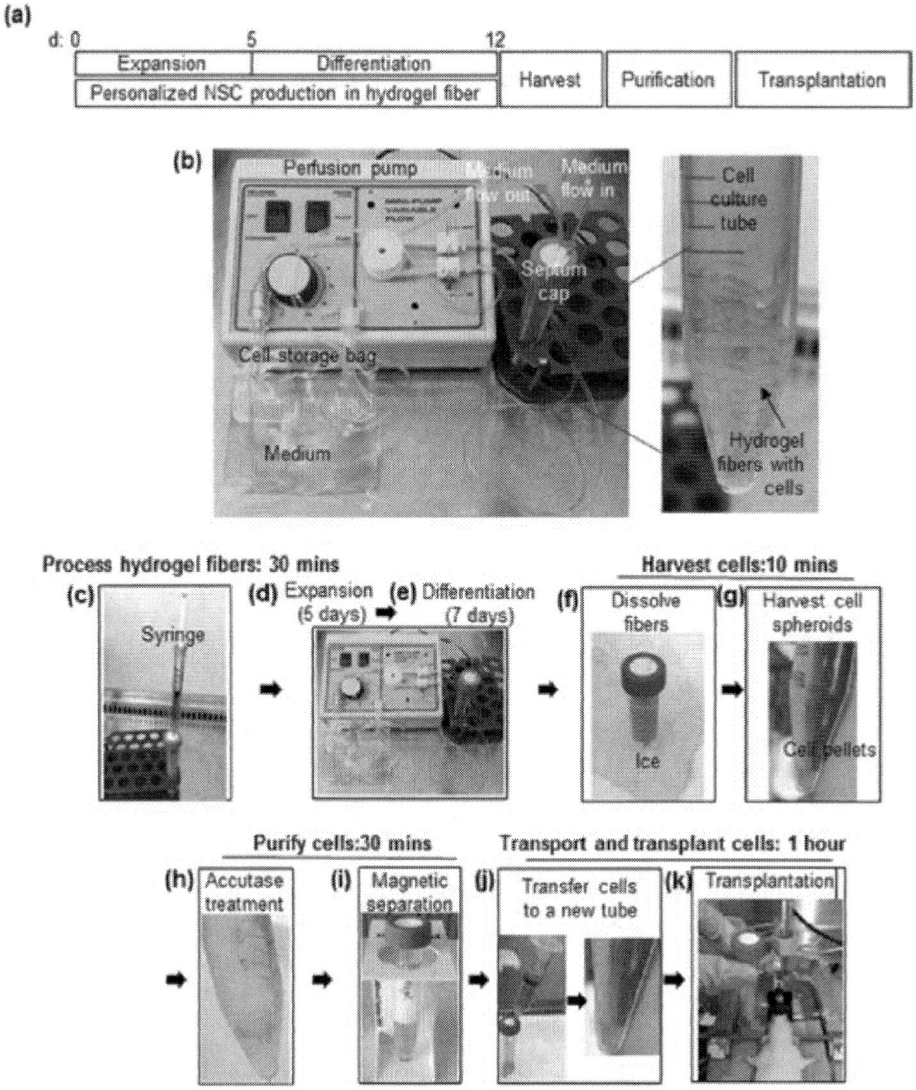

Figure 4. An integrated miniature bioprocessing for iPSC expansion and differentiation into NSCs. (**a**) Illustration of the bioprocessing. (**b**) The miniature cell culture device consists of a pump for medium perfusion, an oxygen-permeable plastic bag for stocking medium and a closed 15 ml conical tube. Fibrous hydrogel fibers with cells were suspended in the tube. (**c**) On day 0, single iPSCs were mixed with 10% PNIPAAm-PEG solution at 4 °C. With a syringe and needle, the mixture was injected into room temperature E8 medium contained in a closed and sterile 15 ml conical tube with a septum cap. Fibrous hydrogels (with diameter <2.0 mm) were instantly formed. (**d,e**) The cells were cultured in E8 medium for 5 days, followed by additional 7 days of neural induction medium in the conical tube. Medium was continuously perfused. (**f**) On day 12, hydrogel scaffolds were liquefied by placing the cell culture tube on ice for 5 minutes. (**g**) Cell spheroids were pelleted by spinning the tube at 100 g for 3 minutes. Medium was removed. (**h**) Cell spheroids were incubated in Accutase at 37 °C for 10 minutes. Removing reagents from the tube and adding reagents to the tube were done with a sterile syringe through the septum cap. (**i**) Magnetic beads coated with anti-SSEA4 antibodies were added into the tube to pull down the undifferentiated SSEA4+ iPSCs with a magnetic cell separator. (**j**) Purified cells in the supernatant were transferred into a new, close tube and transported to the surgical room. (**k**) NSCs were transplanted to the rat brain with a stereotactic injector.

Statistical analysis. Statistical analyses were done using the statistical package Instat (GraphPad Software, La Jolla, CA). For multiple comparisons, the means of triplicate samples were compared using the Tukey multiple comparisons analysis with the alpha level indicated in the figure legend.

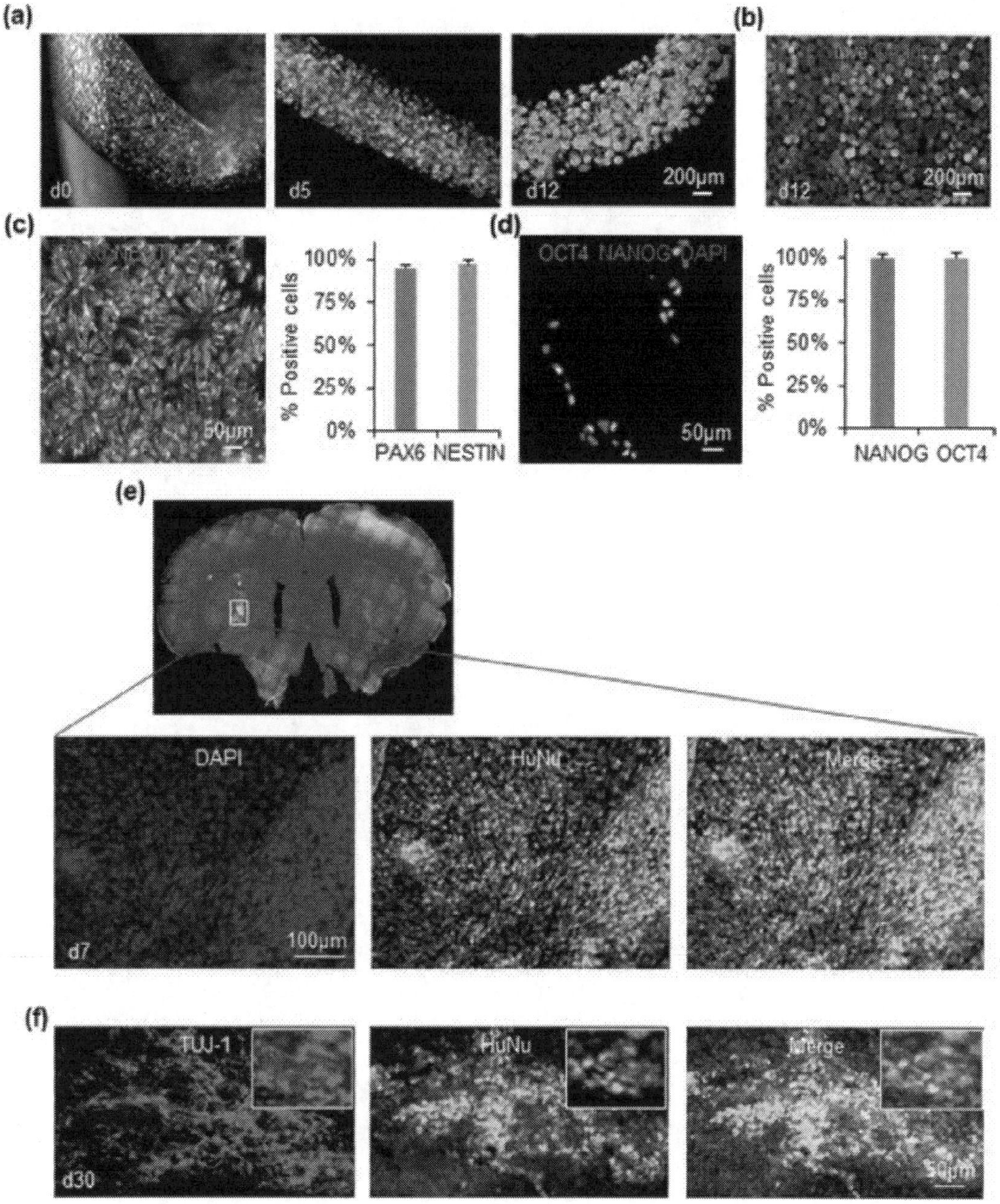

Figure 5. Cells in the miniature bioprocessing. (**a**) Phase images of the hydrogel fibers and cells on day 0, 5 and 12 of the bioprocessing. (**b**) Live/dead staining of cells on day 12. (**c**) ~97% of the purified cell products expressed NSC markers, PAX6 and NESTIN. (**d**) Cells pull down by the magnetic anti-SSEA4 beads were positive for OCT4 and NANOG. (**e**) HuNu+ (human nuclear antigen) NSCs survived well in the rat brain 7 days post-transplantation. (**f**) Substantial amounts of transplanted NSCs became HuNu+ and TUJ-1+ neurons in the rat brain 30 days post-transplantation.

Results

The human iPSCs used for this study were made from human mesenchymal stem cells with retrovirus containing the OKSM (i.e. *OCT4, KLF4, SOX2*, and *c-MYC*) reprogramming factors[47]. They formed compact colonies when cultured on Matrigel-coated cell culture plates in the chemical defined Essential 8 (E8) medium. iPSCs expressed the pluripotency makers, OCT4 and NANOG. They could be differentiated into all the three germ layer cells, such the NESTIN+ ectodermal, α-SMA+ mesodermal and FOXA2+ endodermal cells in the embryoid body (EB) assay (Fig. S1). They formed teratomas after transplanting to the immunodeficient SCID mice. All the three germ layer tissues and cells were found in the teratomas (Fig. S1). To show that results found in this paper is general to human pluripotent stem cells (hPSCs), we also used the H9 hESCs for the research.

Both the iPSCs and H9s could be encapsulated and cultured in the thermoreversible PNIPAAm-PEG hydrogels for long term (Fig. S1)[43,46]. The aqueous 10% PNIPAAm-PEG polymer solution is liquid at low temperature (e.g. below 4 °C) and forms elastic hydrogel with storage modulus around 1000 Pa upon heating above room temperature. With E8 medium supplied with 10 μM ROCK inhibitor Y-27632, single iPSCs could survive and undergo clonal expansion into uniform spherical cell aggregates (spheroids) in the hydrogel (Fig. S1a). When iPSCs were initially seeded at 1.0×10^6 cells/ml, 20-fold expansion could be achieved on day 5, generating ~2.0×10^7 cells/ml. iPSCs could be cultured in the hydrogel scaffold for long term (above 10 passages) (Fig. S1c). The cell growth rate and doubling time (i.e. ~27 hours) were equal to these of cells cultured in Matrigel-coated 2D plates (Fig. S1c). Similar results were found for H9 cells (Fig. S1b,c). These results indicate the 3D PNIPAAm-PEG hydrogel is appropriate for hPSC culture and does not significantly change the cell behavior.

We then studied whether the hPSCs as spheroids could be differentiated into NSCs in the PNIPAAm-PEG hydrogel. Literature research has shown hPSCs cultured in 2D plates could be efficiently induced into neural stem cells (NSCs) by inhibiting the SMAD signaling under defined condition[49]. Our preliminary research confirmed these results. However, cells in the spheroids in the hydrogel face environments that are very different from these of the monolayer cells in the 2D cultures. We asked whether the 3D environments affected cell differentiation and the same differentiation protocol could be used to induce hPSCs (iPSCs and H9s) into NSCs in both 2D and 3D. To test this, we differentiated iPSCs into NSCs in parallel as monolayers on 2D surfaces and as spheroids in the 3D hydrogels. iPSC monolayers and spheroids with diameter around 100 μm were used for the 2D and 3D differentiation, respectively (Fig. 1a,b). To differentiate iPSCs into NSCs, bFGFs and TGFβs were removed from E8 medium and two small molecules LDN193189 (100 nM) and SB431542 (10 μM), which inhibit the SMAD signaling, were supplied to the medium. Very few dead cells were found in both 2D and 3D cultures during the differentiation as shown by the Live/Dead cell staining (Fig. 1a,b). On day 7, ~3.0×10^5 cells/cm^2 and 5.0×10^7 cells/ml were yielded for 2D and 3D culture, respectively. Immunostaining demonstrated that >95% of the starting iPSCs expressed the pluripotency marker, OCT4 and NANOG. On day 7, less than 2% of the cells were positive for OCT4 and NANOG (Fig. 1c). 95% of the day 7 cells expressed high level of NSC markers, NESTIN and PAX6. There were no significant difference for the expressions of protein markers between the 2D and 3D differentiation (Fig. 1d,e).

We further compared the differentiation in 2D and 3D using qPCR (Fig. S2). qPCR data showed the pluripotency marker *OCT4* gradually decreased, while *NANOG* dropped sharply on day 1. NSC markers, such as *PAX6*, *NESTIN*, *FOXG1*, *SOX1* and *SOX2*, increased gradually. While there was no difference for the *PAX6*, *NESTIN* expressions between 2D and 3D differentiation, the 3D environment speeded up and enhanced the expressions of *FOXG1*, *SOX1*, and delayed the expression of *SOX2*. Similar results were found for the H9s (Figs S3 and S4). This comparative study indicates the small molecule-based differentiation protocol works efficiently in both 2D and 3D and the 3D environment does affect the expressions of some genes.

We then studied whether the hPSC-derived NSCs could be cultured *in vitro* to expand their numbers. We cultured the NSCs derived from iPSCs in both the 3D PNIPAAm-PEG hydrogels and 2D plates (Fig. 2 and S5). When NSCs made in the 3D hydrogels were dissociated and further cultured in 3D PNIPAAm-PEG hydrogels, they grew into spheroids and could be passaged every 5 days. However, at passage 4, significant cell death was seen as shown by the live/dead staining (Fig. 2a) and the culture could not be continued. When NSCs made in the 3D hydrogels were cultured on Matrigel-coated plates, cells were could be propagated at least 10 passages without significant cell death (Fig. 2b). Cells expressed the NSC markers, PAX6 and NESTIN, during the long term culture (Fig. 2c). However, qPCR data showed the transcript levels of *PAX6, NESTIN, FOXG1, SOX1* and *SOX2* decreased during the culture (Fig. 2d). NSCs derived from the iPSCs in 2D cultures (Fig. S5), and NSCs differentiated from H9s in both 2D cultures and 3D PNIPAAm-PEG hydrogels showed very similar results (Figs S6 and S7).

We studied whether iPSCs could be differentiated into regional NSCs in the 3D PNIPAAm-PEG hydrogel. Using dual SMAD inhibition without any patterning factors, the produced NSCs were default to be the precursor of cortical neurons[49]. When the day 11 NSCs released from the 3D hydrogels were cultured on Matrigel-coated plates for additional 19 days in neural differentiation medium, 95% of the cells became TUJ-1+ neurons. Among them, ~65% were TBR-1+ cortical neurons (Fig. 3a–c). qPCR showed the significant higher expressions of *TUJ-1* and *TBR-1* on day 30 than day 0 (Fig. 3d). If 0.7 μM CHIR99021 and 2 μM purmorphamine plus 200 ng/ml sonic hedgehog (shh) proteins were added to neural induction medium to activate the canonical Wnt and shh signaling, respectively, the NSCs could be patterned to the ventral midbrain fate as shown by the expression of ventral and midbrain markers: LMX1A and FOXA2 on day 11 (Fig. 3f)[15]. After maturing in the neural differentiation medium in 2D for additional 19 days, 92% of the cells were TUJ-1+ neurons. Among them, ~63% were TH+ midbrain neurons (Fig. 3g–i). Majority of these cells expressed the ventral and midbrain markers: LMX1A and FOXA2 (Fig. 3g). qPCR showed the significant higher expressions of *TUJ-1* and *TH* on day 30 than day 0 (Fig. 3h). These results indicate regional NSCs can be made from iPSCs in the 3D PNIPAAm-PEG hydrogels with pattern factors supplied to the medium.

We then took advantage of the high cell yield in the PNIPAAm-PEG hydrogels to build a prototype miniature bioprocessing for making NSCs from iPSCs for personalized cell therapies (Fig. 4). On day 0, single iPSCs were mixed with 10% PNIPAAm-PEG solution at 4 °C. With a syringe and needle, the mixture was injected into room temperature E8 medium contained in a closed and sterile 15 ml conical tube with a septum cap (Fig. 4c). Fibrous hydrogels (with diameter <2.0 mm) were instantly formed with single iPSCs uniformly distributed in the hydrogels. The cells were cultured in a cell culture incubator at 37 °C and 5% CO_2. Medium stocked in a gas-permeable bag was continuously perfused into the cell culture tube (Fig. 4b). E8 medium was supplied for 5 days (Fig. 4d), followed by additional 7 days of neural induction medium (Fig. 4e). On day 7, hydrogel scaffolds were liquefied by placing the cell culture tube on ice for 5 minutes (Fig. 4f). Cell spheroids were pelleted by spinning the tube at 100 g for 3 minutes (Fig. 4g). Medium was removed. Cell spheroids were incubated in Accutase at 37 °C for 10 minutes (Fig. 4h). Removing reagents from the tube and adding reagents to the tube were done with a sterile syringe through the septum cap. Magnetic beads coated with anti-SSEA4 antibodies were added into the tube to pull down the undifferentiated SSEA4+ iPSCs with a magnetic cell separator (Fig. 4i). Purified cells in the supernatant were transferred into a new, close tube (Fig. 4j) and transported to the surgical room. NSCs were transplanted to the brains of Sprague dawley rats with a stereotactic injector (Fig. 4k).

Single iPSCs in hydrogel fibers grew into iPSC spheroids on day 5, and then became NSC spheroids on day 12 (Fig. 5a). With initial seeding density at 1×10^6 cells/ml, 25-fold expansion and 2.5×10^7 cells/ml hydrogel were achieved on day 12. Total of 1.0×10^8 cells were produced in 4 ml of hydrogel in a 15 ml conical tube. Cell viability was >95% on day 12. 2% of the day 12 cells were SSEA4+. Live/dead cell staining showed no or undetectable dead cells (Fig. 5b). After magnetic separation, the produced cells expressed PAX6 and NESTIN (Fig. 5c) and OCT4+/NANOG+ cells were not detectable. Cells pull down by the magnetic beads expressed both OCT4 and NANOG (Fig. 5d). 7 days post-transplantation, substantial numbers of human cells, as shown by the expression of human nuclear antigen (HuNu+) were found in the rat brain (Fig. 5e). 30 days post-transplantation, substantial numbers of human neurons (e.g. HuNu+ and TUJ-1+) were found in the rat brain (Fig. 5f).

Discussion

Transgene free iPSCs can be readily made from adults cells with the most recent integration free reprogramming vectors[6]. iPSCs can be expanded *in vitro* to generate large numbers of cells. They can also be differentiated presumably all the human cell types. Additionally, there are no ethic debates with using iPSCs derived products. Thus, iPSCs and their derivatives have enormous potential to treat human diseases, either by transplanting them to replace the lost cells or to alter the disease progression through stand by mechanisms, or using them to engineer tissues or organs to replace the dysfunctional ones[1–5].

There are two approaches to use iPSC-derived products. iPSCs and their derivatives can be manufactured in large scale and used as off-shelf products[50]. The advantages of this approach include: (a) off-shelf products can be available when the patients need them, and (b) the large-scale production reduces the biomanufacturing cost. However, allogenic cells usually induce severe immune response. A strategy to minimize the immune response is to generate libraries of iPSCs to provide human leukocyte antigen (HLA)-matched cells to majority of the patients[51–53]. However, the cost and effort to establish these libraries are not trivial[51–53]. In addition, HLA-matched cells may still induce immune response[1,2,54]. The second approach is to make patient-specific iPSC products[1,2,54]. Research has shown patient-specific iPSC products induce no or minimal immune response. However, the cost for biomanufacturing personalized iPSCs and their derivatives with current technologies are extremely high and not affordable for majority of patients[1–5].

Automating the bioprocessing in individualized, closed, computer controlled miniature cell culture devices can significantly reduce the biomanufacturing cost[42]. However, it is very challenging to develop such type of cell culture device with current 2D cell culture (e.g. cell culture dishes) and 3D suspension culture (e.g. spinner flasks) technologies[44]. The low yield and frequent cell passaging make the 2D culture systems not a good choice as personalized cellular biomanufacturing system. In 3D suspension cultures, cells frequently aggregate into large cell agglomerates[43,44,46]. Due to the diffusion limits, the supply of growth factors, nutrients and oxygen to and the depletion of the metabolic waste from the cells located at the center of cell agglomerates became insufficient, leading to slow cell proliferation, apoptosis and uncontrolled differentiation. Stirring or shaking is widely used to reduce the cellular agglomeration and enhance mass transport in suspension cultures[44]. However, it also generated hydrodynamic stress that negatively affects the cell growth and quality[55,56]. High cell density in the culture also promotes spheroid agglomeration[46]. Considering all these factors, cells are generally seeded at low density (e.g. ~5.0×10^5 cells/ml) and stirred at 70 to 120 rpm in suspension cultures. Even under these optimized conditions, cells grow slowly and cell death is common. For instance, iPSCs typically expand 4 times per 4 days to yield around 2.0×10^6 cells/ml in spinner flasks[46]. It is good to note that cells occupy <0.4% of the medium volume with this yield. It would require a 50 ml to 500 ml of 3D suspension cultures to produce 1×10^8 to 1×10^9 cells[44]. This low volumetric yield makes the 3D suspension culture systems not attractive for developing miniature cell culture devices.

A previous study showed hPSCs could be cultured in thermoreversible PNIPAAm-PEG hydrogels at much higher efficiency[46]. The hydrogel scaffolds provided 3D spaces for cell growth and also acted as physical barriers to prevent cell agglomeration and isolate the hydrodynamic stresses. The high cell growth rate, yield and quality makes this hydrogel very attractive for developing personalized cell culture device. In this paper, we demonstrated that personalized NSCs could be made from iPSCs within a miniature cell culture device with the assist of the PNIPAAm-PEG hydrogels (Figs 4 and 5). We showed that: (a) the whole bioprocess including iPSC expansion, iPSC differentiation into NSCs, the product purification, concentration and transport to the surgery room, could be completed with two closed conical tubes. In this study, we depleted the undifferentiated iPSCs with anti-SSEA4 magnetic beads and the purified NSCs were transferred to a second tube (Figs 4 and 5). An alternative approach is to positively isolate the NSCs with magnetic beads coated with antibodies recognizing NSCs. With this approach, the whole bioprocess can be done with one tube; (b) ~1.0×10^8 high purity NSCs could be made with 4 ml hydrogel scaffolds within a 15 ml conical cell culture tube. It is calculated that 1×10^9 cells, which are needed to treat many human diseases, can be made with 40 ml of fibrous hydrogel scaffolds in a 50 ml conical tube. This miniaturized device allows high throughput biomanufacturing of personalized cells for large numbers of patients at the point-of-care; (c) majority of the bioprocessing could be done automatically. In the future, processing fibers, harvesting, purifying and concentrating cells will also be automated. Recent research demonstrated reprogramming human adult cells into iPSCs could be efficiently done in 3D hydrogel scaffolds[57]. And our preliminary research also showed iPSCs could be made from fibroblasts in the 3D PNIPAAm-PEG hydrogels. In the future, we will also integrate the reprogramming into the miniaturized bioprocessing within the conical tube.

iPSCs were differentiated into NSCs to demonstrate the personalized cellular biomanufacturing system in this paper. iPSC-derived NSCs are being widely studied for treating various human neuronal diseases[58,59]. These NSCs could survive, mature and outgrowth long nerve fibers in rodent model of spinal cord injuries[60]. iPSC-derived NSCs patterned to the forebrain cortical fate could survive and significantly improve the function of stroked brains in rodents[33]. iPSC-derived NSCs could migrate toward brain tumors and thus are being studied as gene or drug delivery vehicles for treating brain cancers[61]. NSCs isolated from fetal brains have been in clinical trials for treating a few rare neuronal degenerative diseases[62]. iPSC-derived NSCs have potential to replace these fetal NSCs to avoid the ethical issues and host immune response.

The protocols for differentiating hPSCs into NSCs in the literature are developed in 2D culture systems and may be not applicable for differentiating hPSCs as spheroids in the hydrogel scaffolds. The cell spheroids may have increased amounts of extracellular proteins, cell-cell, cell-matrix interactions[63]. The concentrations of many protein factors in the 3D cell mass may be different from these in the 2D cell monolayers. In addition, gradients may exist in the 3D cell mass and hydrogel. The transport of medium components and metabolic waste in the cell spheroids and hydrogel scaffold are also different from these in the 2D cultures. We systematically compared the iPSC differentiation into NSCs as spheroids in the 3D hydrogel and as monolayers in 2D. In general, the small molecules-based differentiation protocol used in this research worked efficiently in both 3D and 2D cultures (Fig. 1 and S3). However, the transcripts of SOX1 and FOXG1 appeared earlier and stronger in 3D, while the transcript of SOX2 was delayed in 3D (Figs S2 and S4). Literature research has shown that, during the neural differentiation of hPSCs as embryoid body, hPSCs first became primitive neuroepithelia NSCs expressing PAX6 and SOX2, but no SOX1, then became definitive neuroepithlia NSCs expressing PAX6, SOX2 and SOX1[64,65]. Our results showed NSCs generated in 2D cultures and in 3D hydrogel might be mainly primitive and definitive neuroepithelia NSCs, respectively, and the 3D environment promoted forebrain fate NSCs, as shown by the high level of forebrain marker FOXG1 (Fig. S2)[66]. Our research showed that NSCs derived from hPSCs in both 3D and 2D could be cultured for long term in 2D, but not in 3D hydrogels (Fig. 2 and S5). In addition, the transcript level of some genes that are important for NSCs were altered during the long term culture (Fig. 2 and S6).

In summary, we showed that, with the assist of appropriate hydrogel scaffolds, personalized iPSC expansion and differentiation, as well as the subsequent product purification could be integrated into miniature bioprocessing that can be automated. We believe this type of bioprocessing can help the advancement of iPSC-based personalized cell therapies. To our best knowledge, this is the first report on hydrogel-based miniature cell culture device/bioprocessing for personalized iPSC expansion and differentiation.

References

1. Mount, N. M., Ward, S. J., Kefalas, P. & Hyllner, J. Cell-based therapy technology classifications and translational challenges. *Philos. Trans. R. Soc. Lond. B. Biol. Sci.* **370,** 1680–1696 (2015).
2. Neofytou, E., Brien, C. G. O., Couture, L. a & Wu, J. C. Hurdles to clinical translation of human induced pluripotent stem cells. *J. Clin. Invest.* **125,** 2551–2557 (2015).
3. Byrne, J. Overcoming clinical hurdles for autologous pluripotent stem cell-based therapies. *OA Stem Cells* **1,** 3 (2013).
4. Nienow, W., Mccall, M. J., Coopman, K., Kara, B. & Hewitt, C. J. The translation of cell-based therapies : clinical landscape and manufacturing challenges. *Regen. Med.* **10,** 49–64 (2015).
5. Trounson, A. & Mcdonald, C. Review Stem Cell Therapies in Clinical Trials : Progress and Challenges. *Stem Cell* **17,** 11–22 (2015).
6. Okita, K. *et al.* A more efficient method to generate integration-free human iPS cells. *Nat. Methods* **8,** 409–412 (2011).
7. Chen, G. *et al.* Chemically defined conditions for human iPSC derivation and culture. *Nat. Methods* **8,** 424–429 (2011).
8. Kimbrel, E. a. & Lanza, R. Current status of pluripotent stem cells: moving the first therapies to the clinic. *Nat. Rev. Drug Discov.* **14,** 681–692 (2015).
9. Trounson, A. & Dewitt, N. D. Pluripotent stem cells progressing to the clinic. *Nat Rev Mol Cell Biol* **17,** 194–200 (2016).
10. Shi, Y., Kirwan, P., Smith, J., Robinson, H. P. C. & Livesey, F. J. Human cerebral cortex development from pluripotent stem cells to functional excitatory synapses. *Nat. Publ. Gr.* **15,** 477–486 (2012).
11. Shi, Y., Kirwan, P. & Livesey, F. J. Directed differentiation of human pluripotent stem cells to cerebral cortex neurons and neural networks. *Nat. Protoc.* **7,** 1836–1846 (2012).
12. Liu, Y. *et al.* Directed differentiation of forebrain GABA interneurons from human pluripotent stem cells. *Nat. Protoc.* **8,** 1670–1679 (2013).
13. Maroof, A. M. *et al.* Directed Differentiation and Functional Maturation of Cortical Interneurons from Human Embryonic Stem Cells. *Cell Stem Cell* **12,** 559–572 (2013).
14. Hunt, R. F., Girskis, K. M., Rubenstein, J. L. Alvarez-Buylla, A. & Baraban, S. C. GABA progenitors grafted into the adult epileptic brain control seizures and abnormal behavior. *Nat. Neurosci.* **16,** 692–697 (2013).
15. Kriks, S. *et al.* Dopamine neurons derived from human ES cells efficiently engraft in animal models of Parkinson's disease. *Nature* **480,** 547–551 (2011).
16. Kirkeby, A. *et al.* Generation of regionally specified neural progenitors and functional neurons from human embryonic stem cells under defined conditions. *Cell Rep.* **1,** 703–714 (2012).
17. Orlova, V. V. *et al.* Generation, expansion and functional analysis of endothelial cells and pericytes derived from human pluripotent stem cells. *Nat. Protoc.* **9,** 1514–1531 (2014).
18. Prasain, N. *et al.* Differentiation of human pluripotent stem cells to cells similar to cord-blood endothelial colony-forming cells. *Nat. Biotechnol.* **32,** 1151–1157 (2014).
19. Lian, X. *et al.* Efficient Differentiation of Human Pluripotent Stem Cells to Endothelial Progenitors via Small-Molecule Activation of WNT Signaling. *Stem Cell Reports* **3,** 804–816 (2014).
20. Kimbrel, E. a. *et al.* Mesenchymal Stem Cell Population Derived from Human Pluripotent Stem Cells Displays Potent Immunomodulatory and Therapeutic Properties. *Stem Cells Dev.* **23,** 1611–1624 (2014).
21. Nishio, M. & Saeki, K. Differentiation of human pluripotent stem cells into highly functional classical brown adipocytes. *Methods in enzymology* **537,** 177–197 (2014).
22. Burridge, P. W. *et al.* Chemically defined generation of human cardiomyocytes. *Nat. Methods* **11,** 855–860 (2014).
23. Lian, X. *et al.* Robust cardiomyocyte differentiation from human pluripotent stem cells via temporal modulation of canonical Wnt signaling. *Proc. Natl. Acad. Sci. USA* **109,** E1848–E1857 (2012).

24. Lian, X. *et al.* Directed cardiomyocyte differentiation from human pluripotent stem cells by modulating Wnt/β-catenin signaling under fully defined conditions. *Nat. Protoc.* **8,** 162–175 (2013).
25. Si-Tayeb, K. *et al.* Highly efficient generation of human hepatocyte-like cells from induced pluripotent stem cells. *Hepatology* **51,** 297–305 (2010).
26. Hannan, N. R. F., Segeritz, C.-P., Touboul, T. & Vallier, L. Production of hepatocyte-like cells from human pluripotent stem cells. *Nat. Protoc.* **8,** 430–437 (2013).
27. Szkolnicka, D., Farnworth, S. L., Lucendo-Villarin, B. & Hay, D. C. Deriving functional hepatocytes from pluripotent stem cells. *Curr. Protoc. Stem Cell Biol.* **30,** 1G.5.1–1G.5.12 (2014).
28. Rezania, A. *et al.* Reversal of diabetes with insulin-producing cells derived *in vitro* from human pluripotent stem cells. *Nat. Biotechnol.* **32,** 1121–1134 (2014).
29. Pagliuca, F. W. *et al.* Resource Generation of Functional Human Pancreatic β Cells *In Vitro. Cell* **159,** 428–439 (2014).
30. Tabar, V. & Studer, L. Pluripotent stem cells in regenerative medicine: challenges and recent progress. *Nat. Rev. Genet.* **15,** 82–92 (2014).
31. Grealish, S. *et al.* Human ESC-Derived Dopamine Neurons Show Similar Preclinical Efficacy and Potency to Fetal Neurons when Grafted in a Rat Model of Parkinson's Disease. *Cell Stem Cell* **15,** 653–665 (2014).
32. Tong, L. M., Fong, H. & Huang, Y. Stem cell therapy for Alzheimer's disease and related disorders: current status and future perspectives. *Exp. Mol. Med.* **47,** e151 (2015).
33. Tornero, D. *et al.* Human induced pluripotent stem cell-derived cortical neurons integrate in stroke-injured cortex and improve functional recovery. *Brain* **136,** 3561–3577 (2013).
34. Okano, H. & Yamanaka, S. iPS cell technologies: significance and applications to CNS regeneration and disease. *Mol. Brain* **7,** 22 (2014).
35. Nakamura, M. & Okano, H. Cell transplantation therapies for spinal cord injury focusing on induced pluripotent stem cells. *Cell Res.* **23,** 70–80 (2013).
36. Nori, S. *et al.* Grafted human-induced pluripotent stem-cell-derived neurospheres promote motor functional recovery after spinal cord injury in mice. *Proc. Natl. Acad. Sci. USA* **108,** 16825–16830 (2011).
37. Mothe, A. J. & Tator, C. H. Advances in stem cell therapy for spinal cord injury. *Thew J. Clin. Investig.* **122,** 3824–3834 (2012).
38. Wright, L. S., Phillips, M. J., Pinilla, I., Hei, D. & Gamm, D. M. Induced pluripotent stem cells as custom therapeutics for retinal repair: Progress and rationale. *Exp. Eye Res.* **123,** 161–172 (2014).
39. Zhong, X. *et al.* Generation of three dimensional retinal tissue with functional photoreceptors from human iPSCs. *Nat. Commun.* **5,** 1–31 (2015).
40. Chow, M. Z., Boheler, K. R. & Li, R. A. Human pluripotent stem cell-derived cardiomyocytes for heart regeneration, drug discovery and disease modeling : from the genetic, epigenetic, and tissue modeling perspectives. *Stem Cell Res. Ther.* **4,** 1 (2013).
41. Takahashi, K. *et al.* Induction of pluripotent stem cells from adult human fibroblasts by defined factors. *Cell* **131,** 861–872 (2007).
42. Trainor, N., Pietak, A. & Smith, T. Rethinking clinical delivery of adult stem cell therapies. *Nat. Biotechnol.* **32,** 729–735 (2014).
43. Lei, Y. & Schaffer, D. V. A fully defined and scalable 3D culture system for human pluripotent stem cell expansion and differentiation. *Proc. Natl. Acad. Sci. USA* **110,** 1–10 (2013).
44. Serra, M., Brito, C., Correia, C. & Alves, P. M. Process engineering of human pluripotent stem cells for clinical application. *Trends Biotechnol.* **30,** 350–359 (2012).
45. Hourd, P., Ginty, P., Chandra, A. & Williams, D. J. Manufacturing models permitting roll out/scale out of clinically led autologous cell therapies: Regulatory and scientific challenges for comparability. *Cytotherapy* **16,** 1033–1047 (2014).
46. Lei, Y., Jeong, D., Xiao, J. & Schaffer, D. V. Developing Defined and Scalable 3D Culture Systems for Culturing Human Pluripotent Stem Cells at High Densities. *Cell. Mol. Bioeng.* **7,** 172–183 (2014).
47. Park, I.-H. *et al.* Reprogramming of human somatic cells to pluripotency with defined factors. *Nature* **451,** 141–146 (2008).
48. Lippmann, E. S. *et al.* Defined Human Pluripotent Stem Cell Culture Enables Highly Efficient Neuroepithelium Derivation Without Small Molecule Inhibitors. *Stem cell* 1032–1042 (2014).
49. Chambers, S. M. *et al.* Highly efficient neural conversion of human ES and iPS cells by dual inhibition of SMAD signaling. *Nat. Biotechnol.* **27,** 275–280 (2009).
50. Baghbaderani, B. A. *et al.* CGMP-manufactured human induced pluripotent stem cells are available for pre-clinical and clinical applications. *Stem Cell Reports* **5,** 647–659 (2015).
51. Rao, M. Forum iPSC Crowdsourcing : A Model for Obtaining Large Panels of Stem Cell Lines for Screening. *Stem Cell* **13,** 389–391 (2013).
52. Turner, M. *et al.* Forum Toward the Development of a Global Induced Pluripotent Stem Cell Library. *Stem Cell* **13,** 382–384 (2013).
53. Taylor, C. J., Peacock, S., Chaudhry, A. N., Bradley, J. A. & Bolton, E. M. Generating an iPSC bank for HLA-matched tissue transplantation based on known donor and recipient hla types. *Cell Stem Cell* **11,** 147–152 (2012).
54. Almeida, P. E. De, Ransohoff, J. D., Nahid, A. & Wu, J. C. Immunogenicity of Pluripotent Stem Cells and Their Derivatives. *Circ. Res.* **112,** 523–533 (2013).
55. Hu, W., Berdugo, C. & Chalmers, J. J. The potential of hydrodynamic damage to animal cells of industrial relevance: Current understanding. *Cytotechnology* **63,** 445–460 (2011).
56. Brindley, D. *et al.* Bioprocess forces and their impact on cell behavior: implications for bone regeneration therapy. *J. Tissue Eng.* **2011,** 620247–620260 (2011).
57. Caiazzo, M. *et al.* Defined three-dimensional microenvironments boost induction of pluripotency. *Nat. Mater.* **15,** 344–352 (2016).
58. Jensen, M. B., Yan, H., Krishnaney-davison, R., Sawaf, A. Al & Zhang, S. Survival and Differentiation of Transplanted Neural Stem Cells Derived from Human Induced Pluripotent Stem Cells in A Rat Stroke Model. *J. Stroke Cerebrovasc. Dis.* **22,** 304–308 (2013).
59. Manley, N. C., Azevedo-pereira, R. L., Bliss, T. M. & Steinberg, G. K. Cell Therapy for Brain Injury: Neural stem cells in stroke: intracerebral approaches. Chapter 7, 91–109 (2015).
60. Lu, P. *et al.* Long-Distance Axonal Growth from Human Induced Pluripotent Stem Cells after Spinal Cord Injury. *Neuron* **2,** 789–796 (2014).
61. Lee, E. X. *et al.* Glioma gene therapy using induced pluripotent stem cell derived neural stem cells. *Mol. Pharm.* **8,** 1515–1524 (2011).
62. Tsukamoto, A., Uchida, N., Capela, A., Gorba, T. & Huhn, S. Clinical translation of human neural stem cells. *Stem Cells Res. Ther.* **4,** 102 (2013).
63. Peerani, R. *et al.* Niche-mediated control of human embryonic stem cell self-renewal and differentiation. *EMBO J.* **26,** 4744–4755 (2007).
64. Zhang, X. & Zhang, S. Differentiation of neural precursors and dopaminergic neurons from human embryonic stem cells. *Methods Mol. Biol.* **584,** 355–366 (2010).
65. Pankratz, M. T. *et al.* Directed Neural Differentiation of Human Embryonic Stem Cells via an Obligated Primitive Anterior Stage. *Stem cells* **25,** 1511–1520 (2007).
66. Yuan, F. *et al.* Efficient generation of region- specific forebrain neurons from human pluripotent stem cells under highly defined condition. *Sci. Rep.* **5,** 18550 (2015).

Acknowledgements
This work was partially funded by Nebraska DHHS Stem Cell Research Project (to Y.L.).

Author Contributions
Y.L. and H.L. conceived the idea and designed the study. H.L. and Q.L. performed experiments and analyzed data. Y.L. and H.L. wrote the manuscript.

Additional Information
Supplementary information accompanies this paper at http://www.nature.com/srep

Competing financial interests: The authors declare no competing financial interests.

How to cite this article: Lin, H. *et al.* An Integrated Miniature Bioprocessing for Personalized Human Induced Pluripotent Stem Cell Expansion and Differentiation into Neural Stem Cells. *Sci. Rep.* 7, 40191; doi: 10.1038/srep40191 (2017).

Publisher's note: Springer Nature remains neutral with regard to jurisdictional claims in published maps and institutional affiliations.

498802 Oligodendrocyte Precursor Cell Intracellular Redox State Is Dependent on 3D Hydrogel Properties

Abstract:

The intracellular redox state of a cell is indicative not only of a cell's proliferation or differentiation capabilities, but the cell's ability to build and maintain cellular structures as well. In the central nervous system, this could correlate to an oligodendrocyte's ability to generate the myelin sheath, the electrically insulating layer around neuronal axons. This myelin sheath enables neurons to send quick and efficient electrical signals to surrounding neurons, and is the main tissue component that is damaged in demyelinating diseases like multiple sclerosis. Biomaterials may be a potential method of repair following insults or injuries to the central nervous system and the myelin sheath, however little is known about what biomaterial factors stimulate functional regrowth and subsequent remyelination of oligodendrocytes from their parent cells: oligodendrocyte precursor cells.

Here we investigate how tuning the biomaterial properties of a polyethylene glycol based hydrogel affects the intracellular redox state and myelin production of an oligodendrocyte precursor cell line. PEG-dimethacrylate hydrogels with storage moduli from 230-1900 Pa were formed by tuning the concentration and molecular weight from 6 to 10% and 4600 to 8000 Da, respectively. When cells are encapsulated in the hydrogels, not only do cells proliferate in a stiffness dependent trend[1], but the cellular redox state appears to be modulated as well. The cellular redox state can be calculated by measuring the amounts of glutathione in the reduced, GSH, and oxidized GSSG form through a simple assay kit. Cells in the more compliant materials are found to have a more reduced intracellular redox state, observed by increased ratios of reduced glutathione, GSH, to total glutathione in reduced and oxidized forms (GSH and GSSG), compared to a more oxidized intracellular redox state in the least compliant, or stiffer materials. After 1 day in 3D culture, cells in the more compliant material had a 40% increase in reduced glutathione compared to the stiffer material, and a 60% increase at 7 days. When poly(lactic acid) is built into the PEG backbone, lactic acid, a powerful antioxidant and metabolite, can be slowly released as the polymer backbone hydrolytically degrades. Doping in soluble lactic acid as well as the release of lactic acid from the polymer backbone were also found to increase the amount of glutathione in the reduced form (40% with 5μmol/mL of soluble lactic acid) as compared to the total amount of glutathione in both the GSH and GSSG form. Further research investigates how the incorporation of electrospun fibers as topographical cues for myelination will further modulate the intracellular redox state and promote subsequent myelin production by differentiated oligodendrocytes in the 3D biomaterial. These results together suggest the potential use of an engineered PEG hydrogel environment to promote OPC growth, maturation, and repair of the myelin sheath.

[1]Russell, LN & Lampe, KJ, *Cells Tissues and Organs*, 2016.

497077 NANOG Restores the Myogenic Differentiation Potential of Senescent Myoblasts

Abstract:

Skeletal muscle loss due to aging, sarcopenia, is a major challenge facing the elderly. Adult skeletal muscle regeneration relies on the activity of resident satellite cells in skeletal muscle niche. However, systemic and intrinsic factors decrease the myogenic differentiation potential of senescent satellite cells. Here we employed a doxycycline (Dox) regulatable system to show that expression of the embryonic transcription factor NANOG restored the impaired myogenic differentiation potential of senescent myoblasts. This reversal in the myogenic differentiation was shown at the functional level by formation of myotubes in 2D and 3D, and at the molecular level by restoring the expression level of key myogenic regulatory factors such as Myf5, Myod, Myogenin, and MRF4 and members of myocyte enhancer factor 2 family. Interestingly, the rejuvenating effects of NANOG on senescent myoblasts were sustained after withdrawal of Dox. The myoblasts maintained their rejuvenated capacity up to a month after NANOG withdrawal, suggesting that NANOG might have imparted epigenetic changes. In conclusion, these results shed light on the potential of NANOG to restore the myogenic differentiation potential of senescent myoblasts and to reverse the loss of muscle regeneration due to aging.

499694 Pseudomonas Sp. ADP Biofilms: Differentiation of Expression in Atrazine-Degrading Genes and Lectin Binding Analysis

Abstract:

Atrazine, the second most used herbicide in the United States, frequently contaminates soil and groundwater. This pollutant exceeds the maximum contaminant level set by the Environmental Protection Agency of 3 parts per billion in many water sources throughout the Midwest. A proficient alternative to physio-chemical remediation methods relies on enzymes within microbes to perform biotransformations on recalcitrant compounds to innocuous chemicals in a process known as bioremediation. Pseudomonas sp. ADP, a strain capable of degrading atrazine, completely mineralizes the contaminant into ammonia and carbon dioxide following a six-step catabolic pathway. Each enzyme, AtzA-AtzF, catalyzes the biotransformation of atrazine into various compounds, including cyanuric acid and urea.

Despite extensive studies on genetic expression and the biocatalytic capabilities of the strain Pseudomonas sp. ADP in the planktonic mode of growth, little has focused on the sessile lifestyle of the same strain as a biofilm. Biofilms demonstrate great promise is remediation efforts due to the presence of a self-secreted extracellular polymeric matrix for protection from the environment, increased genetic variation due to horizontal gene transfer, and high frequency of cell-to-cell communication via quorum sensing. Reverse transcription real time quantitative polymerase chain reaction was used to determine the differential expression of each atrazine-degrading gene in Pseudomonas sp. ADP cells grown as a biofilm and as planktonic cells. Genes *AtzB, AtzC, AtzD, AtzE*, and *AtzF* demonstrated an increase in expression in a sessile lifestyle relative to planktonic cells, conceivably indicating higher metabolic activity of the atrazine-degrading enzymes in the six-step pathway.

Pseudomonas sp. ADP biofilms were grown in a drip-flow reactor at various temperatures to establish differential expression of atrazine-degrading genes. Decreased expression for genes *AtzA, AtzB,* and *AtzC* and increases in expression for genes *AtzD, AtzE,* and *AtzF* were observed for biofilms grown at higher temperatures. These results give an increased understanding on assessing temperature-dependence of gene activity in strains capable of degrading recalcitrant compounds. Additionally, to further characterize the biofilms, the sugar moieties were identified and compositions were determined using fluorescent lectin binding analysis (FLBA) and laser scanning confocal microscopy. The results of these studies will help further knowledge on the application of biofilms in industrial remediation and other biocatalytic processes.

501660 Biocatalytic Production of the High-Value Biochemical Paraxanthine from Caffeine

Abstract:

Paraxanthine (1,7-dimethylxanthine) is a natural product produced during metabolism of caffeine (1,3,7-trimethylxanthine) in humans and animals. Although paraxanthine is the major caffeine metabolite, its retail value is roughly 30,000 times higher than caffeine because there are limited natural sources from which to isolate the compound, and the chemical synthesis is cumbersome, low yielding, and environmentally unfavorable. Paraxanthine has multiple potential applications across various industries (i.e. pharmaceuticals, cosmetics, and food), but these applications have not been fully explored due to the high cost.

In bacteria, caffeine degradation occurs via sequential *N*-demethylation to xanthine. The soil bacterium *Pseudomonas putida* CBB5 metabolizes caffeine to xanthine using three positional-specific *N*-demethylase enzymes. First the N_1-demethylase, NdmA, converts caffeine to theobromine (3,7-dimethylxanthine). Theobromine is further N_3-demethylated by NdmB, resulting in 7-methylxanthine, which then undergoes N_7-demethylation by NdmC to yield xanthine. Although NdmB is active toward the N_3-methyl group, its activity is blocked by the N_1-methyl group present on the caffeine molecule. However, NdmA exhibits a slight promiscuity toward the N_3-methyl group, resulting in small amounts of paraxanthine produced from caffeine.

Recent elucidation of the NdmA and NdmB crystal structures revealed nine key amino acids involved in substrate binding. Notably, only two of the nine amino acids differ between NdmA and NdmB. Mutation of the two unique amino acids in NdmA to mimic the NdmB active site produced a mutant enzyme with a paraxanthine:theobromine ratio of at least 3:1, over a 100-fold improvement from the wild-type ratio (1:39). However, the activity was also reduced, with less than 20% of the caffeine consumed over 2 hours by *E. coli* expressing the mutant enzyme. Additionally, a peptide loop near the active sites also differs between NdmA and NdmB. Mutation of the NdmA loop sequence to match that of the NdmB loop increased the yield of paraxanthine to over 40%. This research confirms that biocatalytic production of paraxanthine from caffeine is achievable

501424 Combinatorial Crispri Expedites Microbial Metabolic Engineering

Abstract:

Metabolic engineering is entering a new era in which addressable synthetic transcription factors like RNA-guided, catalytically inactive CRISPR effector proteins (dCas9, dCpf1, etc.) are opening new possibilities for dynamic regulation and for rapid assessment of multiplexed genetic interventions to alter the metabolic landscape in both workhorse and previously intractable microbes. Toward this end, our group has focused on several CRISPRi-enabled metabolic engineering applications. Previously, we developed a modular assembly method for traditional restriction-ligation cloning of CRISPR array libraries for multiplex, combinatorial dCas9-mediated transcriptional repression. Importantly, we demonstrated for the first time that dCas9 can be utilized for metabolic engineering in

Expanding upon this work, we have developed a novel, streamlined cloning strategy to assemble natural type II-A CRISPR array libraries in a one-pot reaction, enabling simultaneous repression of all genetic target combinations from a defined set in multiple disparate strains over the span of only a few days. We showcase this method by improving production of two distinct classes of flavonoids (naringenin and cyanidin 3-*O*-glucoside) through repression of several novel knockdown targets aimed at co-opting distinct, endogenous co-substrate pools. This ability to rapidly implement combinatorial genetic interventions in a variety of host strains potentially obviates the role of cumbersome combinatorial gene deletions during hypothesis testing stages early in the strain development pipeline.

Furthermore, to facilitate extension of this assembly method toward construction of distinct CRISPR array types that are compatible with the rapidly expanding repertoire of Class 2 CRISPR effectors, we have developed a web-accessible algorithm that outputs a design strategy and list of DNA oligonucleotides required to assemble defined or combinatorial libraries of any natural CRISPR array type/class. We showcase the algorithm by constructing functional CRISPR array libraries for two distinct CRISPR effector types. These studies underscore the immense impact that RNA-programmed synthetic transcription factors will have on the fields of metabolic engineering and synthetic biology.

492703 Metabolic Engineering of Microorganisms for the Efficient Synthesis of Polysaccharides

Abstract:

Heparin and chondroitin sulfate (CS) are glycosaminoglycans with many biological and physiological functions that contribute to their widespread utilization as anticoagulants and anti-inflammatory drugs. These sulfated polysaccharides are primarily extracted and purified from animal tissues, where factors such as source material, manufacturing processes, and the presence of contaminants impact overall safety and biological and pharmacological efficacy. As part of ongoing efforts to separate the food chain from the drug chain, we are engineering microbial metabolism in order to produce chondroitin sulfate in *Escherichia coli,* and heparosan, a valuable precursor to heparin, in *Bacillus megaterium.*

The engineered production of non-sulfated chondroitin in moderate quantities (~1 g/L) in various microorganisms has been reported, but no industrial scale biotechnological process using *E. coli* to produce CS has been established to date. We are engineering *E. coli* for the production of CS through a biotransformation scheme, where the microbial non-sulfated chondroitin backbone is sulfated *in vitro* using multiple *E. coli* strains to generate the required components. The biotransformation scheme involves expression of several active biosynthetic pathway enzymes and requires examination of their specificity and stability. To this end, chondroitin 6-sulfotransferase, which catalyzes the transfer of sulfate to position-6 of *N*-acetylgalactosamine residues of chondroitin, was expressed by *E. coli* in active form. Together with non-sulfated chondroitin produced by co-expression of *E. coli* strain K4 genes *kfoA*, *kfoC*, and *kfoE* in the non-pathogenic strain *E. coli* BL21, two important components of the biotransformation scheme are made available for *in vitro* CS production. This proof of principle represents an important checkpoint on the path toward *de novo* biosynthesis of chondroitin sulfate from simple carbon sources like glucose and glycerol in a single microbial host.

Furthermore, although heparosan production in engineered *Bacillus subtilis* has been previously reported, the larger GRAS organism, *Bacillus megaterium,* is a more attractive alternative for industrial scale production since it possesses the intrinsic favorable properties of low protease activity and high secretion capability. To exploit these attributes, we have engineered *B. megaterium* for heparosan production. The T7 RNA polymerase (T7 RNAP) expression system for *B. megaterium,* which allows tightly regulated and efficient induction of genes of interest, has been co-opted for control of heparosan synthase (*PmHS2*). Specifically, we show that *B. megaterium* MS941 cells co-transformed with pT7-RNAP and pPT7-pmHS2 plasmids are capable of producing heparosan upon induction with xylose, providing an alternative, safe source of heparosan as a precursor for heparin production.

492492 Metabolic Engineering for Bacterial Production of Caffeic Acid Derived Phenethyl Esters and Amides in *Escherichia coli*

Abstract:

Knowledge-based metabolic engineering of natural products in microbes is a sustainable and cost-effective biomanufacturing alternative to direct extraction from plant tissues or chemical synthesis. Especially, genetic amenability makes the well-characterized microbes like *Escherichia coli* suitable industrial microbial platform for production of noncognate natural products from simple carbon source. However, the limited availability of known pathways or functional enzymes sometimes presents a significant challenge for metabolic engineering. In the present research, we established artificial biosynthetic pathways for therapeutically important compounds including caffeic acid derived phenethyl esters and amides, whose natural biosynthetic routes remain unknown in nature. We first identified functional acyltransferases capable of generating caffeic acid derived esters or amides from caffeoyl-CoA with counterpart aromatic alcohols or amines. Then we took advantage of this CoA-dependent trans-esterification pathway and constructed an *E. coli* platform for production of a collection of caffeic acid derived phenethyl esters and amides from glucose. Further production enhancement was achieved by up to 3-fold via host engineering and pathway optimization. This work permitted the bacterial production of natural compounds caffeic acid derived phenethyl esters and amides from simple carbon source, which also expanded the capabilities of microbial production of value-added aromatic esters and amides.

501799 Rerouting Acetyl-CoA and NADPH to Improve Lipid and Oleochemical Production in *Yarrowia lipolytica*

Abstract:

Global demand for lipid fuels and concerns about climate change have stimulated increasing efforts to produce carbon-neutral fuels directly from renewable resources. Microbially derived fatty acid fuels, the petroleum-replica fuels, have emerged as promising alternatives to meet this challenge. This is because fatty acid-based fuels offer several advantages such as high energy density, low hygroscopicity, miscibility with diesel fuels and compatibility with existing infrastructure. Recent development of the oleaginous yeast biorefinery platform has advanced the possibility to upgrade low value carbohydrates to high value fuels and oleochemicals. It is generally believed nitrogen starvation triggers lipid accumulation in oleaginous species. Under nitrogen starvation conditions, TCA metabolic activity is repressed and the overflown acetyl-CoA flux is rerouted to the fatty acid biosynthetic pathway. The central theme on this mechanism is about the acetyl-CoA supply mode: regulation of mitochondrial isocitrate dehydrogenase (ICDH). An ignored part about fatty acid biosynthesis in oleaginous yeast is the supply of NADPH, which provides the reducing equivalent to extend the carbon backbone. To overcome these challenges, we have engineered alternative cytosolic acetyl-CoA pathways to bypass the ICDH regulation, and recycled mitochondrial electrons to push the limit of lipid production in *Yarrowia lipolytica*. We have also been able to capture additional reducing equivalents from nitrogen metabolism and start drawing a clear picture how nitrogen starvation regulates lipogenesis in *Y. lipolytica*. Taken together, the reported strategy represents viable strategies to upgrade low value carbons to high value fuels and oleochemicals.

503875 Engineering Biomolecular Scaffolds for Enhanced Biocatalysis

Abstract:

Proteins are the most versatile among the various biological building blocks and efforts in protein engineering have resulted in many industrial and biomedical applications. However, the strength of proteins - their versatility and specific interactions - also complicates and hinders their systematic design and engineering. One potential approach is to exploit the modular nature of different protein domains in order to design synthetic protein scaffolds that can perform completely new biological functions. In this talk, I will outline several successful examples in connecting exchangeable protein domains for biocatalysis in (1) bioenergy applications, (2) Fuel cell applications, and (3) biosensing.

500890 Challenges of Process Modeling in the Agrifood Industry

Abstract:

For about 20 years, Cargill has been developing a library of process modeling and simulation tools to improve existing and new processes. We simulate parts of, or entire, existing agrifood processing facilities to uncover and quantify opportunities as well as to design solutions to make our existing processes more sustainable. We also apply this same library of tools to better design new processes. We present, in varying levels of detail, some of the tools we have created along with challenges and successes from our equation-oriented system modeling approach.

494689 Effect of Different Short-Chain Carboxylic Acids on the Kinetics of Growth and Production of Natamycin By *Streptomyces Natalensis*

Abstract:

Natamycin is an industrially important tetraene antibiotic produced in submerged culture by different strains of *Streptomyces* sp. Natamycin has a wide spectrum activity against molds and yeasts with low toxicity against mammalian cells. It is used in the treatment of bronchopulmonary aspergillosis, mycotic keratitis, yeast keratomycosis and candidiasis. Natamycin is used in many food industries for food preservation and has been granted the GRAS status (Generally Regarded As a Safe) by FDA. This present work was carried out to investigate the optimization of natamycin production through the addition of different short-chain carboxylic acids (acetic, propionic and butyric) and their effect on cell growth and natamycin biosynthesis. The optimization was performed in shake-flask level as well as 2-L stirred-tank bioreactor. Cultivations have been evaluated after addition of different short-chain carboxylic acids (acetate, propionate, and butyrate). The obtained results showed that the addition of acetic and propionic acids at concentrations below 2 g.L$^{âˆ'1}$Â at cultivation start showed stimulatory effects on natamycin production. Furthermore, the addition of acetic and propionic acids in a mixture (7:1) at a total concentration of 2 g.L^{-1}, increased antibiotic production significantly. Conclusively, the addition of short chain carboxylic acids increased antibiotic yield and decreased the production time.

496683 Prediction of Swelling Behavior of Cross Linked Maize Starch Suspensions

Abstract:

Pasting behavior of starch greatly influences the texture of a variety of food products. Starch consumption is believed to be linked to the occurrence of diabetes and obesity. It is important to characterize the connection between the structure, composition and architecture of the starch granules and its pasting behavior to design modified starch of desirable rate of digestion and texture. In this research, evolution of granule size distribution of maize starch when subjected to different heating temperatures and heating rates were characterized. The effect of extent of cross linking of maize starch on its swelling behavior was examined. The changes in starch architecture during swelling were characterized by cryo TEM. As expected, swelling was more pronounced at higher temperature, higher heating rate and lower extent of cross linking and resulted in a shift of granule size distribution to larger sizes with a corresponding increase in average size from 13 μm to 23-26 μm. Most of the swelling occurred within the first 10 min. Polymer solution theory was applied to predict the evolution of granule size distribution of starch at different heating temperatures, rates and cross linking in terms of its molecular weight, second virial coefficient and granule elasticity and compared with experimental data. Pasting behavior of waxy maize at different temperatures was also characterized from the measurements of G' and G" for different heating times. G' was found to increase with temperature at holding time of 2 min followed by its decrease. This behavior is believed to be due to the predominant effect of swelling at small times. However, G" was insensitive to temperature and holding times.

498778 Thermal and Rheological Characterization of Sugar Cane Honeys for the Modeling of Evaporation Systems

Abstract:

Thermal and rheological characterization of sugar cane honeys for the modeling of evaporation systems

Angela L. Alarcón R.[1], Paulo C. Narváez[1], Alvaro Orjuela[1], Edgar Camilo Camacho Poveda[2]

[1] Chemical and Environmental Engineering Department. Engineering Faculty. Universidad Nacional de Colombia, Bogotá (Colombia) alalarconr@unal.edu.co

[2] Electric and electronics Engineering Department. Engineering Faculty. Universidad Nacional de Colombia, Bogota (Colombia)

Panela (in South America), or jaggery (in India) is the product obtained by the open evaporation of sugar cane juices. It is also known as non-centrifuged sugar or raw-sugar, and it is mainly composed of sucrose, glucose, fructose, vitamins, minerals and proteins. This product is directly consumed as a sweetener and food, or use for bakery and pastry products in several countries. Panela production starts with the extraction of sugar cane juices (*Saccharum Officinarum)* with initial concentration of 18 – 20 °Brix, using a mill. Then, the juices are treated with a natural surfactant to remove natural mucilage and other impurities, followed by pH adjustment to avoid sucrose hydrolysis during processing. Subsequently, juices are concentrated by evaporation to achieve 93 – 96 °Brix, and finally, the mix is cool down in molds to obtain square or round blocks for the market.

As blocks are not convenient for consumers, granulation and powdering processes have been carried out using manual procedures (Grinding and sieving) in order to obtain products of rapid dissolution. These have an increasing market as alternative sweeteners for refine sugar. As technification is required, spray drying might be a useful processing technology to obtain powdered jaggery. However, few reports have studied the thermal and rheological properties of sugar cane juices, which are required for process modeling and up-scaling of the spray drying systems.

Taking into account the above, thermal and rheological properties of sugar cane juices such as thermal conductivity, thermal diffusivity, specific heat capacity, density and viscosity were evaluated. The dependence of such properties with water content and temperature were evaluated and regressed with suitable models. A specific experimental device was constructed for evaluation of thermal conductivity.

Sugar cane juices (CSCJ) were collected in Jaggery production facilities that operated with steam or combustion gases as heating source. Collected samples had a concentration of 30.97, 48.81 and 70.81°Brix, and they were characterized by HPLC to evaluate the sugar distribution. Results indicate that jaggery is mainly composed of sucrose (28.76 - 57.59 %), glucose (1.36 - 2.63 %) and fructose (1.11 – 2.09%). On the other hand, evaluated thermal properties between 30 - 75°C corresponded to thermal conductivity (0.47 – 0.27 W m^{-1} K^{-1}), thermal diffusivity (1.13 x10^{-7} – 7.19 x10^{-8} m^2 s^{-1}), specific heat capacity (3.71 – 2.73 J g^{-1} K^{-1}), and density between 20 – 35°C (1.1356 – 1.3512 g ml^{-1}). Viscosity (3.36 – 36.95 mPas) was evaluated between 35 - 75°C. Finally, mathematical models for thermal and rheological properties were obtained as a function of temperature and concentration.

References

Alarcón, Á., Narváez, P. C., Orjuela, Á., & Osorio, C. (2016). Physicochemical characterization and profile of sugars of honey sugar cane in the production of panela, *34*. http://doi.org/10.15446/agron.colomb.v34n1supl.58414

Alexandre, R., Cabral, F., Orrego-alzate, C. E., Gabas, A. L., & Telis-romero, J. (2007). Rheological and thermophysical properties of blackberry juice, *27*(3), 589–596.

García, H., Albarracín, L., Toscano, A., Santana, N., & Insuasty, O. (2007). *Guía tecnológica para el manejo integral del sistema productivo de caña panelera*. Bogotá: Ministerio de Agricultura y Desarroll Rural; Corpoica. Retrieved from http://www.panelamonitor.org/documents/519/guia-tecnologica-para-el-manejo-integral-del-siste/

García, J. M., Narváez, P. C., Heredia, F. J., Orjuela, Á., & Osorio, C. (2017). Physicochemical and sensory (aroma and colour) characterisation of a non-centrifugal cane sugar ("panela") beverage. Food Chemistry, 228, 7-13.

Guerra, M. J., & Mujica, M. V. (2010). Physical and chemical properties of granulated cane sugar" panelas". Food Science and Technology (Campinas), 30(1), 250-257.

Jaffe, W. R. (2015). Nutritional and functional components of non centrifugal cane sugar: A compilation of the data from the analytical literature. Journal of Food Composition and Analysis, 43, 194-202.

Seguí, L., Calabuig-Jiménez, L., Betoret, N., & Fito, P. (2015). Physicochemical and antioxidant properties of non-refined sugarcane alternatives to white sugar. International Journal of Food Science & Technology, 50(12), 2579-2588.

495471 (Keynote) Development of Bioprocess Platform for Cordyceptin Production By *Cordyceps MilitarisÂ*

Abstract:

Mushrooms have been used since centuries as food and medicine in many ancient cultures and become one of the main components of Traditional Chinese Medicine (TCM). They produce many bioactive compounds with high therapeutic values. *Cordyceps militaris* is well known for production of immunomodulators and anticancer active compounds of different molecular weights. Of these compounds, Cordycepin [CAS#73-03-0] (3â€™-deoxyadenosine) is nucleoside derivative which exhibited significant activity as anticancer and antitumor in both *in vitro* and *in vivo* studies. In this work, complete industrial platform was developed for the production of this bioactive molecule. This start with optimization of growth medium and cultivation conditions to enhance the production of codycepin in both small scale cultivation vessel. After this step, cultivations were conducted in stirred tank bioreactor for the optimization of bioprocessing parameters during cultivation in batch and fed-batch mode. Furthermore, optimization of downstream process using solvent extraction system was carried out for final cordycepin separation in pure form.

490797 Selecting for Small Molecule Production in Competitive Microfluidic Droplets

Abstract:

Evolutionary selections can be powerful tools for metabolic pathway optimization. However, finding an appropriate selective pressure for the production of a desired compound can be difficult. To extend these methods to new classes of compounds, we have developed *in vitro* selections in which the production of a small molecule mediates competition between two microbial strains. We construct systems in which a resistant producer can inhibit the growth of a sensitive competitor by producing the target small molecule. When the two strains are mixed and co-cultured in microfluidic droplets, producers that make more of the target compound have a fitness advantage. We have developed strategies for manipulating the sensitivity of the producer and competitor strains to the target compounds and have characterized the dynamics and selective pressures in the competitive co-cultures. We are applying this selection strategy to the production of fuels and chemicals from sugars and aromatics.

498854 Improving Glucaric Acid Production By Alleviating Oxidative Stress in *E. coli*

Abstract:

Native metabolism continually produces reactive oxygen species (ROS) that promote cellular damage through redox reactions with DNA, proteins, and lipids. Endogenous scavenging enzymes and mechanisms are typically sufficient to reduce ROS to minimal levels, but genetic engineering for new applications can exhaust the cellular antioxidant capacity. While this problem has been reported in the literature, a general framework is needed to identify and address it as part of the metabolic engineering toolkit. To better understand oxidative stress in engineered metabolic pathways, we examined a heterologous pathway from glucose to glucaric acid in *E. coli*, with particular focus on the conversion of *myo*-inositol to glucuronic acid by *myo*-inositol oxygenase (MIOX). Overexpression of the genes encoding endogenous catalase (*katE*) and superoxide dismutases (*sodA* and *sodB*) led to significant increases in glucuronic acid production, suggesting that hydrogen peroxide and superoxide levels are elevated in the pathway and impinge on productivity. The increase in glucuronic acid titers also varied between *E. coli* strains, verifying previously observed differences in native antioxidant capacity. This work provides further evidence that engineered metabolic pathways can produce elevated ROS levels and that overexpressing ROS scavenging enzymes is an effective strategy to improve pathway performance.

500090 Investigation of Metabolite Channeling in Central Bacterial Pathways

Abstract:

Bacterial cytoplasm is a distinctive *in vivo* environment, crowed with macromolecules which creates glass-like properties (1). Intracellular enzymes of central pathways can be organized into a variety of assemblies, forming channels that pass intermediates from one enzyme to the next without release into the cellular medium. Such reaction mechanisms overcome intracellular diffusion limitations, improve reaction thermodynamics, aid low copy number enzymes, stabilize metabolic fluxes, and reduce side products from promiscuous enzymes. Comparing to transcriptional, translational, or allosteric regulations, channeling is an overlooked regulatory mechanism that leads to stability of a cell's fluxome under environmental/genetic perturbations, allowing cells to remember the topology for distributing substrate influxes throughout its metabolic network (2). Understanding of the scope and mechanisms of metabolite channeling will lead to better design and engineering of heterologous pathways for synthetic biology applications. The channeling can be inferred from *in vitro* analysis, reaction thermodynamics, transport/reaction modeling, measurements of molecular diffusion and protein interactions, or steady state/dynamic isotopic labeling approaches. In this study, ^{13}C-pathway tracing and kinetic models demonstrate several evidences for channeling in native pathways: 1. In ^{13}C-labeling experiments, both cyanobacteria and *Escherichia coli* wild type strains demonstrate unsequential labeling order for cascade metabolites through glycolysis, pentose phosphate and TCA pathways; 2. An overexpressed pathway (Entner-Doudoroff)'s flux can be doubled by knocking out the cell's phosphotransferase system (PTS), the proposed first channeled step; 3. Diffusion/reaction models highlight the relationship between diffusion coefficients, enzyme copy numbers, and enzyme distances and support the necessity of channeling for optimal cell metabolism. This work, together with numerous other studies, shows that native enzymes of bacterial pathways are likely more organized than we thought, and thus impede fluxes towards engineered pathways. If the hypothesis is true, metabolic modeling may have to re-considered, which assumes homogenous intracellular environments.

References:

1. Parry, B.R., Surovtsev, I.V., Cabeen, M.T., O'Hern, C.S., Dufresne, E.R., Jacobs-Wagner, C., 2014. The Bacterial Cytoplasm Has Glass-like Properties and Is Fluidized by Metabolic Activity. Cell 156, 183–194. doi:10.1016/j.cell.2013.11.028

2. Hollinshead, W.D., Rodriguez, S., Martin, H.G., Wang, G., Baidoo, E.E.K., Sale, K.L., Keasling, J.D., Mukhopadhyay, A., Tang, Y.J., 2016. Examining Escherichia coli glycolytic pathways, catabolite repression, and metabolite channeling using Δpfk mutants. Biotechnol. Biofuels 9, 212. doi:10.1186/s13068-016-0630-y

501856 Cell-Free Metabolic Engineering for Heterologous *in vivo* Pathway Optimization

Abstract:

Cell-free metabolic engineering methods offer unique methods for rapidly prototyping heterologous pathways. In cases where substrates or intermediates are toxic and where non-native products are desired, optimization of entire heterologous pathways in vivo can be laborious and inefficient. Cell-free prototyping offers an effective alternative. Previous such efforts have focused on the use of lysate-based systems incorporating cell-free protein synthesis, to enable high-throughput testing of large numbers of gene variants and the use of combinatorial methods for optimizing pathway enzyme loadings. This study pursues an alternative approach, where purified enzymes are recombined to create a reconstituted pathway. This approach allows individual and collective characterization, enabling the parameterization of a kinetic model describing the entire pathway. Once validated, the model can be optimized for maximum throughput. In this case, a non-native product pathway in *Pseudomonas putida* KT 2440 has been optimized using this cell-free method, resulting in a significant increase in the titer of several target aromatic catabolic intermediates. This success bolsters the case for the use of *Pseudomonas* spp. to convert lignin monomers to high-value products. The method will be compared with a directed evolution approach, to demonstrate the increased efficacy of the rational cell-free approach compared to the stochastic method. Longer-term potential for the use of cell-free systems as conversion platforms will also be discussed.

488668 Genetically Programmable Assembly of Microbial Communities for Enhanced Biosynthetic Efficiency

Abstract:

Microbial consortia are increasingly used to perform multi-step biosynthetic transformations that are difficult to perform in a single organism. However, in simple co-culture systems, intermediate metabolites are rapidly diluted by the extracellular medium, leading to reduced rates of downstream biosynthetic steps. Close proximity of microbes to one another may enable metabolite exchange between cells before they are diluted by the bulk culture, and may thereby increase the yield of the final product. In this work, we have engineered microbes to display surface proteins that induce the formation of large (up to 100 μm diameter) clusters. Reaction-diffusion modeling of biological transformations within these clusters shows that cluster size is an important tunable parameter for controlling the efficiency of multicellular biosynthetic cascades, and experiments have shown that cluster size can be controlled by the stoichiometry of hetero-associating surface proteins. By understanding substrate diffusion and the kinetics of each biosynthetic step, we are developing design rules to optimize the spatial organization of microbial consortia to maximize substrate conversion and product yield. In the future, the techniques under development in this work may allow reactions that normally must be done under separate conditions to be performed in a single bioreactor.

501362 A Generic Strategy to Maintain Stable, Multi-Member Microbial Consortia

Abstract:

Microbial community is spotlighted recently in metabolic engineering applications, since it has many benefits such as combining strengths of multiple microbial species, accelerating strain development, and reducing metabolic burden on microbial workhorse. Despite the increasing attention from the relevant field, there are still some technical limitations on establishing stable microbial consortia, a feature required for developing industrially relevant processes. For example, current strategies to maintain stable co-culture often require use of certain carbon substrates. Here, we report a generic strategy to maintain stable cell ratio of multi-member co-culture. The qPCR based method we established for quantifying cell number in multi-member co-culture should also be useful to this community.

494906 Protease-Responsive Droplets Engineered from Self-Assembled Disordered Proteins

Abstract:

Certain intrinsically-disordered proteins (IDPs) self-assemble via liquid-liquid phase separation into spherical droplets. These droplets have been discovered to function as membrane-less organelles in living cells. Here, we engineered protease-responsive microspherical droplets from self-assembling recombinant IDPs. We selected a prototypical droplet-forming IDP domain, the arginine- and glycine-rich RGG domain from LAF-1, and designed a variant consisting of two RGG domains in tandem, separated by a thrombin protease cleavage site. This variant, which we refer to as tandem RGG, exhibits phase separation as a function of protein concentration, temperature, and salt concentration. Importantly, due to multivalent interactions, assembly of tandem RGG droplets occurs under physiological conditions at lower protein concentration than is observed with a single RGG domain. We harnessed this property to engineer protease-triggered droplet dissolution. By operating at a protein concentration at which tandem RGG phase separates but a single RGG domain does not, thrombin treatment causes droplets formed by tandem RGG to slowly dissolve, simply by cleaving tandem RGG in half. We measured the kinetics of droplet dissolution by quantitative optical microscopy and interpreted the results with a reaction-diffusion model. Next, we demonstrated two strategies to release passenger proteins from the droplets. In the first approach, the passenger protein is incorporated into tandem RGG as a genetic fusion and is released as the droplets dissolve upon thrombin treatment. In the second approach, tandem RGG is tagged with a receptor and the passenger protein is tagged with a high-affinity ligand to that receptor. This receptor-ligand interaction recruits the passenger protein into the droplets; subsequent protease treatment to cleave off the ligand de-recruits the passenger protein from the droplets, while the droplets remain intact. We anticipate that these recombinant biomolecular materials will have applications as a technology for topical drug delivery to wounds, where thrombin is present, and more broadly as a modular platform for controlled release of peptides and proteins. We also envision applications in engineering the behavior of membrane-less organelles within cells.

498043 Engineering Anti-EGFR Fibronectin Nanorings for Cancer Immunotherapy

Abstract:

Engineered protein scaffolds have gained interest in recent years to overcome the challenges faced with antibody-based drugs. Several scaffolds have shown promise in both preclinical models and clinical trials. One of these promising scaffolds is the tenth type III domain of human fibronectin, which consists of seven Î²-strands connected with flexible loops. Analogous to the complementarity determining regions of antibodies, the amino acid sequences of these loop regions can be engineered to provide target-specific binders with high affinity and specificity. One such target, epidermal growth factor receptor (EGFR) is a well-studied cancer biomarker and a promising target for cancer therapeutics. While some groups have already evolved high affinity EGFR-binding fibronectins using different methods, their use in cancer immunotherapy applications has not yet been explored.

Â Our lab has developed Chemically Self-Assembling Nanorings (CSANs) as a non-genetic approach for cell-directed immunotherapies. Multimeric CSANs are formed when a bis-methotrexate dimerizer induces the oligomerization of dihydrofolate reductase-dihydrofolate reductase ($DHFR^2$) fusion proteins. An advantage of the CSAN platform is that it can be easily expanded through fusion of additional protein domains or by chemical modifications to the dimerizer.

By fusing an EGFR-targeting fibronectin to the $DHFR^2$ subunits, we formed fibronectin CSANs with both high affinity and high avidity for EGFR-overexpressing cancer cells. After investigating the biological outcomes of the fibronectin nanorings on cancer cells, we developed the bispecific nanorings that will target both the cancer cells and the T cells. We have formed bispecific-CSANs by oligomerizing the fibronectin-$DHFR^2$ subunits alongside anti-CD3 scFv-$DHFR^2$ subunits that target T cells. Furthermore, T cells functionalized with EGFR-CD3 bispecific CSANs are able to selectively recognize and eradicate EGFR-overexpressing tumor cells. A unique advantage of this bispecific CSAN system is the ability to disassemble the CSANs *in vitro* and *in vivo* by introducing the FDA-approved antibiotic, trimethoprim, thus providing control over the T cell-directing activity of the CSANs. As the findings here suggest, fibronectin CSANs provide unique features for enhanced cancer immunotherapy.

496681 Engineered Protein-Polymer Conjugates for Drug Delivery to the Central Nervous System

Abstract:

The blood-brain barrier (BBB) protects the central nervous system (CNS) from potentially harmful compounds that circulate in the blood, presenting a major obstacle to delivering drugs intended to treat disorders of the CNS like Alzheimerâ€™s disease, Parkinson's disease, or brain tumors. In order to transport specific molecules that are needed for CNS function into the brain, there are dedicated receptors on the BBB to facilitate active transport from the blood into the CNS. One strategy to address the challenge of drug delivery to the CNS is the use of receptor mediated transport, utilizing native transport systems to carry non-native cargo, such as a therapeutic, into the brain.

In our work, we have engineered proteins that bind to the human transferrin receptor (TfR) and are developing methods to conjugate these proteins to novel polymer systems to create drug delivery complexes. TfR is a BBB receptor whose native function is in iron regulation, and the receptor has previously been shown to effectively transport therapeutic cargo across the BBB into the CNS in animal studies. Despite much progress in pursuing receptor mediated transport into the brain, there are still no FDA approved therapeutics that use receptor mediated drug transport. In previous studies, several essential qualities needed for a successful use of TfR mediated drug transport have been elucidated. The binding affinity of an engineered molecule that targets TfR should optimally be ~100s nanomolar dissociation constant, so that the drug complex can bind to the TfR in brain blood vessels and be released after crossing into CNS tissue. To avoid serious toxic side effects, TfR binding molecules should not interfere with native transferrin ligand binding. Furthermore, if a targeting molecule is to be used as part of a drug delivery complex, the targeting molecule should be suitable for appropriate chemical modifications, such as conjugation to a polymer system.

Using yeast surface display and directed evolution, we have engineered TfR binding proteins based on the 10th domain of human fibronectin type III (Fn3), beginning with a naïve Fn3 library. Our engineered Fn3 proteins bind TfR in the range of the optimal binding affinity to specifically bind TfR and then be released into the CNS. Our results indicate that the candidate drug delivery proteins do not compete with the transferrin ligand binding epitope. The highly stable, single domain structure of the approximately 10 kDa Fn3 proteins are amenable to further chemical modification in a variety of solvent systems. We are currently conjugating our engineered targeting proteins to multifunctional linear polymer and micelle systems towards the ultimate goal of developing a modular drug delivery complex for the central nervous system.

497484 Covalent Heterobivalent Inhibitors for Drug Allergies

Abstract:

Drug allergies are a growing problem in the world today that accounts for over half of the allergy fatalities that occur each year. Patients in already compromised states in hospitals, either from infection, cancer or other disease, are particularly at risk. Often, while drug allergies may not cause fatal reactions, the presence of moderate reactions compels physicians to alter treatment regiments and compromise patient care. In particular, this is a major concern for cancer chemotherapeutics, where drug allergy rates can be as high as 20%. These immediate hypersensitivity reactions are triggered when drug specific immunoglobulin E (IgE) molecules bind drug labeled serum proteins (allergen) and trigger immune responses. In an effort to prevent these reactions, we have developed a specific inhibitor molecule, covalent heterobivalent inhibitors (cHBIs), that can disrupt the allergen-IgE interaction by specifically and permantely binding the IgE antigen binding site. By only targeting allergen specific IgEs through bivalent binding and subsequent covalent inhibition, cHBIs offer potent and long lasting inhibition of allergic reactions to any drug molecule. To date, we have engineered cHBIs to an antibiotic, penicillin G, and a common chemotherapeutic, oxaliplatin, representing both of the major classes of drug allergies. We demonstrate cHBIâ€™s selective IgE targeting and *in vitro* and *in vivo* efficacy for preventing allergic reactions to both penicillin G and oxaliplatin. While, these two cHBI formulations are clinically viable treatments for very common drug allergies, this study also reveals the utility of the cHBI design, which can be adapted to any drug allergy.

499535 Exploring the HIV V1V2 Loop Conformational Landscape with Protein Engineering

Abstract:

Developing an effective HIV vaccine remains an elusive goal; the vast sequence diversity of HIV presents an enormous challenge with respect to identifying vaccine targets for broad protection. Characterization of several classes of broadly neutralizing antibodies isolated from infected individuals has identified several â€˜sites of vulnerabilityâ€™ on the HIV envelope protein such as the V1V2 loops, and vaccine-induced antibody responses to these sites may provide protection. Structure-based vaccine design approaches have sought to utilize structural understanding of these antibodiesâ€™ recognition of their epitopes to inform design of epitope-focusing immunogens. Despite advances in structural knowledge with the crystallization of the full-length HIV envelope trimer, our limited understanding of the more conformationally dynamic V1V2 loops may hinder structure-based immunogen design for this site.

The V1V2 loops had previously been considered as intractable targets for vaccine design because of their sequence variation, yet their structural and functional conservation, their association with vaccine-mediated protection in the RV144 trial, and the isolation of V1V2 loop-specific broadly neutralizing antibodies now point to their relevance to vaccine design. Monoclonal antibodies recognizing distinct conformations of the V1V2 loops suggest that these loops may exist dynamically, and may be presented differentially by monomeric V1V2 and trimeric envelope spikes. That antibodies preferring quaternary and conformational epitopes (i.e. PG9 and 697-30D) also exhibit greater neutralization potency and breadth as compared to antibodies recognizing linear epitopes (i.e. CH58 and CH59) suggests that the former may be more desirable in a vaccination setting. Thus, achieving proper conformational presentation of the V1V2 loops in an epitope-focusing immunogen presents a protein engineering problem. Here, we explore the conformational landscape of the V1V2 loops by characterizing the conformational effects of a novel mutation identified by protein engineering.

Through computational design and directed evolution of a mini-V1V2 antigen, we have identified a novel mutation in the V1V2 loops that may result in a more preferable quaternary-like conformation; inclusion of this substitution in our mini-V1V2 antigen design and in monomeric gp70 V1V2 scaffolds results in enhanced binding to antibodies preferring quaternary and conformational epitopes (PG9, PG16, 697-30D), as well as a reduction in binding to antibodies recognizing linear epitopes (CH58 and CH59) in both yeast and mammalian expression systems. The location of this substitution is distal from known anti-V1V2 epitopes and in a region of the V1V2 loops well conserved in sequence and in structure. Additionally, its effect on several classes of antibodies with distinct binding modes, as well as its phenotypic effect across 4 HIV strains covering 3 clades, suggests that a more global structural modification or stabilization of V1V2 secondary structure may be taking place. As a single point mutation, this substitution may have utility in simplifying minimal V1V2 immunogen design, in which current approaches require many mutations or loop grafting for proper V1V2 conformation. Furthermore, inclusion of this substitution in a soluble V1V2 probe may allow for profiling of anti-V1V2 responses with finer resolution towards conformation-specific responses currently missed by existing state-of-the-art probes. Taken together, this result has potential to further elucidate structural intricacies of the V1V2 loops and corresponding antibody responses, which will enhance future structure-based vaccine design efforts.

492017 A Small Peptide Localizes to a Distinct Region of the Cell Membrane By Sensing Curvature

Abstract:

Many cellular functions rely on the correct subcellular localization of proteins. These proteins often localize by recognizing a pre-localized binding partner. How, then, do the very first proteins localize to its correct destination? The small peptide SpoVM, a critical component in endospore assembly in Bacillus subtilis, was shown to localize to the nascent forespore by sensing positive (convex) curvature within the cell membrane. We investigate how this nanometer-scale peptide can recognize curvature on the micrometer scale, and elucidate the critical features of the 26-amino acid peptide for sensing membrane curvature. We first reconstituted membrane curvature on the micron scale in vitro by creating solid-supported lipid bilayers (SSLBs) on silica beads, and show SpoVM preferentially binds to membranes with higher degrees of curvature. Next, we developed a flow cytometry assay to quantify binding of SpoVM to SSLBs and show that preferential binding can be attributed to a higher kinetic rate of binding, coupled with binding cooperativity. Finally, through the rational design of SpoVM mutants, we show that the curvature-sensing ability of SpoVM is dependent on a critical proline residue at position 9, which delineates the peptide into a flexible N-terminus and an amphipathic alpha-helical C-terminus. As more proteins are being reported to localize to curved membranes, these findings may reveal a common mechanism by which proteins achieve preferential binding to membranes with positive curvature.

500888 Prediction of Tablet Dissolution By Process Parameters in Continuous Manufacturing

Abstract:

Tablet dissolution is an attribute that is critical to the product quality and should be well controlled in order to produce an acceptable pharmaceutical product. Tablet dissolution testing is a lengthy procedure that is time-consuming, expensive, and requires media preparation. In addition, it is a destructive test, and the sample will no longer be available for further characterization. Hence, there has been a growing interest in the development of non-destructive predictive tools for tablet dissolution to enable the real time release testing (RTRt) strategy. Application of such a strategy would not only lower laboratory and personnel costs but also increase product quality. A statistical model for predicting tablet dissolution profiles will enable rapid techniques in continuous assessment of tablets ensuring the desired quality attributes.

In this work, we developed a statistical model to predict tablet dissolution profile in a continuous direct compaction manufacturing line. Tablets consisted of model API, Acetaminophen (APAP), lactose monohydrate as the filler, and magnesium stearate as lubricant. Multivariate effects were analyzed using Quality-by-Design methods. The experiment was based on a fractional factorial design with 30 experiments, including 3 repeated center points, to study the influence of four process variables – drug concentration (%), compaction force (KN), blending speed (rpm), and feed frame speed (rpm) – on the final product. Dissolution profiles of 180 tablets (six tablets for each experimental run) were obtained and fit into the Weibull model (including parameters of α and β), and the values of α and β were calculated. Multi variate analysis of variance (MANOVA) was used to determine the statistical differences between the dissolution profiles. The results indicated that compaction force and API concentration were the main parameters affecting tablet dissolution in this process. Therefore, a predictive model was developed to correlate the dissolution model parameters α and β to the process parameters. The model was verified using an independently manufactured validation set. The values of α and β of the validation set were estimated by the model. The dissolution profiles of these tablets were predicted using the estimated values of α and β. The predicted dissolution profiles obtained from the model were in agreement with the experimental results using the similarity factor f_2 value greater than 50. This work establishes an approach to the real time release testing (RTRt) of pharmaceutical tablets and the possibility of assessing the quality characteristics of tablets non-destructively.

501282 Integrating Sensors for Monitoring Blend Content in a Pharmaceutical Continuous Manufacturing Plant

Abstract:

In a pharmaceutical manufacturing process, Critical Quality Attributes (CQAs), as defined by the FDA, need to be monitored not only for the final product but also for intermediates. Blend uniformity of powders is one such attribute that needs to be measured to ensure the quality of the final product. In-line sensors, to monitor the blend content of the powders, were implemented within a Direct Compaction (DC) continuous tablet manufacturing line. In most cases, since the primary ingredient of interest is the active pharmaceutical ingredient (API), the amount of the API was monitored/predicted over the course of manufacturing. For the calibration model building process, a unique setup involving dynamic powder spectral acquisition method was used. This setup was aimed at mimicking the powder flow characteristics within the manufacturing line. A Raman probe and a portable NIR were used as sensors and they were placed at the exit of the blending process before the tableting stage. The performance of the two sensors and their respective models were evaluated in terms of accuracy, precision, operating range, measurement frequency, placement, reliability, robustness. Additionally, their performances were validated by off-line traditional analytical measurements.

501455 Scientific and Regulatory Considerations for Developing the Control Strategy of a Continuous Manufacturing Process

Abstract:

Advances in manufacturing technology over the last decade have provided the pharmaceutical industry opportunities for adopting innovative ways to develop flexible, efficient and robust processes. In recent years, continuous manufacturing has emerged as a new technology for the manufacture of pharmaceuticals. Compared to traditional batch process, continuous manufacturing has various advantages, such as, shorter processing times, smaller facility footprint, etc. On the other hand, it has unique elements that lead to uncertainty on how to implement a continuous process within the current regulatory framework. The Food and Drug Administration supports the adoption of innovative technology and is communicating with pharmaceutical companies and academia to facilitate the development of continuous processes. This presentation will discuss several scientific, current good manufacturing practice (CGMP) and regulatory considerations for developing the control strategy of a continuous manufacturing process.

500629 Uncertainty Quantification and Global Sensitivity Analysis of Drug Delivery Primary Containers to Understand Process Capability and Key Risk Factors

Abstract:

A simple high-fidelity model was created to predict the plunger position and air gap in a pre-filled syringe primary container to be used as a component in a combination product. The product-container system has six sources of uncertainty and two uncertain variables related to operating conditions. An uncertainty analysis was carried out to compute the joint distribution of plunger position and air gap performance parameters with the goal to understand the operating ranges for these CQA (critical quality attributes). In order to relate these to operational practices, process capability indices (Cpk) were calculated via a naïve approach where the marginal plunger-depth distribution was fit to a normal distribution in order to enable the index (Cpk) calculations. This simplified approach was compared with a full propagation of factor uncertainties via the rigorous nonlinear model, which resulted in a remarkable difference in the CpK values. This indicates that the Gaussian approximation was unsatisfactory and more refined techniques are required. As a final step, a global sensitivity analysis based on the Sobolâ€™ indices technique was carried out in order to identify the key factors affecting process capability and inform continuous process improvement initiatives. Overall, four of the eight factors were found to be unessential (non-key).

This work demonstrates the use of rigorous mathematical process models and advanced uncertainty/sensitivity techniques to shed light on design and operation performance characteristics (QbD and QbC). The model and techniques were implemented in a web-based deployment platform in order to leverage the rigorous models/techniques and share these simple-to-use applications with all users across the organization, regardless of their background.

500701 Nucleation Behavior of Eszopiclone - Butyl Acetate Solutions from Metastable Zone Widths

Abstract:

Experimental data on the critical supercooling ΔT_{max}, a measure of metastable zone width (MSZWs), as a function of saturated temperature T_0 and cooling rate R were analyzed to describe the nucleation behavior of unseeded eszopiclone-butylacetate solutions. The Nývlt's semiempirical model and Sangwal's theory were employed, respectively. And moreover, we propose a modified model based on the Sangwal's theory in which the nucleation parameters are not dependent on saturation temperature T_0 and nucleation temperature T_1 which can fit the experimental data better than Sangwal's theory. Furthermore, the pre-exponential A and solid – liquid interfacial energy γ can be obtained by employing this modified model, thus, sheds light on the relationship between nucleation parameters and the cooling rate R. Meanwhile, based on the Classical Nucleation Theory, with the estimated interfacial energies from the slope in the modified model, the critical nucleus size and critical Gibbs free energy were calculated and discussed. Then based on the Sangwal's theory, the nucleation activation barrier is found to be dependent on ln R with an exponential relationship, hence, the relationship between nucleation barrier and pre-exponential A can be constructed. At last, the effects of agitation rate and working volume on MSZWs were investigated and discussed.

Development, Implementation, and Use of an On-line Laser Diffraction Particle Size Method in a Spray Drying Manufacturing Process

Pharmaceutical Discovery, Development and Manufacturing Forum
2017 AIChE Annual Meeting

Martin Warman, Martin Warman Consultancy Ltd, and Alon Vaisman, Malvern Instruments

Malvern Panalytical

Why is this important - background

- The Biopharmaceutics Classification System (BCS)
  - Class I - High Permeability, High Solubility
  - Class II - High Permeability, Low Solubility
  - Class III - Low Permeability, High Solubility
  - Class IV - Low Permeability, Low Solubility
- In general, product performance for Class I + III compounds is driven by API concentration and tablet disintegration
- However product performance for Class II + IV compounds is driven by many more factors

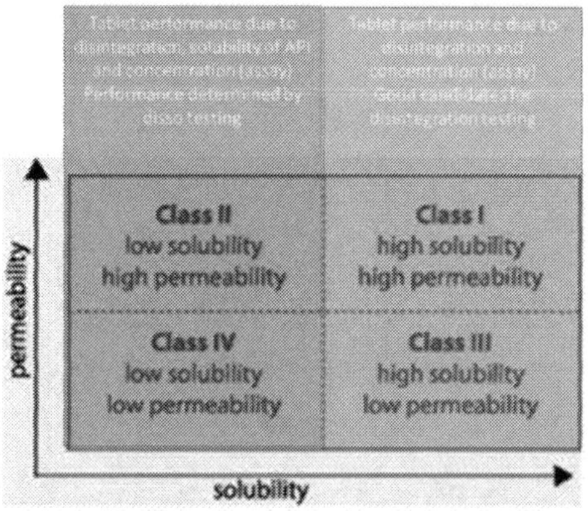

Malvern Panalytical

What is disintegration and weight are not enough

- For BCS II and IV API can't use simple indicators of tablet performance
  - Need to establish 'rate of release' data
  - Disso testing
- However the disso performance is a combination of numerous chemical and physical properties of the matrix
  - Concentration
  - Solubility
  - Hardness
  - Density
  - Porosity
  - Particle size
  - Inter-particulate forces
  - Etc
- Many of these are impacted by the physical API properties but also the intermediate particle properties
  - Spray Dried Dispersion
    - % crystallinity
  - Granulation

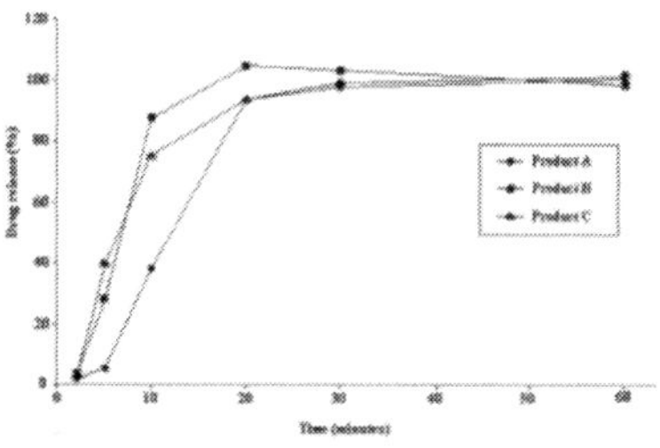

Malvern Panalytical

Case Study - High-level description of Process

- Crystalline API production
  - Reaction steps, crystallization, drying
- Amorphous API spray dried dispersion
  - Spray solution prep, spray drying, secondary drying
- Formulation
  - Pre lube blend, post lube blend
- Compression
  - Compression, film coat
- Described as full QbD, with elements of real-time release

Malvern Panalytical

Identifying steps critical to tablet performance

- For this product we could show the product performance was a result of
  - Tablet compression
  - SDD bulk density and particle size
  - SDD particle size could be controlled byt the spary drying process

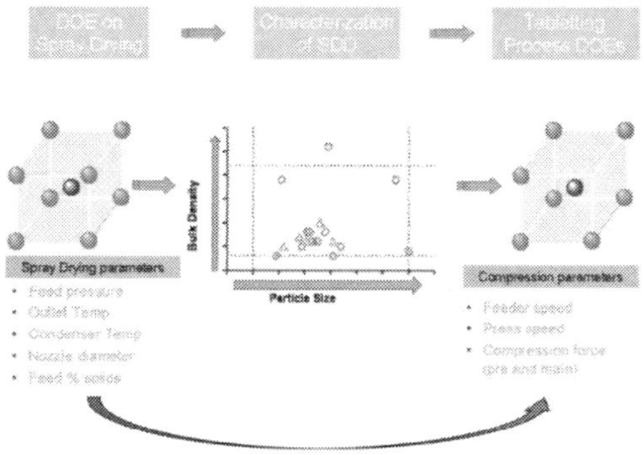

Malvern Panalytical

Identifying control models

- Regression analysis showed
  - The particle size of the wet SDD was determined by nozzle diameter and feed pressure
  - Whereas bulk density related to speed of drying, with droplets that dried fast having symmetrical shape (often with entrained gas bubbles) and droplets that dried slowly having high surface area/higher bulk density
- The key message is these could be controlled independently

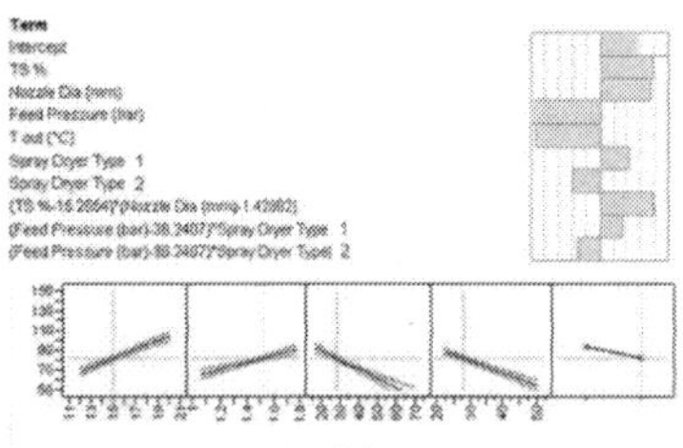

Particle size

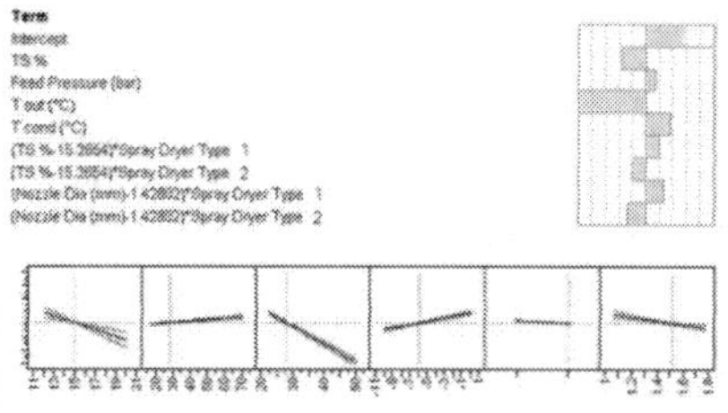

Bulk density

Malvern Panalytical

Graphical Illustration of Design Space

- More so we could demonstrate a three dimensional Design Space for product performance (disso)
- Control strategy complexity pyramid
  - First published by FDA ETT in the AAPS Journal, Vol 16, Number 4, July 2014
  - Included in the paper, Modernizing Pharmaceutical Manufacturing: From Batch to Continuous Production, Journal of Pharmaceutical Innovation, Vol 10, Issue 3, Sept 2015,
- The product filling pre-dates the complexity pyramid by 4 years but in hindsight the product was filled with a level 2 CS, but with the process knowledge to allow level 1 control

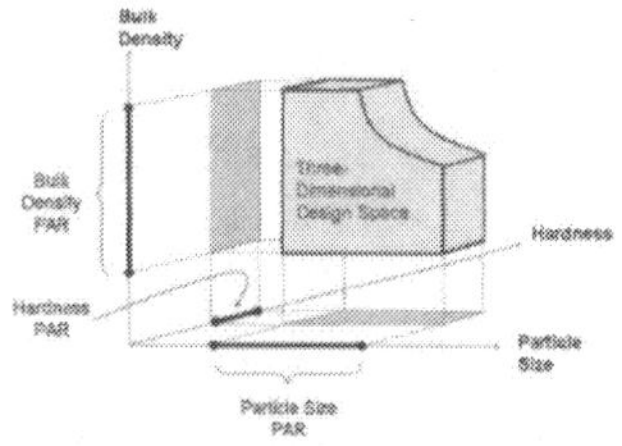

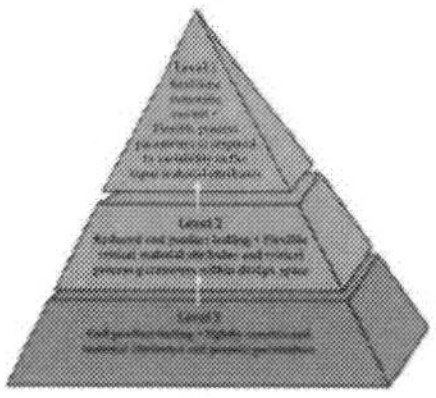

Malvern Panalytical

Control Strategy Implementation

- Dashboard sits on plant control system showing
- Predicted disso ranges shown to SDD process operator
  - If particle size and bulk density stay in the green range, the tablet compression set points are wide
  - If particle size and bulk density are in the yellow range, the tablet compression set point becomes critical
  - If the particle size goes red, there are no approved compression set points that will give acceptable product performance
- In reality the process runs in a NOR where we are always making good product
- Key message is we have real-time indicator of CQA predictions from PAT and parametric data

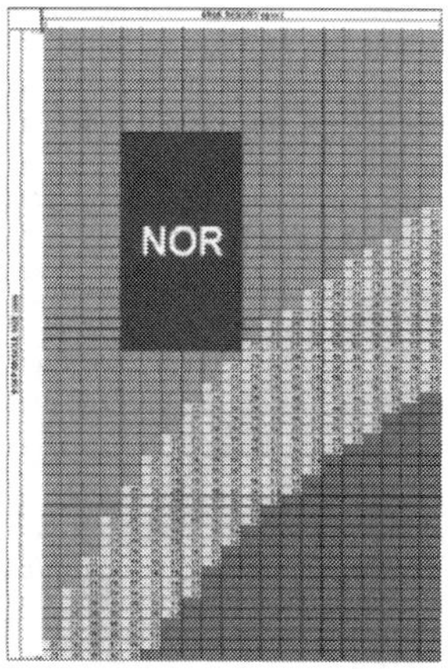

Malvern Panalytical

Key Question
How do we get that data?

Malvern Panalytical

Particle Size Application
Ensuring Representative Measurement

- Appropriate selection of sampling location
- Powder flow is linear (post vortex breaker)
  - However this means it is the 'shadow' of the vortex breaker
- Tried all commercial sampling systems
  - J-type, single funnel, etc
- Poor correlation with off-line data
- Implemented sampling arrangement is the result of empirical testing of various restrictions and funnel configurations
- Has been proven to generate the equivalent results
- Covers 25% of the cross sectional area, i.e. 25% of production runs through the analyser!

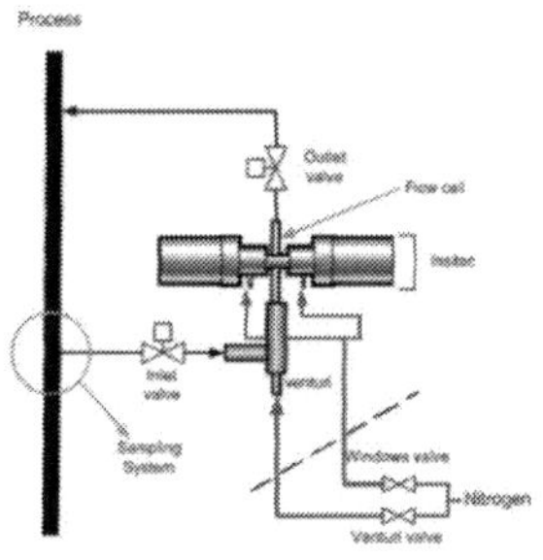

Malvern Panalytical

Prediction models

- First thing to realize is these the regression analysis is to SDD particles which had already been through the secondary dryer
  - Output of spray drying still has some residual solvent
  - Need to predict 'dry' values from 'wet' data
  - Models using process data predict dry particle size and dry bulk density from achieved set-points
- Secondly off-line methods are not the same as PAT methods
  - PAT methods need to predict off-line values as off-line values have been used in the control models
- In terms of particle size
  - Off-line method uses same fundamental measurement technique (laser, scattering, Mastersizer)
    - BUT Mastersizer uses the Fraunhofer approximation, and the Insitec used Mie scattering theory
    - And the off-line method uses a pressure titration to distribute the particles, whereas the PAT method uses flow titration (velocity)
  - PAT method needs to predict off-line values

Malvern Panalytical

Initial correlation between PAT and off-line

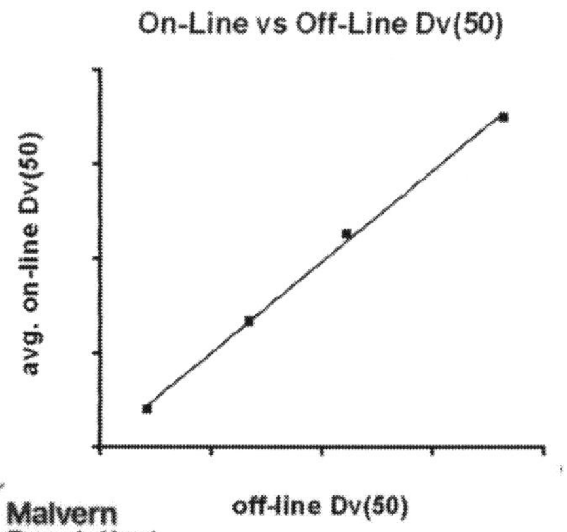

	Parameter	Value
Hardware	Online Laser Diffraction System	Insitec X
	Lens	200 mm (1-400 micron focal length)
	Venturi	90 degree
	Flow rate	9.5 Nm^3/hr
Material Settings	Model	Mie Scattering
	Particulate Refractive Index	1.5 + 0.1i
	Media Refractive Index	1
	Particle Density	1
Data Analysis	Min Particle Diameter (microns)	0.1
	Max Particle Diameter (microns)	600
	Scattering Threshold	8
	Multiple Scattering	Yes
	First Scattering Ring	5
System Suitability	Averaging	8-minute
	Transmission	≥ 80 %
	Counts at Sc(15)	≥ 500

Malvern Panalytical

How do we know this works?

- Example of a process change
- Correlation of process parameters to intermediate attribute means we can change process parameters to cause desired change in attribute
- Make a change expected to move the Dv50 by 3um
    - Dv50 responds immediately and stabilizes in 8mins (one block of data)
- More importantly we can set the press parameters to give the disso performance we require.

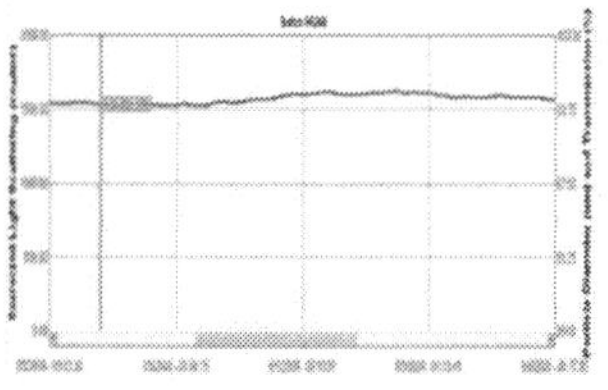

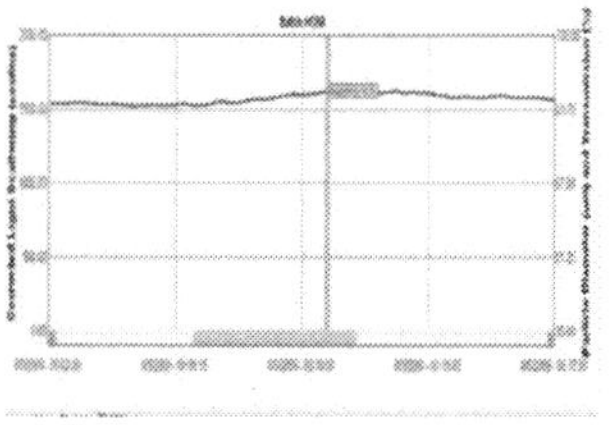

Malvern Panalytical

13

What does this all look like together?

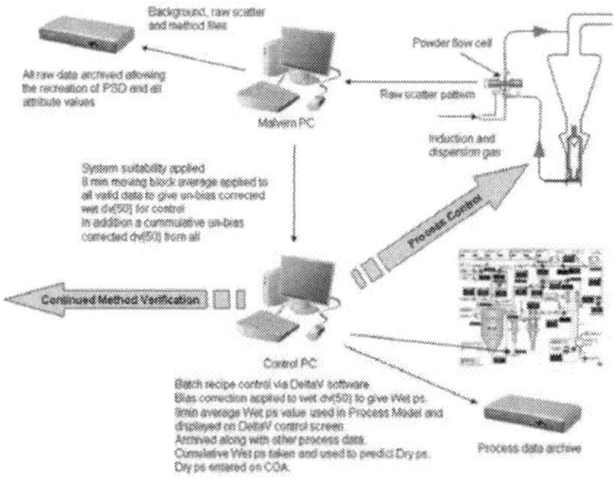

Malvern Panalytical

PAT Method Lifecycle

- Trend analysis of key performance indicators
  - Difference between predicted values (from achieved set-points)
    - ‘parallel testing’ of soft-sensor methods
    - Method updates
  - Difference between PAT measured values to off-line
  - Causes of PAT method system suitability (transmission at Sc15)
  - Predicting how process variance would impact the above
    - Proactive triggers for method updates

Malvern Panalytical

Differences between PAT and off-line methods

- Early PAT prediction methods of dry, off-line data where simple linear regressions (bias adjustments)
- However over time, and by using expanded ranges, we could show that the slope of all three PSD attributes (d(50), d(10) and d(90)) were not 45 degrees
- They are close enough to predict in the operational ranges
  - Fixed bias changed to slope adjustment
  - Fixed was OK, slope was better

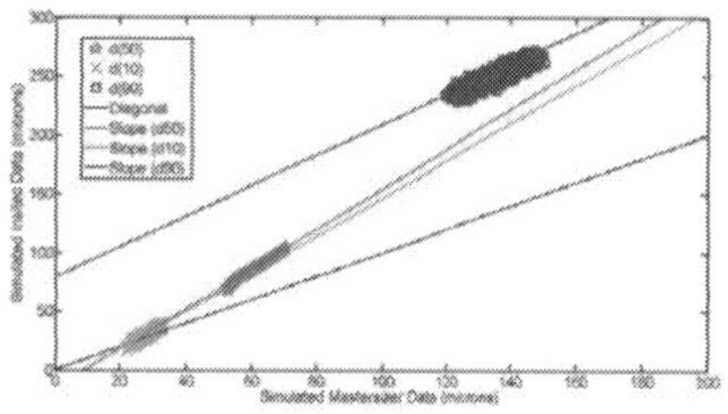

Malvern Panalytical

Characterizing Sc15 and transmission 'failures'

- PAT method uses transmission and counts on Sc15 as a system suitability
- Initially these were simple acceptance values
  - Found to be more sophisticated
- When transmission passes the PSD is mono-modal and SC15 has highest count
- As actually the system suitability should be on signal 'shape'

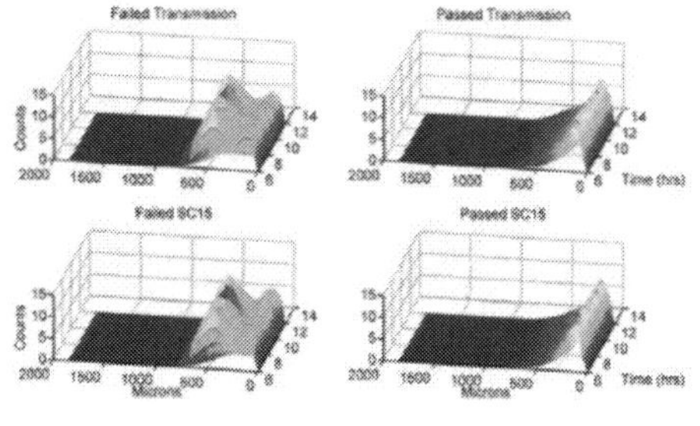

Malvern Panalytical

Impact of viscosity of d(50)

- During development an acceptance range for API/polymer solution was defined
- During production it was found that the viscosity of API/polymer solution varied (within range) and that if it varied the d(50) that would be generated changed
  - Suggested from regression analysis (positive correlation with %TS)
  - Fine if we stay in the narrow operating range, but if the range was expanded then a set-point correction should be applied to correct for cP
- However during the life cycle of the product various factors were identified which impacted viscosity
  - New acceptance tests for raw materials used in spray solution

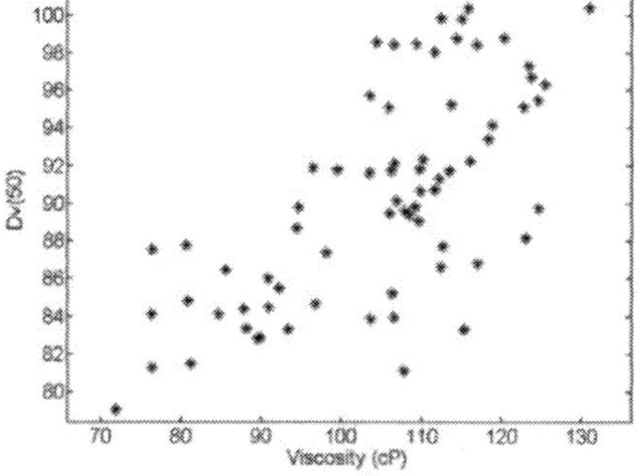

Malvern Panalytical

Continued application development

- Product underwent tech transfer to a bigger spray dryer, in a different facility
- Insitec installation moved upstream
  - Directly below drying tower
  - Before cyclone
- Measurement across entire discharge pipe
- Even then the installation required fine tuning
  - Lens were originally parallel, found to be reflection (multiple bounces) impacting scatter pattern
  - Original installation was across the center of the pipe, Insitec now rotated to look 'up-down' rather than 'left-right'
    - To account for stratified flow in the pipe
  - Method updated to include setting the 'zero' when process is spraying solvent
    - Changes refractive index

Malvern Panalytical

Conclusions

- In general, understand the inputs to tablet performance
  - Fishbone/Ishikawa diagrams are very useful
- Collect the data (either directly or via modelling) to have representative inputs into those diagrams
  - Use high frequency data collection rates to for control models and design space time limits, averaged data to determine product quality attributes
  - Justify data collection and decision frequency
- Make sure the data is weighted appropriately
  - For example in tablets with high drug load, weight variance is more important than low drug load, where variance in concentration has a bigger impact
- If you do, then full real-time release is achievable and has been achieved.

Malvern Panalytical

Acknowledgements

- Vertex
  - Joseph Medendorp, John Bric, Kelly Tolton, Greg Connelly
- Hovione
  - Filipe Gaspar, Nuno Matos, Filipe Neves
- Malvern Panalytical
  - Sensors and automation solutions team

Thank you for your attention.....

Malvern Panalytical

490592 Impurity Rejection in Crystallization â€“Quantification and Mechanisms

Abstract:

Other than isolating the product as a solid, the perhaps most important objective in crystallization of pharmaceuticals is purification. Nevertheless, most of the engineering research in academia and industry has been focused on particle size and polymorph control. Notwithstanding their importance, crystallization as a unit operation is relied on not just in the API step but across the entire synthetic scheme to reject a number of different impurities. These can be reagents, reaction by-products, process impurities, degradants, chiral impurities, inorganic impurities and even genotoxic impurities. Purification via crystallizations is thus critical in producing a drug substance that is suitable for pre-clinical and clinical trials and eventually for market demands. A stringent regulatory oversight is also in place to ensure that the API meets all purity and quality attributes prior to administration to humans.

In Pharma industry, most of the impurity control in crystallization has traditionally been carried out by synthetic chemists who mainly rely on a before and after approach that is highly empirical in nature. Presented herein is the use of engineering principles to accurately quantify rejection of impurities throughout the crystallization. Various mechanisms of impurity rejection are identified and linked to thermodynamic and/or kinetic effects. Case studies from pharmaceutical development are presented that provide a clear understanding on the impurity purge mechanisms and how they are affected by process conditions. This work thus enables simple control measures to be introduced in the crystallization process that are capable of producing materials with targeted purity profile, whether that is for toxicity studies or for clinical trials.

492027 Bi-Directional Control of Glycosylation of Antibodies

Abstract:

Monitoring and control of critical quality attributes (CQAs) is important for biologics manufacturing. CQAs need to be controlled within desirable ranges in order to manufacture a clinically effective product. These ranges can be particularly tight for biosimilars, because similarity to the innovator molecule must be demonstrated. To address these challenges, new flexible methods of monitoring and modulating CQAs are required. Multivariate model predictive control (MPC) can be used to target CQAs within desirable ranges to ensure the delivery of high quality biologic products. In this study, MPC was applied to control the CQA of interest (CQA1) of a recombinant antibody by using two media additives and a cell culture parameter as modulation levers. A partial factorial experiment (n=25) was first conducted in bench-top bioreactors to understand the impact of these levers on CQA1. Both media additives were able to modulate CQA1 but in opposite directions. The cell culture parameter alone was able to modulate CQA1 in both directions. After these levers were identified and characterized, an empirical multivariate model was created to predict and control CQA1 during the cell culture process. Next, the robustness of this control strategy was tested in 3 L bench-top and 100 L pilot-scale bioreactors. For the test, CQA1 was purposefully deviated in either direction (high or low) away from the target using the cell culture parameter as a lever. Our results indicate that the model effectively predicted the amount of media additive required to bring CQA1 back to the target. The target CQA1 range was met for both scales at the end of the cell culture. Thus, the use of media additives to achieve bi-directional model predictive control of CQA1 has been achieved. This study demonstrates that MPC is a robust tool to modulate CQAs within desired targets for cell culture processes.

492107 Bioprocess Optimization for Production of Recombinant Proteins in Chinese Hamster Ovary Cells in Consideration of Metabolic Shifts By Shear Stress

Abstract:

Biopharmaceuticals such as monoclonal antibody drugs have attracted considerable interest in recent years for currently untreatable diseases. A series of large production plants have been built in different parts of the world. Chinese hamster ovary (CHO) cells are widely used as a host cell line and cultured in fed-batch cultivation in a large stirred bioreactor for commercial production of biopharmaceuticals. When a bioreactor is designed for commercial-scale production, it is indispensable to evaluate the impact of shear stress on cell growth and cell viability due to their mechanical fragility. In this study we focused on the impacts of shear stress on cellular metabolism to optimize fed-batch cultivation for production of tissue plasminogen activator (tPA) in CHO cells.

We went through the two steps to optimize a fed-batch cultivation process. The first step is the quantitative evaluation of the effects of shear stress. In the case of a stirred bioreactor, a quantitative evaluation is difficult to be performed due to the distribution of shear stress in a tank. Therefore we introduced a flow chamber which can impart a steady laminar shear stress on the cells for the evaluation of the effects of shear stress and found that metabolism shift occur with increased shear stress. We determined the appropriate range of shear stress for the production from the results. The second step is the optimization of an impeller configuration and feed medium in consideration of the metabolism shift. We designed 3L bioreactors which sustain appropriate shear stress in a tank using computational fluid dynamics (CFD) and feed medium in consideration of metabolism shift. CHO cells producing tPA protein were cultured with our optimized bioprocess. As a result, the production of tPA increased by 1.5 times compared to the conventional fed-batch process.

We found that metabolism shift occur by increasing shear stress and the optimization of a bioprocess taken into account the metabolism shift contributes to increasing productivity.

Using machine learning tools in bioprocess scale up when the number of batches is small

Viktor Konakovsky
Minneapolis, MN, US
Nov 2nd 2017

Topics

- What was happening in my project BioRapid so far
- Today I will talk about using PCA in combination with TOST time series
- The case study will focus on quantifying equivalence between 2L and 200L

My project BioRapid

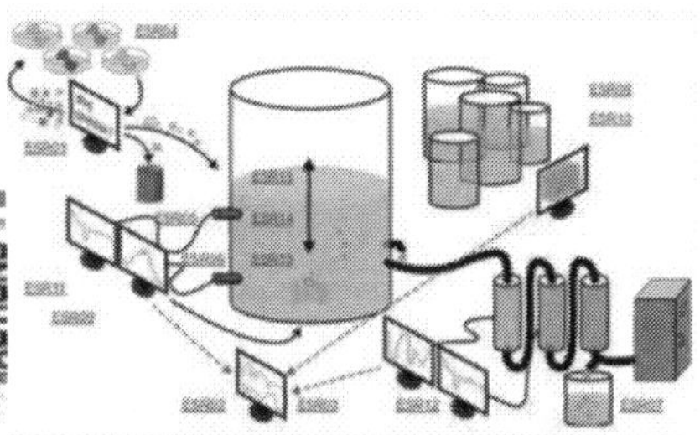

Meet my Team

Bio RAPID

http://www.bio-rapid.eu/

Financial support from the EU-Horizon 2020 Marie Skłodowska-Curie actions (MSCA) ITN project BIORAPID (no. 643056).

Motivation

- This work was initiated while writing a report...

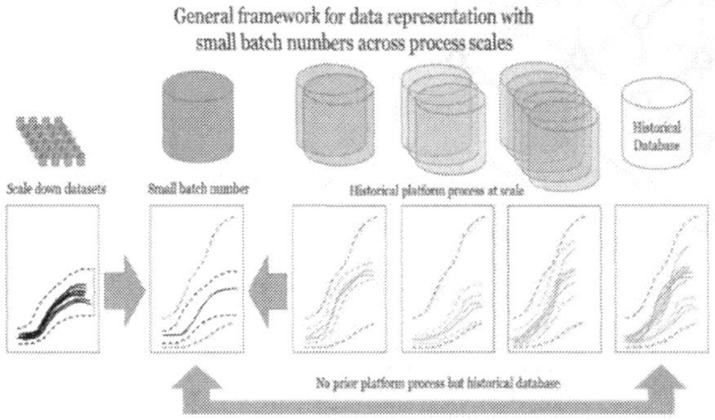

Bio RAPID

"The Rapid Bioprocess Development ITN, effective training framework in rapid development of novel bioactive molecules from the very early stages of potency and efficacy testing to the bio-manufacturing process characterization and monitoring"

EU-Horizon 2020 - Marie Curie ITN project BIORAPID (Grant Agreement n° 643056)

Deliverable D2.5: General framework for data representation with small batch numbers, across process scales

- ...and we are writing up to publish the results soon in Biotechnology Journal

Scientific output

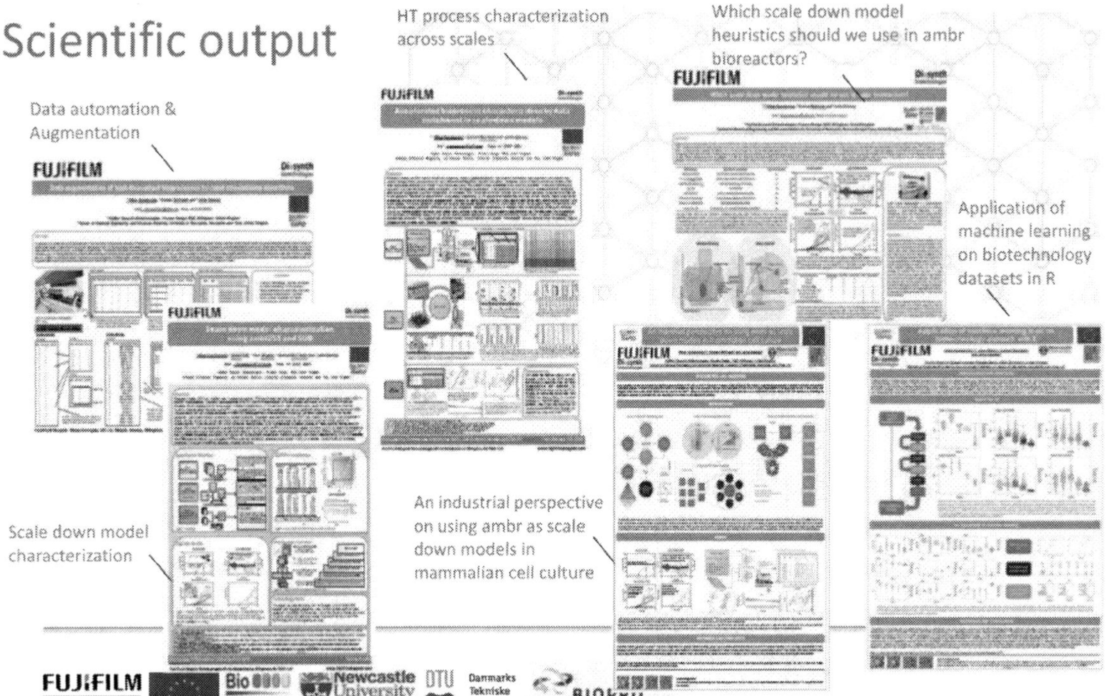

Introduction

My part is to get value out of databases, and to develop predictive modeling and machine learning tools on dedicated industrial projects. Those projects were mostly focused on high throughput upstream data sources in mammalian cell culture but also extend to CLD and to microbial manufacturing

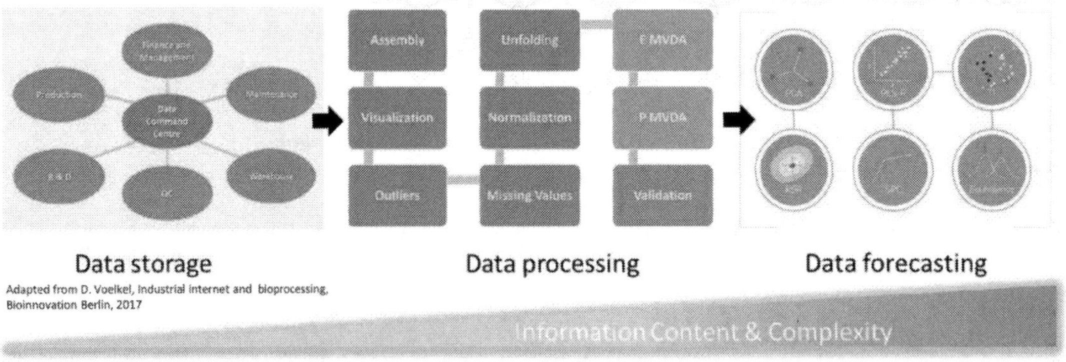

Adapted from D. Voelkel, Industrial internet and bioprocessing, Bioinnovation Berlin, 2017

6

The case

- In HT (high throughput) we have a lot of data, but as we scale up this is not the case any more
- Often at this stage we want to know whether batches are equivalent, or if not, what is different?
- This dataset is a CHO cell culture fed-batch

Scale	*Regime*	*n*
2L	C	6
2L	D	9
10L	E	2
200L	D	1
sum		**18**

The TOST

- Question: Is data from group A equivalent to group B?
- Typically would use TOST (two one sided t-test) to quantify a significant difference
- Statisticians in bioprocessing often use 2 standard deviations (SD) of a reference group to calculate delta

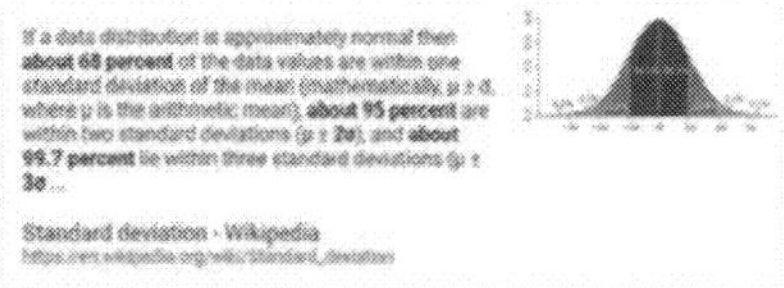

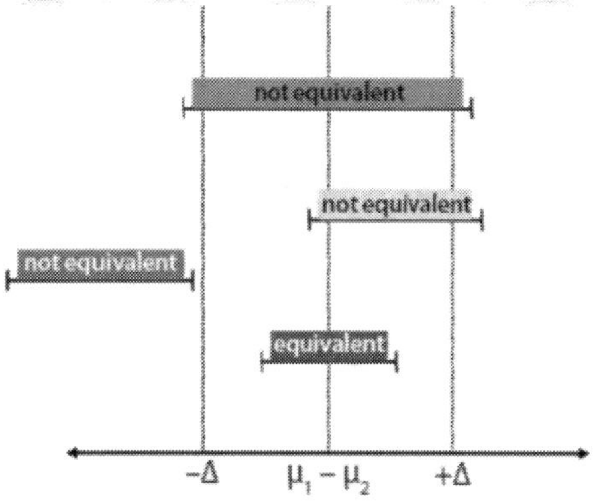

Ref: C. DiCesare et al.: Development, Qualification, and Application of a Bioreactor Scale-Down Process: Modeling Large-Scale Microcarrier Perfusion Cell Culture." *BioProcess International, 2016*

The challenge

- What if the number in one group is just 1? -> Time series TOST testing
- Definition of how many SD are used to form the equivalence interval
- Time series could be problematic, rules for passing/failing equivalence are required
- What if one variable is equivalent, another is not, the next is, the nth variable again is not...

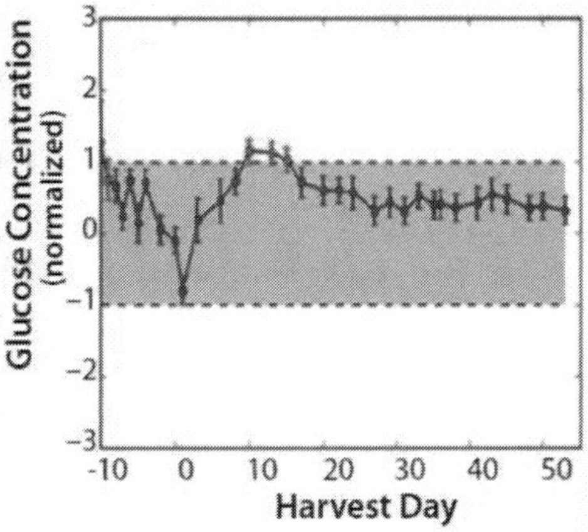

DiCesare used a rule of ‚three in a row' as maximum acceptable equivalence outliers

Ref: C. DiCesare et al.: Development, Qualification, and Application of a Bioreactor Scale-Down Process: Modeling Large-Scale Microcarrier Perfusion Cell Culture." *BioProcess International, 2016*

FUJIFILM Diosynth biotechnologies | BioRAPID | Newcastle University | DTU Danmarks Tekniske Universitet | BIOPRO World Talent Campus

9

The challenge

- Especially relevant for CQAs or CPPs (critical quality attributes or critical process parameters)
- If it was critical to be equivalent, it becomes relevant how close equivalence was missed
- Sometimes those differences can be explained, but often this is means lengthy investigations have to be done

Variable	***Equivalent?***	***Was it a CQA or CPP?***	***P value debate***
V1	Pass	Yes	OK
V2	Pass	Yes	OK
V3	Pass	No	OK
V4	Pass	Yes	OK
V5	Fail	Yes	Clear fail
V6	Pass	Yes	OK
V7	Fail	Yes	Borderline fail
V8	Pass	Yes	OK
V10	Pass	Yes	OK
...			

FUJIFILM Diosynth biotechnologies | BioRAPID | Newcastle University | DTU Danmarks Tekniske Universitet | BIOPRO World Talent Campus

10

The opportunity

- Are there any tools that allow us to capture most of the variance in the process?
- Can we get a quantitative measure of how equivalent the runs are in %, including which variables are equivalent and which are not?
- Today I present how PCA can be combined with a TOST test on time series to do exactly that using the TOSTER package in R

Ref: D Lakens, Two One-Sided Tests (TOST) Equivalence Testing, Package 'TOSTER', 2017

Goal

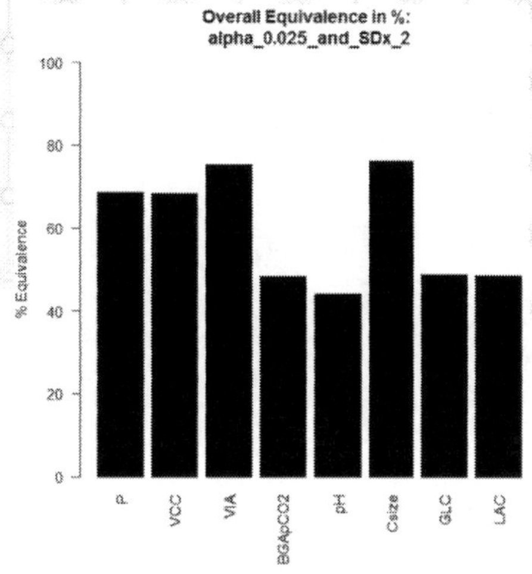

Total Equivalence (TEQ): 59.78%

FUJIFILM Diosynth biotechnologies · BioRAPID · Newcastle University · DTU Danmarks Tekniske Universitet · BIOPRO World Talent Campus

Data

- Prepare dataset by filtering the data into subsets of all batches per day
- Data was mean centered and divided by the SD before extracting the scores with PCA

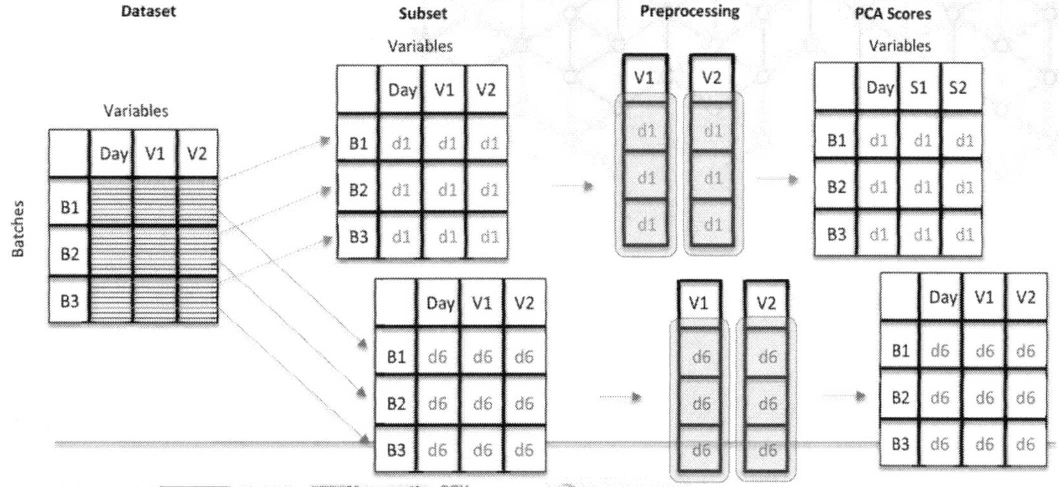

FUJIFILM Diosynth biotechnologies · BioRAPID · Newcastle University · DTU Danmarks Tekniske Universitet · BIOPRO World Talent Campus

Workflow

- Take the scores and run a TOST test on it using the parameters alpha and SD multiplier to define whether these PCs on that day are equivalent or not

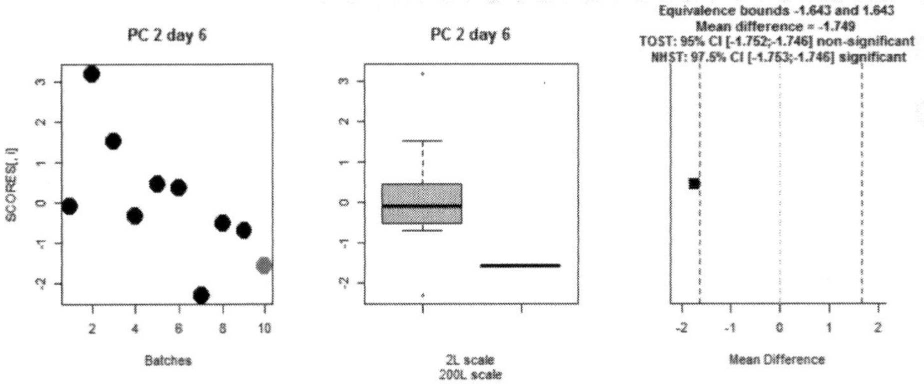

- alpha_0.025_and_SDx_1

FUJIFILM Diosynth biotechnologies | Bio RAPID | Newcastle University | DTU Danmarks Tekniske Universitet | BIOPRO World Talent Campus

13

Workflow

- Take the scores and run a TOST test on it using the parameters alpha and SD multiplier to define whether these PCs on that day are equivalent or not

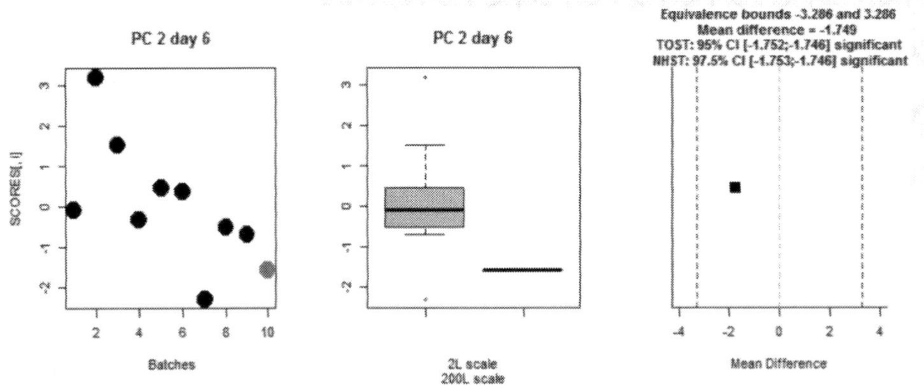

- alpha_0.025_and_SDx_2

FUJIFILM Diosynth biotechnologies | Bio RAPID | Newcastle University | DTU Danmarks Tekniske Universitet | BIOPRO World Talent Campus

14

Workflow

- Take the scores and run a TOST test on it using the parameters alpha and SD multiplier to define whether these PCs on that day are equivalent or not

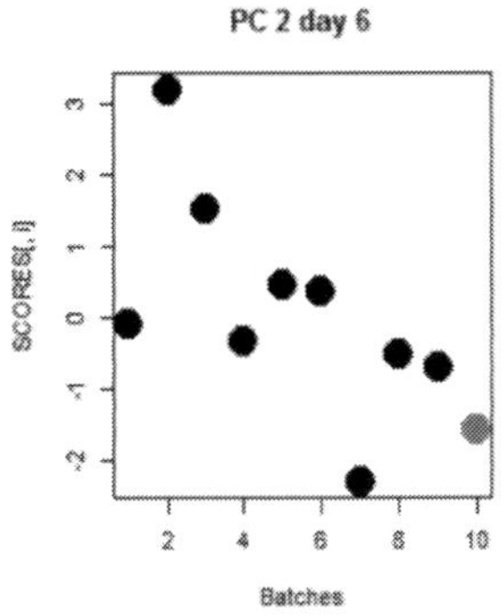

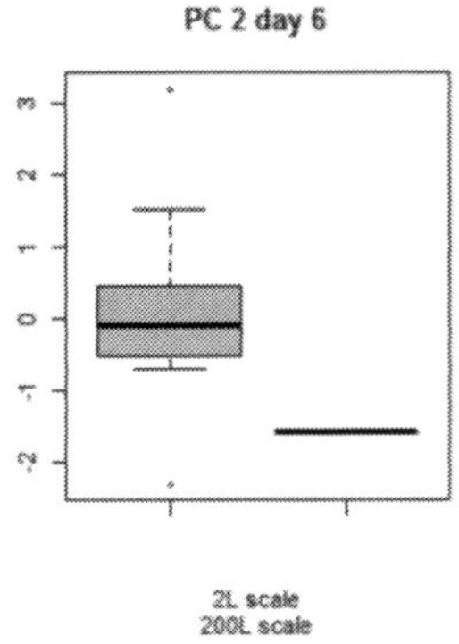

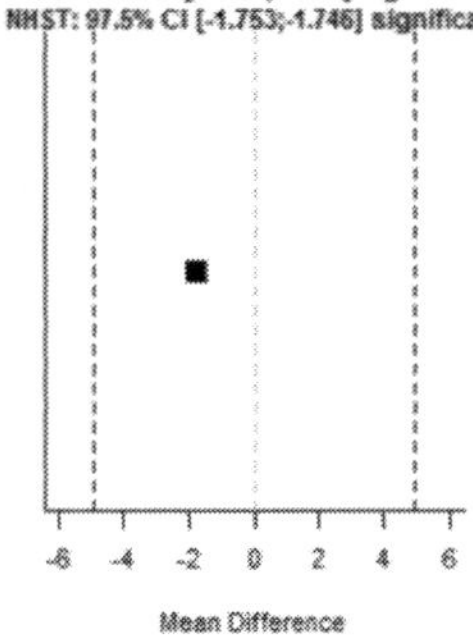

- alpha_0.025_and_SDx_3

Workflow

- The parameter alpha is not relevant in this example as there is only one observation in one of the groups. If would determine if the data point lies within or outside of equivalence.

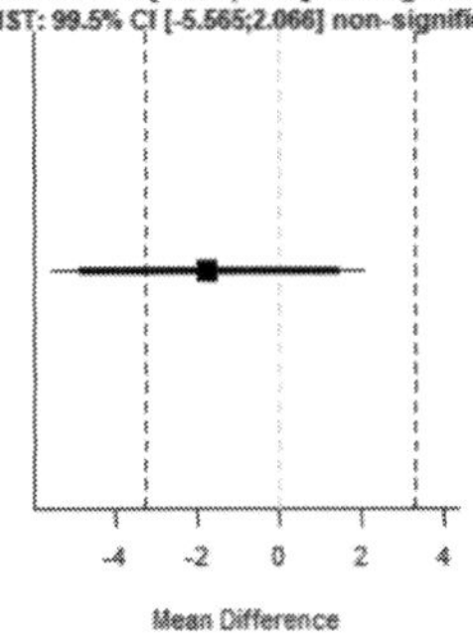

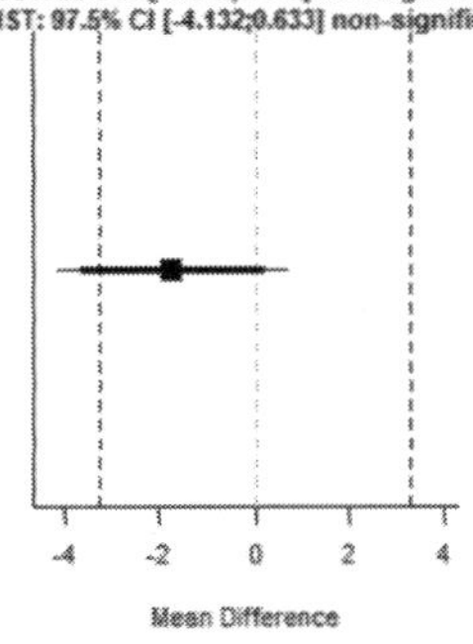

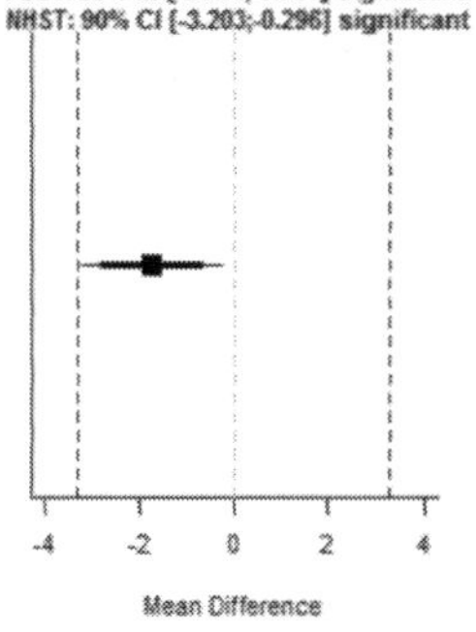

- alpha_0.005_and_SDx_2 alpha_0.025_and_SDx_2 alpha_0.1_and_SDx_2

Workflow

- Different combinations of conditions, alpha and SDmultiplier, are evaluated in a grid search because they have a strong impact on TOST

alpha = 0.005, 0.010, 0.025, 0.050, 0.100
SDmultiplier = 1.0, 1.5, 2.0, 2.5, 3.0

- Create a weight matrix: which PCs were equivalent or not

Name	PC1	PC2	PC3	PC4	PC5	PC6	PC7	PC8
alpha_0.005_and_SDx_1	1	0	0	1	0	0	0	0
alpha_0.01_and_SDx_1	1	0	0	1	0	0	0	0
alpha_0.025_and_SDx_1	1	0	0	1	0	0	0	0
alpha_0.05_and_SDx_1	1	0	0	1	0	0	0	0
alpha_0.1_and_SDx_1	1	0	0	1	0	0	0	0
alpha_0.005_and_SDx_1.5	1	1	0	1	0	0	1	0
alpha_0.01_and_SDx_1.5	1	1	0	1	0	0	1	0
alpha_0.025_and_SDx_1.5	1	1	0	1	0	0	1	0
alpha_0.05_and_SDx_1.5	1	1	0	1	0	0	1	0
alpha_0.1_and_SDx_1.5	1	1	0	1	0	0	1	0
alpha_0.005_and_SDx_2	1	0	0	1	0	1	1	0
alpha_0.01_and_SDx_2	1	1	0	1	0	1	1	0
alpha_0.025_and_SDx_2	1	0	0	1	0	1	1	0
alpha_0.05_and_SDx_2	1	1	0	1	0	1	1	0
alpha_0.1_and_SDx_2	1	1	0	1	0	1	1	0
alpha_0.005_and_SDx_2.5	1	1	0	1	0	1	1	0
alpha_0.01_and_SDx_2.5	1	1	0	1	0	1	1	0
alpha_0.025_and_SDx_2.5	1	1	0	1	0	1	1	0
alpha_0.05_and_SDx_2.5	1	1	0	1	0	1	1	0
alpha_0.1_and_SDx_2.5	1	1	0	1	0	1	1	0
alpha_0.005_and_SDx_3	1	1	0	1	0	1	1	0
alpha_0.01_and_SDx_3	1	1	0	1	0	1	1	0
alpha_0.025_and_SDx_3	1	1	0	1	0	1	1	0
alpha_0.05_and_SDx_3	1	1	0	1	0	1	1	0
alpha_0.1_and_SDx_3	1	1	0	1	0	1	1	0

FUJIFILM Diosynth biotechnologies | BioRAPID | Newcastle University | DTU Danmarks Tekniske Universitet | BIOPRO World Talent Campus

Workflow

- Let us have a look what those PCs represent - weighting of all PCs, on day 6

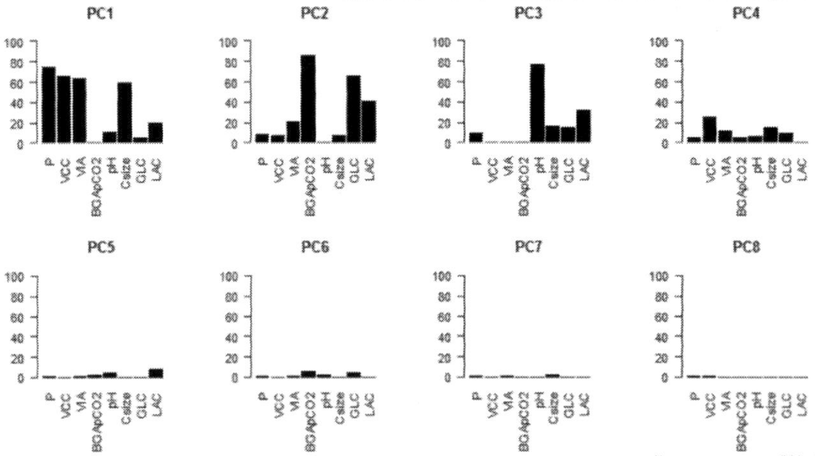

Name	PC1	PC2	PC3	PC4	PC5	PC6	PC7	PC8
SD	1.72	1.54	1.23	0.88	0.41	0.38	0.2	0.13
Var%	37	30	19	10	2	2	1	0
CumVar	37	67	86	95	98	99	100	100

FUJIFILM Diosynth biotechnologies | BioRAPID | Newcastle University | DTU Danmarks Tekniske Universitet | BIOPRO World Talent Campus

Results

- Sum up PCs passing the TOST test
- For every condition
- For every variable
- For every day
- The variance covered in the passing PCs gives a measure of overall equivalence in %

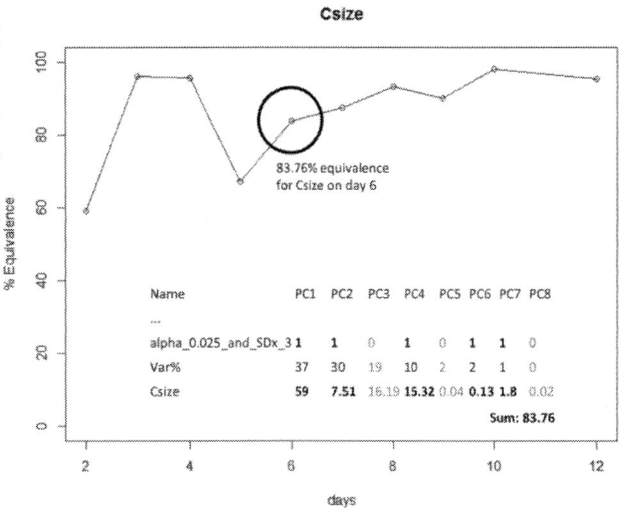

FUJIFILM Diosynth biotechnologies · BioRAPID · Newcastle University · DTU Danmarks Tekniske Universitet · BIOPRO World Talent Campus

19

Results

- Visualize explained equivalence per variable as time profiles, for all days and all variables

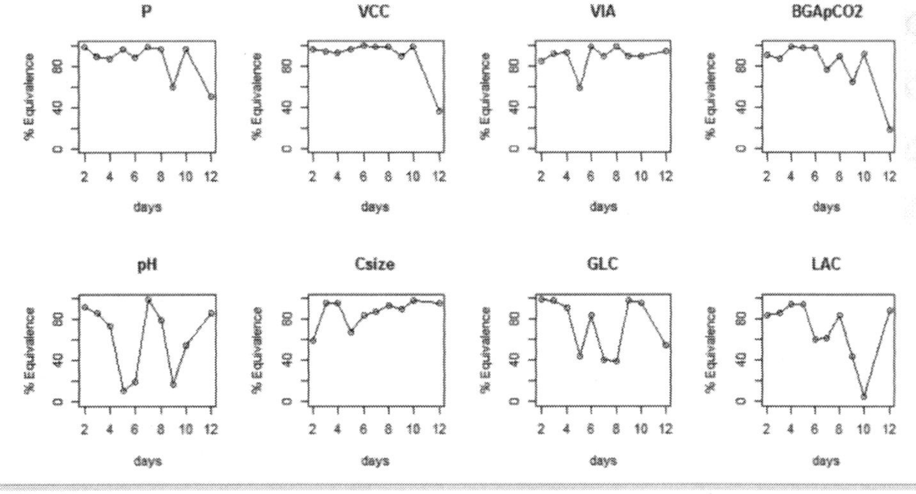

FUJIFILM Diosynth biotechnologies · BioRAPID · Newcastle University · DTU Danmarks Tekniske Universitet · BIOPRO World Talent Campus

20

Results

- Get an overall metric for equivalence with a certain condition of alpha and SDmultiplier

Mean of all equivalent variables over the whole process time:

Total Equivalence (TEQ): 59.78%

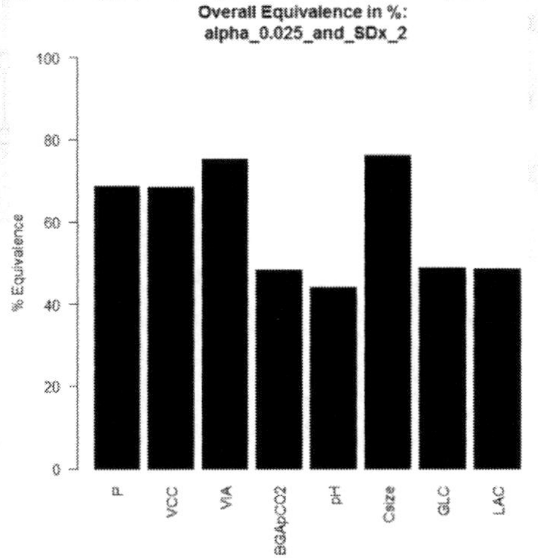

FUJIFILM Diosynth biotechnologies · Bio RAPID · Newcastle University · DTU Danmarks Tekniske Universitet · BIOPRO World Talent Campus

21

Results

- Visualize main differences per variable as time profiles

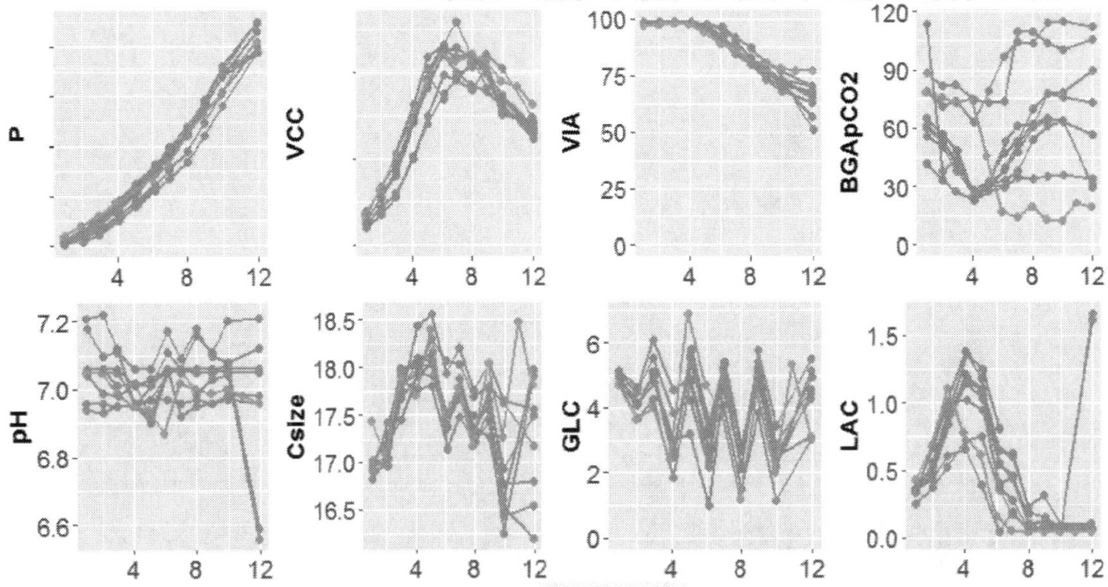

Results

- Get an overall metric for conditions tested at different levels of alpha and SD multipliers

TEQ: 28.17% **TEQ: 59.78%** **TEQ: 79.80%**

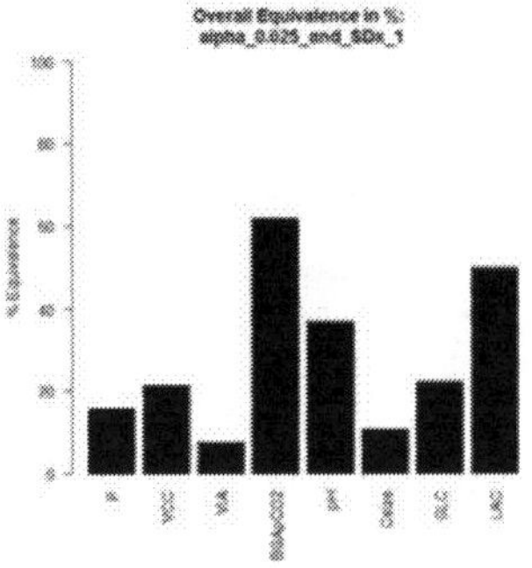

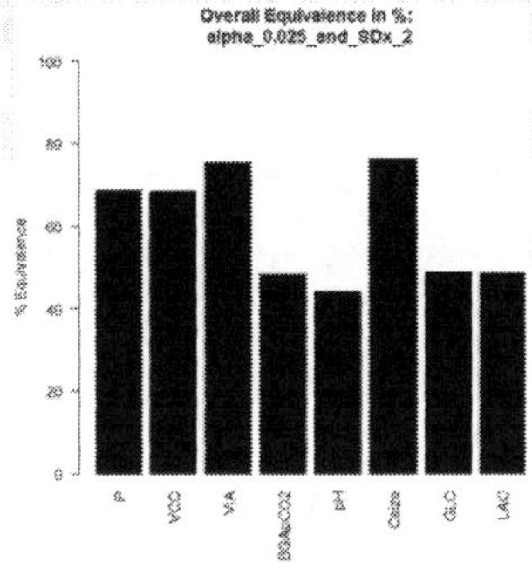

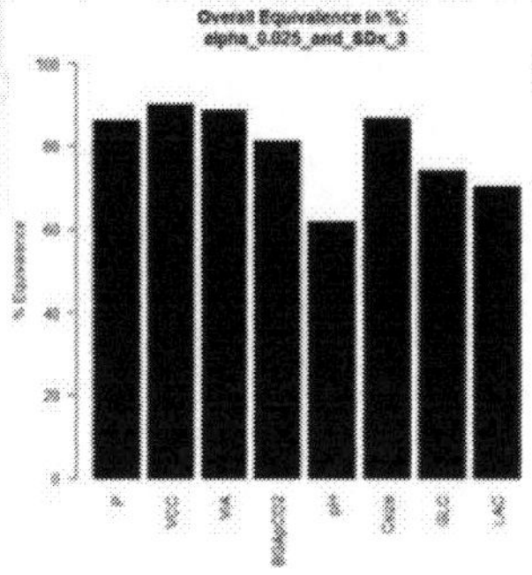

- alpha_0.025_and_SDx_1 alpha_0.025_and_SDx_2 alpha_0.025_and_SDx_3

Conclusions

- A method was presented that enables us to quantify equivalence with a novel metric TEQ
- The metric is based on PCs, which inherently explain the most variance in a dataset, and could be better suited than regular variables to compare groups holisticly using time series TOST
- Two major factors influencing the TOST (alpha and SDmultiplier) were evaluated for the new metric to allow for a better understanding of the TOST results
- This method summarizes which variables were most and least equivalent, which may help to find the root cause of undesireable process behaviour faster
- It is designed for HT datasets, but also worked for small batch numbers of n = 1, when data was scarce during scale up

Acknowledgements

Thanks to my Mentors
Graham McCreath, Head of Process Design, Fujifilm Diosynth Biotechnologies
Jarka Glassey, Professor, University of Newcastle

Thanks for continued support at Fujifilm Diosynth Biotechnologies!

Thanks to my colleagues and friends especially at NCL and DTU!

This project has received funding from the European Union's Horizon 2020 research and innovation programme under the Marie Curie Skłodowska-Curie grant agreement No 643056.

FUJIFILM Diosynth biotechnologies | Bio RAPID | Newcastle University | DTU Danmarks Tekniske Universitet | BIOPRO World Talent Campus

25

Results

- A different (non equivalent) case with more data: Regime C vs Regime D

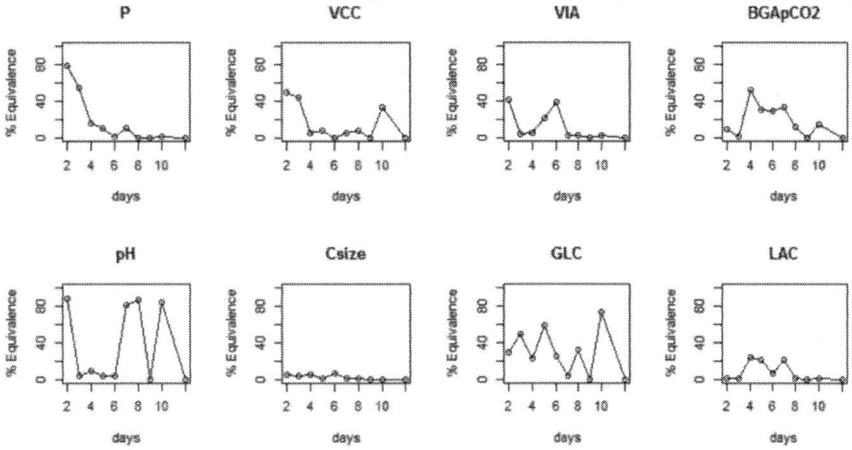

FUJIFILM Diosynth biotechnologies | BioRAPID | Newcastle University | DTU Danmarks Tekniske Universitet | BIOPRO World Talent Campus

27

Results

- A different (non equivalent) case with more data: Regime C vs Regime D

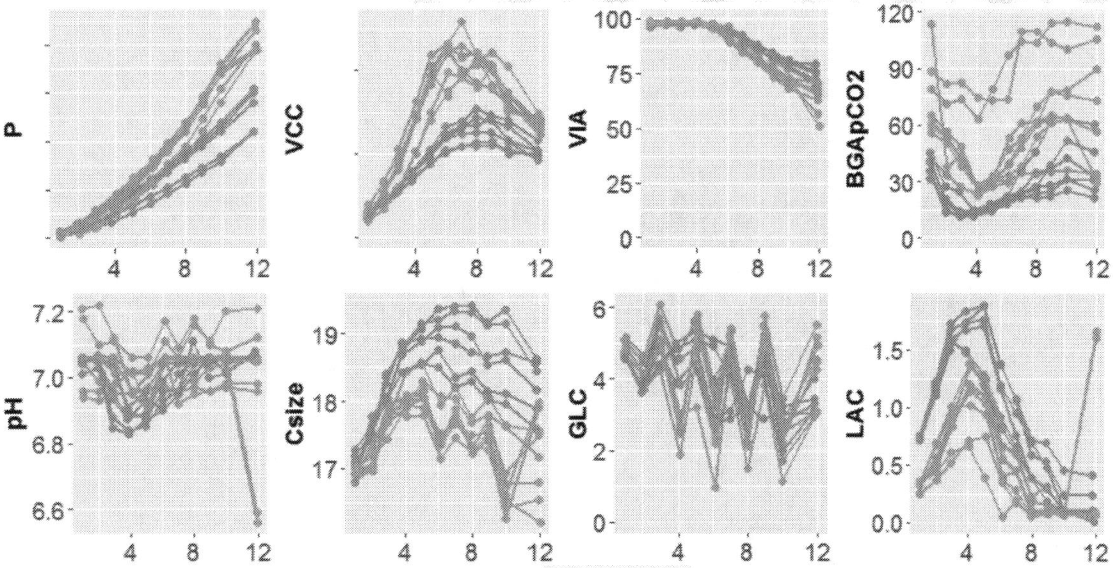

Results

- A different (non equivalent) case with more data: Regime C vs Regime D

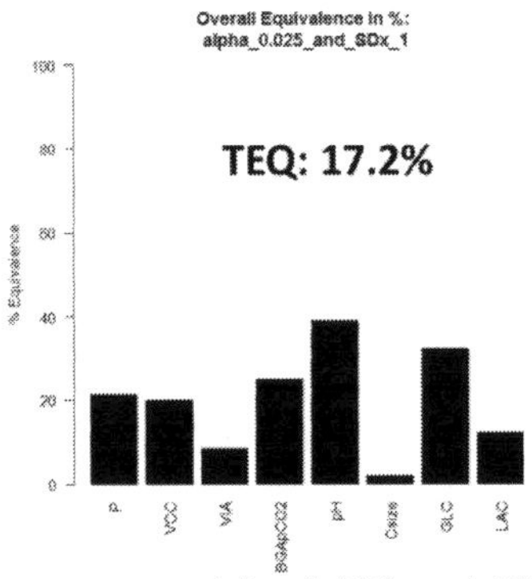

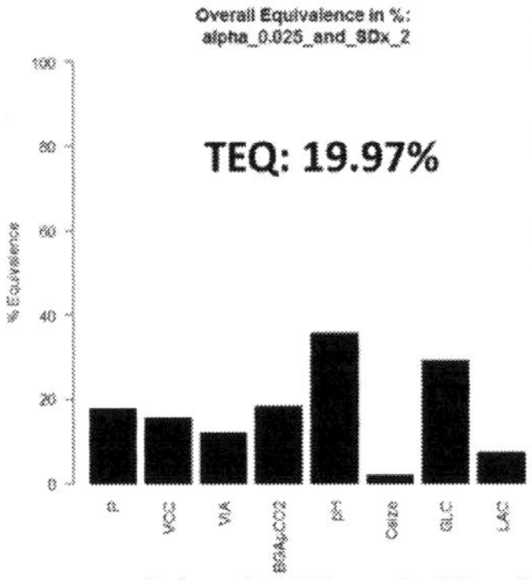

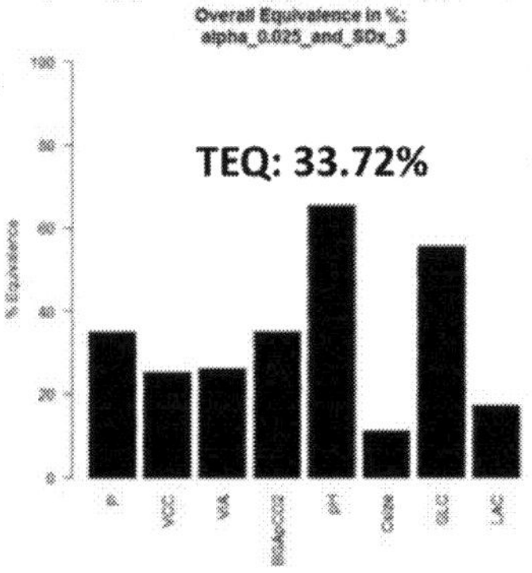

- alpha_0.025_and_SDx_1 alpha_0.025_and_SDx_2 alpha_0.025_and_SDx_3

Workflow

- Combine weighted equivalence matrix with variance covered by individual variables.
- Not all PCs will be used, only those that passed the TOST test.

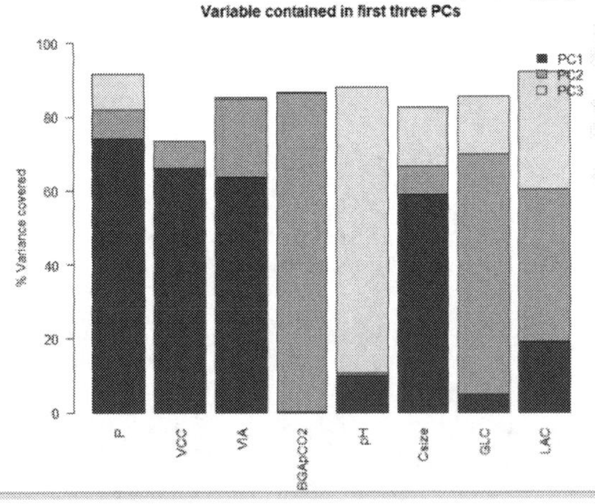

Name	PC1	PC2	PC3	PC4	PC5	PC6	PC7	PC8
...								
alpha_0.025_and_SDx_3	1	1	0	1	0	1	1	0
SD	1.72	1.54	1.23	0.88	0.41	0.38	0.2	0.13
Var%	37	30	19	10	2	2	1	0
CumVar	37	67	86	95	98	99	100	100
P	74.16	7.83	9.67	5.32	1.36	0.57	0.57	0.52
VCC	66.2	7.23	0.07	25.7	0.01	0.02	0.18	0.59
VIA	63.69	21.1	0.45	11.11	1.17	1.07	1.27	0.15
BGApCO2	0.18	86.35	0.12	4.95	2.57	5.81	0	0.01
pH	9.98	0.7	77.25	6.34	4.05	1.66	0.01	0.01
Csize	59	7.51	16.19	15.32	0.04	0.13	1.8	0.02
GLC	4.73	65.23	15.54	9.14	0.22	4.84	0.14	0.16
LAC	19.06	41.18	32.02	0.24	7.27	0.01	0.08	0.15

A workflow-based framework for managing product analytical data and statistical results for lot release

G. Joglekar[1], Q. Cai[2], P. DeLaurentis[1], L. Mockus[1], K. Morris[2], G.V. Reklaitis[1]

[1]Davidson School of Chemical Engineering, Purdue University

[2]Lachman Institute for Pharmaceutical Analysis, Long Island Univ

AIChE Annual Meeting

Paper 665(e)

November 2, 2017

Outline

- Open data requirements
- KProMS system
- The collaborative research project
- KProMS benefits
- Conclusions

Objective

- Demonstrate the use of a workflow-based knowledge management system to support a collaborative research project
  - ✓ Data gathering
  - ✓ Data curation
  - ✓ Data analysis
  - ✓ Multiple groups
  - ✓ Open data

Open Data

- Directive to funding agencies to ensure that public has access to the results of government-funded research (Office of Sc and Tech Policy, 2013)
- The directive requires that data are
  - ✓ Richly described with machine-actionable metadata
  - ✓ Complete, self-explanatory, and accurate
  - ✓ Accessible and preserved beyond the life of a grant
  - ✓ Protected for confidentiality and privacy
- Private funders require that data be released with some degree of openness
- Companies would have similar requirements for internal use of confidential data

Data Sharing

- Even though more data are becoming available, they are not used widely because
  - ✓ Full context is not available
  - ✓ Meta-data is not available
- Data sharing among academic researchers very sporadic
- Tech transfers often time consuming
- The need for well curated data is expected to grow in the future
- There are no guidelines for good practices, or standards for publishing data
  - ✓ Putting it on the cloud is not sufficient

KProMS: Knowledge Provenance Management System

- KProMS framework produces shareable and reusable data by
  - ✓ Capturing the context of data
  - ✓ Streamlining data submission into a repository
  - ✓ Organizing data for machine and manual access over the web

KProMS Architecture

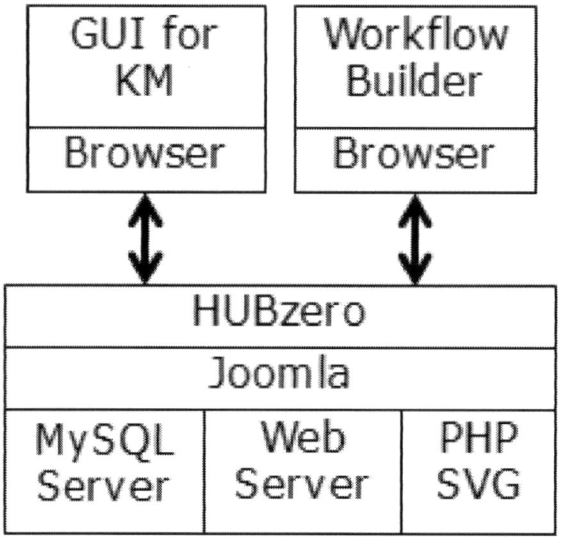

Joglekar, G., Giridhar, A., Reklaitis, G.V. "A Workflow Modeling System for capturing data provenance", Computers & Chemical Engineering (DOI: 10.1016/j.compchemeng.2014.04.006)
Joglekar, G., Giridhar, A., Reklaitis, G.V. "A Novel Knowledge Management System Based on Workflows", Computer Aided Chemical Engineering Series, Vol 33, Elsevier.

Collaborative Project

- Perform analytical tests associated with lot release on drugs purchased in open market
  - ✓ Dissolution
  - ✓ Content uniformity
  - ✓ Physical properties: Weight, Dimension, Mechanical strength
- Perform Bayesian analysis for hypothesis testing
- Funded by the FDA
  - ✓ Long Island University, Purdue University

Project Execution

- Model each experimental procedure as a workflow
- Procure drug to be studied
- Record each experiment as an instance of the associated workflow model
  - ✓ Parameter values
  - ✓ Upload instrument data
- Perform appropriate data checks
- Perform data analysis

Interactions of Working Groups

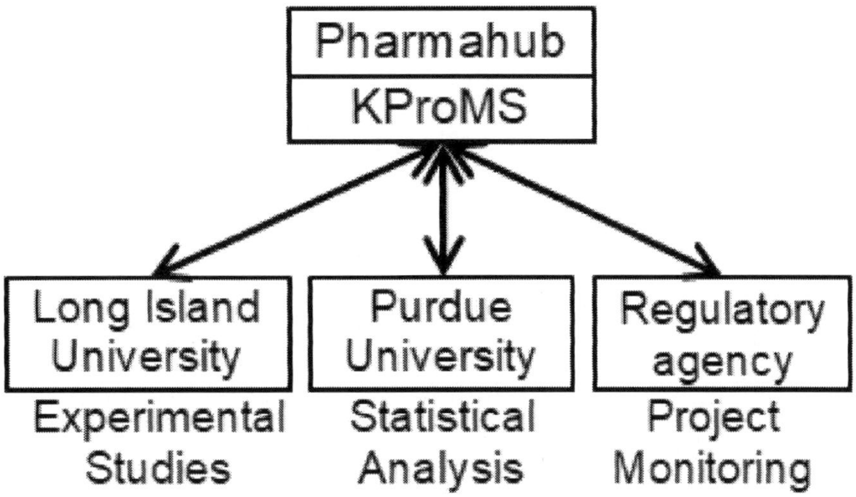

Dissolution Workflow

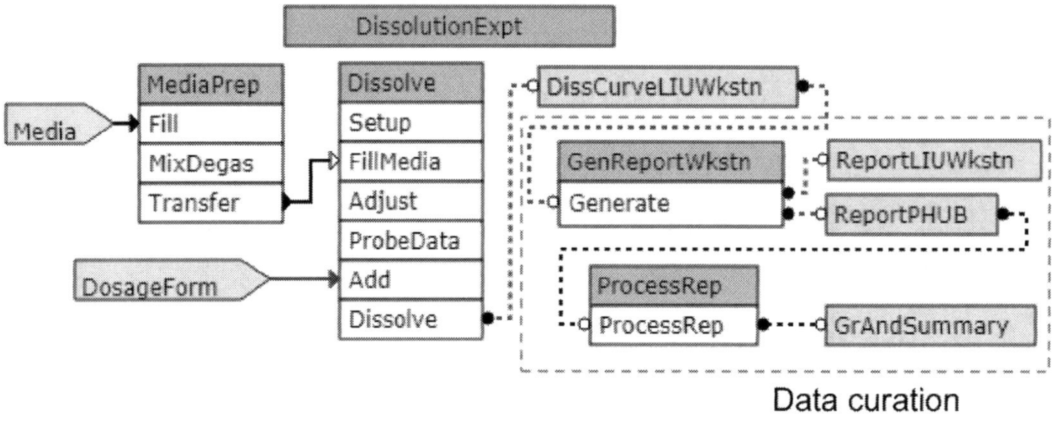

Parameters

		Parameter
Material:Media		Amount, liter
Material:DosageForm		Material lot, ID
Task:MediaPrep	*Subtask*:Fill	pH, number
	Subtask:MixDegas	Degas, Yes/No
Task:Dissolve		Number of vessels, integer
	Subtask:Setup	Bath temperature, C Accessory, Paddle/Basket
	Subtask:FillMedia	Amount per vessel, ml
	Subtask:ProbeData	Exposure Autoset, Yes/No Standard A blank, Yes/No Standard B blank, Yes/No Sample blank, Yes/No Standard A, g/l Standard B, g/l Response factor, Number Wavelength max, nm Path length, mm Calibration curve, Yes/No Baseline correction, Yes/No
	Subtask:Add	Number of units per vessel, number
Task:GenReportWkstn	*Subtask*:Generate	Program name, OPT-DISS
Task:ProcessRep	*Subtask*:ProcessRep	Library program, DigData Input file type, Dissolution report Output data, Pct diss and graph/Zip data

GrAndSummary Data Node

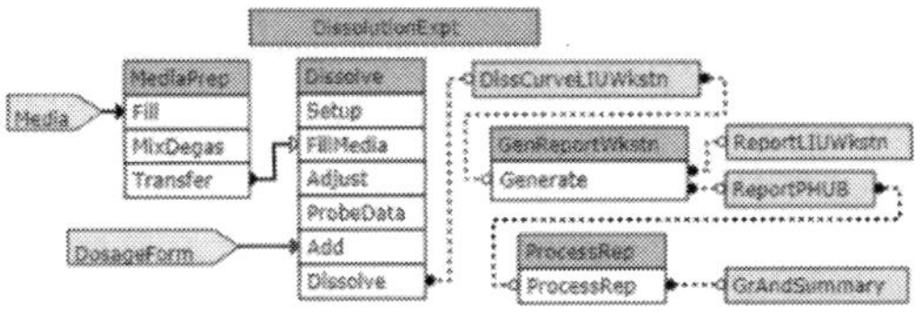

Number of tablets	6					
Dissolution end time	00:30:00					
Pct dissolved at end time	Ch1	Ch2	Ch3	Ch4	Ch5	Ch6
	104.95	100.99	99.90	102.41	102.74	97.26
Av Pct dissolved	101.3750					
Std dev	2.6447					

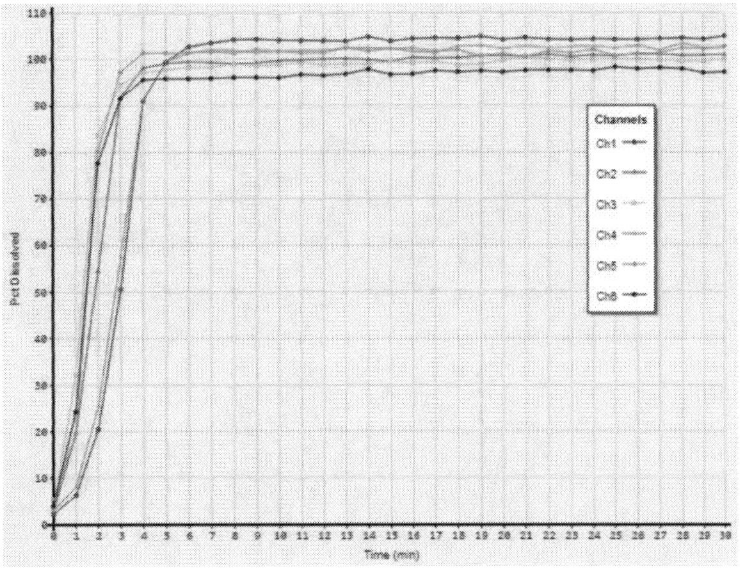

Workflows

Dissolution

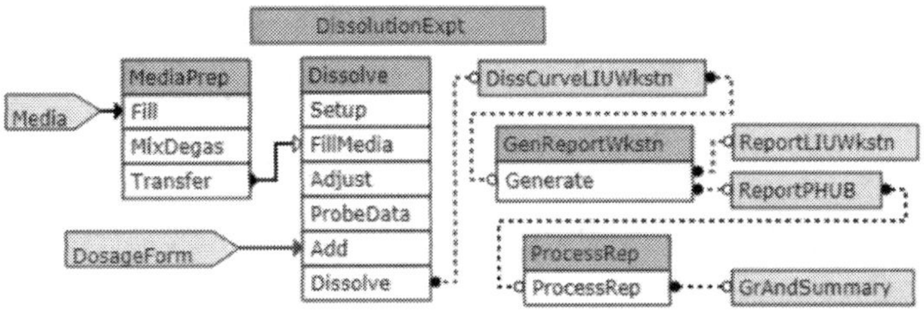

Content uniformity

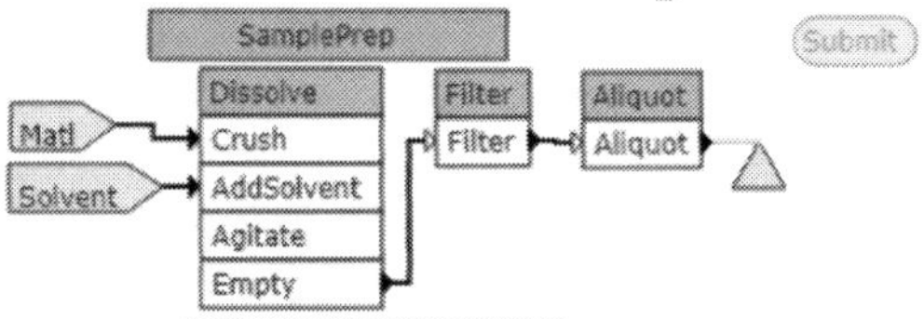

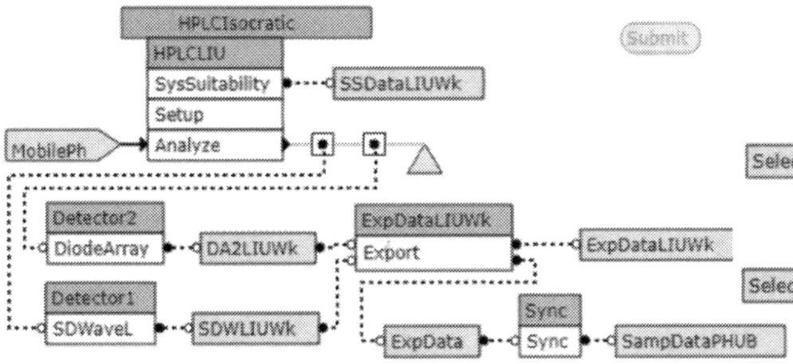

Physical Properties

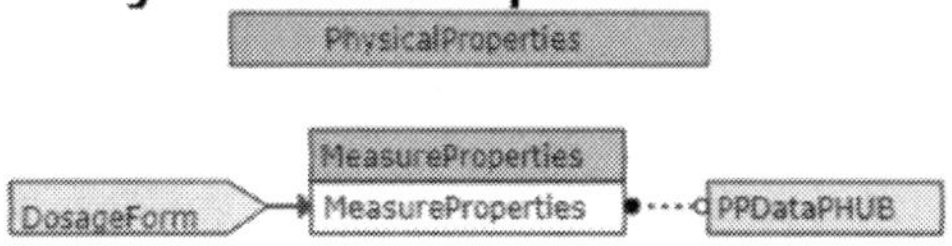

Bayesian Analysis

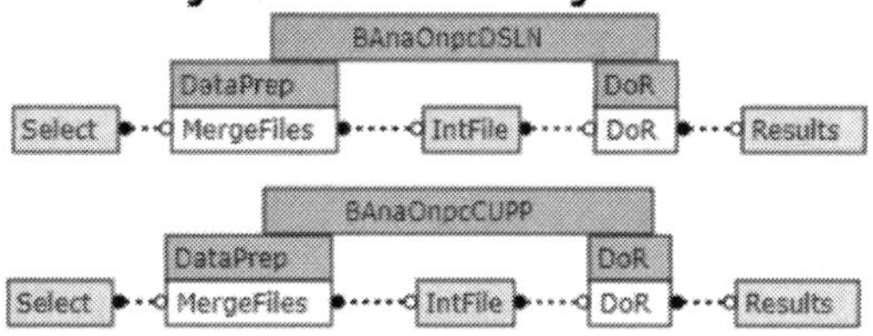

Bayesian Analysis

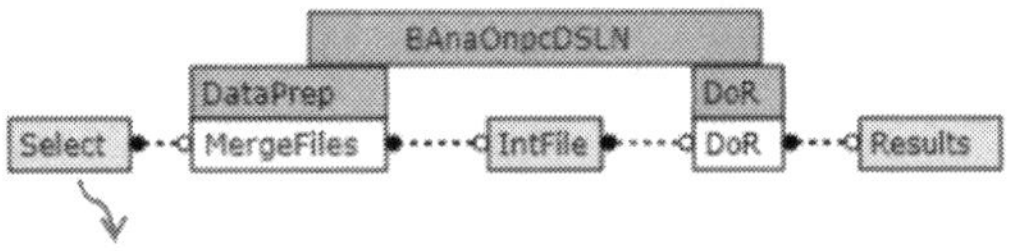

IntFile – Downloaded to PC

Results – Uploaded to repository

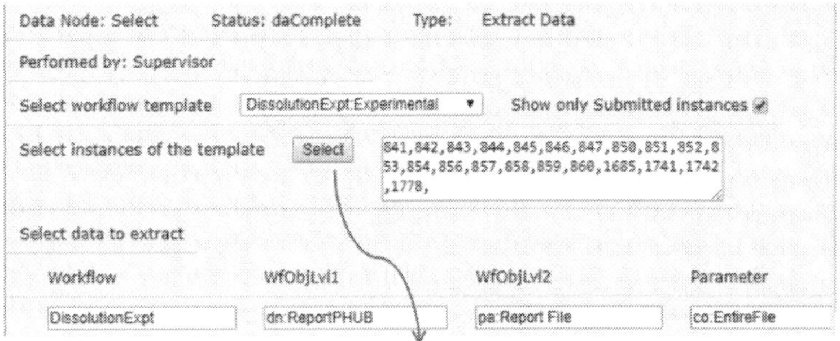

DoR – Only records the R package used. Execution done on PC

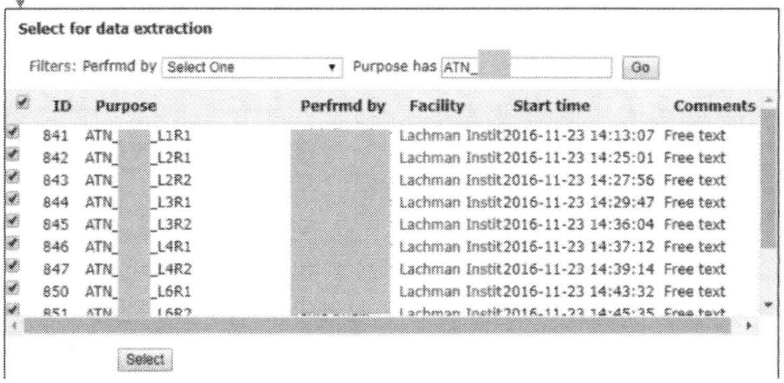

Benefits of Using KProMS

- Workflows tailored to the underlying procedure
- All steps and parameters explicitly defined
- Workflow developed based on input from all collaborators
- Improved quality of data
- Data available on the web as soon as it is submitted
- Facilitates discovery
- Uploaded data is protected

Summary

- KProMS framework provides complete context to the underlying data
- KProMS satisfies the requirement of Open Data archival system

Questions?

495398 Evaluation of Recovery Pathways for Improved Bioavailability of Clofazimine Nanoparticles

Abstract:

Clofazimine is an anti-mycobacterial agent used as part of a multidrug treatment for leprosy for over four decades. Recently clofazimine has shown promising activity against multidrug resistant tuberculosis. Due to its extremely poor aqueous solubility and high lipophilicity, the formulation development has focused on the solubility enhancement. Nanoparticles enhance bioavailability by two mechanisms. First, relative to larger drug crystals, nanoparticles have higher surface-to-volume ratios, which enhances dissolution rates. Second, rapid nanoparticle precipitation can trap drug molecules in higher energy, amorphous states, which have higher solubility. Here we use Flash Nanoprecipication (FNP) to form clofazimine-loaded nanoparticles with hypromellose acetate succinate (HPMC-AS) as polymer stabilizers, and investigate how to recover the clofazimine-loaded nanoparticles from the aqueous dispersions. Dry nanoparticle powders that can be readily reconstituted would enable delivery of drug dispersions to children and infants. Three approaches were used to produce dry powders: direct salt precipitation, spray drying and lyophilization. The yield efficiency, particle morphology, drug loading, dissolution rate and the amorphous/crystalline state of clofazimine in the dry powders are determined and compared among the processes. Based on these results, we evaluate the advantages and disadvantages of each process regarding the maintenance of the bioavailability after drying.

497815 Assessment of the Impact of Interfacial and Shear Stress on Biologics Drug Product Via Mini-Piloting Tools

Abstract:

Assessment of the Impact of Interfacial and Shear Stress on Biologics Drug Product via Mini-piloting Tools

During biologics development, processes such as membrane filtration, filling, and mixing are critical in the drug product manufacturing. However, these processes, as well as turbulent flow conditions, and multiple passes through pumps, valves and â€œpinchâ€ points; impose various stresses on the protein. These stresses can lead to protein instability. Shear, cavitation, and interfacial stress are often attributed as the main causes of protein unfolding and aggregation.

Proper risk analysis needs to be in place to understand the susceptibility of the protein to unfold and aggregate in presence of interfacial and shear stress. There are certain techniques, such as agitation/shaking studies that have been traditionally used to understand the impact of these stresses on the protein physical stability. However, the stresses applied in these systems are convoluted, making it difficult to define the control strategy, (i.e. adjustment in process parameters to reduce foaming/bubble formation, etc.). In this study, we have developed two mini-piloting tools that allow for the isolation of interfacial and shear stress respectively. The interfacial stress is applied to the sample by generating uniform bubbles when recirculating air from the headspace of the vial in a closed loop. The bubbles generated are passed through the solution for finite periods of time. For the shear stress, an estimation of the shear rates present during processing is made by fundamental transport relations and CFD modeling. The small scale shear stress tool simulates the shear experienced by the protein during processing, using a small-scale high pressure pump connected to a fine stainless steel tubing.

These systems help simulate the normal operating ranges as well as proven acceptable ranges for unit operation such as TFF, mixing and filling. Different protein formulations were subjected to controlled interfacial and shear stresses. The quality attributes evaluated after the stress were focused on size variant (particulate matter and HMW species), which is most likely the mechanism associated with interfacial and shear stress. It was possible to identify the susceptibility of the protein to aggregate depending of the stress that was applied. In addition, these tools contributed in formulation screening by evaluating which formulation would better mitigate the impact of the imposed stress.

During development, it is important to identify the protein formulations that have a tendency to aggregate in order to implement appropriate risk mitigation strategies. And also, to have scale down models to validate the design space for the process parameters of key unit operations.

499509 Suitable Raw Material Attributes for Continuous Manufacturing of a Drug Product Via Direct Compression

Abstract:

Continuous manufacturing for small molecule drug products is a rapidly maturing technology for innovator pharmaceutical companies and contract manufacturing organizations, supported by intensive study at academic research institutions. Of particular interest are the raw material properties which predict behavior in the novel unit operations of feeding and continuous blending, as well as their impact on residence time distributions. This study presents a thoughtful approach to the selection and justification of raw material property measurements based on process and product risks, predictive power, laboratory resource utilization, accuracy, robustness, and long-term sustainability. A suitable risk-based raw material monitoring plan for a continuous direct compression product is proposed and discussed.

501486 Wettability of Pharmaceutical Powders of Different Particle Size By Droplet Penetration Technique

Abstract:

We have recently developed a droplet penetration method to characterize the wettability of pharmaceutical powders. This method relies on two key observations during the penetration of a drop deposited on a powder bed: 1) the capillary pressure inside the porous powder is the dominant driving force for the imbibition process and 2) the contact area between the penetrating drop and the powder bed remains constant during most of the process. Under these conditions, we have shown that the penetration process in non-dimensional variables is universal. Hence, using a *reference* liquid it is possible to decouple the properties of the powder bed from the assessment of the contact angle of the *test* liquid under consideration. In addition, measuring the penetration times is sufficient to measure the contact angle, without having to solve the penetration problem.

In this work, we use the droplet penetration method to investigate the effect of particle size distribution on the wetting behavior of pharmaceutical powders. We consider two powder materials, an excipient and an active pharmaceutical ingredient (API). The excipient is lactose monohydrate, and was sieved to obtain particles in the following size ranges: 38-45, 45-53, 53-63, 63-75, 75-90, 90-106 mm. The API used is caffeine anhydrous with particle size distribution characterized by d50 = 15, 19 and 30 mm, obtained by jet milling. In all cases, the test liquid is deionized water and the reference liquid used is silicone oil. Lactose powders show a relatively constant contact angle, except for the largest particles. In the case of caffeine, we observe that larger particles yield smaller contact angles. The observed differences could be attributed to differences between milling and sieving processes. Other possible mechanisms leading to the observed dependence of the contact angle on particle size will be discussed.

492184 Investigation of Microsphere Strength for Spray Drying Applications By Means of Acoustic Levitation

Abstract:

Investigation of Microsphere Strength for Spray Drying Applications by Means of Acoustic Levitation

M. Kreimer[1], I. Aigner[1], S. Sacher[1], M. Krumme[2], T. Mannschott[2], P. van der Wel[3], A. Kaptein[3], H. Schröttner[4,5], G. Brenn[6], J.G. Khinast[1,7]

[1] Research Center Pharmaceutical Engineering (RCPE) GmbH, Inffeldgasse 13, 8010 Graz, Austria

[2] Novartis Pharma AG, Novartis Campus, 4056 Basel, Switzerland

[3] Hosokawa Micron B.V., Gildenstraat 26, 7005 BL Doetinchem, Netherlands

[4] Graz University of Technology, Institute for Electron Microscopy and Nanoanalysis, Steyrergasse 17, 8010 Graz, Austria

[5] Austrian Centre for Electron Microscopy and Nanoanalysis (FELMI-ZFE), Steyrergasse 17, 8010 Graz, Austria

[6] Graz University of Technology, Institute of Fluid Mechanics and Heat Transfer, Inffeldgasse 25 F, 8010 Graz, Austria

[7] Graz University of Technology, Institute for Process and Particle Engineering, Inffeldgasse 13, 8010 Graz, Austria

Growing demands on flexibility and agility in the production of pharmaceutical compounds have led to a paradigm shift, away from batch processing towards advanced continuous manufacturing. Continuous production of solid oral dosage forms starts with API synthesis, continuing with crystallization, purification, filtration, drying, blending, granulation and typically ends via tableting/capsule filling.

While tableting is a continuous process, several other unit operations currently exist mainly in batch mode. One example is the drying of crystals. However, spray drying is a commonly used technology to transform a fluid feed stream into dry powders. An advantage of spray drying is that it can be applied even to heat-sensitive products, since drying occurs close to the wet bulb temperature. Moreover, solutions and suspensions can be dried by evaporation of the liquid phase in a gaseous drying medium.

In the case of spray drying of suspensions, the objective is not to change material properties, such as particle size distribution. However, after crystallization, washing and filtration, different amounts of dissolved solids are present in the suspension slurry. During spray drying of the slurry, the dissolved material will solidify and form solid bridges between the suspension particles. Clearly, the amount of the dissolved material has a crucial impact on the final particle properties. The more material is dissolved in the slurry, the larger and stronger the bridges between primary particles will be. This has significant impact on the final particle properties and the associated processability.

This work highlights the influence of suspension composition on the material properties after drying. Drying of a model compound in a binary-liquid mixture was studied via acoustic levitation. This method enabled observations of the drying behavior of single droplets on a microscopic scale. Changes in the suspension composition had a strong impact on the mechanical properties and morphology of the final particles (microspheres). The morphology varied from very loose agglomerates to densely packed microspheres with crust formation. Hardness of single microspheres was measured via a compression strength test and correlated to the suspension composition.

494080 Nanostructural Interfaces of Additives on the API Surfaces Influencing Process Performance

Abstract:

The objective of this research is to detect and quantify the % extent of nanosmearing distribution of hydrophobic functional groups on multiple pharmaceutical particle surfaces during high shear mixing conditions of pharmaceutical formulations affecting the critical blend flow properties. Two sets of pharmaceutical blends were prepared by mixing 90% fast flo lactose as excipient, 9% micronized acetaminophen (Acetyl Para Amino Phenol - APAP) as Active Pharmaceutical Ingredient (API) and 1% hydrophobic Magnesium Stearate (MgSt) as lubricant at two shear rates of 100 rpm and 500 rpm in a v-blender. In the first stage, APAP particles from the two blend samples were studied under Raman Spectroscopy to detect and confirm the presence of hydrophobic functional groups of MgSt nanosmears and then to map the nanosmeared area distribution on the surface of multiple APAP particles. In the second stage, using Image-J histogram analysis, the Raman images of MgSt mapping on APAP particles were analyzed and the % nanosmeared area covered by the hydrophobic MgSt nanosmears on multiple APAP particles were quantified. For the first time in the field of particle technology, Raman Spectroscopy Mapping method was employed on the organic particulate matter to develop a novel methodology to quantify the presence and % extent of nanosmeared area distribution of hydrophobic MgSt additive on multiple particles. The results indicated that the applied process shear has a significant effect on the percentage area covered by MgSt nanosmears. The % nanosmeared area of MgSt on multiple APAP particles increased from 22% to 41% as the applied process shear during mixing increased from 100 rpm to 500 rpm. The increased nanosmeared area improved the powder blend flow properties. The nanosmears on APAP particles were found to be non-uniform and partially coated. A direct correlation between % nanosmearing area with applied process shear and powder cohesion was found.

501114 Rational Strategies for Physical Characterization of Amorphous Solid Dispersions during Drug Product Development and Manufacture

Abstract:

As amorphous solid dispersions become more widely exploited in pharmaceutical products to enhance active ingredient solubility, strategies to more efficiently confirm the absence of undesired crystallinity during manufacture and shelf life become necessary. In recent years, rapid improvement in the application of solid state characterization techniques, such as calorimetry, spectroscopy, XRPD and ssNMR, has permitted order-of-magnitude increases in method sensitivity and development of reliable quantitative limit tests to justify control strategy selection. Verification of these strategies by transfer of solid state techniques to traditional quality control laboratories remains a challenge. Companies seeking to avoid extensive use of solid state techniques in routine manufacturing release testing may instead choose to leverage such techniques in well-designed development experiments based upon fundamentals.

Using several recent cases of drug products (prepared with hot melt extrudate or spray dried intermediates), we outline efficient approaches for identification of credible phase transition risks and development of methods to explore the propensity for such risks to occur. We find methods can often achieve limits of detection of ~5% amorphous to crystalline conversion on an active ingredient basis (particularly when the drug loading exceeds 10wt%). Experience from both marketed and late stage development programs are used to illustrate a rational, risk-based strategy can be demonstrated during process validation. In contrast to extensive, multi-dimensional stability studies and measurements aiming to anticipate the infinite set of excursion scenarios that might be anticipated during the product lifecycle, we instead advocate a more fundamental exploration of physical stability boundaries. These are often associated with solid dispersion glass transition temperatures. We emphasize the use of well-designed product characterization studies to understand the failure modes of solid dispersion intermediates and drug products, and compare a more limited set of experiential measurements to predictions from nucleation and growth theories. Specialist solid state methodologies can then be confined mainly to development laboratories, yet manufacturing facilities are still provided with robust amorphous control strategies and the knowledge to react to manufacturing excursions comprehensively and efficiently.

494868 Investigation of Compression Induced Amorphization of Crystalline API

Abstract:

Active Pharmaceutical Ingredient (API) often exists in several solid forms (crystalline or amorphous). The different solid forms are known to influence the physical, chemical and mechanical properties of the API. Despite of significant API form screening effort, the selected and thoroughly studied API form may still undergo phase transformations during typical pharmaceutical processing. Among the various pharmaceutical process induced phase transforms, crystalline to amorphous phase transformation and its underlying mechanism has been rarely reported. This study will discuss the recent findings on compression induced crystalline to amorphous phase transformation. The effect of axial and shear stresses as well as excipient and formulation will be discussed.

496787 Steadycom: Modeling Microbial Communities Under Steady-State Growth

Abstract:

Genome-scale metabolic modeling has become widespread for analyzing microbial metabolism. Extending this established paradigm to more complex microbial communities is emerging as a promising way to unravel the interactions and biochemical repertoire of these omnipresent systems. While several modeling techniques have been developed for microbial communities, little emphasis has been placed on the need to impose a time-averaged constant growth rate across all members for a community to ensure co-existence and stability. In the absence of this constraint, the faster growing organism will ultimately displace all other microbes in the community. This is particularly important for predicting steady-state microbiota composition as it imposes significant restrictions on the allowable community membership, composition and phenotypes. In this study, we introduce the SteadyCom optimization framework for predicting metabolic flux distributions consistent with the steady-state requirement. SteadyCom can be rapidly converged by iteratively solving linear programming (LP) problem whose number is independent of the number of organisms. A significant advantage of SteadyCom is compatibility with flux variability analysis. SteadyCom is first demonstrated for a community of four E. coli double auxotrophic mutants and is then applied to a gut microbiota model consisting of nine species, with representatives from the phyla Bacteroidetes, Firmicutes, Actinobacteria and Proteobacteria. In contrast to the direct use of FBA, SteadyCom is able to predict the change in species abundance in response to changes in diets with minimal additional imposed constraints on the model. By randomizing the uptake rates of microbes, an abundance profile with a good agreement to experimental gut microbiota is inferred. SteadyCom provides an important step towards the cross-cutting task of predicting the composition of a microbial community in a given environment.

501587 Intracellular Metabolic Circuits Shape Intercellular Interactions in Multi-Species Microbial Games

Abstract:

Metabolite exchanges in microbial communities give rise to ecological interaction networks that govern ecosystem diversity and stability. These exchanges are mediated by complex intracellular pathways thus raising the question of whether ecological interactions are inferable from genomes. We address this question by integrating genome-scale models of metabolism, to compute the fitness of interacting microbes, with evolutionary game theory, which uses these fitness values to infer evolutionarily stable interactions in multi-species microbial â€œgamesâ€. After validating our approach using data on sucrose hydrolysis by *S. cerevisiae*, we performed over 80,000 *in silico* experiments to evaluate the rise of unidirectional and cross-feeding metabolic dependencies in populations of *Escherichia coli* secreting 189 amino acid pairs. We found that, despite the diversity of exchanged amino acids, most pairs conform to general patterns of inter-species interactions. However, several amino acid pairs deviate from these patterns due to pleiotropy and epistasis in metabolic pathways. We also identified evolutionary paths that could lead to obligate cross-feeding by performing targeted *in silico* invasion experiments,. Overall, our study provides mechanistic insights into the rise of evolutionarily stable inter-species dependencies, with important implications for biomedicine and microbiome engineering.

494596 Elucidating Parameters in Dynamic Microbial Community Models Using Nonlinear Programming

Abstract:

The dynamic behavior of microbial communities can be described by Lotka-Volterra models, which involve highly nonlinear sets of differential and algebraic equations. Estimating parameters for such types of models is computationally challenging. In this work, we demonstrate that state-of-the-art nonlinear programming tools can tackle this class of parameter estimation models [1]. In our framework, we use JuMP (a mathematical programming modeling language) to implement a discrete-time version of the microbial community model, which enables us to compute exact derivative information and to exploit algebraic sparsity. We solve the resulting nonlinear programming problem by using interior-point solvers. We combine these computational capabilities with real experimental data to estimate parameters for a microbial community model. The model under study involves 12 different species and we use over 100 experimental data sets to perform the estimation. The resulting estimation problems contain over 1,000,000 variables and constraints and can be solved in around 10 minutes by using IPOPT. We show that these times can be reduced further by using the parallel solver PIPS-NLP. These favorable solution times enable us to consider additional analysis tasks such as identification of the level sets of the parameter posterior distribution [2]. The proposed computational framework can also be applied more broadly to high-dimensional and nonlinear biological systems models including signaling networks, metabolic pathways, and ecological systems.

[1] Zavala, V. M., Laird, C. D., & Biegler, L. T. (2008). Interior-point decomposition approaches for parallel solution of large-scale nonlinear parameter estimation problems. Chemical Engineering Science, 63(19), 4834â€“4845.Â

[2] Bardsley, J. M., Solonen, A., Haario, H., & Laine, M. (2014). Randomize-Then-Optimize: a Method for Sampling From Posterior Distributions in Nonlinear Inverse Problems. Siam Journal on Scientific Computing, 36(4), A1895â€“A1910.

491980 Constraint-Based Community Modeling Reveals Condition-Dependent Alternate Interactions

Abstract:

Phototrophs and heterotrophs were combined as a novel sustainable symbiotic platform for the production of biofuel precursors. Synthetic communities of cyanobacterium-bacterium and fungus-algae pairs were studied through genome-scale metabolic modeling. Six co-culture metabolic models (CMM) were constructed and used to predict growth phenotypes. The CMM enabled the systematic characterization of communities and the elucidation of the nature of specific interactions between individual members (e.g. mutualism, competition, etc).

The CMM of *Chlorella vulgaris* and *Saccharomyces cerevisiae* was reconstructed using the toolbox Constraint-Based Reconstruction and Analysis of Communities, which was developed in the framework of this project. Based on experimental observations, we corroborated our predictions about growth rates, culture medium, and genetic modifications for this co-culture pair. While yeast uptakes O_2 and provides CO_2 when nitrate is the nitrogen source in medium, both members split the available glucose and grow as mutualists. By using the reconstructed CMM, we predicted and validate that when NH_4 was added to the culture medium yeast dominated the co-culture and outcompeted Chlorella, changing the interaction type. The growth of this co-culture is mediated by exchange of ten metabolites; validation by targeted metabolomics is under way. We also evaluated the effect of 1,748 single gene deletions for monocultures and the co-culture. The experimental growth rates match with predictions, e.g. yeastâ€™s mutant alters both Chlorella and yeast growth rates. In some cases, deleting a gene can offer a growth advantage in co-culture, including an improved growth phenotype for the participating species. These examples demonstrate how community metabolic models can accurately predict the behavior of heterogeneous co-culture pairs thus improving production phenotype for bioproduction.

487822 Engineered Biofilms for Isobutanol Production: *In silico* Modeling of a Two-Species Bacterial Community

Abstract:

Most efforts to engineer microbial consortia for biotechnological applications have focused on well mixed systems. For example, well-mixed microbial communities offer the potential to achieve consolidated bioprocessing (CBP) of soluble lignocellulosic materials where different microbes are responsible for biomass degradation, conversion of the released sugars to desired products and consumption of toxic metabolic byproducts. In nature, the majority of microbes grow as biofilms in diverse communities to compete for and efficiently utilize available nutrients. While considerable progress towards engineering synthetic communities has been realized, existing approaches are largely based on well mixed planktonic cultures which fundamentally lack the three-dimensional structure of naturally occurring microbial biofilms. We believe that engineered biofilm communities hold great promise for diverse applications including biofuels production.

Our long-term goal is to engineer a biofilm community for conversion of cellulose to the second generation liquid fuel isobutanol. As a first step towards this goal, we have developed a biofilm metabolic model of a two-species system comprised of the facultative sugar utilizer *Escherichia coli* engineered for aerobic isobutanol synthesis and the anaerobic organic acid oxidizer/iron(III) reducer *Geobacter metallireducens*. Glucose and oxygen are supplied at the top of the biofilm to support *E. coli* growth and isobutanol synthesis. Organic acids and iron(III) are supplied at the bottom of the biofilm to support *Geobacter metallireducens* growth and mimic organic acid (acetate, ethanol) synthesis by unmodeled cellulose degraders. Our model predicts that the two species will spatially organize along gradients of the electron donors and acceptors. *E. coli* located in the aerobic region synthesizes isobutanol and acetate, which diffuses into anaerobic region where the acetate is consumed by *G. metallireducens*. We predict that the *E. coli-G. metallireducens* biofilm community will produce higher levels of isobutanol than *E. coli* alone due to *G. metallireducens* detoxification of environment. This result holds even when *G. metallireducens* only consumes acetate produced by *E. coli*, suggesting that the two species biofilm community may serve as a modular platform for development of more complex biofuel producing systems.

490927 Microbiome-Virome Interactions in Bovine Rumen: The Role of Auxiliary Metabolic Genes in Relaxing Metabolic Bottlenecks

Abstract:

The complex microbial ecosystem of the rumen has a significant impact on the ruminant hostâ€™s health and productivity through the breakdown and fermentation of feed into volatile fatty acids. Although advancements in high-throughput sequencing are providing access to the vast diversity and functions of this complex microbial ecosystem, our understanding of factors shaping rumen microbial communities is in its infancy. In addition to the complex syntrophic, mutualistic, and sometimes competitive interactions within the rumen microbiome, viruses have been shown to impact microbial populations through a myriad of processes, including cell lysis and reprogramming of host metabolism via Auxiliary Metabolic genes (AMGs). However, little is known about the processes shaping the distribution of rumen viruses or modulation of microbe-driven processes in the rumen. It is well established that bovine dietary variations significantly alter the ruminal microbiome. However, a comprehensive and systematic computational study to decipher the effect of diet utilization and viral AMGs on the microbiome is still missing. With a gradual increase in computational capability and abundance of in silico genome-scale metabolic reconstruction tools, metabolic networks combined with constraint-based modeling provides opportunities to identify microbe-microbe and phage-microbe interactions. In this work, we investigated how rumen microbial and viral community structure and function respond in varied dietary treatments and investigated the distribution and ecological drivers of rumen viruses alongside their bacterial counterparts.

Utilizing four diets with varying degrees of total digestible nutrients (TDN), we investigated the effect of dietary perturbations of the rumen ecosystem by metagenome sequencing of the total microbiome and the viral fraction. For building a simplified rumen community metabolic model, we selected Prevotella ruminicola, Methanobrevibacter gottschalkii, and Ruminococcus flavefaciens as representative organisms that represent starch and protein digesters, methanogens and fiber digesters within the rumen, respectively. We have completed draft genome-scale models for P. ruminicola (467 genes, 1129 metabolites, 1043 reactions), M. gottschalkii (318 genes, 977 metabolites, 840 reactions), and R. flavefaciens (461 genes, 1100 metabolites, 981 reactions) from ModelSEED and are in the process of automatic and manual curation of the models. A microbiome community model will be developed by integrating the three models via interspecies constraints, using existing multi-level and multi-objective modeling frameworks[1-3]. The community model will be validated through experimental observations on metabolite secretion profiles and community compositions as a function of diet and host-specific variations. The identified functions of viral AMGs will be incorporated into the model as regulatory information. The metabolic hubs and bottlenecks in the community will be identified and the metabolite pools for important energy currencies in the community (i.e. ATP, NAD(P) etc.) will be predicted.

Our experimental results show rumen viruses have affected the structure of the previously identified core rumen microbiota and also the microbial metabolism through a vast array of AMGs. While viral communities displayed large shifts following dietary perturbations, 38 viral groups were shared across all host-diet combinations, suggesting the

presence of a core rumen virome largely comprised of novel viruses. Analysis of viral AMGs shows glycosidic hydrolase activity to augment the breakdown of complex carbohydrates, redirecting carbon flux to pentose phosphate pathway, and boost viral replication. We showed that viral community dynamics is highly correlated with microbial density and dietary factors instead of host-specificity. This model serves to answer key ecological questions of ruminant nutrition through diet-virome-microbiome interactions, discover unidentified metabolite transactions, and promises to develop novel strategies for methane mitigation and increasing nutritional efficiency of domesticated bovine species.

References:
1. Zhuang K, Ma E, Lovley DR, Mahadevan R. The design of long-term effective uranium bioremediation strategy using a community metabolic model. Biotechnology and bioengineering. 2012;109(10):2475-2483.
2. Zomorrodi AR, Maranas CD. OptCom: a multi-level optimization framework for the metabolic modeling and analysis of microbial communities. PLoS computational biology. Feb 2012;8(2):e1002363.
3. Zomorrodi AR, Islam MM, Maranas CD. d-OptCom: Dynamic Multi-level and Multi-objective Metabolic Modeling of Microbial Communities. ACS synthetic biology. Apr 18 2014;3(4):247-257.

505501 Development and Application of Integrated Pipeline for the Modeling and Analysis of Microbial Communities in the Doe Systems Biology Knowledgebase

Abstract:

Understanding the molecular interactions and ecological principles that guide the behavior of microbial communities is critical to numerous applications in medicine, in the environment, and in industry. Toward this end, we have developed a pipeline within the DOE Systems Biology Knowledgebase (KBase) that supports the development and analysis of community metabolic models starting from metagenomic data. This pipeline integrates 11 analysis steps including: metagenome assembly; binning assembled contigs by species; assessment of assembly and genome quality and completeness; genome annotation; species metabolic model reconstruction; merging of models into a community model; mapping RNA-seq reads to individual species; model gapfilling; flux balance analysis; comparison of flux and expression profiles; and simulation of growth phenotypes. All eleven steps are available within a user-friendly point-and-click interface at http://narrative.kbase.us.

We applied this pipeline to understand the ecology and trophic interactions occurring within three microbiome-based datasets: (i) a lab-constructed community comprised of the cyanobacterium *Thermosynechococcus elongatus* supporting the heterotrophic bacterium, *Meiothermus ruber*; (ii) ten naturally occurring highly coupled communities comprised of 2-3 species each; and (iii) a larger 18-species natural community comprising an epsomitic phototrophic microbial mat in Hot Lake, Washington. We demonstrate how the tools in KBase can integrate community Biolog profiles and transcriptomic profiles in our community models of these systems. Further, we show how our platform can predict the trophic interactions that occur among species in each of these systems. Overall, we find significant trophic interactions in all communities, involving numerous essential metabolites, including amino-acids, vitamins, and cofactors. We also find numerous cases of metabolic hand-off, particularly in our lab constructed and Hot Lake communities, which both include an autotrophic organism that performs carbon fixation functions for all other species in the community. We see how species form ecological networks of interactions where each species derives chemical energy from the nutrients in the environment, while stability is preserved through essential trophic interactions. We also see how changes in the nutrient content of the environment can disrupt these webs of trophic interactions.

495113 A Versatile Synthetic Biology Platform for High-Throughput Structure and Activity Screening of Ribosomally-Synthesized and Post-Translational Modified Peptides (RiPPs)

Abstract:

Ribosomally synthesized and post-translationally modified peptides (RiPPs) form a major class of natural products that are ubiquitous in the currently sequenced genomes, and exhibit antimicrobial, anticancer, anti-protease, and antiviral activities. As the mature product is synthesized from a ribosomal peptide, combinatorial variants can be readily generated by mutagenesis of the precursor gene, providing a rich source for studying structure-activity relationship and engineering improved medicinal properties. However, most RiPP precursor peptides consist of an N-terminal leader region, which is important for recognition by biosynthetic machinery, export, and self-resistance. Leader removal in the final step of biosynthesis is critical to obtain both bioactivity and high-resolution mass spectra, but is difficult to achieve in heterologous hosts such as *Escherichia coli*, due to the challenges in functional reconstitution of proteases or transporters responsible for leader cleavage.

In this work, we sought to overcome this limitation via protease compartmentalization coupled with inducible self-lysis, which enables *in situ* production of mature, active RiPP variants from *E. coli* colonies. Briefly, the protease is directed to the periplasm using a signal peptide so that it is physically separated from the precursor peptide and biosynthetic enzymes that are expressed in the cytosol. Such compartmentalization is necessary as leader attachment is required for installation of post-translational modifications, after which expression of a phage protein is induced by a temperature cue to lyse *E. coli*. Self-lysis not only allows contact between the protease and the modified precursor peptide in cell lysate to initiate proteolytic leader cleavage, but also release intracellular mature products for direct mass spectrometry (MS) analysis and antimicrobial assays to be performed directly using *E. coli* colonies.

As a proof-of-concept, we screened the analog library of a two-component lanthipeptide, haloduracin, which is natively produced in *Bacillus halodurans*. We refactored the haloduracin pathway to incorporate the above-mentioned synthetic biology approaches. With optimized genetic design and process parameters, production of mature, active haloduracin from *E. coli* colonies was confirmed via both MS analysis and antimicrobial assays. We then created and screened the site-saturation mutagenesis libraries targeting all non-cyclized residues in one of the precursors. Relative antibiotic activities of analogs were estimated using the sizes of inhibition zones in an agar overlay assay towards an indicator *Lactococcus lactis* strain, and amino acid substitutions were rapidly assigned using predicted mass shifts in mass spectra. Within one week, the structure-activity relationship at 16 residues was characterized by analyzing >3,000 *E. coli* colonies, achieving >95% probability in full coverage of 20 canonical amino acids at each residue. Notably, a conserved Pro16 residue common to all related peptides was found to be essential for activity, whereas a non-conserved Pro4 exhibited much higher tolerance to mutations. We also obtained mature, active nisin from *E. coli* colonies, suggesting our method could be generally applicable. We are in the process of purifying three analogs exhibiting substantially larger inhibition zones relative to the wild-type haloduracin during screening to analyze their specific activities.

Separately, we developed a high-throughput MS method for structure-based screening of *E. coli* colonies producing RiPP variants. Instead of using liquid handling robots or manual processes to pattern colonies into arrays on MS target plates, we utilized machine vision for automatic MS analysis at randomly distributed colonies prepared using standard microbiological methods. Computational tools using multivariate analyses were developed to enumerate structural diversity and cluster colonies with isobaric residue substitutions. Moreover, as the MS screening only consumes ~30% of the analytes, follow-up structural elucidation using high-resolution MS and tandem MS on colonies of interest can be performed on the same sample subsequent to screening. In addition to haloduracin, we screened a substrate library of plantazolicin, which belongs to another RiPP class called linear azol(in)e-containing peptides. We identified both known and new analogs, and all MS assignment of amino acid mutations using predicted mass shifts was consistent with DNA sequencing results.

Together, we successfully developed a synthetic biology platform for structure- and activity-based screening of RiPPs from *E. coli* colonies. Future improvements include ultra-high-throughput screening of combinatorial mutagenesis targeting more than one residue, differentiation of isobaric ions with different modification states (L- vs D-amino acids), and integration with more types of bioactivity assays such as anticancer activities.

494588 Hydroclassified Combinatorial Saturation Mutagenesis for Evolving Stereoselectivity and Cosubstrate Specificity of a Diaryl Alcohol Dehydrogenase

Abstract:

Optically active diaryl alcohols is an important class of chemical chiral compounds which could be applied in the synthesis of pharmaceuticals, agrochemicals and natural products, such as antihistamine, hydragogue, and antitussive drugs. Among them, chiral (4-chlorophenyl)-(pyridine-2-yl)-methanol (CPMA) is a vital intermediate for anti-allergic drug Betahistine and could be synthesized by asymmetric reduction of its corresponding (4-chlorophenyl)-(pyridine-2-yl)-ketone (CPMK). Due to the strong steric hindrance of the diaryl groups, very few alcohol dehydrogenases have been reported with desirable activity and enantioselectivity. Previously, a diaryl alcohol dehydrogenase from *Kluyveromyces polyspora* (*Kp*ADH) was identified by genome hunting, however with a moderate enantioselectivity of 81%. Inspired by the smallest amino acid alphabets strategy used in the saturation mutagenesis of an epoxide hydrolase[1], we proposed a hydroclassified combinatorial saturation mutagenesis (HCSM) strategy for construction of smart libraries of *Kp*ADH with reduced library size and hydroclassified amino acids. Additionally, a highly sensitive carbonyl group-dependent colorimetric method was developed using 2,4-dinitrophenylhydrazine (DNPH) for high-throughput screening of smart libraries of *Kp*ADH toward various ketones including diaromatic ketones, aliphatic ketones/diketones, and aromatic ketones[2]. Hot spots in the substrate-binding pocket were predicted based on molecular docking analysis. After site-directed and combinatorial mutagenesis, and the enantioselectivity of *Kp*ADH was enhanced from 81% to 99.7%*ee* (*R*). Additionally, the stereoselectivity of *Kp*ADH was inverted to 97.8%*ee* (*S*) by reshaping substrate binding pocket based on hydrogen bonds networks. Consequently, our study provides carbonyl reductases with promising applications in the preparation of optically active diaryl alcohols.

[1] Sun, Z.; Lonsdale, R; Kong X. D.; Xu J. H.; Zhou J.; Reetz M. T. Angew. Chem. Int. Ed. **2015**, *12*, 12410-12415.

[2] Zhou, J.; Xu, G.; Han R.; Dong, J.; Zhang W.; Zhang, R.; Ni, Y. Catal. Sci. Technol. **2016**, *6*, 6320-6327.

493400 Facilitating Electron Channeling in a Self-Assembling Metabolon Containing Two Dehydrogenases and a Nife-Hydrogenase

Abstract:

To accelerate the electron transport rate for cascade enzymatic production of hydrogen gas, we constructed a self-assembling metabolon (complexed enzyme) containing a three cohesin-containing mini-scaffoldin, two dockerin-containing dehydrogenases, one His-tagged NiFe-hydrogenase, and one bifunctional peptide. This bifunctional peptide had one terminus containing a modified dockerin binding the mini-scaffoldin and the other after chemical modification binding a His-tagged hydrogenase. This new strategy expanded the cohesin-dockerin strategy by using His-tagged proteins directly. The metabolon exhibited an initial reaction rate 12.3 times that of the enzyme mixture at the same enzyme loading based on reduction of oxidized benzyl viologen. The hydrogen generation rate catalyzed by the metabolon was approximately 5.6 times of that of the enzyme mixture. Such reaction rate enhancements suggested strong electron channelling among the adjacent redox enzymes of the metabolon.

500620 Beyond Iron: Iridium-Containing P450 for Selective Cyclopropanations of Alkenes

Abstract:

Enzymes catalyze organic transformations with exquisite levels of selectivity, including chemoselectivity, stereoselectivity, and substrate selectivity, but the types of reactions catalyzed by enzymes are more limited than those of chemical catalysts. Thus, the convergence of chemical catalysis and biocatalysis can enable enzymatic systems to catalyze abiological reactions with high selectivity. Recently, we disclosed artificial enzymes constructed from the apo form of heme proteins and iridium porphyrins that catalyze the insertion of carbenes into a Câ€"H bond. We postulated that the same type of Ir(Me)-PIX enzymes could catalyze the cyclopropanation of a broad range of alkenes with control of multiple modes of selectivity. Here, we report the evolution of artificial enzymes that are highly active and highly stereoselective for the addition of carbenes to a wide range of alkenes. These enzymes catalyze the cyclopropanation of terminal and internal, activated and unactivated, electron-rich and electron-deficient, conjugated and nonconjugated alkenes. In particular, Ir(Me)-PIX enzymes derived from CYP119 catalyze highly enantio- and diastereoselective cyclopropanations of styrene with Â±98% ee, >70:1 dr, >75% yield, and â^¼10,000 turnovers (TON), as well as 1,2-disubstituted styrenes with up to 99% ee, 35:1 dr, and 54% yield. Moreover, Ir(Me)-PIX enzymes catalyze cyclopropanation of internal, unactivated alkenes with up to 99% stereoselectivity, 76% yield, and 1300 TON. They also catalyze cyclopropanation of natural products with diastereoselectivities that are complementary to those attained with standard transition metal catalysts. Finally, Ir(Me)-PIX P450 variants react with substrate selectivity that is reminiscent of natural enzymes; they react preferentially with less reactive internal alkenes in the presence of more reactive terminal alkenes. Together, the studies reveal the suitability of Ir-containing P450s to combine the broad reactivity and substrate scope of transition metal catalysts with the exquisite selectivity of enzymes, generating catalysts that enable reactions to occur with levels and modes of activity and selectivity previously unattainable with natural enzymes or transition metal complexes alone.

499833 Bioprospecting to Discover Keto-Aryl Reductases with Enhanced Specificity Towards Longer-Chain, Aliphatic Substrates

Abstract:

Medium-chain length aliphatic compounds (C4-C10) have promising uses as polymeric precursors, biofuels, flavors and fragrances, and pharmaceutical precursors. The reverse beta-oxidation (RBO) cycle, where aliphatic CoA molecules are condensed and then reduced, is commonly used to make these chemicals. Cycles of condensation result in longer molecules, and precursor feeds can be varied to alter terminal R-groups (i.e. adding a branched methyl group). Implementing this method has been successful in making numerous molecules; however, products suffer from low titers and are made as heterogeneous mixtures due to the promiscuity of several pathway enzymes, leading to costly downstream separations. The reduction of keto-aryl-CoA is thought to be an excellent step to add specificity to the RBO pathway due to the equilibrium for the condensation reaction being heavily favored in the reverse direction, which will permit only selective reductases to form the desired reduced product. We generated a sequence similarity network of 5,000 putative reductases based on the often-used keto-aryl reductase PhaB from *Cupriavidus necator*. From this network, we selected 15 reductases located in various protein clusters and at different phylogenetic distances from PhaB to screen *in vivo* for higher specificity for 3-oxo-pentanoyl-CoA (C5) over 3-oxo-butanoyl-CoA (C4) substrates. From this initial panning experiment, we identified two reductases that showed at least a two-fold increase in specificity for the C5 to C4 compounds compared to PhaB. Using this knowledge, we decided to test the enzymes that cluster near those with high specificity as well as those that cluster near PhaB to see if these enzymes have similar specificities or if a more catalytically active variant could be found. Enzymes near PhaB in the similarity network all showed lower specificity towards C5 to C4 products than the previously identified high-specificity reductases or PhaB. Searching the clusters of enzymes containing the high-specificity reductases identified ones that had similar or higher specificity to the C5 compounds. *In vitro* activity of these enzymes is currently being investigated as are potential mechanisms for enhanced specificity. We hope this method of searching through enzyme sequence can be applied to other enzyme classes to enable identification of clusters with desired activities.

493032 Microbial Synthesis of Novel Terpenoid-Based Inhibitors

Abstract:

Terpenoids represent an important, pharmaceutically relevant class of natural products; unfortunately, their low natural abundance and pronounced chemical complexity often hinder their development into drugs. In this talk, we introduce an approach for building terpenoid-based inhibitors, or "leads", that can be readily synthesized in microbial hosts. Starting with a simple diterpenoid pathway, we use directed evolution to build inhibitors of an enzyme implicated in diabetes. The resulting platform, which accommodates thermodynamic, enzymological, and metabolic considerations, addresses two problems often examined separately—the discovery of natural-product based leads and the synthesis of those leads—by supplying an approach to lead discovery in which microbial "synthesizability" is a search constraint.

503876 Extreme Makeover: Engineering the Thermostable Alcohol Dehydrogenase D (AdhD) Protein Scaffold for New Applications

Abstract:

The thermostable alcohol dehydrogenase D (AdhD) from *Pyrococcus furiosus* is an NAD(H)-dependent enzyme that oxidizes a range of sugars and alcohols but is most active with secondary alcohols. It is monomeric, and a member of the aldo-keto reductase (AKR) superfamily of proteins. We have set out to engineer just about every aspect of this enzyme to improve its performance and develop it for use in several new applications.

First, we have added alpha-helical appendages to both termini of the enzyme to create a bifunctional chimeric protein that can self-assemble into hydrogels. The impact of these modifications on the enzymatic activity is negligible. And the resultant hydrogel is robust and active.[1]

Next we rationally altered the cofactor specificity of the enzyme via two site-directed mutations (K249G/H255R) in the cofactor-binding pocket. This significantly increased activity with NAD(H) and improved the activity with NADP(H) by an order of magnitude. A pre-steady state kinetic analysis was also performed and we observed a wide array of conformational behavior indicating that the previously established kinetic mechanism in the AKRs may need to be reevaluated.[2] We have recently expanded this work through the identification of an amino acid location distal to the cofactor binding pocket that can also be mutated to alter cofactor specificity by altering transition state binding energy as opposed to cofactor ground state binding energy.[3]

We demonstrated the utility of an engineered AdhD for use in enzymatic biofuel cells. The performance of systems employing nicotinamide cofactor-dependent enzymes can be limited by the diffusion of the cofactor between the enzyme and the electrode. The cofactor-specificity mutant (K249G/H255R) fortuitously exhibits activity with nicotinamide mononucleotide (NMN(H)) but with activity that is several orders of magnitude lower that what is observed with the native cofactor. When the enzyme was incorporated into an enzymatic biofuel cell, similar power densities were obtained, and the maximum current density for the NMN(H) electrode was higher than that of the NAD(H). Thus, although there was a significant decrease in kinetic activity of the enzyme, the improvement in the diffusion of the truncated cofactor led to improved performance of the enzymatic biofuel cell.[4]

We also modulated the substrate specificity of the enzyme by making chimeric enzymes between AdhD and the mesostable NADP(H)-dependent human aldose reductase (hAR). Active-site loops were swapped to transfer hAR activity to the thermostable AdhD scaffold, including the reversed cofactor-specificity.[5]

One of the loops that was altered did not ablate AdhD activity, and so we chose this site for the introduction of an RTX domain from the adenylate cyclase protein from *Bordatella pertussis*. This peptide is disordered and forms the beta roll secondary structure motif upon calcium addition. The insertion of the RTX domain into AdhD reduced overall activity. However, the conformational change of the RTX domain affected NAD^+-dependent activity more than $NADP^+$-dependent activity such that calcium could be used as a â€œrheostat switchâ€ to control cofactor selectivity in the enzyme.[6]

Finally, there is a need to develop sensors for explosive molecules such as TNT and RDX and antibody-based systems suffer from high cost, and poor stability. We have used ribosome display to screen randomized AdhD libraries over immobilized RDX. Mutants with affinity for RDX have been identified and characterized by isothermal titration calorimetry (ITC). Thus we have converted an enzyme into a thermostable antibody replacement.[7]

References

1. Wheeldon IR, Campbell E, & Banta S (2009) A chimeric fusion protein engineered with disparate functionalities-enzymatic activity and self-assembly. *J Mol Biol* 392(1):129-142.

2. Campbell E, Wheeldon IR, & Banta S (2010) Broadening the cofactor specificity of a thermostable alcohol dehydrogenase using rational protein design introduces novel kinetic transient behavior. *Biotechnol Bioeng* 107(5):763-774.

3. Solanki K, Abdallah W, & Banta S (2017) Engineering the cofactor specificity of an alcohol dehydrogenase via single mutations or insertions distal to the 2'-phosphate group of NADP(H). *Protein Eng Des Sel* 30(5):373-380.

4. Campbell E, Meredith M, Minteer SD, & Banta S (2012) Enzymatic biofuel cells utilizing a biomimetic cofactor. *Chemical communications* 48(13):1898-1900.

5. Campbell E, Chuang S, & Banta S (2013) Modular exchange of substrate-binding loops alters both substrate and cofactor specificity in a member of the aldo-keto reductase superfamily. *Protein engineering, design & selection : PEDS* 26(3):181-186.

6. Abdallah W, Solanki K, & Banta S (2017) Insertion of a calcium-responsive beta roll domain into a thermostable alcohol dehydrogenase enables tunable control over cofactor selectivity. *(Submitted)*.

7. Bulutoglu B, Haghpanah J, Campbell E, & Banta S (2017) Directed evolution of a thermostable alcohol dehydrogenase to bind RDX for biosensor applications. *(Submitted)*.

499121 Solvent Production Using a CO_2 - Fixing, Synthetic, and Syntrophic *Clostridium* Co-Culture

Abstract:

The demand for energy is steadily growing, and despite the current low oil prices, the stockpiles of petroleum reserves are quickly being depleted. As a result, new technologies capable of utilizing low value feedstocks and waste gases, such as H_2, CO and CO_2, must be developed to meet our growing need for liquid fuels. One interesting solution is to utilize the capabilities of the organisms found in nature. As such, solventogenic *Clostridium acetobutylicum* can utilize a wide variety of monosaccharides (glucose, fructose, xylose and mannose) and polysaccharides (starch, hemicellulose, and molasses) to produce acetone, ethanol and butanol in a process known as the ABE fermentation. During the solvent production, *Cac* releases excess electrons as H_2 gas, and approximately one third of the sugar substrate as CO_2 waste. In comparison, acetogenic *C. ljungdahlii* is limited in its ability to consume sugars (notably glucose) and instead can efficiently consume CO_2 gas in the presence of electron donors, such as H_2 or CO, using the Wood-Ljungdahl pathway. In the process *C. ljungdahlii* converts two molecules of CO_2 into acetyl-CoA, which is then converted into acetate and ethanol. Many previous studies focused on genetic engineering, and optimization of a single organism to carry out multiple reactions and produce commodity chemicals. In nature, most organisms live in microbial consortia, where the metabolic burden is divided among multiple organisms to produce molecules necessary for survival. Therefore, this study examined a synthetic co-culture of *C. acetobutylicum* and *C. ljungdahlii.* In the co-culture *C. acetobutylicum* converts the sugar substrate into solvents, while releasing CO_2 and H_2 gases as waste. At the same time *C. ljungdahlii,* which is unable to consume the sugar substrate, feeds on gaseous waste released by *C. acetobutylicum* to survive, and produce additional acetate. The extra acetate is then utilized by *C. acetobutylicum* to produce additional solvents. In the process, *C. ljungdahlii* improves the carbon balance of the process by recapturing the waste released by *C. acetobutylicum.* The synthetic co-culture discussed here was found to produce high titers of 3-C isopropanol which cannot be produced by either organism individually, and 2,3-butanediol, which can be produced by *C. ljungdahlii* only at trace levels. Here we present data from our efforts to engineer a synthetic *Clostridium* co-culture capable of utilizing glucose as a substrate together with inorganic waste gases in order to improve the carbon balance of the fermentation process, and maximize the solvent production.

SUPPORTED by the National Science Foundation through the NSF grant (Award No. CBET-1511660) and the IGERT fellowship (Award No. 1144726).

501369 Engineering *E. coli* to Consume Methanol

Abstract:

Due to the volatility of sugar prices, growing concerns in the "food vs fuel" debate, and the recent availability of natural gas, methanol has emerged as a promising feedstock for bio-based manufacturing of fuels and chemicals. Organisms that natively metabolize methanol are not generally amenable to metabolic engineering, owing to poorly developed genetic tools, and an incomplete understanding of their physiology, biochemistry and regulation. By contrast, *E. coli*, the workhorse of metabolic engineering, does not natively consume methanol. Here we present progress in engineering *E. coli* to grow and produce a variety of target molecules toward from methanol. We take a rational approach that combines methodologies of metabolic engineering and of chemical biology to develop new techniques to experimentally probe key limitations in the pathway. Our approach yielded two main targets: poor kinetics of the enzyme catalyzing the first step of the pathway, and "flux leakage" from the RuMP pathway that ablates the pathway driving force by leading to the build-up of formaldehyde. Successful experimental strategies to overcome these obstacles were developed. These results lay the foundation for the development of an *E. coli* strain that can be "dropped in" to existing bioprocesses to enable the utilization of the alternative substrate, and underscore the value of developing interdisciplinary methodologies for the rapid identification of pathway bottlenecks to speed up the design-build-test cycle in metabolic engineering.

491759 Removal of Ribosome Stalling Motifs to Improve Oxygenation of Key Intermediate in the Taxol Pathway

Abstract:

Taxol is an effective natural product chemotherapeutic applicable for the treatment of multiple cancers. However, production of Taxol has proven to be a significant challenge for society, as bark extraction, chemical synthesis, and semisynthesis using plant cell cultures all have issues with either yield or scalability. To address this problem, synthetic biologists have generated a strategy to produce Taxol precursors in rapid-growing microbial hosts by the heterologous expression of Taxol pathway enzymes. Although this approach has been successful in generating initial intermediates of the pathway, there is a distinct bottleneck that occurs at the first oxygenation of taxadiene in *Eschericha coli*. Previous work indicated that only minor changes in expression of this P450 enzyme (CYP725A4) severely undermined overall cellular protein production, resulting in poor pathway productivity. Global protein expression effects hinted that some property of this enzyme was inhibiting protein synthesis, perhaps at the level of the ribosome. Recently, XPPX motifs, two consecutive prolines in an amino acid sequence, have been found to induce ribosome stalling. Indeed, CYP725A4 contains three such motifs. Here, we mutate away these XPPX motifs to probe their impact on protein expression and oxygenation efficiency.

492875 Engineering *Yarrowia Lipolytica* for the Production of Triacetic Acid Lactone

Abstract:

Yarrowia lipolytica is an industrially attractive oleaginous yeast with a high innate capacity to produce acyl-CoA molecules, especially lipids. Recently, we sought to divert the flux of these precursors away from native products like fatty acids and toward the proposed biorenewable platform chemical triacetic acid lactone (TAL). To this end, TAL has been proposed as an industrially interesting molecule as it can be easily converted into existing downstream products including sorbic acid, which has a large global demand. Previous efforts to produce TAL in hosts such as *E. coli* and *S. cerevisiae* have been limited by the availability of acyl-CoA precursors. Here, we sought to use the 2-pyrone synthase enzyme isolated from *Gerbera hybrida* to establish TAL production in *Y. lipolytica*. This enzyme performs the iterative condensation of two malonyl-CoAs with an acetyl-CoA starter unit to ultimately form TAL. Initial, simple overexpression of this heterologous synthase was sufficient to produce modest titers. To further improve production, we performed a series of strain engineering approaches to increase flux toward acetyl-CoA and malonyl-CoA intermediates. Final strains were optimized in bioreactor fermentations to improve titer, rate, and yield. Ultimately, we engineered a strain of *Y. lipolytica* capable of titers far exceeding that of any previously published study. This talk will highlight these efforts of strain engineering leading to this unmatched microbial production of TAL.

501650 Engineering a Novel Omega-3 Fatty Acid Biosynthesis Pathway in Yarrowia Lipolytica

Abstract:

Omega-3 long chain fatty acids are essential for human health and development. The stagnant production of omega-3 fatty acids from fish oils no longer meet the demands of the fast growing aquaculture and nutraceutical market. A sustainable and economic source of omega-3 fatty acids is urgently needed. Here we report a systematic approach to metabolic engineering non-conventional yeast *Y. lipolytica* for eicosapentaenoic acid (EPA) production from non-sugar substrates such as cheap plant oils and rendered animal fats. We first established an alternative EPA biosynthesis pathway into *Y. lipolytica* by heterologous expression of a set of elongases and desaturases utilizing acyl-CoA instead of phospholipids, resulting in a high level of EPA production, while avoiding accumulation of undesired intermediate fatty acids. We also enhanced the assimilation and synthesis pathways of triacylglycerides, and disrupted its degradation pathway to improve the efficiency of utilizing exogenous lipids as building block for EPA production. By redirecting carbon flux towards fatty acid elongation and desaturation, we further increased the amount of EPA that was incorporated into triacylglycerol. Additional strategies to increase acyl-CoA flux will be discussed. Our results illustrate that the acyl-CoA pathway circumvents bottlenecks in the conventional phosphatidylcholine dependent desaturase pathways.

494303 Establishing a Novel and Efficient Biosynthetic Pathway for Anticoagulant Precursor 4-Hydroxycoumarin in Engineered *Escherichia coli*

Abstract:

4-hydroxycoumarin (4HC) serves as the building block for synthesis of various anticoagulants, such as warfarin, dicoumarol, phenprocoumon and acenocoumarol, which have great pharmaceutical significance in treatment of thromboembolic diseases. Currently, petroleum-derived chemical synthesis is the major approach for production of 4HC, which is expensive, unsustainable and harm to the environment. Notably, the natural 4HC biosynthetic route is not yet understood. In this work, we constructed an artificial pathway to achieve microbial production of 4HC from renewable carbon source for the first time. First, we proposed a novel pathway to produce 4HC from salicylate via salicoyl-CoA. Identifying enzymes suitable for catalyzing the sequential reactions enabled 2.3 mg/L of 4HC produced from salicylate. Second, a salicylate over-producer was constructed by screening the best candidates that capable of efficiently converting chorismate into salicylate. Third, the rate-limiting step in this pathway was eliminated by function-based enzyme bioprospecting. The replacement of an initially used plant biphenyl synthase with a microbial FabH-like quinolone synthase leads to high efficiency on the 4HC-forming reaction. Remarkably, this effort combined with pathway enzymes modular optimization yielded 108.9 mg/L 4HC from glucose. Then, this titer was further increased to around 500 mg/L in shake flasks by enhancing the availability of chorismate and malonyl-CoA. Furthermore, we stepwise identified 16 thioesterases candidates in *E. coli* and measured their enzymes activities for degradation of the pathway key intermediate salicoyl-CoA by both *in vivo* and *in vitro* assays. Among them, YdiI was found to be the dominant contributor to the salicoyl-CoA degradation. Remarkably, deletion of *ydiI* in host strain further improved the 4HC titer to 935 mg/L, which is the highest titer reported to date. Finally, we explored *in situ* semi-synthesis of warfarin by feeding with the other precursor benzyldeneacetone and the catalyst (S,S)-1,2-diphenylethylenediamine, 43.7 mg/L of warfarin was generated via sonication. This work shows the potential for scaled-up microbial production of 4HC and opens up the possibility of biosynthesizing various coumarin molecules with pharmaceutical significance.

504097 Metabolic Engineering and/or Synthetic Biology for Improved Biotechnological Production: Promises and Realities

Abstract:

Metabolic engineering (ME) emerged as an autonomous research activity in the mid to late 1980s, at the point where genetic manipulation of cellular pathways became possible. Both experimental and computational tools were developed over the next 20 years to the point that one could confidently undertake the development of an engineered organism to produce a novel chemical or improve the production of a native metabolite. Industrial processes slowly emerged, and the field achieved increased recognition and respect. ME should not be confused with the development of genetically modified animal cells to express a transgene for producing therapeutic or diagnostic proteins, an activity that developed in parallel with ME, but led to the enormous, in impact and value, biopharmaceuticals industry. Some 15-20 years later, in the early 2000s, Synthetic Biology (SB) emerged as another enterprise based on the ability to synthesize large pieces of DNA and modify (the new word now was "edit") DNA. In the beginning, there was a marked confusion: was SB different from ME, and if so, how and why? As the field moved forward, a combination of "political" and scientific advances gave SB a distinct character, but yet with large overlap with ME. SB became more sophisticated on engineering complex traits based on engineered regulatory components and focusing largely in non-pathway dependent traits, but also on generating precise DNA editing tools. I will discuss what is the present reality in terms of practical applications.

497182 Inhibition of Bacterial Toxin Activity Using Receptor-Based Peptides

Abstract:

New approaches to the treatment of bacterial illnesses are desperately needed, as the number of antibiotic-resistant strains of bacteria increases and the pipeline of new antibiotics dries up. In addition, traditional antibiotics kill both pathogenic and commensal organisms, leading to detrimental effects to the microbiota. Our lab is interested in developing novel, targeted approaches to this problem; to do so, we focus on bacterial toxins, which are secreted by pathogenic bacteria to destroy host cells, thereby allowing the bacteria to colonize and survive within the host. Our goal is to understand the initial interaction of bacterial toxins with target cell membranes, with the objective of inhibiting the activity of the toxins to prevent bacterial survival.

In this project, we designed targeted receptor-based peptides to inhibit the activity of the leukotoxin (LtxA), secreted by the bacterium *Aggregatibacter actinomycetemcomitans*. This toxin is a member of the RTX (repeats-in-toxin) family of toxins and targets leukocytes through its recognition of the lymphocyte function-associated antigen-1 (LFA-1) integrin, which is only expressed by human white blood cells, as well as its strong affinity for cholesterol in the plasma membrane of the target cell. We designed a panel of small peptides, based on the b-propeller region of the CD11a subunit of LFA-1, which has been implicated in LtxA binding, and characterized their capability to inhibit LtxA binding to LFA-1 and subsequent cytotoxicity in human immune cells. We found that four of the five peptides, specifically those corresponding to sequential b-strands in the b-propeller domain, inhibited LtxA activity, demonstrating the effectiveness of this approach. These peptides specifically inhibit the ability of LtxA to bind LFA-1, thus preventing internalization and subsequent cytotoxicity. A panel of peptides based on the murine form of LFA-1, for which LtxA has no affinity, was used as a negative control. These peptides were less effective than the human-based peptides in inhibiting LtxA activity, demonstrating that this inhibition is specific to the LtxA-LFA-1 interaction. Together, these results demonstrate the possibility of using receptor-based peptides to inhibit LtxA activity. We expect that a similar approach could be used to hinder the activity of other RTX toxins.

497491 Novel Inclusion Complexes for Cancer Treatment. an Approach Based on Energetics Metabolism

Abstract:

Cancer is considered as one of the illness that causes a great amount of deaths each year in developing countries. From several years, cancer treatment has been focused on drugs that interact with DNA or other proteins that are involved in cell division and replication. However, the main drawback of that treatment was the no-selectivity of the treatment, affecting equally healthy and tumor cells. Nevertheless, differences in how tumor cells produce energy has been discovered from the study of their metabolism during last years. Those differences have been named as Warburg effect. It has been proved that tumor cells produces energy preferentially by pentose phosphate pathway, which is a way of minor importance in healthy cells. Based on those differences, a new way for treatment can be opened if pentose phosphate pathway is blocked selectively.

In this work, an inclusion complex based on cyclodextrin and a steroid is prepared so as to block that pathway. Specifically, it reduces the activity of the first enzyme in the pathway (glucose-6-phosphate dehydrogenase). Inclusion complex was prepared successfully with a yield greater than 80% and its effect was tested â€œin vitroâ€ in 8 different cell lines: 6 tumor cell lines (liver, lung, breast, ovary, colorectal and prostate) and 2 healthy cell lines (fibroblast and kidney). Results show that the complex designed is able to reduce significantly cell viability (measured by MTT assay) in all tumour cell lines, and without significant effect in healthy ones. That selectivity suggests a new alternative for tumour treatment. Moreover, other experiments were carried out so as to quantify the effect on cell metabolism, and results show a reduction in G6PD activity and a modification in the level of ROS (reactive oxygen species) inside the cell. Furthermore, the differences between effectiveness along the different types of cells lines was also related with expression levels of sugar transports and the relative expression of the enzyme inhibited. That correlation may allow setting a dose of the treatment depending on the tissue.

Summing up, a new system for cancer treatment was developed, with a wide range of action, and with high levels of selectivity so as to reduce the undesirable side effects.

495091 Colony Formation within Hydrogel Microdroplets Enabled High Throughput Yeast Colony RNA-Seq

Abstract:

Yeasts have long been used as a model system for biological and biomedical research and played vital roles in biotechnology. While measuring genome-wide expression profile in a high throughput of yeasts is essential for both deepening our understanding of yeasts based biological and biomedical research and accelerating the characterization of yeasts in synthetic biology, such method has not been developed. This is because that bulk yeast culture RNA-seq is limited by the high cost and low throughput, and on the other hand, single yeast RNA-seq is technically challenging due to the low RNA content and difficulty of lysis. To circumvent these challenges, we developed a novel droplet microfluidic based high throughput single yeast colony RNA-seq workflow to enable the characterization of thousands of yeast colonies in parallel. Specifically, we first encapsulated single yeast cells into agarose hydrogel microdroplets using a flow focusing-splitter design followed by cultivating them into individual colonies. We then captured the RNA of individual colonies with barcoded RNA capture beads (Drop-seq beads) followed by reverse transcription, library preparation and next generation sequencing to collect single colony genome-wide expression profiles. The colony formation within hydrogel microdroplets amplifies the RNA content and enables effective yeast cell handling and lysis. The barcoded RNA capture with droplets microfluidic enables high throughput process and analysis. We first demonstrated the usage of this workflow by characterizing heterogeneity of genome-wide expression profile of over one thousand *S. cerevisiae* colonies. We further applied this workflow to analyze opaque-white switch of pathogenic yeast *C. albican* at single colony level. This method can be widely applied to enhance the understanding of yeasts based biological and biomedical research and also to accelerate the characterization of yeasts for synthetic biology.

495382 Microfluidic ChIP-Seq Device for Rapid and Parallel Analysis of Histone Modifications

Abstract:

Epigenetic changes, such as histone modifications, have been implicated in a number of different disorders such as heart disease, cancer, and autism. Therefore, it is imperative that we develop a methodology to effectively study epigenetic modifications in an efficient and rapid manner. Furthermore, a variety of histone modifications (i.e. tens of histone marks), are routinely involved in gene regulation. This creates the need for parallel analysis of samples with high throughput in order to understand the complete picture of epigenomes. In this paper we demonstrate an automated and high throughput microfluidic platform capable of producing multiple sets of high quality ChIP-seq (Chromatin immunoprecipitation followed by next generation sequencing) data in a single run that takes as little as 1 hour.

The microfluidic system consists of 4 parallel units, which are controlled using a bank of computer controlled solenoid valves and syringe pumps. Each run generates 4 ChIP DNA samples (which can be prepared into 4 sequencing libraries). The total on-chip operation takes as little as 1 h, with a hands-on working time of less than 20 min. We demonstrate the testing of chromatin samples from as few as 100 cells per assay and there was high reproducibility among 4 datasets generated by the same device giving Pearson correlation coefficients of 0.956, 0.970 and 0.870 for 1,000, 300, and 100 cell samples respectively.

The development of this automated ChIP assay lays the groundwork for large scale multiplexing which will allow fast, simultaneous analysis of multiple samples and marks, which will aid in clinical analysis of disease progression.

496076 Development of Autonomous Microfluidic Chemotaxis Assay Platform for Tip-Growing Plant Cell Using Cell Elongation Assisted Capillary Driven Flow

Abstract:

Chemotaxis is a phenomenon that cells exhibit directional movement in response to external chemical cues. Chemotactic behavior of various cells and microorganisms such as bacterial cells, endothelial cells, neurons, and *Caenorhabditis elegans* have been examined on microfluidic platforms which can offer well-controlled concentration gradient of chemoattractants or repellents with multiplexing capabilities.

Typically, microfluidic-based chemotaxis assay is operated by first introducing chemotactic cells to the micro-channels/chambers, and a chemical environment with either a steady or non-steady state concentration gradient is created around the cells upon the addition of reagents to the reservoirs. For most cases, cells migrate in a particular direction under a certain chemical environment, and researchers evaluate their chemotactic response by quantitating the number of cells in a specific location assigned in the micro-channels/chambers. However, such an end-point analysis may not be applicable for some cases where chemotactic cells do not follow its concentration gradient preciously for a long distance.

A pollen tube (PT) can be categorized in such a hard-to-analyze chemotactic cell *in vitro*. A PT is a tip-growing plant cell germinated from a single pollen grain and responsible for carrying sperm cells to an egg cell for fertilization. PTs elongate a very long distance and eventually reach to their target (egg cell) by sensing multiple guidance cues from their surroundings.

In this work, we have designed a microfluidic channel network which can guide a single PT to an individually separated reaction chamber where a chemoattractant is autonomously released upon the entrance of a PT in the particular zone.

This idea was realized by creating two regions in the device: (1) PT growth chamber filled with a liquid culture medium, and (2) microchannels filled with chemoattractant in agarose gel. Also, these two regions are connected by an empty channel network. While PTs are initially cultured in region (1), some of them randomly elongate through the empty channel and eventually reach to region (2). During this process, a PT drags the liquid culture medium from region (1) with it, and at the same time, the empty channel is instantly filled with the liquid medium by capillary driven flow. As soon as the liquid reach to region (2), the chemoattractant which has been compartment by air (empty channel) start diffusing in this region, creating a concentration gradient at the tip of a PT. In this situation, when the PT responds to the attractant, it changes its growth direction toward the microchannel filled with the chemical. Therefore, the evaluation of the PTâ€™s chemotactic response can be made by the presence of a PT in the particular channel. As a proof of concept, this chemotactic assay device was demonstrated using *Torenia fournieri* and its PT attractant peptide, LURE.

500935 High-Throughput, Single Cell Analysis of Peptide Uptake and Deubiquitinating Enzyme Activity Using a Microfluidic Droplet Trapping Array

Abstract:

Recent advances in molecularly-targeted therapeutics, such as the proteasome inhibitors Bortezomib and Carfilzomib, have shown significant promise in the treatment of multiple myeloma. However, challenges in drug efficacy frequently occur due to tumor heterogeneity, drug resistance, and personalized responses. In the case of multiple myeloma, recent efforts have demonstrated that increased activity of deubiquitinating enzymes (DUBs) facilitate a developed resistance to proteasome-based therapeutics. As such, attention has shifted towards personalized medicine consisting of multiple agent therapies. This has created a need for high-throughput biochemical assays capable of quantifying enzyme activity in intact single cells to examine the potential effectiveness of multiplexed therapeutics. The goal of this project was to develop a microfluidic droplet trapping array capable of dynamic, high-throughput, single cell analysis of DUB activity in intact single cells. The microfluidic platform can generate aqueous droplets at a rate of 200 Hz which are immediately trapped in a downstream trapping array with ~99% trapping efficiency. Single cells were encapsulated in ~40-50% of the trapped droplets. As a proof-of-concept, the droplet trapping array was first used to assess cell penetrating peptide (CPP) uptake across a population of cancer cells. These experiments demonstrated a significant heterogeneity in CPP uptake with ~20% of the population of cells exhibited no observable uptake. Next, the microfluidic platform was used to quantify DUB activity in a population of myeloma-derived cell lines. Here, the microfluidic platform was combined with a novel cell permeable, long lived, fluorescent, DUB specific, peptide-based reporter developed by our group. Preliminary findings demonstrate significant heterogeneity in DUB activity across the population of cells in addition to a variable response to the commercially available DUB inhibitor PR-619. Finally, a statistical analysis was performed across cells challenged with the DUB inhibitor to identify drug resistant subpopulations. The work presented here serves a technology first step for ex vivo diagnostics to help in developing treatment protocols using molecularly-targeted therapeutics towards DUBs.

505037 Protein Engineering Approaches to Dissect Specificity in Cell Signaling

Abstract:

Cells are packed with a high density of proteins, whose interactions drive most physiological processes. Aberrant protein-protein interactions are very often implicated in many debilitating or fatal diseases such as diabetes, neurodegenerative diseases, and cancer. Cell signaling emerges from an ensemble of dynamic, transient, protein-protein interactions in crowded cellular environments. Established structural and cell biological techniques are mostly limited to addressing the function of stable protein complexes, and do not address emergent behavior stemming from multiple transient interactions. My research program centers on the concept that protein interaction networks in cells have evolved to harness multi-protein interactions, requiring *in cell* biochemical measurements of cellular processes and *in vitro* reconstitution of transient multi-protein assemblies. We use a genetically encoded technique called Systematic Protein Affinity Strength Modulation that allows us to control protein interactions both *in vitro* and in cells to understand the molecular mechanisms that dictate context-dependent specificity in G protein-coupled receptors (GPCRs) and kinases. I will present recent work highlighting the synergistic effects of G proteins on GPCR signaling, and the biochemical basis of substrate-specificity in protein kinases.

500173 A Detailed Model of Electroenzymatic Glutamate Biosensors Aids Applications *in vivo*

Abstract:

Electroenzymatic glutamate microbiosensors have proven useful for enabling local measurements of concentration transients in the brain, but due to the diffusional and kinetic resistances of the biosensor's permselective and immobilized enzyme coatings, the sensor response cannot generally be assumed to be a true representation of actual glutamate signals. To model the biosensor's ability to track changing concentrations and to optimize its ability to do so, a mathematical model describing transport through sensor coatings, ping-pong enzyme kinetics, and H_2O_2 electrooxidation kinetics was constructed and solved using a finite element solver (COMSOL). Optimization studies showed that reducing permselective Nafion thickness to 1 µm from ~10 µm, and the enzyme layer thickness to 3 µm from ~20 µm, could increase sensitivity several-fold to ~400 nA/µM/cm^2 and bring response times down from ~0.7 s to ~30 ms. Simulated biosensors with ms response times were also shown to accurately detect glutamate concentration peaks with standard deviations as small as <0.25 s, while slower simulated sensors reflective of the state-of-the-art underestimated the glutamate in such peaks substantially. The modeling results also showed that changing the relative proportions of glutamate oxidase and bovine serum albumin in the crosslinked enzyme layer had little effect on biosensor performance unless the enzyme layer was <5 µm thick. Finally, oxygen depletion within the enzyme layer at high (>5 mM) glutamate concentrations explains experimental observations *in vitro* of apparent K_m values for the immobilized enzyme that deviate significantly from the intrinsic. These modeling results suggest that the performance of electroenzymatic glutamate biosensors can be improved dramatically, which will contribute to better resolution of glutamate transients *in vivo* and better correlation of glutamate signaling with observed behavior.

491439 Role of Interfacial Charge Concentration on Biosensing By Electrolyte Gated TransistorsÂ

Abstract:

Electrolyte gated transistors (EGTs) have been identified as promising transducers for biosensing applications due to fast and sensitive signal transduction, low voltage operation, and facile fabrication. A floating gate EGT (FGT) is a derivative of this device utilizing a floating gate to physically separate and electronically couple the active sensing area with the transistor. This eliminates the degradation and optimization issues associated with interactions of the semiconductor or the dielectric with the analyte medium, typically seen in traditional transistor based detection systems. Previous work has demonstrated the utility of FGTs for the fast and reliable detection of ssDNA and ricin in media such as orange juice, milk and aqueous buffer.

While the FGT strategy has yielded promising results, questions remain regarding its fundamental operating mechanism for detecting biomolecules. In this work, we implement well-established self-assembled monolayer (SAM) chemistry on the sensing area to create a charged surface and hence characterize the role that interfacial charge concentration plays in relation to potentiometric changes. Carboxylic acid terminated SAMs were exposed to buffers of different pH and the response of the device was studied. The results agree with theoretical predictions from Grahameâ€™s equation from conventional double layer theory, thereby rationalizing the nonlinear responses obtained at high analyte concentrations in previous work with this platform. Our study further elucidates an important effect in the sensing mechanism of floating gate EGTs, which are seen as portable, low cost biosensors that could perform well in distributed systems, bringing the sensitivity and functionality of conventional biodetection methods to field applications.

500951 An Inexpensive, Point-of-Care Urine Test for Bladder Cancer in Patients

Abstract:

An inexpensive, point-of-care urine test for bladder cancer in patients

A. P. Acharya[1], K. M. Theisen[6], A. Correa[6], T. Meyyappan[1], A. Apfel[8], T. Sun[8], T. V. Tarin[6], S. R. Little[1, 2, 3, 4, 5, 7*]

1 Department of Chemical and Petroleum Engineering, University of Pittsburgh, PA

2 Department of Bioengineering, University of Pittsburgh, PA

3 Department of Pharmaceutical Sciences, University of Pittsburgh, PA

4 Department of Immunology, University of Pittsburgh, PA

5 Department of Ophthalmology, University of Pittsburgh, PA

6 Department of Urology, University of Pittsburgh Medical Center, PA

7 McGowan Institute for Regenerative Medicine, University of Pittsburgh, PA

8 Department of Biostatistics, Graduate School of Public Health, University of Pittsburgh, PA, USA

Hematuria is the most common symptom of bladder cancer (BC), but only 5-30% of the evaluated patients (via expensive tests) are positive for BC, which makes BC one of the most expensive cancers to treat. Therefore, a cost-effective, office-based test that reliably rules out BC in hematuria evaluation is greatly needed. Herein a urinary assay for a BC biomarker matrix metalloproteinases-2 and -9 (MMP2/9) ("Ammps") is presented, which follows WHO's ASSURED guidelines, are non-invasive, cost-effective, and based on the MMP2/9 activity on the Ammps substrates. Ammps substrates, are composed of gelatin-crosslinked-alginate nanoparticles chelating large amounts of Fe(II), which identify MMP2/9 activity by initiating an autocatalytic Fenton's reaction. In a study of 88 patients, Ammps had 100% sensitivity, 85% specificity, and a negative predictive value (NPV) of 100%. This high NPV can be useful in ruling out BC in hematuria evaluation and can have a broad impact in cancer detection.

Introduction: Bladder cancer (BC) patients require continuous monitoring due to its high recurrence rate. State of the art monitoring of BC is performed using cystoscopy, which is an expensive and invasive procedure. Moreover, since monitoring remains essential throughout the life of the patient, BC is one of the most expensive malignancies to treat from detection to death.

Materials and Methods: Ammps were performed on urine donated by 88 patients (on the same day of donation) suffering from different diseases including bladder cancer, prostate cancer, benign prostate hyperplasia, urinary tract infection among others. The experiments were performed according to the IRB protocol approved by UPMC.

Results and Discussion: Ammps follow the ASSURED guidelines as shown by the table here (above). Moreover, Ammps detect 100% of untreated bladder cancer patients, and generate a range of color for treated bladder and prostate cancer patients.

Conclusions: Ammps, a point-of-care urine test that may be utilized for the detection and surveillance of bladder cancer. This test can be cost-effective, is instrument free, and can be performed in a clinical setting. Ammps has a high negative predictive value and sensitivity for detection of bladder cancer, and thus also has the potential to serve as a screening test in high risk populations. With its capacity for bladder cancer diagnostics at the bedside and ability to satisfy the WHO's ASSURED criteria, Ammps can also have an impact on monitoring of, and possibly screening for, bladder cancer in resource-poor settings. A larger prospective trial is underway to confirm these results.

Figure 1. (A) The mechanism of Ammps. **(B)** Ammps show 100% negative predictive value for bladder cancer in hematuria positive population. **(C)** Ammps follow ASSURED criteria set by WHO.

499729 A Viscosity-Based Measurement System for Pathogen Detection

Abstract:

There is a great need to develop rapid, sensitive, and accurate infectious disease diagnostics. Current gold standard pathogen detection often involves fluorescence-based DNA amplification readings or antibody-based assays to confirm disease presence in either patient or environmental samples. Here we introduce an alternative method, particle diffusometry, where the presence of amplified DNA is detected by measuring change in solution viscosity. The presence of *V. cholerae*, *S. aureus*, and *K. pneumoniae* with specialized oligonucleotide primers allows for amplification using loop-mediated isothermal amplification (LAMP). This amplification increases the length of polynucleotides in the solution, which in turn makes the fluid sample more viscous. We suspend 200 nm particles in a fluid sample and image their Brownian motion with fluorescence microscopy. In the presence of amplified pathogen DNA, these particles exhibit slower Brownian motion, indicating a change in the particle diffusion coefficient, and in turn the solution viscosity. Particle diffusometry calculates the diffusion coefficient of the particles by correlating successive particle images (at time Dt) to one another (cross-correlation, s_c) and the particle image with itself (autocorrelation, s_a) at a magnification (M):

From the diffusion coefficient, the viscosity (η) of the solution is calculated by the Stokes-Einstein equation, where k is the Boltzmann's constant, T is absolute temperature, and a is the hydrodynamic radius of the 200 nm particle.

Using particle diffusometry we successfully determine the presence of amplified *V. cholerae* DNA in blinded studies involving (A) *V. cholerae* DNA with oligos that underwent LAMP, and the following controls: (B) no *V. cholerae* DNA but with oligos that underwent LAMP, (C) *V. cholerae* DNA with oligos that did not undergo LAMP, and (D) no *V. cholerae* DNA but with oligos that did not undergo LAMP. Particle diffusometry measurements are also comparable to qPCR (fluorescence signal). We compare both techniques using a concentration range of pure DNA that is equivalent to what is found in samples containing 1-10,000 cells/mL. Further, we determine the change in viscosity of DNA amplification in the presence of whole cells for *V. cholerae*, *K. pneumoniae*, and *S. aureus*. Finally, we can measure the presence of DNA amplified *V. cholerae* cells in environmental pond water sources, indicating the feasibility of particle diffusometry for the point of care. From these studies, we find that particle diffusometry has a lower limit of detection of 100 cells/mL, a relevant concentration for the detection of *V. cholerae* in water sources. We will also discuss our current work on the implementation of particle diffusometry onto a smartphone-based platform as a quantitative point of care pathogen detection device.

500154 Switch-like DNA Amplification for Biomarker Detection

Abstract:

Detection of low concentration biomolecules is time and labor intensive, particularly when found in complex matrices such as bodily fluids. A switch-like chemistry for signal amplification is advantageous for simple, robust biomolecule detection. An "ON" or "OFF" signal is not affected by drifting signal or changes in reaction kinetics. We have developed a switch-like DNA amplification system that responds to short DNA oligonucleotide reporter molecules. It is well known in literature that small DNA oligonucleotide "reporters" can be transduced from proteins, miRNA, and genomic DNA. Our switch-based chemistry can amplify the oligonucleotide reporter in a biphasic fashion; during the initial phase the reporter molecule is produced at a low rate, and eventually the production of the reporter significantly slows. When the reporter molecule reaches a threshold concentration, the chemistry produces a high-gain burst of reporter DNA molecules. If the initial reaction rate or reporter concentration is high enough, the amplification moves directly into the high-gain phase. The DNA amplification chemistry thus provides a built-in switch function, with low-output DNA amplification serving as an "OFF" phase, and high-output DNA amplification serving as an "ON" phase. The high-gain amplification phase shows ultrasensitive, Hill-type kinetics; researchers have noted the utility of Hill-type amplification kinetics in biosensors, genetic logic gates, and other materials that are required to be highly responsive. The kinetics of each phase and the reporter DNA threshold concentration can be controlled by thermodynamics of DNA complexes or the reaction solution conditions. Ultimately, we will use this chemistry for simple, robust detection of biologically relevant molecules.

502042 Rapid Diagnosis and Discrimination of Healthy and Breast Cancer Tissues Using Classical and Imaging FTIR

Abstract:

Rapid characterization of normal and malignant breast tissues at the molecular level was aimed using classical and imaging Fourier Transform Infrared (FTIR) spectroscopy. Differences in spectral response due to compositional changes in breast tissue after cancer development constitute the basis of this study. Macroscopic size samples from 17 patients, who diagnosed with invasive ductal or lobular carcinoma at Acibadem Maslak Hospital, Istanbul, Turkey, cut in cubes and spectral responses collected from each face of cubes in duplicate by attenuated internal reflection (ATR) mode operating FTIR spectroscopy. Spectral investigation of breast carcinoma showed that three major responses can be obtained depending on the fat content of any spot contacted with the diamond ATR crystal (0.5 m diameter). We divided the spectrum into three groups as, Type I , Type II and Type II according to the intensity of fat related IR bands at 2800-300 cm^{-1} of lipid CH_3 chains and at 1750 cm^{-1} C=O of fatty acid ester functional groups (Figure 1). Spectra exhibiting strong fat related IR peaks were found to be typical of normal breast tissue and denoted as spectrum Type I. Mid to low intensity fatty IR bands rich spectra were nominated as spectrum type II which was spectral trend of some healthy and a few malign tissue samples. No fat bands containing spectra were classified as spectrum type III and this was the common observation for almost all malignant tissues.

The talk will cover spectral and chemometric investigation of IR data to develop a classification tool to discriminate tumors from healthy tissues. For this, spectral responses collected from healthy and malignant tissues were first treated with multivariate statistical tools to develop a model to distinguish healthy and malignant tissues. Spectroscopic data was first reduced into factors and loading applying partial least square (PLS) and factors of PLS were then used to develop discrimination models to separate healthy and tumor tissues on a 2D canonical variate plane using canonical variate analysis (CVA). This hybrid PLS-CVA analysis was employed to fingerprint, DNA/RNA or Amide I&II IR absorption bands portions of the original FTIR spectra separately. Since the fingerprint region covers DNA/RNA and Amide absorption bands regions in the 100-1650 cm^{-1} wavenumber, the best discrimination results were obtained using from this region. Figure 2 shows discrimination of 38 healthy and 66 tumor tissues in three groups.

After successful classification of breast tissues into healthy and tumor groups, discrimination of tumors with respect to their clinical histopathologic evaluations was performed. Dicrimination of tumors based on their pathologic grading (Figure 3), staging with respect to tubulus formation (T score), spread to lymph nodes (N score) and metastasis of cancer (M score). Discrimination of tumor based on histopathological information will be introduced and successful discrimination models will be shown.

Finally, IR image of a breast tissue will be introduced and tumorsâ€™ compositional differentiation depended spectral response changes will be presented to prove the rapid discrimination nature of the proposed technique using molecular level information.

504094 Genetic Circuit Design

Abstract:

I will describe the construction of genetically-encoded sensors that respond to diverse input signals (colors, chemicals, metabolites) and the circuits to respond to these sensors. To do this, we have developed the first programming language for cells (Cello). To compose a design, the user specifies the genetic sensors, circuitry, and actuators (cellular functions) using a descriptive language borrowed from electrical engineering (Verilog). The organism/strain, gate technology, and valid operating conditions are specified in a modular âۥuser constraint fileâ€ (UCF). The software compiles the DNA sequence using a library of precise and highly-insulated genetic parts. I will show work to expand beyond the model organism Escherichia coli to bacteria that live in the human gut (*Bacteroides*) and soil environment (*Rhizobium and Pseudomonas*) and eukaryotes, including mammalian cells. As such, Cello has the potential to serve as a platform to engineer cells for a broad range of medical, materials, and sensing applications.

499423 Oral Formulations: Perspectives from Additive Manufacturing

Abstract:

Selection of oral formulation strategy plays a significant role in development of new drug candidates throughout the development and commercial product stages; requirement for oral bioavailability must be balanced against additional complexity in the manufacturing process[1,2]. Formulation strategies may present manufacturing challenges, particularly for poorly water-soluble drugs, resulting from the necessity of particle size reduction and multiple active excipients to achieve bioavailability targets[3]. As an alternative to established manufacturing strategies, much recent work has focused on additive manufacturing, partially due to its capacity for accurate and precise delivery of drug amounts in the micro- to milli-gram range[4]. The technology also exhibits potential as a flexible platform capable of producing doses in a number of forms from solvent- and melt-based forumulations[5,6], which may include colloidal particles[7].

The manufacture of multiple capsule-based formulations using non-colloidal particulate suspensions is presented in this study. Particles selected from micronized and non-micronized powder grades are characterized in terms of particle size and bulk powder properties, which, in concert with rheology experiments are used to guide printing fluid formulations. Quantitative measures of product physical properties and respective production rates are presented for four active substances in 1-100 milligram amounts, delivered by hydro-philic/phobic carrier fluid platforms. Based on these results, perspectives are given on the ability of the platforms to complement current formulation selection strategies for BCS Class I and Class II drugs and on the potential for the technology serve at both the personalized medicine and industrial manufacturing scales.

References

1. M. Kuentz, R. Holm, and D. P. Elder, 2016, Methodology of oral formulation selection in the pharmaceutical industry, European Journal of Pharmaceutical Sciences, 87, 136-163.
2. M. Ierapetritou, F. Muzzio, and G. Reklaitis, 2016, Perspectives on the Continuous Manufacturing of Powder-Based Pharmaceutical Processes, AIChE Journal, 62, 1846-1862.
3. H.D. Williams, N.L. Trevaskis, S.A. Charman, R.M. Shanker, W.N. Charman, C.W. Pouton, and C.J.H. Porter, 2013, Strategies to Address Low Drug Solubility in Discovery and Development, Pharmacoligical Reviews, 65, 315-499.
4. R. Daly, T.S. Harrington, G.D. Martin, and I.M. Hutchings, 2015, Inkjet printing for pharmaceutics â€" A review of research and manufacturing, International Journal of Pharmaceutics, 494, 554-567.
5. L. Hirshfield, A. Giridhar, L.S. Taylor, M.T. Harris, and G.V. Reklaitis, 2014, Dropwise Additive Manufacturing of Pharmaceutical Products for Solvent-Based Dosage Forms, Journal of Pharmaceutical Sciences, 103, 496-506.
6. E. IÃ§ten, A. Giridhar, L.S. Taylor, Z.K. Nagy, and G.V. Reklaitis, 2015, Dropwise Additive Manufacturing of Pharmaceutical Products for Melt-Based Dosage Forms, Journal of Pharmaceutical Sciences, 104, 1641-1649.
7. J. Pardeike, D.M. Strohmeier, N. Schrödl, C. Voura, M. Gruber, J.G. Khinast, and A. Zimmer, Nanosuspensions as advanced printing ink for accurate dosing of poorly soluble drugs in personalized medicines, 2011,

International Journal of Pharmaceutics, 420, 93-100.

487351 Predicting the Free Energy of Solvation: A Hybrid QSPR Model for Organic Solute/Solvent Pairs

Abstract:

The free energy of solvation is a useful quantity in the journey of a pharmaceutical molecule from discovery through to commercial process development and product design. In this paper, we review some of the roles played by this property across the moleculeâ€™s life cycle, including in the early assessment of the quantities such as the octanol/water partition coefficient or on the assessment of solvent effects on reaction energetics for screening purposes. Initial efforts to propose quantitative structure property relationships (QSPR) and predict the Gibbs free energy of solvation for a wide range of solute/solvent pairs are presented here. They include the use of partial least square (PLS) and multivariate linear regression (MLR) methods. A dataset consisting of 303 solutes in 215 solvents with total number of 1800 Gibbs free energy data points has been built. The proposed models combine experimental descriptors with quantum mechanical (QM) descriptors. For the solvents, experimental descriptors are proposed, as these provide a good description of the bulk properties of solvents and the range of solvents considered is usually relatively narrow; twelve of these descriptors are used in the proposed model (surface tension, dipole moment, octanol-water partition coefficient, dielectric constant, refractive index, boiling point, molecular weight, enthalpy of vaporization, molar volume, critical temperature, critical pressure, and critical volume). For the solutes, on the other hand, QM descriptors are used, making it possible to model solutes that have not been synthesized or that are short-lived/unstable; nine QM solute descriptors are proposed (polarizability, energy of the highest occupied molecular orbital, energy of the lowest unoccupied molecular orbital, dipole moment, most positive partial charge on hydrogen atom in molecule, most negative atomic partial charge in molecule, molecular van der Waals volume, total energy at 0 K, and entropy at 298 K). The accuracy of these initial models is found to be acceptable for fast calculation of the Gibbs free energy of solvation of a wide range of solutes in different solvents and to compare favourably with existing methods. Building on these encouraging results further work to explore whether this level of accuracy is suitable for a range of activities, including screening and process design will also be of interest.

492658 *In silico* Prediction of Absolute Rates of Crystal Growth from Solution

Abstract:

Predicting rates of crystal growth from solution *in silico* remains extremely challenging, but presents a lucrative engineering opportunity to more efficiently design crystals for pharmaceutical, catalytic, and other purposes. Mechanistic crystal growth models [1-3] capture key thermodynamic and kinetic phenomena, and they have successfully predicted steady-state crystal shapes. [4,5] These models can facilitate *in silico* crystal growth rate predictions if rates of attachment to and detachment from kink sites are known; kinks are favorable surface sites for attachment of growth units (atoms, molecules, etc.). Complex computational techniques are currently the only approach to calculate these rates. Moreover, these methods are necessary to gain mechanistic insight to aid future development of a model to predict attachment/detachment rates.

In this work, we demonstrate that *in silico* predictions of absolute crystal growth rates can yield favorable agreement with experimentally measured growth rates. We examine a proof-of-concept system â€“ NaCl in water â€“ which provides valuable insights that will aid future investigations of organic molecules and organic salts. We compute the rates of Na^+ and Cl^- attachment and detachment at kink sites and also elucidate the mechanistic role of water molecules during these events. [6] These rates are implemented in the mechanistic spiral growth model [7,8] to make *in silico* NaCl crystal growth rate predictions, which are in approximate agreement with experimental measurements. [9-11] These results promote the investigation of additional systems and, ultimately, the development of a model to predict attachment and detachment rates. It will be interesting to explore the role of rotational degrees of freedom and functional groups during kink attachment/detachment of organic molecules.

[1]Â J. Li, C.J. Tilbury, S.H. Kim, M.F. Doherty, *Prog. Mater. Sci.* **2016**, *82*, 1-38.

[2]Â M.A. Lovette, A.R. Browning, D.W. Griffin, J.P. Sizemore, R.C. Snyder, M.F. Doherty, *Ind. Eng. Chem. Res.* **2008**, *47*, 9812-9833.

[3] P. Dandekar, Z.B. Kuvadia, M.F. Doherty, *Annu. Rev. Mater. Res.* **2013**, *43*, 359-386.

[4] S.H. Kim, P. Dandekar, M.A. Lovette, M.F. Doherty, *Cryst. Growth Des.* **2014**, *14*, 2460-2467.

[5] C.J. Tilbury, D.A. Green, W.J. Marshall, M.F. Doherty, *Cryst. Growth Des.* **2016**, *16*, 2590-2604.

[6] M.N. Joswiak, M.F. Doherty, B. Peters, *in prep.*

[7] W.K. Burton, N. Cabrera, F.C. Frank, *Phil. Trans. R. Soc. Lond. A* **1951**, *243*, 299-358.

[8] R. Snyder, M.F. Doherty, *P. Roy. Soc. A-Math. Phy.* **2009**, *465*, 1145-1171.

[9] H. Offerman, G. von Brachel, A. Al-Sabbagh, F. Farelo, *Cryst. Res. Technol.* **1995**, *30*, 651-658.

[10] S.B. Zhang, J.J. Yuan, H.A. Mohameed, J. Ulrich, *Cryst. Res. Technol.* **1996**, *31*, 19-25.

[11] S. Al-Jibbouri, J. Ulrich, *J. Cryst. Growth* **2002**, *234*, 237-246.

501156 Mathematical Modeling and Simulation of Magnetophoresis. Application to Drug Delivery Using Magnetic-Field

Abstract:

The effects of magnetic field on paramagnetic species such as ions, biological molecules and solid particles have been investigated theoretically[1-4] and experimentally.[5,6] These paramagnetic species in the presence of an applied magnetic field experiences three kinds of body forces such as paramagnetic gradient force due to concentration gradient, field gradient force due to magnetic field gradient, and Lorentz force due to interaction of moving charged species with the magnetic field.[3,7] The flux of a population of paramagnetic species due to these magnetic body forces have been previously modeled using four different approaches- (A) the magnetic forces are included as the external body forces in the Navier-Stokes equation and the resulting velocity is added to the convective term of the flux expression,[2,3,8] (B) the velocity of paramagnetic particles is approximated as a product of mobility and a net magnetic force, which is then included in the convective term of the flux expression,[4,9,10] (C) the mobility in the migration term of the Nernst-Planck equation is modeled as a matrix containing Hall and Transversal mobilities, which are the functions of magnetic fields,[1] and (D) the diffusion coefficients are modified to include the effects from Lorentz force and the velocity in the convective term is modeled as a product of mobility and field gradient force.[11] The approaches A-D have a few common limitations that these models are not applicable to the concentrated solutions of paramagnetic particles, and they do not account for all magnetic forces including magnetic dipole interactions between particles. Here we propose a novel model for the flux of paramagnetic species that overcomes all the limitations of the preceding approaches. Moreover, the proposed model is based on the first principles that accounts for fluxes due all three magnetic forces with inter-particle interactions.

References:

(1) Aaboubi, O.; Chopart, J.; Douglade, J.; Olivier, A.; Gabrielli, C.; Tribollet, B. Journal of the Electrochemical Society 1990, 137, 1796.

(2) Waskaas, M.; Kharkats, Y. I. The Journal of Physical Chemistry B 1999, 103, 4876.

(3) Hinds, G.; Coey, J.; Lyons, M. Electrochemistry communications 2001, 3, 215.

(4) Furlani, E. Journal of Applied Physics 2006, 99, 024912.

(5) Lueshen, E.; Venugopal, I.; Soni, T.; Alaraj, A.; Linninger, A. Journal of biomedical nanotechnology 2015, 11, 253.

(6) Venugopal, I.; Pernal, S.; Fusinatto, T.; Ashkenaz, D.; Linninger, A. Nano Biomedicine & Engineering 2016, 8.

(7) Weston, M. C.; Gerner, M. D.; Fritsch, I. Analytical chemistry 2010, 82, 3411.

(8) Waskaas, M.; Kharkats, Y. I. Journal of Electroanalytical Chemistry 2001, 502, 51.

(9) Furlani, E.; Ng, K. Physical review E 2006, 73, 061919.

(10) Furlani, E. P.; Xue, X. Microfluidics and nanofluidics 2012, 13, 589.

(11) Kinouchi, Y.; Tanimoto, S.; Ushita, T.; Sato, K.; Yamaguchi, H.; Miyamoto, H. Bioelectromagnetics 1988, 9, 159.

486626 Polyalkoxylated Alcohols As Excipients for Poorly Soluble Drugs

Abstract:

Polyalkoxylated Alcohols as Excipients for Poorly Soluble Drugs

Thomas H. Kalantar[1], Jin Zhao[2], Mladen Ladika[1], Keith Harris[1], Chris Tucker[1], Michael Tulchinsky[1], TC Kuo[1], Joe Kiefer[1], Xiaoyun Chen[1], Robert Krystosek[1]

(1) The Dow Chemical Company, Herbert D. Doan R&D Center, Midland, Michigan, 48674, USA

(2) Pharmaceutical Excipients R&D, Dow Food, Pharma & Medical Films, Midland, MI 48674

There has been a steady increase in the number of active pharmaceutical ingredients (APIs) that are characterized with low aqueous solubility. It has been estimated that 40 to 80% of new drug candidates suffer from poor aqueous solubility and/or slow dissolution, reducing their bioavailability and efficacy, and even preventing their general use [1]. This study explored a solubilization approach using low-T_g alkylated polyoxyalkylene copolymers to enhance drug solubility and dissolution rate[2], where the drug was combined with the polymeric excipient to form an amorphous solid dispersion at the molecular level. Such low-T_g excipients yield liquid or semi-solid drug/excipient formulations.

In this study, a novel class of alkylated polyols was synthesized using renewably sourced alkyl initiators, with a specific emphasis on terpene alcohol precursors. These alcohols were first alkoxylated with butylene oxide then further ethoxylated with ethylene oxide. The effect of solubilization by these copolymers was studied using four model drugs (ketoprofen, danazol, griseofulvin, and probucol). Raman spectroscopy suggested that several copolymers from this study were effective in stabilizing ketoprofen in its amorphous form. The aqueous solubility and dissolution rate of drugs in presence of these excipients were studied using nephelometry and optical microscopy. Several drug/excipient samples showed a dramatic reduction in turbidity relative to the turbidity of drugs alone in water. Optical microscopy of those drug/excipient samples in water also indicated an increase in drug dissolution. Drug solubility was also measured by an HPLC method in which a capsule filled with probucol/excipient mixture was placed in phosphate buffer solution for 30 minutes after capsules were disintegrated.

Medium to high enhancement of solubility were observed with blends of common commercial polymeric excipients with three drugs: phenytoin, griseofulvin, and probucol. Probucol solubility was significantly enhanced when mixed with excipient polymers at a molecular level. When mixtures of probucol and novel alkyl-BO-EO excipients were loaded into capsules and subjected to ambient conditions for one week and accelerated aging at 40 °C and 75% RH for 4 months, the observed increases in probucol solubility were significant (up to 8,000-fold) under both sets of storage condition.

Many polyoxyalkylene copolymers can be used as solubilization-excipients for poorly soluble drugs; however the structure of alkyl-BO-EO copolymers can be optimized to specific drug structures to enable otpimization of solubilization.

References

Â

1. Benet, L. Z., The role of BCS (biopharmaceutics classiï¬cation system) and BDDCS (biopharmaceutics drug disposition classiï¬cation system) in drug development. *J Pharm Sci.* **2013,** *102* (1), 34-42.

2. (a) Tang, J.-L.; Sun, J.; He, Z.-G., *Current Drug Therapy* **2007,** *2,* 85-93; (b) Nielsen, F. S.; Petersen, K. B.; Mullertz, A., *Eur. J. Pharm. Biopharm.* **2008,** *69,* 553-562; (c) Kohli, K.; Chopra, S.; Dhar, D.; Arora, S.; Khar, R. K., *Drug Dicsovery Today* **2010,** *15,* 958-965; (d) Holm, R.; Jorgensen, E. B.; Harborg, M.; Larsen, R.; Holm, P.; Mullertz, A.; Jacobsen, J., *Eur. J. Pharm. Sci.* **2011,** *42,* 416-422.

503694 2D Printed Product Design of Patient Centric Dosage Forms for Adaptive Clinical Trials

Abstract:

Adaptive clinical trial design has been encourage by both FDA and EMA as a flexible clinical conduct which allows the sponsors to bring successful drugs to the market faster while saving money and resources. As a consequence, for exploratory clinical studies, the conventional, rigid structure for manufacture, warehousing and distribution of Investigational Medicine Products (IMPs) can be advantageously replaced by manufacturing technologies which can tailor the dosage strength with real time adjustment in due course alongside with the open clinical trial and minimize the waste of drug compounds.

In this work, we investigate the potential of industrial inkjet printing technology as a flexible and scalable approach of producing clinical supplies and also overcoming biopharmaceutics challenges of difficult-to-deliver drugs such as poor solubility. A systematic development of poorly soluble model drug was accomplished through (1) rational design for amorphous solid dispersions (ASDs) formulation, (2) inkjet printing process understanding, development and optimization for printing ASD producing ink solution and (3) product design for ASD-printed film that could be filled in hard capsules (PFC).

We will present the results of our ink development approach which selects the desired drug-polymer-solvent systems based on systematic solid-state screening, jetting/printing performance and product key attributes (e.g. dosage size and drug load) and substrate selection. PFC prototypes were produced using a R&D inkjet system and preliminary *in vitro* dissolution and short-term stability studies were carried out. Both the potential and remaining challenges of inkjet printing technology in the context of adaptive drug development will be discussed and summarized.

496170 Identifying Molecular Targets of Drugs Using an Integrative Network Analysis of Protein-Protein, Protein-DNA and Transcriptomics Data

Abstract:

The identification of the molecular targets of compounds has great importance in drug discovery and drug re-purposing, not only for understanding mechanism of action, but also for elucidating hidden phenotypes and off-target effects. Gene expression or transcriptomics analysis provides a common means to obtain genome-wide information of cellular response to drug's actions and to identify genes with altered expressions due to a drug treatment. Computational algorithms for such transcriptomics data have been developed and applied with the purpose of determining the direct molecular targets of drugs, as opposed to indirect (downstream) targets. A family of these algorithms uses a network of gene regulatory interactions to predict the direct drug targets by viewing the drug action as a form of network perturbations. For example, causal-reasoning methods [1,2] and Master Regulatory Inference algorithm (MARINa) [3] employ literature-based and reverse-engineered network, respectively, to score a potential target based on statistical enrichment of the downstream molecules among the differentially expressed genes. More recent and advanced methods leverage multiple sources of information on the networks of protein and gene interactions, including protein-protein interaction (PPI) and transcription factor binding site (TFBS). The-state-of-the-art method, called Detecting Mechanism of Action by Network Dysregulation (DeMAND) [4] scores drug targets based on transcriptional dysregulations of the downstream genes using a protein-gene network, combining PPI and an inferred gene transcriptional network.

In this work, we developed Systems Analysis and Learning for inferring Modifiers of Networks (SALMON), a novel method for identifying drug targets. SALMON uses a protein-gene network (PGN) that is constructed using a combination of TFBS and PPI datasets. The strengths and signs of the protein-gene interactions in the PGN are inferred from the gene transcriptomics data by solving the linear regression model described in our previous network inference method called DeltaNeTS [5] with ridge regression strategy. To determine the targets of each drug, each candidate protein is scored based on the dysregulation of the protein-gene interactions in the drug treatment sample. A positive score indicates that the drug enhances the gene regulatory activity of a particular protein, and vice versa, a negative score implies an attenuation of the regulatory activity of the protein target.

We tested the performance of SALMON, using three drug treatment datasets in mammalian cells: (A) NCI-DREAM drug synergy challenge using human lymphoma cells [6], which comprises time-series gene expression profiles of human lymphoma cells from 14 compound treatment experiments, (B) genotoxic compound study using human liver cancer cells [7], which consists of time-series profiles from 62 genotoxic and non-genotoxic compound treatment experiments, and (C) drug treatment study using mouse pancreas cells [8], which includes time-series samples from 29 chromatin-targeting compound treatments. For the human cells, we used tissue-specific TFBS from RegulatoryCircuit database [9] and human PPI from ENRICHR [10] and STRING [11] databases. For the mouse cells, we used mouse-pancreas-cell-specific

TFBS from CellNet [12] and mouse PPI from STRING.

We compared the predictions of SALMON to those from DeMAND, our previous strategy DeltaNeTS [5], and differential gene expression analysis (DE). SALMON significantly outperformed DeMAND, DeltaNeTS and DE analysis in terms of AUROC (area under the receiver operating characteristic curve of true positive rate vs. false positive rate) for all three datasets (see Table 1). Moreover, SALMON was able to more accurately differentiate multiple classes of drugs based on their known activities, such as genotoxicity and histone deacetylation, than the other methods. For example, in NCI-DREAM drug synergy dataset, SALMON was able to specifically detect the action of mitomycin – a DNA crosslinking agent – in activating the DNA crosslink repair, while DeMAND failed to reveal.

The superior performance of SALMON in predicting drug targets is a result of its ability to integrate multiple sources of information on the protein-gene network and to account for the strengths and signs of the protein-gene interactions. Because SALMON uses a kinetic model of the gene transcriptional network, the method is able to naturally accommodate time-series trancriptomics dataset, a common type of data in drug treatment studies. Finally, SALMON gives not only the prediction of the drug targets, but also the signs of the drug action (i.e. enhancement and attenuation). Our test cases demonstrated that SALMON is a highly robust algorithm, capable of providing high prediction accuracy over three different mammalian cells, without the need for exhaustive parameter tuning.

Table 1. Averaged AUROC over compounds in three data sets by SALMON, DeMAND, DeltaNeTS, and DE

	SALMON	DeMAND	DeltaNeTS	DE
Drug synergy challenge	0.807	0.694	0.627	0.625
Genotoxic study	0.768	0.734	0.675	0.636
Mouse cell data	0.900	0.749	0.618	0.677

[1] Chindelevitch, L., D. Ziemek, A. Enayetallah, R. Randhawa, B. Sidders, C. Brockel, and E. S. Huang. 2012. Causal reasoning on biological networks: Interpreting transcriptional changes. Bioinformatics 28 (8): 1114–1121.

[2] Martin, F., T. M. Thomson, A. Sewer, D. a Drubin, C. Mathis, D. Weisensee, D. Pratt, J. Hoeng, and M. C. Peitsch. 2012. Assessment of network perturbation amplitudes by applying high-throughput data to causal biological networks. BMC systems biology 6: 54.

[3] Lefebvre, C., P. Rajbhandari, M. J. Alvarez, P. Bandaru, W. K. Lim, M. Sato, K. Wang, et al. 2010. A human B-cell interactome identifies MYB and FOXM1 as master regulators of proliferation in germinal centers. Molecular systems biology 6 (377): 377.

[4] Woo, J. H., Y. Shimoni, W. S. Yang, P. Subramaniam, A. Iyer, P. Nicoletti, M. Rodríguez Martínez, et al. 2015. Elucidating Compound Mechanism of Action by Network Perturbation Analysis. Cell 162 (2): 441–51.

[5] Noh, H., H. Ziyi, and R. Gunawan. 2016. Inferring Causal Gene Targets from Time Course Expression Data. IFAC-PapersOnLine 49 (26): 350–356.

[6] Bansal, M., et al., A community computational challenge to predict the activity of pairs of compounds, Nature Biotechnology, 2014, pp.1213-1222.

[7] Magkoufopoulou, C., S. M. H. Claessen, M. Tsamou, D. G. J. Jennen, J. C. S. Kleinjans, and J. H. M. Van

delft. 2012. A transcriptomics-based in vitro assay for predicting chemical genotoxicity in vivo. Carcinogenesis 33 (7): 1421–1429.

[8] Kubicek, S., J. C. Gilbert, D. Fomina-yadlin, A. D. Gitlin, and Y. Yuan. 2012. Chromatin-targeting small molecules cause class-speci fi c transcriptional changes in pancreatic endocrine cells.

[9] Marback, D., et al., Tissue-specific regulatory circuits reveal variable modular perturbations across complex diseases, Nature Methods, 2016.

[10] Kuleshov MV, Jones MR, Rouillard AD, Fernandez NF, Duan Q, Wang Z, Koplev S, Jenkins SL, Jagodnik KM, Lachmann A, McDermott MG, Monteiro CD, Gundersen GW, Ma'ayan A. Enrichr: a comprehensive gene set enrichment analysis web server 2016 update. Nucleic Acids Research. 2016; gkw377.

[11] Szklarczyk, D., A. Franceschini, S. Wyder, K. Forslund, D. Heller, J. Huerta-Cepas, M. Simonovic, et al. 2015. STRING v10: Protein-protein interaction networks, integrated over the tree of life. Nucleic Acids Research 43 (D1): D447–D452.

[12] Cahan, P., H. Li, S. A. Morris, E. Lummertz Da Rocha, G. Q. Daley, and J. J. Collins. 2014. CellNet: Network biology applied to stem cell engineering. Cell 158 (4): 903–915.

499086 Integrative -Omics Analysis of Cancer Protein Secretion

Abstract:

Early detection and diagnosis of cancer is critical to the successful treatment of the disease, as later events such as metastasis can severely reduce the chances of survival. Although a number of screening methods exist for different cancers, many require an invasive surgical procedure or suffer from poor sensitivity and/or specificity. This has motivated efforts to identify biomarkers present in biofluids (e.g., blood or urine) that could accurately predict the presence or progress of cancer within an individual without the need for more invasive approaches. However, the heterogeneity of the disease has made such efforts nontrivial, exhibiting substantial variation across different individuals and tissues-of-origin, and even among cells comprising the same tumor. A systems-level approach is therefore necessary to account for the governing biological networks contributing to such behavior, and to identify core patterns for the development of highly specific cancer biomarkers.

To identify potential biomarkers likely to be detectable in patient biofluids, we focused specifically on the subset of proteins processed by the secretory pathway, termed the â€œsecretome.â€ Secretome members were selected based on their annotation in the Human Protein Atlas (HPA), and the presence of an N-terminal signal peptide. Transcriptomic (RNA-Seq) data from solid tumors and matched normal tissues spanning over 20 different cancer types were retrieved from The Cancer Genome Atlas (TCGA), and analyzed using a combination of statistical and integrative approaches to predict candidate biomarkers. A differential expression analysis was performed for each individual cancer type between tumor and matched normal tissue, to reduce tissue-specific effects unrelated to cancer. These results were combined with absolute transcript abundance profiles spanning all normal tissue samples, and cross-referenced with the human plasma proteome to identify a subgroup of upregulated secretome genes as potential biomarkers for each cancer type. The resulting collection of candidate biomarkers included a number of previously-identified proteins known to be involved in malignant transformation or tumor development, as well as many new candidates whose role in cancer had not yet been explored. Beyond providing a collection of potential biomarkers to improve cancer detection and diagnostics, we elucidated cancer-specific and pan-cancer patterns in candidate protein features, such as combinations of post translational modifications, and their connectivity in protein-protein interaction networks. These results provide a deeper understanding of the cancer-specific protein secretory program, which is a critical step toward improving the detection and treatment of the disease.

501500 A Cybernetic Approach to Modeling Lipid Metabolism in Mammalian Cells

Abstract:

The goal-oriented control policies of cybernetic models have been used to predict metabolic phenomena ranging from complex substrate uptake patterns and dynamic metabolic flux distributions to the behavior of gene knockout strains. The primary logic to this approach is to understand the dynamics of how a system evolves in a regulatory manner towards an objective. Cybernetic theory builds on the perspective that metabolic regulation is organized towards achieving goals relevant to an organism's survival or performing specific biological functions. For example, in multi-substrate cultures of E. coli, cells invest resources into enzymes for pathways which maximize their growth rate. While goal-oriented control has successfully yielded the prediction of numerous metabolic phenomena, the control policy itself has not been compared with data for metabolic regulation.

In this work, we build on preliminary work done by Ramkrishna *et al.* in using cybernetic control variables weighed against gene expression data (an indicator of cellular regulation) in bacterial systems and apply that to mammalian systems. We model the dynamic behavior of prostaglandin formation from arachidonic acid in the mouse macrophage RAW 264.7 cells stimulated by lipopolysaccharide resulting in inflammation and/or compactin, a statin drug. Prostoglandins are a well characterized set of inflammatory lipids derived from arachidonic acid. They are widely studied due to their influence on inflammation and related functions. Statin drugs are used by 93% of adults on cholesterol-lowering medication in the United States[1], and while intended to reduce cholesterol levels in human physiology, the consequences to inflammatory metabolism are largely unknown making this system a good case study.

1. Gu Q, Paulose-Ram R, Burt VL, Kit BK (2014) Prescription cholesterol-lowering medication use in adults aged 40 and over: United States, 2003–2012. NCHS Data Brief: 1–8

496987 A Constraint-Based Method for Integrating Omics Datasets to Improve Flux Predictions

Abstract:

Transcriptomic and/or proteomic data has been integrated into constraint-based models to enhance flux predictions. However, it has been observed that for *E.coli* and *S. cerevisiae*, the predictions obtained by parsimonious flux balance analysis (pFBA) (*1*) are as good or better than those integrating transcriptomic and/or proteomic data (*2*). We developed a constraint-based method called Linear Bound Flux Balance Analysis (LBFBA). The method integrates transcriptomic and/or proteomic data to predict the flux distribution. It requires a training set containing transcriptomic and/or proteomic and fluxomic data under different experimental conditions. These data in the training set are used to further tighten the upper and lower bounds for reactionsâ€™ flux.

We applied LBFBA to *E.coli* and *S. cerevisiae* and compared its predictions to those from pFBA. For *E.coli*, LBFBA predictions (normalized errors around 20%) were more accurate than pFBA predictions (normalized errors around 40%), demonstrating LBFBAâ€™s power to more accurately predict the flux distribution. For *S. cerevisiae,* LBFBA improved the flux prediction and had a smaller error range for every condition, indicating smaller flux variability across alternate optimal solutions. LBFBA predicts flux distribution more precisely since it uses reaction-specific bounds to limit individual fluxes instead of using the same bound for every flux like pFBA or using a global assumption for all reactions (e.g. highly expressed reactions have high flux). Sensitivity analysis on training set size showed that four or five experimental conditions with measured omics datasets was enough to improve LBFBA predictions. LBFBA is the first constraint-based method which has been shown to improve flux predictions by integrating proteomic and/or transcriptomic data.

1. N. E. Lewis *et al.*, Omic data from evolved E. coli are consistent with computed optimal growth from genome-scale models. *Mol. Syst. Biol.* **6**(2010), doi:10.1038/msb.2010.47.

2. D. Machado, M. Herrgård, Systematic Evaluation of Methods for Integration of Transcriptomic Data into Constraint-Based Models of Metabolism. *PLoS Comput. Biol.* **10**, e1003580 (2014).

499901 Supported Biomembrane Microenvironments of Controlled Phase and Cholesterol Content for Gamma-Secretase Substrate Cleavage Assays

Abstract:

The intramembrane protease gamma-secretase is a current target of therapeutic intervention, with pivotal pathological functions within Alzheimer's disease and cancer. The biomembrane microenvironment and local phase compostion is thought to strongly influence the function of this enzyme. Our primary objective is to expand methods for *in vitro* studies of the intramembrane proteases (IMPs) with the development of a microsphere-supported (proteolipobead) and planar biomembrane platforms with phase compostion control. IMPs are found throughout all branches of life and their functions are extremely broad. Despite extensive studies, understanding of IMP regulation and catalysis has been hampered by membrane-associated enzymology. Rhomboids and Secretases are polytopic membrane proteases that are widely conserved in all organisms. The precise reaction mechanisms of the intermembrane proteases remain to be elucidated and furthermore, rigorous analysis of the kinetics of interfacial catalysis in these systems has not yet been undertaken.

High throughput of screening of drug candidates has been carried out in bulk assay systems using cell membrane fragments, solubilized enzymes and is underway in proteoliposomes. A critical barrier to further progress in the study and HTS of gamma-secretase is that such bulk systems do not allow for the direct in situ quantification of enzyme, substrates, or inhibitors or their relative distributions within the structures under assay. We have expanded in vitro models of gamma-secretase to include supported biomembranes, enabling: 1) characterization and verification of biomembrane loading of substrates, inhibitors and protein effectors, 2) studies of co-localization, phase-partitioning and lateral mobility of membrane-bound assay constituents, and 3) the development of flow cytometry-based assays of cleavage. In this work we have characterized the structures, phase localization and compositions of gamma-secretase proteolipobead and planar systems by employing correlative optical and surface microscopy from the micro- to the nanoscale. We have probed the formation of Lo domains in situ, monitored using 3D FRET phase detection. We have probed the phase partitioning of enzyme, substrates and cleavage products in PLB systems with superresolution microscopy in tandem with atomic force microscopy. A combination of biomembrane mobility methods are being used to probe the diffusivities and mobile fractions of lipid, substrates and gamma-secretase. The results of phase partioning and diffusivity measurements are being integrated into bottom-up spatial diffusion-reaction models to correlate microenvironment with enzyme function. As new drug targets in cancer and infectious disease involved in regulated intramembrane proteolysis (RIP) based cell signaling are uncovered, it is expected that other IMP enzymes can be subsequently studied and probed at high throughput with this platform.

499580 Continuous Microfluidic Fabrication of Synthetic Asymmetric Vesicles As Biological Membrane Model Systems

Abstract:

Engineered synthetic vesicles (*i.e.* asymmetric liposomes) support comprehensive *in vitro* studies on membrane organization and associated biological processes (*e.g.* signaling, adhesion, protein transport, endocytosis, etc.). Membrane vesicles are spherical structures consisting of a single lipid bilayer enclosing an aqueous lumen. In nature, biological membranes possess many different lipid compositions with some degree of lipid asymmetry across the bilayer. A robust strategy for preparing customizable asymmetric vesicles will have a significant impact on basic and applied research in membrane biology because they more accurately represent the complexity of natural membranes. Here, we report a novel microfluidic-based fabrication process to build synthetic asymmetric vesicles while simultaneously controlling membrane unilamellarity, monodispersity, vesicle size, and luminal content.

Our microfluidic device consists of a triangular post region and two flow-focusing regions that can generate synthetic vesicles at high-throughput. The vesicle fabrication process has four steps: (1) generate highly monodispersed water-in-oil (w/o) emulsions in an oil/inner-leaflet-lipid solution that serve as precursors to form asymmetric vesicles based on the spontaneous assembly of lipid molecules; (2) replace the oil/inner-leaflet-lipid solution that surrounds the w/o emulsions with an oil/outer-leaflet-lipid solution inside the triangular post region; (3) form water-in-oil-in-water (w/o/w) double emulsions; and (4) extract excess oil/outer-leaflet-lipid solution from the double emulsions. Bilayer membrane asymmetry and unilamellarity are assessed by conducting a fluorescence quenching assay and an Î±-hemolysin (Î±-HL) protein insertion assay, respectively. Our approach addresses many of the deficiencies found in current technologies for building vesicles. This method enables us to observe the vesicle formation process on-chip (*i.e.* due to the transparency of the polydimethylsiloxane microfluidic device) and conduct off-chip experiments immediately after vesicle formation (*i.e.* since the last step for forming the vesicles occurs off-chip). Over 80% of the vesicles remained stable for at least 6 weeks and the membrane asymmetry was maintained for over 30 hours. The asymmetric vesicles built using this strategy are used to investigate fundamental properties of membranes, lipid-protein interactions, and can be used as drug delivery vehicles.

501151 Understanding the Metabolic Shift of *Scheffersomyces Stipitis* from Aerobic Growth to Oxygen-Limited Fermentation at Genome-Scale

Abstract:

Understanding the metabolic shift of *Scheffersomyces stipitis* from aerobic growth to oxygen-limited fermentation at genome-scale

Matthew Hilliard, Thomas Jeffries, Q. Peter He and Jin Wang

Xylose fermentation is essential in making lignocellulosic biofuel processes economically feasible. *Scheffersomyces stipitis* is a strain with the highest native capability of converting xylose to ethanol by fermentation. Because *s. stipitis* is a Crabtree negative yeast strain, oxygen availability is one of the key factors that affect ethanol production due to redox (im)balance in its xylose assimilation pathways. Many studies have examined the effects of oxygenation conditions on ethanol yield and productivity. As most of them were conducted under batch conditions without accurate control of oxygen supply, different conclusions have been drawn regarding the optimal oxygenation condition for ethanol production. In addition, although the cellular transcriptomic profiles of *s. stipitis* were compared for conditions under aerobic growth and oxygen-limited fermentation, up to now, none of the studies have examined the transcriptomic shifts at genome-scale during the transition stage from aerobic growth to oxygen-limited fermentation.

In order to gain fundamental understandings on the metabolic shift and regulatory mechanism of *s. stipitis* during the transition from aerobic growth to oxygen-limited fermentation, transcriptomic profiles of time-series samples collected during the transition period were obtained through RNAseq at a sampling interval of 10 minutes. Specifically, 4 batches of experiments were conducted under controlled chemostat condition to achieve aerobic growth and 2 samples were taken for the aerobic growth condition; then, oxygen supply was reduced to achieve a predetermined carbon-to-oxygen ratio (C/O) that optimizes ethanol production. At the same time, dilution rate was adjusted correspondingly to maintain constant cell density within the bioreactor. Once the oxygen supply was reduced, samples were taken every 10 minutes, with the aim to capture the metabolic shift during the transition via transcriptomic profiling. In addition, xylose, acetic acid, and xylitol concentration, as well as biomass density, oxygen consumption, and CO2 production were measured for each sampling point.

To the best of our knowledge, the obtained transcriptomic data set is the very first time-series data set that captures the transition process of *s. stipitis* from aerobic growth to oxygen-limited condition. Such a data set could provide important understanding on how cells adapt to oxygen-limited condition, and offer key insights on how to design mutants with improved ethanol production performance. Common bioinformatics tools[1,2] will be applied to analyze the data, but more importantly, as cellular metabolism shares many similarities with chemical plants, this work will aim to adapt some of the process engineering tools[3,4], developed for monitoring complex dynamic chemical plants, to analyze the data set. Through these analyses, we expect to identify the key pathways that are up- or down-regulated during the transition, as well as identify pathways that are co-regulated. Such information will be used to generate hypotheses on gene-regulatory mechanisms. Our preliminary results indicate that during the transition, cells induce heat shock response although the temperature was maintained as a constant. In addition, certain genes show initial up-regulation, and subsequently, down-regulation. This behavior suggests that a dynamic model is necessary to capture the cellular metabolic shift during the transition.

[1] Li, B., Dewey, C. N., RSEM: accurate transcript quantification from RNA-Seq data with or without a reference genome, BMC Bioinformatics, (2011).

[2] Robinson, M. D., McCarthy, D. J., Smyth, G. K. edgeR: a Bioconductor package for differential expression analysis of digital gene expression data, Bioinformatics (2010).

[3] G.H. Dunteman, Principal component analysis (Quantitative applications in the social sciences), (1989).

[4] Hua, S., Sun, Z., Support vector machine approach for protein subcellular localization prediction, Bioinformatics, (2001).

499531 Isolation and Metagenome Analysis of Single Microdroplet-Cultivated Microbes

Abstract:

Droplet microfluidics is a powerful tool for investigating biological systems including microbial communities, featuring high throughput and small material requirement. However, isolation and analysis of single droplets become very challenging after generation and incubation of these mini-bioreactors. In particular, the amount of material, such as DNA, from droplets is often not enough for analytical techniques such as sequencing. In this work, we have developed a microfluidic device utilizing a pneumatic closed valve to space and dispense single droplets from a large pool. We achieved up to 83% accuracy in dispensing a single droplet. After droplet dispensing, we successfully amplified the genomic DNA from *E. coli* cells cultured in a single droplet via Multiple Displacement Amplification (MDA). We verified the fidelity of the amplified genome by RT-PCR and whole genome sequencing. We also amplified DNA from droplets containing two species of bacteria and used RT-PCR to confirm both genomes were amplified by MDA. Our technology platform is capable of isolating single droplets for downstream DNA amplification and analysis, widely applicable to the investigation of numerous microbial communities. Applying this technology to a real microbiome, our ongoing work is examining microbial interactions in the human gut by droplet co-cultivation, MDA, and metagenomic sequencing.

488219 Controlled Liquid-Liquid Phase Separation of Recombinant Oleosin

Abstract:

Protein engineering enables the creation of materials with designer functionality and responsiveness. In this work, we designed a protein that phase separates into liquid droplets upon oxidation of a cysteine residue. The phase transition temperature can be tuned via the location of the cysteine residue in the protein backbone. Our work is based on the naturally occurring plant protein, oleosin. The protein is amphiphilic, with distinct hydrophilic and hydrophobic domains: an N-terminal hydrophilic segment, a highly hydrophobic central core, and a C-terminal hydrophilic segment. From wild type oleosin, a truncated version was engineered that reduced the hydrophobic core from 87 amino acids to 30 and added 5 glycines. Circular dichroism indicates that this mutant has less secondary structure than the wild type and approximately 65% of the protein is disordered. A single amino acid mutation was made to introduce a cysteine residue in the hydrophilic arm. Oxidation of this residue results in a dimer. Placing the cysteine closer to the N-terminus raises the transition temperature. The phase behavior of the protein can be eliminated by reducing the protein. Because the transition temperature varies with the location of the added cysteine, this protein provides a novel way to control protein liquid droplet formation.

500753 Evaluation of the Procedure to Obtain Chitosan Based Gels with Potential Use As Bone Adhesive on Clinical Settings

Abstract:

Comminuted fractures are defined as fractures in which an anatomical region is broken into multiple pieces. Medical treatments today have problems dealing with this kind of fractures because of the complications associated with these injuries. Among the techniques used nowadays to treat such wounds are the closed reduction, external fixation of pins, internal fixation with wires and internal reduction and fixation with the use of plates, screws, and wires[1]. Some of these treatments end up in a wrong fixation of fragments in the fracture [2]. Furthermore, in most cases, there are fragments so small that they cannot be fixated by any of the aforementioned techniques and are therefore discarded[3]. This only worsens the treatment, since the fact of having missing pieces may require a shortening in the bone to make an appropriate fixation with traditional methods and avoid an improper fixation[4].

Acknowledging this issue, the use of synthetic osseous adhesives can be considered as an alternative. Such adhesives usually present a great adhesion power, but they fail on being biocompatible [5–8]. Due to the lack of biodegradation and biocompatibility properties, the use of synthetic adhesives is restricted to big fragments of bone. On the other side, some biological based adhesives have been under research, but none of them have been able to reach the desired strength to hold through the stress a bone is constantly under [5, 9–11]. Therefore, this work proposes the use of chitosan as the main component for a bio adhesive that allows fixation of small bone fragments. Additionally, this adhesive is designed to allow cellular growth and biodegradation in the human body, which allows faster recovery without harming the patient.

Previous works have achieved to formulate this bio-adhesive, using three other components apart of chitosan. Two of the components are biocompatible agents that allow the adhesive to be more similar to the bone extracellular matrix, giving it a better chance of cellular growth and differentiation [12–14]. The third component is used as a cross linker to link the polymer chains and form a gel[15]. Due to the chemical interaction of the components with the polymer, the preparation method design is in stages. This allows the different components to interact with chitosan, and to restrict the crosslinking process giving properties to the hydrogels intended to work as bone adhesives (including wanted gelation time, malleability, wettability and mechanical strength.) [14, 15]. Accordingly, the method developed has two mixing stages. Each of these stages uses a different set of stirrers and impellers. The first stage is mixed with a DISPERMAT® LC30 and a Lightweight stainless steel dispersion impeller of 25mm. The second stage is done with a Mechanical stirrer EUROSTAR power control-visc 6000 (IKA) and a three-blade boat propeller of 35mm[14, 15].

While performing preclinical trials, we identified the need to modify the preparation process of the osseous bio adhesive. This was done since the equipment used to produce the gels on laboratory settings could not be sterilized and transported to a clinical setting. Therefore, in this study, a throughout evaluation of chemical and mechanical properties of adhesives prepared with different agitation methods is conducted. In it, different small and easy to sterilize home appliances' motors with fixed velocities and impellers were tested. Additionally, the most viable motor was tested with the original impellers (dispersion and three blade boat) to analyze what the critical factors in the production process are. The analysis of these gels has demonstrated that rheological properties do not reflect the difference in mechanical strength as believed before this work. Nevertheless, the change of the equipment used during the preparation of the adhesive changed the mechanical properties of it, decreasing the adhesive strength up to 62%. Additionally, it has been proven that one of the agitation stages has a bigger impact on the adhesive properties of the product. The mixing velocity resulted to have a more significant impact on the properties of the bioadhesive compared to the impeller used. Based on the given results, we propose a novel mixing method to produce gels on clinical settings and decrease the difference in the mechanical properties comparing the adhesive prepared on clinical and laboratory settings.

The results of the chemical and mechanical tests on the gels produced using both the laboratory and the clinical set of equipment will be presented in detail. Furthermore, we present the novel formulation procedure for the osseous chitosan-based bio-adhesive.

References

1. Holz U (2015) The elements of fracture fixation. Indian J Orthop 49:682. doi: 10.4103/0019-5413.168763

2. Shah JP, Patel SG, Singh B, Shah JP (2012) Jatin Shah's head and neck surgery and oncology. Elsevier/Mosby

3. Schatzker J, Tile M (1987) The rationale of operative fracture care. Springer-Verlag

4. Herscovici D, Scaduto JM (2014) Assessing Leg Length After Fixation of Comminuted Femur Fractures. Clin Orthop Relat Res 472:2745–2750. doi: 10.1007/s11999-013-3292-0

5. Porter JR, Ruckh TT, Popat KC (2009) Bone tissue engineering: A review in bone biomimetics and drug delivery strategies. Biotechnol Prog 25:NA-NA. doi: 10.1002/btpr.246

6. Arora M, Chan EK, Gupta S, Diwan AD (2013) Polymethylmethacrylate bone cements and additives: A review of the literature. World J Orthop 4:67–74. doi: 10.5312/wjo.v4.i2.67

7. Toriumi DM, Raslan WF, Friedman M, Tardy ME (1990) Histotoxicity of Cyanoacrylate Tissue Adhesives: A Comparative Study. Arch Otolaryngol - Head Neck Surg 116:546–550. doi: 10.1001/archotol.1990.01870050046004

8. Fedak PWM, Kolb E, Borsato G, Frohlich DEC, Kasatkin A, Narine K, Akkarapaka N, King KM (2010) Kryptonite Bone Cement Prevents Pathologic Sternal Displacement. Ann Thorac Surg 90:979–985. doi: 10.1016/j.athoracsur.2010.05.009

9. Farrar DF (2012) Bone adhesives for trauma surgery: A review of challenges and developments. Int J Adhes Adhes 33:89–97. doi: 10.1016/j.ijadhadh.2011.11.009

10. Donkerwolcke M, Burny F, Muster D (1998) Tissues and bone adhesives—historical aspects. Biomaterials 19:1461–1466. doi: 10.1016/S0142-9612(98)00059-3

11. Heiss C, Kraus R, Schluckebier D, Stiller A-C, Wenisch S, Schnettler R (2006) Bone Adhesives in Trauma and Orthopedic Surgery. Eur J Trauma 32:141–148. doi: 10.1007/s00068-006-6040-2

12. Swetha M, Sahithi K, Moorthi A, Srinivasan N, Ramasamy K, Selvamurugan N (2010) Biocomposites containing natural polymers and hydroxyapatite for bone tissue engineering. Int J Biol Macromol 47:1–4. doi: 10.1016/j.ijbiomac.2010.03.015

13. Fujihara K, Kotaki M, Ramakrishna S (2005) Guided bone regeneration membrane made of polycaprolactone/calcium carbonate composite nano-fibers. Biomaterials 26:4139–4147. doi: 10.1016/j.biomaterials.2004.09.014

14. Pinzon LM, Cedano FJ, Castro CI, Briceño JC, Casas JP, Tabima DM, Salcedo F (2017) Formulation and Characterization of Chitosan-Based Biocomposites with Potential Use for Bone Adhesion. Int J Polym Mater Polym Biomater 00914037.2016.1263948. doi: 10.1080/00914037.2016.1263948

15. Cedano Serrano FJ, Pinzón LM, Narváez DM, Castro Paéz CI, Moreno-Serrano CL, Tabima DM, Salcedo F, Briceno JC, Casas-Rodriguez JP (2017) Evaluation of a water-resistant and biocompatible adhesive with potential use in bone fractures. J Adhes Sci Technol 31:1480–1495. doi: 10.1080/01694243.2016.1263055

486710 Deep Space Drug Shielding

Abstract:

Space medications help prevent astronauts from being ill and provide acute treatment in medical emergency situation for them to fulfill the primary goal of a successful long duration space mission: maintaining astronauts' health in a unique, isolated, and extreme space environment. However, recently, National Aeronautics and Space Administration (NASA) have reported a shorter shelf life of space medications caused by chronic ionizing radiation and space medications' long-term efficacy in space became questionable. Hence, to help extend the space medications' stability and protect pharmaceuticals from the indirect radiation damage, Trolox, free radical scavenging antioxidant, were bio-conjugated with PLL (Poly-l-lysine) on the surface of drug encapsulated PLGA (Poly-lactic-co-glycolic acid) nanoparticles (NPs) using carbodiimide crosslinking chemistry. In this study, Melatonin, a sleep aid medication used in ISS, was chosen as a model drug. Radiation induced hydroxyl radicals, highly reactive free radicals, will be oxidized by Trolox before it could react with the pharmaceuticals in PLGA-PLL-Trolox composites' core. The mean size of PLGA-PLL-Trolox composites were 491.3 ± 33.09 nm with zeta potential value of 26.4 ± 10.18 mV. A successful conjugation of PLL-Trolox on the surface of melatonin encapsulated PLGA NPs were determined by FT-IR and H^1-NMR analysis. Hydroxyl radical scavenging capacity assay (HOSC) was performed to test Trolox's free radical quenching ability. These results suggest a new molecular approach to shield pharmaceuticals from indirect radiation damage using antioxidant conjugated biocompatible polymer.

501173 Synthesis of Designer Lipids Using "Click" Chemistries

Abstract:

Liposomes have great potential in the fields of drug delivery, imaging and in simplified cell membrane studies. However, to achieve their full potential in these applications, it frequently becomes necessary to incorporate synthetic constructs and new functionalities into the membrane architecture. We have created a simplified method for incorporation of functional groups of interest into the phospholipid structure through the covalent attachment of a second aliphatic tail onto a lysolipid backbone. This addition is now possible through the Copper-catalyzed Azide-Alkyne Cycloaddition (CuAAC), thiol-acrylate and/or thiol-Michael reactions using lysolipids functionalized with an alkyne, thiol, or acrylate group, respectively. This wide range of chemistries enables more simplified synthesis of lipids containing variable chemical moieties and enables the covalent attachment of these functionalized tails using a range of catalysts. In the end, this approach will ease the testing of liposome systems containing new behaviors and characteristics.

501181 Loading and Mobility of RNA in Porous Silica Nanoparticles for Â Delivery to Insects

Abstract:

RNA interference (RNAi) technology, which enables the suppression of targeted genes, is of wide interest to insect pest management through the silencing the genes that relate to the insect fitness or mortality. Identifying efficient methods to deliver the dsRNA to insects is the key challenge to achieve effective RNAi. Porous silica nanoparticles (pSNPs), which have tunable pore morphology, large surface area, biocompatibility and ease of surface functionalization, are potential carriers for biomolecules. The goal of this work is to demonstrate the loading, release, and protection of RNA in porous silica nanoparticles. pSNPs with pore sizes appropriate for biomolecule loading have been successfully synthesized to encapsulated the dsRNA within the pores. The silica particles were further functionalized with aminosilanes, providing amine groups with positive charge on the surface necessary to load negatively charged nucleic acids. To investigate the RNA interaction with the particles, direct visualization of the loading of RNA on micron-sized silica particles with similar pore sizes to the pSNPs was performed under confocal laser scanning microscopy. The interaction of dsRNA with the pSNP and the accessibility of the pores of the pSNP can be described from the diffusivity (mobility) of the dsRNA in RNA-loaded pSNPs. Using Fluorescence Recovery after Photobleaching (FRAP), the effect of both pore size (nonporous, 3.6 nm, 7.4, and 11.8 nm diameter) and length of dsRNA (84 bp and 282 bp of fragments within the inhibitor of apoptosis (IAP) genes of *Spodoptera frugiperda*) on the diffusivity at the surface and in the core of dsRNA-loaded pSNPs has been quantified. The importance of pores to a mobile dsRNA network is demonstrated by the lack of measurable mobility for both lengths of RNA on nonporous materials. In addition, when the dsRNA could not penetrate the pores, dsRNA mobility is also not measurable at the surface of the particle. Thus, the pores seem to serve as a â€œsinkâ€ in providing a mobile network of dsRNA on the surface of the particle. The mobility of dsRNA in the nanopores of the pSNPs is expected to have a functional effect on delivery of dsRNA to insects and the delivery of the IAP genes to *Spodoptera frugiperda* using pSNPs is interpreted from loading and mobility as a function of pore size. This work successfully demonstrates the loading of RNA on functionalized pSNPs and identified factors that affects RNA loading and releasing, which provides basis for the delivery of RNA-loaded silica particles *in vivo*.

505175 Engineering and Functionalizing Protein-Based Materials

Abstract:

The development of materials with diverse mechanical and functional properties enables a broad range of applications. For materials composed of protein, well-established molecular biology techniques can theoretically be used to alter the mechanical and functional properties of materials. However, repetition of amino acid sequences in the constituent proteins can complicate engineering the mechanical properties. Furthermore, most protein-based materials are assembled under denaturing conditions, thus precluding incorporation of full-length proteins in an active state. Our lab discovered that the *Drosophila melanogaster* transcription factor Ultrabithorax (Ubx) self-assembles into materials. These materials are stabilized by specific dityrosine bonds that spontaneously form during assembly. The location of the participating tyrosines in unique sequence motifs enables the reliable tuning of the materials by mutagenesis to be either weaker or stronger. Because Ubx materials rapidly form in mild, aqueous buffers, a surprisingly wide variety of proteins can be incorporated via gene fusion without harming either materials assembly or the function of the appended protein. Indeed, when Ubx is fused to angiogenic growth factors, the resulting fibers control cell signaling and cell behavior *in vitro*, and can instigate and guide blood vessel formation *in vivo*. Finally, Ubx proteins in materials retain the ability to bind specific DNA sequences with high affinity. Ubx materials provide an inexpensive and facile platform for customizing materials for a variety of applications.

496520 Lab Automation for Industrial Biotechnology: High-Throughput Cloning for Genome-Scale over-Expression Library Creation and Parts Assembly

Abstract:

In the post-genomic era, the capability of a regular lab could not meet the growing demand for high-throughput cloning and multiple parts assembly in the field of industrial biotechnology research. The development of laboratory automation make it possible to perform complex molecular biology procedures in large-scale with minimal human labor, error and cost. Here, we have developed and optimized a series of automation procedures with high-throughput robotic platform for cloning all 3057 genes of *Corynebacterium glutamicum* individually into a controllable over-espression plasmid within a month. The single round cloning accuracy is ~85%. Such genome-scale over-expression library would be an extremely valuable tool for research related to this industrial workhorse strain. Then we applied our automation platform in DNA assembly process and constructed 96 plasmids in one experiment with full combinations of 6 different kinds of biological parts, including 6 replication origins, 4 antibiotic resistance genes and 4 promoters. In the highly competitive industrial biotechnology filed, with further development and optimization, our lab automation platform would significantly reduce labor requirement and enable researchers to design and test novel ideas in a scale that was not possible before.

497897 Non-Growth Metabolism in Synthetic Auxotrophic Knockouts of *Escherichia coli* for Chemical Production

Abstract:

Decoupling growth and production phases in microbial cell factories to limit carbon loss to biomass is of substantial interest to improve product yields. However, *E. coli* that stop growing due to a stress (e.g. nitrogen limitation) have drastically reduced overall metabolism. Therefore, deregulating non-growth metabolism is essential to achieving efficient stationary phase chemical production. We have evaluated the effect of metabolic gene deletions on central carbon metabolic regulation. We have investigated non-growth conditions that have enhanced glucose uptake rates compared to stationary phase cells. We want to further understand these metabolic changes to identify regulatory bottlenecks in central carbon metabolism for future strategies to increase non-growth metabolism and divert carbon to valuable products. To do so, a robust analysis pipeline must be in place including: rapidly quenching metabolism, derivatization of metabolites for GC-MS analysis, design of quality control procedures, and benchmarking available software packages to correctly identify and quantify each metabolite. We have used this platform to characterize synthetic auxotrophic knockouts of *E. coli*, and have investigated chemical production in select strains. This metabolomics platform is generally useful to inform strain engineering strategies to improve yields and productivities for the biosynthesis of a wide range of different chemicals and fuels.

Engineering of a Highly Efficient *Escherichia coli* Strain for Mevalonate Fermentation through Chromosomal Integration

Jilong Wang
Kechun Zhang Lab
University of Minnesota

November 2, 2017

UNIVERSITY OF MINNESOTA
Driven to Discover

Many Challenges Facing Chemical Industry

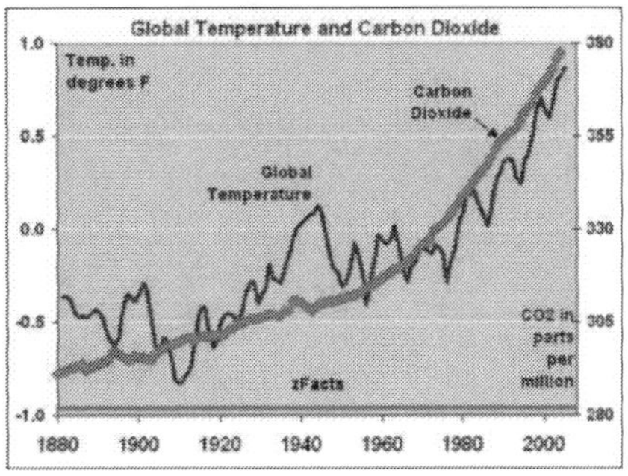

UNIVERSITY OF MINNESOTA
Driven to Discover

2

Zhang Lab Approach to Expanding Metabolism

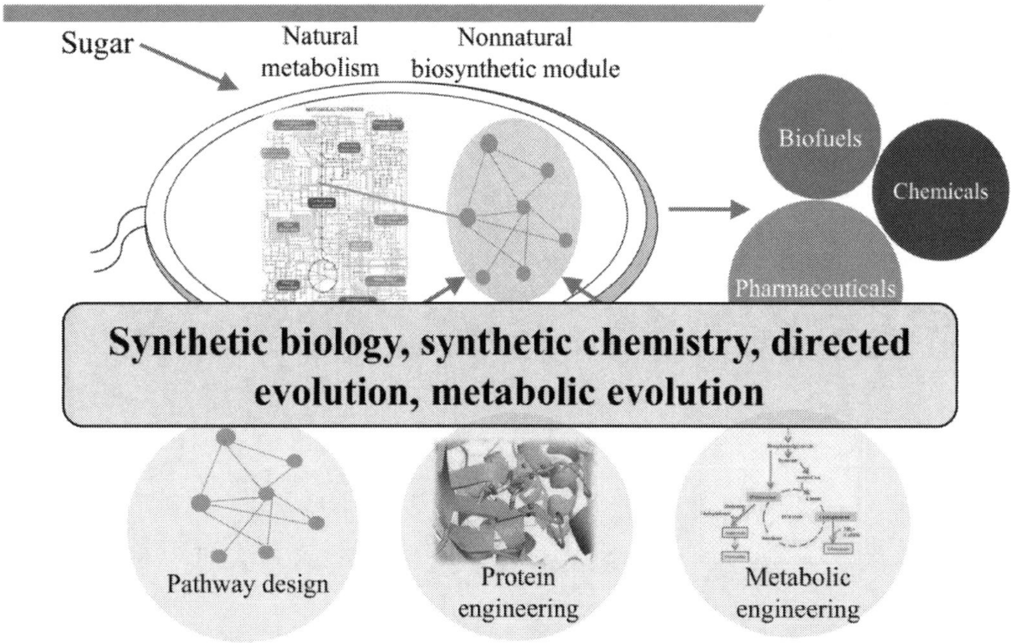

Metabolic Engineering for Large Scale Fermentation

Disadvantages

- Plasmid-based fermentation
  - ➢ Metabolic burden
  - ➢ Segregational instability
- Inducible promoters
  - ➢ Inducer cost
  - ➢ transcriptional heterogeneity

Solutions

- Technologies for chromosome manipulation
  Chromosomal integration of desired genes
- Inducer free promoters
  - ➢ Constitutive promoters
  - ➢ Temperature-sensitive
  - ➢ Anaerobic-inducible

Mevalonate production as an example

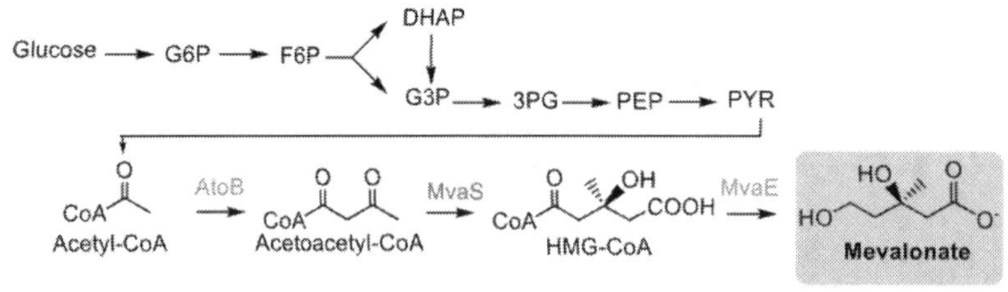

Application of Mevalonate

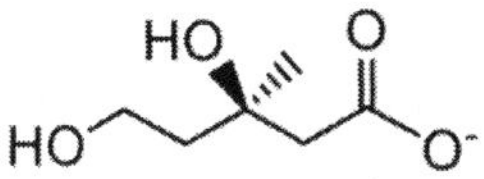

Mevalonate was discovered as growth factor in 1956.

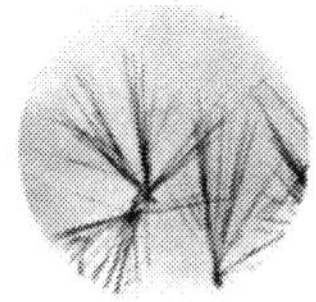

It has been used in:

- Cosmetics
- Sustainable polymers

Precursor to:

- Drugs
- Food additives
- Isoprenoids
- β-methyl-δ-valerolactone (MVL)

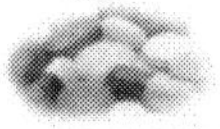

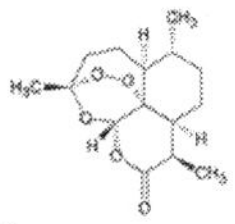

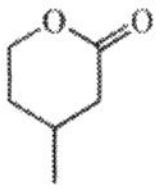

Designed Pathway for MVL

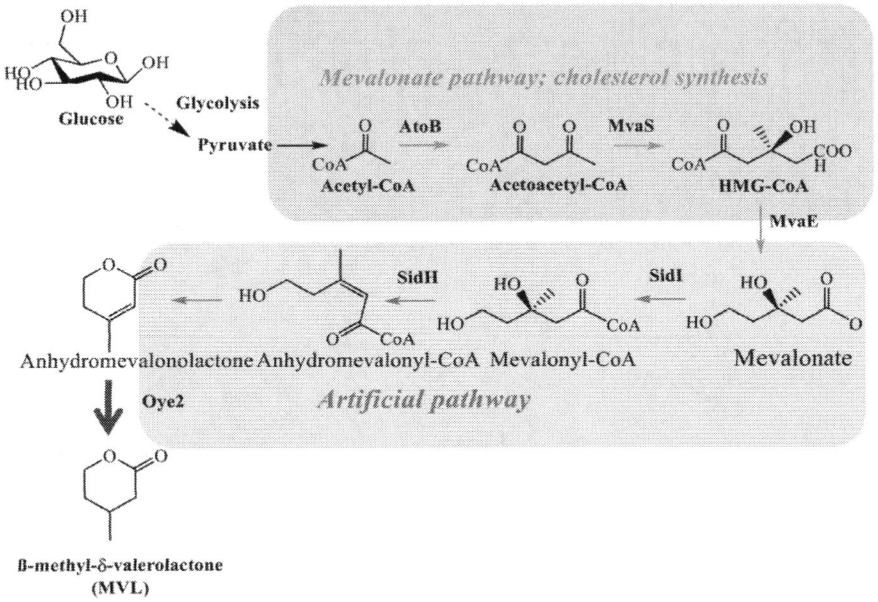

Xiong *et al*. PNAS, 2014 ;111(23):8357-62.

Mechanically Tunable Block Copolymers

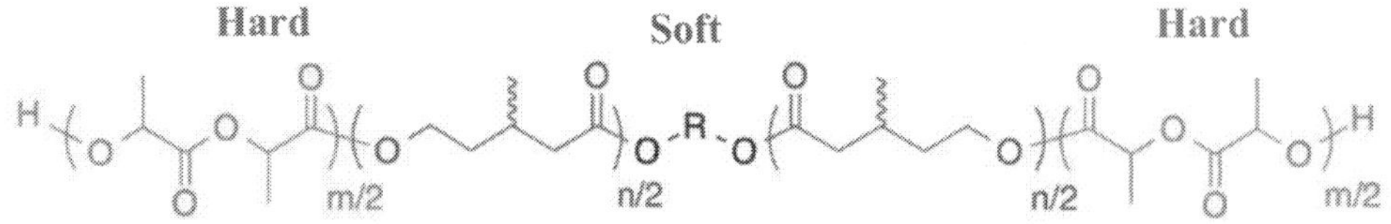

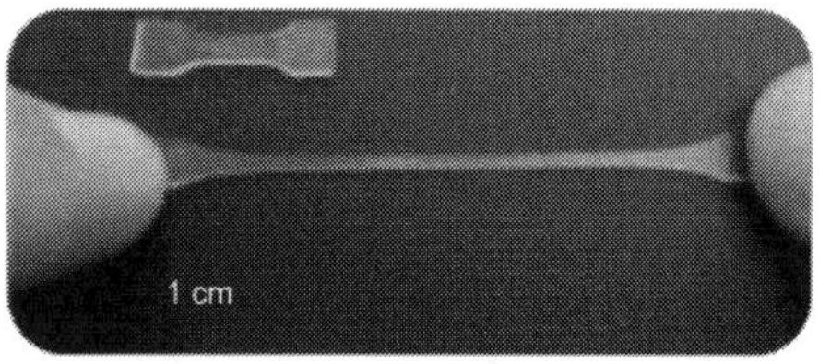

Polymer synthesized by Debbie Schneiderman

Characteristics:
E: 1.9 Mpa
σ_b: 9 Mpa

Stretches to 18 times its original length without breaking

UNIVERSITY OF MINNESOTA
Driven to Discover

Xiong *et al.* PNAS, 2014 ;111(23):8357-62.

Using Biodegradable Monomers for Block Co-Polymers

Using multi-blocks:

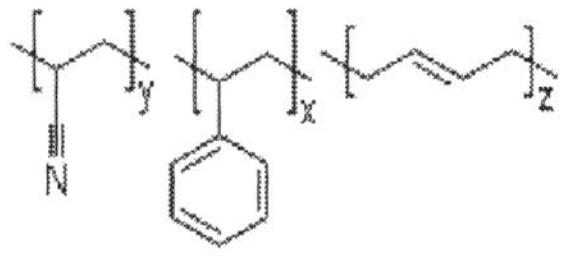

ABS polymer

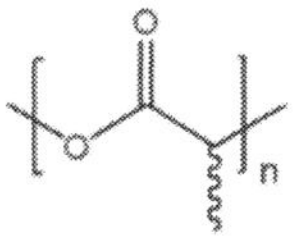

$T_g = 58$ °C

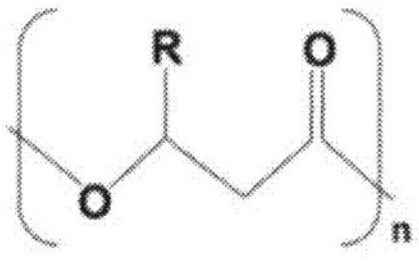

$T_g = 5$ °C, but semicrystalline

Make products like:

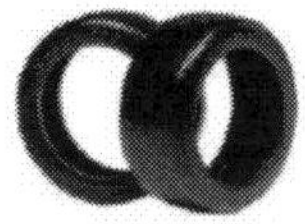

Multiblock polymers: panacea or Pandora's box? Bates *et al.* *Science.* 2012 336(6080):434-40.

UNIVERSITY OF MINNESOTA
Driven to Discover

Chromosomal Integration for Mevalonate Production

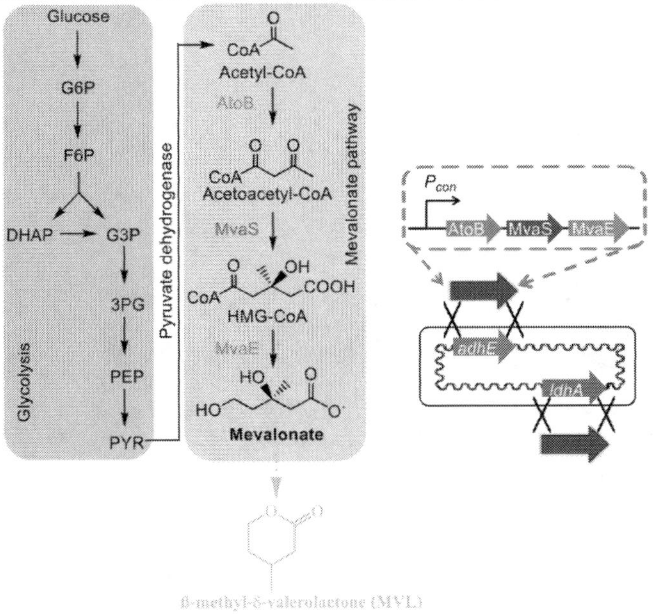

UNIVERSITY OF MINNESOTA
Driven to Discover

Wang *et al.* Appl Environ Microbiol. 2016;82(24):7176-7184.

Chromosomal Integration for Mevalonate Production

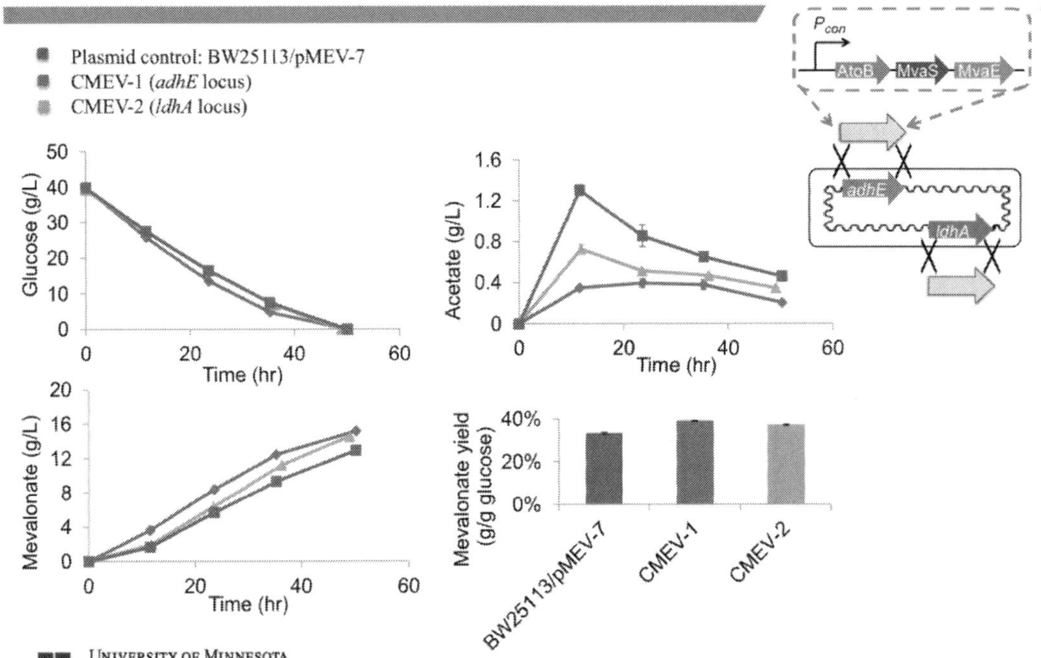

UNIVERSITY OF MINNESOTA
Driven to Discover

Wang *et al.* Appl Environ Microbiol. 2016;82(24):7176-7184.

Improvement of The Production Rate

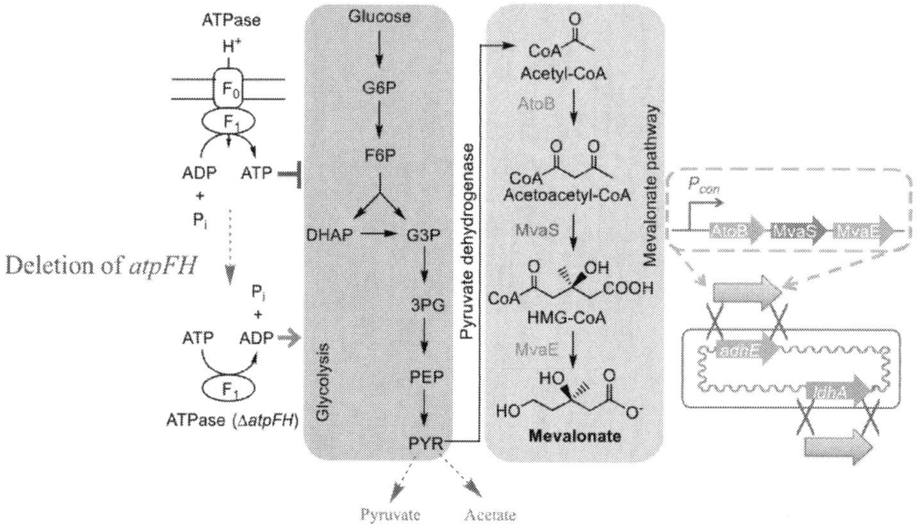

Wang *et al*. Appl Environ Microbiol. 2016;82(24):7176-7184.

Improvement of The Production Rate

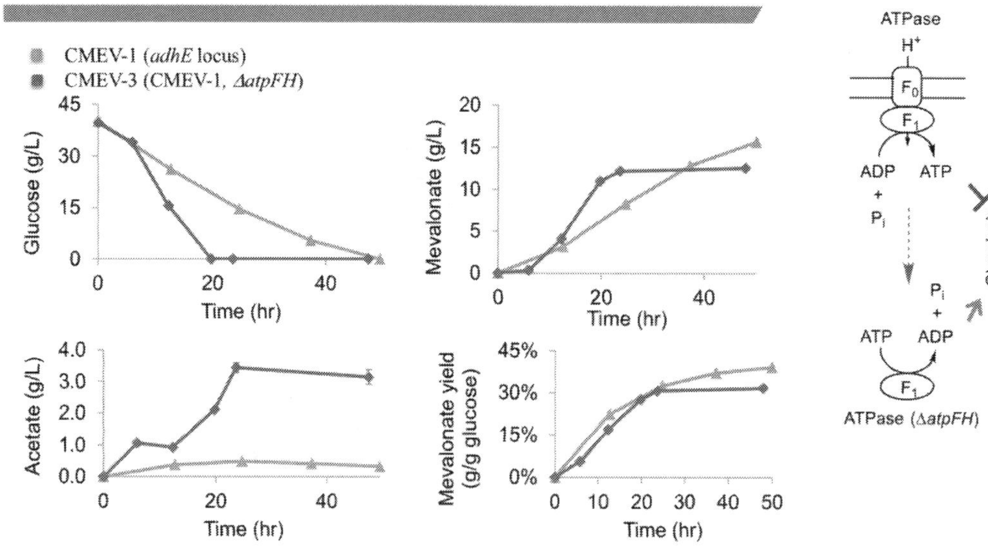

Production rate increased from 0.43 to 0.92 g/L/hr

Wang *et al*. Appl Environ Microbiol. 2016;82(24):7176-7184.

Enhancing Carbon Flux into Mevalonate Synthesis.

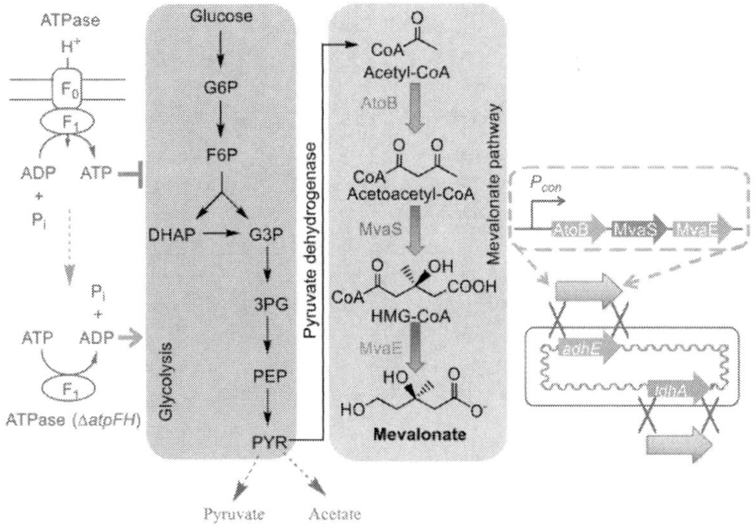

UNIVERSITY OF MINNESOTA
Driven to Discover

Wang *et al.* Appl Environ Microbiol. 2016;82(24):7176-7184.

Enhancing Carbon Flux into Mevalonate Synthesis.

- CMEV-3 (*adhE* locus, *ΔatpFH*) 1 copy
- CMEV-6 (*adhE* and *ldhA* loci, *ΔatpFH*) 2 copies

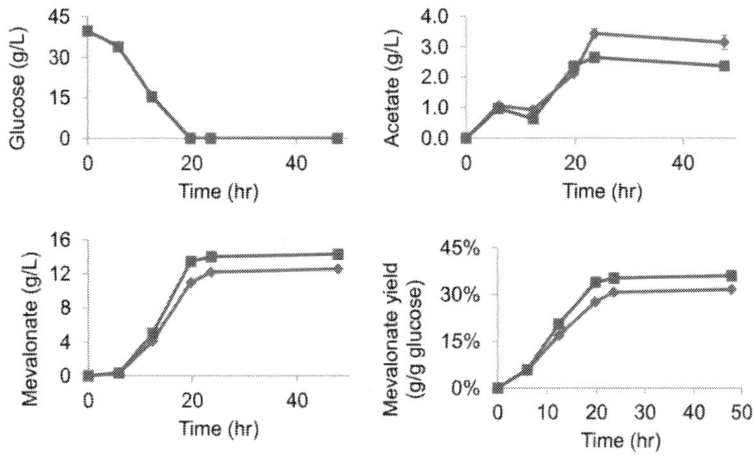

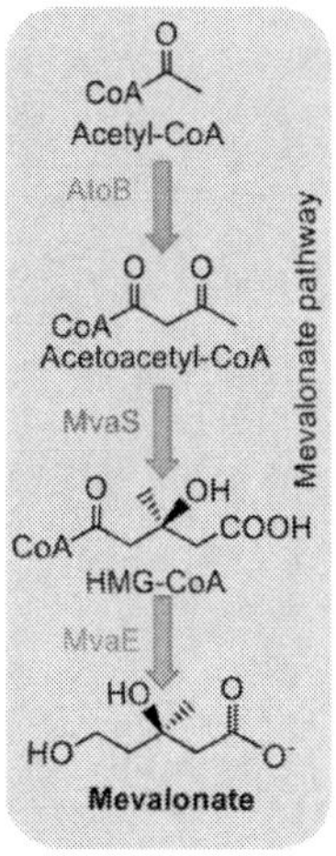

UNIVERSITY OF MINNESOTA
Driven to Discover

Wang *et al.* Appl Environ Microbiol. 2016;82(24):7176-7184.

Fed-batch Fermentation Study

- CMEV-3 (*adhE* locus, *ΔatpFH*) 1 copy
- CMEV-6 (*adhE* and *ldhA* loci, *ΔatpFH*) 2 copies

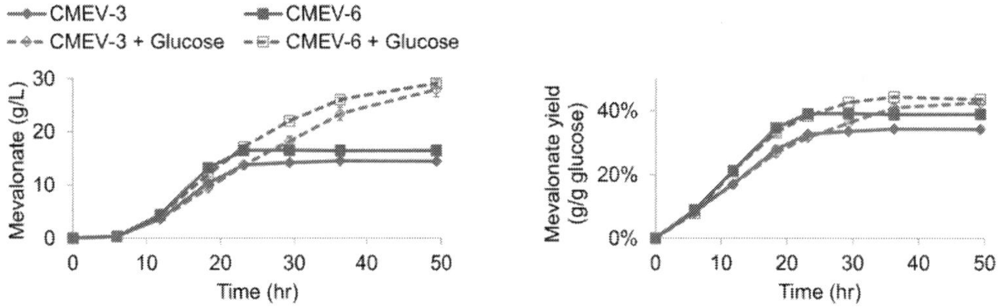

Wang *et al*. Appl Environ Microbiol. 2016;82(24):7176-7184.

Deletion of *sucA*

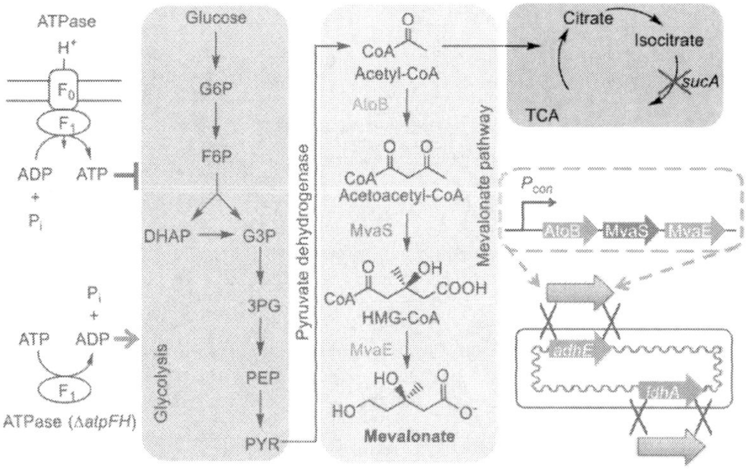

Wang *et al*. Appl Environ Microbiol. 2016;82(24):7176-7184.

Deletion of *sucA*

- CMEV-6 (*adhE* and *ldhA* loci, Δ*atpFH*)
- CMEV-7 (*adhE* and *ldhA* loci, Δ*atpFH*, Δ*sucA*)

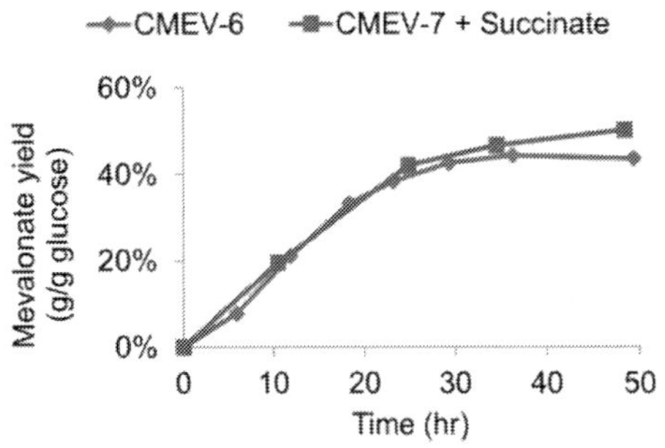

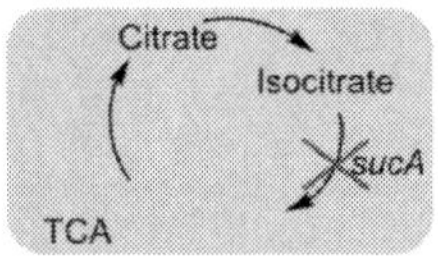

Summary

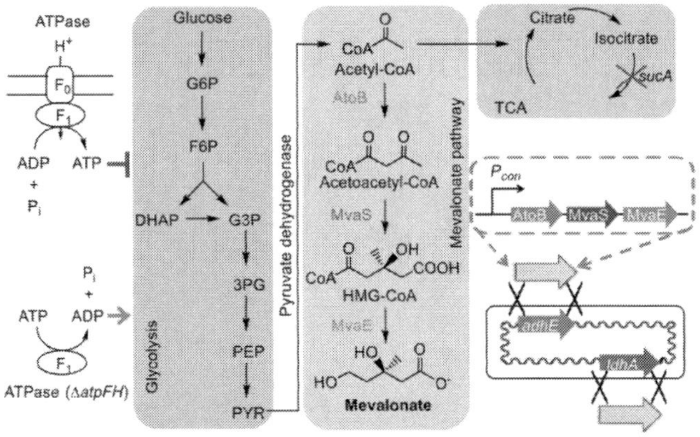

Optimization strategy	Titer (g/L)	Productivity (g/L/hr)	Yield (% theoretical yield)
Chromosomal integration	12.9 to 15.2	0.34 to 0.43	60.6 to 71.2
Δ*atpFH*	15.2 to 27.9	0.43 to 0.92	71.2 to 77.6
Integration of two copies	27.9 to 29.1	0.92 to 1.08	77.6 to 79.4
Δ*sucA*	29.1 to 30.8	1.08 to 1.01	79.4 to 91.6

The Effect of Yeast Extract

Yeast extract (g/L)	OD_{600}	Acetate (g/L)	Mevalonate (g/L)	Yield g/g glucose	Yield (% theoretical yield)
0	2.93 ± 0.04	1.05 ± 0.08	5.3 ± 0.3	0.472 ± 0.009	86.1 ± 1.7
0.5	4.61 ± 0.17	1.25 ± 0.28	8.8 ± 0.2	0.462 ± 0.016	84.3 ± 2.9
1	6.37 ± 0.59	1.08 ± 0.30	13.2 ± 1.1	0.453 ± 0.017	82.6 ± 3.2
5	11.61 ± 0.96	0.21 ± 0.10	23.3 ± 1.2	0.520 ± 0.026	94.9 ± 4.7

Conclusions and Future Directions

- The chromosomal integration is plasmid free and inducer free.

- Δ*atpFH* strain increases the acetyl-CoA flux
  - ❖ Productivity: 0.43 to 0.92 g/L/hr
  - ❖ Titer: 15.2 to 27.9 g/L/hr

- Development of an efficient platform by chromosomal integration

  Large scale production of acetyl-CoA-, pyruvate-, or mevalonate-derived chemicals. (Shake flask experiment: ~ 30 g/L after 48 hrs; ~1 g/L/hr)

Acknowledgements

Zhang Group:

Current:

- Maria McClintock
- Qiuge Zhang
- Zhengyang Liu
- Yuxiao Wu

Former:

- Dr. Mingyong Xiong
- Dr. Jingyu Wang
- Suthamat Niyompanich
- Yishu Tai
- Prof. Wenqin Bai
- Pooja Jambunathan

UNIVERSITY OF MINNESOTA
Driven to Discover

Thanks to the funding support from NSF CHE-1413862

21

497099 Enabling Glucose-Xylose Co-Utilization in Yeast through Expression of Xylose-Specific Transporters

Abstract:

Co-utilization of glucose and xylose is highly desired for cost-effective industrial biofuel and biochemical production from lignocellulosic biomass. However, in yeast, xylose uptake is significantly repressed by glucose, which impedes the implementation of yeast as an efficient platform for biorenewables production. Most of the previous studies focused on engineering yeast hexose transporters to alleviate glucose inhibition as well as increase xylose affinity to improve co-sugar utilization. In contrast to *Saccharamyces cerevisiae*, which cannot naturally metabolize xylose, nearly all the yeasts from CUG clade possess the inherent capability to assimilate xylose. Taking advantage of this property, we were able to identify three xylose-specific transporters from CUG clade yeasts, which could efficiently transport xylose even in the presence of glucose. Among these three, the best transporter allowed simultaneous co-sugar utilization, and was able to assimilate 30 g/L xylose in the mixture of 70 g/L glucose and 40 g/L xylose. The transporters were characterized in an industrially relevant CUG clade strain, *Scheffersomyces stipitis,* which possesses a superior capability of producing ethanol from xylose. The performance of the best newly identified transporter was compared with the best reported transporters in *S. stipitis*. Our work demonstrates that besides reshaping hexose transporters, seeking naturally existing high-performance xylose-specific transporters free from glucose repression is practical. Our results further enhance the potential of *S. stipitis* as a platform strain for commodity and specialty chemicals production in industry.

500829 Deciphering the Elusive Role of Branched-Chain Amino Acid Transaminases in the Production of Branched-Chain Higher Alcohols in *Saccharomyces cerevisiae*

Abstract:

The yeast *Saccharomyces cerevisiae* is an attractive host for the production of branched-chain higher alcohols (BCHAs), including isobutanol, isopentanol, and 2-methyl-1-butanol, which are promising advanced biofuels. Yeast is known to naturally produce these alcohols as degradation products of valine, leucine, and isoleucine, respectively. Metabolic engineering of BCHA biosynthesis in yeast is complicated by the separation of the upstream (branched-chain amino acid biosynthesis) and downstream (Ehrlich degradation) biosynthetic pathways between the mitochondria and the cytosol, respectively. Moreover, each of these compartments contains distinct branched-chain amino acid transaminases (BCATs), encoded by *BAT1* (the mitochondrial BCAT) and *BAT2* (the cytosolic BCAT). Previous studies examining the effects of deleting or overexpressing *BAT1* and/or *BAT2* have yielded inconsistent results, which have obscured the role of these key enzymes and further complicated our ability to engineer optimal BCHA biosynthetic pathways in yeast. In this work, we resolve these previous conflicting results and conclusively elucidate the roles of BCATs in the biosynthesis of BCHAs. We achieve this by taking a rational and combinatorial approach to metabolic engineering and pathway compartmentalization. With this new knowledge, and by manipulating transamination activity alone, we were able to increase isobutanol production by more than 15-fold over wild type strains. Further addressing the intracellular transport bottleneck due to transamination activity in different compartments, we obtained an additional doubling of BCHA production. This study answers the longstanding question of the role of transamination activity in BCHA production in yeast, and develops valuable strains for future optimization of these advanced biofuels.

494347 Metabolic Engineering for Terpenoids Overproduction and Discovery

Abstract:

Terpenoids comprise the largest family of natural products with wide applications, and many of them have been engineered for commercial use. Our strategy for engineering terpenoids is based on the instruction of our *in vitro* reconstitution assay of mevalonate (MVA) pathway. The information from this system guided us to establish an efficient precursor-producing platform for terpenoids. Base on this well optimized platform, a high titer of farnesene was obtained in a short time with little mutant constructed[1]. Subsequently, we directly extend the application of this strategy and platform to realize a high titer production of lycopene[2], astaxanthin[3] and taxadiene[4] production in *E. coli, S. cerevisiae* and filamentous fungi quickly and efficiently.

Besides the production of terpenoids with important industrial applications, efficient heterologous terpenoids overproduction platforms also provide the perfect solution for us to activate the silent genes, gene clusters. Or resolve the problem that the production of target compounds is too low to be characterized, which may frequently encounter for natural product discovery. With the help of well-established terpenoids overproduction platforms in *E. coli*[5] and *S. cerevisiae*, we have characterized a large number of sesquiterpene, diterpene and sesterterpene synthase, which were usually silent or with low expression level in filamentous fungi. Some new terpene skeletons and new compounds that were usually difficult to obtain were discovered by this strategy. It's reasonable for us to believe that the our technical platform and strategy can promote the progress of natural products discovery greatly.

References:

1. Zhu F, Zhong X, Hu M, *et al*. (2014). *Biotechnol Bioeng*, 111, 1396-1405.

2. Zhu F, Lu L, Fu S, *et al*. (2015). *Process Biochem*, 50, 341-346.

3. Ma T, Zhou Y, Li X, *et al*. (2015). *Biotechnol J*, 11, 228-237.

4. Bian G, Yuan Y, Tao H, *et al*. (2017). *Biotechnol J*, doi: 10.1002/biot.201600697.

5. Bian G, Han Y, Hou A, *et al*. (2017). *BioRxiv*, http://biorxiv.org/content/early/2017/02/25/105247.

503869 Industrialization of Biology: Making Nature Accessible and Affordable

Abstract:

Plants produce a variety of rare natural products which are used in our daily lives as flavors, fragrances, food ingredients, cosmetics, vitamins, pharmaceuticals and agricultural chemicals. Despite their intrinsic value, however, sourcing remains a bottleneck to more widespread use due to their low abundance in nature. Manus Bio has established an innovative platform technology for engineering cell factories grounded in modular and data-driven design and developed a commercial organism which can produce a myriad of typically ultra-rare and costly ingredients used in our daily lives. These cellular factories are capable of converting carbon feedstock to product at high yields and have been adapted to produce a mature pipeline of products, or "BioAssemblyLine."

496472 Use of a Trickled Bed Reactor to Improve the Commercial Feasibility of the Hydrogenation of a Nitro-Compound

Abstract:

Use of a Trickled Bed Reactor to Improve the Commercial Feasibility of the Hydrogenation of a Nitro-Compound

Carla Luciani, Jonas Buser, Michael Laurila, Richard Cope, Kevin Cole, Bradley Campbell, Justin Burt, Martin Johnson, Joseph Martinelli, David Mitchell

Lilly Research Laboratories, Lilly Corporate Center, Indianapolis, IN-46285, USA

The use of continuous processing to manufacture intermediates and drug substances has grown significantly over the last few years. While the major reasons for the rapid adoption of continuous technologies in the pharmaceutical industry are typically linked to productivity and safety as well as quality benefits, there are cases where the use of continuous processing is required long term to ensure the process robustness required for commercial manufacturing.

This work investigates the Pd-catalyzed reduction of a nitro compound to produce an aniline during the second step of the proposed registered sequence. Major challenges resulting from the complexity of the reaction matrix were overcome by the use of advanced monitoring techniques (flow-NMR and on-line HPLC). Those monitoring techniques produced exceptionally detailed kinetic profiles that were used to fit kinetic models. The kinetic models were used to simulate the reaction profiles using both continuous and batch reactor types. In addition to the typical manipulated variables investigated for the batch option, such as temperature, pressure, catalyst loading, and catalyst poisons entering the process, several other process parameters were investigated to understand the trickle bed reactor performance (e.g., temperature and concentration profiles in both radial and axial direction, flowrates, wetted catalyst surface, catalyst deactivation mechanism, etc.). The model was used to compare the resulting feasible operating regions/design spaces for both batch and flow modes. Finally, the model was used to investigate re-packing strategies and optimal temperature profiles during operation.

500963 Use of Process Analytical Technology to Gain Mechanistic Insights into Spherical Agglomeration of Active Pharmaceutical Ingredients

Abstract:

Crystallization is a solid-liquid separation process widely used in the pharmaceutical industry to purify and establish physical properties of the drug substance. The performance of subsequent downstream unit operations often depends on the quality of crystals generated in the crystallization. Drug substance attributes (*e.g.*, crystal size, shape, etc.) directly impact the filtration and drying performance in addition to drug product manufacturing processes.

Although pharmaceutical manufacturing processes such as crystallization are well established, more intensified downstream unit operations are being pursued as alternative manufacturing strategies. Spherical agglomeration is a process intensification unit operation that combines crystallization, milling, and granulation into a single unit operation. Historically, growth dominant crystallization processes have been used to generate large drug substance crystals for improved filterability, impurity control, and process robustness. However, in some cases, the product crystals must then be milled to in order to achieve the desired bioavailability. The powder properties that result from the milling operation are often non-ideal for use in the drug product, since the powders can be poor flowing, cohesive, and electrostatic. A powder granulation step is often required to help mitigate these short comings. Spherical agglomeration might have the ability to address these short comings and combine three unit operations (crystallization, milling, granulation/blending) into a single step process, and deliver a powder with desired drug product manufacturing attributes.

In the present work, Process Analytical Technology[1] (PAT) is used to gain mechanistic insight into the effect of various process parameters on spherical agglomeration. Agglomeration of the drug substance has been carried out using a variant of spherical agglomeration, the Quasi Emulsion Solvent Diffusion (QESD) technique. In this case, a quasi-emulsion is formed and used to agglomerate drug substance crystals inside emulsion droplets. It has been shown in the open literature that changing key process parameters such as the turbulent energy dissipation rate, the concentration of solute in the initial solution, and the solvent to anti-solvent ratio changes the dynamics of quasi-emulsion that is formed and hence the quality of agglomerates that are produced. This contribution proposes a mechanistic understanding of QESD through the utilization of PAT, and provides insight into the fundamental rate processes that are required in order to design QESD processes with tailored powder properties.

References:

[1] Y. Yang *et al.*, â€œApplication of Ultra-Performance Liquid Chromatography as an Online Process Analytical Technology Tool in Pharmaceutical Crystallization.â€, *Cryst. Growth Des., 2016, 16, 7074â"7082*

489889 Process Dynamics of Continuous Cooling Crystallization of Carbamazepine in a Mixed Suspension Mixed Product Removal (MSMPR) System

Abstract:

Current pharmaceutical manufacturing is mainly based on batch processing which has inherent drawbacks and disadvantages. To overcome these drawbacks, emerging technologies such as continuous manufacturing (CM) have been highlighted in the industry in the past few years [1, 2]. As the final purification and isolation step, crystallization has a significant impact on the final physicochemical properties of drug substance and is considered one of the critical process steps in the continuous manufacturing of drug substances. The ability to monitor process dynamics of a continuous process is critical for successful control of crystal quality attributes, such as size distribution. PAT tools such as Focused Beam Reflectance Measurement (FBRM) and spectroscopy based system (i.e. NIR, FTIR, and Raman) can be used to monitor the crystallization process. In this work, a systematic framework to evaluate and understand process dynamics using PAT tools is presented.

The continuous cooling crystallization of carbamazepine (CBZ) in ethanol in a multi-stage MSMPR system is used as the case study. FBRM and Raman spectroscopy are implemented for on-line monitoring. Various startup conditions are evaluated to determine the optimal scenario for the continuous system of interest. The impact of disturbances on the process dynamics and state of operation are evaluated by implementing pulse or step changes to the temperature, flowrate, feed solution concentration, and polymorphic form. Slurry samples for all experiments are obtained and characterized by off-line HPLC, laser diffraction and XRPD before and after the disturbances. The on-line and off-line data obtained are used to characterize the effect on crystal properties (i.e. size and form) and process parameters (i.e. solution concentration). Moreover, simulation studies for the crystallization process are conducted using population balance models to further understand the impact of process disturbances on quality attributes for the continuous cooling crystallization of CBZ.

1. Lee, S.L., et al., *Modernizing pharmaceutical manufacturing: from batch to continuous production.* Journal of Pharmaceutical Innovation, 2015. **10**(3): p. 191-199.

2. Poechlauer, P., et al., *Continuous processing in the manufacture of active pharmaceutical ingredients and finished dosage forms: an industry perspective.* Organic Process Research & Development, 2012. **16**(10): p. 1586-1590.

501155 Monte Carlo Stepwise Regression for More Accurate Selection of Critical Process Parameters during Process Characterization

Abstract:

Design of experiments (DOE) is commonly used during process characterization (PC) to define control strategies for the manufacture of drug substance. Although the use of sophisticated DOE techniques in establishing the conditions for PC is common, the analysis of the results is often done with a stepwise regression reliant on p-value thresholds. This technique can often produce artificially large models with multiple false positives, leading to unnecessary work to examine spurious variable relationships and potential implementation of overly complex layers of control in the manufacturing process. This work describes the use of Monte Carlo cross validation (MCCV) simulations to aid in the selection of critical process parameters (CPP). Once models were selected defining the relationship between CPPs and critical quality attributes (CQA), MCCV was used to assess the accuracy of those models. Monte Carlo simulations were then used to establish an appropriate control strategy with defined operating ranges (OR) predicted to meet quality standards.

Three recombinant monoclonal antibody cell culture processes were evaluated. The process parameters that were studied included dissolved oxygen, pH, temperature, inoculation density, and feeding strategy. The quality attributes that were evaluated included aggregate levels and charge variants. Each PC was carried out using a D-optimal design performed in 2L bioreactors. The resulting data was analyzed using MATLAB. Stepwise regression with backward elimination was performed to establish baseline models. MCCV stepwise regression with both forward and backward elimination was then performed to identify process parameters which may have been inappropriately included by the baseline method. When available, large-scale data was used to validate the accuracy of the models along with MCCV. ORs were established by Monte Carlo simulation by modifying CPP ranges until appropriate quality thresholds were met.

A comparison of stepwise regression with other model building techniques was also performed. The simplified interpretation of stepwise regression presented here has the advantage of clearly defining CPPs, despite the modest increases in predictive accuracy afforded by more advanced and complex methods, such as regularization (LASSO, ridge) and dimensionality reduction (PCA) . Applying these techniques to a PC program can lead to more robust interpretation of data, higher statistical confidence in CPP selection, and establishment of more reliable ORs.

Delivering a Design Space for a Continuous Drug Substance Manufacturing Process Using Simulation

Neil Hodnett
GlaxoSmithKline R&D

Agenda

1. Introduction
   - Overview of continuous manufacturing process
   - Motivation for this work
2. Approach
3. Results
   - Specific example of impurity control
4. Summary
   - Key learnings
   - Future opportunities

Introduction

Process overview

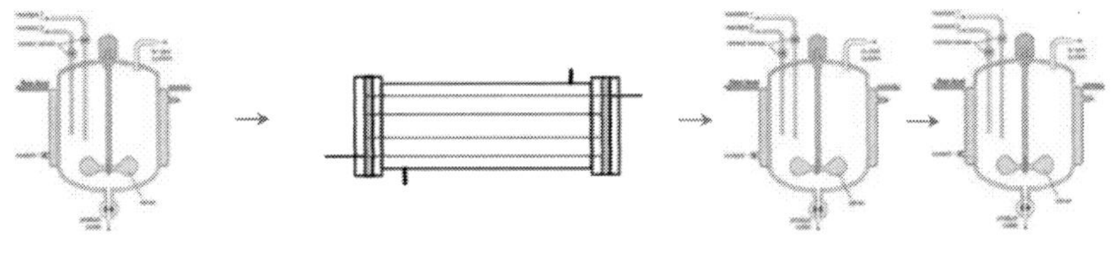

- 5 telescoped transformations
- 13 process parameters

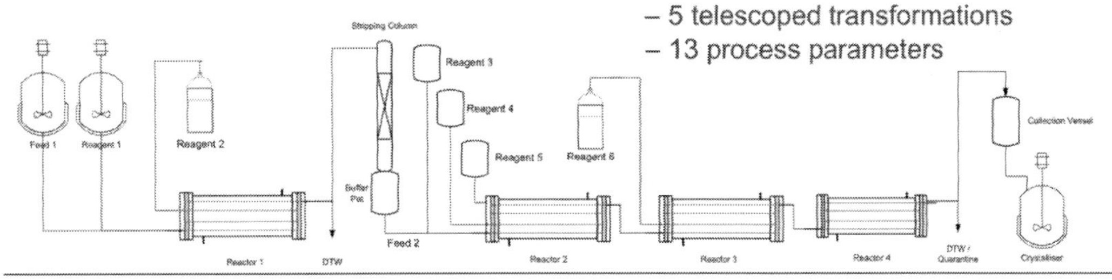

Motivation

Objective: Identify conditions that are most forcing wrt 9 specific process-derived impurities to run in R&D Pilot Facility (laboratory confirmation of Design Space)

- DoE approach unsuitable due to:
  - Challenge in translating lab scale batch/flow results to R&D Pilot and Commercial Manufacturing
  - High material and resource requirements for running DoE in R&D Pilot facility

- Alternative approach based on reaction kinetics and process simulation
  1. Develop a deep understanding of chemistry and capture in kinetic models
  2. Characterize equipment impact and couple with chemistry in process model
  3. Identify most forcing conditions and interactions for each impurity via simulation '*in silico* DoE'

Approach

Process Simulation Approach to Design Space

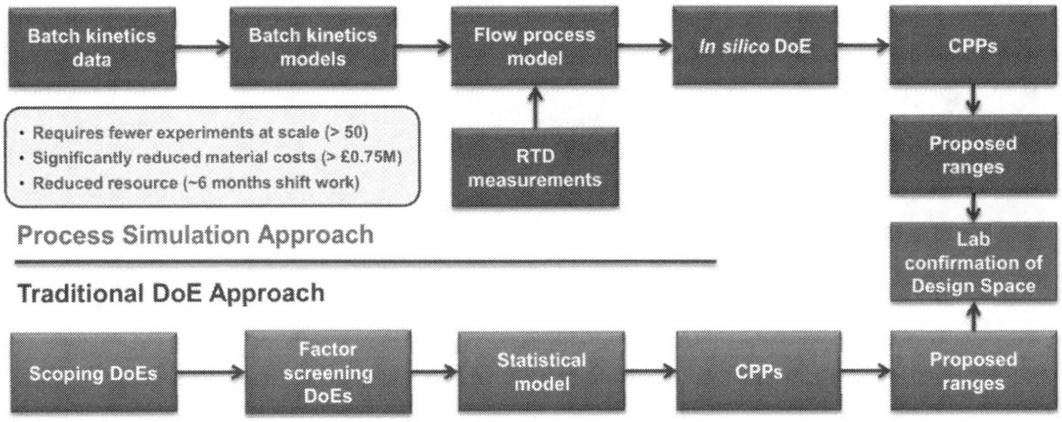

Generating the Batch kinetics data

- Structured sequence of batch experimentation for each chemistry transformation:

1. Profile at ~centre-point conditions
2. Repeat run (reproducibility)
3. Different stoichiometry (less reactant c.f. 1)
4. Different temperature (colder c.f. 1)

- Profiles fitted to proposed mechanisms (DynoChem)
- Outputs are rate constant (k) and activation energy (E_a) for each step

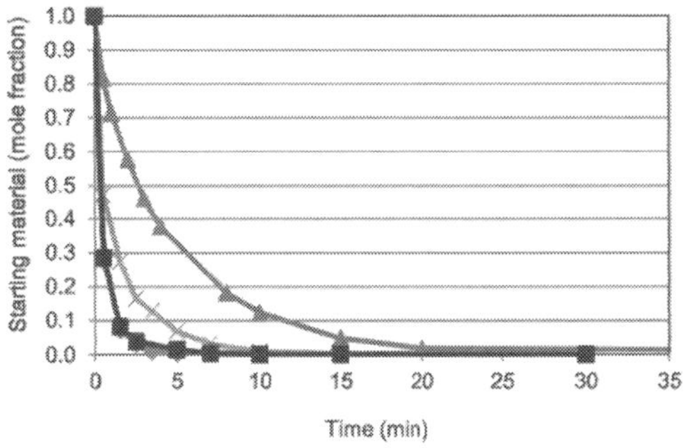

Developing the Batch kinetics models

- Chemistry captured as a network of coupled 1st and 2nd order equations
- Complex mixture at final stage (potentially >30 components)
  - Excess reagents
  - By-products
  - Impurities

Simplifying assumption: Protonation equilibria excluded from model due to buffering and minimal impact on rate expected within proposed ranges

Origin of process derived impurities:

Residual intermediates	Derived impurities	Alternative pathways	Over-reaction
4	3	2	0

Developing the Process model

- Characterization of continuous processing equipment allows calculation of true system volume
- Characterisation combined with kinetics parameters to model output concentrations of components
  - This model is used to simulate the continuous process (ideal plug flow)
  - Combines existing data (no additional data generated)

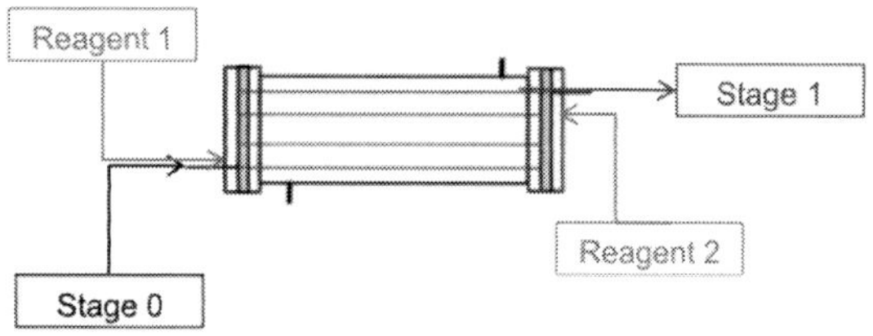

Developing the Process model

- Characterization of continuous processing equipment allows calculation of true system volume
- Characterisation combined with kinetics parameters to model output concentrations of components
  - This model is used to simulate the continuous process (ideal plug flow)
  - Combines existing data (no additional data generated)

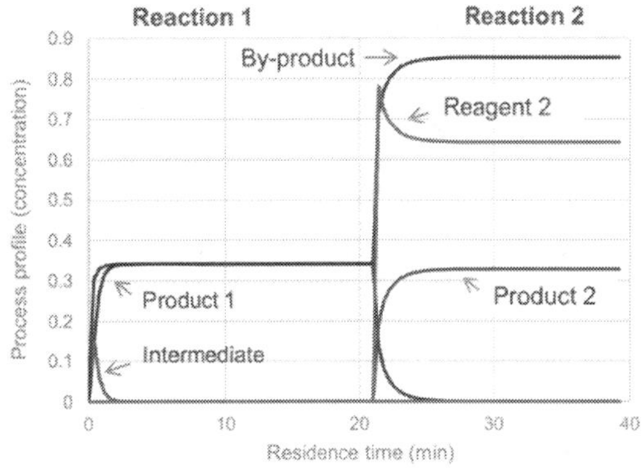

RTD Measurements

- RTD measurements were carried out on the R&D Pilot and commercial manufacturing equipment
- In both cases, all reactors can be approximated as ideal Plug Flow Reactors (PFRs)
- Data justifies the use of a process model where axial dispersion is ignored
  - Simulation demonstrates measured dispersion is unimportant to kinetics

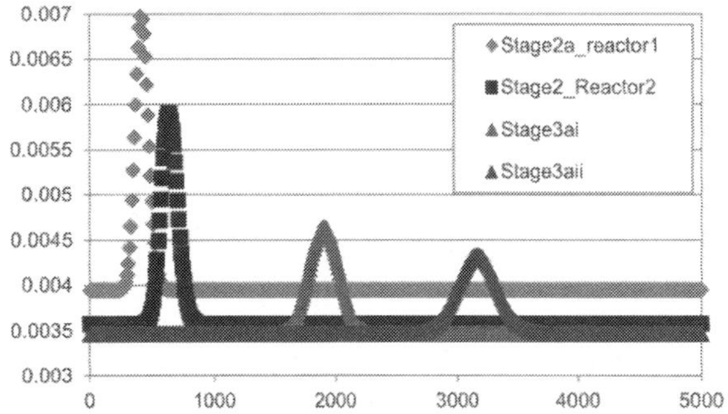

Evolution of the process model

- Process initially modelled as four discrete PFRs (ideal)
  - Gas-stripper excluded
  - Steady state simulations
  - Transcription intensive
- Later moved to single flowsheet model
  - Simpler to use
  - Impact of axial dispersion modeled
  - Dynamic simulation possible

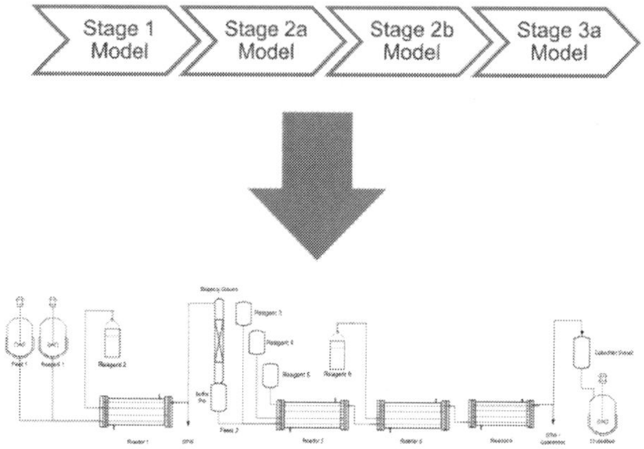

In silico DoE

- Impact of varying each process parameter on predicted levels of process impurities was investigated using a 2-level factorial design
- Enables visualisation of simulation results in a familiar format
- Most forcing conditions identified and ran in R&D Pilot Facility
  - Compared with optimisation results

Run	Parameter 1	Parameter 2	Parameter 3
1	+ 1	+ 1	+ 1
2	− 1	+ 1	+ 1
3	+ 1	− 1	+ 1
4	− 1	− 1	+ 1
5	+ 1	+ 1	− 1
6	− 1	+ 1	− 1
7	+ 1	− 1	− 1
8	− 1	− 1	− 1
9	0	0	0

2^{13} = 8192 possible combinations!

In silico DoE

- Impact of varying each process parameter on predicted levels of process impurities was investigated using a 2-level factorial design
- Enables visualisation of simulation results in a familiar format
- Most forcing conditions identified and ran in R&D Pilot Facility
  - Compared with optimisation results

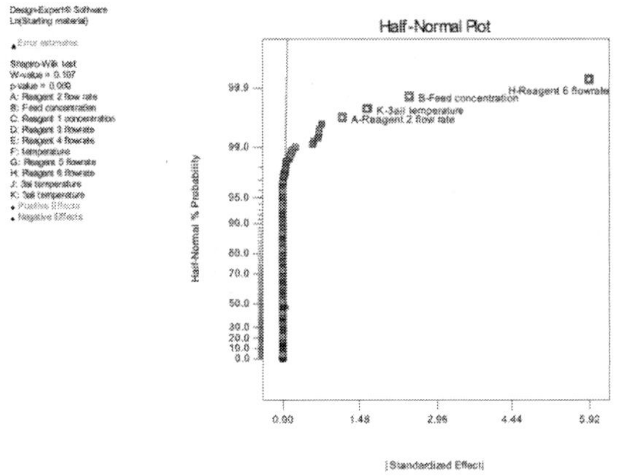

Results

Simulation output

- Output from process simulation is a matrix of most forcing conditions wrt each process derived impurity
- Conditions ran in R&D Pilot Facility

✓ Operating within Design Space delivers API that meets proposed specification

✓ Model-designed experiment successful in 8 / 10 cases

✓ Trends well predicted

  - Differences attributed to equipment performance and calibration errors

Process parameter	Impurity 1	Impurity 2
Concentration of Feed 1 solution	H	H
Concentration of Reagent 1 solution	L	L
Stage 1 residence time	L	L
Flow rate of Reagent 1 solution	L	L
Stage 1 temperature	L	L
Flow rate of Reagent 2	L	L
Flow rate of Reagent 3	H	L
Flow rate of Reagent 4	L	H
Stage 2 temperature	L	H

Impurity 1 = residual Stage 0, Impurity 2 = derived from Stage 0

Stage 3a

Control of process impurities

- Stage 3 input requires tight control in continuous process due to poor purging downstream
- Impurity formed by competing parallel pathway
- Options investigated via process simulation:
  - Increased residence time
  - Increased amount of Reagent 6
  - Increased Stage 3 temperature
- Simulation results highlighted need to tighten & shift process parameter ranges to manufacturing

$$\frac{d[\text{Stage 3a}]}{dt} = k_1[\text{Stage 2b}][\text{Reagent 6}]$$

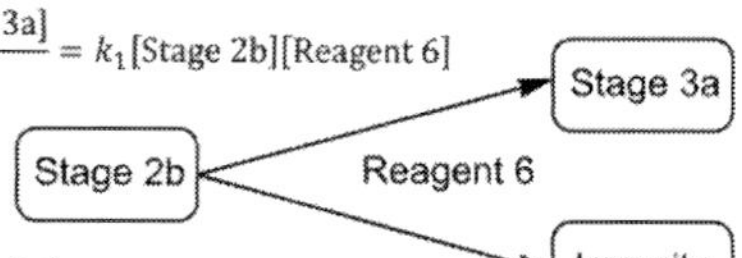

$$\frac{d[\text{Impurity}]}{dt} = k_2[\text{Stage 2b}][\text{Reagent 6}]$$

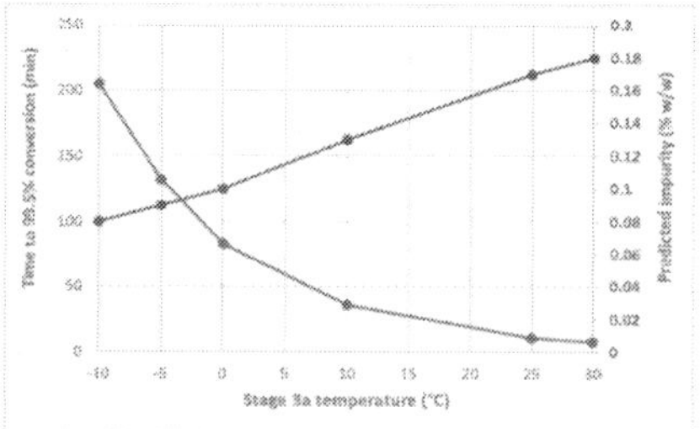

Stage 3a

Control of impurities

- **Two** reactors used: one cold, one warmer
- Compromise between selectivity and conversion in acceptable time frame

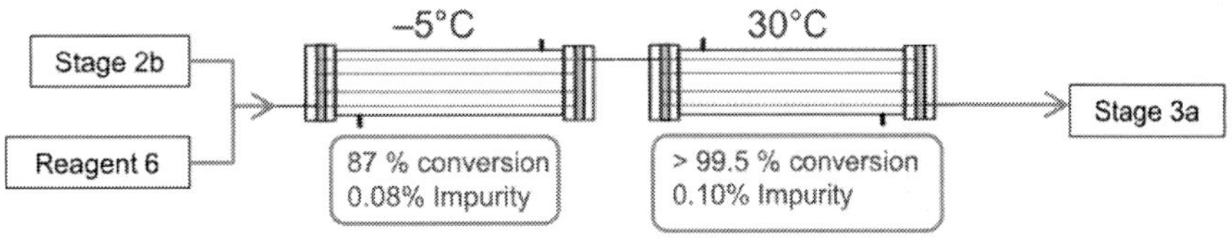

- Reactor temperature set points result from optimization using process model
- Optimized temperatures successfully demonstrated in R&D Pilot Facility

Input to Risk Assessment

Steady state risk assessment

- Process simulation results used to pre-populate risk assessment scoring
- Further simulations of interest were ran 'live'

Disturbance risk assessment

- Process simulation in dynamic mode
- Assess the impact of univariate process disturbances
- Output used to inform Divert to Waste strategy

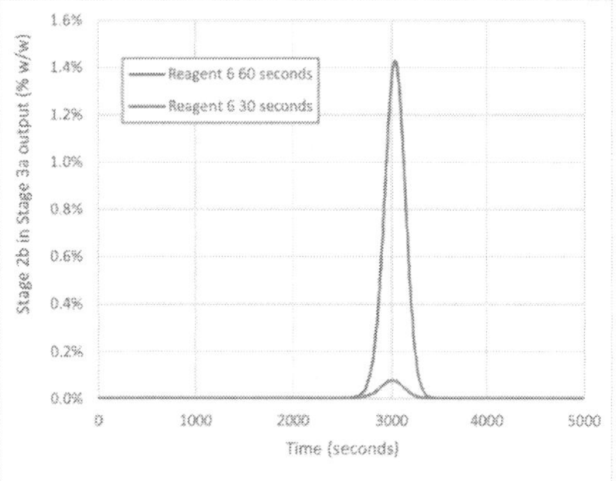

Summary

Summary

- A development strategy based on reaction kinetics and process simulation was used to identify the Design Space for a highly telescoped continuous API manufacturing process
- This approach resulted in significant savings in time, resource and materials
- Process model was a valuable tool for risk assessment - particularly for disturbance handling
- Visualizing the output from multiple simulations remains challenging

Acknowledgements

Malcolm Berry
Peter Shapland
Hannah Robinson
Peter Clements
Mark Hughes
Sabri Ukuser
Peter Hamilton
Irene Areri
Chris Clarke
Lesley Wong

Gary Breen
Augustine Ochen
Markus Andersson
Andrew Richards
Melanie Dumarey
Mohammad Yahyah

Andrew Bird
Joe Hannon

495685 Estimating NRTL-SAC Conceptual Segments of Molecules Using Sigma Profile

Abstract:

The Non-Random Two Liquid Segment Activity Coefficient (NRTL-SAC) model[1] as a solubility model is now routinely used in the industry[2-3]. In this model, the conceptual segments are defined to describe molecular surface characteristics, i.e., hydrophobicity, hydrophilicity, polarity. In practice, the conceptual segments of molecules must be identified from experimental phase equilibrium data with other solvents, and that limits the model applicability to molecules with sufficient phase equilibrium data. In this work, a methodology is presented to estimate the conceptual segments from the sigma profile, which is the molecular surface charge density distribution calculated from computational quantum chemistry, and is used in COSMO-based models[4-5]. We present the rationale and algorithm of segment numbers estimation from sigma profiles. Therefore, the applicability of NRTL-SAC model can be expanded to any components with known structures. The predicted phase equilibria, e.g., VLE, LLE, SLE, are compared with experimental data and predictions from other activity coefficient models to validate the model performance. The proposed approach allows the NRTL-SAC model to predict solubility in unconventional solvents, excipients, or polymers, where phase equilibrium data are insufficient to identify their segment numbers.

1. Chen, C.-C.; Song, Y., Solubility modeling with a nonrandom two-liquid segment activity coefficient model. *Industrial & Engineering Chemistry Research* **2004,** *43* (26), 8354-8362.

2. Gani, R.; JimÃ©nez-GonzÃ¡lez, C.; ten Kate, A.; Crafts, P. A.; Jones, M.; Powell, L.; Atherton, J. H.; Cordiner, J. L., A modern approach to solvent selection. *Chemical Engineering* **2006,** *113* (3), 30-43.

3. Tung, H. H.; Tabora, J.; Variankaval, N.; Bakken, D.; Chen, C. C., Prediction of pharmaceutical solubility Via NRTLâ€SAC and COSMOâ€SAC. *Journal of Pharmaceutical Sciences* **2008,** *97* (5), 1813-1820.

4. Klamt, A., Conductor-like screening model for real solvents: a new approach to the quantitative calculation of solvation phenomena. *The Journal of Physical Chemistry* **1995,** *99* (7), 2224-2235.

5. Lin, S.-T.; Sandler, S. I., A priori phase equilibrium prediction from a segment contribution solvation model. *Industrial & Engineering Chemistry Research* **2002,** *41* (5), 899-913.

498449 Discrete Element Modeling to Predict Pharmaceutical Powder Flow at Different Humidity Conditions

Abstract:

Pharmaceutical powder flow can alter significantly based on the exposed humidity conditions. Lack of predictive models to understand the flow variability at different humidity conditions may undermine its subsequent impediments. A Discrete Element Model (DEM) has been developed to predict the effects of humidity on powder flow, altering the cohesive forces based on granular bond numbers in simple hopper geometries. Experiments similar to the simulations are further performed for different pharmaceutical excipients at different humidity conditions. To predict cohesive properties for powders, those did not flow under gravity; vertical vibration has been introduced in both experiments and simulations. The equivalent DEM bond numbers are in good accordance with the experimental results. Pharmaceutical powders with lower moisture sorption show more variability in flow properties in comparison to cohesive powders having higher moisture contents. The equivalent DEM parameters suggest the same and are being implemented to predictpowder flow variability in different geometries and humidity conditions.

491765 Long-Term Coating Predictions for in a Wurster Fluidized-Bed Coater with Using a Combined CFD-DEM and Markov-Chain Monte Carlo Approach

Abstract:

Coating of particles, in particular, microspheres, using the Wurster fluid-bed coater is a common unit operation in the pharmaceutical industry. Such coatings are typically used to layer active drugs or other ingredients that control drug release, and also for taste/odor masking and enteric barrier coatings that prevent dissolution/disintegration. To ensure control over the drug-release rate and to achieve a consistent inter-particle drug loading (potency), a high degree of coating uniformity is essential. However, relating the coating uniformity of the beads to the process parameters (such as fluidization air-flow rate, atomizing air-flow rate, Wurster-gap height, and bead size) is not trivial.

Experimental methods to characterize bead-flow dynamics inside Wurster coaters are difficult for large systems; moreover, sampling techniques are insufficient to quantify the coating uniformity of millions of beads. In such situations, coupled CFD-DEM simulations (computational-fluid dynamics coupled with discrete-element method) can model the complex particle and airflow in a fluidized bed. However, due to the high computational costs (several days per process-second for an industrial system), simulating several hours of a coating process using CFD-DEM is infeasible. Therefore, a multi-scale model is required to predict the long-term coating uniformity, based on inputs from the CFD-DEM simulations. The Markov-chain Monte-Carlo-based compartment model is one such approach to develop a reduced-order model.

The compartment model subdivides the coater into non-overlapping regions, i.e., compartments. The exchange of particles between adjoining compartments is modeled as stochastic process described by a probability matrix *P*, where P_{ij} describes the probability of a particle moving from compartment *i* to compartment *j* during some small compartment-model time step Î"t_{CM}. This probability matrix *P* is calculated from the detailed CFD-DEM particle trajectories; therefore, the exact positions of the particles are not tracked in the compartment model. Changes in the process parameters affect the flow behavior of the beads which is reflected as changes in the probability-matrix entries P_{ij}.

Certain compartments near the spray nozzle are designated as the spray zone; every particle that visits these spray compartments gains some coating mass, based on the spray rate, particle size, and the compartment-model time stepÂ Î"t_{CM}. The probability matrix *P* is then used to update the positions of the particles in various compartments. By tracking the flow of a large, representative sample of beads, the evolution of the coating mass distribution can be tracked with time - these predictions can be generated in minutes for hours of real process time using the reduced-order Markov-chain Monte Carlo method.

Results from the compartment model for a given set of process conditions were successfully validated against

experimental results of particle diameters obtained using Sympatec QicPic and coating thickness from optical-coherence tomography (OCT). After validation, a virtual design of experiments (DoE) study with varying process conditions has been performed. The results show that a larger Wurster-gap height reduces the variation of the coating mass and thickness but, owing to the initial size distribution of the uncoated beads, the variation in the coating potency increases. Depending on the application, it may be more desirable to optimize the coating potency (to maintain a constant coating-mass fraction for coatings that contain active drug), or optimize the coating thickness (e.g., for protective coatings where a predefined barrier thickness is needed). The present Markov-chain Monte Carlo approach, in combination with CFD-DEM simulations, can be used to design the optimal process parameters for both applications.

496208 Microfluidization as an Enabling Technology for Solubility Enhancement

Abstract:

Microfluidization as an enabling technology for solubility enhancement

Iris Duarte, Clara Sa Couto, Tiago Porfirio, João Vicente and Marcio Temtem

R&D Drug Product Development, Hovione Farmacência S.A, Sete Casas, 2674-506 Loures, Portugal

INTRODUCTION

Emerging drugs are essentially poorly water-soluble and such limited aqueous solubility has been one of the major hurdles in the development of oral-dosage forms, as poor solubility typically hinders oral bioavailability. This challenge has been addressed with different formulation strategies and enabling technologies, including: complexation with cyclodextrines (CD), production of amorphous solid dispersions (ASD) or particle-size reduction, such as the production of nanosuspensions and/or nanocrystals [1].

This work describes a technique which encompasses a microfluidization followed by a spray drying step, to produce nano- to microparticles of crystalline drug-alone, cyclodextrins complexes or amorphous solid dispersions. Three different case-studies will be detailed in this work. The manufacturing process is scalable to industrial scales and can operate either in batch or continuous manufacturing mode as benchmarked in the current work.

The microfluidization step is defined as a dynamic high-pressure process where two liquid streams, solution or suspension, pass through micro-channels toward an impingement area through which the fluids flow and interact [2]. The microfluidization step renders in either a liquid solution or liquid suspension that is typically followed by a spray drying step to remove the solvents and produce the dried particulate system. A schematic illustration of the process train is shown in Figure 1. This process was successfully demonstrated using fluticasone propionate as drug alone, itraconazole (ITZ) / Eudragit L-100 amorphous solid dispersions and ITZ / cyclodextrins inclusion complexes.

Case study I: production of micro / nanocrystals using microfluidization: Fluticasone propionate was dissolved in acetone. As anti-solvent, a mass of deionized water corresponding to 10 times that of the solvent was used. The two solutions were fed and mixed using PureNano^{TM} in a single passage. The particle size of the isolated product was characterized through scanning electron microscopy, SEM (Figure 2). Roughly, particles with nano-scale were obtained [3].

Case study II: complexation of a poorly water soluble API using cyclodextrines: ITZ was dissolved in DCM and SBE-ß-CD dissolved in water. The two solutions were fed and mixed using PureNano^{TM} in a single passage. Two different trials at different temperatures were done, a filtration step was done prior to the spray dryer to remove the

API not complexed. 3% was the maximum drug load obtained in the dried product [4].

Case study III: co-precipitation of nano-solid dispersions: ITZ and Eudragit L-100, in a mass proportion 1:9 and 40:60, were dissolved in a mixture of ethanol / acetone in a volume proportion of 1:1. Acidified cold water (pH=2) was used as anti-solvent with a ratio solvent:anti-solvent of 1:10 w/w. The results indicated that amorphization of ITZ was successful when using the present method (Figure 4) [5].

CONCLUSION

This work demonstrates that the technique herein described, consisting in a microfluidization step followed by an isolation step via spray drying is able to produce drug-alone nano-crystals, cyclodextrins complexes and amorphous solid dispersions. The process is scalable to industrial scales.

Figure 1 - Experimental set-up for the batch and continuous controlled precipitation process.

Figure 2 - SEM of spray dried particles of fluticasone propionate.

Figure 3 - i) X-ray diffractograms of ITZ solid dispersion at two different ratios and its crystalline form and ii) SEM of spray dried dispersion of ITZ.

[1] Perrie Y., Rades T. "Pharmaceutics - Drug Delivery and Targeting", Pharmaceutical Press, 2010;

[2] Rakesh P. Patel, Ashok H Baria, Nikunjana A Patel, An overwiew of size reduction technologies in the field of pharmaceutical manufacturing, Asian Journal of Pharmaceutics October-December (2008) 216-220.

[3] Fonseca, T., Duarte, I., Vicente, J., Temtem, M. "Continuous production of particles", Patent No. WO2016156841 A1, 2016.

[4] Lisboa, H., Santos, F., Vicente, J., Temtem, M. "Continuous complexation of active pharmaceutical ingredients", Portuguese Patent No. 109117, 2016.

[5] Temtem, M., Pereira, R., Vicente, J., Duarte, I., Gaspar, F., Duarte, I. "A method of preparing amorphous solids dispersions in submicron range by co-precipitation", Patent No. WO2016016665 A1, 2016.

491353 Droplet-Templated Anti-Solvent Spherical Crystallization of Hydrophilic and Hydrophobic Drugs with an *in situÂ* Formed Binder

Abstract:

Conventional pharmaceutical crystallization processes in batch reactors often lead to broad crystal size distributions with frequent batch-to-batch variations. These crystals subsequently undergo several secondary formulation processes for crystal and dosage uniformity. To consolidate these processes, spherical crystallization is a promising method, which combines crystallization, milling, and granulation into a single unit operation. Droplet-templated spherical crystallization methods have been developed recently to achieve highly monodisperse crystal microspheres with droplet microfluidics and evaporation techniques. We present here a novel droplet-templated anti-solvent spherical crystallization method applicable to both hydrophilic and hydrophobic drugs. In both cases, an alginate hydrogel binder forms in situ simultaneously, effectively encapsulating the drug crystals into highly monodisperse spheres. We study in detail the crystallization and gelation processes with mass transfer modelling, and the characterization of the obtained alginate/drug spheres. While glycine and carbamazepine are used as model hydrophilic and hydrophobic drugs, this method can be generalized to other drugs as well. In particular, this method offers additional benefits such as minimal thermal impact, fast crystallization rates, high drug-binder loading ratios, and high selectivity towards metastable polymorphs.

Tailoring CSD & Shape of Ibuprofen to Improve Flow Properties in Batch & Continuous Crystallization

Yunliang He, Brian Glennon, Yuan Gao
SSPC, School of chemical & bioprocess engineering, University College Dublin
Anne-Marie Healy, Zelalem Worku
SSPC, School of pharmacy & pharmaceutical sciences, Trinity College Dublin

- The significance of powder flow properties in pharmaceutical industry

In pharmaceutical industry, most API and excipients are manufactured or processed in the form of powders.
Flow properties are critical for handling, storing and formulation of powders
borderline flow properties for high speed tableting
flow factor: 6.74(±0.27) and bulk density and tap density is 0.353 and 0.476 g/ml respectively.

- The effect of size distribution & shape on flow properties

Common belief
Big particles show good flow properties, particles >300 microns are free-flowing
Plates and needles are always poor-flowing

- Tailor crystal size distribution and shape of ibuprofen

micronization
Granulation
Batch Crystallization
Continuous Crystallization

Is it possible to prepare ibuprofen for direct compression from crystallization?

www.sspc.ie INNOVATION THROUGH COLLABORATION

Batch & continuous crystallization of ibuprofen

❖ Fast crystallization to prepare fine crystals

10~30g ibuprofen/100g 50%m Ethanol-Water or 20~30g ibuprofen/100ml hexane was heated above 50℃ to dissolve and crash out by quick cooling to 5 ℃(in 20 mins).

❖ Slow crystallization to prepare big crystals

10~30g ibuprofen/100g 50%m Ethanol-Water or 25g ibuprofen/100ml hexane was heated above 50C to dissolve and then cool to about 40 cool slowly (2~6 ℃/h) from about 40C to about 5C

❖ Continuous crystallization to prepare narrow distribution big crystals

MSMPR(mixed suspension mixed product removal)
Feeding solution: 15g per 100g 50%m ethanol-water
Feeding rate:1~2ml/s
Working volume: 80ml
Residence time:40~80min
Generally, after 6~8 residence time, crystallization will reach steady state

Control of crystal size based on metastable zone

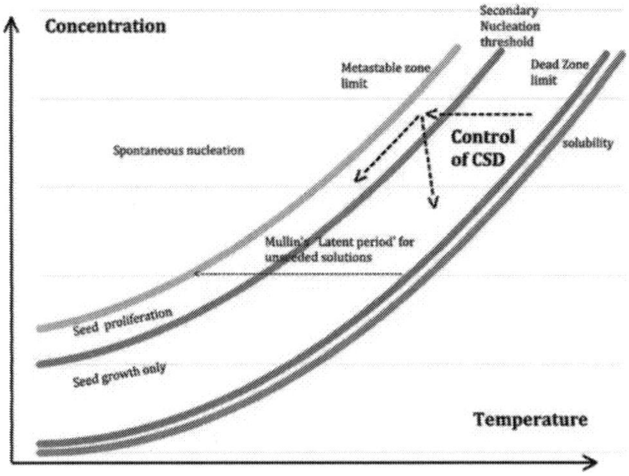

Primary and secondary metastable zone

Threlfall and Coles, Cryst. Eng. Comm. 2016, 18, 369

For batch crystallization

If crystallization was restricted in secondary metastable zone, there will be only crystal growth and no nucleation. The counts and average size R could be controlled by seed counts N.

$$R = \sqrt[3]{\frac{mass}{density * N * shape\ factor}}$$

For continuous crystallization

Crystallization must be operated in primary metastable zone. Continuous secondary nucleation will provide enough seeds to maintain steady state during continuous suspension removal. Crystal size is controlled by supersaturation and residence time.

Control of crystal size in batch crystallization

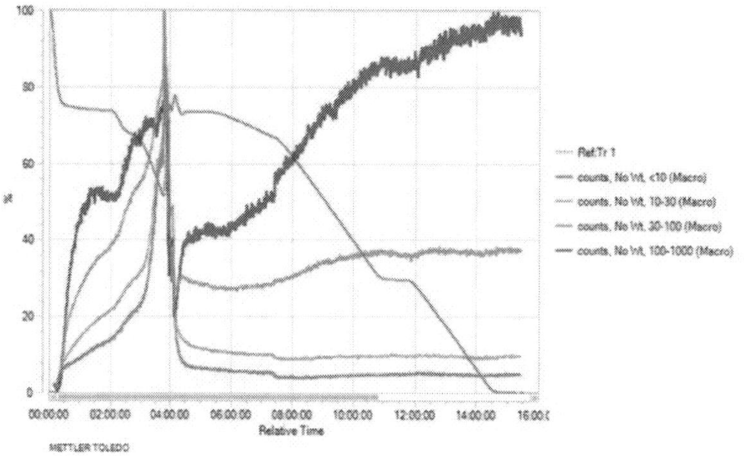

Failed Control of Crystal Size in Batch Crystallization

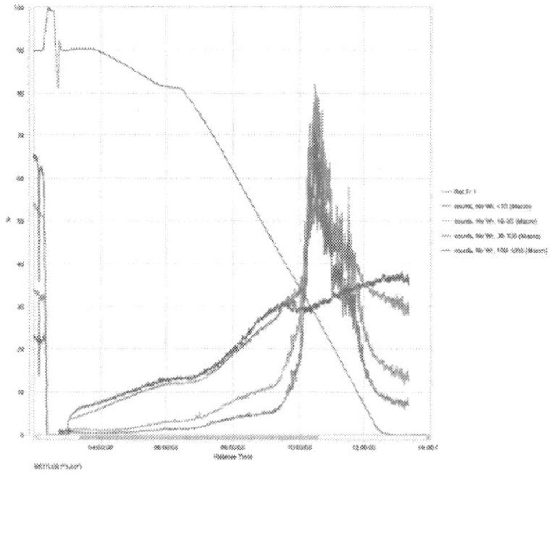

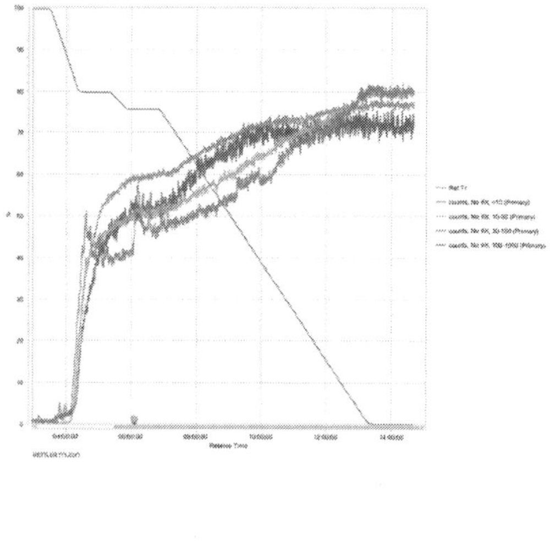

Control of crystal size in continuous crystallization

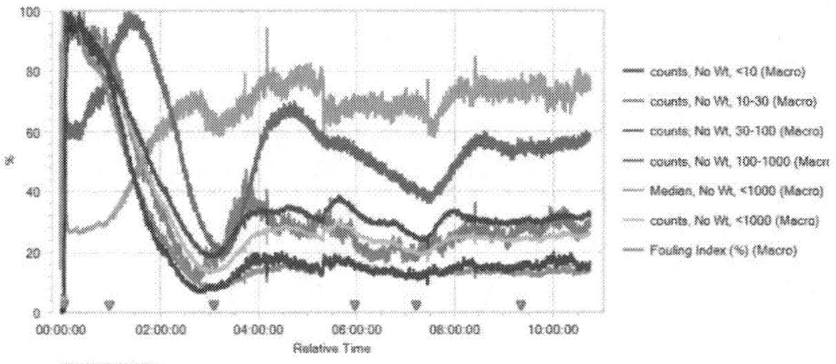

METTLER TOLEDO

Crystal growth mode and control of shape

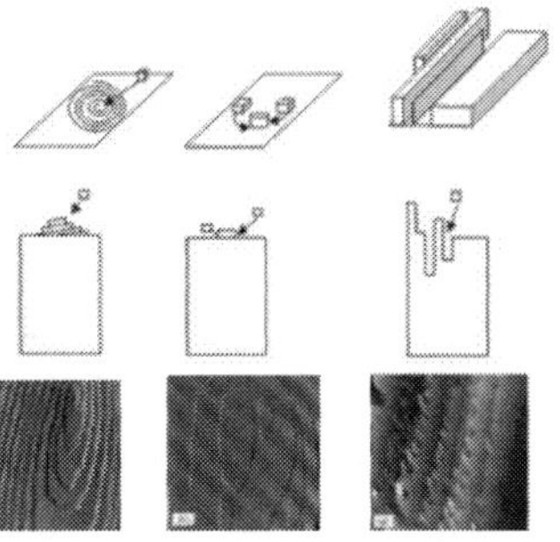

Dislocation 2D nucleation Continuous

Y. Matsuzuki et al. J. Cryst. Growth 242(1-2):192

Factors on crystal shape
- ◆Intrinsic crystal anisotropy
- ◆Interfacial energy
- ◆Supersaturation

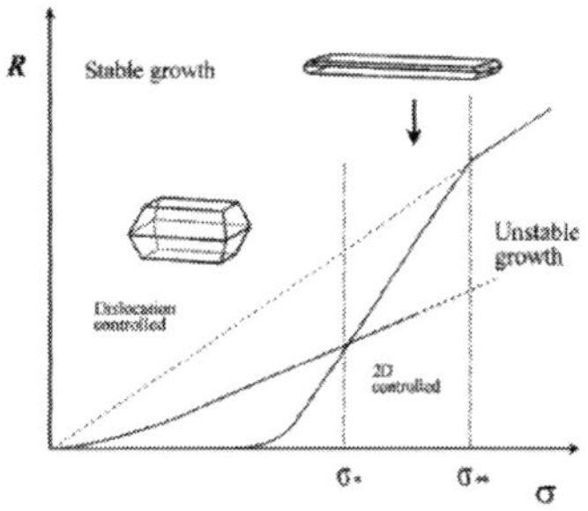

X. Liu and P. Bennema, Physical Review B 53(5):2314

Ibuprofen Crystal Shape From Controlled Crystallization

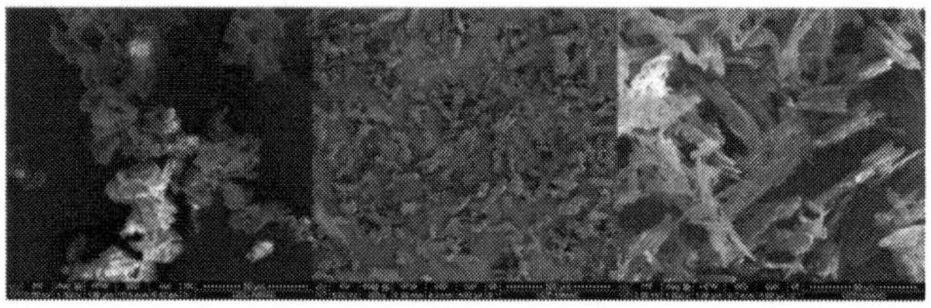

fast batch product from E-W(left & middle) and hexane(right)

continuous product from E-W

slow batch product from E-W

slow batch product from hexane

Comparison of CSD from batch & continuous crystallization

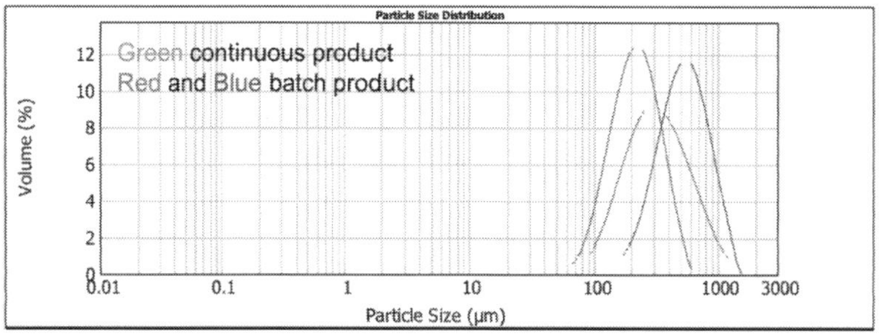

sample		Span	D [4, 3]	D [3, 2]	d (0.1)	d (0.5)	d (0.9)
continuous	Green	1.26	229.85	169.06	110.23	212.59	378.74
slow batch 1	red	1.88	364.50	163.45	123.72	303.34	693.37
slow batch 2	blue	1.40	542.40	277.64	218.59	506.72	929.39

ibuprofen crystal size distribution

Table 1 size distribution of batch and continuous crystallization product in ethanol-water

sample	Span	D [4, 3]	D [3, 2]	d (0.1)	d (0.5)	d (0.9)
Continuous1	1.42	373.07	259.46	156.62	341.15	639.69
Continuous2	1.44	275.20	194.24	119.19	247.18	475.34
Continuous3	1.32	200.34	143.47	90.89	184.74	335.40
Continuous4	1.39	252.26	166.16	108.05	231.41	430.38
Continuous5	1.18	215.49	164.80	109.80	200.61	346.29
Continuous6	1.26	229.85	169.06	110.23	212.59	378.74
Fast batch1	8.25	511.40	17.38	11.57	164.65	1370.03
Fast batch2	2.41	117.41	59.04	32.45	88.51	245.58
Fast batch3	1.57	79.73	49.34	31.46	69.95	141.52
Slow batch1	1.40	542.40	277.64	218.59	506.72	929.39
Slow batch2	1.88	364.50	163.45	123.72	303.34	693.37
Slow batch3	1.22	326.10	171.05	145.53	315.02	530.69
Slow batch4	1.18	160.11	114.33	77.85	140.03	243.60
Slow batch5	1.53	178.78	112.27	74.00	159.14	317.84

Table 2 Size distribution of product from n-hexane

sample	Span	D [4, 3]	D [3, 2]	d (0.1)	d (0.5)	d (0.9)
Fast batch	3.29	44.07	14.40	7.19	18.31	67.43
slow batch1	3.57	354.63	105.74	67.86	219.16	849.73
Slow batch2	2.43	193.91	100.92	59.74	144.71	410.69
Slow batch3	2.06	288.17	134.67	85.25	236.76	573.11

FT4 Powder Rheometer

FT4

Accessories

Shear Cell Test

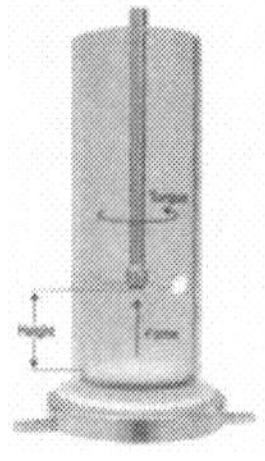

Stability &VFR

www.sspc.ie

INNOVATION THROUGH COLLABORATION

Shear cell test

FF vs D[3,2] and D[4,3]

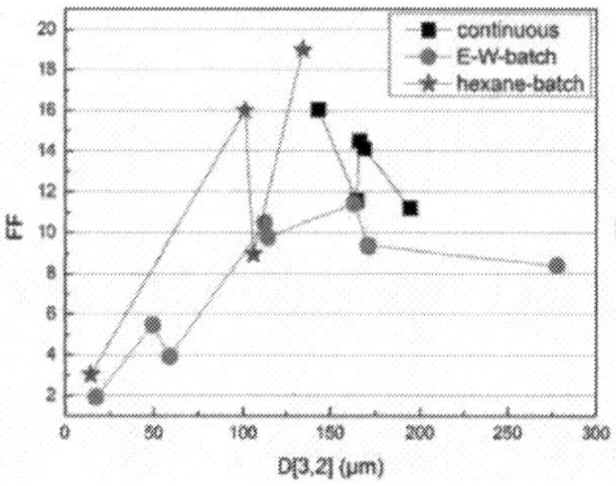

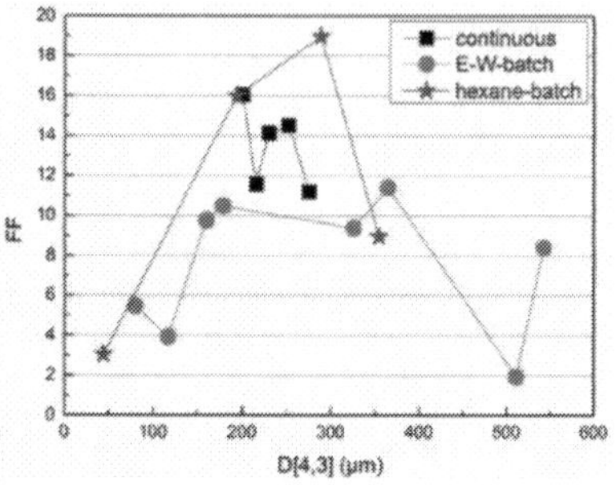

Type of flow	Flow Function Value
Free-flowing	10 < FF
Easy-flowing	4 < FF < 10
Cohesive	2 < FF < 4
Very cohesive and non-flowing	FF < 2

cohesion vs D[3,2] and D[4,3]

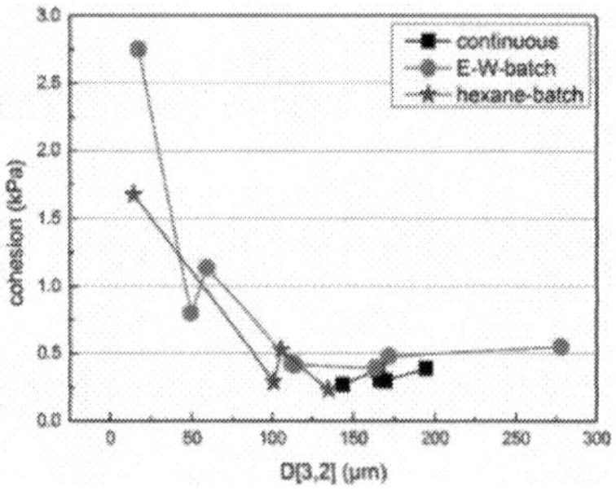

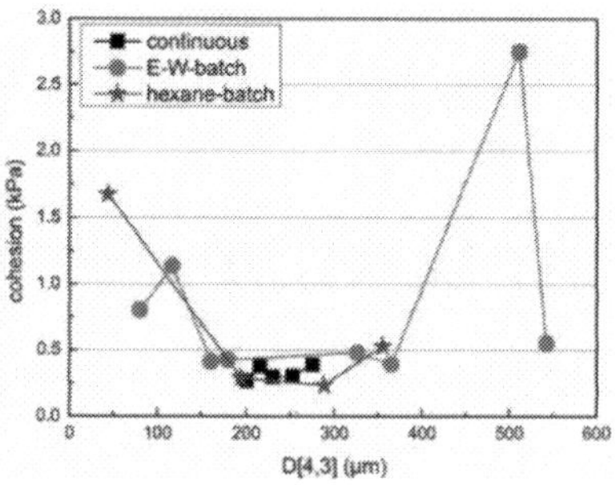

Density in shear cell vs D[3,2] and D[4,3]

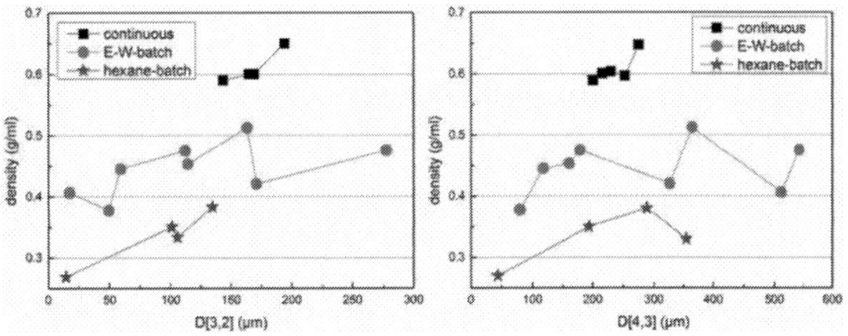

Stability & VFR test

BFE vs D[3,2] and D[4,3]

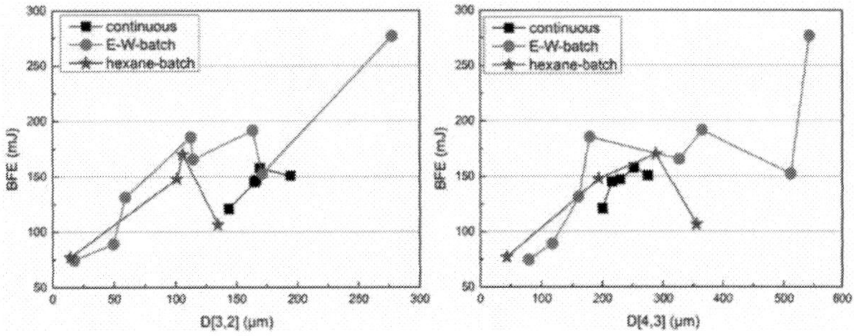

SE vs D[3,2] and D[4,3]

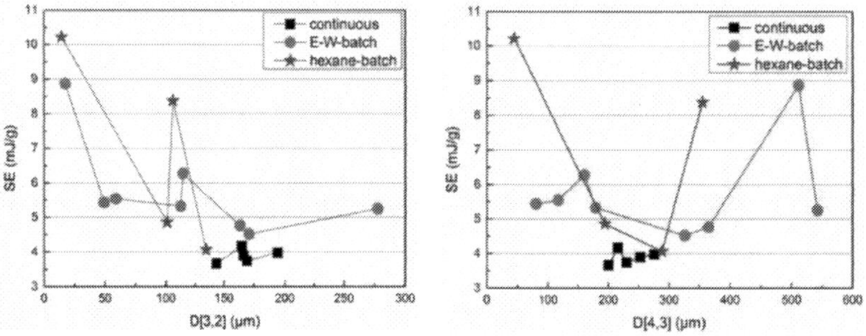

CBD vs D[3,2] and D[4,3]

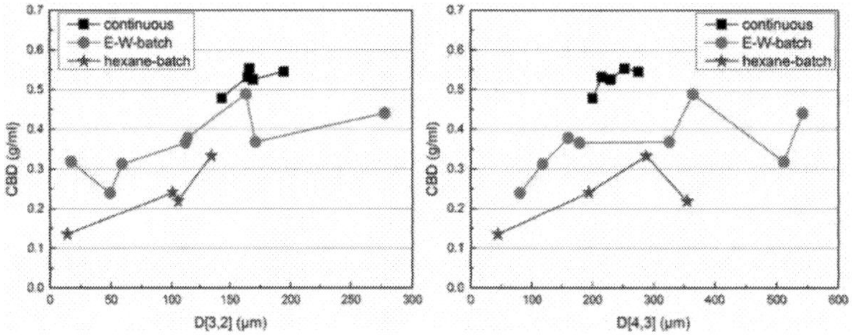

Conclusion

- Continuous products show equant shape, free-flowing, high density and small variance. Continuous crystallization is promising in preparing free-flow powders.
- Flow properties are better correlated with D[3,2] than D[4,3]
- Flow function is mainly determined by size while density is more influenced by shape than size. Rod crystals could be free-flowing by increasing size above 100 μm.
- Flowability(FF) doesn't simply increase with size. Conglomerate and shape also play critical roles in flowability.

Further study

- Investigate the direct compression and tableting behavior of tailor-made ibuprofen
- Set up a continuous manufacture process from crystallization to tableting

Thanks for your attention!

Any Questions?

SSPC

www.sspc.ie

INNOVATION THROUGH COLLABORATION

497920 Formation and Dissolution Characteristics of Paracetamol Granules

Abstract:

The ability of active pharmaceutical ingredients (APIs) to affectively influence the targeted region is of great importance. Therefore, there is a need to offer lasting solutions and improve upon the current methods of formulation. This research analyzes physical and dissolution characteristics to determine the bonding tendencies of paracetamol granules formed using the drop method. Paracetamol is a common API and is considered a Category III drug. Initially, a series of experiments were conducted to form similarly sized granules. The process involved creating a paracetamol powder bed and dropping liquid binder onto the bed. Granules were separated from the powder bed using sieve trays. The cohesiveness of the granules was tested by dropping the granules onto a flat surface from various heights. The weights of each granule was measured before and after dropping to determine a percent attrition. The final investigation involved the ability of the granule to dissolve over time. This testing method involved the use of a USP Apparatus 3 to prevent coning seen in Apparatus 2. Further testing was conducted using UV Vis. Various binders were considered for this research including natural and synthetic polymers. Results show that using 8 drops of binder to form granules will deliver close to the normal dosage (325 â€“ 500 mg) of the API. Synthetic binders produced more cohesive granules during the drop testing. Granules formed using polyvinylpyrrolidone (PVP) binder had the smallest percent attrition (0.1%) when dropped from a height of 24 inches. The rate of dissolution differed for each material examined; however, polyethylene glycol had a very consistent dissolution rate and reached a percent dissolution of 85% in one hour.

492697 Size Reduction through Flash Nanoprecipitation to Improve Solubility, Dissolution, and Bioavailability of Clofazimine

Abstract:

Clofazimine is a poorly soluble riminophenazine antibiotic (logP= 7.66) approved by Food and Drug Administration (FDA) with good safety record. Thus, a reliable, efficacious, and fast-release formulation of clofazimine is highly desired. To improve the drug bioavailability and dissolution rate, clofazimine was encapsulated in nanoparticles prepared with three different surface stabilizers, hypromellose acetate succinate (HPMCAS), lecithin, and zein, through the Flash Nanoprecipitation (FNP) process with high encapsulation efficiency. To convert liquid colloidal suspensions to powders, lyophilization and spray drying, were applied to solidify nanoparticles. Dissolution tests were carried out to evaluate release kinetics of clofazimine nanoparticles prepared by different surface coatings, comparing to raw clofazimine powder and its currently market product Lamprene®. Remarkably improved dissolution rate and supersaturation levels were observed with all clofazimine nanoparticles despite different release kinetics. Additionally, calorimetry technique was explored to study the crystallinity of clofazimine in nanoparticles. The results strongly suggest that the new clofazimine nanoparticles prepared with affordable materials can be used as viable cryptosporidiosis therapeutics.

496138 Formulation Strategies for Solid Dispersions Containing Tablets

Abstract:

Formulation strategies for solid dispersions containing tablets

S. Doktorovova, E. Voney, J. Henriques

R&D Drug Product Development, Hovione Farmaciência S.A, Sete Casas, 2674-506 Loures, Portugal

In this work we present the development of formulation strategies for amorphous solid dispersions containing hypromellose acetate-succinate (HPMC-AS-M). Solid dispersions, especially spray dried, frequently have poor tableting properties due to low bulk densities and small particle size [1]. Tablet formulation then relies on excipients with capacity to improve compactability and blend flow. Three strategies are outlined: i) using a plastic filler/binder that greatly improves blend compactability and assure good mechanical strength of the tablets, ii) using a free flowing filler to improve powder flow properties and guarantee low mass and content variation iii) use a combination of a plastic and brittle excipient. The later should assure a good balance between acceptable powder flow and compactability. For the purpose of this work, a simple formulation compromising the solid dispersion (here represented by sieved HPMC-AS-M), a diluent or combination of two diluents, a disintegrant and a lubricant was considered.

The strategy based on formulating HPMC-AS-M with free flowing diluent largely improved powder flow properties of the blend without compromising compactability, but the improvement of powder flow properties did not translate into good tableting properties. Formulation of HPMC-AS-M with plastic diluent/binder resulted in excellent mechanical strength of the produced tablets but increasing ratio of the diluent compromised mass uniformity. A combination of a plastic and a well flowing diluent was therefore required to achieve a compromise between mechanical strength and mass variability. Fine tuning of the required ratio of diluents depended on the percentage of polymer present in the blend, tablet tensile strength being impacted to greater extent by increased polymer fraction than mass variability.

A model using bulk properties of excipients and polymer was set up as a tool to optimize the formulation powder flow properties of blends and quality attributes of tablets. Various mixture models were used for each property, the one that showed the best correlation between predicted and observed values were selected for the model. Prediction of blend powder flow and tablet properties using the model will enable sparing material in solid dispersion tablet development.

Figure 1: Modeling of a) powder flow properties using specific energy (SE, mJ/g) and c) tableting properties using tablet tensile strenght (TS, MPa) at 12 kN compression force and b) a sweet spot between acceptable powder flow and tableting properties.

References

[1] Henriques, J., Valente, P. & Winters, C. "Formulating Amorphous Solid Dispersions: Bridging Particle Engineering and Formulation", American Pharmaceutical Review, Nov/Dec. 2016.

493167 Modeling the Effects of Material Properties on Tablet Compaction a Case Study for Development

Abstract:

Interest in pharmaceutical continuous manufacturing (PCM) is rapidly growing. As it has been extensively reported PCM can deliver enormous advantages, including faster product development using less material, smaller equipment, superior process control, optimized performance, and more reliable quality performance [1-5]. However, to achieve these advantages in full, we need to develop accurate models of the process able of accounting for effects of material properties. Such models can be used to support faster and more economical process design as well as better process and quality control. As the pharmaceutical industry modernizes its manufacturing practices and incorporates more efficient processing approaches, it is important to reevaluate which process design elements affect product quality and the means to study these systems. The purpose of this presentation is to provide insight on a methodology to correlate and link the effect of raw material properties to equipment and process performance from using both data-driven and semi-empirical models. Such methodology aims at providing modeling tools to predict process performance *in silico*, minimizing the amount of experiments and/or narrowing the experimental scope. The presentation will include a case study where this protocol was applied to the characterization of tablet compression.

Lubricated blends of three (3) commonly used in the pharmaceutical industry: Microcrystalline Cellulose (Avicel) 101, Microcrystalline Cellulose (Avicel) 301, and Milled Lactose. Blends where made as formulations with varying levels of the lubricant, Magnesium Stereate, ranging from 0.25-1.5%. Material characterization was performed for all twelve (12) blends. The characterization included compressibility, permeability, density, particle size, and cohesion measurements. The characterization was part of the development of a reduce size Raw Material Properties Database (RMPD), which was later used to develop correlations between semi-empirical unit operation model parameters and material properties. Using Principal Component Analysis, the raw material properties were collapsed to two (2) major principle components, in order to understand the operating material space. All blends were then compressed using a two by three (2x3) experimental design, varying the fill depth and tablet thickness of the tablets produced, respectively. Tablet properties (e.g., weight, hardness, thickness, and compression force) were collected for all tablets and collected in a results matrix.

Results from the PCA analysis showed that raw material properties were slightly affected by the addition of the lubricant to the three (3) blends of Microcrystalline Cellulose (Avicel) 101, Microcrystalline Cellulose (Avicel) 301, and Milled Lactose. From the results matrix, we regressed the constant model coefficients for the Kuentz Equation, which is a model relating tablet hardness to changes in porosity. Two coefficients (i.e., the maximum tensile strength and the critical relative density) were regressed for each formulationâ€™s tablet compression profile. We evaluated several empirical models to correlate the regressed coefficient values to the original blend material properties from the RMPD. Models evaluated included Regression PCA (R-PCA), Partial Least Square (PLS), and simple linear regressions. From the analysis, we found that each of the coefficients regressed related to a material property from the RMPD. The models will be presented in this presentation. The empirical model between blend material properties and

the regressed constant values for the Kuentz equations showed less than 8% error for the formulations. The parity plots indicated a coefficient of variability (R-Squared) over 95% for both empirical model regressions along with p-values under 0.05 for all coefficients in the regression.

The results from this study indicate that correlations between material properties and semi-empirical equation regression constants are sufficiently accurate. This proof of concept of developing correlations between raw material properties and unit operation models can aid in process development, especially in the area of design space characterization and robustness analysis. Overall, we have shown with this approach that it is possible to develop these correlations and have predictive capabilities of process performance based on material property values.

References:

1. Plumb, K., *Continuous Processing in the Pharmaceutical Industry: Changing the Mind Set.* Chemical Engineering Research and Design, 2005. **83**(6): p. 730-738.

2. Ooi, S.M., et al., *Continuous processing and the applications of online tools in pharmaceutical product manufacture: developments and examples.* Therapeutic Delivery, 2013. **4**(4): p. 463-470.

3. Rantanen, J. and J. Khinast, *The Future of Pharmaceutical Manufacturing Sciences.* Journal of Pharmaceutical Sciences, 2015: p. n/a-n/a.

4. Lee, S., et al., *Modernizing Pharmaceutical Manufacturing: from Batch to Continuous Production.* Journal of Pharmaceutical Innovation, 2015. **10**(3): p. 191-199.

5. Mascia, S., et al., *End-to-End Continuous Manufacturing of Pharmaceuticals: Integrated Synthesis, Purification, and Final Dosage Formation.* Angewandte Chemie International Edition, 2013. **52**(47): p. 12359-12363.

496597 Quality By Design Approaches for Dry Powder Inhalation Products: A Regulatory Perspective

Abstract:

FDA has advocated companies to implement Quality by Design (QbD) into product lifecycles, with particular emphasis during product development. The goal of this presentation is to discuss the benefit and advantage the QbD approach brings to the design and manufacture of orally inhaled drug product, specifically the dry powder inhaler (DPI) variety. Since DPI is also a combination product, its successful development requires thorough understanding of the complex interaction between the powder formulation and the delivery device. We will illustrate how the QbD paradigm can aid in this process through case studies, including device-metered (reservoir) DPI and pre-metered (blister) DPI. We will start by identifying some key Quality Target Product Profile as well as Critical Quality Attributes (CQA). Through risk assessment and experimentation during development, Critical Material Attributes and Critical Process Parameters can be linked to the CQAs, thus allowing the design and implementation of an adequate control strategy, and reducing the need of intensive in-process testing. This is critical for DPI manufacturing since the standard assessment methods (Aerodynamic Particle Size Distribution and Delivered Dose Content Uniformity) have low testing efficiency. We will conclude with a discussion of regulatory submission expectations with respect to in-process control strategy based on commonly seen deficiencies in New Drug Applications and Abbreviated New Drug Applications.

497742 Effect of Bulk Properties of the Tracer on the Measurement of Residence Time Distributions in Continuous Powder Based Unit Operations

Abstract:

In continuous pharmaceutical manufacturing environment, the development of process mixing models is paramount to characterize the impact of upstream disturbances on the product, establish traceability of raw materials, and assess the process's performance and robustness. Upstream disturbance dissipation is key to determine whether material will remain is specification given a process deviation. The number of potentially affected products by a disturbance can be quickly quantified based on the disturbance's dissipation (i.e., back-mixing with in-specification material). Tracing raw material composition allows for product lot-to-lot delineation based on time and lots of raw material used. Process performance and robustness evaluation can allow us to determine potential failure modes and aid in the selection of appropriate process control action to correct deviations.

A commonly used tool in the characterization of mixing systems as well as mixing model development is Residence Time Distribution (RTD). Introduced in the 1950s by Danckwerts, RTD serves as a mean to describe non-ideal mixing in unit operations by characterizing the in-and-out time probability distribution function of a particle entering the unit at steady state [1]. To perform an RTD evaluation, a set of base and tracer materials need to be selected. The base material can be a single or multicomponent ingredient set, which resembles the actual process stream. The tracer is a material that can be quantified at the outlet using visual or chemical methods. Two major methods can be used for introducing a tracer into the system: pulse (or impulse) and step change. The selection of the methodology is very important for both the evaluation and characterization of the mixing behavior. Pulse experiments are effective at keeping the tracer material requirements low, while step changes often require large amounts of material to effectively produce a change in concentration at the system's outlet. Nonetheless, pulse experiments are more difficult to correctly perform due to the nature of the assumption. Failure to introduce pulse materials *evenly* and *instantaneously* may lead to misrepresentations of the system's mixing behavior [2].

RTD methods have been extensively used in liquid systems, and served the purpose of determining the mixing behavior inside of the unit. Over the last decade, RTD methods have been used to characterize and understand mixing in continuous powder (i.e., granular material) units, such as blenders and feed frames [1, 3, 4]. However, as we study in closer detail how the application of RTD methods compares between liquid and solids systems, we see a clear are of research that requires further evaluation: the selection of a tracer material. Similarly, to the selection of materials for performing unit characterization experiments, tracers are often selected based on a formulation, *ad hoc*, or based on the availability of materials with easy means of detection (e.g., easy to recognize spectral bands on an Near Infrared Spectra). Although this latter property is a key requirement for tracer materials, it is important to remember that in order to accurately measure the RTD, tracers introduced either through a pulse or step, *must* not alter the dynamic behavior of the bulk material within the unit operation [5].

Selecting the appropriate tracer requires understanding of the system material properties. Properties of a good tracer include: (1) an ability to be detectable from the bulk, (2) similar physical properties to the process stream, (3) no or minimal adherence to the equipment's walls or surfaces, (4) can be well mixed in the system, and (5) is representative of the materials involved in the system [1]. Tracers whose material properties are vastly different from the bulk can alter granular dynamics within the unit, resulting in inaccurate measure of the RTD. Similarly, a step change, which causes substantial change in the behavior of the bulk powder within the unit operation, may result in an inaccurate characterization of the RTD. Tracer materials in several published works for the characterization of mixers in pharmaceutical manufacturing include pure active pharmaceutical ingredients (API), which tend to have poor flowing characteristics and adhere to the walls and surfaces of vessels [4, 6-9]. Nevertheless, the selection of API is often the most relevant given the fact that it is the only material whose composition is measured in the product.

This work constitutes the development of a scientific framework to select appropriate tracers (i.e., meet all five criterions) for measuring RTDs. The framework is based on the concept of quantifying disparity between bulk and tracer physical material properties using multivariate analysis tools, such as Principal Component Analysis (PCA) [10]. Using the orthogonal decomposition of PCA, several material property values were collapsed into a reduced set of principal components that express material vast set of properties. Materials were placed on a multi-dimensional principal component space, where their weighted Euclidian distance from each other was measured. The goal of creating this space was to determine the level of similarity (or dissimilarity) between materials. Greater distances between two materials on the multi-dimensional space indicated a greater disparity between material property values and henceforth, the materials themselves. Our work proposes an acceptable multi-dimensional principal component region in which materials that lie within it are considered as acceptable tracers.

Both pulse and step change RTD experiments were performed to assess the effectiveness of our tracer selection based on the multivariate material property strategy. A single bulk material was run through a continuous blender (Gericke GCM-250) at a fixed mass flowrate. The impeller speed, blade arrangement and angle of the exit weir were kept constant. Nine unique materials, at variable distances from the bulk material on the PCA scores plot, and consequently exhibiting variable disparity between its own properties and those of the bulk, were selected as tracers. RTD curves were obtained from each of these nine materials. Each experiment was repeated in triplicate. Statistical methods were used to compare the RTD curves to the '*true*' RTD of the process. The true RTD was obtained by using dyed bulk material as the tracer. Materials, which resulted in curves that were statistically similar to the true RTD, were labelled as acceptable tracers, and vice versa. An acceptable Euclidian radius was proposed based on these findings. Step changes of increasing magnitude were performed for a binary mixture, and their RTD curves were obtained. The properties of the material post step were characterized and mapped on the PCA space. Step changes resulting in curves which were statistically similar to the true RTD were identified, and their Euclidean distance noted. An acceptable Euclidean radius and thus, an acceptable magnitude of the step change were proposed.

495948 Assessing Suspension Homogeneity Using CFD-DEM for Enhanced Content Uniformity of Spray Dried Intermediates

Abstract:

Ensuring a uniform concentration of API in the feed suspension/slurry of a spray drying process is a crucial step towards achieving a uniform dosage of the drug product intermediate. To accomplish this, an adequate feed tank design and process conditions are required to avoid heterogeneity of the suspension driven by, e.g., density differences between the solid and fluid phases. However, this heterogeneity can only be assessed *a posteriori* and the process optimization can only be achieved by trial and error. Here we present an alternative to this iterative and cost-ineffective development strategy by employing computational tools such as computational fluid dynamics coupled to discrete element methods (CFD-DEM) that allow offline experiments to be performed for various process conditions. These tools can also enable optimization of the agitation elements, such as number of impellers and distance to the bottom.

Here we focus on a spray drying case study where the feed suspension consists on a solid phase (the active pharmaceutical ingredient, or API) and the fluid phase is a solvent with a dissolved binder and other excipients. The process is run in batch mode while controlling key operation parameters, such as reactor wall temperature and agitator rotating speed, to obtain a homogeneous distribution of particles in the volume of the tank. Mixing in stirred tank reactors is enabled by fluid convection (at a larger scale) and turbulent exchanges (at a smaller scale) [1,2]. Therefore detailed information on the vessel, baffles and impeller (see Figure 1a), which determined the flow patterns, is required [2], [3].

A computational fluid dynamics (CFD-DEM) methodology (see Figure 1b), based on the OpenFOAM® framework, is used to aid process development and foresee potential issues related to accumulation zones within the reactor.

Upon choosing the reactor, its configuration and operating conditions, a steady-state simulation of the turbulent fluid flow, modeled via the Reynolds averaged Navier-Stokes (RANS) equations and shear stress transport (SST) turbulence model is conducted. The selected transient solver then uses this pre-calculated velocity field to update particles velocities considering the effect the forces exerted on them by fluid drag; gravity; and Coriolis acceleration. Furthermore, other capabilities such as particle-wall and particle-particle collision are available and may be incorporated into the solution (at the cost of extra computational effort). This way, the present workflow offers insight on how the interchangeable parts and operating conditions of the reactor might influence the steady-state particle concentration distribution.

[1] A. Ochieng and M. S. Onyango, "CFD simulation of solids suspension in stirred tanks: Review," *Hem. Ind.*, vol. 64, no. 5, pp. 365–374, 2010.

[2] G. R. Kasat, A. R. Khopkar, V. V. Ranade, and A. B. Pandit, "CFD simulation of liquid-phase mixing in solid-liquid stirred reactor," *Chem. Eng. Sci.*, vol. 63, no. 15, pp. 3877–3885, 2008.

[3] P. K. Biswas, S. C. Dev, K. M. Godiwalla, and C. S. Sivaramakrishnan, "Effect of some design parameters on the suspension characteristics of a mechanically agitated sand–water slurry system," *Mater. Des.*, vol. 20, no. 5, pp. 253–265, 1999.

492410 Advanced Model Predictive Control of Powder Level in Continuous Pharmaceutical Manufacturing Pilot-Plant

Abstract:

Currently, pharmaceutical companies are going under paradigm shift from batch to continuous manufacturing. In continuous pharmaceutical manufacturing (CPM), real-time effective control of Critical Process Parameters (CPPs) and Critical Quality Attributes (CQAs) is necessary to consistently produce a very high quality product as mandated by regulatory authority and needed for drug efficiency and patient safety [1]. Efficient automated coordination among different unit operations involved in continuous manufacturing is prerequisite to achieve consistent flow rate, inlet composition, line holdup, and thereby consistent product quality using QbD principles. In CPM, blender is connected with tablet press through a transfer pipe (chute). The chute also provides a suitable interface to integrate sensors (e.g. NIR, Raman) for real time monitoring and control. Maintaining consistent powder level in chute is needed to assure the presence of powder in front of sensors, to avoid over flow and under flow of the powder and to achieve the consistent product quality. However, different level of disturbances, change in raw materials and product specifications and startup, shutdown, speedup, slowdown processes could affect the powder level in chute significantly and thereby process operation, monitoring and product quality. Therefore, real time monitoring and control of powder level is necessary for continuous pharmaceutical manufacturing.

In this work, an advanced model predictive control strategy as well as a PID based control strategy for powder level control in a chute placed in between blender and tablet press unit operation of continuous tablet manufacturing process has been developed, implemented and evaluated. This work has been accomplished in three phases. In **first phase**, the *Insilco* study employing the process and sensor model has been performed to identify the best sensing and control strategy for powder level. Two types of the sensors, the one that gives discrete signal (on/off) and the one that gives continuous signal can be used for real time monitoring of powder level in chute. Discrete and continuous control strategy has been developed specifically for these two types of the sensors. The performance of two types of the sensor for real time monitoring and control of powder level has been compared *Insilco*. The performance of advanced model predictive controller (MPC) for powder level has been also compared with PID controller. We found that a continuous control strategy is more effective than a discrete control strategy, and that a MPC control strategy is more effective than a PID control strategy. In **second phase**, a novel noninvasive technique based on change in electric field concept has been developed, integrated and evaluated for real time monitoring of powder level in continuous manufacturing pilot-plant. The sensor has been integrated with a control panel (DeltaV (Emerson)) through relays and charms. The control panel then send the signal (4-20 mA) to control platform where the control strategy, a program to operate the sensor and a program to convert the sensor signal into powder level have been implemented. The performance of electric field based sensor for real time monitoring of powder level has been compared with a webcam based sensor. In **third phase**, an advanced model predictive controller as well as PID controller have been implemented into our continuous pharmaceutical manufacturing pilot-plant facility. The performance of both control strategies have been practically evaluated for set point tracking and disturbance rejection via utilizing electric filed based sensor. The MPC performs better than PID.

The objective of this presentation is two-fold; first to highlight the *Insilco* studies for identification of best sensing and control strategy for power level and then demonstrate the performance of best sensing and control strategies into our continuous pharmaceutical manufacturing pilot-plant facility

Reference

1. Singh, R., Román-Ospino, A. D., Romañach, R. J., Ierapetritou, M., Ramachandran, R. (2015). Real time monitoring of powder blend bulk density for coupled feed-forward/feed-back control of a continuous direct compaction tablet manufacturing process. International Journal of Pharmaceutics, 495, 612-625. http://dx.doi.org/10.1016/j.ijpharm.2015.09.029

AUTHOR INDEX

AUTHOR INDEX

AUTHOR INDEX

AUTHOR INDEX

AUTHOR INDEX

AUTHOR INDEX

AUTHOR INDEX

AUTHOR INDEX

AUTHOR INDEX

AUTHOR INDEX

AUTHOR INDEX

AUTHOR INDEX

AUTHOR INDEX

AUTHOR INDEX

AUTHOR INDEX

AUTHOR INDEX

AUTHOR INDEX

AUTHOR INDEX

AUTHOR INDEX

AUTHOR INDEX

AUTHOR INDEX